Toward a Sustainable Agriculture Through Plant Biostimulants

Toward a Sustainable Agriculture Through Plant Biostimulants: From Experimental Data to Practical Applications

Editors

Youssef Rouphael
Giuseppe Colla

MDPI • Basel • Beijing • Wuhan • Barcelona • Belgrade • Manchester • Tokyo • Cluj • Tianjin

Editors
Youssef Rouphael
University of Naples Federico II
Italy

Giuseppe Colla
University of Tuscia
Italy

Editorial Office
MDPI
St. Alban-Anlage 66
4052 Basel, Switzerland

This is a reprint of articles from the Special Issue published online in the open access journal *Agronomy* (ISSN 2073-4395) (available at: https://www.mdpi.com/journal/agronomy/special_issues/plant_biostimulants).

For citation purposes, cite each article independently as indicated on the article page online and as indicated below:

LastName, A.A.; LastName, B.B.; LastName, C.C. Article Title. *Journal Name* **Year**, *Volume Number*, Page Range.

ISBN 978-3-0365-0028-7 (Hbk)
ISBN 978-3-0365-0029-4 (PDF)

Cover image courtesy of Giuseppe Colla.

Contents

About the Editors

Youssef Rouphael is Associate Professor at the University of Naples Federico II, Italy. He is Editor-in-Chief of *Agronomy* MDPI and has served as Guest Editor of numerous international journals based on his expertise in biostimulants (*Scientia Horticulturae, Frontiers in Plant Science,* and *Agronomy* MDPI). He is a member of the Biostimulant.com scientific committee. He is internationally recognized for his research in horticultural science.

Giuseppe Colla is Professor of Vegetable Production, Floriculture and Greenhouse Crop Management at the University of Tuscia, Italy. He serves on the editorial boards of several leading journals in horticultural science. Professor Colla has an international reputation for his work on plant nutrition and plant biostimulants and has coordinated numerous projects to improve horticultural production.

agronomy

MDPI

Editorial

Toward a Sustainable Agriculture Through Plant Biostimulants: From Experimental Data to Practical Applications

Youssef Rouphael [1,*] and Giuseppe Colla [2]

1 Department of Agricultural Sciences, University of Naples Federico II, 80055 Portici, Italy
2 Department of Agriculture and Forest Sciences, University of Tuscia, 01100 Viterbo, Italy; giucolla@unitus.it
* Correspondence: youssef.rouphael@unina.it; Tel.: +39-081-2539134

Received: 18 September 2020; Accepted: 22 September 2020; Published: 24 September 2020

Abstract: Modern agriculture increasingly demands an alternative to synthetic chemicals (fertilizers and pesticides) in order to respond to the changes in international law and regulations, but also consumers' needs for food without potentially toxic residues. Microbial (arbuscular mycorrhizal and plant growth promoting rhizobacteria: *Azotobacter*, *Azospirillum* and *Rizhobium* spp.) and non-microbial (humic substances, silicon, animal- and vegetal-based protein hydrolysate and macro- and micro-algal extracts) biostimulants represent a sustainable and effective alternative or complement for their synthetic counterparts, bringing benefits to the environment, biodiversity, human health and economy. The Special Issue "Toward a sustainable agriculture through plant biostimulants: from experimental data to practical applications" compiles 34 original research articles, 4 review papers and 1 brief report covering the implications of microbial and non-microbial biostimulants for improving seedling growth and crop performance, nutrient use efficiency and quality of the produce as well as enhancing the tolerance/resistance to a wide range of abiotic stresses in particular salinity, drought, nutrient deficiency and high temperature. The present compilation of high standard scientific papers on principles and practices of plant biostimulants will foster knowledge transfer among researchers, fertilizer and biostimulant industries, stakeholders, extension specialists and farmers, and it will enable a better understanding of the physiological and molecular mechanisms and application procedure of biostimulants in different cropping systems.

Keywords: humic substances; protein hydrolysates; silicon; arbuscular mycorrhiza; plant growth promoting rhizobacteria; macroalgae; microalgae; abiotic stresses; nutrient use efficiency; physiological mechanisms

---

1. Biostimulants in Agriculture: Rationale

Modern agriculture needs to review and broaden its practices and business models, by integrating opportunities coming from different adjacent sectors and value chains, including the biobased industry, in a fully circular economy strategy [1–3]. Farmers need to operate as managers of the countryside, valorizing their own by-products and using agricultural products with improved environmental profile. Therefore, searching for new technologies and approaches to boost crop productivity under optimal and sub-optimal conditions and to improve resources use efficiency (water and fertilizers) is crucial to ensure food security, while preserving soil quality and providing opportunities of business for farmers [4]. Biobased products such as biostimulants represent a sustainable, efficient technology or complement to their synthetic counterparts (i.e., agrochemicals) to improve nutrient use efficiency and secure yield stability of agricultural and horticulture crops under optimal and sub-optimal conditions [5,6]. Recently, under the new Regulation (EU) 2019/1009, plant biostimulants were defined based on four agricultural functional claims as follow: "*EU fertilising product the function of which is to stimulate plant nutrition processes independently of the product's nutrient content with the sole aim of improving one or more of the following characteristics of the plant and/or the plant rhizosphere: (1) nutrient use efficiency, (2) tolerance resistance to (a)biotic stress, (3) quality characteristics, or (4) availability of confined nutrients in the soil or rhizosphere*" [7]. Many diverse natural substances and chemical derivatives of natural or synthetic compounds as well as beneficial microorganisms are catalogued as plant biostimulants including: (i) humic substances; (ii) vegetal- or animal-based protein hydrolysates; (iii) macro- and micro-algal extracts; (iv) silicon; (v) arbuscular mycorrhizal fungi (AMF); and (vi) plant growth promoting rhizobacteria (PGPR) belonging to the genus *Azotobacter*, *Azospirillum* and *Rizhobium* spp. [8–16].

Plant biostimulants were initially used in organic production, but now they are adopted in several cropping systems such as conventional and integrated crop production [17]. Microbial and non-microbial plant biostimulants are usually used for open field and greenhouse crops including fruit trees, berry crops, grapevines, vegetables, ornamentals, cereals and turfs [18–21]. The biostimulants market is increasing year by year; as a matter of fact, the market of active ingredient biostimulants (amino acids, seaweed extracts, humic substances and microbial amendments) is estimated to account for 2.6 billion dollars in 2019 and is projected to reach almost 5 billion dollars by 2025, at a compound annual growth rate of 11.2% during the forecast period [7,22]. Moreover, more than 1000 scientific papers published in the last 10 years (2010–2020) were found by searching the term "plant biostimulants" and many more articles are available on the Scopus database using related words/terms (i.e., humic substances, seaweed extracts, microalgae, silicon, AMF or PGPR) (www.scopus.com).

The current Special Issue collects 39 scientific contributions (34 research papers, 4 reviews and 1 brief report) covering the different aspects of the agronomic and horticultural crops response to microbial and non-microbial biostimulants application. We highly believe that the current Special Issue: (i) will foster knowledge transfer among scientists, commercial enterprises, stakeholders and farmers; and (ii) will shed light on the cellular, molecular and physiological mechanisms as well as the application procedure of biostimulants in different cropping systems including organic farming.

2. The Role of Non-Microbial and Microbial Biostimulants in Morpho-Anatomical, Biochemical and Physiological Traits of Crops

Applications of non-microbial and microbial plant biostimulants have been shown to enhance plant growth and development, as well as macro- and micronutrient uptake and translocation in several agronomic and horticultural crops resulting in increased biomass production and yield [3]. The stimulation of seedling growth and crop productivity in response to application of non-microbial and microbial plant biostimulants is attributed to the action of bioactive substances on the primary and/or secondary metabolisms, leading to a wide array of biochemical, physiological and molecular responses [3]. Seven combinations of soy flour, diatomaceous earth, concentrated vermicompost extract (liquid) and micronized vermicompost were investigated in laboratory experiments to assess their potential biostimulant action to improve cover crops (red clover and perennial ryegrass) germination and seedling growth [23]. In their research, the authors reported that coated treatments affected in a species-specific manner the germination rate and uniformity, with a significant improvement in total germination rate recorded in red clover, while a reduction was observed in perennial ryegrass. Interestingly, the application of soy flour:diatomaceous earth at a rate of 30:70 boosted the seedlings performance in terms of shoot and root growth as well as dry matter percentage in both tested species. The authors concluded that soy flour provided a sustained source of key amino acids, thus positively influencing N uptake and transplant quality. Furthermore, Ben-Jabeur et al. [24] conducted a three-year experiment on durum wheat aiming to assess the effect of coating wheat seeds with thyme essential oil or *Paraburkholderia phytofirmans* PsJN strain on yield and resistance/tolerance to spetoria leaf botch. The two tested biostimulants were able to alleviate the Septoria leaf botch and to enhance yield in terms of number of spikes per square meter as well as straw and grain yields. The dual beneficial effect (i.e., biocontrol and biostimulant action) was also observed on tomato, where the application of four commercial biostimulants: neem seed cake, sesame oil, quillay extract and seaweeds significantly mitigated the parasitism of root-knot nematodes by reducing eggs and galls on tomato roots with the best results recorded on neem seed cake and sesame oil treatments [25]. The authors also demonstrated that the four tested biostimulants triggered shoot and root biomass production compared to untreated control. The dual beneficial effect was also recorded on tomato, since Allaga et al. [26] reported that a composite bioinoculant containing beneficial fungi and bacteria (*Trichoderma*, *Azotobacter* and *Streptomyces*) was an efficient biocontrol agent, as well as an efficient biostimulant able to improve growth and photosynthetic activity of tomato.

Ertani et al. [27] carried out a short-term trial on hydroponically grown maize to assess the physiological responses to leonardite-humate- and lignosulfonate-based biostimulants. The biostimulants application in particular lignosulfonates boosted root and leaf growth by 51–140% and 5–35%, respectively. The authors concluded that a putative mechanism involved in the biostimulant action of these products might be the stimulation of N metabolism in the belowground organs (i.e., roots) according to the increased activity of key enzymes such as glutamine synthetase and glutamate synthase [27]. Moreover, Kim et al. [28], elucidated the hormonal effects of a commercial vegetal-based biostimulants containing amino acids, lateral root promoting peptide, lignosulfonates and micronutrients on cuttings of basil, tomato and chrysanthemum, characterized by different relative root ability: easy, moderate and difficult, respectively. Thanks to the combination of morphological, biochemical and metabolomics approaches, the authors demonstrated that the vegetal-based biostimulant exerted similar effects to the synthetic hormone (i.e., auxin) by improving adventitious rooting responses. Finally, the authors shed light for the first time onto hormonal regulation of vegetal-based biostimulant and the crucial role of brassinosteroids in adventitious root formation.

Different amino acids (L-methionine, L-glycine and L-tryptophan at 20, 210 and 220 mg/L, respectively) were applied separately on hydroponically grown butterhead lettuce to assess their stimulators role [29]. In their study, L-methionine boosted lettuce growth parameters, whereas a negative effect was observed when L-glycine and L-tryptophan were applied. Based on the results of the first experiment, Khan and co-workers conducted a second experiment with five increasing concentrations of L-methionine (0.02, 0.2, 2.2, 22, 220 and 2220 mg/L). The authors concluded that L-methionine at a concentration of 0.2 mg/L exhibited the best effect of lettuce growth parameters. In fact, it is well established that key amino acids are rapidly absorbed by the crops and act as a stable source of molecule precursors to be integrated into plant metabolism [30]. This was demonstrated by the former authors, who reported that foliar application of glutamate to creeping bentgrass foliage was rapidly absorbed and directly utilized as a precursor to synthesize gamma-aminobutyric acid and proline, two important metabolites with well-known roles in plant stress adaptation.

Bákonyi et al. [31] and Kisvarga et al. [32] reported that alfalfa brown juice could be considered a potential growth stimulator. In their studies, Celosia seedlings where sprayed at five increasing rates of fermented brown juice (0.5%, 1.0%, 1.5%, 2.0% or 2.5%), while basil was sprayed at three different increasing doses (0.5%, 1.0% or 2.5%). Water was adopted in both experiments as an untreated control. The application of alfalfa brown juice at a rate of 0.5% boosted plant growth parameters in both tested species due to the modulation of the anatomical and biochemical responses, in particular increasing the antioxidant activity of key enzymes (catalase and peroxidase) and photosynthetic pigments (chlorophyll a and b) as well as reducing the content of malondialdehyde. Moreover, Niewiadomska et al. [33] carried out a three-year experiment on white lupine cultivation, where two commercial biostimulants and six foliar fertilizers were tested. The commercial biostimulants and fertilizers were able to boost some of the biochemical activity of the soil. The authors attributed the better performance of treated-white lupine to a higher uptake, translocation and assimilation of macro- and microelements.

Seaweed extracts, also known as macroalgae, are considered an important category of non-microbial plant biostimulants due to their use on several agronomic and horticultural crops under both conventional and organic farming systems [34]. Several authors reported that macroalgae such as *Ascophyllum nodosum*, *Ecklonia maxima* or *Pterocladia capillacea* can: (i) improve the agronomic performance of soybean and bean [35,36], potato [37], and Jew's mallow [38]; and (ii) enhance fruit setting in eggplant [39]. In addition to seaweed extracts, the use of PGPR such as *Bacillus thuringiensis* was also considered an efficient approach to boost yield in a sustainable manner. Jo and co-workers [40] inoculation of *Bacillus thuringiensis* KNU-07 incurred a significant increase of total growth biomass of pepper seedlings. The beneficial effect recorded on inoculated pepper plants was associated with a strong modulation of the soil bacterial community even quantitatively or qualitatively.

3. The Role of Non-Microbial and Microbial Biostimulants in Enhancing Nutrient Uptake and Efficiency

Non-microbial and microbial plant biostimulants may positively influence nutrient use efficiency (NUE), in particular nitrogen (N) by enhancing root system architecture and soil exploration as well as increasing macro- and micronutrient solubilization that can result in an increase in NUE [17,41]. Di Mola et al. [42] demonstrated that foliar application of vegetal- (protein hydrolysates or tropical plant extract) and seaweed extract-based biostimulants (brown macroalgae: *Ecklonia maxima*) is considered a sustainable approach to increase greenhouse baby lettuce productivity and NUE in low-input cropping systems. In their study, the authors reported that the application of legume-derived protein hydrolysates and especially seaweed extract elicited important increases in fresh yield under sub-optimal and optimal N conditions (0 and 10 kg ha^{-1}) compared to the untreated and tropical plant extract-treated plants, but the beneficial effect of plant biostimulants was not apparent under luxurious N fertilization conditions (20 and 30 kg ha^{-1}). Similar results were also observed by the same research group [43] on two other important greenhouse leafy vegetables, namely baby spinach and lamb's lettuce, treated with a legume-derived protein hydrolysates and grown under optimal and sub-optimal N regimes. Interestingly, the foliar application of vegetal-based biostimulants incurred a significant increase in N uptake and N use efficiencies in both leafy vegetables (19% and 18%, respectively, for baby spinach and 50% and 73%, respectively, for lamb's lettuce). The authors concluded that improved agronomical performance and use efficiency of baby lettuce, baby spinach and lamb's lettuce was associated with a better photosynthetic activity and biochemical status (higher content of chlorophyll a, b and total and carotenoids) [42,43]. The synergistic biostimulant action through the application of microbial (*Trichoderma virens*) and non-microbial biostimulant (vegetal biopolymer containing amino acids, peptides and vitamins) was demonstrated on greenhouse lettuce grown with three N conditions: sub-optimal, optimal and supra-optimal (0, 70 and 140 kg ha^{-1}) [44]. Lettuce grown under non-fertilized conditions showed an increase in marketable yield when inoculated with *T. virens* alone (45%) and a greater increase with both microbial and non-microbial biostimulant (67%). The beneficial effect of plant biostimulant was less pronounced under optimal N condition and absent under luxurious N conditions. Rouphael and co-workers concluded that, based on the improved fresh yield and NUE in greenhouse lettuce plants, treatment with plant biostimulants improved not only the chlorophyll synthesis and mineral status but also the synthesis and accumulation of antioxidant metabolites that were responsible for reactivating the photosynthetic activity and consequently the agronomic performance.

Concerning floricultural species, Leoni et al. [45] investigated the application of chemical fertilization and integrated nutrient management on yield, quality attributes and NUE of two chrysanthemum cut flower cultivars. Integrated nutrient management based on 50% synthetic fertilizers plus seaweed extract (*A. nodosum*) and microbial consortium (*Glomus* sp. and *Bacillus* sp.) was able to boost yield, quality parameters and NUE compared to the untreated control treatment.

4. The Role of Non-Microbial and Microbial Biostimulants in Abiotic Stresses Tolerance/Resistance

Abiotic stresses, in particular drought, salinity, heat stress, hypoxia and nutrient deficiency, are responsible for 60–70% of yield gap, dictated by global climate changes [46]. To overcome the detrimental effects of sub-optimal conditions on agronomic and horticultural crops, plant biostimulants have been proposed as an efficient agronomic tool to improve tolerance/resistance to unfavorable environment and soil conditions [47]. In their review paper, Bulgari and co-workers summarized the biostimulants literature (humic substances, seaweed extracts, protein hydrolysates, amino acids and beneficial microorganisms) regarding their use on vegetables, focusing on their application and mode of actions to counteract the most common abiotic stresses: cold/chilling stress, heat, salinity, drought stress and nutrient deficiency. In addition to the categorized plant biostimulants, Arnao and Hernández-Ruiz [48] proposed the dual use of melatonin (N-acetyl-5-methoxytryptamine) as plant protector and biostimulant. In their review paper, they discussed the different legal aspects to categorize this natural substance as potential biostimulant at the European level. Arnao and Hernández-Ruiz [48] summarized studies of different responses of melatonin in different plant species and under diverse stress conditions by reporting the observed effects/mechanisms.

The application of four commercial biostimulants containing protein hydrolysates, humic acid and especially brown seaweed extracts (*A. nodosum*) were found to mitigate the negative effects of water stress (70% or 50% of the container substrate capacity) on potted mint by increasing the antioxidant activity of key enzymes such as catalase and superoxide dismutase and by reducing the H_2O_2 accumulation in leaf tissue [49]. The physiological and biochemical effects of β-(1,3)-glucan (paramylon) purified from the microalga *Euglena gracilis* on water-stress Micro-Tom were also assessed by an Italian research group [50]. The eco-physiological approach adopted in this study allowed the identification of several physiological and biochemical mechanisms of improved water stress tolerance, following the application of paramylon nanofibers, for example: (i) increasing of the photosynthetic rate; and (ii) reducing the sensitivity of photosystem II to potential dehydration damages. Moreover, Petropoulos et al. [51] showed that the application of four commercial microbial biostimulants containing AMF, *Trichoderma* and rhizosphere symbiotic bacteria enriched with amino acids or seaweed extracts were able to increase the pods and seeds yield as well as nutritional value and chemical composition of common bean under both optimal and sub-optimal water regimes. In the study by Mannino et al. [52], the impacts of four microbial biostimulants, namely AMF mono fungal inoculum, AMF multi fungal inoculum, PGPB and AMF + PGPB, on molecular and physiological responses of water-stressed tomato were evaluated. Different physiological and molecular responses of tomato to water limitation were recorded depending on microbial inocula, confirming the importance to characterize the optimal plant/beneficial microorganism genotype combination(s) to enhance plant resilience to water stress condition. Non-microbial plant biostimulants such as amino acids/peptides-based product and protein hydrolysates can also be considered an effective tools to improve the tolerance to a wide range of abiotic stresses: heat, hypotoxic, nutrient and salt stresses as well as combined environmental stresses [53,54]. The application of biostimulant based on plant and yeast extracts and containing amino acids, soluble peptides and vitamins improved the heat stress tolerance of four tomato landraces grown under Mediterranean conditions. The biostimulant effects were associated to physiological and biochemical mode of actions, for example: (i) stronger antioxidant defense system; and (ii) maximal photochemical efficiency (F_v/F_m) in leaves of the four tested tomato landraces [53]. Finally, Trevisan et al. [54] demonstrated in a short-term trial that the application of a protein hydrolysates-based biostimulant was able to mitigate the detrimental effects of single (hypoxia, salt or nutrient deficiency) and multiple (nutrient stress + hypoxia or nutrient stress + salinity) stresses of hydroponically grown maize. Root development in terms of biomass and architecture (length and density) was strongly influenced by protein hydrolysates, by upregulating the expression of key genes involved in nitrate transport and reactive oxygen species detoxification and consequently inducing a significant boost of shoot biomass.

5. The Role of Non-Microbial and Microbial Biostimulants in Improving Quality Traits

Pre-clinical and clinical studies have demonstrated the functional (i.e., health-promoting) effects of fruit and vegetables consumption in supporting human health and longevity [55]. In their review paper, Drobek et al. [56] gave an overview on how the application of microbial and non-microbial plant biostimulants can modulate the primary and secondary metabolisms of horticultural species, leading to the synthesis and accumulation of lipophilic and hydrophilic antioxidant molecules also known as phytochemicals [3,15]. The application of vegetal-based biostimulants, in particular tropical plant extract and legume-derived protein hydrolysates, in two important leafy and fruit vegetables induced significant increase in lettuce and tomato nutritional and functional quality [57,58]. Weekly foliar application of tropical plant extract incurred a significant increase of hydrophilic antioxidant activity and total ascorbic acid in lettuce compared to untreated control [57]. Similar results were also recorded in tomato fruits, where tropical plant extract and protein hydrolysates resulted in higher bioactive compounds (total phenols and vitamin C) and lipophilic antioxidant activity than those observed in the non-treated control [58]. Concerning berry fruits, Soppelsa et al. [59] investigated the application of ten commercial biostimulants belonging to almost all the categories including: alfalfa hydrolysate, humic acids, macro-seaweed, extract and microalgal hydrolysate, amino acids alone or in combination with micronutrient (zinc), B-group vitamins, chitosan and a commercial product containing silicon. Biostimulant products based on chitosan had a major impact on strawberry pulp firmness, whereas biostimulant products based on alfalfa hydrolysate, macro-seaweed extract and microalgal hydrolysate induced an improvement in phenolic compounds compared to the remaining treatments. Moreover, in three varieties of winter rape, the application of three biostimulants with the following active substances improved the content of crude fiber and fat: titanium, sodium orto nitrophenol, sodium para nitrophenol, sodium 5-nitroguaiacolate and silicon [60].

Concerning the implications of microbial plant biostimulants on improving produce quality, Chandrasekaran et al. [61] reported that the inoculation of PGPR strain, *Bacillus subtilis* CBR05 induced a significant increase in tomato quality in terms of carotenoids profile (β-carotene and lycopene). Finally, Caser et al. [62,63] showed that the inoculation of soilless-grown saffron with *Rhizophagus intraradices* and to a lesser extent with a mixture of *R. intraradices* and *Funneliformis mosseae* boosted significantly the synthesis and accumulation of health-promoting molecules such as anthocyanins, polyphenols and vitamin C; antioxidant activity; and important bioactive compounds in saffron, such as crocin II, picrocrocin and quercitrin.

6. Conclusions and Looking Forward

In the coming few years, we can expect that plant biostimulants including both natural and synthetic substances, as well as microbial inoculants, will not only make a significant contribution to ecologically and economically sustainable crop production systems within more resilient agro-ecosystems, but will also lay the cornerstone for a future large-scale sustainable agriculture catalyzed by the biobased industry. Although plant biostimulants appear to be a novel and potential category of agricultural inputs complementing synthetic fertilizers, there is an urgent need among the research community and fertilizer industries to elucidate the molecular and physiological mechanisms which will definitely facilitate the diffusion of these bio-products in the agricultural sector. Briglia et al. [64] demonstrated that the combination of phenomic (high-throughput plant phenotyping) and genomic (Next Generation Sequencing) tools opens new perspectives to release effective biostimulant formulations to meet the emerging needs of crops. Finally, Giovannini et al. [65] suggested that, in the near future, transcriptomics research should be adopted as an integrated tool to identify the best synergistic combinations of AMF and associated bacterial communities able to enhance resources use efficiency, plant resilience and boosting nutraceutical compounds in plant species.

Author Contributions: Conceptualization, Y.R. and G.C.; writing—original draft preparation, Y.R. and G.C.; and writing—review and editing, Y.R. and G.C. All authors have read and agreed to the published version of the manuscript.

Funding: This research received no external funding.

Conflicts of Interest: The authors declare no conflict of interest.

References

1. Colla, G.; Rouphael, Y. Biostimulants in horticulture. *Sci. Hortic.* **2015**, *196*, 1–2. [CrossRef]

2. Rouphael, Y.; Colla, G. Synergistic Biostimulatory Action: Designing the Next Generation of Plant Biostimulants for Sustainable Agriculture. *Front. Plant Sci.* **2018**, *9*, 9. [CrossRef] [PubMed]
3. Rouphael, Y.; Colla, G. Editorial: Biostimulants in Agriculture. *Front. Plant Sci.* **2020**, *11*, 40. [CrossRef] [PubMed]
4. Jägermeyr, J. Agriculture's historic twin-challenge toward sustainable water use and food supply for all. *Front. Sustain. Food Syst.* **2020**, *4*, 35. [CrossRef]
5. Fiorentino, N.; Ventorino, V.; Woo, S.L.; Pepe, O.; De Rosa, A.; Gioia, L.; Romano, I.; Lombardi, N.; Napolitano, M.; Colla, G.; et al. Trichoderma-Based Biostimulants Modulate Rhizosphere Microbial Populations and Improve N Uptake Efficiency, Yield, and Nutritional Quality of Leafy Vegetables. *Front. Plant Sci.* **2018**, *9*, 9. [CrossRef]
6. Rouphael, Y.; Spíchal, L.; Panzarová, K.; Casa, R.; Colla, G. High-Throughput Plant Phenotyping for Developing Novel Biostimulants: From Lab to Field or From Field to Lab? *Front. Plant Sci.* **2018**, *9*, 1197. [CrossRef]
7. E.U. Regulation of the European Parliament and of the Council Laying Down Rules on the Making Available on the Market of EU Fertilising Products and Amending Regulations (EC) No 1069/2009 and (EC) No 1107/2009 and Repealing Regulation (EC) No 2003/2003. 2019. Available online: https://eur-lex.europa.eu/legal-content/EN/TXT/?uri=OJ:L:2019:170:TOC (accessed on 10 September 2020).
8. Battacharyya, D.; Babgohari, M.Z.; Rathor, P.; Prithiviraj, B. Seaweed extracts as biostimulants in horticulture. *Sci. Hortic.* **2015**, *196*, 39–48. [CrossRef]
9. Canellas, L.P.; Olivares, F.L.; Aguiar, N.O.; Jones, D.L.; Nebbioso, A.; Mazzei, P.; Piccolo, A. Humic and fulvic acids as biostimulants in horticulture. *Sci. Hortic.* **2015**, *196*, 15–27. [CrossRef]
10. López-Bucio, J.; Pelagio-Flores, R.; Herrera-Estrella, A. Trichoderma as biostimulant: Exploiting the multilevel properties of a plant beneficial fungus. *Sci. Hortic.* **2015**, *196*, 109–123. [CrossRef]
11. Ruzzi, M.; Aroca, R. Plant growth-promoting rhizobacteria act as biostimulants in horticulture. *Sci. Hortic.* **2015**, *196*, 124–134. [CrossRef]
12. Savvas, D.; Ntatsi, G. Biostimulant activity of silicon in horticulture. *Sci. Hortic.* **2015**, *196*, 66–81. [CrossRef]
13. Colla, G.; Nardi, S.; Cardarelli, M.; Ertani, A.; Lucini, L.; Canaguier, R.; Rouphael, Y. Protein hydrolysates as biostimulants in horticulture. *Sci. Hortic.* **2015**, *196*, 28–38. [CrossRef]
14. Colla, G.; Hoagland, L.; Ruzzi, M.; Cardarelli, M.; Bonini, P.; Canaguier, R.; Rouphael, Y. Biostimulant Action of Protein Hydrolysates: Unraveling Their Effects on Plant Physiology and Microbiome. *Front. Plant Sci.* **2017**, *8*, 2202. [CrossRef] [PubMed]
15. Rouphael, Y.; Franken, P.; Schneider, C.; Schwarz, D.; Giovannetti, M.; Agnolucci, M.; De Pascale, S.; Bonini, P.; Colla, G. Arbuscular mycorrhizal fungi act as biostimulants in horticultural crops. *Sci. Hortic.* **2015**, *196*, 91–108. [CrossRef]
16. Chiaiese, P.; Corrado, G.; Colla, G.; Kyriacou, M.C.; Rouphael, Y. Renewable Sources of Plant Biostimulation: Microalgae as a Sustainable Means to Improve Crop Performance. *Front. Plant Sci.* **2018**, *9*, 9. [CrossRef]
17. De Pascale, S.; Rouphael, Y.; Colla, G. Plant biostimulants: Innovative tool for enhancing plant nutrition in organic farming. *Eur. J. Hortic. Sci.* **2018**, *82*, 277–285. [CrossRef]
18. Basile, B.; Rouphael, Y.; Colla, G.; Soppelsa, S.; Andreotti, C. Appraisal of emerging crop management opportunities in fruit trees, grapevines and berry crops facilitated by the application of biostimulants. *Sci. Hortic.* **2020**, *267*, 109330. [CrossRef]
19. Rouphael, Y.; Kyriacou, M.C.; Petropoulos, S.A.; De Pascale, S.; Colla, G. Improving vegetable quality in controlled environments. *Sci. Hortic.* **2018**, *234*, 275–289. [CrossRef]
20. Rouphael, Y.; Lucini, L.; Miras-Moreno, B.; Colla, G.; Bonini, P.; Cardarelli, M. Metabolomic Responses of Maize Shoots and Roots Elicited by Combinatorial Seed Treatments With Microbial and Non-microbial Biostimulants. *Front. Microbiol.* **2020**, *11*, 664. [CrossRef]
21. Rouphael, Y.; Colla, G.; Graziani, G.; Ritieni, A.; Cardarelli, M.; De Pascale, S. Phenolic Composition, Antioxidant Activity and Mineral Profile in Two seed-Propagated Artichoke Cultivars as Affected by Microbial Inoculants And Planting Time. *Food Chem.* **2017**, *234*, 10–19. [CrossRef]
22. Biostimulants Market by Active Ingredient (Humic Substances, Amino Acids, Seaweed Extracts, Microbial Amendments), Crop Type (Fruits & Vegetables, Cereals, Turf & Ornamentals), Application Method, Form, and Region—Global Forecast to 2025. Available online: https://www.marketsandmarkets.com/Market-Reports/b

iostimulant-market-1081.html?gclid=CjwKCAjw4_H6BRALEiwAvgfzq1LVX47L4C4O0v0leN5GfYGuk0xW2oF25JDZhWGs03E3I2rL1kEwGxoCnsAQAvD_BwE (accessed on 10 September 2020).
23. Qiu, Y.; Amirkhani, M.; Mayton, H.; Chen, Z.; Taylor, A. Biostimulant Seed Coating Treatments to Improve Cover Crop Germination and Seedling Growth. *Agronomy* **2020**, *10*, 154. [CrossRef]
24. Ben-Jabeur, M.; Kthiri, Z.; Harbaoui, K.; Belguesmi, K.; Serret, M.D.; Araus, J.; Hamada, W.; Jabeur, B. Seed Coating with Thyme Essential Oil or Paraburkholderia phytofirmans PsJN Strain: Conferring Septoria Leaf Blotch Resistance and Promotion of Yield and Grain Isotopic Composition in Wheat. *Agronomy* **2019**, *9*, 586. [CrossRef]
25. D'Addabbo, T.; Laquale, S.; Perniola, M.; Candido, V.; Addabbo, D. Biostimulants for Plant Growth Promotion and Sustainable Management of Phytoparasitic Nematodes in Vegetable Crops. *Agronomy* **2019**, *9*, 616. [CrossRef]
26. Allaga, H.; Bóka, B.; Poor, P.; Nagy, V.D.; Szűcs, A.; Stankovics, I.; Takó, M.; Manczinger, L.; Vágvölgyi, C.; Kredics, L.; et al. A Composite Bioinoculant Based on the Combined Application of Beneficial Bacteria and Fungi. *Agronomy* **2020**, *10*, 220. [CrossRef]
27. Ertani, A.; Nardi, S.; Francioso, O.; Pizzeghello, D.; Tinti, A.; Schiavon, M. Metabolite-Targeted Analysis and Physiological Traits of *Zea mays* L. in Response to Application of a Leonardite-Humate and Lignosulfonate-Based Products for Their Evaluation as Potential Biostimulants. *Agronomy* **2019**, *9*, 445. [CrossRef]
28. Kim, H.-J.; Ku, K.-M.; Choi, S.; Cardarelli, M. Vegetal-derived Biostimulant Enhances Adventitious Rooting in Cuttings of Basil, Tomato, and Chrysanthemum via Brassinosteroid-mediated Processes. *Agronomy* **2019**, *9*, 74. [CrossRef]
29. Khan, S.; Yu, H.-J.; Li, Q.; Gao, Y.; Sallam, B.N.; Wang, H.; Liu, P.; Jiang, W. Exogenous Application of Amino Acids Improves the Growth and Yield of Lettuce by Enhancing Photosynthetic Assimilation and Nutrient Availability. *Agronomy* **2019**, *9*, 266. [CrossRef]
30. McCoy, R.M.; Meyer, G.W.; Rhodes, D.; Murray, G.C.; Sors, T.G.; Widhalm, J.R. Exploratory Study on the Foliar Incorporation and Stability of Isotopically Labeled Amino Acids Applied to Turfgrass. *Agronomy* **2020**, *10*, 358. [CrossRef]
31. Bákonyi, N.; Kisvarga, S.; Barna, D.; Tóth, I.O.; El-Ramady, H.; Abdalla, N.; Kovács, S.; Rozbach, M.; Fehér, C.; Elhawat, N.; et al. Chemical Traits of Fermented Alfalfa Brown Juice: Its Implications on Physiological, Biochemical, Anatomical, and Growth Parameters of Celosia. *Agronomy* **2020**, *10*, 247. [CrossRef]
32. Kisvarga, S.; Barna, D.; Kovács, S.; Csatári, G.; Tóth, I.O.; Fári, M.; Makleit, P.; Veres, S.; Alshaal, T.; Bákonyi, N. Fermented Alfalfa Brown Juice Significantly Stimulates the Growth and Development of Sweet Basil (*Ocimum basilicum* L.) Plants. *Agronomy* **2020**, *10*, 657. [CrossRef]
33. Niewiadomska, A.; Sulewska, H.; Wolna-Maruwka, A.; Ratajczak, K.; Waraczewska, Z.; Budka, A. The Influence of Bio-Stimulants and Foliar Fertilizers on Yield, Plant Features, and the Level of Soil Biochemical Activity in White Lupine (*Lupinus albus* L.) Cultivation. *Agronomy* **2020**, *10*, 150. [CrossRef]
34. Carillo, P.; Ciarmiello, L.F.; Woodrow, P.; Corrado, G.; Chiaiese, P.; Rouphael, Y. Enhancing Sustainability by Improving Plant Salt Tolerance through Macro- and Micro-Algal Biostimulants. *Biology* **2020**, *9*, 253. [CrossRef] [PubMed]
35. Kocira, S.; Szparaga, A.; Kuboń, M.; Czerwinska, E.; Piskier, T. Morphological and Biochemical Responses of *Glycine max* (L.) Merr. to the Use of Seaweed Extract. *Agronomy* **2019**, *9*, 93. [CrossRef]
36. Kocira, A.; Lamorska, J.; Kornas, R.; Nowosad, N.; Tomaszewska, M.; Leszczyńska, D.; Kozłowicz, K.; Tabor, S. Changes in Biochemistry and Yield in Response to Biostimulants Applied in Bean (*Phaseolus vulgaris* L.). *Agronomy* **2020**, *10*, 189. [CrossRef]
37. Wadas, W.; Dziugieł, T. Changes in Assimilation Area and Chlorophyll Content of Very Early Potato (*Solanum tuberosum* L.) Cultivars as Influenced by Biostimulants. *Agronomy* **2020**, *10*, 387. [CrossRef]
38. Ashour, M.; El-Shafei, A.A.; Khairy, H.M.; Abd-Elkader, D.Y.; Mattar, M.; Alataway, A.; Hassan, S. Effect of Pterocladia capillacea Seaweed Extracts on Growth Parameters and Biochemical Constituents of Jew's Mallow. *Agronomy* **2020**, *10*, 420. [CrossRef]
39. Pohl, A.; Grabowska, A.; Kalisz, A.; Sękara, A. Biostimulant Application Enhances Fruit Setting in Eggplant—An Insight into the Biology of Flowering. *Agronomy* **2019**, *9*, 482. [CrossRef]
40. Jo, H.; Tagele, S.B.; Pham, H.Q.; Kim, M.-C.; Choi, S.; Park, Y.-J.; Ibal, J.C.; Park, G.-S.; Shin, J.-H.; Kim, M.-J. Response of Soil Bacterial Community and Pepper Plant Growth to Application of Bacillus thuringiensis KNU-07. *Agronomy* **2020**, *10*, 551. [CrossRef]

41. Carillo, P.; Colla, G.; Fusco, G.M.; Dell'Aversana, E.; El-Nakhel, C.; Giordano, M.; Pannico, A.; Cozzolino, E.; Mori, M.; Reynaud, H.; et al. Morphological and Physiological Responses Induced by Protein Hydrolysate-Based Biostimulant and Nitrogen Rates in Greenhouse Spinach. *Agronomy* **2019**, *9*, 450. [CrossRef]
42. Di Mola, I.; Cozzolino, E.; Ottaiano, L.; Giordano, M.; Rouphael, Y.; Colla, G.; Mori, M. Effect of Vegetal- and Seaweed Extract-Based Biostimulants on Agronomical and Leaf Quality Traits of Plastic Tunnel-Grown Baby Lettuce under Four Regimes of Nitrogen Fertilization. *Agronomy* **2019**, *9*, 571. [CrossRef]
43. Di Mola, I.; Cozzolino, E.; Ottaiano, L.; Nocerino, S.; Rouphael, Y.; Colla, G.; El-Nakhel, C.; Mori, M. Nitrogen use and uptake effciency and crop performance of baby spinach (*Spinacia oleracea* L.) and lamb's lettuce (*Valerianella locusta* L.) grown under variable cub-optimal N regimes combined with plant-based biostimulant application. *Agronomy* **2020**, *10*, 278. [CrossRef]
44. Rouphael, Y.; Carillo, P.; Colla, G.; Fiorentino, N.; Sabatino, L.; El-Nakhel, C.; Giordano, M.; Pannico, A.; Cirillo, V.; Shabani, E.; et al. Appraisal of Combined Applications of *Trichoderma virens* and a Biopolymer-Based Biostimulant on Lettuce Agronomical, Physiological, and Qualitative Properties under Variable N Regimes. *Agronomy* **2020**, *10*, 196. [CrossRef]
45. Leoni, B.; Loconsole, D.; Cristiano, G.; De Lucia, B.; De Lucia, B. Comparison between Chemical Fertilization and Integrated Nutrient Management: Yield, Quality, N, and P Contents in *Dendranthema grandiflorum* (Ramat.) Kitam. Cultivars. *Agronomy* **2019**, *9*, 202. [CrossRef]
46. Rouphael, Y.; Kyriacou, M.C.; Colla, G. Vegetable Grafting: A Toolbox for Securing Yield Stability under Multiple Stress Conditions. *Front. Plant Sci.* **2018**, *8*, 2255. [CrossRef]
47. Bulgari, R.; Franzoni, G.; Ferrante, A. Biostimulants Application in Horticultural Crops under Abiotic Stress Conditions. *Agronomy* **2019**, *9*, 306. [CrossRef]
48. Arnao, M.B.; Hernández-Ruiz, J. Melatonin as a Chemical Substance or as Phytomelatonin Rich-Extracts for Use as Plant Protector and/or Biostimulant in Accordance with EC Legislation. *Agronomy* **2019**, *9*, 570. [CrossRef]
49. Elansary, H.O.; Mahmoud, E.A.; El-Ansary, D.O.; Mattar, M. Effects of Water Stress and Modern Biostimulants on Growth and Quality Characteristics of Mint. *Agronomy* **2019**, *10*, 6. [CrossRef]
50. Barsanti, L.; Coltelli, P.; Gualtieri, P. Paramylon Treatment Improves Quality Profile and Drought Resistance in *Solanum lycopersicum* L. cv. Micro-Tom. *Agronomy* **2019**, *9*, 394. [CrossRef]
51. Petropoulos, S.A.; Fernandes, Â; Plexida, S.; Chrysargyris, A.; Tzortzakis, N.; Barreira, J.C.M.; Barros, L.; Ferreira, I.C. Biostimulants Application Alleviates Water Stress Effects on Yield and Chemical Composition of Greenhouse Green Bean (*Phaseolus vulgaris* L.). *Agronomy* **2020**, *10*, 181. [CrossRef]
52. Mannino, G.; Nerva, L.; Gritli, T.; Novero, M.; Fiorilli, V.; Bacem, M.; Bertea, C.M.; Lumini, E.; Chitarra, W.; Balestrini, R. Effects of di different microbial inocula on tomato tolerance to water deficit. *Agronomy* **2020**, *10*, 170. [CrossRef]
53. Francesca, S.; Arena, C.; Mele, B.H.; Schettini, C.; Ambrosino, P.; Barone, A.; Rigano, M.M. The Use of a Plant-Based Biostimulant Improves Plant Performances and Fruit Quality in Tomato Plants Grown at Elevated Temperatures. *Agronomy* **2020**, *10*, 363. [CrossRef]
54. Trevisan, S.; Manoli, A.; Quaggiotti, S. A Novel Biostimulant, Belonging to Protein Hydrolysates, Mitigates Abiotic Stress Effects on Maize Seedlings Grown in Hydroponics. *Agronomy* **2019**, *9*, 28. [CrossRef]
55. Kyriacou, M.C.; Rouphael, Y. Towards a new definition of quality for fresh fruits and vegetables. *Sci. Hortic.* **2018**, *234*, 463–469. [CrossRef]
56. Drobek, M.; Frąc, M.; Cybulska, J. Plant Biostimulants: Importance of the Quality and Yield of Horticultural Crops and the Improvement of Plant Tolerance to Abiotic Stress—A Review. *Agronomy* **2019**, *9*, 335. [CrossRef]
57. Cozzolino, E.; Giordano, M.; Fiorentino, N.; El-Nakhel, C.; Pannico, A.; Di Mola, I.; Mori, M.; Kyriacou, M.C.; Colla, G.; Rouphael, Y. Appraisal of Biodegradable Mulching Films and Vegetal-Derived Biostimulant Application as Eco-Sustainable Practices for Enhancing Lettuce Crop Performance and Nutritive Value. *Agronomy* **2020**, *10*, 427. [CrossRef]
58. Caruso, G.; Giordano, M.; Cozzolino, E.; Cuciniello, A.; Cenvinzo, V.; Bonini, P.; Colla, G.; Rouphael, Y. Yield and Nutritional Quality of Vesuvian Piennolo Tomato PDO as Affected by Farming System and Biostimulant Application. *Agronomy* **2019**, *9*, 505. [CrossRef]
59. Soppelsa, S.; Kelderer, M.; Casera, C.; Bassi, M.; Robatscher, P.; Matteazzi, A.; Andreotti, C. Foliar Applications of Biostimulants Promote Growth, Yield and Fruit Quality of Strawberry Plants Grown under Nutrient Limitation. *Agronomy* **2019**, *9*, 483. [CrossRef]

60. Gugała, M.; Sikorska, A.; Zarzecka, K.; Findura, P.; Malaga-Toboła, U.; Toboła, M. Chemical Composition of Winter Rape Seeds Depending on the Biostimulators Used. *Agronomy* **2019**, *9*, 716. [CrossRef]
61. Chandrasekaran, M.; Chun, S.C.; Oh, J.-W.; Paramasivan, M.; Saini, R.K.; Sahayarayan, J.J. Bacillus subtilis CBR05 for Tomato (*Solanum lycopersicum*) Fruits in South Korea as a Novel Plant Probiotic Bacterium (PPB): Implications from Total Phenolics, Flavonoids, and Carotenoids Content for Fruit Quality. *Agronomy* **2019**, *9*, 838. [CrossRef]
62. Caser, M.; Demasi, S.; Victorino, Í.M.M.; Donno, D.; Faccio, A.; Lumini, E.; Bianciotto, V.; Scariot, V. Arbuscular Mycorrhizal Fungi Modulate the Crop Performance and Metabolic Profile of Saffron in Soilless Cultivation. *Agronomy* **2019**, *9*, 232. [CrossRef]
63. Caser, M.; Victorino, Í.M.M.; Demasi, S.; Berruti, A.; Donno, D.; Lumini, E.; Bianciotto, V.; Scariot, V. Saffron Cultivation in Marginal Alpine Environments: How AMF Inoculation Modulates Yield and Bioactive Compounds. *Agronomy* **2018**, *9*, 12. [CrossRef]
64. Briglia, N.; Petrozza, A.; Hoeberichts, F.A.; Verhoef, N.; Povero, G. Investigating the Impact of Biostimulants on the Row Crops Corn and Soybean Using High-Efficiency Phenotyping and Next Generation Sequencing. *Agronomy* **2019**, *9*, 761. [CrossRef]
65. Giovannini, L.; Palla, M.; Agnolucci, M.; Avio, L.; Sbrana, C.; Turrini, A.; Giovannetti, M. Arbuscular Mycorrhizal Fungi and Associated Microbiota as Plant Biostimulants: Research Strategies for the Selection of the Best Performing Inocula. *Agronomy* **2020**, *10*, 106. [CrossRef]

agronomy

MDPI

Article

Saffron Cultivation in Marginal Alpine Environments: How AMF Inoculation Modulates Yield and Bioactive Compounds

Matteo Caser [1], Íris Marisa Maxaieie Victorino [2,3,4], Sonia Demasi [1], Andrea Berruti [2], Dario Donno [1], Erica Lumini [2], Valeria Bianciotto [2] and Valentina Scariot [1,*]

1 Department of Agricultural, Forest and Food Sciences, University of Torino, Largo Paolo Braccini 2, 10095 Grugliasco, Italy; matteo.caser@unito.it (M.C.); sonia.demasi@unito.it (S.D.); dario.donno@unito.it (D.D.)

2 Institute for Sustainable Plant Protection, National Research Council (CNR), Viale Mattioli 25, 10125 Torino, Italy; irisvictorino@gmail.com (Í.M.M.V.); berruti.andrea@gmail.com (A.B.); erica.lumini@unito.it (E.L.); valeria.bianciotto@ipsp.cnr.it (V.B.)

3 Department of Life Sciences and Systems Biology, University of Turin, Viale Mattioli 25, 10125 Torino, Italy

4 Biological Science Department, Science Faculty, Eduardo Mondlane University (UEM), Av. Julius Nyerere-Campus Universitário, Maputo 3453, Mozambique

* Correspondence: valentina.scariot@unito.it; Tel.: +39-011-670-8932

Received: 3 December 2018; Accepted: 30 December 2018; Published: 31 December 2018

Abstract: Arbuscular mycorrhizal fungi (AMF) establish mutualistic symbiotic associations with plant roots and act as biofertilizers by enhancing plant nutrient and water uptake. Moreover, AMF colonization may influence the biosynthesis of plant bioactive compounds in medicinal and aromatic plants. There is limited information on AMF associations with *Crocus sativus* L. (saffron) roots and their effect on crop performances and spice quality. In the present work we verified the efficiency of root mycorrhization in potted conditions, and then we evaluated the yield and quality of the saffron produced in two Alpine sites during two cultivation cycles with the application of AMF. Two inocula were applied, either a single-species (*Rhizophagus intraradices*) or a multispecies mixture (*R. intraradices* and *Funneliformis mosseae*). The trial conducted in potted conditions confirmed that both AMF commercial inocula established symbiotic relationships with saffron roots. The multispecies inoculation yielded the highest content of arbuscules in colonized portions of the root (100%), while the single-species was slightly less (82.9%) and no AMF were recorded in untreated control corms. In open-field conditions, AMF colonization of the root systems, flower production, and saffron yields were monitored, and bioactive compounds contents and antioxidant activity in the dried spice were analyzed using spectrophotometry and high performance liquid chromatography. Overall, the saffron produced was high quality (ISO category) and had high contents of bioactive compounds, with very high total polyphenol content and elevated antioxidant activity. The use of arbuscular mycorrhizal symbionts as biostimulants positively affected saffron cultivation, improving the crop performances and the content of important nutraceutical compounds. In particular, the inoculum composed by *R. intraradices* and *F. mosseae* increased flower production and the saffron yield. *R. intraradices* alone enhanced the spice antioxidant activity and the content of bioactive compounds such as picrocrocin, crocin II, and quercitrin. Since saffron is the world's highest priced spice, the increases in yield and quality obtained using AMF suggests that farms in marginal areas such as alpine sites can increase profitability by inoculating saffron fields with arbuscular mycorrhiza.

Keywords: *Crocus sativus* L.; biofertilization; arbuscular mycorrhizal fungi; antioxidant activity; crocin; picrocrocin; polyphenols; safranal

1. Introduction

Saffron (*Crocus sativus* L.) is a triploid herbaceous geophyte that is reproduced by means of replacement corms and is cultivated in environments with very different soil characteristics [1–3] for its red scarlet stigmas that are used worldwide as a spice and natural dye [4]. Origin, abiotic stresses, agronomical practices, and processing methods (stigma separation, drying, and storage) can influence both the plant and the saffron spice yield, composition, and quality [5,6]. The spice's organoleptic properties are ascribed to the relative percentage of peculiar secondary metabolites—crocin, picrocrocin, and safranal—which provide the unique color, bitter taste, and aroma, respectively. The concentrations of these constituents combine to determine the saffron spice quality, as defined by the International Organization for Standardization [7]. Studies related to saffron quality are expanding mainly due to the antioxidant properties of this spice and their positive influence on human health [8]. Antitumor and cancer-preventive properties are mainly attributed to the high carotenoids content [9].

Reproductive, vegetative, and dormancy are the main phenological stages [10]. Saffron flower induction is a very complicated mechanism directly related to ecological conditions and field management [11,12]. As in most geophyte plants, both seasonal and daily thermoperiodism are involved as the main environmental factors [11]. Flower induction requires an incubation of the corms at high temperature (23–27 °C), followed by a period of exposure at moderately low temperature (17 °C) for flower emergence. In Mediterranean environments, flower induction occurs from early spring to mid summer, while flower emergence occurs from early- to late-autumn. Differences in the time required for flower initiation have mostly been attributed to the corm size [13]. In addition, Molina et al. [14] reported that air and soil temperatures might be responsible for differential flower induction and duration of up to two months. Flowering is followed by a vegetative stage throughout the winter and formation of replacement corms at the base of shoots. At the end of spring, the leaves reach the highest length, start to senesce, and wither, and the bulbs go into dormancy [14].

Due to its unique biological, physiological, and agronomic traits, saffron is able to exploit marginal land and is included in low-input cropping systems, even if high amount of skilled labour is required [11]. In Italy, saffron cultivation is gaining increasing attention as an alternative crop for sustainable agriculture systems [11,15], where it could represent a valid mean for increasing incomes of multifunctional farms, with a positive impact on the recovery and economy of these areas [15,16]. Since saffron is the world's highest-priced spice due to the intensive hand labour required for daily flower picking and stigma separation [14], small increases in the yield and/or quality can connote a large increase in profitability. In this context, the adoption of sustainable cultivation techniques such as the use of biostimulants may represent further help in both the increase in spice yield and active ingredients accumulation [17].

Recent research has focused on the benefits of soil organisms to crops, especially to promote plant nutrient uptake and assimilation [18,19]. Indeed the soil is not only the location of plant life cycle stages, but also the main reservoir for a wide range of plant biostimulants (PBs), including arbuscular mycorrhizal fungi (AMF) [19–21]. Ubiquitous and abundant, AMF are obligate endosymbionts living inside most plant roots present in diverse environments, including productive agricultural systems [22–25]. When colonizing roots, hyphae extend root limits, improving water and inorganic nutrient acquisition from the soil, mainly phosphorus (P) and other minerals, in exchange for photosynthetic products. The use of AMF has a demonstrated economic impact on agriculture and horticulture and they may also confer pathogen protection by altering plant physiological parameters, and improving soil nutrition and aggregation under different growing conditions [26–28].

Mounting evidence indicates that AMF may induce changes in primary and secondary metabolism of host plants, increasing polyphenols, flavonoids, and phytohormone dynamics [29,30]. Such metabolic changes may be ascribed to a transient activation of host defence reactions in colonized roots [20,31]. The role of AMF symbiosis in flowering date and flower production is fragmented [32].

In medicinal and aromatic plants (MAPs), such as *Arnica montana* L., *Coriandrum sativum* L., and *Anethum graveolens* L., AMF colonization influenced bioactive compound biosynthesis such as ascorbic acid, flavonoids, polyphenols, carotenoids, and vitamins [33–36]. Inoculation with *Funneliformis mosseae* Gerd. & Trappe and *G. versiforme* P. Karst. improved plant growth and enhanced the glycyrrhizin concentration in *Glycyrrhiza uralensis* Fisch plants [26]. Moreover, under low P availability, a mix of AMF increased the production of root biomass and of pseudohypericin and hypericin content in flowers of *Hypericum perforatum* L. [32]. Although widely applied, evidence for AMF symbiosis efficacy and persistence is scant, incomplete, or lacking [37,38] and the use of AMF in crop production is facing some limitations due to product costs, producer awareness levels, and variability in mycorrhizal inoculum quality [21,27]. Many factors can affect the success of inoculation and AMF persistence, including environmental and cultivation conditions, species compatibility, degree of spatial competition with other soil organisms, and the time of inoculation. However, once AMF inoculation is restored and well established in soil, the AMF community will persist through time. If detrimental practices are minimized before and after cultivation, biodiverse mycorrhizal hyphal networks will remain unaltered and infective in the field [27]. Hence, it is important to assess the effects of AMF on crop traits both as early application and as residual persistence in the following crop cultivation seasons.

Incidence of AMF, alone or in combination with plant growth promoting bacteria (PGPB), was reported in corms of *C. sativus* [39–44]. Different authors report that well-established AMF colonization of saffron roots results in increased corm P content, chlorophyll, fresh and dry corm mass, and leaf matter, and greater soil P and nitrogen assimilation [43–45]. Shajari et al. [44] indicated a significant effect of AMF in corm growth and mineral assimilation during the second cultivation season, supporting their effective residual effects in saffron cultivation. However, little is known about the effects of AMF on spice yield, and phytochemical profiles in open field cultivation [46,47].

The possibility that AMF can enhance the economic value of saffron by increasing yield and quality is even more interesting if we consider the worldwide increase in use of biocompounds in the food and pharmaceutical industries. Thus, the aims of the present study were (1) to preliminarily verify the constitutive association of AMF with saffron roots in sterile pot conditions, and (2) to assess the AMF symbiosis in open field conditions and its effects on saffron plant growth, productivity, and bioactive compounds content in Alpine open field conditions.

2. Materials and Methods

2.1. AMF Inoculation in Pot

Saffron corms with horizontal diameters of 1.3 to 2.8 cm were sown in pots (4 L; 1 corm per pot) in the last ten days of August 2016. Pots were filled with sterile quartz sand (3 L per pot) on a layer of sterilized expanded clay (1 L per pot). Corms were treated with two inocula (MycAgro Lab, Breteniére, FR), one composed of a single fungus *Rhizophagus intraradices* (Ri) and one of *R. intraradices* and *Funneliformis mosseae* (Ri + Fm). Ten grams of each inoculum were placed under each corm in order to guarantee the contact between the inoculum and the roots and therefore to favor the symbiosis between AMF and roots. Saffron corms used as controls were not inoculated (AMF-). Corms were not treated against fungal pathogens. A randomized block design was used with a total of 48 pots displayed in two experimental plot units (24 pots per unit) and three treatments (8 pots per treatment). Cultivation lasted for one cycle (August 2016–April 2017) in a heated glasshouse of the Department of Agricultural Forest and Food Sciences (DISAFA) of the University of Torino (Italy, 45°06′23.21″ N Lat, 7°57′82.83″ E Long; 293 m a.s.l.), with an average temperature of 22 °C during the day and 16 °C in the night. Irrigation water (pH 7.4, EC 505 µS cm) was added weekly (250 mL per pot) with a drip system. The corms were fertilized by fertigation (VIGORFLOR, AL.FE. srl, MN, Italy) every two weeks starting from the emergence of the spate, in quantities of 1.5 g L^{-1} of water. No flowering occurred because of the small size of the corms.

2.2. AMF Inoculation in Open Field

Saffron corms with horizontal diameters of 2.5 to 3.5 cm were planted in the last ten days of August 2016 in two Alpine experimental sites located in the municipality of Morgex (45°45′35.1″ N; 7°02′37.3″ E; 1000 m a.s.l.) and Saint Cristophe (45°45′06.9″ N; 7°20′37.0″ E; 700 m a.s.l.) in Italy and cultivation lasted for two cycles (2016–2017 and 2017–2018). Both sites were cultivated with saffron for at least the previous three years. Before starting the experiment both fields were milled. To assess the effects of AMF inocula on saffron cultivation and production, the same treatments used in the pot trial were applied (Ri, Ri + Fm or AMF-). A randomized block design was used, with three experimental plot units (blocks). Each plot unit consisted of 56 corms, planted in a 1.44 m^2 area (39 corms m^{-2}). Inter-row planting distance was of 7 cm, while between-row distance was 25 cm. Plots were separated from each other with at least 4 m distance. Before planting, 10 g of inoculum was placed under the corms to ensure contact between plant and the treatment. Irrigation was provided when needed and hand weeding control was conducted during cultivation, while no preplanting fertilization, tillage, or treatments against pathogens were applied. The two Alpine sites were characterized by semicontinental climate, with a long and cold winter (Supplementary Figure S1). In general, both sites had a sandy-loam texture according to the USDA classification and similar chemical characteristics (Supplementary Table S1).

2.3. AMF Evaluation

At the end of the vegetative phase in both pot (February 2017) and open field experiments (April 2017 and 2018), saffron roots were harvested, rid of topsoil, cleaned and stained with 0.1% (*w/v*) cotton blue in 80% lactic acid overnight, then destained 3 times with lactic acid for 18 h, cut into 1-cm-long segments and placed on microscope slides for further morphological analysis. Approximately 25 fragments were observed under light microscope for each replicate for a total of 300 root fragments. Fungal colonization was determined and calculated as described by Trouvelot et al. [48].

2.4. Plant Performance and Saffron Yield in Open Field

The daily number of picked flowers per corm (Supplementary Figure S2) and the yield of spice (i.e., stigmas dried at 40 °C for 8 h in an oven) were measured at flowering (November 2016 and 2017). When leaves were fully expanded (April 2017 and 2018), 50 mg of fresh leaves per treatment were used to determine chlorophyll and carotenoids content as described by Caser et al. [49]. Simultaneous with leaf sampling, the Chlorophyll Meter SPAD-502 (Konica Minolta Sensing Inc., Osaka, Japan) was used to determine the relative quantity of chlorophyll present in 27 randomly selected plants per treatment in the field.

At the end of full plant development (April 2017 and 2018), the leaves length of all corms was measured. Then, 27 plants per treatment were lifted, and corms rid of topsoil, cleaned, and detunicated. The wilted rate as the ratio between the number of wilted corms and the total number of sown corms, the shoot caliber size, and the number, the size and the weight of replacement corms were determined.

2.5. Saffron Extract Preparation and Quality

The saffron aqueous extracts were prepared according to Gresta et al. [11]. Fifty mg of powdered saffron from each treatment and both cultivation years were put into 5 mL of deionized water. After stirring for 1 h at room temperature (circa 21 °C) in the dark, the solution was filtered with polytetrafluoroethylene (PTFE, VWR international, Milano, Italy) filters of 25 mm diameter and 0.45 μm pore size. The saffron extract obtained was diluted 1:10 with deionized water (1 mg mL^{-1}). Saffron extracts were analyzed with a spectrophotometer (Ultrospec 2100 Pro, GE Healthcare, UK Ltd., Little Chalfont, Buckinghamshire, UK) to determine the amount of picrocrocin, crocin, and safranal, according to ISO 3632 [7].

2.6. Total Phenols

The content of total phenols (TPC) was measured by using the Folin–Ciocalteu's phenolic method and determined as reported by Donno et al. [50]. Five hundred µl of saffron extract was added and mixed with 30 mL of deionized water, 2.5 mL of Folin–Ciocalteu's reagent (diluted 1:10), and, after eight minutes, 10 mL of 7.5% (*w/v*) saturated sodium carbonate solution. The solution was incubated at room temperature for 2 h in the dark and the absorbance was detected at 765 nm with a spectrophotometer (Ultrospec 2100 Pro, GE Healthcare, UK Ltd., Little Chalfont, Buckinghamshire,UK). The results were expressed as mg of gallic acid equivalents (GAE) per 100 g of fresh weight (FW).

2.7. Total Anthocyanins

The total anthocyanins content (TAC) was determined using the pH-differential method [50]. Saffron extracts were added to a pH 1 and pH 4.5 buffer solutions. Absorbance of samples was determined at 515 nm and 700 nm after a 15 min equilibration. The formula for calculating TAC is as follows

$$\text{TAC (mg L}^{-1}) = (\text{A} \times \text{sample dilution factor} \times 1000)/(\text{molar absorptivity} \times 1) \quad (1)$$

where A is (Absorbance 515 nm—Absorbance 700 nm) at pH 1.0—(Absorbance 515 nm Absorbance 700 nm) at pH 4.5. The results were expressed as milligrams of cyanidin 3-O-glucoside (C3G) per 100 gof fresh weight (mg of C3G 100 g^{-1} FW).

2.8. Antioxidant Activity

The antioxidant activity (AOA) was determined using the ferric reducing antioxidant power (FRAP) method as reported by Caser et al. [51] and the 2′-azinobis (3-ethylbenzothiazoline-6-sulfonic acid) (ABTS) method as described by Urbani et al. [52].

For the FRAP method, a total of 30 µL of saffron extract was added and mixed with 90 µL of deionized water and 900 µL of the FRAP reagent. After incubation at 37 °C for 30 min, the absorbance of the solutions was measured at 595 nm using a spectrophotometer (Ultrospec 2100 Pro, GE Healthcare, UK Ltd., Little Chalfont, Buckinghamshire, UK). Results were expressed as millimoles of ferrous iron (Fe^{2+}) equivalents per kilogram of fresh weight.

The ABTS radical cation (ABTS +) was obtained by the reaction of 7.0 mM ABTS stock solution with 2.45 mM potassium persulfate solution. After the incubation for 12–16 h before use in the dark and at room temperature, the solution was diluted with distilled water to obtain an absorbance of 0.70 (±0.02) at 734 nm. After addition of 0.6 mL of diluted ABTS·+ solution to 180 µL of saffron extract, the reaction was left in the dark at room temperature for six min. The absorbance was then measured at 734 nm using a spectrophotometer (Ultrospec 2100 Pro, GE Healthcare, UK Ltd., Little Chalfont, Buckinghamshire, UK). The antioxidant activity was expressed as µmol of Trolox equivalents per gram of dry weight (µmol TE g^{-1} DW).

All analyses were performed in three replicates.

2.9. Identification and Quantification of Bioactive Compounds

The chromatographic analysis of saffron extracts (Supplementary Table S2) was conducted with an Agilent 1200 high-performance liquid chromatograph coupled to a diode array detector (HPLC-DAD; Agilent Technologies, Santa Clara, CA, USA), according to established methods [53]. Different chromatographic methods were used for analysis: benzoic acids (ellagic and gallic acids), catechins ((+) catechin and (−) epicatechin), cinnamic acids (caffeic, chlorogenic, coumaric, and ferulic acids), flavonols (hyperoside, isoquercitrin, quercetin, quercitrin, and rutin), carotenoids (crocin I and II and safranal), and vitamin C (ascorbic + dehydroascorbic acids).

Four chromatographic methods were used to separate the bioactive molecules on a Kinetex C18 column (4.6 × 150 mm, 5 µm, Phenomenex, Torrance, CA, USA). Several mobile phases were used for bioactive compound identification and ultraviolet (UV) spectra were recorded at different wavelengths, based on HPLC methods, previously tested and validated [4], with some modifications. UV spectra were recorded at 330 nm (α), 280 nm (β), 310 and 441 nm (χ), and 261 and 348 nm (δ).

All single compounds were identified in samples by comparison and combination of their retention times and UV spectra with those of authentic standards analyzed with the same chromatographic conditions.

2.10. Chemicals and Reagents

All the chemicals and reagents used for the AMF evaluation, phenols, anthocyanins, FRAP, and ABTS assays and bioactive quantification were purchased from Sigma-Aldrich (Saint Louis, MO, USA).

2.11. Statistical Analysis

Arcsin transformation was performed on all percent incidence data before statistical analysis in order to improve homogeneity of variance (Levene test). All the analyzed data were checked for normality of variance. For all indices analyzed in the greenhouse assay, mean differences were computed using a one-way analysis of variance (ANOVA) with Tukey's post hoc test ($p \leq 0.05$). Data from open field were analyzed by means of a linear mixed effect models considering AMF treatments as a fixed factor, year as a repeated measure, and sites and blocks as random factors. The following interactions (year × AMF treatment) were included in the model. Pairwise comparisons (according to sequential Bonferroni post hoc tests) were used to separate means when a treatment was significantly affecting the variable at a $p \leq 0.05$. All presented values are means of untransformed data. All computations were conducted with SPSS statistical package (version 25.0; SPSS Inc., Chicago, IL, USA).

3. Results

3.1. Assessment of Saffron Mycorrhization at Pot and Open Field Scale

Values concerning intensity of colonization in the root system and abundance of arbuscules or coils in the saffron roots in potted conditions are shown in Table 1. Corms treated with Ri + Fm reached the highest level of mycorrhization (M%) (93.33%), however, high levels were also obtained with Ri inoculum (71.37%). The Ri + Fm treatment also had the highest occurrence of arbuscules (a%) in the mycorrhizal portions (100%), being significantly higher than Ri (82.99%) and AMF-(0%).

Table 1. AMF colonization indices (intensity in the whole root system, M; intensity of the mycorrhizal portions, m; presence of arbuscules in the whole root system, A; presence of arbuscules in the mycorrhizal portions, a) of *Crocus sativus* L. roots treated with the inoculum composed by *Rhizophagus intraradices* and *Funneliformis mosseae* (Ri + Fm), *R. intraradices* alone (Ri), or the control (AMF-) in the saffron pot cultivation.

	Index (%)			
Treatment	**M**	**m**	**A**	**a**
Ri + Fm	93.33 a	93.33 a	93.33 a	100.00 a
Ri	71.37 b	80.28 b	58.98 b	82.99 b
AMF-	0.07 c	0.33 c	0.00 c	0.00 c
p	***	***	***	***

Mean values with the same letter are not statistically different at $p \leq 0.05$ according to Tukey's post-hoc tests. The statistical relevance is provided (*** $p < 0.001$).

In open field conditions, the AMF root colonization measurements in *C. sativus* treated with Ri + Fm or with Ri alone during the two cultivation cycles are presented in Table 2. In general both the presence of arbuscules in the mycorrhizal portions (a%) and in the whole root system (A%) indices were affected by the inoculum composition only in the first cultivation year, while control plants

(AMF-) were not colonized. In the second year, low root colonization was observed and no differences among the treated and untreated corms were detected.

Table 2. AMF colonization intensity in open field conditions after the first and second cultivation year of the whole root system (M) and of the mycorrhizal portions (m), and presence of arbuscules in the whole root system (A) and in the mycorrhizal portions (a) of *Crocus sativus* roots treated with inoculum composed of *Rhizophagus intraradices* and *Funneliformis mosseae* (Ri + Fm), *R. intraradices* alone (Ri), or the control (AMF-).

Effect	Index (%)			
Year 1	**M**	m	**A**	a
Ri + Fm	11.6 a	11.7	4.0 a	26.6 a
Ri	13.8 a	14.2	6.9 a	38.1 a
AMF-	1.7 b	3.4	0.0 b	0.0 b
p	*	ns	***	***
Year 2				
Ri + Fm	7.0	8.5	0.8	12.5
Ri	16.1	16.5	1.6	8.31
AMF-	4.73	6.1	2.5	18.8
p	ns	ns	ns	ns
Year × Treatment (*p*)	*	ns	*	*

Values with the same letter denote no significant differences. The statistical relevance is provided (ns, not significant; * $p < 0.05$; *** $p < 0.001$).

3.2. *Impact of AMF Symbiosis on Saffron Productivity and Qualitative Traits in Open Field*

Significant differences between the two cultivation years emerged for several studied parameters. In general, the wilting rate, all the main productivity traits (number of flowers m^{-2}, number of flowers per corm, mg of saffron m^{-2}, saffron per flower, and the number of replacement corms), and the content of leaf chlorophyll and carotenoids significantly increased after the second year of cultivation (Table 3); a reduction in leaf length, SPAD unit, and shoot size was also observed.

Table 3. Effects of cultivation seasons (Year 1 and Year 2), AMF treatments (Ri + Fm was composed of *Rhizophagus intraradices* and *Funneliformis mosseae*, Ri of *R. intraradices* alone, and AMF-was the uninoculated control), and their interaction (Year × AMF treatment) on saffron plant growth and productivity based on linear mixed-effects models considering AMF treatments as a fixed factor, year as a repeated measure, and sites and blocks as random factors.

	Growing Seasons			**AMF Treatments**				**Year × AMF**
Traits	**Year 1**	**Year 2**	*p*	**Ri + Fm**	**Ri**	**AMF-**	*p*	*p*
Wilting rate (%)	39.0	54.4	***	44.3	50.0	45.8	ns	ns
Flower (n m^{-2})	49.8	101.7	***	91.8 a	61.9 b	66.4 b	*	*
Flower/corm (n)	1.5	4.2	***	5.1 a	3.8 b	3.9 b	*	*
Saffron yield (mg m^{-2})	278.0	700.0	***	645.3 a	377.4 b	477.2 b	*	*
Saffron/flower (mg)	6.0	7.0	**	7.1 a	5.8 b	7.3 a	*	*
Leaf length (cm)	36.8	24.1	***	31.4	30.3	29.9	ns	ns
SPAD unit	74.8	45.7	***	60.0	61.1	59.7	ns	ns
Shoot size (mm)	5.3	4.1	**	5.5 a	3.3 b	4.2 ab	**	*
Corm size (mm)	21.1	20.2	ns	19.8	20.0	22.2	ns	ns
Replacement corm (n)	2.2	3.7	*	2.8	3.4	2.7	ns	ns
Corm weight (g)	7.7	6.5	ns	7.8	7.4	6.3	ns	ns
Chlorophyll (μg mg^{-1})	1.6	4.1	***	2.9	2.9	2.7	ns	ns
Carotenoids (μg mg^{-1})	0.6	2.2	***	1.4	1.5	1.4	ns	ns

Values with the same letter denote no significant differences. The statistical relevance is provided (ns, not significant; * $p < 0.05$; ** $p < 0.01$; *** $p < 0.001$).

In Table 3, the productivity and growth traits influenced by the AMF treatments are also reported. Particularly, the inoculum composed of the mix of *R. intraradices* and *F. mosseae* significantly increased the number of flowers m^{-2} (+138.2%), the number of flowers $corm^{-1}$ (+130.8%), and the mg of saffron m^{-2} (+135.2%) in comparison to other treatments. In contrast, the mg of saffron $flower^{-1}$ and the shoot size were significantly reduced (−20% and −40%, respectively) by the inoculum of *R. intraradices* alone in comparison to Ri + Fm and AMF-. Significant interaction between cultivation year and AMF treatments resulted for the number of flowers m^{-2}, the number of flowers $corm^{-1}$, saffron yield, saffron $flower^{-1}$, and shoot size.

Regarding the synthesis of bioactive molecules in the studied saffron spice, differences between the two cultivation seasons occurred (Table 4). Overall, the saffron produced at the two experimental sites belonged to the quality category I for the picrocrocin, safranal, and crocins analysis [7] with a significant increase after the second cultivation year. On the contrary, different bioactive compounds (isoquercitirin, quercitrin, ellagic acid, safranal, and total vitamin C) were significantly reduced. Very few differences were observed among AMF treatments (Table 4). Both Ri + Fm and Ri positively affected the antioxidant activity (FRAP assay) of the saffron produced. While, the effect of the Ri inoculum significantly increased the absorbance value of picrocrocin (ISO 3632) and the content of quercitrin in comparison to Ri + Fm, and the content of crocin II compared to AMF- (Table 4). A significant interaction between cultivation seasons and AMF treatments resulted for picrocrocin (ISO), quercitrin, crocin II, and antioxidant activity (FRAP assay).

Table 4. Effects of cultivation seasons (Year 1 and Year 2), AMF treatments (Ri + Fm was composed of *Rhizophagus intraradices* and *Funneliformis mosseae*, Ri of *R. intraradices* alone, and AMF- was the uninoculated control), and their interaction (year × AMF treatment) on bioactive compounds, total polyphenol content (TPC), anthocyanins, quality traits as defined by ISO 3632 [7], and antioxidant activity of the produced saffron based on liner mixed-effects models considering AMF treatments as a fixed factor, year as a repeated measure, and sites and blocks as random factors.

	Growing Seasons			AMF Treatments				Year × AMF
Traits	Year 1	Year 2	*p*	Ri + Fm	Ri	AMF-	*p*	*p*
ISO 3632 [7]								
Picrocrocin	131.4	135.0	*	130.2 b	138.7 a	136.1 a	*	*
Safranal	38.8	44.2	**	39.9	43.8	40.8	ns	ns
Crocins	207.1	368.5	***	275.9	303.5	284.1	ns	ns
Bioactive compounds (mg 100 g^{-1} dry weight)								
Coumaric acid	23.6	23.5	ns	23.6	23.5	23.7	ns	ns
Isoquercitrin	2.6	2.5	*	2.5	2.5	2.6	ns	ns
Quercitrin	22.8	16.0	***	17.0 b	22.3 a	18.9 ab	*	*
Gallic acid	5.0	4.9	ns	4.9	4.9	5.1	ns	ns
Ellagic acid	2.7	0.8	***	2.0	2.1	1.3	ns	ns
Catechin	3.4	3.1	ns	2.7	3.0	4.3	ns	ns
Epicatechin	6.1	8.3	ns	6.4	6.3	9.0	ns	ns
Safranal	4.4	4.0	***	4.2	4.3	4.2	ns	ns
Crocin I	32.5	67.9	**	49.2	37.7	63.8	ns	ns
Crocin II	31.1	36.6	*	35.0 ab	38.8 a	27.7 b	*	*
Total Vitamin C	76.5	67.0	**	71.4	70.1	73.2	ns	ns
TPC (mg_{GAE} 100 g^{-1} DW)	1340.7	2355.5	ns	1906.1	1868.8	1819.5	ns	ns
Antocyanins (mg_{C3G} 100 g^{-1} DW)	1866.5	1633.6	ns	964.1	2418.8	1867.3	ns	ns
Antioxidant activity								
FRAP (mmol Fe^{2+} kg^{-1})	408.9	1937.1	***	424.8 a	463.8 a	338.2 b	***	***
ABTS (μmol_{TE} g^{-1})	4.2	4.6	ns	4.3	4.5	4.6	ns	ns

Values with the same letter denote no significant differences. The statistical relevance is provided (ns, not significant; * $p < 0.05$; ** $p < 0.01$; *** $p < 0.001$).

4. Discussion

4.1. AMF Colonization

In the literature only some studies report AMF colonization of *C. sativus* roots. In the present study, their presence in potted cultivation was detected in *C. sativus* roots subjected to both AMF treatments (Ri + Fm and Ri). Saffron root fragments showed extensive AM fungal colonization, characterized by a moderate to high intensity of colonization and arbuscule formation. Saffron root colonization in the present pot cultivation trial was markedly superior to the results obtained in the open field test. This could be due to the antagonistic action of the naturally occurring fungi in the soil that compete with the AMF and by the different cultivation substrate used. However, to the best of our knowledge, this is the first report clearly indicating and measuring successful symbiosis between *C. sativus* roots and AMF under pot cultivation conditions.

Our open field data are equal or lower than findings obtained in other open field trials as reported by Aimo et al. [40] and Lone et al. [43]. Applying the percent colonization method for root AMF evaluation, these authors reached a maximum of 30% and 60% mycorrhizal colonization in saffron roots in Italy and Kashmir cultivation fields when using a mix of AMF belonging to the genus Glomus, respectively. In a similar study conducted in Iranian fields, the percentage of root colonization of saffron was of 39% [42], while in a field in Kashmir, ranged between 15 and 90% on the basis of the season [43]. As reported in Supplementary Table S1, P Olsen values measured at the experimental sites are high (>69.2 mg kg^{-1}), indicating the potential for a detrimental effect of P on AMF colonization in our experiment. As the cost of the symbiosis to the plant outweighs the benefit of access to P via the fungal pathway, plants reduce fungal access to carbohydrate [54]. Similar data were reported also in other species such as *Zea mays* L. in which the AMF root colonization was reduced with a soil P content of 90 mg kg^{-1} [55]. In other geophyte plants, such as *Allium tricoccum* Aiton., a low level of AMF symbiosis was observed in the absence of leaves and photosynthetic activity. However, once leaves elongate in early spring, root colonization increases rapidly. This is similar to the pattern of *Maianthemum racemosum* L., where AMF colonization peaked during vegetative growth [56]. Here, AMF sampling was performed during maximum leaf elongation, and therefore, the detection of low colonization is likely more related to soil characteristics than to other physiological or biochemical parameters.

Taken together, all these findings indicated that under open field conditions in alpine environments, AMF colonization was substantially lower than under pot conditions as already indicated in literature. This is in agreement with the meta-analysis of Berruti et al. [27], in which successful outcomes of AMF inoculation were more often found in controlled (greenhouse and growth chamber) conditions. In this condition, environmental extremes and variation are minimized or absent [38]. Moreover, one of the most important confounding factors in pot or field experiments is the effect of root temperature on the AMF growth [57]. The higher temperatures typical of greenhouse conditions favor greater growth and superior colonization by AMF [58].

4.2. AMF Modulate Crop Performance and Spice Quality

Flower yield is a difficult parameter to forecast in saffron since it is influenced by a combination of agronomic, biological, and environmental factors [11]. Generally, a saffron field may produce from 200 to 3000 mg m^{-2} of spice, depending on the cultivation factors [11] and obviously, by the planting density, which may vary considerably. By planting at a 55 corms m^{-2} density in southern Italy (Sicily), Gresta et al. [3] obtained more than 1200 mg m^{-2}. In the area of Navelli (central Italy) [59], with a similar corm density, the average yield ranged between 1000 and 1600 mg m^{-2}. In Iranian fields with a density of 150 and 100 corms m^{-2}, Mollafilabi et al. [60] and Koocheki et al. [61] obtained 740 and 370 mg m^{-2} of saffron, respectively. Recently, the path coefficient analysis conducted by Bayat et al. [62] highlighted that fresh stigma weight, flower number, dry stigma and flower weight, leaf size, and number and size of replacement corms have the highest positive correlation with saffron yield.

Arbuscular mycorrhizal fungi are known to be beneficial to several important plants, including some medicinal plants [30]. Unfortunately, very scarce reports of the effective role of AMF in saffron yield are available. Only, Aimo et al. [40] indicate an increase in flower production m^{-2} (equal to 68%, compared to control) using a mix of AMF species belonging to the genus Glomus. Our results are generally more supportive of the benefits of AMF inoculation with an increase of flower production m^{-2} of circa 140%. Taken together, these findings suggest a beneficial effect of AMF inoculation with a mixture of *R. intraradices* and *F. mosseae* on saffron yield performance.

Few spices are able to provide the combination of color, taste, and aroma to the foods and possess several nutraceutical properties for human health as saffron. Most of the beneficial effects of saffron, recognized since ancient times, are due mainly to its total phenolic content (TPC) and antioxidant activity (FRAP and ABTS assays). *R. intraradices* alone was found to induce an increase in secondary metabolite contents, such as terpenes and phenolics, in *Salvia officinalis* L. [63] and *Echinacea purpurea* L. [64]. Overall, the saffron produced in the studied alpine areas had very high TPC (ranges between 1340.7 and 2355.5 mg GAE 100 g^{-1} DW), which was more than saffron cultivated in different areas of Lebanon (160 mgGAE 100 g^{-1} DW) [65], and is much greater when compared with other common food additives and spices, such as *Eugenia caryophylate* (Thunb.), *Lavandula* spp., *Curcuma domestica* Val, and *Curcuma longa* L. (0.26, 0.22, 23, and 36 mg GAE 100 g^{-1} DW, respectively) (Table 4) [66,67]. Results of ABTS and FRAP assays also demonstrated elevated antioxidant activity (Table 4). ABTS assay values were comparable to what was found in Greek saffron by Ordoudi et al. [68]. FRAP assay values (between 408.9 and 1937.1 mmol Fe^{2+} kg^{-1}) were generally higher in comparison to the Iranian samples (circa 570 mmol Fe^{2+} kg^{-1}) analyzed by Karimi et al. [69]. The saffron produced in the west Italian Alps also had different bioactive compounds (Table 4) known for their health-promoting activity, that is, cinnamic acids, flavonols, benzoic acids, catechins, and carotenoids [50]. Other studies report that water-soluble carotenoids such as crocins have antioxidant effects superior to α-tocopherol [67]. It was recently observed in a clinical study that high crocin I and crocin II contents (4000 and 1000 mg, respectively) inhibit β-amyloid and tau aggregation [70]. Apart from crocins, Asdaq and Inamdar [71] suggest that flavonols are responsible for the synergistic antihyperlipidemic and antioxidant potential of saffron. Amin et al. [72] indicated that a concentration of 1 mg of safranal attenuated the behavioral symptoms of neuropathic pain. Our data indicate that the saffron produced presented high crocin II content (27.7–38.8 mg 100 g^{-1} DW), almost in line with the saffron produced in Sardinia (Italy, DOP Zafferano di Sardegna) [73], while also presenting a higher content of gallic acid compared to what was found in Iranian and Greek saffron (2 mg and 1.2 100 g^{-1} DW) by Karimi et al. [69] and Proestos et al. [74], respectively. Thus, the saffron obtained could be of particular interest for its elevated antioxidant properties.

5. Conclusions

Saffron quality may vary greatly by site on the basis of several factors, among which are climatic conditions and cultivation techniques. We hereby provide data indicating the production of high quality saffron in marginal alpine areas, thus confirming that this crop is a strategic resource and good alternative for mountainous areas building multifunctional economies. Besides the phytochemical profile highlighted, the crop had many bioactive compounds. The use of arbuscular mycorrhizal symbionts as biostimulants positively affected saffron cultivation, mainly by increasing crop productivity, and partially by increasing the content of important nutraceutical compounds. Specifically, the inoculum composed by *R. intraradices* and *F. mosseae* was particularly effective in increasing flower production and saffron yield, while *R. intraradices* alone increased the content of some bioactive compounds—picrocrocin, quercitrin, crocin II—as well as antioxidant activity. Since saffron is the world's highest priced spice, the increases in yield and quality obtained using AMF should allow for an increase in profitability.

Furthermore, a new perspective can be envisaged. Since AMF symbiosis was more effective under soilless pot cultivation, this system may be a valuable alternative for saffron production and further work is underway to assess the potential of AMF inocula in saffron soilless cultivation.

Supplementary Materials: The following are available online at http://www.mdpi.com/2073-4395/9/1/12/s1, Figure S1: Climatic conditions of the Alpine experimental sites, Figure S2: Effects of AMF inoculum composed by *Rhizophagus intraradices* and *Funneliformis mosseae* (Ri + Fm), *R. intraradices* alone (Ri), or control (AMF-) on flower production m-2 during the first (a) and second (b) cultivation cycle, Table S1: Physical and chemical properties of the soils collected in the three saffron experimental fields located in the municipality of Saint Cristophe and Morgex (north west Italy), Table S2: Characteristics of the HPLC methods applied to analyse the bioactive compounds present in the studied saffron samples.

Author Contributions: M.C., A.B., E.L., V.B., and V.S. contributed to the experimental design. M.C., Í.M.M.V., S.D., D.D., A.B., and V.S. acquired and interpreted data. M.C. drafted the manuscript. V.B. and V.S. conceived, coordinated the work, and critically revised the manuscript.

Funding: This research was funded by the project titled 'SaffronALP—Lo zafferano di montagna: tecniche sostenibili per una produzione di qualità'—Fondazione Cassa di Risparmio di Torino (RF = 2017.1966) and by the program Interreg V-A Francia Italia Alcotra "Attività innovative per lo sviluppo della filiera transfrontaliera del fiore edule—Antea' n. 1139.

Acknowledgments: The authors acknowledge Dario Sacco for statistical assistance, Alessandro Putzolu for technical assistance, and Azienda Agricola La Branche di Diego Bovard and Azienda Agricola Rosset for providing plant material and hosting cultivation.

Conflicts of Interest: The authors declare no conflict of interest.

References

1. Molina, R.; García-Luis, A.; Coll, V.; Ferrer, C.; Valero, M.; Navarro, Y.; Guardiola, J. Flower formation in the saffron crocus (*Crocus sativus* L.). The role of temperature. *Acta Hortic.* **2004**, *650*, 39–47. [CrossRef]
2. Mollafilabi, A. *Production and Modern Agronomic of Saffron*; Scientific and Industrial Research Organization of Khorasan: Mashhad, Iran, 2004; pp. 22–35.
3. Gresta, F.; Santonoceto, C.; Avola, G. Crop rotation as an effective strategy for saffron (*Crocus sativus* L.) cultivation. *Sci. Hortic.* **2016**, *211*, 34–39. [CrossRef]
4. D'Archivio, A.; Giannitto, A.; Maggi, M.; Ruggieri, F. Geographical classification of Italian saffron (*Crocus sativus* L.) based on chemical constituents determined by high-performance liquid-chromatography and by using linear discriminant analysis. *Food Chem.* **2016**, *212*, 110–116. [CrossRef] [PubMed]
5. Carmona, M.; Zalacain, A.; Pardo, J.; López, E.; Alvarruiz, A.; Alonso, G. Influence of different drying and aging conditions on saffron constituents. *J. Agric. Food Chem.* **2005**, *18*, 3974–3979. [CrossRef] [PubMed]
6. Tong, Y.; Yan, Y.; Zhu, X.; Liu, R.; Gong, F.; Zhang, L.; Wang, P. Simultaneous quantification of crocetin esters and picrocrocin changes in Chinese saffron by high-performance liquid chromatography-diode array detector during 15 years of storage. *Pharmacogn. Mag.* **2015**, *11*, 540–545. [PubMed]
7. ISO 3632-1:2011. *Spices—Saffron (Crocus sativus L.). Food Products SC 7, Spices, Culinary Herbs and Condiments*; International Organization for Standardization: Geneva, Switzerland, 2011.
8. Shahi, T.; Assadpour, E.; Jafari, S. Main chemical compounds and pharmacological activities of stigmas and tepals of 'red gold'; saffron. *Trends Food Sci. Technol.* **2016**, *58*, 69–78. [CrossRef]
9. Bagur, M.; Salinas, G.; Jiménez-Monreal, A.; Chaouqi, S.; Llorens, S.; Martinez-Tomé, M.; Alonso, G. Saffron: An Old Medicinal Plant and a Potential Novel Functional Food. *Molecules* **2018**, *23*, 30. [CrossRef]
10. De Juan, J.A.; Córcoles, H.L.; Muñoz, R.M.; Picornell, M.R. Yield and yield components of saffron under different cropping systems. *Ind. Crops Prod.* **2009**, *30*, 212–219. [CrossRef]
11. Gresta, F.; Lombardo, G.; Siracusa, L.; Ruberto, G. Saffron an alternative crop for sustainable agricultural systems. A review. *Agron. Sustain. Dev.* **2008**, *28*, 95–112. [CrossRef]
12. Yarami, N.; Kamgar-Haghighi, A.A.; Sepaskhah, A.R.; Zand-Parsa, S. Determination of the potential evapotranspiration and crop coefficient for saffron using a water-balance lysimeter. *Arch. Agron. Soil Sci.* **2011**, *57*, 727–740. [CrossRef]
13. Maggio, A.; Raimondi, G.; Martino, A.; De Pascale, S. Soilless cultivation of saffron in mediterranean environment. *Acta Hortic.* **2006**, *718*, 515–522. [CrossRef]

14. Molina, R.; Valero, M.; Navarro, Y.; Guardiola, J.; García-Luis, A. Temperature effects on flower formation in saffron (*Crocus sativus* L.). *Sci. Hortic.* **2005**, *103*, 361–379. [CrossRef]
15. Giorgi, A.; Pentimalli, D.; Giupponi, L.; Panseri, S. Quality traits of saffron (*Crocus sativus* L.) produced in the Italian Alps. *Open Agric.* **2017**, *2*, 52–57. [CrossRef]
16. Manzo, A.; Panseri, S.; Bertoni, D.; Giorgi, A. Economic and qualitative traits of Italian Alps saffron. *J. Mt. Sci.* **2015**, *12*, 1542–1550. [CrossRef]
17. De Pascale, S.; Rouphael, Y.; Colla, G. Plant biostimulants: Innovative tool for enhancing plant nutrition in organic farming. *Eur. J. Hortic. Sci.* **2017**, *82*, 277–285. [CrossRef]
18. De Ponti, T.; Rijk, B.; Van Ittersum, M. The crop yield gap between organic and conventional agriculture. *Agric. Syst.* **2012**, *108*, 1–9. [CrossRef]
19. Colla, G.; Nardi, S.; Cardarelli, M.; Ertani, A.; Lucini, L.; Canaguier, R.; Rouphael, Y. Protein hydrolysates as biostimulants in horticulture. *Sci. Hortic.* **2015**, *196*, 28–38. [CrossRef]
20. Rouphael, Y.; Cardarelli, M.; Colla, G. Role of arbuscular mycorrhizal fungi in alleviating the adverse effects of acidity and aluminium toxicity in zucchini squash. *Sci. Hortic.* **2015**, *188*, 97–105. [CrossRef]
21. Mercy, L.; Lucic-Mercy, E.; Nogales, A.; Poghosyan, A.; Schneider, C.; Arnholdt-Schmitt, B. A Functional Approach towards Understanding the Role of the Mitochondrial Respiratory Chain in an Endomycorrhizal Symbiosis. *Front. Plant Sci.* **2017**, *8*, 417. [CrossRef]
22. Błaszkowski, J.; Kovács, G.; Balázs, T.; Orłowska, E.; Sadravi, M.; Wubet, T.; Buscot, F. *Glomus africanum* and *G. iranicum*, two new species of arbuscular mycorrhizal fungi (Glomeromycota). *Mycologia* **2010**, *102*, 1450–1462. [CrossRef]
23. Johnson, N.; Gehring, C. Mycorrhizas: Symbiotic mediators of rhizosphere and ecosystem processes. *Rhizosphere* **2007**, 73–100. [CrossRef]
24. Berruti, A.; Borriello, R.; Lumini, E.; Scariot, V.; Bianciotto, V.; Balestrini, R. Application of laser microdissection to identify the mycorrhizal fungi that establish arbuscules inside root cells. *Front. Plant Sci.* **2013**, *4*, 135. [CrossRef] [PubMed]
25. Borriello, R.; Berruti, A.; Lumini, E.; Della Beffa, M.T.; Scariot, V.; Bianciotto, V. Edaphic factors trigger diverse AM fungal communities associated to exotic camellias in closely located Lake Maggiore (Italy) sites. *Mycorrhiza* **2015**, *25*, 253–265. [CrossRef] [PubMed]
26. Liu, J.; Wu, L.; Wei, S.; Xiao, X.; Su, C.; Jiang, P.; Song, Z.; Wang, T.; Yu, Z. Effects of arbuscular mycorrhizal fungi on the growth, nutrient uptake and glycyrrhizin production of licorice (*Glycyrrhiza uralensis* Fisch). *Plant Growth Regul.* **2007**, *52*, 29–39. [CrossRef]
27. Berruti, A.; Lumini, E.; Balestrini, R.; Bianciotto, V. Arbuscular mycorrhizal fungi as natural biofertilizers: Let's benefit from past successes. *Front. Microbiol.* **2016**, *6*, 1559. [CrossRef] [PubMed]
28. Bianciotto, V.; Victorino, I.; Scariot, V.; Berruti, A. Arbuscular mycorrhizal fungi as natural biofertilizers: Current role and potential for the horticulture industry. *Acta Hortic.* **2018**, *1191*, 207–215. [CrossRef]
29. López-Ráez, J.; Verhage, A.; Fenández, I.; García, J.; Azcón-Aguilar, C.; Flors, V.; Pozo, M. Hormonal and transcriptional profiles highlight common and differential host responses to arbuscular mycorrhizal fungi and the regulation of the oxylipin pathway. *J. Exp. Bot.* **2010**, *61*, 2589–2601. [CrossRef]
30. De Sousa Oliveira, M.; da Silva Campos, M.; de Albuquerque, U.P.; da Silva, F.S.B. Arbuscular mycorrhizal fungi (AMF) affects biomolecules content in *Myracrodruon urundeuva* seedlings. *Ind. Crops Prod.* **2013**, *50*, 244–247. [CrossRef]
31. Bona, E.; Lingua, G.; Manassero, P.; Cantamessa, S.; Marsano, F.; Todeschini, V.; Copetta, A.; D'Agostino, G.; Massa, N.; Avidano, L.; et al. AM fungi and PGP pseudomonads increase flowering, fruit production, and vitamin content in strawberry grown at low nitrogen and phosphorus levels. *Mycorrhiza* **2015**, *25*, 181–193. [CrossRef]
32. Lazzara, S.; Militello, M.; Carrubba, A.; Napoli, E.; Saia, S. Arbuscualr mycorrhizal fungi altered the hypericin, pseudohypericin, and hyperforin conten in flowers of Hypericum perforatum grown under contrasting P availability in a highly organic substrate. *Mycorrhiza* **2017**, *27*, 345–354. [CrossRef]
33. Jurkiewicz, A.; Ryszka, P.; Anielska, T.; Waligórski, P.; Bialońska, D.; Góralska, K.; Tsimilli-Michael, M.; Turnau, K. Optimization of culture conditions of *Arnica montana* L.: Effects of mycorrhizal fungi and competing plants. *Mycorrhiza* **2010**, *20*, 293–306. [CrossRef] [PubMed]

34. Zeng, Y.; Guo, L.; Chen, B.; Hao, Z.; Wang, J.; Huang, L.; Yang, G.; Cui, X.; Yang, L.; Wu, Z.; et al. Arbuscular mycorrhizal symbiosis and active ingredients of medicinal plants: Current research status and prospectives. *Mycorrhiza* **2013**, *23*, 253–265. [CrossRef] [PubMed]
35. Giovannetti, M.; Avio, L.; Sbrana, C. Improvement of nutraceutical value of food by plant symbionts. In *Natural Products, Phytochemistry, Botany and Metabolism of Alkaloids, Phenolics and Terpenes*; Ramawat, K.G., Merillon, J.M., Eds.; Springer: Berlin, Germany, 2013; pp. 2641–2662.
36. Rydlová, J.; Jelínkova, M.; Dušek, K.; Dušková, E.; Vosátka, M.; Püschel, D. Arbuscular mycorrhiza differentially affects synthesis of essential oils in coriander and dill. *Mycorrhiza* **2016**, *26*, 123–131. [CrossRef] [PubMed]
37. Hart, M.M.; Antunes, P.M.; Chaudhary, V.B.; Abbott, L.K. Fungal inoculants in the field: Is the reward greater than the risk? *Funct. Ecol.* **2017**, *32*, 126–135. [CrossRef]
38. Ryan, M.H.; Graham, J.H. Little evidence that farmers should consider abundance or diversity of arbuscular mycorrhizal fungi when managing crops. *New Phytol.* **2018**, *220*, 1092–1107. [CrossRef] [PubMed]
39. Berg, G. Plant-microbe interactions promoting plant growth and health: Perspectives for controlled use of microorganisms in agriculture. *Appl. Microbiol. Biotechnol.* **2009**, *84*, 11–18. [CrossRef] [PubMed]
40. Aimo, S.; Gosetti, F.; D'Agostino, G.; Gamalero, E.; Gianotti, V.; Bottaro, M.; Gennaro, M.; Berta, G. Use of arbuscular mycorrhizal fungi and beneficial soil bacteria to improve yield and quality of saffron (*Crocus sativus* L.). *Acta Hortic.* **2010**, *850*, 159–162. [CrossRef]
41. Venkateshwaran, M.; Volkening, J.; Sussman, M.; Ané, J. Symbiosis and the social network of higher plants. *Curr. Opin. Plant Biol.* **2013**, *16*, 118–127. [CrossRef]
42. Mohebi-Anabat, M.; Riahi, H.; Zanganeh, S.; Sadeghnezhad, E. Effects of arbuscular mycorrhizal inoculation on the growth, photosynthetic pigments and soluble sugar of *Crocus sativus* L. (saffron) in autoclaved soil. *Int. J. Agron. Agric. Res.* **2015**, *6*, 296–304.
43. Lone, R.; Shuab, R.; Koul, K. AMF association and their effect on metabolite mobilization, mineral nutrition and nitrogen assimilating enzymes in saffron (*Crocus sativus*) plant. *J. Plant Nutr.* **2016**, *39*, 1852–1862. [CrossRef]
44. Shajari, M.A.; Moghaddam, P.R.; Ghorbani, R.; Koocheki, A. Increasing saffron (*Crocus sativus* L.) corm size through the mycorrhizal inoculation, humic acid application and irrigation managements. *J. Plant Nutr.* **2018**, *41*, 1047–1064. [CrossRef]
45. Moghaddam, P.R.; Koocheki, A.; Molafilabi, A.; Seyyedi, M. Effect of biological and chemical fertilizers on replacement corm and flower yield of saffron (*Crocus sativus* L.). *Iran. J. Crop Sci.* **2013**, *15*, 234–246.
46. Cusano, E.; Consonni, R.; Petrakis, E.A.; Astraka, K.; Cagliani, L.R.; Polissiou, M.G. Integrated analytical methodology to investigate bioactive compounds in *Crocus sativus* L. flowers. *Phytochem. Anal.* **2018**, *29*, 476–486. [CrossRef]
47. Vahedi, M.; Kabiri, M.; Salami, S.A.; Rezadoost, H.; Mirzaie, M.; Kanani, M.R. Quantitative HPLC-based metabolomics of some Iranian saffron (*Crocus sativus* L.) accessions. *Ind. Crops Prod.* **2018**, *118*, 26–29. [CrossRef]
48. Trouvelot, A.; Kough, J.; Gianinazzi-Pearson, V. Mesure du taux de mycorhization VA d'un système radiculaire. Recherche de méthodes d'estimation ayant une signification fonctionnelle. In *Mycorrhizae: Physiology and Genetics*; Gianinazzi-Pearson, V., Gianinazzi, S., Eds.; INRA Press: Paris, France, 1986; pp. 217–221.
49. Caser, M.; D'Angiolillo, F.; Chitarra, W.; Lovisolo, C.; Ruffoni, B.; Pistelli, L.; Pistelli, L.; Scariot, V. Ecophysiological and phytochemical responses of *Salvia sinaloensis Fern.* to drought stress. *Plant Growth Regul.* **2018**, *84*, 383–394. [CrossRef]
50. Donno, D.; Beccaro, G.; Mellano, M.; Cerutti, A.; Bounous, G. Goji berry fruit (*Lycium* spp.): Antioxidant compound fingerprint and bioactivity evaluation. *J. Funct. Foods* **2015**, *18*, 1070–1085. [CrossRef]
51. Caser, M.; D'Angiolillo, F.; Chitarra, W.; Lovisolo, C.; Ruffoni, B.; Pistelli, L.; Pistelli, L.; Scariot, V. Water deficit regimes trigger changes in valuable physiological and phytochemical parameters in *Helichrysum petiolare* Hilliard & B.L. Burtt. *Ind. Crops Prod.* **2016**, *83*, 680–669.
52. Urbani, E.; Blasi, F.; Simonetti, M.; Chiesi, C.; Cossignani, L. Investigation on secondary metabolite content and antioxidant activity of commercial saffron powder. *Eur. Food Res. Technol.* **2016**, *242*, 987–993. [CrossRef]

53. Donno, D.; Mellano, M.G.; Prgomet, Z.; Beccaro, G.L. Advances in *Ribes* x *nidigrolaria* Rud. Bauer & A. Bauer fruits as potential source of natural molecules: A preliminary study on physico-chemical traits of an underutilized berry. *Sci. Hortic.* **2018**, *237*, 20–27.
54. Gosling, P.; Mead, A.; Proctor, M.; Hammond, J.P.; Bending, G.D. Contrasting arbuscular mycorrhizal communities colonizing different host plants show a similar response to a soil phosphorus concentration gradient. *New Phytol.* **2013**, *198*, 546–556. [CrossRef]
55. Watt-Williams, S.J.; Smith, F.A.; Jakobsen, I. Soil phosphorus availability is a driver of the responses of maize (*Zea mays*) to elevated CO_2 concentration and arbuscular mycorrhizal colonization. *Symbiosis* **2018**, 1–10. [CrossRef]
56. Brundrett, M.; Kendrick, B. The roots and mycorrhizas of herbaceous woodland plants. II. Structural aspects of morphology. *New Phytol.* **1990**, *114*, 469–479. [CrossRef]
57. Gavito, M.E.; Olsson, P.A.; Rouhier, H.; Medina-Peñafiel, A.; Jakobsen, I.; Bago, A.; Azcón-Aguilar, C. Temperature constraints on the growth and functioning of root organ cultures with arbuscular mycorrhizal fungi. *New Phytol.* **2005**, *168*, 179–188. [CrossRef] [PubMed]
58. Gavito, M.E.; Azcón-Aguilar, C. Temperature stress in arbuscular mycorrhizal fungi: A test for adaptation to soil temperature in three isolates of *Funneliformis mosseae* from different climates. *Agric. Food Sci.* **2012**, *21*, 2–11. [CrossRef]
59. Tammaro, F. Saffron (*Crocus sativus* L.) in Italy. In *Saffron: Crocus sativus L.*; Negbi, M., Ed.; Harwood Academic Publishers: Amsterdam, NL, 1999; pp. 53–62.
60. Mollafilabi, A.; Koocheki, A.; Moeinerad, H.; Kooshki, L. Effect of plant density and weight of corm on yield and yield components of saffron (*Crocus sativus* L.) under soil, hydroponic and plastic tunnel cultivation. *Acta Hortic.* **2013**, *997*, 51–58. [CrossRef]
61. Koocheki, A.; Seyyedi, S. How irrigation rounds and mother corm size control saffron yield, quality, daughter corms behavior and phosphorus uptake. *Sci. Hortic.* **2016**, *213*, 132–143. [CrossRef]
62. Bayat, M.; Amirnia, R.; Rahimi, M. Phenotypic and genotypic relationships between traits in saffron (*Crocus sativus* L.) as revealed by path analysis. *J. Appl. Res. Med. Aromat. Plants* **2017**, *5*, 33–40. [CrossRef]
63. Geneva, M.; Stancheva, I.; Boychinova, M.; Mincheva, N.; Yonova, P. Effects of foliar fertilization and arbuscular mycorrhizal colonization on *Salvia officinalis* L. growth, antioxidant capacity, and essential oil composition. *J. Sci. Food Agric.* **2010**, *90*, 696–702.
64. Araim, G.; Saleem, A.; Amason, J.; Charest, C. Root colonization by an arbuscular mycorrhizal (AM) fungus increases growth and secondary metabolism of purple coneflower, *Echinacea purpurea* (L.) Moench. *J. Agric. Food Chem.* **2009**, *57*, 2255–2258. [CrossRef]
65. Makhlouf, H.; Saksouk, M.; Habib, J.; Chahine, R. Determination of antioxidant activity of saffron taken from the flower of *Crocus sativus* grown in Lebanon. *Afr. J. Biotechnol.* **2011**, *10*, 8093–8100.
66. Gülçin, I.; Güngör, Ş.; Beydemir, Ş.; Elmastas, M.; İrfan, K. Comparison of antioxidant activity of clove (*Eugenia caryophylata* Thunb) buds and lavender (*Lavandula stoechas* L.). *Food Chem.* **2004**, *87*, 393–400. [CrossRef]
67. Chen, Y.; Zhang, H.; Tian, X.; Zhao, C.; Liu, Y.; Jia, L.; Yin, H.X.; Chen, C. Antioxidant potential of crocins and ethanol extracts of *Gardenia jasminoides* ELLIS and *Crocus sativus* L. A relationship investigation between antioxidant activity and crocin contents. *Food Chem.* **2008**, *109*, 484–492. [CrossRef]
68. Ordoudi, S.; Befani, C.; Nenadis, N.; Koliakos, G.; Tsimidou, M. Further Examination of Antiradical Properties of *Crocus sativus* Stigmas Extract Rich in Crocins. *J. Agric. Food Chem.* **2009**, *57*, 3080–3086. [CrossRef] [PubMed]
69. Karimi, E.; Oskoueian, E.; Hendra, R.; Jaafar, H. Evaluation of *Crocus sativus* L. stigma phenolic and flavonoid compounds and its antioxidant activity. *Molecules* **2010**, *15*, 6244–6256. [CrossRef] [PubMed]
70. Inoue, E.; Shimizu, Y.; Masui, R.; Hayakawa, T.; Tsubonoya, T.; Hori, S.; Sudoh, K. Effects of saffron and its constituents, crocin-1, crocin-2, and crocetin on α-synuclein fibrils. *J. Nat. Med.* **2018**, *72*, 274–279. [CrossRef] [PubMed]
71. Asdaq, S.; Inamdar, M. Potential of garlic and its active constituent, S-allyl cysteine, as antihypertensive and cardioprotective in presence of captopril. *Phytomedicine* **2010**, *17*, 1016–1026. [CrossRef] [PubMed]
72. Amin, S.; Parker, M.; Armbrust, E. Interactions between diatoms and bacteria. *Microbiol. Mol. Boil. Rev.* **2012**, *76*, 667–684. [CrossRef] [PubMed]

73. Del Campo, P.; Carmona, M.; Maggi, L.; Kanakis, C.; Anastasaki, E.; Tarantilis, P.; Polissiou, M.; Alonso, G. Picrocrocin Content and Quality Categories in Different (345) Worldwide Samples of Saffron (*Crocus sativus* L.). *J. Agric. Food Chem.* **2010**, *58*, 1305–1312. [CrossRef]
74. Proestos, C.; Chorianopoulos, N.; Nychas, G.; Komaitis, M. RP-HPLC Analysis of the Phenolic Compounds of Plant Extracts. Investigation of Their Antioxidant Capacity and Antimicrobial Activity. *J. Agric. Food Chem.* **2005**, *53*, 1190–1195. [CrossRef]

agronomy

MDPI

Article

A Novel Biostimulant, Belonging to Protein Hydrolysates, Mitigates Abiotic Stress Effects on Maize Seedlings Grown in Hydroponics

Sara Trevisan [1,2,*], Alessandro Manoli [1] and Silvia Quaggiotti [1]

1. Department of Agronomy, Food, Natural Resources, Animals and Environment (DAFNAE), University of Padova, viale dell'Università 16, 35020 Legnaro (PD), Italy; alessandro.manoli@hotmail.it (A.M.); silvia.quaggiotti@unipd.it (S.Q.)
2. Department of Land, Environment, Agriculture and Forestry (TESAF), University of Padova, viale dell'Università 16, 35020 Legnaro (PD), Italy

* Correspondence: sara.trevisan@unipd.it; Tel.: +39-049-827-2913

Received: 28 November 2018; Accepted: 7 January 2019; Published: 9 January 2019

Abstract: The main challenge to agriculture worldwide is feeding a rapidly growing human population, developing more sustainable agricultural practices that do not threaten human and ecosystem health. An innovative solution relies on the use of biostimulants, as a tool to enhance nutrient use efficiency and crop performances under sub-optimal conditions. In this work a novel biostimulant (APR®, ILSA S.p.A., Arzigano VI, Italy), belonging to the group of protein hydrolysates, was supplied to maize seedlings in hydroponic and its effects were assessed in control conditions and in the presence of three different kinds of stresses (hypoxia, salt and nutrient deficiency) and of their combination. Our results indicate that APR® is soluble and is able to influence root and shoot growth depending on its concentration. Furthermore, its effectiveness is clearly increased in condition of single or combination of abiotic stresses, thus confirming the previously hypothesised action of this substance as enhancer of the response to environmental adversities. Moreover, it also regulates the transcription of a set of genes involved in nitrate transport and ROS metabolism. Further work will be needed to try to transfer this basic knowledge in field experiments.

Keywords: Maize; biostimulant; root; stress; growth; gene expression

1. Introduction

Global population is expecting to increase to nine billion by 2050 [1] and agriculture will need to push crop production accordingly in order to sustain the greater demand for food. This is especially true in developing countries where high rates of population growth are associated to an increased urbanization, leading to changes in income levels and food preferences [2].

Moreover, climate change leading to abiotic pressures, such as rising droughts and other stresses correlated to higher temperature, are predicted to escalate in their severity and frequency [3,4] thus seriously compromising crop productivity [5]. In fact, abiotic stress can reduce crop yields by more than 60% for major crops [6–8].

New crop protection solutions able to mitigate the main abiotic stresses represent a substantial opportunity to contribute to secure, higher and more stable yields. These innovations span across conventional breeding to biotechnology solutions [9] and also encompass new generations of agrochemicals [10]. The global crop protection market attained US $56.7 billion in 2014. However, there are only limited solutions currently available to mitigate abiotic stresses.

In recent years, the use of natural-derived biostimulants is proposed as an innovative solution to address the challenges of sustainable agriculture, by ensuring optimal nutrient uptake, crop yield, quality, and tolerance to abiotic stress [11].

An innovative technology with promising application potential entails the use of a particular class of biostimulants, the protein hydrolysates (PHs). PHs are mixtures of polypeptides, oligopeptides, and free amino acids derived by chemical or enzymatic hydrolysis of plant residues or animal connective tissues. The protein hydrolysates have been demonstrated to stimulate root growth and leaf biomass of several crops. Du Jardin [11] reviewed various effects resulting from the application of these compounds to crops and Van Oosten [12] reported several studies demonstrating the role of PHs in abiotic stress response.

Although the effects of protein hydrolysates on crop performance have been documented by several studies [11,12] the scientific basis of their action has only partially been elucidated mainly due to the complex nature of these products [13]. However, the synthesis of the enzymatic hydrolysis of protein has been an advantageous, ecologically safe strategy to produce biostimulant [13], and more studies are needed to improve protein hydrolysates production techniques and to ensure a low- cost product for consumption and a high use efficiency [14].

In an earlier study we demonstrated a role for a new-synthetized PH (APR®, ILSA S.p.A.) in regulating the expression level of a thousand of genes in maize roots, and hypothesised that it could act by improving the plant responses to various environmental stresses [15]. Based on the results therein obtained APR® has been proposed to enhance plant response to stress. However, this preliminary work has tested APR® on plants grown in not adverse conditions and APR® was applied directly to the soil mixture as solid granules. The chemical composition of this compound (identified also as AA309) is reported in subsequent study by Ertani et al. [16] which also performed Fourier transform infrared (FTIR). The chemical analysis revealed the presence of several amino acids, as lysine, phenylalanine, glycine, aspartate, and isoleucine. The present substance is still under study and its dossier is expected to be definitively completed by the next three months.

Due to the economic importance of maize and the limitations in fertilizer applications imposed during its development, it is important to dissect the effects of the biostimulant on the initial growth of maize seedlings, at both morphological and transcriptomic levels. For this reason, most of the present works on APR® are focused on its effects on this species.

Our previous work [15] suggested that APR® is at least in part soluble and reach root through the soil solution. Furthermore, it seems to act as a stress tolerance enhancer, by modulating the transcription of a wide set of genes involved in ROS detoxification and nutrient acquisition. However, no results on its effects in abiotic stress conditions were gained until now. Our various results with this species indicate that maize is able to sense and rapidly respond to nutritional fluctuations already after hours or minutes of treatment [17–20]. Therefore in the present work, we tried to deepen the effects of APR®, supplied in in hydroponic, in affecting the early response of maize seedlings to abiotic stresses. To this aim we first aimed to assess the APR® activity by measuring its effects on plant growth and identified the optimal concentration to be used in further experiments. Subsequently, to study the effectiveness of APR® as an enhancer of plant tolerance to abiotic stress we grown maize seedlings in the presence of different single and combined abiotic stresses and supplying them with APR®. Our results on root and shoot growth and on the expression profiles of a number of previously identified genes [16] provide further evidence of the APR® biostimulant activity, which early induce root to elongate and affects gene expression, especially increased in conditions of environmental limitations.

2. Materials and Methods

2.1. Maize Seedlings Growth

Seeds of maize (*Zea mays* L.), inbred line B73, were washed in distilled water and germinated on wet filter paper at 25 °C in the dark. After three days, maize seedlings were transferred in a controlled environmental chamber in 500 ml tanks containing a nutrient medium which was constantly aerated and composed as previously described in Quaggiotti [21] and changed every two days. Plants were grown in a growth chamber with an 8-h photoperiod under 200 µmol m^{-2} s^{-1}

of photosynthetically active radiation (PAR; daylight and warm white 1:1, LF-40W) at day/night temperatures of 21/18 °C [21]. The pH of the medium was checked during the growth period and remained at a stable level of around pH 6.5. For analysis of RNA root samples were frozen in liquid nitrogen and stored at −80 °C.

2.2. Set up of the Novel Biostimulant Concentration to be Supplied to Stressed Maize Plants

Maize seedlings were hydroponically grown for three days in distilled water containing different APR® concentrations, resulting in a variable nitrogen content which ranged from 1% to 10% of the amount of nitrogen supplied by the Hoagland-modified nutrient solution previously described. APR® granules were added to tanks 2 h before putting plants into the water and constantly stirred until all product has dissolved.

This series of concentrations was selected basing on their relative content of nitrogen, paying attention to keep it to a sub-nutritional level. To evaluate their effects the root length, root and shoot fresh weight were measured. Data are expressed as the average of three replicates ($n = 10$) ± standard error. For statistical analysis, we compared morphological data derived from the corresponding four different APR® concentrations with those of control plants.

2.3. Stress Application and Morphological Analyses

To try to assess the effect of APR® on maize tolerance to abiotic stress three single stress (hypoxia, salt and nutrient starvation) and two stress combination (hypoxia plus nutrient starvation, salt plus nutrient starvation) were imposed to seedlings for three days. Comparisons were made among non-stressed and stressed plants, which were then compared with plants supplied also with 5% of APR®.

Hypoxic stress conditions were achieved by not bubbling air through the liquid solution for the entire experiment. For salt stress, a 25 mM NaCl concentration, which corresponds to mild salt stress in maize was employed [22–24]. For nutritional stress, seedlings were grown in distilled water only. Each treatment was performed in three biological replicates.

After 3 days, roots and shoots of control and APR® treated plants were harvested. For the morphological analyses, 10 randomly selected seedlings for biological replicate were used.

The remaining plants were immediately frozen in liquid nitrogen and kept at −80 °C for subsequent RNA extraction.

2.4. RNA Extraction, and cDNA Synthesis

Total RNA was extracted from root tissues using TRIzol reagent (Invitrogen, Carlsbad, CA, USA) as previously described by Trevisan et al., [17]. DNAse digestion was performed whit RQ1 RNAse-free DNAse (Promega, Madison, WI, USA) on an aliquot of total RNA as described by Trevisan et al. [17]. RNA was quantified using a Nanodrop 1000 (Thermo Scientific, Nanodrop Products, Wilmington, DE, USA) and its quality further validated by sterile agarose gel electrophoresis. cDNA was synthesized from 500 ng of total RNA mixed with 1 µl of 10 µM oligo-dT, as described by Manoli et al. [25].

2.5. Real Time qPCR

Relative quantification of transcripts by RT-qPCR was performed in a StepOne Real-Time PCR System (Applied Biosystems, Foster City, CA, USA). Reactions were performed using SYBR Green chemistry (Applied Biosystems fast SYBR Green Master Mix, Thermo Fisher Scientific, Waltham, MA, USA), following the manufacturer's instructions. Reverse-transcribed RNA (2.5 ng) was used as template in each reaction as indicated by Manoli [26]. Three technical replicates were performed for each thesis using the conditions described by Trevisan et al. [17]. Melting-curve analysis confirmed the absence of multiple products and primer dimers. Data were exported and analysed according to the method of Livak and Schmittgen [27] and MIQE guidelines [28]. Only transcripts showing amplification with quantification cycle (Cq) < 35 were selected for subsequent gene expression analysis.

All of the primers used in these assays are listed in Table S1.

2.6. Data Analysis

The expression levels of the analysed genes were normalized via comparison to the expression of the internal reference gene (MEP, membrane proteinPB1A10.07c, primers: forward 5′-TGTACTCGGCAATGCTCTTG-3′ and reverse 5′-TTTGATGCTCCAGGCTTACC-3′), as the reference gene [25]. The standard error was calculated from the standard deviation and the variation coefficient of the reference gene and of the genes under assessment.

For statistical analysis, we compared stress condition plants with its own control. Data represent means ± SD of 3 independent experiments performed in triplicate. For the gene expression levels analyses and the choice of the APR® concentration, multiple comparison statistics were calculated using the software RStudio (https://www.rstudio.com/) Version 1.1.453. differences among samples were verified with either ANOVA (normality and homogeneous variances) or Welch's one-way ANOVA (normality and non-homogeneous variances) followed by post hoc LSD or Waller-Duncan test, respectively, and with Kruskal-Wallis (non-normality and homogeneous variances) or Friedman test (non-normality and nonhomogeneous variances). For all statistics a p-value threshold of 0.05 was adopted. For the subsequent growth analyses, one-way ANOVA test followed by Tukey's HSD test was performed. Asterisks indicate significant differences (*: $p < 0.05$; **: $p < 0.02$). One-way ANOVA, Tukey's HSD test.

3. Results

3.1. Choice of the Novel Biostimulant Concentration to be Used for Subsequent Treatments in Hydroponics

In order to assess the most effective APR® concentration four different APR® concentrations (1%, 2%, 5% and 10%) were used and their effects were observed in comparison to those measured for seedlings grown in distilled water for the same period (Ctrl) (Figure 1).

APR® application induced a significant increment of root length when supplied in hydroponic at a concentration of 5%The 10% dose showed a reduction on root length respect to the 5%.

In the case of root and shoot weight no statistically significant differences were observed (Figure 1B,C). According to these data, we decided to use a 5% concentration of APR® for all subsequent analyses.

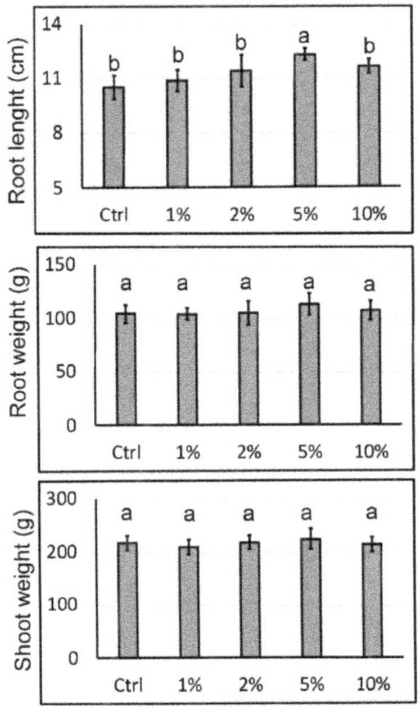

Figure 1. Effects of different APR® concentrations on the physiological growth parameters of *Zea mays* L. seedlings. The graphs represent the root length, the root weight and the shoot weight of maize seedlings hydroponically grown for three days at increasing concentrations of APR®. The values represented in the graphs were calculated from three independent experiments ($n = 10$) and represent the mean ± the standard error. Significantly different values ($p < 0.05$) are evidenced by different letters (One-way Anova, LSD post-hoc test).

3.2. Biostimulant Effects on Root Length in the Presence of Different Stress

When maize seedlings were subjected to hypoxic stress (H) the root length showed values 12% lower if compared to control plants (Figure 2). However, when APR® was supplied to the nutrient solution a significant increase of root elongation was measured, with values 10% higher than those observed for hypoxic plants and similar to those noticed for control plants.

A similar pattern was observed when plants were subjected to salt stress (S), which triggered a visible reduction in primary root length. However, the provision of APR® triggered a significant increment of root length, thus restoring the phenotype of control plants.

Also in the case of nutrient deprivation (N) the supply of APR® significantly induced the primary root to elongate.

The positive effect of APR® provision was even more marked in the case of combined stress. In fact, when hypoxia was associated to nutritional stress (N/H) primary root length was visibly in comparison to the control, but the presence of APR® markedly and significantly restrained this negative effect leading to a phenotype comparable with that observed in not stressed plants. A positive influence of the biostimulant was observed also in the case of the combination between salt stress and nutritional stress (N/S), which inhibited the primary root growth, but the provision of APR® led to a root elongation 40% higher than that measured for stressed plants.

The provision of APR® to control plants did not induce significant effects on root elongation.

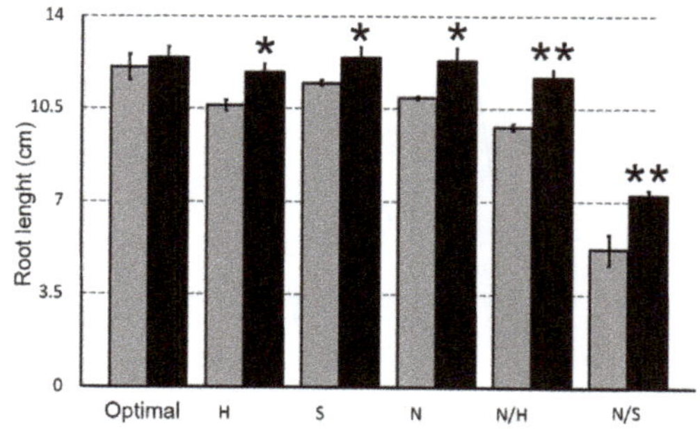

Figure 2. APR® counteracts the negative effects of single and combined abiotic stress on root length. Maize seedlings were subjected for 3 days to several abiotic stresses in absence (grey bars) or in presence (black bars) of 5% APR®. The applied single stresses were: hypoxic stress (H), salt stress (S), nutritional stress (N). The single stresses were combined as: hypoxic stress plus nutritional stress (N/H) and salt stress plus nutritional stress (N/S). The values of root length (cm) are represented in the graphs (mean ± SE) and were calculated from three independent experiments (n = 10). Significantly different values are evidenced by * (*: $p < 0.05$; **: $p < 0.02$; One-way ANOVA, Tukey's HSD test).

3.3. Biostimulant Effects on Root and Shoot Weight in the Presence of Different Stress

To verify if the increments observed in terms of primary root length were associated to an increase of total root weight, these parameters was measured in the same conditions described above (Figure 3). When maize seedlings were subjected singularly to the three different stresses (hypoxia, salt, nutritional deficiency) or to the hypoxia and nutritional deficiency (N/H) combination the root weight did not evidence significant differences in comparison to the control plants nor in response to the biostimulant. On the contrary, a significant increase of root weight in response to APR® provision was measured when plants were subjected to the combination of nutritional and salt stress (N/S).

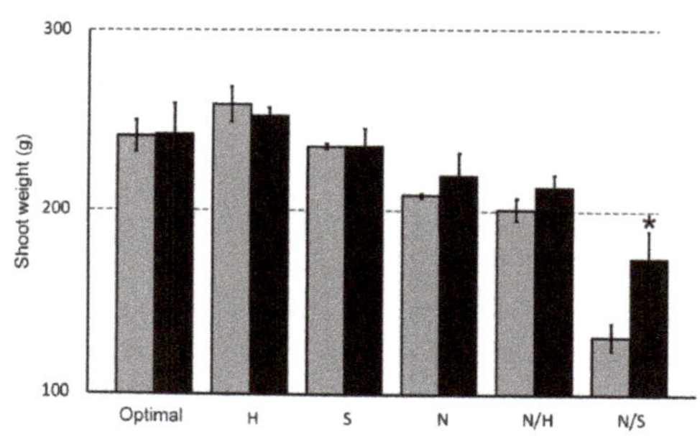

Figure 3. Effects of APR® on root weight in response to different single and combined abiotic stresses. Maize seedlings were subjected for 3 days to several abiotic stresses in absence (grey bars) or in presence (black bars) of 5% APR®. The applied single stresses were: hypoxic stress (H), salt stress (S), nutritional stress (N). The single stresses were combined as: hypoxic stress plus nutritional stress (N/H) and salt stress plus nutritional stress (N/S). The values of root weight (g) are represented in the graphs (mean ± SE) and were calculated from three independent experiments (n = 10). Significantly different values are evidenced by * (*: $p < 0.05$; **: $p < 0.02$; One-way ANOVA, Tukey's HSD test).

As far as the shoot weight was concerned (Figure 4) no differences were observed in response to stress, nor providing APR®, except in the case of the contemporary presence of nutritional starvation and salt stress (N/S). In fact, the co-presence of these two stresses highly inhibited shoot weight accumulation, which was significantly induced in response to APR®.

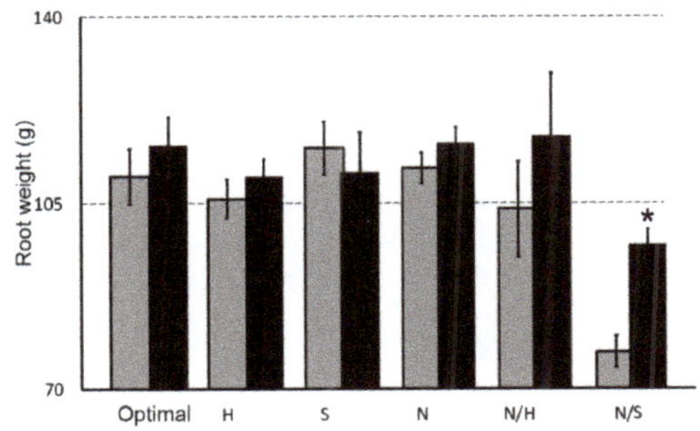

Figure 4. Effects of APR® on shoot weight in response to different single and combined abiotic stresses. Maize seedlings were subjected for 3 days to several abiotic stresses in absence (grey bars) or in presence (black bars) of 5% APR®. The applied single stresses were: hypoxic stress (H), salt stress (S), nutritional stress (N). The single stresses were combined as: hypoxic stress plus nutritional stress (N/H) and salt stress plus nutritional stress (N/S). The values of root weight (g) are represented in the graphs (mean ± SE) and were calculated from three independent experiments ($n = 10$). Significantly different values are evidenced by * (*: $p < 0.05$; **: $p < 0.02$; One-way ANOVA, Tukey's HSD test).

3.4. Biostimulant Regulation of Gene Expression

A number of genes belonging to the group of nitrate transporters and of ROS metabolism were selected basing both on previous results (Trevisan et al. 2017 [20]) and on their putative physiological role (Table 1).

Table 1. Description and classification of the targets genes studied in qPCR expression analysis. The expression levels of genes belonging to nitrate transport system (HATS and LATS) and related to reactive oxygen species (ROS) generation and homeostasis were analyzed.

	Maize GDB Accession ID	Code	Description
HATS	Zm00001d054057	ZmNRT2.1	High affinity nitrate transporter
	Zm00001d054060	ZmNRT2.2	High affinity nitrate transporter
	Zm00001d014976	ZmNRT2.3	High affinity nitrate transporter
	Zm00001d017095	ZmNAR2.1	High affinity nitrate transporter
	Zm00001d003287	ZmNAR2.2	High affinity nitrate transporter
LATS	Zm00001d024587	ZmNRT1.1	Nitrate transporter
	Zm00001d029932	ZmNRT1a	Dual-affinity nitrate transporter
	Zm00001d036941	ZmNRT1b	Nitrate transporter
	Zm00001d017666	ZmNRT1.5	Nitrate transporter
	Zm00001d007785	ZmNRT	Nitrate and chloride transporter
ROS	Zm00001d042961	ZmRbohA	Respiratory burst oxidase protein A
	Zm00001d043543	ZmRbohB	Respiratory burst oxidase protein B
	Zm00001d038762	ZmRbohC	Respiratory burst oxidase protein C
	Zm00001d052653	ZmRbohD	Respiratory burst oxidase protein D
	Zm00001d031908	ZmSOD	Superoxide dismutase [Cu-Zn]
	Zm00001d027511	ZmCAT2	Catalase 2

The nitrate transporters genes include five genes encoding putative high affinity components of nitrate transport (three *ZmNRT2* and two *ZmNAR2* genes respectively, Figure 5) and five encoding members of the NRT1 gene family which is involved in the low affinity nitrate transport system (Figure 6). As far as the ROS genes were concerned this group comprise four genes encoding

NADPHoxidase (*ZmRBOH a, b, c* and *d*), one encoding Catalase2 (*ZmCAT2*) and a gene encoding a Cu-Zn Superoxide dismutase (*ZmSOD*).

The specific effects of the biostimulant on the different groups of genes in conditions of different stress are discussed below.

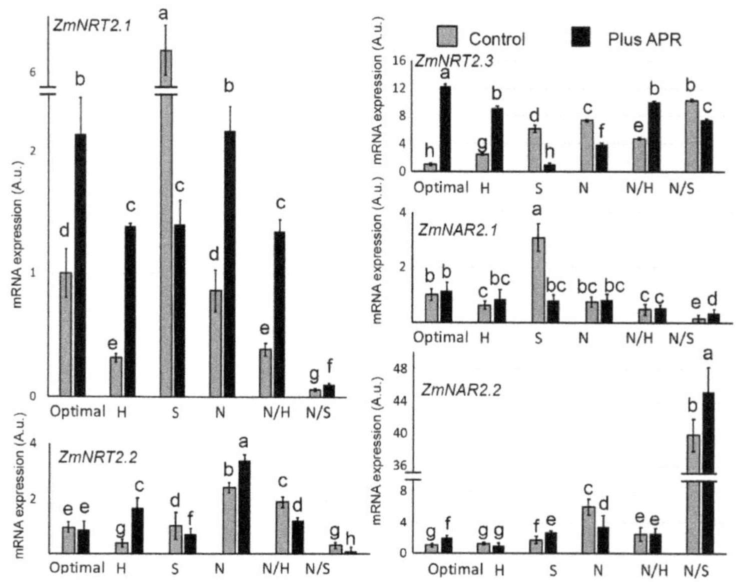

Figure 5. The gene expression patterns of nitrate transporters belonging to the High Affinity Transport Systems (HATS) maize gene family in response to different single and combined abiotic stress are influenced by the presence of APR®. Q-PCR analyses were carried out on root of stressed (H, S, N, N/H, N/S) or unstressed (optimal) maize seedlings grown for 3 days in absence (grey bars) or in presence (black bars) of 5% APR®. Relative mRNA level represents data normalized to control (Optimal = 1). The values shown are means of three biological replicates ± SE. Significantly different values ($p < 0.05$) are evidenced by different letters (One-way Anova, LSD post-hoc test).

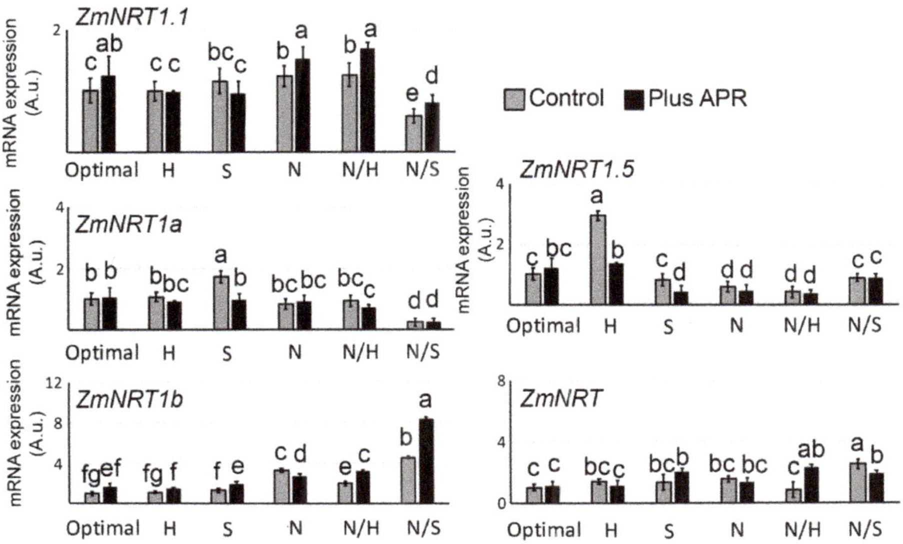

Figure 6. The gene expression patterns of nitrate transporters belonging to the Low Affinity Transport Systems (LATS) maize gene family in response to different single and combined abiotic stress are influenced by the presence of APR ®. Q-PCR analyses were carried out on root of stressed (H, S, N, N/H, N/S) or unstressed (optimal) maize seedlings grown for 3 days in absence (grey bars) or in presence (black bars) of 5% APR®. Relative mRNA level represents data normalized to control (Optimal = 1). The values shown are means of three biological replicates ± SE. Significantly different values ($p < 0.05$) are evidenced by different letters (One-way Anova, LSD post-hoc test).

3.5. Biostimulant Effects on ZmNRT2 Genes Expression

In general, the most striking effects of APR® on gene transcription regulation in roots were observed in conditions of stress and for the group of genes operating in the high affinity nitrate transport system. In hypoxic conditions, the expression of *ZmNRT2.1*, *ZmNRT2.2*, *ZmNRT2.3* and *ZmNAR2.1* was down-regulated in response, but when APR® was supplied a significant increase of their transcription was observed (Figure 5).

The same group of genes were, on the contrary, up-regulated in response to salt stress, but the provision of APR® significantly counteracted this effect, leading to restore the phenotype of un-stressed roots (Figure 5).

In the case of nutrient starvation, instead, unique behaviours were observed for each gene belonging to the high affinity nitrate transport group, with *ZmNRT2.1* and *ZmNRT2.2* being significantly up-regulated and, *ZmNRT2.3* and *ZmNAR2.2*, being down-regulated as a consequence of APR® provision (Figure 5).

When seedlings were subjected to a combination of hypoxia and nutritional stress the transcription of *ZmNRT2.1* and *ZmNRT2.3* were clearly induced, whilst *ZmNRT2.2* expression was down-regulated (Figure 5).

The co-presence of nutritional deficiency and salt triggered for all these genes significant dysregulation of transcription. However, APR® provision restrained this outcome for all of them, except for *ZmNAR2.3* which was further induced by APR®. Except for this situation, the transcription of *ZmNAR2.3* evidenced always minor changes in response to both stress conditions and APR® provision (Figure 5).

In the case of *ZmNRT2.1*, *ZmNRT2.3* and *ZmNAR2.2* a significant up-regulation of expression was noticed also in control condition (un-stressed plants),

3.6. Biostimulant Effects on NRT1 Genes Expression

The transcription of genes implicated in the low affinity transport apparatus was less affected by both stress conditions and APR®, if compared to that of high affinity constituents. Hypoxic conditions induced an increase of the transcription of *ZmNRT1.5* which was significantly counteracted when APR® was provided to the solution (Figure 6). Salt stress triggered an increased transcription of *ZmNRT1a*, which was then inhibited by APR®. The supply of APR® to nutritional starved roots triggered significant change of transcription for *ZmNRT1.1*, *ZmNRT1b*, *ZmNRT* (Figure 6).

APR® significantly affected the expression of these genes, except for *ZmNRT1.5* in plants subjected to a combination of hypoxia and nutrient deficiency (Figure 6). When the combination of nutritional starvation and salt was applied to plants and APR® was supplied to roots significant changes of transcription were noticed, except for *ZmNRT1a* and *ZmNRT1.5* (Figure 6).

3.7. Biostimulant Effects on ROS Genes Expression

As observed for *NRT1* genes also in this case no regulation of expression was noticed upon APR® supply on unstressed seedlings (Figure 7). A more appreciable effect of the biostimulant was observed upon stress conditions. As reported in Figure 7 *ZmSOD1A* transcription was induced upon APR® treatment in hypoxia, nutritional deficiency, association between hypoxia and nutritional stress and also in the case of nutritional and salt combined stresses. On the contrary no evident alterations of expression were measured for *ZmCAT2* neither in response to stress nor in response to APR® (Figure 7). As far as *ZmRboh* genes were concerned their expression was regulated by APR® in response to single and combined stress, even if to a lower extent in compared to *ZmSOD1A* (Figure 7).

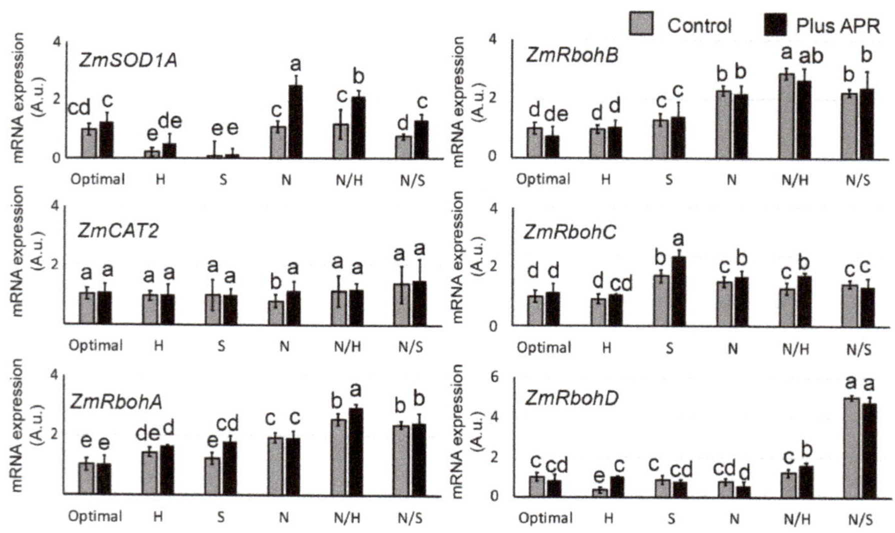

Figure 7. The expression patterns of ROS-related genes in response to different single and combined abiotic stress are influenced by the presence of APR ®. Q-PCR analyses were carried out on root of stressed (H, S, N, N/H, N/S) or unstressed (optimal) maize seedlings grown for 3 days in absence (grey bars) or in presence (black bars) of 5% APR®. Relative mRNA level represents data normalized to control (Optimal = 1). The values shown are means of three biological replicates ± SE. Significantly different values ($p < 0.05$) are evidenced by different letters (One-way Anova, LSD post-hoc test).

4. Discussion

Protein hydrolysates are defined as 'mixtures of polypeptides, oligopeptides and amino acids that are manufactured from protein sources using partial hydrolysis' [29]. Their positive effects on plant performance have encouraged an increasing interest for their use in a more sustainable model of agriculture [30], this leading likewise to a promising solution to the issue of waste disposal [29–33].

Recently a trascriptomic approach was used to study the molecular effects of a collagen derived protein thermal hydrolysate (APR®) produced by Ilsa S.p.A. (Arzignano) on maize roots grown in a solid medium and supplied with localized patches of APR® [16]. Globally the results allowed to recognize a complex APR® action on physiological pathways involved in the stress response and in nutrient acquisition, which seems likely to prime the plant to better tolerate environmental adversities. In the present work the effectiveness of this biostimulant in modulating and improving the maize tolerance to environmental constrains was tested by growing seedlings in different specific abiotic stress conditions and supplying APR® in hydroponics. Overall our data indicate that this compound is soluble in an aqueous solution, suggesting the idea that in soil it can likely move toward roots through mass flow and diffusion and not only being intercepted as a nutrient patch by root growth.

To choose the most effective APR® concentration on plant development, the root and shoot growth were assessed by determining their fresh weights and the root length upon four different concentrations, chosen on the basis of our previous results [16]. The most accepted scientific definition for biostimulants is: "a plant biostimulant is any substance or microorganism applied to plants with the aim to enhance nutrition efficiency, abiotic stress tolerance and/or crop quality traits, regardless of its nutrients content", as reported in du Jardin [11]. According to this, we tested different solutions containing four APR® amounts to which corresponded four different N sub-nutritional concentrations (1%, 2%, 5% and 10% respect to the control Hoagland solution). The most remarkable effect was observed for primary root growth which was stimulated in response to APR® concentrations ranging from 1 to 5% and then inhibited in the presence of a 10% concentration. Detrimental effects of high concentrations of various protein hydrolysates have been observed also by other authors depending on the crop, the typology of biostimulant and the conditions of application [34].

Furthermore, the present results showed that APR® affects root elongation and gene expression in particular when seedlings were subjected to different kind of stresses, confirming the hypothesis put forward by Trevisan and co-authors [16] and thus supporting the suggestion that biostimulants could act as plant protectors able to improve stress tolerance [11], likely by activating the main signalling pathways underlying the response to adverse conditions. Other reports showed that protein hydrolysates modulate plant growth, increase yield and alleviate the impact of abiotic stress on crops [35,36]. The present results, together with those of Trevisan [16] further suggest that this action could involve the molecular regulation of definite genes.

In general, the combination of two or more abiotic stresses has a detrimental impact on crops that is not predictable from that of each of the stresses composing the combination if applied individually. In recent years stress combination has been acknowledged as a novel state of stress and as a major cause of crop loss worldwide [37–39]. For this reason, we decided to assess the APR® potentiality in alleviating stress impact also in condition of stress combination.

As expected the most striking effect of APR® on growth re-establishment in conditions of abiotic stress was observed for roots, which are the main target for hypoxia, salt and nutrient deprivation stresses. The plastic control of the root development throughout time and in response to endogenous and exogenous stimuli allows plants to efficiently adapt to environmental constraints [40,41]. Root apex is highly responsive to external stimuli and rapidly adjusts its growth to efficiently adapt to environmental constraints and resources availability [19,26,42–47]. In this work a clear induction of primary root growth upon APR® treatment was noticed in all the conditions examined, with the most prominent effect in the case of combination of stresses. The simultaneous presence of nutritional deficiency and salt stress led to the most relevant arrest of growth which was, however, at least partially prevented when plants were supplied with APR®. In this case a similar behaviour was observed also in shoot, leading to hypothesise that APR® is able to act also as a systemic clue, firstly perceived by root cells, but likewise triggering a phenotypic response in shoots. This systemic action could be the outcome of the already hypothesised function of APR® as activator of the stress tolerance [16].

Moreover, it could depend on the protein hydrolysates ability to interfere with hormonal signaling, due to the presence of bioactive peptides (for a review Colla [48]) or aminoacids, as confirmed by the chemical composition of this same compound described by Ertani et al., [15]. Recent transcriptomic findings which highlighted the regulation of hormonal key elements by APR® [16] and a different study aimed to characterize the metabolomic regulation by biostimulant [49] reinforces this hypothesis.

Protein hydrolysates seem to improve nutrient uptake through modifications of root architecture (density, length and number of lateral roots), as well as through complexation of nutrients by peptides and amino acids, and also enhancing microbial activity thus increasing the nutrient availability in soil [11,34]. Moreover, a recent paper [50]. demonstrated that protein hydrolysates modulate plant growth and the expression of key genes in N assimilation (including Nitrate and ammonia transporters) in tomato. However only few information has been obtained on protein hydrolysates regulation of nutrient transport system. To better decipher this last aspect, a number of previously identified by Trevisan et al. [16] target genes involved in nitrate transport were chosen as markers for evaluating the transcriptional effects of the treatment.

Our results evidenced a marked regulation of the transcription of genes encoding members of the high affinity nitrate transport system (HATS, NRT2 and NAR genes), which was particularly relevant in condition of abiotic stresses. The impact of APR® supply on the molecular regulation of the Low Affinity Transport System was less evident, leading to suppose that the provision of APR® mainly affects the functioning of the uptake of nitrate in the range of the High Affinity System, which are recognised to play a crucial role in determining the global Nitrogen Use Efficiency (NUE) in condition of limited nutritional inputs [51].

Trevisan et al. [16] also hypothesised that APR® could activate tolerance pathways, by mimicking the plant responses to environmental stresses, thus priming them against unfavourable conditions through the regulation of enzymes involved in the pathway governing the response to oxidative stress. To deepen this hypothesis the analyses of the expression of six genes involved in ROS signalling and defence was assessed, in condition of stress and in the presence of APR®. Only *SOD1A* showed a clear regulation in response to APR® which almost in all the conditions analysed induced its expression, whilst for the other five genes no significant differences were evidenced upon APR® supply.

Superoxide dismutases (SODs) are key enzymes functioning as the first line of antioxidant defence by virtue of the ability to catalyse the enzymatic dismutation of superoxide to H_2O_2 [52]. The present result reinforces the hypothesis that APR might preventively prepare plants to oxidative stresses, by enhancing their own detoxifying tools.

In conclusion, basing on the more acknowledged definition of biostimulant [11], our results confirm the effectiveness of APR® as an enhancer of abiotic stress tolerance, thus allowing to definitely include it among the category of biostimulants (Figure 8). Moreover, present results strengthen the importance of root as a target for APR®, which has been proven to affect both root development and transcription of genes involved in Nitrogen Use Efficiency and ROS detoxification. Both these actions could lead to an improved tolerance to abiotic stresses, as nutritional starvation, salt and hypoxia which take place in the soil environment.

These preliminary knowledges should be in the future transferred in field experiment to further assess the APR® usefulness in agriculture.

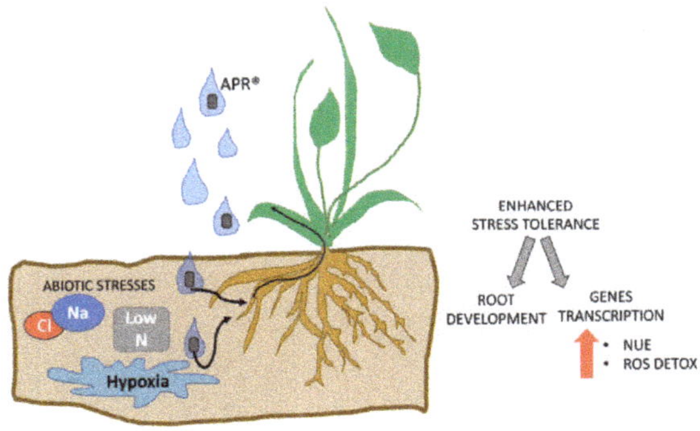

Figure 8. Direct and indirect effects of APR on plant growth in response to single and combined abiotic stresses. APR is perceived by the roots and modulates the root length by balancing the expression of genes involved in nitrate transport and ROS detoxification. APR could be systemically transported to the upper part of the plant, inducing a growth response in the shoot.

Supplementary Materials: The following are available online at http://www.mdpi.com/2073-4395/9/1/28/s1, Table S1: List of primers used in gene expression analysis.

Author Contributions: Conceptualization, S.Q. and A.M.; Methodology, S.T. and A.M.; Data Curation, S.T.; Writing—Original Draft Preparation, S.Q. and S. T.; Writing—Review & Editing, S.Q. and S.T.; Supervision, S.Q.; Project Administration, S.Q.; Funding Acquisition, S.Q.

Funding: This work has been supported by ILSA S.p.A and by University of Padova (DOR 2017). S.T. was financed by a grant from ILSA S.p.A.

Acknowledgments: The authors are thankful to F. Favaron (Department of Land, Environment, Agriculture and Forestry, University of Padova) for the constructive comments in improving this study. Special thanks to C. Franceschi (ILSA S.p.A.) and C. Manoli (ILSA S.p.A.) who assisted in this work.

Conflicts of Interest: The authors declare no conflict of interest. The funders had no role in the design of the study; in the collection, analyses, or interpretation of data; in the writing of the manuscript; or in the decision to publish the results.

References

1. UN Department of Economic and Social Affairs. Population Division. 2017. World Population Prospects: The 2017 Revision. Key Findings and Advance Tables. Working Paper No. ESA/P/WP/248. Available online: https://www.compassion.com/multimedia/world-population-prospects.pdf (accessed on 7 January 2019).
2. UNPD Department of Economic and Social Affairs. World Urbanization Prospects: The 2014 Revision. United Nations Development Programme; 2014: Food and Agriculture Organization of the United Nations (FAO). 2014. The Future of the Food and Agriculture: Trends and Challenges. FAO: Rome, Italy, 2017. Available online: http://www.fao.org/publications/fofa/en/ (accessed on 2 July 2018).
3. Rosenzweig, C.; Elliott, J.; Deryng, D.; Ruane, A.C.; Müller, C.; Arneth, A.; Boote, K.J.; Folberth, C.; Glotter, M.; Khabarov, N.; et al. Assessing agricultural risks of climate change in the 21st century in a global gridded crop model intercomparison. *Proc. Natl. Acad. Sci. USA* **2014**, *111*, 3268–3273. [CrossRef] [PubMed]
4. Myers, S. Planetary health: Protecting human health on a rapidly changing planet. *Lancet* **2017**, *390*, 860–2868. [CrossRef]
5. Coleman-Derr, D.; Tringe, S.G. Building the crops of tomorrow: Advantages of symbiont-based approaches to improving abiotic stress tolerance. *Front. Microbiol.* **2014**, *5*, 283. [CrossRef] [PubMed]
6. Bray, E.A.; Bailey-Serres, J.; Weretilnyk, E. Responses to abiotic stresses. In *Biochemistry and Molecular Biology of Plants*; Buchanan, B.B., Gruissem, W., Jones, R.L., Eds.; American Society of Plant Physiologists: Rockville, MD, USA, 2000; pp. 1158–1203.
7. Grover, A.; Kapoor, A.; Lakshmi, O.S.; Agarwal, S.; Sahi, C.; Agarwal, K.; Agarwal, M.; Dubey, H. Understanding molecular alphabets of the plant abiotic stress responses. *Curr. Sci.* **2001**, *80*, 206–216.

8. Singhal, P.; Jan, A.T.; Azam, M.; Haq, Q.M.R. Plant abiotic stress: A prospective strategy of exploiting promoters as alternative to overcome the escalating burden. *Front. Life Sci.* **2016**, *1*, 52–63. [CrossRef]
9. Lamaoui, M.; Jemo, M.; Datla, R.; Bekkaoui, F. Heat and Drought Stresses in Crops and Approaches for Their Mitigation. *Front. Chem.* **2018**, *19*, 6–26. [CrossRef]
10. Antoniou, C.; Savvides, A.; Christou, A.; Fotopoulos, V. Unravelling chemical priming machinery in plants: The role of reactive oxygen-nitrogen-sulfur species in abiotic stress tolerance enhancement. *Curr. Opin. Biotechnol.* **2016**, *33*, 101–107. [CrossRef] [PubMed]
11. du Jardin, P. Plant biostimulants: Definition; concept; main categories and regulation. *Sci. Hortic.* **2015**, *196*, 3–14. [CrossRef]
12. Van Oosten, M.J.; Pepe, O.; De Pascale, S.; Silletti, S.; Maggio, A. The role of biostimulants and bioeffectors as alleviators of abiotic stress in crop plants. *Chem. Biol. Technol. Agric.* **2017**, *4*, 5. [CrossRef]
13. da Silva, R.R. Bacterial and fungal proteolytic enzymes: Production; catalysis and potential applications. *Appl. Biochem. Biotechnol.* **2017**, *183*, 1–19. [CrossRef] [PubMed]
14. da Silva, R.R. Comment on mRNA-Sequencing Analysis Reveals Transcriptional Changes in Root of Maize Seedlings Treated with Two Increasing Concentrations of a New Biostimulant. *J. Agric. Food. Chem.* **2018**, *66*, 2061–2062. [CrossRef] [PubMed]
15. Ertani, A.; Francioso, O.; Ferrari, E.; Schiavon, M.; Nardi, S. Spectroscopic-Chemical Fingerprint and Biostimulant Activity of a Protein-Based Product in Solid Form. *Molecules* **2018**, *23*, E1031. [CrossRef] [PubMed]
16. Trevisan, S.; Manoli, A.; Ravazzolo, L.; Franceschi, C.; Quaggiotti, S. mRNA-Sequencing Analysis Reveals Transcriptional Changes in Root of Maize Seedlings Treated with Two Increasing Concentrations of a New Biostimulant. *J. Agric. Food. Chem.* **2017**, *65*, 9956–9969. [CrossRef]
17. Trevisan, S.; Manoli, A.; Begheldo, M.; Nonis, A.; Enna, M.; Vaccaro, S.; Caporale, G.; Ruperti, B.; Quaggiotti, S. Transcriptome analysis reveals coordinated spatiotemporal regulation of hemoglobin and nitrate reductase in response to nitrate in maize roots. *New Phytol.* **2011**, *192*, 338–352. [CrossRef] [PubMed]
18. Trevisan, S.; Begheldo, M.; Nonis, A.; Quaggiotti, S. The miRNA-mediated post-transcriptional regulation of maize response to nitrate. *Plant Signal. Behav.* **2012**, *7*, 822–826. [CrossRef] [PubMed]
19. Trevisan, S.; Manoli, A.; Ravazzolo, L.; Botton, A.; Pivato, M.; Masi, A.; Quaggiotti, S. Nitrate sensing by the maize root apex transition zone: A merged transcriptomic and proteomic survey. *J. Exp. Bot.* **2015**, *66*, 3699–3715. [CrossRef]
20. Trevisan, S.; Trentin, A.R.; Ghisi, R.; Masi, A.; Quaggiotti, S. Nitrate affects transcriptional regulation of UPBEAT1 and ROS localisation in roots of *Zea mays* L. *Physiol. Plant.* **2018**. [CrossRef] [PubMed]
21. Quaggiotti, S.; Ruperti, B.; Borsa, P.; Destro, T.; Malagoli, M. Expression of a putative high-affinity NO_3- transporter and of an H+-ATPase in relation to whole plant nitrate transport physiology in two maize genotypes differently responsive to low nitrogen availability. *J. Exp. Bot.* **2003**, *54*, 1023–1031. [CrossRef] [PubMed]
22. Farooq, M.; Hussain, M.; Wakeel, A.; Siddique, K.H.M. Salt stress in maize: Effects; resistance mechanisms; and management. A review. *Agron. Sustain. Dev.* **2015**, *35*, 461–481. [CrossRef]
23. Henry, C.; Bledsoe, S.; Griffiths, C.A.; Paul, M.J.; Kollman, A.; Sakr, S.; Lagrimini, M. Differential role for trehalose metabolism in salt stressed maize. *Plant Physiol.* **2015**, *169*, 1072–1089. [CrossRef] [PubMed]
24. Zörb, C.; Mühling, K.H.; Kutschera, U.; Geilfus, C.M. Salinity Stiffens the Epidermal Cell Walls of Salt-Stressed Maize Leaves: Is the Epidermis Growth-Restricting? *PLoS ONE* **2015**, *10*, e0118406. [CrossRef]
25. Manoli, A.; Sturaro, A.; Trevisan, S.; Quaggiotti, S.; Nonis, A. Evaluation of candidate reference genes for qPCR in maize. *J. Plant Physiol.* **2012**, *169*, 807–815. [CrossRef] [PubMed]
26. Manoli, A.; Begheldo, M.; Genre, A.; Lanfranco, L.; Trevisan, S.; Quaggiotti, S. NO homeostasis is a key regulator of early nitrate perception and root elongation in maize. *J. Exp. Bot.* **2014**, *65*, 185–200. [CrossRef] [PubMed]
27. Livak, K.J.; Schmittgen, T.D. Analysis of relative gene expression data using real-time quantitative PCR and the 2-ΔΔCT method. *Methods* **2001**, *25*, 402–408. [CrossRef] [PubMed]
28. Bustin, S.A.; Nolan, T. Pitfalls of quantitative real-time reverse-transcription polymerase chain reaction. *J. Biomol. Tech.* **2004**, *15*, 155–166. [PubMed]
29. Schaafsma, G. Safety of protein hydrolysates; fractions thereof and bioactive peptides in human nutrition. *Eur. J. Clin. Nutr.* **2009**, *63*, 1161–1168. [CrossRef] [PubMed]

30. du Jardin, P. The Science of Plant Biostimulants—A bibliographic analysis. Ad hoc Study Report to the European Commission DG ENTR 2012. Available online: https://orbi.uliege.be/handle/2268/169257 (accessed on 7 January 2019).
31. Kasparkova, V.; Kolomaznik, K.; Burketova, L.; Sasek, V.; Simek, L. Characterization of low-molecular weight collagen hydrolysates prepared by combination of enzymatic and acid hydrolysis. *J. Am. Leather Chem. Assoc.* **2009**, *104*, 46–51.
32. Pecha, J.; Fürst, T.; Kolomaznik, K.; Friebrova, V.; Svoboda, P. Protein biostimulant foliar uptake modeling: The impact of climatic conditions. *AIChE J.* **2012**, *58*, 2010–2019. [CrossRef]
33. Baglieri, A.; Cadili, V.; Monterumici, C.M.; Gennari, M.; Tabasso, S.; Montoneri, E.; Nardi, S.; Negre, M. Fertilization of bean plants with tomato plants hydrolysates. Effect on biomass production; chlorophyll content and N assimilation. *Sci. Hortic.* **2014**, *176*, 194–199. [CrossRef]
34. Colla, G.; Nardi, S.; Cardarelli, M.; Ertani, A.; Lucini, L.; Canaguier, R.; Rouphael, Y. Protein hydrolysates as biostimulants in horticulture. *Sci. Hortic.* **2015**, *96*, 28–38. [CrossRef]
35. Calvo, P.; Nelson, L.; Kloepper, J.W. Agricultural uses of plant biostimulants. *Plant Soil.* **2014**, *383*, 3–41. [CrossRef]
36. Yakhin, O.I.; Lubyanov, A.A.; Yakhin, I.A.; Brown, P.H. Biostimulants in plant science: A global perspective. *Front. Life Sci.* **2017**. [CrossRef] [PubMed]
37. Mittler, R. Abiotic stress; the field environment and stress combination. *Trends Plant Sci.* **2006**, *11*, 15–19. [CrossRef] [PubMed]
38. Mittler, R.; Blumwald, E. Genetic engineering for modern agriculture: Challenges and perspectives. *Annu. Rev. Plant Biol.* **2010**, *61*, 443–462. [CrossRef] [PubMed]
39. Suzuki, N.; Rivero, R.M.; Shulaev, V.; Blumwald, E.; Mittler, R. Abiotic and biotic stress combinations *New Phytol.* **2014**, *203*, 32–43. [CrossRef] [PubMed]
40. Malamy, J.E. Intrinsic and environmental response path- ways that regulate root system architecture. *Plant Cell Environ.* **2005**, *28*, 67–77. [CrossRef]
41. Hodge, A.; Berta, G.; Doussan, C.; Merchan, F.; Crespi, M. Plant root growth; architecture and function. *Plant Soil* **2009**, *321*, 153–187. [CrossRef]
42. Kong, X.; Liu, G.; Liu, J.; Ding, Z. The Root Transition Zone: A Hot Spot for Signal Crosstalk. *Trends Plant Sci.* **2018**, *23*, 403–409. [CrossRef]
43. Alarcon, M.V.; Salguero, J. Transition zone cells reach G2 phase before initiating elongation in maize root apex. *Biol. Open* **2017**, *6*, 909–913. [CrossRef]
44. Yang, Z.B.; Geng, X.; He, C.; Zhang, F.; Wang, R.; Horst, W.J.; Ding, Z. TAA1-regulated local auxin biosynthesis in the root-apex transition zone mediates the aluminum-induced inhibition of root growth in Arabidopsis. *Plant Cell* **2014**, *26*, 2889–2904. [CrossRef]
45. Baluska, F.; Volkmann, D.; Barlow, P.W. Specialized zones of development in roots: View from the cellular level. *Plant Physiol.* **1996**, *112*, 3–4. [CrossRef] [PubMed]
46. Baluska, F.; Mancuso, S.; Volkmann, D.; Barlow, P.W. Root apex transition zone: A signalling-response nexus in the root. *Trends Plant Sci.* **2010**, *15*, 402–408. [CrossRef] [PubMed]
47. Baluska, F.; Mancuso, S. Root apex transition zone as oscillatory zone. *Front. Plant Sci.* **2013**, *4*, 354. [CrossRef] [PubMed]
48. Colla, G.; Hoagland, L.; Ruzzi, M.; Cardarelli, M.; Bonini, P.; Canaguier, R.; Rouphael, Y. Biostimulant Action of Protein Hydrolysates: Unraveling Their Effects on Plant Physiology and Microbiome. *Front. Plant Sci.* **2017**, *8*, 2202. [CrossRef]
49. Lucini, L.; Rouphael, Y.; Cardarelli, M.; Bonini, P.; Baffi, C.; Colla, G. A Vegetal Biopolymer-Based Biostimulant Promoted Root Growth in Melon While Triggering Brassinosteroids and Stress-Related Compounds. *Front. Plant Sci.* **2018**, *9*, 472. [CrossRef] [PubMed]
50. Sestili, F.; Rouphael, Y.; Cardarelli, M.; Pucci, A.; Bonini, P.; Canaguier, R.; Colla, G. Protein Hydrolysate Stimulates Growth in Tomato Coupled With N-Dependent Gene Expression Involved in N Assimilation. *Front. Plant Sci.* **2018**, *9*, 1233. [CrossRef]

51. Plett, D.C.; Holtham, L.R.; Okamoto, M.; Garnett, T.P. Nitrate uptake and its regulation in relation to improving nitrogen use efficiency in cereals. *Semin. Cell Dev. Biol.* **2018**, *74*, 97–104. [CrossRef] [PubMed]
52. Wang, W.; Xia, M.X.; Chen, J.; Yuan, R.; Deng, F.N.; Shen, F.F. Gene Expression Characteristics and Regulation Mechanisms of Superoxide Dismutase and Its Physiological Roles in Plants under Stress. *Biochemistry* **2016**, *81*, 465–480. [CrossRef] [PubMed]

agronomy

MDPI

Article

Vegetal-Derived Biostimulant Enhances Adventitious Rooting in Cuttings of Basil, Tomato, and Chrysanthemum via Brassinosteroid-Mediated Processes

Hye-Ji Kim [1,*], Kang-Mo Ku [2], Seunghyun Choi [1] and Mariateresa Cardarelli [3]

1 Department of Horticulture and Landscape Architecture, Purdue University, West Lafayette, IN 47907, USA; choi510@purdue.edu
2 Division of Plant and Soil Sciences, West Virginia University, Morgantown, WV 26505, USA; kangmo.ku@mail.wvu.edu
3 Consiglio per la ricerca in agricoltura e l'analisi dell'economia agraria, Centro di Ricerca Orticoltura e Florovivaismo, 84098 Pontecagnano, Italy; mteresa.cardarelli@crea.gov.it
* Correspondence: hjikim@purdue.edu

Received: 23 December 2018; Accepted: 5 February 2019; Published: 10 February 2019

Abstract: Plant-derived protein biostimulants exhibit hormone-like activities promoting plant growth and yield, yet detailed investigations on hormonal function have remained limited. This study was conducted to investigate the effects of vegetal-derived-biostimulant on morphological and metabolic changes in cuttings of three herbaceous species demonstrating different rooting ability, basil (*Ocimum basilicum* L.), tomato (*Solanum lycopersicum* L.), and chrysanthemum (*Chrysanthemum indicum* L.), in comparison to auxin. Unrooted cuttings were applied with or without biostimulant (100, 1000, 5000, and 10,000 mg L^{-1}) or auxin [1% indole-3-butyric acid (IBA) plus 0.5% 1-naphthaleneacetic acid (NAA); 100, 200, 300, and 500 mg L^{-1}] as a basal quick-dip, stuck into inert media, and evaluated at 20 days after placement under intermittent mist. Both compounds increased adventitious rooting in all cuttings. Biostimulant required a significantly higher threshold for a series of adventitious rooting responses than auxin, and the maximum effectiveness was achieved at 5000 mg L^{-1} for biostimulant and 100, 200, and 300 mg L^{-1} for auxin in basil, tomato, and chrysanthemum, respectively. Adventitious rooting responses (dry mass and length) to biostimulant showed a gradual logarithmic rise as a function of increasing dosages, which was not in agreement with biphasic dose-response of auxin. Biostimulant significantly increased or tended to increase fine roots in all tested cuttings, which was not consistent with auxin. Relatively high levels of endogenous brassinosteroids (BRs) were present in non-treated cuttings of basil, tomato, and chrysanthemum in decreasing order. Both compounds had no effects or concomitantly increased or decreased BR levels in plant tissues, with fewer effects on basil and tomato, containing high BR levels, but more prominent effects on chrysanthemum, containing relatively low BR levels. Contrasting effects of biostimulant and auxin were found in antioxidant activities, which were promoted by biostimulant but inhibited by auxin either in roots or shoots. These results indicate that the hormonal effects of vegetal-derived biostimulant are primarily exerted by BR-mediated processes while involving interaction with auxin. Both the biostimulant-derived BRs and auxin were suggested to modulate endogenous BR pool via overlapping and interdependent regulatory functions, inducing morphological and metabolic changes during adventitious rooting of cuttings in a plant species-specific manner.

Keywords: stem cuttings; propagation; root morphology traits; indole-3-acetic acid (IAA); indole-3-butyric acid (IBA); gibberellins; phenolic compounds

1. Introduction

Plant-derived biostimulants represent a well-known group of biostimulants and have been proposed as an innovative tool to address the sustainability challenges facing horticulture and to ensure high yield and quality of horticultural commodities [1–3]. Manufactured from plant protein sources using partial hydrolysis, plant biostimulants are considered as a subgroup of growth regulators and bioregulators which are composed of a mixture of polypeptides, oligopeptides, and amino acids [4]. Plant-derived biostimulants are reported to be more effective than animal-derived biostimulants as they contain a higher concentration of amino acids and soluble peptides, with peptides being the principal active compounds [5–7]. Plant-derived biostimulants are defined as materials other than fertilizers that promote plant growth when applied in small quantities or metabolic enhancers [8]. They are available on the market as various forms, including liquid products, soluble powder or in granular form, and were demonstrated to be effective as a seed treatment, foliar spray, and soil drench for crop production [9–11]. When applied as a foliar spray or soil-drench, biostimulants can induce a series of physiological responses in crops changing their phenotypic characteristics and promoting plant growth [10,12].

It has been proposed that such plant responses induced by biostimulants are derived from hormone-like activities and the production of secondary metabolites [13]. Auxin- and gibberellin-like activities were demonstrated in corn coleoptiles and tomato cuttings [5,6], particularly due to the presence of bioactive peptides [14,15]. Peptides are known to be involved in cell differentiation, protease inhibitor induction, cell division, and pollen self-incompatibility response [16,17]. Similar results were reported in degraded soybean meal products, which had promotive effects on root hairs in *Brassica rapa* and tomato cuttings [7].

The positive effects of biostimulant on plant growth and yield have been demonstrated in many studies. The application of biostimulant not only enhanced the growth of corn seedlings [5,6,18] and stem cuttings of tomato [5], but also improved nutrient status, yield, and quality of herbaceous and woody plants, including corn, bean, tomato, sweet yellow pepper, strawberry, banana, papaya, and red grape [5,13,19–22]. It also enhanced tolerance to a wide range of abiotic stresses, such as drought [23], salinity [9], extreme temperatures [24], nutrient deficiency [25], and adverse soil pH [26]. The application of biostimulant increased root morphology, such as root dry mass, total root length, and root surface area, which was associated with improved nitrogen status [5,6]. However, it is not clear how such morphological and physiological changes are induced by the biostimulant.

Adventitious rooting involves significant cellular metabolic activities, leading to the formation of new roots at the base of stem cuttings. Auxin plays a pivotal role in promoting cell growth, cell division, and adventitious root formation in cuttings [27–29] and its mode of action on adventitious rooting is well-elucidated [27,30–32]. Rooting compounds commonly contain indole-3-butyric acid (IBA), 1-naphthaleneacetic acid (NAA), or a combination of the two compounds. The application of auxin to unrooted cuttings promotes adventitious rooting at relatively low doses.

Meanwhile, the mode of action of plant-derived biostimulants on adventitious rooting is largely unknown. An auxin-signaling mediated pathway was proposed to be involved in adventitious rooting of tomato cuttings and their improved nitrogen status, as represented by higher Soil Plant Analysis Development (SPAD) values, following the biostimulant Trainer® treatment [5]. They also found that biostimulant applications increased the shoot elongation rate of dwarf pea plants, prompting the idea of gibberellin involvement in regulating their shoot growth. Unlike auxins, gibberellins are known to inhibit the production of adventitious roots [33–35]. Therefore, stem cuttings provide an ideal experimental system with which mechanistic investigations on hormonal regulations associated with plant-derived-biostimulant can be undertaken. The system eliminates: (1) gibberellic acid as a potential candidate for biostimulant effects due to their antagonistic nature on adventitious rooting and (2) nutritional effects of biostimulant because nutrients are not required for initial stages of adventitious root formation. Nevertheless, carbohydrates play important roles in adventitious rooting, not only by providing energy and carbon chains for biosynthetic processes of new meristems and roots, but also by affecting gene expression, in collaboration with auxin [32].

Recent metabolomic investigations of the hormonal profile on greenhouse melon demonstrated that the application of biostimulant induced upregulation of metabolites related to brassinosteroids (BRs) and their interactions with other phytohormones were postulated to play a critical role in plant growth responses [36]. Similarly, transcriptomic profiles of lateral roots of maize seedlings demonstrated the involvement of BR signal transduction when treated with biostimulant [37]. Meanwhile, the effects of biostimulants were varied by plant species and/or cultivars, growing seasons, and the application method and concentration of the product [38] although the causes for these variations are not clear.

The objectives of the present study were: (i) to examine the hormonal effects of a plant-derived-biostimulant on adventitious rooting in cuttings of three herbaceous plant species with different rooting ability, (ii) to determine dose responses of stem cuttings to biostimulant and auxin, and (iii) to characterize morphological and metabolic changes induced by biostimulant. Cuttings of basil, tomato, and chrysanthemum were treated with biostimulant by a basal quick-dip, and morphological, physiological, and metabolic changes were evaluated to elucidate the hormonal regulation of biostimulant involved in adventitious rooting formation.

2. Materials and Methods

2.1. Plant Materials

Based on our preliminary observations on adventitious root formation of herbaceous plants, cuttings of basil (*Ocimum basilicum* L. cv. Genovese), tomato (*Solanum lycopersicum* L. cv. Washington Cherry), and chrysanthemum (*Chrysanthemum indicum* L. cv. Hollister) were chosen in this study for differences in their relative rooting ability: easy-to-root, moderate-to-root, and difficult-to-root, respectively. In general, herbaceous plant species can produce adventitious roots without application of exogenous auxin; however, auxin application is of commercial importance in cutting propagation, because the endogenous level of auxin is critical to increase the ease during root induction period [34]. The experiment was carried out in summer 2017 to spring 2018, in a glass greenhouse situated at Purdue University, West Lafayette, IN (lat. 40N, long. 86W; altitude 188m above sea level).

Seeds of tomato and basil were acquired from a commercial source (Johnny's Selected Seeds, Albion, ME, USA). The seeds were sown and grown in a growth room for 2 to 3 weeks at Plant Growth Facilities. Meanwhile, unrooted cuttings of chrysanthemum 'Hollister' were obtained from a commercial source (Syngenta Flowers, LLC., Gilroy, CA, USA). Immediately upon receipt, a box of stem cuttings was kept in a refrigerator maintained at 5 °C. The cuttings were applied with auxin within three days and then stuck into the media and propagated as described below to produce stock plants. Uniform seedlings of basil and tomato and rooted cuttings of chrysanthemum were randomly chosen and transplanted into 2 L plastic containers filled with a commercial potting mix (Fafard 2P Mix; Conrad Fafard, Agawam, MA, USA). Plants were fertigated with acidified water supplemented with a combination of two water-soluble fertilizers (3:1 mixture of 15N–2.2P–12.5K and 21N–2.2P–16.6K, respectively; Everris NA Inc., Dublin, OH, USA) to provide the following (in mg L^{-1}): 150 nitrogen (N), 20 phosphorous (P), 122 potassium (K), 38 calcium (Ca), 15 magnesium (Mg), 0.8 iron (Fe), 0.4 manganese (Mn) and zinc (Zn), 0.2 copper (Cu) and boron (B), and 0.1 molybdenum (Mo). Nitrate form was 76% of nitrogen provided. Irrigation water was supplemented with 93% sulfuric acid (Brenntag, Reading, PA, USA) at 0.08 mL L^{-1} to reduce alkalinity to 100 mg L^{-1} calcium carbonate ($CaCO_3$) and pH to a range of 5.8 to 6.2. The stock plants were grown in a glass-glazed greenhouse with exhaust fan and evaporative-pad cooling, radiant hot water heating, and retractable shade curtains controlled by an environmental computer (Maximizer Precision 10; Priva Computers, Vineland Station, ON, Canada). The average day and night temperatures were 23.8 ± 0.8 and 20.3 ± 0.9 °C, respectively. The photoperiod was 14-h (0800 to 2200 HR) consisting of natural day lengths with supplemental lighting using high-pressure sodium (HPS) lamps. A supplemental photosynthetic photon flux (*PPF*) was measured using a quantum sensor (LI-250A light meter; LI-COR Biosciences, Lincoln, NE, USA) and was approximately 150 μmol m^{-2} s^{-1} at canopy height. The relative humidity inside the greenhouse ranged from 50% to 70% during the study.

Cuttings were taken from the tips of mature stock plants grown in a greenhouse for 2 to 3 months. The cuttings were prepared to have four apical leaves by removing extra leaves from the basal node and trimmed to be uniform in length.

2.2. Biostimulant and Auxin Treatments and Propagation Conditions

The commercial plant-derived biostimulant Quik-link® (Italpollina S.p.a, Rivoli Veronese, Italy) was used in this study. It contains trace elements (10 g kg^{-1} Fe; 7 g kg^{-1} Mn; 3 g kg^{-1} Zn; 1 g kg^{-1} Cu; 0.2 g kg^{-1} Mo) and organic compounds biologically active like vegetal amino acids and peptides. The aminogram (expressed as percentage of the total amino acids) is: Ala(4.5), Arg(6.7), Asp(12.7), Cys(1.1), Glu(20.2), Gly(4.5), His(3.0), Ile(4.9), Leu(8.3), Lys(6.8), Met (1.5), Phe (5.6), Pro (5.6), Thr (4.1), Trp (1.1), Tyr (4.1), Val (5.3). The product also contains the Root Hair Promoting Peptide (RHPP) which is a signaling peptide stimulating root growth [36].

Quik-link is allowed in organic agriculture according to the Council Regulation (EC) No. 834/2007 of 28 June 2007 [36], and is manufactured by Italpollina USA Inc. (Anderson, IN, USA). The biostimulant was prepared in five concentrations of 0 (control), 1000, 3000, 5000, and 10,000 mg L^{-1}. Meanwhile, a commercial formulation of indole-3-butyric acid (IBA) and 1-naphthaleneacetic acid (NAA) (Dip'N Grow, Inc., Clackamas, OR, USA) was used for auxin treatment, since indole-3-acetic acid (IAA), a naturally occurring compounds, can be easily degraded in the presence of light and is susceptible to destruction in the plant by IAA-oxidase [27–29]. IBA and NAA are more effective than the naturally occurring or synthetic IAA for rooting, and therefore, are the most widely used auxins for rooting stem cuttings [34]. The formulation was prepared in five concentrations of 0, 100, 200, 300, and 500 mg L^{-1}, providing IBA and NAA concentrations at 0, 492 μM IBA + 537 μM NAA, 984 μM IBA + 1074 μM NAA, 1476 μM IBA + 1611 μM NAA, and 2460 μM IBA + 2685 μM NAA, respectively.

Stem base of unrooted cuttings were dipped into a solution of either biostimulant or auxin using a basal quick dip method for 3 s to a depth of 2 cm. The stems were quickly stuck into polystyrene cell packs (300 cm^3 soil volume per cell) filled with inert media (1:1 (v/v) perlite and vermiculite mixture). The cell packs were then placed into polystyrene trays and placed under an intermittent mist, providing bottom heat and overhead mist for 10 s every 20 min during daylight hours with 76 to 98% relative humidity at canopy height for a rooting period of 21 days. The photoperiod was 14-h (0800 to 2200 HR) consisting of natural day lengths with supplemental lighting using high-pressure sodium (HPS) lamps. A supplemental photosynthetic photon flux (*PPF*) was approximately 75 μmol m^{-2} s^{-1} at canopy height and daily maximum/minimum temperatures in the greenhouse were 23.3 ± 0.8 /22.6 ± 0.7 °C.

2.3. Plant Growth Measurements

When the maximum rooting was observed at day 20, stem length was measured from the stem end to the apical growing point and the number of leaves were recorded. The Soil Plant Analysis Development (SPAD) value, an index of chlorophyll content per unit leaf area, was measured using the SPAD chlorophyll meter (Minolta Corporation, Ltd., Osaka, Japan) on three newly expanded leaves and three fully matured leaves separately, and averaged at each group.

At the end of the rooting experiment, plant parts were separated into leaves, stems and roots. The fresh mass of each part was determined immediately after harvest and were dried in a forced-convection oven at 70 °C (Heratherm OMH400, Thermo Scientific Inc., Waltham, MA, USA) for 3 days until a constant weight was reached. Shoot dry mass was calculated as the sum of aerial vegetative parts, and total dry mass was calculated as the shoot and root dry mass. The root-to-shoot ratio was calculated based on the dry mass of roots and shoots. Total plant dry mass was calculated by adding the dry mass of each plant part. Shoot and root dry mass were analyzed by regression as a continuous response to log [concentration] and by analysis of variance to the treatment. The dried plant tissues were ground in a Wiley mill to pass through a 20-mesh screen, and 0.1 mg samples were weighed and subsequently analyzed for the nitrogen content using a Flash EA elemental analyzer (Thermo Scientific, Waltham, MA, USA).

2.4. Measurements of Root Morphological Traits

Cuttings were subjected to root morphological analysis at day 20. The number of adventitious roots were counted manually at harvest when the roots were separated from the stems using a razor blade. Entire roots were carefully rinsed and scanned using the Epson Expression 11000XL scanner (Epson America Inc., Long Beach, CA, USA). The debris removal filter was set to discount objects less than 1 cm^2 with a length/width ratio less than 4. The scanned images were then used to determine root morphological traits, such as total root length, root surface area, average root diameter, and root volume, using WinRHIZO Pro software (Regent Instrument Inc., Quebec City, QC, Canada). After root images were taken, the roots were weighed and dried in an oven set at 75 °C until the samples were completely dry to weigh dry mass. Diameter class length (root length within a diameter class) were generated in the images of adventitious roots acquired from WinRHIZO. The roots were divided into 26 diameter classes at 0.25 mm intervals and root length per each root diameter class was calculated. The root diameter class distribution was computed based on the proportion of the root length in each root diameter class compared to the total root length.

2.5. Primary Metabolite Extraction and Qualitative Analysis from Biostimulant

Primary metabolites were extracted following published protocols with modifications of extraction solvent volume. Quik-link (0.5 mL) were weighed into 2 mL microcentrifuge tubes, followed by the addition of 0.2 mL of water. To fractionate non-polar compounds, 0.375 mL of cold chloroform (−20 °C) and 0.7 mL methanol were added. After vigorous up-and-down mixing by hand (50 times), the extracts were centrifuged at 12,000 × *g* for 4 min, 100 μL supernatant (water soluble metabolites) and organic phase (lipid soluble metabolites) were transferred to 1.5 mL microcentrifuge tubes, respectively. The extracts were dried using Vacufuge concentrator (Eppendorf, Thermo Fisher Scientific, Waltham, MA, USA) with 20 μL of methanol to facilitate water evaporation. For water soluble metabolites, dried extracts were derivatized with 50 μL methoxyamine hydrochloride (40 mg ml^{-1} in pyridine) for 90 min at 37 °C, then with 100 μL MSTFA + 1% TMCS at 50 °C for 20 min. For lipid soluble metabolites, dried extracts were derivatized with 200 μL n,o-bis(trimethylsilyl)trifluoroacetamide with 1% of trimethylchlorosilane at 75 °C for 30 min. Metabolites were analyzed using a gas chromatography-mass spectrometry (GC-MS) (Trace 1310 GC, Thermo Fisher Scientific, Waltham, MA, USA) coupled to an MS detector system (ISQ QD, Thermo Fisher Scientific, Waltham, MA, USA) and an autosampler (Triplus RSH, Thermo Fisher Scientific, Waltham, MA, USA). A capillary column (Rxi-5Sil MS, Restek, Bellefonte, PA, USA; 30 m × 0.25 mm × 0.25 μm capillary column w/10 m Integra-Guard Column) was used to detect polar metabolites. For water-soluble metabolite analysis, after an initial temperature hold at 80 °C for 2 min, the oven temperature was increased to 330 °C at 15 °C min^{-1} and held for 5 min. For lipid-soluble metabolite analysis, after an initial temperature hold at 150 °C for 1 min, the oven temperature was increased to 320 °C at 12 °C min^{-1} and held for 7 min. Injector and detector temperatures were set at 250 °C and 250 °C, respectively. An aliquot of 1 μL was injected with the split ratio of 70:1. The helium carrier gas was kept at a constant flow rate of 1.2 mL min^{-1}. The mass spectrometer was operated in positive electron impact mode (EI) at 70.0 eV ionization energy at m/z 40–500 scan range. Metabolite identification was based on the National Institute of Standards and Technology (NIST) library.

2.6. Quantification of BRs in Plant-Derived Biostimulant and Plant Samples

Campesterol, stigmasterol, and beta-sitosterol were quantified based on GC-MS. Quik-link (0.5 mL) were weighed into 2 mL microcentrifuge tubes, followed by the addition of 0.2 mL of water. To fractionate non-polar compounds, 0.375 mL of cold chloroform (–20 °C) and 0.7 mL methanol were added. After vigorous up-and-down mixing by hand (50 times), the extracts were centrifuged at 12,000× *g* for 4 min, and 187.5 µL chloroform layer were transferred to 1.5 mL microcentrifuge tubes. For the quantification of BRs from plant samples, 100 µL of organic phase from the primary metabolite analysis above session was used. The extracts were dried using Vacufuge concentrator (Eppendorf, Thermo Fisher Scientific, Waltham, MA, USA). Dried extracts were derivatized with 200 µL n,o-bis(trimethylsilyl)trifluoroacetamide with 1% of trimethylchlorosilane at 75 °C for 30 min. BRs were analyzed using a GC-MS (Trace 1310 GC, Thermo Fisher Scientific, Waltham, MA, USA) coupled to an MS detector system (ISQ QD, Thermo Fisher Scientific, Waltham, MA, USA) and an autosampler (Triplus RSH, Thermo Fisher Scientific, Waltham, MA, USA). A capillary column (Rxi-5Sil MS, Restek, Bellefonte, PA, USA; 30 m × 0.25 mm × 0.25 µm capillary column w/10 m Integra-Guard Column) was used to detect polar metabolites. After an initial temperature hold at 150 °C for 1 min, the oven temperature was increased to 320 °C at 12 °C min^{-1} and held for 7 min. Injector, MS detector temperatures were set at 250 °C, 250 °C, and 300 °C respectively. An aliquot of 1 µL was injected with the splitless mode. The helium carrier gas was kept at a constant flow rate of 1.2 mL min^{-1}. The mass spectrometer was operated in positive electron impact mode (EI) at 70.0 eV ionization energy at m/z 45–600 scan range. Metabolite identification was based on standard compounds in comparison with the mass spectra and retention time. The standard BRs were injected from 25 ng mL^{-1} to 1000 ng mL^{-1} concentrations.

2.7. Amino Acid Quantification of Vegetal-Biostimulant

To quantify the free amino acid content in the sample, EZ:faast free amino acid for GC-MS kit (Phenomenex, Torrance, CA, USA) was utilized to extract and measure the amino acid concentration. 75 mg of the sample was incubated with 1.5 mL water overnight to extract the free amino acid from the sample. After the 24 h incubation, samples were centrifuged at 12,000 × g for 3 min. Amino acid purification and derivatization were conducted on EZ:faast instruction. The analysis of amino acid was carried out in a gas chromatograph (Trace 1310 GC, Thermo Fisher Scientific, Waltham, MA, USA) coupled to a flame ionization detector (FID), and an autosampler (Triplus RSH, Thermo Fisher Scientific, Waltham, MA, USA). A capillary column (ZebronTM EZ-AAA amino acid GC, Phenomenex, Torrance, CA, USA; 10 m, 0.25 mm) was used. The injection ratio was set at 1:15 and the injection temperature was 250 °C. The injection volume was 1.5 µL. The carrier gas was helium and the flow rate was 1.1 mL min^{-1}. The column oven was set at 110 °C and increased 30 °C per minute to 320 °C. FID temperature was set at 220 °C and the air flow 450 mL min^{-1} and the hydrogen flow was 45 mL min^{-1}.

2.8. Total Phenolic Content and Antioxidant Capacity

Total phenolic content and antioxidant capacity were analyzed using methanol extracts that described above primary metabolite analysis based on the published methods [39]. Freeze-dried samples (20 mg) were extracted in 1.4 mL of 100% methanol at 60 °C for 10 min. After centrifuge the supernatants were used for the total phenolic content, 2,2′-azino-bis(3-ethylbenzothiazoline-6-sulphonic acid) (ABTS), and 2,2-diphenyl- 1-picryl-hydrazyl-hydrate (DPPH) antioxidant capacity analyses. Various concentrations of vitamin C were used as standard curves for ABTS and DPPH assays [39]. For the DPPH assay, reaction mixtures containing test samples (10 µL) and 190 µL of a 200 µM DPPH in ethanol were incubated at room temperature for 30 min in 96-well plates. The absorbance of the DPPH free radical was measured at 515 nm using an Epoch 2 plate reader (Biotek Instruments Inc., Winooski, VT, USA). Antioxidant data were expressed as vitamin C equivalent concentration (µg g^{-1} DW). For the ABTS assay, 7 mM ABTS ammonium salt was dissolved in a potassium phosphate buffer (pH 7.4) and treated with 2.45 mM potassium persulfate. The mixture was then allowed to stand at room temperature for 12–16 h for full color development (dark blue). The solution was then diluted with potassium phosphate buffer until absorbance reached 1.0 ± 0.02 at 735 nm using an Epoch 2 plate reader (Biotek Instruments Inc., Power Wave XS, Winooski, VT, USA).

Subsequently, 190 µL of this solution was mixed with 10 µL of the sample extracts. The absorbance was recorded at room temperature after 6 min. Antioxidant data were expressed as vitamin C equivalent concentration (µg g^{-1} DW). For total phenolic content, Folin-ciocalteu reagent was used to determine total phenolic content [39]. Each sample (10 µL) was mixed with (100 µL) of Folin-Ciocalteu reagent (0.2 N) followed by 3 min of incubation at room temperature. Then, 90 µL of sodium carbonate (7.5%) was added. After 60 min of incubation in the dark at room temperature, absorbance was obtained at 735 nm. The total phenolic concentration was determined based on a standard curve of gallic acid.

2.9. Experimental Design and Statistical Analysis

Treatments were arranged in a completely randomized block design. The procedure was repeated at three different time blocks, and each block consisted of 9 treatments and 10 replicates per treatment, amounting to a total of 270 cuttings (90 samples per each plant species) per each time block. All data were subjected to analysis of variance using JMP for Windows, Version 13.2 (SAS Institute Inc., Cary, NC, USA). Polynomial contrasts were used to compare the treatment effects of biostimulant and auxin. Mean separation within each measured parameter was performed by Tukey's honestly significant difference (HSD) test at $p < 0.05$. Regression analysis was carried out to look for trends in response to the concentration for each treatment. Results from the three experiments showed similar trends and the data sets were consistent with each other. However, because the error variance was not homogeneous between experiments, statistical analyses were conducted separately for each experiment, and data from the two trials were pooled and presented here.

3. Results

3.1. The Effects of Biostimulant on Adventitious Rooting in Cuttings of Basil, Tomato, and Chrysanthemum

All cuttings achieved 100% rooting regardless of plant species and treatment. The average number of adventitious roots in untreated cuttings of basil, tomato, and chrysanthemum were 39, 22, and 16, respectively, demonstrating genetic variations in rooting ability. Total root length was higher in the order of tomato, basil, and chrysanthemum (Table 1).

Meanwhile, both biostimulant and auxin increased adventitious rooting in a dose-dependent manner: The number of adventitious roots, root dry mass, and total root length in all plant species increased or showed an increasing trend by higher concentrations of biostimulant and auxin, with exception of overdoses (IBA + NAA0.5) in auxin (Table 1). However, the response level of plant species varied significantly by the treatments. Rooting response increased more prominently by auxin than biostimulant. An optimal level of auxin to induce rooting was highly plant-species specific and maximum root length was achieved nearly at concentrations of 100, 200, and 300 mg L^{-1} in basil, tomato, and chrysanthemum, respectively (Figure 1f). When compared to auxin, biostimulant was required approximately 15- to 50-time higher concentrations to induce the onset of adventitious root formation. In general, the application of biostimulant at a concentration of 5000 mg L^{-1} increased total root length in all tested cuttings (Table 1). An overdose of auxin tended to negatively affect the dry mass of adventitious roots in basil and tomato to the levels of unrooted cuttings but not in chrysanthemum. Such response was contrasting to biostimulant, where higher concentrations of biostimulant tended to increase or gradually increased adventitious rooting, and even the highest concentration at 10,000 mg L^{-1} did not negatively affect root morphological characteristics (Table 1). The root dry mass was positively correlated with the total root length of cuttings treated with either biostimulant or auxin (Figure S1); however, the relationship between dry mass and total root length slightly varied among plant species and between the treatments (Figure S1), indicating that dry mass and/or total root length do not precisely predict response levels of cuttings to biostimulant and auxin applications.

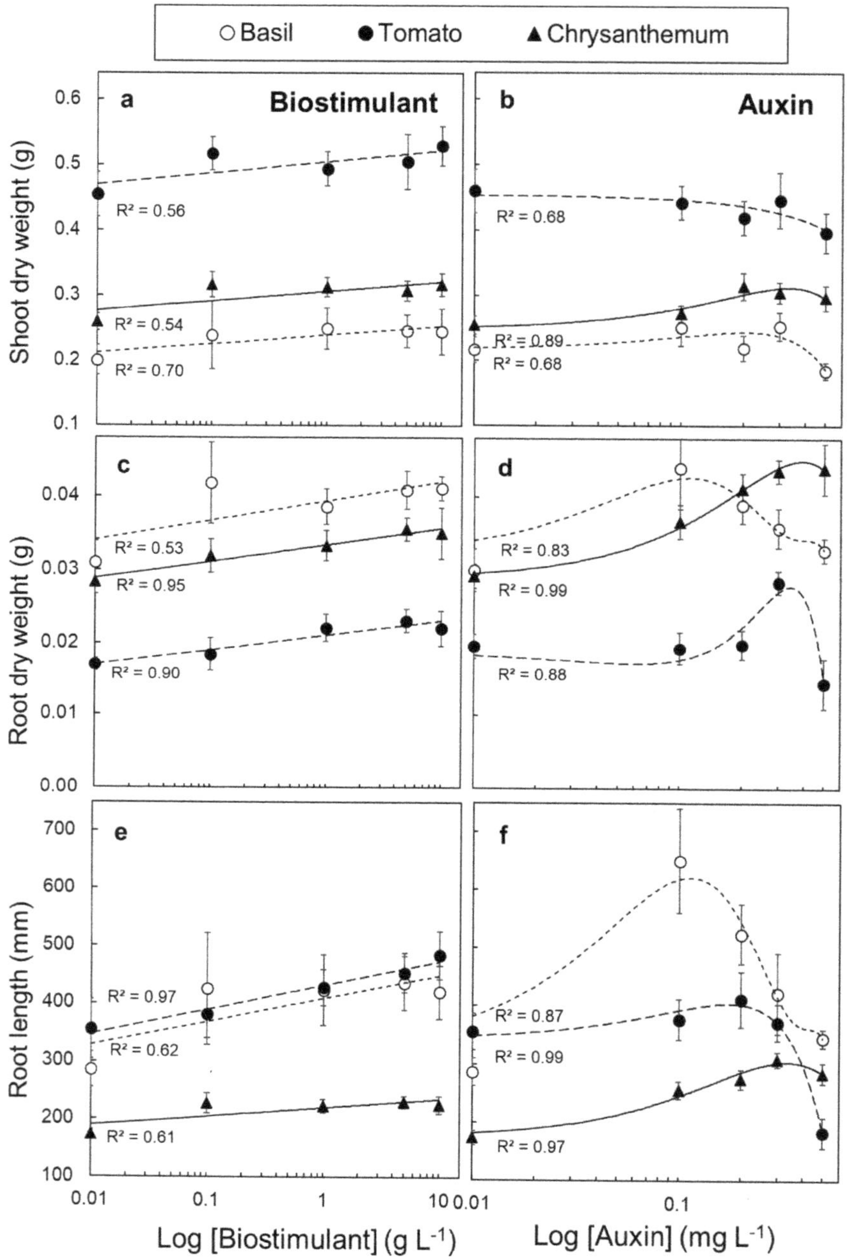

Figure 1. Dose response curves showing the effects of biostimulant or auxin applications on shoot and root dry mass, and total length of adventitious roots in cuttings of basil, tomato, and chrysanthemum. Each data point is the mean ± SE of 20 replicates.

Table 1. Effects of biostimulant (B) or auxin (IBA + NAA) applications on adventitious root characteristics in cuttings of basil, tomato, and chrysanthemum. Cuttings were treated with or without biostimulant (100, 1000, 5000 (B_5), and 10,000 (B_{10}) mg L^{-1}) or auxin (100, 200 (IBA + $NAA_{0.2}$), 300, and 500 (IBA + $NAA_{0.5}$) mg L^{-1}) at the stem base as quick-dip, stuck into inert media, and evaluated at day 20 after placement under intermittent mist. Note that two concentrations per each treatment were presented here.

Treatment	Root Number	Root Dry Mass (g $plant^{-1}$)	Total Root Length (mm)	Root Surface Area (mm^2)	Root Volume (mm^3)	Root Diameter (mm)
			Basil			
Control	39.4	0.033 a	286 b	48.7 b	0.67 b	0.55
B_5	52.0	0.041 a	436 a	69.0 a	0.88 a	0.51
B_{10}	51.9	0.041 a	420 ab	66.4 a	0.84 a	0.52
IBA+$NAA_{0.2}$	42.4	0.034 a	322 ab	60.1 ab	0.89 a	0.60
IBA+$NAA_{0.5}$	49.8	0.033 a	346 ab	60.9 ab	0.87 a	0.57
Significance	ns	ns	*	*	*	ns
			Tomato			
Control	22.2 c	0.017 ab	355 b	41.8 a	0.39	0.38 b
B_5	36.9 a	0.023 a	454 ab	50.2 a	0.44	0.35 bc
B_{10}	36.2 a	0.022 a	484 a	50.7 a	0.42	0.33 c
IBA+$NAA_{0.2}$	28.3 b	0.021 ab	362 ab	42.4 a	0.40	0.38 b
IBA+$NAA_{0.5}$	16.2 c	0.014 b	183 c	28.0 b	0.34	0.48 a
Significance	***	*	***	***	ns	***
			Chrysanthemum			
Control	16.4 c	0.029 d	173 c	35.0 c	0.57 b	0.63
B_5	20.1 bc	0.033 cd	223 b	45.7 b	0.75 a	0.65
B_{10}	21.1 bc	0.036 bc	230 b	47.0 b	0.77 a	0.66
IBA+$NAA_{0.2}$	39.8 a	0.041 ab	275 ab	53.4 ab	0.83 a	0.63
IBA+$NAA_{0.5}$	25.1 b	0.046 a	285 a	57.1 a	0.92 a	0.64
Significance	***	***	***	***	***	ns

ns, *, **, and *** indicate non-significant, or significant at $p < 0.05$, 0.01, and 0.001, respectively. Different letters within each column indicate significant differences according to Tukey's HSD test ($p = 0.05$). Data are means of 20 replicates.

3.2. The Effects of Biostimulant on Root Diameter Class Distribution

Average diameters of adventitious roots varied among plant species. Tomato cuttings had relatively fine roots with an average root diameter of 0.38 mm, while the roots of basil and chrysanthemum composed of coarser roots with average root diameters of 0.55 and 0.63 mm, respectively (Table 1). In tomato, most of the roots were in the finer root class ranging from 0.0 to 0.50 mm, accounting for 75% of total root length (Table 2). The roots of basil consisted of a mixture of finer root diameter classes: about 43% of the total roots were in the finer root class (0.0 to 0.50 mm) and about 41% were intermediate root class (0.50 to 0.75 mm) (Table 2). On the other hand, chrysanthemum produced a wide range of root diameter classes (0.0 to 3.0 mm). About 40% of the total roots were in the finer root class, while the rest of the roots were composed of coarser roots (>0.5 mm). Unlike basil and tomato, where the roots thicker than 1 mm were only a small fraction among the roots (less than 3 and 0.5%, respectively) and were primarily proximal near the stem, more than 10% of the total roots of chrysanthemum were in the coarse root class (>1.0 mm).

Root diameter class distribution analyses revealed treatment differences even within the same plant species (Table 2). In basil and tomato, increasing biostimulant concentrations promoted fine roots (0.0 to 0.25 mm). These results were contrasting to auxin-treated cuttings, in which higher auxin concentrations had an increasing trend of promoting coarser roots (0.50 to 1.00 mm). The response of chrysanthemum roots was quite different from those of basil and tomato: Auxin had more pronounced effects on changing root morphological traits in chrysanthemum, and an optimal auxin concentration (IBA + NAA0.2) significantly promoted the formation of fine roots (0 to 0.25 mm) while decreasing coarser roots (>0.75 mm). Likewise, vegetal-biostimulant tended to promote finer roots in chrysanthemum, but to a lesser degree than auxin.

Table 2. Root diameter class (mm) and relative diameter class length (%) of cuttings of basil, tomato, and chrysanthemum. Cuttings were treated with or without biostimulant (100, 1000, 5000 (B_5), and 10,000 (B_{10}) mg L^{-1}) or auxin (100, 200 (IBA + $NAA_{0.2}$), 300, and 500 (IBA + $NAA_{0.5}$) mg L^{-1}) at the stem base as quick-dip, stuck into inert media, and evaluated at day 20 after placement under intermittent mist. Note that two concentrations per each treatment were presented here. Percentage values at each diameter class are given.

Treatment	Root diameter class (mm)				
	0–0.25	0.25–0.50	0.50–0.75	0.75–1.00	>1.00
	Relative root diameter class length (%)				
	Basil				
Control	15.7 b	27.0	41.0	13.8	2.5
B_5	22.7 a	28.5	40.0	11.7	3.8
B_{10}	21.6 a	30.3	38.2	11.4	4.5
IBA+$NAA_{0.2}$	20.6 ab	26.8	33.4	11.3	3.1
IBA+$NAA_{0.5}$	15.5 b	25.3	32.2	15.5	3.6
Significance	**	ns	ns	ns	ns
	Tomato				
Control	35.2 b	40.1	22.4 b	1.2 b	0.3 b
B_5	39.2 ab	42.3	17.3 bc	1.0 b	0.2 b
B_{10}	43.4 a	41.6	13.8 c	1.0 b	0.2 b
IBA+$NAA_{0.2}$	33.1 b	45.7	18.6 bc	1.9 b	0.7 b
IBA+$NAA_{0.5}$	21.6 c	38.1	34.3 a	4.0 a	2.1 a
Significance	***	ns	***	***	***
	Chrysanthemum				
Control	10.4 b	30.1 a	29.9 b	19.1 a	10.4 ab
B_5	11.5 ab	29.8 a	27.2 b	20.1 a	11.4 a
B_{10}	11.9 ab	26.9 ab	30.4 b	19.3 a	11.5 a
IBA+$NAA_{0.2}$	14.0 a	27.2 ab	32.0 b	17.7 ab	8.9 b
IBA+$NAA_{0.5}$	12.8 ab	24.8 b	38.7 a	15.0 b	8.6 b
Significance	*	**	***	**	**

ns, *, **, and *** indicate non-significant, or significant at $p < 0.05$, 0.01, and 0.001, respectively. Different letters within each column indicate significant differences according to Tukey's HSD test ($p = 0.05$). Data are means of 20 replicates.

3.3. *The Effects of Biostimulant on Shoot Growth*

Consistently with adventitious rooting, response levels of shoots to biostimulant and auxin slightly varied among plant species (Table 3; Tables S1–S3). Increasing concentrations of biostimulant increased or showed an increasing trend of shoot dry mass (Table 3). Biostimulant increased shoot dry mass in cuttings by 10 to 20% at a concentration of 5000 mg L^{-1}, which was somewhat associated with the increase in leaf or stem dry mass of the cuttings. Contrarily, auxin did not affect shoot dry mass of cuttings in basil and tomato with exception of chrysanthemum. SPAD index measured on newly expanded leaves and three fully matured leaves were not significantly different among the treatments, and therefore, pooled for comparisons. The results showed that SPAD index increased only in chrysanthemum when applied with auxin at an optimum level (IBA + NAA0.2), but there were no differences in total nitrogen (N) concentration among treatments (Table 3).

Table 3. Effects of biostimulant (B) or auxin (IBA + NAA) applications on stem length, leaves, stems, and shoot dry mass, the Soil Plant Analysis Development (SPAD) index, total nitrogen (N), and root-to-shoot ratio of basil, tomato, and chrysanthemum cuttings. Cuttings were treated with or without biostimulant (100, 1000, 5000 (B_5), and 10,000 (B_{10}) mg L^{-1}) or auxin (100, 200 (IBA + $NAA_{0.2}$), 300, and 500 (IBA + $NAA_{0.5}$) mg L^{-1}) at the stem base as quick-dip, stuck into inert media, and evaluated at day 20 after placement under intermittent mist. Note that two concentrations per each treatment were presented here.

Treatment	Dry Mass (g plant^{-1})				Stem Length (cm)	SPAD Index	Total N (%)	Root-to-Shoot Ratio
	Total	Shoots	Leaves	Stems				
					Basil			
Control	0.228 ab	0.200 ab	0.145	0.055	6.0	35.1	1.42	0.144
B_5	0.287 a	0.246 a	0.185	0.061	6.3	35.1	1.39	0.170
B_{10}	0.285 a	0.244 a	0.177	0.059	6.6	31.7	1.56	0.177
IBA+$NAA_{0.2}$	0.234 ab	0.200 ab	0.115	0.050	5.8	35.5	1.44	0.175
IBA+$NAA_{0.5}$	0.218 b	0.185 b	0.137	0.049	5.9	32.5	1.45	0.183
Significance	**	*	ns	ns	ns	ns	ns	ns
					Tomato			
Control	0.473 bc	0.457 bc	0.323 ab	0.133 b	7.2 ab	41.5	2.53	0.038 b
B_5	0.526 ab	0.505 ab	0.360 a	0.146 ab	7.5 ab	41.4	2.54	0.046 a
B_{10}	0.551 a	0.530 a	0.358 a	0.173 a	7.8 a	40.8	2.47	0.042 a
IBA+$NAA_{0.2}$	0.414 c	0.399 c	0.283 bc	0.111 b	7.0 b	39.5	2.22	0.046 a
IBA+$NAA_{0.5}$	0.412 c	0.398 c	0.262 c	0.134 b	6.8 b	39.2	2.38	0.036 b
Significance	**	**	**	*	*	ns	ns	ns
					Chrysanthemum			
Control	0.287 b	0.259 b	0.182	0.077 b	6.9 b	34.1 c	3.34	0.104 c
B_5	0.346 a	0.313 a	0.209	0.099 a	7.6 ab	35.4 bc	3.00	0.109 c
B_{10}	0.343 ab	0.307 ab	0.206	0.101 a	8.3 a	34.6 bc	3.02	0.118 bc
IBA+$NAA_{0.2}$	0.356 a	0.315 a	0.212	0.103 a	6.9 b	37.6 a	2.93	0.136 b
IBA+$NAA_{0.5}$	0.343 ab	0.297 ab	0.210	0.087 ab	7.3 ab	36.4 ab	2.97	0.159 a
Significance	*	*	ns	***	***	***	ns	***

ns, *, **, and *** indicate non-significant, or significant at $p < 0.05$, 0.01, and 0.001, respectively. Different letters within each column indicate significant differences according to Tukey's HSD test ($p = 0.05$). Data are means of 20 replicates.

3.4. Tissue Responsiveness to Biostimulant and Auxin

Cuttings of basil, tomato, and chrysanthemum differently responded to biostimulant in comparison to auxin. Biostimulant induced gradual and progressive changes on shoot and root dry mass, as shown in the logarithmic curves (Figure 1a,c), and there were no detrimental effects or phytotoxicity caused by higher doses of biostimulant. This was contradictory to auxin-treated cuttings where higher doses had negative effects on rooting responses, especially in basil and tomato (Figure 1b,d). The rooting responses, as expressed as root dry mass and total root length, were less dramatically influenced by biostimulant than by auxin in all plant species tested.

Basil cuttings were more responsive to a lower concentration of auxin compared to tomato and chrysanthemum, rapidly increasing adventitious roots (Figure 1d,f). At an optimal concentration, auxin-treated cuttings produced similar or higher root biomass relative to biostimulant-treated cuttings (Figure 1c,d). Regression analyses showed that auxin responsiveness of plant species increased in a biphasic manner with increasing concentrations (Figure 1d,f). The response pattern of adventitious rooting to auxin showed that basil and tomato were highly responsive. Basil responds to a lower threshold for rooting followed by a rapid polynomial decay, while tomato required a higher threshold than basil (Figure 1d). Chrysanthemum responded to a lower threshold for rooting, but displayed a gradual polynomial rise to a wide range of auxin, possibly followed by a gradual polynomial fall to a higher concentration of auxin. This rooting response was associated with increased total root length in all the plant species tested (Figure S1). However, a universal scenario of increased root dry mass and/total root length accompanied by biostimulant treatment does not explain the root morphological changes as represented by the proliferation of fine roots, as such subtle changes contribute less to dry mass. Shoots were slightly less responsive to biostimulant applications than roots as characterized by a gentle slope in a logarithmic plot (Figure 1a). Shoot dry mass of basil and tomato did not increase even when a wide range of auxin was applied; however, that of chrysanthemum increased with higher doses of auxin, indicating that chrysanthemum had different responsiveness to auxin from basil and tomato (Figure 1b).

3.5. BRs in Roots and Shoots of Cuttings

Metabolic analyses demonstrated that biostimulant contained precursors of BRs, such as campesterol and stigmasterol (or β-sitosterol) at 11.87 and 28.86 ng mL^{-1} in 5000 mg L^{-1} solution. In addition, a large profile of various compounds, including sugars (ribofuranose, arabinose, and galactose), organic acids (lactic, oxalic acid, glycolic acid, butanoic acid, tartaric acid, and gluconic acid), glucono-1,4-lactone, and fatty acids (palmitic acid and stearic acid) were found to be present as major compounds in the biostimulant (data not shown).

Metabolic profiling of cuttings elucidated that relatively high levels of BRs were present in non-treated cuttings of basil, tomato, and chrysanthemum in decreasing order. Total sterol levels were higher in roots than shoots by 3.4-, 5.3-, and 1.4-times in basil, tomato, and chrysanthemum, respectively) with the highest concentration in roots of basil, tomato, and chrysanthemum in decreasing order, which averaged at 1126, 397, and 213 μg g^{-1} dry weight, respectively (Table 4). There were three major phytosterols present in these plant species: stigmasterol, beta-sitosterol, and campesterol (Table 4). Overall, the combined proportions of stigmasterol and sitosterol were more than 80% of the total sterols, and the proportion of campesterol was less than 20%.

Notably, biostimulant and auxin treatments concomitantly increased or decreased BR levels in plant tissues or had no effects on the levels. Stigmasterol levels in roots tended to be affected by the treatments in all the tested crops ($p \leq 0.12$); however, in a different manner. For example, in basil and chrysanthemum, the optimum level of biostimulant tended to increase stigmasterol levels in roots, while, biostimulant significantly ($p < 0.01$) reduced the levels in tomato. Auxin had similar effects as biostimulant on stigmasterol levels in roots. Correlation relationships were determined between BRs and growth parameters of basil, tomato, and chrysanthemum, i.e., root dry mass, total root length, length of fine roots (0.00 to 0.25 mm) and shoot dry mass. Overall, total sterol levels were not correlated or weakly correlated with root growth parameters in basil and tomato, but moderately correlated (e.g., root dry mass: $r^2 = 0.51$, $p < 0.001$; root length: $r^2 = 0.27$, $p < 0.01$) in chrysanthemum.

Total sterol levels in shoots were the highest in basil, chrysanthemum, and tomato in decreasing order, and averaged at 327, 155, and 75 μg g^{-1} dry weight, respectively (Table 4). Both biostimulant and auxin treatments appeared to have similar increasing or decreasing effects on the levels of BRs as observed in roots. Sitosterol levels were significantly increased in shoots of tomato and chrysanthemum cuttings by biostimulant. Auxin had similar increasing effects on the levels of sitosterol in shoots of those cuttings.

Table 4. Sterol profiles of roots and shoots in cuttings of basil, tomato, and chrysanthemum at day 20 after treatment with or without biostimulant at 5000 (B_5) and 10,000 mg L^{-1} (B_{10}), or auxin at 200 mg L^{-1} (IBA + $NAA_{0.2}$).

Plant Tissue	Treatment	Stigmasterol	ß-Sitosterol	Campesterol	Total
			(μg g^{-1} DW)		
			Basil		
Roots	Control	440 ± 10	454 ± 19	232 ± 10	1126 ± 36
	B_5	474 ± 27	430 ± 17	231 ± 90	1136 ± 51
	B_{10}	410 ± 9	389 ± 32	214 ± 18	1013 ± 52
	IBA+$NAA_{0.2}$	420 ± 13	402 ± 15	217 ± 10	1039 ± 31
	Significance	ns[a]	ns	ns	ns
Shoots	Control	59 ± 40	204 ± 13	64 ± 60	327 ± 23
	B_5	59 ± 60	158 ± 14	55 ± 6	272 ± 25
	B_{10}	47 ± 20	164 ± 40	53 ± 2	264 ± 5
	IBA+$NAA_{0.2}$	47 ± 3	166 ± 16	53 ± 4	261 ± 22
	Significance	ns	ns	ns	ns
			Tomato		
Roots	Control	270 ± 16 a	97 ± 12	30 ± 3 a	397 ± 28 a
	B_5	197 ± 17 b	79 ± 9	22 ± 3 ab	298 ± 27 b
	B_{10}	182 ± 5 b	69 ± 6	20 ± 1 ab	270 ± 90b
	IBA+$NAA_{0.2}$	155 ± 17 b	71 ± 6	17 ± 2 b	243 ± 20 b
	Significance	***	ns	*	**
Shoots	Control	55 ± 8	17 ± 3 b	3.0 ± 0.6	75 ± 12
	B_5	54 ± 40	22 ± 2 ab	5.7 ± 1.8	81 ± 6
	B_{10}	62 ± 90	31 ± 5 a	4.0 ± 1.9	97 ± 15
	IBA+$NAA_{0.2}$	48 ± 50	32 ± 2 a	3.1 ± 1.9	82 ± 7
	Significance	ns	**	ns	ns
			Chrysanthemum		
Roots	Control	83 ± 7	111 ± 9	19.5 ± 2.3	213 ± 18
	B_5	92 ± 5	115 ± 6	20.9 ± 1.5	228 ± 10
	B_{10}	104 ± 6	111 ± 5	20.7 ± 0.8	235 ± 7
	IBA+$NAA_{0.2}$	103 ± 6	114 ± 8	21.0 ± 1.9	238 ± 15
	Significance	ns[b]	ns	ns	ns
Shoots	Control	89 ± 5	65 ± 5 b	1.3 ± 0.5 b	155 ± 10
	B_5	95 ± 2	71 ± 2 ab	1.8 ± 0.2 b	168 ± 40
	B_{10}	99 ± 4	86 ± 4 a	2.0 ± 0.4 b	169 ± 7
	IBA+$NAA_{0.2}$	104 ± 7	78 ± 4 ab	6.5 ± 1.6 a	188 ± 13
	Significance	ns[b]	*	**	ns

ns, *, **, and *** indicate non-significant, or significant at $p < 0.05$, 0.01, and 0.001, respectively. Different letters within each column indicate significant differences according to Tukey's HSD test ($p = 0.05$). Data shown are means ± SE of five replicates. a Significant at $p \leq 0.12$. a Significant at $p \leq 0.1$.

3.6. *Antioxidant Capacities and Total Phenolic Content of Cuttings*

The radical scavenging activities of roots and shoots in cuttings of basil, tomato, and chrysanthemum were examined as estimated by the DPPH (Figure 2a,d,g) and ABTS assays (Figure 2b,e,h). The DPPH method is one of the most frequently used and inexpensive antioxidant assays; however, pH sensitivity is a major disadvantage of the assay [40]. In order to generate robust results, we used the two different scavenging radical assays in this study. The antioxidant capacities obtained from DPPH assay were in accordance with those obtained from ABTS assays regardless of plant species and tissue type (basil: roots $r^2 = 0.94$, shoots $r^2 = 0.79$; tomato: roots $r^2 = 0.75$, shoots $r^2 = 0.82$; chrysanthemum: roots $r^2 = 0.94$, shoots $r^2 = 0.91$).

The results showed large variations in antioxidant capacities among plant species and tissues. Roots of basil showed the highest antioxidant activities (3.6 mg ascorbic acid equivalent per g DW) followed by chrysanthemum (2.3 mg) and tomato (0.4 mg) (Figure 2). In basil, antioxidant capacities were three-times higher in roots than shoots and were less affected by the treatment (Figure 2a,b), while tomato and chrysanthemum was approximately two-times lower in roots than shoots and were either positively or negatively affected by the treatment (Figure 2d,e,g,h).

Biostimulant significantly increased the scavenging activities of roots of chrysanthemum, and such increases were strongly correlated with the concentrations of total phenolic acids (r^2 = 0.77 for DPPH and r^2 = 0.78 for ABTS) (Figure 2g–i). Meanwhile, shoots of chrysanthemum behaved differently and demonstrated significantly higher antioxidant activities concomitantly by both biostimulant and auxin compared to control. There was a trend of concentration-dependent increase in radical scavenging activities by biostimulant in shoots and roots of basil and chrysanthemum (Figure 2). Overall, biostimulant had stimulatory effects on antioxidant activities of adventitious roots in cuttings (based on DPPH and ABTS assays), and such results were contradictory to those induced by auxin.

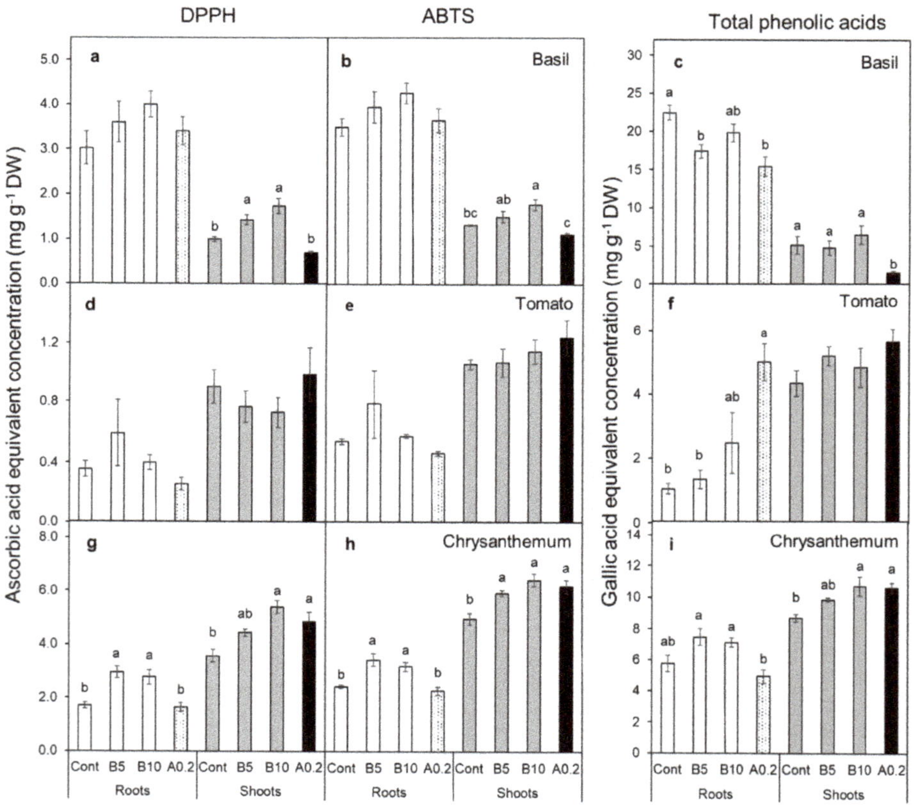

Figure 2. Antioxidant capacity and total phenolic compounds of roots and shoots in cuttings of basil (**a**–**c**), tomato (**d**–**f**) and chrysanthemum (**g**–**i**) at day 20 after treatment with either control (Cont), biostimulant at 5000 (B5) or 10,000 mg L^{-1} (B10), or IAA+NAA at 200 mg L^{-1} (A0.2). Antioxidant capacity was estimated by the DPPH and ABTS assays. The antioxidant capacity and total phenolic compounds of the aqueous extracts are equivalent to indicated concentrations of water-soluble standard antioxidant ascorbic acid (mg g^{-1} DW) and gallic acid (mg g^{-1} DW), respectively. Different letters indicate significant differences within each plant part (roots or shoots) according to Tukey's HSD test (p = 0.05). Data shown are means of five replicates.

4. Discussion

4.1. Biostimulant Promotes Adventitious Rooting Responses of Stem Cuttings Similar to Auxin, but to a Lesser Extent

It is well established that auxin promotes the formation of adventitious roots [41,42] and lateral roots [43–45]. Previous studies have reported on root morphological changes induced by biostimulant applications; however, the changes have been focused primarily on the increases in root biomass, total root length, and root surface area [5,6] without sufficient information on root characteristics. Further, no detailed investigations have been made on hormonal effects of biostimulant in promoting plant growth and yield although these aspects have been demonstrated in many studies.

First, we measured morphological responses of stem cuttings, including the number of adventitious roots, root dry mass, total root length, and average root diameter, as a function of the exogenous concentration of either biostimulant or auxin. While most of these variables exhibited plant species-specific responses, partly due to the genetics of differential tissue responsiveness, it was clear that biostimulant was effective in promoting adventitious rooting of basil and tomato, easy-to-root types, as well as chrysanthemum, moderate-to-root type.

Dose-response analyses were used to evaluate the relationship between compound dosage and plant response (Figure 1), and it was found that rooting responses to biostimulant not only were considerably less compared to auxin but also did not fully in agreement with those to auxin. Biostimulant promoted adventitious rooting leading to a gradual logarithmic rise as a function of increasing dosages, while auxin induced a biphasic dose response characterized by rapid polynomial rise and fall in a plant species-specific manner. A significantly higher threshold was required for biostimulant to induce a series of responses compared to auxin. One of the reasons for the mild changes over a wide range of concentrations induced by biostimulant in cuttings is partly due to a basal quick dip method employed in this study. It was confirmed that such an approach eliminates the possibility of biostimulant as a nutrient source, since there were no differences in total nitrogen level regardless of treatments (Table 3). Changes in endogenous auxin pool may be another possibility because biostimulant Quik-link we used in this study contained about 4.1% tryptophan, as well as other amino acids. As a precursor for auxin biosynthesis pathways in plants, tryptophan might have exerted a weak auxin-induced process, which was postulated in maize seedlings treated with animal-based biostimulant [6]. We did not quantify IAA and other auxin derivatives from plant samples, and therefore, it was not possible to determine how biostimulants interact with endogenous auxin in adventitious rooting formation. Nevertheless, a similar but somewhat unique behavior of biostimulant-treated cuttings observed in our study cannot be justified solely by auxin-mediated activities, opening the possibility of other hormonal regulation in this process.

Variations in adventitious rooting responses were also observed in root architectural traits. An adventitious root system has two major components of root: long, relatively thick roots arising either from the cut stem end or the lower part of the stem that forms its framework and shorter, fine lateral roots arising either directly from these framework roots or indirectly as higher-order lateral roots. Since the complete physical separation of lateral roots from adventitious roots was not possible, particularly in basil and tomato due to fibrous nature of their roots, we performed root diameter class distribution analyses to differentiate these root components by carefully manipulating parameters. This method has been proven to be effective in separating different root types [46], and commonly used in root studies. The results revealed that the roots examined in our study are actually classified into very fine (<0.5) to fine (0.5–2 mm) [47] and we further classified them into multiple categories within the range. The adventitious roots of untreated tomato cuttings were composed primarily of finer roots (average root diameter: 0.38 mm) with about 76% of the total roots within 0 to 0.50 mm diameter class, whereas those of basil cuttings consisted primarily of fine to intermediate roots with about 70% of the roots within 0.25 to 0.75 mm diameter class (average root diameter: 0.55 mm). More than 30% of the total roots in chrysanthemum were composed of coarse roots (>0.75 mm) (average root diameter: 0.63 mm) with a wide range of root diameter classes (Table 2).

Interestingly, biostimulant applications significantly increased or tended to increase finer root classes in these plant species, providing direct evidence that biostimulant stimulates proliferation of lateral roots in cuttings. This is in agreement with a recent study in maize seedlings, in which protein hydrolysates increased length and surface area of lateral roots by about 7 and 1.5 times compared to inorganic nitrogen and free amino acids, respectively [48]. Fine roots are considered to be the most permeable part of a root system and play the key role in the acquisition of water and nutrients and root adaptation to extreme environments, particularly in herbaceous plants [49] and such developmental changes may confer significant advantages on long-term plant growth and survival, particularly under suboptimal water and nutrient conditions.

4.2. Biostimulant Induces Adventitious Rooting of Stem Cuttings Primarily via BR-Mediated Processes

In this study, we measured metabolic responses, including BR levels in plant tissues, antioxidant capacities, and total phenolic compounds in roots and shoots of basil, tomato, and chrysanthemum. We found that biostimulant negatively or positively affects BR biosynthesis in plant tissues and increases antioxidant activities and total phenolic compounds in both roots and shoots of cuttings. Consistently with morphological traits, these metabolic responses were not fully in agreement with auxin.

As a group of steroidal plant hormones, BRs are known to mediate modulation of various components of the antioxidant defense system in plants under abiotic stresses, including drought, salinity, and temperature extremes [50]. BRs were reported to be involved in mitigating the adverse effect of high temperature stress on snap bean plants by increasing total free amino acids in leaves and total phenolic acids in the pod [51]. Increases in antioxidative capacities and phenolic compounds were also found in BR-treated *Brassica junica* seedlings under lead toxicity [52]. The BR-mediated antioxidant system was also demonstrated to modulate root growth as the *Arabidopsis det2-9* mutant defective in BR biosynthesis exhibited inhibited root growth and accumulated more reactive oxygen species than the wild type [53]. We found that stigmasterol, sitosterol, and campesterol were the major phytosterols in cuttings of plant species tested. These phytosterols serve as precursors for BR biosynthesis and are integral membrane components which regulate the permeability and fluidity of membranes [54], and phytoserol composition in the plasma membrane affects the proper functioning of auxin transporters [55]. Campesterol influences the level of active BR, and regulates a number of physiological activities in plant development, such as cell elongation, xylem differentiation, and stress tolerance [54].

Herein, we postulate that biostimulant induces adventitious root formation primarily via BR-mediated processes while interacting with auxin-mediated mechanisms and that native BR pool in plant tissues influences adventitious rooting responses to biostimulant. There are at least six pieces of evidence to support this view: (1) endogenous auxin plays the key role in adventitious rooting formation in these cuttings as adventitious roots were produced in cuttings that did not receive any treatment, (2) endogenous BRs also play a critical role in adventitious rooting formation in these cuttings as relatively high levels of native BRs were present in cuttings that did not receive any treatment, (3) both biostimulant and auxin influenced endogenous BR levels in most cuttings, (4) biostimulant exerted weaker effects on adventitious rooting of cuttings than did auxin treatment, (5) adventitious rooting responses to biostimulant was most prominent in chrysanthemum cuttings that have relatively low native BR levels and are less responsiveness to auxin, and (6) antioxidant activities in adventitious roots tended to be increased by biostimulant but decreased by auxin.

As discussed earlier, the extent to which the increased induction and formation of adventitious rooting varied greatly in response to the compound, with more prominent effects by auxin than biostimulant (Figure 2). For example, the optimal levels of auxin and biostimulant increased the dry mass of adventitious roots by 54% and 20% in chrysanthemum, 67% and 26% in tomato, and 42% and 26% in basil, compared to untreated cuttings. Further, major differences between auxin and biostimulant existed not only in the patterns of dose response curve, but also in the absolute amount of the compounds required for promoting adventitious rooting (Figure 1).

Clouse et al. [56] noted that measurable effects on cell elongation induced by BR required much longer treatment time compared with the rapid effects caused by auxin. Nemhauser et al. [57] elucidated that auxin-response element ARFAT is the crucial intersection point of BR and auxin pathways, which is BR responsive and requires BR biosynthesis for normal expression. These findings are consistent with our observations and support our interpretations that the hormonal effects induced by biostimulant is more likely to be related to BRs than auxin and that auxin and BRs interact in controlling BR pool in plant tissues and work coordinately in fine-tuning adventitious rooting responses of cuttings.

4.3. Biostimulant-Induced BRs and Auxin Have Overlapping Functions in Adventitious Root Formation

We found that roots and shoots of cuttings produce relatively high levels of endogenous BRs which were increased or decreased concomitantly by biostimulant and auxin. This similar effect of both compounds demonstrates that their overlapping role in BR biosynthesis. Although auxin was not quantified in this study, there is no doubt that auxin plays an important role in adventitious rooting formation in these tested plant species. Auxin and BRs are two important phytohormones and are known to exert some similar physiological effects exclusively or through their functional interaction, which include cell division and expansion, vascular differentiation, root growth, and senescence [58]. It was reported that a shared auxin and BR pathway is required for seedling growth, and response from one pathway requires the function of the other, and this interdependence occurs at gene expression level [57]. Consistently, auxin-treated cuttings in our study showed increased levels of BRs, indicating that auxin treatments in cuttings also involve BR biosynthesis. The extent of increased response levels and more remarkable effects on rooting responses induced by auxin indicate that auxin triggers cellular and molecular responses of adventitious rooting synergistically and interdependently from BRs. While such synergistic and interdependent interactions of auxin and BRs have been demonstrated in other plant systems and was well reviewed by Tian et al. [59], this is the first time demonstrating the interactions between biostimulant-induced BRs and auxin in adventitious rooting responses of cuttings. The interaction also includes the lateral root formation of an adventitious root system. BRs are required for lateral root development in Arabidopsis and act synergistically with auxin to promote lateral root formation by increasing acropetal auxin transport [58,60]. BRs mainly function at the lateral root primordia initiation while auxin is required for both initiation and emergence stages of lateral root formation [43,58].

Based on this view, various responses of plant species to biostimulant and auxin can be explained by endogenous BR pools of plant species. The application of biostimulant and auxin had negative effects on BR levels in basil and tomato, highly responsive plant species containing a higher level of native BRs, but had positive effects on BR levels in chrysanthemum, less responsive plant species containing lower BR levels (Table 4). We also demonstrated that a high level of auxin has an inhibitory effect on antioxidant capacities and phenolic compounds in chrysanthemum cuttings (Figure 2). A similar inhibitory effect of increased auxin levels on BR-induced growth responses was observed in auxin-overproducing *yucca* mutants [57]. Thus, it is likely that different plant species have a different level of BR-pool which restricts plant growth response to these compounds, and that increased auxin levels saturate the BR-pool, significantly reducing BR-effects on regulatory changes.

The induction phase in cuttings or detached organs, such as leaves, is generally marked by the immediate consequences of the wounding response caused by severance. It encompasses the first hours after cutting removal, with a local increase in jasmonate, phenolic compounds and auxin at the cutting base [32]. Phenolic compounds exert antioxidant properties against oxidative stress [61], and were demonstrated to promote adventitious roots of stem slices from apple microshoots by protecting IAA from decarboxylation and the tissue from oxidative stress caused by wounding [62], contributing to the auxin stability for adventitious root induction [62]. The high positive correlation ($p < 0.001$)

between antioxidant capacities and total phenolic content indicates that phenolic compounds are a major contributor to the antioxidant activities of these plants. Phenolic compounds act as antioxidants protecting auxins from decarboxylation and the tissue from oxidative stress, allowing more auxin is available to induce roots [62].

In addition to BR-related proteins, other plant hormone related proteins were identified when maize seedlings were exposed to protein hydrolysate [37]. Metabolomics studies of greenhouse melons treated with biopolymer-based biostimulant as substrate drench demonstrated that BRs interact with other hormones in the leaves, possibly via translocation from roots, as the compounds related to other hormones were observed in the leaves [36], and this translocation may explain the lower level of BRs in shoot of cuttings observed in our study. These findings suggest that there are cross-talks among hormones during the adventitious rooting process. BRs positively regulate lateral root formation whereas cytokinin and abscisic acid negatively regulate the event, and ethylene has positive and negative roles during lateral root formation [60]. On the contrary, the root growth-stimulating effect of BRs was proposed to be independent of auxin and gibberellin action, in which processes genes related to other phytohormones did not show changes, suggesting that the stimulatory effect of BRs on root growth is an autonomous effect rather than cross-talks with other phytohormones [63].

Relatively little is known about the effects of BRs on growth and development of adventitious roots. There are only a few reports showing that BRs mainly inhibit adventitious root development in cuttings of tomato and mung bean at low concentrations (0.1 μM), but the effects mainly occur on the shoot [64,65]. BRs have shown to be involved in jasmonate signaling and exert a mild negative regulation of jasmonate-induced inhibition of root growth [66]. Increase in adventitious root formation in geranium stem cuttings were observed in the treatment with BRs which also improved shoot growth of coleus cuttings [67,68].

Our results provide evidence that adventitious rooting responses of cuttings treated with biostimulant involve BR biosynthesis and their overlapping function with auxin, leading to the morphological and metabolic changes occurring during adventitious root formation. Due to the short-term investigations on adventitious rooting processes, we did not find subsequent effects of morphological changes occurring in roots and shoots. Yet, it is expected that such developmental changes improve crop performance and resource acquisition under suboptimal water and nutrient environment and confer significant advantages on long-term plant growth and survival, particularly under abiotic stresses.

5. Conclusions

To elucidate the hormonal effects of plant-derived-biostimulant, adventitious rooting responses of cuttings were examined after a basal quick-dip treatment with various concentrations of biostimulant in comparison to auxin. This approach allows detailed investigations on the hormonal function of biostimulant as auxin is known to play a key role in adventitious rooting process and eliminates potential nutrient effects of the compound. Biostimulant exerted similar effects as auxin increasing adventitious rooting responses. Dose-response analyses revealed that biostimulant showed a gradual logarithmic rise as a function of increasing dosages, contrary to a typical biphasic dose response of auxin, and required a significantly higher threshold than auxin. Metabolic profiles showed that BRs were highly present in non-treated cuttings of basil, tomato, and chrysanthemum in decreasing order, and both biostimulant and auxin had fewer effects in basil and tomato, high BR producers, and greater effects in chrysanthemum, less BR producer, indicating that native BR-pools of plant species influence adventitious rooting responses to biostimulant, as well as auxin. Biostimulant promoted antioxidant activities and phenolic compounds in cuttings, particularly in chrysanthemum, while auxin inhibited these metabolic responses. The inhibitory effect of auxin is likely due to the saturation of BR-pool, significantly reducing BR-effects. These provide evidence that biostimulant has overlapping functions with auxin in adventitious root formation, while exerting distinctive and independent contributions. We demonstrate for the first time that biostimulant induces adventitious rooting responses of cutting via BR-mediated processes while interacting with auxin and that there are interdependent effects of BRs and auxin on antioxidant activities of cuttings. Our results provide new insight into the hormonal regulation of biostimulant and a fine-tuning role of BRs in adventitious root formation.

Supplementary Materials: The following are available online at http://www.mdpi.com/2073-4395/9/2/74/s1, Figure S1. The relationship between root dry mass and total root length of basil, tomato, and chrysanthemum cuttings as affected by biostimulant and auxin applications, Table S1. Polynomial contrasts on the means of adventitious root number, root dry mass, total root length, total root surface area, root volume and average root diameter of basil, tomato, and chrysanthemum cuttings as affected by biostimulant and auxin applications, Table S2. Polynomial contrasts on the means of root diameter class (mm) and relative diameter class length (%) of basil, tomato, and chrysanthemum cuttings as affected by biostimulant and auxin applications. Percentage values at each diameter class are given, Table S3. Polynomial contrasts on the means of on stem length, leaves, stems, and shoot dry mass, SPAD index, and root-to-shoot ratio of basil, tomato, and chrysanthemum cuttings as affected by biostimulant and auxin applications. Percentage values at each diameter class are given.

Author Contributions: H.K. coordinated and supervised the research, provided intellectual inputs for defining the experimental design, data analysis, and interpretation, and wrote the entire manuscript. K.K. performed metabolic analysis and contributed on metabolomics results, S.C. conducted experiments, collected data, assisted in data analysis and interpretation, and implemented the manuscript. M.C. gave support in the experimental design, data analysis, and interpretation, and implemented the manuscript.

Funding: This research was supported by the USDA National Institute of Food and Agriculture, Hatch/Multi State project NE-1335 Resource Management in Commercial Greenhouse Production; Purdue University Research Funds.

Acknowledgments: We are grateful to Rob Eddy, Michael Russell, Nathan Deppe, and Dan Little for their technical assistance; Hye Su Lee and Gaotian Zhu for their help with crop management and data collection; and Hélène Reynaud for the helpful discussions and support on the research.

References

1. Calvo, P.; Nelson, L.; Kloepper, J.W. Agricultural uses of plant biostimulants. *Plant Soil.* **2014**, *383*, 3–41. [CrossRef]
2. Colla, G.; Rouphael, Y. Biostimulants in horticulture. *Sci. Hortic.* **2015**, *196*, 1–2. [CrossRef]
3. du Jardin, P. Plant biostimulants: Definition, concept, main categories and regulation. *Sci. Hortic.* **2015**, *196*, 3–14. [CrossRef]
4. Schaafsma, G. Safety of protein hydrolysates, fractions thereof and bioactive peptides in human nutrition. *Eur. J. Clin. Nutr.* **2009**, *63*, 1161–1168. [CrossRef] [PubMed]
5. Colla, G.; Rouphael, Y.; Canaguier, R.; Svecova, E.; Cardarelli, M. Biostimulant action of a plant-derived protein hydrolysate produced through enzymatic hydrolysis. *Front. Plant Sci.* **2014**, *5*, 448. [CrossRef] [PubMed]
6. Ertani, A.; Cavani, L.; Pizzeghello, D.; Brandellero, E.; Altissimo, A.; Ciavatta, C.; Nardi, S. Biostimulant activity of two protein hydrolyzates in the growth and nitrogen metabolism of maize seedlings. *J. Plant Nutr. Soil Sci.* **2009**, *172*, 237–244. [CrossRef]
7. Matsumiya, Y.; Kubo, M. *Soybean Peptide: Novel Plant Growth Promoting Peptide from Soybean*; InTech Europe Rijeka, Croatia, 2011; pp. 215–230.
8. Kauffman, G.L.; Kneivel, D.P.; Watschke, T.L. Effects of a biostimulant on the heat tolerance associated with photosynthetic capacity, membrane thermostability, and polyphenol production of perennial ryegrass. *Crop. Sci.* **2007**, *47*, 261–267. [CrossRef]
9. Lucini, L.; Rouphael, Y.; Cardarelli, M.; Canaguier, R.; Kumar, P.; Colla, G. The effect of a plant-derived biostimulant on metabolic profiling and crop performance of lettuce grown under saline conditions. *Sci. Hortic.* **2015**, *182*, 124–133. [CrossRef]
10. Colla, G.; Cardarelli, M.; Bonini, P.; Rouphael, Y. Foliar applications of protein hydrolysate, plant and seaweed extracts increase yield but differentially modulate fruit quality of greenhouse tomato. *Hortscience* **2017**, *52*, 1214–1220. [CrossRef]
11. Colla, G.; Hoagland, L.; Ruzzi, M.; Cardarelli, M.; Bonini, P.; Canaguier, R.; Rouphael, Y. Biostimulant action of protein hydrolysates: Unraveling their effects on plant physiology and microbiome. *Front. Plant Sci.* **2017**, 8. [CrossRef]

12. Rouphael, Y.; Colla, G.; Giordano, M.; El-Nalchel, C.; Kyriacou, M.C.; De Pascale, S. Foliar applications of a legume-derived protein hydrolysate elicit dose-dependent increases of growth, leaf mineral composition, yield and fruit quality in two greenhouse tomato cultivars. *Sci. Hortic.* **2017**, *226*, 353–360. [CrossRef]
13. Ertani, A.; Pizzeghello, D.; Francioso, O.; Sambo, P.; Sanchez-Cortes, S.; Nardi, S. *Capsicum chinensis* L. growth and nutraceutical properties are enhanced by biostimulants in a long-term period: Chemical and metabolomic approaches. *Front. Plant Sci.* **2014**, 5. [CrossRef] [PubMed]
14. Ito, Y.; Nakanomyo, I.; Motose, H.; Iwamoto, K.; Sawa, S.; Dohmae, N.; Fukuda, H. Dodeca-CLE peptides as suppressors of plant stem cell differentiation. *Science* **2006**, *313*, 842–845. [CrossRef] [PubMed]
15. Kondo, T.; Sawa, S.; Kinoshita, A.; Mizuno, S.; Kakimoto, T.; Fukuda, H.; Sakagami, Y. A plant peptide encoded by CLV3 identified by in situ MALDI-TOF MS analysis. *Science* **2006**, *313*, 845–848. [CrossRef] [PubMed]
16. Ryan, C.A.; Pearce, G. Polypeptide hormones. *Plant Physiol.* **2001**, *125*, 65–68. [CrossRef]
17. Ryan, C.A.; Pearce, G.; Scheer, J.; Moura, D.S. Polypeptide hormones. *Plant Cell* **2002**, *14*, S251–S264. [CrossRef]
18. Schiavon, M.; Ertani, A.; Nardi, S. Effects of an alfalfa protein hydrolysate on the gene expression and activity of enzymes of the tricarboxylic acid (TCA) cycle and nitrogen metabolism in *Zea mays* L. *J. Agric. Food Chem.* **2008**, *56*, 11800–11808. [CrossRef]
19. Baglieri, A.; Cadili, V.; Monterumici, C.M.; Gennari, M.; Tabasso, S.; Montoneri, E.; Nardi, S.; Negre, M. Fertilization of bean plants with tomato plants hydrolysates. Effect on biomass production, chlorophyll content and N assimilation. *Sci. Hortic.* **2014**, *176*, 194–199. [CrossRef]
20. Paradikovic, N.; Vinkovic, T.; Vrcek, I.V.; Zuntar, I.; Bojic, M.; Medic-Saric, M. Effect of natural biostimulants on yield and nutritional quality: An example of sweet yellow pepper (Capsicum annuum L.) plants. *J. Sci. Food Agric.* **2011**, *91*, 2146–2152.
21. Parrado, J.; Escudero-Gilete, M.L.; Friaza, V.; Garcia-Martinez, A.; Gonzalez-Miret, M.L.; Bautista, J.D.; Heredia, F.J. Enzymatic vegetable extract with bio-active components: Influence of fertiliser on the colour and anthocyanins of red grapes. *J. Sci. Food Agric.* **2007**, *87*, 2310–2318. [CrossRef]
22. Zodape, S.T.; Gupta, A.; Bhandari, S.C.; Rawat, U.S.; Chaudhary, D.R.; Eswaran, K.; Chikara, J. Foliar application of seaweed sap as biostimulant for enhancement of yield and quality of tomato (*Lycopersicon esculentum* Mill.). *J. Sci. Ind. Res.* **2011**, *70*, 215–219.
23. de Vasconcelos, A.C.F.; Zhang, X.Z.; Ervin, E.H.; Kiehl, J.D. Enzymatic antioxidant responses to biostimulants in maize and soybean subjected to drought. *Sci. Agric.* **2009**, *66*, 395–402. [CrossRef]
24. Botta, A. Enhancing plant tolerance to temperature stress with amino acids: An approach to their mode of action. In *I World Congress on the Use of Biostimulants in Agriculture*; Silva, S.S., Brown, P., Ponchet, M., Eds.; International Society for Horticultural Science: Leuven, Belgium, 2013; Volume 1009, pp. 29–35.
25. Colla, G.; Svecova, E.; Cardarelli, M.; Rouphael, Y.; Reynaud, H.; Canaguier, R.; Planques, B. Effectiveness of a plant-derived protein hydrolysate to improve crop performances under different growing conditions. In *I World Congress on the Use of Biostimulants in Agriculture*; Silva, S.S., Brown, P., Ponchet, M., Eds.; International Society for Horticultural Science: Leuven, Belgium, 2013; Volume 1009, pp. 175–179.
26. Rouphael, Y.; Cardarelli, M.; Bonini, P.; Colla, G. Synergistic action of a microbial-based biostimulant and a plant derived-protein hydrolysate enhances lettuce tolerance to alkalinity and salinity. *Front. Plant Sci.* **2017**, *8*, 131. [CrossRef] [PubMed]
27. Epstein, E.; Ludwigmuller, J. Indole-3-butyric acid in plants—Occurrence, synthesis, metabolism and transport. *Physiol. Plant.* **1993**, *88*, 382–389. [CrossRef]
28. Ludwig-Muller, J. Indole-3-butyric acid in plant growth and development. *Plant Growth Regul.* **2000**, *32*, 219–230. [CrossRef]
29. Nordstrom, A.C.; Jacobs, F.A.; Eliasson, L. Effect of exogenous indole-3-acetic-acid and indole-3-butyric acid on internal levels of the respective auxins and their conjugation with aspartic-acid during adventitious root-formation in pea cuttings. *Plant Physiol.* **1991**, *96*, 856–861. [CrossRef] [PubMed]
30. Husen, A.; Pal, M. Metabolic changes during adventitious root primordium development in Tectona grandis Linn. f. (teak) cuttings as affected by age of donor plants and auxin (IBA and NAA) treatment. *New Forests.* **2007**, *33*, 309–323. [CrossRef]
31. De Klerk, G.J.; Van der Krieken, W.; De Jong, J.C. Review—The formation of adventitious roots: New concepts, new possibilities. *In Vitro Cell. Dev. Biol. Plant.* **1999**, *35*, 189–199. [CrossRef]

32. da Costa, C.T.; de Almeida, M.R.; Ruedell, C.M.; Schwambach, J.; Maraschin, F.S.; Fett-Neto, A.G. When stress and development go hand in hand: Main hormonal controls of adventitious rooting in cuttings. *Front. Plant Sci.* **2013**, 4. [CrossRef]
33. Fabijan, D.; Taylor, J.S.; Reid, D.M. Adventitious rooting in hypocotyls of sunflower (*Helianthus annuus*) seedlings.2. Action of gibberellins, cytokinins, auxins and ethylene. *Physiol. Plant.* **1981**, *53*, 589–597. [CrossRef]
34. Hartmann, H.T.; Kester, D.E.; Davies, F., Jr.; Geneve, R.L. *Plant Propagation: Principles and Practices*, 8th ed.; Prentice-Hall: Upper Saddle River, NJ, USA, 2011.
35. Liu, J.H.; Reid, D.M. Adventitious rooting in hypocotyls of sunflower (*Helianthus annuus*) seedlings.4. The role of changes in endogenous free and conjugated indole-3-acetic-acid. *Physiol. Plant.* **1992**, *86*, 285–292. [CrossRef]
36. Lucini, L.; Rouphael, Y.; Cardarelli, M.; Bonini, P.; Baffi, C.; Colla, G. A vegetal biopolymer-based biostimulant promoted root growth in melon while triggering brassinosteroids and stress-related compounds. *Front. Plant Sci.* **2018**, 9. [CrossRef] [PubMed]
37. Trevisan, S.; Manoli, A.; Ravazzolo, L.; Franceschi, C.; Quaggiotti, S. mRNA-sequencing analysis reveals transcriptional changes in root of maize seedlings treated with two increasing concentrations of a new biostimulant. *J. Agric. Food Chem.* **2017**, *65*, 9956–9969. [CrossRef] [PubMed]
38. Colla, G.; Nardi, S.; Cardarelli, M.; Ertani, A.; Lucini, L.; Canaguier, R.; Rouphael, Y. Protein hydrolysates as biostimulants in horticulture. *Sci. Hortic.* **2015**, *196*, 28–38. [CrossRef]
39. Ku, K.M.; Choi, J.N.; Kim, J.; Kim, J.K.; Yoo, L.G.; Lee, S.J.; Hong, Y.S.; Lee, C.H. Metabolomics analysis reveals the compositional differences of shade grown tea (*Camellia sinensis* L.). *J. Agric. Food Chem.* **2010**, *58*, 418–426. [CrossRef] [PubMed]
40. Huang, D.; Ou, B.; Prior, R.L. The chemistry behind antioxidant capacity assays. *J. Agric. Food Chem.* **2005**, *53*, 1841–1856. [CrossRef] [PubMed]
41. Falasca, G.; Zaghi, D.; Possenti, M.; Altamura, M.M. Adventitious root formation in *Arabidopsis thaliana* thin cell layers. *Plant Cell Rep.* **2004**, *23*, 17–25. [CrossRef] [PubMed]
42. Sorin, C.; Bussell, J.D.; Camus, I.; Ljung, K.; Kowalczyk, M.; Geiss, G.; McKhann, H.; Garcion, C.; Vaucheret, H.; Sandberg, G.; Bellini, C. Auxin and light control of adventitious rooting in Arabidopsis require ARGONAUTE1. *Plant Cell.* **2005**, *17*, 1343–1359. [CrossRef]
43. Casimiro, I.; Marchant, A.; Bhalerao, R.P.; Beeckman, T.; Dhooge, S.; Swarup, R.; Graham, N.; Inze, D.; Sandberg, G.; Casero, P.J.; et al. Auxin transport promotes *Arabidopsis* lateral root initiation. *Plant Cell.* **2001**, *13*, 843–852. [CrossRef]
44. Malamy, J.E.; Benfey, P.N. Down and out in *Arabidopsis*: The formation of lateral roots. *Trends Plant Sci.* **1997**, *2*, 390–396. [CrossRef]
45. Woodward, A.W.; Bartel, B. Auxin: Regulation, action, and interaction. *Ann. Bot.* **2005**, *95*, 707–735. [CrossRef]
46. Gu, D.X.; Zhen, F.X.; Hannaway, D.B.; Zhu, Y.; Liu, L.L.; Cao, W.X.; Tang, L. Quantitative classification of rice (*Oryza sativa* L.) root length and diameter using image analysis. *PloS ONE* **2017**, *12*, e0169968. [CrossRef] [PubMed]
47. Zobel, R.W.; Waisel, Y. A plant root system architectural taxonomy: A framework for root nomenclature. *Plant Biosyst.* **2010**, *144*, 507–512. [CrossRef]
48. Santi, C.; Zamboni, A.; Varanini, Z.; Pandolfini, T. Growth stimulatory effects and genome-wide transcriptional changes produced by protein hydrolysates in maize seedlings. *Front. Plant Sci.* **2017**, 8. [CrossRef] [PubMed]
49. McCully, M.E. Roots in soil: Unearthing the complexities of roots and their rhizospheres. *Ann. Rev. Plant Physiol. Plant Mol. Biol.* **1999**, *50*, 695–718. [CrossRef]
50. Vardhini, B.V.; Anjum, N.A. Brassinosteroids make plant life easier under abiotic stresses mainly by modulating major components of antioxidant defense system. *Front. Environ. Sci.* **2015**, *2*, 1–16. [CrossRef]
51. El-Bassiony, A.M.; Ghoname, A.A.A.; El-Awadi, M.E.; Fawzy, Z.F.; Gruda, N. Ameliorative effects of brassinosteroids on growth and productivity of snap beans grown under high temperature. *Gesunde Pflanz.* **2012**, *64*, 175–182. [CrossRef]

52. Kohli, S.K.; Handa, N.; Sharma, A.; Gautam, V.; Arora, S.; Bhardwaj, R.; Alyemeni, M.N.; Wijaya, L.; Ahmad, P. Combined effect of 24-epibrassinolide and salicylic acid mitigates lead (Pb) toxicity by modulating various metabolites in *Brassica juncea* L. seedlings. *Protoplasma.* **2018**, *255*, 11–24. [CrossRef] [PubMed]
53. Lv, B.; Tian, H.; Zhang, F.; Liu, J.; Lu, S.; Bai, M.; Li, C.; Ding, Z. Brassinosteroids regulate root growth by controlling reactive oxygen species homeostasis and dual effect on ethylene synthesis in *Arabidopsis*. *PLoS Genet.* **2018**, *14*, e1007144. [CrossRef]
54. Clouse, S.D. Brassinosteroids. *The arabidopsis book* **2011**, *9*, e0151. [CrossRef]
55. Men, S.; Boutte, Y.; Ikeda, Y.; Li, X.; Palme, K.; Stierhof, Y.-D.; Hartmann, M.-A.; Moritz, T.; Grebe, M. Sterol-dependent endocytosis mediates post-cytokinetic acquisition of PIN2 auxin efflux carrier polarity. *Nat. Cell. Biol.* **2008**, *10*, 237–244. [CrossRef] [PubMed]
56. Clouse, S.D.; Zurek, D.M.; McMorris, T.C.; Baker, M.E. Effect of brassinolide on gene-expression in elongating soybean epicotyls. *Plant Physiol.* **1992**, *100*, 1377–1383. [CrossRef] [PubMed]
57. Nemhauser, J.L.; Mockler, T.C.; Chory, J. Interdependency of brassinosteroid and auxin signaling in *Arabidopsis*. *PLoS Biol.* **2004**, *2*, 1460–1471. [CrossRef] [PubMed]
58. Bao, F.; Shen, J.J.; Brady, S.R.; Muday, G.K.; Asami, T.; Yang, Z.B. Brassinosteroids interact with auxin to promote lateral root development in Arabidopsis. *Plant Physiol.* **2004**, *134*, 1624–1631. [CrossRef] [PubMed]
59. Tian, H.Y.; Lv, B.S.; Ding, T.T.; Bai, M.Y.; Ding, Z.J. Auxin-BR interaction regulates plant growth and development. *Front. Plant Sci.* **2018**, *8*, 1–8. [CrossRef] [PubMed]
60. Fukaki, H.; Tasaka, M. Hormone interactions during lateral root formation. *Plant Mol. Biol.* **2009**, *69*, 437–449. [CrossRef] [PubMed]
61. Gill, S.S.; Tuteja, N. Reactive oxygen species and antioxidant machinery in abiotic stress tolerance in crop plants. *Plant Physiol. Biochem.* **2010**, *48*, 909–930. [CrossRef] [PubMed]
62. De Klerk, G.J.; Guan, H.Y.; Huisman, P.; Marinova, S. Effects of phenolic compounds on adventitious root formation and oxidative decarboxylation of applied indoleacetic acid in Malus 'Jork 9'. *Plant Growth Regul.* **2011**, *63*, 175–185. [CrossRef]
63. Mussig, C.; Shin, G.H.; Altmann, T. Brassinosteroids promote root growth in Arabidopsis. *Plant Physiol.* **2003**, *133*, 1261–1271. [CrossRef]
64. Guan, M.; Roddick, J.G. Epibrassinolide-inhibition of development of excised, adventitious and intact roots of tomato (*Lycopersicon esculentum*)-comparison with the effects of steroidal estrogens. *Physiol Plant.* **1988**, *74*, 720–726. [CrossRef]
65. Guan, M.; Roddick, J.G. Comparison of the effects of epibrassinolide and steroidal estrogens on adventitious root-growth and early shoot development in mung bean cuttings. *Physiol. Plant* **1988**, *73*, 426–431. [CrossRef]
66. Ren, C.M.; Han, C.Y.; Peng, W.; Huang, Y.; Peng, Z.H.; Xiong, X.Y.; Zhu, Q.; Gao, B.D.; Xie, D.X. A leaky mutation in *DWARF4* reveals an antagonistic role of brassinosteroid in the inhibition of root growth by jasmonate in Arabidopsis. *Plant Physiol.* **2009**, *151*, 1412–1420. [CrossRef] [PubMed]
67. Swamy, K.N.; Seeta Ram Rao, S. Influence of brassinosteroids on rooting and growth of geranium (*Pelargonium* sp.) stem cuttings. *Asian J. Plant Sci.* **2006**, *5*, 619–622.
68. Swamy, K.N.; Seeta Ram Rao, S. Effect of brassinosteroids on rooting and early vegetative growth of Coleus [*Plectranthus forskohlii* (Wild.) Briq.] stem cuttings. *Indian J. Plant Sci.* **2010**, *1*, 68–73.

agronomy

MDPI

Article

Morphological and Biochemical Responses of *Glycine max* (L.) Merr. to the Use of Seaweed Extract

Sławomir Kocira [1,*], Agnieszka Szparaga [2], Maciej Kuboń [3], Ewa Czerwińska [4] and Tomasz Piskier [2]

1 Department of Machinery Exploitation and Management of Production Processes, University of Life Sciences in Lublin, Akademicka 13, 20-950 Lublin, Poland
2 Department of Agrobiotechnology, Koszalin University of Technology, Racławicka 15–17, 75-620 Koszalin, Poland; agnieszka.szparaga@tu.koszalin.pl (A.S.); tomasz.piskier@tu.koszalin.pl (T.P.)
3 Institute of Agricultural Engineering and Informatics, Agricultural University Krakow, Balicka 116B, 30-149 Krakow, Poland; maciej.kubon@ur.krakow.pl
4 Department of Biomedical Engineering, Koszalin University of Technology, Śniadeckich 2, 75-453 Koszalin, Poland; ewa.czerwinska@tu.koszalin.pl
* Correspondence: slawomir.kocira@up.lublin.pl; Tel.: +48-815319735

Received: 21 December 2018; Accepted: 15 February 2019; Published: 18 February 2019

Abstract: Currently, modern agriculture aims to improve the quantity and quality of crop yield, while minimizing the negative impact of treatments on the natural environment. One of the methods to increase plant yield and quality, especially after the occurrence of both abiotic or biotic stress factors, is the application of biostimulants. The aim of the study was to determine the effect of *Ecklonia maxima* extract on plant growth, and the yield, nutritional, and nutraceutical properties of soybean seeds. A field experiment was conducted in three growing seasons (2014–2016). Soybean seeds of Atlanta cultivar were sown in the third 10-day period of April. *Ecklonia maxima* extract was applied in the form of single or double, spraying in the concentrations of 0.7% and 1.0%. Determinations were conducted for: biometric traits, seed yield, seed number, thousand seeds weight, contents of lipids, and proteins in seeds. Further analyses included the contents of total polyphenols, flavonoids, anthocyanins, and reducing power. The number of seaweed extract applications and its concentration modified biometric traits, yield, and quality of crop, while also also altering the nutraceutical and antioxidative potential of soybean. The application of this preparation improved the growth and yield of soybean without any negative effect on the nutritive value of seeds.

Keywords: antioxidant activity; growth; nutrients; nutraceutical potential; soybean; yield

1. Introduction

Soya (*Glycine max* (L.) Merrill.) is one of the most important leguminous plants that are cultivated around the world because it is a precious source of both protein and fat [1,2]. Its use for production of food, oil, and fodder means that the demand for this plant is continuously growing [3]. Due to its broad use, it is called a "wonderful crop" [4]. However, this plant is sensitive to unfavourable climatic conditions [5]. Thus, to ensure its effective protection against biotic and abiotic factors, it is recommended to use it in the cultivation of biostimulants, which may improve the biochemical, morphological, and physiological processes that take place in a plant [6–8].

Biostimulants as plant-growth promoters were defined for the first time in the world literature by Kaufman [9]. In turn, Du Jardin [10] claims that "a plant biostimulant is any substance or microorganism, in the form in which it is supplied to the user, applied to plants, seeds or the root environment with the intention to stimulate natural processes of plants to benefit their nutrient use efficiency and/or their tolerance to abiotic stress, regardless of its nutrients content, or any combination of such substances and/or microorganisms intended for this use". The European Biostimulants Industry Council (EBIC) was established in order to develop legal regulations regarding the registration of biostimulants, according to the specificity of their action. However, currently, the registration of these preparations is still based on legal regulations that are set for fertilizers and plant protection products [11–13]. According to Colla et al. [14,15] and Battacharyya et al. [16], among the entire group of those preparations, extracts from seaweed and protein hydrolysates constitute the two most-important categories of substances of natural biostimulants. According to Aguilar, brown algae are the most often used in agriculture [17]. The most popular are, among others, *Ecklonia maxima* (Osbeck) Papenfuss and *Ascophyllum nodosum* (L.) Le Jolis. Brown-algae extracts include various phytohormones, such as auxins, gibberellins, cytokininis, abscisic acid, ethylene, betaine, and polyamins, and other growth promoters as well as trace elements and microelements [18]. Seaweeds also include a varied range of organic compounds, among others, aminoacids, such as asparaginic acid, glutamine acid, and alanine. While, alginic acid, laminarin, and mannitol constitute almost half of the total content of carbohydrates in such biostimulating preparations. Seaweeds also contain a wide range of vitamins that can be used by plants, such as C, B2, B12, D3, E, K, niacin, panthotenic acid, and folic acid. Although, vitamin A does not occur in algae extracts, the presence of its precursor—carotene and another possible precursor, fucoxnathin—was determined [18–20].

According to Ecoforce [21] and Van Oosten et al. [22], seaweed extracts that are used as biostimulants increase the yield and its quality in two ways: a. they stimulate hormone synthesis, influence absorption, and translocation of nutrients; b. they condition the soil, improving its ability to retain moisture and stimulate the activity of favourable microorganisms. Medjdoub estimates that the use of biostimulants, which include extracts from seaweeds, has greater meaning in agriculture. That is because plant growth and development is controlled by plant hormones, which directly or indirectly control the course of various physiological reactions and their integration with the total metabolism [23]. Many studies on seaweed extract indicate that they may increase: a. plant growth, b. activity of photosynthesis, c. resistance to fungi, bacteria and viruses, d. tolerance to ground frost, drought, and salt content, and e. yield and productivity of many cultivations [24–26], mainly by activation of protective mechanisms of plants [27]. Foliar application of biostimulants that are based on seaweeds is an agrotechnical treatment that brought many advantages in numerous cultivations, including grapevine, watermelon, strawberry, apple, tomato, spinach, onion, bean, pepper, carrot, potato, wheat, corn, barley, rice, and turf grass. The results show that plants treated with lower concentrations of extract indicated a stronger growth, higher yield, and higher mineral and nutritive elements content relative to the control [28–33]. Positive reactions also included an improved flowering and fructification ability, product quality and efficiency, and resistance to abiotic stress [31,34,35]. Studies that were performed on a wide group of crops proved that the application of sea-algae-based biostimulants stimulates the primary and secondary metabolisms in plants through the absorption and assimilation of nutrients [36–44]. Growth of productivity of crops induced by the use of such biostimulants in optimal and suboptimal conditions may be related to several direct and indirect mechanisms, including the stimulation of enzymatic activities that are related to carbon, nitrogen metabolism, Krebs cycle, and glycolysis. Such use may also induce activity similar to hormones, especially the one that is assigned to auxins and gibberellins, and improve the nutrition of treated plants by the modulation of the root system [14–16,45].

However, other results indicate that the application of such biostimulants, despite its numerous advantages, like faster germination and earlier growth [46,47], may inhibit the growth and development of many plants. This calls for greater care in the use of seaweeds extracts [48]. Therefore, the inhibition of plant growth after application of biostimulants is a potential problem in plant production. The concentration of these products is an important factor in this regard [49]. This issue may be caused directly by elements that are included in the extract [50], or it might be a consequence of modifications in the regular physiological growth of the plant [51,52]. Improvement of commercial-product formulas, and knowing the mechanisms of active substances in plants and their persistence, should mitigate such negative effects. The exact effect of various elements (e.g., nutrients, betaines, oligomers, polymers) from seaweeds on improving plant growth, vigour, and fractioning of extracts is not fully known. Therefore, a detailed analysis of composition and the fractioning of elements on plant physiology, together with a better ability to monitor the impact of such extracts on those variables and on the expression of genes, would shed light on some of the performance mechanisms [53,54].

Among biostimulants that include seaweeds, Kelpak is particularly interesting. It is extracted from the species *Ecklonia maxima* Osbeck and then harvested along the shores of Africa. Kelpak contains phytohormones, such as auxins (11 mg dm^{-3}) and cytokinins (0.031 mg dm^{-3}), and also alginates (1.5 g L^{-1}), amino acids (total 441.3 mg 100 g^{-1}), mannitol (2261 mg L^{-1}), neutral sugars (1.08 g L^{-1}), and small amounts of macro- (mean composition: N 0.09%, P 90.7 mg kg^{-1}, K 7163.3 mg kg^{-1}, Ca 190.4 mg kg^{-1}, Mg 337.2 mg kg^{-1}, Na 1623.7 mg kg^{-1}) and microelements (mean composition: Mn 17.3 mg kg^{-1}, Fe 40.7 mg kg^{-1}, Cu 13.5 mg kg^{-1}, Zn 17.0 mg kg^{-1}, B 33.0 mg kg^{-1}) [55,56]. Moreira Sisalema [57] indicates that Kelpak, due to the unique extraction process, contains a very-high auxins-to-cytokinins ratio. The dominance of auxins stimulates the dynamic growth of plant roots, which increases the absorption of indispensable nutrients and minerals, and consequently, of plant production. Activity of this biostimulant may also increase plant resistance to drought and enable faster plant regeneration after water stress. Cals [58] shows that Kelpak should be used in leguminous plants before flowering in the dose of 2.0 L ha^{-1}, in relation to the condition of plant nutrition. The concentration of these preparations in foliar applications is usually from 0.2% to 1% and it rarely exceeds these concentrations. Depending on the specific cultivation, cultivar, and climatic conditions, farmers are usually recommended to use these biostimulants in the form of two-week spraying in the stage of intense plant growth [28,59,60].

Because of the varied reactions of many plants to the application of biostimulants from seaweeds, and due to small number of studies on their influence in soya cultivation, a three-year log field study was carried out. Its main aim was to assess the impact of the application of *Ecklonia maxima* extract (Kelpak) on plant growth, yield size, and the quality and nutraceutical potential of genetically non-modified seeds of Atlanta cultivar. The initial hypothesis was that the introduction of agrotechnical treatment to soya cultivation in the form of plant spraying with Kelpak preparation would modify the plant growth, yield, and chemical composition of soya seeds. To test this hypothesis, the yield and structural elements of soya cropping were assessed, as well as the protein, fat, and anti-oxidant potential of seeds in relation with the applied doses and concentrations of the tested preparation. In order to know the morphological and biochemical plant reaction on seaweed extract performance, the responses of treated and untreated control plants in the same environmental conditions were compared. It was expected that the observation of plant reaction would considerably increase knowledge regarding the manner of seaweed extract performance, particularly in leguminous plants cultivation, which are sensitive to biotic and abiotic stresses. The present work is a concrete step towards broadening the understanding of the advantages of the application in agricultural practice of *Ecklonia maxima* seaweed extracts for the improvement of the size and quality of crops.

2. Materials and Methods

2.1. Plant Materials and Growth Conditions

The field experiment was carried out in 2014–2016 in Perespa (50°66′ N; 23°63′ E, Poland). It was established in a randomized block design in four replications on experimental plots with an area of 10 m^2. Soybean was cultivated on the soil belonging to the Gleyic Phaeozems, which was characterized by alkaline pH (pH in 1M KCl: 7.4–7.5). The soil content in the assimilable nutrients was at the medium level, as follows: P (12.6–14.2 mg P_2O_5 in 100 g soil), K (15.3–17.1 mg K_2O in 100 g soil), and Mg (6.2–6.8 mg Mg in 100 g soil). Each year, winter wheat was used as a forecrop. Soybean seeds (*Glycine max* (L.) Merr.) of Atlanta cultivar (Agroyoumis, Poland) were sown on the 25 of April in 2014 and 2015, and 23 of April in 2016 in rows every 30 cm at a raw spacing of

3.5 cm. The weeds were mechanically and manually removed. No pesticides were used (pests did not exceed the thresholds of harmfulness). In the growing season, the plants were sprayed with biostimulant (water solutions) that was based on the *Ecklonia maxima* extract (Kelpak). Kelpak contains phytohormones (mostly auxins 11 mg kg^{-1}, cytokinins 0.03 mg kg^{-1}, and auxin: cytokinin ratio 367:1), carbohydrates (16.9 g kg^{-1}), amino acids (2.5 g kg^{-1}), vitamin B1 (0.9 mg kg^{-1}), B2 (0.1 mg kg^{-1}), C (20 mg kg^{-1}), and E (0.7 mg kg^{-1}). The elemental profile of the biostimulant is: N 3.6 g kg^{-1}, P 8.2 g kg^{-1}, K 7.2 g kg^{-1}, Ca 0.8 g kg^{-1}, Mg 0.2 g kg^{-1}, Fe 13.6 mg kg^{-1}, Mn 8.4 mg kg^{-1}, B 0.24 mg kg^{-1}, Zn 4.2 mg kg^{-1}, and Cu 0.2 mg kg^{-1} [14,61]. The scheme of doses, developmental stages of plants, and terms of spraying are presented in Table 1.

Table 1. Plant developmental stages and dates of biostimulants application.

Biostimulant	Number of Sprays and Plant Developmental Stages in Which the Biostimulants were Applied	Concentration	Volume of Wrking Solution/ Working Pressure	Date of Spraying		
				2014	2015	2016
Kelpak SL	Single spraying BBCH 13-15 (LSS)	0.7%	300 l·ha^{-1}/ 0.30 MPa	June 21	June 20	June 7
	Double spraying BBCH 13-15, BBCH 61 (LDS)	0.7%		June 21, July 5	June 20, July 3	June 7, June 23
	Single spraying BBCH 13-15 (HSS)	1.0%		June 21	June 20	June 7
	Double spraying BBCH 13-15, BBCH 61 (HDS)	1.0%		June 21, July 5	June 20, July 3	June 7, June 23

Plants sprayed with water served as the control. The biostimulant (or water) was sprayed with a GARLAND FUM 12B battery field sprayer (Lechler LU 120-03) at a pressure of 0.30 MPa, using 300 l liquid per hectare. The average temperature and rainfalls in the soybean growing season are shown in Table 2.

Table 2. Temperature (T) and rainfalls during the soybean growing season 2014–2016.

Month	Year						Average from 2002 to 2013	
	2014		2015		2016			
	T (°C) Average (min/max)	Rainfall (mm)	T (°C) Average (min/max)	Rainfall (mm)	T (°C) Average (min/max)	Rainfall (mm)	T (°C)	Rainfall (mm)
IV	9.4 (−6.0/22.7)	36.5	8.2 (−1.7/24.3)	30.1	9.2 (−1.2/22.6)	68.4	8.5	41.2
V	13.7 (0.5/27.7)	208.3	12.7 (1.5/24.9)	108.6	13.8 (2.6/26.7)	61.3	12.7	63.4
VI	16.1 (6.7/28.9)	67.1	17.4 (6.6/30.5)	14.1	18.1 (4.2/31.5)	97.1	17.7	68.6
VII	20.3 (10.0/31.0)	104.2	19.6 (8.4/33.4)	59.2	19.5 (8.8/31.2)	107.6	18.9	79.1
VIII	18.2 (6.3/34.0)	115.4	21.6 (5.6/35.5)	23.4	18.2 (7.1/30.7)	95.3	19.4	71.8
IX	13.7 (3.7/25.8)	89.4	15.1 (4.2/34.5)	137.6	15.2 (1.6/28.7)	41.2	14.1	69.2
Average/Total	15.1	620.9	15.8	373.0	17.1	470.9	15.2	393.3

2.2. *Plant Growth, Yield, and Nutritional Value Determination*

After the pods have matured, when the seeds have obtained a typical color and hardness (BBCH 89), the plant height, the internode number on the main shoot, and the first pod height were recorded. In addition, after harvesting, the number of pods per plant, the number of seeds per 1 m^2, the weight of seeds, and the weight of thousand seeds were determined. Subsequently, the seeds were dried and then grinded. The flour was used for further analysis.

Protein content was determined with the Kjeldahl method, whereas the content of lipids was based on the acid hydrolysis method [62].

2.3. Nutraceutical Potential

Seed extract was prepared following the methodology that was proposed by Świeca et al. [63]. The ground soybean seeds were extracted with a mixture of acetone, water, and hydrochloric acid (70:29:1; v/v/v). Afterwards, the samples were centrifuged for 10 min (6800× *g*) and the resulting supernatant was collected and then used for further analyses.

2.3.1. Phenolics Determination

Determination of Total Phenolic Compounds (TPC)

The content of total phenolic compounds (TPC) was determined with the method of Singleton and Rossi using the Folin–Ciocalteau reagent [64]. Absorbance of the samples was measured with a UV-vis spectrophotometer at a wavelength of 725 nm, then TPC was computed and expressed as gallic acid equivalents (GAE) in mg per g of dry matter (DM).

Determination of Flavonoid Content (TFC)

The total content of flavonoids was determined acc. to the method that was presented by Lamaison and Carnet [65]. The prepared soybean extract was mixed with a methanolic solution of $AlCl_3 \times 6H_2O$. After incubation, the absorbance was measured with a UV-vis spectrophotometer at the wavelength of 430 nm. The total flavonoid content was expressed as quercetin equivalents (QE) in mg per g DM.

Determination of Anthocyanins (TAC)

Using the method that was proposed by Fuleki and Francis using potassium chloride and sodium acetate buffer at two pH values (1.0 and 4.5), the content of anthocyanins was assayed [66]. After 15 min, absorbance of each sample was measured at wavelengths of 520 nm and 700 nm. Subsequently, anthocyanin content was calculated as cynidin-3-glucoside equivalents (Cy3-GE) in mg per g DM.

2.3.2. Reducing Power

Reducing power was measured following the method that was provided by Pulido et al. [67]. The soybean extract was mixed with a phosphate buffer (200 mM, pH 6.6) and 1% solution of $K_3[Fe(CN_6)]$. Next, the samples were incubated at 50 °C for 20 min. The reaction was stopped with trichloroacetic acid and the samples were centrifuged (6800× *g*, 10 min). The resulting supernatant was mixed with distilled water and $FeCl_3$. Afterwards, absorbance was measured at the wavelength of 700 nm. Reducing power was expressed as Trolox equivalents in mg per g DM.

2.4. The Index of Biostimulant Effect

The index of biostimulant effect (ABT-C) was determined as the difference between the mean result that was obtained after biostimulant application (ABT) and the control (C), which enabled the evaluation of the effect of biostimulant type on the analyzed traits. The mean value for each treatment has been obtained clustering the means of lower concentration single spraying (LSS), lower concentration double spraying (LDS), single application of the higher concentration (HSS), and higher concentration double spraying (HDS) from different years all together. The standard deviation value (SD) was determined for all reported mean values of ABT-C [5].

2.5. *Statistical Analysis*

The obtained results were statistically elaborated with Statistica 13 software (StatSoft, Inc.). The materials were collected over three seasons (2014–2016). Laboratory analyses were performed in triplicate. Normality of data distribution was assessed with the Shapiro–Wilk test. The significance of differences between the evaluated mean values was estimated with the Tukey test at a significance level of $p < 0.05$.

3. Results

3.1. *Effect of Biostimulants on Biometric Traits*

3.1.1. Plant Height

The single application of the higher concentration (HSS) of Kelpak biostimulant ensured better effects in increasing soybean plant height (increased by 35% as compared to the control) (Table 3). The highest plants were obtained in the growing season 2016 after their single spraying with the higher concentration of Kelpak. In contrast, the smallest plants were produced in the 2015 season and their height differed significantly from the values noted in seasons 2014 and 2016. The biostimulant increased the height of plants, which was indicated by a value of the Kelpak effect index (ABT-C) of 28.2 cm for this trait (Table 4).

Table 3. Effect of *Ecklonia maxima* extract (Kelpak) treatment on biometric traits of soybean (average from 2014-2016).

Parameters	*Ecklonia maxima* Treatment	Season			Average from 2014 to 2016
		2014	2015	2016	
Plant height (cm)	C	85.4[a]	81.9[a]	88.1[a]	85.1[a]
	LSS	114.9[b]	107.7[b]	112.6[b]	111.7[b]
	LDS	117.5[b]	108.0[b]	117.0[b]	114.1[b]
	HSS	118.8[b]	106.4[b]	120.0[b]	115.0[b]
	HDS	114.9[b]	108.0[b]	114.4[b]	112.4[b]
	AS	110.3[b]	102.4[a]	110.3[b]	
Number of internodes in the main shoot	C	11.2[a]	10.1[a]	9.6[a]	10.3[a]
	LSS	10.4[a]	8.6[a]	10.2[a]	9.7[a]
	LDS	9.9[a]	9.3[a]	9.0[a]	9.4[a]
	HSS	10.0[a]	9.8[a]	10.3[a]	10.0[a]
	HDS	9.9[a]	8.8[a]	11.1[a]	9.9[a]
	AS	10.3[b]	9.3[a]	10.0[ab]	
Location height of the first pod (cm)	C	12.5[a]	11.1[a]	11.7[a]	11.7[a]
	LSS	13.0[a]	14.2[a]	12.2[a]	13.2[ab]
	LDS	13.8[a]	14.0[a]	13.3[a]	13.7[b]
	HSS	12.0[a]	12.5[a]	12.2[a]	12.2[ab]
	HDS	13.0[a]	12.7[a]	13.3[a]	13.0[ab]
	AS	12.8[a]	12.9[a]	12.5[a]	
Number of pods (per plant)	C	15.2[a]	14.7[a]	16.3[a]	15.4[a]
	LSS	20.9[b]	22.5[cd]	21.0[b]	21.5[bc]
	LDS	22.4[b]	23.4[d]	21.5[b]	22.4[c]
	HSS	19.9[b]	21.4[bc]	21.1[b]	20.8[b]
	HDS	20.4[b]	20.3[b]	20.8[b]	20.5[b]
	AS	19.8[a]	20.4[a]	20.1[a]	

Abbreviations: C, control; LSS, lower concentration single spraying; LDS, lower concentration double spraying; HSS higher concentration single spraying; HDS, higher concentration double spraying. Means in the columns, concerning the selected traits, followed by different small letters are significantly different at $p < 0.05$.

Table 4. The index of biostimulant effect (ABT-C).

Parameters	Kelpak
Plant height (cm)	28.2
Number of nodes in the main shoot	−0.5
Location height of the first pod (cm)	1.3
Number of pods (per plant)	5.9
Number of seeds (per m^{-2})	622
Seed yield (t ha^{-1})	0.824
1000 seed weight (g 1000^{-1})	−10.7
Total protein (% DM)	0.35
Total fat (% DM)	−1.56
Total phenols (mg g^{-1} DM)	2.53
Total flavonoids (mg g^{-1} DM)	1.23
Anthocyanins (mg g^{-1} DM)	0.01
Reducing power (mg TE g^{-1} DM)	0.15

3.1.2. Number of Internodes in the Main Shoot

Internode number decreased regardless of the Kelpak concentration and the number of its applications, although the differences were insignificant (Table 3). The highest number of internodes on the main shoot was obtained in the first and third season. The highest number of internodes on the main shoot was obtained in the first season and it differed significantly from the number that was determined in 2015. The value of the ABT-C index computed for Kelpak was negative (Table 4).

3.1.3. Location Height of the First Pod

Biostimulant treatment increased the height of the first pod as compared to the control. Significant differences were observed between the double application of the lower concentration of Kelpak and the control (increased by 17%) (Table 3). The tallest heights of the first pods were observed in the 2015 season, however they did not significantly differ from the values that were reported in the two other seasons. Values of the ABT-C index demonstrate that the height of the first pod was larger with the application of Kelpak preparation (Table 4).

3.1.4. Number of Pods per Plant

Double foliar application of the lower concentration of Kelpak permitted achieving the highest number of pods per plant (increased by 45% as compared to the control) (Table 3). The study demonstrated that the mean number of pods determined in particular growing seasons was at a similar level and did not significantly differ among seasons. In turn, biostimulant increased the pod number per plant because the value of Kelpak effect index was 5.9 pods/plant after spraying with this preparation (Table 4).

3.2. *Effect of Biostimulants on Soybean Yield*

3.2.1. Number of Seeds

Double spraying soybean plants with the higher concentrations (HDS) of Kelpak had the largest effect on the increase in seed number per m2 (increased by 43% as compared to the control) (Table 5). The analysis of growing seasons demonstrated the largest value of this trait in 2016 and the smallest one in 2015 (lower by 5% than that noted in 2016). The application of seaweed extract increased this number, which was indicated by values of the ABT-C index that were calculated for this trait (Table 4).

Table 5. Effect of *Ecklonia maxima* extract (Kelpak) treatment on yield and nutritional properties of soybean (average from 2014–2016).

Parameters	*Ecklonia maxima* Treatment	Season			Average from 2014 to 2016
		2014	2015	2016	
Number of seeds (per m^{-2})	C	1793^{a}	1581^{a}	1907^{a}	1760^{a}
	LSS	2255^{b}	2210^{b}	2337^{b}	2267^{b}
	LDS	2340^{b}	2376^{bc}	2406^{bc}	2374^{b}
	HSS	2344^{b}	2350^{b}	2401^{bc}	2365^{b}
	HDS	2466^{c}	2528^{c}	2576^{c}	2524^{c}
	AS	2240^{a}	2209^{a}	2326^{b}	
Seed yield (t ha^{-1})	C	3.267^{a}	2.664^{a}	3.262^{a}	3.064^{a}
	LSS	3.677^{b}	3.636^{b}	3.767^{b}	3.693^{b}
	LDS	3.805^{b}	3.876^{b}	3.852^{bc}	3.844^{b}
	HSS	3.758^{b}	3.874^{b}	3.907^{bc}	3.846^{b}
	HDS	4.137^{c}	4.171^{c}	4.198^{c}	4.169^{c}
	AS	3.729^{a}	3.644^{a}	3.797^{b}	
1000 seed weight (g)	C	182.2^{b}	168.5^{a}	171.0^{b}	173.9^{b}
	LSS	163.1^{a}	164.6^{a}	161.2^{a}	162.9^{a}
	LDS	162.6^{a}	163.2^{a}	160.1^{a}	161.9^{a}
	HSS	160.3^{a}	164.9^{a}	162.7^{a}	162.3^{a}
	HDS	167.8^{a}	165.0^{a}	163.1^{a}	165.3^{a}
	AS	167.2^{b}	165.2^{ab}	163.6^{a}	
Total protein (% DM)	C	36.8^{a}	46.5^{d}	35.9^{a}	39.7^{a}
	LSS	37.7^{b}	45.7^{c}	38.6^{d}	40.7^{a}
	LDS	37.7^{b}	47.4^{e}	36.3^{b}	40.5^{a}
	HSS	38.0^{b}	42.7^{b}	38.9^{d}	39.9^{a}
	HDS	39.1^{c}	40.9^{a}	38.1^{c}	39.4^{a}
	AS	37.9^{b}	44.6^{c}	37.6^{a}	
Total fat (% DM)	C	17.5^{d}	15.0^{d}	16.6^{c}	16.4^{b}
	LSS	14.5^{a}	15.5^{e}	14.5^{a}	14.8^{a}
	LDS	15.4^{b}	12.8^{a}	15.1^{b}	14.4^{a}
	HSS	15.5^{b}	13.8^{c}	15.1^{b}	14.8^{a}
	HDS	15.7^{c}	13.3^{b}	16.4^{c}	15.2^{ab}
	AS	15.7^{c}	14.1^{a}	15.5^{b}	

Abbreviations: C, control; LSS, lower concentration single spraying; LDS, lower concentration double spraying; HSS higher concentration single spraying; HDS, higher concentration double spraying. Means in the columns, concerning the selected traits, followed by different small letters are significantly different at $p < 0.05$.

3.2.2. Seed Yield

The most positive response of plants to the use of biostimulant was observed after double spraying with the higher concentration of Kelpak preparation, as indicated by their seed yield increase by 36% when compared to the control (Table 5). The highest mean seed yield for Atlanta cv. was obtained in 2016. In contrast, the seed yield of 2015 season turned out to be the lowest among the studied seasons (lower by 4% than that noted in 2016). Foliar application of Kelpak increased the seed yield of soybean of Atlanta cv., which was indicated by positive values of the ABT-C index that were calculated for this trait (Table 4).

3.2.3. Thousand Seed Weight

Foliar application of Kelpak decreased 1000 seed weight. Its lowest value was determined after double application of Kelpak in the lower concentration (decrease by 7% as compared to the control) (Table 5). The least decrease of 1000 seed weight was achieved after double plant spraying with the higher concentration of Kelpak biostimulant. The highest mean 1000 seed weight was reported in the 2014 growing season. The values of the biostimulant effect index calculated for this trait were negative, which points to the negative impact of Kelpak preparation on 1000 seed weight (Table 4).

3.3. *Effect of Biostimulant on the Nutritional Properties*

3.3.1. Total Protein in Soybean Seeds

Depending on concentration and number of applications, Kelpak increased or decreased the protein content in a dry matter of seeds. However, the statistical analysis demonstrated that differences in the effects of biostimulant on this trait were insignificant. Increased protein content was determined in seeds of plants single-sprayed with the lower concentrations of Kelpak (Table 5). Concerning growing seasons, the highest protein content of seeds was noted in 2015. Values of the ABT-C index that were calculated for this trait were positive for this preparations (Table 4).

3.3.2. Total Fat in Soybean Seeds

Regardless of the number of sprayings and concentration of biostimulant, its use decreased the fat content in dry matter of soybean seeds, with the greatest decrease (by 14% as compared to the control) being noted after double spraying the plants with the lower concentration of Kelpak (Table 5). In contrast, the smallest decrease in fat content of the seeds as compared to the control was determined after double spraying with the higher concentrations of Kelpak. The highest fat content of soybean seeds was noted in season 2014 and the lowest in 2015. The values of the ABT-C index that were calculated for this preparation were negative (Table 4), which is indicative of its negative effect on fat content in of Atlanta cv. soybean seeds.

3.4. *Effect of Biostimulants on the Antioxidant Potential in Soybean Seeds*

3.4.1. Total Phenolic Content

The use of Kelpak in soybean cultivation caused changes in contents of total polyphenols (TPC) in seeds (Table 6), which varied depending on both the number of applications and the concentration of this preparation. The use of the biostimulant based on *Ecklonia maxima* extract caused an increase in phenolics compounds content in soybean seeds. However, significant differences were only demonstrated in plants that were single-sprayed with 1% Kelpak (HSS). The TPC content that was determined for this combination was over twofold higher, when compared to the control combination. This nutraceutical property of soybean was influenced by meteorological conditions that occurred in a given growing season. The highest significant differences were observed in 2014 and 2016. A positive value of the difference between contents of phenolics in combinations that were treated with Kelpak biostimulant and the control samples (ABT-C) was calculated for soybean seeds (Table 4).

3.4.2. Total Anthocyanins Content

The presence of anthocyanins was detected in seven out of the 15 analyzed combinations of Kelpak biostimulant use in soybean cultivation. These compounds were not detected in the control samples in any of the growing seasons studied.

The use of Kelpak affected the content of anthocyanins in soybean seeds. However, their presence was only detected in 17% of the analyzed combinations. The number of applications and concentration of the biostimulant were the factors that determined anthocyanins content. The highest value of which was noted after plants spraying with the higher concentration of Kelpak. In this case, significant differences were also observed as influenced by conditions that occurred during the plant growth stage (Table 6). The values of biostimulant effect ABT-C index calculated for this trait were positive (Table 4).

Table 6. Effect of *Ecklonia maxima* extract (Kelpak) treatment on the antioxidant potential in soybean seeds (average from 2014–2016).

Parameters	*Ecklonia maxima* Treatment	Season			AA
		2014	2015	2016	
Total phenols ($mg\ g^{-1}$ DM)	C	5.77^{a}	4.50^{a}	5.77^{b}	5.35^{a}
	LSS	7.36^{b}	5.85^{e}	7.74^{c}	6.98^{a}
	LDS	8.56^{c}	4.70^{b}	8.40^{d}	7.22^{a}
	HSS	15.02^{d}	5.05^{c}	15.20^{e}	11.76^{b}
	HDS	5.78^{a}	5.26^{d}	5.54^{a}	5.53^{a}
	AS	8.50^{b}	5.07^{a}	8.53^{b}	
Total flavonoids ($mg\ g^{-1}$ DM)	C	1.99^{a}	1.44^{a}	1.99^{a}	1.81^{a}
	LSS	1.87^{a}	1.92^{c}	1.92^{a}	1.90^{a}
	LDS	2.64^{b}	1.84^{b}	2.59^{b}	2.36^{a}
	HSS	5.10^{d}	2.93^{e}	5.15^{d}	4.39^{b}
	HDS	4.18^{c}	2.08^{d}	4.21^{c}	3.49^{b}
	AS	3.16^{b}	2.04^{a}	3.17^{b}	
Anthocyanins ($mg\ g^{-1}$ DM)	C	0.00^{a}	0.00^{a}	0.00^{a}	0.000^{a}
	LSS	0.00^{a}	0.00^{a}	0.00^{a}	0.000^{a}
	LDS	0.00^{a}	0.02^{b}	0.00^{a}	0.007^{a}
	HSS	0.00^{a}	0.04^{c}	0.00^{a}	0.013^{ab}
	HDS	0.04^{b}	0.00^{a}	0.05^{b}	0.030^{b}
	AS	0.008^{a}	0.012^{b}	0.010^{ab}	
Reducing power (mg TE g^{-1} DM)	C	0.15^{a}	0.10	0.15^{a}	0.13^{a}
	LSS	0.30^{bc}	0.22	0.33^{c}	0.28^{b}
	LDS	0.21^{ab}	0.14	0.28^{b}	0.21^{ab}
	HSS	0.45^{d}	0.08	0.42^{e}	0.31^{b}
	HDS	0.38^{cd}	0.16	0.37^{d}	0.30^{b}
	AS	0.30^{b}	0.14^{a}	0.31^{b}	

Abbreviations: C, control; LSS, lower concentration single spraying; LDS, lower concentration double spraying; HSS higher concentration single spraying; HDS, higher concentration double spraying. Means in the columns, concerning the selected traits, followed by different small letters are significantly different at $p < 0.05$.

3.4.3. Total Flavonoid Content

Flavonoid content analysis showed a significant effect of the application of biostimulant on its values. The foliar application of Kelpak resulted in the increased content of flavonoids in seeds. Significantly, the highest content of these compounds was noted after plant spraying with 1% solution of this preparation, regardless of the number of applications.

The analysis of the effect of biostimulants with different composition revealed that their foliar application resulted in an increased content of flavonoids when compared to the control samples (a positive value of the ABT-C difference) (Table 4).

3.4.4. Reducing Power

The evaluation of the effect of applying biostimulants with different compositions on the antioxidant activity of soybean included the determination of the reducing power, the value of which was increased by almost all combinations of this biostimulant.

Significant differences in reducing power values were observed upon the application of Kelpak biostimulant (Table 6). A tendency for an increase of reducing potential was noted after the application of this preparation in the higher concentration and after single spraying the plants with its 0.7% solution. In the second study year, the value of reducing power was the lowest when compared to the other analyzed years (over twofold decrease of RP value). Foliar application of *Ecklonia maxima* extract increased its reducing power values, which was indicated by positive values of the ABT-C index that were calculated for this trait (Table 4).

4. Discussion

Biostimulants induce the growth and development of plants, from seed germination throughout the entire ontogenesis. They affect the metabolic processes that occur in the plant by enhanced activity and synthesis of phytohormones, by stimulating the growth of the root system, and by improving the uptake, translocation, and retention of nutrients, which determines quantity and quality of crop yield [6,68].

Our study demonstrates a significant increase in the growth of soybean plants after the foliar application of biostimulant that is based on *Ecklonia maxima* extract. An earlier study also showed the growth stimulation of soybean treated with a biostimulant (Fylloton) containing *Ascophyllum nodosum* extract and free amino acids [8]. The marked growth responses in soybean plants are possibly due to *Ecklonia maxima* extract (Kelpak) composition, especially the PGRs (Plant Growth Regulator) that were identified (cytokinins, auxins, polyamines, gibberellins, brassinosteroids) and the mineral content in this biostimulant [55,69,70]. Additionally, the stimulatory role of Kelpak in the production of phytohormones has been demonstrated. For example, it increased the content of cytokinins in *Eucomis autumnalis* [70].

However, despite the observed favourable effects related to the application of biostimulants, including seaweeds, the precise mechanism of their activity still remains mostly unknown [26]. It should be emphasised that a full explanation of the principle or principles of their operation may cause a potential increase in the use of these preparations. According to Crouch and Van Staden [71] and Craigie [72], a wide scope of reported physiological responses of plants in cultivations where seaweed extracts were used is related to the fact that those products include numerous active compounds. Cytokinins, auxins, gibberellins, brassinosteroids, and other activating particles, like, for example, oligomers and polysaccharides are included [73]. According to Depuydt and Hardtke [74], cytokinines, together with auxins, function as regulators of various physiological processes, including those that are related to plant growth and development [75–79]. Thus, each change in the concentration of endogenic cytokinines influences the regulation of many physiological processes and as a result impacts the growth of the entire plant [80,81]. Studies by Aremu et al. [70] proved that the total content of cytokinines increases in plants after the application of the Kelpak preparation. The qualitative composition of the listed compounds is changed, which is related to their functional and physiological role, in particular, during plant morphogenesis. According to Strnad [76], isoprenic cytokinines determine the growth processes that include a continuation of the cell cycle. On the other hand, aromatic cytokinines model growth processes, such as morphogenesis and ageing. Aremu et al. [70] even assumed that the quantification of the endogenic content of cytokinins might provide information regarding possible physiological mechanisms that are related to the application of Kelpak biostimulant. However, researchers stress that, due to numerous active substances and compounds that are contained in Kelpak, the observed, favourable impact on the growth and development of plants may not only be assigned to cytokinins, but instead be the result of possible cross reactions of those compounds with other biologically active particles that are included in seaweed biostimulants. Therefore, further research concerning those fields is indispensable in order to obtain a full explanation for the Kelpak performance [70].

Still, in the literature, the prevailing hypothesis is that the majority of responses of plants that were cultivated with biostimulants' application, including seaweeds, results from the presence of compounds from the group of plant hormones, namely cytokinins [18]. The assumption stems from the fact that these compounds, isolated from seaweed extracts and individually tested in cultivations, mitigates the stress that is caused by free radicals through direct capturing and the prevention of reactive oxygen forms (ROS). This is done through the inhibition of xanthine oxidation [18,34,82–84].

Khan et al. [85] and Panda et al. [18] additionally indicate that extracts from seaweeds, such as Kelpak, support plant tolerance to stress, influencing the increase of K^+ capture in plants.

Additionally, in the literature, we may find hypotheses regarding modelling the growth and development of plants through the application of biostimulants from seaweeds as an effect of the presence of substances that are similar to gibberellins [86]. Research by Stephenson [19] indicates that these extracts include at least two compounds that behave like gibberellins (GA3 and GA7). However, they also show the presence of terpenoids and α-tokopherol, the performance of which may imitate gibberellins' activity on plants [18,87,88].

Recent theories indicate that the activity of seaweed extracts may be the result of their content of betains. Panda et al. shows that these compounds have a similar impact on plants as the aforementioned cytokinins [18]. It was proved that seaweed extracts include, among others, gamma aminobutyric acid betaine, 6-amino valeric acid betaine, and glyco betaine. Mancuso et al. [89] suggested that, due to the presence of those active compounds, extracts influence the mitigation of the osmotic and oxidization stress in plants, which may lead to the damage of DNA lipids, carbohydrates, and proteins, and also disturb correct cell signalling. Genard et al. [90] and Blunden et al. [91] even assigned an improvement of plant yield to the presence of betaines in extracts, since that led to the increased concentration of chlorophyll. According to Naidu et al. [92], betaines constitute a source of nitrogen when they are provided in low doses, or act as osmolites in higher concentrations. Many studies also show that betaines play a role in the correct formation of somatic germs from cotyledons tissues and mature seeds [18,93,94].

The stimulation of plant growth and development may also result from the occurrence of polyamins in biostimulants based on seaweed extracts, since these compounds may act as plant growth regulators. However, it should be emphasised that they are not classified as plant hormones. Several amino groups that usually replace hydrogen in the alkaline chain (putrescine, spermidine, and spermine) are characteristic of polyamins' structure. Research by Haman et al. [95] proves that polyamins determine the stability of various RNA and DNA conformation states. These compounds are often related to important stages of the cell division cycle. They also ensure the stability of a membrane to various cell membranes. Thus, due to the fact that polyamins affect a wide scope of physiological growth processes, their occurrence in biostimulating products that are made of seaweed may influence plant growth [18].

In our study, the use of biostimulant increased the fat and protein content in soybean seeds. The stimulating effect of biostimulants on the nutritional composition of various plants is mainly due to the number of PGR contained in the solutions [89]. In addition, the increased nutritional content in *C. triloba* is probably related to the ability of biostimulators to improve the slow release of nutrients and their uptake by plants [68,96]. A stimulating activity of seaweeds extract is also found in the presence of abscisic acid (ABA) [97,98]. However, the ABA function remains not fully characterized. Nevertheless, it is known that this acid induces protein synthesis, which are needed by plants in dealing with stress factors during water deficiencies [99,100]. Davies [101] shows that, during drought, this compound in plants caused numerous physiological reactions, including the closing of stoma, increase of a trend for accumulation of protein in seeds, gene transcription for proteinase inhibitors, as well as inhibition of sprouts growth or initiation of some states of seeds dormancy.

Unfortunately, the precise mechanisms that are activated by those biostimulants are still difficult to be identified, despite even greater knowledge on the composition of extracts from seaweeds. This is also due to the fact that these preparations constitute an abundance of many biologically active chemical compounds. They also include bioactive secondary metabolites, vitamins, and vitamin precursors [102,103]. Many authors underline the meaning of their synergetic cooperation, which stimulates that growth of plants assuming a mechanism that has not been fully known yet [24,72,104]. One of the main components of seaweed extracts are polysaccharides, including alginians, fucoidans, and laminarans [85,105]. Fucoidans have various structures due to a varied degree of methylation, sulphurization, and branching [72]. Alginians are polymers of mannuronic and galuronic acid, with a confirmed activity of plant-growth promotion [106]. Finally, laminarins that are included in extracts are registered compounds that increase plant resistance to fungi and bacteria pathogens [107].

Improvement of a nutritive value of soya seeds observed in our research, as expressed in protein and fat content, could also be caused by the fact that seaweeds are rich in phenolic compounds (complex chloroclucinol, eckol, and dieckol polymers). Phenols belong to secondary metabolites synthesized in plants under the influence of stress. Their task is to protect cells and the components of cell nuclei [108,109]. The ability to chelate metal ions [110] is a significant role of these compounds, besides their antioxidant activity. Research by Raj et al. [60] confirms that phenolic compounds with dihydroxybenzene or trihydroxybenzene groups show strong chelating activity. Rengasamy et al. [111] also shows that eckol belonging to phenolic compounds proves to have a strong auxinosimilar activity. According to the authors, the impact of these polyphenols, included in seaweed extracts, on the endogenic content of auxins is a key element that is necessary for understanding basic mechanisms of these preparations. Korasick et al. [112] reached similar conclusions. They conclude that auxins have a strong impact on many important stages of physiological growth in the life cycle of a plant. Researchers emphasise that maintaining proper concentration of active auxin in cells is of key significance for controlling almost all aspects that are related to the plant growth. The concentration of cell auxin is affected by the speed of anabolism, catabolism, transport, and conjugation [113]. In relation to the type of concentration, polyphenols may inhibit or stimulate the development of vegetative plants. It mainly takes place due to their abilities to modulate the metabolism and concentration of active auxin forms in plants [114–117]. Gaspar et al., [118] proves that the phenolic inhibitors of oxidase IAA, such as chlorogenic acid, influence the activity of auxins. Some of the mentioned compounds even constitute alternative substrates for the oxidizing enzyme, which in turn is related to the protection of auxins before oxygen decomposition. Wilson and Van Staden [119] prove that some of the phenolic acids that protect auxins before decarboxilation increase the concentration of active forms of auxins, which are indispensable for the stimulation of growth and development of crop roots. However, according to Arem et al., attempts to explain mechanisms that are responsible for a positive response of plants to application of extracts from seaweeds should take into account the possible cross reactions between phytohormones included therein and the quantitative concentration of auxins, which may justify the observed morphological differences [120].

In the literature, one may find hypotheses that assume that the increased growth and yield of plants that are treated with seaweed extracts resulted from a positive impact of those preparations on the activity of esterase enzymes. This enzyme is considered to be a marker of plant growth processes due to its role in organogenesis. It also works as an index of somatic embryogenesis [121–124]. According to Aremu et al., a higher activity of esterase in plants that were treated with seaweed extracts indicated their stimulating impact on the increase of plant biomass production [120].

Plant metabolism may be modelled through the use of biostimulants. According to Nardi et al., this group of active preparations affects most of all carbon and nitrogen metabolism, which is associated with an enhanced activity of enzymes participating in, among others, the process of glycolysis, Krebs cycle, or nitrogen assimilation [125]. Oboh at al. and Ertani et al. demonstrated that biostimulants application yielded metabolic pathways that are linked with secondary metabolites, like e.g. phenolic compounds [126,127]. It should also be emphasized that the synthesis of secondary metabolites proceeds as an element of chemical defense [128]. Already, in 1959, these compounds were no longer treated as ballast substances [129]. Today, they are believed to play a significant role in plant protection against adverse factors [130]. The most common indicator of plants resistance to biotic factors is the content of phenolic compounds [131], which are precursors of more complex phenolic structures, like flavonoids or lignins [132].

In our study, the foliar application of Ecklonia maxima extract (Kelpak) caused a significant increase in polyphenols content. Ertani et al. and Lakhdar et al. showed that the application of biostimulants in plant cultivation enhanced the synthesis of antioxidative compounds, which are indicators of increased plant resistance to biotic and abiotic stress factors [38,133]. The physiological response of plants to the use of biostimulants results from the presence of active substances in them, such as phytohormones, amino acids, proteins, phenols, or triacontanol [38,127,134].

A positive impact of compounds that are included in seaweed extracts on the total content of phenolic compounds in soya seeds has a significant meaning in the attempt to explain the mechanisms of operation of those biostimulants. The antioxidant potential of plants is inseparable from the amount and quality of phenolic compounds [135]. In real environmental conditions, the regulation of phytochemical synthesis includes a range of advanced mechanisms that enable a precise control of production of specific particles in a suitable place and time, and also in response to outside signals [136]. Such compounds include those in seaweed extracts that may activate specific biochemical pathways that are responsible for the synthesis of secondary metabolites in plants [136–138]. According to Cheynier et al., eckol that is contained in biostimulants influences the phenylpropanoid pathway in the biochemical synthesis of phenolic acids [136]. Researchers assume that this impact is caused by the regulation of the enzymatic activity of ammonia lyase of pheyloalanin and chalconic sythase. However, only the approach that is based on the genetic analysis will enable the observation of gene regulation engaged in those pathways [16,120]. Research results that were carried out by Jannin et al. proved that cysteine protease, related to the process of synthesis of phenolic compounds, were regulated downwards, while the expression of genes that are related to photosynthesis, cell metabolism, response to stress, and nitrogen metabolism were significantly raised in the case of plants treated with seaweed extracts [139]. Roupahel et al. observed an increased concentration of phenolic compounds after the application of such biostimulants assigned to their main components, such as polysaccharides (alginians, fucoidans, and laminarins) [140]. These compounds influence endogenic hormonal homeostasis [63,64]. Additionally, processes of synthesis and accumulation of secondary metabolites may be related to the activity of enzymatic groups that are engaged in phytochemical homeostasis (the so-called direct effect) [127,140]. They also depend on the plant nutrition condition and potassium and magnesium concentration (direct effect) [41]. Roupahel et al. [140] also search for the growth of the concentration of bioactive compounds in the activation of key enzymes, such as chalkone isomerase, which is engaged in the biosynthesis of flavon precursors [141].

According to Azcona et al. [142] and Ertani et al. [127], the high effectiveness of biostimulants in plant crops is also influenced by the number of treatments at the appropriate stages of plant development. The first treatment of plant with these preparations resulted most of all in the increased number and weight of leaves, which is referred to as "short-time effect". Another dose of biostimulants, applied at the plant blooming stage, led to the long-term effect, which was manifested by changes in crop size and quality. In the case of fruits, it results in, among others, an increase in their number and weight when compared to control samples that were not treated with biostimulants [125,127]. The increased content of polyphenols in the crop may indeed result from the use of biostimulants at the appropriate growth stages of plants. Experiments that were conducted by Oboh et al. [126] and by Zhang and Hamauzu [143] confirmed that the first application of these preparations led to an increased content of phenolic compounds in leaves, and that this increase was smaller after the second application of biostimulants. According to Ertani et al. [127], changes in the total polyphenolics content resulting from different numbers of applications of biostimulants are also linked with changes in contents of individual phenolic acids.

This was since the increasing total content of phenolic acids led to an increased number of their functional groups, which are sequesters of free radicals [144]. It must be emphasized that the increased content of polyphenols in plant tissues, as evoked by the action of biostimulants, is a beneficial phenomenon, not only because of the increased plant resistance to stress factors, but also because of significant importance to consumers, since such plant products are rich sources of antioxidative compounds being valuable to the human body [145,146]. Phenolic acids, such as caffeic, gallic, and ferulic, are claimed to exhibit anticarcinogenic and antimicrobial activities [147,148].

To sum up, biostimulants that contain seaweed extracts enable many opportunities to improve plant growth. Their use in agriculture is considered to be favourable for cropping. However, the operation mechanism of such products is not completely described [149]. That is because the impact of the application of growth regulators on plants is not only a consequence of their direct ability to control metabolic pathways, since their activity may be multidirectional. Limited knowledge on the mechanisms of the preparations' activity are still mainly based on assumptions and hypotheses highlights the need of further research within this scope [5]. So far it has been proved that seaweed extracts influence the plant physiology through changes in their general profile of transcriptome, and also in metabolism [139,150]. Research by Fan et al., concerning the analysis of gene expression, expanded the understanding of the possible mechanisms that regulate the activity of these preparations [141]. Researchers indicate that, after the application of extracts from seaweed, increases in the amount of transcripts of regulatory enzymes that are related to the nitrogen metabolism (cytolase glutamine synthesis), antioxidant ability (glutathione reductase), and glycine betaine synthesis (betaine aldehyde dehydrogenesis and choline monooxigenesis) were observed [16,18,22,72].

Although biostimulants are extensively used in agricultural practice, presently the most significant research on those preparations requires a better understanding of the mechanism of their influence [22,151]. According to Van Osten et al. [22] and Povero et al. [152], only after obtaining a complete explanation of those mechanisms, can the design and production of new generation biostimulants can take place. Due to their complex composition and interactions between particular compounds, mechanisms of operation of preparations based on seaweed extracts are slowly and successively discovered, with applications of molecular biology, metabolomics, and genomics techniques. However, according to many researchers, observed favourable biological effects of extracts activity is caused by the activity of small organic particles, as well as polymers that are included in products that have an ability to regulate genes' operation responsible for ensuring and modelling plant resistance systems [16,71,139].

5. Conclusions

The number of biostimulant applications and its concentration modified the biometric traits, crop size, and yield, as well as the nutraceutical and antioxidative potential of soybean seeds. The study demonstrated that the foliar application of *Ecklonia maxima* extract improved the growth and yield of soybean without any negative effect on the nutritive value of its seeds. Our experiment showed a positive effect of double foliar application of the higher concentrations of this biostimulant on soybean seed number and yield. The application of *Ecklonia maxima* extract increased the antioxidative activity of soybean seeds, and content of total phenolic compounds, flavonoids, and anthocyanins. The results of our study indicate the need for continuing investigations and extending their scope with the aim to identify responses of different cultivable plants on the use of biostimulants that are based on various biologically-active compounds.

Author Contributions: S.K. and A.S. conceived and designed the research. S.K., A.S. and E.C. performed the experiments. S.K., A.S., E.C., M.K. and T.P. prepared the materials. S.K., A.S., and M.K. analyzed the data. S.K. and A.S. wrote the paper. M.K. and T.P. revised the manuscript. All authors read and approved the final manuscript.

Funding: This research received no external funding.

Conflicts of Interest: The authors declare no conflict of interest.

References

1. Bezerra, A.R.G.; Sediyama, T.; Borém, A.; Soares, M.M. Importância econômica. In *Soja: Do plantio à colheita*; Sediyama, T., Silva, F., Borém, A., Eds.; UFV: Viçosa, Brazil, 2015; pp. 9–26.
2. Kocira, S. Effect of amino acid biostimulant on the yield and nutraceutical potential of soybean. *Chil. J. Agric. Res.* **2019**, *79*, 17–25. [CrossRef]

3. Stein, H.H.; Berger, L.L.; Drackley, J.K.; Fahey, G.C., Jr.; Hernot, D.C.; Parsons, C.M. Nutritional properties and feeding values of soybeans and their coproducts. In *Soybeans, Chemistry, Production, Processing, and Utilization*; Johnson, L.A., White, P.J., Galloway, R., Eds.; AOCS Press: Urbana, IL, USA, 2008; pp. 613–660.
4. Lodhi, K.K.; Choubey, N.K.; Dwivedi, S.K.; Pal, A.; Kanwar, P.C. Impact of seaweed saps on growth, flowering behavior and yield of soybean [*Glycine max* (L.) Merrill.]. *Bioscan* **2015**, *10*, 479–483.
5. Szparaga, A.; Kocira, S.; Kocira, A.; Czerwińska, E.; Świeca, M.; Lorencowicz, E.; Kornas, R.; Koszel, M.; Oniszczuk, T. Modification of growth, yield, and the nutraceutical and antioxidative potential of soybean through the use of synthetic biostimulants. *Front. Plant Sci.* **2018**, *9*. [CrossRef] [PubMed]
6. Basak, A. Biostimulators. Definitions, classification and legislation. In *Biostimulators in Modern Agriculture, General Aspects*; Gawronska, H., Ed.; Editorial Housen Wies Jutra: Warszawa, Poland, 2008; pp. 7–17.
7. Paradiković, N.; Vinkovic, T.; Vinkovic, V.I.; Zuntar, I.; Bojic, M.; Medic, M. Effect of natural biostimulants on yield and nutritional quality: An example of sweet yellow pepper (*Capsicum annuum* L.) plants. *Sci. Food Agric.* **2011**, *91*, 2146–2152. [CrossRef] [PubMed]
8. Kocira, S.; Szparaga, A.; Kocira, A.; Czerwińska, E.; Wójtowicz, A.; Bronowicka-Mielniczuk, U.; Koszel, M.; Findura, P. Modeling biometric traits, yield and nutritional and antioxidant properties of seeds of three soybean cultivars through the application of biostimulant containing seaweed and amino acids. *Front. Plant Sci.* **2018**, *9*, 388. [CrossRef] [PubMed]
9. Kaufmann, G.L., III; Kneivel, D.P.; Watschke, T.L. Effects of a biostimulant on the heat tolerance associated with photosynthetic capacity, membrane thermostability, and polyphenol production of perennial ryegrass. *Crop Sci.* **2007**, *47*, 261–267. [CrossRef]
10. Du Jardin, P. The Science of Plant Biostimulants—A bibliographic analysis. Ad hoc Study Report to the European Commission DGENTR. 2012. Available online: https://orbi.uliege.be/bitstream/2268/169257/1/Plant_Biostimulants_final_report_bio_2012_en.pdf (accessed on 1 February 2019).
11. Traon, D.; Amat, L.; Zotz, F.; Du Jardin, P. *A Legal Framework for Plant Biostimulants and Agronomic Fertiliser Additives in the EU. Report for the European Commission Enterprise & Industry Directorate—General*; Publications Office of the European Union: Luxembourg, 2014.
12. Matyjaszczyk, E. The introduction of biostimulants on the Polish market. The present situation and legal requirements. *Przem. Chem.* **2015**, *10*, 1841–1844. [CrossRef]
13. Du Jardin, P. Plant biostimulants: Definition, concept, main categories and regulation. *Sci. Hortic.* **2015**, *196*, 3–14. [CrossRef]
14. Colla, G.; Cardarelli, M.; Bonini, P.; Rouphael, Y. Foliar applications of protein hydrolysate, plant and seaweed extracts increase yield but differentially modulate fruit quality of greenhouse tomato. *HortScience* **2017**, *52*, 1214–1220. [CrossRef]
15. Colla, G.; Hoagland, L.; Ruzzi, M.; Cardarelli, M.; Bonini, P.; Canaguier, R.; Rouphael, Y. Biostimulant action of protein hydrolysates: Unraveling their effects on plant physiology and microbiome. *Front. Plant Sci.* **2017**, *8*, 2202. [CrossRef] [PubMed]
16. Battacharyya, D.; Babgohari, M.Z.; Rathor, P.; Prithiviraj, B. Seaweed extracts as biostimulants in horticulture. *Sci. Hortic.* **2015**, *196*, 39–48. [CrossRef]
17. Aguilar, J. Algas marinas para la agricultura de alto rendimiento. Available online: http://www.interempresas.net/Horticola/Articulos/136576-Algas-marinas-para-la-agricultura-de-alto-rendimiento.html (accessed on 1 February 2019).
18. Panda, D.; Pramanik, K.; Nayak, B.R. Use of sea weed extracts as plant growth regulators for sustainable agriculture. *Int. J. Bioresour. Stress Manag.* **2012**, *3*, 404–411.
19. Stephenson, W.A. *Seaweed in Agriculture and Horticulture*; Faber and Faber: London, UK, 1968.
20. Crouch, I.J.; van Staden, J. Evidence for the presence of plant growth regulators in commercial seaweed products. *Plant Growth Regul.* **1993**, *13*, 21–29. [CrossRef]
21. Ecoforce, A. Extracto de algas para agricultura ecológica—Fertilizantes ecoforce. Available online: http://fertilizantesecoforce.es/es/blog/index/list/cat/agricultura-ecologica/?p=19 (accessed on 1 February 2019).
22. Van Oosten, M.J.; Pepe, O.; De Pascale, S.; Silletti, S.; Maggio, A. The role of biostimulants and bioeffectors as alleviators of abiotic stress in crop plants. *Technol. Agric.* **2017**, *4*, 5. [CrossRef]
23. Medjdoub, R. *La algas marinas y la agricultura*; Adiego: Zaragoza, Spain; pp. 1–3. Available online: http://catsaigner.adiego.com/sites/default/files/las_algas_marinas.pdf (accessed on 1 February 2019).

24. Norrie, J.; Keathley, J.P. Benefits of *Ascophyllum nodosum* marine-plant extract applications to 'Thompson seedless' grape production. *Acta Hortic.* **2006**, *727*, 243–247. [CrossRef]
25. Gajc-Wolska, J.; Spiżewski, T.; Grabowska, A. The effect of seaweed extracts on the yield and quality parameters of broccoli (*Brassica oleracea* var. *cymosa* L.) in open field production. *Acta Hortic.* **2013**, *1009*, 83–89. [CrossRef]
26. Sharma, H.S.S.; Fleming, C.; Selby, C.; Rao, J.R.; Martin, T. Plant biostimulants: A review on the processing of microalgae and use of extracts for crop management to reduce abiotic and biotic stresses. *J. Appl. Phycol.* **2014**, *26*, 465–490. [CrossRef]
27. Guerrero, J. La aplicación de las algas marinas para la fertilización Hortalizas. Available online: http://www.hortalizas.com/nutricion-vegetal/la-aplicacion-de-las-algas-marinas-para-la-fertilizacion/ (accessed on 1 February 2019).
28. Jayaraj, J.; Norrie, J.; Punja, Z.K. Commercial extract from the brown seaweed *Ascophyllum nodosum* reduces fungal diseases in greenhouse cucumber. *J. Appl. Phycol.* **2011**, *23*, 353–361.
29. Araujo, I.B.; Peruch, L.A.M.; Stadnik, M.J. Efeito do extrato de alga e da argila silicatada na severidade da alternariose e na produtividade da cebolinha comum (*Allium fistulosum* L.). *Trop. Plant Pathol.* **2012**, *37*, 363–367. [CrossRef]
30. Uppal, A.K.; El Hadrami, A.; Adam, L.R.; Tenuta, M.; Daayf, F. Biological control of potato Verticillium wilt under controlled and field conditions using selected bacterial antagonists and plant extracts. *Biol. Control* **2008**, *44*, 90–100. [CrossRef]
31. Zodape, S.T.; Gupta, A.; Bhandari, S.C.; Rawat, U.S.; Cahudhary, D.R.; Eswara, K.; Chikara, J. Foliar application of seaweed sap as biostimulant for enhancement of yield and field quality of tomato (*Lycopersicon esculentum* Mill.). *J. Sci. Ind. Res.* **2011**, *70*, 215–219.
32. El Modafar, C.; Elgadda, M.; El Boutachfaiti, R.; Abouraicha, E.; Zehhar, N.; Petit, E.; El Alaoui-Talibi, Z.; Courtois, B.; Courtois, J. Induction of natural defence accompanied by salicylic acid-dependant systemic acquired resistance in tomato seedlings in response to bioelicitors isolated from green algae. *Sci. Hortic.* **2012**, *138*, 55–63. [CrossRef]
33. Shah, M.T.; Zodape, S.T.; Chaudhary, D.R.; Eswaran, K.; Chikara, J. Seaweed sap as an alternative to liquid fertilizer for yield and quality improvement of wheat. *J. Plant Nutr.* **2013**, *36*, 192–200. [CrossRef]
34. Zhang, X.Z.; Ervin, E.H. Impact of seaweed extract-based cytokinins and zeatin riboside on creeping bentgrass heat tolerance. *Crop Sci.* **2008**, *48*, 364–370. [CrossRef]
35. Jayaraj, J.; Wan, A.; Rahman, M.; Punja, Z.K. Seaweed extract reduces foliar fungal diseases on carrot. *Crop Prot.* **2008**, *10*, 1360–1366. [CrossRef]
36. Botta, A. Enhancing plant tolerance to temperature stress with amino acids: An approach to their mode of action. *Acta Hortic.* **2013**, *1009*, 29–35. [CrossRef]
37. Colla, G.; Svecova, E.; Rouphael, Y.; Cardarelli, M.; Reynaud, H.; Canaguier, R.; Rouphael, Y. Effectiveness of a plant-derived protein hydrolysate to improve crop performances under different growing conditions. *Acta Hortic.* **2013**, *1009*, 175–179. [CrossRef]
38. Ertani, A.; Schiavon, M.; Muscolo, A.; Nardi, S. Alfalfa plant-derived biostimulant stimulate short-term growth of salt stressed *Zea mays* L. plants. *Plant Soil* **2013**, *364*, 145–158. [CrossRef]
39. Lucini, L.; Rouphael, Y.; Cardarelli, M.; Canaguier, R.; Kumar, P.; Colla, G. The effect of a plant-derived protein hydrolysate on metabolic profiling and crop performance of lettuce grown under saline conditions. *Sci. Hortic.* **2015**, *182*, 124–133. [CrossRef]
40. Xu, C.; Leskovar, D. Effects of *A. nodosum* seaweed extracts on spinach growth, physiology and nutrition valued under drought stress. *Sci. Hortic.* **2015**, *183*, 39–47. [CrossRef]
41. Rouphael, Y.; Colla, G.; Giordano, M.; El-Nakhel, C.; Kyriacou, M.C.; De Pascale, S. Foliar applications of a legume-derived protein hydrolysate elicit dose-dependent increases of growth, leaf mineral composition, yield and fruit quality in two greenhouse tomato cultivars. *Sci. Hortic.* **2017**, *226*, 353–360. [CrossRef]
42. Rouphael, Y.; Cardarelli, M.; Bonini, P.; Colla, G. Synergistic action of a microbial-based biostimulant and a plant derived-protein hydrolysate enhances lettuce tolerance to alkalinity and salinity. *Front. Plant Sci.* **2017**, *8*, 131. [CrossRef] [PubMed]
43. Rouphael, Y.; De Micco, V.; Arena, C.; Raimondi, G.; Colla, G.; De Pascale, S. Effect of *Ecklonia maxima* seaweed extract on yield, mineral composition, gas exchange and leaf anatomy of zucchini squash grown under saline conditions. *J. Appl. Phycol.* **2017**, *29*, 459–470. [CrossRef]

44. Di Stasio, E.; Rouphael, Y.; Colla, G.; Raimondi, G.; Giordano, M.; Pannico, A.; El-Nakhel, C.; De Pascale, S. The influence of *Ecklonia maxima* seaweed extract on growth, photosynthetic activity and mineral compositionof *Brassica rapa* L. subsp. *sylvestris* under nutrient stress conditions. *Eur. J. Hortic. Sci.* **2017**, *82*, 286–293. [CrossRef]
45. Colla, G.; Nardi, S.; Cardarelli, M.; Ertani, A.; Lucini, L.; Canaguiere, R.; Rouphael, J. Protein hydrolysates as biostimulants in horticulture. *Sci. Hortic.* **2015**, *196*, 28–38. [CrossRef]
46. Möller, M.; Smith, M.L. The significance of the mineral component of seaweed suspension on lettuce (*Lactuca sativa* L.) seedling growth. *J. Plant Physiol.* **1998**, *153*, 658–663. [CrossRef]
47. Farooq, M.; Aziz, T.; Basra, S.M.A.; Cheema, M.A.; Rehman, H. Chilling tolerance in hybrid maize induced by seed treatments with salicylic acid. *J. Agron. Crop Sci.* **2008**, *194*, 161–168. [CrossRef]
48. Sivritepe, N.; Sivritepe, H.Ö.; Türkan, I.; Bor, M.; Özdemir, F. NaCl pre-treatments mediate salt adaptation in melon plants through antioxidative system. *Seed Sci. Technol.* **2008**, *36*, 360–370. [CrossRef]
49. Zodape, S.T.; Kawarkhe, V.J.; Patolia, J.S.; Warade, A.D. Effect of liquid seaweed fertilizer on yield and quality of okra (*Abelmoschus esculentus* L.). *J. Sci. Ind. Res. India* **2008**, *67*, 1115–1117.
50. Aitken, J.E.; Senn, J.I. Seaweed products as fertilizers and soil conditioners. *Bot. Mar.* **1965**, *8*, 144–147. [CrossRef]
51. Rayorath, P.; Narayanan, J.M.; Farid, A.; Khan, W.; Palanisamy, R.; Hankins, S.; Critchley, A.T.; Prithiviraj, B. Rapid bioassays to evaluate the plant growth promoting activity of *Ascophyllum nodosum* (L.) Le Jol. using a model plant, *Arabidopsis thaliana* (L.) Heynh. *J. Appl. Phycol.* **2008**, *20*, 423–429. [CrossRef]
52. Subramanian, S.; Sangha, J.S.; Gray, B.A.; Singh, R.P.; Hiltz, D.; Critchley, A.T.; Prithiviraj, B. Extracts of the marine brown macroalga, *Ascophyllum nodosum*, induce jasmonic acid dependent systemic resistance in *Arabidopsis thaliana* against *Pseudomonas syringae* pv. *Tomato* DC3000 and *Sclerotinia sclerotiorum*. *Eur. J. Plant Pathol.* **2011**, *131*, 237–248. [CrossRef]
53. Pereira, L.; Gheda, S.F.; Ribeiro-Claro, P.J.A. Analysis by vibrational spectroscopy of seaweed polysaccharides with potential use in food, pharmaceutical, and cosmetic industries. *Int. J. Carbohydr. Chem.* **2013**, *7*. [CrossRef]
54. Ordog, V.; Stirk, W.A.; van Staden, J.; Novak, O.; Strand, M. Endogenous cytokinins in the three genera of microalgae from the Chlorophyta. *J. Phycol.* **2004**, *40*, 88–95. [CrossRef]
55. Stirk, W.A.; Tarkowská, D.; Turečová, V.; Strnad, M.; van Staden, J. Abscisic acid, gibberellins and brassinosteroids in Kelpak®, a commercial seaweed extract made from Ecklonia maxima. *J. Appl. Phycol.* **2014**, *26*, 561–567. [CrossRef]
56. Lötze, E.; Hoffman, E.W. Nutrient composition and content of various biological active compounds of three south African-based commercial seaweed biostimulants. *J. Appl. Phycol.* **2016**, *28*, 1379–1386. [CrossRef]
57. Moreira Sisalema, J.M. Comportamiento agronómico del cultivo de soya (*Glycine max* L.), a la aplicación de tres extractos de algas marinas, en la zona de Puebloviejo. Available online: http://dspace.utb.edu.ec/bitstream/49000/5048/1/TE-UTB-FACIAG-ING%20AGROP-000018.pdf (accessed on 1 February 2019).
58. Ficha Técnica. Available online: https://www.calstiendavirtual.cl/html5/Fichas/6CASA%20JARD%C3%8DN/0126756%20KELPAK.pdf (accessed on 1 February 2019).
59. Ali, N.; Farrell, A.; Ramsubhag, A.; Jayaraman, J. The effect of *Ascophyllum nodosum* extract on the growth, yield and fruit quality of tomato under tropical conditions. *J. Appl. Phycol.* **2016**, *28*, 1353–1362. [CrossRef]
60. Raj, T.; Nishanthi, P.; Hane Graff, K.; Ann Suji, H. Seaweed extract as a biostimulant and a pathogen controlling agent in plants. *Int. J. Trop. Agric.* **2018**, *36*, 563–580.
61. Rouphael, Y.; Giordano, M.; Cardarelli, M.; Cozzolino, E.; Mori, M.; Kyriacou, M.C.; Bonini, P.; Colla, G. Plant- and seaweed-based extracts increase yield but differentially modulate nutritional quality of greenhouse spinach through biostimulant action. *Agronomy* **2018**, *8*, 126. [CrossRef]
62. AOAC. *Official Methods of Analysis of AOAC International*, 17th ed.; Horwitz, W., Ed.; AOAC International: Gaithersburg, MD, USA, 2000.
63. Świeca, M.; Gawlik-Dziki, U.; Kowalczyk, D.; Złotek, U. Impact of germination time and type of illumination on the antioxidant compounds and antioxidant capacity of Lens culinaris sprouts. *Sci. Hortic.* **2012**, *140*, 87–95. [CrossRef]
64. Singleton, V.; Rossi, J. Colorimetry of total phenolics with phosphomolybdic-phosphotungstic acid reagents. *Am. J. Enol. Viticult.* **1965**, *16*, 144–158.

65. Lamaison, J.L.C.; Carnet, A. Teneurs en principaux flavonoids des fleurs de *Crataegeus monogyna* Jacq et de *Crataegeus laevigata* (Poiret D. C) en fonction de la vegetation. *Pharm. Acta Helv.* **1990**, *65*, 315–320. [CrossRef]
66. Fuleki, T.; Francis, F.J. Quantitative methods for anthocyanins. 1. Extraction and determination of total anthocyanin in cranberries. *J. Food Sci.* **1968**, *33*, 72–77. [CrossRef]
67. Pulido, R.; Bravo, L.; Saura-Calixto, F. Antioxidant activity of dietary polyphenols as determined by a modified ferric reducing/antioxidant power assay. *J. Agric. Food Chem.* **2000**, *48*, 3396–3402. [CrossRef] [PubMed]
68. Calvo, P.; Nelson, L.; Kloepper, J.W. Agricultural uses of plant biostimulants. *Plant Soil* **2014**, *383*, 3–41. [CrossRef]
69. Papenfus, H.B.; Stirk, W.A.; Finnie, J.F.; Van Staden, J. Seasonal variation in the polyamines of *Ecklonia maxima*. *Bot. Mar.* **2012**, *55*, 539–546. [CrossRef]
70. Aremu, A.O.; Plačková, L.; Gruz, J.; Bíba, O.; Novák, O.; Stirk, W.A.; Doležal, K.; Van Staden, J. Seaweed-derived biostimulant (Kelpak®) influences endogenous cytokinins and bioactive compounds in hydroponically grown Eucomis autumnalis. *J. Plant Growth Regul.* **2016**, *35*, 151–162. [CrossRef]
71. Crouch, I.J.; van Staden, J. Commercial seaweed products as biostimulants in horticulture. *J. Home Consum. Hortic.* **1994**, *1*, 19–76. [CrossRef]
72. Craigie, J.S. Seaweed extract stimuli in plant science and agriculture. *J. Appl. Phycol.* **2011**, *23*, 371–393. [CrossRef]
73. Stadnik, M.J.; de Freitas, M.B. Algal polysaccharides as source of plant resistance inducers. *Trop. Plant Pathol.* **2014**, *39*, 111–118. [CrossRef]
74. Depuydt, S.; Hardtke, C.S. Hormone signalling crosstalk in plant growth regulation. *Curr. Biol.* **2011**, *21*, 365–373. [CrossRef] [PubMed]
75. Horgan, R. Present and future prospects for cytokinin research. In *Physiology and Biochemistry of Cytokinins in Plants*; Kamĭnek, M., Mok, D.W.S., Eds.; SPB Academic Publishing: Hague, The Netherlands, 1992; pp. 3–13.
76. Strnad, M. The aromatic cytokinins. *Physiol. Plant* **1997**, *101*, 674–688. [CrossRef]
77. Sakakibara, H. Cytokinins: Activity, biosynthesis, and translocation. *Ann. Rev. Plant Biol.* **2006**, *57*, 431–449. [CrossRef] [PubMed]
78. Santner, A.; Calderon-Villalobos, L.I.A.; Estelle, M. Plant hormones are versatile chemical regulators of plant growth. *Nat. Chem. Biol.* **2009**, *5*, 301–307. [CrossRef] [PubMed]
79. Frébort, I.; Kowalska, M.; Hluska, T.; Frébortová, J.; Galuszka, P. Evolution of cytokinin biosynthesis and degradation. *J. Exp. Bot.* **2011**, *62*, 2431–2452. [CrossRef] [PubMed]
80. Stirk, W.A.; Van Staden, J. Flow of cytokinins through the environment. *Plant Growth Regul.* **2010**, *62*, 101–116. [CrossRef]
81. Zalabák, D.; Pospısilová, H.; Šmehilová, M.; Mrızová, K.; Freébort, I.; Galuszka, P. Genetic engineering of cytokinin metabolism: Prospective way to improve agricultural traits of crop plants. *Biotechnol. Adv.* **2013**, *31*, 97–117. [CrossRef] [PubMed]
82. McKersie, B.D.; Leshem, Y.Y. *Stress and Stress Coping in Cultivated Plants*; Kluwer Academic Publishers: Dordrecht, The Netherlands, 1994.
83. Fike, J.H.; Allen, V.G.; Schmidt, R.E.; Zhang, X.; Fontenot, J.P.; Bagley, C.P.; Ivy, R.L.; Evans, R.R.; Coelho, R.W.; Wester, D.B. Tasco-Forage: I. Influence of a seaweed extract on antioxidant activity in tall fescue and in ruminants. *J. Anim. Sci.* **2001**, *79*, 1011–1021. [CrossRef] [PubMed]
84. Ervin, E.H.; Zhang, X.; Fike, J. Ultraviolet-B Radiation Damage on Kentucky Bluegrass II: Hormone Supplement Effects. *HortScience* **2004**, *39*, 1471–1474. [CrossRef]
85. Khan, W.; Rayirath, U.P.; Subramanian, S.; Jithesh, M.N.; Rayorath, P.; Hodges, D.M.; Critchley, A.T.; Craigie, J.S.; Norrie, J.; Prithiviraj, B. Seaweed extracts as biostimulants of plant growth and development. *J. Plant Growth Regul.* **2009**, *28*, 386–399. [CrossRef]
86. Jennings, R.C. Gibberellins as endogenous growth regulators in green and brown algae. *Planta* **1968**, *80*, 34–42. [CrossRef]
87. Gopala, R.P. Gibberellin-like behaviour of α-tocopherol in green gram *Vigna radiata*. *Geobios* **1984**, *11*, 21–25.
88. Jensen, A. Tocopherol content of seaweed and seaweed meal. I. Analytical methods and distribution of tocopherols in benthic algae. *J. Sci. Food Agric.* **1969**, *20*, 449–453. [CrossRef]
89. Mancuso, S.; Azzarello, E.; Mugnai, S.; Briand, X. Marine bioactive substances (IPA extract) improve foliar iron uptake and water tolerance in potted *Vitis vinifera* plants. *Adv. Hort. Sci.* **2006**, *20*, 156–161.

90. Genard, H.; Le Saos, J.; Billard, J.P.; Tremolieres, A.; Boucaud, J. Effect of salinity on lipid composition, glycine betaine content and photosynthetic activity in chloroplasts of *Suaeda maritima*. *Plant Physiol. Biochem.* **1991**, *29*, 421–427.
91. Blunden, G.; Jenkins, T.; Liu, Y. Enhanced leaf chlorophyll levels in plants treated with seaweed extract. *J. Appl. Physiol.* **1997**, *8*, 535–543. [CrossRef]
92. Naidu, B.P.; Jones, G.P.; Paleg, L.G.; Poljakoff-Mayber, A. Proline analogues in *Melaleuca* species: Response of *Melaleuca lanceolata* and *M. uncinata* to water stress and salinity. *Aust. J. Plant Physiol.* **1987**, *14*, 669–677. [CrossRef]
93. Wachira, F.; Ogada, J. In vitro regeneration of *Camellia sinensis* (L.) O. Kuntze by somatic embryogenesis. *Plant Cell Rep.* **1995**, *14*, 463–466. [CrossRef] [PubMed]
94. Akula, A.; Akula, C.; Bateson, M. Betaine: A novel candidate for rapid induction of somatic embryogenesis in tea (*Camellia sinensis* [L.] O. Kuntze). *Plant Growth Regul.* **2000**, *30*, 241–246. [CrossRef]
95. Hamana, K.; Matsuzaki, S.; Niitsu, M.; Samejima, K.; Nagashima, H. Polyamines of inicellular thermoacidophillic red alga, *Cyanidium caldarium*. *Phytochemistry* **1990**, *29*, 377–380. [CrossRef]
96. Rathore, S.S.; Chaudhary, D.R.; Boricha, G.N.; Ghosh, A.; Bhatt, B.P.; Zodape, S.T.; Patolia, J.S. Effect of seaweed extract on the growth, yield and nutrient uptake of soybean (*Glycine max*) under rainfed conditions. *S. Afr. J. Bot.* **2009**, *75*, 351–355. [CrossRef]
97. Hirsch, R.; Hartung, W.; Gimmler, H. Abscisic acid content of algae under stress. *Bot. Acta* **1989**, *102*, 326–334. [CrossRef]
98. Boyer, G.L.; Dougherty, S.S. Identification of abscisic acid in the seaweed *Ascophyllum nodosum*. *Phytochemistry* **1988**, *27*, 1521–1522. [CrossRef]
99. Rensing, S.A.; Lang, D.; Zimmer, A.D.; Terry, A.; Salamov, A.; Shapiro, H.; Nishiyama, T.; Perroud, P.F.; Lindquist, E.A.; Kamisugi, Y.; et al. The *Physcomitrella* genome reveals evolutionary insight into the conquest of land by plants. *Science* **2008**, *319*, 64–69. [CrossRef] [PubMed]
100. Verslues, P.E.; Agarwal, M.; Katiyar-Agarwal, S.; Zhu, J.; Zhu, J.K. Methods and concepts in quantifying resistance to drought, salt and freezing, abiotic stresses that affect water plant status. *Plant J.* **2006**, *45*, 523–539. [CrossRef] [PubMed]
101. Davies, P.J. *Plant Hormones. Biosynthesis, Signal Transduction, Action!* 3rd ed.; Kluwer Academic Publishers: Dordrecht, The Netherlands, 2004; p. 750.
102. Berlyn, G.P.; Russo, R.O. The use of organic biostimulants to promote root growth. *Belowground Ecol.* **1990**, *2*, 12–13.
103. Blunden, G.; Currie, M.; Mathe, J.; Hohmann, J.; Critchley, A.T. Betaine yields from marine algal species utilized in the preparation of seaweed extracts used in agriculture. *Nat. Prod. Commun.* **2010**, *5*, 581–585. [PubMed]
104. Craigie, J.; MacKinnon, S.; Walter, J. Liquid seaweed extracts identified using H NMR profiles. *J. Appl. Phycol.* **2008**, *20*, 665–671. [CrossRef]
105. Rayirath, P.; Benkel, B.; Hodges, D.M.; Allan-Wojtas, P.; MacKinnon, S.; Critchley, A.; Prithiviraj, B. Lipophilic components of the brown seaweed, *Ascophyllum nodosum*, enhance freezing tolerance in *Arabidopsis thaliana*. *Planta* **2009**, *230*, 135–147. [CrossRef] [PubMed]
106. Yabur, R.; Bashan, Y.; Hernandez-Carmona, G. Alginate from the macroalgae *Sargassum sinicola* as a novel source for microbial immobilization material in waste water treatment and plant growth promotion. *J. Appl. Phycol.* **2007**, *19*, 43–53. [CrossRef]
107. Mercier, L.; Laffite, C.; Borderies, G.; Briand, X.; Esquerre-Tugaye, M.T.; Fournier, J. The algal polysaccharide carrageenans can act as an elicitor of plant defence. *New Phytol.* **2011**, *149*, 43–51. [CrossRef]
108. Nakamura, T.; Nagayama, K.; Uchida, K.; Tanaka, R. Antioxidant activity of phlorotannins isolated from the brown alga *Eisenia bicydis*. *Fish. Sci.* **1996**, *62*, 923–926. [CrossRef]
109. Wang, T.; Jonsdottir, R.; Olafsdottir, G. Total phenolic compounds, radical scavenging and metal chelation of extracts from Icelandic seaweeds. *Food Chem.* **2009**, *116*, 240–248. [CrossRef]
110. Andjelkovic, M.; Van Camp, J.; De Meulenaer, B.; Depaemelaere, G.; Socaciu, C.; Verloo, M.; Verhe, R. Iron-chelation properties of phenolic acids bearing catechol and galloyl groups. *Food Chem.* **2006**, *98*, 23–31. [CrossRef]

111. Rengasamy, K.R.R.; Kulkarni, M.G.; Stirk, W.A.; Van Staden, J. Eckol Improves Growth, Enzyme Activities, and Secondary Metabolite Content in Maize (*Zea mays* cv. Border King). *J. Plant Growth Regul.* **2015**, *34*, 410–416. [CrossRef]
112. Korasick, D.A.; Enders, T.A.; Strader, L.C. Auxin biosynthesis and storage forms. *J. Exp. Bot.* **2013**, *64*, 2541–2555. [CrossRef] [PubMed]
113. Ljung, K. Auxin metabolism and homeostasis during plant development. *Development* **2013**, *140*, 943–950. [CrossRef] [PubMed]
114. Peer, W.A.; Murphy, A.S. Flavonoids and auxin transport: Modulators or regulators? *Trends Plant Sci.* **2007**, *12*, 556–563. [CrossRef] [PubMed]
115. Wu, H.C.; Du Toit, E.S.; Reinhardt, C.F.; Rimando, A.M.; Van der Kooy, F.; Meyer, J.J.M. The phenolic, 3,4-dihydroxybenzoic acid, is an endogenous regulator of rooting in *Protea cynaroides*. *Plant Growth Regul.* **2007**, *52*, 207–215. [CrossRef]
116. De Klerk, G.J.; Guan, H.; Huisman, P.; Marinova, S. Effects of phenolic compounds on adventitious root formation and oxidative decarboxylation of applied indoleacetic acid in Malus 'Jork 9'. *Plant Growth Regul.* **2011**, *63*, 175–185. [CrossRef]
117. Teixeira da Silva, J.; Dobranszki, J.; Ross, S. Phloroglucinol in plant tissue culture. *In Vitro Cell. Dev. Biol. Plant* **2013**, *49*, 1–16. [CrossRef]
118. Gaspar, T.; Kevers, C.; Penel, C.; Greppin, H.; Reid, D.; Thorpe, T. Plant hormones and plant growth regulators in plant tissue culture. *In Vitro Cell. Dev. Biol. Plant* **1996**, *32*, 272–289. [CrossRef]
119. Wilson, P.J.; Van Staden, J. Rhizocaline, rooting co-factors, and the concept of promoters and inhibitors of adventitious rooting—A review. *Ann. Bot.* **1990**, *66*, 479–490. [CrossRef]
120. Aremu, A.O.; Nqobile, A.; Kannan, M.; Rengasamy, R.R.; Amoo, S.O.; Gruz, J.; Bíba, O.; Šubrtová, M.; Pênčík, A.; Novák, O.; et al. Physiological role of phenolic biostimulants isolated from Brown seaweed *Ecklonia maxima* on plant growth and development. *Planta* **2015**, *241*, 1313–1324. [CrossRef] [PubMed]
121. Chibbar, R.N.; Schyluk, J.; Georges, F.; Mallard, C.S.; Constabel, F. Esterase isozymes as markers of somatic embryogenesis in cultured carrot cells. *J. Plant Physiol.* **1988**, *133*, 367–370. [CrossRef]
122. Pedersen, S.; Andersen, S.B. Developmental expression of isoenzymes during embryiogenesis in barley anther culture. *Plant Sci.* **1993**, *91*, 75–86. [CrossRef]
123. Krsnik-Rasol, M.; Cipcic, H.; Hagege, D. Isoesterases related to cell differentiation in plant tissue culture. *Chem. Biol. Interact.* **1999**, *120*, 587–592. [CrossRef]
124. Balen, B.; Krsnik-Rasol, M.; Simeon-Rudolf, V. Isoenzymes of peroxidase and esterase related to morphogenesis in *Mammillaria gracillis* Pfeiff. tissue culture. *J. Plant Physiol.* **2003**, *160*, 1401–1406. [CrossRef] [PubMed]
125. Nardi, S.; Carletti, P.; Pizzeghello, D.; Muscolo, A. Biological activities of humic substances, in biophysico-chemical processes involving natural nonliving organic matter in environmental systems. In *Fundamentals and Impact of Mineral-Organic-Biota Interactions on the Formation, Transformation, Turnover, and Storage of Natural Nonliving Organic Matter (NOM)*, 3rd ed.; John Wiley: Hoboken, NJ, USA, 2009; pp. 305–339.
126. Oboh, G.; Puntel, R.L.; Rocha, J.B.T. Hot pepper (*Capsicum annuum*, Tepin and *Capsicum chinese*, Habanero) prevents Fe^{2+}-induced lipid peroxidation in brain-*in vitro*. *Food Chem.* **2007**, *102*, 178–185. [CrossRef]
127. Ertani, A.; Pizzeghello, D.; Francioso, O.; Sambo, P.; Sanchez-Cortes, S.; Nardi, S. *Capsicum chinensis* L. growth and nutraceutical properties are enhanced by biostimulants in a long-term period: Chemical and metabolomic approaches. *Front. Plant Sci.* **2014**, *5*, 375. [CrossRef] [PubMed]
128. Złotek, U.; Świeca, M.; Jakubczyk, A. Effect of abiotic elicitation on main health-promoting compounds, antioxidant activity and commercial quality of butter lettuce (*Lactuca sativa* L.). *Food Chem.* **2014**, *148*, 253–260. [CrossRef] [PubMed]
129. Karolewski, P.; Jagodziński, A.M. Udział węgla w związkach obronnych przed czynnikami biotycznymi u roślin drzewiastych. *Sylwan* **2013**, *157*, 831–841.
130. Czerwińska, E.; Szparaga, A. Antibacterial and antifungal activity of plant extracts. *Rocznik Ochrona Środowiska* **2015**, *17*, 209–229.
131. Stevens, M.T.; Lindroth, R.L. Induced resistance in the interminate growth of aspen (*Populus tremuloides*). *Oecologia* **2005**, *145*, 298–306. [CrossRef] [PubMed]

132. Kocira, S.; Szparaga, A.; Kocira, A.; Czerwińska, E.; Depo, K.; Erlichowska, B.; Deszcz, E. Effect of applying a biostimulant containing seaweed and amino acids on the content of fiber fractions in three soybean cultivars. *Legum. Res.* **2018**. [CrossRef]
133. Lakhdar, A.; Iannelli, M.A.; Debez, A.; Massacci, A.; Jedidi, N.; Abdelly, C. Effect of municipal solid waste compost and sewage sludge use on wheat (Triticum durum): Growth, heavy metal accumulation, and antioxidant activity. *J. Sci. Food Agric.* **2010**, *90*, 965–971. [CrossRef] [PubMed]
134. Jindo, K.; Martim, S.A.; Navarro, E.C.; Pérez-Alfocea, F.; Hernandez, T.; Garcia, C.; Aguiar, N.O.; Canellas, L.P. Root growth promoting by humic acids from composted and non-composted urban organic wastes. *Plant Soil* **2012**, *353*, 209–220. [CrossRef]
135. Gülçin, I. Antioxidant activity of food constituents: An overview. *Arch. Toxicol.* **2012**, *86*, 345–391. [CrossRef] [PubMed]
136. Cheynier, V.; Comte, G.; Davies, K.M.; Lattanzio, V.; Martens, S. Plant phenolics: Recent advances on their biosynthesis, genetics, and ecophysiology. *Plant Physiol. Biochem.* **2013**, *72*, 1–20. [CrossRef] [PubMed]
137. Pavarini, D.P.; Pavarini, S.P.; Niehues, M.; Lopes, N.P. Exogenous influences on plant secondary metabolite levels. *Anim. Feed Sci. Technol.* **2012**, *176*, 5–16. [CrossRef]
138. Moyo, M.; Amoo, S.O.; Aremu, A.O.; Gruz, J.; Šubrtová, M.; Doležal, K.; Van Staden, J. Plant regeneration and biochemical accumulation of hydroxybenzoic and hydroxycinnamic acid derivatives in *Hypoxis hemerocallidea* organ and callus cultures. *Plant Sci.* **2014**, *227*, 157–164. [CrossRef] [PubMed]
139. Jannin, L.; Arkoun, M.; Etienne, P.; Laîné, P.; Goux, D.; Garnica, M.; Fuentes, M.; San Francisco, S.; Baigorri, R.; Cruz, F.; et al. *Brassica napus* growth is promoted by *Ascophyllum nodosum* (L.) Le Jol. Seaweed extract:microarray analysis and physiological characterization of N, C, and S metabolisms. *J. Plant Growth Regul.* **2013**, *32*, 31–52. [CrossRef]
140. Rouphael, Y.; Kyriacou, M.C.; Petropoulos, S.A.; De Pascale, S.; Colla, G. Improving vegetable quality in controlled environments. *Sci. Hortic.* **2018**, *234*, 275–289. [CrossRef]
141. Fan, D.; Hodges, D.M.; Critchley, A.T.; Prithiviraj, B. A commercial extract of brown macroalga (*Ascophyllum nodosum*) affects yield and the nutritional quality of Spinach in vitro. *Commun. Soil Sci. Plant Anal.* **2013**, *44*, 1873–1884. [CrossRef]
142. Azcona, I.; Pascual, I.; Aguirreolea, J.; Fuentes, M.; García-Mina, J.M.; Sánchez-Díaz, M. Growth and development of pepper are affected by humic substances derived from composted sludge. *J. Plant Nutr. Soil Sci.* **2011**, *174*, 916–924. [CrossRef]
143. Zhang, D.; Hamauzu, Y. Phenolic compounds, ascorbic acid, carotenoids and antioxidant properties of green, red and yellow bell peppers. *J. Food Agric. Environ.* **2003**, *1*, 22–27.
144. Pantelidis, G.E.; Vasilakakis, M.; Manganaris, G.A.; Diamantidis, G.R. Antioxidant capacity, phenol, anthocyanin and ascorbic acid contents in raspberries, blackberries, red currants, gooseberries and cornelian cherries. *Food Chem.* **2007**, *102*, 777–783. [CrossRef]
145. Świeca, M.; Sęczyk, Ł.; Gawlik-Dziki, U. Elicitation and precursor feeding as tools for the improvement of the phenolic content and antioxidant activity of lentil sprouts. *Food Chem.* **2014**, *161*, 288–295. [CrossRef] [PubMed]
146. Złotek, U.; Szychowski, K.A.; Świeca, M. Potential in vitro antioxidant, anti-inflammatory, antidiabetic, and anticancer effect of arachidonic acid-elicited basil leaves. *J. Funct. Foods* **2017**, *36*, 290–299. [CrossRef]
147. Raina, K.; Rajamanickam, S.; Deep, G.; Singh, M.; Agarwal, R.; Agarwal, C. Chemopreventive effects of oral gallic acid feeding on tumor growth and progression in TRAMP mice. *Mol. Cancer Ther.* **2008**, *7*, 1258–1267. [CrossRef] [PubMed]
148. Chen, J.Y.; Wen, P.F.; Kong, W.F.; Pan, Q.H.; Zhan, J.C.; Li, J.M.; Wan, S.B.; Huang, W.D. Effect of salicylic acid on phenylpropanoids and phenylalanine ammonia-lyase in harvested grape berries. *Postharvest Biol. Technol.* **2006**, *40*, 64–72. [CrossRef]
149. Polo, J.; Mata, P. Evaluation of a biostimulant (Pepton) based in enzymatic hydrolyzed animal protein in comparison to seaweed extracts on root development, vegetative growth, flowering, and yield of gold cherry tomatoes grown under low stress ambient field conditions. *Front. Plant Sci.* **2018**, *8*, 2261. [CrossRef] [PubMed]
150. Nair, P.; Saveetha, K.; Zhang, J.; Ji, X.; Chris, K.; Bernhard, B.; Mark, D.H.; Alan, T.C.; David, H.; Balakrishnan, P. Transcriptional and metabolomics analysis of *Ascophyllum nodosum* mediatel freezing tolerance in *Arabidopsis thaliana*. *BMC Genom.* **2012**, *13*, 643. [CrossRef] [PubMed]

151. Biesaga-Kościelniak, J.; Kościelniak, J.; Filek, M.; Marcińska, I.; Krekule, J.; Machackova, I.; Kuboń, M. The effect of plant growth regulators and their interaction with electric current on winter wheat development. *Acta Physiol. Plant* **2010**, *32*, 987–995. [CrossRef]
152. Povero, G.; Mejia, J.F.; Tommaso, D.D.; Piaggesi, A.; Warrior, P. A systematic approach to discover and characterize natural plant biostimulants. *Front. Plant Sci.* **2016**, *7*, 435. [CrossRef] [PubMed]

agronomy

MDPI

Article

Comparison between Chemical Fertilization and Integrated Nutrient Management: Yield, Quality, N, and P Contents in *Dendranthema grandiflorum* (Ramat.) Kitam. Cultivars

Beniamino Leoni, Danilo Loconsole *, Giuseppe Cristiano and Barbara De Lucia

Department of Agro-Environmental and Territorial Science (DISAAT), University of Bari, "Aldo Moro", Italy Via Amendola 165/A, 70120 Bari, Italy; beniamino.leoni@uniba.it (B.L.); giuseppe.cristiano@uniba.it (G.C.); barbara.delucia@uniba.it (B.D.L.)

* Correspondence: danilo.loconsole@hotmail.it; Tel.: +39-080-544-3039

Received: 27 February 2019; Accepted: 17 April 2019; Published: 19 April 2019

Abstract: To assess the effects of a new integrated nutrient management protocol on yield and cut stem quality, root morphology, N accumulation, nitrogen utilization efficiency (NUE), and P content in tissue, a biennial (2011 and 2012) chrysanthemum cut flower cultivation was carried out. In both years, two nutrition management (CNM: conventional NM and INM: integrated NM) treatments and two *Dendranthema grandiflorum* (Ramat.) Kitamura cultivar ("White CV_1" and "Yellow CV_2") treatments were compared. The treatments were arranged in a split-plot design with three replicates. CNM was fertilized using a recommended dose fertilization of mineral NPK; INM treatment was fertilized using a half dose (50%) of CNM plus a combined usage of N organic fertilizer, seaweed extract (*Ascophyllum nodosum*), and microrganism consortium (*Glomus* sp. and *Bacillus* sp.). Yield at harvest (+19%), number of leaves (+33%), leaf area (+46%), number of flower heads (+27%), and total aboveground dry weight (+40%) were significantly increased by the INM application compared to the control. In terms of the root system, the increase was evident in terms of length (+174%), volume (+167%), projected area (+166%), and surface area (+165%), tips (+175%), forks (+285%), and crossings (+464%). The greatest N accumulation, in both years, was registered by INM treatment at harvest: +94% in 2011 and +55% in 2012. Differences in the NM were evident in the NUE, which was highest in CNM (on average 162) compared to INM (on average 142). In both years the P content in above-ground chrysanthemum tissues was in the order of head > leaves > stems, which was maintained in both INM and CNM treatments. A higher yield (138 stems m^{-2}) was obtained in "CV_2 Yellow" compared to "CV_1 White" (120 stems m^{-2}). Based on our findings, applying INM to chrysanthemum improves yield, cut flower quality, and plant nutrient uptake, in an agro–environmentally sustainable way. A basic economic analysis on fertilizers, cost gross production, and takings difference obtained, was carried out.

Keywords: N organic fertilizer; seaweed extract; mycorrhizal inoculants; phosphate-solubilizing microorganisms; biofertilizers; microorganism consortium

1. Introduction

Fertilization is essential for optimizing crop productivity [1]. Mineral fertilizers, particularly nitrogen (N) and phosphorus (P), are important for plant nutrition [2,3]. However, when used in overly large doses they are also a potential source of environmental pollution [4–6]. Nutrient overapplication has introduced major challenges in terms of soil infertility [7], N and P runoff [8,9], environmental degradation [10], and climate change [11,12].

Today there is an increasing need for a balanced fertilization strategy, minimizing the use of mineral fertilizers to enhance both crop production and quality and nutrient uptake under low input conditions [13]. Mineral fertilizers can be replaced by organic fertilizers [14], plant biostimulants [15], and beneficial microbial inoculants [16].

Possible interventions in conservation agriculture include the combined use of inorganic and organic fertilizers, as well as biostimulants and biofertilizers in order to increase a balanced nutrient supply [17]. Integrated nutrition management (INM) focusing on the optimization of the biological potential improves fertilizer input efficiency, reduces environmental risks, and increases crop productivity, through root/rhizosphere management [18].

In terms of biostimulants, seaweed extracts are used in sustainable agriculture in order to increase growth, quality, and shelf life [19–21]. Many studies have demonstrated the positive effects of seaweed extracts on a wide range of crops, including cereals [22], ornamental and flowering plants [23], vegetables [24], and field crops [25].

Biofertilizers are also an important alternative source of plant nutrients and are key components of integrated nutrient management in crop production. The use of microbial inoculants with P solubilizing activities in soils is an environmental-friendly alternative to further applications of chemical-based P fertilizers [26,27]. Various studies have examined the potential of different bacterial species to solubilize inorganic phosphate compounds. *Bacillus* spp., and in particular *B. subtilis* and *B. megaterium*, may provide the available forms of P to plants, thus considerably improving plant growth performance [28–31].

Other microbial inoculants, such as arbuscular mycorrhiza fungi (AMF), increase the P availability through the expansion of the root surface area by extraradical hyphae formation [32,33].

The various benefits of AMF include increased growth and nutrient uptake (especially N, P, and K) and crop yields [34–38]. The AMF also produce a heat-stable protein called glomalin, which is a glycoprotein that enhances soil aggregation and helps in soil carbon sequestration. Together, glomalin and mycorrhizal hyphae lead to a stable soil structure.

The combined use of N organic fertilizers, biostimulants, and biofertilizers is therefore a new approach that has not been widely investigated in ornamentals, which entails developing many efficient formulations with low mineral inputs, with positive impacts on crops and environment.

Chrysanthemum (*Dendranthema grandiflorum* (Ramat.) Kitamura) is a commercial cut flower, belonging to the *Asteraceae* family, with nearly 200 cultivars. It is one of the top ten elite cut flowers globally, due to its different shapes, dazzling colors, varying sizes, and excellent vase life. In Italy, where our research was carried out, there is a considerable demand in both domestic and export markets.

Extracts of the plants (stems and flowers) have many potential medicinal properties, including anti-HIV, antibacterial, and antimycotic [39]. N, P, and K play a vital role in the production of good quality flowers. N is essential for the creation of biomass as well as for the biosynthesis of enzymes in chrysanthemum leaves [40].

The N requirements of chrysanthemums are known to be higher during the first seven weeks of growth, and during this time, deficiencies are more difficult to correct than in later stages of development [41]. Chrysanthemums take up N at an even rate from the time of planting until the flower bud differentiation stage where after N uptake decreases [42]. In chrysanthemums, the need for P is significantly lower than that of nitrogen [43]. K requirements are high, and its presence in the plant favorably affects growth and flower color [44].

To the best of our knowledge, there are no available data on how the INM system based on mineral and organic N fertilizers, seaweed extracts, plus a consortium between AMF (arbuscular mycorrhiza fungi) and PSB (Phosphate solubilizing bacteria), affects yield and quality in chrysanthemums.

The goal of this research was to evaluate the effects of an innovative INM compared to conventional nutrient management, in chrysanthemum cut flower cultivation, on: (1) yield and cut stem quality, (2) N concentration, accumulation, and utilization efficiency and P uptake, (3) root architecture, and (4) soil fertility.

2. Materials and Methods

2.1. Experimental Conditions

Two field experiments were carried out in 2011 and 2012, from August to December, at a floricultural farm located in Sannicandro di Bari (southern Italy: 40°59′24″ N, 16°47′01″E, 181 m a.s.l.). The local climatic conditions are characterized by hot dry summers and mild rainy autumns and winters, typical of the Mediterranean climate. During the plant growth period under natural photoperiod, the mean air temperature was 17.2 °C and 18.2 °C in 2011 and 2012, respectively; minimum air temperature was 3.7 °C in December 2011 and 5.4 °C in December 2012; maximum air temperature was 32.4 °C in August 2011, and 32.6 °C in August 2012.

Seasonal chrysanthemum cuttings (Minstreel Serie, Straathof Plants BV, The Netherlands), ideal for blooming from November to late December, were obtained from a local commercial propagator, with the following characteristics: stem length, 11.6 cm; number of leaves, 8; leaf area, 81.1 cm^2; plant fresh weight, 3.1 g; and plant dry weight, 0.3 g. In both years, plants were transplanted on 6 August into an uncovered tunnel. In the first week of October in both years, the tunnel was covered by ethylene vinyl acetate (EVA) film.

The main soil characteristics (taken from 0 to 25 cm depth) are described in Table 1. Soil pH was determined with a pH meter (P9991, Hanna Instruments, Italy) in a settling suspension on a 60 g sample mixed with 150 mL of deionized water, after shaking for 60 min at room temperature (22 °C). The soil used for our experiment was slightly sub-alkaline (pH = 7.34, near to neutrality) and it was representative of Apulian soils in which chrysanthemum was cultivated with remarkable production results. Chrysanthemum plants generally grow with a pH ranging between 6 and 7.2 [45].

The electric conductivity (EC) was measured on water extract (1:5 *v/v*) with a conductivity meter (HI 4321, Hanna Instruments, Italy). Soil organic carbon (SOC) was determined by wet oxidation. Based on USDA classification, experimental soil was classified as clay loam soil. Experimental soil was moderately provided with organic matter and CEC was also classified as moderate [46].

The total Kjeldahl N (TKN) was measured using 1 g samples of both growing media and plant tissues using the Kjeldahl method after 96% H2SO4 hot digestion. Total phosphorus was determined (P) by the colorimetric molybdovanadate phosphoric acid method. Exchangeable K, Ca, and Mg were determined using 0.2 g of dry sample (105 °C for 24 h) after acid digestion in a microwave oven (CEM Mars Xpress, Cologno al Serio, IT). Substrate digests were filtered, diluted, and analyzed by atomic absorption spectrometry (Perkin-Elmer Aanalyst 200, Waltham, MA, USA). The analyses were carried out in triplicate.

The soil was sandy clay with a slightly alkaline pH of 7.3 (IUSS), EC of 1.77 dS m^{-1}, and moderately high CEC (cation exchange capacity) of 23.8 Meq 100 g^{-1}.

Table 1. Initial soil physico–chemical characteristics (mean ± standard error). Data are the means of three samples.

Parameter	Value
pH (soil:H_2O ratio 1:2.5)	7.34 ± 0.2
Electric Conductivity (EC) (soil:H_2O ratio 1:5) (dS m^{-1})	1.77 ± 0.08
Cation exchange capacity (CEC) (Meq100g^{-1})	23.8
Sand (%)	52 ± 3
Silt (%)	16 ± 2
Clay (%)	32 ± 4
Total C (g kg^{-1})	12.54 ± 1.2
Organic matter (g kg^{-1})	21.61 ± 1.9
T Kjeldahl-N (g kg^{-1})	1.15 ± 0.13
P (mg kg^{-1})	71.25 ± 0.9
Available K (mg kg^{-1})	579 ± 10.1
Available Ca (mg kg^{-1})	2160 ± 22
Available Mg (mg kg^{-1})	495 ± 31

2.2. Treatments and Experimental Design

In both years, four treatments in total consisting of two nutrition management (NM) and two *Dendranthema grandiflorum* (Ramat.) Kitamura cultivar (CV) were compared as follows:

1. Conventional NM (CNM or control) and integrated NM (INM);
2. "White CV_1" and "Yellow CV_2".

Treatments were carried out using a split-plot design with three replicates, with NM as the main plot and CV as the subplot. The surface of each experimental plot measured 2.2 m^2.

CNM treatment was applied through a fertigation system using a recommended dose of mineral NPK: 17 g m^{-2} N, 16 g m^{-2} P_2O_5, and 17 g m^{-2} K_2O plus microelements, starting one week after transplanting, every week, for 12 weeks, the last one during the second week of November (flower bud differentiation).

INM treatment was applied by fertigation at a half dose (50%) of CNM plus a mixture of an N organic fertilizer, seaweed extract and microorganism consortium as shown in Table 2, starting from transplantation. Commercial products were applied at the manufacturer's recommended rates.

NPK doses added with INM fertilization were the following: 11.8 g m^{-2} N, 8 g m^{-2} P_2O_5, and 12 g m^{-2} K_2O. N organic fertilizer added to the mineral NPK dose mentioned above, was derived by hydrolyzed animal epithelium, beet molasses extract, and brown seaweed extract.

In the second year, the same treatments were repeated.

In both years, the growing density was 34 plants m^{-2}.

Table 2. Combined use of N organic fertilizer, seaweed extract, and microorganism consortium applied in two experiments (2011 and 2012).

Type and Commercial Product (*)	Content	Total Rate (g/100m^2)	Weeks of Applications (*n*) (**)
N organic fertilizer (Euroflorid)	N = 5% *w/w* (N = 0.5 g/m^2)	75	I, II, III, IV, V
N organic fertilizer (Amminostim-bio)	N = 6% *w/w* (N = 0.9 g/m^2)	25	VI, VII, VIII
Seaweed extract (Euroalg)	*Ascophyllum nodosum* (L.) Le Jol. 32% *w/w*, N = 1.5% *w/w* (N = 0.9 g/m^2), K20 = 5.0% *w/w*	58	I, II, III, IV, V, VI, VII, VIII, IX, X, XI, XII
Microrganism consortium (Micotric L)	*Glomus mosseae* and *G. intraradices* (2 spores g^{-1}) and *Bacillus megaterium var. Phosphaticum* 6 × 107 CFUs g^{-1}	90	I, II, III, IV
Microrganism consortium (Europlus)	*Glomus mosseae* (2 spores g^{-1}) + *Trichoderma viride* and *Bacillus megaterium var. Phosphaticum* 6 × 107 CFUs g^{-1}	162	I, II, III, IV

(*) by Eurovix SpA, Entratico (BG), Italy; (**) fertigation from transplant (week 1) to flower bud differentiation (week 12).

During the experiments, all field management procedures (e.g., irrigation and pest control) were the same among treatments. The irrigation system was a micro drip; each drip line was placed between two plants rows with an emitter (pressure compensating) discharge rate of 2.0 L h^{-1}. Except for nutrition, production was carried out using the grower's standard practices. Cut flowers were harvested when 50% of flower heads had opened.

Morpho–biometric measurements were carried out at the Department of Agro–environmental and Territorial Sciences (DISAAT), University of Bari, Italy. Plants were sampled for aboveground and ground biomass and N and P content (%) at 55, 93, and 131 DAT (days after transplant) in both growing periods.

The growth and yield observations were recorded on twelve randomly selected plants from each treatment.

In both years, at harvest (second ten days of December), the soil was washed from roots, and aboveground plants were divided into stems, leaves, and flowers, which were oven dried at 70 °C until they reached a constant mass to measure the respective dry weights.

At flower harvest, the measurements involved: yield (secondary branches = stems m^{-2}), stem length (cm), inflorescence (*n* and diameter, cm), leaves (*n*), and leaf area (cm^2), Chlorophyll SPAD

(Single-photon avalanche diode) index (Minolta Chlorophyll Meter SPAD-502), dry and fresh weight of leaves, stems, inflorescences, roots, and whole plants. In order to perform root morphology analysis, only in the first year, roots were spread out, washed, and then scanned at 300 dpi on an HP DeskScan II scanner (HEWLETT PACKARD C6261A, Palo Alto, CA, USA). Root analysis was performed using the WinRHIZO® image analysis system (V 4.1c Régent Instruments, Quebec, Canada); measurements involved total root length, average root diameter, projected and surface area, tips, forks, and crossings.

The total Kjeldahl N (TKN) content was measured, both in the first and second years, using 1 g samples of foliar and radical tissues, using the Kjeldahl method after 96% H_2SO_4 hot digestion. On the other hand, the P-Olsen measurement was only used during the first year.

Nitrogen utilization efficiency (NUE) was estimated by the ratio of dry biomass to plant N accumulation at harvest.

2.3. Economic Analysis

A basic economic analysis about fertilizer costs (for CNM and INM), gross sealable production, and profit raised was developed.

2.4. Statistical Analysis

The data were analyzed by three-way ANOVA using CoStat-Statistics Software. Treatment means were separated with Student–Newman–Keuls (SNK) ($p \leq 0.05$).

3. Results

The overall aims of this research were to evaluate the effects of an innovative INM compared to CNM, in a biennial chrysanthemum cut flower cultivation, on (i) yield and cut stem quality; (ii) root morphology; and (iii) N accumulation, NUE, and P content in plant tissue.

The main effect of NM was found to be highly significant for most of the parameters investigated.

Yield at harvest, as determined by the harvestable number of cut stems per plant (Table 3), increased significantly in INM (140 stems m^{-2}, +19%) compared to those under CNM (118 stems m^2).

Genotype influenced marketable yield: CV_2 registered the highest value (138 stems m^{-2}), surpassing that of CV1 by 15% (on average 120 stems m^{-2}).

Concerning the Y factor, the yields were not different (133 stems m^{-2} on average).

Table 3. Main effects of nutrient management, cultivar on yield, stem height, leaf number, leaf area, chlorophyll index, and number of flower heads in chrysanthemum plants over the two years of application.

Treatments	Yield (no. stems m^{-2})	Stem Height (cm)	Leaves (no./plant)	Leaf Area (cm^{-2})	Chl. Index (SPAD)	Flower Heads (no. stems^{-1})
Nutrition Management (NM)						
Conventional NM (CNM)	118a	103	60b	2064b	44.8	6.6b
Integrated NM (INM)	140b	106	80a	3017a	46.9	8.4a
Significance	*	ns	*	**	ns	**
Cultivar (CV)						
CV 1	120b	105	64	2416	45.1	8.2
CV 2	138a	104	76	2665	46.6	6.9
Significance	*	ns	ns	ns	ns	ns
Year						
2011	133	116a	82	2795	46.6	8.1
2012	133	92b	59	2486	45.1	7.3
Significance	ns	*	*	ns	ns	ns
Interaction						
NM × CV	*	ns	ns	ns	ns	*
NM × Year	ns	ns	ns	ns	ns	ns
CV × Year	ns	ns	ns	ns	ns	ns
NM × CV × year	ns	ns	ns	ns	ns	ns

Different letters within each column indicate significant differences according to SNK test ($p \leq 0.05$). NS not significant * $p < 0.05$ and ** 0.01, indicate level of significance.

Table 3 also shows the influence of the treatments on the commercial quality parameters of the cut stems at harvest. The stem height is an important parameter that is used for the classification of the stems for marketing and sales, and in fact, customers often prefer flowers with a longer stem. Stem height was not found to be significant between both NM and CV treatments; however, it showed significant differences among Y: in 2012 it was 20% lower (92 cm) than 2011 (116 cm).

Regarding the number of leaves per plant, the INM treatment led to an increase of 33% (80 leaves/stem) compared to CNM (60 leaves/stem); in 2012 the number of leaves (59) showed an average decrease of 28% (82 leaves) compared to 2011.

The INM treatment also showed a significant increase of 46% (3017 cm^2) in the leaf area value compared to CNM (2064 cm^2).

The chlorophyll index SPAD was not significant in any of the treatments.

The number of flower heads per stem was highest (8.4) with an increase of +27% when plants were treated with INM, compared to CNM (6.6). No differences were found between the cultivars and years.

Concerning the leaves, stems, flower heads, and aboveground dry weight, Table 4 shows the statistically significant differences in favor of INM compared to CNM. Leaf values showed a 38% increase, stem value a 37% increase, and flower heads a 55% increase, which were reflected in the increase of aboveground dry weight (+40%). No difference was found between the cultivars.

Table 4 also shows that 2011 had the highest aboveground dry weight value, which decreased to 25% during 2012.

Table 4. Main effects of nutrient management, cultivar on dry weight of various organs, and above-plant on chrysanthemum over the two years of application at harvest time.

Treatments	Dry Weight (g)			
	Leaves	Stem	Heads	Above-plant
Nutrition Management (NM)				
CNM	8.30 b	14.61 b	5.65 b	28.56 b
INM	11.10 a	20.05 a	8.75 a	39.90 a
Significance	**	**	**	**
Cultivar (CV)				
CV 1	9.40	17.10	8.05	34.55
CV 2	10.00	16.60	7.95	34.55
Significance	ns	ns	*	ns
Year				
2011	10.80	19.80	8.70 a	39.30 a
2012	8.60	14.40	6.50 b	29.5 b
Significance	*	**	*	**
Interaction				
NM × CV	ns	ns	ns	ns
NM × Year	*	ns	*	*
CV × Year	ns	ns	ns	ns
NM × CV × year	ns	ns	ns	ns

Different letters within each column indicate significant differences according to SNK test ($p \leq 0.05$). NS not significant * $p < 0.05$ and ** 0.01, indicate level of significance.

In 2011 the root morphology (Table 5) was evaluated. Parameter values for the plants under INM treatment were higher than CNM as follows: root length (+174%), area projection (+166%), surface area (165%), root volume (+167%), tips (+175%), forks (+285%), and crossings (+464%).

Regarding the CV, the best performing root system was White (CV_1) compared to Yellow (CV_2): root length (+63%), area projection (+37%), surface area (+38%), root volume (+19%), tips (+100%), forks (+109%), and crossings (+197%).

Table 5. Main effects of nutrient protocol management and cultivar on total root length (TRL), area projection (AP), surface area (SA), root volume (RV), root tips (RT), root forks (RF), and root crossings (RC) at 2011 harvest period in chrysanthemum plants.

Treatments	TRL (cm)	AP (cm^{-2})	SA (cm^{-2})	RV (cm^3)	RT (no.)	RF (no.)	RC (no.)
Nutrition Management (NM)							
CNM	382.1 b	21.2 b	66.7 b	0.9 b	1264.8 b	948.4 b	34.7 b
INM	1049.2 a	56.3 a	177.0 a	2.4 a	3486.6 a	3655.4 a	195.6 a
Significance	**	**	**	**	**	**	*
Cultivar (CV)							
CV 1	975.0 a	48.7 a	153.0 a	1.9 a	3535.2 a	3501.8 a	208.2 a
CV 2	597.4 b	35.5 b	110.8 b	1.6 b	1768.3 b	1676.0 b	70.0 b
Significance	*	*	*	*	**	**	**
Interaction							
NM × CV	*	*	*	*	**	**	**

Different letters within each column indicate significant differences according to SNK test ($p \leq 0.05$). NS not significant * $p < 0.05$ and **0.01, indicate level of significance.

Concerning the plant N accumulation (gm^{-2}) at every DAT in both years (Table 6), the maximum value was obtained under INM, which was the result of a simultaneous increase in dry weight (Table 4). The highest N accumulation, in both years, was at harvest (131 DAT), in 2011 with an increase of 94%, and in 2012 with an increase of 55%. No significant difference was found between the CVs, except for the flower head value at 131 DAT in both years.

Table 6. Main effects of nutrient management and cultivar on N accumulation ($g\ m^{-2}$) at three different days after transplant (DAT) in chrysanthemum plants over the two years of application.

Treatments	DAT		
	55	93	131
		First Year	
Nutrition Management (NM)			
CNM	4.33 b	5.47 b	6.20 b
INM	6.28 a	9.13 a	10.39 a
Significance	*	*	*
Cultivar (CV)			
CV 1	5.55	7.20	8.99
CV 2	5.61	7.68	8.48
Significance	ns	ns	ns
Interaction			
NM × CV	ns	ns	ns
		Second Year	
Nutrition Management (NM)			
CNM	2.67	5.05	6.37 b
INM	3.31	6.12	9.90 a
Significance	*	*	**
Cultivar (CV)			
CV 1	2.93	5.42	8.47 a
CV 2	3.03	5.72	7.57 b
Significance	NS	NS	*
Interaction			
NM × CV	ns	ns	ns

Different letters within each column indicate significant differences according to the SNK test ($p \leq 0.05$). NS not significant * $p < 0.05$ and **0.01, indicate level of significance.

Table 7 shows that in both years CNM treatment statistically influenced N accumulation (gm^{-2}) in all plant epigeal organs and on all sample dates. In the first year, the highest N accumulation was observed compared to CNM in the leaves (+48%) at 55 DAT, stems at 93 DAT (+85%), and flower buds (+79%) at 131 DAT. Regarding INM, in the second year, the highest value was recorded in leaves (+28%) at 55 DAT, stems (+46%), and flower buds (+117%) at 131 DAT. In both years the CVs did not influence N accumulation.

Table 7. Main effects of nutrient management and cultivar on N accumulation (g m^2) in different organs at three different DAT in chrysanthemum plants over the two years of application.

Treatments	DAT						
	55		93		131		
	Leaves	Stems	Leaves	Stems	Leaves	Heads	Stems
				First Year			
Nutrition Management (NM)							
CNM	3.05	1.28	3.31	2.16	2.83	1.87	1.50
INM	4.52	1.76	5.14	3.99	4.50	3.34	2.55
Significance	**	*	**	**	**	**	**
Cultivar (CV)							
CV 1	4.09	1.46	4.08	3.13	3.52	3.48	2.00
CV 2	4.04	1.57	4.53	3.26	3.75	2.73	2.00
Significance	ns	ns	ns	ns	ns	*	ns
Interaction							
NM × CV							
				Second Year			
Nutrition Management (NM)							
CNM	1.84	0.83	3.07	1.88	2.75	1.28	1.75
INM	2.35	0.96	3.57	2.55	3.30	2.78	2.57
Significance	*	*	*	*	*	**	*
Cultivar (CV)							
CV 1	2.07	0.87	3.23	2.19	3.09	1.71	2.54
CV 2	2.11	0.91	3.52	2.20	2.95	1.97	1.77
Significance	ns	ns	ns	ns	ns	ns	*
Interaction							
NM × CV							

Different letters within each column indicate significant differences according to SNK test ($p \leq 0.05$). NS not significant * $p < 0.05$ and **0.01, indicate level of significance.

Figure 1 shows that in both years the N utilization efficiency (NUE) value was highest in CNM (on average 162) compared to INM (on average 142); no significant difference was found between the CVs.

In both years, the P content (%), at harvest, in above-ground vegetative tissues (leaves, stems, and heads) of INM plants was higher than those of CNM plants (Table 8). In the first year, the increase in INM compared to CNM was 11% in the leaves, 20% in the stems, and 21% in the flower heads. In the second year, the increase in P content in the leaves under INM was similar to that recorded in the first year (12%), while it was lower for stems (+12%) and flower heads (+14%). In both years the P content in above-ground vegetative tissues were in the order of head > leaves > stems, which was maintained in both INM and CNM treatments.

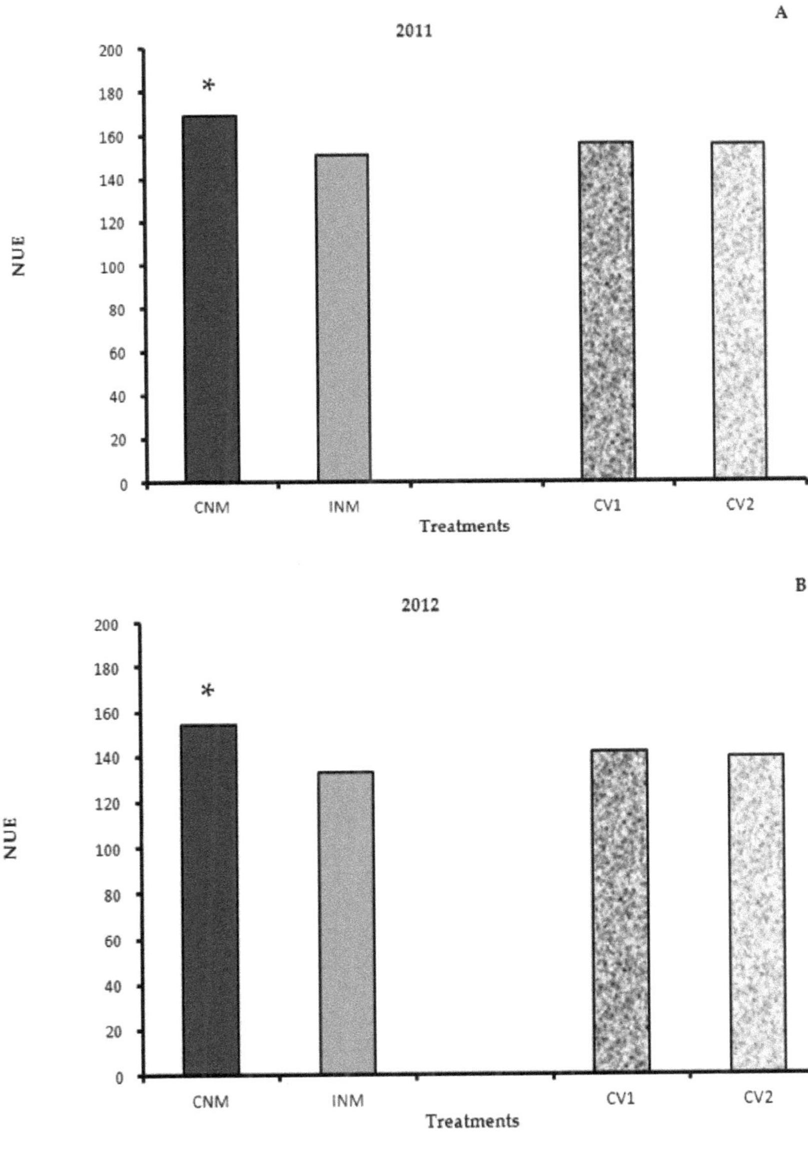

Figure 1. Main effects of nutrient management and cultivar on N utilization efficiency (NUE) in first (A) and second (B) year in chrysanthemum plants at harvest time (* indicates the level of significance at $p < 0.05$).

Table 8. Main effects of nutrient management and cultivar on phosphorus content (%) in different organs at harvest in chrysanthemum plants over the two years of application.

TMTS	Leaves	Stem	Flower Heads
		First Year	
Nutrition Management (NM)			
CNM	0.19 b	0.15 b	0.33 b
INM	0.21 a	0.18 a	0.40 a
Significance	*	*	*
Cultivar (CV)			
CV 1	0.20	0.16	0.32
CV 2	0.21	0.17	0.31
Significance	ns	ns	ns
Interaction			
NM × CV	ns	ns	ns
		Second Year	
Nutrition Management (NM)			
CNM	0.17 b	0.16 b	0.29 b
INM	0.19 a	0.18 a	0.33 a
Significance	*	*	*
Cultivar (CV)			
CV 1	0.18	0.19	0.31
CV 2	0.18	0.18	0.30
Significance	ns	ns	ns
Interaction			
NM × CV	ns	ns	ns

Different letters within each column indicate significant differences according to SNK test ($p \leq 0.05$). NS not significant * $p < 0.05$ and **0.01, indicate level of significance.

Concerning an economic point of view, the increased yield obtained with INM (140 stems m^{-2}) compared with CNM (118 stems m^{-2}) led consequently to an increase of gross production of exactly €50,600.00 (takings difference). This amount was much greater than the cost increases needed for INM compared to CNM (Table 9).

Table 9. Basic economic analysis of fertilizers cost, gross production, and takings difference obtained.

	Fertilizer Cost	Yield	Gross Production	Takings Difference
	(€ ha^{-1})	(stems m^{-2})	(€ ha^{-1}) *	(€)
CNM	1945.00	118	271,400.00	
INM	3144.00	140	322,000.00	+50,600.00

* Chrysanthemum price calculated at 2012–2013: 0.23 €/stem (ISMEA/2012–2013).

4. Discussion

In our study, mineral nutrient management (CNM) and integrated nutrient management (INM) were compared in chrysanthemum cultivation. The INM protocol, which combined the application of half the rate of CNM and seaweed extract, organic and biofertilizer (AMF + PSB), improved yield, cut stem quality traits, root morphology, as well as N accumulation and P content in tissues. Based on other research about INM practices [47–49], this protocol seems to be suitable in order to obtain advantages on profits and sustainability. Our aim was to verify a new mixture in order to reduce mineral fertilizer application, making chrysanthemum cultivation more sustainable, as well as highly profitable.

Compared to CNM, the INM protocol led to a significantly higher yield in terms of the number of secondary branches per m^{-2} (Table 3). This could be attributed to a better nutrient translocation in the plant, which led to the production of a greater number of axillary buds and therefore of secondary axes, in line with Kale et al. [50] in *Salvia* and Nethra [51] in the Chinese aster. In other studies regarding biostimulant applications, yield also increased in seaweed treated plants influenced by cytokinin content, which enhances nutrient mobilization in plant organs [52].

Regarding cut stem quality traits (Table 3), our results are in agreement with Verma et al. [53], who applied an INM on chrysanthemum CV Roja. The treatment that consisted of *Azospirillum*, PSB, vermicompost, and 50% of recommended mineral NPK recorded the highest plant height, number of branches, and flowers per plant. Similar results were reported in *Crossandra* [54] and Dahlia [55]. The combination of biofertilizers with the recommended NPK dose yielded a higher flower production in *Limonium* [56] and *Calendula* [57].

In our study, a higher leaf area value was found in chrysanthemum plants under INM. According to De Lucia and Vecchietti [58], the application of seaweed extract (*A. nodosum*) in *Lilium* CV Brindisi, greatly affected these parameters (12.3 cm^2 of treated plants, compared with 10.3 cm^2 of untreated plants). This was potentially due to the direct effect of the biostimulant containing betaine. The nutrient concentration present in both the N organic fertilizer and seaweed extract biostimulant cannot on its own explain the positive response as an increase in aerial organ dry weight (Table 4). In fact cytokinins have a considerable influence on nutrient mobilization in vegetative and reproductive organs [59].

Microbial inoculants are also good supplement with half the recommended mineral dose of fertilizer. Wu et al. [60] reported that *G. mossae* plus *B. megaterium* on maize increased plant growth and NPK assimilation. As regards the effect of applying INM, the chrysanthemum root development exhibited a remarkable increase in all parameters compared to CNM (Table 5). The root growth promoting activity has been observed in snapdragon, when a biostimulant was applied [61]. Previous research has shown that the brown seaweed extract, rich in auxin, improved lateral root formation when applied to mung bean [62]. A study carried out by Mancuso et al. [63] on potted *Vitis vinifera* under seaweed extracts, showed an increase in the total volume of the root system.

Concerning the nutrient uptake, Biswas et al. [64] and Adesemoye et al. [65] showed that PGPR (plant growth-promoting rhizobacteria) also influences this parameter through a more pronounced development of the root surface area. The INM seems to encourage a better uptake of mineral nutrients by plants, which results in a higher number of branches as well as leaf area, and more flowers [66].

The N uptake by chrysanthemum plants may be enhanced by the use of biofertilizers, possibly because they stimulate better root architecture or due to the influence of growth hormones contained in seaweed extracts. These substances can increase the ability of nutrient absorption as well as enzymatic activity, in agreement with Kumari et al. [67].

The N accumulation (g m^{-2}) value in the INM treatment could be caused by the better availability and uptake of nutrients facilitated by the application of both mineral and organic fertilizers, biostimulants, and biofertilizers (Tables 6 and 7). Mahadik et al. [68] showed that the increase in N and P uptake by chrysanthemum plants was the highest with the application of *Azotobacter* plus PSB, 50% of RDF (Recommended Dose of Fertilizers) (100:100:100 kg ha^{-1} NPK), and 10 t ha^{-1} of vermicompost. Regarding the P content in plant tissue (Table 8), our findings are validated by similar results found in a number of earlier studies on bacteria.

Shirmardi et al. [69] reported that PGPR solubilizes the inorganic phosphate and produces IAA, thus improving plant growth by increasing P-uptake from the soil and its transport to plant shoots. A significant increase in sunflower growth parameters, including plant P content, was found in inoculated plants after inoculation with *Bacillus* sp., possibly due to the P-solubilizing, IAA-synthesizing, and root-colonizing of these strains [70], which increase nutrient uptake.

Richardson (1994) and Rodríguez and Fraga [71] studied the influence of several soil bacteria on the supply of P to plants as a consequence of their capacity for inorganic or organic P solubilization and, therefore, for improving plant growth performance. In addition, in a 1994 study, Garbaye [72] postulated that some PSB behave like mycorrhizal helper bacteria with a synergistic interaction.

Compared to the non-treated control, the combined application of mycorrhizal fungus and rhizobacteria significantly increased growth parameters, i.e., total fresh weight, aerial dry weight, shoot length, and leaf area, in bananas. The leaf mineral content, i.e., N, P, and K, also increased significantly following the combined application of both microorganisms [73].

Finally, integrated nutrient management practices could be viable for sustainable floriculture on a commercial and profitable scale. Our data on the economics of chrysanthemum flowers are in agreement with those Verma, who showed that the cost of fertilizer can be saved with inoculation of both Azospirillum and PSB, obtaining higher flower yield compared to CNM. Angadi too, carried out a study that shed light on the combination of Azospirillum, PSB, 50% vermicompost, and 1/2 recommended NPK dose, giving the maximum net returns per euro invested.

5. Conclusions

The quality and quantity of fertilizers are the key factor affecting the growth, yield, and quality of cut flowers. Since chrysanthemum is an energy-intensive ornamental crop with a very high input of fertilizers, several experiments have been aimed at using alternative methods, reducing mineral fertilizers, and in particular the INM.

Our results shows that the INM protocol, 50% mineral RDF with N organic fertilizer plus biostimulant (seaweed extract) plus biofertilizer (microbial consortium of *Glomus* sp. and *Bacillus* sp.), is effective in enhancing yield, quality, root morphology, and nutrient uptake compared to RDF. This indicates the possibility of the sustainable, eco-friendly cultivation of chrysanthemum. In order to discern the influence of each component of INM mixture on yield and quality traits, future research is needed.

Author Contributions: B.L.: substantial contribution to the experimental work and data collection; D.L.: bibliographic research and support for drafting; G.C.: support with data analysis, interpretation, and critical revision of the manuscript; B.D.L.: experimental design and coordination of the work, interpretation of the data, and critical revision of the manuscript.

Funding: This research was funded by Chimicaverde Flor (Province of Bari).

Conflicts of Interest: The authors declare no conflict of interest.

References

1. Yousaf, M.; Li, J.; Lu, J.; Ren, T.; Cong, R.; Fahad, S.; Li, X. Effects of fertilization on crop production and nutrient-supplying capacity under rice-oilseed rape rotation system. *Sci. Rep.* **2017**, *7*, 1270. [CrossRef] [PubMed]
2. Stellacci, A.M.; Cristiano, G.; Rubino, P.; De Lucia, B.; Cazzato, E. Nitrogen uptake, nitrogen partitioning and N-use efficiency of container-grown Holm oak (*Quercus ilex* L.) under different nitrogen levels and fertilizer sources. *J. Food Agric. Environ.* **2013**, *11*, 990–994.
3. Mantovani, J.R.; Silveira, L.G.D.; Landgraf, P.R.C.; Santos, A.R.D.; Costa, B.D.S. Phosphorus rates and use of cattle manure in potted gerbera cultivation. *Ornam. Hortic.* **2017**, *23*, 412–418. [CrossRef]
4. Vitousek, P.M.; Aber, J.D.; Howarth, R.W.; Likens, G.E.; Matson, P.A.; Schindler, D.W.; Schlesinger, W.H.; Tilman, D.G. Human alteration of the global nitrogen cycle: Sources and consequences. *Ecol. Appl.* **1997**, *7*, 737–750. [CrossRef]
5. Tilman, D. The greening of the green revolution. *Nature* **1998**, *396*, 211–212. [CrossRef]
6. Berg, M.; Köpke, U.; Haas, G. Nitrate leaching: Comparing conventional, integrated and organic agricultural production systems. *Intern. Assoc. Hydrol. Sci.* **2002**, *273*, 131–136.
7. Chen, J.H. The combined use of chemical and organic fertilizers and/or biofertilizer for crop growth and soil fertility. In Proceedings of the International Workshop on Sustained Management of the Soil-Rhizosphere System for Efficient Crop Production and Fertilizer Use, Bangkok, Thailand, 16–20 October 2006; Volume 16, p. 20.
8. Hansen, B.; Thorling, L.; Schullehner, J.; Termansen, M.; Dalgaard, T. Groundwater nitrate response to sustainable nitrogen management. *Sci. Rep.* **2017**, *7*, 8566. [CrossRef] [PubMed]

9. Shu, J.; Wu, H.; Chen, M.; Peng, H.; Li, B.; Liu, R.; Liu, Z.; Wang, B.; Huang, T.; Hu, Z. Fractional removal of manganese and ammonia nitrogen from electrolytic metal manganese residue leachate using carbonate and struvite precipitation. *Water Res.* **2019**, *153*, 229–238. [CrossRef]
10. Torbert, H.A.; King, K.W.; Harmel, R.D. Impact of Soil Amendments on Reducing Phosphorus Losses from Runoff in Sod. *J. Environ. Qual.* **2005**, *34*, 1415. [CrossRef] [PubMed]
11. Bouwman, L.; Goldewijk, K.K.; Van Der Hoek, K.W.; Beusen, A.H.W.; Van Vuuren, D.P.; Willems, J.; Rufino, M.C.; Stehfest, E. Exploring global changes in nitrogen and phosphorus cycles in agriculture induced by livestock production over the 1900–2050 period. *Proc. Natl. Acad. Sci. USA* **2013**, *110*, 20882–20887. [CrossRef]
12. Zhang, X.; Davidson, E.A.; Mauzerall, D.L.; Searchinger, T.D.; Dumas, P.; Shen, Y. Managing nitrogen for sustainable development. *Nature* **2015**, *528*, 51–59. [CrossRef]
13. Murillo-Amador, B.; Morales-Prado, L.E.; Troyo-Diéguez, E.; Córdoba-Matson, M.V.; Hernández-Montiel, L.G.; Rueda-Puente, E.O.; Nieto-Garibay, A. Changing environmental conditions and applying organic fertilizers in *Origanum vulgare* L. *Front. Plant Sci.* **2015**, *6*, 549. [CrossRef]
14. Szparaga, A.; Kocira, S. Generalized logistic functions in modelling emergence of *Brassica napus* L. *PLoS ONE* **2018**, *13*, e0201980. [CrossRef] [PubMed]
15. Alori, E.T.; Babalola, O.O. Microbial Inoculants for Improving Crop Quality and Human Health in Africa. *Front. Microbiol.* **2018**, *9*, 2213. [CrossRef] [PubMed]
16. Fiorentino, N.; Ventorino, V.; Woo, S.L.; Pepe, O.; De Rosa, A.; Gioia, L.; Romano, I.; Lombardi, N.; Napolitano, M.; Colla, G.; et al. Trichoderma-based biostimulants modulate rhizosphere microbial populations and improve N uptake efficiency, yield, and nutritional quality of leafy vegetables. *Front. Plant Sci.* **2018**, *9*, 743. [CrossRef] [PubMed]
17. Wani, S.A.; Chand, S.; Wani, M.A.; Ramzan, M.; Hakeem, K.R. Azotobacter chroococcum—A Potential Biofertilizer in Agriculture: An Overview. In *Soil Science: Agricultural and Environmental Prospectives*; Springer International Publishing: Cham, Switzerland, 2016; pp. 333–348, ISBN 9783319344515.
18. Gruhn, P.; Goletti, F.; Yudelman, M. Integrated nutrient management, soil fertility, and sustainable agriculture: Current issues and future challenges. *Food Agric. Environ. Discuss. Pap.* **2000**, 32.
19. Ertani, A.; Francioso, O.; Tinti, A.; Schiavon, M.; Pizzeghello, D.; Nardi, S. Evaluation of Seaweed Extracts From Laminaria and *Ascophyllum nodosum* spp. as *Biostimulants* in *Zea mays* L. Using a Combination of Chemical, Biochemical and Morphological Approaches. *Front. Plant Sci.* **2018**, *9*, 428. [CrossRef]
20. De Pascale, S.; Rouphael, Y.; Colla, G. Plant biostimulants: Innovative tool for enhancing plant nutrition in organic farming. *Eur. J. Hortic. Sci.* **2018**, *82*, 277–285. [CrossRef]
21. Chiaiese, P.; Corrado, G.; Colla, G.; Kyriacou, M.C.; Rouphael, Y. Renewable Sources of Plant Biostimulation: Microalgae as a Sustainable Means to Improve Crop Performance. *Front. Plant Sci.* **2018**, *9*, 1782. [CrossRef]
22. Nelson, W.R.; van Staden, J. Effect of seaweed concentrate on the growth of wheat. *South Afr. J. Sci.* **1986**, *82*, 199–200.
23. Abetz, P.; Young, C.L. The Effect of Seaweed Extract Sprays Derived from Ascophyllum nodosum on Lettuce and Cauliflower Crops. *Bot. Mar.* **1983**, *26*, 487–492. [CrossRef]
24. Crouch, I.J.; Van Staden, J. Effect of seaweed concentrate on the establishment and yield of greenhouse tomato plants. *J. Appl. Phycol.* **1992**, *4*, 291–296. [CrossRef]
25. Burchett, S.; Fuller, M.P.; Jellings, A.J. Application of seaweed extract improves winter hardiness of winter barley cv Igri. In Proceedings of the Society for Experimental Biology, Annual Meeting, The York University, Toronto, ON, Canada, 22–27 March 1998; Springer: Berlin/Heidelberg, Germany, 1998.
26. Zaidi, A.; Khan, M.S.; Ahemad, M.; Oves, M.; Wani, P.A. Recent Advances in Plant Growth Promotion by Phosphate-Solubilizing Microbes. In *Microbial Strategies for Crop Improvement*; Springer: Berlin/Heidelberg, Germany, 2009; pp. 23–50, ISBN 9783642019791.
27. Rafi, M.M.; Krishnaveni, M.S.; Charyulu, P.B.B.N. Phosphate-Solubilizing Microorganisms and Their Emerging Role in Sustainable Agriculture. In *Recent Developments in Applied Microbiology and Biochemistry*; Buddolla, V., Ed.; Academic Press: Dordrecht, The Netherlands, 2019; Chapter 17; pp. 223–233, ISBN 9780128163283. [CrossRef]
28. Canbolat, M.Y.; Bilen, S.; Çakmakçı, R.; Şahin, F.; Aydın, A. Effect of plant growth-promoting bacteria and soil compaction on barley seedling growth, nutrient uptake, soil properties and rhizosphere microflora. *Biol. Fertil. Soils* **2006**, *42*, 350–357. [CrossRef]

29. Ipsilantis, I.; Karamesouti, M.; Gasparatos, D. Beneficial Microorganisms for the Management of Soil Phosphorus. In *Sustainable Agriculture Reviews*; Lichtfouse, E., Ed.; Springer: Cham, Switzerland, 2018; Volume 32.
30. Rajkumar, M.; Nagendran, R.; Lee, K.J.; Lee, W.H.; Kim, S.Z. Influence of plant growth promoting bacteria and Cr6+ on the growth of Indian mustard. *Chemosphere* **2006**, *62*, 741–748. [CrossRef] [PubMed]
31. Yasmin, S.; Rahman Bakar, M.A.; Malik, K.A.; Hafeez, F.Y. Isolation, characterization and beneficial effects of rice associated plant growth promoting bacteria from Zanzibar soils. *J. Basic Microbiol.* **2004**, *44*, 241–252. [CrossRef]
32. Stewart, L.I.; Hamel, C.; Hogue, R.; Moutoglis, P. Response of strawberry to inoculation with arbuscular mycorrhizal fungi under very high soil phosphorus conditions. *Mycorrhiza* **2005**, *15*, 612–619. [CrossRef]
33. Li, H.; Smith, S.E.; Holloway, R.E.; Zhu, Y.; Smith, F.A. Arbuscular mycorrhizal fungi contribute to phosphorus uptake by wheat grown in a phosphorus-fixing soil even in the absence of positive growth responses. *New Phytol.* **2006**, *172*, 536–543. [CrossRef]
34. Estaun, V.; Camprubi, A.; Joner, E.J. Selecting arbuscular mycorrhizal fungi for field application. In *Mycorrhizal Technology in Agriculture*; Gianinazzi, S., Schuepp, H., Barea, J.M., Haselwandter, K., Eds.; Birkhauser Verlag: Berlin, Germany, 2002; pp. 249–259.
35. Wang, C.; Li, X.; Zhou, J.; Wang, G.; Dong, Y. Effects of arbuscular mycorrhizal fungi on growth and yield of cucumber plants. *Commun. Soil Sci. Plant Anal* **2008**, *39*, 499–509. [CrossRef]
36. Kapulnik, Y.; Tsror (Lahkim), L.; Zipori, I.; Hazanovsky, M.; Wininger, S.; Dag, A. Effect of AMF application on growth, productivity and susceptibility to Verticillium wilt of olives grown under desert conditions. *Symbiosis* **2010**, *52*, 103–111. [CrossRef]
37. Ortas, I. Effect of mycorrhiza application on plant growth and nutrient concentration in cucumber production under field conditions. *Span. J. Agric. Res.* **2010**, *8*, 116–122. [CrossRef]
38. Mwangi, A.M.K.; Kahangi, E.M.; Ateka, E.; Onguso, J.; Mukhongo, R.W.; Mwangi, E.K.; Jefwa, J.M. Growth effects of microorganisms based commercial products inoculated to tissue cultured banana cultivated in three different soils in Kenya. *Appl. Soil Ecol.* **2013**, *64*, 152–162. [CrossRef]
39. Lim, T.K. Chrysanthemum morifolium. In *Edible Medicinal and Non-Medicinal Plants*; Springer Netherlands: Dordrecht, The Netherlands, 2014; pp. 250–269.
40. Liu, D.; Liu, W.; Zhu, D.; Geng, M.; Zhou, W.; Yang, T. Nitrogen effects on total flavonoids, chlorogenic acid, and antioxidant activity of the medicinal plant *Chrysanthemum morifolium*. *J. Plant Nutr. Soil Sci.* **2010**, *173*, 268–274. [CrossRef]
41. Lunt, O.R.; Kofranek, A.M. Nitrogen and potassium nutrition of chrysanthemum. *Proc. Am. Soc. Hort. Sci* **1958**, *72*, 487–497.
42. Yoon, H.S.; Goto, T.; Kageyama, Y. Developing a Nitrogen Application Curve for Spray Chrysanthemums grown in Hydroponic System and its Practical Use in NFT System. *Engei Gakkai zasshi* **2000**, *69*, 416–422. [CrossRef]
43. Li, Y.; Jiang, Y.; Li, Z.; Zhao, L. A study of phosphorus uptake kinetic characteristic of four terrestrial economic plants. In Proceedings of the 2009 Beijing International Environmental Technology Conference, Beijing, China, 16–19 October 2009; pp. 140–145.
44. Vanek, V.; Balik, J.; Cerny, J.; Pavliik, M.; Pavlikova, D.; Tlustos, P.; Valtera, J. The Nutrition of Horticultural Plants. Academic Press: Prague, Czech Republic, 2012.
45. Tesi, R. Colture Protette, Ortoflorovivaismo. *Colt. Protette* 1987.
46. Ceccon, P.; Fagnano, M.; Grignani, C.; Monti, M.; Orlandini, S. (Eds.) *Agronomia*; SES: Napoli, Italy, 2017.
47. Nammidevi, M.; Shalini, B.; Bhongle, S.A.; Khiratkar, S.D. Effect of bio-inoculants with graded doses of NPK on flowering, yield attributes and economics of annual chrysanthemum. *J. Soils Crop.* **2008**, *18*, 217–220.
48. Deshmukh, P.G.; Khiratkar, S.D.; Badge, S.A.; Bhongle, S.A. Effect of bio-inoculants with graded doses of NPK on growth and yield of gaillardia. *J. Soils Crop.* **2008**, *18*, 212–216.
49. Sunitha, H.M.; Hunje, R.; Vyakaranahal, B.S.; Bablad, H.B. Effect of plant population and integrated nutrient management on growth, seed yield and quality of African marigold (*Tagetes erecta* L.). *Karnataka J. Agric. Sci* **2010**, *23*, 783–786.
50. Kale, R.D.; Bano, K.; Sreenivasa, M.N.; Bagyaraj, D.J. Influence of worms cast on the grown and mycorrhizal colonization of two ornamental plants. *South Indian Hortic.* **1987**, *35*, 433–437.

51. Nethra, N.N. Effect PH, Organic and Inorganic Fertilizer on Growth, Yield and Post Harvest Life of China Aster (*Callistephns Chinensis* (L.) Nexx.). Ph.D. Dissertation, University of Agricultural Sciences, GKVK, Karnataka, India, 1996.
52. Gersani, M.; Kende, H. Studies on cytokinin-stimulated translocation in isolated bean leaves. *J. Plant Growth Regul.* **1982**, *1*, 161–171.
53. Verma, S.K.; Angadi, S.G.; Patil, V.S.; Mokashi, A.N.; Mathad, J.C.; Mummigatti, U.V. Growth, yield and quality of chrysanthemum (*Chrysanthemum morifolium* Ramat.) cv. Raja as influenced by integrated nutrient management. *Karn J. Agric. Sci.* **2011**, *24*, 681–683.
54. Narsimha Raju, S.; Haripriya, K. Integrated nutrient management in cossandra (*Crossandra infundibuliformis* L.) Cv. Dindigul local. *South Indian Hortic.* **2001**, *49*, 318–322.
55. Warade, A.P.; Golliwar, V.J.; Chopde, N.; Lanje, P.W.; Thakre, S.A. Effect of organic manures and bio-fertilizers on growth, flowering and yield of dahlia. *J. Soils Crop.* **2007**, *17*, 354–357.
56. Gajathri, H.N.; Jayaprassed, K.V.; Narayanaswamy, P. Response of bio-fertilezers and their combined application with different levels of inorganic fertilizers in statice (*Limonium caspia*). *J. Ornam. Hortic.* **2004**, *7*, 70–74.
57. Shashikanth, G. Effect of Different Sources of Nutrients on Growth and Yield of Tall Marigold. Master's Thesis, University of Agricultural Sciences, Dharwad, India, 2005.
58. De Lucia, B.; Vecchietti, L. Type of bio-stimulant and application method effects on stem quality and root system growth in L.A. Lily. *Eur. J. Hortic. Sci.* **2012**, *77*, 10.
59. Khan, W.; Rayirath, U.P.; Subramanian, S.; Jithesh, M.N.; Rayorath, P.; Hodges, D.M.; Critchley, A.T.; Craigie, J.S.; Norrie, J.; Prithiviraj, B. Seaweed Extracts as Biostimulants of Plant Growth and Development. *J. Plant Growth Regul.* **2009**, *28*, 386–399. [CrossRef]
60. Wu, S.C.; Cao, Z.H.; Li, Z.G.; Cheung, K.C.; Wong, M.H. Effects of biofertilizer containing N-fixer, P and K solubilizers and AM fungi on maize growth: A greenhouse trial. *Geoderma* **2005**, *125*, 155–166. [CrossRef]
61. Cristiano, G.; Pallozzi, E.; Conversa, G.; Tufarelli, V.; De Lucia, B. Effects of an Animal-Derived Biostimulant on the Growth and Physiological Parameters of Potted Snapdragon (*Antirrhinum majus* L.). *Front. Plant Sci.* **2018**, *9*, 861. [CrossRef]
62. Crouch, I.J.; Van Staden, J. Evidence for rooting factors in a seaweed concentrate prepared from Ecklonia maxima. *J. Plant Physiol.* **1991**, *137*, 319–322. [CrossRef]
63. Mancuso, S.; Azzarello, E.; Mugnai, S.; Briand, X. Marine bioactive substances (IPA extract) improve foliar ion uptake and water stress tolerance in potted Vitis vinifera plants. *Adv. Hortic. Sci.* **2006**, *20*, 1000–1006.
64. Biswas, J.C.; Ladha, J.K.; Dazzo, F.B. Rhizobia Inoculation Improves Nutrient Uptake and Growth of Lowland Rice. *Soil Sci. Soc. Am. J.* **2000**, *64*, 1644. [CrossRef]
65. Adesemoye, A.O.; Torbert, H.A.; Kloepper, J.W. Enhanced plant nutrient use efficiency with PGPR and AMF in an integrated nutrient management system. *Can. J. Microbiol.* **2008**, *54*, 876–886. [CrossRef] [PubMed]
66. Patanwar, M.; Sharma, G.; Banjare, C.; Chandravanshi, D.; Sahu, E. Growth and development of chrysanthemum (*Dendranthema grandiflora* Tzvelev) as influenced by integrated nutrient management. *Ecoscan* **2014**, *6*, 459–462.
67. Kumari, A.; Goyal, R.K.; Choudhary, M.; Sindhu, S.S. Effect of different nitrogen levels and biofertilizers on growth, yield and nutrient content of chrysanthemum. *Ann. Agric. Res. New Ser.* **2014**, *35*, 156–163.
68. Mahadik, M.K.; Dalal, S.R.; Patil, S.W.U. Dry matter production and nutrients composition of chrysanthemum in relation to integrated nutrient management. *J. Pharmacogn. Phytochem.* **2017**, *6*, 1879–1882.
69. Shirmardi, M.; Savaghebi, G.R.; Khavazi, K.; Akbarzadeh, A.; Farahbakhsh, M.; Rejali, F.; Sadat, A. Effect of Microbial Inoculants on Uptake of Nutrient Elements in Two Cultivars of Sunflower (*Helianthus annuus* L.) in Saline Soils. *Notulae Scientia Biologicae* **2010**, *2*, 57. [CrossRef]
70. Shahid, M.; Hameed, S.; Tariq, M.; Zafar, M.; Ali, A.; Ahmad, N. Characterization of mineral phosphate-solubilizing bacteria for enhanced sunflower growth and yield-attributing traits. *Ann. Microbiol.* **2015**, *65*, 1525–1536. [CrossRef]
71. Rodríguez, H.; Fraga, R. Phosphate solubilizing bacteria and their role in plant growth promotion. *Biotechnol. Adv.* **1999**, *17*, 319–339. [CrossRef]

72. Garbaye, J. Tansley Review No. 76 Helper bacteria: A new dimension to the mycorrhizal symbiosis. *New Phytol.* **1994**, *128*, 197–210. [CrossRef]
73. Rodríguez-Romero, A.S.; Guerra, M.S.P.; Del Carmen Jaizme-Vega, M. Effect of arbuscular mycorrhizal fungi and rhizobacteria on banana growth and nutrition. *Agron. Sustain. Dev.* **2005**, *25*, 395–399. [CrossRef]

agronomy

MDPI

Article

Arbuscular Mycorrhizal Fungi Modulate the Crop Performance and Metabolic Profile of Saffron in Soilless Cultivation

Matteo Caser [1,†,*], Sonia Demasi [1,†], Íris Marisa Maxaieie Victorino [2,3,4], Dario Donno [1], Antonella Faccio [2], Erica Lumini [2], Valeria Bianciotto [2] and Valentina Scariot [1,2]

1 Department of Agricultural, Forest and Food Sciences, University of Torino. Largo Paolo Braccini 2, 10095 Grugliasco (TO), Italy; sonia.demasi@unito.it (S.D.); dario.donno@unito.it (D.D.); valentina.scariot@unito.it (V.S.)
2 Institute for Sustainable Plant Protection, National Research Council, Viale Mattioli 25, 10125 Torino, Italy; rismarisamazaieie.victorino@unito.it (Í.M.M.V.); antonella.faccio@ipsp.cnr.it (A.F.); erica.lumini@ipsp.cnr.it (E.L.); valeria.bianciotto@ipsp.cnr.it (V.B.)
3 Department of Life Sciences and Systems Biology, University of Turin. Viale Mattioli 25, 10125 Torino, Italy
4 Biological Science Department, Science Faculty, Eduardo Mondlane University (UEM), Av. Julius Nyerere-Campus Universitário, Maputo 3453, Mozambique
* Correspondence: matteo.caser@unito.it; Tel.: +39-011-670-9835
† Both authors contributed equally to this work.

Received: 7 March 2019; Accepted: 7 May 2019; Published: 8 May 2019

Abstract: Saffron (*Crocus sativus* L.) is cultivated worldwide. Its stigmas represent the highest-priced spice and contain bioactive compounds beneficial for human health. Saffron cultivation commonly occurs in open field, and spice yield can vary greatly, from 0.15 to 1.5 g m^{-2}, based on several agronomic and climatic factors. In this study, we evaluated saffron cultivation in soilless systems, where plants can benefit from a wealth of nutrients without competition with pathogens or stresses related to nutrient-soil interaction. In addition, as plant nutrient and water uptake can be enhanced by the symbiosis with arbuscular mycorrhizal fungi (AMF), we also tested two inocula: a single species (*Rhizophagus intraradices*) or a mixture of *R. intraradices* and *Funneliformis mosseae*. After one cultivation cycle, we evaluated the spice yield, quality (ISO category), antioxidant activity, and bioactive compound contents of saffron produced in soilless systems and the effect of the applied AMF inocula. Spice yield in soilless systems (0.55 g m^{-2}) was on average with that produced in open field, while presented a superior content of several health-promoting compounds, such as polyphenols, anthocyanins, vitamin C, and elevated antioxidant activity. The AMF symbiosis with saffron roots was verified by light and transmission electron microscopy. Inoculated corms showed larger replacement corms (+50% ca.). Corms inoculated with *R. intraradices* performed better than those inoculated with the mix in terms of spice quality (+90% ca.) and antioxidant activity (+88% ca.). Conversely, the mixture of *R. intraradices* and *F. mosseae* increased the polyphenol content (+343% ca.). Thus, soilless systems appeared as an effective alternative cultivation strategy for the production of high quality saffron. Further benefits can be obtained by the application of targeted AMF-based biostimulants.

Keywords: biostimulants; *Crocus sativus*; *Funneliformis mosseae*; glasshouse; protected cultivation; *Rhizophagus intraradices*; substrate

1. Introduction

Crocus sativus L. (saffron) is a flowering plant belonging to the Iridaceae family [1], grown for its red scarlet stigmas that represent the world's highest-priced spice. The market price for high quality saffron can reach 15,000–20,000 € kg^{-1} [2]. This species is widely cultivated in several countries, such as Iran, Italy, Spain, Morocco, France, Greece, China, India and Mexico [3], with an annual spice production that exceeds 220,000 kg [4].The importance and notoriety of saffron, used since ancient times as a dye, ingredient for the preparation of spirits, and condiment for food, is due to the substances contained in the spice, primarily crocins, picrocrocin and safranal [5,6]. These compounds confer the saffron's unique colour, taste, and aroma, and can also have positive biological effects. Saffron active constituents, such as carotenoids (i.e., crocins), polyphenols, and vitamins showed significant antioxidant activity [7–12]. Furthermore, saffron extracts exhibit anti-carcinogenic, anti-depressive, anti-hyperglycemic, hypoglycemic, and memory-enhancing effects [3,13]. *Crocus sativus* is a highly hand labour-intensive crop, mainly during flower harvesting and stigma separation. It is traditionally cultivated in small and flat plots, wherein mechanisation is not economically sustainable due to the harvest type and short flowering period [5,8]. Five hundred hand labour hours are needed to obtain 1 kg of dried saffron [4,5]. Saffron cultivation can be carried out on an annual or multi-year cycle [14,15]. Annual cultivation guarantees the effective control of plant diseases with a more accurate corm selection. On the contrary, in a multi-year cycle (e.g., 3–4 years in Spain, 4–5 years in Italy, and 6–8 years in India and Greece) [14], corm multiplication and the size of replacement corms in the ground can decrease drastically over the third year [15]. Environment and cultivation management affect flower induction in *C. sativus* [5,16–18]. In Mediterranean environments, flower induction occurs from early spring to mid-summer, while flower emergence occurs from early- to late-autumn. Differences in the time required for flower initiation have mostly been attributed to the corm size [19]. To produce flowers, the *C. sativus* corm diameter needs to be greater than 1 cm [20]. As the corm increases, flowering increases [16,21] and occurs in advance [22]. Commercially, a 2.5–3.5 cm diameter corm appears to be the most common size used to have full flowering already during the first cultivation cycle [23]. To increase saffron yield and quality, and to reduce production costs, flowering modulation through cultivation in soilless systems has been proposed [6,19,24]. In this cultivation system, plants are grown without the use of soil as a rooting medium and are supplied of inorganic nutrients via the irrigation water [25], and thus can benefit of a wealth of nutrients without competition with pathogens or stresses related to nutrient-soil interaction [26]. However, at present, only limited and controversial reports of saffron soilless cultivation under protected conditions are present in the literature. Molina et al. [18] reported that, in a glasshouse, temperatures may be responsible for production differences in terms of flower induction and flowering duration. Maggio et al. [19] showed that, in southern Italy, cultivation in a cold glasshouse on vermiculite and perlite-based substrates positively affected the yield and number of replacement corms. Similarly, Helal Beigi et al. [27] found that cocopeat and perlite substrates enhanced corm dry weight. While Souret and Weathers [28] and Mollafilabi et al. [24] concluded that soilless cultivation in experiments carried out in France and Iran, respectively significantly decreased the spice yield, in comparison to open field cultivation.

Plant performance in soilless systems may be improved through use of biostimulants, i.e., any natural substance or microorganism applied to plants with the aim to enhance nutrition efficiency, abiotic stress tolerance and/or crop quality traits, regardless of its nutrients content [29], with a consequent decrease of chemicals and increase of sustainability of the production system [30]. Soil microorganisms such as arbuscular mycorrhizal fungi (AMF) are collecting growing interest as biostimulants. They can form mutualistic symbiosis with about 80% of land plant species, including several crops [31]. Across the interface between the plant and the fungus, carbohydrates and mineral nutrients (i.e., N, P, Zn and B) are exchanged [32]. Thus, AMF can alleviate the limitation in plant growth caused by an inadequate nutrient supply and can improve tolerance to biotic and abiotic stress [33]. Additionally, there is evidence to indicate that AMF symbiosis may have a positive impact on crop quality [34]. Increased yield of essential oils, terpenes and polyphenols, and enhanced antioxidant activity were induced by AMF symbiosis in several medicinal and aromatic plants (MAPs) [12,35–38]. This higher concentration of bioactive molecules makes AMF-hosted plants generally more attractive for the pharmaceutical and food industries [39].

The positive effects of AMF on corm growth, spice yield, and the nutraceutical compound content of *C. sativus* have already been reported in open field trials [12,40–42]. However, so far little is known of the proper saffron AMF inocula application and effects in soilless conditions, where plants are cultivated in pots filled with sterilised substrates that are free of AM fungal propagules or highly reduced in AMF diversity [43]. In the meta-analysis performed by Berruti et al. [31], it has been observed that the fungal colonization gain in inoculated plants was significantly more frequent in the greenhouses than in the open-field conditions, even if the effectiveness of AMF inoculation on shoot biomass and yield was equally successful.

Thus, in the literature, saffron cultivation on soilless systems has been proposed for spice production, but no comparison with open field has been reported. While, the effects of AMF-based biostimulants have been investigated only in open fields. To evaluate if saffron cultivation in soilless systems and AMF application may improve crop performance, spice yield and quality, and modulate bioactive compounds content, we cultivated saffron on soilless systems, applying two AMF inocula, and we compared results with those obtained in a previous open field-based trial [12].

2. Materials and Methods

2.1. Plant Material and Soilless Cultivation

Crocus sativus corms with horizontal diameters of 2.5–3.5 cm, provided by the Azienda agricola "Les épices Vda" di Alessandro Putzolu (Chatillon, AO, Italy), were planted during the last 10 days of August 2017 in the experimental heated glasshouse of the Department of Agricultural Forest and Food Sciences (DISAFA) of the University of Torino (Italy, 45°06′23.21″N Lat, 7°57′82.83″E Long; 300 m a.s.l.). Corms were cultivated in pots (4 L, 14 cm diameter and 17 cm height; two corms per pot; density of 91 corms m^{-2}) filled with sterile quartz sand (2 L per pot; bulk density of 1.2 kg m^{-3}) on a layer of sterilised expanded clay (1 L per pot; bulk density of 300 kg m^{-3}) for a total weight of about 1.5 kg. During the flowering period, the average temperatures were 22 °C during the day and 14 °C during the night.

Two inocula (MycAgro Lab, Breteniére, FR) were used in this experiment: one composed of a single fungus *Rhizophagus intraradices* (Ri) and one composed of *R. intraradices* and *Funneliformis mosseae* (Ri+Fm). Both inocula consisted of AMF spores and inorganic substrate (calcined clay, vermiculite and zeolite). Inocula treatments were compared to a control without any formulation (AMF-). Ten grams of each inoculum were inserted into each vase. The treatment was placed under each corm in order to guarantee contact between the inoculum and the roots, therefore, favouring mutualistic symbiosis. Corms were not treated for fungal pathogens and cultivation lasted one cycle (August 2017–April 2018).

A complete randomised block design was used, with a total of 48 pots in two experimental plot units (24 pots per unit) and three treatments (8 pots per treatment). Irrigation water (pH 7.4, EC 505 µS cm) was added weekly (250 mL per pot) with a drip system. The corms were fertilised by fertigation (N:K 13:46; VIGORFLOR, AL.FE. srl, MN, Italy) every 2 weeks starting from the emergence of the spate, in quantities of 1.5 g L^{-1} of water.

2.2. Determination of Flower Production, Stigma Yield and Corm Growth

At flowering (October and November 2017), the number of flowers produced daily per corm and the yield of spice (i.e., stigmas dried at 40 °C for 8 h in an oven) were measured. The spice yield was calculated by weighting the mg of saffron produced per pot (area equal to 196 cm^2) and comparing the values to g of spice per square meter (m^2). At the end of the vegetative period (April 2018), corms were lifted, rid of topsoil, cleaned and de-tunicated, then the number, size and weight of replacement corms were determined.

2.3. Preparation of the Saffron Extract

The saffron aqueous extracts were prepared according to Caser et al. [12]. Briefly, 50 mg of powdered saffron were suspended into 5 mL of deionised water. After stirring (1000 rpm) for 1 hour at room temperature (circa 21 °C) in the dark, the solution was filtered with polytetrafluoroethylene (PTFE, VWR International, Milano, Italy) filters with a 25 mm diameter and 0.45 μm pore size. The saffron extract was then diluted 1:10 with deionised water to obtain the working solution. Each sample was prepared in triplicate.

2.4. Determination of Saffron Quality by ISO 3632

Saffron aqueous extracts were analysed with a spectrophotometer (Ultrospec 2100 Pro, Ultrospec 2100 pro, Amersham Biosciences, Uppsala, Sweden) to determine the content of picrocrocin, safranal, and crocin to have the information on the bitterness, the flavouring strength, and the colouring strength [44]. Data were related to the dry matter percentage and expressed as the absorbance of a 1% aqueous solution of dried saffron at 257, 330 and 440 nm respectively, using a 1 cm pathway quartz cell [A1% 1 cm (λ max)] and calculated according to the following formula [45]:

$$\text{A1\%1cm } (\lambda \text{ max}) = \text{D} \times 10000/\text{m} \times (100-\text{wMV}) \quad (1)$$

where D is the specific absorbance; m is the mass of the evaluated solution in grams; and wMV is the moisture expressed as a percentage mass fraction of the sample.

Moisture content (wMV) was determined using the following formula:

$$\text{wMV} = (\text{m0}-\text{m1}) \times (100/\text{m0})\% \quad (2)$$

where m0 is the mass, in grams, of the saffron portion before drying; and m1 is the mass, in grams, of the dry residue after incubation, performed in an oven for 16 h at 103 ± 2 °C.

All analytical steps were conducted in the dark to prevent analyte degradation.

2.5. Determination of Bioactive Compounds by HPLC

Bioactive compounds were determined by means of four high performance liquid chromatography-diode array detection (HPLC–DAD) methods (Table 1; [46]) using an Agilent 1200 High-Performance Liquid Chromatograph coupled to an Agilent UV-Vis diode array detector (Agilent Technologies, Santa Clara, CA, USA). Phytochemical separation was achieved with a Kinetex C18 column (4.6 × 150 mm, 5 μm, Phenomenex, Torrance, CA, USA) using several mobile phases for compound identification and recording UV spectra at different wavelengths, based on HPLC methods, as previously tested and validated [47], with some modifications. UV spectra were recorded at 330 nm (α), 280 nm (β), 310 and 441 nm (γ), and 261 and 348 nm (δ). All single compounds were identified by a comparison and combination of their retention times and UV spectra with those of authentic standards under the same chromatographic conditions.

Table 1. Characteristics of the HPLC methods applied to analyse the bioactive compounds present in the studied saffron samples.

HPLC Method	Class	Standard	Stationary Phase	Mobile Phase	Flow (mL min^{-1})	Time (min)
α	Cinnamic acids	Caffeic acid Chlorogenic acid Coumaric acid Cerulic acid	KINETEX–C18 column (4.6 × 150 mm, 5 μm)	A: 10mM KH_2PO_4/H_3PO_4, pH = 2.8 B: CH_3CN	1.5	20 + 2 (CT)
	Flavanols	Hyperoside Isoquercitrin Quercetin Wuercitrin Rutin				
β	Benzoic acids	Ellagic acid Gallic acid		A: $H_2O/CH_3OH/HCOOH$ (5:95:0.1 v/v/v), pH = 2.5 B: $CH_3OH/HCOOH$ (100:0.1 v/v)	0.6	23 + 2 (CT)
	Catechins	Catechin Epicatechin				
γ	Carotenoids	Crocin I Crocin II Safranal		A: H_2O B: CH_3CN	0.6	35 + 10 (CT)
δ	Vitamin C	Ascorbic acid Dehydroascorbic acid		A: 5 mM $C_{16}H_{33}N(CH_3)_3Br$/50 mM KH_2PO_4, pH=2.5 B: CH_3OH	0.9	10 + 5 (CT)

CT = conditioning time; Method α—gradient analysis: 5% B to 21% B in 17 min + 21% B in 3 min + 2 min of conditioning time—wavelength: 330 nm; Method β—gradient analysis: 3% B to 85% B in 22 min + 85% B in 1 min + 2 min of conditioning time—wavelength: 280 nm; Method γ—gradient analysis: 5% B to 95% B in 30 min + 95% B to 5% B in 5 min + 10 min of conditioning time—wavelengths: 310 nm + 441 nm; Method δ—isocratic analysis: 10 min + 5 min of conditioning time—wavelengths: 261 nm + 348 nm.

2.6. Phytochemical Characterisation

The phytochemical characterisation of each sample was performed as previously described by Caser et al. [48,49]. Briefly, the total anthocyanin content (TAC) was determined using the pH-differential method. Saffron solution was added to pH 1 and pH 4.5 buffer solutions. The absorbance of samples was determined at 515 and 700 nm after 15 min of equilibration. The results were expressed as milligrams of cyanidin 3-O-glucoside (C3G) per 100 grams of dry weight (mg_{C3G} $100g^{-1}$ DW). The total phenol content (TPC) was measured using the Folin–Ciocalteau phenolic method at 765 nm. The results were expressed as mg of gallic acid equivalents (GAE) per 100 g of dry weight (DW; mg_{GAE} $100g^{-1}$ DW). The antioxidant activity (AOA) was determined at 595 nm using the ferric reducing antioxidant power (FRAP) method and at 734 nm using the 2,2′-azinobis (3-ethylbenzothiazoline-6-sulphonic acid; ABTS) method. Results were expressed as millimoles of ferrous iron (Fe^{2+}) equivalents per kilogram of dry weight (mmol Fe^{2+} kg^{-1} DW) and as μmol of Trolox equivalents per gram of dry weight (μmol TE g^{-1} DW), respectively. All analyses were performed in three replicates and the absorbances were read using a spectrophotometer (Ultrospec 2100 Pro, Ultrospec 2100 pro, Amersham Biosciences, Uppsala, Sweden).

2.7. AMF Evaluation

On the base of saffron highly mycorrhization level (70 to 90% mycorrhizal intensity) previously reported [12], we randomly selected saffron roots in April 2018. Then, the root segments were processed for observation in light and under transmission electron microscopy. Root segments were excised under a stereomicroscope and quickly fixed in 2.5% glutaraldehyde in 0.1 M cacodilate buffer (pH 7.2) for 2 hours at room temperature and overnight at 4 °C. The samples were then post-fixed in 1% OsO4 in the same buffer and dehydrated in an ascending series of ethanol to 100%, incubated in two changes of absolute acetone and infiltrated in Epon-Araldite resin [50]. The resin was polimerised for 24 h at 60 °C. Semi-thin (1 μm) sections were then stained with 1% toluidine blue and ultra-thin (70 nm) sections were counter-stained with uranyl acetate and lead citrate [51], and used for electron microscopy analyses under a Philips CM10 transmission electron microscope.

2.8. Chemicals and Reagents

Sodium carbonate, Folin–Ciocalteu phenol reagent, sodium acetate, citric acid, hydrochloric acid, iron (III) chloride hexahydrate, 2,4,6-tripyridyl-S-triazine (TPTZ) and 1,2-phenylenediamine dihydrochloride (OPDA) were purchased from Sigma Aldrich (St. Louis, MO, USA), whereas acetic acid was purchased from Fluka Biochemika (Buchs, Switzerland). Ethylenediaminetetraacetic acid (EDTA) disodium salt was purchased from AMRESCO (Solon, OH, USA), whereas sodium fluoride was purchased from Riedel-de Haen (Seelze, Germany). Ethanol, acetone, sodium citrate and lead nitrate were purchased from Fluka Biochemika. Analytic HPLC grade solvents, methanol and formic acid were purchased from Sigma Aldrich and Fluka Biochemika, respectively; potassium dihydrogen phosphate, ammonium dihydrogen phosphate and phosphoric acid were also purchased from Sigma Aldrich. Milli-Q ultrapure water was produced by Sartorius Stedium Biotech mod. Arium (Sartorius, Goettingen, Germany). Cetyltrimethylammonium bromide (cetrimide) was purchased from Extrasynthése (Genay, France), whereas 1,2-phenylenediamine dihydrochloride (OPDA) was purchased from Sigma Aldrich. All polyphenolic and terpenic standards were purchased from Sigma Aldrich. The organic acids were purchased from Fluka Biochemika, whereas ascorbic acid and dehydroascorbic acid were purchased from Extrasynthése. All chemicals specific for electron and optical microscopy were purchased from Electron Microscopy Sciences (Newark, PA, USA), i.e., glutharaldeyde, cacodylate buffer, osmium tetroxide, epon/araldite resin, toluidine "O" and uranyl acetate.

2.9. Statistical Analysis

An arcsin transformation was performed on all percentage incidence data before statistical analysis in order to improve the homogeneity of the variance (Levene test). All the analysed data were checked for the normality of variance. For all the analysed parameters, mean differences were computed using a one-way ANOVA with a Tukey *post hoc* test ($p \leq 0.05$). Mean comparisons between data obtained in soilless and those from the first growing season of a previous work conducted in open field [12] cultivations were performed using an independent samples t-test. All analyses were performed using SPSS 24.0 Inc. software (SPSS Inc., Chicago, IL, USA).

3. Results and Discussion

3.1. Crop Performance, Quality and Secondary Metabolite Content of Saffron in Soilless Cultivation

Soilless cultivation in a glasshouse has been recently proposed as an alternative method to open field cultivation for saffron. Maggio et al. [19] and Gresta et al. [6] reported that, by controlling growth conditions, flowering could be modulated, extended and considerably increased, compared with open field cultivation. In the present study, under protected conditions, flowering had the same duration (ca. 22 days) compared to cultivation of the same corms planted on the same days in a northwestern Italian open field [12], but the saffron flowering moved forward about 20 days (from 5 October 2017 to 23–30 October 2017), in agreement with Gresta et al. [6]. Since, for the flower emergence, corms required to be transferred from 23–27 °C to 17 °C [18], the most likely reason for this results is related to the fact that, in a glasshouse, the lowering of seasonal temperatures takes place more slowly than in an open field. In addition to the temperature lowering, Gresta et al. [52] indicated the soil water content as another environmental component that can trigger flowering. However, as in these two studies object of comparison, the cultivation occurred in different substrates (quartz sand vs soil), it appears not possible to make speculations.

Saffron yield can vary from 0.15 to 1.5 g m^{-2}, based on planting density, plantation age (from one to six year crop cycles), and climatic conditions during the crop season [1]. In this study, an average of 0.55 g m^{-2} was obtained, indicating a profitable production already during the first year. This yield was similar to what obtained cultivating the same corms at a density of 39 corms m^{-2} in a northwestern Italian open field [12] and superior to that obtained in south Italy under similar glasshouse conditions by Gresta et al. [6] (corm density equal to 40 corms m^{-2}; 0.46 g m^{-2}) with corms coming from Sardinia (Italy). With similar corm density to our work, Cavusoglu and Erkel [53] and Maggio et al. [19] obtained much higher yields (0.88 g m^{-2} and 2.34 g m^{-2}, respectively) in glasshouses located in Turkey and south Italy. In Iranian open fields, at a corm density similar to our study, Mollafilabi et al. [24] and Koocheki and Seyyedi [54] obtained an average spice yield of 0.48 g m^{-2}. As affirmed by Gresta et al. [52], to trigger saffron flowering, a not yet fully understood combination of temperature and soil water content is needed.

In addition to the spice yield, another economically important attribute of saffron is the number of replacement corms. The obtained values (2.63 replacement corms $corm^{-1}$) are lower of those obtained by Maggio et al. [19] in soilless cultivation in a cold glasshouse in south Italy, by using peat and perlite (1:1) substrates, where corms produced from 3.0 to 4.5 replacement corms per corm. In addition to a different substrate, these authors also incubated corms in the dark for 83 days before planting. Thus, the combination of these two factors could have guarantee a superior result. Comparing to open field experiments that used corms with similar size to our study, results were in agreement with those from our trial in northwest Italy [12], and the trials performed by Turhan et al. [55] in Turkey (2.32 replacement corms $corm^{-1}$), while superior to those obtained by Koocheki and Seyyedi [54] in Iranian fields (1.32 replacement corms $corm^{-1}$).

Guidelines for the analyses of the main compounds that contribute to the sensory profile of saffron have been established by ISO 3632 regulations [44]. These regulations define procedures to determine these compounds by spectrophotometric analyses and have established the limits by which saffron quality is classified into three different categories (first, second and third). Specifically, the saffron produced under soilless conditions belongs to the highest quality, i.e., first category, for all the studied parameters.

The evaluation of antioxidant activity is generally considered as an important method to evaluate the nutraceutical properties of food, as indicated in other previous studies [30]. Apart from crocins, Karimi et al. [56] and Asdaq and Inamdar [57] highlighted that phenols and flavonols are responsible for the antioxidant potential of saffron. Overall, the saffron produced in soilless systems showed a very high TPC (4445.4 mg_{GAE} $100g^{-1}$ DW), more than the saffron cultivated in other sites in the Alps (range between 1340 and 2355 mg_{GAE} $100g^{-1}$ DW) [12], Lebanon (160 mg_{GAE} $100g^{-1}$ DW) [58], and India (828 mg_{GAE} $100g^{-1}$ DW) [8]. In terms of antioxidant activity, FRAP values were superior to those of Iranian and Italian samples (circa 570 and 1250 mmol Fe^{2+} kg^{-1}) [12,56] and ABTS values were comparable to those found in Italian and Greek saffron by Caser et al. [12] and Ordoudi et al. [59].

3.2. AMF Colonisation

In our study, the presence of AMF and their colonisation of saffron roots were confirmed by observations using light microscopy (Figure 1) and transmission electron microscopy (TEM; Figure 2) on semi-thin and thin sections, respectively. Observations on semi-thin sections, stained in blue, show that the saffron roots are mycorrhised when inoculated with both inocula (Figure 1A–C), confirming the mycorrhizal intensity described in Caser et al. [12]. At the level of the cortical root parenchyma, the typical mycorrhizal arbuscular fungal structures have been highlighted (insets Figure 1A,C). Figure 1 shows the presence of intercellular and intracellular hyphae (Figure 1C) and arbuscules (Figure 1A,B). No fungal structures were found in the roots of the control treatments (Figure 1D).

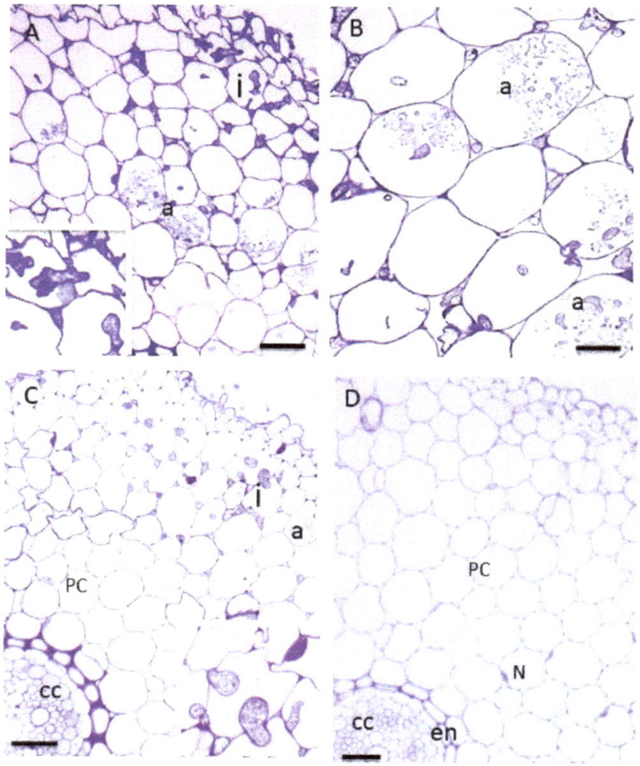

Figure 1. Light microscope images of semi-thin sections of *Crocus sativus* roots inoculated with *Rhizophagus intraradices* and *Funneliformis mosseae* (Ri+Fm, **A**), *R. intraradices* alone (Ri, **B** and **C**) or the control (arbuscular mycorrhizal fungi (AMF)-, **D**), stained with toluidine blue. At the level of the cortical cells, note the presence of intercellular and intracellular hyphae (i) and arbuscules (a). Magnification in insets A and C shows details of the intracellular hyphae. Cortical parenchyma (PC) cells with nucleus (N) are indicated. No fungal structure is present between and inside the root cells in AMF-roots (D). Note the central cylinder (cc) and the endodermide (en). Bars are 20 μm in A, C and D, and 10 μm in B.

Here, the host plasma membrane invaginates and proliferates around all the developing intracellular fungal structures, and cell wall material is laid down between this membrane and the fungal cell surface. The exchange of molecules between the fungal and plant cytoplasm takes place both through their plasma membranes and their cell walls; a functional compartment, known as the symbiotic interface, is thus defined. At the electron microscope level, as seen in Figure 2A,C (arrows), this new apoplastic space, based on membrane proliferation, is evident around the intracellular and arbusculated hyphae of the AMF penetrated inside the saffron root cortical cells. On the basis of TEM observations, we can conclude that the mycorrhizae, formed between saffron roots and the two species of AM fungi in the inocula used in pot experiments, are alive and functionally active.

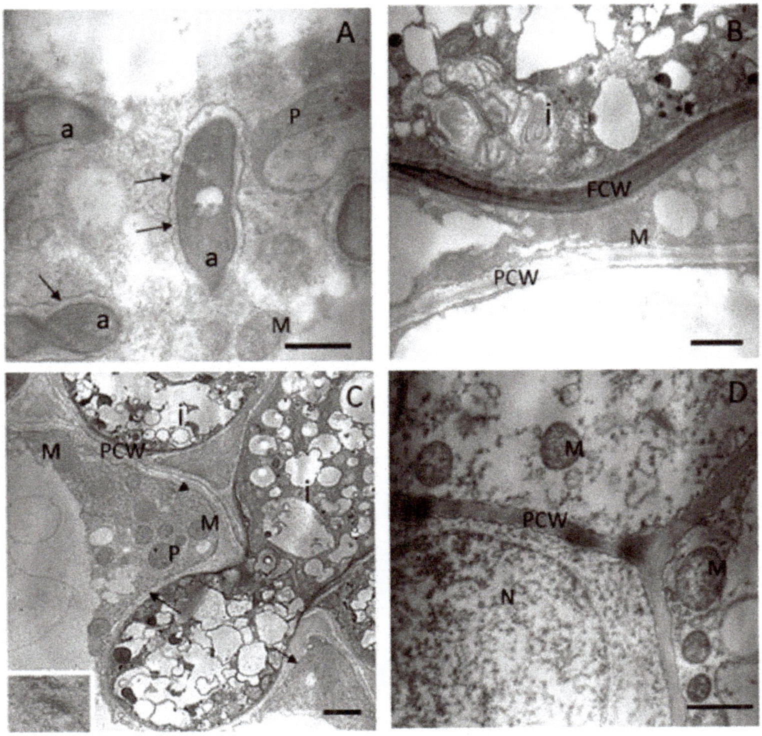

Figure 2. Transmission electron microscopy images of thin sections of saffron roots colonised by *Rhizophagus intraradices* and *Funneliformis mosseae* (Ri+Fm, **A**), *R. intraradices* alone (Ri, **B** and **C**) or the control (AMF-, **D**). In details, a: fungal arbuscule; N: nucleus; M: mithocondria; P: plastids; i: fungal hyphae; PCW: plant cell wall; FCW: fungal cell wall; arrow: plant plasmamembrane; arrowhead and inset: Golgi apparatus. The bar is 1 μm in A, B, C and D.

3.3. Impact of AMF on Saffron in Soilless Cultivation

3.3.1. Crop Performance and Quality Classification

In the present study, slight differences in flowering time and production were detected between treated corms (Figure 3 and Table 2). Both applied inocula (Ri and Ri+Fm) anticipated saffron flowering time of one week, compared to untreated corms (AMF-; 23 October vs. 30 October), whereas the flowering peaks and end of flowering occurred in about the same number of days (6–9 November and 11–13 November, respectively).

No significant differences were observed between the treatments in terms of the number of flowers corm^{-1} and the obtained mg of spice flower^{-1} (Table 2). Very few reports about the effective role of AMF in saffron flowering and yield are available in the literature, and only under open field conditions. Aimo et al. [40] and Caser et al. [12] indicated a positive role of AMF on the saffron productive performance, with an increase in flower production (+68% and +138%, respectively, compared to the untreated corms) using AMF species belonging to the genus *Glomus*.

Table 2. Effects of AMF inoculum composed of *Rhizophagus intraradices* alone (Ri), *R. intraradices* and *Funneliformis mosseae* (Ri+Fm) or the control (AMF-) on yield performances (flower corm^{-1} and saffron flower^{-1}), growth (number of replacement corms corm^{-1}; replacement corm size and weight variation between the end and beginning of the trial) and mean absorbance values for picrocrocin, safranal and crocin of saffron samples obtained during glasshouse cultivation.

	Yield		**Replacement corm**			**Quality category (ISO3632 [44])**		
Treatment	**Flower corm^{-1} (n)**	**Saffron flower^{-1} (mg)**	**Size (%)**	**Corm^{-1} (n)**	**Weight (%)**	**Picrocrocin ($A^{1\%}_{1cm}$ (λ257)**	**Safranal ($A^{1\%}_{1cm}$ (λ330)**	**Crocin ($A^{1\%}_{1cm}$ (λ440)**
Ri	0.84 ± 0.62	6.8 ± 1.3	45.8 ± 4.6a	2.71 ± 1.53	7.8 ± 5.6	143.8 ± 4.6(I)^{β}a	61.0 ± 5.3(I)a	422.6 ± 4.1(I)a
Ri+Fm	0.66 ± 0.60	6.0 ± 1.4	54.6 ± 6.2a	2.25 ± 0.95	8.6 ± 3.8	124.3 ± 3.9(I)c	30.7 ± 3.4(I)c	164.2 ± 3.8(III)c
AMF-	0.97 ± 0.53	6.6 ± 0.4	33.1 ± 6.8b	2.63 ± 1.06	12.6 ± 5.1	135.9 ± 3.4(I)b	54.3 ± 6.7(I)b	324.7 ± 5.9(I)b
p	ns	ns	***	ns	ns	***	***	***

Mean values with the same letter are not statistically different at $p \leq 0.05$, according to a Tukey *post hoc* test. The statistical relevance of 'Between-Subjects Effects' tests (***$p < 0.001$, ns = not significant). ^{β} The quality category (ISO3632) is indicated in brackets. The limits for the first (I) quality category are: picrocrocin>70; safranal 20–50; crocins>200. ISO3632 limits for the second (II) quality category are: picrocrocin >55; safranal 20–50; crocins >170. ISO3632 limits for the third (III) category are: picrocrocin>40; safranal 20–50; crocins>120.

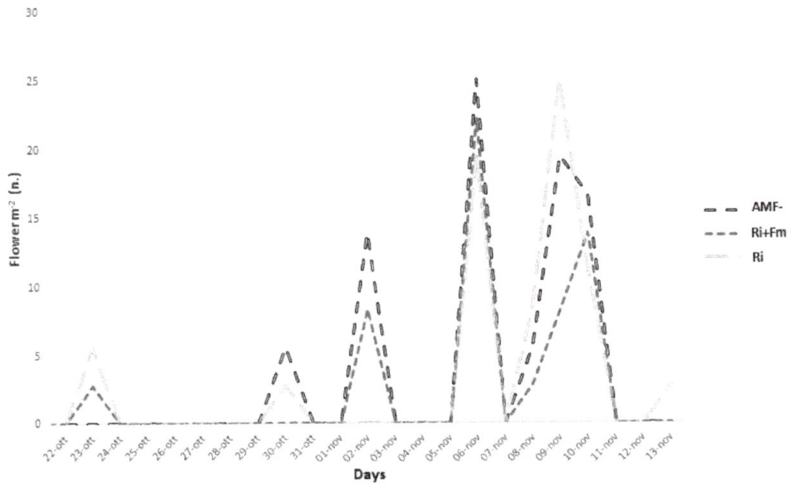

Figure 3. Effects of AMF inoculum composed of *Rhizophagus intraradices* alone (Ri), *R. intraradices* and *Funneliformis mosseae* (Ri+Fm) or the control (AMF-) on the flowering calendar of *Crocus sativus* corms and the daily number of picked flowers m^{-2} during soilless cultivation.

Both of the AMF inocula increased the size of replacement corms in comparison to untreated corms (Table 2), suggesting a positive effect on flower production for the following cultivation cycle, in agreement with Aimo et al. [40] and Mohebi-Anabat et al. [39]. Corm size is indeed a major factor in bulbous plants to determine the flowering capacity and production of new replacement corms [5,42].

Saffron quality greatly depends on the growing conditions [12,60]. In the present study, among the AMF inocula, *R. intraradices* alone significantly increased the content of picrocrocin (bitterness), safranal (flavouring strength) and crocins (colouring strength), in comparison to the other treatments. On the contrary, Ri+Fm significantly reduced the content of these molecules and, thus the quality of the spice, in particular by lowering the crocin content to the third category of ISO 3632. To the best of our knowledge, this is the first report indicating the effect of AMF on the quality (ISO) of saffron obtained by soilless cultivation. The positive role of Ri on the increase of the saffron quality, especially on the content of picrocrocin, was highlighted also in northwestern Italian open field [12]. Thus, the corm inoculation with Ri could further increase the already high quality saffron produced in the Italian Alps [45,61].

3.3.2. Saffron Metabolic Profiling Comparing to Other Foods

In addition to the peculiar organoleptic characteristics, the stigmas of the *C. sativus* flower contain many secondary metabolites with demonstrated pharmacological effects [3,11,62–64]. The identification and quantification of bioactive compounds in saffron and the evaluation of their biological activities are important to gauge their potential efficacy in food and pharmaceutical industries [65]. The range of all chemicals can vary greatly as a result of growing conditions, such as in response to the application of biostimulants [63]. Inoculation with AMF is known to alter the production of secondary metabolites in MAPs, both in roots, shoots, and flowers, even if is not consistent among plant organs [66]. The effects of AMF inocula on the biosynthesis of secondary metabolites in saffron are presented in Table 3. This more in-depth analysis confirmed the results obtained by assessing the spice quality according to ISO3632 guidelines. The single species inoculum Ri significantly increased the content of crocins (crocin I and II), whereas the mix Ri+Fm decreased it; these findings are in agreement with those obtained by Caser et al. [12] under field conditions in a temperate mountain area (north-west Italy), where the saffron obtained by corms inoculated with Ri resulted in superior quality (i.e., quality compared to the ISO standards). Regarding antioxidant activity (AOA), inoculation with Ri resulted in superior values in both used methods (FRAP and ABTS). The AMF inoculum composed of Ri+Fm significantly increased the contents of isoquercitrin and the total phenolic (TPC) compared to Ri, while of ellagic acid in comparison to Ri and AMF-. Differences in results according to the AMF inoculum composition were also observed in other plant species cultivated on different substrates. Among the reviewed studies, it has been found that the single inoculation of *R. intraradices* tend to be more successful for bioactive compounds increase than inoculation experiments with more than one species applied at the same time. In *Echinacea purpurea* Moench. [67] cultivated in a sand and soil (1:1) substrate, *R. intraradices* alone increased more the content of polyphenols than the mixed inoculum, while in *Cynara cardunculus* L. cultivated in sandy soil [68] and *Lactuca sativa* L. cultivated in a mixture of peat, sandy loam soil and calcinated clay (1:1:1) [69] *R. intraradices* enhanced more the antioxidant activity. However, it has not been observed any effect on the accumulation of polyphenols in *Ocimum basilicum* L. cultivated in a sterilised sand and soil (3:1) substrate [70] and in *Salvia officinalis* L. in sand, soil, and expanded clay (1:1:1) [71,72].

Table 3. Bioactive compounds, anthocyanins, total polyphenol content and antioxidant activity (ferric reducing antioxidant power (FRAP) and 2,2′-azinobis(3-ethylbenzothiazoline-6-sulfonic acid) (ABTS); antioxidant activity (AOA)) of the saffron produced via glasshouse cultivation with AMF inocula composed of *Rhizophagus intraradices* alone (Ri), *R. intraradices* and *Funneliformis mosseae* (Ri+Fm) or a control (AMF-).

Class	Compound (mg $100g^{-1}$ DW)	Ri	Ri+Fm	AMF-	*p*
Cinnamic acids	Coumaric acid	23.4 ± 3.5	23.7 ± 2.6	23.7 ± 3.1	ns
Flavonols	Isoquercitrin	1.9 ± 0.3b	2.6 ± 0.2a	2.3 ± 0.3ab	**
	Quercitrin	17.8 ± 4.6	11.6 ± 4.1	19.1 ± 3.6	ns
Benzoic acids	Gallic acid	4.5 ± 1.5	5.1 ± 1.3	4.9 ± 1.4	ns
	Ellagic acid	1.9 ± 0.5b	3.2 ± 0.3a	1.0 ± 0.4b	**
Catechins	Catechin	1.9 ± 0.4	1.6 ± 0.3	1.8 ± 0.3	ns
	Epicatechin	9.8 ± 2.9	5.9 ± 2.1	9.6 ± 2.5	ns
Carotenoids	Safranal	4.0 ± 0.9	4.0 ± 1.2	4.0 ± 0.7	ns
	Crocin I	104.2 ± 8.6a	22.1 ± 6.5c	55.5 ± 8.4b	***
	Crocin II	42.7 ± 9.6a	16.4 ± 3.8b	38.7 ± 12.9ab	**
Vitamin C	Dehydroascorbic acid	28.8 ± 6.5	30.2 ± 4.1	31.8 ± 6.9	ns
	Ascorbic acid	31.1 ± 9.5	36.3 ± 6.7	41.7 ± 4.8	ns
	Total vitamin C	59.9 ± 10.2	66.5 ± 5.9	73.6 ± 8.4	ns
TAC	Anthocyanin (mg_{C3G} $100g^{-1}$ DW)	640.7 ± 84.6b	146.4 ± 29.8c	1654.5 ± 68.4a	*
	Methods				
TPC	Folin–Ciocalteu (mg_{GAE} $100g^{-1}$ DW)	816.5±152.7b	3619.0±400.2a	4445.4±450.2a	***
AOA	FRAP (mmol Fe^{2+} kg^{-1} DW)	3133.9±1524.3a	1383.0±589.7ab	379. 7±128.4b	**
	ABTS (μmol_{TE} g^{-1} DW)	5.4±0.8a	3.6±0.4c	4.5±0.7ab	**

Mean values with the same letter are not statistically different at $p \leq 0.05$, according to a Tukey *post hoc* test. The statistical relevance of 'Between-Subjects Effects' tests (* $p < 0.05$, ** $p < 0.01$, *** $p < 0.001$, ns = not significant).

Karimi et al. [56] and Rahaiee et al. [63] indicated that the antioxidant capacities of saffron might be due to the presence of total phenolics and flavonoids. Based on the obtained results, the content of the bioactive compounds detected in saffron could be compared to other commonly eaten fruits with highly advantageous nutritive properties. Saffron had a higher total phenol content (TPC) and antioxidant activity (AOA) than fresh *Ribes nigrum* L. berries (circa +1000% and +493%, respectively), and fresh (circa +2000% and +1800%, respectively) and dried (circa +900% and +1650%, respectively) *Lycium* spp. fruits [65,73], analysed with the same method. Since saffron showed an antioxidant activity superior than 500 mg_{GAE} $100g^{-1}$ it could be also listed within the health beneficial fruits such as *Rubus glaucus* Berth. and *Prunus serotina* var. Capulì as suggested by Vasco et al. [74]. Its content of vitamin C was similar to what found in *Actinidia deliciosa* (A.Chev.) C.F.Liang & A.R.Ferguson and *Citrus sinensis* (L.) Osb., and even higher than in *Lycium* spp. (+150%) and *Vaccinium* spp. (+580%). Also, the coumaric acid content was superior (+85%) than in *Morus nigra* L. fruits [75] while lower than in *Lycium* spp. fruits, that showed also higher content of gallic acid, ellagic acid, catechin, and epicatechin was generally lower in saffron (on average circa −75%, −70%, −92%, and −95%, respectively) [73,75]. Lastly, the content of anthocyanins, that are suggested to have neuroprotective properties [76], was up to 11654.5 mg_{C3G} $100g^{-1}$ DW, i.e., a value very high in comparison to fresh fruit extracts from *Morus nigra*, *Rubus idaeus* L., and *Fragaria ananassa* D. (80.0, 33.7, and 35.2 mg_{C3G} $100g^{-1}$, respectively) [75].

3.3.3. Soilless Cultivation vs. Open Field

Saffron root colonisation by AMF could be affected by the cultivation conditions related to the substrate composition, root temperature or the presence of antagonistic fungi naturally occurring in the soil [31,40,41,76]. In our recent studies, AM fungal colonisation was noted in *C. sativus* roots inoculated with Ri and Ri+Fm, both in soilless (Figures 1 and 2) and in open field conditions [12]. Figures 4 and 5 report the comparisons of the results obtained by these studies. Compared to open field, in soilless conditions not-inoculated corms (AMF-) showed similar spice yields but with higher quality while, referring to AMF treatments, Ri-inoculated corms produced less spice but with a higher quality, whereas Ri+Fm inoculated corms produced less spice, with a lower quality (i.e., reduction in crocin content).

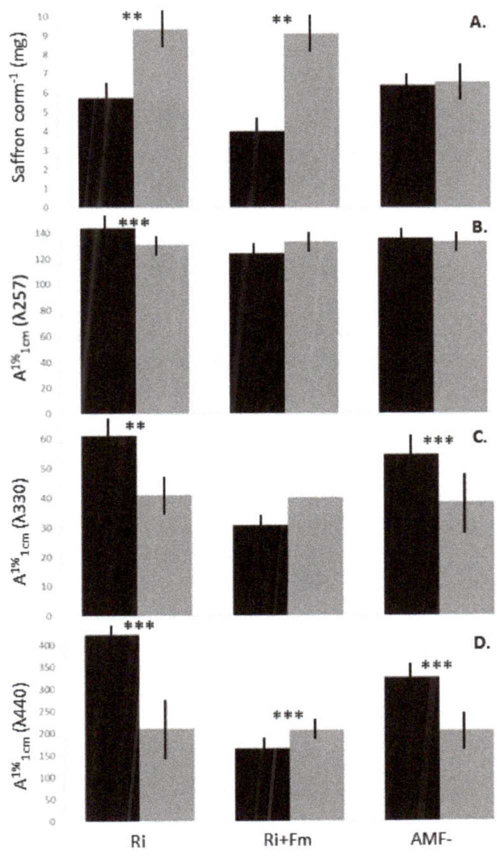

Figure 4. Effects of AMF inoculum consisting of *Rhizophagus intraradices* alone (Ri), *R. intraradices* and *Funneliformis mosseae* (Ri+Fm) or a control (AMF-) on (**A.**) mg of saffron $corm^{-1}$, (**B.**) picrocrocin, (**C.**) safranal, and (**D.**) crocin of *Crocus sativus* corms cultivated in soilless (black bars) and open field (grey bars, [12]) conditions. Mean comparisons of each treatment in the two cultivation types were performed using an independent samples t-test.

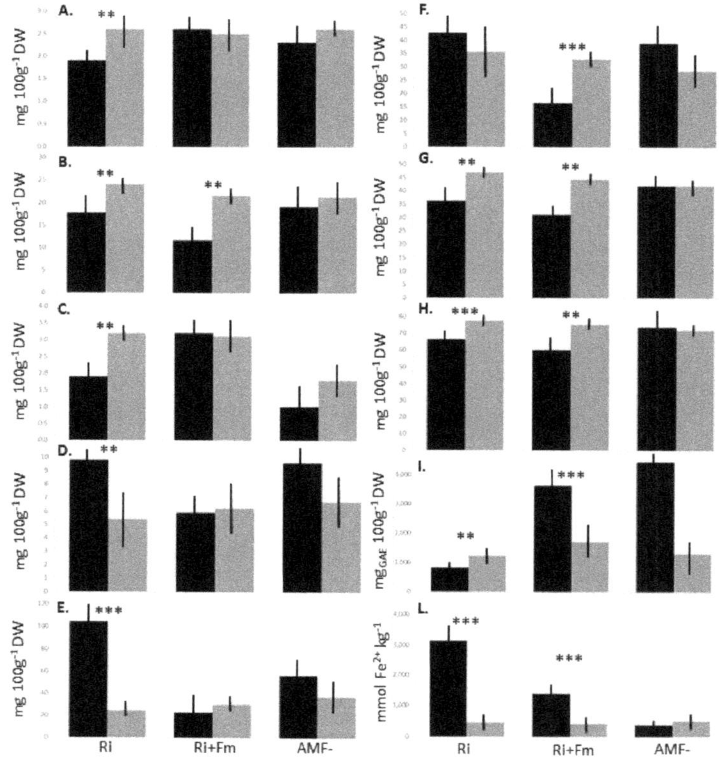

Figure 5. Effects of AMF inoculum consisting of *Rhizophagus intraradices* alone (Ri), *R. intraradices* and *Funneliformis mosseae* (Ri+Fm) or a control (AMF-) on the content of (**A.**) isoquercitrin, (**B.**) quercitrin, (**C.**) ellagic acid, (**D.**) epicatechin, (**E.**) crocin I, (**F.**) crocin II, (**G.**) ascorbic acid, (**H.**) vitamin C, (**I.**) total polyphenol content (TPC), and (L.) antioxidant activity (FRAP assay) of saffron produced in soilless (black bars) and open field (grey bars, [12]) conditions. Mean comparisons of each treatment in the two cultivation types were performed using an independent samples t-test.

With respect to the nutraceutical compounds, the comparisons are presented in Figure 5. No differences were reported between the untreated corms (AMF-), whereas the application of Ri in the soilless condition induced an increase in the contents of epicatechin, crocin I, and antioxidant activity (+80%, +435%, and +675%, respectively), while a decrease in the contents of isoquercitrin, quercitrin, ellagic acid, ascorbic acid, vitamin C, and TPC. Fewer differences were induced by Ri+Fm, which positively stimulated both the total phenolic content and antioxidant activity (+210% and +325%, respectively), but caused a decrease in quercitrin, crocin II, ascorbic acid, and vitamin C.

4. Conclusions

Soilless cultivation in a glasshouse appeared as an effective strategy for the cultivation of saffron with a first-year cultivation spice yield that is comparable with open field production sites. Moreover, the high quality saffron produced via soilless cultivation presented an elevated content of several health-promoting compounds with highly advantageous nutritive properties, such as polyphenols and elevated antioxidant activity. Further studies are needed to define better the methodologies to modulate time and duration of flowering, to improve yield, and to efficiently schedule harvest practices.

Arbuscular mycorrhizal-based products have received great interest in agriculture for their potential to improve crop productivity, nutritional quality, as well as resistance to plant pathogens and numerous environmental stresses. The literature highlights that AMF must be chosen by evaluating different aspects, such as the inoculum type, host plants, and the environmental and growing conditions.

Here, AMF successfully colonised *C. sativus* roots; their effects varied on the basis of inoculum type and cultivation conditions. Among the studied AMF inocula, *R. intraradices* appeared to give more benefits to *C. sativus* than the mix of *R. intraradices* and *F. mosseae*. Specifically, the *R. intraradices* inoculation appeared successful in open field to increase spice yields while in soilless systems to increase the spice quality.

Thus, soilless systems appeared as an effective alternative cultivation strategy for the production of high quality saffron. Further benefits can be obtained by the application of targeted AMF-based biostimulants. A cost-benefit analysis should be performed to assess the economic sustainability.

Author Contributions: M.C., E.L., V.B. and V.S. contributed to the experimental design. M.C., S.D., Í.M.M.V., D.D., A.F. and V.S. acquired and interpreted the data. M.C. drafted the manuscript. V.B. and V.S. conceived the study, coordinated the work and critically revised the manuscript.

Funding: This research was funded by the project titled 'SaffronALP—Lo zafferano di montagna: tecniche sostenibili per una produzione di qualità' - Fondazione Cassa di Risparmio di Torino (RF=2017.1966) and by the program Interreg V-A Francia Italia Alcotra "Attività innovative per lo sviluppo della filiera transfrontaliera del fiore edule–Antea' n. 1139.

Acknowledgments: The authors acknowledge Dario Sacco for statistical assistance and Alessandro Putzolu for technical assistance.

Conflicts of Interest: The authors declare no conflict of interest.

References

1. Gresta, F.; Santonoceto, C.; Avola, G. Crop rotation as an effective strategy for saffron (*Crocus sativus* L.) cultivation. *Sci. Hortic.* **2016**, *211*, 34–39. [CrossRef]
2. Maleki, F.; Kazemi, H.; Siahmarguee, A.; Kamkar, B. Development of a land use suitability model for saffron (*Crocus sativus* L.) cultivation by multi-criteria evaluation and spatial analysis. *Eco Eng.* **2017**, *106*, 140–153. [CrossRef]
3. Giorgi, A.; Scheurer, T. Alpine resources: Assets for a promising future-conclusions from the forum alpinum. *Mount. Res. Dev.* **2015**, *35*, 414–415. [CrossRef]
4. Husaini, A.; Shafiq, A.; Parvaze, S.; Rather, G.; Parray, A.; Shikari, A.; Javid, M. Bioinformatics for saffron (*Crocus sativus* L.) improvement. *Comm. Biomet. Crop. Sci.* **2009**, *4*, 3–8.
5. Gresta, F.; Lombardo, G.; Siracusa, L.; Ruberto, G. Saffron an alternative crop for sustainable agricultural systems. A review. *Agron. Sustain. Dev.* **2008**, *28*, 95–112. [CrossRef]
6. Gresta, F.; Avola, G. Hypothesis on the relation between meteorological data and flowering peaks in saffron (*Crocus sativus L.*) crop. *Acta Hortic.* **2017**, *1184*, 189–194. [CrossRef]
7. Elgazar, A.; Rezq, A.; Bukhari, H. Anti-Hyperglycemic Effect of Saffron Extract in Alloxan-Induced Diabetic Rats. *Europ. J. Biol. Sci.* **2013**, *5*, 14–22.
8. Baba, S.; Mohiudin, T.; Basu, S.; Swarnkar, M.; Malik, A.; Abbas, N.; Singh, A.; Ashraf, N. Comprehensive transcriptome analysis of *Crocus sativus* for discovery and expression of genes involved in apocarotenoid biosynthesis. *BMC Genom.* **2015**, *16*, 698. [CrossRef] [PubMed]
9. Lahmass, I.; Touria, L.; Delporte, C.; Sikdar, S.; Antwerpen, P.; Saalaoui, E.; Mégalizzi, V. The waste of saffron crop, a cheap source of bioactive compounds. *J. Funct. Foods* **2017**, *35*, 341–351. [CrossRef]
10. Cusano, E.; Consonni, R.; Petrakis, E.A.; Astraka, K.; Cagliani, L.R.; Polissiou, M.G. Integrated analytical methodology to investigate bioactive compounds in *Crocus sativus* L. flowers. *Phytochem. Anal. Phytochem. Analysis* **2018**, *29*, 1–11. [CrossRef]
11. Menghini, L.; Leporini, L.; Vecchiotti, G.; Locatelli, M.; Carradori, S.; Ferrante, C.; Zengin, G.; Recinella, L.; Chiavaroli, A.; Leone, S.; et al. *Crocus sativus* L. stigmas and byproducts: Qualitative fingerprint, antioxidant potentials and enzyme inhibitory activities. *Food Res. Int.* **2018**, *109*, 91–98. [CrossRef]
12. Caser, M.; Maxaieie Victorino, Í.M.; Demasi, S.; Berruti, A.; Donno, D.; Lumini, E.; Bianciotto, V.; Scariot, V. Saffron cultivation in marginal Alpine environments: How AMF inoculation modulates yield and bioactive compounds. *Agronomy* **2019**, *9*, 12. [CrossRef]

13. Talaei, A.; Moghadam, H.; Tabassi, S.; Mohajeri, S. Crocin, the main active saffron constituent, as an adjunctive treatment in major depressive disorder: A randomized, double-blind, placebo-controlled, pilot clinical trial. *J. Affect. Disord.* **2015**, *174*, 51–56. [CrossRef]
14. Landi, R. *Lo zafferano: Tradizione e qualità. Camera di commercio: Industria, Artigianato e Agricoltura di Firenze*; CCIAA (Camera di Commercio: Industria, Artigianato e Agricoltura): Firenze, Italy, 2007; Volume 4, p. 65.
15. Lachaud, C. *La Bible du Safranier. Tout savoir sur le Crocus sativus et sur le Safran*, 1st ed.; La Paradisière: Correze, France, 2012; ISBN 978-2-7466-4412-0.
16. De Juan, J.A.; Córcoles, H.L.; Muñoz, R.M.; Picornell, M.R. Yield and yield components of saffron under different cropping systems. *Ind. Crops Prod.* **2009**, *30*, 212–219. [CrossRef]
17. Yarami, N.; Kamgar-Haghighi, A.A.; Sepaskhah, A.R.; Zand-Parsa, S. Determination of the potential evapotranspiration and crop coefficient for saffron using a water-balance lysimeter. *Arch. Agr. Soil Sci.* **2011**, *57*, 727–740. [CrossRef]
18. Molina, R.; Valero, M.; Navarro, Y.; Guardiola, J.; García-Luis, A. Temperature effects on flower formation in saffron (*Crocus sativus* L.). *Sci. Hortic.* **2005**, *103*, 361–379. [CrossRef]
19. Maggio, A.; Raimondi, G.; Martino, A.; De Pascale, S. Soilless cultivation of saffron in mediterranean environment. *Acta Hortic.* **2006**, *718*, 515–522. [CrossRef]
20. Benschop, M. *Crocus*; De Hertogh, A., Le Nard, M., Eds.; The Physiology of Flower Bulbs, Elsevier Science Publishers: Amsterdam, The Netherlands, 1993; pp. 257–272.
21. De Mastro, G.; Ruta, C. Relation between corm size and saffron (*Crocus sativus* L.) flowering. *Acta Hortic.* **1993**, *334*, 512–517. [CrossRef]
22. Douglas, M.H.; Smallfield, B.M.; Wallace, A.R.; McGimpsey, J.A. Saffron (*Crocus sativus* L.): The effect of mother corm size on progeny multiplication, flower and stigma production. *Sci. Hortic.* **2014**, *166*, 50–58. [CrossRef]
23. Kumar, R. Calibration and validation of regression model for non-destructive leaf area estimation of saffron (*Crocus sativus* L.). *Sci. Hortic.* **2009**, *122*, 142–145. [CrossRef]
24. Mollafilabi, A.; Koocheki, A.; Moeinerad, H.; Kooshki, L. Effect of plant density and weight of corm on yield and yield components of saffron (*Crocus sativus* L.) under soil, hydroponic and plastic tunnel cultivation. *Acta Hortic.* **2013**, *997*, 51–58. [CrossRef]
25. Putra, P.A.; Yuliando, H. Soilless culture system to support water use efficiency and product quality: A review. *Agric. Agricult. Sci. Proc.* **2015**, *3*, 283–288. [CrossRef]
26. Augé, R.M.; Toler, H.D.; Saxton, A.M. Arbuscular mycorrhizal symbiosis alters stomatal conductance of host plants more under drought than under amply watered conditions: A meta-analysis. *Mycorrhiza* **2015**, *25*, 13–24. [CrossRef] [PubMed]
27. Helal Beigi, Y.; Khoshgoftar, M.; Shamsi, A.; Zamani, N. Possibility of saffron corm growth in different bed of cultivation in soilless system. In *Proceedings of 1st Congress of Hydroponic in Greenhouse Production*; Isfahan University of Technology: Isfahan, Iran, 2009.
28. Sourell, F.; Weathers, P. The growth of saffron (*Crocus sativus L.*) in aeroponics and hydroponics. *J. Herb. Spices Med. Plants* **2000**, *7*, 113–127.
29. Du Jardin, P. Plant biostimulants: Definition, concept, main categories and regulation. *Sci. Hortic.* **2015**, *196*, 3–14. [CrossRef]
30. Rouphael, Y.; Franken, P.; Schneider, C.; Schwarz, D.; Giovannetti, M.; Agnolucci, M.; De Pascale, S.; Bonini, P.; Colla, G. Arbuscular mycorrhizal fungi acta s biostimulants in horticultural crops. *Sci. Hortic.* **2015**, *196*, 91–108. [CrossRef]
31. Berruti, A.; Lumini, E.; Balestrini, R.; Bianciotto, V. Arbuscular mycorrhizal fungi as natural biofertilizers: Let's benefit from past successes. *Front. Microbiol.* **2016**, *6*, 1559. [CrossRef]
32. Smith, S.E.; Read, D.J. *Mycorrhizal Symbiosis*; Academic Press: London, UK, 1997; Volume 40, p. 154.
33. Todeschini, V.; Lahmidi, N.A.; Mazzucco, E.; Marsano, F.; Gosetti, F.; Robotti, E.; Bona, E.; Massa, N.; Bonneau, L.; Marengo, E.; et al. Impact of beneficial microorganisms on strawberry growth, fruit production, nutritional quality, and volatilome. *Front. Plant. Plant Sci.* **2018**, *9*, 1611. [CrossRef] [PubMed]
34. Pellegrino, E.; Bedini, S. Enhancing ecosystem services in sustainable agriculture: Biofertilization and biofortification of chickpea (*Cicer arietinum* L.) by arbuscular mycorrhizal fungi. *Soil Biol. Biochem.* **2014**, *68*, 429–439. [CrossRef]

35. Zeng, Y.; Guo, L.; Chen, B.; Hao, Z.; Wang, J.; Huang, L.; Yang, G.; Cui, X.; Yang, L.; Wu, Z.; et al. Arbuscular mycorrhizal symbiosis and active ingredients of medicinal plants: Current research status and prospectives. *Mycorrhiza* **2013**, *23*, 253–265. [CrossRef] [PubMed]
36. Giovannetti, M.; Avio, L.; Sbrana, C. *Improvement of nutraceutical value of food by plant symbionts*; Ramawat, K.G., Merillon, J.M., Eds.; Natural Products, Phytochemistry, Botany and Metabolism of Alkaloids, Phenolics and Terpenes; Springer: Berlin/Heidelberg, Germany, 2013; pp. 2641–2662.
37. Swamy, M.K.; Akhtar, M.S.; Sinniah, U.R. *Response of PGPR and AM fungi toward growth and secondary metabolite production in medicinal and aromatic plants*; Hakeem, K., Akhtar, M., Eds.; Plant, Soil and Microbes; Springer: Heidelberg, Germany, 2016.
38. Avio, L.; Turrini, A.; Giovannetti, M.; Sbrana, C. Designing the ideotype mycorrhizal symbionts for the production of healthy food. *Front. Plant Sci.* **2018**, *9*, 1089. [CrossRef] [PubMed]
39. De Sousa Oliveira, M.; da Silva Campos, M.; de Albuquerque, U.P.; da Silva, F.S.B. Arbuscular mycorrhizal fungi (AMF) affects biomolecules content in *Myracrodruon urundeuva* seedlings. *Ind. Crops Prod.* **2013**, *50*, 244–247. [CrossRef]
40. Aimo, S.; Gosetti, F.; D´Agostino, G.; Gamalero, E.; Gianotti, V.; Bottaro, M.; Gennaro, M.; Berta, G. Use of arbuscular mycorrhizal fungi and beneficial soil bacteria to improve yield and quality of saffron (*Crocus sativus* L.). *Acta Hortic.* **2010**, *850*, 159–162. [CrossRef]
41. Lone, R.; Shuab, R.; Koul, K. AMF association and their effect on metabolite mobilization, mineral nutrition and nitrogen assimilating enzymes in saffron (*Crocus sativus*) plant. *J. Plant Nutr.* **2016**, *39*, 1852–1862. [CrossRef]
42. Shajari, M.A.; Moghaddam, P.R.; Ghorbani, R.; Koocheki, A. Increasing saffron (*Crocus sativus* L.) corm size through the mycorrhizal inoculation, humic acid application and irrigation managements. *J. Plant Nutr.* **2018**, *41*, 1047–1064. [CrossRef]
43. Mohebi-Anabat, M.; Riahi, H.; Zanganeh, S.; Sadeghnezhad, E. Effects of arbuscular mycorrhizal inoculation on the growth, photosynthetic pigments and soluble sugar of *Crocus sativus* L. (saffron) in autoclaved soil. *Int. J. Agron. Agric. Res.* **2015**, *6*, 296–304.
44. ISO 3632-1:2011. *Spices—Saffron (Crocus sativus L.)*; Food products SC 7, Spices, culinary herbs and condiments; International Organization for Standardization: Geneva, Switzerland, September 2011.
45. Giorgi, A.; Pentimalli, D.; Giupponi, L.; Panseri, S. Quality traits of saffron (*Crocus sativus* L.) produced in the Italian Alps. *Open Agriculture* **2017**, *2*, 52–57. [CrossRef]
46. Donno, D.; Mellano, M.G.; Prgomet, Z.; Beccaro, G.L. Advances in *Ribes* x *nidigrolaria* Rud. Bauer & A. Bauer fruits as potential source of natural molecules: A preliminary study on physico-chemical traits of an underutilized berry. *Sci. Hortic.* **2018**, *237*, 20–27.
47. Donno, D.; Mellano, M.G.; Prgomet, Z.; Cerutti, A.K.; Beccaro, G.L. Phytochemical characterization and antioxidant activity evaluation of Mediterranean medlar fruit (*Crataegus azarolus* L.): Preliminary study of underutilized genetic resources as a potential source of health-promoting compound for food supplements. *J. Food Nutrit. Res.* **2017**, *56*, 18–31.
48. Caser, M.; D'Angiolillo, F.; Chitarra, W.; Lovisolo, C.; Ruffoni, B.; Pistelli, L.; Pistelli, L.; Scariot, V. Ecophysiological and phytochemical responses of *Salvia sinaloensis* Fern. to drought stress. *Plant Growth Regul.* **2018**, *84*, 383–394. [CrossRef]
49. Caser, M.; Chitarra, W.; D'Angiolillo, F.; Perrone, I.; Demasi, S.; Lovisolo, C.; Pistelli, L.; Pistelli, L.; Scariot, V. Drought stress adaptation modulates plant secondary metabolite production in *Salvia dolomitica* Codd. *Ind. Crop Prod.* **2019**, *129*, 85–96. [CrossRef]
50. Hoch, H.C. *Freeze-substitution of fungi*; Aldrich, H.C., Todd, W.J., Eds.; Ultrastructure techniques of microorganisms; Plenum Press: New York, NY, USA, 1986; pp. 183–211.
51. Reynolds, E.S. The use of lead citrate at high pH as an electron-opaque stain in electron microscopy. *J. Cell. Biol.* **1963**, *17*, 208–212. [CrossRef]
52. Gresta, F.; Avola, G.; Lombardo, G.M.; Siracusa, L.; Ruberto, G. Analysis of flowering, stigmas yield and qualitative traits of saffron (*Crocus sativus* L.) as affected by environmental conditions. *Sci. Hortic.* **2009**, *119*, 320–324.
53. Cavusoglu, A.; Erkel, E. The effect of different planting areas and corm size on yield and harvest period of saffron (*Crocus sativus* L.) in Kocaeli province. *J. Akdeniz Univ. Faculty Agric.* **2005**, *18*, 179–184.

54. Koocheki, A.; Seyyedi, S. How irrigation rounds and mother corm size control saffron yield, quality, daughter corms behavior and phosphorus uptake. *Sci. Hortic.* **2016**, *213*, 132–143. [CrossRef]
55. Turhan, H.; Kahriman, F.; Egesel, C.; Kemal Gul, M. The effects of different growing media on flowering and corm formation of saffron (*Crocus sativus* L.). *Afr. J. Biotech.* **2007**, *6*, 2328–2332.
56. Karimi, E.; Oskoueian, E.; Hendra, R.; Jaafar, H. Evaluation of Crocus sativus L. stigma phenolic and flavonoid compounds and its antioxidant activity. *Molecules* **2010**, *15*, 6244–6256. [CrossRef]
57. Asdaq, S.; Inamdar, M. Potential of garlic and its active constituent, S-allyl cysteine, as antihypertensive and cardioprotective in presence of captopril. *Phytomedicine* **2010**, *17*, 1016–1026. [CrossRef] [PubMed]
58. Makhlouf, H.; Saksouk, M.; Habib, J.; Chahine, R. Determination of antioxidant activity of saffron taken from the flower of *Crocus sativus* grown in Lebanon. *Afr. J. Biotech.* **2011**, *10*, 8093–8100.
59. Ordoudi, S.; Befani, C.; Nenadis, N.; Koliakos, G.; Tsimidou, M. Further Examination of Antiradical Properties of *Crocus sativus* Stigmas Extract Rich in Crocins. *J. Agric. Food Chem.* **2009**, *57*, 3080–3086. [CrossRef]
60. Siracusa, L.; Gresta, F.; Avola, G.; Lombardo, G.M.; Ruberto, G. Influence of corm provenance and environmental condition on yield and apocarotenoid profiles in saffron (*Crocus sativus* L.). *J. Food Comp. Anal.* **2010**, *23*, 394–400. [CrossRef]
61. Manzo, A.; Panseri, S.; Bertoni, D.; Giorgi, A. Economic and qualitative traits of Italian Alps saffron. *J. Mt. Sci.* **2015**, *12*, 1542–1550. [CrossRef]
62. Sánchez-Vioque, R.; Rodríguez-Conde, M.F.; Reina-Ureña, J.V.; Escolano-Tercero, M.A.; Herraiz-Peñalver, D.; Santana-Méridas, O. In vitro antioxidant and metal chelating properties of corm, tepal and leaf from saffron (*Crocus sativus* L.). *Ind. Crop Prod.* **2012**, *39*, 149–153. [CrossRef]
63. Rahaiee, S.; Moini, S.; Hashemi, M.; Shojaosadati, S. Evaluation of antioxidant activities of bioactive compounds and various extracts obtained from saffron (*Crocus sativus* L.): A review. *J. Food Sci. Tech.* **2015**, *52*, 1881–1888. [CrossRef] [PubMed]
64. Shahi, T.; Assadpour, E.; Jafari, S. Main chemical compounds and pharmacological activities of stigmas and tepals of 'red gold'; saffron. *Trends Food Sci. Technol.* **2016**, *58*, 69–78. [CrossRef]
65. Donno, D.; Mellano, M.G.; Raimondo, E.; Cerutti, A.K.; Prgomet, Z.; Beccaro, G.L. Influence of applied drying methods on phytochemical composition in fresh and fried goji fruits by HPLC fingerprint. *Eur. Food Res. Technol.* **2016**, *242*, 1961–1974. [CrossRef]
66. Heidari, M.; Karami, V. Effects of different mycorrhiza species on grain yield, nutrient uptake and oil content of sunflower under water stress. *J. Saudi Soc. Agric. Sci.* **2014**, *13*, 9–13. [CrossRef]
67. Araim, G.; Saleem, A.; Arnason, J.T.; Charest, C. Root colonization by arbuscular mycorrhizal (AM) fungus increases growth and secondary metabolism of purple coneflower, *Echinacea purpurea* (L.) Moench. *J. Agric. Food Chem.* **2009**, *57*, 2255–2258. [CrossRef]
68. Ceccarelli, N.; Curadi, M.; Martelloni, L.; Sbrana, C.; Picciarelli, P.; Giovannetti, M. Mycorrhizal colonization impacts on phenolic content and antioxidant properties of artichoke leaves and flower heads two years after field transplant. *Plant Soil* **2010**, *335*, 311–323. [CrossRef]
69. Avio, L.; Sbrana, C.; Giovannetti, M.; Frassinetti, S. Arbuscular mycorrhizal fungi affect total phenolics content and antioxidant activity in leaves of oak leaf lettuce varieties. *Sci. Hortic.* **2017**, *224*, 265–271. [CrossRef]
70. Toussaint, J.P.; Smith, F.A.; Smith, S.E. Arbuscular mycorrhizal fungi can induce the production of phytochemicals in sweet basil irrespective of phosphorus nutrition. *Mycorrhiza* **2007**, *17*, 291–297. [CrossRef]
71. Nell, M.; Vötsch, M.; Vierheilig, H.; Steinkellner, S.; Zitterl-Eglseer, K.; Franz, C.; Novak, J. Effect of phosphorus uptake on growth and secondary metabolites of garden (*Salvia officinalis* L.). *J. Sci. Food Agric.* **2009**, *89*, 1090–1096. [CrossRef]
72. Geneva, M.P.; Stancheva, I.V.; Boychinova, M.M.; Mincheva, N.H.; Yonova, P.A. Effects of foliar fertilization and arbuscular mycorrhizal colonization on *Salvia officinalis* L. growth, antioxidant capacity, and essential oil composition. *J. Sci. Food Agric.* **2010**, *90*, 696–702.
73. Donno, D.; Beccaro, G.; Mellano, M.; Cerutti, A.; Bounous, G. Goji berry fruit (*Lycium* spp.): Antioxidant compound fingerprint and bioactivity evaluation. *J. Funct. Foods* **2015**, *18*, 1070–1085. [CrossRef]
74. Vasco, C.; Ruales, J.; Kamal-Eldin, A. Total phenolic compounds and antioxidant capacities of majpr fruits from Ecuador. *Food Chem.* **2008**, *111*, 816–823. [CrossRef]

75. Donno, D.; Cerutti, A.K.; Prgomet, I.; Mellano, M.G.; Beccaro, G.L. Foodomics for mulberry fruit (Morus spp.): Analytical fingerprint as antioxidants' and health properties' determination tool. *Food Res. Int.* **2015**, *69*, 179–188. [CrossRef]
76. Ma, H.; Johnson, S.L.; Liu, W.; DaSilva, N.A.; Meschwitz, S.; Dain, J.A.; Seeram, N.P. Evaluation of polyphenol anthocyanin-enriched extracts of blackberry, black raspberry, blueberry, cranberry, res raspberry, and strawberry for free radical scavenging, reactive carbonyl species trapping, anti-glycation, anti-β-amyloid aggregation, and microglial neuroprotective effects. *Int. J. Mol. Sci.* **2018**, *19*, 461.

agronomy

MDPI

Article

Exogenous Application of Amino Acids Improves the Growth and Yield of Lettuce by Enhancing Photosynthetic Assimilation and Nutrient Availability

Shumaila Khan [1,†], Hongjun Yu [1,†], Qiang Li [1,†], Yinan Gao [2], Basheer Noman Sallam [1], Heng Wang [1], Peng Liu [1] and Weijie Jiang [1,*]

1 Institute of Vegetables and Flowers, Chinese Academy of Agricultural Sciences (CAAS), No.12 Zhongguancun South Street, Haidian District, Beijing 100081, China; shumaila_khan@yahoo.com (S.K.); liqiang05@caas.cn (Q.L.); yuhongjun@caas.cn (H.Y.); basir_zen@yahoo.com (B.N.S.); 82101181122@caas.cn (H.W.); 82101179107@caas.cn (P.L.)

2 College of Horticulture and Landscape, Tianjin Agricultural University, Tianjin 300000, China; ilyzp1@gmail.com

* Correspondence: jiangweijie@caas.cn; Tel.: +86-010-8210-8797

† The authors equally contributed to this work.

Received: 19 April 2019; Accepted: 16 May 2019; Published: 26 May 2019

Abstract: As natural plant growth stimulators, amino acids are widely used to improve the yield and quality of crops. Several studies have illustrated the effects of different amino acids on lettuce plant parts. However, the effects of applying single amino acids on root growth remain elusive. The objective of this study was to evaluate the effect of root application of L-methionine on the growth of lettuce. In this study, two successive experiments on butterhead lettuce were conducted under hydroponic conditions. Three amino acids, L-methionine (20 mg/L), L-glycine (210 mg/L), and L-tryptophan (220 mg/L), were applied separately. L-methionine significantly increased the growth performance by 23.60%, whereas growth using L-tryptophan and L-glycine decreased by 98.78% and 27.45%, respectively. Considering the results of the first experiment, a second experiment was established with different concentrations of L-methionine (2200 mg/L, 220 mg/L, 22 mg/L, 2.2 mg/L, 0.2 mg/L, and 0.02 mg/L). The plants were allowed to grow for four weeks. Leaf width, plant area, leaf area, chlorophyll contents, etc., were evaluated. The results show that plant growth significantly improved by applying L-methionine at the lowest concentrations of 0.2 mg/L and 0.02 mg/L, which can, therefore, improve hydroponic production of lettuce and, accordingly, human nutrition.

Keywords: L-methionine; L-tryptophan; L-glycine; lettuce; nitrogen

1. Introduction

Lettuce (*Lactuca sativa* L.) is one of the main vegetable crops widely cultivated in China and consumed as a salad throughout the world. Due to its high nutritional value provided by mineral elements, vitamins, and folate, which play significant roles in human nutrition and diet, lettuce has become the focal point of several studies [1–4].

Nitrogen is an essential element for lettuce plants, an integral component of protein, phospholipid, and chloroplast [5,6]. Nitrogen uptake, assimilation, and utilization play essential roles in plant growth and development [7,8]. The plants mainly take up nitrogen in the form of nitrate (NO_3^-) and ammonium (NH_4^+) or N_2 from the atmosphere through nitrogen-fixing bacteria [9–11]. The application of a large amount of chemical fertilizer to ensure high crop yield causes serious issues for agricultural products [12] and the environment [13,14]. Hence, there is a need to look for sustainable horticultural practices to counteract chemical-based agribusiness. In this respect, the application of amino acids, as a type of growth-promoting substance, supplies plant nutrients but also improves plant quality, which ultimately boosts the yield and commercial output of crops [15]. Therefore, it has become popular in sustainable agriculture [16–20]

Amino acids as biostimulants (substances that promote plant growth, improve nutrient availability, and enhance plant quality) [20,21] are not only getting popular for mitigating injuries caused by abiotic stresses [22] but also serve as hormone precursors [20,23–25]; signaling factors of different physiological progressions, such as glutamate receptors (GRLs) [24–26]; and regulators of nitrogen uptake [27], root development [25,28–30], and antioxidant metabolism [25,30–32]. Better root development supported by the addition of amino acids can boost nitrogen fixation, which induces an enhanced root surface for nutrient uptake [29,33].

Direct application of amino acids and their products could modulate N uptake and assimilation; this phenomenon is mediated by enzymes engaged with N assimilation [15,25,30]. In addition, it could be a follow-up to the flagging pathway that controls N securing amino acids in roots, which are mostly accessible as supplements [30]. Additionally, application of amino acids was also found to increase K^+ in plants both in the presence of salt stress and without salt application [25,30,32]

A recent study [31] showed that seed treatment or foliar application of amino acids had different effects on soybean crops. An amino acid applied individually acts as a signaling component, i.e., increases antioxidant enzyme activity and causes efficient nutrient uptake [25,34,35]. Different investigations have demonstrated a positive impact of foliar application of amino acid mixtures on plants, for example, increased production in *Solanum lycopersicum* L. [36] and accumulation of dry matter, chlorophyll [37], starch, and polysaccharides in *Vicia faba* L. [38].

Previous studies have indicated an association of L-methionine with the biosynthesis of growth regulating substances such as cytokinins, auxins, and brassinosteroids in plants [39,40]. L-methionine functions as a precursor of a significant number of essential biomolecules such as vitamins, polyamine, cofactors [41], and antioxidants such as glutathione, which are considered to be significant determinants of cellular redox homeostasis and many defense compounds [19]. All of these biomolecules contain sulfur moieties that act as functional groups and are derived from L-methionine. In plants, L-methionine biosynthesis plays a central role in fixing inorganic sulfur from the environment, providing the only metabolic sulfide donor for the generation of glutathione, phytochelatins, iron–sulfur clusters, vitamin cofactors, and multiple secondary metabolites [42,43].

However, there is little data on the impact of isolated amino acids, particularly in root application. Additionally, the majority of investigations have been carried out utilizing a mixture of amino acids and other methods of application, such as foliar application and seed treatment. Hence, this study is based on the hypothesis that the root application of individual amino acids can improve the uptake of nitrogen and other growth-related factors, which can lead to increased productivity of lettuce plants. Therefore, the objective of the present work was to evaluate the effect of the separate application of L-tryptophan, L-glycine, and L-methionine in nutrient solution on the growth, yield, and physiology of lettuce plants.

2. Materials and Methods

The research study was carried out at the Vegetables and Flowers Institute, Chinese Academy of Agricultural Sciences, Beijing, China, in 2017–2018 to determine the regulation of lettuce plant growth response under different amino acids and concentrations.

2.1. Plant Material and Growth Conditions

Sowing of butterhead lettuce seeds was done under controlled conditions using a mixture of peat moss with an average of 2–3 seeds per hole. All cultural practices were maintained in order to have a good plant stand. Average minimum and maximum monthly temperatures were set to 24 °C and 34 °C. Plants were provided with natural sunlight with a light intensity of approximately

900–1000 μmoles m^{-2}/s. At pre-emergence stages, the nutrient solution was applied once a week. Plants with at least 2 fully expanded leaves 30 days after sowing were transferred to a closed-loop hydroponic system (Figure 1).

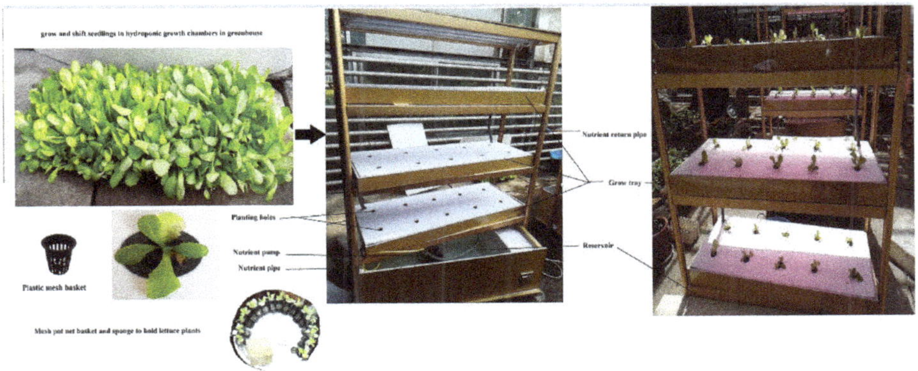

Figure 1. The hydroponic system, mesh basket, and other materials used to grow lettuce plants.

Briefly, the hydroponic system consisted of 3 growing nutrient trays (used as 3 replicates) with 10 holes each, with a distance of 21.2 × 20.5 cm between the holes. The height of the hydroponic growing stand was 160 cm from the ground, while the length and width of the growing tray were 102 cm and 38 cm, respectively. The capacity of the water reservoir was 80 L with 6 cm depth and recyclable. Plastic mesh used to cover the plants was 6 cm long and 2 cm wide.

A pH of 6.0–6.3 and electrical conductivity (EC) of 1.5–2.0 mS cm^{-1} of nutrient solution were maintained regularly for optimal plant growth.

Plantlets were grown with 75% strength Hoagland nutrient solution containing the following nutrients (in mg/L), as previously described [44]: $Ca(NO_3)_2$ = 1122; KNO_3 = 910; KH_2PO_4 = 272; NH_4NO_3 = 40; $MgSO_4$ = 247; EDTA (Ethylenediamine tetraacetic acid Ferric Sodium Salt) = 16.80; $ZnSO_4$ = 1.20; $Na_2B_4O_7$ = 0.28; Na_2MoO_4 = 0.20; $CuSO_4$ = 0.10; and $MnSO_4$ = 0.86.

2.2. *Application of Three Amino Acids on Lettuce*

The experiment was conducted from December 2017 to February 2018 with 3 replications. The concentration of 3 amino acids, L-methionine, L-tryptophan, and L-glycine, was kept at 20 mg/L, 210 mg/L, and 220 mg/L, respectively. This gave 4 treatment combinations (3 amino acids and 1 control treatment). The amino acid treatment was started 8 days after transplanting into the nutrient solution to prevent plants from undergoing nutrient shock. Data were recorded every week and the crop was harvested after 30 days of treatment.

2.3. *Application of L-Methionine Concentrations on Lettuce*

The second experiment was conducted from January to March 2018 with a single amino acid, L-methionine (selected from experiment 1), in 6 concentrations, as 3 treatments and 3 replications. Plants were transplanted to the nutrient solution and treated with the amino acid after 8 days. The concentration of L-methionine applied was 2200 mg/L, 220 mg/L, and 22 mg/L for the treatment and 2.2 mg/L, 0.2 mg/L, and 0.02 mg/L for the control. All other experimental conditions were the same as in the first experiment. The tanks of nutrient solution were refreshed weekly.

2.4. Data Collection and Analysis

Data were recorded for the following morphological and physiological parameters: root length, leaf length, leaf width, leaf area, plant area, chlorophyll content, and fresh and dry mass of root and shoot.

2.4.1. Vegetative Growth Parameters

The number of leaves, plant height, plant diameter, and leaf area were measured every 7 days following standard procedures as proposed in [45–47].

$$AF\left(cm^2\right) = 0.7 \times Length\,(cm) \times Width\,(cm) - 2.4$$

Leaf length and width were measured by using a measuring tape/scale.

Root length was measured by separating roots from plants and placing them on paper and blotting them, then using a measuring tape (recorded in cm). Fresh weight per plant was square root transformed to normalize the error distribution before the analysis, as described [48], using an electronic balance (S = 0.1 g) (Acculab V-1200). The harvested plants were rinsed with distilled water, then the roots were blotted on filter paper and dried completely in an oven at 60–65 °C to determine dry weight [49]. The following formulas were used to calculate the index of growth traits [45].

Relative growth rate (RGR) was calculated by the following formula:

$$\mathrm{RGR} = (\mathrm{lnW_2} - \mathrm{lnW_1})(\mathrm{t_2} - \mathrm{t_1})$$

where W_2 and W_1 denote the plant's dry mass (g) at time t_2 and t_1, respectively.

The net assimilation rate was calculated by the following formula:

$$\mathrm{NAR} = \mathrm{dW}/(\mathrm{A} \times \mathrm{dt})$$

where A is the area of assimilation organs (cm^2), dW is the dry mass increment (g), and dt is the time of cultivation (days). Root mass ratio (RMR; root mass per unit total plant mass) was calculated as described in [49].

2.4.2. Physiological Measurements

Total chlorophyll content was estimated by using a portable The Soil Plant Analysis Development (SPAD) chlorophyll apparatus (SPAD-502 Plus, Konica Minolta, Tokyo, Japan). The leaf net photosynthesis rate for 3 independent lettuce seedlings per experimental replicate was determined using a portable LI-6400 photosynthesis system (Li-Cor 6400-18, Lincoln, NE, USA) [40,50]. The set values used were as follows: photosynthetic photon flux density, 500 $\mu mol \cdot m^{-2} \cdot s^{-1}$; air flow rate inside the sample chamber, 400 $\mu mol \cdot s^{-1}$.

The nutrient contents were measured in dried leaves ground by an electric mortar (multipurpose high-speed disintegrator, Dingia), sieved, weighed out to 0.2 g, and digested by concentrated nitric acid (HNO_3, 5–6 mL) carefully under a laminar flow hood cabinet. All nutrients were analyzed using an optical emission spectrometer (Optima 5300 DV Spectrometer, Shelton, CT, USA).

The total N was determined by the Kjeldahl method [51].

2.5. Statistical Analysis

The recorded data were subjected to analysis of variance (ANOVA) and fixed-factor models [52], and Duncan's multiple range test was used to assess the significance of treatment differences by means of IBM SPSS Statistics for Windows (version 20.0, IBM Corp., Armonk, NY, USA).

3. Results

3.1. Application of Three Amino Acids on Lettuce

The effects of different amino acids were studied for vegetative growth. The plants grown in the modified nutrient solution (Hoagland and amino acids) showed varied responses for metric traits.

The vegetative indicators responded positively and significantly to all applied L-methionine concentrations (Figure 2A–F). Leaf length, width, and the number increased in response to L-methionine application, and decreased with L-glycine and L-tryptophan. The leaf length of L-methionine treated plants increased by 11.41% compared to control plants, while it decreased by 13.76% and 61.92% in L-glycine and L-tryptophan treated plants, respectively (Figure 2A). A significant increase in leaf width (17.46%) was also found with L-methionine, but there was an 18.25% and 63.49% decrease in response to L-glycine and L-tryptophan, respectively (Figure 2B). Similarly, leaf area and leaf numbers also increased under L-methionine treatment (31.41% and 50.4%, respectively), while leaf area decreased under L-glycine and L-tryptophan (29.67% and 86.25%, respectively) and leaf numbers decreased under L-tryptophan (50.36%) compared to control (Figure 2C,D).

Furthermore, plant height and area also had an encouraging response to L-methionine application (Figure 2E,F). The results revealed that there was an abrupt change in plant height and area, which were reduced by 82.91% and 90.78%, respectively, upon L-tryptophan application.

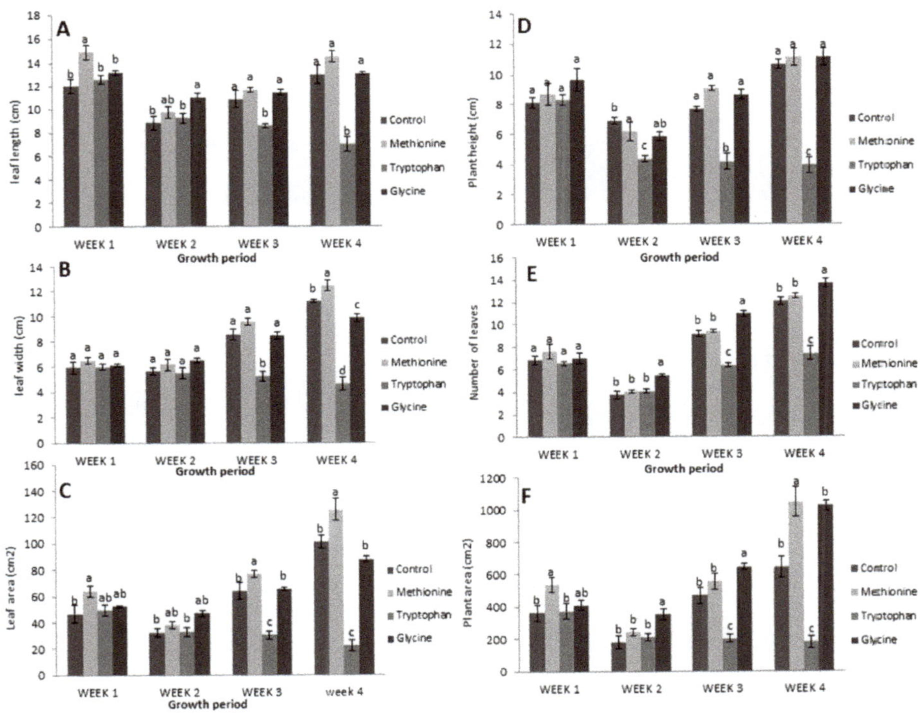

Figure 2. (**A**) Leaf length, (**B**) leaf width, (**C**) leaf area, (**D**) plant height, (**E**) leaf number, and (**F**) plant area with L-methionine (20 mg/L), L-glycine (220 mg/L), and L-tryptophan (210 mg/L). Means followed by the same lowercase letters do not differ significantly from each other in the comparison between amino acid treatments each week using Duncan's multiple range test ($p < 0.05$).

Likewise, root length, shoot-to-root ratio, relative water content, net assimilation rate, and fresh and dry biomass were positively affected by L-methionine treatment and negatively by L-tryptophan and L-glycine (Figure 3). The results revealed that lettuce plants showed a significant increase in root length (Figure 3A), relative water content (Figure 3D), and net assimilation rate (Figure 3C) in response to L-methionine application, and a decrease with the other two amino acids. Interestingly, the shoot-to-root ratio was found to be higher in response to L-tryptophan, which suggests that amino acids other than L-methionine also have an important role in plant growth. Moreover, a relative increase in fresh and dry biomass was observed with L-methionine application, by 20.88% and 15.71%, respectively, and a decrease with L-tryptophan and L-glycine treated plants (Figure 3D). Taken together, these results indicate that biostimulants, specifically the amino acid L-methionine, play a critical role in the growth and development of lettuce plants.

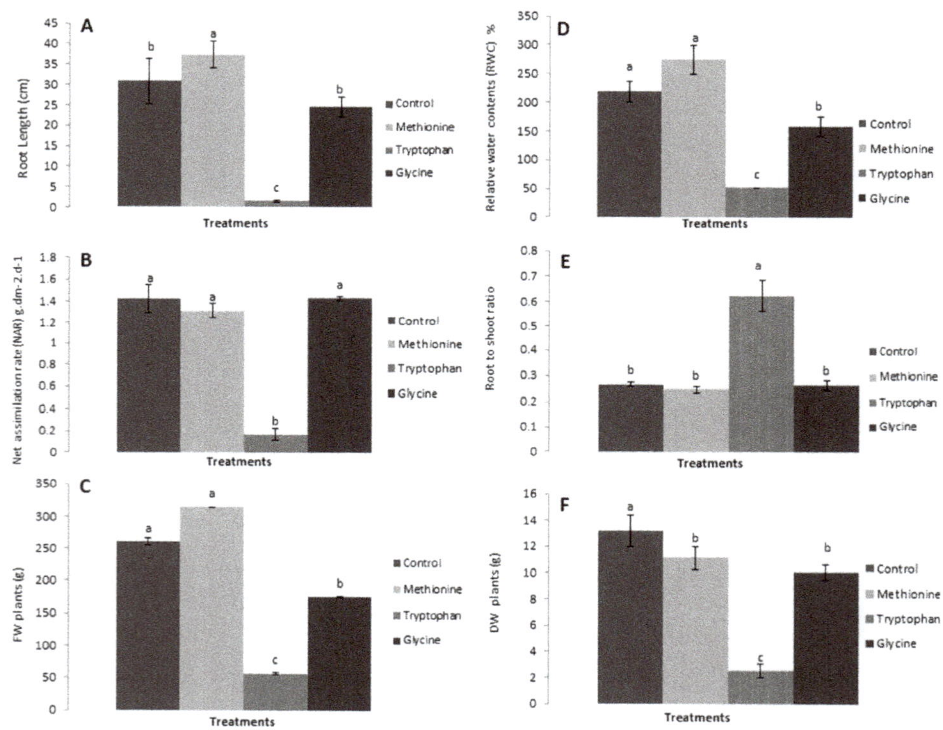

Figure 3. Effects of amino acids on (**A**) root length, (**B**) net assimilation rate (NAR), (**C**) plant fresh weight (FW), (**D**) relative water content (RWC), (**E**) shoot-to-root ratio (S:R), and (**F**) plant dry weight (DW). Means followed by the same lowercase letters do not differ significantly from each other according to Duncan's multiple range test ($p < 0.05$).

3.1.1. Photosynthetic Measurements

Photosynthetic measurements of lettuce leaves (Table 1) included amino acid application effect, rate of net photosynthesis, stomatal conductance, and transpiration rate. They were significantly affected by applied amino acids compared to control, except for total chlorophyll content. Transpiration rate and intracellular CO_2 were relatively higher among all amino acid treated plants. The high levels of photosynthesis and chlorophyll content suggest that amino acids are important regulators of photosynthesis in lettuce, ultimately leading to higher yield and biomass.

Table 1. Effects of amino acids on physiological indicators of lettuce plants.

Treatments	Net Photosynthesis Rate (Pn) (μmol CO_2 m^{-2} s^{-1})	Stomatal Conductance (mol m^{-2} s^{-1})	Ci (μmol/mol)	Transpiration Rate (Tr) (mol H_2O m^{-2} s^{-1})	Total Chlorophyll Content (SPAD Values)
Control	7.31 a	0.07 a	335.78 b	1.86 a	16.5 n.s.
L-methionine (Meth)	4.33 b	0.03 b	369.55 ab	0.97 bc	16.7 n.s.
L-tryptophan (Try)	3.08 c	0.03 b	426.67 a	0.83 c	15.6 n.s.
L-glycine (Gly)	4.63 b	0.04 b	390.34 ab	1.53 b	17.4 n.s.

L-Meth (20 mg/L), L-Try (220 mg/L), and L-Gly (210 mg/L). n.s., not significant. Means followed by the same letters are not statistically different from each other, according to Duncan's multiple range test ($p < 0.05$).

In addition, relative growth parameters including leaf area index, leaf dry matter content, root mass ratio, specific leaf area, and leaf area ratio were higher in response to L-tryptophan treatment (Table 2). The relative growth rate was found to be higher with L-glycine application. L-methionine and L-glycine treated plants had a nonsignificant association with all other traits compared to control plants. These results show that along with L-methionine, L-tryptophan is also an important player, especially for relative growth enhancement in lettuce.

Table 2. Effects of amino acids on growth indices of lettuce plants.

Treatments	LAI (cm^2 cm^{-2})	LDMC (g g^{-1})	RMR (g g^{-1})	SLA (cm^2 g^{-1})	LAR (cm^2 g^{-1})	RGR (g g^{-1} d^{-1})
Control	2.73 b	0.05 b	0.01 b	0.07 n.s.	22.67 bc	4.44 b
L-methionine	2.06 b	0.03 b	0.01 b	0.04 n.s.	6.38 c	1.18 c
L-tryptophan	26.18 a	0.58 a	0.15 a	0.10 n.s.	63.65 a	1.34 c
L-glycine	3.85 b	0.05 b	0.01 b	0.07 n.s.	35.78 b	7.31 a

L-Methionine (20 mg/L), L-Tryptophan (220 mg/L), and L-Glycine (210 mg/L). LAI, leaf area index; LDMC, leaf dry matter content; RMR, root mass ratio; SLA, specific leaf area; LAR, leaf area ratio; RGR, relative growth rate; n.s. not significant. Means followed by the same letters are not statistically different from each other, according to Duncan's multiple range test ($p < 0.05$).

However, vitamin C content was not significantly affected by the applied concentrations, although there was a tendency for it to be higher in L-methionine and L-glycine compared to control (Table 3). In contrast, dry matter percentages were higher in L-tryptophan treated plants.

Table 3. Effects of amino acids on vitamin C content and dry matter percentage of lettuce leaves.

Treatments	Vitamin C (mg 100 g^{-1})	Dry Matter (%)
Control	0.25 n.s.	13.4 bc
L-methionine	0.3 n.s.	10.1 c
L-tryptophan	0.2 n.s.	94.5 a
L-glycine	0.3 n.s.	24.7 b

L-Methionine (20 mg/L), L-Tryptophan (220 mg/L), and L-Glycine (210 mg/L). n.s., not significant. Means followed by the same letters are not statistically different from each other, according to Duncan's multiple range test ($p < 0.05$).

3.1.2. Nutrient Contents

Nutrient content analysis (Table 4) revealed a dynamic change in response to all amino acid concentrations. The content of essential elements such as nitrogen, phosphorus, and potassium varied among treatments.

Table 4. Effects of amino acids on essential elements of lettuce.

Treatments	N	P	K	S	Ca	Mg	Fe	Cu	Na	Zn	Al
	(% DW)	(mg/g)									
Control	1.4 b	12.3 d	208.1 b	9.8 b	93.7 a	22.9 a	6.8 a	0.1 c	18.3 a	0.2 b	13.1 a
L-methionine	4.3 a	32.2 b	420.7 a	10.1 b	66.3 c	12.6 b	2.1 b	0.06 d	4.48 b	0.4 a	7.1 b
L-tryptophan	3.7 a	36.8 a	421.4 a	13.1 a	81.5 b	13.4 b	2.8 b	0.14 b	3.95 b	0.4 a	9.0 ab
L-glycine	4.3 a	21.7 c	230.8 b	13.3 a	44.6 d	12.6 b	6.9 a	0.23 a	3.40 b	0.3 a	9.6 ab

L-methionine (20 mg/L), L-tryptophan (220 mg/L), and L-glycine (210 mg/L); Macronutrients: N, nitrogen; P, phosphorus; K, potassium; S, sulfur; Ca, calcium; Mg, magnesium. Micronutrients: Fe, iron; Cu, copper; Mo, molybdenum; Na, sodium; Zn, zinc; Al, aluminum. Means followed by the same letters are not statistically different from each other, according to Duncan's multiple range test ($p < 0.05$).

3.2. Application of Different L-Methionine Concentrations on Lettuce

In the second trial, L-methionine at a higher concentration had a reduced effect on plant growth and physiology. The results show that lower levels of L-methionine significantly contributed to enhancing the number of leaves, plant height, and leaf length and width (Figure 4). Remarkably, 0.22 mg/L concentration of L-methionine resulted in a gradual increase in vegetative growth compared to control plants during all weeks. In contrast, higher levels were negatively associated with the corresponding measurements. Lettuce plants had short stature and fewer leaves in response to 2200 mg/L and 220 mg/L of L-methionine (Figure 4B,D). Concomitantly, more than 80% and 50% decreases in these two traits were found with increasing amino acid levels.

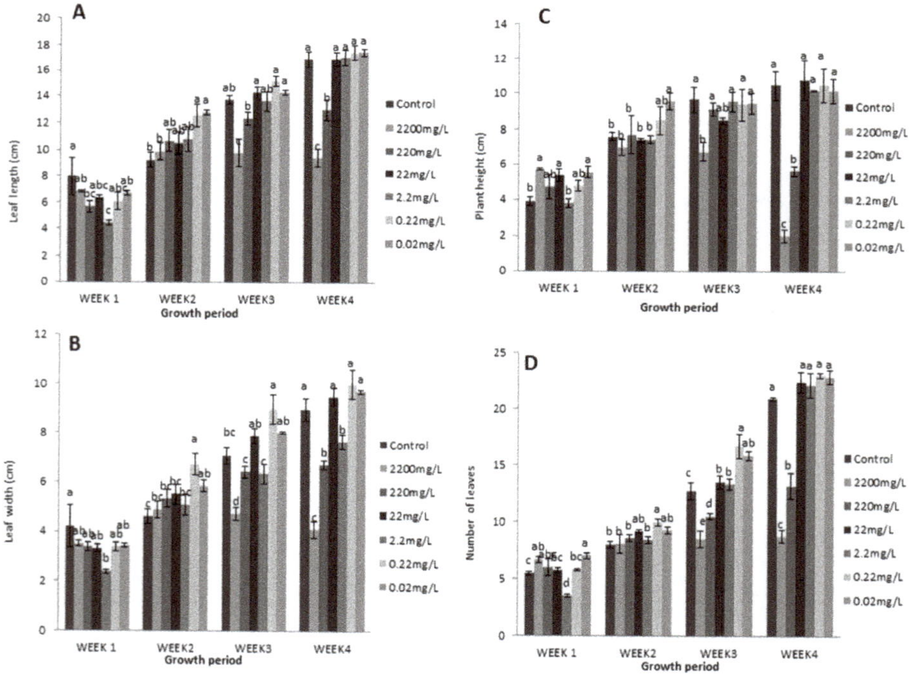

Figure 4. Effects of L-methionine concentrations on (**A**) leaf length, (**B**) plant height, (**C**) leaf width, and (**D**) the number of leaves per plant. Means followed by the same letters are not statistically different from each other, according to Duncan's multiple range test at ($p < 0.05$).

Similarly, relatively decreased leaf length (42.24% and 23.3%, respectively) and width (54.5% and 25.23%, respectively) were observed with increased treatment levels (Figure 4A,C). A strong increasing trend was also found for both traits in response to 0.2 mg/L of L-methionine. Overall, these results indicate that higher levels of L-methionine have an inhibiting effect on plant growth.

Leaf and plant area, root length, and fresh and dry weight of lettuce plants were improved by lower L-methionine concentrations (especially 0.2 mg/L) in advanced growth stages (Figure 5). However, mixed growth patterns were also present with different amino acid levels at each plant stage. A maximum reduction in leaf and plant area (75.5% and 74.653%, respectively) was found in 2200 mg/L treated plants, and a minimum (15.6% and 4.03%, respectively) in 0.2 mg/L plants (Figure 5A,B). Moreover, root length was found to be reduced at all levels except 0.2 mg/L, which caused a relative increase by 14.8% (Figure 5C). At the same time, the fresh and dry weight of roots were also lower with more concentrated treatment (Figure 5D), which suggests that amino acids are essential elements required as micronutrients for plant growth. An increased concentration leads to restricted plant biomass. Moreover, the decreased fresh and dry weight of lettuce plants indicates that nutrient stress and reduced photosynthetic activity were responsible for the lower accumulation of leaf organic material and growth rate.

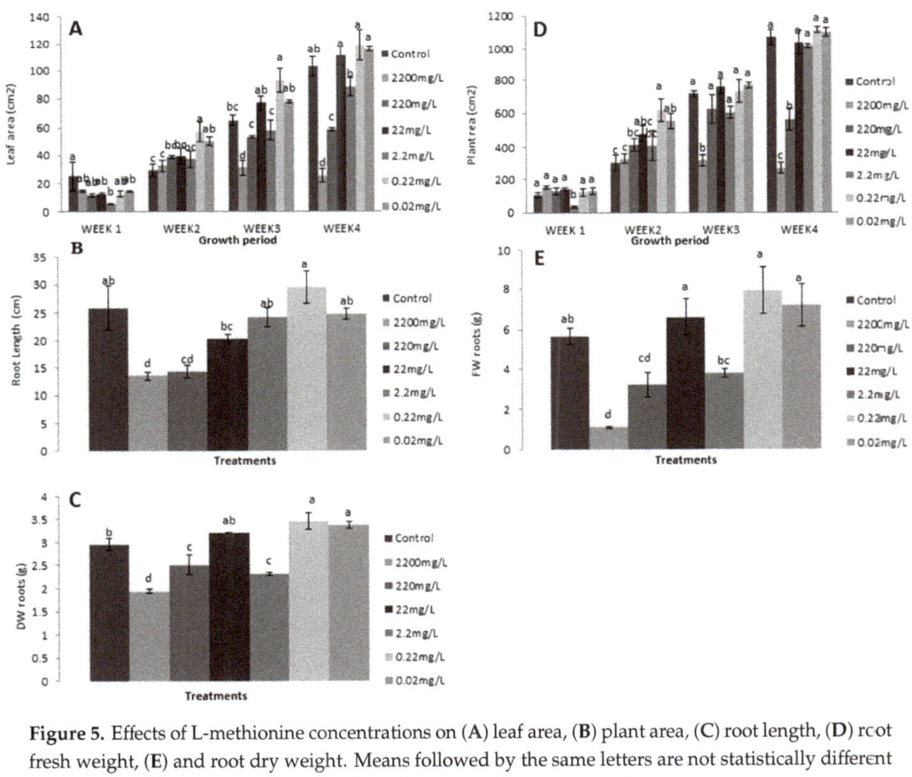

Figure 5. Effects of L-methionine concentrations on (**A**) leaf area, (**B**) plant area, (**C**) root length, (**D**) root fresh weight, (**E**) and root dry weight. Means followed by the same letters are not statistically different from each other, according to Duncan's multiple range test at ($p < 0.05$).

3.2.1. Relative Growth Measurements

Relative growth parameters measured in the second trial revealed that most of the traits were substantially increased by decreased L-methionine concentrations (Table 5). Surprisingly, root mass and leaf area ratios increased ($p < 0.05$) under higher concentration, which shows that plants can respond to a stress environment by maintaining their growth patterns. However, all other measured parameters had a decreasing tendency under a nutrient stress environment.

Table 5. Effects of concentrations of L-methionine on growth indices of lettuce plants.

Treatment (mg/L)	LAI (cm^2 cm^{-2})	RMR (g g^{-1})	SLA (cm^2 g^{-1})	LAR (cm^2 g^{-1})	RGR (g g^{-1} d^{-1})	RWC (%)
Control	0.24 a	0.04 b	1679.50 a	51.5 ab	0.33 c	91.2 a
2200	0.06 c	0.34 a	1259.19 ab	76.6 a	0.34 c	11.11 c
220	0.17 b	0.06 b	572.02 b	42.2 ab	0.23 c	72.22 b
22	0.25 a	0.02 b	1118.38 ab	36.9 b	0.27 c	93.72 a
2.2	0.24 a	0.02 b	1937.15 a	46.1 ab	0.52 c	92.96 a
0.22	0.26 a	0.01 b	1496.32 a	34.5 b	1.6 b	93.28 a
0.02	0.27 a	0.01 b	1641.48 a	45.5 ab	3.7 a	90.77 a

LAI, leaf area index; RMR, root mass ratio; SLA, specific leaf area; LAR, leaf area ratio; RGR, relative growth rate; RWC, relative water content. Means followed by the same letters are not statistically different from each other, according to Duncan's multiple range test ($p < 0.05$).

3.2.2. Photosynthetic Measurements

Enhanced photosynthetic activity, transpiration, and total chlorophyll content (Table 6) at lower L-methionine concentrations suggest that plants require this nutrient only in small amounts. More moderate transpiration activity occurred with all applied treatments, and reduced accumulation of chlorophyll in leaves causes restricted photosynthetic activity. There was no significant difference observed for stomatal conductance but it was relatively higher with 22 mg/L of L-methionine.

Table 6. Effects of L-methionine levels on net physiological and growth indicators of lettuce leaves.

Treatment (mg/L)	Net Photosynthesis Rate (Pn) (μmol CO_2 m^{-2} s^{-1})	Stomatal Conductance (mol m^{-2} s^{-1})	Ci ($\mu mol/mol$)	Transpiration Rate (Tr) (mol H_2O m^{-2} s^{-1})	Total Chlorophyll Content (SPAD)
Control	5.36 b	0.10 abc	363.22 bc	1.94 b	24.7 cd
2200	1.67 e	0.03 c	379.51 ab	0.64 d	23.7 d
220	2.37 de	0.05 bc	398.91 a	1.11 c	29.9 a
22	3.07 d	0.15 a	395.1 a	1.35 c	26.2 bc
2.2	4.47 c	0.06 bc	347.86 c	1.27 c	27.3 b
0.22	7.16 a	0.12 ab	369.23 bc	2.41 a	28.1 ab
0.02	6.61 a	0.09 abc	349.91 c	1.92 b	27.3b

Means followed by the same letters are not statistically different from each other, according to Duncan's multiple range test ($p < 0.05$).

3.2.3. Nutrient Contents

Data in Table 7 describe the effects of L-methionine on macro- and microelements. The increasing trend of essential element accumulation including N, P, and K at reduced L-methionine levels indicates that these elements affect plant metabolism and help to adapt to modified environmental cues, which directly or indirectly affects plant metabolism. For example, a significant increase in nitrogen (N) content in leaf tissues increases photosynthesis efficiency, which is key to increasing crop yield. Plant metabolism is maintained by these elements with lower fractions of amino acids to regulate plant growth and development. Mixed fractions of other elements at different concentrations signify their importance in plant health regulation. For example, in addition to essential elements, S, Mg, Fe, Cu, Mn, and Na accumulation was higher at all levels.

Any change in amino acid concentration leads to stress conditions, and plants respond differently at different levels by changing their growth patterns.

Moreover, a significant ($p < 0.05$) decrease in vitamin C content and leaf dry matter content and percentage also highlights the importance of micronutrients in plant metabolism.

Table 8 shows that vitamin C content decreased significantly ($p < 0.05$) with 2200 mg/L, but increased by 14.21% with 0.22 mg/L as compared to control. In contrast, a significant increase in leaf dry matter content was found with 2200 mg/L and in dry matter percentage ($p < 0.05$) with L-methionine application of 2200 mg/L and 220 mg/L. Moreover, a decrease in the fresh and dry weight of lettuce plants indicates that reduced photosynthetic activity was responsible for the lower accumulation of leaf organic material and reduced growth.

Table 7. Effects of L-methionine concentrations on essential macro- and microelements of lettuce leaves.

Treatment (mg/L)	N	P	K	S	Ca	Mg	Fe	Cu	Na	Zn	Al
	(% DW)	(mg/g)									
Control	5.3 c	17.5 d	128.3 c	9.7 d	67.1 c	11.3 d	2.5 c	0.03 c	4.3 d	0.5 c	7.6 b
2200	2.1 e	27 bc	164.1 c	23.8 a	106.1 ab	19.2 abc	5.5 a	0.08 a	15.4 a	0.8 b	9.9 ab
220	2.3 e	27.1 bc	233.1 b	19.6 b	127.9 a	24.5 a	4.4 ab	0.04 bc	11 b	0.4 c	10.4 a
22	4.2 d	25.2 c	254.3 b	12.4cd	83.2 bc	16.27 cd	3.2 bc	0.04 bc	6 c	0.4 c	8.7 ab
2.2	5.7 bc	39.0 a	351.8 a	14.1 c	108.8 bc	22.7 ab	3.0 c	0.05 b	11 b	1.1 a	9.3 ab
0.22	6.7 a	34.9 ab	336.4 a	13 cd	107.1 ab	19.8 abc	3.2 bc	0.04 bc	9.4 b	0.4 c	9 ab
0.02	6.1 b	29.5 bc	246.5 b	12.5 cd	99.2 abc	17.3 bc	3.4 bc	0.05 b	6.6 b	0.4 c	9.1 ab

Macronutrients: N, nitrogen; P, phosphorus; K, potassium; S, sulfur; Ca, Calcium; Mg, magnesium. Micronutrients: Fe, iron; Cu, copper; Mo, molybdenum; Na, sodium; Zn, zinc; Al, aluminum. Means followed by the same letters are not statistically different from each other, according to Duncan's multiple range test at ($p < 0.05$).

Table 8. Effects of L-methionine concentrations on vitamin C content, leaf dry matter content (LDMC), dry matter percentage (DM%), fresh weight (FW), and dry weight (DW) of lettuce leaves.

Treatment (mg/L)	Vitamin C (mg 100 g^{-1})	LDMC (g g^{-1})	DM (%)	Mean Fresh Weight (g)	Mean Dry Weight (g)
Control	0.21 a	0.13 b	4.7 b	37.97 bc	4.8 c
2200	0.11 b	0.49 a	48.8 a	0.86 d	0.8 d
220	0.19 a	0.12 b	11.9 ab	10.89 cd	2.3 d
22	0.19 a	0.06 b	5.69 b	58.89 ab	5.4 bc
2.2	0.21 a	0.06 b	5.62 b	44.29 ab	7.97 a
0.22	0.23a	0.05 b	5.29 b	70.94 a	8.7 a
0.02	0.21 a	0.04 b	4.23 b	60.89 ab	7.3 ab

Means followed by the same letters are not statistically different from each other, according to Duncan's multiple range test ($p < 0.05$).

As indicated by the outcomes shown in Tables 3 and 8, there was no significant change ($p < 0.05$) observed in plants with or without amino acid treatment. It was observed that all individual amino acid treatments, except for one L-methionine concentration, led to no significant ($p < 0.05$) impact on vitamin C content. The special case was L-methionine at 2200 mg L^{-1}, which prompted a decrease in vitamin C content, essentially contrasting with the control plants ($p < 0.05$).

4. Discussion

Amino acid application is a common practice for horticultural crops worldwide, with the majority of treatments making use of biostimulants with a mixture of amino acids [53]. In our study, we checked the activity of a single amino acid that regulates the nutrient contents involved in growth variables.

Previous investigations have demonstrated that plant developmental cues respond distinctively to the provision of amino acids [54–56]. It is likely that the effects of amino acids on plants rely on the kind of amino acids supplied [56] and the plant cultivars [57]

From our results, we can presume that amino acids (L-methionine, L-tryptophan, and L-glycine) not only make nutrients available to plants but also act as signal transducing molecules [31], as small doses are sufficient for plant development response, while these molecules can act as signals of several beneficial plant physiological processes. Studies demonstrate that amino acids in the form of a foliar spray on plants is a promising technique [38]. In this manner, L-methionine induces more prominent absorption of sulfur and nitrogen in plants, which also depends on the amount applied [26,58–60].

Plants utilize amino acids according to their nutritional needs and genetic background, as well as environmental and developmental cues [61,62]. This might be the reason why amino acid reactions were not consistent in both experiments. Therefore, we can assume that the decreasing effect of L-tryptophan on yield might be due to the inhibitory impact of auxin on vegetative growth. In this association, the reduction of lettuce yield per plant caused by the inhibition effect might be due to the detrimental effect of auxin accumulation stress on growth, the aggravation of mineral nutrient uptake, and the improvement of plant respiration [41]. The distinctive response was reported by Abbas et al. [54], in an investigation on L-tryptophan applied to chickpea at rates of 10^{-1} M, 10^{-2} M, and 10^{-3} M. They found random results with different parameters: root length was increased only in the control compared to the three treatments, the number of nodules increased only with 10^{-1} M, and nodule fresh and dry weight decreased with 10^{-3} M treatment and increased with the other two compared to control, while control remained nonsignificant. The most pods and highest plant weight were shown with 10^{-2} M, but pod weight per plant was significantly affected by all treatments due to the production of phytochromes suitable to chickpea. This experiment may provide evidence for the substantiating inconsistency of lettuce observed in our previous experiment.

However, our results contrast those described in [63], in which numbers of strawberry leaves per plant were significantly higher with the application of L-tryptophan than control.

A few reasons can clarify the positive effects of L-methionine. First, it has a role in maintaining the structure of proteins required for cell division, cell differentiation, and growth. Second, it provides sufficient sulfur and nitrogen according to plant needs. Third, the ability of L-methionine to be converted into polyamines and enlarge by entering the hormone structures [64] allows nitrogen movement between cells and organs [65]. It also functions as a buffer and behaves as a source of carbon and energy [66], and as a precursor of spermidine and gibberellin biosynthesis [43,67], growth regulators, and many secondary metabolites [43]. L-methionine also acts as a growth regulator of cytokinin, brassinosteroids, and auxin, increasing the initiation of roots; helps with the absorption of more nutrients by the plant [39,67], which may stimulate endogenous hormone homeostasis [68,69]; and is required for the development of hairy roots [67] at optimum levels.

Increased L-methionine levels influence phytohormones, which ultimately increases the chlorophyll content and chloroplast development or cytokinins [70,71]. An expected requirement for the prompting of L-methionine application might be the proximity of phytohormones (e.g., auxins and cytokinin). The phytohormones and signaling compounds may improve the photosynthetic activity, leading to better yield. Another possible mechanism involved with the amino acid effect could be related to the stimulation of root growth of treated plants, which may improve water and nutrient uptake capability, leading to yield productivity [68,69], as well as enhanced cell formation and increased fresh and dry matter [72], with increased growth behavior [69,70,73].

Our results demonstrate that high L-methionine concentration reduced plant growth due to damage to the photosynthetic apparatus [61] and blocking of nutrient uptake. Higher levels of this nutrient cause blockage of photosynthesis in stressed environments [65]. Padgett (1996) applied L-methionine to the root zones of chrysanthemum plants, producing a physiological disorder called methionosis, with the typical pattern of a metabolite–antimetabolite relationship [47]. It is thought that in this case, L-methionine, especially because of the large amounts applied, may function as an antimetabolite that interferes with normal amino acid metabolism. In other words, amino acids should be meticulously applied, as they could reduce the percentage of dry weight because they cause swollen, water-filled tissues due to depressed vegetative growth [65,74].

Thus, we speculate that application of L-methionine with 0.22 mg/L in the nutrient solution was sufficient, but other concentrations were too high and might have been a source of stress. Our results are consistent with previous studies [42,75] proposing that the improvement of novel "bio-sound items" ought to continue based on a foundational approach established in chemical synthesis, natural chemistry or biochemistry, and biotechnology connected to genuine plant physiological, agrarian, and environmental constraints. It was proposed that these items should work at low dosages, be biologically and ecologically friendly, and have reproducible advantages in horticultural plant development. The high amount of L-methionine also reduced vitamin C content. Therefore, we conclude that plant metabolism is affected by external N and thus can reflect the changes in N absorption, transport, and metabolism [76]. Similar findings have been reported in Chinese cabbage and lettuce [77]. The optimal required concentration is essential for optimal growth. It was confirmed from the previous study that when amino acids are added alone, care must be taken, as they can inhibit cell growth [63]. In general, the use of amino acids on plants can improve their capacity of transporting mineral components [66].

In view of the synthetic pathway of vitamin C and the synthesis of ascorbic acid requiring plant climatic changes and the conditions of plant sustenance, it may be hypothesized that whatever factor builds the sugar (or glucose) content in plant tissues can thus increase the vitamin C content [78]. It has been reported that amino acids [43] and nitrogen fertilizers do not impact the vitamin C content in broccoli. Conversely, in cauliflower, when nitrogen fertilizers are extended from 80 kg to 120 kg per ha, the ascorbic acid content is reduced by 7% [79].

Optimizing the amino acid content can bring about different morphogenetic responses; higher concentrations generally inhibit growth in *Cicer arietinum* [80]. The available information from various studies suggests that optimal levels of various amino acids may be species- or genotype-dependent, which needs to be determined before recommending their use [81]. Increasingly, plant ecologists working at all levels have become interested in the role of amino acid nutrition in the lives of plants and determining the proper amount of amino acids suitable for plant growth [82].

The depressive effect of L-tryptophan and high amounts of L-methionine on yield may be attributed to the inhibitory effect of auxin accumulation on vegetative growth, the disturbance in mineral uptake, and/or the enhancement of plant respiration [83].

5. Conclusions

Taking into consideration the discussion above, it can be inferred that L-methionine increases the chlorophyll content of plants and contributes to the saving of energy, thus boosting the plant yield. L-methionine led to significant increases in observed physiological factors in lettuce leaves at lower concentrations because at high concentrations it affects auxin uptake, which can kill plants. In brief, L-methionine at a concentration of 0.2 mg/L showed the best effect on the growth of lettuce plants. Therefore, we can say that L-methionine can contribute as a suitable substitute for fertilizers to increase crop yield. Future research should concentrate on assessing the mechanisms of how amino acids can influence the genetic transcription of various parameters, including supplement transporters, hormone production, and antioxidant metabolism. Along these lines, it will be possible to acquire the best understanding of the role of amino acids as biostimulants in lettuce plants.

Author Contributions: Conceptualization, W.J.; methodology, Q.L. and Y.G.; software, S.K.; validation, S.K.; formal analysis, S.K. and B.N.S.; investigation, S.K.; data curation, S.K.; writing—original draft preparation, S.K.; writing—review and editing, H.W., P.L., H.Y., and Q.L.; visualization, H.Y.; S.K.; supervision, W.J.; project administration, Y.G.; funding acquisition, W.J.

Funding: Demonstration of innovative and resource-efficient urban agriculture for multiple benefits in China (National key R&D program of China, (2017YFE0118500).

Conflicts of Interest: The authors declare no conflict of interest.

References

1. Samuolienė, G.; Urbonavičiūtė, A.; Duchovskis, P.; Bliznikas, Z.; Vitta, P.; Žukauskas, A. Decrease in nitrate concentration in leafy vegetables under a solid-state illuminator. *HortScience* **2009**, *44*, 1857–1860. [CrossRef]
2. Son, K.H.; Oh, M.M. Leaf shape, growth, and antioxidant phenolic compounds of two lettuce cultivars grown under various combinations of blue and red light-emitting diodes. *HortScience* **2013**, *48*, 988–995.
3. Christopoulou, M.; Wo, S.R.C.; Kozik, A.; McHale, L.K.; Truco, M.J.; Wroblewski, T.; Michelmore, R.W. Genome-wide architecture of disease resistance genes in lettuce. *G3 Genes Genomes Genet.* **2015**, *5*, 2655–2669. [CrossRef] [PubMed]
4. Amoozgar, A.; Mohammadi, A.; Sabzalian, M. Impact of light-emitting diode irradiation on photosynthesis, phytochemical composition and mineral element content of lettuce cv. Grizzly. *Photosynthetica* **2017**, *55*, 85–95. [CrossRef]

5. Nawaz, M.A.; Wang, L.; Jiao, Y.; Chen, C.; Zhao, L.; Mei, M.; Yu, Y.; Bie, Z.; Huang, Y. Pumpkin rootstock improves nitrogen use efficiency of watermelon scion by enhancing nutrient uptake, cytokinin content, and expression of nitrate reductase genes. *Plant Growth Regul.* **2017**, *82*, 233–246. [CrossRef]
6. Nawaz, M.A.; Chen, C.; Shireen, F.; Zheng, Z.; Sohail, H.; Afzal, M.; Ali, M.A.; Bie, Z.; Huang, Y. Genome-wide expression profiling of leaves and roots of watermelon in response to low nitrogen. *BMC Genom.* **2018**, *19*, 456. [CrossRef] [PubMed]
7. Scheible, W.-R.; Morcuende, R.; Czechowski, T.; Fritz, C.; Osuna, D.; Palacios-Rojas, N.; Schindelasch, D.; Thimm, O.; Udvardi, M.K.; Stitt, M. Genome-wide reprogramming of primary and secondary metabolism, protein synthesis, cellular growth processes, and the regulatory infrastructure of *Arabidopsis* in response to nitrogen. *Plant Physiol.* **2004**, *136*, 2483–2499. [CrossRef]
8. Yuan, L.; Yuan, Y.; Du, J.; Sun, J.; Guo, S. Effects of 24-epibrassinolide on nitrogen metabolism in cucumber seedlings under $Ca(NO_3)_2$ stress. *Plant Physiol. Biochem.* **2012**, *61*, 29–35. [CrossRef] [PubMed]
9. Amalfitano, C.; Gomez, L.D.; Frendo, P.; De Pascale, S.; Pepe, O.; Simister, R.; Ventorino, V.; Agrelli, D.; Borrelli, C.; McQueen-Mason, S.J.; et al. Plant–Rhizobiumsymbiosis, seed nutraceuticals, and waste quality for energy production of *Vicia faba* L. as affected by crop management. *Chem. Biol. Technol. Agric.* **2018**, *5*, 15. [CrossRef]
10. Jämtgård, S.; Näsholm, T.; Huss-Danell, K. Nitrogen compounds in soil solutions of agricultural land. *Soil Biol. Biochem.* **2010**, *42*, 2325–2330. [CrossRef]
11. Marschner, H. *Marschner's Mineral Nutrition of Higher Plants*; Academic Press: New York, NY, USA, 2011.
12. Caruso, G.; Conti, S.; La Rocca, G. Influence of crop cycle and nitrogen fertilizer form on yield and nitrate content in different species of vegetables. *Adv. Hort. Sci.* **2011**, *25*, 81–89.
13. Silva, L.F.; Hower, J.C.; Izquierdo, M.; Querol, X. Complex nanominerals and ultrafine particles assemblages in phosphogypsum of the fertilizer industry and implications on human exposure. *Sci. Total Environ.* **2010**, *408*, 5117–5122. [CrossRef]
14. Verger, P.J.; Boobis, A.R. Reevaluate pesticides for food security and safety. *Science* **2013**, *341*, 717–718. [CrossRef]
15. du Jardin, P. Plant biostimulants: Definition, concept, main categories and regulation. *Sci. Hortic.* **2015**, *196*, 3–14. [CrossRef]
16. Paschalidis, K.A.; Roubelakis-Angelakis, K.A. Spatial and temporal distribution of polyamine levels and polyamine anabolism in different organs/tissues of the tobacco plant. Correlations with age, cell division/expansion, and differentiation. *Plant Physiol.* **2005**, *138*, 142–152. [CrossRef] [PubMed]
17. Liu, H.; Su, X.; Li, X.; Zhao, X.; Zang, L.; Pan, W. Development of prolonged release microspheres of metformin hydrochloride using ion exchange resins. *J. Chin. Pharm. Sci.* **2006**, *15*, 155.
18. Pang, J.; Ross, J.; Zhou, M.; Mendham, N.; Shabala, S. Amelioration of detrimental effects of waterlogging by foliar nutrient sprays in barley. *Funct. Plant Biol.* **2007**, *34*, 221–227. [CrossRef]
19. Paungfoo-Lonhienne, C.; Thierry, G.; Lonhienne, A.; Doris, R.; Nicole, R.; Michael, C.; Richard, I.W.; Gamage, H.K.; Bernard, J.C.; Peer, M.S.; et al. Plants can use protein as a N source without assistance from other organisms. *Proc. Natl. Acad. Sci. USA* **2008**, *105*, 4524–4529. [CrossRef] [PubMed]
20. Rouphael, Y.; Colla, G. Synergistic Biostimulatory Action: Designing the Next Generation of Plant Biostimulants for Sustainable Agriculture. *Front. Plant Sci.* **2018**, *9*, 1655. [CrossRef]
21. Rouphael, Y.; Spíchal, L.; Panzarová, K.; Casa, R.; Colla, G. High-Throughput Plant Phenotyping for Developing Novel Biostimulants: From Lab to Field or From Field to Lab? *Front. Plant Sci.* **2018**, *9*, 1197. [CrossRef]
22. Kowalczyk, K.; Zielony, T.; Gajewski, M. Effect of Aminoplant and Asahi on yield and quality of lettuce grown on rockwool. *Biostimulators Mod. Agric. Veg. Crops* **2008**, 35–43.
23. Zhao, Y. Auxin biosynthesis and its role in plant development. *Annu. Rev. Plant Biol.* **2010**, *61*, 49–64. [CrossRef] [PubMed]
24. Maeda, H.; Dudareva, N. The shikimate pathway and aromatic amino acid biosynthesis in plants. *Annu. Rev. Plant Biol.* **2012**, *63*, 73–105. [CrossRef] [PubMed]
25. Calvo, P.; Nelson, L.; Kloepper, J. Agricultural uses of plant biostimulants. *Plant Soil* **2014**, *383*, 3–41. [CrossRef]
26. Forde, B.G.; Roberts, M.R. Glutamate receptor-like channels in plants: A role as amino acid sensors in plant defence? *F1000Prime Rep.* **2014**, *6*, 37. [CrossRef]

27. Miller, A.J.; Fan, X.; Shen, Q.; Smith, S.J. Amino acids and nitrate as signals for the regulation of nitrogen acquisition. *J. Exp. Bot.* **2007**, *59*, 111–119. [CrossRef]
28. Walch-Liu, P.; Forde, B.G. L-Glutamate as a novel modifier of root growth and branching: What's the sensor? *Plant Signal. Behav.* **2007**, *2*, 284–286. [CrossRef]
29. Weiland, M.; Mancuso, S.; Baluska, F. Signalling via glutamate and GLRs in *Arabidopsis thaliana*. *Funct. Plant Biol.* **2016**, *43*, 1–25. [CrossRef]
30. Halpern, M.; Bar-Tal, A.; Ofek, M.; Minz, D.; Muller, T.; Yermiyahu, U. The Use of Biostimulants for Enhancing Nutrient Uptake. In *Advances in Agronomy*; Sparks, D.L., Ed.; Academic Press: New York, NY, USA, 2015; Volume 130, pp. 141–174.
31. Teixeira, W.F.; Fagan, E.B.; Soares, L.H.; Umburanas, R.C.; Reichardt, K.; Neto, D.D. Foliar and seed application of amino acids affects the antioxidant metabolism of the soybean crop. *Front. Plant Sci.* **2017**, *8*, 327. [CrossRef]
32. Ertani, A.; Pizzeghelio, D.; Altissimo, A.; Nardi, S. Use of meat hydrolyzate derived from tanning residues as plant biostimulant for hydroponically grown maize. *J. Plant Nutr. Soil Sci.* **2013**, *176*, 287–296. [CrossRef]
33. Hildebrandt, T.M.; Nesi, A.N.; Araújo, W.L.; Braun, H.-P. Amino acid catabolism in plants. *Mol. Plant* **2015**, *8*, 1563–1579. [CrossRef] [PubMed]
34. Persson, J.; Högberg, P.; Ekblad, A.; Högberg, M.N.; Nordgren, A.; Näsholm, T. Nitrogen acquisition from inorganic and organic sources by boreal forest plants in the field. *Oecologia* **2003**, *137*, 252–257. [CrossRef]
35. Gioseffi, E.; de Neergaard, A.; Schjørring, J.K. Interactions between uptake of amino acids and inorganic nitrogen in wheat plants. *Biogeosciences* **2012**, *9*, 1509–1518. [CrossRef]
36. Koukounaras, A.; Tsouvaltzis, P.; Siomos, A.S. Effect of root and foliar application of amino acids on the growth and yield of greenhouse tomato in different fertilization levels. *J. Food Agric. Environ.* **2013**, *11*, 644–648.
37. El-Aal, M.A.; Eid, R.S. Effect of foliar spray with lithovit and amino acids on growth, bioconstituents, anatomical and yield features of soybean plant. *Plant Biotechnol.* **2018**, 187–201.
38. SH SADAK, M.; Abdelhamid, M.T.; Schmidhalter, U. Effect of foliar application of aminoacids on plant yield and some physiological parameters in bean plants irrigated with seawater. *Acta Biol. Colomb.* **2014**, *20*, 141–152.
39. El-Awadi, M.; El-Bassiony, A.; Fawzy, Z.; El-Nemr, M. Response of snap bean (*Phaseolus vulgaris* L) plants to nitrogen fertilizer and foliar application with methionine and tryptophan. *Nat. Sci.* **2011**, *9*, 87–94.
40. Yong, J.W.H.; Letham, S.D.; Wong, C.S.; Graham, D.F. Rhizobium-induced elevation in xylem cytokinin delivery in pigeon pea induces changes in shoot development and leaf physiology. *Funct. Plant Biol.* **2014**, *41*, 1323–1335. [CrossRef]
41. Paciorek, T.; Zazímalová, E.; Ruthardt, N.; Petrásek, J.; Stierhof, Y.D.; Kleine-Vehn, J.; Morris, D.A.; Emans, N.; Jürgens, G.; Geldner, N.; et al. Auxin inhibits endocytosis and promotes its own efflux from cells. *Nature* **2005**, *435*, 1251–1256. [CrossRef]
42. Yakhin, O.I.; Aleksandr, A.L.; Ildus, A.Y.; Patrick, H.B. Biostimulants in Plant Science: A Global Perspective. *Front. Plant Sci.* **2017**, *7*, 2049. [CrossRef]
43. Shekari, G.; Javanmardi, J. Effects of Foliar Application Pure Amino Acid and Amino Acid Containing Fertilizer on Broccoli (*Brassica oleracea* L. var. italica) Transplant. *Adv. Crop Sci. Tech.* **2017**, *5*, 280. [CrossRef]
44. Jiang, W.J. *New Techniques for Soilless Cultivation of Vegetables*, rev. ed.; Beijing Jindun Publishing House: Beijing, China, 2007; pp. 197–200.
45. Hunt, R. Plant growth curves. In *The Functional Approach to Plant Growth Analysis*; Edward Arnold Ltd.: London, UK, 1982.
46. Yoshida, S.; Kitano, M.; Eguchi, H. growth of lettuce plants (*Lactuca sativa* L.) under control of dissolved O_2 concentration in hydroponics. *Biotronics* **1997**, *26*, 39–45.
47. Woltz, S.; Jackson, C. Production of yellow strapleaf of chrysanthemum & similar disorders by amino acid treatment. *Plant Physiol.* **1961**, *36*, 197. [PubMed]
48. León, A.P.; Martín, J.P.; Chiesa, A. Vermicompost application and growth patterns of lettuce (*Lactuca sativa* L.). *Agric. Tropica Subtrop.* **2012**, *45*, 134–139. [CrossRef]
49. Medek, D.E.; Ball, M.C.; Schortemeyer, M. Relative contributions of leaf area ratio and net assimilation rate to change in growth rate depend on growth temperature: Comparative analysis of subantarctic and alpine grasses. *New Phytol.* **2007**, *175*, 290–300. [CrossRef] [PubMed]

50. Gao, Y.; Yu, H.; Liu, P.; Ma, C.; Li, Q.; Jiang, W. Ending composting during the thermophilic phase improves cultivation substrate properties and increasing winter cucumber yield. *Waste Manag.* **2018**, *79*, 260–272. [CrossRef]
51. Ahn, Y.S. Plant analysis for evaluating plant nutrition. In *International Training Workshop on Soil Test and Plant Analysis*; RDA& FFTC/ASPAC, 1987.
52. Snedecor, G.; Cochran, W. *Statistical Method*, 7th ed.; The Iowa State University Press: Ames, IA, USA, 1980; pp. 39–63.
53. Colla, G.; Nardi, S.; Cardarelli, M.; Ertani, A.; Lucini, L.; Canaguier, R.; Rouphael, Y. Protein hydrolysates as biostimulants in horticulture. *Sci. Hortic.* **2015**, *196*, 28–38. [CrossRef]
54. Liu, X.Q.; Chen, H.Y.; Ni, Q.X.; Kyu, S.L. Evaluation of the role of mixed amino acids in nitrate uptake and assimilation in leafy radish by using 15N-labeled nitrate. *Agric. Sci. China* **2008**, *7*, 1196–1202. [CrossRef]
55. Tsouvaltzis, P.; Koukounaras, A.; Siomos, A.S. Application of amino acids improves lettuce crop uniformity and inhibits nitrate accumulation induced by the supplemental inorganic nitrogen fertilization. *Int. J. Agric. Biol.* **2014**, *16*, 951–955.
56. Wang, H.J.; Wu, L.H.; Wang, M.Y.; Zhu, Y.H.; Tao, Q.N.; Zhang, F.S. Effects of amino acids replacing nitrate on growth, nitrate accumulation, and macroelement concentrations in pakchoi (*Brassica chinensis* L.). *Pedosphere* **2007**, *17*, 595–600. [CrossRef]
57. Mobini, M.; Khoshgoftarmanesh, A.H.; Ghasemi, S. The effect of partial replacement of nitrate with arginine, histidine, and a mixture of amino acids extracted from blood powder on yield and nitrate accumulation in onion bulb. *Sci. Hortic.* **2014**, *176*, 232–237. [CrossRef]
58. Vincill, E.D.; Bieck, A.M.; Spalding, E.P. Ca2+ conduction by an amino acid-gated ion channel related to glutamate receptors. *Plant Physiol.* **2012**, *159*, 40–46. [CrossRef] [PubMed]
59. Santi, C.; Zamboni, A.; Varanini, Z.; Pandolfini, T. Growth stimulatory effects and genome-wide transcriptional changes produced by protein hydrolysates in maize seedlings. *Front. Plant Sci.* **2017**, *8*, 433. [CrossRef] [PubMed]
60. Price, C.A. Leaf Gui: Analyzing the geometry of veins and areoles using image segmentation algorithms. In *High-Throughput Phenotyping in Plants*; Springer: New York, NY, USA, 2012; pp. 41–49.
61. Singh, R.; Parihar, P.; Prasad, S.M. Sulfur and calcium simultaneously regulate photosynthetic performance and nitrogen metabolism status in As-challenged *Brassica juncea* L. seedlings. *Front. Plant Sci.* **2018**, *9*, 772. [CrossRef] [PubMed]
62. Sebastian, A.; Prasad, M. Vertisol prevent cadmium accumulation in rice: Analysis by ecophysiological toxicity markers. *Chemosphere* **2014**, *108*, 85–92. [CrossRef]
63. El-Shiaty, O.H.; El-Sharabasy, S.F.; Abd El-Kareim, A.H. Effect of some amino acids and biotin on callus and proliferation of date palm (*Phoenix dactylifera* L.) Sewy cultivar. *Arab. J. Biotechnol.* **2004**, *7*, 265–272.
64. Kakkar, R.; Nagar, P.; Ahuja, P.; Rai, V. Polyamines and plant morphogenesis. *Biol. Plant.* **2000**, *43*, 1–11. [CrossRef]
65. Padgett, P.E.; Leonard, R.T. Free amino acid levels and the regulation of nitrate uptake in maize cell suspension cultures. *J. Exp. Bot.* **1996**, *47*, 871–883. [CrossRef]
66. Shafeek, M.; Helmy, Y.; Magda, A.; Shalaby, F.; Nadia, M.O. Response of onion plants to foliar application of sources and levels of some amino acid under sandy soil conditions. *J. Appl. Sci. Res.* **2012**, *8*, 5521–5527.
67. Davies, P.J. *Plant Hormones: Biosynthesis, Signal Transduction, Action!*; Springer: New York, NY, USA, 2004.
68. Colla, G.; Cardarelli, M.; Bonini, P.; Rouphael, Y. Foliar Applications of Protein Hydrolysate, Plant and Seaweed Extracts Increase Yield but Differentially Modulate Fruit Quality of Greenhouse Tomato. *HortScience* **2017**, *52*, 1214–1220. [CrossRef]
69. Rouphael, Y.; Colla, G.; Giordano, M.; El-Nakhel, C.; Kyriacou, M.C.; De Pascale, S. Foliar applications of a legume-derived protein hydrolysate elicit dosedependent increases of growth, leaf mineral composition, yield and fruit quality in two greenhouse tomato cultivars. *Sci. Hortic.* **2017**, *226*, 353–360. [CrossRef]
70. Bahari, A.; Pirdashti, H.; Yaghubi, M. The effects of amino acid fertilizers spraying on photosynthetic pigments and antioxidant enzymes of wheat (*Triticum aestivum* L.) under salinity stress. *Int. J. Agron. Plant Prod.* **2013**, *4*, 787–793.
71. Anne, C.; Thomas, S. Regulation of chloroplast development and function by cytokinin. *J. Exp. Bot.* **2015**, *66*, 4999–5013.

72. Fawzy, Z.; El-Shal, Z.; Yunsheng, L.; Zhu, O.; Sawan, O.M. Response of garlic (*Allium Sativum* L.) plants to foliar spraying of some bio-stimulants under sandy soil condition. *J. Appl. Sci. Res.* **2012**, *8*, 770–776.
73. Romero, I.; Téllez, J.; Yamanaka, L.E.; Steindel, M.; Romanha, A.J.; Grisard, E.C. Transsulfuration is an active pathway for cysteine biosynthesis in *Trypanosoma rangeli*. *Parasites Vectors* **2014**, *7*, 197. [CrossRef] [PubMed]
74. Abbas, S.H.; Sohail, M.; Saleem, M.; Mahmood, T.; Aziz, I.; Qamar, M.; Majeed, A.; Arif, M. Effect of L-tryptophan on plant weight and pod weight in chickpea under rainfed conditions. *SciTech Dev.* **2013**, *32*, 277–280.
75. Herve, J.J. Biostimulants, a new concept for the future; prospects offered by the chemistry of synthesis and biotechnology. *C. R. Acad. Agric. Fr.* **1994**, *80*, 91–102.
76. Yunna, Z.; Gang, L.; Houcheng, L.; Guangwen, S.; Riyuan, C.; Shiwei, S. Effects of partial replacement of nitrate with different nitrogen forms on the yield, quality and nitrate content of Chinese kale. *Commun. Soil Sci. Plant. Anal.* **2018**, 1384–1393.
77. Chen, G.L.; Gao, X.R.; Zhang, X.B. Effect of partial replacement of nitrate by amino acid and urea on nitrate content of non-heading Chinese cabbage and lettuce in hydroponic condition. *Agric. Sci. China* **2002**, 444–449.
78. Valpuesta, V.; Botella, M. Biosynthesis of L-ascorbic acid in plants: New pathways for an old antioxidant. *Trends Plant Sci.* **2004**, *9*, 573–577. [CrossRef]
79. Podsedek, A. Natural antioxidants and antioxidant capacity of Brassica vegetables: A review. *LWT-Food Sci. Technol.* **2007**, *40*, 1–11. [CrossRef]
80. John, S.J.; Guha-Mukherjee, S. *Plant Molecular Biology and Biotechnology*, 1st ed.; Narosa Pub. House: New Delhi, India, 1997; pp. 17–28.
81. El-sharabasy, S.; Fatma, I.; Gehan, H.; El-Dawayaty, M. Effect of different amino acids at different concentrations on multiplication and rooting stage of in vitro propagation of strawberries (Fragaria X Ananassa Duch cv. Chandler). *Egypt. J. Genet. Cytol.* **2015**, *44*, 31–34.
82. Rosati, A.; Day, K.; DeJong, T. Distribution of leaf mass per unit area and leaf nitrogen concentration determine partitioning of leaf nitrogen within tree canopies. *Tree Physiol.* **2000**, *20*, 271–276. [CrossRef] [PubMed]
83. Thimann, K.V. Auxins and the inhibition of plant growth. *Biol. Rev.* **1939**, *14*, 314–337. [CrossRef]

agronomy

MDPI

Review

Biostimulants Application in Horticultural Crops under Abiotic Stress Conditions

Roberta Bulgari, Giulia Franzoni * and Antonio Ferrante

Department of Agricultural and Environmental Sciences, Università degli Studi di Milano, via Celoria 2, 20133 Milano, Italy; roberta.bulgari@unimi.it (R.B.); antonio.ferrante@unimi.it (A.F.)
* Correspondence: giulia.franzoni@unimi.it; Tel.: +39-02-503-16593

Received: 10 April 2019; Accepted: 10 June 2019; Published: 12 June 2019

Abstract: Abiotic stresses strongly affect plant growth, development, and quality of production; final crop yield can be really compromised if stress occurs in plants' most sensitive phenological phases. Additionally, the increase of crop stress tolerance through genetic improvements requires long breeding programmes and different cultivation environments for crop performance validation. Biostimulants have been proposed as agronomic tools to counteract abiotic stress. Indeed, these products containing bioactive molecules have a beneficial effect on plants and improve their capability to face adverse environmental conditions, acting on primary or secondary metabolism. Many companies are investing in new biostimulant products development and in the identification of the most effective bioactive molecules contained in different kinds of extracts, able to elicit specific plant responses against abiotic stresses. Most of these compounds are unknown and their characterization in term of composition is almost impossible; therefore, they could be classified on the basis of their role in plants. Biostimulants have been generally applied to high-value crops like fruits and vegetables; thus, in this review, we examine and summarise literature on their use on vegetable crops, focusing on their application to counteract the most common environmental stresses.

Keywords: plant biostimulant; environmental stress; vegetables

1. Abiotic Stresses

Plants are continuously subjected to a multitude of stressful events, from seed germination through to the whole life cycle. These stresses are commonly divided into two categories—biotic and abiotic stresses—depending on the nature of the trigger factor. The first are caused by other living organisms, including insects, bacteria, fungi, and weeds that affect plant development and productivity. The second are generally linked with the climatic, edaphic, and physiographic components of the environment, when they are limiting factors of plant growth and survival. The most important abiotic stresses limiting agricultural productivity, almost all over the world, are drought, salinity, non-optimal temperatures, and low soil fertility. Among these, drought, and nutrient deficiencies are major problems, mostly in developing countries where the incomes of rural people depend on agriculture [1]. Actually, in "The State of Food and Agriculture 2007", FAO reported that only 3.5% of the global land area is not affected by some environmental constraints. In 1982, Boyer estimated that yield losses caused by unfavourable environments were as much as 70% [2,3]. Farooq et al. [4] reported that drought induced a reduction of yield between 13% and 94% in several crops, depending on the intensity and duration of the stress. Afterwards, Cramer et al. [5] estimated the impacts of different abiotic stresses on crop production in terms of the percentage of global land area affected, considering the 2000 and 2007 FAO reports. They also referred to the increasing number of publications focused on this topic between 2001 and 2011. The exact impact of these changes on agricultural systems is extremely difficult to predict and it depends on numerous parameters that are all not always included in predictive models. Even if some projections show that positive and negative outcomes on crop production could be balanced in the medium term, several studies agree that in the long term, the negative ones will prevail [6,7]. Based on future scenarios, adaptation and mitigation are essential to increase the resilience capacity of agricultural systems and to ensure crops yield and quality. Since environmental conditions cannot be controlled, several strategies on different levels are required, such as agronomical techniques or breeding of more tolerant cultivars [8].

In 2010, at the society's annual conference, Vegetable Breeding and Stress Physiology working groups of the American Society for Horticultural Sciences focused particularly on the "Improvement of Horticultural Crops for Abiotic Stress Tolerance" considering the effects of climate change [9]. Up to now, most studies on climate change impacts focus on major crops, and only few papers pay attention to fruit and vegetable in terms of production, quality, and supply chain [10,11]. An important aspect to take into consideration is the effect of the combination of different stressful factors. Most of the time, crops are subjected to several abiotic stresses that occur simultaneously in the field. In these situations, studying the stresses separately is not enough because plant response is unique and cannot be predicted by the reply obtained when each factor is applied individually [12–14]. Moreover, biotic and abiotic components typically interact in an ecosystem. For instance, environmental conditions affect plant-pest interaction in different ways, by decreasing plant tolerance or increasing the risk of pathogen infection [15,16].

Focusing on horticultural species, the tolerance to abiotic stresses is an important trait because their cash value is usually higher than field crops, they require more resources for farming and because they provide a source of many nutrients, fibre, minerals, and carbohydrates, which are essential in a healthy diet [17]. Food and Agriculture Organization (FAO) reports that about 90% of essential vitamin C and 60% of vitamin A for human comes from vegetables. Indeed, low fruit and vegetable intake is a major contributing risk factor to several widespread and debilitating nutritional diseases. According to the Global Burden of Disease Study, 3.4 million deaths can be attributed to low consumption of fruit and 1.8 million to low vegetables diets worldwide [18]. Therefore, growing high-quality vegetables becomes one of the most important goals of current agriculture, in order to meet the needs of the population and the increasing demand for fruit and vegetables. Abiotic stresses do not only affect the yield but also the quality of these products, triggering morphological, physiological and biochemical changes that can alter the visual appearance and/or the nutraceutical value in a way that the product could become unmarketable [19]. Bisbis et al. [11] investigated the double effects of elevated temperature and increased CO_2 on the physiology of different vegetables. They observed several responses according to plant species and severity of the stress, taking into consideration the possible adaptation strategies that could be implemented in order to mitigate the effects of climate change. Nonetheless, these mechanisms are still under-researched and should be studied in depth, because not only different species but different cultivars also could respond differently to the same environmental stress. For example, cultivars with low levels of antioxidants are particularly vulnerable to oxidative stress compared to those with high antioxidant activity [20–23]. This aspect has a particular importance as selection criterion in the choice of appropriate cultivars for a specific situation. Oxidative stress is a common phenomenon caused by several adverse conditions; it generally occurs when the balance between the production of reactive oxygen species (ROS) and the quenching activity is upset by a stressful event [24]. Low levels of ROS are normally produced by different reactions during physiological metabolisms like photosynthesis or respiration, and they play an important signaling role in plant growth and development. Their amount dramatically increases under abiotic stress conditions and, if not controlled could result in cellular damage and death. Besides their toxicity to proteins, lipids or nucleic acids, the increased production of ROS under stressful conditions plays a key role in the complex signaling network of plants stress responses. Their concentration is maintained at non-toxic levels by the activity of the antioxidant system: a wide range of enzymatic or non-enzymatic antioxidant molecules are accumulated in plant tissues to quench ROS induced by stress [25–28]. Moreover, the maintenance of this equilibrium is also dependent on numerous factors, such as the timing of stress application, its intensity and duration. Indeed, moderate or controlled stress conditions could have a positive effect on quality traits of several crops [29]. For example, water deprivation might be a useful crop management strategy to improve the quality of lettuce and fleshy fruits in terms of nutritive and health-promoting value and taste, by stimulating the secondary metabolism and concentration of different phytochemicals such as α-tocopherol, β-carotene, flavonoid and so on [30,31]. Besides the production of ROS scavenging compounds, plants also increase the biosynthesis and accumulation of compatible solutes with an osmoprotective role, like sugars and proline.

Plants generally reply to non-optimal environmental conditions both with short- and long-term adaptation strategies, by the activation and regulation of the expression of specific stress associated genes [32,33].

Since plants are sessile organisms and they have to cope with adverse external conditions: all these mechanisms are essential for their survival. These strategies are effective if they are activated in time, in order to set a defense response and anticipate the environmental changes that might affect plant growth irreversibly. The trade-off between growth and acclimation metabolisms results in a sort of fitness cost for plants, since energy and nutrients normally destined to growth and production are intended for stress responsive mechanisms [34].

Agronomic management conducted in order to enhance plant tolerance towards abiotic stresses evolved over the centuries due to the technologic progress, climate change, scientific knowledge, and farmers' experiences. The choice of the correct cultivar, the best growing period, the sowing density, and the amount of water or fertilizers are some of the most common strategies applied to mitigate the negative effects of abiotic stresses [8]. Protected cultivation is a cropping technique adopted to preserve plants from unfavourable outdoor conditions. It is mainly suited to vegetables and floriculture production in a non-optimal environment, through the control of temperatures, radiation or atmospheric composition. Another agronomical strategy, especially applied in vegetable crops, is soilless cultivation. This approach allows controlling of water and nutrients, avoiding the use of soil for cultivation and all the problems related to it, like poor quality or contamination.

Grafting is an additional tool adopted to counteract environmental stresses and increase tolerance in vegetable crops. This technique is applied especially to high-yielding fruits and vegetables such as cucurbits and solanaceous to enhance tolerance against saline soil, nutrient or water deficiency, heavy metals or pollutants toxicity [35–37].

Agronomical strategies are essential in mitigating the negative effect of several abiotic stresses, but sometimes their application is not enough. Moreover, current experiments aim to transfer one or more genes involved in signaling or regulatory pathways, or genes encoding to molecules, such as osmolytes and antioxidants, conferring tolerance to a specific abiotic stress [38]. Several functional and regulatory genes involved in abiotic stress tolerance have been identified and studied. Results of these studies can be exploited for genetic improvement aiming to introduce tolerance traits in cultivated crops. Since different physiological traits related to stress tolerance are under multigenic control, the manipulation of a single gene generally is not enough. Hence, scientists have paid more attention to regulatory genes, including transcription factors, due to their ability to regulate a vast array of downstream stress-responsive genes at a time [39–41].

However, the huge existing genetic variability among vegetable species, the lack of knowledge about minor cultivars genome, the complex responses triggered by abiotic stress conditions and the limited strategies currently available make genetic improvement really difficult and often inefficient. Moreover, besides the wide diversity of germplasms available, plant tolerance to stress depends both on stress features such as duration, severity, and frequency, as well as the affected tissues and development stages of crops [24,42–44].

Additionally, the increase of crop tolerance through genetic improvements requires many years of work and different cultivation environments that cannot be always taken into consideration. As a result, several new cultivars that can be used by the growers are released each year.

Another technique widely used for developing stress tolerance in plants is in vitro selection. This culture-based tool allows better understanding of several plants' physiological and biochemical responses to adverse environmental conditions. It has been applied specially to obtain salt/ and drought/tolerant lines in a wide range of plant species, including vegetables [45]. In vitro selection is based on the induction of a genetic variation among cells, tissues or organs, their exposure to a stressor, and the subsequent regeneration of the whole organism starting from the surviving cells [46]. Even if in vitro selection is a less expensive and time-saving approach compared with classic molecular engineering, some limitations, mostly concerning the stability of the selected traits and epigenetic adaptation, still exist.

In addition to these strategies, it has been observed that stress tolerance can also be induced by biostimulants or specific bioactive compounds, if they are applied on vegetable crops when they really need to be protected [47–49]. Biostimulant application on horticultural crops under environmental stress conditions will be discussed in detail below.

2. Biostimulants

Biostimulant products have been considered innovative agronomic tools as demonstrated by the increase of scientific publications and by the constant expansion of their market [50]. France, Italy, and Spain are the leading EU countries in the production of biostimulants [51]. According to a new report by Grand View Research, Inc., the biostimulant market size is expected to reach USD 4.14 billion by 2025 [52]. The complex nature of the composition of these products and the wide range of molecules contained makes it complicated to understand and define which compounds are the most active. The isolation and study of a single component is almost impossible and the efficacy of a biostimulant is not due to a single compound but is the consequence of the synergistic action of different bioactive molecules. Moreover, the application rules and time are not always clear. For all these reasons, the European Commission developed a proposal for a new regulatory framework and a draft for a new fertilizer regulation was prepared in 2016. The amendments to the proposal of the European Commission were adopted by the European Parliament in October 2017, while the legislative resolution on the proposal was approved on 27 March 2019 [53–55].

Plant biostimulants are defined as products obtained from different organic or inorganic substances and/or microorganisms, that are able to improve plant growth, productivity and alleviate the negative effects of abiotic stresses [56,57]. Mineral elements, vitamins, amino acids, and poly- and oligosaccharides, trace of natural plant hormones are the most known components. However, it is important to underline that the biostimulant activity must not depend on the product's nutrients or natural plant hormones content. The mechanisms activated by biostimulants are often difficult to identify and are still under investigation [58]. High-throughput phenotyping and omic technologies seem to be useful approaches to understand biostimulants activity and hypothesize a mode of action [59–61]. They can act directly on plant physiology and metabolism by improving soil conditions [62,63]. They are able to modify some molecular processes that allow to improve water and nutrient use efficiency of crops, stimulate plant development, and counteract abiotic stresses [47] by enhancing primary and secondary metabolism [55,61,63].

One of the key points of the discussion is about the application of these products in stressful conditions and their role as nutrients, not with a curative function. In particular, if a product has a direct effect against biotic stresses, it should not be included in the biostimulant category but should be registered as plant protection products.

2.1. Classification of Biostimulants in Categories

During the years, different authors have proposed several categorizations of biostimulant products on the basis of their main component or mode of action. In many countries outside the European Union, both kinds of information must be reported on the label in order to register these products [55]. The current classification is based on source of raw material, even if this choice does not always provide the correct information about the biological activity of the product [56]. Thus, biostimulants are classified as these major groups:

Humic substances (*HSs*): they include humic acids, fulvic acids and humins. *HSs* are natural constituents of soil organic matter, resulting from the decomposition processes of plants, animals, and microbial residues, but also from the metabolic activity of soil microbes [57]. It has been observed that treatments with humic substances stimulate plants root growth and development [64,65]. This is reflected in a better uptake of nutrients and water, and enhanced tolerance to environmental stresses, [66,67]. How the *HSs* affect plant physiology is not fully understood. This is due to the molecular complexity of these substances and to the abundance and diversity of plants responses altered by their application. Moreover, a strong relationship between medium properties and *HSs* bioactivity has been reported [68]. The positive effects exerted by these complex aggregates could be ascribed both to the hormone-like activity of some of their component and also to IAA-independent mechanisms [69]. For example, like auxins, *HSs* are able to promote plant growth and induce H^{+}ATPase activity in plasma membrane [70–72].

Seaweed extracts: seaweeds are a vast group of macroscopic, multicellular marine algae that can be brown, red, and green. They are an important source of organic matter and fertilizer nutrients. Seaweed extracts have been used in agriculture as soil conditioners or plant stimulators. They are applied as foliar spray and are able to enhance plant growth, abiotic stresses tolerance, photosynthetic activity, and resistance to fungi, bacteria and virus, improving yield and productivity of several crops [73–75]. Seaweeds used for biostimulant production contain cytokinins and auxins or other hormone-like substances [76]. They also contain many active mineral and organic compounds, including complex polysaccharides such as laminarin, fucoidan, alginates and plant hormones that contribute to plant growth [77]. Recently the potential application of micro-algae as plant biostimulants has been considered [78–80].

Hydrolysed proteins and amino acids containing products: hydrolysed proteins are a mixture of amino acids, peptides, polypeptides and denatured proteins that can be obtained by chemical, enzymatic and thermal hydrolysis of proteins (or by combining these different hydrolysis types) from both plant and animal sources [67,81]. Studies reported that the applications of some commercial protein hydrolysate products from animal origin were phytotoxic, having negative effects on plant growth when compared to a commercial protein hydrolysate of plant origin [82,83]. In another study, Botta et al. [84] observed that lettuce plants treated with an animal-based protein hydrolysed had a higher fresh and dry weight compared with the control. Generally, they can induce plant defense responses and increase plant tolerance to many abiotic stresses, as reported by several authors [85–88].

Microorganisms: this group includes bacteria, yeast, filamentous fungi, and micro-algae. They are isolated from soil, plants, water, and composted manures or other organic materials. They are applied to soil to increase crop productivity through metabolic activities. They enhance the uptake of nutrients through nitrogen fixation and the solubilization of nutrients, they modify a hormonal status by inducing plant hormones biosynthesis such as auxins, cytokinins, etc.; they also enhance tolerance to abiotic stresses and produce volatile organic compounds (VOCs), which may also have a direct effect on plants. Plant growth-promoting rhizobacteria (PGPR) are able to ameliorate plant responses to abiotic stresses stimulating physical, chemical and biological activities [89,90]. Positive effects are given by microorganisms that form a protective biofilm on root surface enhancing nutrient and water uptake.

Another category of biostimulants includes those derived from extracts of food waste or industrial waste streams, composts and compost extracts, manures, vermicompost, aquaculture residues and waste streams, and sewage treatments among others [91]. Biostimulants derived from agro-industrial by-products were reported to be effective in improving plant productivity, increasing the synthesis of secondary compounds involved in several plant physiological responses, and enhancing the activity of the enzyme phenylalanine ammonia lyase (PAL E.C. 4.3.1.5) [92]. The effect of biostimulant application on PAL activity and on the expression of genes encoding for this enzyme was observed by several authors [56,88,89] and references therein, even if at present it is not possible to define if this is a direct or indirect effect. Because of the diversity of source materials and extraction technologies, the mode of action of these products is not easily determined [55]. The use of by-products as raw material that can be transformed into fertilizing products is the idea underlying the new fertiliser regulation and the Circular Economy Action Plan, which is focused on reaching a sustainable agriculture. The guidelines for fertiliser regulation, the need to produce in a more environmentally friendly cultivation system maintaining good crop yield and quality, the increase in price of synthetic fertilizer, the withdrawn of several agrochemicals and the multifaceted effects on plants or soil of biostimulants are favouring the expansion of this market.

A new category of biostimulant products, including nanoparticles and nanomaterials, has been recently proposed by Juárez-Maldonado et al. [93]. Nanoparticles and nanomaterials are usually defined as particles with dimensions between about 1 nm and 100 nm that show properties that are not found in their bulk form. They are able to modify the quality of the production and the tolerance to abiotic stresses when applied in small quantities as foliar spray or in nutrient solution, also in vegetable crops [94–97]. Their biostimulant properties seems to be associated with the structure and nature of the materials. The interaction between plant and nanoparticles and nanomaterials surfaces can positively affect ions and metabolites transport and receptors activity by modifying the surrounding environment in terms of energy and charges. This activity is not dependent on chemical composition. Moreover, nanoparticles and nanomaterials release chemical elements like iron or carbon that could be useful for plant when are metabolised.

A study showed that application of zinc oxide nanoparticles on tomato as soil amendment or by foliar spray increased plant height, chlorophyll and total soluble protein content [98].

2.2. *Effect of Biostimulants on Chlorophyll Content, Photosynthesis and Growth in Vegetables*

Biostimulants can be used in vegetable cultivation to improve productivity and yield, and to enhance plant health and tolerance to stress factors. Indeed, they have positive effects on plant metabolism, both in optimal and sub-optimal environmental conditions.

Many authors have observed that plant based biostimulants and seaweed extracts often increase the colour of leaves by stimulating chlorophyll biosynthesis or reducing its degradation [99,100]. Leaf colour is an important quality parameter in vegetable crops because it contributes to the visual appearance of the product, especially in leafy vegetables for which the greenness influences the consumer's appeal. In addition, a higher chlorophyll content also allows for a greater photosynthetic activity of leaves. High concentration of leaf pigments (chlorophyll and carotenoids) has been observed after biostimulant treatments in rocket [101,102], in lettuce, and endive by Bulgari et al. [103]. Amino acids or seaweed extract application had positive effects on photosynthetic pigments, P and K content, fresh and dry weight of celeriac leaves [104]. Similar results have been observed after root inoculation with several plant growth promoting bacteria (PGPR) in broccoli (*Brassica oleracea* 'italica') using *Bacillus cereus*, *Brevibacillus reuszeri*, and *Rhizobium rubi* [105], and tomato under non-stressful conditions treated with PGPRs belonging to the genera *Bacillus*, *Pseudomonas* and *Azotobacter* [106], in strawberry (*Fragaria ananassa*) with five PGPRs (*Bacillus subtilis*, *Bacillus atrophaeus*, *Bacillus spharicus* subgroup, *Staphylococcus kloosii*, and *Kocuria erythromyxa*) [107] and also in lettuce grown under salt stress after inoculation with *Serratia* sp., *Rhizobium* sp., and *Azospirillum* [108,109]. Brown seaweeds are widely used as a biostimulant products to improve plant growth, and recently a phenolic compound isolated from *Ecklonia maxima* showed stimulatory effects in cabbage plants, improving photosynthetic pigments concentration, phytochemicals and myrosinase activity [110].

Abdalla [111] reported that moringa leaf extracts increased vegetative growth, chlorophyll content, total sugars, phenols, ascorbic acid, and photosynthetic rate of rocket salad. Similar effects have been observed in fennel [112,113] and squash under water stress condition (plants under a deficit irrigation of 80% or 60% ETc) [114]. In tomato plants it led to a greater fruit weight, volume and firmness, and enhanced titratableacidity, chlorophyll and ascorbic acid content [115].

Luziatelli et al. [116] recently found that different vegetal-derived bioactive compounds significantly increased the chlorophyll content and fresh weight of lettuce. Kulkarni et al. [117] investigated the promoting effect of bioactive molecules derived from smoke and seaweed in spinach and they observed that morphological, physiological and biochemical parameters including growth, chlorophyll and carotenoids content were positively improved.

Broccoli plants were significantly affected by two different products: Goemar BM86 and Seasol. The content of micro- and macro-nutrients increased, and also the leaf area, stem diameter and biomass, as reported by Gajc-Wolska et al. [74] and Mattner et al. [118].

Paradiković et al. [119] studied the effect of four different commercial biostimulants (Radifarm, Megafol, Viva, and Benefit), containing amino acid, polysaccharides and organic acids as active compounds on pepper plants and observed an increase in both yield and fruit quality. Radifarm and Viva treatments also affected tomato plants, stimulating the root apparatus in optimal and drought condition, respectively [120,121].

Recently, a sago bagasse hydrolysate was tested on tomato plants. The product showed a growth promoting ability as observed by the higher seed germination and protein and sugar content compared to the control. Moreover, the expression of the genes related to carbon and nitrogen metabolisms increased [122].

2.3. *Biostimulants and Crop Tolerance to Abiotic Stresses*

Table 1 is a summary of biostimulant products or bioactive molecules from different origins that have been evaluated for amelioration of abiotic stresses in several vegetables species. The biostimulants effectiveness to counteract the stressful condition depends on several factors, such as timing of application and their mode of action. The application of biostimulants can be carried out with different timings: before the stress affects the cultivation, during the stress, or even after. They could be applied on seeds, when plants are in early stages of growth, or when crops are fully developed, depending on the desired results [123]. As general consideration, biostimulants that contain anti-stress compounds, such as proline or glutamic acid, can be applied when the stress occurs or during stress conditions. On the contrary, those that are involved in the activation of bioactive compounds biosynthesis must be applied before the stress occurs. Proper timing of application during crop development differs from species to species and it also depends on the most critical phases for crop productivity. Thus, the identification of the right time of biostimulant application is as important as the determination of the exact dose, in order to avoid waste of product, high production costs, and unexpected results. Biostimulants can be applied as foliar spray or to the roots, at sowing for protecting the seedling in the early development stages, in a floating system nutrient solution or during blooming or fruit setting. There is no general recipe that works for a crop species and in each stress situation.

The protective role of biostimulants on plants has been increasingly studied. These products are able to counteract environmental stress such as water deficit, soil salinization, and exposure to sub-optimal growth temperatures in several ways [47,56,124,125]: They improve plant performance, enhance plant growth and productivity, interact with several processes involved in plant responses to stress, and increase the accumulation of antioxidant compounds that allow decrease in plant stress sensitivity.

More recent results of interest on vegetable crops tolerance have been obtained after the application of different exogenous treatments. Cao et al. [126] reported that a lower red to far-red ration improved tomato seedling tolerance to salt stress, acting on phytochrome activity. Mertinez et al. [127] showed positive results obtained after the application of exogenous melatonin in tomato plants grown under a combination of salinity and heat. Another interesting approach to induce tolerance to abiotic stresses is soaking plant seeds with different compounds, synthetic or natural. This strategy is generally called seed priming and has been deeply reviewed by Asharaf et al. [128].

2.3.1. Biostimulants and Cold or Chilling Stress

Low temperatures reduce plant metabolism and delay physiological responses. A reduced metabolism, consequent to cold stress, leads to an inhibition of the activity of photosystem II, called photoinhibition. Cold induces damages to cell membranes with destabilization of the phospholipid layers.

In tomato, cold tolerance has been enhanced by the application of psychrotolerant soil bacteria. Several strains have been isolated from soil during winter conditions and used as a cold protectant.

Tomato treated with these psychrotolerant bacteria showed higher seeds germination, reduced membrane damage, and antioxidant systems activation when exposed to chilling temperatures [129,130]. These soil bacteria can be considered as putative biostimulants for protecting plants against cold stress. Since low temperature causes stress to plant, especially during transplant, Marfà et al. [131] studied the effect of an enzymatic hydrolysates obtained from animal haemoglobin on strawberry plants in the firsts growing stages. They observed an increase in roots biomass and in the early production of fruit. The same product was also tested on lettuce plants subjected to cold stress and an increase in fresh weight, dry weight, specific leaf area, and relative growth rate was observed [132].

External applications of an amino acid biostimulant (Terra-Sorb® Foliar) on lettuce plants grown in different cold situations led to an increase in fresh weight and to an higher stomatal conductance [84]. A typical plants response to stress is the accumulation of compatible osmolytes, such as amino acids, which confer tolerance. The exogenous application of amino acids has the benefit of avoiding protein breakdown and saving energy resources in plants, even if the exact mechanism of action is not fully understood. Pepper (*Capsicum annuum*) seedlings were treated with 5-aminolevulinic acid in order to improve chilling tolerance through three different methods—soaking the seeds, spraying the leaves or drenching the soil. All the applications showed good effects in terms of stress tolerance. Fresh biomass, proline, sucrose, and water content were significantly higher while membrane permeability was reduced [133].

Positive effects on coriander plant grown in cold vegetative chambers have been observed in response to Asahi SL or Goemar Gateo (Arysta Life Science) treatments [124]. Results obtained by the study of stress indicators such as antioxidant activity, photosynthetic pigment concentration and activity, hydrogen peroxide and malondialdehyde amount showed that biostimulant application affected different metabolic pathways in a positive way, leading stressed plants to a phase of acclimation to low temperature. The biostimulant action against cold stress usually increases the accumulation of osmotic molecules by stimulating the biosynthetic pathways that lead to the cold protectant substances. These biostimulants also increase membrane thermostability, reducing the chilling injury.

2.3.2. Biostimulants and Heat Stress

Global warming and the projection of a rising temperature have a negative impact on agriculture [134,135]. High temperatures could induce several damages to plant cells, disturbing proteins synthesis and activity, inactivating enzymes and damaging membranes. The range between 30 °C and 45 °C is the optimal temperature for structural integrity and enzymal activity, which are irreversibly denatured when temperature increases above 60 °C. As a consequence, physiological activities like photosynthesis or respiration are affected. An overproduction of toxic compounds, like reactive oxygen species, causing oxidative stress, is one of the most frequent throwbacks [136]. As response, plants start synthesizing compatibles solutes in order to maintain cell homeostasis and turgor, organize proteins, and cellular structures. Moreover, they generally close stomata and increase the number of trachomatous, in order to prevent water loss. Also, at the molecular level there is a variation of the expression of genes involved in the synthesis or activity of antioxidant enzymes related to ROS scavenging, osmolytes or transporters. Temperature above optimum inhibits seeds germination and retards plant growth. Heat stress could negatively affect the yield by interfering with the reproductive phase, decreasing pollen vitality and germination, inhibiting flower differentiation and development and reducing fruit set, which ultimately reduces growth and yield.

Tomato is considered one of the most sensitive species to non-optimal temperatures, and heat stress often results in long style lengths and in a decreased fruit set [137]. There is little information in the literature about treatments specifically applied to vegetable crops exclusively against high temperature since, most of the time, heat stress is combined with drought or salinity. The application of brassinosteroids on tomato [138] and snap bean [139] has resulted in a higher biomass accumulation and net photosynthesis rate, increased growth and quality of snap bean pod in terms of NPK content and the total free amino acids levels in leaves. This might be due to the protective role of brassinosteroids on the photosynthetic apparatus from oxidative stress, increasing the ability to regenerate RuBP and carboxylation efficiency.

Nahar et al. [140] investigated the effect of exogenous application of glutathione against heat stress. Mung bean seedlings treated before their exposition to high temperature, showed a reduced oxidative stress and methylglyoxal content, a reactive compound that damages cells. This results in a more efficient antioxidant defense system. Pre-treatment with glutathione enhanced tolerance to short-term heat stress, improving plant physiological adaptation. For example, leaf relative water content and turgidity, which usually decreases under high temperature, were protected. Positive effect on mung bean has been observed in response to the application of nitric oxide [141] and ascorbic acid [142]. Nitric oxide treatment resulted in a promotion of photosynthetic activity, increasing the quantum maximum efficiency of PS2. It also affected electrolyte leakage, leading to a better cell membrane integrity. Oxidative stress, lipid peroxidation, and H_2O_2 content were decreased and antioxidant enzyme activity was restored. Similar results have been obtained after the application of proline and abscisic acid on chickpea [143,144]. Chickpea is sensitive to high temperature that generally leads to yield and quality losses. After treatments, membrane damage, measured as electrolyte leakage, MDA and H_2O_2 levels was decreased, while leaf water content was increased. These effects might be related with the osmoprotectant role of proline and with the accumulation of osmolytes after ABA treatments. Treated plants also showed a high chlorophyll content and this result, which has already been seen in other experiment with exogenous proline, could be related to membrane stability. The activity of oxidative metabolism was enhanced in treated plants, as expected also by the less oxidative damage of cells.

As discussed above, melatonin treatment exerts a positive effect to counteract chilling stress in coriander plants; otherwise, Martinetz et al. [127] found that melatonin treatments also have a protective role against the combination of heat and salt stress in tomato plants. Biostimulant treatments used against heat stress protect cell membranes by increasing their stability and reduce or avoid the accumulation of ROS.

2.3.3. Biostimulants and Salinity Stress

Among abiotic stresses, salinity is one of the main damaging factors affecting plant growth and metabolism as an effect of osmotic stress caused by salt. Sodium chloride (NaCl) is the more abundant salt presents in saline environments and is toxic in higher concentrations [145]. It happens especially near the coasts, where crops are frequently irrigated with saline water [85,146]. In many Mediterranean areas, the problem of seawater intrusion may cause a reduction of 50% of yield in lettuce cultivation, as reported by Miceli et al. [147]. A significant reduction of both fresh weight and chlorophyll content is a typical effect of salinity condition on plants and was observed also in spinach [148], in bean [149] and other crops [150]. Besides, chlorophyll content is a central parameter of the product quality particularly in green leafy vegetable, not only in terms of plant physiology status but also from a market point of view. This is a huge problem for vegetable crops where the edible parts are leaves, sprouts or flower buds. Consumers choices, in fact, are guided mostly by the visual appearance of products, hence a less green leafy vegetable or a malformed fruit are generally not accepted.

Salt stress causes a nutrient imbalance due to the limited uptake of the nutrients from the soil, threatening the nutritional quality of horticultural crops. Nutrient availability is compromised by salinity that causes several disorders such as competitive uptake with other ions like Ca^{2+}, P and K, mobility problems within the plant and a reduced water potential [151–155]. The solubility of micronutrients such as Cu, Fe, Mn, Mo and Zn is also affected by the pH of the soil solution, and in saline condition their availability is very low. Bano et al. [156] reported an important reduction of total phenolics, total soluble proteins and a suppressed activity of catalase, superoxide dismutase and peroxidase in carrot under saline condition. Salt stress could also alter several metabolic processes in plants, such as photosynthesis [157,158], respiration [159], phytohormone regulation, protein biosynthesis, nitrogen assimilation [160], and can also generate secondary oxidative stress [146,161]. It generally leads to a decrease of production and to a lower quality of the final product, due to an inhibition of leaves and roots growth and a change in leaf colour [17]. To verify the effects deriving from the applications of biostimulants, several trials on lettuce plants under salt stress were performed, since this crop is considered moderately sensitive to salinity.

Lucini et al. [85] showed that a plant-derived protein hydrolysate improved tolerance to salinity in lettuce plants, increasing yield and dry weight. Treated plants also have a higher performance and an increased maximum quantum efficiency of PS2 compared to the control. Similar results have been recently observed in lettuce plants in response to the application of an organic commercial biostimulant named Retrosal® [162].

Several experiments have been carried out using different PGPR that are able to enhance abiotic stress tolerance. Inoculation with *Azospirillum brasilense* showed positive results on lettuce [163,164], sweet pepper [165], chickpea and faba beans [166] grown under salty environment. Lettuce fresh weight, dry weight, ascorbic acid content, and germination percentage were increased; also, the visual appearance of the final product was better because of higher chlorophyll levels. In chickpeas and faba beans, the inoculation relieved the stress caused by salinity, increasing the root and shoot growth compared with the non-inoculated plants. Sweet pepper is a salt-sensitive crop and inoculation showed positive effect mitigating deleterious effects of NaCl. Dry weight, indeed, was higher than non-inoculated plants under several salt concentrations. Moreover, the inoculation also increased the CO_2 assimilation rate. A similar result has been obtained by Cordovilla et al. [167] applying two different *Rhizobium* strain on faba bean and pea plants. Pea plants inoculated with tolerant strain showed no reduction by salt stress condition in shoot and roots dry weight. The same strain was, however, not effective on faba beans. These results highlight the variation existing inter and intra species, and the difficulty in improving tolerance through selection and breeding. A comparable experiment has been carried out by Mayak et al. [168] on tomato seedling. They tested several strains of *rhizobacterium* and found that plants inoculated with *Achromobacter piechaudii* and irrigated with saline water had a higher fresh and dry weights and an increased water use efficiency. Yildirim et al. [169] obtained similar results in squash with the application of several biological products based on the *Bacillus* and *Trichoderma* species.

It is known that humic acids have a lot of beneficial effect stimulating shoot and root growth and improving environmental stress tolerance even if the exact mechanism of action is not completely clear. These activities were confirmed in several vegetable crops like sweet pepper [170], beans [171] and cucumber [172] grown under different salt stress conditions.

Bioactive compounds present in seaweed extracts are able to improve plant tolerance against abiotic stresses too. Two seaweed-based plant biostimulants containing *Ascophyllum nodosum* named Super Fifty® and Acadian were applied respectively on lettuce [173] and strawberry [174] and were associated with a significant increase in yield and root dry weight, despite the adverse salinity condition.

Sulphated exopolysaccharides extracted from the microalgae *Dunaliella salina* were applied on tomato plants to investigate their potential effect alleviating salt stress damages. Results obtained showed that treatment enhance plant growth, antioxidant enzymes activities and several metabolic mechanisms related to jasmonic acid pathway [175].

The application of seaweed extracts from *Sargassum muticum* and *Jania rubens* significantly alleviated the negative effects of salt through regulation of amino acids metabolism, ionic content balanced and improved antioxidant defence in chickpeas plants. Amino acids such as serine, threonine, proline and aspartic acid were identified in roots as responsible for salt stress amelioration [176].

Besides lettuce and pepper, bean is also considered a salt sensitive plant but in most developing countries it is cultivated in saline conditions. Several plant extracts based on licorice root, *Moringa oleifera* or maize grain have been tested on common bean by Egyptian researchers [177–181]. They observed that soaking seeds in propolis or maize grain extract improves seed germination percentage, stability of cell membrane and relative water potential under saline conditions. Antioxidant system activity was increased while lipid peroxidation and electrolyte leakage were reduced compared with the control plants. *Moringa oleifera* leaf extract, used alone or in combination with salicylic acid, and administered as foliar spray or as seed soaking, improved several physiochemical parameters as chlorophyll and carotenoids concentration, total soluble sugars and ascorbic acid content. A very similar trial has been carried out with licorice root extract and best results have been recorded integrating seed soaking and foliar spray applications.

A recent study highlighted the ability of a bee-honey based biostimulant to improve the tolerance of onion plants to salinity stress. Indeed, treated plants showed higher biomass, bulb yield, and photosynthetic pigments. Moreover, the osmoprotectans content as proline, soluble sugars and total free amino acids, the membrane stability index and the enzymatic and non-enzymatic antioxidant activity were enhanced [182]. Hence, biostimulants applied in case of salinity stress induce the accumulation of osmolytes, in order to enhance the cell osmotic potential and the level of protective molecules against oxidative stress.

2.3.4. Biostimulants and Drought Stress

Abiotic stresses are closely connected with the problem of resources availability and farmers are frequently forced to work in suboptimal conditions. A more sustainable use of resources also concerns water availability, a critical growing factor. The increasing use of aquifer-based irrigation by farmers worldwide poses a serious threat to the long-term sustainability of the agricultural system. Over-utilization of this dwindling water supply is leading to an ever-enlarging area in which productive farming itself has ceased or is threatened. Moreover, the increase of irrigation leads to a higher risk of soil salinization. Scientists generally agree with the perspective that several regions could become arid due to the negative impacts of global climate change on water resources [183]. Since one of the main effects of biostimulants is to improve water use efficiency, their application could be a possible strategy to reduce the amount of water added to crops [184]. Drought stress strongly influences plant gas exchange changing photosynthetic and transpiration rates, which are directly linked to yield. Application of *Ascophyllum nodosum* on broccoli [185] and spinach [186] enhanced gas exchange through the reduction of stomatal closure, resulting in increased plant resistance to water stress. Leaf yellowing is another common symptom of drought stress due to chlorophyll degradation during leaf senescence and is used as reliable indicator of metabolic and energetic imbalance in plants under stress. Biostimulant treatments with *A. nodosum* increased total chlorophyll content in tomato leaves [187]. A reduction of water loss, wilting damages and 3-carbon dialdehyde MDA after biostimulant applications were observed. Similar results have been obtained by Petrozza et al. [188] in responses to Megafol treatments in tomato plants. The results revealed that treated plants were healthier than non-treated ones in terms of biomass and chlorophyll fluorescence. Moreover, plants treated with the biostimulant product were able to recover more quickly when they had access to water. The expression of two drought stress marker genes was analysed and the results obtained showed that treated plants were experiencing a low level of water stress.

Sometimes, water stress in plants is caused by bacterial infection clogging xylem vessels and preventing water flow. Romero et al. [189] demonstrated that treatments with *Azospirillum brasilense*, a strain isolated in arid environments, delayed wilting of tomato plants. Treated plants, indeed, showed a high xylem vessels area, resulting in a more efficient water transport from the soil to the leaves. On the other hand, there are several strains of bacteria populating soil promoting plant growth through its metabolic activities and plant interactions. They produce exopolysaccharides, phytohormones, 1-aminocyclopropane-1-carboxylate (ACC) deaminase, volatile compounds, inducing several metabolic plant responses as accumulation of osmolytes and antioxidants, or up or down regulation of stress responsive genes and alteration in root morphology leading to a tolerance of water stress [190,191]. Some examples are reported below. Tomato seedlings treated with *Achromobacter piechaudii* were stimulated to accumulate biomass during the stress period and, the amount of ethylene that usually has negative effects on membrane status was lower than control [168].

Arshad et al. [192] investigated the growth of two plants promoting rhizobacteria on pea (*Pisum sativum*) crop grown under drought stress condition in different phenological phases. They observed that PGPR containing ACC-deaminase, a precursor of ethylene, significantly decreased the stress effects on growth and yield too. Positive results in terms of antioxidant and photosynthetic pigments activity have been collected in basil plants treated with *Pseudomonas sp.* under water stress conditions [193].

Seaweed extracts are already largely used for cultivated plant treatments and most of them contain plant growth hormones, auxins, abscisic acid, cytokinins, gibberellins, polyamines, oligosaccharides, betaines and brassinosteroids. A micro-algae-based biostimulant with known composition was tested on water stressed tomato plants. Results revealed that biostimulant application reduced the damaging effects of stress, increased plant height, root length, and enhanced the number and area of the leaves [78]. Biostimulants are capable of reducing drought injures, are able to enhance the biosynthesis of osmolytes and antioxidants against ROS, such as observed for salinity stress, and of plant hormones, like abscisic acid, regulating transpiration and avoiding excessive water losses.

2.3.5. Biostimulants and Nutrient Deficiency

One of the roles ascribed to biostimulant products is the ability to increase nutrient uptake [53] through different strategies. For instance, they are able to change soil structure or nutrient solubility, modify roots morphology directly or ameliorate nutrient transport in plants [194]. Their application might be really useful in poor soil conditions and in low input horticultural cultivation systems [195]. Indeed, soil nutrient imbalance is an increasing problem for farmers that spend a lot of money every year on fertilizers to resume soil fertility. All these mechanisms result in better nutrient use efficiency for both micro- and micro-nutrients.

Several experiments have been performed to investigate if the application of biostimulants allows a reduction of fertilizers without affecting crop yield and quality.

Koleška et al. [196] showed that the application of a biostimulant product named Viva® on tomato plants, growing under reduced NPK nutrition, help counteract the negative effects of nutrient deficiency. For example, lycopene and chlorophyll content that is usually affected by the availability of macronutrients was preserved in treated plants grown with NPK reduction. Moreover, biostimulant application helped maintain cell homeostasis and prevent oxidative stress. A similar experiment was performed by Anjum et al. [197] on garlic plants grown with half of the recommended dose of nutrients. Garlic growth and yield were positively affected by the biostimulant application in combination with a low dose of macronutrients.

A seaweed-based product (Kelpak®) has been tested on okra seedlings grown with different nutrient deficiencies [198]. Treatments were applied three times a week and were compared with a polyamine solution treatment. Plants treated with the biostimulant showed an increase in growth parameters, such as shoot length, stem thickness, leaves and roots numbers, and fresh weight under phosphorous and potassium deficiency. Kelpak® efficacy might be due to the combination of auxins, cytokinins and polyamines contained in the product.

Spinelli et al. [199] measured the effects of another commercial seaweed extract, named Actiwave® on the vegetative and productive performance of strawberry plants grown on an iron deficient substrate. They found that vegetative growth, chlorophyll content, stomatal density and photosynthetic rate were enhanced after biostimulant treatment. Fruit production and weight were also increased. Nutrient uptake might have been positively influenced by the more developed root system of treated plants. Treatment also contrasted the negative effects of iron chlorosis and this could be linked to betaine contained in this product.

The positive effects of seaweed extracts are usually ascribed to their polysaccharide content that helps the soil structure; nevertheless, Vernieri et al. [102] obtained good results by applying Actiwave in a hydroponic system with different concentrations of nutrient solutions. Yield and leaf area were higher in rocket plants grown with the lowest nutrient concentration, indicating a better nutrient use efficiency.

Most of the biostimulant contains a mixture of different amino acids and short peptides that are usually called protein hydrolysates. They have a positive effect on plant growth and protection against several stresses. The Cerdán et al. [200] study showed that amino acids origin might influence the efficacy of the product. Tomato plants grown under iron deficiency conditions and treated with two products containing amino acids from plant and animal origin showed different responses. Plant-derived amino acids promoted growth and chlorophyll content both in controlled and iron deficiency conditions. This effect might be ascribed to glutamic acid content. Indeed, this amino acid plays an important role in nitrogen metabolism [201] and chlorophyll biosynthesis [202].

Nutrient imbalance might be the cause of several disorders during plant growth and development. Blossom-end rot in pepper is usually caused by a local calcium deficiency in young fruits. Parađiković et al. [203] tested four different biostimulant products for their effects on yield and BER incidence on pepper. They also evaluated the application as foliar spray or in a nutrient solution of the same products. The results obtained revealed that biostimulants applications helped to reduce the occurrence of BER and increase yield. Moreover, nutrient accumulation in fruits and leaves was promoted by the treatments.

These experiments revealed that biostimulant products cannot totally replace fertilizers but could be really useful to reduce the amount of mineral nutrition or help in nutrient deficiency and imbalanced situations. For example, in the floating system cultivation of baby leaf such as rocket, the nutrient solution can be reduced by 75% of Hoagland's solution [101].

The biostimulants that help reduce nutrient deficiencies usually improve crops nutrient uptake by increasing root biomass, nutrient transport/translocation, and enzyme activities involved in nutrient assimilation.

3. Conclusions and Future Prospects

This review reports the progress on the recent development of biostimulant products with special emphasis on their effects, improving tolerance to abiotic stresses in vegetable crops. During their life cycle, crops are often exposed to abiotic stresses, acting individually or in combination, which could dramatically reduce the yield and quality of products. Biostimulants could represent an effective and sustainable tool to enhance plant growth and productiveness, improving tolerance against abiotic stresses. In fact, biostimulants have been successfully applied for:

- improving nutrients and water use efficiency of crops;
- enhancing tolerance against salinity, water stress, cold, high temperature, etc.;
- increasing yield and quality of agricultural crops.

It is important to consider that the complex and variable nature of raw materials used for their production and the heterogeneous mixture of components of the final product can make it difficult to attribute a specific mode of action to each biostimulant. The situation is further complicated by the high number of plants, bacteria and in general, substances included into the category of plant biostimulants. For example, two products obtained by two different plants would fall in the same category, but their effects and their mode of action might be completely different. Moreover, the opposite situation may occur; the same product may produce different effects when applied on different plants. This could be related to the genetic variability among species, variety or cultivars. In addition, the biostimulant activity of a product may also depend on the nature and severity of the abiotic stress.

It must also be considered that trying to link a specific mode of action only to the main component of a product might be a mistake because it would be like excluding the effect of the molecules that are presents in small quantities or in traces, but it is known that the efficacy of biostimulant products is the result of a synergistic or antagonistic effect of many components. Furthermore, our understanding of the mode of action also depends on the amount of information provided by scientific papers, on the numbers of analyses performed, and on their investigation level. The availability of innovative research tools will surely improve the knowledge of biostimulant composition, but this information will not be exhaustive. Therefore, the biostimulant mode of action can be understood through plant responses at the physiological, biochemical, and molecular levels.

Table 1. Examples of biostimulant products or substances with a biostimulant effect on horticultural crops to counteract abiotic stress conditions.

ABIOTIC STRESS	SEVERITY AND TIME OF ExPOUSURE	BIOSTIMULANT PRODUCT OR SUBSTANCES WITH A BIOSTIMULANT EFFECT	DOSE	APPLICATION METHODS AND NUMBER OF TREATMENTS	CROP	BENEFICIAL EFFECTS	REFERENCE
Chilling or cold stress	6 °C for 6 days	Asahi SL (Sodium para-nitrophenolate, sodium ortho-nitrophenolate, sodium 5-nitroguaiacolate) / Goëmar Goteo (Composition (w/v): organic substances 1.3–2.4%, phosphorus (P_2O_5). 24.8%, potassium (K_2O) .4.75%)	0.1%	Foliar spray (3×)	*Coriandrum sativum L.*	↓electrolyte leakage ↑Chlorophyll *a* and carotenoids ↑Fv/Fm ↑E ↑gs ↓Ci	[124]
	10, 12 °C for 7 days / 15 °C for 7, 10 days	*Flavobacterium glaciei, Pseudomonas frederiksbergensis, Pseudomonas vancouverensis*	-	Seed inoculation	*Solanum lycopersicum*	↑shoot height ↑root length ↑biomass accumulation ↓electrolyte leakage ↓lipid peroxidation ↑proline accumulation ↑SOD, CAT, APX, POD, GR activity	[129,130]
	−6 °C for 5 nights	Pepton 85/16 (enzymatic hydrolysates obtained from animal haemoglobin. L-α amino acids (84.83%) and free amino acids (16.52%), organic-nitrogen content (12%), mineral-nitrogen content (1.4%), potassium content (4.45%), iron content (4061 ppm), very low heavy-metal content)	2 L ha^{-1}, 4 L ha^{-1}	Injection into the soil (5×)	*Fragaria × ananassa*	↑new roots ↑flowering ↑fruit weight	[131]
	−3 °C for 4 h	Pepton 85/16	0.4, 0.8, 1.6 g L^{-1}	Soil application (1×)	*Lactuca sativa L.*	↑fresh and dry weight ↑SLA ↑RGR	[132]
	4 °C for 8 days or nights /6 °C for 8 days only to the roots	Terra-Sorb® Foliar (Free amino acids (ASP, SER, GLU, GLY, HIS, ARG, THR, ALA, PRO, CIS, TYR, VAL, MET, LYS, ILE, LEU, PHE, TRP) 9,3% (w/w), Total amino acids 12% (w/w), Total nitrogen (N) 2,1% (w/w), Organic Nitrogen (N) 2,1% (w/w), Boron (B) 0,02% (w/w), Manganese (Mn) 0,05% (w/w), Zinc (Zn) 0,07% (w/w), Organic matter 14,8% (w/w))	3 mL L^{-1}	Foliar spray (3×)	*Lactuca sativa L. var. capitata*	↑roots fresh weight ↑green cover %	[84]
	3 °C for 48 h	5-aminolevulinic acid	0, 1, 10, 25, 50 ppm (15 mL for seed soaking and 25 mL for soil drench)	Seed soaking/ foliar spray/soil drench (1×)	*Capsicum annuum*	↓visual injuring ↑chlorophyll ↑RWC ↑gs ↓membrane permeability ↑shoot and root mass ↑SOD activity	[133]

Table 1. *Cont.*

ABIOTIC STRESS	SEVERITY AND TIME OF EXPOUSURE	BIOSTIMULANT PRODUCT OR SUBSTANCES WITH A BIOSTIMULANT EFFECT	DOSE	APPLICATION METHODS AND NUMBER OF TREATMENTS	CROP	BENEFICIAL EFFECTS	REFERENCE
Drought stress	Occlusion of xylem vessels	*Azospirillum brasilense (BNM65)*	-	Seed inoculation	*Solanum lycopersicum*	↑height plants ↑dry weight ↑xylem vessel area	[189]
	No irrigation for 5 days	Megafol® (Composition (w/v): total nitrogen (N) 3.0% (36.6 g L^{-1}); organic nitrogen (N) 1.0% (12.2 g L^{-1}); ureic nitrogen (N) 2.0% (24.4 g L^{-1}); potassium oxide (K_2O) soluble in water 8.0% (97.6 g); organic carbon (C) of biological origin 9.0% (109.8 g L^{-1}))	2 mL L^{-1}	Foliar spray (1×)	*Solanum lycopersicum*	↑leaf area ↑RLWC	[188]
	50% ET	*Ascophyllum nodosum*	0.50%	Foliar spray and drench	*Spinacia oleracea*	↑RLWC ↑leaf area ↑fresh and dry weight ↑SLA ↑gas exchange	[186]
	No irrigation until symptoms of wilting appear	Pseudomonas spp. (*P. putida P. fluorescens*)	-	Seed inoculation	*Pisum sativum*	↑grain yield ↑root growth ↑shoot length ↑number of pods per plant ↑chlorophyll	[192]
	No irrigation for 12 days	*Achromobacter piechaudii* (ARV8)	-	Seedling inoculation	*Solanum lycopersicum*	↑fresh and dry weight of seedling ↑plant growth ↓ethylene	[168]
	No irrigation for 12 days	*Achromobacter piechaudii* (ARV8)	-	Seedling inoculation	*Capsicum annuum*	↑ fresh and dry weight of seedling ↑plant growth	[168]
	No irrigation for 7 days	*Ascophyllum nodosum*	0.33%	Foliar spray (2×)	*Solanum lycopersicum*	↑RWC ↑plant growth ↑foliar density ↑chlorophyll ↓lipid peroxidation ↑proline ↑soluble sugars	[187]
	No irrigation for 2 days	*Ascophyllum nodosum* + amino acids	-	Soil application (1×)/ foliar spray (3×)	*Brassica oleracea var. italica*	↑Pn ↑gs ↑chlorophyll	[185]
	40, 70% field capacity	Gibbrellic acid and titanium dioxide	250, 500 ppm (GA3) 0.01, 0.03% (titanium nanoparticles)	Stems and foliar spray (2×)	*Ocimum basilicum*	↑CAT activity ↓lipid peroxidation ↑LRWC	[95]
	No irrigation	VIVA®	-	2×	*Solanum lycopersicum*	↑plant biomass ↑roots biomass	[120]
	60, 40% field capacity	*Pseudomonades, Bacillus lentus, Azospirillum brasilens*	-	Seed inoculation	*Ocimum basilicum*	↑CAT, GPX activity ↑chlorophyll	[193]
	60, 40% ET	Moringa leaf extract	3%	Foliar spray (2×)	*Cucurbita pepo*	↑growth ↑HI ↑WUE ↑Fv/Fm ↑PI ↑soluble sugars ↑free proline ↓electrolyte leakage ↑membrane stability	[114]

Table 1. *Cont.*

ABIOTIC STRESS	SEVERITY AND TIME OF ExPOUSURE	BIOSTIMULANT PRODUCT OR SUBSTANCES WITH A BIOSTIMULANT EFFECT	DOSE	APPLICATION METHODS AND NUMBER OF TREATMENTS	CROP	BENEFICIAL EFFECTS	REFERENCE
Heat stress	35 °C	Nano-TiO_2	0.05, 0.1, 0.2 g L^{-1}	Foliar spray (1×)	*Solanum lycopersicum*	↑gs ↑E ↑ Pn	[94]
	40/30 °C for 8 days	Brassinosteroids	0.01, 0.1, and 1.0 mg L^{-1}	Foliar spray (1×)	*Solanum lycopersicum*	↑antioxidant enzyme activities ↓H_2O_2 ↓MDA ↑shoot weight	[138]
	35.2 °C (Tmax)	Brassinosteroids	25, 50, 100 ppm	Foliar spray (2×)	*Phaseolus vulgaris*	↑plant length ↑number of leaves, branches and shoots per plant ↑fresh and dry weight ↑pod weight ↑N, P, K in bean pods	[139]
	45 °C for 90 min	Nitric oxide	150 μM	Immersion of leaf disks	*Phaseolus radiatus*	↑Fm ↓electrolyte leakage	[141]
	35/25 40/30 45/35 °C	Ascorbic acid	50 μM	In a nutrient solution	*Phaseolus radiatus*	↑% germination ↑seedling growth ↓electrolyte leakage ↑TTC reduction ability ↑RLWC ↓MDA ↓H_2O_2 ↑antioxidant activity ↑ascorbic acid ↑GSH ↑proline	[142]
	35/25 40/30 45/35 °C	Proline	5, 10, 15 μM	In a nutrient solution	*Cicer arietinum*	↑% germination ↑shoot and root length ↓electrolyte leakage ↑chlorophyll ↑RLWC ↓lipid peroxidation ↓H_2O_2 ↑GSH ↑proline	[143]
	35/25 40/30 45/35 °C for 10 days	Abscisic acid	2.5 μM	In a nutrient solution	*Cicer arietinum*	↑shoot length ↑osmolytes ↑chlorophyll ↑cellular oxidizing ability	[144]
	42 °C for 48 h	Glutathione	0.5 mM	-	*Vigna radiata L.*	↑RLWC ↑chlorophyll ↑proline ↓MDA ↓ H_2O_2 ↓O_2^- ↓LOX activity ↑ascorbate ↓GSSG	[140]
Heat and salt stress	35 °C and 75 mM NaCl for 15 days	Melatonin	100 μM	Foliar spray (5×)	*Solanum lycopersicum*	↑biomass ↑Pn ↑gs ↑E ↑chlorophyll a ↑carotenoids ↑Fv/Fm ↑efficiency of PSII ↑ETR ↑antioxidant capacity ↓H_2O_2 ↓lipid peroxidation ↓protein oxidation	[127]

Table 1. *Cont.*

ABIOTIC STRESS	SEVERITY AND TIME OF E×POUSURE	BIOSTIMULANT PRODUCT OR SUBSTANCES WITH A BIOSTIMULANT EFFECT	DOSE	APPLICATION METHODS AND NUMBER OF TREATMENTS	CROP	BENEFICIAL EFFECTS	REFERENCE
Iron deficiency	-	Actiwave® (*Ascophyllum nodosum*)(Composition (w/v): total nitrogen (N) 3.0% (38.7 g L^{-1}); organic nitrogen (N) 1.0% (12.9 g L^{-1}); ureic nitrogen (N) 2.0% (25.8 g L^{-1}); potassium oxide (K_2O) soluble in water 7.0% (90.3 g L^{-1}); organic carbon (C) of biological origin 12% (154.8 g L^{-1}); iron (Fe) soluble in water 0.5% (6.45 g L^{-1}); iron (Fe) chelated by ethylenediaminedi (2-hydroxy-5-sulfophenylacetic) acid (EDDHSA) 0.5% (6.45 g L^{-1}); zinc (Zn) soluble in water 0.08% (1.03 g L^{-1}); zinc (Zn) chelated by Ethylenediaminetetraacetic acid (EDTA) 0.08% (1.03 g L^{-1}))	10 mL in 20 mL tap water	In a nutrient solution	*Fragaria ananassa*	↑vegetative growth ↑chlorophyll ↑stomatal density ↑photosynthetic rate ↑ fruit production ↑berry weight	[199]
	-	Amino acids	0.1, 0.2 mL L^{-1} / 0.2, 0.7 mL L^{-1}	Root application/foliar spray (4×)	*Solanum lycopersicum*	↑plant growth ↑root and leaf ferrum chelate reductase activity ↑chlorophyll ↑leaf Fe ↑Fe_2:Fe ratio	[200]
Reduced NPK	NPK reduced of 40%	VIVA® (Composition (w/v): total nitrogen (N) 3.0% (37.2 g L^{-1}); organic nitrogen (N) 1.0% (12.4 g L^{-1}); ureic nitrogen (N) 2.0% (24.8 g L^{-1}); potassium oxide (K_2O) soluble in water 8.0% (99.2 g L^{-1}); organic carbon (C) of biological origin 8.0% (99.2 g L^{-1}); iron (Fe) soluble in water 0.02% (0.25 g L^{-1}); iron (Fe) chelated by EDDHSA 0.02% (0.25 g L^{-1}))	10.5 mL /plant	Foliar spray	*Solanum lycopersicum*	↑yield ↑ascorbic acid ↑lycopene ↑chlorophyll ↑carotenoids	[196]
	NPK deprivation	Kelpak (*Ecklonia maxima*, containing polyamine, cytokinins and auxins, putrescine, spermine)	0.40%	In a nutrient solution (twice per week for 8 weeks)	*Abelmoschus esculentus*	↑number of leaves ↑number of roots ↑stem thickness ↑shoot weight ↑root weight ↑leaf area	[198]
	NPK reduced of 50%	Bio-Cozyme (concentrated micro-biological biostimulant and soil inoculants. Total Nitrogen (N) 0.20%, Soluble Potash (KO) 5.00%, Magnesium (Mg) 1.40%, Boron (B) 0.20%, Copper (Cu) 0.50%, Iron (Fe) 3.00%, Manganese (Mn)1.00%, Molybdenum (Mo) 0.0.25%, Zinc (Zn) 2.00%, Humic Acid, humates & derivatives 8.00%, Vitamins, E, C, B Complex, organic acids, natural sugars carbohydrates, amino acids 1.40%)	2 kg ha^{-1}	Foliar application (4×)	*Allium sativum*	↑bulb yield ↑plant height ↑NPK in leaves	[197]

Table 1. *Cont.*

ABIOTIC STRESS	SEVERITY AND TIME OF E×POUSURE	BIOSTIMULANT PRODUCT OR SUBSTANCES WITH A BIOSTIMULANT EFFECT	DOSE	APPLICATION METHODS AND NUMBER OF TREATMENTS	CROP	BENEFICIAL EFFECTS	REFERENCE
Salt stress	30, 50, 80 mol m^{-3} NaCl for 30 days / 40, 80, 120 mol m^{-3} NaCl	*Azospirillum brasilense*	-	Seed inoculation	*Lactuca sativa*	↑germination % ↑total fresh and dry weight ↑biomass partition ↑plantlets number ↑plantlets dry weight ↑total leaf fresh weight ↑leaf area ↑leaves number ↑chlorophyll ↑root dry weigh ↑ascorbic acid ↑plant survival after transplant	[163,164]
	40, 80, 120 mM NaCl	*Azospirillum brasilense/Pantoea dispersa*	-	Inoculation	*Capsicum annuum*	↑plant dry weight ↑K^+:Na^+ratio ↑gs ↑relative growth rate ↑net assimilation rate ↓ Cl^- accumulation ↑NO_3^- concentration ↑CO_2 assimilation	[165]
	714 mg·L^{-1} NaCl	*Azospirillum brasilense (ATCC 29,729)*	-	Soil inoculation	*Cicer arietinum*	↑nodule formation ↑shoot dry weight	[166]
	100 mmol L^{-1} NaCl	*Rhizobium leguminosarum (GRA19–GRL19)*	-	Seedling inoculation	*Vicia faba / Pisum sativum*	↑plant growth	[167]
	50, 100 mM NaCl	*Bacillus species, Bacillus pumilis, Trichoderma harzannum, Paenibacillus azotoformans and polymyxa*	-	Seed treatment/ watering	*Cucurbita pepo*	↑fresh weight ↑potassium uptake ↓sodium uptake ↑ K^+:Na^+ ratio	[169]
	30, 60, 120 mM (NaCl, Na_2SO_4, $CaCl_2$, $CaSO_4$, KCl, K_2SO_4, $MgCl_2$, $MgSO_4$) for 60 days	Humic acid	0.05, 0.1%	Soil application	*Phaseolus vulgaris*	↑plant nitrate, nitrogen and phosphorus ↓soil electricity conductivity ↓proline ↓electrolyte leakage ↑plant root and shoot dry weight	[171]
	-	Acadian (*Ascophyllum nodosum*)	-	Soil application	*Fragaria ananassa*	↑yield ↑growth ↑root length ↑surface area, volume and number of tips ↑numbers of crowns	[174]
	80 mM NaCl	Super Fifty® (*Ascophyllum nodosum*)	0.4, 1, 2.5, 10 mL L^{-1}	In the nutrient solution	*Lactuca sativa*	↑root, stem, total plant weight	[173]
	25 mM NaCl	Protein hydrolysates	2.5 mL L^{-1}	Foliar spray/soil application	*Lactuca sativa*	↑fresh yield ↑dry biomass ↑root dry weight ↑plant nitrogen metabolism ↑Fv/Fm ↓oxidative stress ↑osmolytes ↑glucosynolates	[85]

Table 1. *Cont.*

ABIOTIC STRESS	SEVERITY AND TIME OF EXPOUSURE	BIOSTIMULANT PRODUCT OR SUBSTANCES WITH A BIOSTIMULANT EFFECT	DOSE	APPLICATION METHODS AND NUMBER OF TREATMENTS	CROP	BENEFICIAL EFFECTS	REFERENCE
	0.8, 1.3, and 1.8 dS/m NaCl	Retrosal® (organic mix with high concentration of carboxylic acids, containing calcium oxide (CaO) 8.0% (w/w) soluble in water and 1.4% complexed by ammonium ligninsulfonate, Zinc (Zn) 0.2% (w/w) soluble in water and 0.2% (w/w) chelated by EDTA.)	0.1 or 0.2 mL/plant	Soil application (4×)	*Lactuca sativa*	↑fresh weight ↑chlorophyll Pn ↑ gas exchange ↓proline ↓ABA	[162]
	43, 207 mM NaCl for 7 weeks	*Achromobacter piechaudii*	-	Seedling inoculation	*Solanum lycopersicum*	↑fresh and dry weights of tomato seedlings ↓ethylene ↑uptake phosphorous and potassium ↑WUE	[204]
	200 mM NaCl	Nano-TiO_2	5, 10, 20 and 40 mg L^{-1}	Foliar spray	*Solanum lycopersicum*	activities of carbonic anhydrase, nitrate reductase, SOD and POX ↑proline ↑glycinebetaine ↑growth ↑yield	[97]
	28, 56 mmol kg^{-1}	*Ascophyllum nodosum*	1, 2 g kg^{-1}	Soil application	*Cucumis sativus*	↑fruit yield ↑Pn	[172]
	7.15, 7.2 dSm^{-1}	Licorice root extract	0.50%	Seed soaking /foliar spray	*Phaseolus vulgaris*	↑plant growth ↑yield ↑RWC ↑chlorophylls ↑free proline ↑total soluble carbohydrates ↑total soluble sugars ↑nutrients ↑selenium ↑K^+:Na^+ ratio ↑membrane stability index ↑activities of all enzymatic antioxidants ↓electrolyte leakage ↓MDA ↓Na^+ ↓H_2O_2 ↓O_2^-	[181]
	100 mM NaCl	Propolis and maize grain extract	1, 2%	Soaking seed	*Phaseolus vulgaris*	↑% germination ↑seedling growth ↑cell membrane stability index ↑RWC ↑free proline ↑total free amino acids ↑total soluble sugars ↑indole-3-acetic acid ↑gibberellic acid ↑activity of the antioxidant system ↓lipid peroxidation ↓electrolyte leakage ↓ABA	[178]

Table 1. *Cont.*

ABIOTIC STRESS	SEVERITY AND TIME OF E×POUSURE	BIOSTIMULANT PRODUCT OR SUBSTANCES WITH A BIOSTIMULANT EFFECT	DOSE	APPLICATION METHODS AND NUMBER OF TREATMENTS	CROP	BENEFICIAL EFFECTS	REFERENCE
	6.23–6.28 dS m^{-1}	Salycilic acid and *Moringa oleifera*	0.30%	Seed soaking /foliar spray	*Phaseolus vulgaris*	↑shoot length ↑number and area of leaves ↑ plant dry weight ↑RWC ↑chlorophyll ↑carotenoid ↑total soluble sugars ↑free proline ↑ascorbic acid ↑N, P, K and Ca, ↑ratios of K/Na and Ca/Na ↑green pod and dry seed yields	[179]
	100 mM NaCl	*Moringa oleifera*	crude extract	Soaking seed	*Phaseolus vulgaris*	↑shoot and root lengths ↑plant dry mass ↑total soluble sugars ↑proline ↑K^+, Na^+ and Cl^- ↑ascorbic acid ↑total glutathione ↓MDA ↓ H_2O_2 ↓O_2^- ↑SOD, APX, GR	[177,180]
	50, 150 mM NaCl	*Sargassum muticum* and *Jania rubens*	1%	Foliar spray (2×)	*Cicer arietinum*	↑plant growth ↑chlorophyll ↑carotenoid ↑soluble sugars ↑phenols ↓Na^+ ↑ K^+ ↓H_2O_2 ↑CAT, SOD, POD, APX activity ↓MDA	[176]
	3, 6 g L^{-1}	*Dunaliella salina* exopolysaccharides	0.1 g L^{-1}	Foliar spray (2×)	*Solanum lycopersicum*	↑chlorophyll ↑protein ↓proline	[175]
	8.81 dS m^{-1}	Bee-honey based biostimulant	25–50 g L^{-1}	Foliar spray	*Allium cepa*	↑biomass ↑bulb yield ↑WUE ↑photosynthetic pigments ↑osmoprotectants ↑membrane stability index ↑RWC ↑enzymatic and non-enzymatic antioxidants	[182]
	8 mM NaCl	phosphorus / humic acid	50, 100, 150 mg kg^{-1} (P)/750, 1500 mg kg^{-1} (humic acid)	Soil application	*Capsicum annuum*	↑fresh and dry weight of shoot and root ↓membrane damage ↑nutrient uptake	[170]
UV-stress	300–340nm illumination for 15 min	Nano-anatase	0.25%	Soaking seed and foliar spray	*Spinacia oleracea*	↓O_2 ↓H_2O_2 ↓MDA ↑ SOD, CAT, APX, GPX activity	[96]

Fv/Fm maximum quantum efficiency of Photosystem II; Pn net photosynthetic rate; E transpiration rate; gs stomatal conductance; Ci sub stomatal CO_2 concentration; SLA specific leaf area; RGR relative growth rate; RLWC relative leaf water content; RWC relative water content; WUE water use efficiency; PI performance index; MDA malondialdehyde; TTC 2,3,5-triphenyltetrazolium chloride; GSH reduced glutathione; GSSG oxidized glutathione; LOX lipoxygenase; CAT catalase; SOD superoxide dismutase; APX ascorbate peroxidase; POX peroxidase; GR glutathione reductase; HI harvest index; ABA abscisic acid; ETR electron transport rate. The symbol ↑ means an increase or ↓ a decrease of the parameter measured. The symbol × represents how many times the treatment was applied.

Author Contributions: Conceptualization, R.B., G.F., A.F.; writing—original draft preparation, R.B. G.F.; writing—review and editing, A.F.

Funding: This research received no external funding.

Conflicts of Interest: The authors declare no conflict of interest.

References

1. Verma, A. Abiotic Stress and Crop Improvement: Current Scenario. *Adv. Plants Agric. Res.* **2016**, *4*, 345–346.
2. Boyer, J.S. Plant Productivity and Environment. *Science (80-)* **1982**, *218*, 443–448. [CrossRef] [PubMed]
3. La Peña, R.D.; Hughes, J. Improving Vegetable Productivity in a Variable and Changing Climate. *J. SAT Agric. Res.* **2007**, *4*, 666–675.
4. Farooq, M.; Wahid, A.; Kobayashi, N.; Fujita, D.; Basra, S.M.A. Plant drought stress: Effects, mechanisms and management To cite this version: Review article. *Agron. Sustain. Dev.* **2009**, *29*, 185–212. [CrossRef]
5. Cramer, G.R.; Urano, K.; Delrot, S.; Pezzotti, M.; Shinozaki, K. Effects of abiotic stress on plants: A systems biology perspective. *BMC Plant Biol.* **2011**, *11*, 163. [CrossRef] [PubMed]
6. Global Warming of 1.5 °C. Available online: https://www.ipcc.ch/sr15/ (accessed on 8 March 2019).
7. FAO. The State of Food and Agriculture (SOFA). Available online: http://www.fao.org/3/a-i6030e.pdf (accessed on 8 March 2019).
8. Mariani, L.; Ferrante, A. Agronomic Management for Enhancing Plant Tolerance to Abiotic Stresses—Drought, Salinity, Hypoxia, and Lodging. *Horticulturae* **2017**, *3*, 52. [CrossRef]
9. Mou, B. Improvement of horticultural crops for abiotic stress tolerance: An introduction. *HortScience* **2011**, *46*, 1068–1069. [CrossRef]
10. Parajuli, R.; Thoma, G.; Matlock, M.D. Environmental sustainability of fruit and vegetable production supply chains in the face of climate change: A review. *Sci. Total Environ.* **2019**, *650*, 2863–2879. [CrossRef]
11. Bisbis, M.B.; Gruda, N.; Blanke, M. Potential impacts of climate change on vegetable production and product quality—A review. *J. Clean. Prod.* **2018**, *170*, 1602–1620. [CrossRef]
12. Mittler, R. Abiotic stress, the field environment and stress combination. *Trends Plant Sci.* **2006**, *11*, 15–19. [CrossRef]
13. Rivero, R.M.; Mestre, T.C.; Mittler, R.; Rubio, F.; Garcia-Sanchez, F.; Martinez, V. The combined effect of salinity and heat reveals a specific physiological, biochemical and molecular response in tomato plants. *Plant Cell Environ.* **2014**, *37*, 1059–1073. [CrossRef] [PubMed]
14. Choudhury, F.K.; Rivero, R.M.; Blumwald, E.; Mittler, R. Reactive oxygen species, abiotic stress and stress combination. *Plant J.* **2017**, *90*, 856–867. [CrossRef] [PubMed]
15. Pandey, P.; Irulappan, V.; Bagavathiannan, M.V.; Senthil-Kumar, M. Impact of Combined Abiotic and Biotic Stresses on Plant Growth and Avenues for Crop Improvement by Exploiting Physio-morphological Traits. *Front. Plant Sci.* **2017**, *8*, 537. [CrossRef] [PubMed]
16. Atkinson, N.J.; Urwin, P.E. The interaction of plant biotic and abiotic stresses: From genes to the field. *J. Exp. Bot.* **2012**, *63*, 3523–3543. [CrossRef] [PubMed]
17. Shannon, M.C.; Grieve, C.M. Tolerance of vegetable crops to salinity. *Sci. Hortic. (Amst.)* **1998**, *78*, 5–38. [CrossRef]
18. Fruit and Vegetables for Health Initiative. Available online: http://www.fao.org/3/a-i6807e.pdf (accessed on 8 March 2019).
19. Rao, N.K.S.; Laxman, R.H.; Shivashankara, K.S. Physiological and Morphological Responses of Horticultural Crops to Abiotic Stresses. In *Abiotic Stress Physiology of Horticultural Crops*; Rao, N.K.S., Shivashankara, K.S., Laxman, R.H., Eds.; Springer: New Delhi, India, 2016; pp. 3–7, ISBN 978-81-322-2723-6.
20. Sánchez-Rodríguez, E.; Rubio-Wilhelmi, M.; Cervilla, L.M.; Blasco, B.; Rios, J.J.; Rosales, M.A.; Romero, L.; Ruiz, J.M. Genotypic differences in some physiological parameters symptomatic for oxidative stress under moderate drought in tomato plants. *Plant Sci.* **2010**, *178*, 30–40. [CrossRef]
21. Mittova, V.; Tal, M.; Volokita, M.; Guy, M. Up-regulation of the leaf mitochondrial and peroxisomal antioxidative systems in response to salt-induced oxidative stress in the wild salt-tolerant tomato species *Lycopersicon pennellii*. *Plant Cell Environ.* **2003**, *26*, 845–856. [CrossRef] [PubMed]
22. Abedi, T.; Pakniyat, H. Antioxidant enzyme changes in response to drought stress in ten cultivars of oilseed rape (*Brassica napus* L.). *Czech J. Genet. Plant Breed.* **2010**, *46*, 27–34. [CrossRef]

23. Venkateswarlu, B.; Shanker, A.K.; Shanker, C.; Maheswari, M. *Crop Stress and Its Management: Perspectives and Strategies*; Springer: Berlin/Heidelberg, Germany, 2012; ISBN 978–94–007–2219–4.
24. Kusvuran, S.; Kiran, S.; Ellialtioglu, S.S. Antioxidant Enzyme Activities and Abiotic Stress Tolerance Relationship in Vegetable Crops. In *Abiotic and Biotic Stress in Plants—Recent Advances and Future Perspectives*; InTechOpen: London, UK, 2016.
25. You, J.; Chan, Z. ROS Regulation during Abiotic Stress Responses in Crop Plants. *Front. Plant Sci.* **2015**, *6*, 1092. [CrossRef]
26. Suzuki, N.; Koussevitzky, S.; Mittler, R.; Miller, G. ROS and redox signalling in the response of plants to abiotic stress. *Plant Cell Environ.* **2012**, *35*, 259–270. [CrossRef]
27. Das, K.; Roychoudhury, A. Reactive oxygen species (ROS) and response of antioxidants as ROS-scavengers during environmental stress in plants. *Front. Environ. Sci.* **2014**, *2*, 53. [CrossRef]
28. Mittler, R.; Vanderauwera, S.; Suzuki, N.; Miller, G.; Tognetti, V.B.; Vandepoele, K.; Gollery, M.; Shulaev, V.; Van Breusegem, F. ROS signaling: The new wave? *Trends Plant Sci.* **2011**, *16*, 300–309. [CrossRef] [PubMed]
29. Wang, Y.; Frei, M. Stressed food—The impact of abiotic environmental stresses on crop quality. *Agric. Ecosyst. Environ.* **2011**, *141*, 271–286. [CrossRef]
30. Oh, M.-M.; Carey, E.E.; Rajashekar, C.B. Regulated Water Deficits Improve Phytochemical Concentration in Lettuce. *J. Am. Soc. Hortic. Sci.* **2019**, *135*, 223–229. [CrossRef]
31. Ripoll, J.; Urban, L.; Staudt, M.; Lopez-Lauri, F.; Bidel, L.P.R.; Bertin, N. Water shortage and quality of fleshy fruits-making the most of the unavoidable. *J. Exp. Bot.* **2014**, *65*, 4097–4117. [CrossRef]
32. Hasanuzzaman, M.; Nahar, K.; Alam, M.M.; Roychowdhury, R.; Fujita, M. Physiological, biochemical, and molecular mechanisms of heat stress tolerance in plants. *Int. J. Mol. Sci.* **2013**, *14*, 9643–9684. [CrossRef]
33. Xu, Z.; Jiang, Y.; Zhou, G. Response and adaptation of photosynthesis, respiration, and antioxidant systems to elevated CO_2 with environmental stress in plants. *Front. Plant Sci.* **2015**, *6*, 701. [CrossRef]
34. Bechtold, U.; Field, B. Molecular mechanisms controlling plant growth during abiotic stress. *J. Exp. Bot.* **2018**, *69*, 2753–2758. [CrossRef]
35. Schwarz, D.; Rouphael, Y.; Colla, G.; Venema, J.H. Grafting as a tool to improve tolerance of vegetables to abiotic stresses: Thermal stress, water stress and organic pollutants. *Sci. Hortic. (Amst.)* **2010**, *127*, 162–171. [CrossRef]
36. Al-Harbi, A.R.; Al-Omran, A.M.; Alharbi, K. Grafting improves cucumber water stress tolerance in Saudi Arabia. *Saudi J. Biol. Sci.* **2018**, *25*, 298–304. [CrossRef]
37. Kumar, P.; Rouphael, Y.; Cardarelli, M.; Colla, G. Vegetable Grafting as a Tool to Improve Drought Resistance and Water Use Efficiency. *Front. Plant Sci.* **2017**, *8*, 1130. [CrossRef] [PubMed]
38. Wang, W.; Vinocur, B.; Altman, A. Plant responses to drought, salinity and extreme temperatures: Towards genetic engineering for stress tolerance. *Planta* **2003**, *218*, 1–14. [CrossRef] [PubMed]
39. Zhuang, J.; Zhang, J.; Hou, X.L.; Wang, F.; Xiong, A.S. Transcriptomic, Proteomic, Metabolomic and Functional Genomic Approaches for the Study of Abiotic Stress in Vegetable Crops. *Crit. Rev. Plant Sci.* **2014**, *33*, 225–237. [CrossRef]
40. Wang, H.; Wang, H.; Shao, H.; Tang, X. Recent Advances in Utilizing Transcription Factors to Improve Plant Abiotic Stress Tolerance by Transgenic Technology. *Front. Plant Sci.* **2016**, *7*, 67. [CrossRef] [PubMed]
41. Singhal, P.; Jan, A.T.; Azam, M.; Haq, Q.M.R. Plant abiotic stress: A prospective strategy of exploiting promoters as alternative to overcome the escalating burden. *Front. Life Sci.* **2016**, *9*, 52–63. [CrossRef]
42. Sade, N.; Del Mar Rubio-Wilhelmi, M.; Umnajkitikorn, K.; Blumwald, E. Stress-induced senescence and plant tolerance to abiotic stress. *J. Exp. Bot.* **2018**, *69*, 845–853. [CrossRef] [PubMed]
43. Ahmad, P.; Prasad, M.N.V. *Abiotic Stress Responses in Plants: Metabolism, Productivity and Sustainability*; Springer Science & Business Media: Berlin/Heidelberg, Germany, 2011; ISBN 146140634X.
44. Shah, L.R.; Sharma, A.; Nabi, J.; Prasad, J. Breeding approaches for abiotic stress management in vegetable crops. *J. Pharmacogn. Phytochem.* **2018**, *7*, 1023–1028.
45. Pérez-Clemente, R.M.; Gomez-Cadenas, A. In vitro Tissue Culture, a Tool for the Study and Breeding of Plants Subjected to Abiotic Stress Conditions. In *Recent Advances in Plant In Vitro Culture*; InTechOpen: London, UK, 2012; p. 13.
46. Rai, M.K.; Kalia, R.K.; Singh, R.; Gangola, M.P.; Dhawan, A.K. Developing stress tolerant plants through in vitro selection-An overview of the recent progress. *Environ. Exp. Bot.* **2011**, *71*, 89–98. [CrossRef]

47. Van Oosten, M.J.; Pepe, O.; De Pascale, S.; Silletti, S.; Maggio, A. The role of biostimulants and bioeffectors as alleviators of abiotic stress in crop plants. *Chem. Biol. Technol. Agric.* **2017**, *4.1*, 5. [CrossRef]
48. Yamauchi, Y. Integrated Chemical Control of Abiotic Stress Tolerance Using Biostimulants. In *Plant, Abiotic Stress and Responses to Climate Change*; InTechOpen: London, UK, 2018; pp. 133–143.
49. Colla, G.; Rouphael, Y. Biostimulants in horticulture. *Sci. Hortic. (Amst.)* **2015**, *196*, 1–2. [CrossRef]
50. Povero, G.; Mejia, J.F.; Di Tommaso, D.; Piaggesi, A.; Warrior, P. A Systematic Approach to Discover and Characterize Natural Plant Biostimulants. *Front. Plant Sci.* **2016**, *7*, 435. [CrossRef] [PubMed]
51. Traon, D.; Amat, L.; Zotz, F.; du Jardin, P. *A Legal Framework for Plant Biostimulants and Agronomic Fertiliser Additives in the EU*; Report for the European Commission; Enterprise & Industry Directorate—General: Brussels, Belgium, 2014; p. 133.
52. Biostimulants Market Size Worth $4.14 Billion By 2025|CAGR: 10.2%. Available online: https://www.grandviewresearch.com/press-release/global-biostimulants-market (accessed on 8 March 2019).
53. Ebic Welcomes Compromise Reached At Trilogue Meeting on Fertilising Products Regulation, First Step Towards Eu-Wide Market Creation for Biostimulants. Available online: http://www.biostimulants.eu/2018/11/ebic-welcomes-compromise-reached-at-trilogue-meeting-on-fertilising-products-regulation-first-step-towards-eu-wide-market-creation-for-biostimulants/ (accessed on 5 June 2019).
54. Ep Approves Fpr: A Giant Leap for Biostimulants. Available online: http://www.biostimulants.eu/2019/03/ep-approves-fpr-a-giant-leap-for-biostimulants/ (accessed on 5 June 2019).
55. Yakhin, O.I.; Lubyanov, A.A.; Yakhin, I.A.; Brown, P.H. Biostimulants in Plant Science: A Global Perspective. *Front. Plant Sci.* **2017**, *7*, 2049. [CrossRef] [PubMed]
56. Du Jardin, P. Plant biostimulants: Definition, concept, main categories and regulation. *Sci. Hortic. (Amst.)* **2015**, *196*, 3–14. [CrossRef]
57. Rouphael, Y.; Colla, G. Synergistic Biostimulatory Action: Designing the Next Generation of Plant Biostimulants for Sustainable Agriculture. *Front. Plant Sci.* **2018**, *9*, 1655. [CrossRef]
58. Paul, K.; Sorrentino, M.; Lucini, L.; Rouphael, Y.; Cardarelli, M.; Bonini, P.; Reynaud, H.; Canaguier, R.; Trtílek, M.; Panzarová, K.; et al. Understanding the Biostimulant Action of Vegetal-Derived Protein Hydrolysates by High-Throughput Plant Phenotyping and Metabolomics: A Case Study on Tomato. *Front. Plant Sci.* **2019**, *10*, 47. [CrossRef]
59. Bulgari, R.; Morgutti, S.; Cocetta, G.; Negrini, N.; Farris, S.; Calcante, A.; Spinardi, A.; Ferrari, E.; Mignani, I.; Oberti, R.; et al. Evaluation of Borage Extracts As Potential Biostimulant Using a Phenomic, Agronomic, Physiological, and Biochemical Approach. *Front. Plant Sci.* **2017**, *8*, 935. [CrossRef]
60. Dalal, A.; Bourstein, R.; Haish, N.; Shenhar, I.; Wallach, R.; Moshelion, M.; Sciences, P.; Sciences, W. A High-Throughput Physiological Functional Phenotyping System for Time- and Cost-Effective Screening of Potential Biostimulants. *bioRxiv* **2019**, 525592. [CrossRef]
61. Nardi, S.; Carletti, P.; Pizzeghello, D.; Muscolo, A. Biological activities of humic substances. In *Biophysico-Chemical Processes Involving Natural Nonliving Organic Matter in Environmental Systems*; John Wiley & Sons, Inc.: Hoboken, NJ, USA, 2009; pp. 305–340, ISBN 0470494948, 9780470494943.
62. Bulgari, R.; Cocetta, G.; Trivellini, A.; Vernieri, P.; Ferrante, A. Biostimulants and crop responses: A review. *Biol. Agric. Hortic.* **2015**, *31*, 1–17. [CrossRef]
63. Caradonia, F.; Battaglia, V.; Righi, L.; Pascali, G.; La Torre, A. Plant Biostimulant Regulatory Framework: Prospects in Europe and Current Situation at International Level. *J. Plant Growth Regul.* **2018**, 1–11. [CrossRef]
64. Canellas, L.P.; Olivares, F.L.; Okorokova-Facanha, A.L.; Facanha, A.R. Humic Acids Isolated from Earthworm Compost Enhance Root Elongation, Lateral Root Emergence, and Plasma Membrane H+-ATPase Activity in Maize Roots. *Plant Physiol.* **2002**, *130*, 1951–1957. [CrossRef]
65. Trevisan, S.; Francioso, O.; Quaggiotti, S.; Nardi, S. Humic substances biological activity at the plant-soil interface: From environmental aspects to molecular factors. *Plant Signal. Behav.* **2010**, *5*, 635–643. [CrossRef] [PubMed]
66. Canellas, L.P.; Olivares, F.L.; Aguiar, N.O.; Jones, D.L.; Nebbioso, A.; Mazzei, P.; Piccolo, A. Humic and fulvic acids as biostimulants in horticulture. *Sci. Hortic. (Amst.)* **2015**, *196*, 15–27. [CrossRef]
67. Nardi, S.; Pizzeghello, D.; Schiavon, M.; Ertani, A. Scientia Agricola Plant biostimulants: Physiological responses induced by protein hydrolyzed-based. *Sci. Agric.* **2016**, *73*, 18–23. [CrossRef]

68. Canellas, L.P.; Dobbss, L.B.; Oliveira, A.L.; Chagas, J.G.; Aguiar, N.O.; Rumjanek, V.M.; Novotny, E.H.; Olivares, F.L.; Spaccini, R.; Piccolo, A. Chemical properties of humic matter as related to induction of plant lateral roots. *Eur. J. Soil Sci.* **2012**, *63*, 315–324. [CrossRef]
69. Trevisan, S.; Botton, A.; Vaccaro, S.; Vezzaro, A.; Quaggiotti, S.; Nardi, S. Humic substances affect Arabidopsis physiology by altering the expression of genes involved in primary metabolism, growth and development. *Environ. Exp. Bot.* **2011**, *74*, 45–55. [CrossRef]
70. Nardi, S.; Pizzeghello, D.; Reniero, F.; Rascio, N. Chemical and Biochemical Properties of Humic Substances Isolated from Forest Soils and Plant Growth. *Soil Sci. Soc. Am. J.* **2000**, *64*, 639. [CrossRef]
71. Nardi, S.; Pizzeghello, D.; Muscolo, A.; Vianello, A. Physiological effects of humic substances on higher plants. *Soil Biol. Biochem.* **2002**, *34*, 1527–1536. [CrossRef]
72. Dobbss, L.B.; Canellas, L.P.; Olivares, F.L.; Aguiar, N.O.; Peres, L.E.P.; Azevedo, M.; Spaccini, R.; Piccolo, A.; Façanha, A.R. Bioactivity of chemically transformed humic matter from vermicompost on plant root growth. *J. Agric. Food Chem.* **2010**, *58*, 3681–3688. [CrossRef]
73. Norrie, J.; Keathley, J.P. Benefits of *Ascophyllum Nodosum* Marine-Plant Extract Applications to 'Thompson Seedless' Grape Production. *Acta Hortic.* **2006**, 243–248. [CrossRef]
74. Gajc-Wolska, J.; Spiewski, T.; Grabowska, A. The Effect of Seaweed Extracts on the Yield and Quality Parameters of Broccoli (*Brassica oleracea* var. *cymosa* L.) in Open Field Production. *Acta Hortic.* **2013**, *1009*, 83–90. [CrossRef]
75. Sharma, H.S.S.; Fleming, C.; Selby, C.; Rao, J.R.; Martin, T. Plant biostimulants: A review on the processing of macroalgae and use of extracts for crop management to reduce abiotic and biotic stresses. *J. Appl. Phycol.* **2014**, *26*, 465–490. [CrossRef]
76. Hamza, B.B.; Suggars, A. Biostimulants: Myths and Realities. *TurfGrass Trends* **2001**, *8*, 6–10.
77. Battacharyya, D.; Babgohari, M.Z.; Rathor, P.; Prithiviraj, B. Seaweed extracts as biostimulants in horticulture. *Sci. Hortic. (Amst.)* **2015**, *196*, 39–48. [CrossRef]
78. Oancea, F.; Velea, S.; Mincea, C.; Ilie, L. Micro-algae based plant biostimulant and its effect on water stressed tomato plants. *Rom. J. Plant Prot.* **2013**, *VI*, 104–117.
79. Mógor, Á.F.; Ördög, V.; Lima, G.P.P.; Molnár, Z.; Mógor, G. Biostimulant properties of cyanobacterial hydrolysate related to polyamines. *J. Appl. Phycol.* **2018**, *30*, 453–460. [CrossRef]
80. Chiaiese, P.; Corrado, G.; Colla, G.; Kyriacou, M.C.; Rouphael, Y. Renewable Sources of Plant Biostimulation: Microalgae as a Sustainable Means to Improve Crop Performance. *Front. Plant Sci.* **2018**, *9*, 1782. [CrossRef] [PubMed]
81. Colla, G.; Nardi, S.; Cardarelli, M.; Ertani, A.; Lucini, L.; Canaguier, R.; Rouphael, Y. Protein hydrolysates as biostimulants in horticulture. *Sci. Hortic. (Amst.)* **2015**, *196*, 28–38. [CrossRef]
82. Cerdán, M.; Sánchez-Sánchez, A.; Oliver, M.; Juárez, M.; Sánchez-Andreu, J.J. Effect of foliar and root applications of amino acids on iron uptake by tomato plants. *Acta Hortic.* **2009**, *830*, 481–488. [CrossRef]
83. Lisiecka, J. The effect of animal protein hydrolysate on quantity and quality of strawberry daughter plants cv. 'Elsanta'. *Acta Sci. Pol. Hortorum Cultus* **2011**, *10*, 31–40.
84. Botta, A. Enhancing plant tolerance to temperature stress with amino acids: An approach to their mode of action. *Acta Hortic.* **2012**, *1009*, 29–36. [CrossRef]
85. Lucini, L.; Rouphael, Y.; Cardarelli, M.; Canaguier, R.; Kumar, P.; Colla, G. The effect of a plant-derived biostimulant on metabolic profiling and crop performance of lettuce grown under saline conditions. *Sci. Hortic. (Amst.)* **2015**, *182*, 124–133. [CrossRef]
86. Rouphael, Y.; Cardarelli, M.; Bonini, P.; Colla, G. Synergistic Action of a Microbial-based Biostimulant and a Plant Derived-Protein Hydrolysate Enhances Lettuce Tolerance to Alkalinity and Salinity. *Front. Plant Sci.* **2017**, *8*, 131. [CrossRef] [PubMed]
87. Colla, G.; Hoagland, L.; Ruzzi, M.; Cardarelli, M.; Bonini, P.; Canaguier, R.; Rouphael, Y. Biostimulant Action of Protein Hydrolysates: Unraveling Their Effects on Plant Physiology and Microbiome. *Front. Plant Sci.* **2017**, *8*, 2202. [CrossRef] [PubMed]
88. Trevisan, S.; Manoli, A.; Quaggiotti, S. A Novel Biostimulant, Belonging to Protein Hydrolysates, Mitigates Abiotic Stress Effects on Maize Seedlings Grown in Hydroponics. *Agronomy* **2019**, *9*, 28. [CrossRef]
89. Ruzzi, M.; Aroca, R. Plant growth-promoting rhizobacteria act as biostimulants in horticulture. *Sci. Hortic. (Amst.)* **2015**, *196*, 124–134. [CrossRef]

90. Turan, M.; Yildirim, E.; Kitir, N.; Unek, C.; Nikerel, E.; Ozdemir, B.S.; Güneş, A.; Mokhtari, N.E.P. Beneficial Role of Plant Growth-Promoting Bacteria in Vegetable Production Under Abiotic Stress. In *Microbial Strategies for Vegetable Production*; Zaidi, A., Khan, M.S., Eds.; Springer International Publishing: Cham, Switzerland, 2017; pp. 151–166, ISBN 978-3-319-54401-4.
91. Xu, L.; Geelen, D. Developing Biostimulants from Agro-Food and Industrial By-Products. *Front. Plant Sci.* **2018**, *9*, 1567. [CrossRef] [PubMed]
92. Ertani, A.; Schiavon, M.; Altissimo, A.; Franceschi, C.; Nardi, S. Phenol-containing organic substances stimulate phenylpropanoid metabolism in *Zea mays*. *J. Plant Nutr. Soil Sci.* **2011**, *174*, 496–503. [CrossRef]
93. Juárez-Maldonado, A.; Ortega-Ortíz, H.; Morales-Díaz, A.; González-Morales, S.; Morelos-Moreno, Á.; Cabrera-De la Fuente, M.; Sandoval-Rangel, A.; Cadenas-Pliego, G.; Benavides-Mendoza, A. Nanoparticles and Nanomaterials as Plant Biostimulants. *Int. J. Mol. Sci.* **2019**, *20*, 162. [CrossRef]
94. Qi, M.; Liu, Y.; Li, T. Nano-TiO_2 Improve the Photosynthesis of Tomato Leaves under Mild Heat Stress. *Biol. Trace Elem. Res.* **2013**, *156*, 323–328. [CrossRef]
95. Kiapour, H.; Moaveni, P.; Habibi, D.; Sani, B. Evaluation of the application of gibbrellic acid and titanium dioxide nanoparticles under drought stress on some traits of basil (*Ocimum basilicum* L.). *Int. J. Agron. Agric. Res* **2015**, *6*, 138–150.
96. Lei, Z.; Mingyu, S.; Xiao, W.; Chao, L.; Chunxiang, Q.; Liang, C.; Hao, H.; Xiaoqing, L.; Fashui, H. Antioxidant Stress is Promoted by Nano-anatase in Spinach Chloroplasts Under UV-B Radiation. *Biol. Trace Elem. Res.* **2008**, *121*, 69–79. [CrossRef] [PubMed]
97. Khan, M.N. Nano-titanium Dioxide (Nano-TiO_2) Mitigates NaCl Stress by Enhancing Antioxidative Enzymes and Accumulation of Compatible Solutes in Tomato (*Lycopersicon esculentum* Mill.). *J. Plant Sci.* **2016**, *11*, 1–11. [CrossRef]
98. Raliya, R.; Nair, R.; Chavalmane, S.; Wang, W.-N.; Biswas, P. Mechanistic evaluation of translocation and physiological impact of titanium dioxide and zinc oxide nanoparticles on the tomato (*Solanum lycopersicum* L.) plant. *Metallomics* **2015**, *7*, 1584–1594. [CrossRef] [PubMed]
99. Abbas, S.M.; Akladious, S.A. Application of carrot root extract induced salinity tolerance in cowpea (*Vigna sinensis* L.) seedlings. *Pak. J. Bot.* **2013**, *45*, 795–806.
100. Khan, W.; Rayirath, U.P.; Subramanian, S.; Jithesh, M.N.; Rayorath, P.; Hodges, D.M.; Critchley, A.T.; Craigie, J.S.; Norrie, J.; Prithiviraj, B. Seaweed Extracts as Biostimulants of Plant Growth and Development. *J. Plant Growth Regul.* **2009**, *28*, 386–399. [CrossRef]
101. Vernieri, P.; Borghesi, E.; Ferrante, A. Application of Biostimulants in Floating System for Improving Rocket Quality. *J. Food Agric. Environ.* **2005**, *3*, 86.
102. Vernieri, P.; Borghesi, E.; Tognoni, F.; Serra, G.; Ferrante, A.; Piaggesi, A. Use of biostimulants for reducing nutrient solution concentration in floating system. *Acta Hortic.* **2006**, *718*, 477–484. [CrossRef]
103. Bulgari, R.; Podetta, N.; Cocetta, G.; Piaggesi, A.; Ferrante, A. The effect of a complete fertilizer for leafy vegetables production in family and urban gardens. *Bulg. J. Agric. Sci.* **2014**, *20*, 1361–1367.
104. Shehata, S.M.; Abou El-Yazied, A. Effect of Foliar Spraying with Amino Acids and Seaweed Extract on Growth Chemical Constitutes, Yield and its Quality of Celeriac Plant Effect of Bio-stimulants on Yield and Quality of Head Lettuce Grown Under Two Sources of Nitrogen View project. *Eur. J. Sci. Res.* **2011**, *58*, 257–265.
105. Yildirim, E.; Karlidag, H.; Turan, M.; Dursun, A.; Goktepe, F. Promotion of Broccoli by Plant Growth Promoting Rhizobacteria. *HortScience* **2011**, *46*, 932–936. [CrossRef]
106. Narendra Babu, A.; Jogaiah, S.; Ito, S.; Kestur Nagaraj, A.; Tran, L.-S.P. Improvement of growth, fruit weight and early blight disease protection of tomato plants by rhizosphere bacteria is correlated with their beneficial traits and induced biosynthesis of antioxidant peroxidase and polyphenol oxidase. *Plant Sci.* **2015**, *231*, 62–73. [CrossRef] [PubMed]
107. Karlidag, H.; Yildirim, E.; Turan, M.; Pehluvan, M.; Donmez, F. Plant growth-promoting rhizobacteria mitigate deleterious effects of salt stress on strawberry plants (*Fragaria* × *ananassa*). *HortScience* **2013**, *48*, 563–567. [CrossRef]
108. Han, H.S.; Lee, K.D. Plant Growth Promoting Rhizobacteria Effect on Antioxidant Status, Photosynthesis, Mineral Uptake and growth of lettuce under soil salinity. *Res. J. Agric. Biol. Sci.* **2005**, *1*, 210–215.
109. Fasciglione, G.; Casanovas, E.M.; Yommi, A.; Sueldo, R.J.; Barassi, C.A. Azospirillum improves lettuce growth and transplant under saline conditions. *J. Sci. Food Agric.* **2012**, *92*, 2518–2523. [CrossRef] [PubMed]

110. Rengasamy, K.R.R.; Kulkarni, M.G.; Pendota, S.C.; Van Staden, J. Enhancing growth, phytochemical constituents and aphid resistance capacity in cabbage with foliar application of eckol—A biologically active phenolic molecule from brown seaweed. *New Biotechnol.* **2016**, *33*, 273–279. [CrossRef] [PubMed]
111. Mona, M.A. The potential of *Moringa oleifera* extract as a biostimulant in enhancing the growth, biochemical and hormonal contents in rocket (*Eruca vesicaria subsp.* sativa) plants. *Int. J. Plant Physiol. Biochem.* **2013**, *5*, 42–49. [CrossRef]
112. Abou-Sreea, A.I.; Matter, F.M. Using moringa leaf extract as biostimulant and giberrellic acid for enhancing fennel (*Foeniculum vulgare* var. *azoricum Mill.*) Growth and oil yield. *Acta Sci. Intellectus* **2016**, *2410*, 9738.
113. Abdel-Rahman, S.S.A.; Abdel-Kader, A.A.S. Response of Fennel (*Foeniculum vulgare*, Mill) plants to foliar application of moringa leaf extract and benzyladenine (BA). *S. Afr. J. Bot.* **2019**. [CrossRef]
114. Abd El-Mageed, T.A.; Semida, W.M.; Rady, M.M. Moringa leaf extract as biostimulant improves water use efficiency, physio-biochemical attributes of squash plants under deficit irrigation. *Agric. Water Manag.* **2017**, *193*, 46–54. [CrossRef]
115. Ullah, A.; Ali, S.; Shah, S.M. Influence of foliar application of bio-stimulants on growth, yield and chemical composition of tomato Influence of foliar application of bio-stimulants on growth, yield and chemical composition of tomato. *Int. J. Biosci.* **2019**, *14*, 309–316.
116. Luziatelli, F.; Ficca, A.G.; Colla, G.; Baldassarre Švecová, E.; Ruzzi, M. Foliar Application of Vegetal-Derived Bioactive Compounds Stimulates the Growth of Beneficial Bacteria and Enhances Microbiome Biodiversity in Lettuce. *Front. Plant Sci.* **2019**, *10*, 60. [CrossRef] [PubMed]
117. Kulkarni, M.G.; Rengasamy, K.R.R.; Pendota, S.C.; Gruz, J.; Plačková, L.; Novák, O.; Doležal, K.; Van Staden, J. Bioactive molecules derived from smoke and seaweed *Ecklonia maxima* showing phytohormone-like activity in *Spinacia oleracea* L. *New Biotechnol.* **2019**, *48*, 83–89. [CrossRef] [PubMed]
118. Mattner, S.W.; Wite, D.; Riches, D.A.; Porter, I.J.; Arioli, T. The effect of kelp extract on seedling establishment of broccoli on contrasting soil types in southern Victoria, Australia. *Biol. Agric. Hortic.* **2013**, *29*, 258–270. [CrossRef]
119. Paradiković, N.; Vinković, T.; Vinković Vrček, I.; Žuntar, I.; Bojić, M.; Medić-Šarić, M. Effect of natural biostimulants on yield and nutritional quality: An example of sweet yellow pepper (*Capsicum annuum* L.) plants. *J. Sci. Food Agric.* **2011**, *91*, 2146–2152. [CrossRef] [PubMed]
120. Petrozza, A.; Summerer, S.; Di Tommaso, G.; Di Tommaso, D.; Piaggesi, A. An evaluation of tomato plant root development and Morpho- Physiological response treated with VIVA® by image analysis. *Acta Hortic.* **2013**, *1009*, 155–160. [CrossRef]
121. Petrozza, A.; Summerer, S.; Di Tommaso, G.; Di Tommaso, D.; Piaggesi, A. Evaluation of the Effect of RADIFARM® treatment on the Morpho—Physiological characteristics of root systems via image analysis. *Acta Hortic.* **2013**, *1009*, 149–154. [CrossRef]
122. Kumar, S.; Chinnannan, K.; Thamilarasan, S.K.; Seralathan, M.; Shanmuganathan, R.; Padikasan, I.A. Enzymatically hydrolysed sago bagasse improves physiological, biochemical and molecular attributes of Solanum lycopersicum. *Biocatal. Agric. Biotechnol.* **2019**, *17*, 499–506. [CrossRef]
123. Kunicki, E.; Grabowska, A.; Sękara, A.; Wojciechowska, R. The effect of cultivar type, time of cultivation, and biostimulant treatment on the yield of spinach (*Spinacia oleracea* L.). *Folia Hortic.* **2010**, *22*, 9–13. [CrossRef]
124. Pokluda, R.; Sękara, A.; Jezdinský, A.; Kalisz, A.; Neugebauerová, J.; Grabowska, A. The physiological status and stress biomarker concentration of *Coriandrum sativum* L. plants subjected to chilling are modified by biostimulant application. *Biol. Agric. Hortic.* **2016**, *32*, 258–268. [CrossRef]
125. Le Mire, G.; Nguyen, M.L.; Fassotte, B.; du Jardin, P.; Verheggen, F.; Delaplace, P.; Jijakli, M.H. Implementing plant biostimulants and biocontrol strategies in the agroecological management of cultivated ecosystems. A review. *Biotechnol. Agron. Soc. Environ.* **2016**, *20*, 299–313.
126. Cao, K.; Yu, J.; Xu, D.; Ai, K.; Bao, E.; Zou, Z. Exposure to lower red to far-red light ratios improve tomato tolerance to salt stress. *BMC Plant Biol.* **2018**, *18*, 10–15. [CrossRef] [PubMed]
127. Martinez, V.; Nieves-Cordones, M.; Lopez-Delacalle, M.; Rodenas, R.; Mestre, T.; Garcia-Sanchez, F.; Rubio, F.; Nortes, P.; Mittler, R.; Rivero, R. Tolerance to Stress Combination in Tomato Plants: New Insights in the Protective Role of Melatonin. *Molecules* **2018**, *23*, 535. [CrossRef] [PubMed]
128. Ashraf, M.A.; Akbar, A.; Askari, S.H.; Iqbal, M.; Rasheed, R.; Hussain, I. Recent Advances in Abiotic Stress Tolerance of Plants through Chemical Priming: An Overview. In *Advances in Seed Priming*; Rakshit, A., Singh, H.B., Eds.; Springer: Singapore, 2018; pp. 51–79, ISBN 978-981-13-0032-5.

129. Subramanian, P.; Kim, K.; Krishnamoorthy, R.; Mageswari, A.; Selvakumar, G.; Sa, T. Cold stress tolerance in psychrotolerant soil bacteria and their conferred chilling resistance in tomato (*solanum lycopersicum* Mill.) under low temperatures. *PLoS ONE* **2016**, *11*, e0161592. [CrossRef] [PubMed]
130. Subramanian, P.; Mageswari, A.; Kim, K.; Lee, Y.; Sa, T. Psychrotolerant Endophytic Pseudomonas sp. Strains OB155 and OS261 Induced Chilling Resistance in Tomato Plants (*Solanum lycopersicum* Mill.) by Activation of Their Antioxidant Capacity. *Mol. Plant-Microbe Interact.* **2015**, *28*, 1073–1081. [CrossRef] [PubMed]
131. Marfà, O.; Cáceres, R.; Polo, J.; Ródenas, J. Animal protein hydrolysate as a biostimulant for transplanted strawberry plants subjected to cold stress. *Acta Hortic.* **2009**, *842*, 315–318. [CrossRef]
132. Polo, J.; Barroso, R.; Ródenas, J.; Azcón-Bieto, J.; Cáceres, R.; Marfà, O. Porcine hemoglobin hydrolysate as a biostimulant for lettuce plants subjected to conditions of thermal stress. *Horttechnology* **2006**, *16*, 483–487. [CrossRef]
133. Korkmaz, A.; Korkmaz, Y.; Demirkiran, A.R. Enhancing chilling stress tolerance of pepper seedlings by exogenous application of 5-aminolevulinic acid. *Environ. Exp. Bot.* **2010**, *67*, 495–501. [CrossRef]
134. Challinor, A.J.; Watson, J.; Lobell, D.B.; Howden, S.M.; Smith, D.R.; Chhetri, N. A meta-analysis of crop yield under climate change and adaptation. *Nat. Clim. Chang.* **2014**, *4*, 287–291. [CrossRef]
135. Fahad, S.; Bajwa, A.A.; Nazir, U.; Anjum, S.A.; Farooq, A.; Zohaib, A.; Sadia, S.; Nasim, W.; Adkins, S.; Saud, S.; et al. Crop Production under Drought and Heat Stress: Plant Responses and Management Options. *Front. Plant Sci.* **2017**, *8*, 1147. [CrossRef]
136. Hasanuzzaman, M.; Nahar, K.; Fujit, M. Extreme Temperature Responses, Oxidative Stress and Antioxidant Defense in Plants. In *Abiotic Stress—Plant Responses and Applications in Agriculture*; InTechOpen: London, UK, 2013; pp. 169–205.
137. Camejo, D.; Rodríguez, P.; Morales, M.A.; Dell'Amico, J.M.; Torrecillas, A.; Alarcón, J.J. High temperature effects on photosynthetic activity of two tomato cultivars with different heat susceptibility. *J. Plant Physiol.* **2005**, *162*, 281–289. [CrossRef]
138. Ogweno, J.O.; Song, X.S.; Shi, K.; Hu, W.H.; Mao, W.H.; Zhou, Y.H.; Yu, J.Q.; Nogués, S. Brassinosteroids Alleviate Heat-Induced Inhibition of Photosynthesis by Increasing Carboxylation Efficiency and Enhancing Antioxidant Systems in Lycopersicon esculentum. *J. Plant Growth Regul.* **2008**, *27*, 49–57. [CrossRef]
139. El-Bassiony, A.M.; Ghoname, A.A.; El-Awadi, M.E.; Fawzy, Z.F.; Gruda, N. Ameliorative Effects of Brassinosteroids on Growth and Productivity of Snap Beans Grown Under High Temperature. *Gesunde Pflanz.* **2012**, *64*, 175–182. [CrossRef]
140. Nahar, K.; Hasanuzzaman, M.; Alam, M.M.; Fujita, M. Exogenous glutathione confers high temperature stress tolerance in mung bean (*Vigna radiata* L.) by modulating antioxidant defense and methylglyoxal detoxification system. *Environ. Exp. Bot.* **2015**, *112*, 44–54. [CrossRef]
141. Yang, J.D.; Yun, J.Y.; Zhang, T.H.; Zhao, H.L. Presoaking with nitric oxide donor SNP alleviates heat shock damages in mung bean leaf discs. *Bot. Stud.* **2006**, *47*, 129–136.
142. Kumar, S.; Kaur, R.; Kaur, N.; Bhandhari, K.; Kaushal, N.; Gupta, K.; Bains, T.S.; Nayyar, H. Heat-stress induced inhibition in growth and chlorosis in mungbean (*Phaseolus aureus* Roxb.) is partly mitigated by ascorbic acid application and is related to reduction in oxidative stress. *Acta Physiol. Plant.* **2011**, *33*, 2091. [CrossRef]
143. Kaushal, N.; Gupta, K.; Bhandhari, K.; Kumar, S.; Thakur, P.; Nayyar, H. Proline induces heat tolerance in chickpea (*Cicer arietinum* L.) plants by protecting vital enzymes of carbon and antioxidative metabolism. *Physiol. Mol. Biol. Plants* **2011**, *17*, 203–213. [CrossRef]
144. Kumar, S.; Kaushal, N.; Nayyar, H.; Gaur, P. Abscisic acid induces heat tolerance in chickpea (*Cicer arietinum* L.) seedlings by facilitated accumulation of osmoprotectants. *Acta Physiol. Plant.* **2012**, *34*, 1651–1658. [CrossRef]
145. Viégas, R.A.; da Silveira, J.A.G.; Lima, A.R.D., Jr.; Queiroz, J.E.; Fausto, M.J.M. Effects of NaCl-salinity on growth and inorganic solute accumulation in young cashew plants. *Rev. Bras. Eng. Agrícola Ambient.* **2006**, *5*, 216–222. [CrossRef]
146. Colla, G.; Rouphael, Y.; Leonardi, C.; Bie, Z. Role of grafting in vegetable crops grown under saline conditions. *Sci. Hortic. (Amst.)* **2010**, *127*, 147–155. [CrossRef]

147. Miceli, A.; Moncada, A.; D'Anna, F. Effect of salt stress in lettuce cultivation. *Acta Hortic.* **2003**, *609*, 371–375. [CrossRef]
148. Kaya, C.; Higgs, D.; Sakar, E. Response of two leafy vegetables grown at high salinity to supplementary potassium and phosphorus during different growth stages. *J. Plant Nutr.* **2002**, *25*, 2663–2676. [CrossRef]
149. Taïbi, K.; Taïbi, F.; Ait Abderrahim, L.; Ennajah, A.; Belkhodja, M.; Mulet, J.M. Effect of salt stress on growth, chlorophyll content, lipid peroxidation and antioxidant defence systems in *Phaseolus vulgaris* L. *S. Afr. J. Bot.* **2016**, *105*, 306–312. [CrossRef]
150. Yarsi, G.; Sivaci, A.; Dasgan, H.Y.; Altuntas, O.; Binzet, R.; Akhoundnejad, Y. Effects of salinity stress on chlorophyll and carotenoid contents and stomata size of grafted and ungrafted galia C8 melon cultivar. *Pak. J. Bot.* **2017**, *49*, 421–426.
151. Munns, R.; Termaat, A. Whole-plant responses to salinity. *Aust. J. Plant Physiol.* **1986**, *13*, 143–160. [CrossRef]
152. Blaylock, A.D. *Soil Salinity, Salt Tolerance, and Growth Potential of Horticultural and Landscape Plants*; B-988; University of Wyoming, Cooperative Extension Service: Laramie, WY, USA, 1994.
153. Maas, E.V.; Grattan, S.R. Crop yields as affected by salinity. *Agronomy* **1999**, *38*, 55–110.
154. Shaheen, S.; Naseer, S.; Ashraf, M.; Akram, N.A. Salt stress affects water relations, photosynthesis, and oxidative defense mechanisms in *Solanum melongena* L. *J. Plant Interact.* **2013**, *8*, 85–96. [CrossRef]
155. Xu, C.; Mou, B. Responses of Spinach to Salinity and Nutrient Deficiency in Growth, Physiology, and Nutritional Value. *J. Am. Soc. Hortic. Sci.* **2016**, *141*, 12–21. [CrossRef]
156. Bano, S.; Ashraf, M.; Akram, N.A.; Al-Qurainy, F. Regulation in some vital physiological attributes and antioxidative defense system in carrot (*Daucus carota* L.) under saline stress. *J. Appl. Bot. Food Qual. Angew. Bot.* **2012**, *85*, 105.
157. Agastian, P.; Kingsley, S.J.; Vivekanandan, M. Effect of salinity on photosynthesis and biochemical characteristics in mulberry genotypes. *Photosynthetica* **2000**, *38*, 287–290. [CrossRef]
158. Sayyad-Amin, P.; Jahansooz, M.R.; Borzouei, A.; Ajili, F. Changes in photosynthetic pigments and chlorophyll-a fluorescence attributes of sweet-forage and grain sorghum cultivars under salt stress. *J. Biol. Phys.* **2016**, *42*, 601–620. [CrossRef] [PubMed]
159. Moud, A.M.; Maghsoudi, K. Salt Stress Effects on Respiration and Growth of Germinated Seeds of Different Wheat (*Triticum aestivum* L.) Cultivars. *World J. Agric. Sci.* **2008**, *4*, 351–358.
160. Flores, P.; Botella, M.Á.; Cerdá, A.; Martínez, V. Influence of nitrate level on nitrate assimilation in tomato (*Lycopersicon esculentum*) plants under saline stress. *Can. J. Bot.* **2004**, *82*, 207–213. [CrossRef]
161. Breusegem, F. Van Reactive oxygen species in plant cell death. *Plant Physiol.* **2006**, *141*, 384–390. [CrossRef] [PubMed]
162. Bulgari, R.; Trivellini, A.; Ferrante, A. Effects of Two Doses of Organic Extract-Based Biostimulant on Greenhouse Lettuce Grown Under Increasing NaCl Concentrations. *Front. Plant Sci.* **2019**, *9*, 1870. [CrossRef]
163. Barassi, C.A.; Ayrault, G.; Creus, C.M.; Sueldo, R.J.; Sobrero, M.T. Seed inoculation with Azospirillum mitigates NaCl effects on lettuce. *Sci. Hortic. (Amst.)* **2006**, *109*, 8–14. [CrossRef]
164. Fasciglione, G.; Casanovas, E.M.; Quillehauquy, V.; Yommi, A.K.; Goñi, M.G.; Roura, S.I.; Barassi, C.A. Azospirillum inoculation effects on growth, product quality and storage life of lettuce plants grown under salt stress. *Sci. Hortic. (Amst.)* **2015**, *195*, 154–162. [CrossRef]
165. Del Amor, F.M.; Cuadra-Crespo, P. Plant growth-promoting bacteria as a tool to improve salinity tolerance in sweet pepper. *Funct. Plant Biol.* **2012**, *39*, 82–90. [CrossRef]
166. Hamaoui, B.; Abbadi, J.M.; Burdman, S.; Rashid, A.; Sarig, S.; Okon, Y. Effects of inoculation with Azospirillum brasilense on chickpeas (*Cicer arietinum*) and faba beans (*Vicia faba*) under different growth conditions. *Agronomie* **2003**, *21*, 553–560. [CrossRef]
167. Del Pilar Cordovilla, M.; Berrido, S.I.; Ligero, F.; Lluch, C. Rhizobium strain effects on the growth and nitrogen assimilation in Pisum sativum and Vicia faba plant growth under salt stress. *J. Plant Physiol.* **1999**, *154*, 127–131. [CrossRef]
168. Mayak, S.; Tirosh, T.; Glick, B.R. Plant growth-promoting bacteria that confer resistance to water stress in tomatoes and peppers. *Plant Sci.* **2004**, *166*, 525–530. [CrossRef]
169. Yildirim, E.; Taylor, A.G.; Spittler, T.D. Ameliorative effects of biological treatments on growth of squash plants under salt stress. *Sci. Hortic. (Amst.)* **2006**, *111*, 1–6. [CrossRef]
170. Mesut Çimrin, K.; Türkmen, Ö.; Turan, M.; Tuncer, B. Phosphorus and humic acid application alleviate salinity stress of pepper seedling. *Afr. J. Biotechnol.* **2010**, *9*, 5845–5851.

171. Aydin, A.; Kant, C.; Turan, M. Humic acid application alleviate salinity stress of bean (*Phaseolus vulgaris* L.) plants decreasing membrane leakage. *Afr. J. Agric. Res.* **2012**, *7*, 1073–1086. [CrossRef]
172. Demir, K.; Günes, A.; Inal, A.; Alpaslan, M. Effects of Humic Acids on the Yield and Mineral Nutrition of Cucumber (*Cucumis Sativus*, L.) Grown with Different Salinity Levels. *Acta Hortic.* **1999**, *492*, 95–104. [CrossRef]
173. Guinan, K.J.; Sujeeth, N.; Copeland, R.B.; Jones, P.W.; O'Brien, N.M.; Sharma, H.S.S.; Prouteau, P.F.J.; O'Sullivan, J.T. Discrete Roles for Extracts of *Ascophyllum Nodosum* in Enhancing Plant Growth and Tolerance To Abiotic and Biotic Stresses. *Acta Hortic.* **2013**, *1009*, 127–135. [CrossRef]
174. Ross, R.; Holden, D. Commercial extracts of the brown seaweed *Ascophyllum nodosum* enhance growth and yield of strawberries. *HortScience* **2010**, *45*, S141.
175. EL Arroussi, H.; Benhima, R.; Elbaouchi, A.; Sijilmassi, B.; EL Mernissi, N.; Aafsar, A.; Meftah-Kadmiri, I.; Bendaou, N.; Smouni, A. Dunaliella salina exopolysaccharides: A promising biostimulant for salt stress tolerance in tomato (*Solanum lycopersicum*). *J. Appl. Phycol.* **2018**, *30*, 2929–2941. [CrossRef]
176. Abdel Latef, A.A.H.; Srivastava, A.K.; Saber, H.; Alwaleed, E.A.; Tran, L.S.P. Sargassum muticum and Jania rubens regulate amino acid metabolism to improve growth and alleviate salinity in chickpea. *Sci. Rep.* **2017**, *7*, 1–12. [CrossRef]
177. Rady, M.M.; Varma, B.; Howladar, S.M. Common bean (*Phaseolus vulgaris* L.) seedlings overcome NaCl stress as a result of presoaking in *Moringa oleifera* leaf extract. *Sci. Hortic. (Amst.)* **2013**, *162*, 63–70. [CrossRef]
178. Semida, W.M.; Rady, M.M. Presoaking application of propolis and maize grain extracts alleviates salinity stress in common bean (*Phaseolus vulgaris* L.). *Sci. Hortic. (Amst.)* **2014**, *168*, 210–217. [CrossRef]
179. Rady, M.M.; Mohamed, G.F. Modulation of salt stress effects on the growth, physio-chemical attributes and yields of *Phaseolus vulgaris* L. plants by the combined application of salicylic acid and *Moringa oleifera* leaf extract. *Sci. Hortic. (Amst.)* **2015**, *193*, 105–113. [CrossRef]
180. Howladar, S.M. A novel *Moringa oleifera* leaf extract can mitigate the stress effects of salinity and cadmium in bean (*Phaseolus vulgaris* L.) plants. *Ecotoxicol. Environ. Saf.* **2014**, *100*, 69–75. [CrossRef] [PubMed]
181. Rady, M.M.; Desoky, E.S.M.; Elrys, A.S.; Boghdady, M.S. Can licorice root extract be used as an effective natural biostimulant for salt-stressed common bean plants? *S. Afr. J. Bot.* **2019**, *121*, 294–305. [CrossRef]
182. Semida, W.M.; Abd El-Mageed, T.A.; Hemida, K.; Rady, M.M. Natural bee-honey based biostimulants confer salt tolerance in onion via modulation of the antioxidant defence system. *J. Hortic. Sci. Biotechnol.* **2019**, 1–11. [CrossRef]
183. Elliott, J.; Deryng, D.; Müller, C.; Frieler, K.; Konzmann, M.; Gerten, D.; Glotter, M.; Flörke, M.; Wada, Y.; Best, N.; et al. Constraints and potentials of future irrigation water availability on agricultural production under climate change. *Proc. Natl. Acad. Sci. USA* **2014**, *111*, 3239–3244. [CrossRef] [PubMed]
184. Calvo, P.; Nelson, L.; Kloepper, J.W. Agricultural uses of plant biostimulants. *Plant Soil* **2014**, *383*, 3–41. [CrossRef]
185. Kałuzewicz, A.; Krzesiński, W.; Spizewski, T.; Zaworska, A. Effect of biostimulants on several physiological characteristics and chlorophyll content in broccoli under drought stress and re-watering. *Not. Bot. Horti Agrobot. Cluj-Napoca* **2017**, *45*, 197–202. [CrossRef]
186. Xu, C.; Leskovar, D.I. Effects of A. nodosum seaweed extracts on spinach growth, physiology and nutrition value under drought stress. *Sci. Hortic. (Amst.)* **2015**, *183*, 39–47. [CrossRef]
187. Goñi, O.; Quille, P.; O'Connell, S. Ascophyllum nodosum extract biostimulants and their role in enhancing tolerance to drought stress in tomato plants. *Plant Physiol. Biochem.* **2018**, *126*, 63–73. [CrossRef] [PubMed]
188. Petrozza, A.; Santaniello, A.; Summerer, S.; Di Tommaso, G.; Di Tommaso, D.; Paparelli, E.; Piaggesi, A.; Perata, P.; Cellini, F. Physiological responses to Megafol® treatments in tomato plants under drought stress: A phenomic and molecular approach. *Sci. Hortic. (Amst.)* **2014**, *174*, 185–192. [CrossRef]
189. Romero, A.M.; Vega, D.; Correa, O.S. Azospirillum brasilense mitigates water stress imposed by a vascular disease by increasing xylem vessel area and stem hydraulic conductivity in tomato. *Appl. Soil Ecol.* **2014**, *82*, 38–43. [CrossRef]
190. Vurukonda, S.S.K.P.; Vardharajula, S.; Shrivastava, M.; SkZ, A. Enhancement of drought stress tolerance in crops by plant growth promoting rhizobacteria. *Microbiol. Res.* **2016**, *184*, 13–24. [CrossRef] [PubMed]
191. Kumar, A.; Verma, J.P. Does plant—Microbe interaction confer stress tolerance in plants: A review? *Microbiol. Res.* **2018**, *207*, 41–52. [CrossRef] [PubMed]

192. Arshad, M.; Shaharoona, B.; Mahmood, T. Inoculation with Pseudomonas spp. Containing ACC-Deaminase Partially Eliminates the Effects of Drought Stress on Growth, Yield, and Ripening of Pea (*Pisum sativum* L.). *Pedosphere* **2008**, *18*, 611–620. [CrossRef]
193. Heidari, M.; Golpayegani, A. Effects of water stress and inoculation with plant growth promoting rhizobacteria (PGPR) on antioxidant status and photosynthetic pigments in basil (*Ocimum basilicum* L.). *J. Saudi Soc. Agric. Sci.* **2012**, *11*, 57–61. [CrossRef]
194. Halpern, M.; Bar-Tal, A.; Ofek, M.; Minz, D.; Muller, T.; Yermiyahu, U. The Use of Biostimulants for Enhancing Nutrient Uptake. In *Advances in Agronomy*; Academic Press: Cambridge, MA, USA, 2015; Volume 130, pp. 141–174, ISBN 9780128021378.
195. Toscano, S.; Romano, D.; Massa, D.; Bulgari, R.; Franzoni, G.; Ferrante, A. Biostimulant applications in low input horticultural cultivation systems. *Italus Hortus* **2019**, *25*, 27–36.
196. Koleška, I.; Hasanagić, D.; Todorović, V.; Murtić, S.; Klokić, I.; Paradiković, N.; Kukavica, B. Biostimulant prevents yield loss and reduces oxidative damage in tomato plants grown on reduced NPK nutrition. *J. Plant Interact.* **2017**, *12*, 209–218. [CrossRef]
197. Anjum, K.; Ahmed, M.; Baber, J.K.; Alizai, M.A.; Ahmed, N.; Tareen, M.H. Response of Garlic Bulb Yield To Bio-Stimulant (Bio-Cozyme) Under Calcareous Soil. *Life Sci. Int. J.* **2014**, *8*, 3058–3062.
198. Papenfus, H.B.; Kulkarni, M.G.; Stirk, W.A.; Finnie, J.F.; Van Staden, J. Effect of a commercial seaweed extract (Kelpak®) and polyamines on nutrient-deprived (N, P and K) okra seedlings. *Sci. Hortic. (Amst.)* **2013**, *151*, 142–146. [CrossRef]
199. Spinelli, F.; Fiori, G.; Noferini, M.; Sprocatti, M.; Costa, G. A novel type of seaweed extract as a natural alternative to the use of iron chelates in strawberry production. *Sci. Hortic. (Amst.)* **2010**, *125*, 263–269. [CrossRef]
200. Cerdán, M.; Sánchez-Sánchez, A.; Jordá, J.D.; Juárez, M.; Sánchez-Andreu, J. Effect of commercial amino acids on iron nutrition of tomato plants grown under lime-induced iron deficiency. *J. Plant Nutr. Soil Sci.* **2013**, *176*, 859–866. [CrossRef]
201. Miflin, B.J.; Lea, P.J. The pathway of nitrogen assimilation in plants. *Phytochemistry* **1976**, *15*, 873–885. [CrossRef]
202. Porra, R.J. Invited Review Recent Progress in Porphyrin and Chlorophyll Biosynthesis. *Regulation* **1997**, *65*, 492–516.
203. Paradiković, N.; Vinković, T.; Vinković Vrček, I.; Tkalec, M. Natural biostimulants reduce the incidence of BER in sweet yellow pepper plants (*Capsicum annuum* L.). *Agric. Food Sci.* **2013**, *22*, 307–317. [CrossRef]
204. Mayak, S.; Tirosh, T.; Glick, B.R. Plant growth-promoting bacteria confer resistance in tomato plants to salt stress. *Plant Physiol. Biochem.* **2004**, *42*, 565–572. [CrossRef] [PubMed]

agronomy

MDPI

Review

Plant Biostimulants: Importance of the Quality and Yield of Horticultural Crops and the Improvement of Plant Tolerance to Abiotic Stress—A Review

Magdalena Drobek, Magdalena Frąc and Justyna Cybulska *

Institute of Agrophysics, Polish Academy of Sciences, Doświadczalna 4, 20-290 Lublin, Poland; m.drobek@ipan.lublin.pl (M.D.); m.frac@ipan.lublin.pl (M.F.)

* Correspondence: j.cybulska@ipan.lublin.pl; Tel.: +48-81-744-5061

Received: 26 April 2019; Accepted: 20 June 2019; Published: 24 June 2019

Abstract: Biostimulants are among the natural preparations that improve the general health, vitality, and growth of plants and protect them against infections. They can be successfully used in both agri- and horticultural crops. The main active substances used in such preparations are humic and fulvic acids, protein hydrolysates, compounds containing nitrogen, seaweed extracts, beneficial fungi, and bacteria. Biostimulant formulations may be single- or multi-component, but the synergic action of several different components has been observed. Many groups of biostimulants have been distinguished through their method of application (soil, foliar), the material from which they were produced (plant, animal), or the process by which they were created (hydrolysis, fermentation, extraction). Natural soil stimulants can induce the development of beneficial soil organisms that provide substrates for plant growth. The use of natural preparations that are not harmful to the environment is particularly important in connection with the progressive processes of soil degradation and atmospheric pollution. This review gives an overview of the importance and influence of different natural plant biostimulants on both the yield and quality of crops.

Keywords: fruit quality; plants biostimulants; yielding

---

1. Introduction

The quality and quantity of crops are influenced by both biotic and abiotic factors. Quality may be defined as a set of agronomic (e.g., fruit size, yield, resistance to bacteria and fungi) and organoleptic (e.g., colour, shape, firmness) properties as well as nutrient and vitamin content [1]. The abiotic factors include soil composition, extreme salinity, acidity, high and low temperatures, drought, pollution, humidity, rain, wind, or ultraviolet radiation. Stress caused by unfavourable stimuli can significantly reduce harvest yields because plants respond by using their energy reserves to fight stress instead of concentrating on yielding. Biotic factors include various bacteria, fungi, or viruses that are the cause of numerous plant diseases. Fungal and bacterial infections may not only reduce yield but may also lead to the loss of the entire harvest. To prevent this, various types of plant protection products are used. In accordance with the recommendations of the European Union [2], chemical and mineral plant protection agents are intended to be slowly replaced by natural preparations. The reason for this is the adverse influence of chemical and mineral plant protection agents on the natural environment, as well as on the health benefits of plant crops. Moreover, artificial fertilizers are responsible for the eutrophication of many bodies of water. This results in the formation of dead zones devoid of living organisms. The Baltic Sea alone is distinguished by having oxygen-free zones making up around 60,000 km^2 of area caused by water pollution due to fertilizers. This area constitutes, on average, 3.5% of the catchment area of the Baltic Sea [3]. The effects of fertilizers have an unfavourable effect on algae, plants, animals, and people. Due to the fact that man is a higher-order consumer, people are particularly severely exposed to the harmful effects of fertilizer compounds accumulated at the lower levels of the food chain. Harmful compounds from fertilizers may weaken enzymes or interfere with protein production or vitamin absorption in the human body [3]. Natural preparations called biostimulants increase the efficiency of nutrient utilization and tolerance to abiotic stress and improve the quality of crops [4]. Biostimulants include organic and non-organic substances and/or microorganisms [5]. Farmers who manage organic farms are also eager to use natural stimulants to improve crop quality [6]. Increasing consumer awareness concerning healthy food favours the enhancement of the significance of organic farming [7].

The effects of the stimulators may be multifaceted. The effects of their activities vary depending on the type of biostimulant used and the plant variety. However, it should be noted that most of them have a beneficial effect on crops [8].

2. Definitions and Classification of Biostimulants

Biostimulants can be treated as an additive to fertilizers and support the uptake of nutrients, promote plant growth, and increase tolerance to abiotic stress [9]. The definition of biostimulants is wide and not sufficiently precise. However, there are two main features that distinguish biostimulants from other growth and plant-protection agents. A biostimulant may be any substance or mixture of substances of natural origin or microorganism which improves the condition of crops without causing adverse side effects [10].

Enzymes, proteins, amino acids, micronutrients, and other compounds may be used as biostimulants. Natural stimulants are often included under the term biostimulants, including phenols, salicylic acid, humic and fulvic acids, or protein hydrolases [10,11]. An important group of plant biostimulants are organisms including fungi and bacteria that change the species composition of organisms found in the soil or plants. Their presence may accelerate the rate of degradation processes or limit the number of specific fungal and bacterial groups [12,13]. Popular fungi used as biostimulants include *Glomus intraradices* [14], *Trichoderma atroviride* [14], *Trichoderma reesei*, and *Heteroconium chaetospira* [10,15–19]. Useful bacteria include *Arthrobacter* spp., *Enterobacter* spp., *Acinetobacter* spp., *Pseudomonas* spp., *Ochrobactrum* spp., *Bacillus* spp., and *Rhodococcus* spp. [18,19].

Biostimulants cannot be defined as fertilizers because they do not provide nutrients directly to plants. Biostimulants may facilitate the acquisition of nutrients by supporting metabolic processes in the soil and plants. An example of such an activity is the facilitation of the development of arbuscular mycorrhizal fungi that transport nutrients to the host plant [20].

3. Sources of Biostimulants

Biostimulants are preparations made from natural raw materials. Some of them are plant extracts such as rosemary, which stimulates the growth of tomato plants with a concentration of 1000 ppm. Rosemary oil contributes to improved nutrient uptake and increases the fresh mass of roots [21]. Plant and animal biostimulants are formed, for example, as a result of chemical or enzymatic hydrolysis. The products of hydrolysis are mixtures of peptides and amino acids (protein hydrolysates). Chemical acid or alkaline hydrolysis is used to produce biostimulants of animal origin, from raw materials such as hen feathers, bone meal, casein, collagen from skins, animal tissue or fish waste (Table 1). Biostimulants of plant origin are produced using enzymatic hydrolysis. In the production of plant biostimulants, for example, alfalfa hay, pulses, and vegetable or fruit waste may be used [13,22]. Protein hydrolysates contain amino acids, peptides, and non-protein compounds. Protein hydrolysates stimulate plant growth, reduce the use of inert fertilizers, and are environmentally friendly [22]. The solution, which at the same time allows for a reduction in the amount of organic waste and the creation of biostimulating preparations, is actually a fermentation process. Biostimulants may also be the products of anaerobic digestion (Table 1). Dissolved organic matter is formed in fermentation chambers and has stimulating properties. The source of dissolved organic matter is usually plant, animal, and lignin biomass [23]. Biopreparations from marine algae may contain low-molecular polypeptides and amino acids, vitamins, enzymes, phytohormones, sugars, and antioxidants. These compounds activate the processes of rhizogenesis and lead to positive morphological and anatomical changes in plants (Table 1). Two applications of a biopreparation derived from algae enhanced the development of *Cornus alba* "Aurea" roots by 80% when compared to the control. Such biopreparations can be used for rooting young seedlings or improving the process of adult rhizogenesis [24].

Literature data indicate the positive effect of seaweed extracts as plant biostimulants. The extracts from *Ascophyllum nodosum* are listed as the most frequently used [25,26]. Equally popular are *Solanum lycopersicum* L. [27], *Ecklonia maxima*, *Sargassum* spp. [12], *Laminaria* spp., *Durvillaea potatumum*, *Ulva lactuca*, *Caulerpa sertularioides*, *Padina gymnospora*, *Sargassum liebmannii*, and *Sargassum johnstonii* [25,28].

The group of biostimulants also includes consortia of beneficial fungi or bacteria. Among the fungi used in the cultivation of plants, the following species are noted: *Glomus intraradices*, *Trichoderma atroviride*, *Trichoderma reesei*, and *Heteroconium chaetospira* (Table 1) [10,14–17]. Symbiotic arbuscular mycorrhizal fungi have a positive effect on crop quality. Arbuscular mycorrhizal fungi, along with *Rhizoglomus irregular*, promote the growth of *Stevia rebaudiana* Bertoni [20]. Plant growth bacteria include *Arthrobacter* spp., *Enterobacter* spp., *Acinetobacter* spp., *Pseudomonas* spp., *Ochrobactrum* spp., *Bacillus* spp., and *Rhodococcus* spp. (Table 1) [18,19]. The largest group of beneficial bacteria includes *Rhizobium* spp. and plant growth-promoting rhizobacteria [10]. The plant growth-promoting rhizobacteria group includes *Streptomyces* spp., *Pseudomonas* spp., and *Bacillus* spp. The literature results indicate that *Streptomyces* spp. protect tomato plants against putrefactive bacteria *Pectobacterium carotovorum* subsp. brasiliensis (Pcb). In addition, the volatiles produced by six *Streptomyces* spp. isolates stimulate the growth of tomato roots. The largest fresh root mass obtained due to the volatile substances was 138.2 ± 16.1 mg in comparison to a control mass of 111.5 ± 10.3 mg. The volatile substances also positively affect the dry matter content and root volume as well as the dry and fresh mass of the shoot [29]. Zhao et al. [30] managed to isolate 276 endophytic bacteria from soybean root nodules that protected soybean roots against fungal infections of *Phytophthora sojae*. The antagonistic bacteria included *Enterobacter* spp., *Acinetobacter* spp., *Pseudomonas* spp., *Ochrobactrum* spp., and *Bacillus* spp. Also, lactic acid bacteria, e.g., *Lactobacillus plantarum*, and *Lactobacillus paracasei* are used to promote plant growth and indirectly control diseases [31].

Table 1. Activity of different types of biostimulants.

Source of Biostimulant	Example	Main Activity
Hydrolysis Product	Enzymatic (alfalfa hay, pulses, and vegetable or fruit waste) and chemical (feathers, bone meal, casein, collagen from skins, animal tissue, or fish waste)	Increase in yield [32] Increase in nitrogen and phosphorus content in leaves and macro- and micronutrients [33,34] Increase in protein content in cereal grains [33] Protection against biotic and abiotic stresses [35] Increased soil fertility through the development of soil microorganisms [36]
Anaerobic Digestion Product	Plant, animal, and lignin biomass	Call of the auxin-like effect [23,37] Improving the availability of nutrients [31]
Biopreparations from Marine Algae	*Ascophyllum nodosum*, *Sargassum wightii*, *Ecklonia maxima*, *Enteromorpha intestinalis*, *Gelidium pectinutum*	Antioxidant potential and ability to capture free radicals [12] Chelating effect [12] Increase in plant resistance to fungal and bacterial infections [12] Extension of the shelf-life of fruit for consumption [12] Improve the thermal resistance of plants [12] Protection against drought stress [38]
Consortia of Beneficial Fungi	*Rhizophagus intraradices*, *Rhymbocarpus aggregatus*, *Glomus viscosum*, *Glomus etunicatum*, *Glomus claroideum*, *Trichoderma* sp., *Heteroconium chaetospira*	Increase in the growth and yield of plants alone and in symbiosis with bacteria, e.g., from *Azotobacter* spp. [39] Plant protection against oxidative stress [40]

4. Application Method of Biostimulants

Biostimulants may be used in the form of soil preparations (powders, granules, or solutions added to the soil) or as liquid foliar application products [41]. Biostimulants containing humic substances and nitrogen compounds are often applied directly onto the soil, whereas various types of extracts from plants and seaweed are used in the form of foliar applications. Biostimulants can be introduced into the irrigation system and taken up by plants along with water. One example is the Kelpak SL (*Ecklonia maxima* extract) biostimulant which was sprayed in an aqueous solution of *Phaseolus vulgaris* L. [41]. Biostimulants are used regularly during the whole vegetative period or proactively, i.e., once during the decline of vital forces of the plant. In this case, the biostimulants were administered once during the occurrence of a strong stress factor, e.g., frost [42–44]. The results show that the soil application of the biostimulant was not as effective as foliar application. The foliar application of a biostimulant obtained from sewage sludge increased the level of macro- and micronutrients in the leaves of maize. The nitrogen content in maize leaves increased by 26% (dose 3.6 L/ha) and 46% (dose 7.2 L/ha) [45]. Biostimulants can also be used in the form of biomass or meal from seaweed, however, this method has some limitations. Biomass and meal may be used in areas located close to the source of seaweed acquisition due to transport problems. Biomass or meal is applied directly to the soil long before planting in order to enrich the substrate with nutrients. Agro-technical measures such as ploughing are used to mix biomass or meal with the topsoil [12].

Common forms in which biostimulants occur are ready-to-use extracts or powder to make an aqueous solution. Soil biostimulants often affect the structure of the root, increasing, among other factors, its ability to absorb nutrients. Foliar extracts protect the plant against biotic and abiotic stresses. The circadian rhythm of plants should be taken into consideration. Biostimulants should be applied in the morning when the stomata are open and the assimilation rate is at its peak [12,38]. Biostimulants are also applied directly onto harvested fruits. It was noted that biostimulants containing a combination of extracts from *Sargassum* spp., *Laminaria* spp., and *A. nodosum* (Table 1) significantly extended the shelf-life and the storage life of oranges. After using the biostimulant, the fruits became more resistant to mechanical damage and putrefaction which allowed for an extension of the storage time and suitability for consumption. The extract used produced a better effect than, for example, the calcium chloride normally used in the industry to protect fruit against putrefactive bacteria [12,38].

5. Effect of Biostimulants on Yielding

The popularity of biostimulants in agriculture is associated with the possibility of obtaining higher yields without the need to discontinue the production of ecological crops. According to numerous scientific studies, biostimulants have a positive effect on yielding plants [36]. The yield is usually determined as the amount of fruit obtained from one plant or plot. The yield depends on the type of biostimulant used, the dose, the method of application, and the plant variety. Increased yield is often associated with improving the quality of vegetables or fruit. This is particularly important in organic farming, where artificial fertilizers cannot be used [37,45]. The quality of fruits and vegetables is shaped from the moment of plant growth to the time of harvesting of fruits and vegetables and it consists of the taste and the content of nutrients. The quality is influenced by genetic and agro-environmental factors [46].

The positive influence of biostimulants based on humic, fulvic, and carboxylic acids on the yielding of apricot fruits has been proven [8]. Control trees showed a yield of 12 kg fruit/tree and after the application of humic and fulvic acids together and carboxylic acids in a separate experiment, the yield of the trees increased to 21 kg of fruit/tree and 19 kg of fruit/tree, respectively (Table 2). However, this relationship was observed only in the second year of using the biostimulant. During the first growing season, the yield of the control trees was higher than that of the trees that were treated with biostimulants containing humic and fulvic acids. The biostimulant based on polysaccharides turned out to be ineffective with this variety of apricot trees, which showed a yield comparable to the control in both growing seasons [8]. Preparations containing all amino acids allowed for an increase in mango yield. With a dose of 3 L/ha, the yield increased by 18% compared to the control. At the same time, the biostimulants caused a 15% decrease in fruit weight. The authors explained this phenomenon through the competition of fruit for nutrients [32]. Biostimulants containing phenolic compounds such as sodium para-nitrophenolan, sodium orto-nitrophenolan, and sodium 5-nitroguajakolan proved to be good preparations for raspberry bushes. As a result of the foliar application of the phenolic compounds, a 20% (Table 2) increase in raspberry yield was obtained [47]. The highest yield was achieved when a biostimulant containing phenol compounds was applied to the "Polka" raspberry variety (yield = 23.03 kg/plot), while the yield in the control was 18.28 kg/plot. Already in the first year after the use of biostimulants, the amount of fruit collected from one bush increased. The best results were produced by 6-benzyladenine in a dose of 100 mg/L, which caused an increase in yield in the first year by about 0.5 kg. In the second year of using the biostimulant 6-benzyladenine at a dose of 100 mg/L and α-naphthaleneacetic acid at a dose of 20 mg/L, the yield increased by more than 1 kg from a single shrub [37]. Strawberry yielding significantly increased after using biostimulants containing herbal and marine plant extracts. In this case, the study proved that soil biostimulants are a source of nitrogen compounds. Moreover, foliar biostimulant application did not produce results as good as biostimulants added to the soil. Biostimulants added to the soil caused a significant increase in the amount of fruit and also improved the condition of the plants. Plants were more resistant to weather conditions and pathogens [42]. Extracts of "Moscatel" vine-shoots improved the yielding of the grapevine variety "Airén". The literature results indicate that two foliar biostimulant variants were prepared, non-toasted and toasted vine-shoots. A significantly higher yield was achieved in the case of two preparations, which were the non-toasted vine-shoots extract (3.09 ± 0.05 kg/plant) and the toasted vine-shoots extract (3.57 ± 0.05 kg/plant) in comparison to the control (2.54 ± 0.03 kg/plant) (Table 2) [48].

A mixture of four biostimulants containing amino acids, polysaccharides, vitamins, humic acids, organic carbon, and enzymatic proteins caused a comparable increase in the yield of two varieties of yellow pepper [49]. The Blondy F1 variety produced a crop at a level of 5.98a 0.23 kg/plot (yield = 5.24 ± 0.30 kg/plot). The Century F1 variety produced a yield of 5.76 ± 0.20 kg/plot (yield in the control = 5.06 ± 0.28 kg/plot) (Table 2). The peptides and amino acids contained in these formulations demonstrated a protective action against excessively high temperatures in the summer season and induced root growth and development, while vitamins and humic acids resulted in fruit growth [49]. Horseradish extract increased pumpkin yield by 12.5% [50] and the beneficial effect of fungal species *Glomus intraradices* and *Trichoderma atroviride* positively influenced the yielding of zucchini, resulting in a yield increase of 0.39 kg per plant on average, probably by increasing the effectiveness of nutrients uptake [14]. A 3% *Moringa oleifera* extract in combination with 0.6% $ZnSO_4$ and 0.25% K_2SO_4 increased the yield of "Kinnow" mandarin plants by 65% (Table 2) compared to the control [51].

The results of the study underline the positive effect of humic acids on the yielding of fruit trees [8]. The use of phenolic compounds [47] resulted in the increased yielding of fruit bushes. A high yield of vegetables may be obtained by using a mixture of amino acids, polysaccharides, vitamins, humic acids, and other compounds. Each of the substrates of the mixture affects another biochemical process occurring in the soil and plant, which allows for the achievement of the desired effect [49].

Table 2. Effect of selected biostimulants on quality of fruit and vegetables.

Biostimulant	Dose	Plant	Type of Claims	Benefitial Effect
Humic and fulvic acids	500 g/100 L	Apricot (*Prunus armeniaca* L.)	Yield	Increase by 75% in the second year compared to the control [8]
Carboxylic acids	200 mL/100 L	Apricot (*Prunus armeniaca* L.)	Yield	Increase 58% in the second year compared to the control [8]
All amino acids	3 L/ha	Mango (*Mangifera indica*)	Yield	Increase by 18% compared to the control [32]
6-benzyladenine	100 mg/L	Blueberry (*Vaccinium corymbosum* L.)	Yield	Increase by 0.5 kg/tree in the first year of using the biostimulant compared to the control [37]
α-naphthaleneacetic acid and 6-benzyladenine	100 mg/L (α-naphthaleneacetic acid) 20 mg/L (6-benzyladenine)	Blueberry (*Vaccinium corymbosum* L.)	Yield	Increase by 1 kg/tree in the second year of using the biostimulant compared to the control [37]
Humic acids	3g/L	Cucumber (*Cucumis sativus* L.)	Diameter	Increase in vegetable diameter by 1.23 cm (first season) and 1.55 cm (second season) compared to the control [52]
Nitrogen, amino acids, auxins	0.45 cm/L	Cucumber (*Cucumis sativus* L.)	Length	Lengthening vegetables by 3.85 cm (first season) and 3.49 cm (second season) compared to the control [52]
Biostimulant	**Dose**	**Plant**	**Type of Claims**	**Benefitial Effect**
Carboxylic acids	200 mL/100 L	Apricot (*Prunus armeniaca* L.)	Diameter	Increase diameter by 2.6 mm (second growing season) compared to the control [8]
Arbuscular mycorrhizal fungi and *Pseudomonas fluorescens* C7	The bacterial inoculum was replicated, watering each plant with 200 mL of bacterial suspension (density about 108 colony forming unit CFU/mL)	Tomato (*Solanum lycopersicum* L.)	Weight	Increase in tomato fruit mass by 6.9 g compared to control [53]
6-benzyladenine	100 mg/L	Blueberry cv. Duke and Bluecrop	Weight	Increase in the weight of blueberry fruit by about 32.4% (first season) and 33.6% (second season) for the blueberry cultivar Duke and 43.5% (first season) and 33.1% (second season) for the blueberry cultivar Bluecrop compared to the control [53]
α-naphthaleneacetic acid	20mg/L	Blueberry cv. Duke and Bluecrop	Weight	Increased the weight of the blueberry cultivar Duke fruit by 41.9% (first season) and 20.0% (second season) and the blueberry cultivar Bluecrop by 55.0% (first season) and 25.4% (second season) compared to the control [37]
Humic Acids	3g/L	Cucumber (*Cucumis sativus* L.)	Length	Increase in vegetable diameter by 1.23 cm (first season) and 1.55 cm (second season) compared to the control [52]
Nitrogen, amino acids, auxins	0.45 cm/L	Cucumber (*Cucumis sativus* L.)	Length	Lengthening vegetables by 3.85 cm in the first and 3.49 cm in the second growing season compared to the control [52]

Table 2. *Cont.*

Carboxylic acids	200 mL/100 L	Apricot (*Prunus armeniaca* L.)	Diameter	Widening of fruit by 2.6 mm on average in the second growing season compared to the control [8]
Arbuscular mycorrhizal fungi and *P. fluorescent* C7	The bacterial inoculum was replicated, watering each plant with 200 mL of bacterial suspension (density about 108 CFU/mL)	Tomato (*Solanum lycopersicum* L.)	Weight	Increase in tomato fruit mass by 6.9 g compared to control [53]
6-Benzyladenine	100 mg/L	Blueberry cv. Duke and Bluecrop	Weight	Increase in the weight of blueberry fruit by about 32.4% (first season) and 33.6% (second season) for the blueberry cultivar Duke and 43.5% (first season) and 33.1% (second season) for the blueberry cultivar Bluecrop compared to the control [53]
α-naphthaleneacetic acid	20 mg/L	Blueberry cv. Duke and Bluecrop	Weight	Increase in weight of the blueberry cultivar Duke fruit by 41.9% (first season) and 20.0% (second season) and the blueberry cultivar Bluecrop by 55.0% (first season) and 25.4% (second season) compared to the control [46]
Biostimulant including, among others, nitrogen, amino acids, auxins	0.45 cm/L	Cucumber (*Cucumis sativus* L.)	Plant height	Increase in plant height of 14.5 cm (in the first growing season) and 19.75 cm (in the second growing season) in comparison with the control plants [52]
Moringa oleifera leaf extract	3% treatments were replicated three times	Pumpkin (*Cucurbita pepo* L.)	Chlorophyll content	34.6% increase in the chlorophyll content compared to the control [50]
Salicylic acid–chitosan nanoparticles	(concentration 0.01%–0.16%)	Maize CV. Surya local	Chlorophyll content	Chlorophyll content in the control was 10.72 mg/g, the chlorophyll content in the maize leaves treated with the biostimulant (concentration 0.01%–0.16%) was in the range of 16.43 to 25.88 mg/g on average [54]
Ascophyllum nodosum Seaweed Extract	1.5 kg/ha	"Sangiovese" grapes	Phenolic content	Phenolic content increased by 1.063 mg/cm^2 compared to the control [55]
A. nodosum Seaweed Extract	3kg/ha	"Sangiovese" grapes	Phenolic content	Phenolic content increased by 0.951 mg/cm^2 compared to the control [55]
Biostimulant (chicken feathers) with Fertilizer (300 kg N/ha + 120 kg K/ha)	3.6 L/ha	Maize (*Zea mays* L. cv PR32W86 Pioneer)	Nitrogen content	Nitrogen content of corn leaves increased by 14.4% (first season) 15% (second season) compared to the control [33]
Biostimulant (chicken feathers) with fertilizer (300 kg N/HA + 120 kg K/HA)	7.2 L/ha	Maize (*Zea mays* L. cv PR32W86 Pioneer)	Nitrogen content	Increase nitrogen content by 39.1% (first season) and 33.3% (second season) compared to the control [33]
Protein hydrolysate	5.0 and 2.5 mL/L	Tomato (*Solanum lycopersicum* L.)	Content of lycopene	Increased by 34.9% and 18.0%, respectively, compared to the control [3]
Protein hydrolysate	2.5 mL/L	Tomato (*Solanum lycopersicum* L.)	Ascorbic acid	Increase ascorbic acid content by 27.3% compared to the control [3]
3% corn seed extract	Soaking and praying 1 mM mg plants	Sunflower seed (*Helianthus annuus* L.)	Enzymatic activity of superoxide dismutase, catalase, and peroxidase	Enzymatic activity of superoxide dismutase, catalase and peroxidase increased by 65.5%, 77.8%, and 84.6%, respectively, as compared to the controls [56]

6. Effect of Biostimulants on the Growth and Size of Plants

The way in which biostimulants work may be defined as multifaceted. The literature describes the positive effect of biostimulants on the growth of fruits and vegetables. At the same time, there are studies in which no effect of biostimulants on fruit size was found. The lack of biostimulant effects is explained by the use of a biostimulant unsuitable for the tested cultivar [37,53].

Fruit producers are interested in biopreparations that allow for the attainment of the largest and healthiest-looking fruit that draw consumers' attention [52]. An increase in the average length and diameter of cucumbers was attained after the use of humic acids and a mixture of nitrogen, amino acids, and auxins. Humic acids in a concentration of 3g/L increased the average length of the fruit in the first and second season by 9.9 cm and 12.2 cm, respectively. The same concentration of humic acids increased the diameter of the cucumbers in the first and second season by an average of 1.23 cm and 1.55 cm, respectively [52]. Three tested biostimulators caused an elongation and increase in the diameter of the vegetables. One biostimulant (containing nitrogen, amino acids, and auxins) led to the elongation of the cucumbers by 3.85 cm in the first and 3.49 cm in the second growing season. The diameter of the fruit increased by 1.12 cm and 1.56 cm, respectively, in the first and second growing season under the influence of this biostimulant. The application of humic acids and biostimulants containing auxins in particular makes it possible to obtain elongated and thickened cucumbers [52]. The use of biostimulants containing humic and fulvic acids as well as carboxylic acids led to a tenfold enlargement of the apricot fruit. The greatest influence on the size of the fruit was carboxylic acids, which contributed to the widening of fruit by 2.6 mm on average in the second growing season [8].

Studies show that the perfect biostimulants that cause the growth of fruits and vegetables are consortia of microorganisms. Examples are arbuscular mycorrhizal fungi and plant growth-promoting bacteria, the use of which resulted in increased tomato weight. A positive effect was obtained through the combination of arbuscular mycorrhizal fungi containing fungi of the following species: *Rhizophagus* spp., *Rhizophagus aggregatus*, *Septoglycus viscosum*, *Claroideoglomus etunicatum*, *Claroideoglomus claroideum* and various types of plant growth-promoting bacteria. All of the biostimulants based on microorganisms (arbuscular mycorrhizal fungi + *Pseudomonas* sp. Strain 19Fv1T, arbuscular mycorrhizal fungi + *Pseudomonas fluorescens* C7, arbuscular mycorrhizal fungi + *Pseudomonas* sp. 19 Fv1T and *Pseudomonas fluorescens* C7) caused an increase in tomato fruit mass, but the most effective result was demonstrated by biostimulants including arbuscular mycorrhizal fungi and *P. fluorescent* C7, which caused an increase in tomato mass to 71.3 ± 0.6 g (weight of control tomatoes = 64.4 ± 0.9 g). The microorganisms used caused a slight elongation of tomatoes as well. The length of the fruit in the controls ranged from 5.49 ± 0.03 cm to 5.81 ± 0.03 cm, while in combination with the biostimulants used, it rose to 5.88 ± 0.03 cm to 6.05 ± 0.02 cm. There was also a slight increase in fruit diameter from 4.24 ± 0.02 cm to 4.62 ± 0.03 cm to 4.64 ± 0.03 cm to 4.78 ± 0.02 cm [53].

The application of 6-benzyladenine as a biostimulant at a dose of 100 mg/L resulted in an increase in the weight of blueberry fruit by about 32.4% (first season) and 33.6% (second season) for the blueberry cultivar Duke and 43.5% (first season) and 33.1% (second season) for the blueberry cultivar Bluecrop compared to the control. The literature results showed that α-naphthaleneacetic acid at a dose of 20mg/L was also an effective biostimulant, which increased the weight of the blueberry cultivar Duke fruit by 41.9% (first season) and 20.0% (second season) and the blueberry cultivar Bluecrop by 55.0% (first season) and 25.4% (second season). As demonstrated in the study, one inefficient biostimulant was gibberellic acid at a dose of 200 mg/L. Gibberellic acid increased the weight of blueberry cultivar Duke fruit by 4.7% (first season) and 14.3% (second season) as well as Bluecrop blueberry cultivar by 0.8% (first season) and 11.5% (second season) compared to the control. For this reason, gibberellic acid is not recommended as a biostimulant for soft fruits [37].

The titanium compounds with which raspberries were treated caused an increase in fruit weight from 4.44 g (control) to an average of 5.4 g, but only at the beginning of the harvest season. At the end of the harvest, a 57% decrease in raspberry weight was observed [47]. The same relationship was observed for phenolic compounds. At the beginning of the harvest, the mass of raspberries treated with phenolic compounds was 5.04 g on average, while at the end of the harvest there was a 44.4% loss in fruit weight. The fact is that the raspberry collection is characterized by the loss of fruit mass at the end of the harvest, but if these three biostimulants were used, the resulting losses were greater than in the control. The control fruits were characterized by a 42.3% weight loss, while the fruits treated with biostimulants showed a mass loss at the end of the harvest in the range from 44.4% to 57.0% [47]. The increase in fruit weight after the use of biostimulants compounds was also observed in the case of cherries. The use of salicylic acid with the addition of calcium resulted in a 15% increase in sour cherries "Sweetheart" (2015) and "Skeena" (2016) [57].

Regarding the multifaceted effect of biostimulants, it should be emphasized that these formulations may affect many of the characteristics of the plant, e.g., fruit size, plant height, and root length [52]. The use of humic acids in a concentration of 3g/L and other biostimulants caused an increase in cucumber plant height, the number of leaves, and the number of stems in both growing seasons [52]. In the first growing season, the height of the control plants was 78.13 cm. Plants treated with 3 g/L of humic acids were 14.25 cm taller on average. In the second growing season plants treated with humic acids were taller by 13.25 cm. Similar dependencies may be observed after using other biostimulators. In the first and second growing season, this biostimulant (including, among others, nitrogen, amino acids, and auxins) was the most effective, causing an increase in plant height of 14.5 cm (in the first growing season) and 19.75 cm (in the second growing season) in comparison with the control plants. The biostimulant including, among others, naphthyl acetic acid was the least effective, and resulted in an increase in the growth of plants by 4.38 cm and 7.92 cm, respectively to the first and second growing seasons. Similar dependencies were noted in the number of leaves and new stems [52].

7. Impact of Biostimulants on Physical Characteristics

Biostimulants also have an influence on mechanical properties, i.e., the firmness of fruits or vegetables. Depending on the type, biostimulants may cause the stiffening of cell walls, thereby reducing their extensibility [8]. Biostimulants that increase the flexibility of cell walls at the same time extend the shelf-life of fruits and vegetables for consumption and facilitate their storage. Biostimulants based on carboxylic, humic, and fulvic acids and also the biopolymers of polysaccharides increased the mechanical strength of apricot fruits during two years of biostimulant use [8]. In turn, biostimulants containing phenolic compounds or chitosan resulted in the loss of fruit firmness of the three raspberry varieties studied. The use of biostimulants based on titanium compounds did not alter the fruit firmness, which was comparable to the firmness in the control test [47]. The use of spic cytozyme containing essential plant nutrients and growth biostimulants in the amount of 4 mL/L significantly reduced the cracking of pomegranate fruit [58]. In addition to improving mechanical properties, biostimulants change the shape and colour of fruits and vegetables. Fruit with larger length and diameter, as well as the right colours, are preferred by consumers [8]. However, consumer preferences are subject to dynamic changes.

An important visual feature that proves the quality of fruit is colour. The colour of the fruit is substantially influenced by the content of anthocyanins. Weber et al. [59] examined the content of anthocyanins in strawberries treated with *Ascophyllum nodosum* extract with silicon. Fruits treated with a biostimulator were characterized by a higher content of anthocyanins in the initial fruiting period, therefore, they were more red than the control fruits [59]. "Sweetheart" cherries treated with glycine and betaine were characterized by a darker skin than the control fruits. Although the mechanism of action of betaine and glycine on the formation of anthocyanins is not fully understood, it is known that the darker colour of the fruit was caused by a higher content of antioxidants [57]. Tarantino et al. (2018) in the second year of using biostimulants obtained apricot fruits with a lighter skin compared to the first year. This could be due to the higher concentration of biostimulants used in the first growing season. There were significant differences in the colour of the fruit. In the second year of the experiment the colour of the apricots was redder than in the first year of fruiting. There were no significant differences in the colour of the fruit produced by the three biostimulants used (1—biopolymers of polysaccharides; 2—humic and fulvic acids; 3—carboxylic acids) [8].

It should be emphasized that the increased mechanical strength or fruit colour change results from the good condition of fruit plants, which in turn is a result of their proper nutrition [50]. Biostimulants are supplied externally, which indirectly, e.g., induced by the photosynthesis process, plays an important role in the nutrition of plants. Indirect induction, for example, consists in increasing the leaf area. Leaves are the main organs in which photosynthesis takes place, therefore increasing the leaf area leads to an increase in photosynthesis. Intensively photosynthetic plants are better nourished. Increasing the leaf area also leads to an increase in the transpiration surface. This phenomenon has the especially important function of protecting the plant from overheating. *Moringa oleifera* leaf extract increased the surface area of the *Cucurbita pepo* L. by 9.7% and simultaneously led to a 34.6% increase in the chlorophyll content of the leaves compared to the control [50]. The positive effect of the *A. nodosum* extract in a period of drought on the growth of spinach has been proven. Seaweed extract increased the relative water content of the leaves from 76% to 82%. The surface area of the leaf was also increased by 16% (foliar spray), 21% (biostimulant in the irrigation system), and 38% (biostimulant in the irrigation system and in a spray). Increasing the area and turgor of the leaf led to an increase in the intensity of photosynthesis and improved the conditions for growing spinach under stress conditions [25].

8. Effect of Biostimulants on Chemical Composition

Biostimulants can affect a number of the chemical properties of fruits and vegetables, including dry mass, acidity or vitamin content. The chemical composition of the fruit directly affects their palatability. It is assumed that fruits with a content of dissolved solids (SSC) above 12°Brix are characterized by an excellent taste [47]. In the first year of using biostimulants containing the biopolymers of polysaccharides, humic and fulvic acids as well as carboxylic acids, the average value of SSC in apricots stood at 10.7°Brix. In the second year of using these biostimulants, fruit taste values improved significantly, as evidenced by the increase in the SSC level to an average of 14.1°Brix [8]. Biostimulants containing phenol compounds or chitosan reduced the dissolved solids content in the fruits of the three raspberry varieties (Pokusa, Polka, and Poranna Rosa). The opposite effect was produced by biostimulants based on titanium compounds, the use of which resulted in an increase in the content of dissolved solids in the raspberry fruit [47]. The quality of the fruit is also demonstrated by the ratio content of dissolved solids to their titratable acidity. Fruit quality is defined as good if the ratio content of dissolved solids to titratable acidity is within the range of 10 to 15. The treatment of fruit trees with biostimulants containing biopolymers of polysaccharides (16.7) and humic and fulvic acids (16.1) leads to an increase in the ratio content of dissolved solids to titratable acidity in relation to the control (14.0) and thus negatively influenced the sensory quality of fruit [8].

It is important to grow fruit that has an appropriate level of acidity but it is difficult to say, however, whether changes in acidity at the level of several percent have a significant impact on the fruit taste, because it is based on the subjective impression of the consumer. Although the literature data present studies on the effect of biostimulants on fruit acidity, there is no explicit interpretation of the results. It is not clear whether the changes in fruit acidity should be understood in terms of the positive or negative effects of the biostimulants used. The use of phenolic compounds and titanium compounds as a biostimulant in the cultivation of raspberries led to an increase in fruit acidity to 2.26% and 2.18%, respectively (control, 2.08%) [47]. A decrease in apricot acidity was noted after the use of biostimulants containing polysaccharides, humic and fulvic acids, and carboxylic acids [8]. In the second year after using these biostimulants, fruit acidity was reduced from an average of 3.45 (control pH) to an average of 3.7–3.8 (pH after using biostimulants) [8].

An important health-related feature of fruits and vegetables is the content of vitamin C and nitrogen compounds. The role of nitrogen in plants results from its influence on growth and development. It is a component of nucleic acids, it participates in the process of photosynthesis, and it builds amino acids that form a part of plant proteins [60]. The content of vitamin C and nitrogen compounds in the fruit depends mainly on the plant variety [47]. The use of selected biostimulants (phenolic compounds, chitosan, and titanium compounds) increased both the levels of ascorbic acid and nitrates. It turned out that the three biostimulators tested positively influenced the level of nitrates in raspberries. It was found that phenolic compounds contained in one of the biostimulants increased the level of vitamin C most effectively [47]. Assuming that phenolic compounds increased the content of vitamin C in fruit, it was necessary to determine how different biostimulators affect the content of phenolic compounds. Zarzecka et al. [61] studied the effect of herbicides and biostimulators on the polyphenol content of potato tubers. The experiment was conducted over a period of two years. Three potato varieties were treated with different substances: Harrier herbicide 295 ZC, Harrier 295 ZC + Kelpak SL growth regulator, and Sencor 70 WG herbicide, 5 Sencor 70 WG + Asahi growth regulator. In the case of Asahi, the active substances were phenolic compounds, while for Kelpak SL, auxins and cytokinins were the active substances. The applied biostimulants and herbicides caused an increase in the polyphenol content in tubers of all potato varieties (on average, 159.8–161.3 mg/kg) compared to the control (average of 156.0 mg/kg). The use of biostimulants and herbicides increased the content of polyphenols in the leaves of the potato to an average of 289.2–291.2 mg/kg compared to the control (287.8 mg/kg). The content of polyphenols in tubers is of particular importance for humans. Polyphenols reduce the risk of numerous diseases, e.g., blocking carcinogenic compounds [61]. It was observed that after using a biostimulant containing seaweed *A. nodosum* and silicon, the content of phenolic compounds in strawberries was slightly lower. Phenols are also defined as compounds produced by plants under stressful conditions, hence the conclusion about the positive effect of the tested biostimulant on strawberries [59]. The foliar spraying of the "Airén" grapevines by non-toasted and toasted biostimulants increased the content of phenolic compounds. In this case, the biostimulants were extracts from the "Moscatel" vine shoots. An important group of phenolic grape buds are hydroxycinnamic acids (trans-caffeic and trans-p-coumaric), which affect the taste of wine. Both non-toasted biostimulants (14.10 ± 0.13) and toasted biostimulants (11.26 ± 0.27) led to a trans-p-coumaric acid increase relative to the control (8.60 ± 0.03). The non-toasted biostimulant (1.14 ± 0.01) and toasted biostimulant (0.95 ± 0.03) also led to a trans-caffeic acid growth compared to the control (0.92 ± 0.03). The effect of the higher content of, among other compounds, hydroxycinnamic acid is a better quality wine [48]. A biostimulant containing *A. nodosum* seaweed extract increased the phenol content of "Sangiovese" grapes. A 1.5 kg/ha dose of biostimulant increased the phenolic content to 1.063 mg/cm^2, while the biostimulant in a dose of 3 kg/ha increased the phenolic content to 0.951 mg/cm^2. The phenol content of the control was 0.753 mg/cm^2. The results were statistically significant [55].

The positive effect of biostimulants is also based on increasing the content of chlorophyll in leaves and thus increasing the efficiency of the process of photosynthesis. Salicylic acid-chitosan nanoparticles used as a biostimulant led to an increase in the content of chlorophyll in leek corn [54]. While the chlorophyll content in the control was 10.72 mg/g, the chlorophyll content in the maize leaves treated with the biostimulant (concentration 0.01%–0.16%) was in the range of 16.43 to 25.88 mg/g on average. In plants treated only with chitosan and salicylic acid, a decrease in the chlorophyll content to an average of 9.24 mg/g and 9.79 mg/g was observed [54]. After the foliar application of the *Moringa oleifera* leaf extract, a 34.6% increase in the chlorophyll content of *Cucurbita pepo* L. leaves was recorded compared to the control (plants sprayed only with water) [50]. The increase in the chlorophyll content combined with the increased intensity of the photosynthesis process was noted during the cultivation of Hibiscus treated with biostimulants formed in the process of the hydrolysis of waste. A 15% increase in the chlorophyll content of leaves resulted in a 24% increase in the photosynthesis rate compared to the control [62].

During the tests to determine the chemical composition of fruits, the content of glucose, fructose, sucrose, ascorbate, proteins, and macro- and micro-elements is often determined. Plants treated with one of the tested biostimulators (arbuscular mycorrhizal fungi + *Pseudomonas* sp. 19 Fv1T and *P. fluorescens* C7) showed an increase in the concentration of glucose in tomatoes at 11.83 g/kg, while in controls the content of glucose was 10.45–11.0 g/kg. After using this biostimulant, the fructose content also increased to about 12.86 g/kg, while in the controls it was 10.77–11.14 g/kg. After using a biostimulator based on *A. nodosum* seaweed extract and a silicon extract, a slight increase in the level of sugars in the strawberry fruit was observed. The most common sugars were glucose and fructose. Sucrose accounted for 11% of total sugars [59]. The use of an extract of *M. oleifera* leaves increased the total soluble sugar content in pumpkin by about 80.6% [50]. An interesting relationship was observed in the case of ascorbate. The use of biostimulant containing arbuscular mycorrhizal fungi, *Pseudomonas* sp. 19 Fv1T and *P. fluorescens* C7 led to an increase in the ascorbate content (10.75 mg/100 g), whereas the use of a biostimulant containing arbuscular mycorrhizal fungi and *P. fluorescens* C7 reduced the ascorbate content in tomatoes (4.30 mg/100 g). In the controls, tomatoes contained about 5.47–7.12 mg/100 g ascorbate. In plants treated with the biostimulant containing arbuscular mycorrhizal fungi and *P. fluorescens* C7, an increase in β-carotene in tomatoes was observed (controls: 2.117–2.224 μg/100 g fresh weight; β-carotene content in the biostimulator study: 2.829 μg/100 g fresh weight). β-carotene may be converted into vitamin A and can protect against the adverse effects of free radicals [53]. The protein content is particularly important in the case of grain plants. The foliarly used biostimulator containing sewage sludge caused an increase in protein content in maize grains by about 30% in both growing seasons [45]. The biostimulator formed as a by-product of the two-stage process of pressing olive oil led to an increase in the protein content of maize grains by 19% [63].

Combining the biostimulant (chicken feathers) with a fertilizer gave a better quality of maize yield than using only the fertilizer. Three combinations of the agent used for spraying the corn were used. In the first variant, only fertilizer was used (300 kg N/ha + 120 kg K/ha). For the second variant, fertilizer was used (300 kg N/ha + 120 kg K/ha) in combination with a biostimulant (3.6 L/ha). In the third variant, fertilizer (300 kg N/ha + 120 kg K/ha) was used in combination with a biostimulant (7.2 L/ha). The treatment was applied during two seasons. The highest level of nitrogen was obtained after the application of a biostimulant containing fertilizer in combination with the highest dose of the biostimulant (7.2 L/ha). The nitrogen content of corn leaves increased by 14.4% (fertilizer with the biostimulant of 3.6 L/ha) and 39.1% (fertilizer with the biostimulant of 7.2 L/ha) in the first vegetative season in comparison to the nitrogen content of maize leaves treated only with fertilizer. In the second growing season, the nitrogen content increased by 15% (fertilizer with the biostimulant of 3.6 L/ha) and 33.3% (fertilizer with the biostimulant of 7.2 L/ha) respectively. The use of a biostimulant in combination with the fertilizer also resulted in an increase in phosphorus content in the leaves of maize. The treatment increased *p* levels by 32.8% (fertilizer with a biostimulant of 3.6 L/ha) and 52.2% (fertilizer with a biostimulant 7.2 L/ha) in the first season and by 43.5% (fertilizer with a biostimulant of 3.6 L/ha) and 51.1% (fertilizer with a biostimulant of 7.2 L/ha) in the second season compared to the control [33]. An effective biostimulant for the "Kinnow" mandarin trees proved to be *Moringa oleifera* extract. The 3% *Moringa oleifera* extract foliar application with 0.6% $ZnSO_4$ and 0.25% K_2SO_4 resulted in a 1.35-fold (first season) and 1.42-fold (second season) increase in nitrogen content compared to the control. Trees sprayed only with 3% *Moringa oleifera* extract showed a 1.09 times (first season) and 1.07 times (second season) higher phosphorus content compared to control trees [51].

The improvement in the chemical properties of fruits may increase not only their pro-health values, but also lead to an improvement in their sensory values. One example may be guaiacol, which was applied in a foliar way to improve the quality of wine. Guaiacol was shown to increase the amount of glycosylated aromatic compounds in "Microvine" grapes. These compounds have influenced the improvement of wine quality in the final step of wine formation. The guaiacol-treated fruits were characterized by a higher aglycone content (534.25 μg/g) compared to the control (157.52 μg/g). Treatment with guaiacol also increased the content of monomethyl alcohols from 2.94 μg/g in control fruits to 170.30 μg/g in guaiacol-treated fruits [64].

Biostimulants are becoming a viable option for solving the problem of the ineffective uptake of nutrients from fertilizers by plants. The fact is that a large proportion of fertilizer nutrients are not taken up by plants. Reducing the amount of mineral fertilizers introduced into the soil limits environmental degradation. It is thought that the development of certain biostimulants has the potential to increase the amount of nutrients taken up by plants [65]. The increase in the amount of nutrients taken up by plants may be achieved through the use of fertilizers and biostimulants in combination [33]. An increased nitrogen content was obtained thanks to the use of a biostimulant formed in the process of the hydrolysis of chicken feathers in combination with nitrogen fertilizer, while the increase in phosphorus content was the result of using a biostimulant formed in the process of the hydrolysis of chicken feathers with phosphate fertilizer [33] and an extract of *Moringa oleifera* as a biostimulant [51].

9. Effect of Biostimulants on Antioxidant Properties

Antioxidant activity is an often-studied property of fruits and vegetables. Antioxidants are listed as compounds that inhibit tumour cell proliferation and protect against oxidative stress caused by excess free radicals. The result of oxidative stress may be, among other factors, damage to DNA, cell membranes, or enzymes [66]. It was shown that the use of biostimulants in plant breeding can change the activity of enzymes and affect the antioxidant properties. Lycopene, ascorbic acid, phenolic compounds and others have antioxidant properties. Reactive oxygen molecules, e.g., OH, O^{2-}, and H_2O_2, are detoxified by antioxidant compounds (e.g., phenols, ascorbic acid) and enzymes (e.g., catalase, peroxidase, superoxide dismutase) [67].

Protein hydrolysate applied as a biostimulant to tomatoes had no effect on the level of phenolic compounds, while its effect on the content of ascorbic acid and lycopene was noted. After using biostimulant doses of 5.0 and 2.5 mL/L, the content of lycopene increased by 34.9% and 18.0%, respectively, compared to the control. The dose of 2.5 mL/L biostimulant increased the content of ascorbic acid by 27.3% [68]. The use of biostimulants on apricot fruit trees increased the antioxidant capacity of fruits. In the first season (average 76.8 mg/100 g), after using the stimulants, the antioxidant capacity of fruit was higher than in the second season (average 66.5 mg/100 g). The observed differences in antioxidative abilities between the two seasons were explained by changes in climatic conditions [8].

Also, a biostimulant based on salicylic acid and chitosan nanoparticles (SA-CS NPs) had an effect on the enzyme and antioxidant activity in maize leaves. The enzyme activity in leaves treated with chitosan, salicylic acid, and a control was comparable. After two days of treating the plants with the biostimulant, the activity of superoxide dismutase increased by two times compared to plants treated with salicylic acid. After three days of treating plants with a biostimulant, superoxide dismutase activity was 3.2 times higher than for plants treated with only salicylic acid. Peroxidase activity in plants treated with a biostimulant was 7.7 (after two days) and 5.2 (after three days) times higher than for plants treated with only salicylic acid. Catalase activity, phenylalanine ammonia lyase, and polyphenol oxidase increased by 2.9, 2.3, and 1.5-fold, respectively, after the second day of treatment with nanoparticles compared to salicylic acid treatment. It should be emphasized that the enzyme activity occurring in the leaves of plants treated with a biostimulant increased during the first three days. After the fourth day of treatment, the enzyme activity decreased in all variants of the experiment. The content of hydrogen peroxide in leaves treated with SA-CS NPs was 1.7 (first day of treatment), 3.6 (second day of treatment), and 1.7 (third day of treatment) times higher than in plants treated with salicylic acid [54].

The use of *M. oleifera* extract as a biostimulant resulted in a decrease in the activity of the antioxidant enzymes (catalase, peroxidase, and superoxide dismutase) in rocket plants (*Eruca vesicaria* subsp. *Sativa*). At the same time, the content of phenol and ascorbic acid was higher with increasing concentrations of the biostimulant [67]. Aqueous garlic extract improved tomato oxidation properties. Superoxide dismutase activity increased in proportion to the aqueous garlic extract concentration. The highest activity of this enzyme was observed with the foliar application of the biostimulant in a volume of 200 μg/L; also, the peroxidase activity was highest after using the biostimulant at this concentration. A lower aqueous garlic extract concentration (50 μg/L) did not affect the activity of these enzymes [69]. Soaking sunflower seed *Helianthus annuus* L. in a 3% corn seed extract and spraying 1 mM Mg plants stimulated the sunflower's antioxidant system. The enzymatic activity of superoxide dismutase, catalase, and peroxidase increased by 65.5%, 77.8%, and 84.6%, respectively, as compared to the controls. The increased level of antioxidant enzymes was related to the foliar application of Mg ions, the use of which also increased the intensity of the photosynthesis process [56].

Biostimulants increased the phenylalanine ammonia lyase enzyme activity. While the phenylalanine ammonia lyase level in the control was 7.9 ± 0.22 IU/mL × min (0.4% *E. maxima extract*), after using biostimulants it increased to 9.0 ± 0.01 and 9.7 ± 0.01 IU/mL × min (10−6 M *E. maxima* extract). Phenylalanine ammonia lyase is an enzyme catalyzing the first step in the synthesis of phenyl compounds. An increased production of phenolic compounds is observed during plant stress. From this, it may be concluded that biostimulants can induce plant stress to increase the production of secondary metabolites [70]. Biostimulatory properties also reveal many components of compost. Depending on the raw materials and methods used for the composting process, the compost may contain, among others, polysaccharides, amino acids, and organic nitrogen. Compost can be used to replace peat in greenhouse cultures. In addition, it may be produced from organic waste, such as wood, plant residue, or other residues. Compost which is considered by the European Union to be ecological must consist solely of natural raw materials, characterized by a limited content of heavy metals and hazardous elements (Se, Mo, S). The product of the composting process must be free of pathogenic agents (*Salmonella* sp. and *Escherichia coli*) [71]. Agroindustrial compost proved to be an alternative to peat in the cultivation of red lettuce. The compost increased the content of antioxidant compounds in lettuce leaves. In the autumn season, lettuce leaves cultivated in the compost showed 1.5 times more antioxidant activity than lettuce grown in compost in the summer season and it was also higher that of lettuce grown in peat in the autumn season [72].

10. Conclusions

Biostimulants are preparations of natural origin that support the pro-ecological cultivation of vegetables and fruits. Although for several years a positive effect of biostimulants has been widely reported, they are rarely introduced into standard cultivation technologies. This is connected with the insufficient knowledge of farmers on functions and usage of biostimulants what results in a fear of an increase in the cost of cultivation and a reduction in the quality and quantity of plants, which would affect the profitability of crops. The problem is also the multitude of preparations and the need to select a proper biostimulant for a specific plant variety in order to obtain the highest and the best quality yields. The market requires the development of preparations with a broad spectrum of functionality, which is easy to apply and has the possibility of combination with other agents.

The use of biostimulants on a commercial scale would limit the amount of mineral fertilizers introduced into the environment, thus reducing the pollution of soils, water, and air. This is especially important in the case of global warming. Global agriculture accounts for an average of 21% [73] of the global greenhouse effect, of which around 13% [74] is concerned with the effect of artificial fertilizers. The newly developed technologies of biopreparations may constitute a significant contribution to environmental protection, but primarily they are closely linked with sustainable agricultural and horticultural production with the aim of obtaining cheap, easily available, and high quality food. The effect of biostimulants depends on many factors, from the raw material and the process as a result of which they arose to the plant varieties, application method, and climate. The positive effect of consortia of microorganisms and plant hydrolysates on growth and yield of crops plants should be particularly emphasized. It is also important to increase the antioxidant potential of plants treated with biostimulants containing algae. A positive impact on crop quality and performance, no negative or harmful impact on people, animals, or the environment, increased biodiversity of beneficial microorganisms, and improvement of soil properties are the main advantages of biostimulants. However, the nature of their positive influence is not fully understood, therefore their mechanisms of action are, in some cases, still a challenge and need to be recognized. For this reason, biostimulants are among the hot topics in agriculture and still require detailed research.

Author Contributions: M.D. collected the data, interpreted the results and wrote the manuscript. M.F. provided ideas and performed reviews and corrections of the manuscript. J.C. designed the study, helped draft the manuscript, interpreted results and corrected the text. All authors read and approved the final manuscript.

Funding: The National Centre for Research and Development, Poland, project BIOSTRATEG, contract number BIOSTRATEG3/344433/16/NCBR/2018.

Conflicts of Interest: The authors declare no conflict of interest.

References

1. Di Vittori, L.; Mazzoni, L.; Battino, M.; Mezzetti, B. Pre-Harvest Factors Influencing the Quality of Berries. *Sci. Hortic.* **2018**, *233*, 310–322. [CrossRef]
2. European Union. II Non-Legislative Acts. DECISIONS: Commission Implementing Decision (EU) 2017/1442 of 31 July 2017 Establishing Best Available Techniques (BAT) Conclusions, under Directive 2010/75/EU of the European Parliament and of the Council, for Large Combustion Pl. *Off. J. Eur. Union. L* **2017**, *60*, 1–88.
3. Møller, A.P.; Laursen, K. Reversible Effects of Fertilizer Use on Population Trends of Waterbirds in Europe. *Biol. Conserv.* **2015**, *184*, 389–395. [CrossRef]
4. Rouphael, Y.; Spichal, L.; Panzarova, K.; Casa, R.; Colla, G. High-Throughput Plant Phenotyping for Developing Novel Biostimulants: From Lab to Field or from Field to Lab? *Front. Plant Sci.* **2018**, *9*, 1–19. [CrossRef] [PubMed]
5. Rouphael, Y.; Colla, G. Synergistic Biostimulatory Action: Designing the Next Generation of Plant Biostimulants for Sustainable Agriculture. *Front. Plant Sci.* **2018**, *9*, 1–24. [CrossRef]
6. Bradshaw, T.L.; Berkett, L.P.; Griffith, M.C.; Darby, H.M.; Moran, R.E.; Garcia, M.E. Assessment of Kelp Extract Biostimulants on Tree Growth, Yield, and Fruit Quality in a Certified Organic Apple Orchard. In *II International Organic Fruit Symposium 1001*; International Society for Horticultural Science: Brabant, Belgium, 2013; pp. 191–198. [CrossRef]
7. Kyriacou, M.C.; Rouphael, Y. Towards a New Definition of Quality for Fresh Fruits and Vegetables. *Sci. Hortic.* **2018**, *234*, 463–469. [CrossRef]
8. Tarantino, A.; Lops, F.; Disciglio, G.; Lopriore, G. Effects of Plant Biostimulants on Fruit Set, Growth, Yield and Fruit Quality Attributes of 'Orange Rubis®' Apricot (*Prunus armeniaca* L.) Cultivar in Two Consecutive Years. *Sci. Hortic.* **2018**, *239*, 26–34. [CrossRef]
9. The European Parliament and the Council of the European Union. *Regulation (Eu)/Laying down Rules on the Making Available on the Market of EU Fertilising Products and Amending Regulations (EC) No 1069/2009 and (EC) No 1107/2009 and Repealing Regulation (EC) No 2003/2003*; European Commission: Brussels, Belgium, 2019; Volume 0270.
10. du Jardin, P. Plant Biostimulants: Definition, Concept, Main Categories and Regulation. *Sci. Hortic.* **2015**, *196*, 3–14. [CrossRef]
11. Chiaiese, P.; Corrado, G.; Colla, G.; Kyriacou, M.C.; Rouphae, Y. Renewable Sources of Plant Biostimulation: Microalgae as a Sustainable Means to Improve Crop Performance. *Front. Plant Sci.* **2018**, *9*, 1–6. [CrossRef]
12. Battacharyya, D.; Babgohari, M.Z.; Rathor, P.; Prithiviraj, B. Seaweed Extracts as Biostimulants in Horticulture. *Sci. Hortic.* **2015**, *196*, 39–48. [CrossRef]
13. Colla, G.; Nardi, S.; Cardarelli, M.; Ertani, A.; Lucini, L.; Canaguier, R.; Rouphael, Y. Protein Hydrolysates as Biostimulants in Horticulture. *Sci. Hortic.* **2015**, *196*, 28–38. [CrossRef]
14. Colla, G.; Rouphael, Y.; Di Mattia, E.; El-Nakhel, C.; Cardarelli, M. Co-Inoculation of Glomus Intraradices and Trichoderma Atroviride Acts as a Biostimulant to Promote Growth, Yield and Nutrient Uptake of Vegetable Crops. *J. Sci. Food Agric.* **2015**, *95*, 1706–1715. [CrossRef] [PubMed]
15. Mukherjee, P.K.; Horwitz, B.A.; Herrera-Estrella, A.; Schmoll, M.; Kenerley, C.M. *Trichoderma* Research in the Genome Era. *Annu. Rev. Phytopathol.* **2013**, *51*, 105–129. [CrossRef] [PubMed]
16. Nicolás, C.; Hermosa, R.; Rubio, B.; Mukherjee, P.K.; Monte, E. Trichoderma Genes in Plants for Stress Tolerance-Status and Prospects. *Plant Sci.* **2014**, *228*, 71–78. [CrossRef] [PubMed]
17. Usuki, F.; Narisawa, K. A Mutualistic Symbiosis between a Dark Septate Endophytic Fungus, Heteroconium Chaetospira, and a Nonmycorrhizal Plant, Chinese Cabbage. *Mycologia* **2007**, *99*, 175–184. [CrossRef] [PubMed]

18. Gaiero, J.R.; McCall, C.A.; Thompson, K.A.; Day, N.J.; Best, A.S.; Dunfield, K.E. Inside the Root Microbiome: Bacterial Root Endophytes and Plant Growth Promotion. *Am. J. Bot.* **2013**, *100*, 1738–1750. [CrossRef] [PubMed]
19. Zhao, D.; Zhao, H.; Zhao, D.; Zhu, X.; Wang, Y.; Duan, Y.; Xuan, Y.; Chen, L. Isolation and Identification of Bacteria from Rhizosphere Soil and Their Effect on Plant Growth Promotion and Root-Knot Nematode Disease. *Biol. Control* **2018**, *119*, 12–19. [CrossRef]
20. Tavarini, S.; Passera, B.; Martini, A.; Avio, L.; Sbrana, C.; Giovannetti, M.; Angelini, L.G. Plant Growth, Steviol Glycosides and Nutrient Uptake as Affected by Arbuscular Mycorrhizal Fungi and Phosphorous Fertilization in Stevia Rebaudiana Bert. *Ind. Crops Prod.* **2018**, *111*, 899–907. [CrossRef]
21. Souri, M.K.; Bakhtiarizade, M. Scientia Horticulturae Biostimulation Effects of Rosemary Essential Oil on Growth and Nutrient Uptake of Tomato Seedlings. *Sci. Hortic.* **2019**, *243*, 472–476. [CrossRef]
22. Ugolini, L.; Cinti, S.; Righetti, L.; Stefan, A.; Matteo, R.; D'Avino, L.; Lazzeri, L. Production of an Enzymatic Protein Hydrolyzate from Defatted Sunflower Seed Meal for Potential Application as a Plant Biostimulant. *Ind. Crops Prod.* **2015**, *75*, 15–23. [CrossRef]
23. Scaglia, B.; Pognani, M.; Adani, F. The Anaerobic Digestion Process Capability to Produce Biostimulant: The Case Study of the Dissolved Organic Matter (DOM) vs. Auxin-like Property. *Sci. Total Environ.* **2017**, *589*, 36–45. [CrossRef] [PubMed]
24. Pacholczak, A.; Szydo, W.; Jacygrad, E.; Federowicz, M. Effect of auxins and the biostimulator algaminoplant on rhizogenesis in stem cuttings of two dogwood cultivars (*Cornus alba* "Aurea" and 'Elegantissima'). *Hortorum Cultus* **2012**, *11*, 93–103.
25. Xu, C.; Leskovar, D.I. Effects of A. nodosum Seaweed Extracts on Spinach Growth, Physiology and Nutrition Value under Drought Stress. *Sci. Hortic.* **2015**, *183*, 39–47. [CrossRef]
26. Spann, T.M.; Road, E.S.; Alfred, L.; Little, H.A. Applications of a Commercial Extract of the Brown Seaweed Ascophyllum Nodosum Increases Drought Tolerance in Container-Grown 'Hamlin' Sweet Orange Nursery Trees. *HortScience* **2011**, *46*, 577–582. [CrossRef]
27. Hernández-Herrera, R.M.; Santacruz-ruvalcaba, F.; Ruiz-lópez, M.A.; Norrie, J.; Hernández-Carmona, G. Effect of Liquid Seaweed Extracts on Growth of Tomato Seedlings (*Solanum lycopersicum* L.). *J. Appl. Phycol.* **2014**, 619–628. [CrossRef]
28. Kumari, R.; Kaur, I.; Bhatnagar, A.K. Effect of Aqueous Extract of Sargassum Johnstonii Setchell & Gardner on Growth, Yield and Quality of Lycopersicon Esculentum Mill. *J. Appl. Phycol.* **2011**, 623–633. [CrossRef]
29. Dias, M.P.; Bastos, M.S.; Xavier, V.B.; Cassel, E.; Astarita, L.V.; Santarém, E.R. Plant Growth and Resistance Promoted by Streptomyces Spp. in Tomato. *Plant Physiol. Biochem.* **2017**, *118*, 479–493. [CrossRef] [PubMed]
30. Zhao, L.F.; Xu, Y.J.; Lai, X.H. Antagonistic Endophytic Bacteria Associated with Nodules of Soybean (*Glycine max* L.) and Plant Growth-Promoting Properties. *Braz. J. Microbiol.* **2018**, *49*, 269–278. [CrossRef]
31. Lamont, J.R.; Wilkins, O.; Bywater-Ekegärd, M.; Smith, D.L. From Yogurt to Yield: Potential Applications of Lactic Acid Bacteria in Plant Production. *Soil Biol. Biochem.* **2017**, *111*, 1–9. [CrossRef]
32. Morales-Payan, J.P. Influence of Foliar Sprays of an Amino Acid Formulation on Fruit Yield of "Edward" Mango. *Acta Hortic.* **2015**, *1075*, 157–160. [CrossRef]
33. Tejada, M.; Rodríguez-Morgado, B.; Paneque, P.; Parrado, J. Effects of Foliar Fertilization of a Biostimulant Obtained from Chicken Feathers on Maize Yield. *Eur. J. Agron.* **2018**, *96*, 54–59. [CrossRef]
34. Ramírez-Pérez, L.J.; Morales-Díaz, A.B.; Benavides-Mendoza, A.; De-Alba-Romenus, K.; González-Morales, S.; Juárez-Maldonado, A. Dynamic Modeling of Cucumber Crop Growth and Uptake of N, P and K under Greenhouse Conditions. *Sci. Hortic.* **2018**, *234*, 250–260. [CrossRef]
35. Panfili, I.; Bartucca, M.L.; Del Buono, D. The Treatment of Duckweed with a Plant Biostimulant or a Safener Improves the Plant Capacity to Clean Water Polluted by Terbuthylazine. *Sci. Total Environ.* **2018**, *646*, 832–840. [CrossRef] [PubMed]
36. García-Martínez, A.M.; Díaz, A.; Tejada, M.; Bautista, J.; Rodríguez, B.; Santa María, C.; Revilla, E.; Parrado, J. Enzymatic Production of an Organic Soil Biostimulant from Wheat-Condensed Distiller Solubles: Effects on Soil Biochemistry and Biodiversity. *Process Biochem.* **2010**, *45*, 1127–1133. [CrossRef]
37. Milić, B.; Tarlanović, J.; Keserović, Z.; Magazin, N.; Miodragović, M.; Popara, G. Bioregulators Can Improve Fruit Size, Yield and Plant Growth of Northern Highbush Blueberry (*Vaccinium corymbosum* L.). *Sci. Hortic.* **2018**, *235*, 214–220. [CrossRef]

38. Goñi, O.; Quille, P.; O'connell, S. Ascophyllum Nodosum Extract Biostimulants and Their Role in Enhancing Tolerance to Drought Stress in Tomato Plants. *Plant Physiol. Biochem.* **2018**. [CrossRef]
39. Xavier, L.J.C.; Boyetchko, S.M. Arbuscular Mycorrhizal Fungi as Biostimulants and Bioprotectants of Crops. *Appl. Mycol. Biotechnol.* **2002**, *2*, 311–340. [CrossRef]
40. Wu, Q.S.; Zou, Y.N.; Fathi Abd-Allah, E. *Mycorrhizal Association and ROS in Plants*; Elsevier Inc.: Amsterdam, The Netherlands, 2014. [CrossRef]
41. Kocira, A.; Świeca, M.; Kocira, S.; Złotek, U.; Jakubczyk, A. Enhancement of Yield, Nutritional and Nutraceutical Properties of Two Common Bean Cultivars Following the Application of Seaweed Extract (*Ecklonia maxima*). *Saudi J. Biol. Sci.* **2018**, *25*, 563–571. [CrossRef]
42. Filipczak, J.; Żurawicz, E.; Paszt, L.S. Wpływ wybranych biostymulatorów na wzrost i plonowanie roślin truskawki "Elkat" influence of selected biostimulants on the growth and yielding of "Elkat" strawberry plants. *Zesz. Naukowe Inst. Ogrod.* **2016**, *24*, 43–58.
43. Kocira, A.; Kocira, S.; Świeca, M.; Złotek, U.; Jakubczyk, A.; Kapela, K. Effect of Foliar Application of a Nitrophenolate–Based Biostimulant on the Yield and Quality of Two Bean Cultivars. *Sci. Hortic.* **2017**, *214*, 76–82. [CrossRef]
44. Kocira, S.; Sujak, A.; Kocira, A.; Wójtowicz, A.; Oniszczuk, A. Effect of Fylloton Application on Photosynthetic Activity of Moldavian Dragonhead (*Dracocephalum moldavica* L.). *Agric. Agric. Sci. Procedia* **2015**, *7*, 108–112. [CrossRef]
45. Tejada, M.; Rodríguez-Morgado, B.; Gómez, I.; Franco-Andreu, L.; Benítez, C.; Parrado, J. Use of Biofertilizers Obtained from Sewage Sludges on Maize Yield. *Eur. J. Agron.* **2016**, *78*, 13–19. [CrossRef]
46. Rouphael, Y.; Kyriacou, M.C. Quality and Safety of Fresh Fruits and Vegetables at Harvest. *Sci. Hortic.* **2018**, *239*, 78–79. [CrossRef]
47. Grajkowski, J.; Ochmian, I. Influence of three biostymulants on yielding and fruit quality of three primocane raspberry cultivars. *Hortorum Cultus* **2007**, *6*, 29–36.
48. Sánchez-Gómez, R.; Zalacain, A.; Pardo, F.; Alonso, G.L.; Salinas, M.R. Moscatel Vine-Shoot Extracts as a Grapevine Biostimulant to Enhance Wine Quality. *Food Res. Int.* **2017**, *98*, 40–49. [CrossRef] [PubMed]
49. Paradiković, N.; Vinković, T.; Vinković Vrček, I.; Žuntar, I.; Bojić, M.; Medić-Šarić, M. Effect of Natural Biostimulants on Yield and Nutritional Quality: An Example of Sweet Yellow Pepper (*Capsicum annuum* L.) Plants. *J. Sci. Food Agric.* **2011**, *91*, 2146–2152. [CrossRef] [PubMed]
50. Abd El-Mageed, T.A.; Semida, W.M.; Rady, M.M. Moringa Leaf Extract as Biostimulant Improves Water Use Efficiency, Physio-Biochemical Attributes of Squash Plants under Deficit Irrigation. *Agric. Water Manag.* **2017**, *193*, 46–54. [CrossRef]
51. Nasir, M.; Sattar, A.; Basra, S.M.A.; Ullah, A. Foliar Application of Moringa Leaf Extract, Potassium and Zinc Influence Yield and Fruit Quality of 'Kinnow' Mandarin. *Sci. Hortic.* **2016**, *210*, 227–235. [CrossRef]
52. Fawzy, Z.F. Response of Growth and Yield of Cucumber Plants (*Cucumis sativus* L.) to Different Foliar Applications of Humic Acid and Bio-Stimulators. *Int. Res. J. Appl. Basic Sci.* **2012**, *6*, 630–637.
53. Bona, E.; Todeschini, V.; Cantamessa, S.; Cesaro, P.; Copetta, A.; Lingua, G.; Gamalero, E.; Berta, G.; Massa, N. Combined Bacterial and Mycorrhizal Inocula Improve Tomato Quality at Reduced Fertilization. *Sci. Hortic.* **2018**, *234*, 160–165. [CrossRef]
54. Kumaraswamy, R.V.; Kumari, S.; Choudhary, R.C.; Sharma, S.S.; Pal, A.; Raliya, R.; Biswas, P.; Saharan, V. Salicylic Acid Functionalized Chitosan Nanoparticle: A Sustainable Biostimulant for Plant. *Int. J. Biol. Macromol.* **2019**, *123*, 59–69. [CrossRef] [PubMed]
55. Frioni, T.; Sabbatini, P.; Tombesi, S.; Norrie, J.; Poni, S.; Gatti, M.; Palliotti, A. Effects of a Biostimulant Derived from the Brown Seaweed Ascophyllum Nodosum on Ripening Dynamics and Fruit Quality of Grapevines. *Sci. Hortic.* **2018**, *232*, 97–106. [CrossRef]
56. ur Rehman, H.; Alharby, H.F.; Alzahrani, Y.; Rady, M.M. Magnesium and Organic Biostimulant Integrative Application Induces Physiological and Biochemical Changes in Sun Flower Plants and Its Harvested Progeny on Sandy Soil. *Plant Physiol. Biochem.* **2018**, *126*, 97–105. [CrossRef] [PubMed]
57. Queirós, F.; Ribeiro, C.; Vilela, A.; Aires, A.; Barros, A.I.; Schouten, R.; Paula, A.; Gonçalves, B. Scientia Horticulturae Effects of Calcium and Growth Regulators on Sweet Cherry (*Prunus avium* L.) Quality and Sensory Attributes at Harvest. *Sci. Hortic.* **2019**, *248*, 231–240. [CrossRef]
58. Aziz, R.A.; Naira, A.; Moieza, A. Effect of Plant Biostimulants on Fruit Cracking and Quality Attributes of Pomegranate Cv. Kandhari Kabuli. *Sci. Res. Essays* **2013**, *8*, 2171–2175. [CrossRef]

59. Weber, N.; Schmitzer, V.; Jakopic, J.; Stampar, F. First Fruit in Season: Seaweed Extract and Silicon Advance Organic Strawberry (*Fragaria* × *Ananassa* Duch.) Fruit Formation and Yield. *Sci. Hortic.* **2018**, *242*, 103–109. [CrossRef]
60. Jämtgård, S.; Näsholm, T.; Huss-Danell, K. Nitrogen Compounds in Soil Solutions of Agricultural Land. *Soil Biol. Biochem.* **2010**, *42*, 2325–2330. [CrossRef]
61. Zarzecka, K.; Gugała, M.; Sikorska, A.; Mystkowska, I.; Baranowska, A.; Niewęgłowski, M.; Dołęga, H. The Effect of Herbicides and Biostimulants on Polyphenol Content of Potato (*Solanum tuberosum* L.) Tubers and Leaves. *J. Saudi Soc. Agric. Sci.* **2018**. [CrossRef]
62. Massa, D.; Prisa, D.; Montoneri, E.; Battaglini, D.; Ginepro, M.; Negre, M.; Burchi, G. Scientia Horticulturae Application of Municipal Biowaste Derived Products in Hibiscus Cultivation: Effect on Leaf Gaseous Exchange Activity, and Plant Biomass Accumulation and Quality. *Sci. Hortic.* **2016**, *205*, 59–69. [CrossRef]
63. Alberola, C.; Lichtfouse, E.; Navarrete, M.; Debaeke, P.; Souchère, V. Agronomy for Sustainable Development. *Ital. J. Agron.* **2008**, *3*, 77–78. [CrossRef]
64. Sánchez-gómez, R.; Torregrosa, L.; Zalacain, A.; Ojeda, H.; Bouckenooghe, V.; Schneider, R.; Alonso, G.L.; Salinas, M.R. Scientia Horticulturae Behavior of Glycosylated Aroma Precursors in Microvine Fruits after Guaiacol Foliar Application. *Sci. Hortic.* **2019**, *246*, e1–e8. [CrossRef]
65. Halpern, M.; Bar-Tal, A.; Ofek, M.; Minz, D.; Muller, T.; Yermiyahu, U. *The Use of Biostimulants for Enhancing Nutrient Uptake*; Elsevier Inc.: Amsterdam, The Netherlands, 2015; Volume 130. [CrossRef]
66. Han, M.; Li, G.; Liu, X.; Li, A.; Mao, P.; Liu, P.; Li, H.; Journal, H.P.; Li, G.; Liu, X.; et al. Phenolic Profile, Antioxidant Activity and Anti-Proliferative Activity of Crabapple Fruits. *Hortic. Plant J.* **2019**. [CrossRef]
67. Abdalla, M.M. Boosting the Growth of Rocket Plants in Response to the Application of Moringa oleifera Extracts as a Biostimulant. *Life Sci. J.* **2014**, *11*, 1097–8135.
68. Rouphael, Y.; Colla, G.; Giordano, M.; El-Nakhel, C.; Kyriacou, M.C.; De Pascale, S. Foliar Applications of a Legume-Derived Protein Hydrolysate Elicit Dose-Dependent Increases of Growth, Leaf Mineral Composition, Yield and Fruit Quality in Two Greenhouse Tomato Cultivars. *Sci. Hortic.* **2017**, *226*, 353–360. [CrossRef]
69. Hayat, S.; Ahmad, H.; Ali, M.; Ren, K.; Cheng, Z. Aqueous Garlic Extract Stimulates Growth and Antioxidant Enzymes Activity of Tomato (*Solanum lycopersicum*). *Sci. Hortic.* **2018**, *240*, 139–146. [CrossRef]
70. Kulkarni, M.G.; Rengasamy, K.R.R.; Pendota, S.C.; Gruz, J.; Plačková, L.; Novák, O.; Doležal, K.; Van Staden, J. Bioactive Molecules Derived from Smoke and Seaweed Ecklonia Maxima Showing Phytohormone-Like Activity in *Spinacia oleracea* L. *New Biotechnol.* **2019**, *48*, 83–89. [CrossRef] [PubMed]
71. López-López, N.; López-Fabal, A. Scientia Horticulturae Compost Based Ecological Growing Media According EU Eco-Label Requirements. *Sci. Hortic.* **2016**, *212*, 1–10. [CrossRef]
72. Giménez, A.; Fernández, J.A.; Pascual, J.A.; Ros, M. Scientia Horticulturae an Agroindustrial Compost as Alternative to Peat for Production of Baby Leaf Red Lettuce in a Floating System. *Sci. Hortic.* **2019**, *246*, 907–915. [CrossRef]
73. FAO. *The State of Food And Agriculture*; FAO: Rome, Italy, 2016.
74. FAO. *Greenhouse Gas Emissions from Agriculture, Forestry and Other Land Use*; FAO: Rome, Italy, 2014.

agronomy

MDPI

Article

Paramylon Treatment Improves Quality Profile and Drought Resistance in *Solanum lycopersicum* L. cv. Micro-Tom

Laura Barsanti [1], Primo Coltelli [2] and Paolo Gualtieri [1,*]

[1] Istituto di Biofisica, Consiglio Nazionale delle Ricerche (CNR), Via Moruzzi 1, 56124 Pisa, Italy
[2] Istituto Scienza e Tecnologie dell'Informazione, CNR, Via Moruzzi 1, 56124 Pisa, Italy
* Correspondence: paolo.gualtieri@pi.ibf.cnr.it; Tel.: +39-050-621-3026

Received: 15 May 2019; Accepted: 14 July 2019; Published: 17 July 2019

Abstract: Tomatoes, the most cultivated vegetables worldwide, require large amounts of water and are adversely affected by water stress. *Solanum lycopersicum* L., cv. Micro-Tom was used to assess the effects of β-(1,3)-glucan (paramylon) purified from the microalga *Euglena gracilis* on drought resistance and fruit quality profile. Plants were grown in an aeroponic system under three cultivation conditions: optimal water regimen, water scarcity regimen, and water scarcity regimen coupled with a root treatment with paramylon. Eco–physiological, physicochemical and quality parameters were monitored and compared throughout the lifecycle of the plants. Drought stress caused only a transient effect on the eco–physiological parameters of paramylon-treated plants, whereas physicochemical and biochemical parameters underwent significant variations. In particular, the fruits of paramylon-treated plants reached the first ripening stage two weeks before untreated plants grown under the optimal water regime, while the fruits of stressed untreated plants did not ripe beyond category II. Moreover, antioxidant compounds (carotenoids, phenolic acid, and vitamins) of fruits from treated plants underwent a two-fold increase with respect to untreated plants, as well as soluble carbohydrates (glucose, fructose, and sucrose). These results show that paramylon increases plant resistance to drought and highly improves the quality profile of the fruits with respect to untreated plants grown under drought stress.

Keywords: Biostimulants; *Euglena gracilis*; algal polysaccharide; β-glucan; water stress; tomato; aeroponics

---

1. Introduction

Paramylon is the storage product of the unicellular alga *Euglena gracilis*. This polysaccharide is a β-(1,3)-glucan endogenously synthetized as 1–2 μm granules consisting of 100% glucose [1]. The granules are composed of concentric segments, which possibly indicates the successive deposition of unbranched linear β-(1,3)-glucan chains on a central nucleus [2–4]. Wild type (WT) photosynthetic cells can accumulate paramylon up to 60% of cell dry weight (DW) [2], while the WZSL mutant of *E. gracilis* (spontaneous, non-chloroplastic, osmotrophic mutant; W describes the white color of the cells; Z means *E. gracilis* Klebs, Z strain; S means spontaneous mutant, and L means light grown parent culture) [5] can accumulate large amounts of it (up to 95% DW) when grown in the dark with an adequate carbon source [2].

Β-glucans are PAMPs (pathogen-associated molecular patterns) recognized by specific membrane receptors (pattern recognition receptors, PRRs) which trigger the activation of the innate immune system [6,7]. Both the molecular structure and degree of polymerization affect the strength and efficacy of β-glucans recognition by PRRs, as well as their successive reactions [2,8–10].

Linear β-(1,3)-glucans bind preferentially to Dectin-1, a C-type lectin receptor expressed on most cells of the innate immune system [11,12]. A minimum of 10 units of glucose is necessary to trigger an immune response [8,13,14].

Evaluations of the effective potential of linear β-(1,3)-glucans have been often made by testing preparations from plant/algae/fungal sources that are always contaminated by pigments, proteins, and membranes resulting in non-specific immunoresponses. β-(1,3)-glucan purified from paramylon synthetized by the WZSL mutant lacks any kind of contaminations from cellular components, which are always present in the paramylon extracted from WT cells. β-1-3-glucan purified from the WZSL mutant is further processed to produce linear nanofibers suited for binding to Dectin-1 receptors of target cell membranes. The effect of these nanofibers has been already investigated in our laboratory on tomato plants, animals, and humans [4,15–17]. In tomatoes, paramylon nanofibers modulate conductance to carbon dioxide (CO_2) diffusion from air to the carboxylation sites by regulating hormone levels and water-use efficiency, leading to an increase of plant defense capacity against drought [16].

Tomatoes are the most cultivated vegetable worldwide, being one of the most nutritionally and economically important crops. They require large amounts of water and are adversely affected by drought, which limits photosynthesis and, consequently, plant growth and yield worldwide [18]. Hence, we investigated the role of β-1,3-glucan nanofibers as elicitors of tomato plants response to drought to understand their physiological and photosynthetic responses to this stress.

In this study, tomato cv. Micro-Tom was chosen because of its small size (10–20 cm in height) and short life cycle of about 3 months. Plants were grown in an aeroponic system under three cultivation conditions: optimal water regimen, water scarcity regimen (drought), and water scarcity regimen (drought) coupled with a root treatment with paramylon to monitor and compare eco–physiological (leaf water potential, CO_2 assimilation rate, stomatal conductance, internal CO_2 concentration, photosystem II (PSII) photochemical efficiency, actual photon yield of PSII, and photochemical quenching of PSII), physicochemical (dry biomass, ashes, dry matter, moisture, microelements, weight, and size), and quality parameters (antioxidant compositions and activities, as well as soluble carbohydrates) throughout the lifecycle of the plants.

2. Materials and Methods

2.1. Aeroponic Culture System

The aeroponic cultivation system used was the Nutriculture Twin Amazon 16 (Nutriculture DGS, UK), which consists of a 100 L reservoir tank (160 × 75 × 46 cm), a root chamber housing the delivery system and a molded plastic lid holding sixteen 75 mm mesh pots (Figure 1). The Maxi Jet 1000 pump (14 W, flow rate 1000 Lh^{-1}, max head height 142 cm, NEWA Tecno Industria SRL, Italy) supplies a powerful spray creating a miniature rainstorm inside the chamber through eight 360 degree sprinklers that shower the whole root area, leaving no blind spots. Since the aeroponic container can be assimilated to a closed system, the condensation of growth medium eventually balances its evaporation, reducing the amount of evapotranspiration.

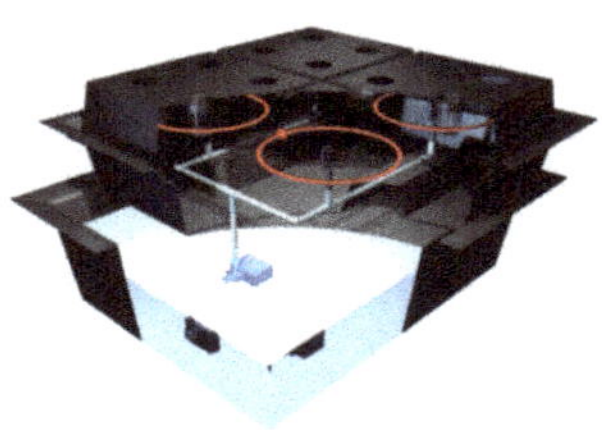

Figure 1. The aeroponic cultivation system.

2.2. *Paramylon Nanofibers Preparation*

Paramylon granules were extracted and purified from two-days-old cultures of *Euglena gracilis* WZSL mutant according to Barsanti et al. [2]. Nanofibers were obtained by alkaline degradation of the granules [2]. Assuming a 20% loss during the procedure, the final concentration of β-1,3-glucan nanofibers was about 0.8% w/v.

2.3. *Plant Material, Growth Conditions and Paramylon Treatment*

Seeds of *Solanum lycopersicum* L., cv. Micro-Tom [19] were surface sterilized with a bleach solution (commercial bleach 30% v/v, Triton X-100 0.02% v/v) for 15 min, washed 3 times with sterile water, and placed overnight in the dark at 4 °C. Seeds were then sown on filter paper covering the bottom of 150 mm diameter Petri dishes (about 20 seeds per plate); the paper was wetted with sterile demineralized water, the dishes were wrapped with Parafilm, and they were placed in the dark for about 2 days to aid germination. Upon germination, the cover was removed to allow the development of the roots (2–3 days). Seedlings were then transferred to rock wool (Grodan® Pro Plug) plugs, one seed per plug, to grow under controlled climate conditions (16/8 h light/dark; 400 $\mu molm^{-2}\ s^{-1}$ PAR, Photosynthetic Active Radiation; 22 °C) for approximately 2 weeks in a plastic tray containing half strength nutrient solution [$Ca(NO_3)_2\ 4H_2O$, 0.944 gL^{-1}; KNO_3, 0.808 gL^{-1}; $MgSO_4\ 2H_2O$, 0.492 gL^{-1}; $NH_4H_2PO_4$, 0.152 gL^{-1}; microelements, 0.050 gL^{-1}; pH 5.5] according to the protocol by Motohashi et al. [20] in a non-circulating hydroponic method. Upon appearance of the first true leaf, seedlings selected for uniform development were transferred to 3 aeroponic cultivation systems, 16 plants per system, in a glasshouse with a mean temperature of 24 °C, a 14/10 h light/dark photoperiod, and 800 $\mu molm^{-2}$ s^{-1} PAR irradiance. Plants were transferred into the greenhouse on May 14, 2018, 39 days after germination (dag). Each aeroponic tank was filled with 100 L of half strength nutrient solution, average pH (±SE) 5.50 ± 0.02.

Three cultivation conditions were tested: optimal water regimen (WW_P$^-$), water scarcity regimen (drought, WS_P$^-$), and water scarcity regimen (drought) coupled with root treatment with paramylon (WS_P$^+$).

Paramylon was added to the nutrient solution of one of the 3 tanks to a final concentration of 500 mg L^{-1}. The paramylon concentration was chosen according to the results of previous experiments [16]. The atomization spray time and interval time were 3-s on/5 min off in the control system (optimal water regimen), and 3-s on/120 min off in the other two systems (water scarcity regimen with and without paramylon). As a consequence, the irrigation supply was 8.64 $Lplant^{-1}\ d^{-1}$ in WW_P$^-$ and 0.36 $Lplant^{-1}d^{-1}$ in both WS_P$^+$ and WS_P$^-$. The parameters of the aeroponics system used in the experiment are shown in Table 1.

The optimal water regimen was chosen according to Johnstone et al. [21], who used an aeroponic system for nutritional studies on *Lycopersicon esculentum* Mill. cv Cannery Row and established a misting regimen to plant roots of about 10 Ld^{-1} $plant^{-1}$. The water scarcity regimen (0.36 Ld^{-1} $plant^{-1}$) was assessed previously by a watering threshold experiment and by checking the representative features of the plants for yellowing and wilting symptoms in 100 tomato plants 30 dag in early spring. We want to stress that the irrigation supply did not correspond to the water needed by the plants.

The levels of nutrient solution and pH were monitored daily; a fresh solution was added in order to maintain a volume of 100 L in each tank. Controlled conditions were maintained throughout the experiment (14/10 h light/dark; 800 $\mu molm^{-2}\ s^{-1}$ PAR; 24 °C; pH (±SE) 5.5 ± 0.022).

Table 1. Parameters of the aeroponic system used in the experiment.

Parameters	Units	WW_P⁻	WS_P⁻ and WS_P⁺
Root available space	m^{-3}	0.012	0.012
Resident water	L	100	100
Emitter capacity	mLs^{-1}	20	20
Emitters per system		8	8
Plants per system		16	16
Irrigation duration	sh^{-1}	36	1.5
Irrigation supply per emitter	mLh^{-1}	720	30
Irrigation supply per system	Lh^{-1}	5.76	0.24
Irrigation supply per plant	Ld^{-1}	8.64	0.36

2.4. *Scanning Electron Microscopy (SEM) Preparations*

Root samples were fixed in 100% methanol for 20 min and then transferred in 100% dry ethanol for 30 min with a further change into fresh 100% ethanol overnight. After dehydration, samples were dried in a critical-point dryer apparatus, coated with gold and viewed using a Philips-SEM 505 microscope (Eindhoven, The Netherlands).

2.5. *Water Potential, Gas Exchanges and Chlorophyll a Fluorescence*

Preliminary measurements of water potential, gas exchanges and chlorophyll *a* fluorescence were done to assess whether there were variations of these parameters in the different leaves of each plant. Since no variation was detected, measurement was performed on a single leaf per plant, using the same leaf for all the measurement of the experiment.

The first measure of all the eco–physiological parameters was performed upon the transfer of the plants to the aeroponic system, i.e., 39 dag. The predawn leaf water potential (Ψ_w) was measured on one fully expanded mature leaf per plant (n = 16), using a Scholander-type pressure chamber (model 600, PMS Instrument, Albany, OR, USA) and N_2 for the application of pressure, following the precautions suggested by Turner and Long [22].

Leaf gas exchanges and chlorophyll *a* fluorescence measurements were determined between 10:00 AM and 1:00 PM (solar time) on one fully-expanded mature leaf plant (n = 16). Instantaneous measurements of steady state photosynthetic carbon dioxide (CO_2) assimilation rate (A), stomatal conductance (g_s), and internal CO_2 concentration (C_i) were performed using an LI-6400 portable photosynthesis system (Li-Cor, Lincoln, NE, USA), according to Scartazza et al. [16]. The modulated chlorophyll *a* fluorescence and the status of the electron transport of photosystem II (PSII) were measured with a PAM-2000 fluorometer (Walz, Effeltrich, Germany) on the same leaves used for gas exchange after 40 min of dark-adaptation. Fluorescence in light-adapted leaves was induced according to Scartazza et al. [16] and Schreiber et al. [23]. The maximum efficiency of PSII photochemistry was calculated as $F_v/F_m = (F_m - F_0)/F_m$; the actual photon yield of PSII photochemistry (Φ_{PSII}) was calculated as $(F'_m - F')/F'_m$; the photochemical quenching qP was calculated as $(F'_m - F_t)/(F'_m - F'_0)$. The used variables in dark-adapted state were as follows: F_v was the variable fluorescence, F_m was the maximal fluorescence, and F_0 was the minimum fluorescence. The used variables in light-adapted state were: F′ was the fluorescence at the actual state of PSII reaction centers, F'_m was the maximal fluorescence, F'_0 was the minimal fluorescence, and F_t was the transient fluorescence.

2.6. *Physicochemical Parameters and Mineral Content*

At the end of the experiment, all plants were sampled and separated in fruits, leaves, stems, and roots. Fruits were harvested at the ripening category intermediate between VI and VII [24]. All the fruits produced by each plant were gathered and samples (16 fruits) for the analysis were taken from this pool. Several quality attributes were determined on the plants: fresh and dry weights, dry matter content, size (using Vernier calipers), number of fruits, and total ash. The roots, stems, leaves, and fruits of all the plants were dried at 40 °C for 96 h in a ventilated oven. One gram of dry matter was ashed at 550 °C for 6 h. Total micronutrient content (Ca, Cu, Fe, K, Mg, Mn, Na and Zn) was determined in a sub-sample after digestion with HNO_3 and $HClO_4$ [25]. The digests were analyzed by means of a Varian AA240 FS (Varian, USA) hydride generation atomic absorption spectrophotometer equipped with flow vapor generation accessory VGA 77 (Agilent Technologies, USA). Eight independent replicates were used for chemical analysis, and data are presented as mean ± standard deviation (n = 8).

2.7. Antioxidant Compounds, Total Antioxidant Capacity and Carbohydrates

Retinol content was obtained using standard conversion formula (1 µg retinol = 1 retinol equivalent (RE) method used; 1 µg β-carotene = 0.167 µg RE), according to Aremu and Nweze [26]. Lycopene and β-carotene contents were measured spectrophotometrically according to Georgé et al. [27]. Dried tomatoes (500 mg) were homogenized in a mortar with 100 mL of hexane/acetone/ethanol (50/25/25, v/v/v) in the dark, ultrasonically disrupted, and centrifuged at 12,000× g for 20 min at 4 °C. The supernatants were filtered through 0.2 µm Minisart SRT 15 filters and transferred into a separating funnel. The organic phase was washed three times with 20 mL of distilled water (in order to remove acetone and ethanol). The aqueous phase was discarded, and the remaining water in the organic phase was removed by adding anhydrous sodium sulphate. The final volume was made up to 50 mL with hexane. The reaction mixture absorbance was measured at 436 and 450 for β-carotene determination and 503 nm for lycopene determination.

Reduced (ascorbic acid, AsA) and oxidized (dehydroascorbate, DHA) ascorbate contents were measured spectrophotometrically according to Kampfenkel et al. [28]. Fresh fruit samples (250 mg) were homogenized in a mortar with 0.5 mL of 2% (w/v) phosphoric acid and centrifuged at 12,000× g for 15 min at 4 °C. The AsA assay mixture contained 50 µL of sample extract, 150 µl of K/P 200 mM (pH 7.4), 50 µl of trichloroacetic acid (TCA) 6% (w/v) and 50 µL of water. The total ascorbate (AsA + DHA) assay mixture contained 50 µl of sample extract, 150 µl of K/P 200 mM (pH 7.4), 50 µl of TCA 6% (w/v) and 50 µL of dithiothreitol (DTT) 10 mM. The reaction mixture was left at room temperature for 15 min; 50 µL of N-ethylmaleimide 0.5% (w/v) were added after the reduction of DHA to AsA. The color was developed in both assays by adding the reagents in the following sequence: 250 µL of TCA 10% (w/v), 200 µL of ortho-phosphoric acid 42% (v/v), 200 µL of 2,2-dipyridyl 4.0% (w/v) in ethanol 70% (v/v) and 100 µL of $FeCl_3$ 3% (v/v) to a final volume of 1 mL. Controls were also run, and the solution was allowed to stand at 40 °C for an Fe^{2+}-bathophenanthroline complex to develop. The DHA levels were estimated on the basis of the difference between total ascorbate and AsA values. A standard calibration curve covering 0–10 nM of AsA or DHA range was used.

Tocopherols were determined by HPLC according to Döring et al. [29]. Fresh fruit samples (250 mg) were homogenized in a mortar with 0.4 mL of 100% HPLC-grade methanol and incubated overnight at 4 °C in the dark. The supernatant was filtered through 0.2 µm Minisart SRT 15 filters and immediately analyzed at room temperature with a reverse-phase Dionex column (Acclaim 120, C18, 5 µm particle size, 4.6 mm internal diameter × 150 mm length). Tocopherols were eluted at a flow rate of 1 mL min^{-1} using 100% solvent A (acetonitrile/methanol, 75/25, v/v) for the first 14 min, followed by a 3 min linear gradient to 100% solvent B (methanol/ethylacetate, 68/32, v/v) and 15 min with 100% solvent B. Tocopherols were detected at 280 nm. Authentic standards (Sigma-Aldrich, Italy) were used to quantify the tocopherols content of each sample.

The content of carbohydrates in fruits was determined spectrophotometrically according to Aguiar et al. [30] and quantified using a K-SUFRG commercial kit (Megazyme, Wicklow, Ireland), following the manufacturer's protocol.

The antioxidant properties of the fruits were assessed spectrofluorimetrically by the oxygen radical absorption capacity (ORAC) and hydroxyl radical antioxidant capacity (HORAC) assays [31,32].

Fresh fruit samples (10 mg) were added to 0.75 mL of 100% ethanol/methanol/water/formic acid (35:35:28:2, v/v/v/v) and centrifuged at 12,000× *g* for 10 min at 4 °C. The supernatant was collected, and 10 μL were mixed with 170 μL of 48 nM fluorescein (FL). The reagents were transferred into the main reagent wells (OptiPlate 96 F plates, Perkin Elmer, Waltham, MA, USA) and incubated at 37 °C for 20 min before recording the initial fluorescence (excitation/emission = 485/527 nm). After incubation, 20 μL of the 2,2′-azobis(2-methylpropionamidine) dihydrochloride reagent (51.5 mM final concentration) were added, and fluorescence readings were taken every minute for 60 min. A phosphate buffer (75 mM, pH 7.4) was used as a blank, and a Trolox solution (0.78–25 μM) was used as a standard. The final ORAC values were calculated by using a regression equation between the Trolox concentration and the net area under the FL decay curve and expressed as μmol Trolox equivalents (TE) per gram of fresh weight (FW). In the HORAC assay, 10 μL of supernatant were mixed with 170 μL of 48 nM fluorescein (605 mM final concentration) and incubated at 37 °C for 10 min, before recording the initial fluorescence (excitation/emission = 485/520 nm). After incubation, 10 μL of H_2O_2 (27.5 mM final concentration) and 10 μL of Co(II) (230 μM final concentration) solutions were added, and fluorescence readings were taken every minute for 60 min. A phosphate buffer was used as a blank, and a gallic acid solution (100–600 μM) was used as a standard. The final HORAC values were calculated using a regression equation between the standard antioxidant concentration and the net area under the curve. One HORAC unit was assigned to the net protection area provided by 1 μM gallic acid, and the activity of the sample was expressed as μmol gallic acid equivalent (GAE) per gram of fresh weight (FW).

The content of total phenolic compounds was determined spectrophotometrically according to Waterhouse [33]. Fresh fruit samples (100 mg) were homogenized in a mortar with 5 mL of methanol acidified with 1 % HCl (v/v) for 20 h in the dark at 4 °C. Extracts were centrifuged for 15 min at 12,000× *g* at 4 °C, and the supernatants were filtered through 0.2 μm Minisart SRT 15 filters and stored in test tubes at −20 °C. Fifty μL of a 4-times diluted extract was mixed with 2.45 mL of distilled water and 250 μL of Folin–Ciocalteu's phenol reagent. After incubation at room temperature for 6 min, 750 μL sodium carbonate 7.5% (w/v) and 500 μL of deionized water were mixed. After 120 min incubation at room temperature, the reaction mixture absorbance was measured at 760 nm. A calibration curve was prepared using a standard solution of gallic acid (range 0–1 mg mL^{-1}). Eight independent replicates were used for chemical analysis, and data are presented as mean ± standard deviation ($n = 8$).

2.8. Statistical Analysis

The statistical analysis was performed using JMP 12 (SAS institute, Cary, NC, USA). The normality of the data was preliminary tested by the Shapiro–Wilk W test. If measurements were carried out for more than two time points, data were analyzed using one-way repeated measures ANOVA, and comparison among means was determined by Tukey's HSD (honestly significant difference) multiple comparison test ($p < 0.05$). All the other data were analyzed by Student's *t* test.

3. Results and Discussion

Aeroponics is a soil-less cultivation system considered a specialized version of hydroponics [34]. It is an air-water system in which the roots of the plant extend and grow inside a closed container in the dark, are exposed to air, and directly sprayed with a nutrient-water mix through atomizers (Figure 1). The aerial portions of the plant (leaves, stem and crown) extend above the wet zone separated from the root. Aeroponics systems are very useful for plant root studies under controlled conditions; we chose it to monitor the response of tomato plants to drought and feasibility of paramylon nanofibers, directly applied to the root system, in modulating the response of the whole plant to this stressor.

The representative features of the plants (aerial parts and root system) 60 dag under the three cultivation conditions tested are shown in Figures 2 and 3. Under the optimal water regimen (Figure 2A), plants showed a normal compact growth habit with short internodes, fully expanded leaves, and regular fruit size typical of the determinate growth of Micro-Tom. The root system was fully developed with extremely long roots and plenty lateral roots, indicating a superior growth with respect to the plants under the other cultivation conditions (Figure 3A). Drought stressed plants (Figure 2B) showed wilting symptoms, with yellowing and rolling of the lower leaves, as well as a reduced fruit size. The root system appeared reduced in density and length (Figure 3B). Stressed plants with paramylon root treatment (Figure 2C) did not show any wilting; the growth habit was quite compact, the internode length was greater with respect to plants grown under optimal water regimen, and the fruit sizes were comparable. The root system showed a dramatic reduction of both density and length, coupled with an increase of the lateral rootlets (Figure 3C).

Figure 2. The representative features of the tomato plants (aerial parts) under the three cultivation conditions tested 60 days after germination (dag): (**A**) Optimal water regimen; (**B**) water scarcity regimen; (**C**) water scarcity regimen coupled with paramylon root treatment.

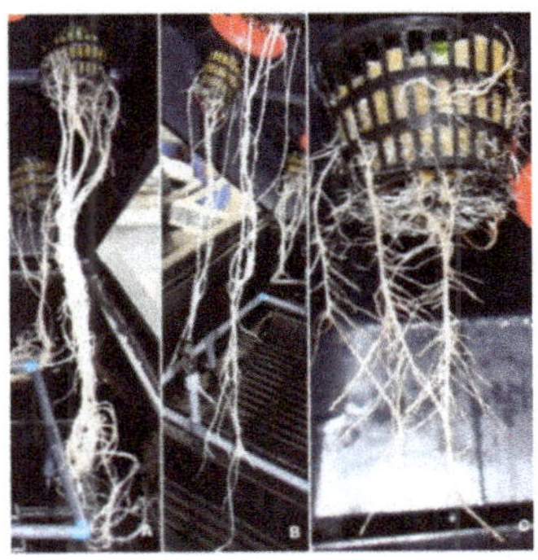

Figure 3. The root system of the tomato plants under the three cultivation conditions tested 60 dag. (**A**) Optimal water regimen, root length about 100 cm; (**B**) water scarcity regimen, root length about 80 cm; (**C**) water scarcity regimen coupled with paramylon root treatment, root length about 10 cm.

Figures 4 and 5 show the variations of the main eco-physiological parameters, monitored to highlight the effect of the paramylon root treatment during the life cycle of the plants.

Figure 4 shows that the leaf water potential (Ψ_w) (Figure 4A), CO_2 assimilation rate (A) (Figure 4B), stomatal conductance (gs) (Figure 4C), and internal CO_2 concentration (Ci) (Figure 4D) had time series with a similar trend in each growth condition tested. In WW_P$^-$ plants, the values of these parameters were almost constant and comparable with the values present in the literature [35] (Figure 4A–D). These data confirm that the chosen water regimen (8.64 Ld^{-1} $plant^{-1}$) was optimal for Micro-Tom growth [21]. In WS_P$^-$ plants, the four parameters underwent a steep decrease after a week of treatment (46 dag), which reached saturation before the end of experiment (Figure 4A–D). The reduction of the stomatal conductance as a reaction to water stress was the cause of this decreasing trend [36,37]. Additionally, in WS_P$^+$ plants, the four parameters underwent a steep decrease after a week of treatment (46 dag), but the values recovered to those of control plants (WW_P$^-$) after one or two weeks (Figure 4A–D). This delayed effect of paramylon nanofibers was mainly due to the time necessary to colonize the root system (Figure 6).

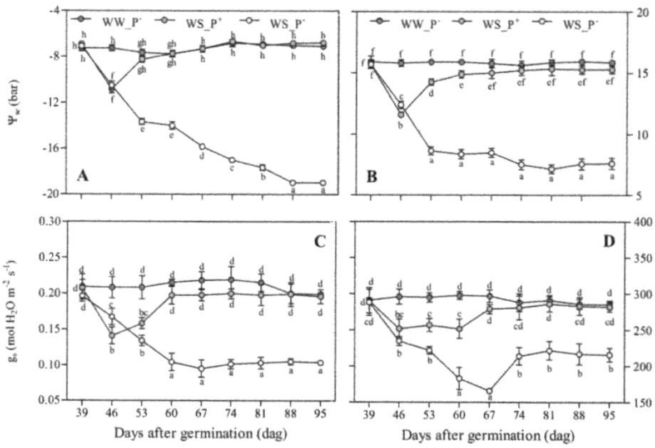

Figure 4. (**A**) Leaf water potential (Ψ_w), (**B**) CO_2 assimilation rate (A), (**C**) stomatal conductance (g_s), and (**D**) internal CO_2 concentration (C_i) in leaves of *Solanum lycopersicum* cv. Micro-Tom grown under well-watered (WW) and water-stressed (WS) conditions coupled (or not) with root treatment with paramylon (P^+ and P^-). Data are shown as mean ± standard deviation ($n = 16$), and measurements were made starting from 39 dag. In each graph, different letters indicate significant differences among treatments ($p < 0.05$, Tukey's HSD post hoc test).

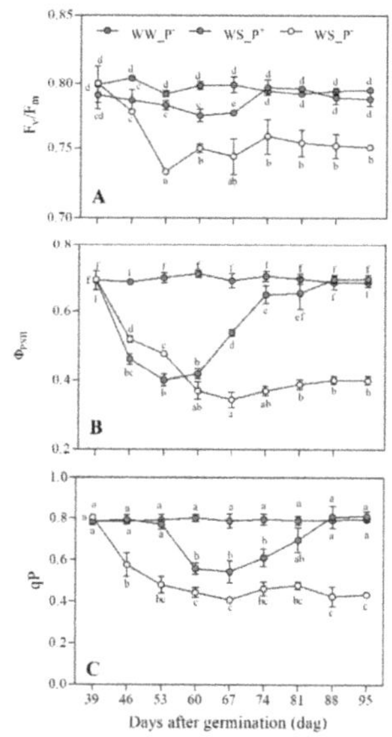

Figure 5. (**A**) Photosystem II (PSII) photochemical efficiency (F_v/F_m), (**B**) actual photon yield of PSII photochemistry (Φ_{PSII}), and (**C**) photochemical quenching state of PSII (qP) in leaves of *Solanum lycopersicum* cv. Micro-Tom grown under well-watered (WW) and water-stressed (WS) conditions coupled (or not) with root treatment with paramylon (P^+ and P^-). Data are shown as mean ± standard deviation ($n = 16$), and measurements were made starting from 39 dag. In each graph, different letters indicate significant differences among treatments ($p < 0.05$, Tukey's HSD post hoc test).

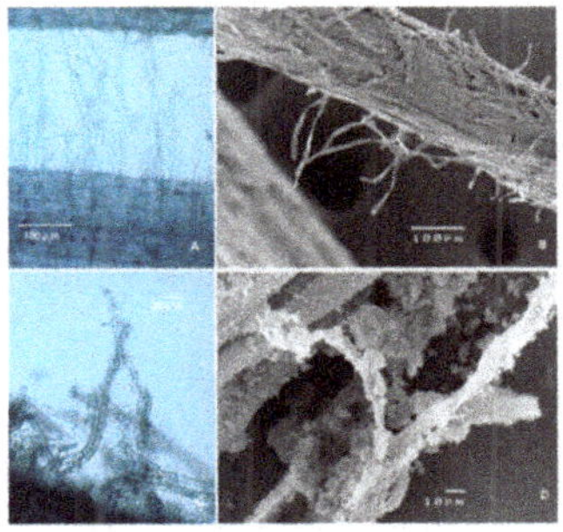

Figure 6. Optical microscopy and scanning electron microscopy images of tomato root hairs 50 dag. (**A** and **B**) Hair surface of non-treated plants. (**C** and **D**) Hair surface of paramylon-treated plants.

Figure 5 shows that the PSII photochemical efficiency (F_v/F_m) (Figure 5A), photon yield of PSII photochemistry (Φ_{PSII}) (Figure 5B), and photochemical quenching of PSII (qP) (Figure 5C) also had time series with a similar trend for each growth condition tested.

In WW_P$^-$ plants, the values of the three parameters were almost constant (Figure 5A–C). In WS_P$^-$ plants, they underwent a steep decrease after a week of treatment (46 dag), which reached saturation before the end of experiment (Figure 5A–C). Additionally, in WS_P$^+$ plants, the three parameters underwent a steep decrease after a week of treatment (46 dag), but their values recovered to those of control plants (WW_P$^-$) after several weeks (Figure 5A–C).

We can say that paramylon nanofibers induced dehydration tolerance and improved intrinsic water use efficiency by influencing stomatal behaviour. Scartazza et al. [16] monitored the effects of water removal on the Ψ_w and Φ_{PSII} of paramylon-treated tomato plants. They suggested that the paramylon caused an increase of CO_2 diffusional constraints but also promoted the ability of tomato plants to reduce water losses and counteract the reduction of Φ_{PSII} caused by the drought. According to these authors [16], β-glucan nanofibers play a potential role in reducing the sensitivity of PSII to potential dehydration damages thanks to a strong stomatal control associated with the transient modified profile of the three major plant hormones content (i.e., abscissic acid, jasmonic acid, salicylic acid) in the xylem sap. Furthermore, they stated that paramylon induced a consistent increase of g_m/g_s ratio (mesophyll conductance/stomatal conductance) and carbon gained per unit water used, which represents a relevant adaptive trait under water-limited conditions [16,36].

All our eco-physiological data confirmed the potentiality of paramylon nanofibers in the buffering water stress effect on both the photosynthetic rate and the PSII photochemical efficiency. The rate of linear electron transport (Φ_{PSII}) returned to the value of the control (WW_P$^-$ plants) with the consequent increase of the proportion of PSII reaction centres that were open (qP). Therefore, PSII photoinhibition and photodamage were counteracted (e.g., a complete recovery of F_v/F_m and Φ_{PSII} values), reducing the sensitivity of PSII to potential dehydration damages.

The values of the root, stem, leaf and fruit dry biomass measured 95 dag on the plants under the three different cultivation conditions are shown in Figure 7. WS_P$^-$ plants showed a significant reduction only in leaf DW (−23% compared with WW_P$^-$) data relative to the fruits are not shown because the fruit did not ripe beyond the II category (mature green) [24]. Drought stress conditions were so severe that all the eco-physiological parameters were deeply altered (Figure 2, Figure 5, and Figure 6) and inhibited the progress of ripening.

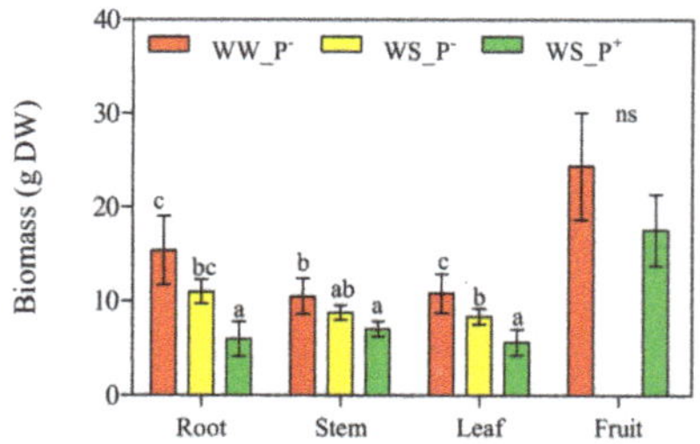

Figure 7. Dry biomass expressed as g dry weight (DW) of leaf, stem, root, and red ripe fruits of *Solanum lycopersicum* cv. Micro-Tom grown under well-watered (WW) and water-stressed (WS) conditions coupled (or not) with root treatment with paramylon (P^+ and P^-). Data are shown as mean ± standard deviation ($n = 8$). Measurements were made 95 days after germination (dag). For each parameter, different letters indicate significant differences among treatments ($p < 0.05$, Tukey's HSD post hoc test).

A significant reduction of root, stem and leaf DW was observed in WS_P$^+$ plants (−61%, −33% and −48% compared with WW_P$^-$). This reduction of growth could be explained as an adaptation strategy to reduce resource spending and allow the plants to divert energy in the fruits. No significant differences ($p > 0.05$) were observed between red ripe fruits DW of WW_P$^-$ and WS_P$^+$ plants. It is clear that WS induced a negative effect on leaf biomass and fruit ripening, as confirmed by the reduction of leaf DW and the alteration of fruit-ripening processes. Our results are not consistent with those by other authors [38–40] but are in accordance with Khan et al. [41]. The differences observed in the WS sensitivity may be due to the severity of drought.

The other physicochemical parameters (ashes, microelements, weight, size and yield) and the quality parameters (antioxidant compositions and activities, soluble carbohydrates) were measured on fruits assigned to a ripening category intermediate between the VI (red ripe) and the VII (red overripe) categories (Figure 8) [24].

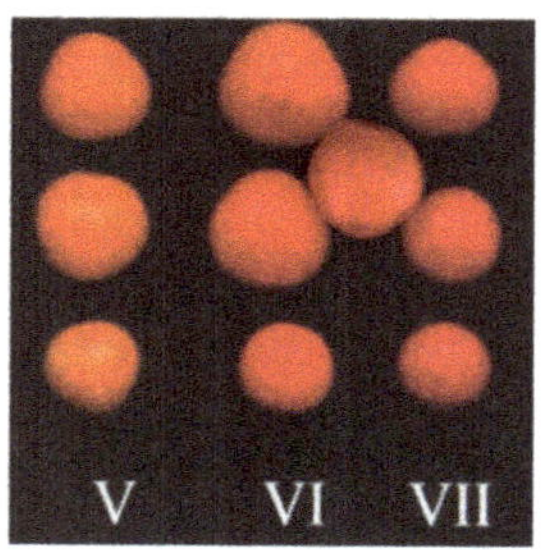

Figure 8. Ripening categories table according to Leide et al. [24]: Fruits used for the measurement of physicochemical and quality parameters belong to a ripening category intermediate between the VI (red ripe) and the VII (red overripe) categories.

No significant differences were observed between the number, weight and dimensions of the fruits of WW_P$^-$ and WS_P$^+$ (data not shown), though paramylon-treated plants showed precocious fruiting and ripening (Figure 9). As already stated, data relative to fruits of the drought stressed plants without paramylon (WW_P$^-$) are not shown because the fruits did not ripe beyond the II category (mature green).

The effects of cultivation conditions on red ripe fruits were monitored by the quantification of biometric parameters (ashes, dry matter, and moisture; Figure 10), microelement content, and bioactive compounds (antioxidant compositions and activities, soluble carbohydrates; Figures 11–13). In fruits harvested from WS_P$^+$ plants, a significant increase of ashes and dry matter was observed compared with WW_P⁻ plants, while moisture showed an opposite trend (Figure 10). These data confirmed that paramylon treatment could improve fruit quality.

Figure 9. Precocious fruiting and ripening of the fruits of WS_P⁺ plants (**A**) with respect to WW_P⁻ plants (**B**) 74 dag.

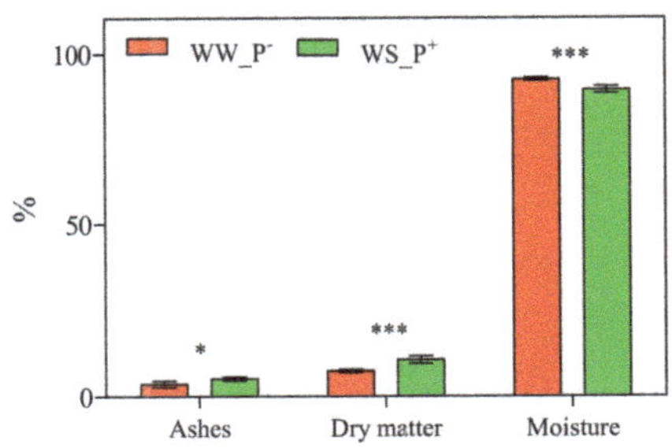

Figure 10. Biometric parameters (dry ash, dry matter and moisture) in red ripe fruits of *Solanum lycopersicum* cv. Micro-Tom grown under well-watered (WW) and water-stressed (WS) conditions coupled (or not) with root treatment with paramylon (P^+ and P^-). Data are shown as mean ± standard deviation (n = 8). Measurements were made 95 dag. For each parameter, the data were analyzed by Student's t test. The significant differences are for: *** = $p < 0.001$ and * = $p > 0.05$

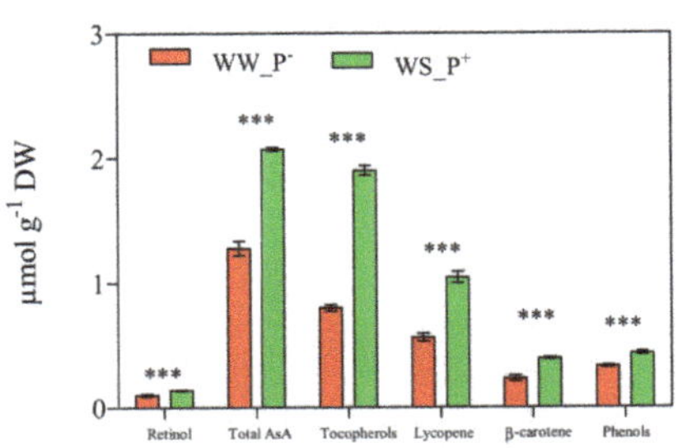

Figure 11. Antioxidant compounds in red ripe fruits of *Solanum lycopersicum* cv. Micro-Tom grown under well-watered (WW) and water-stressed (WS) conditions coupled (or not) with root treatment with paramylon (P^+ and P^-). Data are shown as mean ± standard deviation (n = 8). Measurements were made 95 dag. For each parameter, the data were analyzed by Student's t test. The significant differences are for: *** = $p < 0.001$. Abbreviations: DW, dry weight; Total AsA, total ascorbate (reduced and oxidized forms).

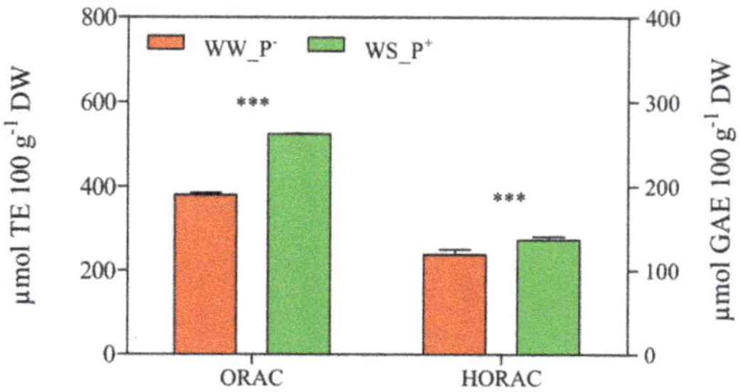

Figure 12. Antioxidant capacity expressed as oxygen radical absorbance capacity (ORAC) and hydroxyl radical antioxidant capacity (HORAC) in red ripe fruits of *Solanum lycopersicum* cv. Micro-Tom grown under well-watered (WW) and water-stressed (WS) conditions coupled (or not) with root treatment with paramylon (P^+ and P^-). Data are shown as mean ± standard deviation ($n = 8$). Measurements were made 95 dag. For each parameter, the data were analyzed by Student's t test. The significant differences are for: *** = $p < 0.001$. Abbreviations: DW, dry weight; GAE, gallic acid equivalent; TE, trolox equivalent.

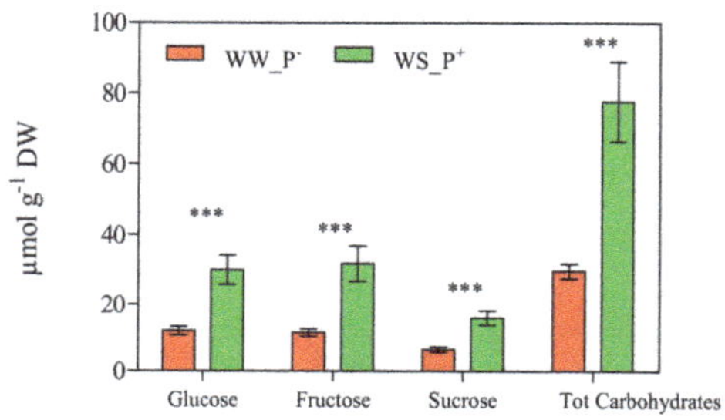

Figure 13. Carbohydrates in red ripe fruits of *Solanum lycopersicum* cv. Micro-Tom grown under well-watered (WW) and water-stressed (WS) conditions coupled (or not) with root treatment with paramylon (P^+ and P^-). Data are shown as mean ± standard deviation ($n = 8$). Measurements were made 95 dag. For each parameter, the data were analyzed by Student's t test. The significant differences are for: *** = $p < 0.001$. Abbreviations: DW, dry weight.

No significant differences were observed between WW_P$^-$ and WS_P$^+$ fruits regarding microelements content (data not shown).

The cultivation conditions significantly affected the amount of antioxidant compounds of the fruits: all biochemical parameters (retinol, i.e., Vitamin A, tot AsA, i.e., Vitamin C, tocopherols, i.e., Vitamin E, lycopene, β-carotene and phenols) deeply increased in fruits harvested from WS_P$^+$ plants (about two-fold higher than those recorded in WW_P$^-$ plants; Figure 11). A high AsA content is an important trait in tomato fruits, as it prevents oxidative stress (especially during fruit ripening) and thus enhances shelf life [42]. Similarly, an increase of lycopene (a highly characteristic phytonutrient of tomatoes) could affect the final nutritional quality and commercial value of tomato fruit [43]. In fact, lycopene greatly enhances fruit quality thanks to its intense antioxidant activity that suppresses cell proliferation and interferes with the growth of cancer cells [44,45]. In addition, a marked increase of the antioxidant activity expressed as the ORAC and HORAC values was observed in fruits harvested from WS_P$^+$ plants (+38% and +15%, compared with WW_P$^-$; Figure 12). This result confirms the involvement of lycopene, β-carotene, and retinol in the antioxidant response of WS_P$^+$ fruits. Tocopherols are non-enzymatic lipid-soluble antioxidants that protect the pigments, proteins, and polyunsaturated fatty acids of the photosynthetic apparatus against reactive oxygen species [46]. It has been reported that the tocopherols content in tomato fruits depends on many factors such as the level of irrigation, light, and NaCl [47]. We found that plants grown under WS_P$^+$ conditions produced fruits with a high tocopherol content, thus indicating an induction of defense mechanisms. Phenolic compounds and

AsA represent the main water-soluble antioxidants in tomatoes [48]. In WS_P$^+$ fruits, a significant increase of total phenols was observed, suggesting that they contribute positively to the antioxidant activity of the tomato water-soluble fraction by reducing the levels of free radicals due to WS (as confirmed by the increase of ORAC and HORAC levels). This response can be considered an adaptive mechanism to water stress that promotes the *de novo* synthesis of these metabolites [40]. At the end of the experiment, the content of carbohydrates showed a similar trend. Glucose, fructose, and sucrose concentrations, as well as total carbohydrates, in the fruits of WS_P$^+$ plants were nearly three-fold higher than in fruits harvested from WW_P$^-$ plants (Figure 13). Carbohydrates are essential for plant growth and survival, as well as maintenance and repair processes, and they are also major sources of cellular energy [49]. These compounds play a key role in regulating overall cellular metabolism, maintaining osmotic equilibrium, and preventing turgor loss in tissue [50,51]. They also can act as scavengers of reactive oxygen species (ROS) and contribute to the protection of membranes and macromolecules [52,53]. Micro-Tom fruits contain the reducing sugars fructose and glucose with trace amounts of sucrose, typical of tomatoes [54]. The carbohydrates composition found here was in agreement with that reported in the literature [55].

4. Conclusions

Drought is by far the most important environmental stressor in agriculture worldwide, and it is expected to contribute to the severe salinization of more than 50% of world arable land by 2050. The research efforts to improve crop productivity under water limiting conditions, focused mainly on natural selection and the breeding activity of tolerant genotypes. Root treatment could be another method to cope with the drought stress. In this paper, we showed the results of the direct application of paramylon on the root system of Micro-Tom tomatoes. Paramylon extracted from the *E. gracilis* WZSL mutant was processed to linear nanofibers that interacted with the Dectin-1 receptors present on the target cell membranes of tomato roots, enhancing the plant defense capacity against drought. Drought tolerance was achieved by influencing stomatal behavior and inducing an effective improvement of water use efficiency, obtained by modulating the conductance to CO_2 diffusion from air to the carboxylation sites through the modulation of hormone levels [16]. We observed that the paramylon treatment allowed the optimal water regimen of about 8.64 L plant^{-1} day^{-1} to be lowered to 0.36 Lplant^{-1} day^{-1} without a detrimental effect on the yield and eco–physiological parameters. The great increase of antioxidant compounds (Vitamin A/C/E, lycopene, β-carotene and phenols) together with the increase of carbohydrates (glucose, fructose and sucrose) in the fruits of paramylon-treated plants improved their nutritional value and sensory quality.

These results confirm the biostimulant activity of paramylon in increasing plant adaptation capacity for abiotic stress.

Author Contributions: Conceptualization, methodology, investigation, writing—L.B. and P.G.; resources, software, and data curation—P.C.

Funding: This research received no external funding.

Conflicts of Interest: The authors declare no conflict of interest.

References

1. Barsanti, L.; Passarelli, V.; Evangelista, V.; Frassanito, A.M.; Gualtieri, P. Chemistry, physico-chemistry and applications linked to biological activities of β-glucans. *Nat. Prod. Rep.* **2011**, *28*, 457–466. [CrossRef] [PubMed]
2. Barsanti, L.; Vismara, R.; Passarelli, V.; Gualtieri, P. Paramylon (β-1,3-glucan) content in wild type and WZSL mutant of *Euglena gracilis*: effects of growth conditions. *J. Appl. Phycol.* **2001**, *13*, 59–65. [CrossRef]
3. Shibakami, M.; Tsuboichi, G.; Nakamura, M.; Hayashi, M. Preparation of carboxylic acid-bearing polysaccharide nanofiber made from euglenoid β-1,3-glucans. *Carbohydr. Polym.* **2013**, *15*, 95–101. [CrossRef] [PubMed]

4. Russo, R.; Barsanti, L.; Evangelista, V.; Frassanito, A.M.; Longo, V.; Pucci, L.; Penno, G.; Gualtieri, P. *Euglena gracilis* paramylon activates human lymphocytes by upregulating pro-inflammatory factors. *Food Sci. Nutr.* **2017**, *5*, 205–214. [CrossRef] [PubMed]
5. Rosati, G.; Barsanti, L.; Passarelli, V.; Giambelluca, A.; Gualtieri, P. Ultrastructure of a novel non-photosynthetic *Euglena* mutant. *Micron* **1996**, *27*, 367–373. [CrossRef]
6. Takeuchi, O.; Akira, S. Pattern recognition receptors and inflammation. *Cell* **2010**, *140*, 805–820. [CrossRef] [PubMed]
7. Thompson, I.J.; Oyston, P.C.F.; Williamson, D.E. Potential of the β-glucans to enhance innate resistance to biological agents. *Expert Rev. Anti. Infect. Ther.* **2010**, *8*, 339–352. [CrossRef] [PubMed]
8. Adams, E.L.; Rice, P.J.; Graves, B.; Ensley, H.E.; Yu, H.; Brown, G.D.; Williams, D.L. Differential high-affinity interaction of dectin-1 with natural or synthetic glucans is dependent upon primary structure and is influenced by polymer chain length and side chain branching. *J. Pharmacol. Exp. Ther.* **2008**, *325*, 115–123. [CrossRef]
9. Fu, C.; Iyer, P.; Herkal, A.; Abdullah, J.; Stout, A.; Free, S.J. Identification and characterization of genes required for cell-to-cell fusion in *Neurospora crassa*. *Eukaryot. Cell* **2011**, *10*, 1100–1109. [CrossRef]
10. Bashir, K.M.I.; Choi, J.-S. Clinical and physiological perspectives of β-glucans: the past, present, and future. *Int. J. Mol. Sci.* **2017**, *18*, 1906. [CrossRef]
11. Dennehy, K.M.; Brown, G.D. The role of the beta-glucan receptor dectin-1 in control of fungal infection. *J. Leukoc. Biol.* **2007**, *82*, 253–258. [CrossRef]
12. Monie, T.P. Immune cells and the process of pattern recognition. In *The innate immune system: A compositional and functional perspective*; Monie, T.P., Ed.; Academic Press: London, UK, 2017; pp. 41–83.
13. Palma, A.S.; Feizi, T.; Zhang, Y.; Stoll, M.S.; Lawson, A.M.; Diaz-Rodríguez, E.; Campanero-Rhodes, A.S.; Costa, J.; Brown, G.D.; Chai, W. Ligands for the beta-glucan receptor, dectin-1, assigned using "designer" microarrays of oligosaccharide probes (neoglycolipids) generated from glucan polysaccharides. *J.Biol. Chem.* **2006**, *281*, 5771–5779. [CrossRef] [PubMed]
14. Tsoni, S.V.; Brown, G.D. β-glucans and dectin-1. *Ann. N. Y. Acad. Sci.* **2008**, *1143*, 45–60. [CrossRef] [PubMed]
15. Vismara, R.; Vestri, S.; Frassanito, A.M.; Barsanti, L.; Gualtieri, P. Stress resistance induced by paramylon treatment in *Artemia* sp. *J. Appl. Phycol.* **2004**, *16*, 61–67. [CrossRef]
16. Scartazza, A.; Picciarelli, P.; Mariotti, L.; Curadi, M.; Barsanti, L.; Gualtieri, P. The role of *Euglena gracilis* paramylon in modulating xylem hormone levels, photosynthesis and water-use efficiency in *Solanum lycopersicum* L. *Physiol. Plant.* **2017**, *161*, 486–501. [CrossRef] [PubMed]
17. Kusmic, C.; Barsanti, L.; Di Lascio, N.; Faita, F.; Evangelista, V.; Gualtieri, P. Anti-fibrotic effect of paramylon nanofibers from the WZSL mutant of *Euglena gracilis* on liver damage induced by CCl_4 in mice. *J. Funct. Foods* **2018**, *46*, 538–545. [CrossRef]
18. Morison, J.I.L.; Baker, N.R.; Mullineaux, P.M.; Davies, W.J. Improving water use in crop production. *Philos. Trans. R. Soc. Lond.* **2008**, *363*, 639–658. [CrossRef] [PubMed]
19. Scott, J.W.; Harbaugh, B.K. *Micro-Tom: A Miniature Dwarf Tomato*; Agricultural Experiment Station, Institute of Food and Agricultural Sciences, University of Florida: Gainesville, FL, USA, 1989; pp. 1–6.
20. Motohashi, R.; Enoki, H.; Fukazawa, C.; Kiriiwa, S. Hydroponic culture of Micro-Tom tomato. *Bio-protocol* **2015**, *5*, e1613. [CrossRef]
21. Johnstone, P.R.; Nichols, M.A.; Fisher, K.J.; Reid, J. Nutritional studies with processing tomato grown in aeroponic. *Acta Horticulture* **2001**, 143–150. [CrossRef]
22. Turner, N.C.; Long, M.J. Errors arising from rapid water loss in the measurement of leaf water potential by the pressure chamber technique. *Aust. J. Plant Physiol.* **1980**, *7*, 527–537. [CrossRef]
23. Schreiber, U.; Schilwa, U.; Bilger, W. Continuous recording of photochemical and non-photochemical chlorophyll fluorescence quenching with a new type of modulation fluorometer. *Photosynth. Res.* **1986**, *10*, 51–62. [CrossRef] [PubMed]
24. Leide, J.; Hildebrandt, H.; Reussing, K.; Riederer, M.; Vogg, G. The developmental pattern of tomato fruit wax accumulation and its impact on cuticular transpiration barrier properties: effects of a deficiency in a β-ketoacyl-coenzyme A synthase (LeCER6). *Plant Physiol.* **2007**, *144*, 1667–1679. [CrossRef] [PubMed]
25. Zasoski, R.J.; Burau, R.G. A rapid nitric-perchloric acid digestion method for multi-element tissue analysis. *Comm. Soil Sci. Plant Anal.* **1977**, *8*, 425–436. [CrossRef]

26. Aremu, S.O.; Nweze, C.C. Determination of vitamin A content from selected Nigerian fruits using spectrophotometric method. *Bangladesh J. Sci. Ind. Res.* **2017**, *52*, 153–158. [CrossRef]
27. Georgé, S.; Tourniaire, F.; Gautier, H.; Goupy, P.; Rock, E.; Caris-Veyrat, C. Changes in the contents of carotenoids, phenolic compounds and vitamin C during technical processing and lyophilization of red and yellow tomatoes. *Food Chem.* **2011**, *124*, 1603–1611.
28. Kampfenkel, K.; Van Montagu, M.; Inzé, D. Extraction and determination of ascorbate and dehydroascorbate from plant tissue. *Anal. Biochem.* **1995**, *10*, 165–167. [CrossRef] [PubMed]
29. Döring, A.S.; Pellegrini, E.; Campanella, A.; Trivellini, A.; Gennai, C.; Petersen, M.; Nali, C.; Lorenzini, G. How sensitive is *Melissa officinalis* to realistic ozone concentrations? *Plant Physiol. Biochem.* **2014**, *74*, 156–164. [CrossRef]
30. Aguiar, F.; Menezes, V.; Rogez, H. Spontaneous postharvest fermentation of acai (*Euterpe oleracea*) fruit. *Postharvest Biol. Technol.* **2013**, *86*, 294–299. [CrossRef]
31. Ou, B.X.; Hampsch-Woodill, M.; Prior, R.L. Development and validation of an improved Oxygen Radical Absorbance Capacity Assay using fluorescein as the fluorescent probe. *J. Agric. Food Chem.* **2001**, *49*, 4619–4626. [CrossRef]
32. Ou, B.X.; Huang, D.J.; Hampsch-Woodill, M.; Flanagan, J.A.; Deemer, E.K. Analysis of antioxidant activities of common vegetables employing oxygen radical absorbance capacity (ORAC) and ferric reducing antioxidant power (FRAP) assays: a comparative study. *J. Agric. Food Chem.* **2002**, *50*, 3122–3128. [CrossRef]
33. Waterhouse, A.L. Polyphenolics: determination of total phenolics. In *Current Protocols in Food Analytical Chemistry*; Wrolstad, R.E., Ed.; Wiley: New York, NY, USA, 2002; pp. 1–4.
34. Lakhiar, I.A.; Gao, J.; Syed, T.N.; Chandio, F.A.; Buttar, N.A. Modern plant cultivation technologies in agriculture under controlled environment: a review on aeroponics. *J. Plant Interact.* **2018**, *13*, 338–352. [CrossRef]
35. Alves, F.R.R.; Melo, H.C.; Crispin-Filho, A.J.; Costa, A.C.; Nascimiento, K.J.T.; Carvalho, R.F. Physiological and biochemical responses of photomorphogenic tomato mutants (cv. Micro-Tom) under water withholding. *Acta Physiol. Plant* **2016**, *38*, 155–170. [CrossRef]
36. Centritto, M.; Lauteri, M.; Monteverdi, M.C.; Serraj, R. Leaf gas exchange, carbon isotope discrimination, and grain yield in contrasting rice genotypes subjected to water deficits during the reproductive stage. *J. Exp. Bot.* **2009**, *60*, 2325–2339. [CrossRef] [PubMed]
37. Flexas, J.; Niinements, U.; Gallé, A.; Barbour, M.M.; Centritto, M.; Diaz-Espejo, A.; Douthe, C.; Galmés, J.; Ribas-Carbo, M.; Rodriguez, P.L.; et al. Diffusional conductances to CO_2 as a target for increasing photosynthesis and photosynthetic water-use efficiency. *Photosynth. Res.* **2913**, *117*, 45–59. [CrossRef] [PubMed]
38. Patanè, C.; Cosentino, S.L. Effects of soil water deficit on yield and quality of processing tomato under a Mediterranean climate. *Agric. Water Manag.* **2010**, *97*, 131–138. [CrossRef]
39. Casa, R.; Rouphael, Y. Effects of partial root-zone drying irrigation on yield fruit quality, and water-use efficiency in processing tomato. *J. Hortic. Sci. Biotechnol.* **2014**, *89*, 389–396. [CrossRef]
40. Guida, G.; Sellami, M.H.; Mistretta, C.; Oliva, M.; Buonomo, R.; De Mascellis, R.; Patanè, C.; Rouphael, Y.; Albrizio, R.; Giorio, P. Agronomical, physiological and fruit quality responses of two Italian long-storage tomato landraces under rain-fed and full irrigation conditions. *Agric. Water Manag.* **2017**, *180*, 126–135. [CrossRef]
41. Khan, S.H.; Khan, A.; Litaf, U.; Shah, A.S.; Khan, M.A.; Bilal, M.; Ali, M.U. Effect of drought stress on tomato cv. Bombino. *J. Food Process. Technol.* **2015**, *6*, 465. [CrossRef]
42. Slimestad, R.; Verheul, M. Review of flavonoids and other phenolics from fruits of different tomato (*Lycopersicon esculentum* L.) cultivars. *J. Sci. Food Agric.* **2009**, *89*, 1255–1270. [CrossRef]
43. Dumas, Y.; Dadomo, M.; Licca, G.D.; Grolier, P. Effects of environmental factors and agricultural techniques of antioxidant content of tomatoes. *J. Sci. Food Agric.* **2003**, *83*, 369–382. [CrossRef]
44. Levi, F.; Pasche, C.; Lucchini, F.; La Vecchia, C. Dietary intake of selected micronutrients and breast-cancer risk. *Int. J. Cancer* **2001**, *91*, 260–263. [CrossRef]
45. Stahl, W.; Sies, H. Bioactivity and protective effects of natural carotenoids. *Biochim. Biophys. Acta* **2005**, *30*, 101–107. [CrossRef] [PubMed]

46. Quadrana, L.; Almeida, J.; Otaiza, S.N.; Duffy, T.; Corrêa da Silva, J.V.; de Godoy, F.; Asís, R.; Bermúdez, L.; Fernie, A.R.; Carrari, F.; et al. Transcriptional regulation of tocopherol biosynthesis in tomato. *Plant Mol. Biol.* **2013**, *81*, 309–325. [CrossRef] [PubMed]
47. Raiola, A.; Tenore, G.C.; Barone, A.; Frusciante, L.; Rigano, M.M. Vitamin E content and composition in tomato fruits: beneficial roles and bio-fortification. *Int. J. Mol. Sci.* **2015**, *8*, 29250–29264. [CrossRef] [PubMed]
48. Fanasca, S.; Colla, G.; Maiani, G.; Venneria, E.; Rouphael, Y.; Azzini, E.; Saccardo, F. Changes in antioxidant content of tomato fruits in response to cultivar and nutrient solution composition. *J. Agric. Food Chem.* **2006**, *54*, 4319–4325. [CrossRef]
49. Kusvuran, S.; Ellialtioglu, S.; Polat, Z. Antioxidative enzyme activity lipid peroxidation, and proline accumulation in the callus tissues of salt and drought tolerant and sensitive pumpkin genotypes under chilling stress. *Hortic. Environ. Biotechnol.* **2013**, *54*, 319–325. [CrossRef]
50. Clifford, S.C.; Arndt, S.K.; Corlett, J.E.; Joshi, S.; Sankhla, N.; Popp, M.; Jones, H.G. The role of solute accumulation, osmotic adjustment and changes in cell wall elasticity in drought tolerance in *Ziziphus mauritania* (Lamk.). *J. Exp. Bot.* **1998**, *49*, 967–977. [CrossRef]
51. Sivakumar, P.; Sharmila, P.; Jain, V.; Pardha-Saradhi, P. Sugars have potential to curtail oxygease acitivity of Rubisco. *Biochem. Biophys. Res. Commun.* **2002**, *298*, 247–250. [CrossRef]
52. Hoekstra, F.A.; Golovina, E.A.; Buitink, J. Mechanisms of plant desiccation tolerance. *Trends Plant Sci.* **2001**, *6*, 431–438. [CrossRef]
53. Mahajan, S.; Tuteja, N. Cold, salinity and drought stresses: an overview. *Arch. Biochem. Biophys.* **2005**, *444*, 139–158. [CrossRef]
54. Malundo, T.M.M.; Shewfelt, R.L.; Scott, J.W. Flavor quality of fresh tomato (*Lycopersicon esculentum* Mill) as affected by sugar and acid levels. *Postharvest Biol. Technol.* **1995**, *6*, 103–110. [CrossRef]
55. Gomez, P.; Ferrer, M.A.; Fernandez-Trujillo, J.P.; Calderon, A.; Artes, F.; Egea-Cortines, M.; Weiss, J. Structural changes, chemical composition and antioxidant activity of cherry tomato fruits (cv Micro-Tom) stored under optimal and chilling conditions. *J. Sci. Food Agric.* **2009**, *89*, 1543–1551. [CrossRef]

agronomy

MDPI

Article

Metabolite-Targeted Analysis and Physiological Traits of *Zea mays* L. in Response to Application of a Leonardite-Humate and Lignosulfonate-Based Products for Their Evaluation as Potential Biostimulants

Andrea Ertani ' Serenella Nardi [1,*], Ornella Francioso [2], Diego Pizzeghello [1], Anna Tinti [2] and Michela Schiavon [1]

1 Department of Agronomy, Food, Natural resources, Animals and Environment (DAFNAE), University of Padua, 35100 Padua, Italy
2 Department of Agricultural and Food Sciences, Alma Mater Studiorum-University of Bologna, Viale Fanin 40, 40127 Bologna, Italy
3 Department of Biomedical and Neuromotor Sciences, Alma Mater Studiorum-University of Bologna, Via Belmeloro 8/2, 40126 Bologna, Italy
* Correspondence: serenella.nardi@unipd.it; Tel.: +39-0498272911

Received: 24 June 2019; Accepted: 9 August 2019; Published: 12 August 2019

Abstract: The main aim of this study is to identify and investigate specific humates (Hs) as potential biostimulants. Five specialty lignosulfonates (LS1-5), one commercial leonardite-humate (PH), and one commercial lignosulfonate (LH), were analyzed for their carbon, nitrogen, and sulfur contents, and the distribution of functional groups using Fourier transform infrared (FTIR) and Raman spectroscopies. Hs were further supplied for two days to *Zea mays* L. in hydroponics to test their capacity to trigger changes in physiological target-responses. LS1, LS2, LS3, and LS5 determined the most pronounced effects on plant growth and accumulation of proteins and phenolics, perhaps because of their chemical and spectroscopic features. Root growth was more increased (+51–140%) than leaf growth (+5–35%). This effect was ascribed to higher stimulation of N metabolism in roots according to the increased activity of N-assimilation enzymes (GS and GOGAT) and high consumption of sugars for energy-dependent processes. Increased values of RuBisCO, SPAD (Soil Plant Analysis Development values), and leaf sugar accumulation refer to enhanced photosynthesis attributed to Hs. We conclude that Hs tested in this study functioned as biostimulants, but the specialty lignosulfonates were more efficient in this role, possibly because of the type of starting material and process used for their production, which may have influenced their chemical properties.

Keywords: *Zea mays* L; lignohumate; lignosulfonate; biological activity; nitrogen metabolism; carbon metabolism; proteins; phenolics; sugars

1. Introduction

Increasing food production for a developing world population and the protection of environmental resources represents a great challenge in the field of agricultural sciences. Traditional agronomical practices especially have negatively impacted a number of environmental aspects and have been in part responsible for soil and water pollution [1]. In addition, the quality of most agricultural soils has long been injured by the thorough application of mineral fertilizers in order to achieve high crop yield requirements [2].

The decline of soil chemical and physical properties is generally accompanied by the decrease of soil fertility, a reduced content of soil organic carbon, and the impoverishment of microbial communities' biodiversity [3]. Therefore, new advances in support of environmentally friendly crop productions are required. Among them, one potential strategy could be the application of biostimulant products during crop cultivation [4]. Biostimulants are "formulated products of biological origin, either with or without plant growth promoting microorganisms (PGPMs), able to stimulate plant productivity at very low dosages by virtue of synergic effects of the different bioactive constituents" [5]. Biostimulants promote plant nutrition and tolerance to environmental stresses [6,7] and, based on their origin and the starting source for their manufacturing, they are divided in different groups, as follows: Humic substances (HS), seaweed extracts, protein hydrolysates, and microbial inoculants, such as mycorrhizal fungi and rhizobacteria, and beneficial elements [8]

Humic substances or humates are regarded as a major category of biostimulants, with a big market share [9]. They represent the most stable and recalcitrant component of soil organic matter and derive from the chemical and microbial degradation of vegetal and animal residues [10,11]. They are useful for improving the quality of soils, as well as the plant metabolism and root morphological traits, via their interaction with a plurality of biochemical mechanisms and physiological processes occurring at the plant-soil interface [10,11]. Specifically, humic substances stimulate plant growth via hormone-like effects and increased photosynthesis efficiency, enhance the respiration rate, and improve root nutrient uptake through an effect, either direct or indirect, on the expression of genes encoding H^+-ATPase isoforms and membrane transporters [7,10–12].

Over the last decades, commercial humic products designed lignohumates have found various applications in environmental technologies and agriculture [13,14] and are commonly used for several industrial purposes [14]. They share similar properties with humic substances in terms of chelation, buffering, and cation exchange capacity because of the great number of carboxylic and phenol groups bonded to the aromatic ring [10,15]. Lignohumates are water soluble anionic polymers containing high and low molecular weight molecules, as well as a large number of charged groups, and are by-products generated from the sulphite process of wood, in which fibers of cellulose are separated from lignin by the action of bisulphite [16,17]. The lignin fraction in wood is sulfonated, degraded and solubilized in water during this procedure [18]. In this way, the production of humates from materials that do not primarily contain them becomes a very fast process, which otherwise would naturally take many years. Researches have only clarified the primary structure of these polymers in part, so far, and only a few studies have investigated their effects on plant growth and metabolism [19–21].

The production of humates derived from different salts of humic acids, such as ammonium humates and potassium (K) humates is increasingly growing. Potassium humates, in particular, are used as biostimulants to ameliorate soil chemical, physical, and biological properties, such as the content of organic matter, water retention capacity, structure, deactivation of toxic metals, and microbiome. In addition, they can increase the efficiency of inorganic fertilizers by prompting plant growth, yield and quality, enhancing nutrient uptake and assimilation, and promoting plant resistance to stress conditions [22–27].

Interestingly, the chemistry and physiological functioning of humates can vary depending on the starting material (e.g., leonardite, wood) from which they originate, extraction processes (KOH extraction for leonardite, wood bisulphite extraction for lignosulfonates) and modification technologies used to obtain the products. Indeed, humates derived from the same source and obtained by the same company can widely differ in composition [28]. On this account and in view of the plant diversity from which humates can be produced, it appears relevant to characterize the marketed products to test their effectivity in agriculture as biostimulants.

In light of such considerations, seven humates were investigated in this study to evaluate their biostimulant potential. The humates included a commercial lignosulfonate-based product (LH, LignoHumate®, produced using a patented oxidation process) consisting of a highly concentrated plant and soil amendment, a commercial humate extracted from leonardite (PH), produced and marketed by Borregaard. The remaining humates (LS1, LS2, LS3, LS4, LS5) were specialty lignosulfonates developed by Borregaard and applying proprietary technology (different from the one used to obtain LH) to modify the starting material. We first assayed differences in their content of main elements (C, N, and S), and in the occurrence and distribution of principal functional groups using two complementary spectroscopic techniques (FT-IR and FT-Raman). Then, we applied these products to *Zea mays* L. plants in order to evaluate differences in their capacity to trigger positive changes in physiological and biochemical traits associated with plant productivity. We chose to test the products on *Zea mays* L. because it is a relevant staple crop for many populations worldwide. One of the novelties of the study is that most of products tested in this study were specialty lignosulfonates developed by Borregaard's company using proprietary technology and, thus, they were supposed to be very different in chemical features from standard lignosulfonates.

2. Results

2.1. Chemical and Spectroscopic Features of Hs

The elemental composition in percent content (w/w) of Hs is reported in Table 1. Carbon (C) content was strongly correlated ($R^2 = 0.83$) with nitrogen (N) content for all Hs and varied from 33.04% (w/w) in LH to 54.56% (w/w) in LS1. Nitrogen content was also maximum in LS1 (2.18% w/w), but minimum in LS4 (1.58% w/w). Sulfur (S) content was low only in PH (1.30%), while it was higher in lignosulfonates, varying from 5.13% (w/w) in LS5 to 7.83% (w/w) in LS4.

Table 1. Elemental analysis of carbon (C), nitrogen (N), and sulfur (S) in the different humates.

Product	C	N	S
		% (*w/w*)	
LS1	54.56 ± 1.02	2.18 ± 0.13	5.56 ± 0.34
LS2	41.08 ± 1.10	1.70 ± 0.15	6.12 ± 0.40
LS3	41.28 ± 1.32	1.95 ± 0.21	5.49 ± 0.12
LS4	37.11 ± 1.15	1.58 ± 0.17	7.83 ± 0.21
LS5	48.15 ± 1.50	2.12 ± 0.13	5.13 ± 0.34
PH	38.05 ± 1.01	1.67 ± 0.20	1.30 ± 0.23
LH	33.04 ± 1.14	1.64 ± 0.22	5.33 ± 0.31

FTIR and Raman analyses were performed to evaluate the main chemical attributes of Hs. The attributions of the main peaks for different functional groups identified in the FTIR and Raman spectra were mainly obtained by references [29]. With respect to FTIR spectra, we decided to display only the peak fitting results obtained in the region from 1800 to 1370 cm^{-1}, because the main differences in variation were observed in this region. The region between 1200 cm^{-1} and 1000 cm^{-1} was heavily dominated by strong bands, probably originated by the SO_3H group vibrations (Figure 1) [30].

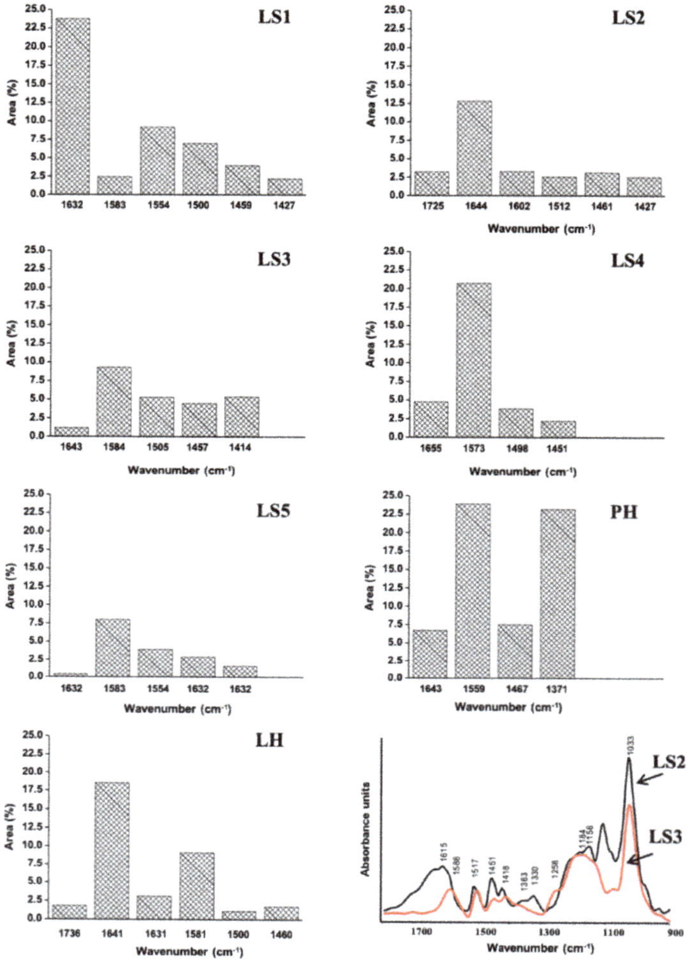

Figure 1. FTIR histograms of humates (Hs) peak areas processes by using curve fitting (from 1800 to 1370 cm^{-1}).

In LS2 and LH, due to C=O bonds of acetyl ester from residual hemicelluloses, a band between 1735 and 1725 cm^{-1} was evident. A very weak band at 1705 cm^{-1} was observed only in LS5. This band, associated with those at 1258 and 1418 cm^{-1}, may be attributed to the C=O stretching of COOH groups, while the other two bands may be due to C (=O)\O stretching vibration and OH in-plane deformation vibrations, respectively (spectra not shown). The appearance of carboxyl acid groups could be related to the removal of hemicellulose in this sample [31]. The bands at 1644 in LS2, LS3, LH, and PH, and at 1632 cm^{-1} in LS1 and LH, were likely associated with H_2O and C=O stretching in conjugated *p*-substituted aryl ketones [32]. In addition, the peak at 1655 cm^{-1} recorded in LS4 could be assigned to C=O in alkyl groups of the lignin side chains, conjugated with the aromatic rings [33]. These bands were completely absent in LS5. Other bands identified between 1600 and 1573 cm^{-1} corresponded to vibration of aromatic rings. The intensity of these bands depends on the number of C-O bonds to the aromatic ring [34]. Intermolecular aromatic C=C bonds may also have contributed to the intensity of these bands. The peaks from 1512 to 1498 cm^{-1} are typical of the skeletal and stretching vibration of aromatic moieties in lignin. Such peaks were present in all products. The bands at around

1460 and 1414 cm^{-1} were attributed to the bending vibration of the methoxyl on benzene rings and methylene groups, respectively. The peak at 1370 cm^{-1}, observed only in PH, may be due to aromatic CH generated by cleavage of ether bonds within the lignin (spectra not shown).

The relative area percentage gave an estimation of the functional group distribution in the Hs (Figure 1). The band at around 1640 cm^{-1} showed a variable distribution among products. For instance, it was dominant in LS1 (24%), LH (18%), LS2 (13%), and totally absent in LS5. The aromatic structure was diversified into different bands at around 1580, 1559, and 1500 cm^{-1}. The first band was dominant in LS4 (21%), LS3 (9.4%), and LH (9.0%), and absent in PH. In the other products, this band ranged from 8% to 2.4%. The second band at 1559 cm^{-1} accounted for 24% in PH and 9% in LS1. The last band at around 1500 cm^{-1} was prevalent in PH (7.5%), LS1 (7%), and LS3 (5%). In other lignosulfonates, it varied from 4% in LS4 and LS5, 2.5% in LS2, and 1% in LH. Finally, the band at 1371 cm^{-1} accounted for 23% in the commercial humate PH.

The Raman spectra of LS2 and LS5 are reported in Figure 2, while the complete attributions of the two lignosulfonates are shown in Table 2. Both spectra display bands at 3490 and 3250 cm^{-1}, attributable to OH stretching free or H-bonded, respectively, and both aliphatic (at 2940 and 2846 cm^{-1}) and aromatic (at 3070 cm^{-1}) CH stretching in the higher wavenumber region. Moreover, the shoulder at about 1670 cm^{-1} could be ascribed to conjugated C=O stretching [35], the bands at 1630, 1604, and about 1500 cm^{-1}, together with that one at 1190 cm^{-1}, were all attributable to phenolic rings, the last one specifically to lignin [35,36]. The peaks at 1460, 1370, and 1330 cm^{-1} corresponded to bending vibrations of O-CH_3, CH, and aliphatic OH in lignin and cellulose, respectively [35]. The peaks at 1284 and 1082 cm^{-1}, together with that recorded at 815 cm^{-1} indicated the presence of sulfated groups [37,38]. Other bands observed in the Raman spectra were less indicative to identify the functional groups present in LS2 and LS5. The relative intensity of the over reported bands is different in the two examined spectra. In particular, for LS5 the bands attributable to aromatic groups (at 3070, 1633, 1604, and 1190 cm^{-1}) displayed a higher intensity compared to LS2, indicating that the aromatic component was higher in LS5. On the contrary, the bands at 1330 and 898 cm^{-1}, both attributable to cellulose, were more intense in LS2, indicating a higher content of this component in LS2 compared to LS5.

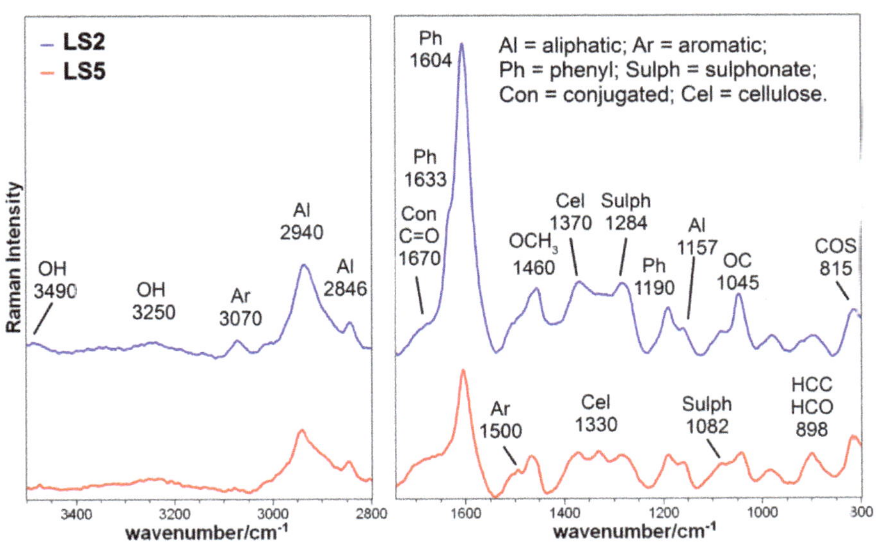

Figure 2. FT-RAMAN spectra of lignosulfonates LS2 and LS5.

Table 2. Main bands observed in the Raman spectra of humates LS2 and LS5. S = strong; m = medium; w = weak; v = very; sh = shoulder.

Attributions	LS2	LS5
OH stretching	3490 w	3490 vw
OH stretching	3250 w	3250 w
Aromatic CH stretching	3070 w	3070 vw
Aliphatic CH stretching	2940 m	2940 w
Aliphatic CH stretching	2846 m–w	2846 m–w
1670 conjugated C=O	1670 w. sh	1670 w. sh
Phenolic peak	1633 s. sh	
Aryl ring stretch, symmetric (lignin); Phenolic peak	1604 vs	1604 s
C_{ar}-H in plane bend, CO(H) str.	About 1500 vw. sh	About 1500 w. sh
CH_3 bending in OCH_3 (lignin and carbohydrates)	1460m–w	1460 vw
C-H bend in R_3C-H (cellulose)	1370 m	1370 m
Aliphatic O-H bend (cellulose)		1330 m
Sulfate group, asymmetric stretching	1284 m	1284 m
Phenol (lignin)	1190 m–w	1190 m–w
C-C skeletal mode OCH_3 loop rocking	1157 w. sh	1157 w. sh
Sulfate group, symmetric stretching	1082 w. sh	1082 m–w. sh
$OC(H_3)$ stretching and rocking	1045 m	1045 m
H-C-C and H-C-O bending at C6 (cellulose)	898 w	898 m
bending of primary C6-O-S	815 m	815 m

2.2. *Effect of Hs on Maize Plant Growth*

The effect of Hs application on maize plant growth is reported in Figure 3. Results indicated that LS5 was the most effective in promoting the leaf (Figure 3A) and root (Figure 3B) dry weight (+140% and +35%, respectively), compared to the untreated plants. The remaining Hs did not substantially improve the leaf biomass produced by plants. However, they all stimulated the root growth appreciably. Specifically, LS3, LS4, and LH increased the root biomass of plants by 51%, 57% and 52%, respectively, while LS2 and PH were by about 85%, and LS1 was by 111%.

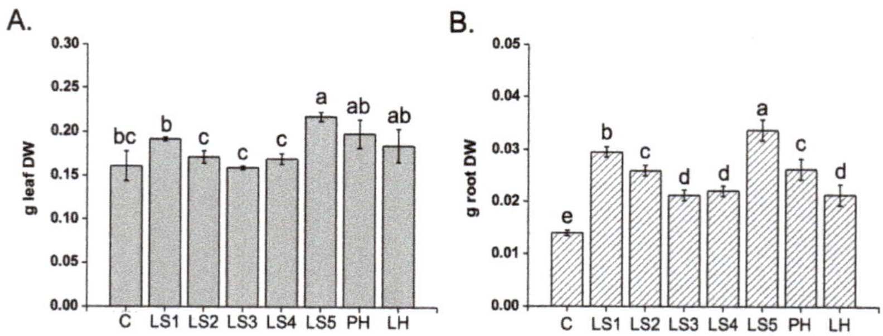

Figure 3. Effect of individual humates (Hs) on leaf (**A**) and root (**B**) dry weight of *Z. mays* L. plants. Twelve-day-old plants were supplied for two days with Hs at 1 mg C L^{-1}. Different letters above bars indicate significant differences at $p < 0.05$, according to the Student–Newman–Keuls test. Data represent the means of three measurements with ten plants in each (±SD). C = control; LH = commercial lignosulfonate-based product; PH = commercial humate extracted from leonardite; LS1 – LS5 = specialty lignosulfonates.

2.3. *Effects of Hs on SPAD, RuBisCO activity, and N-compounds (Proteins and Phenolics)*

The effect of Hs on maize plants was additionally evaluated in terms of photosynthetic efficiency by measuring the SPAD index (Figure 4A) and the activity of the RuBisCO enzyme (Figure 4B). In general, Hs prompted the increase of the SPAD index values of plants to a similar extent (Figure 4A). Analogously, RuBisCO activity was increased by all Hs, but differences in the percent stimulation caused by individual Hs were observed in this case (Figure 4B). LS2, in particular, was the most effective in enhancing the activity of this enzyme (by about 70%), followed by LS1, LS3, LS5, and PH (+30–50%). The other Hs stimulated the RuBisCO activity to a lower extent.

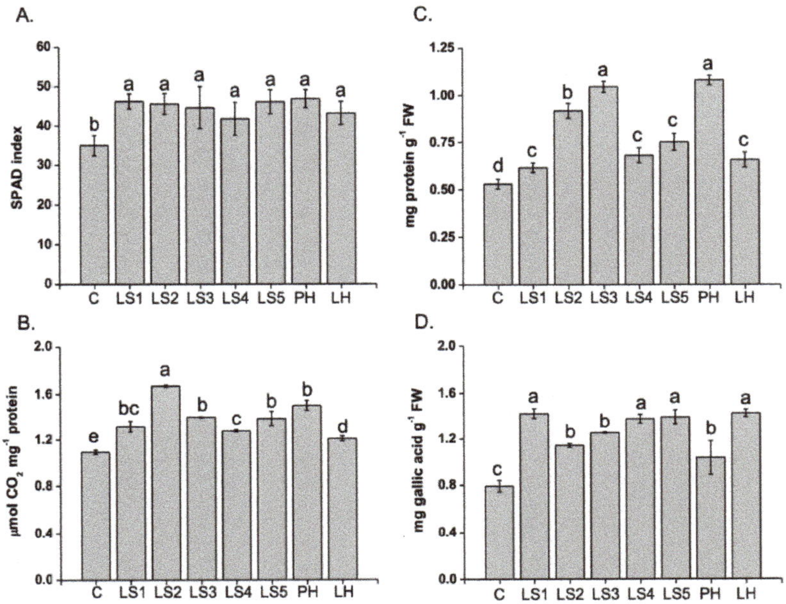

Figure 4. Effect of humates (Hs) on SPAD index (**A**), RuBisCO activity (**B**), protein content (**C**), and total phenolic compounds (**D**) in leaves of *Z. mays* L. plants. Twelve-day-old plants were supplied with Hs at 1 mg C L^{-1} for two days. Different letters above bars indicate significant differences at $p < 0.05$, according to Student–Newman–Keuls test. Data represent the means of three measurements with three plants in each (±SD). C = control; LH = commercial lignosulfonate-based product; PH = commercial humate extracted from leonardite; LS1-LS5= specialty lignosulfonates.

As the SPAD index is associated to the amount of N compounds in plants, the quantification of proteins, total phenols, and individual phenolic acids was performed. It is noteworthy that the content of total N was also measured in the plants (data not shown), but no significant differences were recorded, likely because of the limited duration of the experiment. Protein accumulation was enhanced in leaves of maize plants supplied with Hs (Figure 4C). LS2, LS3, and PH, in particular, induced the most pronounced increases (+74%, +98%, and +104%, respectively). The synthesis of phenol compounds (Figure 4D) was stimulated in leaves of maize plants treated with Hs as well. In this case, however, LS1, LH, LS4, and LS5 were responsible for the greatest increments (by about 80%).

Differential accumulation of individual phenolic acids was also observed between maize plants supplied with Hs and the controls, as well as among plants treated with distinct Hs (Table 3). There were three derivatives of cinnamic acids (caffeic, *p*-coumaric, and ferulic acids), one ester of caffeic acid and (−)-quinic acid (chlorogenic acid), and one derivative of benzoic acid (*p*-hydroxybenzoic acid). In most cases, Hs induced significantly higher accumulation of chlorogenic, caffeic, p-coumaric, ferulic, and p-hydroxybenzoic acids in leaves of maize plants compared to the controls. LS1, LS2, LS3, LS4, and LS5 especially, accounted for the most appreciable effects in this respect. Specifically, very high values of leaf phenolic acid accumulation were measured for chlorogenic and caffeic acids in plants treated with LS2 (+168% and 184%, respectively) and LS4 (+651% and 262%, respectively), for ferulic acid in plants provided with LS1 (+472%), LS2 (328%), LS3 (+222%), and LS4 (+413%), and for *p*-hydroxybenzoic acid in plants given with LS1 (+193%), LS2 (+187%), and LS4 (+202%).

Table 3. Profile of phenolic compounds in leaves and roots of *Z. mays* L. Plants were grown for 12 days in a nutrient solution and supplied with individual humates at 1 mg C L^{-1} for two days. n. d. = not detectable. Values along the same column following by different letters are statistically different at $p < 0.05$ ($n = 3$, ± SD) according to Student–Newman–Keuls test. C= control; LH = commercial lignosulfonate-based product; PH = commercial humate extracted from leonardite; LS1−LS5= specialty lignosulfonates.

	Chlorogenic		Caffeic		*p*-Cumaric		Ferulic		*p*-Hydroxybenzoic	
					Leaves (mg k^{-1} FW)					
C	30.29	± 4.11e	0.52	± 0.02d	1.28	± 0.05d	0.58	± 0.02c	16.32	± 0.13e
LS1	67.65	± 8.54b	1.41	± 0.02c	3.02	± 0.05a	3.32	± 0.03a	47.88	± 3.45a
LS2	81.13	± 12.37a	3.90	± 0.02a	1.33	± 0.03c	2.48	± 0.06a	46.88	± 5.33a
LS3	51.24	± 6.32c	1.34	± 0.05c	2.61	± 0.06b	1.87	± 0.07b	30.04	± 7.34c
LS4	85.98	± 7.10a	1.88	± 0.05b	2.41	± 0.07b	4.13	± 0.05a	49.24	± 6.13a
LS5	57.10	± 10.22c	1.47	± 0.01c	2.35	± 0.05b	1.24	± 0.05b	39.94	± 5.28b
PH	34.09	± 5.13d	0.72	± 0.02d	3.09	± 0.07a	1.22	± 0.05b	40.69	± 6.13b
LH	37.48	± 5.08d	0.92	± 0.03d	2.03	± 0.03c	0.72	± 0.03c	24.78	± 5.13d
					Roots (mg kg^{-1} FW)					
C	3.54	± 0.12c	n.d.		5.71	± 0.30b	0.58	± 0.01c	29.97	± 4.14a
LS1	3.56	± 0.11c	n.d.		1.04	± 0.62c	3.09	± 0.03a	3.65	± 0.84d
LS2	6.05	± 0.13a	n.d.		3.16	± 0.61b	2.86	± 0.05a	11.57	± 3.12c
LS3	3.44	± 0.12c	n.d.		6.29	± 0.82a	0.85	± 0.03c	21.07	± 2.34b
LS4	5.67	± 0.23b	n.d.		5.10	± 0.72b	1.12	± 0.03b	19.46	± 4.13b
LS5	7.61	± 0.14a	n.d.		4.89	± 0.53b	1.69	± 0.05b	24.39	± 6.81b
PH	7.54	± 0.17a	n.d.		6.46	± 0.52a	2.65	± 0.05a	31.12	± 5.68a
LH	3.48	± 0.18c	n.d.		6.15	± 0.51a	0.75	± 0.01c	20.29	± 3.12b

In roots, only chlorogenic and ferulic acids were more accumulated in plants treated with Hs than the controls. The highest values of chlorogenic acid content were observed in roots after plant treatment with LS2 (+71%), LS4 (+60%), LS5 (+115%), PH (+113%). With respect to ferulic acid, maximum accumulation was measured in roots of plants supplied with LS1 (+436%), LS2 (+396%), and PH (+361%).

2.4. Effects of Hs on GS and GOGAT Activities

Further effects of Hs on maize plant metabolism were investigated by measuring the activities of two enzymes (GS and GOGAT) that catalyze key steps in N assimilation (Figure 5). Overall, a greater activity of such enzymes was determined in plants supplied with Hs. The activity of GS in leaves in particular, was increased by LH (+44%), LS4 (+24%), and LS5 (+18%) (Figure 5A), while the activity of GOGAT was stimulated by all Hs applied to plants (Figure 5B). LS3 accounted for the maximum leaf activity of GOGAT (+98%). In roots, the activity of both GS and GOGAT enzymes was enhanced by all Hs (Figure 5C,D). In the case of GS, the highest activity was detected in roots of plants treated with LS2 (Figure 5C), while maximum GOGAT activity was measured in plants supplied with LS1 and LS5 (Figure 5D).

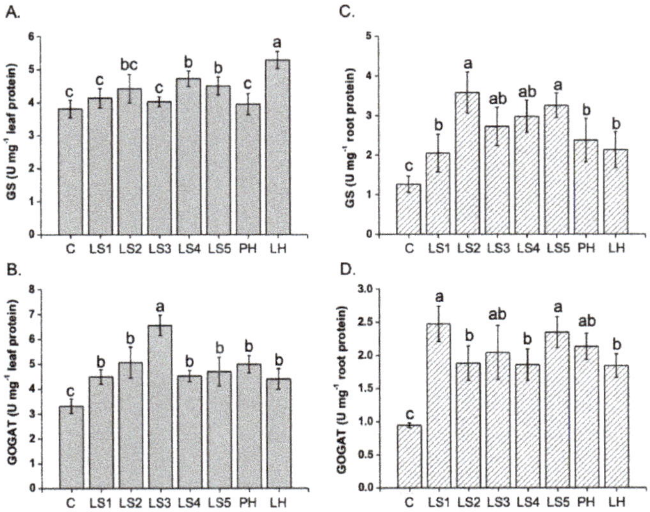

Figure 5. Effect of humates (Hs) on glutamine synthetase (GS) and glutamate synthase (GOGAT) activity in leaves (**A**, **B**, respectively) and roots (**C**, **D**, respectively) of *Z. mays* L. plants. Twelve-day-old plants were supplied with Hs at 1 mg C L^{-1} for two days. Different letters above bars indicate significant differences at $p < 0.05$, according to the Student–Newman–Keuls test. Data represent the means of three measurements with three plants in each (±SD). C = control; LH = commercial lignosulfonate-based product; PH = commercial humate extracted from leonardite; LS1 – LS5 = specialty lignosulfonates.

2.5. *Effects of Hs on Reducing Sugar Accumulation*

The content of soluble reducing sugars (glucose and fructose) was increased in leaves of plants treated with Hs (Figure 6). Precisely, improved glucose accumulation was observed in leaves of maize plants after treatment with LS1, LS2, and LH (+39%, +58%, +41%, respectively, Figure 6A). With respect to fructose, all Hs stimulate its accumulation, with maximum values determined by LS2 and LS3 (+92% and +111%, respectively, Figure 6A). In roots, an opposite trend was evident, as the content of both sugars decreased when plants were treated with Hs, with few exceptions (Figure 6B).

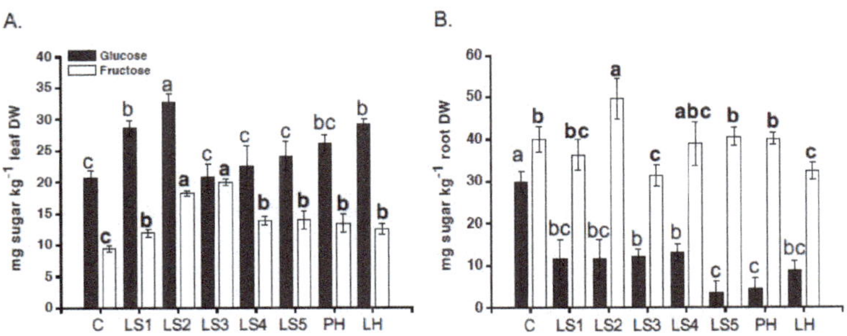

Figure 6. Effect of individual humates (Hs) on glucose and fructose accumulation in leaves (**A**) and roots (**B**) of *Z. mays* L. plants. Twelve-day-old plants were supplied with Hs at 1 mg C L^{-1} for two days. Different letters above bars (un-bolded for glucose and bolded for fructose) indicate significant differences at $p < 0.05$, according to the Student–Newman–Keuls test. Data represent the means of three measurements with three plants in each (±SD). C = control; LH = commercial lignosulfonate-based product; PH = commercial humate extracted from leonardite; LS1 – LS5 = specialty lignosulfonates.

2.6. Statistical Analysis of Data

The correlation analysis evidenced significant relationships between the parameters analyzed in maize plants subjected to treatment with Hs (Table S1). The root dry weight, which was more stimulated than the leaf dry weight by Hs, positively correlated with SPAD, total phenols, GS and GOGAT root activity, and RubisCO activity, whereas it negatively correlated with the content of glucose in roots. SPAD index values displayed a positive correlation with the content of N metabolites (proteins and phenols), the activity of GOGAT, GS (only in roots), RubisCO, and the leaf fructose content. However, SPAD negatively correlated with the root glucose content. Total phenols showed positive correlation with GS activity in leaves and roots and GOGAT activity in roots. The activity of GS in leaves did not show any correlation with the other parameters analyzed, but GS activity in roots positively correlated with the activity of GOGAT in leaves and roots. The activity of both N enzymes also positively correlated with RubisCO activity. The activity of all three enzymes, GS (in roots), GOGAT, and RubisCO, negatively correlated with glucose content in roots. RuBisCO positively correlated with the leaf glucose content and fructose content in both leaves and roots, whereas it negatively correlated with the root glucose.

With respect to PCA analysis, three factors accounted for 91% of the total variance. Factor 1 explained 53.6% of the variance and positively correlated with GS and GOGAT activity in roots, SPAD, total phenols, while it negatively correlated with glucose content in roots. Factor 2 explained 22.7% of the variance and was positively correlated with GOGAT activity in leaves, protein content, and leaf fructose amount. Factor 3 explained the remaining 14.8% of the variance and was correlated with the content of fructose in roots and GS activity in leaves. Plotting data reported in Table S2 according to PC1 and PC2 allowed three clusters to be identified (Figure S1A,C); a main group constituted by plants 1, 2, 3, 4, 5, 6, and 7, corresponding to LS1, LS2, LS4, LS5, PH, and LH, and the other two by control (untreated, 8) and LS3 (3). In particular, LS1, LS2, LS4, LS5, PH, and LH were characterized by high values of GS and GOGAT activity in roots, SPAD, and total phenols, whilst LS3 had high values of GOGAT activity in roots and protein. The control plants had higher values of glucose content in roots. Plotting PC1 and PC2 also revealed that, among plants treated with humates LS1, LS2, LS4, LS5, PH and LH, those treated with LH tended to be at the bottom of the cloud, and PH was at the top, along the axis 2. It should be also noted that plotting PC1 and PC3, LH (7) differed from the other treatments for high GOGAT activity in leaves.

3. Discussion

Humates can differ in composition depending on the source material and process type employed for their production. Therefore, they can show significant variation in biostimulant properties. In this study, we assayed seven humates (a commercial lignosulfonate-based product, a commercial humate extracted from leonardite, and five specialty lignosulfonates provided by Borregaard's company) by determining their elemental content and dissecting the major functional groups occurring in their formulation. Then, in order to determine the plant-growth promoting potential of Hs, we evaluated differences in their capacity to promote plant biomass production, N assimilation into organic compounds (chlorophylls, proteins and phenols), and photosynthesis.

We found that all products were able to stimulate plant growth and the metabolic responses typically triggered by biostimulants. Therefore, untreated plants were different from plants treated with tested Hs in terms of performance, as revealed by PCA analysis. However, LS1, LS2, LS3, and LS5 appeared to be the most effective in this respect, being able to induce the greatest increments (up to 184%) of most physiological parameters (dry weight, root GS activity, GOGAT activity, RuBisCO activity) and targeted-biochemical markers (SPAD, proteins, phenols, fructose content) in maize, compared to the untreated plants. A general overview of such increments is depicted in the heat map of plant-associated parameters influenced by individual humates, reported in Figure S2. LS2 and LS3 contained a similar percent content of total C and N, as well as LS1 and LS5. The spectroscopic characteristic of all samples and especially LS2 and LS5 revealed the presence of cellulose residues and aromatic groups. LS4 and PH contained the highest percentage in aromatic groups according to the deconvolution process of FT-IR spectra, while for LS1 the functional group distribution appeared to be a mixture of the same groups observed in LS4 and PH, but with a considerable hydrophilic feature (see the band at 1632 cm^{-1}). Therefore, the C and N composition and profile in functional groups of specialty products LS1, LS2, LS3, and LS5 could explain their better efficiency as biostimulants compared to the lignosulfonate LH.

Overall, root growth was more stimulated (+51–140%) than leaf growth by all Hs (+5–35%), with more pronounced effects observed in plants treated with LS1 and LS5. These results are in line with the current literature that reports early root growth as a typical response of plants treated with humic substances, while the stimulation of leaf growth is generally recognized as a delayed response [10,17,39]. One possible explanation of this effect is that humic substances can act on root development by influencing the hormonal balance within the plants and nitric oxide distribution, either directly or indirectly, and by modifying the nutrient uptake by plants and the activity of root membrane H^+-ATPase [8,39,40]. Early root development could also be ascribed to the biological properties of humic substances, whose hormone-like activity has been previously described [41,42], and that Hs tested in this study might possess as well. Ertani et al. [17], in particular, reported the auxin-like and gibberellin-like activity of two lignosulfonates, and the gibberellin-like activity of a leonardite humic acid. The hormone-like activity of humic substances and commercial humates are likely due to their content in auxin-like substances, as well as to the presence of phenol-C groups with biological activity [43,44].

Hs were also effective in promoting N metabolism. In particular, LS1, LS2, and LS5 determined the highest increases in the activity of N assimilation enzymes, i.e., glutamine synthetase (GS) and glutamate synthase (GOGAT), in roots. This finding could explain why plants treated with these products developed their roots more. In this respect, the root dry weight of maize plants positively correlated with GS and GOGAT root activity. In general, all Hs enhanced the activity of GS and GOGAT more in roots than in leaves, which may suggest that early root growth stimulation in maize by Hs was also a result of a more pronounced N metabolism enhancement and decreased N storage. Similar findings and hypothesis have been previously reported by Jannin et al. [45]. Higher activity of N enzymes in roots might be due to metabolic changes related to differences in the root/shoot nitrate balance occurring under LH treatment [39]. In leaves, GOGAT activity was significantly stimulated by all Hs, while GS activity was stimulated by only four of them. Such differences could be ascribed to distinct mechanisms of regulation of N enzymes induced by several factors, including N metabolites (e.g., ammonium, glutamine, and glutamate) that are known to exert feedback effects [46–48]. In this respect, those Hs determining the highest increases in leaf protein accumulation were responsible for the least increases in GS leaf activity. Interestingly, they also stimulated the accumulation of phenolic compounds as the other Hs, but to a less extent. This observation seems to suggest that when plants are treated with Hs, two preferential metabolic pathways can be mainly stimulated, i.e., the N primary metabolism that produces proteins and the secondary metabolism involved in the synthesis of phenolics. These two metabolic pathways have been previously identified as principal targets of humic substances and other biostimulants, including lignosulfonate-humates, in maize and other plant species [17,49]. With respect to phenolic compounds, the increase in content of a number of them, especially in leaves, to levels that were not injurious to plants, can be deemed as an important result because these phytochemicals have recognized health beneficial properties, are implied in the plant defense responses against stress conditions, and mediate plant relationships with ecological partners [50–53].

The positive effects of all Hs on plant metabolism was also confirmed by the increased activity of RuBisCO, i.e., the enzyme responsible for CO_2 fixation in the Calvin cycle. Indeed, measuring the RuBisCO activity allowed for knowing whether Hs stimulated the photosynthetic efficiency of plants, because higher activity generally corresponds to higher photosynthetic rates and productivity. The increased activity of RuBisCO in plants under treatment by humic substances could be due to increased number of chloroplasts per cell, as proposed by Jannin et al. [45]. RuBisCO activity positively correlated with the SPAD index values and the leaf content of reducing sugars. Similar results were previously reported by Ertani et al. [17].

In our study, we observed a reduction in glucose and fructose accumulation in roots of maize plants. Glucose is mainly produced in the cytosol from triose-phosphate precursors produced during the Calvin cycle and its accumulation in cells is influenced by different factors, like the photosynthetic rate, the need of glucose for energy-dependent processes, and the metabolic fate of the precursor glutaraldehyde 3-P (including the synthesis of starch). In roots, the level of carbohydrates depends on the source of N they receive (NO_3, NH_4, or amino acids), the rate of transport of photosynthates and the quantity of reserves that are stored in the root tissues. The different distribution of glucose between leaves and roots also depends on the need of the plant to use glucose in a specific organ for a metabolic requirement. The decrease of glucose in the roots, for instance, may be indicative of a high demand for ATP-dependent nutrient transport and other energy-requiring processes in the root cells, including growth processes, and could be associated with the increased need of C-skeleton for the synthesis of N compounds. A similar reasoning can be made for fructose.

4. Materials and Methods

4.1. Elemental Composition and Spectroscopic Analysis of Hs

Seven humates (Hs) were tested in this study for their biostimulant properties. All these products completely dissolved in H_2O without leaving insoluble clumps. The carbon (C), nitrogen (N), and sulfur (S) contents of Hs were determined via dry combustion conducted in the element analyzer vario MACRO CNS (Hanau, Germany).

The Fourier transform infrared (FTIR) spectra of these products were recorded using an ALPHA FTIR spectrometer (Bruker Optics, Ettlingen, Germany) equipped with an ATR (attenuated total reflectance) sampling device containing diamond crystals. The absorbance spectra were recorded between 4000 cm^{-1} and 400 cm^{-1}, at a spectral resolution of 4 cm^{-1}, with 64 scans co-added and averaged. A background spectrum of air was recorded under the same procedure conditions before each series of measurements. Spectra were processed with the Grams/386 spectroscopic software (version 6.00, Galactic Industries Corporation, Salem, NH). Overlapping peaks were resolved using a peak fitting analysis in the spectral region from 1800 to 1000 cm^{-1} by using the Grams/386 spectroscopic software (version 6.00, Galactic Industries Corporation, Salem, NH). The overlapping bands were resolved with a Gaussian function. The best fitting parameters were determined by minimization of the reduced Chi square (χ^2). Good agreement between experimental and calculated profiles was obtained, with coefficients of determination, R^2, ranging from 0.999 to 0.988 and the standard error, SE, from 0.001 to 0.003. All data are expressed as percentage area.

FT-Raman spectra of Hs were recorded in solid state with a Multiram FT-Raman spectrometer (Bruker Optics, Ettlingen, Germany) equipped with a cooled Ge-diode detector. The excitation source was a Nd-YAG laser (1.064 nm, about 30 mW laser power on the sample) in the backscattering (180°) configuration. The low laser power was due to the brown color of the samples, which burned out using a higher laser power. As a consequence of burning, it was possible to record only the spectra of LS2 and LS5.

4.2. Plant Material and Experimental Design

Seeds of *Zea mays* L. (P1921, Pioneer HI-BRED, Italia Sementi S.r.l.) were soaked in distilled water overnight and then surface-sterilized in 5% (v/v) sodium hypochlorite for 10 min while shaking. Seeds were germinated on filter paper wetted with distilled water for 60 h in the dark at 25 °C. Seedlings were then transferred into 3 L pots in the presence of a thoroughly aerated Hoagland solution, with a density of 24 plants per pot. The nutrient solution was renewed every 48 h and contained the following salts (μM): KH_2PO_4 (40), $Ca(NO_3)_2$ (200), KNO_3 (200), $MgSO_4$ (200), FeNaEDTA (10), H_3BO_3 (4.6),

$CuCl_2 \cdot 2H_2O$ (0.036), $MnCl_2 \cdot 4H_2O$ (0.9), $ZnCl_2$ (0.09), and $NaMoO \cdot 2H_2O$ (0.01). Plants were grown inside a chamber with 14 h of light per day, in air temperatures of 21 °C (night) and 27 °C (day), at a relative humidity of 70/85%, and with a photon flux density of 280 mol $m^{-2}s^{-1}$. After twelve days of growth in hydroponics, each Hs was added in a unique application to the nutrient solution at 1 mg C L^{-1} (for each treatment with single Hs, 3 pots were prepared). After 48 h from the addition of Hs, plants were harvested. The choice of this short incubation time was dictated by results obtained in several previous studies, where a period of 24–48 h was found to induce early molecular responses and morpho-physiological changes in both roots and leaves. Plants that were not added with Hs served as controls (3 pots, 24 plants per pot).

At the end of the treatment, plants were randomly harvested and then carefully washed and dried with blotting paper. A sub-sample of the plant material was immediately frozen with liquid nitrogen and kept at −80 °C, to be used for biochemical analyses. For dry weight measurement, 10 plants randomly harvested were used (ten per treatment from each pot). The samples were placed in a drying oven for 2 d at 70 °C and allowed to cool for 2 h inside a closed bell jar. The dry weight of individual roots and leaves was measured for each plant.

4.3. Determination of the SPAD Index

The relative chlorophyll content was determined using a non-destructive method that employed light transmission across a leaf, at two wavelengths, to quantify the greenness and thickness of leaves. The ratio of the transmission of the two wavelengths provides a chlorophyll content index that is also named the SPAD index. The analyses were performed using a SPAD (Soil Plant Analysis Development) chlorophyll meter (SPAD-502 model, Minolta Camera Co, Ltd., Osaka, Japan) and the SPAD index was measured on the last expanded leaf of maize plants. The determination was carried out on 5 measurements per leaf from 10 plants per each treatment.

4.4. Analysis of Soluble Proteins and Reducing Sugars

For protein extraction, frozen foliar tissues (100 mg) of five plants per pot were ground in liquid nitrogen and vortexed in the presence of 5 mL buffer (100 mM Tris-HCl pH 7.5, 1 mM Na_2EDTA, 5 mM DTT) and centrifuged at 14,000 g. The supernatants were mixed with 10% (w/v) trichloroacetic acid and then centrifuged. The pellets were finally re-suspended in 0.1 N NaOH. The protein concentration was determined using the Bradford method through a UV/VIS spectrophotometer (Lambda 1, Perkin-Elmer, Monza, Italy) at $\lambda = 595$ nm. Protein concentration was expressed as mg of protein g^{-1} fresh weight (FW).

For reducing sugar analysis, foliar tissues (100 mg) of five plants per pot were dried for 48 h at 80 °C, ground to obtain a fine powder, and then extracted with 2.5 mL 0.1 N H_2SO_4. Samples were incubated in a heating block for 40 min at 60 °C and then centrifuged at 6000 g for 10 min at 4 °C. Supernatants were filtrated (0.2 µm, Membra-Fil® Whatman Brand, Whatman, Milan, Italy) and further analyzed via HPLC (Perkin Elmer 410). Soluble sugars were separated using a Biorad Aminex 87 C column (300 × 7.8 mm) with H_2O as eluent at a flow rate of 0.6 mL min^{-1}. Sugar concentration was expressed as mg g^{-1} dry weight (DW).

4.5. Analysis of Total and Individual Phenolic Compounds

The content of total phenols in plant samples was quantified using the Folin–Ciocalteu method. For individual phenol detection, extraction from frozen plant material of five plants (1 plant = 1 biological replicate) was performed using water/methanol (1:1 v/v), filtered at 0.45 µm. Phenols were separated via an HPLC 2700 (Thermo Finnigan, San Jose, CA, USA) coupled with an 1806 UV/Vis (Thermo Finnigan, San Jose, CA, USA) detector. The column was a TM-LC 18 (Supelcosil) equipped with pre-column TM-LC 18 (Pelliguard, Supelco). Elution was conducted at a flow rate of 1.2 mL min^{-1} using a mixture of water/ n-butanol/ acetic acid (80.5:18:1.5 v/v) as the mobile phase. The injection volume of each sample was 20 µL. Detection was performed at $\lambda = 275$ nm and the identification of compounds was obtained by comparison of their retention time values with those of corresponding standards. The calibration curve and quantification were performed considering the relationship between peak areas vs. standard concentrations at four concentrations ($n = 4$). A linear fitting with an R squared value of (R^2) = 0.99 was obtained.

4.6. Determination of GS, GOGAT and RuBisCO Activity

For the assay of glutamine synthetase (GS) and glutamate synthase (GOGAT) enzyme activity, fresh root and leaf tissues (1 g) were ground in a mortar with 10 mL of 100 mM Hepes-NaOH solution at pH 7.5, 5 mM $MgCl_2$ solution, and 1 mM dithiothreitol. For the RuBisCO enzyme, the extraction protocol was the same as for GS and GOGAT, but the enzyme activity in this case was measured in leaves only and the ratio of plant material to buffer was 1:3 (w/v). The extracts were filtered through two layers of muslin and centrifuged at 20,000 g for 15 min at 4 °C. The supernatants were used for enzymatic assays.

For the glutamine synthetase (GS EC 6.3.1.2) assay, each mixture contained 90 mM imidazole-HCl (pH 7.0), 60 mM hydroxylamine (neutralized), 20 mM $KAsO_4$, 3 mM $MnCl_2$, 0.4 mM ADP, 120 mM glutamine, and enzyme extract. The assay was performed in a final volume of 750 µL. The enzymatic reaction was developed for 15 min at 37 °C. The α-glutamyl hydroxamate was colorimetrically determined by addition of 250 µL of a mixture (1:1:1) of 10% (w/v) $FeCl_3 \cdot 6H_2O$ in 0.2 M HCl, 24% (w/v) trichloroacetic acid and 50% (w/v) HCl. The optical density was measured at $\lambda = 540$ nm. Enzyme activity was expressed in μmol^{-1} g^{-1} FW, representing the amount of enzyme catalyzing the formation of 1 nmole γ-glutamyl-hydroxamate min^{-1}.

The glutamate synthase (GOGAT EC 1.4.7.1) assay contained 25 mM Hepes-NaOH (pH 7.5), 2 mM L glutamine, 1 mM α-ketoglutaric acid, 0.1 mM NADH, 1 mM Na_2EDTA, and 100 µL of enzyme extract. GOGAT activity was measured spectrophotometrically by monitoring NADH oxidation at $\lambda = 340$ nm. The enzyme activity was expressed in μmol^{-1} g^{-1} FW, representing the amount of enzyme catalyzing the oxidation of 1 nmole NADH min^{-1}.

The activity of RuBisCO (EC 4.1.1.39) was determined spectrophotometrically in a coupled assay by measuring the production of 3-phosphoglycerate following a 5 min period of incubation with 2 mL of 10 mM $MgCl_2$ and 20 mM $NaHCO_3$ [54].

For each enzyme activity assay, analyses were conducted in three biological replicates (1 plant = 1 biological replicate) per treatment and the absorbance in the samples was measured using a JASCO V-530 UV/VIS spectrophotometer.

4.7. Statistical Analysis

For all determinations, the analysis of variance (ANOVA) was performed using the SPSS software version 19.0 (SPSS Inc. 1999), which was followed by pair-wise post hoc analyses (Student–Newman–Keuls test) to determine which means differed significantly at $p < 0.05$ (±SD). The number of biological replicates varied depending on the analysis performed and is indicated in the figure and table legends. Correlations between variables were determined using Pearson's coefficient. To identify the structure of the interdependences between the main parameters, a joint principal component analysis (PCA) was performed on the following variables, considering both untreated plants (control) and plants treated with the different humates: Root dry weight, leaf dry weight, SPAD, proteins, total phenols, leaf GS, root GS, leaf GOGAT, root GOGAT, RuBisCO, leaf glucose, leaf fructose, root glucose, and root glucose. The standardized variables were subjected to PCA. Rotated orthogonal components (varimax method of rotation) were extracted and the relative scores were determined. Only PCs with an eigenvalue > 1 were considered for the discussion. Statistics were performed using SPSS software version 25.0 (SPSS, Chicago, IL).

5. Conclusions

In conclusion, the current study provides clear evidence that all tested products acted as biostimulants. Additionally, the specialty lignosulfonates provided by Borregaard's company were apparently the most effective in this role, likely because of the novel process employed for their production and the products' chemical features (e.g., different C content values and presence of functional groups). These results support the importance of setting up new technologies and advanced industrial processes for the production of novel commercial humates and lignosulfonates with better formulation performance, which can be used as efficient biostimulants during crop cultivation in the framework of sustainable agriculture. Future studies could be performed in field trials and using other crop species, including horticultural crops, to definitely confirm the positive characteristics of these products under varying and/or stress conditions.

Supplementary Materials: The following are available online at http://www.mdpi.com/2073-4395/9/8/445/s1, Table S1: Correlations between variables determined using Pearson's coefficient. Asterisks indicate significant correlation at $p < 0.05$ (*) or $p < 0.01$ (**). r = root, l = leaf, dw = dry weight, TP = total phenols, GS = glutamine synthetase, GOGAT = glutamate synthase, FRU = fructose, GLU = glucose, PROT = proteins, Table S2: Loadings values of the plant variables on the axes identified by principal components (PC) analysis for the different types of treatment and control. r = root, l = leaf, dw = dry weight, TP = total phenols, GS = glutamine synthetase, GOGAT = glutamate synthase, FRU = fructose, GLU = glucose, PROT = proteins. Figure S1: Position of the treated and untreated plants (1 = LS1, 2 = LS2, 3 = LS3, 4 = LS4, 5 = LS5, 6 = PH, 7 = LH, and 8 = control) in the reduced space of the first two principal components (PC1 and PC2) (A) and on PC1 and PC3 (B); variables projected in the plane determined by PC1 and PC2 (D) and PC1 and PC3 (C). r = root, l = leaf, dw = dry weight, TP = total phenols, GS = glutamine synthetase, GOGAT = glutamate synthase, FRU = fructose, GLU = glucose, PROT = proteins, Figure S2: Heat map of plant-associated parameters influenced by individual humates. Different colors indicate different levels of induction/repression (more red more repression, more blue more induction). r = root, l = leaf, dw = dry weight, TP = total phenols, GS = glutamine synthetase, GOGAT = glutamate synthase, FRU = fructose, GLU = glucose, PROT = protein.

Author Contributions: O.F. and A.T. performed the spectroscopic analysis and wrote the relative part in the ms; A.E. performed the physiological analyses, bioassays and chemical analyses of the products and wrote the ms; M.S. wrote the ms; D.P. critically read the ms and performed the statistical analyses of data; S.N. designed the study. All the authors critically reviewed the ms.

Funding: This study was funded by Borregaard (Hjalmar Wessels vei, Sarpsborg, Norway).

Acknowledgments: We would like to thank Federica Zanellato for the precious help in conducting the physiological analyses.

Conflicts of Interest: The authors declare no conflict of interest.

References

1. Karamesouti, M.; Detsis, V.; Kounalaki, A.; Vasiliou, P.; Salvati, L.; Kosmas, C. Land-use and land degradation processes affecting soil resources: Evidence from a traditional Mediterranean cropland. *Catena* **2015**, *132*, 45–55. [CrossRef]
2. Shaviv, A.; Mikkelsen, R.L. Controlled-release fertilizers to increase efficiency of nutrient use and minimize environmental degradation-A review. *Fertil. Res.* **1993**, *35*, 1–12. [CrossRef]
3. Trivedi, P.; Delgado-Baquerizo, M.; Anderson, I.C.; Singh, B.K. Response of soil properties and microbial communities to agriculture: Implications for primary productivity and soil health indicators. *Front. Plant Sci.* **2016**, *7*, 990. [CrossRef] [PubMed]
4. Canellas, L.P.; Olivares, F.L.; Aguiar, N.O.; Jones, D.; Nebbioso, A.; Mazzei, P.; Piccolo, A. Humic and fulvic acids as biostimulants in horticulture. *Sci. Hortic.* **2015**, *196*, 15–27. [CrossRef]
5. Yakhin, O.I.; Lubyanov, A.A.; Yakhin, I.A.; Brown, P.H. Biostimulants in Plant Science: A Global Perspective. *Front. Plant Sci.* **2017**, *7*, 2049. [CrossRef] [PubMed]
6. Colla, G.; Nardi, S.; Cardarelli, M.; Ertani, A.; Lucini, L.; Canaguier, R.; Rouphael, Y. Protein hydrolysates as biostimulants in horticulture. *Sci. Hortic.* **2015**, *196*, 28–38. [CrossRef]
7. Nardi, S.; Pizzeghello, D.; Schiavon, M.; Ertani, A. Plant biostimulants: Physiological responses induced by protein hydrolyzed-based products and humic substances in plant metabolism. *Sci. Agric.* **2016**, *73*, 18–23. [CrossRef]
8. Du Jardin, P. Plant biostimulants: Definition, concept, main categories and regulation. *Sci. Hortic.* **2015**, *196*, 3–14. [CrossRef]
9. Calvo, P.; Nelson, L.; Kloepper, J.W. Agricultural uses of plant biostimulants. *Plant Soil* **2014**, *383*, 3–41. [CrossRef]

10. Nardi, S.; Carletti, P.; Pizzeghello, D.; Muscolo, A. Biological activities of humic substances. In *Biophysico-Chemical Processes Involving Natural Nonliving Organic Matter in Environmental Systems; Part, I. Fundamentals and Impact of Mineral-Organic-Biota Interactions on the Formation, Transformation, Turnover, and Storage of Natural Nonliving Organic Matter (NOM)*; Senesi, N., Xing, B., Huang, P.M., Eds.; John Wiley & Sons: Hoboken, NJ, USA, 2009; Volume 2, pp. 305–339.
11. Nardi, S.; Ertani, A.; Francioso, O. Soil–root cross-talking: The role of humic substances. *J. Plant Nutr. Soil Sci.* **2017**, *180*, 5–13. [CrossRef]
12. Zandonadi, D.B.; Santos, M.P.; Dobbss, L.B.; Olivares, F.L.; Canellas, L.P.; Binzel, M.L.; Okorokova, A.L.; Façanha, A.R. Nitric oxide mediates humic acids-induced root development and plasma membrane H^+-ATPase activation. *Planta* **2010**, *231*, 1025. [CrossRef]
13. Montoneri, E.; Adani, F.; Genevini, P.L.; Ricca, G.; Cherubini, S.; Spitaleri, C. Polyalkylphenyl-sulphonic acids with acid groups of variable strength from animal-vegetable wastes. *Waste Manag.* **2004**, *24*, 513–522. [CrossRef]
14. Gargulak, J.D.; Lebo, S.E. Commercial use of lignin-based materials. In *Lignin: Historical, Biological and Haterials Perspectives*; Glass, W.G., Northey, R.A., Schultz, T.P., Eds.; ACS Symposium Series 742; American Chemical Society: Washington, DC, USA, 2000; Volume 15, pp. 2–559.
15. Lutnaes, B.F.; Myrvold, B.O.; Lauten, R.A.; Endeshaw, M.M. 1H and 13C NMR data of benzylsulfonic acids model compounds for lignosulfonate. *Magn. Reson. Chem.* **2008**, *46*, 299–305. [CrossRef]
16. McGovern, J.N. *Pulp and Paper: Chemistry and Chemical Technology*, 3rd ed.; Casey, J.P., Ed.; Wiley: New York, NY, USA, 1980; Volume 4.
17. Ertani, A.; Francioso, O.; Tugnoli, V.; Righi, V.; Nardi, S. Effect of commercial lignosulfonate-humate on *Zea mays* L. metabolism. *J. Agric. Food Chem.* **2011**, *59*, 11940–11948. [CrossRef]
18. Matsushita, Y. Conversion of technical lignins to functional materials with retained polymeric properties. *J. Wood Sci.* **2015**, *61*, 230–250. [CrossRef]
19. Cieschi, M.T.; Lucena, J.J. Iron and Humic Acid Accumulation on Soybean Roots Fertilized with Leonardite Iron Humates under Calcareous Conditions. *J. Agric. Food Chem.* **2018**, *66*, 13386–13396. [CrossRef]
20. Kovács, K.; Czech, V.; Fodor, F.; Solti, A.; Lucena, J.J.; Santos-Rosell, S.; Hernández-Apaolaza, L. Characterization of Fe-Leonardite complexes as novel natural iron fertilizers. *J. Agric. Food Chem.* **2013**, *61*, 12200–12210. [CrossRef]
21. Aguirre, E.; Leménager, D.; Bacaicoa, E.; Fuentes, M.; Baigiorri, R.; Zamarreño, A.M.; García-Mina, J.M. The root application of a purified leonardite humic acid modifies the transcriptional regulation of the main physiological root responses to Fe deficiency in Fe-sufficient cucumber plants. *Plant Physiol. Biochem.* **2009**, *47*, 215–223. [CrossRef]
22. Kulikova, N.A.; Stepanova, E.V.; Koroleva, O.V. Mitigating activity of humic substances: Direct influence on biota. In *Use of Humic Substances to Remediate Polluted Environments: From Theory to Practice*; Springer: Dordrecht, The Netherlands, 2005; pp. 285–309.
23. Billingham, K. Humic products potential or presumption for agriculture? Can humic products improve my soil? In Proceedings of the 27th Annual Conference of the Grassland Society of NSW Inc., Orange, NSW, Australia, 24–26 July 2012; Harris, C., Lodge, G., Waters, C., Eds.; Grassland Society of NSW Inc.: Wagga, Australia, 2012; pp. 43–50.
24. Mikkelsen, R.L. Humic materials for agriculture. *Better Crop* **2005**, *89*, 6–10.
25. Lobartini, J.C.; Tan, K.H.; Rema, J.A.; Gingle, A.R.; Pape, C.; Himmelsbach, D.S. The geochemical nature and agricultural importance of commercial humic matter. *Sci. Total Environ.* **1992**, *113*, 1–15. [CrossRef]
26. Gaur, A.C.; Bhardwaj, K.K.R. Influence of sodium humate on the crop plants inoculated with bacteria of agricultural importance. *Plant Soil* **1971**, *35*, 613–621. [CrossRef]
27. Mora, V.; Bacaicoa, E.; Baigorri, R.; Zamarreno, A.M.; Garcia-Mina, J.M. NO and IAA key regulators in the shoot growth promoting action of humic acid in *Cucumis sativus* L. *J. Plant Growth Regul.* **2014**, *33*, 430–439. [CrossRef]
28. Iakimenko, O.S. Commercial humates from coal and their influence on soil properties and initial plant development. In *Use of Humic Substances to Remediate Polluted Environments: From Theory to Practice*; Springer: Dordrecht, The Netherlands, 2005; pp. 365–378.
29. Schulz, H.; Baranska, M. Identification and quantification of valuable plant substances by IR and Raman spectroscopy. *Vib. Spectrosc.* **2007**, *43*, 13–25. [CrossRef]

30. Suganuma, S.; Nakajima, K.; Kitano, M.; Yamaguchi, D.; Kato, H.; Hayashiet, S. Hydrolysis of cellulose by amorphous carbon bearing SO3H, COOH, and OH groups. *J. Am. Chem. Soc.* **2008**, *130*, 12787–12793. [CrossRef]
31. Semwal, S.; Gaur, R.; Mukherjee, S.; Chopra, A.; Gupta, R.P.; Kumar, R.; Tuli, D.K. Structural features of dilute acid pretreated *Acacia mangium* and impact of sodium sulfite supplementation on enzymatic hydrolysis. *Sustain. Chem. Eng.* **2016**, *4*, 4635–4644. [CrossRef]
32. Nadji, H.; Diouf, P.N.; Benaboura, A.; Bedard, Y.; Riedl, B.; Stevanovic, T. Comparative study of lignins isolated from Alfa grass (*Stipa tenacissima* L.). *Bioresour. Technol.* **2009**, *100*, 3585–3592. [CrossRef]
33. Hsu, T.C.; Guo, G.L.; Chen, W.H.; Hwang, W.S. Effect of dilute acid pretreatment of rice straw on structural properties and enzymatic hydrolysis. *Bioresour. Technol.* **2010**, *101*, 4907–4913. [CrossRef]
34. Hergert, H.L. *Infrared spectra, In Lignins: Occurrence, Formation, Structure and Reactions*; Sarkanen, K.V., Ludwig, C.H., Eds.; John Wiley & Sons, Inc.: New York, NY, USA, 1971; pp. 267–296.
35. Ozgenc, O.; Durmaz, S.; Hakki Boyaci, I.; Eksi-Kocak, H. Determination of chemical changes in heat-treated wood using ATR-FTIR and FT Raman spectrometry. *Spectrochim. Acta Part A Mol. Biomol. Spectrosc.* **2017**, *171*, 395–400. [CrossRef]
36. Séné, C.; McCann, M.C.; Wilson, R.H.; Crinter, R. Fourier-Transform Raman and Fourier-Transform lnfrared Spectroscopy. *Plant Physiol.* **1994**, *106*, 1623–1631.
37. Synytsya, A.; Bleha, R.; Synytsya, A.; Pohl, R.; Hayashi, K.; Yoshinaga, K.; Nakano, T.; Hayashi, T. Mekabu fucoidan: Structural complexity and defensive effects against avian influenza A viruses. *Carbohydr. Polym.* **2014**, *111*, 633–644. [CrossRef]
38. Marinval, N.; Saboural, P.; Haddad, O.; Maire, M.; Bassand, K.; Geinguenaud, F.; Djaker, N.; Ben Akrout, K.; Lamy de la Chapelle, M.; Robert, R.; et al. Identification of a pro-angiogenic potential and cellular uptake mechanism of a LMW highly sulfated fraction of fucoidan from *Ascophyllum nodosum*. *Mar. Drugs* **2016**, *14*, 185. [CrossRef]
39. Mora, V.; Baigorri, R.; Bacaicoa, E.; Zamarrenõ, A.M.; García-Mina, J.M. The humic acid induced changes in the root concentration of nitric oxide, IAA and ethylene do not explain the changes in root architecture caused by humic acid in cucumber. *Environ. Exp. Bot.* **2012**, *76*, 24–32. [CrossRef]
40. Mora, V.; Bacaicoa, E.; Zamarrenõ, A.M.; Aguirre, E.; Garnica, M.; Fuentes, M.; García-Mina, J.M. Action of humic acid on promotion of cucumber shoot growth involves nitrate-related changes associated with the root-to-shoot distribution of cytokinins, polyamines and mineral nutrients. *J. Plant Physiol.* **2010**, *167*, 633–642. [CrossRef]
41. Pizzeghello, D.; Nicolini, G.; Nardi, S. Hormone-like activity of humic substances in *Fagus sylvaticae* L. forests. *New Phytol.* **2001**, *151*, 647–657. [CrossRef]
42. Muscolo, A.; Sidari, M.; Francioso, O.; Tugnoli, V.; Nardi, S. The auxin-like activity of humic substances is related to membrane interactions in carrot cell cultures. *J. Chem. Ecol.* **2007**, *33*, 115–129. [CrossRef]
43. Pizzeghello, D.; Zanella, A.; Carletti, P.; Nardi, S. Chemical and biochemical characterization of dissolved organic matter from Silver fir and beech forest soils. *Chemosphere* **2006**, *65*, 190–200. [CrossRef]
44. Muscolo, A.; Sidari, M.; Attinà, E.; Francioso, O.; Tugnoli, V.; Nardi, S. Biological activity of humic substances is related to their chemical structure. *Soil Sci. Soc. Am. J.* **2007**, *71*, 75–85. [CrossRef]
45. Jannin, L.; Arkoun, M.; Ourry, A.; Laîné, P.; Goux, D.; Garnica, M.; Fuentes, M.; San Francisco, S.; Baigorri, R.; Cruz, F.; et al. Microarray analysis of humic acid effects on *Brassica napus* growth: Involvement of N, C and S metabolisms. *Plant Soil* **2012**, *359*, 297–319. [CrossRef]
46. Vance, C.P.; Gantt, J.S. Control of nitrogen and carbon metabolism in root nodules. *Physiol. Plant.* **1992**, *85*, 266–274. [CrossRef]
47. Forde, B.G. Nitrate transporters in plants: Structure, function and regulation. *BBA-Biomembr.* **2000**, *1465*, 219–235. [CrossRef]
48. Wang, M.; Shen, Q.; Xu, G.; Guo, S. New insight into the strategy for nitrogen metabolism in plant cells. *Int. Rev. Cell Mol. Biol.* **2014**, *310*, 1–37.
49. Ertani, A.; Schiavon, M.; Nardi, S. Transcriptome-wide identification of differentially expressed genes in *Solanum lycopersicon* L. in response to an alfalfa-protein hydrolysate using microarrays. *Front. Plant Sci.* **2017**, *8*, 1159. [CrossRef]

50. Lattanzio, V.; Lattanzio, V.M.T.; Cardinali, A. Role of phenolics in the resistance mechanisms of plants against fungal pathogens and insects. In *Phytochemistry Advances in Research*; Imperato, F., Ed.; Research Signpost: Kerala, India, 2006; pp. 23–67.
51. Vermerris, W.; Nicholson, R. Phenolic compound biochemistry. In *The Role of Phenols in Plant Defense*; Springer: Dordrecht, The Netherlands, 2006; pp. 211–234.
52. Chen, F.; Tholl, D.; Bohlmann, J.; Pichersky, E. The family of terpene synthases in plants: A mid-size family of genes for specialized metabolism that is highly diversified throughout the kingdom. *Plant J.* **2011**, *66*, 212–229. [CrossRef]
53. Zhang, F.P.; Yang, Q.Y.; Zhang, S.B. Dual effect of phenolic nectar on three floral visitors of *Elsholtzia rugulosa* (Lamiaceae) in SW China. *PLoS ONE* **2016**, *11*, e0154381. [CrossRef]
54. Di Marco, G.; D'Ambrosio, N.; Giardi, M.T.; Massacci, A.; Tricoli, D. Photosynthetic properties of leaves of a yellow green mutant of wheat compared to its wild type. *Photosynth. Res.* **1989**, *21*, 117–122.

agronomy

MDPI

Article

Biostimulant Application Enhances Fruit Setting in Eggplant—An Insight into the Biology of Flowering

Alicja Pohl, Aneta Grabowska, Andrzej Kalisz and Agnieszka Sękara *

Department of Vegetable and Medicinal Plants, University of Agriculture in Krakow, 29-Listopada 54, 31-425 Kraków, Poland

* Correspondence: agnieszka.sekara@urk.edu.pl; Tel.: +48-12-662-52-16

Received: 15 July 2019; Accepted: 23 August 2019; Published: 26 August 2019

Abstract: Eggplant (*Solanum melongena* L.) is a warm climate crop. Its cultivation extends to temperate regions where low temperatures can affect the course of the generative phase, which is primarily sensitive to abiotic stress. The novelty of the present investigation consisted of characterising the heterostyly, pollination, and fertilisation biology of eggplants in field cultivations, which provided a basis for explaining the effect of a protective biostimulant on these processes. We aimed to investigate the flowering biology of three eggplant hybrids treated with Göemar BM-86®, containing *Ascophylum nodosum* extract, to determine the crucial mechanisms behind the increased flowering and fruit set efficiency and the final effect of increased yield. The flower phenotype (long, medium or short styled), fruit setting, and the number of seeds per fruit were recorded during the two vegetation periods. The numbers of pollen tubes and fertilised ovules in ovaries were evaluated during the generative stage of development to characterise the course of pollination and fertilisation for all types of flowers depending on the cultivar and biostimulant treatment. The expression of heterostyly depended on the eggplant genotype, age of the plant, fruit load, and biostimulant treatment. Domination by long-styled flowers was observed, amounting to 41%, 42%, and 55% of all flowers of "Epic" F_1, "Flavine" F_1, and "Gascona" F_1, respectively. This flower phenotype contained the highest number of pollen tubes in the style and the highest number of fertilised ovules. The biostimulant had a positive effect on the flower and fruit set numbers, as well as on the pollination efficiency in all genotypes. *Ascophylum nodosum* extract could be used as an efficient stimulator of flowering and fruit setting for eggplant hybrids in field conditions in a temperate climatic zone.

Keywords: *Ascophyllum nodosum*; *Solanum melongena*; heterostyly; pollination efficiency

1. Introduction

Eggplant is a photoperiodically inert plant with bisexual and partially self-pollinating flowers, although cross-pollination increases the effectiveness of fruit setting [1].

The downward-facing flowers are born solitary or in clusters. The eggplant produces three types of flowers: With a long-style pistil, where the stigma is localised above the anthers; with a medium-style pistil, where the stigma is at the same level as the anthers; and with a short-style pistil, where the stigma is below the anthers (Figure 1). This flower character promotes outcrossing between morphs via delivery and uptake of pollen by pollinators [2,3]. The stamen pores of the long- and medium-styled pistils are localised above or close to the stigma, favouring self-pollination. On the contrary, the stigmas of short-styled pistils are inside the downward-facing anther cone, making self-pollination difficult [4–6].

a b c

Figure 1. Stylar heteromorphism in eggplant: The flower with long-styled pistil of "Flavine" F_1 (**a**), medium-styled pistil of "Gascona" F_1 (**b**), short-styled pistil of "Epic" F_1 (**c**).

Anthers are ready to release pollen and the stigma is receptive from the first opening of the flower (Figure 2). Stigma receptiveness gradually decreases with the plant's age, and by the fifth day of flowering, receptiveness is negligible, and the stigma turns brown [2].

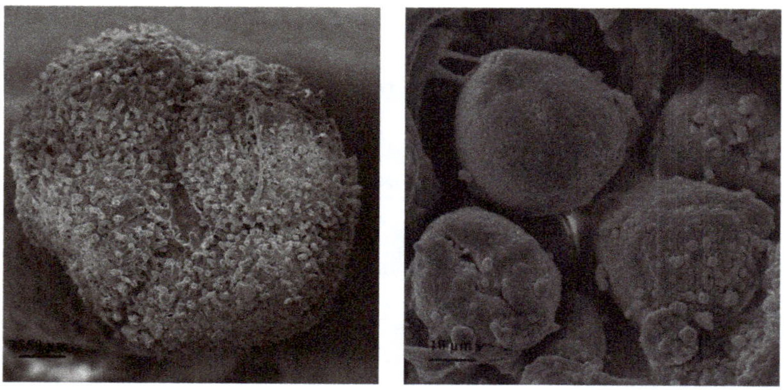

a b

Figure 2. Stigma in eggplant pistil with visible papillae in a receptive phase, (**a**), and pollen grains (**b**) in "Epic" F_1 by scanning electron microscopy.

All types of flowers are found in the same plant and even within the same cluster. The expression of heterostyly depends on the plant genotype, the age of the plant, fruit load, environmental conditions, and growing practices. Generally, domination by long-styled flowers has been reported, amounting to 50–100% of all flowers [7–9]. The higher fruit setting efficiency of this phenotype results from well-developed nodules with high pollen absorption capacity. However, the development of the ovules and their position in the placenta, as well as pollen grain shape, size, and amount in anthers, were nondifferentiated among long-, medium-, and short-styled pistils [6,10], although Wang et al. [11] demonstrated that lower fruit setting from short-styled flowers resulted from stigma-pollen incompatibility. The bumblebee (*Bombus terrestris*) is the most effective eggplant pollinator for plants under covers. Yield increase and better fruit quality are considered to be the major benefits of bumblebee application as compared to self-pollination or inflorescence vibrating [8,12]. Optimisation of eggplant yield in unfavourable conditions could also be achieved by introducing the cultivation of parthenocarpic cultivars [13]. Pollination leading to fruit and seed formation is associated with the production of endogenous growth regulators such as auxins. In this respect, the use of fruit-setting using auxin-based growth regulators has also been recommended to enhance fruit setting under suboptimal temperatures [5,14]. Investigations on the control of eggplant flowering through growth regulators have been successively performed since the end of the 20th century, but their results have been inconclusive [15,16]. Eggplant tolerance to biotic and abiotic stresses can be managed through grafting. The effects of rootstock/scion combinations on eggplant performance were investigated in terms of yield and fruit quality [17,18]. It can be assumed that this technique affects the flowering biology as well, but this issue needs future investigation. In Poland, eggplants are cultivated mainly under unheated foil covers from spring to autumn. To lower costs, cultivation is also performed in open fields where air temperatures may fall below the optimum, causing a reduction in flowering and fruit setting [19]. A promising way to control eggplant generative development could be biostimulant application. Biostimulants have been a focus of global interest of the scientific community since the end of the 20th century, giving promising results in different branches of agriculture as stimulators of crops growth, stress tolerance and yield [20,21]. Seaweed extracts (SWE) are among the main biostimulants, recognised as nontoxic, nonpolluting and nonhazardous to various organisms [22,23]. The majority of the SWE formulations are based on the extract of the brown algae *Ascophyllum nodosum* (L.) Le Jolis. Although seaweed extracts are heterogeneous in nature, the leading companies standardise their chemical composition to ensure consistent product quality [24,25]. Some authors have reported the stimulatory effect of seaweed extracts on eggplant yield [26,27], but there are no references on the flowering biology of this species as affected by SWE biostimulation. SWE action is extremely complex, but interdisciplinary investigation of biostimulant vs plant interactions may shed new light on the effective utilisation of these promising bioproducts in horticulture.

We hypothesise that seaweed extract affects the flowering and fruit setting of eggplant in a multidirectional manner. The reaction of plants to biostimulant treatment depends on the flowering biology of the cultivars, particularly the proportions of different flower phenotypes and their fertility. We aimed to investigate the flowering biology of three eggplant hybrids treated with seaweed extract Göemar BM-86® (Arysta LifeScience North America, LLC) to determine the crucial mechanisms behind the final effect of increased yield.

2. Materials and Methods

2.1. Experimental Arrangement

A two-factorial experiment was set up using randomised blocks in three replications, in the years 2013 and 2015, at the University of Agriculture in Krakow, Poland. The investigated eggplant hybrids, "Epic" F_1 (Seminis Vegetable Seeds), "Flavine" F_1 (Gautier Semences), "Gascona" F_1 (Gautier Semences), were selected on the basis of preliminary studies evaluating their performance in field cultivation under temperate climate conditions [26–28], determined by the earliness, vigour, and yield potential of those plants. Biostimulant Göemar BM-86® (Arysta LifeScience North America, LLC) was applied three times in two week intervals as a foliar application, in a dose of 1.5 $dm^3\ ha^{-1}$. Control plants were sprayed with distilled water. Goemar BM 86® is standardised *Ascophyllum nodosum* (L.) Le Jolis extract, which provides a constant and balanced formulation containing (in %): N, 5.0; Mg, 2.4, S, 3.2, B, 2.07; and Mo, 0.02 [29].

2.2. Cultivation Procedures

Eggplant seeds were sown on 1 March 2013 and 3 March 2015 in seed boxes. After three weeks, the seedlings with one fully developed leaf were transplanted into black 40-cell multipots (VEFI, Norway) with a single cell volume of 0.23 dm^3. Seedlings were grown in a greenhouse, in temperatures of 20/17 ± 2 °C day/night. The growing medium was peat substrate KlasmanTS2 (Klasmann-Deilmann GmbH, Geeste, Germany). The foliar fertiliser Kristalon Green (Yara, Szczecin, Poland) was applied twice in a dose of 10 g dm^{-3} water during seedling production. A gradual decrease in temperature and irrigation was used for the hardening of seedlings seven days before being transplanted to the experimental field (50°04′ N, 19°51′ E) on 7 May 2013 and 15 May 2015, with spacing of 0.75 × 0.60 m. Experimental plots covered 15 plants per treatment for observations of flowering and fruit setting and an additional 15 plants per treatment for flower collection for microscopic observations. Plots were surrounded by shelterbelts. The soil of the experimental field was Fluvic Cambisol (Humic) according to the FAO (Food and Agriculture Organization of the United Nations) classification with a C_{org} level of 2% and pH_{KCl} 6.11. Before the field experiment was established, the soil samples were analysed, and doses of fertilisers were applied to achieve a stable content of nutrients (in mg dm^3): N, 100; P, 90; K, 220; Ca, 1,100; Mg, 70. Cultivation procedures of weeding, irrigation, and plant protection were performed according to the standard recommendations for eggplant cultivated in field conditions in Poland, described by Sękara [30].

2.3. Weather Conditions

The climate of the experimental station is humid continental (Dfb) according to the Köppen's classification. Detailed data concerning the mean air temperature, photosynthetically active radiation (PAR), and the total rainfall during the vegetation seasons in 2013 and 2015 are presented in Table 1. Data were collected from automatic HOBO Pro RH/Temp loggers to assess temperature and a HOBO Weather Station (Onset Comp. Corp., Cape Cod, USA) to assess light characteristics and rainfall at the experimental site. The growing season in 2015 was generally warmer than that in 2013 regarding mean monthly temperatures, with the exception of June. In 2015, precipitation was distributed evenly, while in 2013, 45% of rainfall was recorded in June. A cool September in both years and low PAR caused a continuous decline in eggplant yield (Table 1).

Table 1. Mean monthly temperature, photosynthetically active radiation (PAR) and sum of rainfall in vegetation seasons 2013 and 2015.

Month	2013			2015		
	Temperature (°C)	PAR (μmol m^{-2} s^{-1})	Sum of Rainfall (mm)	Temperature (°C)	PAR (μmol m^{-2} s^{-1})	Sum of Rainfall (mm)
May	14.3	345	83	13.1	357	93
June	17.6	392	188	17.5	401	40
July	19.4	477	28	20.4	489	39
August	18.8	396	51	21.2	357	58
September	12.1	256	63	14.9	283	60

2.4. Procedures for Flowering and Fruit Setting Observations

The observations were conducted on 5 plants per replication (15 plants per treatment and cultivar) during the flowering period, from June to September. Single flowers were labeled according to the order of appearance on each plant. The numbers of flowers of particular phenotypes (with long-styled, medium-styled, or short-styled pistil) were recorded after the opening of petals. Then, the number of fruits set from flowers of a particular phenotype was also recorded about one week after fertilisation, when fruit sets reached 1–2 cm in diameter. Flowers which did not set fruits were naturally aborted. Fruits in a stage of harvest maturity were picked to reflect the standard cultivation conditions and to exclude excessive metabolite sink by ripening fruits. Data were calculated and presented as a sum of flowers and fruits per plant per month for the two experimental years separately.

2.5. Procedures for Microscopic Observations

At full flowering, 20 pollinated flowers of each phenotype per treatment and cultivar were collected in 2013 and 2015. Data are presented as a sum of observations for investigated seasons, $N = 40$. The styles were isolated and fixed in FAA (formalin-acetic-alcohol), according to Martin's method [31] adapted by Sękara [30]. The germination of pollen on stigmas, growth of pollen tubes, and fertilisation of ovules were examined under fluorescence microscopy with the use of SteREO LUMAR V12 microscope (Carl Zeiss AG, Jena, Germany) (Figure 3). The number of pollen tubes in half of the style and the number of fertilised ovules were evaluated. The numbers of pistils having a number of pollen tubes in the ranges 0–100; 100–200; 200–300; 300–400; 400–500; 500–600; 600–700; 700–800, 800–900; 900–1000 were determined. For fertilised ovules, the following ranges were included: 0–50; 50–100; 100–150; 150–200; 200–250; 250–300; 350–400.

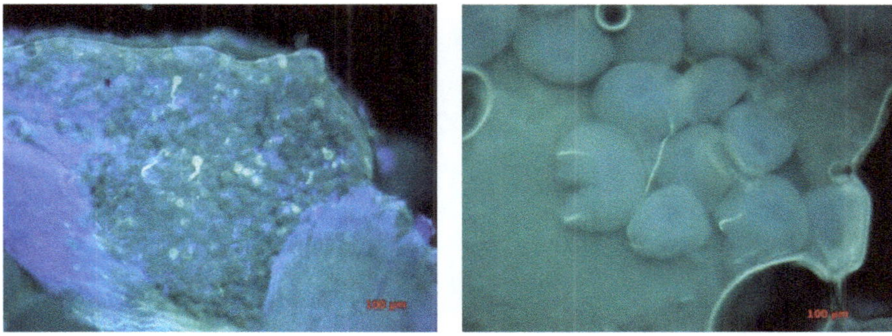

a b

Figure 3. Germination of the pollen on stigmas (**a**) and fertilisation of ovules (**b**) of eggplant observed under fluorescence microscopy after Martin's aniline blue fluorescence technique.

2.6. Statistical Analyses

Statistical analyses were performed using the Statistica 12.0 software package (StatSoft Inc., Tulsa, OK, USA). A three-way analysis of variance followed by Tukey's honest significance test was used to determine the main effects of the type of flower, biostimulant, and time of sampling, as well as interactions between main effects, at the $p \leq 0.05$ significance level. Data shown in the tables and figures are averages of three replicates.

3. Results

In the conditions of the present experiment, the eggplants started flowering at the beginning of June, while the period of the most intensive flowering fell in August. The investigated hybrids showed flower heterostyly—the presence of long-, medium-, and short-styled flowers was observed for all investigated plants. Moreover, heterostyly expression significantly depended on biostimulant treatment and the age of the plants (Table 2).

Table 2. Chosen aspects of flowering and fruit setting of eggplant as depended on fruit type and biostimulant treatment.

Parameter	Year	Type of Flower					
		With Long-Styled Pistil		With Medium-Styled Pistil		With Short-Styled Pistil	
		C *	B	C	B	C	B
"Epic" F_1							
Number of particular types of flowers per plant	2013	12.2 d **	13.4 d	5.6 b	8.2 c	3.4 a	3.5 ab
	2015	12.1 c	13.5 c	6.4 b	7.6 b	3.5 a	3.6 a
Number of fruits per plant set from particular types of flowers	2013	2.4 b	4.2 d	0.8 a	2.0 b	1.0 a	1.4 ab
	2015	3.2 b	5.6 c	1.0 a	3.2 b	1.0 a	1.2 a
Effectiveness of fruit setting as depended on type of flower (%)	2013	20.7	27.6	15.6	24.7	26.7	40.0
	2015	33.7	48.8	18.9	46.7	26.6	36.7
Number of seeds per fruit	2013	332 d	365 e	245 b	289 c	102 a	125 a
	2015	358 e	372 f	255 c	312 d	95 a	138 b
"Flavine" F_1							
Number of particular types of flowers per plant	2013	7.8 cd	10.0 d	6.6 bc	10.0 d	4.2 a	4.8 ab
	2015	8.4 b	11.2 c	6.8 b	11.0 c	3.8 a	4.8 a
Number of fruits per plant set from particular types of flowers	2013	2.0 abc	2.8 b	1.4 ab	2.4 bc	1.2 a	1.6 ab
	2015	1.8 a	3.6 b	1.2 a	3.2 b	1.2 a	2.0 a
Effectiveness of fruit setting as depended on type of flower (%)	2013	26.9	28.7	22.8	26.0	20.0	33.3
	2015	26.0	36.2	16.7	27.2	30.0	41.1
Number of seeds per fruit	2013	312 c	328 c	258 b	325 c	142 a	155 a
	2015	322 c	333 d	289 b	316 c	134 a	143 a

Table 2. *Cont.*

Parameter	Year	Type of Flower					
		With Long-Styled Pistil		With Medium-Styled Pistil		With Short-Styled Pistil	
		C *	B	C	B	C	B
		"Gascona" F_1					
Number of particular types of flowers per plant	2013	7.6 c	8.0 c	5.6 b	8.2 c	3.4 a	3.6 a
	2015	7.6 bc	7.8 c	6.0 b	7.8 c	4.2 a	3.8 a
Number of fruits per plant set from particular types of flowers	2013	2.0 abc	3.2 c	1.6 ab	2.4 bc	1.0 a	0.8 a
	2015	2.4 ab	3.0 b	1.6 a	2.2 ab	1.4 a	1.4 a
Effectiveness of fruit setting as depended on type of flower (%)	2013	26.2	41.5	22.6	29.1	23.3	20.0
	2015	38.7	37.1	11.7	30.6	36.7	33.3
Number of seeds per fruit	2013	289 d	322 e	257 c	285 d	78 a	127 b
	2015	321 e	356 f	269 c	297 d	98 a	159 b

* C, control; B, biostimulant; ** Means within rows, followed by different letters, are significantly different at $p \leq 0.05$, $N = 3$. Comparisons were performed with the use of Tukey's honest significance test.

Among 23 flowers set by "Epic" F_1 plant during one vegetation period, 55% had long-style pistils, 30% had medium-style pistils, and 15% had short-style pistils. "Epic" F_1 plants produced 7 fruits during the vegetation season, on average; 57% of these were from long-styled flowers, 26% from medium-styled flowers, and 17% from short-styled flowers. Biostimulant treatment significantly increased the number of only medium-styled flowers in 2013 and the number of fruits set by long- and medium-styled flowers in both vegetation periods. The most effective in fruit setting were long-styled flowers. The biostimulant positively affected the percentage of fruits set by all flower phenotypes and the number of seeds, with the exception of flowers with short-styled pistils in 2013. The first fruits were collected at the end of July. The highest number of long- and medium-styled flowers was observed in August; the lowest was observed in September (Figure 4, Table 3).

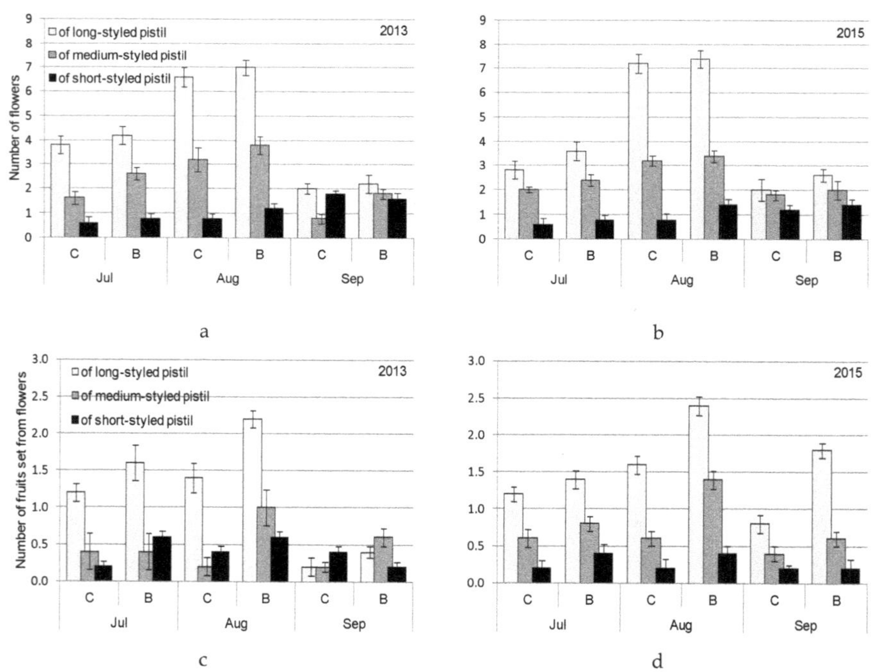

Figure 4. The course of flowering and fruit setting of "Epic" F_1 eggplant as depended on fruit type and biostimulant treatment. C, control; B, biostimulant. Bars represent mean number of flowers per plant in 2013 (**a**), 2015 (**b**) and fruits per plant in 2013 (**c**), and 2015 (**d**) (error bars indicate SE).

Table 3. Results of ANOVA for parameters of flowering and fruit setting of "Epic" F_1 eggplant presented in Figure 4.

ANOVA Source of Variation	"Epic" F_1			
	No of Flowers 2013	No of Fruits 2013	No of Flowers 2015	No of Fruits 2015
Type of flower (F)	***	***	***	***
Biostimulant (B)	**	*	***	***
Month (M)	***	***	*	**
F × B	ns	*	ns	*
F × M	***	***	***	ns
B × M	ns	ns	ns	*

Levels of significance for ANOVA: * $p \leq 0.05$; ** $p \leq 0.01$; *** $p \leq 0.001$; ns, not significant; $N = 3$. Comparisons were performed with the use of Tukey's honest significance test.

The number of short-styled flowers increased in line with aging of the plants. The number of fruits set from long- and medium-styled flowers increased from July to August, then decreased in September. We observed, on average, 106 pollen tubes in the short-styled pistils, 422 in medium-styled pistils, and 610 in long-styled pistils collected from control plants and 129, 490, and 778 pollen tubes, respectively, collected from biostimulant-treated plants (Figure 5). The ovaries of the short-styled flowers contained approximately 36 fertilised ovules, and more fertilised ovules were found in the remaining types of flowers: 199 and 225 in medium- and long-styled flowers, respectively, produced by control plants. The flowers of biostimulant-treated plants contained 39%, 32%, and 36% more fertilised ovules in the ovaries of short-, medium-, and long-styled flowers, respectively.

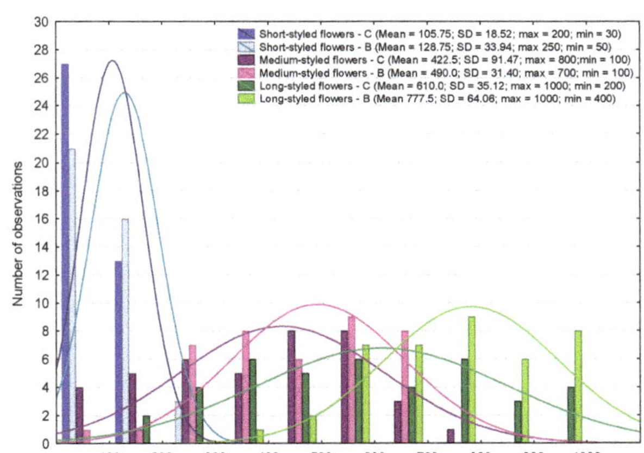

a

Figure 5. *Cont.*

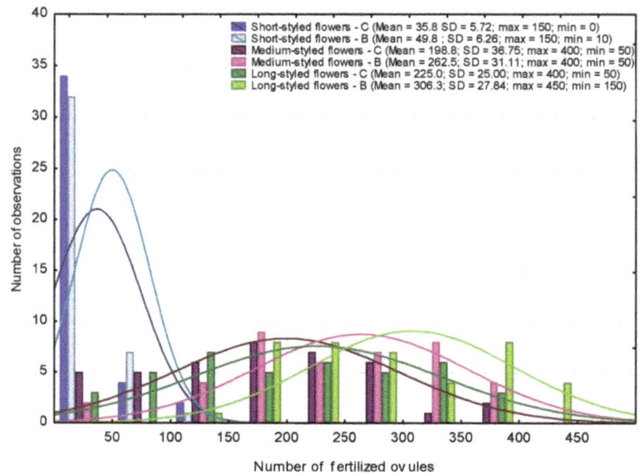

b

Figure 5. Number of pollen tubes (**a**) and fertilised ovules (**b**) in the styles of different flower types in "Epic" F_1.

The "Flavine" F_1 plants produced 22 flowers during the vegetation period; 41% had long-styled pistils, 38% had medium-styled pistils, and 19% had short-styled ones (Table 2). The number of fruits collected from a plant was 6 on average; 42% of these were from long-styled flowers, 34% from medium-styled flowers, and 24% from short-styled flowers. Biostimulant treatment significantly increased the number of long-styled flowers in 2015, the numbers of medium-styled flowers in 2013 and 2015, and the numbers of fruits set by long- and medium-styled flowers in 2015. The most effective in fruit setting were long-styled flowers. The biostimulant positively affected the percentage of fruits set by all flower phenotypes and the number of seeds in fruits born by long-styled flowers in 2015 and by medium-styled flowers in both years of the experiment. The highest number of long- and medium-styled flowers was observed in August; the lowest was in September (Figure 6, Table 4). The number of fruits set from long-styled flowers was the highest in August. We observed, on average, 149 pollen tubes in the middle of the style in the short-styled flowers of control plants, 410 in medium-styled flowers, and 595 in long-styled ones (Figure 7). In biostimulant-treated flowers, a 23% higher number of pollen tubes was observed in short-styled pistils, 16% higher in medium-styled pistils, and 20% higher in long-styled pistils. The ovaries of the short-styled flowers collected from the control plants contained, on average, 54 fertilised ovules; more fertilised ovules were found in the remaining types of flowers: 208 and 243 in medium- and long-styled flowers, respectively. The numbers of fertilised ovules in analogous types of flowers collected from biostimulant-treated plants were 102%, 24%, and 23% higher, respectively.

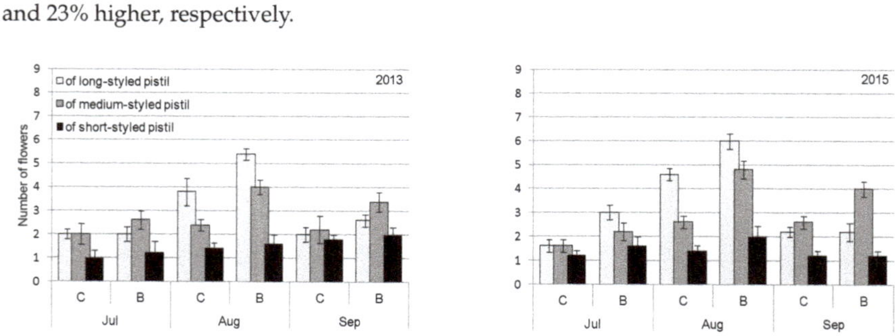

a b

Figure 6. *Cont.*

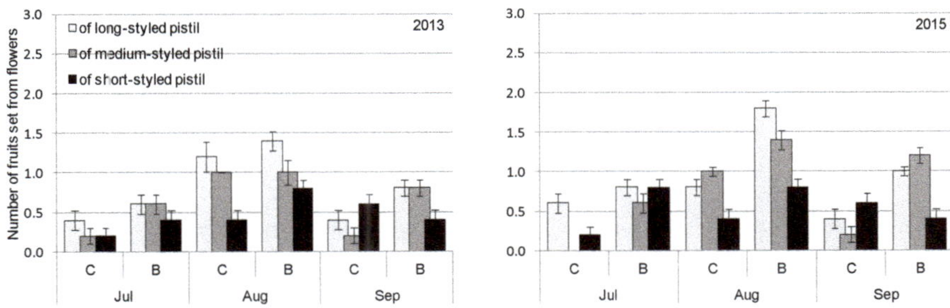

c d

Figure 6. The course of flowering and fruit setting of "Flavine" F_1 eggplant as depended on fruit type and biostimulant treatment. C, control; B, biostimulant. Bars represent mean number of flowers per plant in 2013 (**a**), 2015 (**b**) and fruits per plant in 2013 (**c**), and 2015 (**d**) (error bars indicate SE).

Table 4. Results of ANOVA for parameters of flowering and fruit setting of "Flavine" F_1 eggplant presented in Figure 6.

ANOVASource of Variation	"Flavine" F_1			
	No of Flowers 2013	**No of Fruits 2013**	**No of Flowers 2015**	**No of Fruits 2015**
Type of flower (F)	***	***	***	**
Biostimulant (B)	**	***	***	***
Month (M)	***	***	***	***
F × B	ns	*	*	ns
F × M	***	***	***	ns
B × M	ns	*	*	ns

Levels of significance for ANOVA: * $p \leq 0.05$; ** $p \leq 0.01$; *** $p \leq 0.001$; ns, not significant; $N = 3$. Comparisons were performed with the use of Tukey's honest significance test.

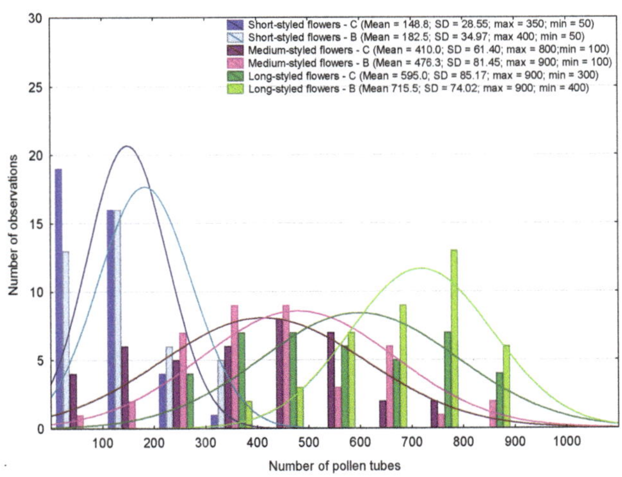

a

Figure 7. *Cont.*

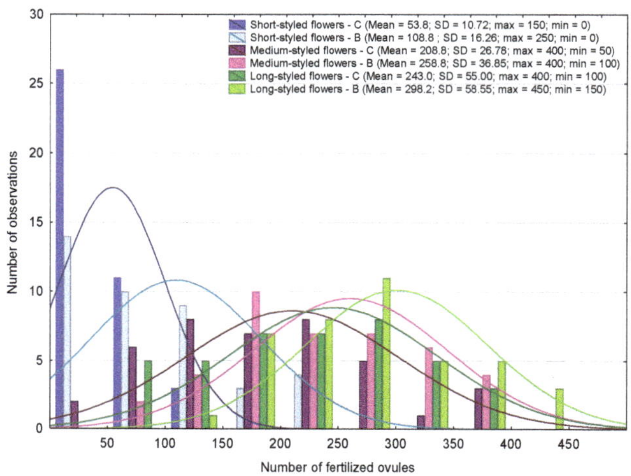

b

Figure 7. Number of pollen tubes (**a**) and fertilised ovules (**b**) in the styles of different flower types in "Flavine" F_1.

"Gascona" F_1 plants produced 18 flowers during the vegetation period, with 42% of these being long styled, 38% medium styled, and 20% short styled (Table 2). The number of fruits collected from a plant was 6, on average; 46% of these were from long-styled flowers, 34% from medium-styled flowers, and 20% from short-styled flowers. Biostimulant treatment significantly increased the number of medium-styled flowers in 2013 and 2015 but did not affect the number of fruits. The most effective in fruit setting were long-styled flowers. The biostimulant positively affected the percentage of fruits set by long-styled flowers in 2015 and medium-styled flowers in both years, but it negatively affected the effectiveness of fruit setting by short-styled flowers. Biostimulant treatment positively affected the seed number (Table 2). The highest number of long-, and medium-styled flowers was observed in August; the lowest was in September (Figure 8, Table 5). The number of fruits set from long- and medium-styled flowers increased from July to August, then decreased in September. Long-, medium-, and short-styled flowers were analysed regarding the number of pollen tubes in the styles. Differences in the course of pollination and fertilisation between investigated cultivars concerned the number of pollen tubes and fertilised ovules in the pistils. For control "Gascona" F_1 plants, we observed, on average, 119 pollen tubes in the middle of the style in the short-styled flowers, 418 in medium-styled flowers, and 595 in long-styled ones (Figure 9). The ovaries of the short-styled flowers contained approximately 0–50 fertilised ovules, while more fertilised ovules were found in the remaining types of flowers: 200–400 in medium- and long-styled flowers. The flowers of control plants contained lower numbers of both pollen tubes and fertilised ovules in all types of flowers.

Table 5. Results of ANOVA for parameters of flowering and fruit setting of "Gascona" F_1 eggplant presented in Figure 8.

ANOVASource of Variation	"Gascona" F_1			
	No of Flowers 2013	No of Fruits 2013	No of Flowers 2015	No of Fruits 2015
Type of flower (F)	***	***	***	*
Biostimulant (B)	*	***	ns	ns
Month (M)	***	ns	***	***
F × B	*	ns	*	ns
F × M	***	ns	***	*
B × M	ns	ns	ns	ns

Levels of significance for ANOVA: * $p \leq 0.05$; *** $p \leq 0.001$; ns, not significant; $N = 3$. Comparisons were performed with the use of Tukey's honest significance test.

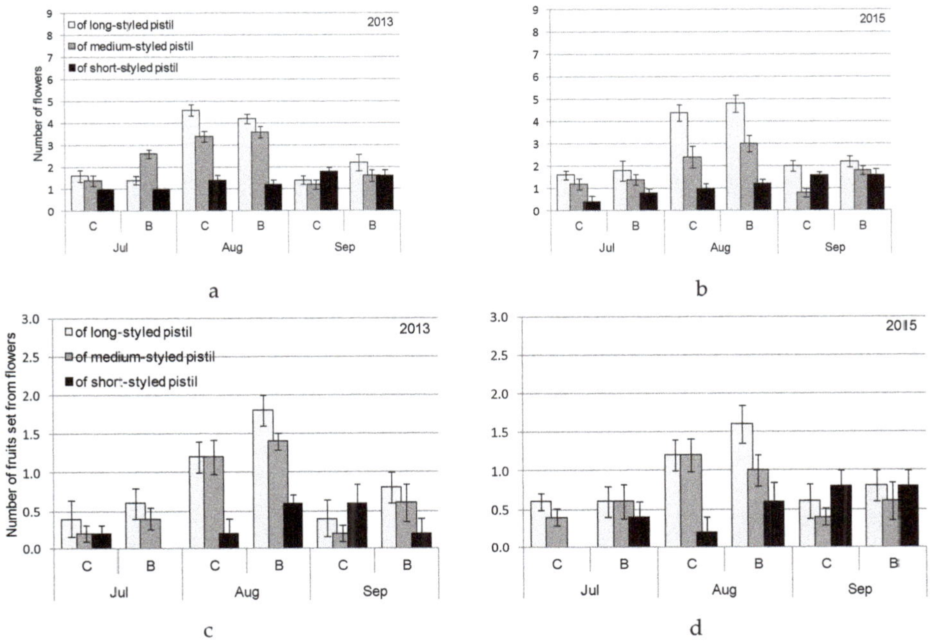

Figure 8. The course of flowering and fruit setting of "Gascona" F_1 eggplant as depended on fruit type and biostimulant treatment. C, control; B, biostimulant. Bars represent mean number of flowers per plant in 2013 (**a**), 2015 (**b**) and fruits per plant in 2013 (**c**), and 2015 (**d**) (error bars indicate SE).

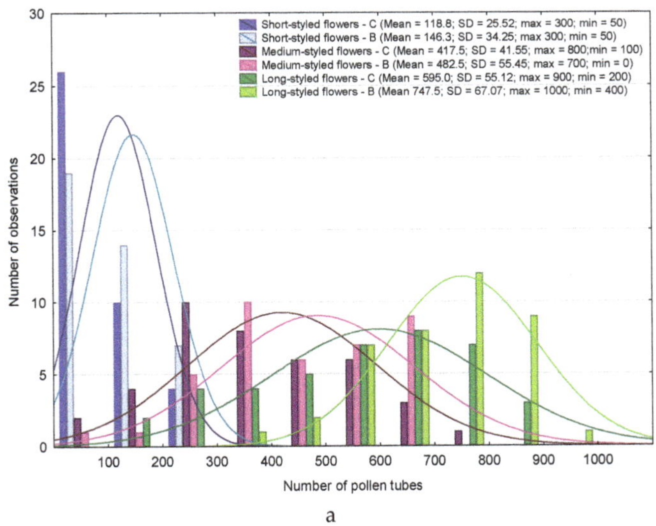

a

Figure 9. *Cont.*

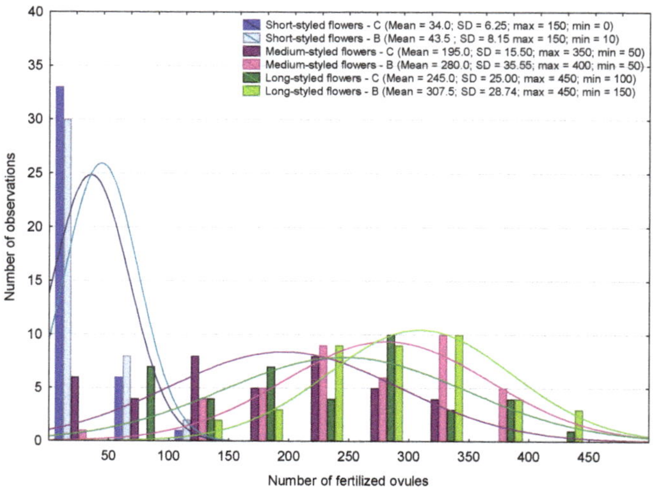

b

Figure 9. Number of pollen tubes (**a**) and fertilised ovules (**b**) in the styles of different flower types in "Gascona" F_1.

4. Discussion

4.1. Heterostyly Expression in Eggplant as Affected by Biostimulant Treatment and Cultivar

The recent research aimed to develop an overview of the heterostyly phenomenon in eggplant, and its implications on fruit setting biology. We demonstrated the presence of three phenotypes of flowers and the differentiated fertility of them, specific to the investigated hybrids. Generally, long-styled flowers dominated, but the fruit setting efficiency was not directly determined by the flower phenotype. A study by Srinivas et al. [32] indicated that for two eggplant hybrids of Indian breeding, with long, green and round, purple fruits, 80% of fruits were set by long-styled flowers, whereas 20% of the fruits were set by medium-styled flowers and no fruit by short-styled flowers. The partial sterility of short-styled flowers demonstrated in the cited research was due to small stigmas with under-developed papillae on which pollen grains failed to germinate. The short-styled flowers of the eggplant hybrids which are the subject of the present investigations were fertile, although the lowest number of pollen tubes and fertilised ovules was observed in this flower phenotype. Despite this fact, "Epic" F_1 and "Gascona" F_1 plants set about 20% of fruits from short-styled flowers. For "Flavine" F_1, the percentage of fruits set by this mentioned flower phenotype was 30%. Sękara and Bieniasz [6] determined that the ovules of short-styled pistils were typically developed, but that their fruit setting efficiency was low. On the contrary, results by Hazra et al. [33] indicated full sterility of short-styled flowers due to some problem related to ovary development. Observations with the use of a fluorescence microscope allowed us to verify the correct growth of pollen tubes in the styles of all types of pistils but their number was significantly affected by flower type and biostimulant treatment and by cultivar to a lesser extent. This observation is contrary to the results of Wang et al. [11], who determined that the structure of the stigmatic surface in short-styled flowers inhibited pollen germination. On the grounds of highly genotype-dependent heterostyly expression in eggplant, results on short-styled pistil performance may be divergent.

Application of Göemar BM-86® caused an increase in the numbers of pollen tubes and fertilised ovules. This phenomenon was common for all types of flowers and is directly attributable to pistil characteristics. Biostimulant-treated and control plants were not isolated, so they both could act as pollen donors. The effect of biostimulants on pollen production and fertility should also be an object of future research. Based on the available literature, we can only conclude that a wide pool of bioactive seaweed extract compounds provided balanced development and enhanced the flowering and fruiting of the investigated eggplant hybrids. The biostimulant-treated plants could be able to develop a better canopy for effective interception of light and—through a significant reduction in interplant competition for solar energy and nutrients—build suitable carbohydrate reserves earlier. Such mechanisms beyond increased flowering and fruit setting in seaweed extract treated plants were proposed by Arthur et al. [34] for bell peppers and by Helaly et al. [35] for tomatos.

The increasing number of short-styled flowers in line with plant aging, in the conditions of the present experiment, could be the result of increasing fruit load with the vegetation season's flow. Having well-developed anthers, short-styled flowers act as pollen donors to provide reproductive success. Araméndiz Tatis et al. [9] demonstrated that short-styled flowers of the "Lilac" eggplant landrace and "Long Purple" increased male fitness and thus produced an imbalance in functioning between male and hermaphrodite flowers. According to Khah et al. [36], fruit load negatively affected style length but not anther cone length in eggplant, even under favourable climatic conditions. The investigated hybrids could reduce energy outlines by creating flowers with reduced pistil and decreased fertility at the end of the vegetation period, but do so while producing pollen in the normally shaped anthers, promoting male behaviour. Short-styled flowers could be borne by fruit-loaded plants as a source of pollen for insects. The construction of the eggplant stamens is an expression of adaptation to pollination through vibrations. Such an adaptation limits the potential pollinators to species that are able to introduce vibrations into anthers, including bumblebees [13]. Bumblebees commonly visited eggplant flowers in the conditions of the presented experiment.

4.2. Biostimulant-Affected Flower and Fruit Set Effectiveness

Biostimulants have shown promising results in promoting flowering and reducing the fruit drop agents in many fruit trees, like apple, avocado, clementine, orange, olive, and pomegranate [37–40]. In this respect, seaweed extracts enriched in microelements are the most effective [41]. Vegetables with edible fruits are characterised by competition between the flowers and fruits at different stages of growth and in different positions in relation to inflowing assimilates [42]. Dropping of flowers, typical for eggplant, could be the mixed effect of lack of pollination or limited inflow of assimilates and the phenomenon of domination of fruit producing growth regulators. In the present research, hybrids treated with seaweed extract bore more flowers and fruits than did untreated ones. More intensive flower setting was elicited either by improved plant growth through seaweed extract application or by endogenous components, especially cytokinins, which enhance nutrient partitioning in vegetative plant organs and increase in the transport of assimilates to the growing fruits. A similar effect was observed for eggplant treated with seaweed extract by Abd El-Gawad and Osman [43]. Under the conditions of the presented experiment, biostimulant application also increased the number of pollen tubes and fertilised ovules in all types of flowers of the investigated cultivars. The overall positive influence of seaweed extracts on the plants resulted in better reproductive effectiveness and increased fruit yield and quality, described in detail by Pohl et al. [26,27]. Gómez-Cadenas et al. [44] investigated the effect of a biostimulant product containing macronutrients on citrus fruit set abscission. The beneficial effects of the biostimulant resulted from an increase in the photosynthetic efficiency which led to better transport of carbohydrates from leaves to fruit sets. Seaweed-treated apple trees also showed higher leaf chlorophyll contents and increased rates of photosynthesis and respiration due to treatment decreasing the oscillations in yield between "on" and "off" years and increasing the average fruit weight on plants affected by too high a crop load [45]. Pollination and fertilisation are very stress-sensitive stages of development [1]. Based on research on tomato, low temperatures, especially during the night, are not detrimental to ovule development but could affect stigma and style function [5]. Pollen viability is the highest at 20–22 °C [14], while the mean temperatures for the flowering period (July–September) were 16.8 and 18.8 °C in 2013 and 2015, respectively. Pollen development and viability depend on carbohydrate supply [5], so the increased photosynthetic performance of biostimulant-treated plants could improve sugar partitioning to developing pollen grains. The bioactive compounds of seaweed extracts enhance the tolerance of eggplant to abiotic stresses [46] and this tolerance can also cover the generative reproduction of this crop in temperate regions.

5. Conclusions

Eggplant is a warm climate crop and is cultivated for fruits, widely used in many world cuisines because of their unique taste and dietetic value. Nonoptimal growing conditions, especially in temperate climatic zones, affect plant flowering and fruit setting. Biostimulant application in the experiments presented herein affected the flowering biology of eggplant cultivars in different ways. Generally, the biostimulant positively affected the percentages of the most fertile medium- and long-styled flowers and the effectiveness of fruit setting by all flower phenotypes. Increased numbers of pollen tubes and fertilised ovules in all types of flowers of the investigated cultivars were noted. The overall positive influence of Göemar BM-86® on the plants resulted in increased reproductive effectiveness. Biostimulant application seems to be a promising solution to improve eggplant flowering and fruit setting in unfavourable growing conditions.

Author Contributions: Conceptualization, A.P. and A.S.; methodology, A.P. and A.S.; software, A.P. and A.S.; validation, A.P. and A.S. and A.K.; formal analysis, A.P.; investigation, A.P. and A.G.; resources, A.P. and A.S.; data curation, A.P. and A.S.; writing—original draft preparation, A.P. and A.S.; writing—review and editing, A.P. and A.S. and A.K.; visualization, A.P. and A.S.; supervision, A.S.

Funding: This research was funded by the Ministry of Science and Higher Education of the Republic of Poland.

Conflicts of Interest: The authors declare no conflict of interest.

References

1. Pandit, M.K.; Thapa, H.; Akhtar, S.; Hazra, P. Evaluation of brinjal genotypes for growth and reproductive characters with seasonal variation. *J. Crop. Weed* **2010**, *6*, 31–34.
2. Jagatheeswari, D. Morphological studies on flowering plants (Solanaceae). *Int. Lett. Nat. Sci.* **2014**, *10*, 36–43. [CrossRef]
3. Keller, B.; Thomson, J.D.; Conti, E. Heterostyly promotes disassortative pollination and reduces sexual interference in Darwin's primroses: Evidence from experimental studies. *Funct. Ecol.* **2014**, *28*, 1413–1425. [CrossRef]
4. Glover, B.J.; Bunnewell, S.; Martin, C. Convergent evolution within the genus *Solanum*: The specialised anther cone develops through alternative pathways. *Gene* **2004**, *331*, 1–7. [CrossRef] [PubMed]
5. Karapanos, I.C.; Mahmood, S.; Thanopoulos, C. Fruit set in solanaceous vegetable crops as affected by floral and environmental factors. *Eur. J. Plant Sci. Biotechnol.* **2008**, *2*, 88–105.
6. Sękara, A.; Bieniasz, M. Pollination, fertilization and fruit formation in eggplant (*Solanum melongena* L.). *Acta Agrobot.* **2008**, *61*, 107–113.
7. Kowalska, G. The influence of heterostyly, pollination method and harmonization on eggplant's (*Solanum melongena* L.) flowering and fruiting. *Acta Agrobot.* **2003**, *56*, 61–76. [CrossRef]
8. Kowalska, G. Eggplant (*Solanum melongena* L.) flowering and fruiting dynamics depending on pistil type as well as way of pollination and flower harmonization. *Folia Hortic.* **2006**, *18*, 17–29.
9. Araméndiz Tatis, H.; Cardona Ayala, C.; Espitia Camacho, M. Floral morphology characterization of two cultivars of eggplant (*Solanum melongena* L.) (Solanaceae). *Rev. Fac. Nac. Agron. Medellin* **2009**, *62*, 5125–5134.
10. Rylski, I.; Nothmann, J.; Arcan, L. Differential fertility in short-styled eggplant flowers. *Sci. Hortic.* **1984**, *22*, 39–46. [CrossRef]
11. Wang, Y.; Liu, A.; Li, W.; Jiang, Y.; Song, S.; Li, Y.; Chen, R. Comparative proteomic analysis of eggplant (*Solanum melongena* L.) heterostylous pistil development. *PLoS ONE* **2017**, *12*, e0179018. [CrossRef] [PubMed]
12. Gemmill-Herren, B.; Ochieng, A.O. Role of native bees and natural habitats in eggplant (*Solanum melongena*) pollination in Kenya. *Agric. Ecosyst. Environ.* **2008**, *127*, 31–36. [CrossRef]
13. Kikuchi, K.; Honda, I.; Matsuo, S.; Fukuda, M.; Saito, T. Stability of fruit set of newly selected parthenocarpic eggplant lines. *Sci. Hortic.* **2008**, *115*, 111–116. [CrossRef]

14. Makrogianni, D.I.; Karapanos, I.C.; Passam, H.C. Seasonal fluctuations in pollen production and viability in eggplant and the quality of seed-containing and seedless (auxin-set) fruits. *J. Plant Growth Regul.* **2018**, *37*, 937–946. [CrossRef]
15. Sharma, M.D. Effect of plant growth regulators on growth and yield of brinjal at Khajura, Banke. *J. Inst. Agric. Anim. Sci.* **2006**, *27*, 153–156. [CrossRef]
16. Netam, J.L.; Sharma, R. Efficacy of plant growth regulators on growth characters and yield attributes in brinjal (*Solanum melongena* L.) cv. Brinjal. 3112. *J. Agric. Veterin. Sci.* **2014**, *7*, 27–30. [CrossRef]
17. Sabatino, L.; Iapichino, G.; D'Anna, F.; Palazzolo, E.; Mennella, G.; Rotino, G.L. Hybrids and allied species as potential rootstocks for eggplant: Effect of grafting on vigour, yield and overall fruit quality traits. *Sci. Hort.* **2018**, *228*, 81–90. [CrossRef]
18. Sabatino, L.; Iapichino, G.; Rotino, G.L.; Palazzolo, E.; Mennella, G.; D'Anna, F. *Solanum aethiopicum* gr. gilo and its interspecific hybrid with *S. melongena* as alternative rootstocks for eggplant: Effects on vigor, yield, and fruit physicochemical properties of cultivar 'Scarlatti'. *Agronomy* **2019**, *9*, 223. [CrossRef]
19. Caruso, G.; Pokluda, R.; Sękara, A.; Kalisz, A.; Jezdinský, A.; Kopta, T.; Grabowska, A. Agricultural practices, biology and quality of eggplant cultivated in Central Europe. A Review. *Hort. Sci.* **2017**, *44*, 21–42. [CrossRef]
20. Grabowska, A.; Kunicki, E.; Sękara, A.; Kalisz, A.; Jezdinský, A.; Gintrowicz, K. The effect of biostimulants on the quality parameters of tomato grown for the processing industry. *Agrochimica* **2015**, *59*, 203–217.
21. Kopta, T.; Pavlikova, M.; Sękara, A.; Pokluda, R.; Marsalek, B. Effect of bacterial-algal biostimulant on the yield and internal quality of lettuce (*Lactuca sativa* L.) produced for spring and summer crop. *Not. Bot. Horti. Agrobot. Cluj-Napoca* **2018**, *46*, 615–621. [CrossRef]
22. DuJardin, P. Plant biostimulants: Definition, concept, main categories and regulation. *Sci. Hortic.* **2015**, *196*, 3–14. [CrossRef]
23. Di Stasio, E.; Van Oosten, M.J.; Silletti, S.; Raimondi, G.; Dell'Aversana, E.; Carillo, P.; Maggio, A. *Ascophyllum nodosum*-based algal extracts act as enhancers of growth, fruit quality, and adaptation to stress in salinized tomato plants. *J. Appl. Phycol.* **2018**, *30*, 1–12. [CrossRef]
24. Sharma, H.S.; Fleming, C.; Selby, C.; Rao, J.R.; Martin, T. Plant biostimulants: A review on the processing of macroalgae and use of extracts for crop management to reduce abiotic and biotic stresses. *J. Appl. Phycol.* **2014**, *26*, 465–490. [CrossRef]
25. Yakhin, O.I.; Lubyanov, A.A.; Yakhin, I.A.; Brown, P.H. Biostimulants in plant science: A global perspective. *Front Plant Sci.* **2017**, *7*, 2049–2080. [CrossRef] [PubMed]
26. Pohl, A.; Grabowska, A.; Kalisz, A.; Sękara, A. Preliminary screening of biostimulative effects of Goëmar BM-86 on eggplant cultivars grown under field conditions in Poland. *Acta Agrobot.* **2018**, *71*, 1752. [CrossRef]
27. Pohl, A.; Grabowska, A.; Kalisz, A.; Sękara, A. The eggplant yield and fruit composition as affected by genetic factor and biostimulant application. *Not. Bot. Horti. Agrobot. Cluj-Napoca* **2019**, *47*. [CrossRef]
28. Sękara, A.; Bączek-Kwinta, R.; Gawęda, M.; Kalisz, A.; Pokluda, R.; Jezdinský, A. Sequential abiotic stress applied to juvenile eggplant modifies the seedlings parameters, plant ontogeny and yield. *Hort. Sci.* **2016**, *43*, 149–157. [CrossRef]
29. UPL Limited, Arysta Lifescience. Available online: https://www.arystalifescience.com/ (accessed on 23 January 2019).
30. Sękara, A. *Biology of the Vegetative and Generative Development of Eggplant (Solanum melongena L.) in the Field Production. Chosen Aspects*; Zesz Nauk UR Wydawnictwo Uniwersytetu Rolniczego: Krakow, Poland, 2010; Volume 459, pp. 19–24.
31. Martin, F.W. Staining and observing pollen tubes in the style by means of fluorescence. *Stain. Technol.* **1959**, *34*, 125–128. [CrossRef]
32. Srinivas, G.; Jayappa, A.H.; Patel, A.I. Heterostyly: A threat to potential fruit yield in brinjal (*Solanum melongena* L.). *Adv. Life Sci.* **2016**, *5*, 1211–1215.
33. Hazra, P.; Mandal, J.; Mukhopadhyay, T.P. Pollination behaviour and natural hybridization in *Solanum melongena* L., and utilization of functional male sterile line in hybrid seed production. *Caps. Eggpl. Newsl.* **2003**, *22*, 143–146.
34. Arthur, G.D.; Stirk, W.A.; Van Staden, J. Effect of a seaweed concentrate on the growth and yield of three varieties of *Capsicum annuum*. *S. Afr. J. Bot.* **2003**, *69*, 207–211. [CrossRef]
35. Helaly, M.N.; Arafa, A.A.; Ibrahim, H.M.; Ghoniem, K.H. Improving growth and productivity of tomato by some biostimulants and micronutrients with or without mulching. *J. Phytol.* **2018**, *10*, 15–23.

36. Khah, E.M.; Antonopoulos, A.; Passam, H.C. Floral behaviour and fruit set in four cultivars of aubergine. *Acta Hortic.* **2002**, *579*, 259–264. [CrossRef]
37. Fornes, F.; Sánchez-Perales, M.; Guardiola, J.L. Effect of a seaweed extract on the productivity of 'de Nules' Clementine mandarin and Navelina orange. *Bot. Mar.* **2002**, *45*, 486–489. [CrossRef]
38. Basak, A. Effect of preharvest treatment with seaweed products, Kelpak® and Goëmar BM 86®, on fruit quality in apple. *Int. J. Fruit Sci.* **2008**, *8*, 1–14. [CrossRef]
39. Abubakar, A.R.; Naira, A.; Moieza, A. Effect of plant biostimulants on flowering, fruit drop, yield and return bloom of pomegranate cv. Kandhari Kabuli. *Asian J. Hortic.* **2012**, *7*, 473–477.
40. Morales-Payan, J.P.; Candelas, C.D. Increasing organic avocado fruit yield using a *Ascophyllum nodosum* biostimulant and fertilization. *Acta Hortic.* **2014**, *1042*, 121–124. [CrossRef]
41. Chouliaras, V.; Tasioula, M.; Chatzissavvidis, C.; Therios, I.; Tsabolatidou, E. The effects of a seaweed extract in addition to nitrogen and boron fertilization on productivity, fruit maturation, leaf nutritional status and oil quality of the olive (*Olea europaea* L.) cultivar Koroneiki. *J. Sci. Food Agric.* **2009**, *89*, 984–988. [CrossRef]
42. Marcelis, L.F.M.; Elings, A.; Bakker, M.J.; Brajeul, E.; Dieleman, J.A.; de Visser, P.H.B.; Heuvelink, E. Modelling dry matter production and partitioning in sweet pepper. *Acta Hortic.* **2006**, *718*, 121–128. [CrossRef]
43. Abd El-Gawad, H.G.; Osman, H.S. Effect of exogenous application of boricacid and seaweed extract on growth, biochemical content and yield of eggplant. *J. Hortic. Sci. Ornam. Plants* **2014**, *6*, 133–143.
44. Gómez-Cadenas, A.; Pérez-Santamarina, R.; Ghorbel, R. Effect of a biostimulant product containing macronutrients and a carboxylic acid (AMEC®) on citrus fruitlet abscission. *Acta Hort.* **2012**, 189–193.
45. Spinelli, F.; Fiori, G.; Noferini, M.; Sprocatti, M.; Costa, G. Perspectives on the use of a seaweed extract to moderate the negative effects of alternate bearing in apple trees. *J. Hort. Sci. Biotechnol.* **2009**, *84*, 131–137. [CrossRef]
46. Van Oosten, M.J.; Pepe, O.; De Pascale, S.; Silletti, S.; Maggio, A. The role of biostimulants and bioeffectors as alleviators of abiotic stress in crop plants. *Chem. Biol. Technol. Agric.* **2017**, *4*, 5. [CrossRef]

agronomy

MDPI

Article

Foliar Applications of Biostimulants Promote Growth, Yield and Fruit Quality of Strawberry Plants Grown under Nutrient Limitation

Sebastian Soppelsa [1,2], Markus Kelderer [2], Claudio Casera [2], Michele Bassi [2], Peter Robatscher [2], Aldo Matteazzi [2] and Carlo Andreotti [1,*]

1 Faculty of Science and Technology, Free University of Bozen-Bolzano, Piazza Università, 5-39100 Bozen-Bolzano, Italy

2 Laimburg Research Centre, Laimburg 6, 39051 Pfatten-Vadena, Italy

* Correspondence: carlo.andreotti@unibz.it

Received: 11 July 2019; Accepted: 22 August 2019; Published: 26 August 2019

Abstract: Biostimulants have been found effective in enhancing plant resistance toward stressful conditions. The aim of the present study was to evaluate the efficacy of selected biostimulants to overcome the negative effects of nutrient limitation on the growth performances and on the fruit quality of soilless cultivated strawberry plants. The condition of nutrient limitation was imposed by supplying the plants with only a single fertilization at transplantation and by excluding any further nutrient supply for the entire duration of the experiment (three months, from May to July). Strawberry plants were treated seven times during the period from preflowering up to berry maturation with different classes of biostimulants (humic acids, alfalfa hydrolysate, macroseaweed extract and microalga hydrolysate, amino acids alone or in combination with zinc, B-group vitamins, chitosan, and a commercial product containing silicon) at commercial dosages. The use of alfalfa hydrolysate, vitamins, chitosan, and silicon was able to promote biomass accumulation in roots (four to seven folds) and fruits (+20%) of treated plants, whereas the total leaf area increased by 15%–30%. Nutrient concentrations in leaves and roots showed variations for microelements (e.g., Fe, B, Zn, and Si) in response to biostimulant applications, whereas no significant differences were observed for macronutrient contents among treatments. Final berry yield was found around 20% higher in chitosan- and silicon-treated plants. Chitosan treatment significantly increased pulp firmness (by 20%), while a high nutritional value (e.g., phenolic compounds concentration) was observed in alfalfa- and seaweed-treated fruits (+18%–20% as compared to control). The overall outcomes of the present experiment show that selected biostimulants can be considered as a valid agronomic tool able to contrast the negative consequence of growing crops under insufficient nutritional conditions.

Keywords: soilless conditions; abiotic stress; alfalfa hydrolysate; chitosan; zinc; ascorbic acid; *Fragaria x ananassa*

1. Introduction

The nutritional status of strawberry plants is of major relevance for the achievement of the expected levels of productivity and overall fruit quality [1]. Growth, yield, and quality parameters were found to be positively correlated with the rate of mineral nutrients (macro and micro) used during the strawberry production cycle [1–3]. The current world strawberry production is often conducted under covered systems (greenhouses and plastic tunnels) and in combination with soilless technologies [4]. Soilless production systems require a particularly fine control of nutrient supply, but this control is difficult to be achieved especially when the use of mineral fertilizers is banned (i.e., in organic farming).

Under such conditions, nutrient deficiencies/imbalances often become a severe limiting factor for the overall economic sustainability of the cultivation [5,6].

With the aim to help both organic and integrated growers to overcome the problem of the insufficient nutritional status of soilless cultivated crops, different technical solutions are currently available on the market of fertilizer products, including the vast group of biostimulant compounds. These compounds were recently considered in the new EU regulation of fertilizer products and defined as products able to improve one or more of the following characteristics: (i) nutrient use efficiency; (ii) tolerance toward abiotic stresses; (iii) quality traits and (iv) the availability of confined nutrients in soil or the rhizosphere [7].

During the last few years, the international scientific literature has largely documented the beneficial effects of the use of biostimulants on several different crops. As for the use of biostimulant to contrast abiotic stresses, Goñi et al. [8] and Di Stasio et al. [9] recently reported the use of seaweed extracts from *Ascophyllum nodosum* to enhance the tolerance of tomato plants to drought and salinity stress, respectively. This positive effect was observed also with other biostimulants, such as protein hydrolysate on lettuce [10], chitosan on white clover [11], and silicon on tomato [12]. Regarding the stress caused by nutrient deficiency, biostimulants can contribute to soil nutrient availability, uptake, and assimilation of nutrients to the plants [13]. Mechanisms by which biostimulants affect nutrient uptake are generally related to an improvement of physicochemical properties of soil, nutrient solubility, root morphology, and root colonization by arbuscular mycorrhizal fungi [13,14]. Mattner et al. [15] found an increased root length and density in strawberry plants treated with seaweed extracts (*Duvillaea potatorum* and *A. nodosum*), which implies a more efficient use of nutrients by the crop. Extracts from other seaweed species (e.g., *Ecklonia maxima* and *Kappaphycus alvarezii*) also increased the uptake of several macro- and micronutrients, leading to a higher biomass accumulation and final yield in lettuce, soybean, and zucchini squash [16–18]. Spinelli et al. [19] observed the efficacy of seaweed extract in increasing root growth and enhancing tolerance to iron deficiency on strawberry plants, demonstrating that *A. nodosum* might be a valid substitute of Fe–chelate compounds (e.g., sequestrene). In addition to seaweed extracts, other biostimulant products based on amino acids (such as arginine, glycine, histidine, and phenylalanine) were also found to be able to reduce the negative effects of iron-induced chlorosis and the incidence of physiological disorders in horticultural products (tomato and apple) [20,21].

Biostimulants were also found to be effective in promoting yield in several vegetable and fruit species, while differently modulating the nutritional and functional properties of their edible products. Foliar applications of different biostimulants (protein hydrolysate, seaweed, and plant extracts) were all effective in increasing the yield of greenhouse tomato but showed different modulating effects on final fruit quality (e.g., protein hydrolysate-treated fruits showed a higher lycopene concentration, whereas treatments with plant extracts reduced the level of undesirable components such as nitrates) [22]. As concerns the fruit crops, the application of seaweed extracts increased the total soluble solids and organic acids content of mango fruits [23], whereas the use of chitosan and humic acids enhanced the strawberry fruit average size and marketability [24,25]. Also well-established is the induction effect of several biostimulants on the phenolic and flavonoid metabolism in horticultural crops. Treatments with *A. nodosum* induced a higher accumulation of total anthocyanins in strawberry fruits [26], whereas the application of plant and seaweed-based extracts increased total phenolics accumulation in spinach leaves [27,28]. Chitosan and alginate applications promoted the accumulation of anthocyanin and phenolic acids such as 3-O-glucosyl-resveratrol in *Vitis vinifera* [29]. Portu et al. [30] and El-Sayed [31] showed that the application of amino acids (e.g., phenylalanine) can enhance the content of anthocyanins, flavonols, and the total phenols in table grape. These biostimulants resulted in being able to be active on the phenylpropanoid pathway of the treated plants, as demonstrated by the enhanced expression of constitutive genes such as PAL, CHS, and CHI [32–37].

The purpose of this study was to understand if the use of selected biostimulant products could be considered to contrast the negative effects of nutrient limitation in soilless cultivated strawberry. With this aim, repeated preharvest applications of biostimulant compounds were performed and the effects on the growth and fruit yield measured. Moreover, changes in strawberry quality and nutritional value (total phenolic and anthocyanin content, antioxidant potential) were evaluated. Some of the selected substances were tested for the first time on strawberry plants and, to the best of our knowledge, this was the first study where the efficacy of several biostimulants was evaluated simultaneously during the same growing cycle.

2. Materials and Methods

2.1. Experimental Site and Biostimulant Applications

The trial was carried out in a greenhouse located at the Laimburg Research Centre, municipality of Vadena/Pfatten (46°22′ N; 11°17′ E; 237 m a.s.l.) in Alto-Adige/South Tyrol, Italy, during the period April–July 2016. Climatic conditions inside the greenhouse (temperature and humidity) were monitored and are reported in Figure 1.

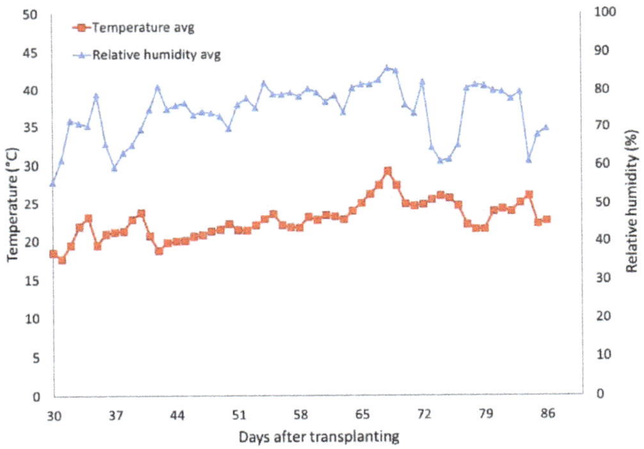

Figure 1. Climatic conditions (average daily temperature and relative humidity) inside the greenhouse during the evaluation period.

A total of 176 cold stored strawberry tray plants (*Fragaria* × *ananassa* Duch.) cv. Elsanta (the most commonly chosen cultivar for protected and open field cultivation in the area), very uniform in size and general conditions (Figure S1, Supplementary Material), were transplanted (4 plants per pot) on the 21st of April 2016 into rectangular plastic pots having the following size: 48 × 23 × 13 cm. A mixture of white peat and natural clay was used as growth medium. A week after the transplantation, plants were fertigated as used commercially for strawberry under greenhouse conditions with a starter fertilizer NPK (24-10-10 and microelements; Peters Excel, Scotts Company, Italy) at a rate of 1 g L^{-1} (EC 1.1 mS cm^{-1}; 0.15 L of solution $plant^{-1}$). Details on the substrate and on the nutrient solution composition are shown in Table 1. The plant material was characterized at the beginning of the experiment (at transplantation) by assessing the dry weight of the different organs (Table 2). For this purpose, four strawberry plants out of the initial group of plants were randomly sampled, washed, and divided into roots, crowns, and leaves. The plant organs were then put in an oven (ED 56, Binder GmbH, Tuttlingen, Germany) at 65 °C until they reached a stable weight and the dry mass recorded (g dry weight (DW) $plant^{-1}$). Moreover, mineral element concentration in leaves and roots was determined as described in Section 2.5 (Table 2). All plants were uniformly irrigated until runoff at two-day intervals using a watering can. During the growing cycle, no chemical pesticides were used, and no further fertilization with mineral nutrients was applied to plants for the whole experiment duration. Starting from the 20th of May (at preflowering stage, 30 DAT—days after transplanting), plants were foliar treated at weekly intervals with the biostimulant products. Seven applications were performed, in total using 10 different biostimulants. Details on the names, abbreviations and physicochemical characteristics of the utilized biostimulants are reported in Table 3. The applied concentrations were those reported in the label for commercial products, whereas those treatments still at prototype stage were applied according the information found in literature for similar compounds. The mechanism by which foliar applied biostimulants penetrate the leaf surface was not investigated in the present research, even though other studies conducted on similar products evidenced that the penetration occurs mainly through the cuticular layer and the stoma [38]. The penetration of these substances into leaves is a passive process and is influenced by several factors, including the concentration and the chemical nature of the applied molecules, the morphology of the leaf cuticle, and the degree of the stomatal opening [39]. Control plants were sprayed only with water. All the plants were sprayed until the runoff point using a hand sprayer (50 mL per plant). Pollination was performed with a brush by taking the pollen from the flower's own anthers. The experiment setup was organized as a completely randomized design with 4 replicates composed by 4 plants per replicate (i.e., 16 plants per treatment and 176 plants in total).

Table 1. Composition of the substrate and of the nutrient solution used for the experiment.

Physico-Chemical Composition of Substrate		Composition of the Nutrient Solution Used at Transplant	
pH	5.1	N (mg L^{-1})	240
Dry matter (%)	42.6	P_2O_5 (mg L^{-1})	100
Humidity (%)	57.4	K_2O (mg L^{-1})	100
Soluble salts (g L^{-1})	2.0	B (mg L^{-1})	0.2
NO_3 (mg L^{-1})	3.3	Cu (mg L^{-1})	0.1
NH_4 (mg L^{-1})	462.7	Fe (mg L^{-1})	1.2
P_2O_5 (mg L^{-1})	37.0	Mn (mg L^{-1})	0.5
K_2O (mg L^{-1})	245.0	Mo (mg L^{-1})	0.1
Mg (mg L^{-1})	121.0	Zn (mg L^{-1})	0.3
Na (mg L^{-1})	13.0		
B (mg L^{-1})	0.26		
Fe (mg L^{-1})	63.0		
Mn (mg L^{-1})	10.2		
Cu (mg L^{-1})	1.2		
Zn (mg L^{-1})	2.1		

Table 2. Characterization of the cold stored strawberry plants at the transplanting: dry weight of the organs and mineral element concentrations in leaves and roots.

Dry Weight of Plant Organs (g DW [a] plant^{-1})		Mineral Element Concentration	Leaves	Roots
Leaves	2.07 ± 0.31	N (%)	3.90 ± 0.28	2.65 ± 0.19
Crown	0.71 ± 0.23	P (%)	0.55 ± 0.10	0.43 ± 0.07
Roots	2.17 ± 0.58	K (%)	3.14 ± 0.20	0.76 ± 0.16
Total weight	4.95 ± 0.97	Ca (%)	1.39 ± 0.20	1.33 ± 0.34
		Mg (%)	0.42 ± 0.03	0.32 ± 0.06
		S (%)	0.52 ± 0.26	0.42 ± 0.13
		B (mg kg^{-1})	35.58 ± 13.20	22.93 ± 1.66
		Cu (mg kg^{-1})	19.98 ± 9.67	115.13 ± 65.82
		Fe (mg kg^{-1})	74.68 ± 32.96	149.83 ± 40.22
		Mn (mg kg^{-1})	168.35 ± 5.84	61.80 ± 23.66
		Zn (mg kg^{-1})	33.35 ± 5.32	85.68 ± 40.17
		Na (mg kg^{-1})	163.59 ± 91.87	299.73 ± 100.71
		Si (mg kg^{-1})	382.35 ± 61.93	181.83 ± 32.92

[a] DW: dry weight.

Table 3. Biostimulant characterization, properties, and concentration applied for the experiment.

Treatment	Active Ingredients	Moisture (%)	Ash (%)	Density (kg dm^{-3})	Organic Matter (%)	pH	Electrical Conductivity (dS m^{-1})	Total Organic C (% w w^{-1})	Total Organic N (% w w^{-1})	Free Amino Acids (% w w^{-1})	Total Amino Acids (% w w^{-1})	Other Characteristics	Concentration
CON	Water												
HAL	Humic acids	-	-	1.1	-	9.2	1.2	7.5	0.1	-	-	-	1.0 g L^{-1}
APH	Alfalfa protein hydrolysate	70.0	7.0	1.2	23.0	5.5	1.6	-	-	1.5	5.1	-	3.0 g L^{-1}
SEA	Macroseaweed extract	84.0	1.5	1.0	14.5	4.5	0.4	3	≤0.1	-	-	From *A. nodosum*	4.0 g L^{-1}
SPI	Microalga hydrolysate	-	-	1.2	-	5.5	1.5	16.8	3.9	6.5	-	From *Spirulina spp* (37% hydrol. degree)	4.0 g L^{-1}
MAA	Mix of amino acids	45.0	5.0	1.2	50.0	5.5	0.8	24.5	9	1.5	55.0	-	3.0 g L^{-1}
PHE	MAA combined with pure phenylalanine	55.0	5.0	1.2	40.0	5.5	-	19.6	7.2	1.8	45.0	Phenylalanine (1%)	3.0 g L^{-1}
ZIN	MAA combined with zinc	55.0	7.0	1.2	38.0	5.5	-	19.6	7.2	0.8	44.0	Zn (2%)	3.0 g L^{-1}
VIT	B-group vitamins (Sigma-Aldrich, USA)	-	-	-	-	-	-	-	-	-	-	B1-thiamine (33.3%), B2-riboflavin (33.3%), B6-pyridoxine (33.3%)	1.0 g L^{-1}
CHI	Chitosan—ChitoPlant Solution® (Agritalia, Italy)	98.3	0.01	-	1.7	5.2	-	-	-	-	-	-	10 mL L^{-1}
SIL	Siliforce® (ILSA S.p.A., Italy)	-	-	1.2	-	2.0	-	-	-	-	-	Si (8 g kg^{-1}), Zn (1.8%), Mo (0.2%)	0.3 mL L^{-1}

2.2. Vegetative Growth and Leaf Gas Exchanges

Plant growth dynamic as affected by biostimulant applications was determined every two weeks from preflowering (about 30 DAT) to the end of harvest (about 90 DAT) by measuring the area of two young trifoliate leaves per strawberry plant with a Li-Cor 3000 Leaf Area Meter (Li-Cor Inc., Lincoln, NE, USA). Biomass accumulation in different plant organs was calculated by subtracting the initial dry weight measured at transplanting stage (0 DAT, Table 2) from the final dry weight reached at the end of the growth cycle (90 DAT). The dry weight of the different plant organs at the end of production cycle was determined on four plants per each treatment as previously described in Section 2.1. Measurements included the determination of the final dry matter of organs that were already present at transplanting (roots, crown and leaves) as well as the weight of the newly formed organs (shoots, stolons, and fruits).

The leaf chlorophyll content was indirectly determined with a SPAD-502 Chlorophyll Meter (Konica Minolta, Tokyo, Japan) at biweekly intervals on two randomly selected trifoliate leaves per plant. Moreover, after the plants had received four applications of all the treatments, the net assimilation (A, μmol m^{-2} s^{-1}) and transpiration (E, mmol m^{-2} s^{-1}) rates of leaves were evaluated using a portable gas exchange analyzer (LCpro ADC, Hoddesdon Bioscientific, Ltd, UK). The gas exchange evaluations were conducted immediately before the fifth application of biostimulants (at 57 DAT), and 24 and 48 h after the application. Measurements were performed on a young, fully expanded single leaf of four randomly selected plants per treatment and were taken under saturating light conditions (1.800 μmol photons m^{-2} s^{-1}), around midday (11:00–13:00 h), using a broad leaf gas chamber with a window size of 6.25 cm^2 and a flow rate of 400 mL min^{-1}.

2.3. Yield and Fruit Quality

Ripe strawberry fruits (fully red color) were harvested every three days during the period from approximately 60 to 80 DAT. Harvested berries were counted and weighted with a digital scale in the laboratory. The cumulated yield per plant was determined by adding the weight of all the strawberries collected from the same plant during the harvesting period. Fruit quality was assessed on four strawberries per plant sampled at an intermediate pick during the harvesting period. Flesh firmness (kg cm^{-2}) was measured with a penetrometer equipped with a 6 mm diameter cap. The total soluble solids (°Brix) was determined with a portable refractometer (PAL-1, ATAGO, Tokyo, Japan), whereas the titratable acidity (g L^{-1} of citric acid) was measured with a titrator (TitroLine easy, SCHOTT, Mainz, Germany) by titrating strawberry pulp to pH 8.2 using 0.1 M NaOH. The external fruit color was assessed with a colorimeter (Minolta, model CR-400, Tokyo, Japan) by measuring the same four fruits at three different positions around the equatorial side of each fruit. The colorimetric coordinates (L*, a*, b*) were used to calculated the chroma and hue angle values with the formulas (1) and (2), respectively [40].

$$C^* = (a^{*2} + b^{*2})^{1/2} \quad (1)$$

$$h° = arctan\ (b^*/a^*) \quad (2)$$

2.4. Biochemical Analysis of Strawberry Fruits

2.4.1. Sample Preparation and Extraction Procedure

Around 20 g of strawberry fruits per replicate was randomly collected at red mature stage and immediately frozen in liquid nitrogen and stored at −80 °C. The extraction was conducted using 25 mg of lyophilized sample, which was homogenized and extracted in 1.8 mL of extraction solution (80% methanol acidified with H_3PO_4, pH 1.0) and in 30 μL of 0.1 M NaF solution for 15 min at 5 °C. The extract was then filtered with PTFE filters (0.45 mm, Thermo Fisher Scientific), and the filtrate was stored at −80 °C until analysis. The extraction procedure specific for ascorbic acid analysis is described in Section 2.4.5.

2.4.2. Total Polyphenols Content (TPC)

Total phenolic content determination was performed on strawberry fruit extracts using the Folin–Ciocalteu assay following the methodology described in Meyers et al. [41] with some modifications. A total of 60 µL of the sample extract was diluted with 250 µL of deionized water. Then, 60 µL of Folin-Ciocalteu reagents was added, and the mixture was allowed to react in the dark for 6 min at 20 °C in agitation at 1500 rpm in the Thermomixer. A total of 630 µL of Na_2CO_3 (7.5% *w/v*) was added and incubated for 90 min at 20 °C always in the dark and under agitation at 1500 rpm. The total polyphenol content was determined at 740 nm using a spectrophotometer Cary 60 UV–Vis (Agilent Technologies, Santa Clara, CA, USA). Gallic acid standard solutions were used to calibrate the method (range 5–500 mg L^{-1}, $r^2 > 0.999$). The content of total polyphenols in each strawberry fruit extract was calculated and expressed as mg of gallic acid equivalents (GAE) per 100 g of dry weight (mg GAE 100 g^{-1} DW).

2.4.3. Determination of Total Anthocyanin Content (TAC)

Total anthocyanin content in strawberry extracts was determined according to Lee et al. [42] using the spectrophotometric pH differential method. Two dilutions of the same sample were prepared by adding 200 µL of extract to 800 µL of potassium chloride (0.25 M, pH 1) and to 800 µL of sodium acetate (0.4 M, pH 4.5), respectively. Absorbance was measured with a Cary 60 UV–Vis spectrophotometer at 520 and 700 nm at pH 1 and 4.5, where A = (A520−A700) pH 1−(A520−A700) pH 4.5. Total anthocyanin content was calculated using the Lambert–Beer law (ε = 26900 L mol^{-1} cm, MW = 449.2 g mol^{-1}) as mg cyanidin 3-glucoside equivalents (CGE) per 100 g of dry weight (mg CGE 100 g^{-1} DW).

2.4.4. Antioxidant Activity (ABTS)

Radical scavenging activity of the strawberry extracts was determined as described by Re et al. [43] with some modifications. ABTS was dissolved in water to a 7 mM concentration; the ABTS radical was obtained from reaction of ABTS stock solution with 2.45 mM potassium persulfate and allowing the mixture to stand in the dark at 4 °C for 16 h before use. The ABTS radical solution was diluted with deionized water to reach absorbance of 0.700 ± 0.02 at 734 nm. A total of 30 µL of each sample extract was added to 1.97 mL of diluted ABTS solution. The absorbance was measured at 734 nm on the Cary 60 UV–Vis spectrophotometer after 10 min in dark conditions. For each sample, the percentage of inhibition after 10 min of reaction was measured, and the concentration of sample (Trolox equivalent antioxidant capacity, TEAC) was calculated using the external calibration curve of the Trolox standard (Trolox, range 15.6–500 mg L^{-1}, $r^2 > 0.999$). The results were expressed as milligrams Trolox equivalents per 100 g of dry weight (mg Trolox 100 g^{-1} DW).

2.4.5. Ascorbic Acid Quantification

The analytical method for the ascorbic acid was based on Bassi et al. [44]. A total of 50 mg of freeze-dried sample was extracted using 1 mL of the extraction solution (700 µL deionized H_2O containing 8% (*v/v*) acetic acid and 3% (*w/v*) metaphosphoric acid added with 300 µL of methanol) [45], mixed at 3200 rpm for about 20 s at room temperature and filtered through a 0.20 µm PTFE filter. An HPLC Agilent 1260 Infinity system (Santa Clara, CA, USA) with a diode array (1260 DAD VL) detector, controlled through the software Agilent ChemStation™ (ver. C.01.03) (Agilent, Santa Clara, CA, USA), was used for the analysis of the ascorbic acid. The separation of the analyte was carried out at 25 °C using a Kinetex 5 µ C18 100 Å column (150 mm × 4.6 mm, 5 µm particle size; Phenomenex, Torrance, CA, USA) equipped with a precolumn (4.6 mm; Security Card, Phenomenex, Torrance, CA, USA). Detection wavelength was 260 nm. The mobile phases used were 5 mM KH_2PO_4, pH 4.8 (solvent A), and methanol (solvent B). The analytical gradient was as follows: 0 min, 100% A; 2.5 min, 100% A; 6 min, 80% A; 8 min, 100% A; and 13 min, 100% A. The flow rate was set at 1.0 mL min^{-1}. The temperature of the autosampler was 4 °C, and injection volume was 5 µL.

2.4.6. Chemicals

Ethanol (96%) was purchased from J.T. Baker (Center Valley, PA, USA). Acetic acid (96%) and potassium chloride were purchased from Merck (Kenilworth, NJ, USA), whereas hydrochloric acid (36%) and sodium acetate (anhydrous) were obtained from Fisher Chemical (Fisher Scientific, MA, USA). Phosphoric acid (≥99%), Folin–Ciocalteu's phenol reagent, sodium carbonate, 2,2′-azino-bis(3-ethylbenzothiazoline-6-sulphonic acid) diammonium salt (ABTS), sodium fluoride, ascorbic acid (99%), and Trolox (6-hydroxy-2,5,7,8-tetramethylchroman-2-carboxylic acid) were acquired from Sigma-Aldrich (St. Louis, MO, USA). Potassium persulfate ($K_2S_2O_8$) and gallic acid (≥99%) were purchased from Carl Roth (Karlsruhe, Germany). Methanol (HPLC-grade) was purchased from VWR Chemicals (Milan, Italy), and meta-phosphoric acid (≥99%) and monopotassium phosphate (≥99%) were acquired from Thermo Fisher Scientific Inc. (Waltham, MA, USA). The ultrapure water was prepared with a Millli-Q-water purification system (EMD Millipore Corporation, Billerica, MA, USA).

2.5. Mineral Element Analysis in Plant Tissues

The analysis of macro- and micronutrients was performed by collecting a 5 g DW sample of leaves and roots from 4 randomly selected plants at 0 DAT and from each of the 4 replicates per treatment at 90 DAT. Nitrogen content was determined with an elemental analyzer (LECO Truspec, LECO Corporation, St. Joseph, MI, USA); method according to Dumas [46]. The other macro- (P, K, Ca, Mg) and microelements (S, Fe, Cu, B, Zn, Mn, Na, Si) were analyzed with microwave-assisted acid digestion (EPA 3052 1996; [47]; Milestone Srl, Sorisole, BG, Italy Mod. UltraWave) using the inductively coupled plasma optical emission spectrometry (ICP-OES; [48]; Agilent, Santa Clara, CA, USA, Model 720).

2.6. Statistical Analysis

Significant differences between treatments were determined with a one-way analysis of variance (ANOVA) after that data were checked for normal distribution (Shapiro–Wilk normality test, $p > 0.05$) and equality of variances (Bartlett's test, $p > 0.05$). Mean separation was performed with the Dunnett test for $p < 0.05$. This procedure is recommended when working with several treatments [49], as in this case. Data expressed in percentage were arcsine-transformed prior to the application of the ANOVA. All analyses were carried out in R v. 3.3.1. (R Development Core Team, Vienna, Austria, 2016). Values were expressed as mean ± standard deviation (SD).

3. Results

3.1. Plant Growth Parameters

Biomass accumulation in strawberry leaves varied approximately between 1 and 5 g DW (Table 4 and Figure S2, supplementary material). Mix of amino acids (MAA)-treated plants showed a significantly lower leaf dry weight accumulation as compared to control. No significant differences were detected in the biomass accumulation in crown, shoots, and stolons of the differently treated plants (Table 4). Root biomass accumulation in untreated plants was generally low (0.6 g DW on average), whereas treatments with alfalfa protein hydrolysate (APH), macroseaweed extract (SEA), microalga hydrolysate (SPI), B-group vitamins (VIT), and chitosan (CHI) significantly enhanced biomass accumulation in the root system (e.g., approximately 7-fold in APH-treated plants). Fruit dry matter accumulation appeared to be significantly reduced in MAA- and amino acids combined with pure phenylalanine (PHE)-treated plants (values around 2 and 3 g DW plant^{-1} respectively, approximately −40% as compared to control) and increased in CHI- and Siliforce® (SIL)-treated plants (approximately by 20%). All biostimulant-treated plants (with the exception of SIL) resulted in being characterized by a higher relative share of dry matter allocation in the root system at the end of the growth cycle (+40%–80%) as compared to control (Figure 2).

Table 4. Accumulated biomass in different plant organs (g DW plant^{-1}) as affected by biostimulant applications.

Treatment	Leaves	Shoots and Stolons	Crowns	Roots	Fruits
CON	3.35 ± 1.32 [1]	0.87 ± 0.38	0.04 ± 0.25	0.59 ± 0.87	4.01 ± 0.19
HAL	3.17 ± 2.36	0.47 ± 0.20	0.07 ± 0.16	2.26 ± 1.72	3.96 ± 0.15
APH	3.47 ± 2.08	0.85 ± 0.30	0.13 ± 0.21	4.19 ± 0.34 ***	4.31 ± 0.15
SEA	2.05 ± 1.16	0.59 ± 0.22	0.08 ± 0.13	3.06 ± 1.49 *	4.26 ± 0.33
SPI	4.85 ± 1.81	0.83 ± 0.21	0.25 ± 0.25	2.91 ± 1.19 *	3.76 ± 0.20
MAA	0.87 ± 0.78 *	0.56 ± 0.39	0.10 ± 0.22	0.79 ± 0.97	1.98 ± 0.58 ***
PHE	1.70 ± 1.47	0.45 ± 0.37	0.00 ± 0.17	2.00 ± 1.70	3.01 ± 0.16 ***
ZIN	2.40 ± 1.00	0.65 ± 0.26	0.01 ± 0.22	1.56 ± 0.64	3.79 ± 0.08
VIT	5.30 ± 1.17	1.16 ± 0.21	0.33 ± 0.39	4.13 ± 2.29 **	4.32 ± 0.12
CHI	4.81 ± 0.85	0.86 ± 0.42	0.58 ± 0.37	3.75 ± 1.85 **	5.05 ± 0.39 ***
SIL	3.14 ± 1.42	0.69 ± 0.44	0.12 ± 0.31	1.49 ± 1.19	4.64 ± 0.57 *

[1] Means ± SD. Values followed by asterisk indicate significant differences between a single treatment group and control group, according to Dunnett's test ($n = 4$). *** $p < 0.001$; ** $p < 0.01$; * $p < 0.05$; values followed by no asterisk indicate absence of any significant differences. Treatments legend: CON, control; HAL, humic acids; APH, alfalfa protein hydrolysate; SEA, macroseaweed extract; SPI, microalga hydrolysate; MAA, mix of amino acids; PHE, MAA combined with pure phenylalanine; ZIN, MAA combined with zinc; VIT, B-group vitamins; CHI, chitosan; SIL, Siliforce®.

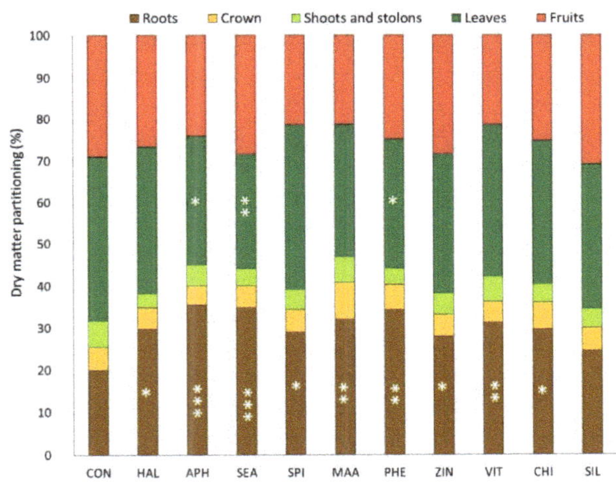

Figure 2. Dry matter distribution in strawberry plant organs treated with different biostimulant products and water (control). $n = 4$. The asterisk indicates significant differences between a single treatment group and control group within each plant organ, according to Dunnett's test. *** $p < 0.001$; ** $p < 0.01$; * $p < 0.05$; no asterisk indicates absence of any significant differences. Treatments legend: CON, control; HAL, humic acids; APH, alfalfa protein hydrolysate; SEA, macroseaweed extract; SPI, microalga hydrolysate; MAA, mix of amino acids; PHE, MAA combined with pure phenylalanine; ZIN, MAA combined with zinc; VIT, B-group vitamins; CHI, chitosan; SIL, Siliforce®.

Relative leaf dry matter allocation was significantly lower in SEA-, APH-, and PHE-treated plants, whereas in fruits, crown, shoots, and stolons, there were no significant differences as compared to the control. At different phases of the experimental cycle, the leaf area of plants treated with APH, VIT, CHI, and SIL was significantly larger than control (+15%–30%), whereas humic acids (HAL) and PHE applications significantly reduced the leaf surface (about 80 cm^2 compared to 100 cm^2 in control plants) at the end of the experiment (Figure 3).

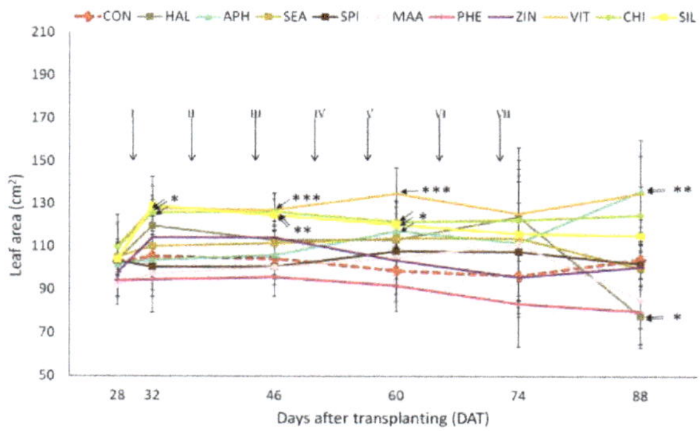

Figure 3. Leaf area dynamics during the period from 28 to 88 days after transplanting (DAT) (end of growth and production) in strawberry plants treated with different biostimulant products and water (control). Arrows indicate when the 7 biostimulant applications were performed. Vertical bars indicate mean ± SD, $n = 4$. The asterisk indicates significant differences according to Dunnett's test. *** $p < 0.001$; ** $p < 0.01$; * $p < 0.05$; no asterisk indicates absence of any significant differences.

3.2. Nutrient Concentration in Plant Tissues and SPAD Values

Macronutrient concentrations in strawberry leaves were not affected by the tested biostimulants (Table 5). Nitrogen content ranged between 2.4% and 2.7% DW, whereas phosphorus concentration was around 0.5% DW. Leaf content for potassium and calcium was approximately 2.6% and 2.2%, respectively. As for the magnesium (Mg) and sulfur (S), their concentrations were found around 0.4% (Mg) and 0.2% (S). The micronutrients zinc (Zn) and silicon (Si) presented higher concentrations in leaves of some of the tested biostimulants (Table 6). In detail, Zn was found significantly higher in leaves of MAA combined with zinc (ZIN)- and SIL-treated plants (+338% and +190% as compared to control, respectively), whereas Si content in strawberry leaves was significantly increased in plants treated with APH, ZIN and SIL, with values higher than 600 mg kg^{-1} DW compared to around 400 mg kg^{-1} DW of the control leaves. No significant differences were observed for the other micronutrient contents.

Macronutrient concentrations were not affected by biostimulant applications also at root level (Table 7). As far as the micronutrient concentrations are concerned, root boron content was found significantly higher (+20–30%) in control plants as compared to the majority of the biostimulant-treated plants (Table 8). Iron (Fe) content in root tissues was promoted by SPI and VIT applications (+273% and +202% as compared to control, respectively). Moreover, the Zn concentration was significantly increased in roots of ZIN-treated plants (140 mg kg^{-1} DW in comparison to 77 mg kg^{-1} DW in control roots). Finally, Si content was found to be higher in roots treated with HAL, SEA, SPI, ZIN, VIT, CHI, and SIL (around +75% as compared to control).

Table 5. Macronutrient concentrations in strawberry leaves at 90 DAT (days after transplanting) as affected by biostimulant applications.

Treatment	N (% DW)	P (% DW)	K (% DW)	Ca (% DW)	Mg (% DW)	S (% DW)
CON	2.59 ± 0.25 [1]	0.51 ± 0.04	2.52 ± 0.18	2.37 ± 0.15	0.38 ± 0.02	0.22 ± 0.03
HAL	2.46 ± 0.04	0.52 ± 0.05	2.72 ± 0.42	2.27 ± 0.27	0.39 ± 0.02	0.22 ± 0.03
APH	2.60 ± 0.27	0.58 ± 0.08	2.64 ± 0.12	2.26 ± 0.22	0.38 ± 0.01	0.22 ± 0.03
SEA	2.56 ± 0.17	0.55 ± 0.05	2.81 ± 0.38	2.25 ± 0.20	0.39 ± 0.03	0.24 ± 0.05
SPI	2.49 ± 0.35	0.56 ± 0.07	2.58 ± 0.23	2.42 ± 0.16	0.42 ± 0.04	0.24 ± 0.07
MAA	2.56 ± 0.09	0.62 ± 0.04	2.59 ± 0.12	2.14 ± 0.13	0.38 ± 0.03	0.20 ± 0.01
PHE	2.74 ± 0.39	0.52 ± 0.11	2.77 ± 0.08	2.42 ± 0.32	0.40 ± 0.03	0.24 ± 0.05
ZIN	2.47 ± 0.16	0.52 ± 0.09	2.75 ± 0.08	2.16 ± 0.24	0.37 ± 0.03	0.21 ± 0.05
VIT	2.38 ± 0.29	0.62 ± 0.02	2.79 ± 0.32	2.24 ± 0.18	0.42 ± 0.02	0.19 ± 0.03
CHI	2.36 ± 0.17	0.54 ± 0.05	2.69 ± 0.15	2.10 ± 0.15	0.39 ± 0.01	0.20 ± 0.01
SIL	2.38 ± 0.21	0.51 ± 0.06	2.74 ± 0.25	2.16 ± 0.15	0.37 ± 0.03	0.18 ± 0.03

[1] Means ± S.D. Values followed by asterisk indicate significant differences between a single treatment group and control group, according to Dunnett's test ($n = 4$). *** $p < 0.001$; ** $p < 0.01$; * $p < 0.05$; no asterisk indicates absence of any significant differences.

Table 6. Micronutrient concentrations in strawberry leaves at 90 DAT (days after transplanting) as affected by biostimulant applications.

Treatment	B (mg kg^{-1} DW)	Fe (mg kg^{-1} DW)	Mn (mg kg^{-1} DW)	Cu (mg kg^{-1} DW)	Zn (mg kg^{-1} DW)	Na (mg kg^{-1} DW)	Si (mg kg^{-1} DW)
CON	66.58 ± 10.93	81.65 ± 4.89	381.88 ± 107.48	7.00 ± 0.00	28.93 ± 3.01	240.12 ± 84.80	416.88 ± 61.75
HAL	58.75 ± 3.78	111.15 ± 28.87	331.58 ± 47.57	10.10 ± 3.59	34.53 ± 10.27	319.56 ± 45.58	538.53 ± 158.67
APH	60.55 ± 7.41	102.68 ± 32.96	355.08 ± 60.70	7.35 ± 0.64	25.98 ± 1.79	251.92 ± 126.22	782.65 ± 68.74 ***
SEA	55.93 ± 3.32	197.75 ± 143.89	300.90 ± 101.74	7.15 ± 0.30	28.73 ± 6.49	363.06 ± 74.44	424.30 ± 18.28
SPI	62.08 ± 16.99	114.60 ± 13.64	378.10 ± 104.30	7.00 ± 0.00	31.75 ± 2.32	327.93 ± 127.08	471.50 ± 33.98
MAA	61.18 ± 5.57	800.55 ± 1348.74	367.60 ± 57.87	9.53 ± 3.18	34.88 ± 7.92	243.16 ± 66.33	457.83 ± 113.69
PHE	61.75 ± 13.01	94.93 ± 16.11	375.75 ± 117.59	7.38 ± 0.75	31.38 ± 4.81	440.30± 262.85	480.68 ± 146.04
ZIN	59.43 ± 6.13	310.85 ± 332.71	285.73 ± 83.13	7.65 ± 1.23	127.43 ± 13.95 ***	270.27 ± 68.68	624.68 ± 98.52 **
VIT	62.40 ± 5.67	91.75 ± 13.21	347.75 ± 59.23	7.58 ± 0.67	36.15 ± 4.11	234.82 ± 79.06	398.65 ± 71.72
CHI	59.50 ± 4.46	115.40 ± 43.02	393.08 ± 38.55	7.13 ± 0.25	30.90 ± 0.85	218.92 ± 20.99	487.33 ± 110.15
SIL	60.40 ± 4.43	93.48 ± 23.08	366.53 ± 87.71	7.23 ± 0.45	83.80 ± 13.61 ***	254.18 ± 126.09	689.13 ± 41.81 ***

[1] Means ± S.D. Values followed by asterisk indicate significant differences between a single treatment group and control group, according to Dunnett's test ($n = 4$). *** $p < 0.001$; ** $p < 0.01$; * $p < 0.05$; no asterisk indicates absence of any significant differences.

Table 7. Macronutrient concentrations in strawberry roots at 90 DAT (days after transplanting) as affected by biostimulant application.

Treatment	N (% DW)	P (% DW)	K (% DW)	Ca (% DW)	Mg (% DW)	S (% DW)
CON	1.44 ± 0.12 [1]	0.15 ± 0.03	0.49 ± 0.18	1.60 ± 0.22	0.21 ± 0.02	0.35 ± 0.03
HAL	1.28 ± 0.17	0.13 ± 0.02	0.46 ± 0.16	1.74 ± 0.17	0.22 ± 0.01	0.37 ± 0.08
APH	1.45 ± 0.14	0.14 ± 0.05	0.44 ± 0.22	1.85 ± 0.29	0.23 ± 0.05	0.45 ± 0.09
SEA	1.41 ± 0.15	0.14 ± 0.03	0.40 ± 0.08	1.97 ± 0.16	0.23 ± 0.02	0.50 ± 0.08
SPI	1.35 ± 0.17	0.13 ± 0.02	0.43 ± 0.09	1.77 ± 0.16	0.22 ± 0.03	0.40 ± 0.07
MAA	1.65 ± 0.18	0.18 ± 0.03	0.49 ± 0.07	1.68 ± 0.21	0.22 ± 0.03	0.46 ± 0.13
PHE	1.39 ± 0.17	0.14 ± 0.02	0.48 ± 0.18	1.86 ± 0.23	0.20 ± 0.02	0.45 ± 0.13
ZIN	1.45 ± 0.08	0.15 ± 0.02	0.50 ± 0.14	1.82 ± 0.17	0.24 ± 0.04	0.43 ± 0.07
VIT	1.44 ± 0.08	0.14 ± 0.02	0.45 ± 0.09	1.91 ± 0.35	0.26 ± 0.01	0.57 ± 0.28
CHI	1.26 ± 0.19	0.13 ± 0.03	0.39 ± 0.03	1.93 ± 0.17	0.23 ± 0.01	0.46 ± 0.12
SIL	1.40 ± 0.07	0.17 ± 0.03	0.59 ± 0.18	1.69 ± 0.25	0.23 ± 0.02	0.41 ± 0.13

[1] Means ± S.D. Values followed by asterisk indicate significant differences between a single treatment group and control group, according to Dunnett's test ($n = 4$). *** $p < 0.001$; ** $p < 0.01$; * $p < 0.05$; no asterisk indicates absence of any significant differences.

Table 8. Micronutrient concentrations in strawberry roots at 90 DAT (days after transplanting) as affected by biostimulant application.

Treatment	B (mg kg^{-1} DW)	Fe (mg kg^{-1} DW)	Mn (mg kg^{-1} DW)	Cu (mg kg^{-1} DW)	Zn (mg kg^{-1} DW)	Na (mg kg^{-1} DW)	Si (mg kg^{-1} DW)
CON	17.28 ± 2.29	484.63 ± 94.84	56,25 ± 23,07	80.08 ± 53.24	77.08 ± 43.06	702.14 ± 81.07	350.25 ± 78.52
HAL	13.10 ± 3.45 *	1227.43 ± 538.00	51,43 ± 22,80	98.13 ± 73.45	67.48 ± 56.12	662.19 ± 118.41	528.63 ± 97.22 *
APH	11.43 ± 2.66 **	620.83 ± 215.52	81,40 ± 12,66	52.65 ± 19.62	44.98 ± 30.09	624.75 ± 77.51	444.93 ± 92.27
SEA	10.90 ± 0.98 ***	724.53 ± 214.65	55,35 ± 38,78	146.40 ± 84.29	42.95 ± 15.41	681.68 ± 81.97	538.88 ± 105.99 *
SPI	12.35 ± 2.64 **	1807.48 ± 1745.77 **	75,45 ± 31,44	68.90 ± 52.69	57.78 ± 38.74	624.80 ± 141.86	683.63 ± 182.23 ***
MAA	13.73 ± 3.36	648.95 ± 150.92	70,58 ± 19,96	207.95 ± 222.50	127.05 ± 94.10	579.94 ± 100.10	515.35 ± 127.19
PHE	12.98 ± 1.78 *	691.78 ± 309.30	97,50 ± 88,66	83.70 ± 113.54	40.78 ± 15.81	586.40 ± 61.30	461.80 ± 117.93
ZIN	12.33 ± 2.46 **	808.08 ± 157.22	70,25 ± 38,06	69.63 ± 52.43	139.78 ± 32.20 *	643.41 ± 191.04	689.03 ± 93.68 ***
VIT	8.53 ± 1.98 ***	1461.73 ± 956.78 *	100,13 ± 73,10	228.33 ± 63.72	95.98 ± 35.03	600.22 ± 90.50	631.18 ± 99.86 **
CHI	10.33 ± 2.76 ***	1079.63 ± 661.56	133,63 ± 39,71	80.23 ± 41.43	38.28 ± 8.27	655.94 ± 19.66	623.28 ± 187.63 **
SIL	12.68 ± 1.91 *	519.48 ± 183.10	55,30 ± 29,18	69.08 ± 50.39	67.98 ± 22.73	565.72 ± 98.53	590.98 ± 57.86 **

[1] Means ± S.D. Values followed by asterisk indicate significant differences between a single treatment group and control group, according to Dunnett's test ($n = 4$). *** $p < 0.001$; ** $p < 0.01$; * $p < 0.05$; no asterisk indicates absence of any significant differences.

SPAD values were not affected by biostimulant applications during most of the growing cycle. Only at the end of the experiment (88 DAT) was a significant higher SPAD value measured in plants treated with APH (+17% as compared to control) (Figure 4).

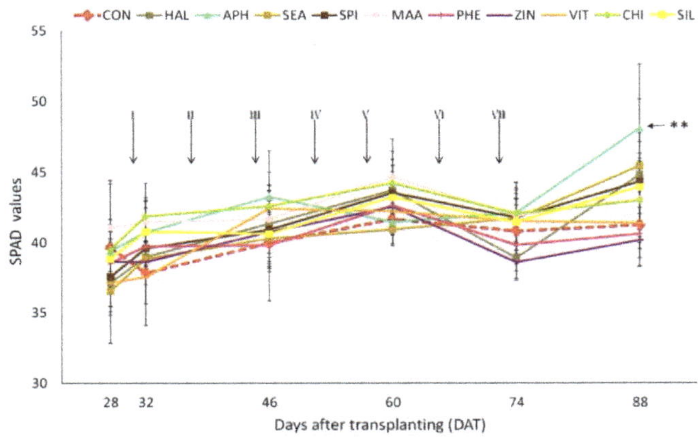

Figure 4. Chlorophyll content (as SPAD values) dynamics from 28 to 88 DAT (end of growth and production) in strawberry plants treated with different biostimulant products and water (control). Arrows indicate when the 7 biostimulant applications were performed. Vertical bars indicate mean ± SD, $n = 4$. The asterisk indicates significant differences according to Dunnett's test. *** $p < 0.001$; ** $p < 0.01$; * $p < 0.05$; no asterisk indicates absence of any significant differences.

3.3. *Leaf Gas Exchanges*

Before the fifth application (exactly at 57 DAT), a significantly higher leaf photosynthetic rate was measured for HAL-, ZIN-, and SIL-treated leaves (Figure 5A). After 24 h from the application, only SIL-treated leaves were still characterized by a significantly higher photosynthetic rate as compared to control (+32%), whereas at 48 h from the treatments, none of the biostimulants tested showed a significantly higher rate than control plants. Biostimulant-treated plants were characterized by a higher intensity of transpiration (almost double as compared to control) immediately before and after the fifth application (Figure 5B). Only HAL-treated plants did not show any significant difference with control 24 and 48 h after the application.

3.4. *Yield, Harvest and Fruit Quality*

CHI- and SIL-treated plants were characterized by a significantly higher average fruit weight as compared to control (Table 9). Their number of fruits per plant did not differ with control, leading to a significant 10–15 g yield increment per plant. Conversely, applications with amino acids (MAA and PHE) decreased the mean fruit weight, the number of fruits per plant, and the final yield (−49% and −22% as compared to control).

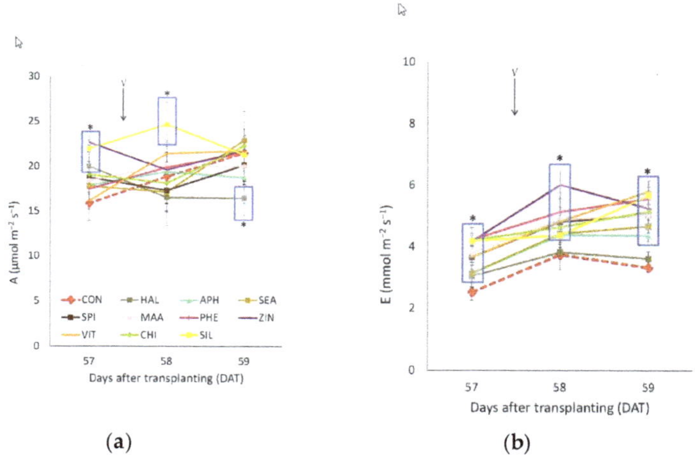

Figure 5. Photosynthetic (**a**) and transpiration (**b**) rates in strawberry leaves treated with different biostimulant products and water (control). 57 DAT: immediately before spray; 58 DAT: 24 h after spray; 59 DAT: 48 h after spray. Arrow indicates the fifth biostimulant application. Vertical bars indicate mean ± SD, $n = 4$. Asterisk (*), and the blue rectangle in the background indicates the group of treatments that significantly differed from control according to Dunnett's test ($p < 0.05$); no asterisk indicates absence of any significant differences.

Table 9. Yield parameters (final yield, number of fruits per plant and mean fruit weight) as affected by biostimulant products.

Treatment	Total Yield (g plant^{-1} FW [a])	Number Fruits Plant^{-1} (N°)	Mean Fruit Weight (g FW)
CON	50.84 ± 3.57 [1]	7.75 ± 0.25	6.55 ± 0.24
HAL	50.95 ± 1.95	7.42 ± 0.72	6.97 ± 0.42
APH	51.67 ± 2.90	7.00 ± 0.25	7.43 ± 0.46 *
SEA	53.57 ± 1.17	8.25 ± 0.87	6.52 ± 0.40
SPI	51.05 ± 7.71	7.17 ± 0.38	7.16 ± 0.07
MAA	25.95 ± 4.69 ***	4.50 ± 0.66 ***	5.74 ± 0.76 *
PHE	39.59 ± 2.73 **	6.58 ± 0.38 *	6.08 ± 0.43
ZIN	47.51 ± 1.89	7.17 ± 0.58	6.82 ± 0.84
VIT	54.58 ± 2.94	8.25 ± 0.25	6.60 ± 0.09
CHI	64.88 ± 3.02 ***	8.42 ± 0.80	7.72 ± 0.22 **
SIL	59.57 ± 6.29 *	7.92 ± 0.88	7.54 ± 0.11 *

[a] FW: fresh weight; [1] Means ± SD. Values followed by asterisk indicate significant differences between a single treatment group and control group, according to Dunnett's test ($n = 4$). *** $p < 0.001$; ** $p < 0.01$; * $p < 0.05$; no asterisk indicates absence of any significant differences.

The majority of strawberries (around 60% of the total yield per plant) were picked during the middle period of the harvesting time, corresponding to the second of a 3-week harvest period (Figure 6). VIT- and MAA-treated plants were characterized by a larger share of fruits picked at the beginning of the harvest, therefore showing an overall earlier ripening of the strawberries. Conversely, SEA-, SPI-, and CHI-treated plants showed a delayed maturation process as suggested by the higher share of fruits that were picked during the last period of the harvest.

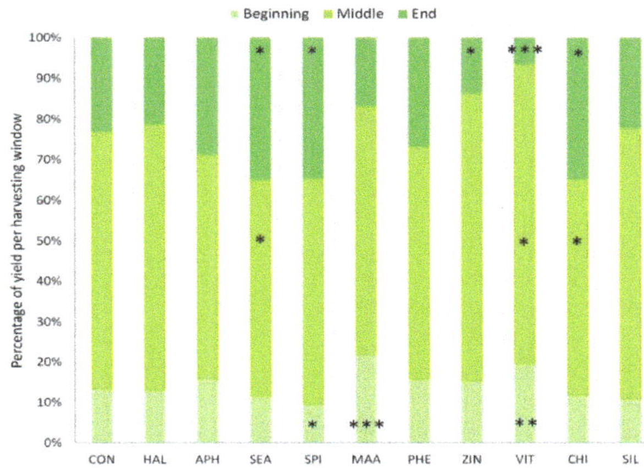

Figure 6. Percentage of the total yield per plant harvested during the three harvesting windows (beginning, middle, and end). $n = 4$. Asterisk indicates significant differences between a single treatment group and control group within a harvesting window, according to Dunnett's test. *** $p < 0.001$; ** $p < 0.01$; * $p < 0.05$; no asterisk indicates absence of any significant differences.

Strawberry flesh firmness ranged between 0.5 and 0.8 kg cm^{-2} (Table 10). Fruits from CHI-treated plants were characterized by significantly firmer pulp as compared to control (+18%), whereas SPI-treated strawberries were significantly softer and sourer (10.1 g L^{-1} TA). Total soluble solids were generally increased by the biostimulant applications, with MAA-treated strawberries being those presenting the highest sugars accumulation (8.7 °Brix TSS, Table 8). Chroma values (C*) were significantly higher in fruits treated with HAL, APH, SEA, and ZIN (around +18% in comparison to control), indicating fruits with a brighter red coloration (Table 10), whereas no significant differences in fruit hue angle were detected.

Total phenolic content varied between 2100 and 2800 mg GAE 100 g^{-1} DW (Table 11). SEA and APH applications significantly enhanced the final total phenolic content of mature strawberry fruits (+20% as compared to control). Total anthocyanin content (TAC) in fruits was not significantly affected by biostimulant applications (Table 11), even though HAL, APH, and SEA were among those treatments characterized by the highest anthocyanin accumulation. The fruit antioxidant potential measured as ABTS was not altered significantly by the treatments, whereas the ascorbic acid concentration was significantly reduced by both biostimulant products containing the zinc mineral element such as ZIN and SIL (−24% and −19% as compared to control, respectively) (Table 11).

Table 10. Fruit quality traits (FF—firmness; TSS—total soluble solids; TA—titratable acidity and color parameters: C*—chroma; h°—hue angle) as affected by biostimulant products.

Treatment	FF (kg cm^{-2})	TSS (° Brix)	TA (g L^{-1})	C *	h°
CON	0.71 ± 0.02 [1]	7.07 ± 0.06	9.11 ± 0.09	27.41 ± 1.36	18.31 ± 3.46
HAL	0.69 ± 0.02	7.73 ± 0.61	9.32 ± 0.08	31.58 ± 2.71 *	17.72 ± 3.90
APH	0.73 ± 0.02	7.07 ± 0.51	9.48 ± 0.18	34.51 ± 2.54 ***	17.28 ± 3.70
SEA	0.72 ± 0.01	7.73 ± 0.50	9.74 ± 0.19	34.72 ± 3.74 ***	18.68 ± 2.01
SPI	0.58 ± 0.09 **	7.03 ± 0.15	10.19 ± 0.99 *	29.89 ± 1.07	16.13 ± 2.84
MAA	0.73 ± 0.04	8.70 ± 0.70 ***	8.66 ± 0.40	30.06 ± 0.90	16.98 ± 0.63
PHE	0.72 ± 0.03	7.23 ± 0.25	9.60 ± 0.29	28.69 ± 3.23	16.23 ± 0.80
ZIN	0.71 ± 0.03	7.97 ± 0.2 *	9.44 ± 0.11	32.86 ± 2.37 **	17.03 ± 4.61
VIT	0.74 ± 0.05	7.80 ± 0.44 *	8.79 ± 0.55	30.25 ± 1.99	14.38 ± 3.18
CHI	0.84 ± 0.01 **	6.97 ± 0.15	8.72 ± 1.00	29.05 ± 1.81	16.94 ± 2.84
SIL	0.73 ± 0.11	7.37 ± 0.15	9.34 ± 0.27	29.75 ± 1.04	17.08 ± 1.55

[1] Means ± SD. Values followed by asterisk indicate significant differences between a single treatment group and control group, according to Dunnett's test ($n = 4$). *** $p < 0.001$; ** $p < 0.01$; * $p < 0.05$; no asterisk indicates absence of any significant differences.

Table 11. Fruit nutraceutical values (TPC—total phenolic content; TAC—total anthocyanin content; ABTS—antioxidant potential; AA—ascorbic acid) as affected by biostimulant products.

Treatment	TPC (mg GAE 100 g^{-1} DW)	TAC (mg CGE 100 g^{-1} DW)	ABTS (mg Trolox 100 g^{-1} DW)	AA (mg AA 100 g^{-1} DW)
CON	2303.95 ± 116.11 [1]	354.89 ± 55.57	2886.04 ± 110.02	556.69 ± 37.18
HAL	2455.53 ± 406.32	435.01 ± 85.90	2865.18 ± 402.51	617.55 ± 102.80
APH	2803.17 ± 488.83 *	461.48 ± 56.26	3420.95 ± 252.11	572.33 ± 12.77
SEA	2734.73 ± 261.74 *	478.59 ± 126.19	3178.61 ± 296.83	548.29 ± 29.01
SPI	2211.79 ± 105.48	408.96 ± 13.62	2799.13 ± 44.49	582.32 ± 53.40
MAA	2327.76 ± 122.79	369.94 ± 139.69	3318.91 ± 463.59	587.54 ± 42.27
PHE	2506.83 ± 240.52	475.10 ± 83.04	2990.51 ± 316.85	539.14 ± 51.51
ZIN	2098.88 ± 152.98	404.44 ± 18.33	3011.75 ± 599.56	422.90 ± 56.38 *
VIT	2114.04 ± 270.33	396.17 ± 100.75	2831.57 ± 55.35	469.69 ± 10.50
CHI	2400.41 ± 40.91	417.22 ± 31.63	3047.21 ± 53.23	599.14 ± 84.73
SIL	2687.08 ± 197.94	307.97 ± 37.55	3220.11 ± 91.69	452.43 ± 94.26 *

[1] Means ± SD. Values followed by asterisk indicate significant differences between a single treatment group and control group, according to Dunnett's test ($n = 4$). *** $p < 0.001$; ** $p < 0.01$; * $p < 0.05$; no asterisk indicates absence of any significant differences.

4. Discussion

Abiotic stresses, including nutrient limitation, are generally the cause of a reduced plant growth and final yield [50]. Under the imposed experimental conditions, soilless cultivated strawberry plants were characterized by vegetative and reproductive performances that were below the standard for similar cultivation systems and cultivar [51,52]. This was also suggested by values of the leaf area to yield ratio that were found below 3 cm^2 g^{-1}, under the threshold considered as indicative of a good vegetative–reproductive balance for strawberry [53]. Despite the limited supply of nutrients to the plants, which consisted of a single fertigation done one week after transplant (Table 1), macronutrient levels in leaves and roots (Tables 5 and 7) were found within the range of sufficiency for strawberry plants as reported in the literature [54,55]. Only nitrogen level was found to be lower (1.4% on average) than the concentration range found in strawberry roots of the cultivar 'Selva´ grown under standard conditions (around 2.4%) [56]. Moreover, none of the applied biostimulants induced significant changes in the final macronutrient concentrations. Differently, some of the micronutrients (e.g., Fe, Zn, and Si) were found at a higher concentration in leaves and roots of plants treated with specific biostimulant compounds (Tables 6 and 8). More in detail, SPI-treated plants presented a higher Fe accumulation in the root system, probably as a consequence of the rhizosphere acidification promoted by selected metabolites included in the seaweed extract (e.g., kahydrin) that resulted in a more efficient mobilization and assimilation of the acid-soluble ions, including iron. A similar mechanism of Fe chelation following seaweed application was described by Spinelli et al. [19] on strawberry. ZIN and SIL applications resulted in higher concentrations of Zn and Si in leaves and roots (Tables 6 and 8). This result can be explained as the consequence of the direct supply of Zn and Si obtained by the application of both biostimulant treatments. Moreover, prior studies have highlighted the importance of amino acids (included in the ZIN formulation) as metal-chelating agents and carriers of micronutrients [57]. Regarding the lower root boron content observed in treated plants, Kaya et al. [58] reported that Si supply to tomato plants reduced boron concentration in plant tissues as a consequence of the B immobilization caused by the formation of boron–silicate complexes in the soil. Moreover, B availability is generally decreased under higher soil pH [59]. Although the substrate pH at the end of the experiment was not measured, biostimulant applications could have slightly enhanced soil pH, therefore limiting B availability for the plants [39,60].

Under the described growing and nutrient conditions, the use of selected biostimulants had a positive effect on strawberry plant growth and fruit yield. APH treatment enhanced biomass accumulation in roots, leaf area, and chlorophyll concentration (Table 4; Figures 3 and 4). These results reflect those of Ertani et al. [61] and Rouphael et al. [62] on pepper and tomato, respectively. The mode of action of the protein hydrolysates is likely to be linked to different plant physiological mechanisms, which include the stimulation of key enzymes involved in both the primary and secondary metabolisms, the hormone-like function of several components of the protein hydrolysate, and the indirect stimulation of the biological activities of the plant-associate microbes [63]. B-group vitamins were also effective in increasing leaf area and root growth (Figure 3 and Table 4). Vitamin B1 (thiamine) is an important cofactor involved in many primary metabolic processes (glycolysis, pentose phosphate pathway, and tricarboxylic acid cycle) [64,65], explaining therefore the enhanced growth performances shown by the treated plants under suboptimal growing conditions. CHI and SIL applications improved both above- and belowground growth and yield performances (Figure 3; Tables 4 and 9). These findings are consistent with those obtained by Mukta et al. [66] and by Hajiboland et al. [67] on chitosan- and silicon-treated strawberry plants and are probably linked with the enhanced stress tolerance shown by the treated plants [68,69]. As shown in Figure 5, biostimulants containing zinc (ZIN and SIL) and HAL were found to be able to enhance leaf photosynthesis. Zinc–metalloenzymes are important for the activity of the carboxylation process key enzyme (Rubisco) [70,71]. Considering that Zn-deficiency reduces the photosynthetic activity of plants [72], the enhanced photosynthetic performances characterizing ZIN- and SIL-treated plants could be therefore explained with their higher zinc concentration in leaves (Table 6).

Biostimulants have been also claimed to improve the quality attributes of horticultural products. The definition of quality of fruit and vegetable is considered extremely dynamic and involves a large number of intrinsic and extrinsic characteristics, including socioeconomic aspects linked to consumers' perceptions and acceptance of the products [73]. The potential role of plant biostimulants in enhancing the quality of greenhouse vegetables has been recently reviewed by Rouphael et al. [74]. In the present experiment, primary fruit quality attributes were slightly affected by the biostimulant applications (Table 10). Chitosan was found to be able to enhance flesh firmness with potential consequences on the shelf life period extension of treated fruits. Similar results were also obtained by Bhaskara Reddy et al. [75] and Hernández-Muñoz et al. [76] on strawberry (different cultivars). Treatments with amino acids and B-group vitamins led to an earlier ripening of the fruits (Figure 6) and to strawberries with higher sugar content (Table 10), a finding that was also described by Khan et al. [77] on grape fruits. Alfalfa protein hydrolysate and seaweed extracts improved the final coloration and the phenolic concentration in strawberry fruit at harvest (Tables 10 and 11). These results confirm the capacity of both biostimulants to interfere with the phenylpropanoid pathway, leading to more colored and stress-resistant crops [21,27,78,79]. Finally, a reduction in ascorbic acid level in fruits treated with biostimulants containing zinc (ZIN and SIL) was recorded (Table 11). This effect might be due to the inhibitory action of selected metals (e.g., zinc and copper) on the enzyme activity of ascorbic acid metabolism as also shown by Olkhovych et al. [80] on other species.

5. Conclusions

The aim of the present research was to examine if the application of different biostimulants could help the growth and yield performances of soilless cultivated strawberry plants under limited nutrient availability. The main findings suggest that selected biostimulants (i.e., alfalfa protein hydrolysate, B-group vitamins, chitosan, and silicon) could be effective in stimulate vegetative growth and final fruit yield. Moreover, biostimulants based on seaweed extracts, protein hydrolysate, and chitosan were found able to improve strawberry commercial (firmness and external color) and nutritional (phenolic compounds) quality. The outcomes of this research allow therefore a positive overall evaluation of biostimulants as agronomic tools able to contrast the negative consequence of growing crops under insufficient nutritional conditions. Further work needs to be done to deepen our understanding of the uptake mechanisms and use efficacy of the biostimulants under the current strawberry industrial production systems, according to their way of application and interaction with the different farming techniques (open field or protected cultivations, fresh or cold stored plants, genotypes, etc.).

Supplementary Materials: The following are available online at http://www.mdpi.com/2073-4395/9/9/483/s1, Figure S1. Cold stored strawberry plants before transplantation (0 DAT). Figure S2. Strawberry plants at the end of the experiment (90 DAT) as affected by the application of the different biostimulants. Treatments legend: CON, control; HAL, humic acids; APH, alfalfa protein hydrolysate; SEA, macroseaweed extract; SPI, microalga hydrolysate; MAA, mix of amino acids; PHE, MAA combined with pure phenylalanine; ZIN, MAA combined with zinc; VIT, B-group vitamins; CHI, chitosan; SIL, Siliforce®.

Author Contributions: S.S. contributed in the set-up of the experimental protocol, performed the experiment, processed the fruit samples, analyzed the data, and participated in the writing of the paper; M.K. contributed to the set-up of the experiment and to the interpretation of the results; C.C. gave technical assistance for the implementation of the study; M.B. contributed to the analysis of phenols, anthocyanins, antioxidant potential, and ascorbic acid; P.R. contributed to the set-up of the analytical work related to the quality indices of the fruits; A.M. contributed to the analysis of mineral elements; C.A. coordinated the research, worked on the statistical analysis, and contributed to the writing of several sections of the manuscript.

Funding: The research was co-founded by the Free University of Bozen-Bolzano (Project BIO_TOOL TN 1B07), Laimburg Research Centre, and ILSA S.p.A. Laimburg Research Centre was funded by the Autonomous Province of Bolzano; financial support for the "Technology Park" (Decision no. 1472, 07.10.2013) is gratefully acknowledged. This work was also supported by the Open Access Publishing Fund of the Free University of Bozen-Bolzano. SS was funded by a Ph.D. grant from the Free University of Bozen-Bolzano.

Conflicts of Interest: ILSA S.p.A. provided part of the biostimulants used for the experiment. The design of the study, the collection, the analysis and interpretation of the data, as well as article writing of the manuscript was carried out by the authors independently. The authors declare that the research was conducted in the absence of any commercial or financial relationships that could be construed as a potential conflict of interest.

References

1. Tagliavini, M.; Baldi, E.; Lucchi, P.; Antonelli, M.; Sorrenti, G.; Baruzzi, G.; Faedi, W. Dynamics of nutrients uptake by strawberry plants (*Fragaria* × *ananassa* Dutch.) grown in soil and soilless culture. *Eur. J. Agron.* **2005**, *23*, 15–25. [CrossRef]
2. Albregts, E.E.; Howard, C.M.; Chandler, C.K. Strawberry responses to K rate on a fine sand soil. *Hort. Sci.* **1991**, *26*, 135–138. [CrossRef]
3. Shirko, R.; Nazarideljou, M.J.; Akbar, M.A.; Naser, G. Photosynthetic reaction, mineral uptake, and fruit quality of strawberry affected by different levels of macronutrients. *J. Plant Nutr.* **2018**, *41*, 1807–1820. [CrossRef]
4. Hukkanen, A.T.; Kokko, H.I.; Buchala, A.J.; McDougall, G.J.; Stewart, D.; Kärenlampi, S.O.; Karjalainen, R.O. Benzothiadiazole induces the accumulation of phenolics and improves resistance to powdery mildew in strawberries. *J. Agric. Food Chem.* **2007**, *55*, 1862–1870. [CrossRef] [PubMed]
5. de Ponti, T.; Rijk, B.; van Ittersum, M.K. The crop yield gap between organic and conventional agriculture. *Agric. Syst.* **2012**, *108*, 1–9. [CrossRef]
6. van Eysinga, J.R.; van Caem, H.E. Nutrition of glasshouse strawberries with nitrogen, phosphorus and potassium. *Sci. Hortic.* **1977**, *7*, 359–368. [CrossRef]
7. European Council. Available online: https://www.consilium.europa.eu/ (accessed on 11 May 2019).
8. Goñi, O.; Quille, P.; O'Connell, S. *Ascophyllum nodosum* extract biostimulants and their role in enhancing tolerance to drought stress in tomato plants. *Plant Physiol. Biochem.* **2018**, *126*, 63–73. [CrossRef]
9. Di Stasio, E.; Van Oosten, M.J.; Silletti, S.; Raimondi, G.; dell'Aversana, E.; Carillo, P.; Maggio, A. *Ascophyllum nodosum*-based algal extracts act as enhancers of growth, fruit quality, and adaptation to stress in salinized tomato plants. *J. Appl. Phycol.* **2018**, *30*, 2675–2686. [CrossRef]
10. Lucini, L.; Rouphael, Y.; Cardarelli, M.; Canaguier, R.; Kumar, P.; Colla, G. The effect of a plant-derived biostimulant on metabolic profiling and crop performance of lettuce grown under saline conditions. *Sci. Hortic.* **2015**, *182*, 124–133. [CrossRef]
11. Li, Z.; Zhang, Y.; Zhang, X.; Merewitz, E.; Peng, Y.; Ma, X.; Huang, L.; Yan, Y. Metabolic pathways regulated by chitosan contributing to drought resistance in white clover. *J. Proteome Res.* **2017**, *16*, 3039–3052. [CrossRef]
12. Shi, Y.; Zhang, Y.; Han, W.; Feng, R.; Hu, Y.; Guo, J.; Gong, H. Silicon enhances water stress tolerance by improving root hydraulic conductance in *Solanum lycopersicum* L. *Front. Plant Sci.* **2016**, *7*, 196. [CrossRef]
13. De Pascale, S.; Rouphael, Y.; Colla, G. Plant biostimulants: Innovative tool for enhancing plant nutrition in organic farming. *Eur. J. Hortic. Sci.* **2018**, *82*, 277–285. [CrossRef]
14. Halpern, M.; Bar-Tal, A.; Ofek, M.; Minz, D.; Muller, T.; Yermiyahu, U. The use of biostimulants for enhancing nutrient uptake. In *Advances in Agronomy*; Sparks, D.L., Ed.; Elsevier Science Publishing Co Inc.: San Diego, CA, USA, 2015; Volume 130, pp. 141–174. ISBN 978-0-12-802137-8.
15. Mattner, S.W.; Milinkovic, M.; Arioli, T. Increased growth response of strawberry roots to a commercial extract from *Durvillaea potatorum* and *Ascophyllum nodosum*. *J. Appl. Phycol.* **2018**, *30*, 2943–2951. [CrossRef]
16. Crouch, I.J.; Beckett, R.P.; Van Staden, J. Effect of seaweed concentrate on the growth and mineral nutrition of nutrient-stressed lettuce. *J. Appl. Phycol.* **1990**, *2*, 269–272. [CrossRef]
17. Rathore, S.S.; Chaudhary, D.R.; Boricha, G.N.; Ghosh, A.; Bhatt, B.P.; Zodape, S.T.; Patolia, J.S. Effect of seaweed extract on the growth, yield and nutrient uptake of soybean (*Glycine max*) under rainfed conditions. *South Afr. J. Bot.* **2009**, *75*, 351–355. [CrossRef]
18. Rouphael, Y.; De Micco, V.; Arena, C.; Raimondi, G.; Colla, G.; De Pascale, S. Effect of *Ecklonia maxima* seaweed extract on yield, mineral composition, gas exchange, and leaf anatomy of zucchini squash grown under saline conditions. *J. Appl. Phycol.* **2017**, *29*, 459–470. [CrossRef]
19. Spinelli, F.; Fiori, G.; Noferini, M.; Sprocatti, M.; Costa, G. A novel type of seaweed extract as a natural alternative to the use of iron chelates in strawberry production. *Sci. Hortic.* **2010**, *125*, 263–269. [CrossRef]
20. Ghasemi, S.; Khoshgoftarmanesh, A.H.; Afyuni, M.; Hadadzadeh, H. Iron(II)–amino acid chelates alleviate salt-stress induced oxidative damages on tomato grown in nutrient solution culture. *Sci. Hortic.* **2014**, *165*, 91–98. [CrossRef]

21. Soppelsa, S.; Kelderer, M.; Casera, C.; Bassi, M.; Robatscher, P.; Andreotti, C. Use of biostimulants for organic apple production: Effects on tree growth, yield, and fruit quality at harvest and during storage. *Front. Plant Sci.* **2018**, *9*, 1342. [CrossRef]
22. Colla, G.; Cardarelli, M.; Bonini, P.; Rouphael, Y. Foliar applications of protein hydrolysate, plant and seaweed extracts increase yield but differentially modulate fruit quality of greenhouse tomato. *Hortscience* **2017**, *52*, 1214–1220. [CrossRef]
23. Mohamed, A.Y.; El-Sehrawy, O.A. Effect of seaweed extract on fruiting of Hindy Bisinnara mango trees. *J. Am. Sci.* **2013**, *9*, 537–544.
24. El-Miniawy, S.; Ragab, M.; Youssef, S.; Metwally, A. Response of strawberry plants to foliar spraying of chitosan. *Res. J. Agric. Biol. Sci.* **2013**, *9*, 366–372.
25. Neri, D.; Lodolini, E.M.; Savini, G.; Sabbatini, P.; Bonanomi, G.; Zucconi, F. Foliar application of humic acids on strawberry (*cv Onda*). In *Proceedings of the Acta Horticulturae*; International Society for Horticultural Science (ISHS): Leuven, Belgium, November 2002; pp. 297–302.
26. Roussos, P.A.; Denaxa, N.-K.; Damvakaris, T. Strawberry fruit quality attributes after application of plant growth stimulating compounds. *Sci. Hortic.* **2009**, *119*, 138–146. [CrossRef]
27. Fan, D.; Hodges, D.M.; Zhang, J.; Kirby, C.W.; Ji, X.; Locke, S.J.; Critchley, A.T.; Prithiviraj, B. Commercial extract of the brown seaweed *Ascophyllum nodosum* enhances phenolic antioxidant content of spinach (*Spinacia oleracea* L.) which protects *Caenorhabditis elegans* against oxidative and thermal stress. *Food Chem.* **2011**, *124*, 195–202. [CrossRef]
28. Rouphael, Y.; Giordano, M.; Cardarelli, M.; Cozzolino, E.; Mori, M.; Kyriacou, M.; Bonini, P.; Colla, G. Plant-and seaweed-based extracts increase yield but differentially modulate nutritional quality of greenhouse spinach through biostimulant action. *Agronomy* **2018**, *8*, 126. [CrossRef]
29. Cai, Z.; Kastell, A.; Mewis, I.; Knorr, D.; Smetanska, I. Polysaccharide elicitors enhance anthocyanin and phenolic acid accumulation in cell suspension cultures of Vitis vinifera. *Plant Cell Tissue Organ Cult. Pctoc* **2012**, *108*, 401–409. [CrossRef]
30. Portu, J.; López-Alfaro, I.; Gómez-Alonso, S.; López, R.; Garde-Cerdán, T. Changes on grape phenolic composition induced by grapevine foliar applications of phenylalanine and urea. *Food Chem.* **2015**, *180*, 171–180. [CrossRef]
31. El-Sayed, M.E.A. Improving fruit quality and marketing of "Crimson Seedless" grape using some preharvest treatments. *J. Hortic. Sci. Ornam. Plants* **2013**, *5*, 218–226.
32. Ertani, A.; Schiavon, M.; Muscolo, A.; Nardi, S. Alfalfa plant-derived biostimulant stimulate short-term growth of salt stressed *Zea mays* L. plants. *Plant Soil* **2012**, *364*, 145–158. [CrossRef]
33. Landi, L.; Feliziani, E.; Romanazzi, G. Expression of defense genes in strawberry fruits treated with different resistance inducers. *J. Agric. Food Chem.* **2014**, *62*, 3047–3056. [CrossRef]
34. Schiavon, M.; Pizzeghello, D.; Muscolo, A.; Vaccaro, S.; Francioso, O.; Nardi, S. High Molecular Size Humic Substances Enhance Phenylpropanoid Metabolism in Maize (*Zea mays* L.). *J. Chem. Ecol.* **2010**, *36*, 662–669. [CrossRef]
35. Vera, J.; Castro, J.; Gonzalez, A.; Moenne, A. Seaweed polysaccharides and derived oligosaccharides stimulate defense responses and protection against pathogens in plants. *Mar. Drugs* **2011**, *9*, 2514–2525. [CrossRef]
36. Xie, L.; Wang, Z.H.; Cheng, X.H.; Gao, J.J.; Zhang, Z.P.; Wang, L.J. 5-Aminolevulinic acid promotes anthocyanin accumulation in Fuji apples. *Plant Growth Regul.* **2013**, *69*, 295–303. [CrossRef]
37. Youssef, K.; Sanzani, S.M.; Ligorio, A.; Ippolito, A.; Terry, L.A. Sodium carbonate and bicarbonate treatments induce resistance to postharvest green mould on citrus fruit. *Postharvest Biol. Technol.* **2014**, *87*, 61–69. [CrossRef]
38. Saa, S.; Olivos-Del Rio, A.; Castro, S.; Brown, P.H. Foliar application of microbial and plant based biostimulants increases growth and potassium uptake in almond (*Prunus dulcis* [Mill.] D. A. Webb). *Front. Plant Sci* **2015**, *6*. [CrossRef]
39. Rouphael, Y.; Colla, G. Synergistic biostimulatory action: Designing the next generation of plant biostimulants for sustainable agriculture. *Front. Plant Sci.* **2018**, *9*, 1655. [CrossRef]
40. Sacks, E.J.; Shaw, D.V. Color change in fresh strawberry fruit of seven genotypes stored at 0C. *HortScience* **1993**, *28*, 209–210. [CrossRef]
41. Meyers, K.J.; Watkins, C.B.; Pritts, M.P.; Liu, R.H. Antioxidant and antiproliferative activities of strawberries. *J. Agric. Food Chem.* **2003**, *51*, 6887–6892. [CrossRef]

42. Lee, J.; Durst, R.W.; Wrolstad, R.E. Determination of total monomeric anthocyanin pigment content of fruit juices, beverages, natural colorants, and wines by the pH differential method: Collaborative study. *J. Aoac Int.* **2005**, *88*, 1269–1278.
43. Re, R.; Pellegrini, N.; Proteggente, A.; Pannala, A.; Yang, M.; Rice-Evans, C. Antioxidant activity applying an improved ABTS radical cation decolorization assay. *Free Radic. Biol. Med.* **1999**, *26*, 1231–1237. [CrossRef]
44. Bassi, M.; Lubes, G.; Bianchi, F.; Agnolet, S.; Ciesa, F.; Brunner, K.; Guerra, W.; Robatscher, P.; Oberhuber, M. Ascorbic acid content in apple pulp, peel, and monovarietal cloudy juices of 64 different cultivars. *Int. J Food Prop.* **2018**, 1–9. [CrossRef]
45. *AOAC Vitamin C in juices and vitamin preparations. Official Method 967.21*; AOAC Official Methods of Analysis, 18th; Association of Official Analytical Chemists: Gaithersburg, MD, USA, 2005; p. 45.1.14.
46. ISO 16634-1:2008. *Food Products—Determination of the Total Nitrogen Content by Combustion According to the Dumas Principle and Calculation of the Crude Protein Content—Part 1: Oilseeds and Animal Feeding Stuffs*; International Organization for Standardization: Geneva, Switzerland, 2008.
47. EPA 3052. *Microwave assisted acid digestion of siliceous and organically based matrics*; U.S. Environmental Protection Agency: Washington, DC, USA, 1996.
48. EPA 6010D. *Inductively coupled plasma–Optical emission spectrometry*; U.S. Environmental Protection Agency: Washington, DC, USA, 2018.
49. Dunnett, C.W. Multiple comparisons between several treatments and a specified treatment. In *Proceedings of the Linear Statistical Inference*; Caliński, T., Klonecki, W., Eds.; Springer: New York, NY, USA, 1985; pp. 39–47.
50. Cramer, G.R.; Urano, K.; Delrot, S.; Pezzotti, M.; Shinozaki, K. Effects of abiotic stress on plants: A systems biology perspective. *Bmc Plant Biol.* **2011**, *11*, 163. [CrossRef]
51. Jarosz, Z.; Konopinska, J. Effect of substrate type and nitrogen fertilization upon yielding and chemical composition of 'Elsanta' strawberry cultivar grown in unheated foil tunnel. *Acta Sci. Pol. Hortorum Cultus* **2010**, *9*, 87–96.
52. Saied, A.S.; Keutgen, A.J.; Noga, G. The influence of NaCl salinity on growth, yield and fruit quality of strawberry cvs. 'Elsanta' and 'Korona'. *Sci. Hortic.* **2005**, *103*, 289–303. [CrossRef]
53. Casierra-Posada, F.; Torres, I.D.; Blanke, M.M. Fruit quality and yield in partially defoliated strawberry plants in the tropical highlands. *Gesunde Pflanz.* **2013**, *65*, 107–112. [CrossRef]
54. Almaliotis, D.; Velemis, D.; Bladenopoulou, S.; Karapetsas, N. Leaf nutrient levels of strawberries (cv Tudla) in relation to crop yield. In *Proceedings of the Acta Horticulturae*; International Society for Horticultural Science (ISHS): Leuven, Belgium, 2002; pp. 447–450.
55. Campbell, C.R.; Miner, G.S. Strawberry, annual hill culture. In *Reference sufficiency ranges for plant analysis in the southern region of the United States*; Southern Cooperative Series Bulletin 394; North Carolina Department of Agriculture & Consumer Services: Raleigh, NC, USA, 2000; ISBN 1-58161-394-6.
56. Jamali, B.; Eshghi, S.; Tafazoli, E. Mineral composition of 'Selva' strawberry as affected by time of application of nitric oxide under saline conditions. *Hortic. Env. Biotechnol.* **2015**, *56*, 273–279. [CrossRef]
57. Popko, M.; Michalak, I.; Wilk, R.; Gramza, M.; Chojnacka, K.; Górecki, H. Effect of the new plant growth biostimulants based on amino acids on yield and grain quality of winter wheat. *Molecules* **2018**, *23*, 470. [CrossRef]
58. Kaya, C.; Tuna, A.L.; Guneri, M.; Ashraf, M. Mitigation effects of silicon on tomato plants bearing fruit grown at high boron levels. *J. Plant Nutr.* **2011**, *34*, 1985–1994. [CrossRef]
59. Tsadilas, C.D.; Kassioti, T.; Mitsios, I.K. Influence of liming and nitrogen forms on boron uptake by tobacco. *Commun. Soil Sci. Plant Anal.* **2005**, *36*, 701–708. [CrossRef]
60. Hellequin, E.; Monard, C.; Quaiser, A.; Henriot, M.; Klarzynski, O.; Binet, F. Specific recruitment of soil bacteria and fungi decomposers following a biostimulant application increased crop residues mineralization. *PLoS ONE* **2019**, *13*, e0209089. [CrossRef]
61. Ertani, A.; Sambo, P.; Nicoletto, C.; Santagata, S.; Schiavon, M.; Nardi, S. The use of organic biostimulants in hot pepper plants to help low input sustainable agriculture. *Chem. Biol. Technol. Agric.* **2015**, *2*, 11. [CrossRef]
62. Rouphael, Y.; Colla, G.; Giordano, M.; El-Nakhel, C.; Kyriacou, M.C.; De Pascale, S. Foliar applications of a legume-derived protein hydrolysate elicit dose-dependent increases of growth, leaf mineral composition, yield and fruit quality in two greenhouse tomato cultivars. *Sci. Hortic.* **2017**, *226*, 353–360. [CrossRef]

63. Colla, G.; Hoagland, L.; Ruzzi, M.; Cardarelli, M.; Bonini, P.; Canaguier, R. Biostimulant action of protein hydrolysates: Unraveling their effects on plant physiology and microbiome. *Front. Plant Sci.* **2017**, *8*, 2202. [CrossRef]
64. Goyer, A. Thiamine in plants: Aspects of its metabolism and functions. *Phytochemistry* **2010**, *71*, 1615–1624. [CrossRef]
65. Rapala-Kozik, M. Vitamin B1 (thiamine): A cofactor for enzymes involved in the main metabolic pathways and an environmental stress protectant. In *Advances in Botanical Research*; Rébeillé, F., Douce, R., Eds.; Academic Press: Cambridge, MA, USA, 2011; Volume 58, pp. 37–91. ISBN 0065-2296.
66. Mukta, J.; Rahman, M.; As Sabir, A.; Gupta, D.R.; Surovy, M.Z.; Rahman, M.; Islam, M.T. Chitosan and plant probiotics application enhance growth and yield of strawberry. *Biocatal. Agric. Biotechnol.* **2017**, *11*, 9–18. [CrossRef]
67. Hajiboland, R.; Moradtalab, N.; Eshaghi, Z.; Feizy, J. Effect of silicon supplementation on growth and metabolism of strawberry plants at three developmental stages. *N. Z. J. Crop Hortic. Sci.* **2018**, *46*, 144–161. [CrossRef]
68. Agostinho, F.; Tubana, B.; Martins, M.; Datnoff, L. Effect of different silicon sources on yield and silicon uptake of rice grown under varying phosphorus rates. *Plants* **2017**, *6*, 35. [CrossRef]
69. Artyszak, A. Effect of silicon fertilization on Crop yield quantity and quality: A literature review in Europe. *Plants* **2018**, *7*, 54. [CrossRef]
70. DiMario, R.J.; Clayton, H.; Mukherjee, A.; Ludwig, M.; Moroney, J.V. Plant carbonic anhydrases: Structures, locations, evolution, and physiological roles. *Mol. Plant* **2017**, *10*, 30–46. [CrossRef]
71. Sturikova, H.; Krystofova, O.; Huska, D.; Adam, V. Zinc, zinc nanoparticles and plants. *J. Hazard. Mater.* **2018**, *349*, 101–110. [CrossRef]
72. Mattiello, E.M.; Ruiz, H.A.; Neves, J.C.L.; Ventrella, M.C.; Araújo, W.L. Zinc deficiency affects physiological and anatomical characteristics in maize leaves. *J. Plant Physiol.* **2015**, *183*, 138–143. [CrossRef]
73. Kyriacou, M.C.; Rouphael, Y. Towards a new definition of quality for fresh fruits and vegetables. *Sci. Hortic.* **2018**, *234*, 463–469. [CrossRef]
74. Rouphael, Y.; Kyriacou, M.C.; Petropoulos, S.A.; De Pascale, S.; Colla, G. Improving vegetable quality in controlled environments. *Sci. Hortic.* **2018**, *234*, 275–289. [CrossRef]
75. Bhaskara Reddy, M.; Belkacemi, K.; Corcuff, R.; Castaigne, F.; Arul, J. Effect of pre-harvest chitosan sprays on post-harvest infection by Botrytis cinerea and quality of strawberry fruit. *Postharvest Biol. Technol.* **2000**, *20*, 39–51. [CrossRef]
76. Hernández-Muñoz, P.; Almenar, E.; Valle, V.D.; Velez, D.; Gavara, R. Effect of chitosan coating combined with postharvest calcium treatment on strawberry (*Fragaria* × *ananassa*) quality during refrigerated storage. *Food Chem.* **2008**, *110*, 428–435. [CrossRef]
77. Khan, A.S.; Ahmad, B.; Jaskani, M.J.; Ahmad, R.; Malik, A.U. Foliar application of mixture of amino acids and seaweed (*Ascophyllum nodosum*) extract improve growth and physicochemical properties of grapes. *Int. J. Agric. Biol.* **2012**, *14*, 383–388.
78. Cluzet, S.; Torregrosa, C.; Jacquet, C.; Lafitte, C.; Fournier, J.; Mercier, L.; Salamagne, S.; Briand, X.; Esquerre'-Tugaye', M.-T.; Dumas, B. Gene expression profiling and protection of *Medicago truncatula* against a fungal infection in response to an elicitor from green algae *Ulva* spp. *Plant Cell Env.* **2004**, *27*, 917–928. [CrossRef]
79. Ertani, A.; Pizzeghello, D.; Francioso, O.; Sambo, P.; Sanchez-Cortes, S.; Nardi, S. *Capsicum chinensis* L. growth and nutraceutical properties are enhanced by biostimulants in a long-term period: Chemical and metabolomic approaches. *Front. Plant. Sci.* **2014**, *5*. [CrossRef]
80. Olkhovych, O.; Volkogon, M.; Taran, N.; Batsmanova, L.; Kravchenko, I. The Effect of copper and zinc nanoparticles on the growth parameters, contents of ascorbic acid, and qualitative composition of amino acids and acylcarnitines in *Pistia stratiotes* L. (Araceae). *Nanoscale Res. Lett.* **2016**, *11*. [CrossRef]

agronomy

MDPI

Article

Yield and Nutritional Quality of Vesuvian Piennolo Tomato PDO as Affected by Farming System and Biostimulant Application

Gianluca Caruso [1,*], Stefania De Pascale [1], Eugenio Cozzolino [2], Antonio Cuciniello [2], Vincenzo Cenvinzo [1], Paolo Bonini [3], Giuseppe Colla [4] and Youssef Rouphael [1]

1 Department of Agricultural Sciences, University of Naples Federico II, 80055 Portici, Italy
2 Council for Agricultural Research and Economics (CREA)–Research Center for Cereal and Industrial Crops, 81100 Caserta, Italy
3 NGA Laboratory, 437162 Tarragona, Spain
4 Department of Agriculture and Forest Sciences, University of Tuscia, 01100 Viterbo, Italy
* Correspondence: gcaruso@unina.it; Tel.: +39-81-2539-104

Received: 7 August 2019; Accepted: 29 August 2019; Published: 2 September 2019

Abstract: Scientific investigations are being increasingly devoted to biostimulant effects on vegetable yield and quality, with the perspective of sustainable crop management. Two farming systems (conventional or organic) in factorial combination with two biostimulant treatments (tropical plant extract (PE); legume-derived protein hydrolysate (PH)) plus a non-treated control were compared in terms of tomato fruit yield, yield components, mineral composition, functional and nutritional indicators. PE- and PH-based biostimulants resulted in higher plant biomass, PH even in higher leaf area index, compared to non-treated control. Marketable yield was not significantly affected by farming system. PH and PE gave higher yield than non-treated control. PH treatment led to higher fruit number than the control, whereas PE incurred significant increase in yield only under organic farming. The mean fruit weight attained the highest value upon PE application under conventional management. Colour component a* (redness) was higher with the conventional system compared to the organic one, whereas an opposite trend was shown by the organic acids malate, oxalate and isocitrate. Irrespective of the farming system, the soluble solids, fruit brightness (L*) and redness as well as the target organic acids malate, oxalate, citrate and isocitrate were significantly higher than untreated plants by 10.1%, 16.1%, 19.8%, 18.9%, 12.1%, 13.5% and 26.8%, respectively, with no significant differences between the PH- and PE-based biostimulants. Higher lipophilic activity and total ascorbic acid concentration but lower lycopene were recorded under organic management. PE and PH application resulted in higher total phenol and ascorbic acid as well as in lycopene content, and lipophilic antioxidant activity than the non-treated control. Biostimulants proved to be an effective sustainable tool for enhancing tomato fruit yield and functional quality both under conventional and organic vegetable systems.

Keywords: antioxidant activity; functional quality; lycopene; organic farming; protein hydrolysate; *Solanum lycopersicum* L.; tropical plant extract

---

1. Introduction

Organic horticulture has been increasing worldwide for the past two decades, as a result of rising demand of consumers for healthy and safer food [1], accounting for 3.5 million ha in 2014, which is almost twofold compared to 2008 [2]. Indeed, this farming management is environmentally-friendly due to food production with minimal harm to ecosystems as well as minimal use of inputs in particular fertilizers and pesticides [3]. However, the lower yield compared to conventional agriculture, i.e.

−20% according to Ponisio et al. [4] and −5% to −34% as reported by Seufert et al. [5], represents a disadvantage of organic farming. The latter yield reduction is mainly associated to higher biotic pressure caused by parasites, pests and pathogens [4,6] and to nutrient limitation in particular N and P [7] which limits production in several organic-based systems [8]. In fact, the rate of major minerals such as nitrogen and phosphorus released from organic fertilizers and crop residues do not often meet the crop demand during the highest rate plant growth, leading to significant yield reduction [9].

Within both conventional and organic farming systems, the use of naturally derived plant biostimulants is a promising sustainable approach [10,11], aiming to enhance (i) plant nutrient availability/uptake/assimilation and use efficiency, (ii) abiotic stress tolerance as well as (iii) product quality [12–14]. Within biostimulants, protein hydrolysates (PHs) are mainly made of amino acids, polypeptides and oligopeptides derived from proteins of animal or plant origin upon partial hydrolysis [15] and can be applied to seeds, leaves or soil in several forms (liquid or granular) [12]. Tropical plant extract (PE) and especially legume-derived protein hydrolysates (PHs) obtained from vegetal origin proteins have been drawing interest in world agricultural areas, compared to animal-derived ones, due to both their higher agronomic value [16] and no use constraints in organic farming. Moreover, PE or PH application to leaves and/or roots reportedly elicit physiological processes, thus resulting in enhancement of growth [17,18], production and quality [18,19], tolerance to abiotic stressors, such as drought, soil and water salinity, extreme temperature, nutrient deficiency, soil acidity and alkalinity [11,20–25]. Notably, PE or PHs also encourage plant activity of key enzymes involved either in N or C metabolism [12,24,26,27]. In addition, PH treatment may boost crop performances, by eliciting auxin- and gibberellin-like activities through bioactive peptides [17,28,29]. PE and PHs also exert indirect effects on plants, as they modify the architecture of roots and increase their hair surface expansion, thus enhancing macro- and microelement uptake [26,28,30–32]. However, limited scientific literature are available with regard to the effect of foliar applications of PH or PE in interaction with either conventional or organic farming on agronomical and fruit quality responses of tomato landraces, in particular the long shelf-life cherry tomato landrace 'Pomodorino del Piennolo del Vesuvio' (PPV), a typical niche product of Campania (Italy) horticultural sector.

In the perspective of the above mentioned topics, a two-year experiment was carried out to assess the response of cherry tomato landrace PPV to foliar applications of a vegetal protein based hydrolysate or a tropical plant extract biostimulant in interaction with organic or conventional crop system, in terms of yield, mineral composition, functional and sensorial quality attributes.

2. Materials and Methods

2.1. Growing Conditions and Experimental Protocol

The experimental research was carried out on open field grown tomato (*Solanum lycopersicum* L.) 'Piennolo del Vesuvio D.O.P.' ecotype Riccia, in Portici (Naples), southern Italy characterized by a typical Mediterranean climate, in 2016 and 2017. The soil was sandy-loam having 77% sand, 14.5% silt, 8.5% clay, with soil electrical conductivity of 342 μS cm^{-1}, 1.6% organic matter, 0.94 g kg^{-1} N, 63.9 mg kg^{-1} P_2O_5, 1.8 g kg^{-1} K_2O. The monthly air temperature (day/night) and rainfall recorded at the plant level, expressed as means of the two research years, were the following: 21.6 °C, 7.9 °C and 47.2 mm in April; 24.5 °C, 11.3 °C and 56.3 mm in May; 29.5 °C, 15.8 °C and 23.7 mm in June; 32.2 °C, 17.1 °C and 17 mm in July.

A factorial combination of biostimulant application (B) and farming system (F) was applied, based on two biostimulant treatments (PH or PE) plus a non-treated control and two farming systems (organic or conventional). The experimental design was a randomized complete-block design with three replications, yielding 18 experimental units (3 B × 2 F × 3 replications). Each experimental unit consisted of an 8 square meter plot. Tomato seedlings were transplanted on 25 and 24 April in the first and second growing season respectively, at a plant density of 4 plants m^{-2}.

The two commercial PH and PE-based biostimulants 'Trainer'®and 'Auxym'®were kindly provided by Italpollina S.p.A., Rivoli Veronese, Italy. The legume-derived PH biostimulant obtained through enzymatic hydrolysis contains 75% of free amino acids and peptides, 22% of carbohydrates and 3% of mineral nutrients. The detailed aminogram of the product along with the phenolics, flavonoids and elemental composition were reported by Rouphael et al. [31] and Paul et al. [25]. The PE biostimulant obtained by fermentation of tropical plants contains 54% of free amino acids and peptide, 17% carbohydrate, 23% mineral nutrients, 6% vitamins and 0.22% phytohormones as reported in detail by Rouphael et al. [33] and Caruso et al. [32].

Cherry tomato plants were sprayed with a solution containing 3 and 2 ml L^{-1} of PH- or PE-based biostimulant, or with water (non-treated control), four times during the growing season at 7-day intervals, starting in coincidence with the early growth of the first fruit truss.

Organic farming practices were performed in compliance with the EC Regulation 834/2007 and related subsequent updates. Both in conventional and organic systems, the fertilization was carried out with 153 kg ha^{-1} of N, 39 kg ha^{-1} of P_2O_5 and 223 kg ha^{-1} of K_2O. Phosphorus was completely supplied at planting, whereas nitrogen and potassium were given both prior to crop establishment (31% and 55% for N and K_2O respectively) and the remainder on dressing. Under the organic management a 6-5-13 Bioilsa organic-mineral fertilizer (based on hydrolyzed collagen and meat flour), N (11%) and N-K (7%–21%) hydrolyzed protein manure were used; ammonium sulphate, potassium sulphate, potassium nitrate and ammonium nitrate were supplied to the conventionally grown crops. Drip irrigation started when the soil available water capacity decreased to 80%. Crop protection was performed against downy mildew, tomato leaf miner, aphids, whitefly, and red spider.

2.2. Yield, Biometric Assessments and Leaf Color Measurements

Harvests of fully ripe fruits were performed from 14 July to 2 August, as an average of the two research years, and the marketable yield, number of fruits per plant and the mean fruit mass were determined on a sub-plot of 4 m^2. Fruits that were deformed or misshaped were considered unmarketable. The final leaf area was measured on 10 plants in each experimental plot using a Licor 3000 electronic area meter (Licor, Lincoln, NE, USA) and then the leaf area index was calculated. A sample of the fresh material was dried at 70 °C for about 3 days until reaching constant weight, to determine dry aboveground biomass.

Cherry tomato color was measured on the two sides of 10 fruits per experimental unit using Minolta CR-300 Chroma Meter (Minolta Camera Co. Ltd., Osaka, Japan) in order to obtain the color space parameters, in particular L* (brightness), a* (redness) and b* (yellowness).

2.3. Juice Total Soluble Solids and Fruit Dry Matter Content

The cherry tomato PPV fruits were homogenized in a blender for 2 min and the homogenate was filtered, then the total soluble solids content was measured using the Bellingham and Stanley digital refractometer (model RFM 81). The tomato fruit dry matter percentage was also determined after drying the fresh material at 70 °C for about 3 days until reaching constant weight. The dried tomato fruit samples were collected for further mineral analysis.

2.4. Mineral and Organic Acids Analysis

The desiccated cherry tomato fruit tissues were ground in a Wiley Mill to pass through an 841 μm screen and used for macro-mineral profile analysis, sodium content and organic acids as described in detail by Rouphael et al. [31] and Kyriacou et al. [34]. Phosphorus, potassium, calcium, magnesium, sulfur, sodium, malate, oxalate, citrate and isocitrate were separated and quantified by ICS-3000 ion chromatography (Dionex, Sunnyvale, CA, USA) coupled to a conductivity detector. Macronutrients, sodium and organic acids concentrations were expressed on a dry weight basis (g kg^{-1} dw).

2.5. Antioxidant Activity Analysis

The lipophylic and hydrophilic antioxidant activities were assessed on extract from freeze-dried cherry tomato PPV fruits (200 mg) added with methanol and distilled water, respectively. The antioxidant activity of the lipophilic and hydrophilic extract fractions were measured with the 2,20-azinobis 3-ethylbenzothiazoline-6-sulfonic acid ABTS [35] and with the N,N-dimethyl-p-phenylenediamine (DMPD) methods [36], respectively. The absorbance of the solutions for LAA and HAA were measured at 734 and 505 nm, respectively. Lipophylic and hydrophilic antioxidant activities were expressed as mmol of Trolox (6-hydroxy-2,5,7,8-tetramethylchro man-2-carboxylic acid) and mmol ascorbic acid per 100 g of dw [36].

2.6. Antioxidant Molecules Analysis

The total ascorbic acid and polyphenols were assessed spectrophotometrically based on the protocol by Kampfenkel et al. [37] and the Folin–Ciocalteau procedure [38], respectively, after slight modifications [34]. For quantification, ascorbate and gallic acid were used as external standards to build calibration curves both for total ascorbic acid and total polyphenols content. The absorbance of the solutions for total ascorbic acid and total polyphenols were measured at 525 and 765 nm, respectively, and the results were expressed as mg ascorbic acid on 100 g fw and mg gallic acid per 100 g dw. Lycopene content was also assessed spectrophotometrically, based on the protocol by Sadler et al. [39], and for the quantification pure lycopene (Sigma, St. Louis, MO) was used to build the calibration curves. The absorbance of the lycopene hexane solution was measured at 472 nm. Lycopene content was expressed in mg 100 g^{-1} fw.

2.7. Statistical Processing

All agronomical and qualitative data were subjected to three-way analysis of variance using the software package SPSS. The means were separated by DMRT test at 0.05 significance level. All the agronomical and quality variables were not significantly affected by the growing season (i.e., year) or its interactions with the two experimental factors applied, and therefore the mean data of the two years were reported.

3. Results and Discussion

3.1. Yield and Morphometric Measurements

As reported in Table 1, the legume-derived protein hydrolysate (PH) resulted in the highest leaf area index and biomass of the vegetative plant parts, though the latter variable was not significantly different from that recorded under the effect of the tropical plant extract (PE).

Marketable yield and its components, fruit number and mean weight, were significantly affected by biostimulant treatment, whereas no differences were recorded between conventional and organic systems. Moreover, fruit number and mean weight were also significantly influenced by the interaction between the two experimental factors (Table 1). The application of PH-based biostimulant resulted in the highest yield but PE biostimulant also gave a significantly higher yield compared to non-treated control (+18.7% and +11.2%, respectively); these outcomes stemmed from the combined effects of fruit number and mean weight (Table 1).

The two latter variables were significantly affected by the interaction between the two studied factors (Table 1). For instance, the fruit number per plant was just connected to the effect of PE-based biostimulant, leading to higher fruit number than the control only under organic system, whereas PH-based biostimulant always showed the best effect. Moreover, the mean fruit weight attained the highest value upon PE application under conventional management, the latter being also higher than that obtained with the organic system, whereas no differences were recorded between the remaining comparisons.

Table 1. Plant growth parameters and yield indicators of 'Piennolo del Vesuvio' cherry tomato as affected by farming system and biostimulant application.

Source of Variance	LAI	Aerial Biomass	Marketable Yield	Marketable Fruits	
				Mean Weight	Number
	(m^2 m^{-2})	(g dw m^{-2})	(t ha^{-1})	(g per Fruit)	No. per Plant
Year (Y)	NS	NS	NS	NS	NS
Farming system (F)	NS	NS	NS	NS	NS
Biostimulant (B)	*	*	*	NS	*
Y × F	NS	NS	NS	NS	NS
Y × B	NS	NS	NS	NS	NS
F × B	NS	NS	NS	*	*
Y × F× B	NS	NS	NS	NS	NS
Year (Y)					
2016	4.92	350.3	14.5	15.5	23.1
2017	5.00	375.5	15.0	15.7	24.3
Farming system (F)					
Organic	5.05	368.0	14.7	15.4	24.0
Conventional	4.87	357.8	14.8	15.8	23.3
Biostimulant formulate (B)					
Control	4.54 c	336.4 b	13.4 c	15.4	21.8 b
Legume-derived protein hydrolysate (PH)	5.31 a	385.5 a	15.9 a	15.4	26.0 a
Tropical plant extract (PE)	5.03 b	366.8 a	14.9 b	16.0	23.2 b

dw, dry weight. ns, * nonsignificant or significant at $p \leq 0.05$, respectively. Different letters within each column indicate significant differences according to Duncan's multiple range test ($p \leq 0.05$).

In contrast with the present research findings, in previous investigation [40] conventional management of different vegetable species led to higher yield than organic one. Consistently with our results, Colla et al. [30] detected growth and yield increase of tomato in greenhouse upon PHs application, which is a whole crop cycle extension of the short-time stimulation effect observed on tomato treated with PH extracts [17,41]. Notably, the effects shown by the applied biostimulant on plants is different from the nutritional input elicited by fertilizers [42]. Indeed, in our research tomato plants showed different patterns of yield components response to the applied substances in interaction with the crop system (Figure 1).

Foliar applications of PE and PHs may have triggered in tomato plants a physiological mechanism linked to the enhanced content of signaling molecules which are the prevailing PE and PH components [12]. In this respect, low-sized molecules such as peptides and free amino acids can regulate plant phenological progress upon their easy absorption through leaves and roots by promoting endogenous biosynthesis of phyto-hormones [43]. Consistently, other authors [17,18,31,33,44] reported that plant growth, fruit setting and yield were enhanced by the auxin- and sometimes gibberellin-like activity of the mentioned biostimulants.

PE- and PH-based biostimulants are likely to boost plant development and yield through: (i) stimulating cell proliferation by signaling molecules such as specific amino acids connected to nitrogen metabolism (i.e., glutamic and aspartic acids) and soluble peptides; (ii) vitamin provision targeted to cell protection from oxidation; (iii) encouraging plant metabolism with micronutrients supply ([26] and references cited therein). Moreover, an important increase in cytokinins content was promoted by biostimulant application in *Spinacea oleracea* [45]. An additional action pattern of PE and legume-derived PH consists of enhancing macronutrient uptake and assimilation through modulation of root biomass, density and lateral root number, as well as microbial activity with the consequent higher availability of soil nutrients [13,30].

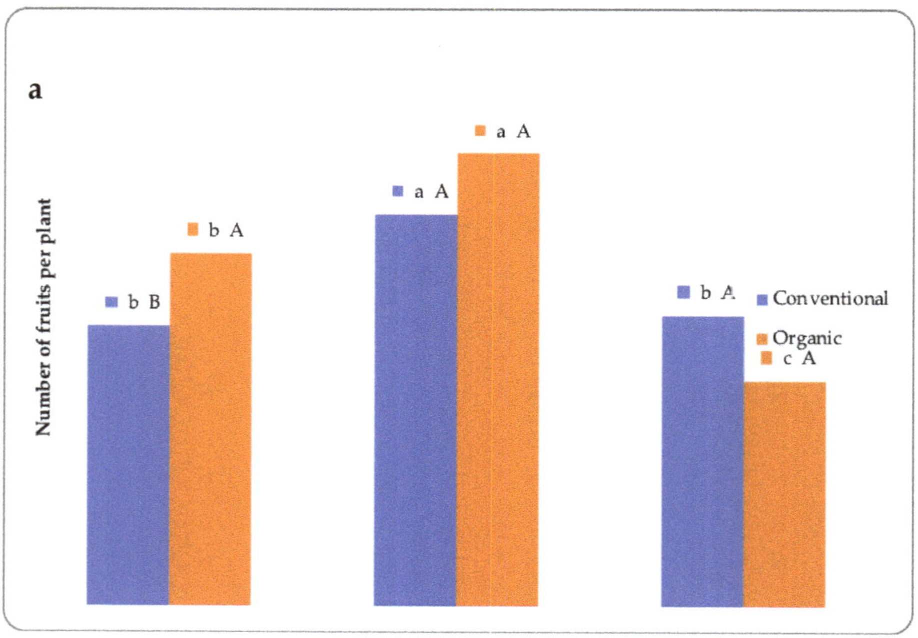

Figure 1. *Cont.*

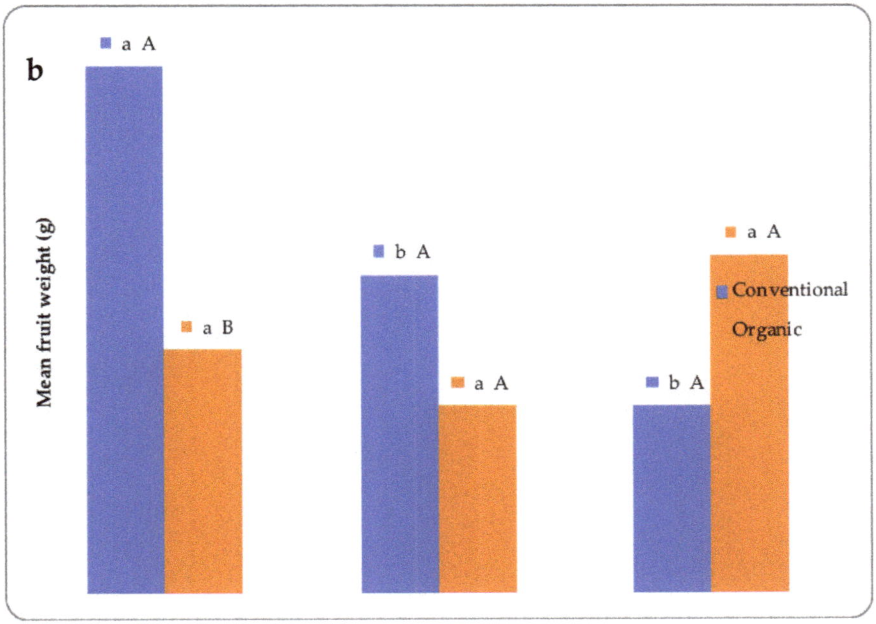

Figure 1. Interaction between farming system and biostimulant application on 'Piennolo del Vesuvio' cherry tomato fruit number per plant (**a**) and mean weight (**b**). Different letters mean significant difference according to Duncan's multiple range test at $p \leq 0.05$. Lowercase letters refer to the comparison between biostimulants, whereas capital letters to the comparison between farming systems within each biostimulant application.

In other experiments, *Lactuca sativa* L. sprayed with PE or PH showed a 11% higher biomass than non-treated control, which may be as a consequence of both the stimulation exerted by the most represented substances such as amino acids and key peptides and of enhancement of cultivable epiphytic bacteria as well as their species richness and diversity [46]. Overall, the direct and/or indirect mode of actions of the applied biostimulants may have boosted both growth and crop productivity of treated cherry tomato plants compared to the non-treated control treatment.

3.2. *Fruit Colorimetry, Nutritional Quality and Mineral Profile*

Farming system significantly affected some target indicators of tomato fruit colorimetry and nutritional quality as well as mineral composition (Tables 2 and 3). Two out of the three variables characterizing the colour (a* and b*) were higher under conventional management compared to the organic one; conversely, the organic acids malate, oxalate and isocitrate attained higher concentrations in the organically grown berries (Table 2). In the present research, both tomato fruit dry matter percentage and soluble solids were not significantly affected by farming management, whereas in previous investigations asparagus spears [47] and leek pseudo-stems [48] organically grown in southern or northern Europe respectively showed higher dry matter and sugar content than those managed conventionally.

Table 2. Flavor compounds and fruit colorimetry of 'Piennolo del Vesuvio' cherry tomato as affected by farming system and biostimulant application.

Source of Variance	Dry Matter	TSS	Fruit Colorimetry			Organic Acids (g kg^{-1} dw)			
	(%)	(°Brix)	L *	A *	B *	Malate	Oxalate	Citrate	Isocitrate
Farming system	NS	NS	NS	*	*	*	*	NS	*
Biostimulant	NS	*	*	*	NS	*	*	*	*
F × B	NS	NS	NS	NS	NS	NS	NS	NS	NS
Farming system									
Organic	8.6	7.3	40.8	31.3	20.4	13.4	1.25	43.8	0.54
Conventional	8.9	7.5	43.6	34.4	23.5	10.9	1.07	40.5	0.43
Biostimulant formulate									
Control	8.4 b	6.9 b	38.1 b	29.0 b	21.1 b	10.8 b	1.07 b	38.7 b	0.41 b
Legume-derived protein hydrolysate	8.9 a	7.6 a	44.9 a	35.1 a	22.1 a	12.4 a	1.22 a	43.3 a	0.53 a
Tropical plant extract	9.0 a	7.6 a	43.6 a	34.4 a	22.6 a	13.3 a	1.18 a	44.6 a	0.51 a

TSS, total soluble solids. ns, * nonsignificant or significant at $p \leq 0.05$, respectively. Different letters within each column indicate significant differences according to Duncan's multiple range test ($p \leq 0.05$).

Table 3. Fruit mineral composition of 'Piennolo del Vesuvio' cherry tomato as affected by farming system and biostimulant application.

Source of Variation	Mineral Composition (g kg^{-1} dw)					
	P	K	S	Ca	Mg	Na
Farming system	NS	NS	*	NS	NS	NS
Biostimulant	*	*	*	NS	*	*
F × B	NS	NS	NS	NS	NS	NS
Farming system						
Organic	0.87	36.43	0.76 a	5.51	1.44	0.31
Conventional	0.93	35.01	0.68 b	6.02	1.52	0.29
Biostimulant formulate						
Control	0.82 b	33.83 b	0.62 c	5.60	1.30 b	0.28 b
Legume-derived protein hydrolysate	1.00 a	36.66 a	0.83 a	5.72	1.59 a	0.30 ab
Tropical plant extract	0.87 ab	36.68 a	0.72 b	5.92	1.56 a	0.32 a

ns, * nonsignificant or significant at $p \leq 0.05$, respectively. Different letters within each column indicate significant differences according to Duncan's multiple range test ($p \leq 0.05$).

Regardless of the farming system, the soluble solids, fruit brightness and redness as well as the target organic acids malate, oxalate, citrate and isocitrate were significantly higher than untreated plants by 10.1%, 16.1%, 19.8%, 18.9%, 12.1%, 13.5% and 26.8%, respectively, with no significant differences between the PH- and PE-based biostimulants (Table 2). The highest fruit juice soluble solids and organic acids obtained in biostimulant-treated plants independently on the formulate could be considered important key quality attributes for consumer satisfaction [49]. Consistently with our findings, Rouphael et al. [31,33], Colla et al. [18] and Ertani et al. [19] reported the increased content of soluble solids, glucose and fructose in greenhouse grown *Solanum lycopersicum* and *Capsicum chinensis* fruits upon the treatment with biostimulants, derived from tropical plant extract fermentation, enzymatic hydrolysis of legume and alfalfa plants or by extraction of red grapes.

Minerals content is essential for the quality of fruit vegetables including tomato. Based on two surveys carried out in Finland and the USA, Levander [50] demonstrated that the contribution of vegetables to dietary intake of phosphorus, potassium, calcium, magnesium and sodium is 7–11%, 31–35%, 5–7%, 18–24% and 11%, respectively. The present work has generated important information regarding the relative abundance of minerals in cherry tomato landrace and its variation range across farming system and biostimulant application. In this respect, K was found by far the most abundant mineral, followed by Ca, P, Mg, S and Na (Table 2).

For all measured minerals no significant interaction between farming system and biostimulant application was observed (Table 3). Neither farming system nor biostimulant application had significant effect of Ca content in fruit (average 5.7 g kg^{-1}). The effect of biostimulant application on tomato fruit mineral profile was much more pronounced than the farming system. K and Mg were positively affected by both biostimulants compared to non-treated control, with no significant difference between them. PH-based biostimulant exhibited a higher content of P; in addition, both commercial biostimulants had a better effect on S content compared to the untreated control, with PH showing the highest values (Table 3). In other investigations, compared to non-treated plants the application of a PH-based biostimulant resulted in better nutritional status: higher K and Mg content in tomato [18,31] and in spinach [33] grown under protected cultivation.

In the present research, the increased concentration of cherry tomato fruit K and Mg induced by the application PH-based biostimulant might have been mediated through several direct/indirect mechanisms involving: (i) enhanced mineral uptake promoted by root growth stimulation encouraging absorption, translocation and accumulation of nutrients [17,51]; (ii) higher nutrient transporter expression in cell membranes [24,52]; (iii) the action of PH biostimulant bioactive compounds (soluble peptides, carbohydrates and free amino acids) in strengthening the sink effect and therefore the movement of nutrients within the plant [42].

3.3. Antioxidant Activity and Bioactive Content

Fruit vegetables in particular tomato are considered good sources of lipophilic and hydrophilic antioxidant molecules such as lycopene, total ascorbic acid and polyphenols. The influence of farming system and biostimulant application on antioxidant activities and bioactive compounds are reported in Table 4. Neither farming system nor biostimulant application had a significant effect on hydrophilic antioxidant activity (average 10.9 mmol ascorbate eq. 100 g^{-1} dw). When averaged over biostimulant application, higher lipophilic activity and total ascorbic acid concentration but lower lycopene were recorded under organic management compared to the conventional one. Moreover, no significant differences between the two farming systems arose with regard to hydrophilic antioxidant activity and phenols content (Table 4). Consistently with our results, in previous research carried out on strawberry in southern Italy [53], organic farming resulted in higher fruit ascorbic acid than the conventional management. As for the biostimulant application, both the PH and PE biostimulants resulted in higher lipophilic antioxidant activity as well as phenols, ascorbic acid and lycopene concentration than non-treated control, with no significant differences between the two commercial biostimulants used (Table 4).

Table 4. Antioxidant activity and bioactive content of 'Piennolo del Vesuvio' cherry tomato as affected by farming system and biostimulant application.

Source of Variation	Antioxidant Capacity		Lycopene	Total Phenols	Total Ascorbic Acid
	Lipophilic (mmol Trolox eq. 100g^{-1} dw)	Hydrophilic (mmol Ascorbate eq. 100g^{-1} dw)	(mg 100g^{-1} fw)	(mg Gallic Acid eq. 100g^{-1} dw)	(mg 100g^{-1} fw)
Farming system	*	NS	*	NS	*
Biostimulant	*	NS	*	*	*
F × B	NS	NS	NS	NS	NS
Farming system					
Organic	8.1 a	11.0	171.0 b	1.9	23.9 a
Conventional	7.7 b	10.8	188.2 a	1.9	18.5 b
Biostimulant formulate					
Control	5.8 b	10.7	150.2 b	1.8 b	14.5 c
Legume-derived protein hydrolysate	9.1 a	11.1	196.3 a	2.0 a	29.9 a
Tropical plant extract	8.7 a	10.9	192.0 a	2.0 a	19.2 b

ns, * nonsignificant or significant at $p \leq 0.05$, respectively. Different letters within each column indicate significant differences according to Duncan's multiple range test ($p \leq 0.05$).

The phytochemical homeostasis requires enzymatic activities leading to a stabilization of the concentration of antioxidants which show an increase both in response to free radical production [19,31] and when K and Mg in the tissues are high [31]. In this respect, the protection against oxidative stresses in maize plants was primed by both protein hydrolysate and plant extract based biostimulant through the expression of superoxide dismutases activity-regulating genes [54], which catalyze the enzymatic dismutation of superoxide to H_2O_2 [55]. The application of protein hydrolysates in greenhouse conditions encouraged the synthesis of ascorbate, p-coumaric, chlorogenic acid, capsaicin and antioxidant activity in *Capsicum chinensis* L. fruits [19], as well vitamin C in tomato fruits [18,31]. Similarly, *Spinacia oleracea* phenolic acids production was enhanced by biostimulant application [45], through the phenylalanine ammonia lyase pathway [56]. Therefore, the foliar application of plant biostimulants such as PH or PE can be instrumental in satisfying increasing consumer standards for the functional quality aspects of fresh cherry tomato PPV landrace [57,58].

4. Conclusions

From research carried out in southern Italy on tomato landrace 'Piennolo del Vesuvio D.O.P.' the effective application of plant biostimulants based on tropical plant extract or legume-derived protein hydrolysate on fruit yield, nutritional and functional attributes arose. Indeed, both formulates overall enhanced production, quality, mineral and antioxidant indicators either under organic or conventional farming systems. Controversial outcomes stemmed from the comparison between the two crop managements, as conventional farming resulted in better colored and lycopene richer fruits, but higher organic acids, ascorbic acid content and lipophilic antioxidant activity was recorded when organic procedures were applied. The present study allows us to draw important conclusions relevant to the significant contribution of biostimulant application in making sustainable even a conventional tomato farming system.

Author Contributions: Conceptualization, G.C. (Gianluca Caruso) and Y.R.; data curation, G.C. (Gianluca Caruso), E.C. and Y.R.; formal analysis, G.C. (Gianluca Caruso); funding acquisition, G.C.(Giuseppe Colla) and P.B.; investigation, G.C. (Gianluca Caruso), E.C., A.C. and V.C.; methodology, G.C. (Gianluca Caruso), S.D.P., E.C. and Y.R.; supervision, Gia.Ca.; writing—original draft, G.C. (Gianluca Caruso) and Y.R.; writing—review and editing, G.C. (Gianluca Caruso), S.D.P., G.C.(Giuseppe Colla) and Y.R.

Funding: This work was partially supported by Italpollina Company (Rivoli Veronese, Italy).

Acknowledgments: The authors wish to thank Roberto Maiello for his assistance with laboratory equipments.

Conflicts of Interest: The authors declare no conflict of interest.

References

1. Dorais, M.; Alsanius, B. Recent advances in organic horticulture technology and management. *Sci. Hortic.* **2016**, *208*, 1–2. [CrossRef]
2. Willer, H.; Lernoud, J. The World of Organic Agriculture—Statistics and Emerging Trends 2016. FiBL-IFOAM Rep. Bonn. Available online: https://shop.fibl.org/fileadmin/documents/shop/1698-organicworld-2016.pdf (accessed on 23 August 2019).
3. Dorais, M. Organic production of vegetables: State of the art and challenges. *Can. J. Plant Sci.* **2007**, *85*, 1055–1066. [CrossRef]
4. Ponisio, L.C.; M'Gonigle, L.K.; Mace, K.C.; Palomino, J.; de Valpine, P.; Kremen, C. Diversification practices reduce organic to conventional yield gap. *Proc. R. Soc. B* **2015**, *282*, 20141396. [CrossRef] [PubMed]
5. Seufert, V.; Ramankutty, N.; Foley, J.A. Comparing the yields of organic and conventional agriculture. *Nature* **2012**, *485*, 229–232. [CrossRef] [PubMed]
6. Orsini, F.; Maggio, A.; Rouphael, Y.; De Pascale, S. 'Physiologically quality' of organically grown vegetables. *Sci. Hortic.* **2016**, *208*, 131–139. [CrossRef]
7. De Ponti, T.; Rijk, B.; Van Ittersum, M.K. The crop yield gap between organic and conventional agriculture. *Agric. Syst.* **2012**, *108*, 1–9. [CrossRef]

8. Berry, P.M.; Sylvester-Bradley, R.; Philipps, L.; Hatch, D.J.; Cuttle, S.P.; Rayns, F.W.; Goslin, P. Is the productivity of organic farms restricted by the supply available nitrogen? *Soil Use Manag.* **2002**, *18*, 248–255. [CrossRef]
9. Lester, G.E.; Saftner, R.A. Organically versus conventionally grown produce: Common production inputs, nutritional quality, and nitrogen delivery between the two systems. *J. Agric. Food. Chem.* **2011**, *59*, 10401–10406. [CrossRef]
10. Colla, G.; Rouphael, Y. Biostimulants in horticulture. *Sci. Hortic.* **2015**, *196*, 1–2. [CrossRef]
11. Rouphael, Y.; Cardarelli, M.; Bonini, P.; Colla, G. Synergistic action of a microbial-based biostimulant and a plant derived-protein hydrolysate enhances lettuce tolerance to alkalinity and salinity. *Front. Plant Sci.* **2017**, *8*, 131. [CrossRef]
12. Colla, G.; Nardi, S.; Cardarelli, M.; Ertani, A.; Lucini, L.; Canaguier, R.; Rouphael, Y. Protein hydrolysates as biostimulants in horticulture. *Sci. Hortic.* **2015**, *196*, 28–38. [CrossRef]
13. Du Jardin, P. Plant biostimulants: Definition, concept, main categories and regulation. *Sci. Hortic.* **2015**, *196*, 3–14. [CrossRef]
14. Rouphael, Y.; Cardarelli, M.; Colla, G. Role of arbuscular mycorrhizal fungi in alleviating the adverse effects of acidity and aluminium toxicity in zucchini squash. *Sci. Hortic.* **2015**, *188*, 97–105. [CrossRef]
15. Schaafsma, G. Safety of protein hydrolysates, fractions thereof and bioactive peptides in human nutrition. *Eur. J. Clin. Nutr.* **2009**, *63*, 1161–1168. [CrossRef] [PubMed]
16. Cerdán, M.; Sánchez-Sánchez, A.; Jordá, J.D.; Juárez, M.; Sánchez-Andreu, J. Effect of commercial amino acids on iron nutrition of tomato plants grown under lime-induced iron deficiency. *J. Plant Nutr. Soil Sci.* **2013**, *176*, 859–866. [CrossRef]
17. Colla, G.; Rouphael, Y.; Canaguier, R.; Svecova, E.; Cardarelli, M. Biostimulant action of a plant-derived protein hydrolysate produced through enzymatic hydrolysis. *Front. Plant Sci.* **2014**, *5*, 448. [CrossRef]
18. Colla, G.; Cardarelli, M.; Bonini, P.; Rouphael, Y. Foliar applications of protein hydrolysate, plant and seaweed extracts increase yield but differentially modulate fruit quality of greenhouse tomato. *Hort. Sci.* **2017**, *52*, 1214–1220. [CrossRef]
19. Ertani, A.; Pizzeghello, D.; Francioso, O.; Sambo, P.; Sanchez-Cortes, S.; Nardi, S. *Capsicum chinensis* L. growth and nutraceutical properties are enhanced by biostimulants in a long-term period: Chemical and metabolomic approaches. *Front. Plant Sci.* **2014**, *5*, 375. [CrossRef]
20. Botta, A. Enhancing plant tolerance to temperature stress with amino acids: An approach to their mode of action. *Acta Hortic.* **2013**, *1009*, 29–35. [CrossRef]
21. Colla, G.; Svecova, E.; Rouphael, Y.; Cardarelli, M.; Reynaud, H.; Canaguier, R.; Planques, B. Effectiveness of a plantderived protein hydrolysate to improve crop performances under different growing conditions. *Acta Hortic.* **2013**, *1009*, 175–179. [CrossRef]
22. Ertani, A.; Schiavon, M.; Muscolo, A.; Nardi, S. Alfalfa plant-derived biostimulant stimulate short-term growth of salt stressed *Zea mays* L. plants. *Plant Soil* **2013**, *364*, 145–158. [CrossRef]
23. Lucini, L.; Rouphael, Y.; Cardarelli, M.; Canaguier, R.; Kumar, P.; Colla, G. The effect of a plant-derived protein hydrolysate on metabolic profiling and crop performance of lettuce grown under saline conditions. *Sci. Hortic.* **2015**, *182*, 124–133. [CrossRef]
24. Sestili, F.; Rouphael, Y.; Cardarelli, M.; Pucci, A.; Bonini, P.; Canaguier, R.; Colla, G. Protein hydrolysate stimulates growth in tomato coupled with N-dependent gene expression involved in N assimilation. *Front. Plant Sci.* **2018**, *9*, 1233. [CrossRef] [PubMed]
25. Paul, K.; Sorrentino, M.; Lucini, L.; Rouphael, Y.; Cardarelli, M.; Bonini, P.; Reynaud, H.; Canaguier, R.; Trtílek, M.; Panzarová, K.; et al. Understanding the biostimulant action of vegetal-derived protein hydrolysates by high-throughput plant phenotyping and metabolomics: A case study on tomato. *Front. Plant Sci.* **2019**, *10*, 47. [CrossRef] [PubMed]
26. Colla, G.; Hoagland, L.; Ruzzi, M.; Cardarelli, M.; Bonini, P.; Canaguier, R.; Rouphael, Y. Biostimulant action of protein hydrolysates: Unraveling their effects on plant physiology and microbiome. *Front. Plant Sci.* **2017**, *8*, 2202. [CrossRef] [PubMed]
27. Schiavon, M.; Ertani, A.; Nardi, S. Effects of an alfalfa protein hydrolysate on the gene expression and activity of enzymes of TCA cycle and N metabolism in *Zea mays* L. *J. Agric. Food Chem.* **2008**, *56*, 11800–11808. [CrossRef] [PubMed]

28. Ertani, A.; Cavani, L.; Pizzeghello, D.; Brandellero, E.; Altissimo, A.; Ciavatta, C.; Nardi, S. Biostimulant activities of two protein hydrolysates on the growth and nitrogen metabolism in maize seedlings. *J. Plant. Nutr. Soil. Sci.* **2009**, *172*, 237–244. [CrossRef]
29. Matsumiya, Y.; Kubo, M. Soybean. In *Soybean and Nutrition*; El-Shemy, H., Ed.; InTech Europe: Rijeka, Croatia, 2011; pp. 215–230. [CrossRef]
30. Colla, G.; Rouphael, Y.; Di Mattia, E.; El-Nakhel, C.; Cardarelli, M. Co-inoculation of *Glomus intraradices* and *Trichoderma atroviride* acts as a biostimulant to promote growth, yield and nutrient uptake of vegetable crops. *J. Sci. Food Agric.* **2015**, *95*, 1706–1715. [CrossRef] [PubMed]
31. Rouphael, Y.; Colla, G.; Giordano, M.; El-Nakhel, C.; Kyriacou, M.C.; De Pascale, S. Foliar applications of a legume-derived protein hydrolysate elicit dose dependent increases of growth, leaf mineral composition, yield and fruit quality in two greenhouse tomato cultivars. *Sci. Hort.* **2017**, *226*, 353–360. [CrossRef]
32. Caruso, G.; Stoleru, V.; De Pascale, S.; Cozzolino, E.; Pannico, A.; Giordano, M.; Teliban, G.; Cuciniello, A.; Rouphael, Y. Production, leaf quality and antioxidants of perennial wall rocket as affected by crop cycle and mulching type. *Agronomy* **2019**, *9*, 194. [CrossRef]
33. Rouphael, Y.; Giordano, M.; Cardarelli, M.; Cozzolino, E.; Mori, M.; Kyriacou, M.; Bonini, P.; Colla, G. Plant and seaweed-based extracts increase yield but differentially modulate nutritional quality of greenhouse spinach through biostimulant action. *Agronomy* **2018**, *8*, 126. [CrossRef]
34. Kyriacou, M.C.; El-Nakhel, C.; Graziani, G.; Pannico, A.; Soterious, G.A.; Giordano, M.; Ritieni, A.; De Pascale, S.; Rouphael, Y. Functional quality in novel food sources: Genotypic variation in the nutritive and phytochemical composition of thirteen microgreens species. *Food Chem.* **2019**, *277*, 107–118. [CrossRef] [PubMed]
35. Re, R.; Pellegrini, N.; Proteggente, A.; Pannala, A.; Yang, M.; Rice-Evans, C. Antioxidant activity applying an improved abts radical cation decolorization assay. *Free Rad. Biol. Med.* **1999**, *26*, 1231–1237. [CrossRef]
36. Fogliano, V.; Verde, V.; Randazzo, G.; Ritieni, A. Method for measuring antioxidant activity and its application to monitoring the antioxidant capacity of wines. *J. Agric. Food Chem.* **1999**, *47*, 1035–1040. [CrossRef] [PubMed]
37. Kampfenkel, K.; Van Montagu, M.; Inzé, D. Extraction and determination of ascorbate and dehydroascorbate from plant tissue. *Anal. Biochem.* **1995**, *225*, 165–167. [CrossRef] [PubMed]
38. Singleton, V.L.; Orthofer, R.; Lamuela-Raventos, R.M. Analysis of total phenols and other oxidation substrates and antioxidants by means of Folin–Ciocalteu reagent. *Methods Enzymol.* **1999**, *299*, 152–178. [CrossRef]
39. Sadler, G.; Davis, J.; Dezman, D. Rapid extraction of lycopene and β-carotene from reconstituted tomato paste and pink grapefruit homogenates. *J. Food Sci.* **1990**, *55*, 1460–1461. [CrossRef]
40. Caruso, G.; Conti, S.; La Rocca, G. Influence of crop cycle and nitrogen fertilizer form on yield and nitrate content in dierent species of vegetables. *Adv. Hortic. Sci.* **2011**, *25*, 81–89.
41. Parrado, J.; Bautista, J.; Romero, E.J.; García-Martínez, A.M.; Friaza, V.; Tejada, M. Production of a carob enzymatic extract: Potential use as a biofertilizer. *Bioresour. Technol.* **2008**, *99*, 2312–2318. [CrossRef]
42. Calvo, P.; Nelson, L.; Kloepper, J.W. Agricultural uses of plant biostimulants. *Plant Soil* **2014**, *383*, 3–41. [CrossRef]
43. Cavani, L.; Ciavatta, C. Attività biostimolante negli idrolizzanti proteici. *L'Informatore Agrario* **2007**, *44*, 46–52.
44. Carillo, P.; Colla, G.; El-Nakhel, C.; Bonini, P.; D'Amelia, L.; Dell'Aversana, E.; Pannico, A.; Giordano, M.; Sifola, M.I.; Kyriacou, M.C.; et al. Biostimulant application with a tropical plant extract enhances *Corchorus olitorius* adaptation to sub-optimal nutrient regimens by improving physiological parameters. *Agronomy* **2019**, *9*, 249. [CrossRef]
45. Kulkarni, M.G.; Rengasamy, K.R.R.; Pendota, S.C.; Gruz, J.; Plačková, L.; Novák, O.; Doležal, K.; Van Staden, J. Bioactive molecules derived from smoke and seaweed *Ecklonia maxima* showing phytohormone-like activity in *Spinacia oleracea* L. *New Biotechnol.* **2019**, *48*, 83–89. [CrossRef] [PubMed]
46. Luziatelli, F.; Ficca, A.G.; Colla, G.; Baldassarre Švecová, E.; Ruzzi, M. Foliar application of vegetal-derived bioactive compounds stimulates the growth of beneficial bacteria and enhances microbiome biodiversity in lettuce. *Front. Plant. Sci.* **2019**, *10*, 60. [CrossRef] [PubMed]
47. Caruso, G.; Villari, G.; Borrelli, C.; Russo, G. Effects of crop method and harvest seasons on yield and quality of green asparagus under tunnel in southern Italy. *Adv. Hort. Sci.* **2012**, *26*, 51–58.

48. Golubkina, N.A.; Seredin, T.M.; Antoshkina, M.S.; Kosheleva, O.V.; Teliban, G.C.; Caruso, G. Yield, quality, antioxidants and elemental composition of new leek cultivars under greenhouse organic or conventional system. *Horticulturae* **2018**, *4*, 39. [CrossRef]
49. Kader, A.A. *Postharvest Technology of Horticultural Crops*; University of California, Division of Agriculture and Natural Resources Publication: Davis, CA, USA, 2002; Volume 3311, p. 535.
50. Levander, O.A. Fruit and vegetable contributions to dietary mineral intake in human health and disease. *Hort. Sci.* **1990**, *25*, 1486–1488.
51. Lucini, L.; Rouphael, Y.; Cardarelli, M.; Bonini, P.; Baffi, C.; Colla, G. A vegetal biopolymer-based biostimulant promoted root growth in melon while triggering brassinosteroids and stress-related compounds. *Front. Plant Sci.* **2018**, *9*, 472. [CrossRef]
52. Ertani, A.; Schiavon, M.; Nardi, S. Transcriptome-wide identification of differentially expressed genes in *Solanum Lycopersicon* L. in response to an alfalfa-protein hydrolysate using microarrays. *Front. Plant Sci.* **2017**, *8*, 1159. [CrossRef]
53. Conti, S.; Villari, G.; Faugno, S.; Melchionna, G.; Somma, S.; Caruso, G. Effects of organic vs. conventional farming system on yield and quality of strawberry grown as an annual or biennial crop in southern Italy. *Sci. Hortic.* **2014**, *180*, 63–71. [CrossRef]
54. Trevisan, S.; Manoli, A.; Quaggiotti, S.A. Novel biostimulant, belonging to protein hydrolysates, mitigates abiotic stress effects on maize seedlings grown in hydroponics. *Agronomy* **2019**, *9*, 28. [CrossRef]
55. Wang, W.; Xia, M.X.; Chen, J.; Yuan, R.; Deng, F.N.; Shen, F.F. Gene expression characteristics and regulation mechanisms of superoxide dismutase and its hysiological roles in plants under stress. *Biochem.* **2016**, *81*, 465–480. [CrossRef]
56. Hyun, M.W.; Yun, Y.H.; Kim, J.Y.; Kim, S.H. Fungal and plant phenylalanine ammonialyase. *Mycobiol.* **2011**, *39*, 257–265. [CrossRef] [PubMed]
57. Rouphael, Y.; Kyriacou, M.C.; Petropoulos, S.A.; De Pascale, S.; Colla, G. Improving vegetable quality in controlled environments. *Sci. Hortic.* **2018**, *234*, 275–289. [CrossRef]
58. Kyriacou, M.C.; Rouphael, Y. Towards a new definition of quality for fresh fruits and vegetables. *Sci. Hortic.* **2018**, *234*, 463–469. [CrossRef]

agronomy

MDPI

Review

Melatonin as a Chemical Substance or as Phytomelatonin Rich-Extracts for Use as Plant Protector and/or Biostimulant in Accordance with EC Legislation

Marino B. Arnao * and Josefa Hernández-Ruiz

Department of Plant Physiology, University of Murcia, 30100-Murcia, Spain; jhruiz@um.es
* Correspondence: marino@um.es; Tel.: +34 868887001

Received: 3 September 2019; Accepted: 19 September 2019; Published: 21 September 2019

Abstract: Melatonin (*N*-acetyl-5-methoxytryptamine) is a ubiquitous molecule present in animals and plants, and also in bacteria and fungi. In plants, it has an important regulatory and protective role in the face of different stress situations in which it can be involved, mainly due to its immobility. Both in the presence of biotic and abiotic stressors, melatonin exerts protective action in which, through significant changes in gene expression, it activates a stress tolerance response. Its anti-stress role, along with other outstanding functions, suggests its possible use in active agricultural management. This review establishes considerations that are necessary for its possible authorization. The particular characteristics of this substance and its categorization as plant biostimulant are discussed, and also the different legal aspects within the framework of the European Community. The advantages and disadvantages are also described of two of its possible applications, as a plant protector or biostimulant, in accordance with legal provisions.

Keywords: biostimulant; fertilizer; melatonin; phytomelatonin; plant protector; plant stress

1. Introduction

Melatonin (*N*-acetyl-5-methoxytryptamine) is a biogenic amine derived from the amino acid tryptophan, which was discovered in 1958 in the cow pineal gland by Lerner and cols. [1]. Two years later, it was detected in humans and its chemical structure was elucidated. This molecule, which was initially only related to changes in the structure of melanocytes in amphibians, fish and reptiles, was soon found to act as a neurohormone in mammals [2,3]. Since its discovery it has become one of the most researched molecules. In animals, it presents a multitude of physiological actions such as a role in the circadian rhythms of several molecules, and its influence on sleep–wake cycles, mood, motor activity and body temperature changes [4–7]. Its influence on food intake and its relationship with metabolic syndrome has also been demonstrated [8–10]. In other more specific situations such as the physiology of the retina, the immune system, sexual behavior and as an anti-cancer effector, melatonin also has a relevant role [11–15]. In addition, interesting and extensive reviews on the role of melatonin in animals and humans can be consulted [16–23].

In 1995, the presence of melatonin in plants was discovered [24–27]. During the following years there was much reluctance on the part of researchers to accept this, since some refused to believe that a neurohormone could be present in plants, and much less that it had any role in their physiology. A key piece was the elucidation of the melatonin biosynthesis route in plants, localized between the mitochondria, chloroplasts and cytoplasm of cells, and which has been studied with great accuracy by K. Back and J. Kong in rice and *Arabidopsis* plants [28–30]. However, it is now fully accepted that melatonin is present in all plant species and that it presents a panoply of interesting actions. Indeed, several studies have demonstrated its role in processes such as seed germination, growth and the development of seedlings, leaves and roots. It takes part in organogenesis processes such as rooting and fruiting, and in processes of leaf and fruit senescence. It acts as a protector of the photosynthetic and stomatic system, and as a regulator of various enzymes of the metabolism of carbohydrates, lipids, amino acids, nitrogen, sulfur and phosphorus. It also has a role in the secondary metabolism, enhancing the synthesis of flavonoids, anthocyanins, and carotenoids, among others. It regulates its own biosynthesis and that of several plant hormones such as auxin, abscisic acid, gibberellins, cytokinins, ethylene, polyamines, jasmonic acid and salicylic acid [31–39].

Of all the aspects investigated, its protective action against stress situations has been the most researched and about which most is known. Melatonin exerts a protective action, mediated by major changes in gene expression, both against abiotic (cold, heat, drought, waterlogging, salinity, alkalinity, acid rain, chemical contamination by heavy metals, UV radiation) and biotic (bacteria, fungi, virus) stressors. As a result, plants are more tolerant and/or resistant to the negative action of such stressors [31,36,40–43] (see below). The term "biostimulant" was first proposed to denote "materials that, in minute quantities, promote plant growth" by Zhang and Schmidt (1997) [44]. Later, the definition was modified by Kaufman et al. (2007) as: "Biostimulants are materials, other than fertilisers, that promote plant growth when applied in low quantities" [45]. According Du Jardin (2015), the following definition is proposed: "A plant biostimulant is any substance or microorganism applied to plants with the aim to enhance nutrition efficiency, abiotic stress tolerance and/or crop quality traits, regardless of its nutrients content", and extended as "plant biostimulants also designate commercial products containing mixtures of such substances and/or microorganisms" [46]. Under the EC (European Community) regulation: "Plant biostimulants will be EC marked as fertilizing products stimulating plant nutrition processes independently of the products' nutrient content with the sole aim of improving one or more of the following characteristics of the plant and the plant rhizosphere or phyllosphere: Nutrient use efficiency, tolerance to abiotic stress, crop quality, availability of confined nutrients in the soil and rhizosphere, humification and degradation of organic compounds in the soil". Extensive revision works on this topic can be consulted [47,48]. The objective of this work is to provide sufficient data to establish the clear protective role of melatonin against adverse environmental situations, and to discuss the possible global use of melatonin as a biostimulant and/or bioprotective agent. Current legislation of the EC, is taken into account and the advantages and disadvantages of its use in plant crops destined for animal and human consumption are analyzed.

2. Melatonin as a Regulator of Plant Stress Physiology

Although there was much evidence in the 1990s that melatonin could exert some role as an antioxidant agent in animal cells and tissues, it was not until 2004 and 2006, in carrot cells and Chinese licorice (*Glycyrrhiza uralensis* Fisch.), that the possible protective role of melatonin in plants was suggested [49–51], although some curious and previous data existed [52]. The initial idea that melatonin in plants, as in animals, could play an important role as an antioxidant was taking shape and results in this regard became ever more plentiful [53–57]. In addition, studies on melatonin as a possible plant regulator were also progressing, especially since the initial studies of Arnao and cols. on the role of melatonin in plant growth and development, and the so-called auxin-like activity [58–64].

It was not until the publication of results on the action of melatonin on changes in gene expression that the extent and potential of melatonin as a regulatory agent of multiple physiological processes in plants became widely known [64–70]. Exceeding previous expectations, melatonin is capable of activating all known molecular stress mechanisms in plants. Thus, gene regulatory factors involved in the response to cold, high temperatures, salinity, drought, chemical toxicity, etc., and also biotic stress, are up-regulated by melatonin [31,38,40,41,43]. Melatonin also regulates the expression of multiple enzymes related to hormonal homeostasis, up-regulating or down-regulating the expression of genes that encode enzymes of the biosynthetic or catabolic pathways of plant hormones including indole-3-acetic acid (auxin), gibberellins such as gibberellin-4 (GA_4), cytokinins, abscisic acid (ABA) and ethylene. It also others regulators such as salicylic acid (SA), jasmonic acid (JA) and polyamines [31–33,35,38,69,71–75]. In general, subjecting plants to a stressful situation—which leads to an increase in endogenous levels of melatonin—or treatment with exogenous melatonin, results in a stress tolerance response mediated by specific stress response factors and changes in the endogenous levels of plant hormones involved in the response [31–36,38,40–43,75–85]. In addition, the recent identification of a melatonin receptor in *Arabidopsis thaliana* has opened new expectations related to its role as a new plant hormone [86]. Figure 1 shows these aspects in a condensed form.

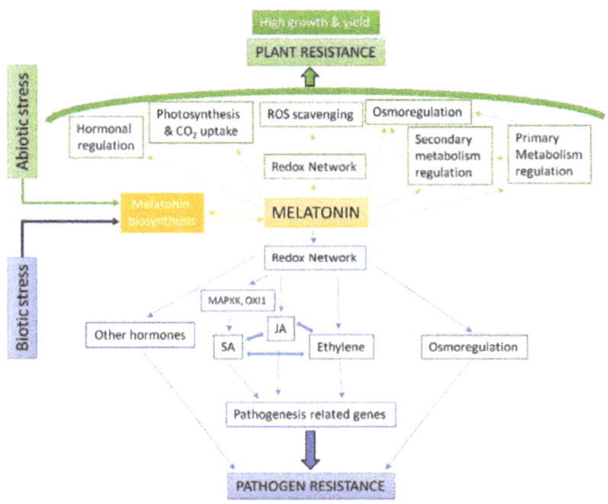

Figure 1. Model of redox network/melatonin action on abiotic and biotic stress responses.

3. Beneficial Responses to Melatonin Treatments in Different Crops in Stress Situations

Studies conducted with melatonin in plants under both abiotic and biotic stress are numerous. Table 1 compiles many of the studies with an agronomic interest since they deal primarily with crop species for human consumption. Table 1 presents studies classified by plant species, where there are many physiological aspects that are investigated in which melatonin exerts some generally beneficial action. These include seed germination, the growth and vegetative development of plants; photosynthesis, its pigments, photorespiration, stomatic conductance and water economy; the yields of seeds and fruits in adverse conditions; osmoregulation, ion exchange and adjustments in osmotic and hydric potentials, and the regulation of the different metabolisms of carbohydrates, lipids, nitrogen compounds, sulfur and phosphorus cycles. In regards to the secondary metabolism, melatonin induces the biosynthesis of flavonoids, anthocyanins and carotenoids, among others; in hormonal homeostasis, it intervenes in the regulation of all plant hormones and its own biosynthesis. It promotes the rooting process of primary, secondary and adventitious roots while during foliar senescence, melatonin regulates the expression of chlorophyll degradation-related and senescence-induced genes. In the postharvest control of fruits, melatonin increases the ethylene and lycopene content, and regulates many enzymes of the cell wall, ethylene biosynthesis, and primary and secondary metabolisms. It also helps preserve cut flowers; in fruiting it induces parthenocarpy. Finally, its role in bacterial, fungal and viral pathogenic infection should be emphasized, slowing damage and stimulating systemic acquired resistance (SAR) to favor crop health.

Obviously, all the above plant physiology aspects are of interest for application in plant production. Indeed, while many of the above studies were at a laboratory level, others have already been put into practice in crops with excellent results.

In general, exogenous melatonin applications are made through the root system, in irrigation water, or by spraying leaves. In the last case, no adjuvant is needed since melatonin is an amphipathic molecule that crosses biological membranes and the waxy cuticles. Melatonin is transported via the xylem from the roots to the rest of the organs of the plant quite effectively [87,88].

Table 1. Studies of different responses to melatonin treatments in different crop species in diverse stress situations.

Plant Species	Stress Type	Melatonin Treatment (μM)	Effects Observed	Reference
Alfalfa	Waterlogging	100	↑ tolerance, growth, photosynthesis, Chls, polyamines, ↓ electrolyte leakage, ROS, ethylene, leaf senescence	[89]
	Metal-Cd	10–200	↑ tolerance, growth, Cd transporters, ↓ Cd in roots, ROS	[90]
	Oxidative	1–100	↑ lateral root formation, cell division	[91]
Apple	Salinity	0.1	↑ shoot height, leaf number, Chls, K^+, ↓ electrolyte leakage, ROS	[92,93]
	Drought	100	↑ tolerance, re-open stomata, water in leaf, photosynthesis, N uptake, N metabolism, growth, ↓ ABA activity, ROS, leaf senescence	[71,94–96]
	Waterlogging	200	↓ chlorosis, wilting of the seedlings, ROS, ↑ tolerance, photosynthesis	[97]
	Alkaline	5	↑ tolerance, root system, redox balance, polyamines	[98]
	Leaf-senescence	10 mM	↓ senescence, ROS burst, ↑ Chls, photosynthesis, sucrose, starch, N	[67,99,100]
	Diplocarpon mali	50–500	↑ resistance to fungal infection, ↓ leaf lesions, cell death, pathogen expansion	[101]
	Apple Replant Disease—ARD	200	↑ growth, photosynthesis, K levels, soil microbial, ↓ ARD effects, ROS	[102]
	Apple Stem Grooving Virus	15	↑ shoot regrowth, 95% shoots virus-free, virus-free area	[103]
Apricot	-	10 ppm	↑ leaf growth, photosynthesis, fruit yield, size and retention, TA, TSS	[104]
Banana	Post-harvest	200–500	↑ shelf life of fruits, ↓ ethylene, ripening, quality sharp changes	[105]
	Anthracnose	10 mM	↑ fruit resistance, banana shelf life, ↓ anthracnose disease	[106]
Barley	Cold, drought	1 mM	↑ photosynthesis efficiency, ABA, water content, ROS	[72]
	Leaf-senescence	0.01–1	↑ Chls, growth, ↓ senescence	[59,107]
Bermudagrass	Cold, salt, drought	20–100	↑ growth, osmoregulation, ↓ ROS burst, cell damage	[108,109]

Table 1. *Cont.*

Plant Species	Stress Type	Melatonin Treatment (μM)	Effects Observed	Reference
Broccoli	-	60 ppm	↑ growth, photosynthetic attributes: LAI, NAR, AGR, CGR, Chls, carotenoids	[110]
Cabbage	Metal-Cu	1–100	↑ germination, growth, ↓ membrane peroxidation	[111]
	-	100–1000	↑ growth, anthocyanins, osmoregulation, redox balance, ↓ ABA, senescence factors, Chls degradation	[112,113]
Cassava	-	100	↓ ROS, postharvest deterioration, starch degradation	[114]
	Xam bacterial bligh		↑ disease resistance, ↓ bacterial propagation in leaves	[115]
Cherry sweet	Rootstocks	0.5–5	↑ number of roots, length, % rooting in 3 cherry rootstocks	[116,117]
			↑ photosynthetic pigments, biomass, total carbohydrates and proline	
	Orchard trees	10	↓ sweet cherries ripening, anthocyanins	[118]
Citrus	Salinity	1	↑ osmoregulation, Chls, ↓ ROS burst, membrane peroxidation	[119]
Coffee	Drought	300	↑ tolerance, root system, photosynthesis, gas exchange, CO_2 fixation, Chls, ASA-GSH cycle, ↓ ROS, MDA	[120]
Cotton	-	20	↑ germination, growth, antioxidant enzymes, GA_3, ↓ ROS, MDA, ABA	[121]
Cucumber	Cold	25–500 mM	↑ germination, ↓ ROS, membrane peroxidation	[122]
	Cold	50–500	↑ GSH pool, ↓ ROS	[123]
	Cold	200	↑ tolerance, photosynthesis, polyamines, ABA, GSH-ASA cycle, ↓ electrolyte leakage, ROS,	[124,125]
	Heat	100	↑ tolerance, N metabol., nitrate, ↓ damage, ammonium	[126]
	N-excess	100	↑ tolerance, growth, NPK balance, Ca, ↓ damage, nitrate, ammonium	[127]
	Salinity	1	↑ germination, GA_4, ↓ ROS, membrane peroxidation, ABA	[69]
		50–150	↑ tolerance, Chls, photosynthesis, GSH-ASA cycle, ↓ ROS	[128]

Table 1. *Cont.*

Plant Species	Stress Type	Melatonin Treatment (μM)	Effects Observed	Reference
		1	↑ germination, protein biosynthesis, lipid and carbohydrate metabol., TCA, ATP	[129]
	Drought	100	↑ germination, root growth	[130]
	Oxidative	50	↑ systemic antioxidant defence, GSH, photosynthesis, ↓ ROS	[131]
	Metal-Cu	0.01	↑ tolerance, growth, Cu-sequestration, TCA, ATP, GSH, ↓ ROS	[132]
	Cinnamic acid	100	↑ tolerance, growth, water and nutrient balance, hormonal balance	[133]
Faba bean	Salinity	100–500	↑ plant height, RWC, photosynthetic pigments, osmolites, phenolic	[134]
Grape	Drought	0.05–0.2	↑ seedling growth, osmoregulation, photosynthesis, ↓ ROS burst	[135]
	Salinity-Rhizobacteria	-	↑ root growth, RWC, melatonin in roots, colonization, ↓ damage, ROS	[136]
	Berry ripening	100	↑ anthocyanins, phenols, flavonoids, proanthocyanidins, resveratrol, ↓ ROS	[137,138]
	Berry/Wine	430	↑ size- and ripening-berries, fruity-, spicy- and sweet-wine	[139]
Kiwifruit	Drought	50–200	↑ tolerance, photosynthesis, CO_2 fixation, growth, biomass, roots, osmoregulation, flavonoids ↓ lipid peroxidation, carotenoid degradation	[140,141]
	Heat	200	↑ tolerance, ASA, proline, antioxidant enzymes, ↓ heat damage, ROS	[142]
Leek	Cold, heat	5	↑ tolerance, germination, growth	[143]
Lychee	Post-harvest	400	↑ redox balance, antioxidant enzymes, ↓ pericarp browning, discoloration, ROS, membrane leakage, loss of phenolics, flavonoids and anthocyanins	[144]
Lupin	Several stress	-	↑ germination, growth, rooting, redox balance, ↓ ROS, foliar senescence	[58,60,62,145]
Maize	Salinity, heat	100	↑ photosynthesis, antioxidant enzymes, ↓ ROS, electrolyte leakage	[146,147]
	-	10–1000	↑ root and stem growth, plant height, leaf surface area, protein, carbohydrates, Chls	[148]

Table 1. *Cont.*

Plant Species	Stress Type	Melatonin Treatment (μM)	Effects Observed	Reference
	Drought	100	↑ tolerance, growth, photosynthesis, stoma conductance, transpiration, RWC, antiox enzymes, ↓ ROS, MDA	[149–151]
	Heat	10–90	↑ tolerance, antioxidant enzymes, osmoregulation, ↓ ROS, MDA, electrolyte leakage	[152]
	Metal-Pb	50–100	↑ tolerance, growth, photosynthesis, Chls, RWC, K, Ca levels, ↓ ROS, MDA	[153]
	-	10	↑ sugar metabolism, photosynthesis, sucrose phloem loading	[154]
	-	50–500	protein synthesis, folding, destination and storage, defence, anti-stresses and detoxifying proteins	[155]
	Aging seeds		↑ viability, growth, antioxidant enzymes, carbohydrate-, secondary-, and amino acid metabol., ↓ ROS, MDA	[156]
Melissa (lemon balm)	Metal-Zn-Cd	1000	↑ tolerance, growth, antioxidant enzymes	[157]
Mung bean	Cold	20	↑ tolerance, growth, plastids, ↓ ROS, lipid peroxidation	[158]
Oat	Salinity, drought	50–100	↑ tolerance, growth, Chls, proline, antioxidant enzymes, ↓ ROS, MDA	[59,159,160]
Onion	Cold, heat	5	↑ tolerance, germination, growth	[143]
Peach (fruit)	Cold	50–200	↑ juice, TSS, polyamines, GABA, proline, ↓ chilling injury, ROS	[161]
	Post-harvest	100	↑ firmness, TSS. ASA, ↓ weight loss, decay incidence, respiration rate,	[162]
Pear (tree)	-	100	↑ photosynthesis, fruit size, TSS, sucrose, sorbitol, starch	[163]
	Parthenocarpy	100	↑ parthenocarpy with expansion, division mesocarp cells, unviable seeds, GAs	[73]
Pear (fruit)	Post-harvest	100	↑ firmness, commercial value, ↓ weight loss, ethylene, softening, core browning	[164,165]
Pepper	Salinity, Fe-low	100	↑ growth, Chls, photosynthesis, fruit yield, Fe, K uptake, antioxidant enzymes	[166]
	Cold	1–5	↑ germination, growth, antioxidant enzymes, ↓ ROS, MDA	[167]

Table 1. *Cont.*

Plant Species	Stress Type	Melatonin Treatment (μM)	Effects Observed	Reference
	Boron-high	1	↑ tolerance, growth, photosynthesis, ↓ B in leaf and fruit, toxicity	[168]
Pea	Oxidative	50–200	↑ photosynthesis efficiency, pigments, water content, ↓ ROS	[169,170]
	Metal-Cu	5	↑ plant survival	[171]
Plum (fruit)	Cold	1–1000	↑ firmness, postharvest life, ASA, phenols, antioxidant activity, ↓ weight loss	[172]
Pomegranate	Cold	100	↑ tolerance, antioxidant enzymes, membrane integrity, phenols, ↓ ROS	[173]
Poplar	Oxidative		↑ redox balance, proline, ↓ ROS, MDA, membrane damage, electrolyte leakage	[174]
Potato	Salinity	0.1–200	↑ tolerance, K^+/Na^+ homeostasis, ATPase, triacylglycerol breakdown, fatty acid β-oxidation, energy turnover	[175]
	Phytophthora infestans (potato late blight)	1–10 mM	↑ plant innate immunity, fungicide resistance and virulence, synergistic anti-fungal effects of melatonin with fungicides	[176]
Radish	Heat	50–300	↑ biomass, quality, antioxidant enzymes, Chls, hormone contents	[177]
Rapeseed	Salinity	0.01–100	↑ tolerance, redox balance, ion homeostasis, ↓ ROS, MDA	[178]
	Drought	500	↑ tolerance, germination, Chls, stoma size, osmoregulation, antioxidant enzymes, ↓ ROS, MDA	[179]
Rice	Cold	20–100	↑ tolerance, growth, photosynthesis, redox balance, ↓ ROS	[180]
	Salinity	10–20	↑ Chls, ↓ senescence, ROS, cell death	[181]
	Metal-Cd		↑ tolerance, growth, photosynthesis, redox balance, panicle number, grain yield	[182,183]
	Bacterial blight	200	↓ bacterial proliferation, motility	[184]
	Salt, cold, Blast fungus	-	↑ tolerance, melatonin induction, hormones, ↓ fungi proliferation	[185]
	-	0.5–1	↑ seminal roots, lateral roots, root growth, root biomass	[186]

Table 1. *Cont.*

Plant Species	Stress Type	Melatonin Treatment (μM)	Effects Observed	Reference
	Soil	10–50	↑ number of lateral roots, root growth, shaping root architecture	[187]
Soybean	Salinity, drought	50–100	↑ tolerance, seedling growth, leaf size, biomass, seed yield	[188]
	Drought	100	↑ RWC, Chls, photosynthetic gas-exchange parameters, osmoregulation, antioxidant enzymes, and seed growth-related indicators	[189]
	Metal-Al	0.1–1	↑ tolerance, root growth, antioxidant enzymes, osmoregulation, ↓ ROS,	[190]
Spinach	Boron	100–300	↑ tolerance, growth, photosynthesis, RWC, CO_2 uptake, sugars, carotenoids, redox balance, ↓ ROS, MDA	[191]
Strawberry	Post-harvest	100	↑ nutritional quality, antioxidant enzymes, anthocyanins, phenols, GABA, ATP ↓ fungal decay	[192]
	Post-harvest	0.1–1	↑ color, firmness, TSS, ASA, flavonoids, ↓ weight loss, senescence, ROS, MDA	[193]
Sunflower	Salinity	15	↑ root, hypocotyl growth, antioxidant potential, antioxidant enzymes, GSH	[194–196]
Tea plant	Salinity, cold, drought	100	↑ photosynthesis, GSH, ASA, antioxidant enzymes, ↓ ROS, MDA	[197,198]
Tobacco	Tobacco mosaic virus	100	↑ tolerance, ↓ virus proliferation, virus-RNA, viral disease	[199]
Tomato	Salinity	50–150	↑ photosynthesis, PSII efficiency, D1 protein turnover, ↓ ROS burst	[200]
	Salinity	20–50	↑ growth, photosynthesis, Rubisco, proline, C-metabol., ASA-GSH cycle, ↓ ROS, MDA	[201]
	Salinity	150	↑ tolerance, photosynthesis, PSII repair, ASA-GSH cycle, ↓ ROS	[202]
	Cold	100	↑ antioxidant enzymes, GSA, ASA, CO_2 uptake, sucrose, proline, Calvin cycle, polyamines, ↓ ROS, MDA, electrolyte leakage	[203]
	Cold	100	↑ tolerance, growth, VAZ cycle, photosynthesis, photosystem efficiency, ↓ ROS, MDA, photoinhibition	[204]
	Cold-fruit	100	↑ tolerance, proline, polyamines, membrane integrity	[205]

Table 1. *Cont.*

Plant Species	Stress Type	Melatonin Treatment (μM)	Effects Observed	Reference
	Heat	10	↑ thermotolerance and cell protection	[206,207]
	Heat-pollen	20	↑ thermotolerance, polen germination, antioxidant enzymes, reproductive development	[208]
	Metal-Cd	25–500	↑ Cd tolerance, phytochelatins, ATPase activity	[209]
	Metal-Cd	-	↑ Cd tolerance, heat-shock factor A1a induction by melatonin	[210]
	Metal-Cd-Se	-	↑ growth, photosynthesis, electrolyte leakage, phytochelatins, GSH, ↓ ROS, Cd leaf,	[211]
	Alkalinity	0.25–1	seedling growth, photosynthesis, ion homeostasis, burst	[212]
	Acid rain	100	↑ tolerance, growth, chloroplast integrity, photosynthesis, antioxidant enzymes, ↓ ROS, MDA	[213]
	Drought	100	↑ tolerance, waxes-cutin leaf, RWC, Chls,	[214]
	Drought	100	↑ tolerance, Chls, antioxidant enzymes, p-coumaric acid, ↓ ROS, MDA	[215]
	S-low	100	↑ S uptake, assimilation, transport and metabolism, peroxiredoxins, redox homeostasis, ↓ ROS, DNA damage	[216]
	Rooting	50	↑ adventitious root formation, auxin, auxin transport and signal transduction	[217]
	On vine-ripening		↑ fruit yield and quality, ASA, citric acid, lycopene, TSS, Ca, P ↓ N, Mg, Cu, Zn, Fe, Mn,	[218]
	Post-harvest	50	↑ fruit ripening, fruit quality, colour, carotenoids, polygalacturonase and related, biosynthesis, perception and signalling of ethylene, anthocyanins, ↓ weight loss	[74,219]
	Mosaic virus	100	↑ tolerance, ↓ virus proliferation, virus-RNA, viral disease	[199]
Valerian	Metal-Zn-Cd	1000	↑ tolerance, growth, antioxidant enzymes	[157]
Watermelon	Cold	150	↑ photosynthesis, ↓ cold-related microRNA	[220]
	Salinity	50–500	↑ tolerance, growth, photosynthesis, antioxidant enzymes, GSH, ASA, ↓ ROS, MDA	[221]

Table 1. *Cont.*

Plant Species	Stress Type	Melatonin Treatment (μM)	Effects Observed	Reference
	Metal-V	0.1	↑ tolerance, growth, photosynthesis, antioxidant enzymes, ↓ V level, V transport, ROS, MDA	[222]
Wheat	Cold	1000	↑ redox balance, Chls, osmoregulation, ↓ ROS	[223]
	Cold	1000	↑ tolerance, growth, Chls, photosynthesis, CO_2 uptake, grain filled	[224]
	Cold	1000	↑ photosynthesis, stomatal conductance, antioxidant enzymes, membrane stability	[225]
	Salinity	1	↑ tolerance, growth, photosynthesis, IAA, polyamines, ↓ ROS	[226]
	Salinity	50–500	↑ growth, yield, antioxidant enzymes, ↓ ROS, MDA	[227]
	Drought	500	↑ tolerance, RWC, photosynthesis, antioxidant enzymes, ASA, GSH, ↓ ROS, membrane damage	[228]
	Metal-Cd	100	↑ tolerance, antioxidant enzymes, ASA, GSH, ↓ ROS	[229]
	Metal-Cd	50–100	↑ tolerance, growth, Chls, photosynthesis, RWC, Ca, K, antioxidant enzymes, ↓ ROS, MDA, Cd	[230]
	Metal-Zn	1000	↑ tolerance, Chls, photosynthesis, Rubisco, ATPase	[231]
	N-low	1	↑ N and nitrate, N absorption, N metabolism, growth, yield, in shoots and roots	[232]

↑, Increased content or increased action. ↓, decreased content or decreased action; ABA, abscisic acid; AGR, absolute growth rate; ASA, ascorbic acid; CGR, crop growth rate; Chls, chlorophylls; CMC, component materials categories of fertilizers; EC, European Community; ECHA, European Chemical Agency; EU, European Union; GA_4, gibberellin-4; GABA, γ-aminobutyric acid; GSH, glutathione; JA, Jasmonic acid; LAI, leaf area index; MDA, malondialdehyde; MAPKK, mitogen-activated protein kinase cascade; NAR, net assimilation rate; OXI1, oxidative signal-inducible1 kinases; PFC, product function categories of fertilizers; ROS, reactive oxygen species; RWC, relative water content; SA, salicylic acid; SAR, systemic acquired resistance; TA, total valuable acidity; TCA, Krebs cycle; TSS, total solid soluble.

4. Melatonin in the Health and Environment of EC

In accordance with the Classification, Labelling and Packaging (CLP, EC-No 1272/2008) regulation, which is based on the United Nations' Globally Harmonized System, which has a purpose to ensure a high level of protection of health and the environment, as well as the free movement of substances, mixtures and articles, the European Chemical Agency (ECHA) classified melatonin (EC No. 200-797-7 (CAS 73-31-4), *N*-(2-(5-methoxyindol-3-yl)-ethyl)-acetamide), as a non-hazardous substance in terms of physical and chemical hazards. With respect to human health, it is classified as a non-hazardous substance in the oral, dermal, inhalation and irritation categories, and in regards to mutagenicity and carcinogenicity. However, melatonin is classified as a health hazard substance (code H-361) in terms of reproductive toxicity because it is suspected of damaging fertility or an unborn child. This classification reflects one of its multiple functions as an animal hormone, in which its participation in the modulation of sexual behavior in mammals has been demonstrated, and also, it is believed, the same of fertility [233,234]. In fact, it is usually applied to sheep as a hormonal regulator of sexual zeal to homogenize the reproductive process in ovine, with demonstrated higher conception and pregnancy rates when applied [235]. Nevertheless, melatonin is classified as non-hazardous in terms of its possible damage to the environment and atmosphere.

5. Melatonin as an Active Substance or as a Plant Biostimulator/Protector in Crops: Concepts and Legal Considerations in EC

After many changes and adaptations, the EC finally seems to have established its policy regarding the authorization, classification, use, distribution, importation, management, etc., of plant protectors and fertilizers, in an attempt to improve agricultural production, while minimizing risks and hazards for humans, animals and the environment. In order to establish the minimum basis for the possible use of melatonin in plant production and post-harvest application, several requirements regarding its human consumption must be taken into account:

(i) Melatonin is a highly studied substance that has given rise to abundant physicochemical and biological data; (ii) there are numerous studies in animals and humans regarding its beneficial effects on health, in aspects as diverse as neurodegenerative, immunological, liver, renal, heart, skin and gastrointestinal diseases, in addition to osteopathy, retinopathy, etc. It also helps in the treatment of various cancers, particularly, chemical and radiological therapies; (iii) in regards to melatonin for human consumption, although it is classified as a drug in the EC, there are some cases in which it does not need a medical prescription, such as those where the amount of melatonin is less than 1 mg. Generally, these are used for jet-lag and sleep disorders. In many other countries (e.g., USA, Canada) melatonin is not treated as a drug, but as a food supplement; (iv) in no case has melatonin been declared as toxic, even at the intake of 1 g/day. Only some slight side effects such as migraine and headache have been described.

The possible use of melatonin in plant production involves particular aspects such as: (i) Melatonin is a molecule that exists in all living things, from bacteria to humans, but also in plants, algae, fungi, etc.; (ii) its action in animals and humans is well known since it has been investigated for many years. In plants, although many physiological effects of melatonin are known, new data are being acquired every day; (iii) in all cases, only positive effects have been described, all beneficial for the development of plants (the same can be said for animals); (iv) little information is available on its effect on bacteria and fungi, especially those that are part of the soil microbiota (rhizosphere); (v) there are also few or no data on its effect on the environment, in particular on agricultural and aquatic fauna; (vi) the levels of melatonin described in plants, and which appear to be effective in pharmacological treatments known to date, are much higher than those described in animals or humans, which may be a cause for caution.

Council Directive 91/414/EEC of 15 July, 1991 concerning the marketing of plant protection products provides rules governing plant protection products and the active substances contained in those products. This old directive has been replaced by two more current ones that are as follows:

- Regulation #1. Regulation EC 1107/2009 of the European Parliament and of the Council of 21 October 2009 concerning the placing of plant protection products on the market and repealing council directives 79/117/EEC and 91/414/EEC, and;
- Regulation #2. Regulation EU 2019/1009 of the European Parliament and of the Council of 5 June 2019 laying down rules on the making available on the market of EU fertilizing products and amending Regulations (EC) No. 1069/2009 and (EC) No. 1107/2009 and repealing Regulation (EC) No 2003/2003.

If we review the actions confirmed so far for melatonin in plants, we find that melatonin exerts a clear action as a plant protector in situations of biotic stress against bacterial, fungal and viral diseases (Regulation #2), but it can also be used as an agent against situations of abiotic stress (Regulation #1). Thus, Regulation #2 says in point 22:

> "Certain substances, mixtures and micro-organisms, referred to as plant biostimulants, are not as such inputs of nutrients, but nevertheless stimulate plants' natural nutrition processes. Where such products aim solely at improving the plants' nutrient use efficiency, tolerance to abiotic stress, quality traits or increasing the availability of confined nutrients in the soil or rhizosphere, they are by nature more similar to fertilising products than to most categories of plant protection products. They act in addition to fertilisers, with the aim of optimising the efficiency of those fertilisers and reducing the nutrient application rates. Such products should therefore be eligible for CE marking under this Regulation and excluded from the scope of Regulation (EC) No 1107/2009".

These two regulations attempt to classify the substances and products applicable to crops into two large groups: Those that are plant protectors (phyto-sanitary) (Regulation #1) and those that can be used as fertilizers (Regulation #2). As we have seen in the previous section, melatonin is classified as a health hazard substance (code H-361) for its reproductive toxicity in ECHA, so its possible authorization as an active substance by regulation EC 1907/2006 of Registration, Evaluation, Authorisation and Restriction of Chemicals (REACH) could be difficult.

Although Regulation #1 on plant protection products extends the concept of an active substance, since it includes microorganisms and preparations (art. 1 point 2): This Regulation shall apply to substances, including micro-organisms having general or specific action against harmful organisms or on plants, parts of plants or plant products, referred to as 'active substances', some interesting restrictions appeared in:

- Art. 23b: "Basic substances shall be approved in accordance with paragraphs 2 to 6. (...) For the purpose of paragraphs 2 to 6, a basic substance is an active substance which (...), (b) does not have an inherent capacity to cause endocrine disrupting, neurotoxic or immunotoxic effects";
- Annex II, Impact on Human Health, 3.6.5: "An active substance, safener or synergist shall only be approved if, on the basis of the assessment of community or internationally agreed test guidelines or other available data and information, including a review of the scientific literature, reviewed by the Authority, it is not considered to have endocrine disrupting properties that may cause adverse effect in humans, unless the exposure of humans to that active substance, safener or synergist in a plant protection product, under realistic proposed conditions of use, is negligible, ... " and in;
- Annex II. Ecotoxicology, 3.8.2. An active substance, safener or synergist shall only be approved if, on the basis of the assessment of community or internationally agreed test guidelines, it is not considered to have endocrine disrupting properties that may cause adverse effects on non-target organisms unless the exposure of non-target organisms to that active substance in a plant protection product under realistic proposed conditions of use is negligible.

Thus, taking into account all this legal information, and ruling out the possibility of using melatonin as an active substance (pure chemical substance) for agronomic application, the possibility of using plant, bacterial, algae, or fungi extracts rich in melatonin would remain. Thus, a good plan might be to use plant (or other) extracts rich in melatonin as a fertilizer, in the category of biostimulants. A biostimulant could also be defined as a formulated product of biological origin that improves plant productivity as a consequence of the emergent properties of its constituents. Thus, biostimulants could be defined by their demonstrated mode of action and origin, or solely by their demonstrated beneficial impact on plant productivity. The challenges in developing a definition are also complicated by the multi-component and largely undefined composition of many biostimulant products and the possibility that the activity of a biostimulant may not be explained by the presence of any individual constituent, but is a result of the interaction of many constituents in the product. Indeed, most biostimulants in use today are complex mixtures of chemicals derived from a biological process or the extraction of biological materials [236].

According Regulation #2 (EU 2019/1009) on fertilizing products, in Annex I, Product Function Categories (PFCs) of EU fertilizing products, in Category 6, two types of plant biostimulant can be developed: Microbial plant biostimulants (subtype A) and non-microbial plant biostimulants (subtype B). In Annex II, it says: "An EU fertilizing product shall consist solely of component materials complying with the requirements for one or more of the CMCs listed in this Annex", where the different component materials categories (CMC) were defined. Of interest are the following:

- CMC2: Plants, plant parts or plant extracts is described as: "An EU fertilizing product may contain plants, plant parts or plant extracts having undergone no other processing than cutting, grinding, milling, sieving, sifting, centrifugation, pressing, drying, frost treatment, freeze-drying or extraction with water or supercritical CO_2 extraction. For the purpose of this point, plants include mushrooms and algae and exclude blue-green algae (cyanobacteria)."
- CMC6: Food industry by-products, point (e): "Plants, plant parts or plant extracts having undergone only heat treatment or heat treatment in addition to processing methods referred to in CMC 2"
- CMC7: Micro-organisms. "An EU fertilising product belonging to PFC 6A may contain micro-organisms, including dead or empty-cell micro-organisms and non-harmful residual elements of the media on which they were produced".

The strategies to obtain melatonin-rich extracts may involve microorganisms (PFC6A) or plants (PFC 6B). At present, there seem to be no data on the production of melatonin by bacteria or fungal cultures. The objective to obtain melatonin-rich plants (CMC2) is ambitious since phytomelatonin levels in plants are usually very low, and less than 5–10 ng per gram of plant. An exhaustive classification of many plants according to their phytomelatonin content can be consulted [37,237,238]. Generally, medicinal plants have high phytomelatonin content, but this tends to vary widely due to the varied origin of plants, technical conditions of growth, variety, post-harvest treatment, etc. Several strategies can be followed: (i) Selecting plant species with high levels of phytomelatonin which can be extracted and concentrated, and (ii) inducing the biosynthesis of phytomelatonin in in vitro cultured pre-selected plant tissues. A discussion on this aspect can be consulted [239]. Our group is developing a formulation where only aromatic/medicinal plants are used to obtain a botanical mixture rich in phytomelatonin through the application of a simple process. A rigorous plant selection protocol and careful management will ensure high phytomelatonin content in the plant extracts generated. The formulation and its protocol are being patented before being made available to interested companies for commercial exploitation. We are currently characterizing it and conducting the appropriate studies and bioassays in plants to confirm its beneficial biological activity related with its high phytomelatonin content.

Figure 2 shows, according to the legislation analyzed, the pros and cons of melatonin (as a chemical substance) and phytomelatonin-rich extracts and its possible regularization as a plant protector or fertilizer (biostimulant).

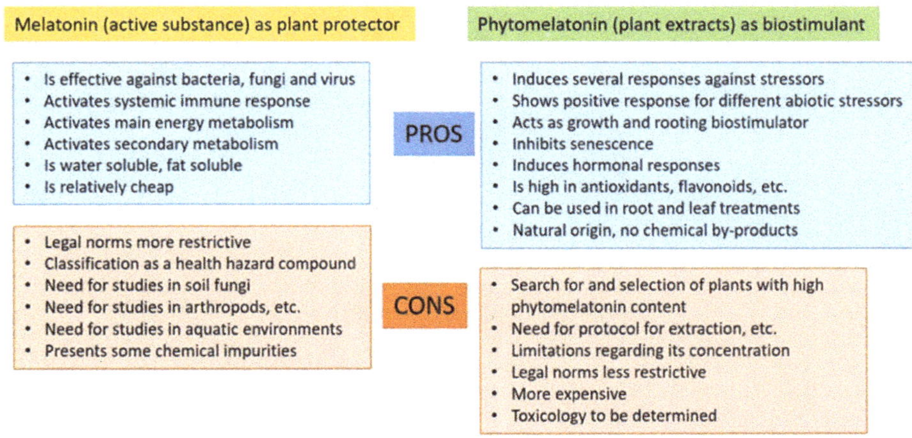

Figure 2. Pros and cons of the possible use of chemical melatonin and rich-phytomelatonin extracts according to EC legislation.

6. Future Prospects

Numerous studies with melatonin have resulted in a set of data that indicate the excellent beneficial effects that this compound has on plants, especially in stress situations. It should not be forgotten that melatonin is a natural compound, endogenous to plants and other organisms including humans. It is this last aspect that makes it more interesting and also more delicate or sensitive, when using it as a plant protective agent or as a biostimulator. However, demonstrating through trials that its use is possible in crops and does not entail risks to human and wildlife health will be the only way forward in this field. The alternative of using phytomelatonin-rich extracts seems more interesting, but also more laborious. The search and selection of plants with high endogenous levels of phytomelatonin is a first requirement for subsequent extraction and preparation. The analysis and study of its potential as a protector against plant stress will throw light on the true effect on crops. However, although many aspects of the mechanism of action of phytomelatonin are already known, there are other relevant aspects to study as: (i) The optimal mode of application, time and rate; (ii) the phenological state; (iii) the effect on rhizosphere; (iv) the persistence in soil or in foliar applications; (v) the synergic or antagonic effects with other plant treatments (pesticides, fertilizers, etc.), among others. Obviously, companies in the phytochemical sector (manufacturers) will need to start field studies and deal with possible legal regularization.

Abbreviations

ABA	abscisic acid
AGR	absolute growth rate
ASA	ascorbic acid
CGR	crop growth rate
Chls	chlorophylls
CMC	component materials categories of fertilizers
EC	European Community
ECHA	European Chemical Agency
EU	European Union
GA_4	gibberellin-4
GABA	γ-aminobutyric acid

GSH	glutathione
JA	jasmonic acid
LAI	leaf area index
MDA	malondialdehyde
MAPKK	mitogen-activated protein kinase cascade
NAR	net assimilation rate
OXI1	oxidative signal-inducible1 kinases
PFC	product function categories of fertilizers
ROS	reactive oxygen species
RWC	relative water content
SA	salicylic acid
SAR	systemic acquired resistance
TA	total valuable acidity
TCA	Krebs cycle
TSS	total solid soluble

Author Contributions: The manuscript was conceived by M.B.A. and written and revised by M.B.A. and J.H.-R.

Funding: Not financial support available for this review.

Conflicts of Interest: The authors declare no conflicts of interest.

References

1. Lerner, A.B.; Case, J.D.; Takahashi, Y.; Lee, T.H.; Mori, W. Isolation of melatonin, a pineal factor that lightens melanocytes. *J. Am. Chem. Soc.* **1958**, *80*, 2587. [CrossRef]
2. Lerner, A.; Case, J.; Mori, W.; Wright, M. Melatonin in peripheral nerve. *Nature* **1959**, *183*, 1821. [CrossRef] [PubMed]
3. Lerner, A.; Case, J.; Heinzelmann, R. Structure of melatonin. *J. Am. Chem. Soc.* **1959**, *81*, 6084–6085. [CrossRef]
4. Xie, Z.; Chen, F.; Li, W.A.; Geng, X.; Li, C.; Meng, X.; Feng, Y.; Liu, W.; Yu, F. A review of sleep disorders and melatonin. *Neurol. Res.* **2017**, *39*, 559–565. [CrossRef] [PubMed]
5. Vadnie, C.A.; McClung, C.A. Circadian rhythm disturbances in mood disorders: Insights into the role of the suprachiasmatic nucleus. *Neural Plast.* **2017**, *2017*, 1504507. [CrossRef]
6. Carpentieri, A.R.; Oliva, C.; Díez-Noguera, A.; Cambras, T. Melatonin administration modifies circadian motor activity under constant light depending on the lighting conditions during suckling. *Chronobiol. Int.* **2015**, *32*, 994–1004. [PubMed]
7. Blume, C.; Angerer, M.; Raml, M.; del Giudice, R.; Santhi, N.; Pichler, G.; Kunz, A.B.; Scarpatetti, M.; Trinka, E.; Schabus, M. Healthier rhythm, healthier brain? Integrity of circadian melatonin and temperature rhythms relates to the clinical state of brain-injured patients. *Eur. J. Neurol.* **2019**, *26*, 1051–1059. [CrossRef]
8. Marková, M.; Adámeková, E.; Kubatka, P.; Bojková, B.; Ahlersová, E.; Ahlers, I. Effect of prolonged melatonin administration on metabolic parameters and organ weights in young male and female Sprague-Dawley rats. *Acta Vet. Brno* **2003**, *72*, 163–173. [CrossRef]
9. Bojková, B.; Marková, M.; Ahlersová, E.; Ahlers, I.; Adámeková, E.; Kubatka, P.; Kassayová, M. Metabolic effects of prolonged melatonin administration and short-term fasting in laboratory rats. *Acta Vet. Brno* **2006**, *75*, 21–32. [CrossRef]
10. Cardinali, D.P.; Hardeland, R. Inflammaging, metabolic syndrome and melatonin: A call for treatment studies. *Neuroendocrinology* **2017**, *104*, 382–397. [CrossRef]
11. Kubatka, P.; Bojková, B.; Mociková, K.; Mníchová, M.; Adámeková, E.; Ahlers, I.; Ahlersová, E.; Cermáková, M. Effects of tamoxifen and melatonin on mammary gland cancer induced by N-methyl-N-nitrosourea and by 7,12-dimethylbenz(a)anthracene, respectively, in female Sprague-Dawley rats. *Folia Biologica (Praha)* **2001**, *47*, 5–10.
12. Di Bella, G.; Mascia, F.; Gualano, L.; Di Bella, L. Melatonin anticancer effects: Review. *Int. J. Mol. Sci.* **2013**, *14*, 2410–2430. [CrossRef]
13. Bojková, B.; Kubatka, P.; Qaradakhi, T.; Zulli, A.; Kajo, K. Melatonin may increase anticancer potential of pleiotropic drugs. *Molecules* **2018**, *19*, 3910. [CrossRef]

14. Nabavi, S.M.; Nabavi, S.F.; Sureda, A.; Xiao, J.; Dehpour, A.R.; Shirooie, S.; Silva, A.S.; Baldi, A.; Khan, H.; Daglia, M. Anti-inflammatory effects of melatonin: A mechanistic review. *Crit. Rev. Food Sci. Nutr.* **2019**, *59*, S4–S16. [CrossRef]
15. Carrascal, L.; Nuñez-Abades, P.; Ayala, A.; Cano, M. Role of melatonin in the inflammatory process and its therapeutic potential. *Curr. Pharm. Des.* **2018**, *24*, 1563–1588. [CrossRef]
16. Carrillo-Vico, A.; Guerrero, J.M.; Lardone, P.J.; Reiter, R.J. A review of the multiple actions of melatonin on the immune system. *Endocrine* **2012**, *27*, 189–200. [CrossRef]
17. Reiter, J.R.; Rosales-Corral, A.S.; Tan, D.X.; cuna-Castroviejo, D.; Qin, L.; Yang, S.F.; Xu, K. Melatonin, a full service anti-cancer agent: Inhibition of initiation, progression and metastasis. *Int. J. Mol. Sci.* **2017**, *18*, 843. [CrossRef]
18. Majidinia, M.; Reiter, R.J.; Shakouri, S.K.; Yousefi, B. The role of melatonin, a multitasking molecule, in retarding the processes of ageing. *Ageing Res. Rev.* **2018**, *47*, 198–213. [CrossRef]
19. Boga, J.A.; Caballero, B.; Potes, Y.; Perez-Martinez, Z.; Reiter, R.J.; Vega-Naredo, I.; Coto-Montes, A. Therapeutic potential of melatonin related to its role as an autophagy regulator: A review. *J. Pineal Res.* **2019**, *66*, e12534. [CrossRef]
20. Luo, C.; Yang, Q.; Liu, Y.; Zhou, S.; Jiang, J.; Reiter, R.J.; Bhattacharya, P.; Cui, Y.; Yang, H.; Ma, H.; et al. The multiple protective roles and molecular mechanisms of melatonin and its precursor N-acetylserotonin in targeting brain injury and liver damage and in maintaining bone health. *Free Radic. Biol. Med.* **2019**, *130*, 215–233. [CrossRef]
21. Arnao, M.B.; Hernández-Ruiz, J. Melatonin: Synthesis from tryptophan and its role in higher plants. In *Amino Acids in Higher Plants*; D' Mello, J.P.F., Ed.; CAB Intern.: Boston, MA, USA, 2015; pp. 390–435.
22. Jan, J.E.; Reiter, R.J.; Wasdell, M.B.; Bax, M. The role of the thalamus in sleep, pineal melatonin production, and circadian rhythm sleep disorders. *J. Pineal Res.* **2009**, *46*, 1–7. [CrossRef]
23. McHill, A.W.; Phillips, A.J.; Czeisler, C.A.; Keating, L.; Yee, K.; Barger, L.K.; Garaulet, M.; Scheer, F.A.; Klerman, E.B. Later circadian timing of food intake is associated with increased body fat. *Am. J. Clin. Nutr.* **2017**, *106*, 1213–1219. [CrossRef]
24. Dubbels, R.; Reiter, R.J.; Klenke, E.; Goebel, A.; Schnakenberg, E.; Ehlers, C.; Schiwara, H.W.; Schloot, W. Melatonin in edible plants identified by radioimmunoassay and by HPLC-MS. *J. Pineal Res.* **1995**, *18*, 28–31. [CrossRef]
25. Hattori, A.; Migitaka, H.; Iigo, M.; Yamamoto, K.; Ohtani-Kaneko, R.; Hara, M.; Suzuki, T.; Reiter, R.J. Identification of melatonin in plants and its effects on plasma melatonin levels and binding to melatonin receptors in vertebrates. *Biochem. Mol. Biol. Int.* **1995**, *35*, 627–634.
26. Kolar, J.; Machackova, I.; Illnerova, H.; Prinsen, E.; van Dongen, W.; van Onckelen, H. Melatonin in higher plant determined by radioimmunoassay and liquid chromatography-mass spectrometry. *Biol. Rhythm Res.* **1995**, *26*, 406–409.
27. van Tassel, D.L.; Roberts, N.; O'Neill, S.D. Melatonin from higher plants: Isolation and identification of N-acetyl-5-methoxytryptamine. *Plant Physiol.* **1995**, *108*, 101.
28. Back, K.; Tan, D.X.; Reiter, R.J. Melatonin biosynthesis in plants: Multiple pathways catalyze tryptophan to melatonin in the cytoplasm or chloroplasts. *J. Pineal Res.* **2016**, *61*, 426–437. [CrossRef]
29. Zheng, X.; Tan, D.X.; Allan, A.C.; Zuo, B.; Zhao, Y.; Reiter, R.J.; Wang, L.; Wang, Z.; Guo, Y.; Zhou, J.; et al. Chloroplastic biosynthesis of melatonin and its involvement in protection of plants from salt stress. *Sci. Rep.* **2017**, *7*, 41236. [CrossRef]
30. Wang, L.; Feng, C.; Zheng, X.; Guo, Y.; Zhou, F.; Shan, D.; Liu, X.; Kong, J. Plant mitochondria synthesize melatonin and enhance the tolerance of plants to drought stress. *J. Pineal Res.* **2017**, *63*, e12429. [CrossRef]
31. Arnao, M.B.; Hernández-Ruiz, J. Melatonin: A new plant hormone and/or a plant master regulator? *Trends Plant Sci.* **2019**, *24*, 38–48. [CrossRef]
32. Arnao, M.B.; Hernández-Ruiz, J. Melatonin in its relationship to plant hormones. *Ann. Bot.* **2018**, *121*, 195–207. [CrossRef]
33. Arnao, M.B.; Hernández-Ruiz, J. The multi-regulatory properties of melatonin in plants. In *Neurotransmitters in Plants*; Roshchina, V.V., Ramakrishna, A., Eds.; Taylor & Francis-CRC: New York, NY, USA, 2018; pp. 71–101.
34. Sharif, R.; Xie, C.; Zhang, H.; Arnao, M.; Ali, M.; Ali, Q.; Muhammad, I.; Shalmani, A.; Nawaz, M.; Chen, P.; et al. Melatonin and its effects on plant systems. *Molecules* **2018**, *23*, 2352. [CrossRef]

35. Arnao, M.B.; Hernández-Ruiz, J. Growth activity, rooting capacity, and tropism: Three auxinic precepts fulfilled by melatonin. *Acta Physiol. Plant* **2017**, *39*, 127. [CrossRef]
36. Arnao, M.B.; Hernández-Ruiz, J. Functions of melatonin in plants: A review. *J. Pineal Res.* **2015**, *59*, 133–150. [CrossRef]
37. Arnao, M.B. Phytomelatonin: Discovery, content, and role in plants. *Adv. Bot.* **2014**, *2014*, e815769. [CrossRef]
38. Wang, Y.; Reiter, R.J.; Chan, Z. Phytomelatonin: A universal abiotic stress regulator. *J. Exp. Bot.* **2018**, *69*, 963–974. [CrossRef]
39. Reiter, R.J.; Rosales-Corral, S.; Tan, D.X.; Jou, M.J.; Galano, A.; Xu, B. Melatonin as a mitochondria-targeted antioxidant: One of evolution´s best ideas. *Cell Mol. Life Sci.* **2017**, *74*, 3863–3881. [CrossRef]
40. Debnath, B.; Islam, W.; Li, M.; Sun, Y.; Lu, X.; Mitra, S.; Hussain, M.; Liu, S.; Qiu, D. Melatonin mediates enhancement of stress tolerance in plants. *Int. J. Mol. Sci.* **2019**, *20*, 1040. [CrossRef]
41. Yu, Y.; Lv, Y.; Shi, Y.; Li, T.; Chen, Y.; Zhao, D.; Zhao, Z. The role of phyto-melatonin and related metabolites in response to stress. *Molecules* **2018**, *23*, 1887. [CrossRef]
42. Shibaeva, T.G.; Markovskaya, E.F.; Mamaev, A.V. Phytomelatonin: A review. *Biol. Bull. Rev.* **2018**, *8*, 375–388. [CrossRef]
43. Arnao, M.B.; Hernández-Ruiz, J. Melatonin and reactive oxygen and nitrogen species: A model for the plant redox network. *Melatonin Res.* **2019**, *2*, 152–168. [CrossRef]
44. Zhang, X.; Schmidt, R. The impact of growth regulators on a-tocopherol status of water-stressed *Poa pratensis* L. *Int. Turfgrass Soc. Res. J.* **1997**, *8*, 1364–1373.
45. Kauffman, G.L.; Kneivel, D.P.; Watschke, T.L. Effects of a biostimulant on the heat tolerance associated with photosynthetic capacity, membrane thermostability, and polyphenol production of perennial ryegrass. *Crop Sci.* **2007**, *47*, 261–267. [CrossRef]
46. du Jardin, P. Plant biostimulants: Definition, concept, main categories and regulation. *Sci. Hort.* **2015**, *196*, 3–14. [CrossRef]
47. Rouphael, Y.; Colla, G. Synergistic Biostimulatory Action: Designing the Next Generation of Plant Biostimulants for Sustainable Agriculture. *Front. Plant Sci.* **2018**, *9*, 1655. [CrossRef]
48. Rouphael, Y.; SpÍchal, L.; Panzarová, K.; Casa, R.; Colla, G. High-Throughput plant phenotyping for developing novel biostimulants: From lab to field or from field to lab? *Front. Plant Sci.* **2018**, *9*, 1197. [CrossRef]
49. Afreen, F.; Zobayed, S.M.A.; Kozai, T. Melatonin in *Glycyrrhiza uralensis*: Response of plant roots to spectral quality of light and UV-B radiation. *J. Pineal Res.* **2006**, *41*, 108–115. [CrossRef]
50. Lei, X.Y.; Zhu, R.Y.; Zhang, G.Y.; Dai, Y.R. Attenuation of cold-induced apoptosis by exogenous melatonin in carrot suspension cells: The possible involvement of polyamines. *J. Pineal Res.* **2004**, *36*, 126–131. [CrossRef]
51. Arnao, M.B.; Hernández-Ruiz, J. The physiological function of melatonin in plants. *Plant Sign. Behav.* **2006**, *1*, 89–95. [CrossRef]
52. Fletcher, R.; Sopher, C. Retardation of leaf senescence by melatonin indicates that it acts as an antioxidant in plants. In Proceedings of the 5th Canadian Pineal and Melatonin Symposium, Guelph, ON, Canada, 13–15 July 1997.
53. Hardeland, R.; Pandi-Perumal, S.; Poeggeler, B. Melatonin in plants: Focus on a vertebrate night hormone with cytoprotective properties. *Funct. Plant Sci. Biotechnol.* **2007**, *1*, 32–45.
54. Reiter, R.J.; Korkmaz, A.; Paredes, S.D.; Manchester, L.C.; Tan, D.X. Melatonin reduces oxidative/nitrosative stress due to drugs, toxins, metals, and herbicides. *Neuro Endocrinol. Lett.* **2008**, *29*, 609–613.
55. Paredes, S.D.; Korkmaz, A.; Manchester, L.C.; Tan, D.X.; Reiter, R.J. Phytomelatonin: A review. *J. Exp. Bot.* **2009**, *60*, 57–69. [CrossRef]
56. Posmyk, M.; Janas, K. Melatonin in plants. *Acta Physiol. Plant* **2009**, *31*, 1–11. [CrossRef]
57. Reiter, R.J.; Paredes, S.D.; Manchester, L.C.; Tan, D.X. Reducing oxidative/nitrosative stress: A newly-discovered genre for melatonin. *Crit. Rev. Biochem. Mol. Biol.* **2009**, *44*, 175–200. [CrossRef]
58. Hernández-Ruiz, J.; Cano, A.; Arnao, M.B. Melatonin: Growth-stimulating compound present in lupin tissues. *Planta* **2004**, *220*, 140–144. [CrossRef]
59. Hernández-Ruiz, J.; Cano, A.; Arnao, M.B. Melatonin acts as a growth-stimulating compound in some monocot species. *J. Pineal Res.* **2005**, *39*, 137–142. [CrossRef]
60. Arnao, M.B.; Hernández-Ruiz, J. Melatonin promotes adventitious and lateral root regeneration in etiolated hypocotyls of *Lupinus albus* L. *J. Pineal Res.* **2007**, *42*, 147–152. [CrossRef]

61. Arnao, M.B.; Hernández-Ruiz, J. Melatonin in plants. More studies are necessary. *Plant Sign. Behav.* **2007**, *2*, 381–382. [CrossRef]
62. Hernández-Ruiz, J.; Arnao, M.B. Melatonin stimulates the expansion of etiolated lupin cotyledons. *Plant Growth Regul.* **2008**, *55*, 29–34. [CrossRef]
63. Park, W.J. Melatonin as an endogenous plant regulatory signal: Debates and perspectives. *J. Plant Biol.* **2011**, *54*, 143–149. [CrossRef]
64. Tan, D.X.; Hardeland, R.; Manchester, L.C.; Korkmaz, A.; Ma, S.; Rosales-Corral, S.; Reiter, R.J. Functional roles of melatonin in plants, and perspectives in nutritional and agricultural science. *J. Exp. Bot.* **2012**, *63*, 577–597. [CrossRef]
65. Byeon, Y.; Park, S.; Kim, Y.S.; Back, K. Microarray analysis of genes differentially expressed in melatonin-rich transgenic rice expressing a sheep serotonin N-acetyltransferase. *J. Pineal Res.* **2013**, *55*, 357–363.
66. Arnao, M.B.; Hernández-Ruiz, J. Melatonin: Plant growth regulator and/or biostimulator during stress? *Trends Plant Sci.* **2014**, *19*, 789–797. [CrossRef]
67. Wang, P.; Sun, X.; Xie, Y.; Li, M.; Chen, W.; Zhang, S.; Liang, D.; Ma, F. Melatonin regulates proteomic changes during leaf senescence in *Malus Hupehensis*. *J. Pineal Res.* **2014**, *57*, 291–307. [CrossRef]
68. Weeda, S.; Zhang, N.; Zhao, X.; Ndip, G.; Guo, Y.D.; Buck, G.; Fu, C.; Ren, S. *Arabidopsis* transcriptome analysis reveals key roles of melatonin in plant defense systems. *PLoS ONE* **2014**, *9*, e93462. [CrossRef]
69. Zhang, H.J.; Zhang, N.; Yang, R.C.; Wang, L.; Sun, Q.Q.; Li, D.B.; Cao, Y.Y.; Weeda, S.; Zhao, B.; Ren, S.; et al. Melatonin promotes seed germination under high salinity by regulating antioxidant systems, ABA and GA_4 interaction in cucumber (*Cucumis sativus* L.). *J. Pineal Res.* **2014**, *57*, 269–279. [CrossRef]
70. Zhang, N.; Zhang, H.J.; Zhao, B.; Sun, Q.Q.; Cao, Y.Y.; Li, R.; Wu, X.X.; Weeda, S.; Li, L.; Ren, S.; et al. The RNA-seq approach to discriminate gene expression profiles in response to melatonin on cucumber lateral root formation. *J. Pineal Res.* **2014**, *56*, 39–50. [CrossRef]
71. Li, C.; Liang, D.; Chang, C.; Jia, D.; Ma, F. Melatonin mediates the regulation of ABA metabolism, free-radical scavenging, and stomatal behavior in two *Malus* species under drought stress. *J. Exp. Bot.* **2015**, *66*, 669–680. [CrossRef]
72. Li, X.; Tan, D.X.; Jiang, D.; Liu, F. Melatonin enhances cold tolerance in drought-primed wild-type and abscisic acid-deficient mutant barley. *J. Pineal Res.* **2016**, *61*, 328–339. [CrossRef]
73. Liu, J.; Zhai, R.; Liu, F.; Zhao, Y.; Wang, H.; Liu, L.; Yang, C.; Wang, Z.; Ma, F.; Xu, L. Melatonin induces parthenocarpy by regulating genes in gibberellin pathways in 'Starkrimson' pear (*Pyrus communis* L.) *Front. Plant Sci.* **2018**, *9*, 946. [CrossRef]
74. Sun, Q.Q.; Zhang, N.; Wang, J.; Zhang, H.J.; Li, D.B.; Shi, J.; Li, R.; Weeda, S.; Zhao, B.; Ren, S.; et al. Melatonin promotes ripening and improves quality of tomato fruit during postharvest life. *J. Exp. Bot.* **2015**, *66*, 657–668. [CrossRef]
75. Hernández-Ruiz, J.; Arnao, B.M. Relationship of melatonin and salicylic acid in biotic/abiotic plant stress responses. *Agronomy* **2018**, *8*, 33. [CrossRef]
76. Erland, L.A.E.; Murch, S.J.; Reiter, R.J.; Saxena, P.K. A new balancing act: The many roles of melatonin and serotonin in plant growth and development. *Plant Sign. Behav.* **2015**, *10*, e1096469. [CrossRef]
77. Reiter, R.J.; Tan, D.X.; Zhou, Z.; Coelho-Cruz, M.H.; Fuentes-Broto, L.; Galano, A. Phytomelatonin: Assisting plants to survive and thrive. *Molecules* **2015**, *20*, 7396–7437. [CrossRef]
78. Tan, D.X. Melatonin and plants. *J. Exp. Bot.* **2015**, *66*, 1–2. [CrossRef]
79. Tan, D.X.; Manchester, C.L.; Esteban-Zubero, E.; Zhou, Z.; Reiter, J.R. Melatonin as a potent and inducible endogenous antioxidant: Synthesis and metabolism. *Molecules* **2015**, *20*, 18886–18906. [CrossRef]
80. Zhang, N.; Sun, Q.; Zhang, H.; Cao, Y.; Weeda, S.; Ren, S.; Guo, Y.D. Roles of melatonin in abiotic stress resistance in plants. *J. Exp. Bot.* **2015**, *66*, 647–656. [CrossRef]
81. Hardeland, R. Melatonin in plants. Diversity of levels and multiplicity of functions. *Front. Plant Sci.* **2016**, *7*, 198. [CrossRef]
82. Nawaz, M.A.; Huang, Y.; Bie, Z.; Ahmad, W.; Reiter, R.J.; Niu, M.; Hameed, S. Melatonin: Current status and future perspectives in plant science. *Front. Plant Sci.* **2016**, *6*, 1230. [CrossRef]
83. Erland, L.A.E.; Saxena, P.K. Melatonin in plant morphogenesis. *In Vitro Cell. Dev. Biol. Plant* **2018**, *54*, 3–24. [CrossRef]

84. Fazal, H.; Abbasi, B.H.; Ahmad, N.; Ali, M. Exogenous melatonin trigger biomass accumulation and production of stress enzymes during callogenesis in medicinally important *Prunella vulgaris* L. (Selfheal). *Physiol. Mol. Biol. Plants* **2018**, *24*, 1307–1315. [CrossRef]
85. Kanwar, M.K.; Yu, J.; Zhou, J. Phytomelatonin: Recent advances and future prospects. *J. Pineal Res.* **2018**, *65*, e12526. [CrossRef]
86. Wei, J.; Li, D.; Zhang, J.; Shan, C.; Rengel, Z.; Song, Z.; Chen, Q. Phytomelatonin receptor PMTR1-mediated signaling regulates stomatal closure in *Arabidopsis thaliana*. *J. Pineal Res.* **2018**, *65*, e12500. [CrossRef]
87. Yoon, H.Y.; Kim, M.; Park, J.W. Foliar accumulation of melatonin applied to the roots of maize (*Zea mays*) seedlings. *Biomolecules* **2019**, *9*, 26. [CrossRef]
88. Li, H.; Chang, J.; Zheng, J.; Dong, Y.; Liu, Q.; Yang, X.; Wei, C.; Zhang, Y.; Ma, J.; Zhang, X. Local melatonin application induces cold tolerance in distant organs of *Citrullus lanatus* L. via long distance transport. *Sci. Rep.* **2017**, *7*, 40858. [CrossRef]
89. Zhang, Q.; Liu, X.; Zhang, Z.; Liu, N.; Li, D.; Hu, L. Melatonin improved waterlogging tolerance in alfalfa (*Medicago sativa*) by reprogramming polyamine and ethylene metabolism. *Front. Plant Sci.* **2019**, *10*, 44. [CrossRef]
90. Gu, Q.; Chen, Z.; Yu, X.; Cui, W.; Pan, J.; Zhao, G.; Xu, S.; Wang, R.; Shen, W. Melatonin confers plant tolerance against cadmium stress via the decrease of cadmium accumulation and reestablishment of microRNA-mediated redox homeostasis. *Plant Sci.* **2017**, *261*, 28–37. [CrossRef]
91. Chen, Z.; Gu, Q.; Yu, X.; Huang, L.; Xu, S.; Wang, R.; Shen, W.; Shen, W. Hydrogen peroxide acts downstream of melatonin to induce lateral root formation. *Ann. Bot.* **2018**, *121*, 1127–1136. [CrossRef]
92. Li, C.; Wang, P.; Wei, Z.; Liang, D.; Jia, D.; Fu, M.; Ma, F. The mitigation effects of exogenous melatonin on salinity-induced stress in *Malus hupehensis*. *J. Pineal Res.* **2012**, *53*, 298–306. [CrossRef]
93. Li, C.; Liang, B.; Chang, C.; Wei, Z.; Zhou, S.; Ma, F. Exogenous melatonin improved potassium content in *Malus* under different stress conditions. *J. Pineal Res.* **2016**, *61*, 218–229. [CrossRef]
94. Wang, P.; Sun, X.; Li, C.; Wei, Z.; Liang, D.; Ma, F. Long-term exogenous application of melatonin delays drought-induced leaf senescence in apple. *J. Pineal Res.* **2013**, *54*, 292–302. [CrossRef]
95. Zuo, B.; Zheng, X.; He, P.; Wang, L.; Lei, Q.; Feng, C.; Zhou, J.; Li, Q.; Han, Z.; Kong, J. Overexpression of MzASMT improves melatonin production and enhances drought tolerance in transgenic *Arabidopsis thaliana* plants. *J. Pineal Res.* **2014**, *57*, 408–417. [CrossRef]
96. Liang, B.; Ma, C.; Zhang, Z.; Wei, Z.; Gao, T.; Zhao, Q.; Ma, F.; Li, C. Long-term exogenous application of melatonin improves nutrient uptake fluxes in apple plants under moderate drought stress. *Environ. Exp. Bot.* **2018**, *155*, 650–661. [CrossRef]
97. Zheng, X.; Zhou, J.; Tan, D.X.; Wang, N.; Wang, L.; Shan, D.; Kong, J. Melatonin improves waterlogging tolerance of *Malus baccata* (Linn.) Borkh. seedlings by maintaining aerobic respiration, photosynthesis and ROS migration. *Front. Plant Sci.* **2017**, *8*, 483. [CrossRef]
98. Gong, X.; Shi, S.; Dou, F.; Song, Y.; Ma, F. Exogenous melatonin alleviates alkaline stress in *Malus hupehensis* Rehd. by regulating the biosynthesis of polyamines. *Molecules* **2017**, *22*, 1542. [CrossRef]
99. Wang, P.; Yin, L.; Liang, D.; Li, C.; Ma, F.; Yue, Z. Delayed senescence of apple leaves by exogenous melatonin treatment: Toward regulating the ascorbate-glutathione cycle. *J. Pineal Res.* **2012**, *53*, 11–20. [CrossRef]
100. Wang, P.; Sun, X.; Chang, C.; Feng, F.; Liang, D.; Cheng, L.; Ma, F. Delay in leaf senescence of *Malus hupehensis* by long-term melatonin application is associated with its regulation of metabolic status and protein degradation. *J. Pineal Res.* **2013**, *55*, 424–434.
101. Yin, L.; Wang, P.; Li, M.; Ke, X.; Li, C.; Liang, D.; Wu, S.; Ma, X.; Li, C.; Zou, Y.; et al. Exogenous melatonin improves *Malus* resistance to *Marssonina* apple blotch. *J. Pineal Res.* **2013**, *54*, 426–434. [CrossRef]
102. Li, C.; Zhao, Q.; Gao, T.; Wang, H.; Zhang, Z.; Liang, B.; Wei, Z.; Liu, C.; Ma, F. The mitigation effects of exogenous melatonin on replant disease in apple. *J. Pineal Res.* **2018**, *65*, e12523. [CrossRef]
103. Chen, L.; Wang, M.R.; Li, J.W.; Feng, C.H.; Cui, Z.H.; Zhao, L.; Wang, Q.C. Exogenous application of melatonin improves eradication of apple stem grooving virus from the infected in-vitro shoots by shoot tip culture. *Plant Pathol.* **2019**, *68*, 997–1006. [CrossRef]
104. Mohamed Abd El-Naby, S.; Ahmed Mohamed, A.; Mohamed El-Naggar, Y. Effect of melatonin, GA_3 and NAA on vegetative growth, yield and quality of *Canino* apricot fruits. *Acta Sci. Pol. Hortorum Cultus* **2019**, *18*, 167–174.

105. Hu, W.; Yang, H.; Tie, W.; Yan, Y.; Ding, Z.; Liu, Y.; Wu, C.; Wang, J.; Reiter, R.J.; Tan, D.X.; et al. Natural Variation in Banana Varieties Highlights the Role of Melatonin in Postharvest Ripening and Quality. *J. Agric. Food Chem.* **2017**, *65*, 9987–9994. [CrossRef]
106. Li, T.; Wu, Q.; Zhu, H.; Zhou, Y.; Jiang, Y.; Gao, H.; Yun, Z. Comparative transcriptomic and metabolic analysis reveals the effect of melatonin on delaying anthracnose incidence upon postharvest banana fruit peel. *BMC Plant Biol.* **2019**, *19*, 289. [CrossRef]
107. Arnao, M.B.; Hernández-Ruiz, J. Protective effect of melatonin against chlorophyll degradation during the senescence of barley leaves. *J. Pineal Res.* **2009**, *46*, 58–63. [CrossRef]
108. Shi, H.; Jiang, C.; Ye, T.; Tan, D.; Reiter, R.J.; Zhang, H.; Liu, R.; Chan, Z. Comparative physiological, metabolomic, and transcriptomic analyses reveal mechanisms of improved abiotic stress resistance in bermudagrass [*Cynodon dactylon* (L). Pers.] by exogenous melatonin. *J. Exp. Bot.* **2015**, *66*, 681–694. [CrossRef]
109. Fan, J.; Hu, Z.; Xie, Y.; Chan, Z.; Chen, K.; Amombo, E.; Chen, L.; Fu, J. Alleviation of cold damage to photosystem II and metabolisms by melatonin in *Bermudagrass*. *Front. Plant Sci.* **2015**, *6*, 925. [CrossRef]
110. Gul, M.; Khan, D.; Wani, A.; Bhat, S.; Mir, S.; Malik, A.; Kumar, A.; Narayan, S.; Stephen, K.; Lone, A. Foliar application of melatonin modulates the growth and photosynthetic pigments in broccoli cv. *Palam Samridhi. SKUAST J. Res.* **2018**, *20*, 193–198.
111. Posmyk, M.M.; Kuran, H.; Marciniak, K.; Janas, K.M. Presowing seed treatment with melatonin protects red cabbage seedlings against toxic copper ion concentrations. *J. Pineal Res.* **2008**, *45*, 24–31. [CrossRef]
112. Zhang, N.; Sun, Q.; Li, H.; Cao, Y.; Zhang, H.; Li, S.; Zhang, L.; Qi, Y.; Ren, S.; Zhao, B.; et al. Melatonin improved anthocyanin accumulation by regulating gene expressions and resulted in high reactive oxygen species scavenging capacity in cabbage. *Front. Plant Sci.* **2016**, *7*, 197. [CrossRef]
113. Tan, X.L.; Fan, Z.; Kuang, J.; Lu, W.; Reiter, R.J.; Lakshmanan, P.; Su, X.; Zhou, J.; Chen, J.; Shan, W. Melatonin delays leaf senescence of Chinese flowering cabbage by suppressing ABFs-mediated abscisic acid biosynthesis and chlorophyll degradation. *J. Pineal Res.* **2019**, *67*, e12570. [CrossRef]
114. Hu, W.; Kong, H.; Guo, Y.; Zhang, Y.; Ding, Z.; Tie, W.; Yan, Y.; Huang, Q.; Peng, M.; Shi, H.; et al. Comparative physiological and transcriptomic analyses reveal the actions of melatonin in the delay of postharvest physiological deterioration of cassava. *Front. Plant Sci.* **2016**, *7*, 736. [CrossRef]
115. Wei, Y.; Chang, Y.; Zeng, H.; Liu, G.; He, C.; Shi, H. RAV transcription factors are essential for disease resistance against cassava bacterial blight via activation of melatonin biosynthesis genes. *J. Pineal Res.* **2018**, *64*, e12454. [CrossRef]
116. Sarropoulou, V.N.; Therios, I.N.; Dimassi-Theriou, K.N. Melatonin promotes adventitious root regeneration in in vitro shoot tip explants of the commercial sweet cherry rootstocks CAB-6P (*Prunus cerasus* L.), Gisela 6 (*P. cerasus x P. canescens*), and MxM 60 (*P. avium x P.mahaleb*). *J. Pineal Res.* **2012**, *52*, 38–46. [CrossRef]
117. Sarropoulou, V.N.; Dimassi-Theriou, K.N.; Therios, I.N.; Koukourikou-Petridou, M. Melatonin enhances root regeneration, photosynthetic pigments, biomass, total carbohydrates and proline content in the cherry rootstock PHL-C (*Prunus avium x Prunus cerasus*). *Plant Physiol. Biochem.* **2012**, *61*, 162–168. [CrossRef]
118. Tijero, V.; Muñoz, P.; Munné-Bosch, S. Melatonin as an inhibitor of sweet cherries ripening in orchard trees. *Plant Physiol. Biochem.* **2019**, *140*, 88–95. [CrossRef]
119. Kostopoulou, Z.; Therios, I.; Roumeliotis, E.; Kanellis, A.K.; Molassiotis, A. Melatonin combined with ascorbic acid provides salt adaptation in *Citrus aurantium* L. seedlings. *Plant Physiol. Biochem.* **2015**, *86*, 155–165. [CrossRef]
120. Campos, C.N.; Avila, R.G.; de Souza, K.R.D.; Azevedo, L.M.; Alves, J.D. Melatonin reduces oxidative stress and promotes drought tolerance in young *Coffea arabica* L. plants. *Agric. Water Manag.* **2019**, *211*, 37–47. [CrossRef]
121. Xiao, S.; Liu, L.; Wang, H.; Li, D.; Bai, Z.; Zhang, Y.; Sun, H.; Zhang, K.; Li, C. Exogenous melatonin accelerates seed germination in cotton (*Gossypium hirsutum* L.). *PLoS ONE* **2019**, *14*, e0216575. [CrossRef]
122. Posmyk, M.M.; Balabusta, M.; Wieczorek, M.; Sliwinska, E.; Janas, K.M. Melatonin applied to cucumber (*Cucumis sativus* L.) seeds improves germination during chilling stress. *J. Pineal Res.* **2009**, *46*, 214–223. [CrossRef]
123. Balabusta, M.; Szafranska, K.; Posmyk, M.M. Exogenous melatonin improves antioxidant defense in cucumber seeds germinated under chilling stress. *Front. Plant Sci.* **2016**, *7*, 575.

124. Zhao, H.; Zhang, K.; Zhou, X.; Xi, L.; Wang, Y.; Xu, H.; Pan, T.; Zou, Z. Melatonin alleviates chilling stress in cucumber seedlings by up-regulation of CsZat12 and modulation of polyamine and abscisic acid metabolism. *Sci. Rep.* **2017**, *7*, 4998. [CrossRef]
125. Zhao, H.; Ye, L.; Wang, Y.; Zhou, X.; Yang, J.; Wang, J.; Cao, K.; Zou, Z. Melatonin increases the chilling tolerance of chloroplast in cucumber seedlings by regulating photosynthetic electron flux and the ascorbate-glutathione cycle. *Front. Plant Sci.* **2016**, *7*, 1814. [CrossRef]
126. Zhao, N.; Sun, Y.; Wang, D.Y.; Zheng, J.X. Effects of exogenous melatonin on nitrogen metabolism in cucumber seedlings under high temperature stress. *Zhiwu Shengli Xuebao/Plant Physiol. J.* **2012**, *48*, 557–564.
127. Zhang, R.; Sun, Y.; Liu, Z.; Jin, W.; Sun, Y. Effects of melatonin on seedling growth, mineral nutrition, and nitrogen metabolism in cucumber under nitrate stress. *J. Pineal Res.* **2017**, *62*, e12403. [CrossRef]
128. Wang, L.Y.; Liu, J.L.; Wang, W.X.; Sun, Y. Exogenous melatonin improves growth and photosynthetic capacity of cucumbers under salinity-induced stress. *Photosynthetica* **2015**, *54*, 1–10. [CrossRef]
129. Zhang, N.; Zhang, H.J.; Sun, Q.Q.; Cao, Y.Y.; Li, X.; Zhao, B.; Wu, P.; Guo, Y.D. Proteomic analysis reveals a role of melatonin in promoting cucumber seed germination under high salinity by regulating energy production. *Sci. Rep.* **2017**, *7*, 503. [CrossRef]
130. Zhang, N.; Zhao, B.; Zhang, H.J.; Weeda, S.; Yang, C.; Yang, Z.C.; Ren, S.; Guo, Y.D. Melatonin promotes water-stress tolerance, lateral root formation, and seed germination in cucumber (*Cucumis sativus* L.). *J. Pineal Res.* **2013**, *54*, 15–23. [CrossRef]
131. Li, H.; He, J.; Yang, X.; Li, X.; Luo, D.; Wei, C.; Ma, J.; Zhang, Y.; Yang, J.; Zhang, X. Glutathione-dependent induction of local and systemic defense against oxidative stress by exogenous melatonin in cucumber (*Cucumis sativus* L.). *J. Pineal Res.* **2016**, *60*, 206–216. [CrossRef]
132. Cao, Y.Y.; Qi, C.D.; Li, S.; Wang, Z.; Wang, X.; Wang, J.; Ren, S.; Li, X.; Zhang, N.; Guo, Y.D. Melatonin alleviates copper toxicity via improving copper sequestration and ROS scavenging in *Cucumber*. *Plant Cell Physiol.* **2018**, *60*, 562–574. [CrossRef]
133. Li, J.; Li, Y.; Tian, Y.; Qu, M.; Zhang, W.; Gao, L. Melatonin has the potential to alleviate cinnamic acid stress in cucumber seedlings. *Front. Plant Sci.* **2017**, *8*, 1193. [CrossRef]
134. Dawood, M.G.; El-Awadi, M.E. Alleviation of salinity stress on *Vicia faba* L. plants via seed priming with melatonin. *Acta Biol. Colomb.* **2015**, *20*, 223–235. [CrossRef]
135. Meng, J.F.; Xu, T.F.; Wang, Z.Z.; Fang, Y.L.; Xi, Z.M.; Zhang, Z.W. The ameliorative effects of exogenous melatonin on grape cuttings under water-deficient stress: Antioxidant metabolites, leaf anatomy, and chloroplast morphology. *J. Pineal Res.* **2014**, *57*, 200–212. [CrossRef]
136. Ma, Y.; Jiao, J.; Fan, X.; Sun, H.; Zhang, Y.; Jiang, J.; Liu, C. Endophytic bacterium *Pseudomonas fluorescens RG11* may transform tryptophan to melatonin and promote endogenous melatonin levels in the roots of four grape cultivars. *Front. Plant Sci.* **2017**, *7*, 2068. [CrossRef]
137. Xu, L.; Yue, Q.; Bian, F.E.; Sun, H.; Zhai, H.; Yao, Y. Melatonin enhances phenolics accumulation partially via ethylene signaling and resulted in high antioxidant capacity in grape berries. *Front. Plant Sci.* **2017**, *8*, 1426. [CrossRef]
138. Xu, L.; Yue, Q.; BIAN, F.; Zhai, H.; Yao, Y. Melatonin Treatment Enhances the Polyphenol Content and Antioxidant Capacity of Red Wine. *Hortic. Plant J.* **2018**, *4*, 144–150. [CrossRef]
139. Meng, J.F.; Xu, T.F.; Song, C.Z.; Yu, Y.; Hu, F.; Zhang, L.; Zhang, Z.W.; Xi, Z.M. Melatonin treatment of pre-veraison grape berries to increase size and synchronicity of berries and modify wine aroma components. *Food Chem.* **2015**, *185*, 127–134. [CrossRef]
140. Liang, D.; Ni, Z.; Xia, H.; Xie, Y.; Lv, X.; Wang, J.; Lin, L.; Deng, Q.; Luo, X. Exogenous melatonin promotes biomass accumulation and photosynthesis of kiwifruit seedlings under drought stress. *Sci. Hort.* **2019**, *246*, 34–43. [CrossRef]
141. Liang, D.; Shen, Y.; Ni, Z.; Wang, Q.; Lei, Z.; Xu, N.; Deng, Q.; Lin, L.; Wang, J.; Lv, X.; et al. Exogenous melatonin application delays senescence of kiwifruit leaves by regulating the antioxidant capacity and biosynthesis of flavonoids. *Front. Plant Sci.* **2018**, *9*, 426. [CrossRef]
142. Liang, D.; Gao, F.; Ni, Z.; Lin, L.; Deng, Q.; Tang, Y.; Wang, X.; Luo, X.; Xia, H. Melatonin improves heat tolerance in kiwifruit seedlings through promoting antioxidant enzymatic activity and glutathione S-transferase transcription. *Molecules* **2018**, *23*, 584. [CrossRef]
143. Hanci, F.; Çingi, M.; AKINCI, H. Influence of L-tryptophan and melatonin on germination of onion and leek seeds at different temperatures. *Turk. J. Agric. Res.* **2019**, *6*, 214–221. [CrossRef]

144. Zhang, Y.; Huber, D.J.; Hu, M.; Jiang, G.; Gao, Z.; Xu, X.; Jiang, Y.; Zhang, Z. Delay of postharvest browning in litchi fruit by melatonin via the enhancing of antioxidative processes and oxidation repair. *J. Agric. Food Chem.* **2018**, *66*, 7475–7484. [CrossRef] [PubMed]
145. Arnao, M.B.; Hernández-Ruiz, J. Growth conditions determine different melatonin levels in *Lupinus albus* L. *J. Pineal Res.* **2013**, *55*, 149–155. [CrossRef] [PubMed]
146. Chen, Y.E.; Mao, J.-J.; Sun, L.-Q.; Huang, B.; Ding, C.-B.; Gu, Y.; Liao, J.-Q.; Hu, C.; Zhang, Z.W.; Yuan, S.; et al. Exogenous melatonin enhances salt stress tolerance in maize seedlings by improving antioxidant and photosynthetic capacity. *Physiol. Plant.* **2018**, *164*, 349–363. [CrossRef] [PubMed]
147. Pehlivan, N.; Guler, N.S. Protective effect of a natural ally on simultaneous mild heat and salt episodes in maize seedlings. *Acta Physiol. Plant* **2018**, *40*, 203. [CrossRef]
148. Erdal, S. Melatonin promotes plant growth by maintaining integration and coordination between carbon and nitrogen metabolisms. *Plant Cell Rep.* **2019**, *38*, 1001–1012. [CrossRef]
149. Ye, J.; Wang, S.; Deng, X.; Yin, L.; Xiong, B.; Wang, X. Melatonin increased maize (*Zea mays* L.) seedling drought tolerance by alleviating drought-induced photosynthetic inhibition and oxidative damage. *Acta Physiol. Plant* **2016**, *38*, 48. [CrossRef]
150. Fleta-Soriano, E.; Díaz, L.; Bonet, E.; Munné-Bosch, S. Melatonin may exert a protective role against drought stress in maize. *J. Agron. Crop Sci.* **2017**, *203*, 286–294. [CrossRef]
151. Huang, B.; Chen, Y.E.; Zhao, Y.Q.; Ding, C.B.; Liao, J.Q.; Hu, C.; Zhou, L.J.; Zhang, Z.W.; Yuan, S.; Yuan, M. Exogenous melatonin alleviates oxidative damages and protects Photosystem II in maize seedlings under drought stress. *Front. Plant Sci.* **2019**, *10*, 677. [CrossRef]
152. Li, Z.G.; Xu, Y.; Bai, L.K.; Zhang, S.Y.; Wang, Y. Melatonin enhances thermotolerance of maize seedlings (*Zea mays* L.) by modulating antioxidant defense, methylglyoxal detoxification, and osmoregulation systems. *Protoplasma* **2018**, *256*, 471–490. [CrossRef]
153. Okant, M.; Kaya, C. The role of endogenous nitric oxide in melatonin-improved tolerance to lead toxicity in maize plants. *Environ. Sci. Pollut. Res.* **2019**, *26*, 11864–11874. [CrossRef]
154. Zhao, H.; Su, T.; Huo, L.; Wei, H.; Jiang, Y.; Xu, L.; Ma, F. Unveiling the mechanism of melatonin impacts on maize seedling growth: Sugar metabolism as a case. *J. Pineal Res.* **2015**, *59*, 255–266. [CrossRef] [PubMed]
155. Kolodziejczyk, I.; Dzitko, K.; Szewczyk, R.; Posmyk, M.M. Exogenous melatonin expediently modifies proteome of maize (*Zea mays* L.) embryo during seed germination. *Acta Physiol. Plant* **2016**, *38*, 146. [CrossRef]
156. Su, X.; Xin, L.; Li, Z.; Zheng, H.; Mao, J.; Yang, Q. Physiology and transcriptome analyses reveal a protective effect of the radical scavenger melatonin in aging maize seeds. *Free Radic. Res.* **2018**, *52*, 1094–1109. [CrossRef] [PubMed]
157. Hodzic, E.; Balaban, M.; Suskalo, N.; Galijasevic, S.; Hasanagic, D.; Kukavica, B. Antioxidative response of Melissa officinalis L. and Valeriana officinalis L. leaves exposed to exogenous melatonin and excessive zinc and cadmium level. *J. Serb. Chem. Soc.* **2018**, *84*, 70. [CrossRef]
158. Szafranska, K.; Glinska, S.; Janas, K.M. Ameliorative effect of melatonin on meristematic cells of chilled and re-warmed *Vigna radiata* roots. *Biol. Plant* **2013**, *57*, 91–96. [CrossRef]
159. Gao, W.; Feng, Z.; Bai, Q.; He, J.; Wang, Y. Melatonin-mediated regulation of growth and antioxidant capacity in salt-tolerant naked oat under salt stress. *Molecules* **2019**, *20*, 1176. [CrossRef] [PubMed]
160. Gao, W.; Zhang, Y.; Feng, Z.; Bai, Q.; He, J.; Wang, Y. Effects of melatonin on antioxidant capacity in naked oat seedlings under drought stress. *Molecules* **2018**, *23*, 1580. [CrossRef]
161. Cao, S.; Song, C.; Shao, J.; Bian, K.; Chen, W.; Yang, Z. Exogenous melatonin treatment increases chilling tolerance and induces defense response in harvested peach fruit during cold storage. *J. Agric. Food Chem.* **2016**, *64*, 5215–5222. [CrossRef]
162. Gao, H.; Zhang, Z.K.; Chai, H.K.; Cheng, N.; Yang, Y.; Wang, D.N.; Yang, T.; Cao, W. Melatonin treatment delays postharvest senescence and regulates reactive oxygen species metabolism in peach fruit. *Postharvest Biol. Technol.* **2016**, *118*, 103–110. [CrossRef]
163. Liu, J.; Yue, R.; Si, M.; Wu, M.; Cong, L.; Zhai, R.; Yang, C.; Wang, Z.; Ma, F.; Xu, L. Effects of exogenous application of melatonin on quality and sugar metabolism in *Zaosu* pear fruit. *J. Plant Growth Regul.* **2019**. [CrossRef]
164. Zhai, R.; Liu, J.; Liu, F.; Zhao, Y.; Liu, L.; Fang, C.; Wang, H.; Li, X.; Wang, Z.; Ma, F.; et al. Melatonin limited ethylene production, softening and reduced physiology disorder in pear (*Pyrus communis* L.) fruit during senescence. *Postharvest Biol. Technol.* **2018**, *139*, 38–46. [CrossRef]

165. Liu, J.; Yang, J.; Zhang, H.; Cong, L.; Zhai, R.; Yang, C.; Wang, Z.; Ma, F.; Xu, L. Melatonin inhibits ethylene synthesis via nitric oxide regulation to delay postharvest senescence in pears. *J. Agric. Food Chem.* **2019**, *67*, 2279–2288. [CrossRef] [PubMed]
166. Kaya, C.; Higgs, D.; Ashraf, M.; Alyemeni, M.; Ahmad, P. Integrative roles of nitric oxide and hydrogen sulfide in melatonin-induced tolerance of pepper (*Capsicum annuum* L.) plants to iron deficiency and salt stress alone or in combination. *Physiol. Plant* **2019**. [CrossRef] [PubMed]
167. Korkmaz, A.; Karakas, A.; Kocacinar, F.; Cuci, Y. The effects of seed treatment with melatonin on germination and emergence performance of pepper seeds under chilling stress. *J. Agric. Sci.* **2017**, *23*, 167–176.
168. Sarafi, E.; Tsouvaltzis, P.; Chatzissavvidis, C.; Siomos, A.; Therios, I. Melatonin and resveratrol reverse the toxic effect of high boron (B) and modulate biochemical parameters in pepper plants (*Capsicum annuum* L.). *Plant Physiol. Biochem.* **2017**, *112*, 173–182. [CrossRef] [PubMed]
169. Szafranska, K.; Reiter, R.J.; Posmyk, M.M. Melatonin application to *Pisum sativum* L. seeds positively influences the function of the photosynthetic apparatus in growing seedlings during paraquat-induced oxidative stress. *Front. Plant Sci.* **2016**, *7*, 1663. [CrossRef]
170. Szafranska, K.; Reiter, R.J.; Posmyk, M.M. Melatonin improves the photosynthetic apparatus in pea leaves stressed by paraquat via chlorophyll breakdown regulation and its accelerated *de novo* synthesis. *Front. Plant Sci.* **2017**, *8*, 878. [CrossRef] [PubMed]
171. Tan, D.X.; Manchester, L.C.; Helton, P.; Reiter, R.J. Phytoremediative capacity of plants enriched with melatonin. *Plant Sign. Behav.* **2007**, *2*, 514–516. [CrossRef]
172. Bal, E. Physicochemical changes in *Santa Rosa* plum fruit treated with melatonin during cold storage. *J. Food Meas. Charact.* **2019**, *13*, 1713–1720. [CrossRef]
173. Jannatizadeh, A. Exogenous melatonin applying confers chilling tolerance in pomegranate fruit during cold storage. *Sci. Hort.* **2019**, *246*, 544–549. [CrossRef]
174. Ding, F.; Wang, G.; Zhang, S. Exogenous melatonin mitigates methyl viologen-triggered oxidative stress in poplar leaf. *Molecules* **2018**, *23*, 2852. [CrossRef] [PubMed]
175. Yu, Y.; Wang, A.; Li, X.; Kou, M.; Wang, W.; Chen, X.; Xu, T.; Zhu, M.; Ma, D.; Li, Z.; et al. Melatonin-stimulated triacylglycerol breakdown and energy turnover under salinity stress contributes to the maintenance of plasma membrane H^+-ATPase activity and K^+/Na^+ homeostasis in sweet potato. *Front. Plant Sci.* **2018**, *9*, 256. [CrossRef] [PubMed]
176. Zhang, S.; Zheng, X.; Reiter, R.J.; Feng, S.; Wang, Y.; Liu, S.; Jin, L.; Li, Z.; Datla, R.; Ren, M. Melatonin attenuates potato late blight by disrupting cell growth, stress tolerance, fungicide susceptibility and homeostasis of gene expression in *Phytophthora infestans*. *Front. Plant Sci.* **2017**, *8*, 1993. [CrossRef] [PubMed]
177. Jia, C.; Yu, X.; Zhang, M.; Liu, Z.; Zou, P.; Ma, J.; Xu, Y. Application of melatonin-enhanced tolerance to high-temperature stress in cherry radish (*Raphanus sativus* L. var. radculus pers). *J. Plant Growth Regul.* **2019**. [CrossRef]
178. Zhao, G.; Zhao, Y.; Yu, X.; Felix, K.; Han, H.; Guan, R.; Wang, R.; Shen, W. Nitric oxide is required for melatonin-enhanced tolerance against salinity stress in rapeseed (*Brassica napus* L.) seedlings. *Int. J. Mol. Sci.* **2018**, *19*, 1912. [CrossRef] [PubMed]
179. Khan, M.N.; Zhang, J.; Luo, T.; Liu, J.; Rizwan, M.; Fahad, S.; Xu, Z.; Hu, L. Seed priming with melatonin coping drought stress in rapeseed by regulating reactive oxygen species detoxification: Antioxidant defense system, osmotic adjustment, stomatal traits and chloroplast ultrastructure perseveration. *Ind. Crop Prod.* **2019**, *140*, 111597. [CrossRef]
180. Han, Q.H.; Huang, B.; Ding, C.B.; Zhang, Z.W.; Chen, Y.E.; Hu, C.; Zhou, L.J.; Huang, Y.; Liao, J.Q.; Yuan, S.; et al. Effects of melatonin on anti-oxidative systems and Photosystem II in cold-stressed rice seedlings. *Front. Plant Sci.* **2017**, *8*, 785. [CrossRef] [PubMed]
181. Liang, C.; Zheng, G.; Li, W.; Wang, Y.; Hu, B.; Wang, H.; Wu, H.; Qian, Y.; Zhu, X.G.; Tan, D.X.; et al. Melatonin delays leaf senescence and enhances salt stress tolerance in rice. *J. Pineal Res.* **2015**, *59*, 91–101. [CrossRef]
182. Lee, K.; Back, K. Overexpression of rice serotonin N-acetyltransferase 1 in transgenic rice plants confers resistance to cadmium and senescence and increases grain yield. *J. Pineal Res.* **2017**, *62*, e12392. [CrossRef]
183. Lee, H.Y.; Back, K. Cadmium disrupts subcellular organelles, including chloroplasts, resulting in melatonin induction in plants. *Molecules* **2017**, *22*, 1791. [CrossRef]
184. Chen, X.; Sun, C.; Laborda, P.; Zhao, Y.; Palmer, I.; Fu, Z.Q.; Qiu, J.; Liu, F. Melatonin treatment inhibits the growth of *Xanthomonas oryzae* pv. *oryzae*. *Front. Microbiol.* **2018**, *9*, 2280. [CrossRef] [PubMed]

185. Wei, Y.; Zeng, H.; Hu, W.; Chen, L.; He, C.; Shi, H. Comparative transcriptional profiling of melatonin synthesis and catabolic genes indicates the possible role of melatonin in developmental and stress responses in rice. *Front. Plant Sci.* **2016**, *7*, 676. [CrossRef] [PubMed]
186. Park, S.; Back, K. Melatonin promotes seminal root elongation and root growth in transgenic rice after germination. *J. Pineal Res.* **2012**, *53*, 385–389. [CrossRef] [PubMed]
187. Liang, C.; Li, A.; Yu, H.; Li, W.; Liang, C.; Guo, S.; Zhang, R.; Chu, C. Melatonin regulates root architecture by modulating auxin response in rice. *Front. Plant Sci.* **2017**, *8*, 134. [CrossRef] [PubMed]
188. Wei, W.; Li, Q.; Chu, Y.-N.; Reiter, R.J.; Yu, X.M.; Zhu, D.H.; Zhang, W.K.; Ma, B.; Lin, Q.; Zhang, J.S.; et al. Melatonin enhances plant growth and abiotic stress tolerance in soybean plants. *J. Exp. Bot.* **2015**, *66*, 695–707. [CrossRef] [PubMed]
189. ZOU, J.N.; JIN, X.J.; WANG, M.X.Y. Effects of melatonin on photosynthesis and soybean seed growth during grain filling under drought stress. *Photosynthetica* **2019**, *57*, 512–520. [CrossRef]
190. Zhang, J.; Zeng, B.; Mao, Y.; Kong, X.; Wang, X.; Yang, Y.; Zhang, J.; Xu, J.; Rengel, Z.; Chen, Q. Melatonin alleviates aluminium toxicity through modulating antioxidative enzymes and enhancing organic acid anion exudation in soybean. *Funct. Plant Biol.* **2017**, *44*, 961–968. [CrossRef]
191. Moussa, H.R.; Algamal, S.M.A. Does Exogenous Application of Melatonin Ameliorate Boron Toxicity in Spinach Plants? *Int. J. Veg. Sci.* **2017**, *23*, 233–245. [CrossRef]
192. Aghdam, M.S.; Fard, J.R. Melatonin treatment attenuates postharvest decay and maintains nutritional quality of strawberry fruits (*Fragaria x anannasa* cv. Selva) by enhancing GABA shunt activity. *Food Chem.* **2017**, *221*, 1650–1657. [CrossRef]
193. Liu, C.; Zheng, H.; Sheng, K.; Liu, W.; Zheng, L. Effects of melatonin treatment on the postharvest quality of strawberry fruit. *Postharvest Biol. Technol.* **2018**, *139*, 47–55. [CrossRef]
194. Mukherjee, S.; David, A.; Yadav, S.; Baluska, F.; Bhatla, S.C. Salt stress-induced seedling growth inhibition coincides with differential distribution of serotonin and melatonin in sunflower seedling roots and cotyledons. *Physiol. Plant* **2014**, *152*, 714–728. [CrossRef] [PubMed]
195. Arora, D.; Bhatla, S.C. Melatonin and nitric oxide regulate sunflower seedling growth under salt stress accompanying differential expression of Cu/Zn SOD and Mn SOD. *Free Radic. Biol. Med.* **2017**, *106*, 315–328. [CrossRef] [PubMed]
196. Kaur, H.; Bhatla, S.C. Melatonin and nitric oxide modulate glutathione content and glutathione reductase activity in sunflower seedling cotyledons accompanying salt stress. *Nitric Oxide* **2016**, *59*, 42–53. [CrossRef] [PubMed]
197. Li, J.; Yang, Y.; Sun, K.; Chen, Y.; Chen, X.; Li, X. Exogenous Melatonin Enhances Cold, Salt and Drought Stress Tolerance by Improving Antioxidant Defense in Tea Plant (*Camellia sinensis* (L.) O. Kuntze). *Molecules* **2019**, *24*, 1826. [CrossRef] [PubMed]
198. Li, J.; Arkorful, E.; Cheng, S.; Zhou, Q.; Li, H.; Chen, X.; Sun, K.; Li, X. Alleviation of cold damage by exogenous application of melatonin in vegetatively propagated tea plant (*Camellia sinensis* (L.) O. Kuntze). *Sci. Hort.* **2018**, *238*, 356–362. [CrossRef]
199. Zhao, L.; Chen, L.; Gu, P.; Zhan, X.; Zhang, Y.; Hou, C.; Wu, Z.; Wu, Y.F.; Wang, Q.C. Exogenous application of melatonin improves plant resistance to virus infection. *Plant Pathol.* **2019**. [CrossRef]
200. Zhou, X.; Zhao, H.; Cao, K.; Hu, L.; Du, T.; Baluska, F.; Zou, Z. Beneficial roles of melatonin on redox regulation of photosynthetic electron transport and synthesis of D1 protein in tomato seedlings under salt stress. *Front. Plant Sci.* **2016**, *7*, 1823. [CrossRef]
201. Siddiqui, H.M.; Alamri, S.; Al-Khaishany, Y.M.; Khan, N.M.; Al-Amri, A.; Ali, M.H.; Alaraidh, A.I.; Alsahli, A.A. Exogenous melatonin counteracts NaCl-induced damage by regulating the antioxidant system, proline and carbohydrates metabolism in tomato seedlings. *Int. J. Mol. Sci.* **2019**, *20*, 353. [CrossRef]
202. Yin, Z.; Lu, J.; Meng, S.; Liu, Y.; Mostafa, I.; Qi, M.; Li, T. Exogenous melatonin improves salt tolerance in tomato by regulating photosynthetic electron flux and the ascorbate-glutathione cycle. *J. Plant Interact.* **2019**, *14*, 453–463. [CrossRef]
203. Ding, F.; Liu, B.; Zhang, S. Exogenous melatonin ameliorates cold-induced damage in tomato plants. *Sci. Hort.* **2017**, *219*, 264–271. [CrossRef]
204. Ding, F.; Wang, M.; Liu, B.; Zhang, S. Exogenous melatonin mitigates photoinhibition by accelerating non-photochemical quenching in tomato seedlings exposed to moderate light during chilling. *Front. Plant Sci.* **2017**, *8*, 244. [CrossRef] [PubMed]

205. Aghdam, M.S.; Luo, Z.; Jannatizadeh, A.; Sheikh-Assadi, M.; Sharafi, Y.; Farmani, B.; Fard, J.R.; Razavi, F. Employing exogenous melatonin applying confers chilling tolerance in tomato fruits by upregulating ZAT2/6/12 giving rise to promoting endogenous polyamines, proline, and nitric oxide accumulation by triggering arginine pathway activity. *Food Chem.* **2019**, *275*, 549–556. [CrossRef] [PubMed]
206. Xu, W.; Cai, S.Y.; Zhang, Y.; Wang, Y.; Ahammed, G.J.; Xia, X.J.; Shi, K.; Zhou, Y.H.; Yu, J.Q.; Reiter, R.J.; et al. Melatonin enhances thermotolerance by promoting cellular protein protection in tomato plants. *J. Pineal Res.* **2016**, *61*, 457–469. [CrossRef] [PubMed]
207. Ahammed, G.J.; Xu, W.; Liu, A.; Chen, S. Endogenous melatonin deficiency aggravates high temperature-induced oxidative stress in *Solanum lycopersicum* L. *Environ. Exp. Bot.* **2019**, *161*, 303–311. [CrossRef]
208. Qi, Z.Y.; Wang, K.X.; Yan, M.Y.; Kanwar, K.M.; Li, D.Y.; Wijaya, L.; Alyemeni, N.M.; Ahmad, P.; Zhou, J. Melatonin alleviates high temperature-induced pollen abortion in *Solanum lycopersicum*. *Molecules* **2018**, *23*, 386. [CrossRef] [PubMed]
209. Hasan, M.; Ahammed, G.J.; Yin, L.; Shi, K.; Xia, X.; Zhou, Y.; Yu, J.; Zhou, J. Melatonin mitigates cadmium phytotoxicity through modulation of phytochelatins biosynthesis, vacuolar sequestration and antioxidant potential in *Solanum lycopersicum* L. *Front. Plant Sci.* **2015**, *6*, 601. [CrossRef] [PubMed]
210. Cai, S.Y.; Zhang, Y.; Xu, Y.P.; Qi, Z.Y.; Li, M.Q.; Ahammed, G.J.; Xia, X.J.; Shi, K.; Zhou, Y.H.; Reiter, R.J.; et al. HsfA1a upregulates melatonin biosynthesis to confer cadmium tolerance in tomato plants. *J. Pineal Res.* **2017**, *62*, e12387. [CrossRef]
211. Li, M.Q.; Hasan, M.; Li, C.X.; Ahammed, G.J.; Xia, X.J.; Shi, K.; Zhou, Y.H.; Reiter, R.J.; Yu, J.Q.; Xu, M.X.; et al. Melatonin mediates selenium-induced tolerance to cadmium stress in tomato plants. *J. Pineal Res.* **2016**, *61*, 291–302. [CrossRef]
212. Liu, N.; Jin, Z.; Wang, S.; Gong, B.; Wen, D.; Wang, X.; Wei, M.; Shi, Q. Sodic alkaline stress mitigation with exogenous melatonin involves reactive oxygen metabolism and ion homeostasis in tomato. *Sci. Hort.* **2015**, *181*, 18–25. [CrossRef]
213. Debnath, B.; Hussain, M.; Irshad, M.; Mitra, S.; Li, M.; Liu, S.; Qiu, D. Exogenous melatonin mitigates acid rain stress to tomato plants through modulation of leaf ultrastructure, photosynthesis and antioxidant potential. *Molecules* **2018**, *23*, 388. [CrossRef]
214. Ding, F.; Wang, G.; Wang, M.; Zhang, S. Exogenous melatonin improves tolerance to water deficit by promoting cuticle formation in tomato plants. *Molecules* **2018**, *23*, 1605. [CrossRef] [PubMed]
215. Karaca, P.; Cekic, F.O. Exogenous melatonin-stimulated defense responses in tomato plants treated with polyethylene glycol. *Int. J. Veg. Sci.* **2019**. [CrossRef]
216. Hasn, M.; Liu, C.X.; Pan, Y.T.; Ahammed, G.J.; Qi, Z.Y.; Zhou, J. Melatonin alleviates low-sulfur stress by promoting sulfur homeostasis in tomato plants. *Sci. Rep.* **2018**, *8*, 10182. [CrossRef] [PubMed]
217. Wen, D.; Gong, B.; Sun, S.; Liu, S.; Wang, X.; Wei, M.; Yang, F.; Li, Y.; Shi, Q. Promoting roles of melatonin in adventitious root development of *Solanum lycopersicum* L. by regulating auxin and nitric oxide signaling. *Front. Plant Sci.* **2016**, *7*, 718. [CrossRef] [PubMed]
218. Liu, J.; Zhang, R.; Sun, Y.; Liu, Z.; Jin, W.; Sun, Y. The beneficial effects of exogenous melatonin on tomato fruit properties. *Sci. Hort.* **2016**, *207*, 14–20. [CrossRef]
219. Sun, Q.; Zhang, N.; Wang, J.; Cao, Y.; Li, X.; Zhang, H.; Zhang, L.; Tan, D.X.; Guo, Y.D. A label-free differential proteomics analysis reveals the effect of melatonin on promoting fruit ripening and anthocyanin accumulation upon postharvest in tomato. *J. Pineal Res.* **2016**, *61*, 138–153. [CrossRef] [PubMed]
220. Li, H.; Dong, Y.; Chang, J.; He, J.; Chen, H.; Liu, Q.; Wei, C.; Ma, J.; Zhang, Y.; Yang, J.; et al. High-throughput microRNA and mRNA sequencing reveals that microRNAs may be involved in melatonin-mediated cold tolerance in *Citrullus lanatus* L. *Front. Plant Sci.* **2016**, *7*, 1231. [CrossRef] [PubMed]
221. Li, H.; Chang, J.; Chen, H.; Wang, Z.; Gu, X.; Wei, C.; Zhang, Y.; Ma, J.; Yang, J.; Zhang, X. Exogenous melatonin confers salt stress tolerance to watermelon by improving photosynthesis and redox homeostasis. *Front. Plant Sci.* **2017**, *8*, 295. [CrossRef]
222. Nawaz, M.A.; Jiao, Y.; Chen, C.; Shireen, F.; Zheng, Z.; Imtiaz, M.; Bie, Z.; Huang, Y. Melatonin pretreatment improves vanadium stress tolerance of watermelon seedlings by reducing vanadium concentration in the leaves and regulating melatonin biosynthesis and antioxidant-related gene expression. *J. Plant Physiol.* **2018**, *220*, 115–127. [CrossRef]

223. Turk, H.; Erdal, S.; Genisel, M.; Atici, O.; Demir, Y.; Yanmis, D. The regulatory effect of melatonin on physiological, biochemical and molecular parameters in cold-stressed wheat seedlings. *Plant Growth Regul.* **2014**, *74*, 139–152. [CrossRef]
224. Li, X.; Brestic, M.; Tan, D.X.; Zivcak, M.; Zhu, X.; Liu, S.; Song, F.; Reiter, R.J.; Liu, F. Melatonin alleviates low PS I-limited carbon assimilation under elevated CO_2 and enhances the cold tolerance of offspring in chlorophyll b-deficient mutant wheat. *J. Pineal Res.* **2018**, *64*, e12453. [CrossRef] [PubMed]
225. Sun, L.; Li, X.; Wang, Z.; Sun, Z.; Zhu, X.; Liu, S.; Song, F.; Liu, F.; Wang, Y. Cold priming induced tolerance to subsequent low temperature stress is enhanced by melatonin application during recovery in wheat. *Molecules* **2018**, *23*, 1091. [CrossRef] [PubMed]
226. Ke, Q.; Ye, J.; Wang, B.; Ren, J.; Yin, L.; Deng, X.P.; Shiwen, W. Melatonin mitigates salt stress in wheat seedlings by modulating polyamine metabolism. *Front. Plant Sci.* **2018**, *9*, 914. [CrossRef] [PubMed]
227. Zafar, S.; Hasnain, Z.; Anwar, S.; Perveen, S.; Iqbal, N.; Noman, A.; Ali, M. Influence of melatonin on antioxidant defense system and yield of wheat (*Triticum aestivum* L.) genotypes under saline condition. *Pak. J. Bot.* **2019**, *51*, 1987–1994. [CrossRef]
228. Cui, G.; Zhao, X.; Liu, S.; Sun, F.; Zhang, C.; Xi, Y. Beneficial effects of melatonin in overcoming drought stress in wheat seedlings. *Plant Physiol. Biochem.* **2017**, *118*, 138–149. [CrossRef] [PubMed]
229. Ni, J.; Wang, Q.; Shah, A.F.; Liu, W.; Wang, D.; Huang, S.; Fu, S.; Wu, L. Exogenous melatonin confers cadmium tolerance by counterbalancing the hydrogen peroxide homeostasis in wheat seedlings. *Molecules* **2018**, *23*, 799. [CrossRef] [PubMed]
230. Kaya, C.; Okant, M.; Ugurlar, F.; Alyemeni, M.N.; Ashraf, M.; Ahmad, P. Melatonin-mediated nitric oxide improves tolerance to cadmium toxicity by reducing oxidative stress in wheat plants. *Chemosphere* **2019**, *225*, 627–638. [CrossRef]
231. Zuo, Z.; Sun, L.; Wang, T.; Miao, P.; Zhu, X.; Liu, S.; Song, F.; Mao, H.; Li, X. Melatonin improves the photosynthetic carbon assimilation and antioxidant capacity in wheat exposed to nano-ZnO stress. *Molecules* **2017**, *22*, 1727. [CrossRef]
232. Qiao, Y.; Yin, L.; Wang, B.; Ke, Q.; Deng, X.; Wang, S. Melatonin promotes plant growth by increasing nitrogen uptake and assimilation under nitrogen deficient condition in winter wheat. *Plant Physiol. Biochem.* **2019**, *139*, 342–349. [CrossRef]
233. Hoffman, R.A.; Reiter, R.J. Pineal gland: Influence on gonads of male hamsters. *Science* **1965**, *148*, 1609–1611. [CrossRef]
234. Choi, D. Effects of dietary supplement containing melatonin on reproductive activity in male golden hamsters. *Dev. Reprod.* **2019**, *23*, 101–110. [CrossRef] [PubMed]
235. deNicolo, G.; Morris, S.T.; Kenyon, P.R.; Morel, P.C.H.; Parkinson, T.J. Melatonin-improved reproductive performance in sheep bred out of season. *Anim. Reprod. Sci.* **2008**, *109*, 124–133. [CrossRef] [PubMed]
236. Yakhin, O.I.; Lubyanov, A.A.; Yakhin, I.A.; Brown, P.H. Biostimulants in Plant Science: A Global Perspective. *Front. Plant Sci.* **2017**, *7*, 2049. [CrossRef] [PubMed]
237. Hernández-Ruiz, J.; Arnao, M.B. Phytomelatonin, an interesting tool for agricultural crops. *Focus Sci.* **2016**, *2*, 1–7. [CrossRef]
238. Arnao, M.B.; Hernández-Ruiz, J. The potential of phytomelatonin as a nutraceutical. *Molecules* **2018**, *23*, 238. [CrossRef] [PubMed]
239. Arnao, M.B.; Hernández-Ruiz, J. Phytomelatonin, natural melatonin from plants as a novel dietary supplement: Sources, activities and world market. *J. Funct. Foods* **2018**, *48*, 37–42. [CrossRef]

agronomy

MDPI

Article

Effect of Vegetal- and Seaweed Extract-Based Biostimulants on Agronomical and Leaf Quality Traits of Plastic Tunnel-Grown Baby Lettuce under Four Regimes of Nitrogen Fertilization

Ida Di Mola [1,*], Eugenio Cozzolino [2,*], Lucia Ottaiano [1], Maria Giordano [1], Youssef Rouphael [1], Giuseppe Colla [3] and Mauro Mori [1]

1 Department of Agricultural Sciences, University of Naples Federico II, Portici 80055, Italy; lucyottaiano@gmail.com (L.O.); maria.giordano@unina.it (M.G.); youssef.rouphael@unina.it (Y.R.); mori@unina.it (M.M.)

2 Council for Agricultural Research and Economics (CREA)-Research Center for Cereal and Industrial Crops, Caserta 81100, Italy

3 Department of Agriculture and Forest Sciences, University of Tuscia, Viterbo 01100, Italy; giucolla@unitus.it

* Correspondence: ida.dimola@unina.it (I.D.M.); eugenio.cozzolino@crea.gov.it (E.C.)

Received: 3 September 2019; Accepted: 20 September 2019; Published: 22 September 2019

Abstract: Nitrogen (N) fertilizers play a crucial role in agriculture, representing a powerful tool for farmers for increasing yields throughout the seasons under both optimal and suboptimal conditions. At the same time, their synthetic/chemical nature could have several influences on ecosystems and human health. For this reason, there is an urgent need to find new and more sustainable means of production to increase plant productivity and optimize nitrogen use. An experiment was conducted in a plastic tunnel to assess the response of baby lettuce crop to the foliar application of three plant biostimulants (PBs): Legume-derived protein hydrolysate (LDPH) 'Trainer®', tropical plant extract (TPE) 'Auxym®' and seaweed extract (SwE) from *Ecklonia maxima* 'Kelpak®' under different N rates of 0, 10, 20 and 30 kg N·ha^{-1}. The responses of baby lettuce plants were assessed in terms of yield, growth parameters and physicochemical composition of the leaves. The fresh yield of baby lettuce in both biostimulant-treated and untreated plants was positively affected by increasing N rates from 0 to 20 kg N·ha^{-1}, reaching a plateau thereafter indicating luxury N conditions at 30 kg N·ha^{-1}. However, high N fertilizer application (20 and especially 30 kg N·ha^{-1}) resulted in undesirable decreases in antioxidant activities and total ascorbic acid (TAA). Under non-fertilized regimens, foliar PBs application boost growth and yield of baby lettuce in comparison to non-treated plants. Foliar spray with LDPH and especially SwE elicited significant increases in marketable fresh yield (averaging 14%, 6% and 7% at 10, 20 and 30 kg N·ha^{-1}, respectively) compared to TPE and untreated plants. Improved agronomical performance of baby lettuce under optimal (10 kg N ha^{-1}) and especially suboptimal N regimens (0 kg N ha^{-1}) was associated with increasing photochemical efficiency and a better activity of photosystem II (higher Soil Plant Analysis Development-SPAD index and chlorophyllous pigments biosynthesis). The application of LDPH enhanced antioxidant capacity and TAA in baby lettuce leaf and did not increased nitrate content as recorded in SwE and TPE treatments. Overall, plant biostimulants may be considered as a sustainable tool of production to increase leafy vegetable productivity in low fertility soils.

Keywords: *Lactuca sativa* L.; legume-derived protein hydrolysate; nitrate; tropical plant extract; seaweed extract

1. Introduction

In recent years, the consumption of fresh-cut leafy vegetables has increased and among them, baby leaf lettuce is very widespread. Baby leaf lettuce is widely cultivated in Italy under both open field and greenhouse conditions [1]. Baby leaf vegetables are characterized by a short cycle but it requires a correct agronomic management to avoid high levels of nitrate accumulation and pesticide residues in the final product [2]. Therefore, there is a paramount interest in enhancing its production and quality, and at the same time reducing the nitrate levels of leafy vegetables, in order to overcome the legal limit for the marketing imposed by the European Community (Reg. n° 1258/2011).

With the aim to boost yield and to contain the risk of nitrate accumulation in the leaves, the research community is focusing on the use of sustainable production technologies, including application of beneficial microorganisms (Plant growth promoting Rhizobacteria, mycorrhiza and *Trichoderma*) and plant biostimulants [3–6]. In function of their origin, non-microbial plant biostimulants can be classified into five categories: (i) Seaweed extracts and microalgae, (ii) protein hydrolysate (PH) and amino acid containing products, (iii) plant extracts, (iv) humic substances and (v) silicon, with the first three groups commanding 75% of the market share [7–10]. Protein hydrolysate and amino acids containing products are normally obtained by enzymatic and/or chemical hydrolysis depending on the organic matrix (animal or vegetal) and are characterized by high percentages of amino acids and peptides, followed by carbohydrates and small amounts of micronutrients [7,8]. Moreover, plant extracts are normally produced through the fermentation of tropical plants and contain amino acids and peptides, carbohydrates but also vitamins and micronutrients with small quantities of phytohormones [8], while seaweed extracts particularly the brown macroalgae are obtained through a process called 'cold cell burst' and contain polysaccharides, osmolytes (proline and betaines), macro- and micro-nutrients, brassinosteroids and phytohormones (auxins, cytokinins and gibberellins; [11,12]).

Recent studies carried out on vegetable crops including leafy greens have demonstrated that foliar and/or root applications of plant or seaweed-based biostimulants elicit several physiological and molecular processes, thus resulting in improvements in growth, productivity, nutritional quality and nutrient use efficiency (NUE) and tolerance to abiotic stressors such as drought, soil and water salinity, nutrient deprivation and extreme temperatures [12–24]. The beneficial effects of vegetal- and seaweed-based biostimulants have been attributed to direct and indirect stimulation mechanisms [9]. The direct stimulation action of biostimulants include: (i) Activity enhancement of key enzymes involved in carbon and nitrogen (N) metabolism [13,20,25], (ii) eliciting hormone-activity in particular auxin- and gibberellin-like activities through bioactive peptides [26–28], (iii) physiological, biochemical and anatomical changes such as the production of antioxidant enzymes, pigments, secondary metabolites and smallest cell guard length and width [7,8,12,29]. In addition to direct mechanisms, indirect modes of actions on agronomical performance and nutrient uptake and assimilation have been also reported when vegetal- or seaweed extract-based biostimulants were applied as substrate drench and/or foliar spray [8,29]. The better nutritional status in biostimulant-treated plants in comparison to untreated plants has been mostly associated with root system modulation (increases in root biomass, root length and diameter and lateral root branching [8,12,29]).

Among the different agronomical claims of plant biostimulants, the capacity to improve NUE in particular, N is one of the most important claims supporting their placement in the market for both economic and environmental reasons [9]. However, limited scientific literature are available regarding the effects of plant biostimulants on vegetable crops under sub-optimal N regimens [20,22,30–32]. For instance, Sestili and co-workers [20] demonstrated that the application of a PH at optimal and sub-optimal N regimens enhanced hydroponically grown tomato performance, especially substrate drench. Interestingly, the same authors observed that protein hydrolysate at low N conditions upregulated gene expression for amino acid transporter and glutamine synthetase, leading to a higher assimilation of N with a positive impact on plant growth. Similarly, Carillo et al. [22] reported that foliar application of PH, especially under suboptimal N fertilization regimes (0 or 15 kg N ha^{-1}) boost marketable yield of greenhouse spinach due to an enhancement of nutrient acquisition and to an increase in total amino acids in plants as well as to an improvement of photochemical efficiency, thus boosting yield.

Since there is ample evidence of species-specific response to plant biostimulants, especially that of leaf biostimulant permeability (through leaf cuticule and stomatal aperture) and thus the efficacy of the biostimulant product is species-dependent [9]; there is an urgent need among researchers to assess the effect of vegetal- and seaweed based-biostimulants on baby lettuce performance at different N fertilization regimes.

Taking into account all the previous considerations, an experiment was conducted in a plastic tunnel to assess the response of baby lettuce crop to the foliar application of the legume-derived PH 'Trainer®', tropical plant extract 'Auxym®' and seaweed extract from *Ecklonia maxima* 'Kelpak®' under different N rates of 0, 10, 20 and 30 kg N·ha^{-1}. The responses of baby lettuce plants were assessed in terms of yield, leaf morphometric parameters and leaf quality traits.

2. Materials and Methods

2.1. Experimental Setting, Plant Material and Design

The experiment was carried out in a unheated plastic tunnel covered by polyethylene during the winter 2018 growing season (January 16—March 12) at Gussone Park, experimental site of the Department of Agricultural Sciences (40°48.870′ N; 14°20.821′ E; 70 m above sea level) located in Portici, southern Italy. The trend of daily maximum and minimum air temperature inside the plastic tunnel is reported in Figure 1. The baby leaf lettuce (*Lactuca sativa* L.) cv. 'Zarina' (ISI Sementi SpA, Parma, Italy) was used as the selected crop.

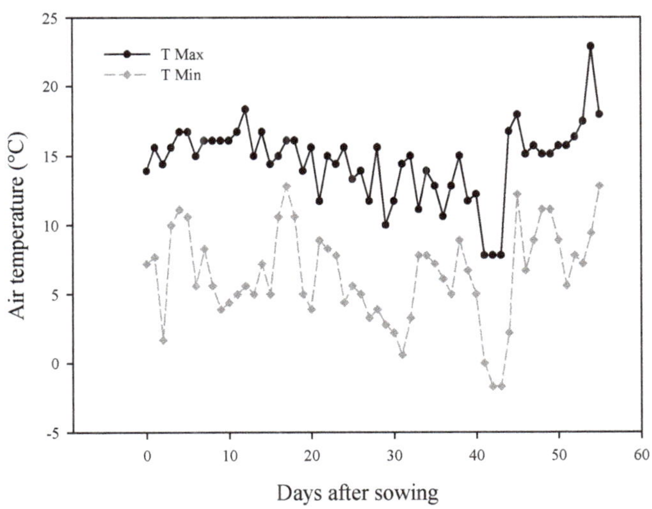

Figure 1. Trend of the maximum and minimum air temperature inside the plastic tunnel during the growing period of baby lettuce.

A factorial combination of N fertilization (N) and biostimulant application (B) was applied based on four increasing N fertilization levels (0, 10, 20 or 30 kg N·ha^{-1}; N0, N10, N20 and N30, respectively) and three plant biostimulants (seaweed extract—SwE, legume-derived protein hydrolysate—LDPH and tropical plant extract—TPE) and a non-treated control. The experimental design was a randomized complete-block design with three replications, yielding 48 experimental units (4N × 4B × 3 replications) established in large lysimeters of reinforced fiber glass with a diameter of 0.70 m and a depth of 0.60 m.

Each experimental unit consisted of one large lysimeters. The lysimeters were filled with a soil having the following chemical and physical characteristics reported in Table 1.

Table 1. Physical and chemical properties of the soil used in this work.

Soil Properties	Units	Mean Values
Texture		
Coarse sand	%	69.1
Fine sand	%	21.9
Silt	%	4.5
Clay	%	4.5
Chemical properties		
pH	-	6.54
Electrical conductivity	$dS \cdot m^{-1}$	0.64
Organic matter	$g \cdot kg^{-1}$	32.4
Total N (Kjeldahl method)	$g \cdot kg^{-1}$	1.2
P_2O_5 (Olsen method)	$mg \cdot kg^{-1}$	312.8
K_2O (Tetraphenylborate method)	$mg \cdot kg^{-1}$	620.7
NO_3-N	$mg \cdot kg^{-1}$	10.0
NH_4-N	$mg \cdot kg^{-1}$	9.0

2.2. *Nitrogen Fertilization Levels, Cultural Practices and Biostimulants Application*

The baby leaf lettuce was hand seeded on January 16 at a plant density of 2500 seeds·m^{-2}. The N was applied as calcium nitrate (26%) in a single operation 14 days after sowing. The calcium nitrate was used based on standard commercial practices used in Italy.

The three commercial SwE, LDPH and TPE-based biostimulants were made by 'Kelpak®' (Kelp Products (Pty) Ltd., Cape Town, South Africa), 'Trainer®' and 'Auxym®' (Italpollina S.p.A., Rivoli Veronese, Italy), respectively.

The SwE obtained through 'cold cell burst' mainly contained phytohormones (auxins and cytokinins with a very high auxin-to-cytokinin ratio), carbohydrates, amino acids, vitamins (B1, B2, C and E) and macro- and micro-nutrients [18,19]. The LDPH-based biostimulant contained free amino acids and peptides (75%), carbohydrates (22%) and mineral nutrients (3%). The detailed aminogram was reported by Paul et al. [33,34]. The TPE biostimulant obtained by fermentation of tropical plants contained 54% of free amino acids and peptide, 17% carbohydrate, 23% mineral nutrients, 6% vitamins and 0.22% phytohormones as reported in detail by Caruso et al. [23,24]. Baby lettuce leaf plants were sprayed with a biostimulant solution containing 3 mL·L^{-1} of SwE and LDPH and 2·mL L^{-1} for TPE-based biostimulant, or with water (non-treated control), five times during the growing season at 7-day intervals, starting three weeks after sowing. The relative doses of the three commercial plant biostimulants were used based on manufacturer recommendations. The volume of the solution used during the five foliar applications was 100 mL per square meter.

2.3. *Plant Growth Parameters, Marketable Yield, Leaf Colorimetry and Sampling*

On March 12, the baby leaf lettuce was harvested in all experimental units. The leaf area was measured using an electronic leaf area meter (Li-Cor3000, Li-Cor, Lincoln, NE, USA) in order to calculate the leaf area index (LAI). The marketable fresh yield was also measured and expressed in tons per ha, and a sub-sample was oven dried at 70 °C for 3 days in order to determine the leaf dry matter percentage, and the dry samples were consequently used for the mineral analysis. Furthermore, the specific leaf weight (leaf dry weight per unit area; mg·cm^{-2}) as well as leaf succulence (leaf water content per unit area; mg·cm^{-2}) were also recorded.

Leaf colorimetry was measured on the upper side of 10 leaves per experimental unit using Minolta CR-300 Chroma Meter (Minolta Camera Co. Ltd., Osaka, Japan) in order to obtain the color space parameters (L^*, a^* and b^*) and a portable chlorophyll meter SPAD-502 (Konica Minolta, Tokyo, Japan) was also used to measure the SPAD (Soil Plant Analysis Development) index.

Batch samples of fresh leaves from each experimental unit were frozen in liquid nitrogen immediately after harvest, lyophilized Christ, Alpha 1–4 (Osterode, Germany) and stored at −80 °C until further analysis.

2.4. Antioxidant Capacity Analysis

The lipophylic and hydrophilic antioxidant capacities were assessed on extract from freeze-dried baby lettuce leaves (200 mg) added with methanol and distilled water, respectively. The antioxidant activity of the lipophilic and hydrophilic extract fractions was measured spectrophotometrically based on the methods of Re et al. [35] and Fogliano et al. [36], respectively. The absorbance of the solutions for lipophilic and hydrophilic extract fractions were measured at 734 and 505 nm, respectively. Lipophylic and hydrophilic antioxidant activities were expressed as mmol of Trolox and mmol ascorbic acid per 100 g of dry weight (dw) [36].

2.5. Chlorophyllous Pigments and Nitrate Analysis

Chlorophyll and carotenoids content of the baby lettuce leaves were also assayed spectrophotometrically after the extraction of the fresh material (500 mg) using pure acetone as described in detail by Lichtentahler and Buschmann [37], whereas the nitrate content was determined based on the method of Sah [38]. The absorbance of the solutions for chlorophyll a and b, carotenoids and nitrate were measured at 662, 645, 470 and 550 nm. The chlorophyllous pigments were expressed as mg g^{-1} fresh weight (fw), while the nitrate content was expressed as mg kg^{-1} fw.

2.6. Total Ascorbic Acid Analysis

The total ascorbic acid (expressed as mg ascorbic acid on 100 g fw) was also assessed spectrophotometrically based on the protocol by Kampfenkel et al. [39]. The absorbance of the solution for total ascorbic acid was measured at 525 nm.

2.7. Statistical Processing

Morphological and qualitative data were statistically analyzed by a 2-way ANOVA using the SPSS 21 software package for Windows 2007. The means were separated by a Duncan's test (significance level 0.05).

3. Results and Discussion

3.1. Effect of N Fertilization Levels and Biostimulant Application on Yield and Growth

The results regarding morphological parameters and marketable yield of baby lettuce are reported in Figure 2; Figure 3 and Table 2. For marketable fresh yield and leaf area index (LAI) significant interaction between fertilization (F) and biostimulant application (B) was observed, whereas leaf succulence and specific leaf weight (SLW) were only influenced by the two tested factors with no significant F × B interaction (Figures 2 and 3, and Table 2). The fresh yield of baby lettuce in both biostimulant-treated and untreated plants was positively affected by increasing N fertilization levels from 0 to 20 kg N·ha^{-1}, reaching a plateau thereafter indicating a luxury N conditions at 30 kg N·ha^{-1} (Figure 2). The marketable fresh yield of baby lettuce at N0 was clearly higher by 19% in biostimulant-treated plants compared untreated plants, with no significant differences between the three plant biostimulants tested (Figure 2). Interestingly, foliar spray with LDPH and especially SwE elicited significant increases (average 14%, 6% and 7% at 10, 20 and 30 kg N·ha^{-1}, respectively) compared to TPE and untreated plants (Figure 2). Similarly to the effects on marketable fresh yield, the leaf area index (LAI) in SwE and LDPH-treated plants at 10, 20 and 30 kg N·ha^{-1} was significantly higher compared to baby lettuce treated with TPE or untreated plants, whereas under non-fertilized conditions LAI was significantly higher in biostimulant compared to untreated plants, irrespective of the commercial biostimulants used (Figure 3).

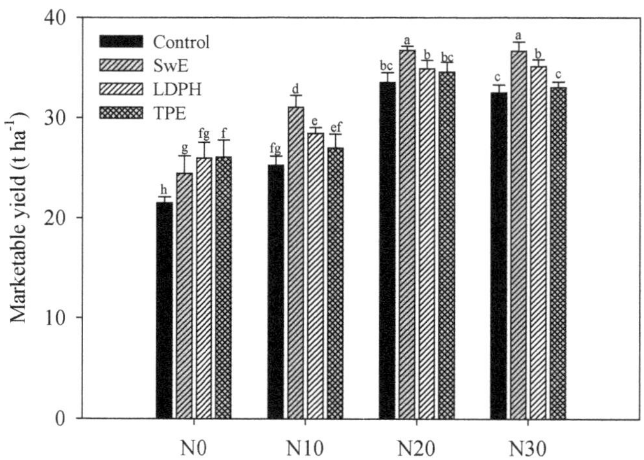

Figure 2. Effects of nitrogen (N) fertilization levels (0, 10, 20 and 30 kg N·ha^{-1}; N0, N10, N20 and N30, respectively) and biostimulant applications (untreated control, SwE: Extract of seaweed *Ecklonia maxima*, LDPH: Legume-derived protein hydrolysate and TPE: Tropical plant extract) on the marketable fresh yield of baby lettuce plants. Different letters indicate significant differences according to the Duncan's test (significance level 0.05).

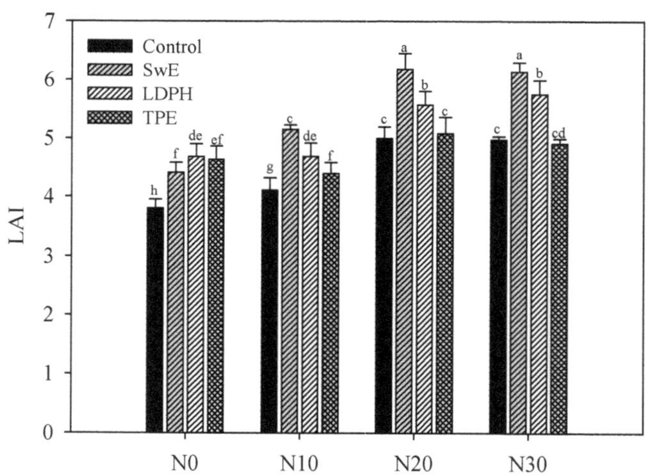

Figure 3. Effects of nitrogen (N) fertilization levels (0, 10, 20 and 30 kg N·ha^{-1}; N0, N10, N20 and N30, respectively) and biostimulant applications (untreated control, SwE: Extract of seaweed *Ecklonia maxima*, LDPH: Legume-derived protein hydrolysate and TPE: Tropical plant extract) on the leaf area index (LAI) of baby lettuce plants. Different letters indicate significant differences according to the Duncan's test (significance level 0.05).

Table 2. Effects of nitrogen (N) fertilization levels (0, 10, 20 and 30 kg N·ha^{-1}; N0, N10, N20 and N30, respectively) and biostimulant applications (untreated control, SwE: Extract of seaweed *Ecklonia maxima*, LDPH: Legume-derived protein hydrolysate and TPE: Tropical plant extract) on leaf succulence and specific leaf weight (SLW) of baby lettuce plants.

Treatments	Succulence	SLW
	(mg H_2O·cm^{-2})	(mg dm·cm^{-2})
Fertilization (F)		
N0	59.9 b	2.99 a
N10	71.5 a	3.26 a
N20	69.9 a	2.66 b
N30	67.7 a	2.60 b
Biostimulant (B)		
Control	60.5 c	2.94 a
SwE	65.2 b	2.56 b
LDPH	70.4 a	2.92 a
TPE	72.9 a	3.10 a
Significance		
F	**	**
B	*	*
F × B	NS	NS

NS, *, and ** indicate non-significant, significant at $p < 0.05$, significant at $p < 0.01$, respectively. Different letters indicate significant differences according to the Duncan's test (significance level 0.05).

When averaged over biostimulant application (F × B = not significant), the leaf succulence increased quadratically by increasing N fertilization levels from 10 to 30 kg N·ha^{-1} with no significant difference among the three N fertilization rates, whereas the SLW declined at 20 and 30 kg N·ha^{-1} (Table 2). Averaged over N fertilization levels, significant differentiation in terms of leaf succulence and SLW was recorded in response to biostimulants application with the higher values of succulence observed with LDPH and TPE followed by SwE as opposed to untreated plants, whereas the lowest values of SLW were recorded in baby lettuce treated with *Ecklonia maxima* extract (Table 2).

The stimulation effect of commercial biostimulants (6%–19%) recorded in the current research is in line with previous studies carried out on greenhouse fresh tomato treated with seaweed extracts of *E. maxima* or *Ascophyllum nodosum*, LDPE and TPE (7%–25%; [12,40]) but far lower than those recorded on greenhouse spinach [19]. The different stimulation effect among tested species indicates a crop-specific differential response to plant biostimulant applications and thus requires additional crop-specific studies to optimize plant biostimulants application, taking into consideration the following factors: environment, management practice and plant morphological traits (e.g., leaf permeability and cuticle thickness [9,28]).

Interestingly, LDPH (at 0 kg N·ha^{-1}) and SwE (at 10, 20 and 30 kg N·ha^{-1}) are likely to boost growth response and crop productivity as a consequence of the presence of bioactive molecules such as amino acids (tryptophan, glutamic and aspartic acids), soluble peptides (in LDPH) and polysaccharides (laminarans, fucoidans and alginates), phenolic compounds, osmolytes (proline, betaine and manitol) and phyohormones (abscisic acid, auxins, brassinosteroids, cytokinins and gibberellins) (in SwE) [8,29]. These former molecules present in seaweed and PH-based biostimulants may have triggered a signal transduction pathway through elicitation of endogenous phytohormone synthesis, thus leading to a higher crop productivity compared to untreated-baby lettuce plants [19,20]. Another possible mechanism of action (indirect mechanism) behind the stimulation of LAI and marketable fresh yield could be the modulation of the root system architecture in terms of root biomass, root volume and length and higher root branching triggered by tryptophan in LDPH and auxins in SwE, which improved nutrient uptake/translocation/assimilation, leading to a higher agronomical performance [12,19,41,42]. Our results are in agreement with those of Carillo and co-workers [22] who reported that foliar application of LDPH at a rate of 4 mL·L^{-1} under suboptimal N fertilization conditions (0 and 15 kg N·ha^{-1}) increased the fresh yield of greenhouse spinach through an increase of the nutritional status (higher macronutrient accumulation), better photosynthetic activity and improving the total acid content.

3.2. Effect of N Fertilization Levels and Biostimulant Application on Leaf Colorimetry and SPAD Index

Among the physical properties that may affect the purchasing decisions of vegetable consumers is product appearance, in particular, the color of the vegetable [43]. In the present study, no significant interaction between N fertilization and biostimulants application was recorded for the three leaf colorimetric parameters lightness (L^*), green color intensity (negative values of a^*) and yellow color intensity (positive values of b^*) (Table 3). The colorimetric CIELAB components L^* and b^* were significantly influenced by the two tested factors, whereas a^* was only affected by N fertilization levels (Table 3). Increasing the N fertilization levels from 0 to 30 kg N·ha^{-1} yielded lighter baby lettuce leaf expression by increasing L^* values, but with a decrease in a^* values (Table 3). Moreover, when averaged over N fertilization levels, the foliar application of SwE and TPE-based biostimulants elicited an increase in L^* values compared to the untreated control, whereas LDPH treatment exhibited intermediate values (Table 3).

Table 3. Effects of nitrogen (N) fertilization levels (0, 10, 20 and 30 kg N·ha^{-1}; N0, N10, N20 and N30, respectively) and biostimulant applications (untreated control, SwE: Extract of seaweed *Ecklonia maxima*, LDPH: Legume-derived protein hydrolysate and TPE: Tropical plant extract) on leaf hunter color parameters of baby lettuce plants.

Table	L^*	a^*	b^*
Fertilization (F)			
N0	55.9 d	−20.8 a	39.8 b
N10	56.7 c	−21.2 a	40.1 b
N20	58.7 b	−22.7 b	41.8 a
N30	59.5 a	−22.7 b	42.4 a
Biostimulants (B)			
Control	56.5 c	−21.5	39.8 b
SwE	58.5 a	−22.2	41.6 a
LDPH	57.5 b	−21.8	41.0 a
TPE	58.3 a	−21.9	41.6 a
Significance			
F	**	*	*
B	*	NS	*
F × B	NS	NS	NS

NS, *, and ** indicate non-significant, significant at $p < 0.05$, or significant at $p < 0.01$, respectively. Different letters indicate significant differences according to the Duncan's test (significance level 0.05).

Interestingly, the foliar application of commercial plant biostimulants improved the SPAD index significantly; this is an important physiological parameter having a crucial role on the primary metabolism of plants. With the exception of under N20, where no significant difference in the SPAD index was observed, between biostimulants-treated and untreated plants, the foliar application with SwE (at 10 and 30 kg N·ha^{-1}) and with the three commercial biostimulants (at 0 kg N·ha^{-1}) incurred a significant increase in the SPAD index (Figure 4). Our findings have been also demonstrated in many leafy vegetable species such as jute, lettuce, and spinach [16,19,22]. The highest SPAD values observed after the application of plant biostimulants in particular extracts from brown macroalgae could be attributed to several putative mechanisms like the following: (i) better translocation of soluble sugars via the phloem, (ii) increases in the biogenesis of chloroplast, as well as (iii) limited chlorophyll degradation, and thus, delayed senescence [29,44,45].

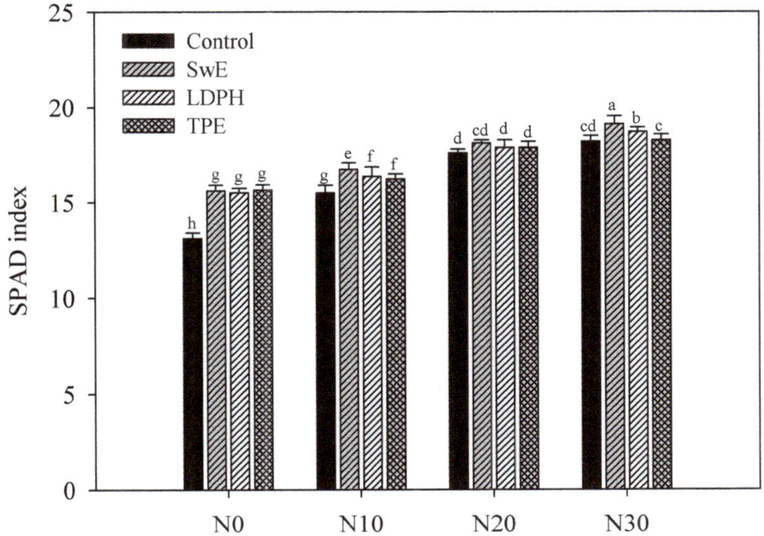

Figure 4. Effects of nitrogen (N) fertilization levels (0, 10, 20 and 30 kg N·ha^{-1}; N0, N10, N20 and N30, respectively) and biostimulant applications (untreated control, SwE: Extract of seaweed *Ecklonia maxima*, LDPH: Legume-derived protein hydrolysate and TPE: Tropical plant extract) on the SPAD index of baby lettuce plants. Different letters indicate significant differences according to the Duncan's test (significance level 0.05).

3.3. Effect of N Fertilization Levels and Biostimulant Application on Nitrate Accumulation and Biochemical Parameters

Nitrate was affected by both N fertilization levels and biostimulant application, without significant F × B interaction (Table 4). The nitrate concentration in baby lettuce leaf was negatively affected by N fertilization levels. Increasing the N fertilization from 0 to 30 kg N·ha^{-1} increased the nitrate accumulation in leaves, especially at 20 and 30 kg N·ha^{-1}, where the content of nitrate was above the upper limits set by the European Union (EU) for safe lettuce marketing (Commission Regulation No. 1258/2011; 3000 to 5000 mg NO_3^-·kg^{-1} of lettuce depending on growing season and cultivation conditions). On the other hand, when averaged over N fertilization levels, the nitrate concentration in LDPH treated plants was significantly lower on average by 21.2% compared to baby lettuce treated with SwE or TPE and it was not significantly different than untreated-baby lettuce plants (Table 4). The capacity of LDPH, which is mainly composed of soluble solids and especially amino acids, to accumulate less nitrate in the leaf tissue, could be attributed to a molecular mechanism such as the up-regulation of genes involved in N metabolism such as nitrate reducatse, and consequently to an augmenting assimilation of nitrates into amino acids [46,47]. Furthermore, other studies conducted by Sady and Smoleń [31] and Smoleń and Sady [32] on carrots and spinach, respectively, reported that after the foliar application of 'Pentakeep V' containing 5-aminolevulinic acid was able to reduce nitrate accumulation in combination with a 50% N dose, whereas an opposite trend was observed in combination with 100% N. The authors concluded that nitrate accumulation in response to biostimulant application may change in relation to several interacting variables including species, variety and N application rates.

Table 4. Effects of nitrogen (N) fertilization levels (0, 10, 20 and 30 kg N·ha^{-1}; N0, N10, N20 and N30, respectively) and biostimulant applications (untreated control, SwE: Extract of seaweed *Ecklonia maxima*, LDPH: Legume-derived protein hydrolysate and TPE: Tropical plant extract) on nitrate, chlorophyll and carotenoids content of baby lettuce plants.

Treatments	Nitrate	Chlorophyll a	Chlorophyll b	Total chlorophyll	Carotenoids
	(mg·kg^{-1} fw)	(mg·g^{-1} fw)	(mg·g^{-1} fw)	(mg·g^{-1} fw)	(μg·g^{-1} fw)
Fertilization (F)					
N0	703.3 d	0.298 b	0.211	0.508 b	156 b
N10	1476.0 c	0.330 a	0.210	0.540 a	178 a
N20	6206.7 b	0.334 a	0.201	0.535 a	170 a
N30	7288.1 a	0.338 a	0.209	0.546 a	170 a
Biostimulants (B)					
Control	3366.5 b	0.302 b	0.191 c	0.493 b	155 c
SwE	4467.6 a	0.319 b	0.192 c	0.511 b	181 a
LDPH	3504.4 b	0.342 a	0.214 b	0.556 a	173 ab
TPE	4426.5 a	0.337 a	0.232 a	0.569 a	164 bc
Significance					
F	**	*	NS	*	*
B	*	*	**	*	*
F × B	NS	NS	NS	NS	NS

NS, *, and ** indicate non-significant, significant at $p < 0.05$, or significant at $p < 0.01$, respectively. Different letters indicate significant differences according to the Duncan's test (significance level 0.05).

One of the beneficial responses of plant biostimulants application is an increase in chlorophyllous pigments such as chlorophyll a, b and total, as well as carotenoids. This was the case in the current research study, since the foliar application of LDPH and TPE incurred a significant increase in chlorophyll a and b and consequently the total chlorophyll compared to SwE and untreated-baby lettuce plants (Table 4). Furthermore, the content of carotenoids was positively affected by the foliar application of SwE and LDPH compared to the control treatment (Table 4). This beneficial effect of vegetal and seaweed extract-based biostimulants on carotenoids and especially chlorophyll content has been recorded also in corn, jute and eggplant [21,22,48,49]. The increase in chlorophyll a and total content in both LDPH and TPE (characterized by the high percentage of free amino acids (75% and 54%, respectively [23,24]) could be attributed to the higher content of primary amino acids in the vegetal-based treated plants as amino acids (e.g., alanine, aspartate, asparagines and glutamate) which help to boost chlorophyll content, and consequently, increase photosynthetic activity as well as the quantum yield of O_2 evolution [22].

3.4. *Effect of N Fertilization Levels and Biostimulant Application on Antioxidant Capacity and Bioactive Content*

Lipophilic (LAA) and hydrophilic (HAA) antioxidant activities as well as total ascorbic acid (TAA) were significantly affected by both factors with a significant F × B interaction (Table 5). Antioxidant scavenging activity was an important functional quality parameter in assessing the nutritional properties of foods including leafy vegetables, since lipophilic (e.g., β-carotene, lutein, α-tocopherol, etc.) and hydrophilic (e.g., vitamin C, caffeic acid, ferulic acid, quercitin, etc.) antioxidant molecules impart beneficial effects to human health, as these bioactive molecules are known to play a primary role in delaying oxidative damage, thus, preventing a wide range of diseases [50–54]. In the current study, LAA, HAA and TAA of the baby lettuce ranged from 19.9 to 32.3 mmol trolox 100·g^{-1} dw, from 3.0 to 8.2 mmol ascorbic acid 100 g^{-1} dw and from 6.8 to 33.4 mg g^{-1}, respectively (Table 5).

Table 5. Effects of nitrogen (N) fertilization levels (0, 10, 20 and 30 kg N ha^{-1}; N0, N10, N20 and N30, respectively) and biostimulant applications (untreated control, SwE: Extract of seaweed *Ecklonia maxima*, LDPH: Legume-derived protein hydrolysate and TPE: Tropical plant extract) lipophilic (LAA) and hydrophilic (HAA) antioxidant activities and total ascorbic acid (TAA) of baby lettuce plants.

Treatments		LAA	HAA	TAA
		mmol Trolox eq. 100 g^{-1} dw	mmol ascorbic acid eq. 100 g^{-1} dw	mg g^{-1} fw
N0	Control	26.2 b	6.7 c	25.0 b
	SwE	27.6 b	6.6 c	33.4 a
	LDPH	32.3 a	8.2 a	18.0 c
	TPE	25.2 b	7.2 bc	18.6 c
N10	Control	21.0 c	7.6 b	17.3 c
	SwE	20.4 c	7.5 b	16.8 c
	LDPH	20.9 c	7.7 b	18.1 c
	TPE	30.9 a	7.6 b	16.9 c
N20	Control	22.0 c	4.9 d	14.0 de
	SwE	22.2 c	4.8 d	14.3 d
	LDPH	19.9 c	4.4 d	12.6 de
	TPE	22.1 c	3.0 e	14.5 d
N30	Control	21.8 c	3.3 e	13.2 de
	SwE	20.1 c	3.2 e	11.7 e
	LDPH	21.9 c	4.5 d	12.6 de
	TPE	21.5 c	3.4 e	6.8 f
Significance				
Fertilization (F)		**	**	**
Biostimulant (B)		*	*	**
F × B		**	**	**

*, ** significant at $p < 0.05$ and 0.01, respectively. Different letters indicate significant differences according to the Duncan's test (significance level 0.05).

High N fertilizer application (20 and especially 30 kg·N ha^{-1}) resulted in undesirable decreases in HAA and TAA of baby lettuce leaves (Table 5), which is in agreement with the results of Wang et al. [55] who reported that high N fertilization levels can result in undesirable changes in the quality attributes of fruit and leafy vegetables such as soluble solids and ascorbic acid leading to a decrease in commercial, nutritional and functional quality.

The vegetal- and seaweed extract-based biostimulants applied to baby lettuce resulted in higher antioxidant capacity and bioactive content depending on the N fertilization levels compared to untreated control treatment. For instance, at N0 the highest antioxidant activities and TAA compared to the untreated control were recorded in baby lettuce treated with LDPH and SwE-based biostimulant plants, respectively, whereas at N10 and N30 the highest LAA and HAA contents were observed in TPE and LDPH treated plants, respectively (Table 5). Our findings on the effect of plant biostimulants on nutritional and functional quality of the product were in line with previous research on vegetal-based biostimulants (protein hydrolysate and plant extract) conducted by Caruso et al. [23], in which foliar application at weekly interval increases the LAA, HAA and TAA contents of perennial wall rocket compared to the non-treated control. Similarly, Vasantharaja et al. [56] demonstrated that the application of seaweed extract-based biostimulant (*Sargassum swartzii*) boosted the antioxidant activity and the bioactive content (e.g., phenols and vitamin C) of cowpea. A mechanistic explanation of the beneficial effect of plant biostimulants, in particular LDPH, on the biosynthesis of antioxidant molecules could be due to: i) the activity stimulation of key enzymes involved in antioxidant homeostasis in cells, and ii) the higher macro- and micro-nutrient assimilation of biostimulant-treated plants which could contribute to the synthesis of amino acids, phenylalanine and tyrosine [7,40].

4. Conclusions

The idea of working with plant biostimulants to increase yield under both optimal and suboptimal conditions is gaining interest among leafy vegetable growers, as well as private companies and researchers for both economic and environmental reasons. The foliar application of vegetal and seaweed extract-based biostimulants, in particular SwE and LDPH enhanced plant growth, and productivity especially under sub-optimal N regimens, and to a lesser extent; at 20 and 30 kg $N \cdot ha^{-1}$. The foliar application of SwE and LDPH was effective in supporting better physiological and biochemical status in terms of the SPAD index, chlorophyll and carotenoids content leading to a higher agronomical performance. Interestingly, the leaf quality traits of baby lettuce leaf can be improved by biostimulation action, especially with LDPH which delivered leaves with high antioxidant activity and total ascorbic acid as well as low nitrate content. The results of the current experiment highlight the benefit of using vegetal and seaweed extract-based biostimulants in baby lettuce to improve productivity under both optimal and especially suboptimal N regimens, bringing benefits to farmers and to the environment.

Author Contributions: Conceptualization, M.M.; methodology, I.D.M., E.C.; software, L.O.; validation, M.M., I.D.M., E.C. and L.O.; formal analysis, L.O.; investigation, I.D.M.; resources, M.M.; data curation, I.D.M. and L.O.; writing—original draft preparation, I.D.M.; writing—review and editing, Y.R.; visualization, M.G. and G.C.; supervision, G.C. and M.M.; project administration, M.M.; funding acquisition, M.M.

Conflicts of Interest: The authors declare no conflicts of interest.

References

1. Colonna, E.; Rouphael, Y.; Barbieri, G.; De Pascale, S. Nutritional quality of leafy vegetables harvested at two light intensities. *Food Chem.* **2016**, *199*, 702–710. [CrossRef] [PubMed]
2. Skovgaard, M.; Renjel Encinas, S.; Jensen, O.C.; Andersen, J.H.; Condarco, G.; Jørs, E. Pesticide residues in commercial lettuce, onion, and potato samples from Bolivia—A threat to public health? *Environ. Health Insights* **2017**, *11*, 1–8. [CrossRef] [PubMed]
3. De Pascale, S.; Rouphael, Y.; Colla, G. Plant biostimulants: Innovative tool for enhancing plant nutrition in organic farming. *Eur. J. Hortic. Sci.* **2017**, *82*, 277–285. [CrossRef]
4. Rouphael, Y.; Franken, P.; Schneider, C.; Schwarz, D.; Giovannetti, M.; Agnolucci, M.; De Pascale, S.; Bonini, P.; Colla, G. Arbuscular mycorrhizal fungi act as biostimulants in horticultural crops. *Sci. Hortic.* **2015**, *196*, 91–108. [CrossRef]
5. Rouphael, Y.; Kyriacou, M.C.; Petropoulos, S.A.; De Pascale, S.; Colla, G. Improving vegetable quality in controlled environments. *Sci. Hortic.* **2018**, *234*, 275–289. [CrossRef]
6. Rouphael, Y.; Spíchal, L.; Panzarová, K.; Casa, R.; Colla, G. High-throughput plant phenotyping for developing novel biostimulants: From lab to field or from field to lab? *Front. Plant Sci.* **2018**, *9*, 1197. [CrossRef] [PubMed]
7. Colla, G.; Nardi, S.; Cardarelli, M.; Ertani, A.; Lucini, L.; Canaguier, R.; Rouphael, Y. Protein hydrolysates as biostimulants in horticulture. *Sci. Hortic.* **2015**, *196*, 28–38. [CrossRef]
8. Colla, G.; Hoagland, L.; Ruzzi, M.; Cardarelli, M.; Bonini, P.; Canaguier, R.; Rouphael, Y. Biostimulant action of protein hydrolysates: Unraveling their effects on plant physiology and microbiome. *Front. Plant Sci.* **2017**, *8*, 2202. [CrossRef]
9. Rouphael, Y.; Colla, G. Synergistic biostimulatory action: Designing the next generation of plant biostimulants for sustainable agriculture. *Front. Plant Sci.* **2018**, *9*. [CrossRef]
10. Chiaiese, P.; Corrado, G.; Colla, G.; Kyriacou, M.C.; Rouphael, Y. Renewable sources of plant biostimulation: Microalgae as a sustainable means to improve crop performance. *Front. Plant Sci.* **2018**, *9*, 1782. [CrossRef]
11. Di Stasio, E.; Rouphael, Y.; Colla, G.; Raimondi, G.; Giordano, M.; Pannico, A.; El-Nakhel, C.; De Pascale, S. The Influence of *Ecklonia maxima* seaweed extract on growth, photosynthetic activity and mineral composition of *Brassica rapa* L. subsp. sylvestris under nutrient stress conditions. *Eur. J. Hortic. Sci.* **2017**, *82*, 286–293. [CrossRef]

12. Rouphael, Y.; De Micco, V.; Arena, C.; Raimondi, G.; Colla, G.; De Pascale, S. Effect of *Ecklonia maxima* seaweed extract on yield, mineral composition, gas exchange and leaf anatomy of zucchini squash grown under saline conditions. *J. Appl. Phycol.* **2017**, *29*, 459–470. [CrossRef]
13. Ertani, A.; Cavani, L.; Pizzeghello, D.; Brandellero, E.; Altissimo, A.; Ciavatta, C.; Nardi, S. Biostimulant activity of two protein hydrolyzates in the growth and nitrogen metabolism of maize seedlings. *J. Plant Nutr. Soil Sci.* **2009**, *172*, 237–244. [CrossRef]
14. Ertani, A.; Schiavon, M.; Muscolo, A.; Nardi, S. Alfalfa plant-derived biostimulant stimulate short-term growth of salt stressed *Zea mays* L. plants. *Plant Soil* **2013**, *64*, 145–158. [CrossRef]
15. Botta, A. Enhancing plant tolerance to temperature stress with amino acids: An approach to their mode of action. *Acta Hortic.* **2013**, *1009*, 29–35. [CrossRef]
16. Lucini, L.; Rouphael, Y.; Cardarelli, M.; Canaguier, R.; Kumar, P.; Colla, G. The effect of a plant-derived biostimulant on metabolic profiling and crop performance of lettuce grown under saline conditions. *Sci. Hortic.* **2015**, *182*, 124–133. [CrossRef]
17. Lucini, L.; Rouphael, Y.; Cardarelli, M.; Bonini, P.; Baffi, C.; Colla, G. A vegetal biopolymer-based biostimulant promoted root growth in melon while triggering brassinosteroids and stress-related compounds. *Front Plant Sci.* **2018**, *9*, 472. [CrossRef]
18. Rouphael, Y.; Colla, G.; Giordano, M.; El-Nakhel, C.; Kyriacou, M.C.; De Pascale, S. Foliar applications of a legume-derived protein hydrolysate elicit dose dependent increases of growth, leaf mineral composition, yield and fruit quality in two greenhouse tomato cultivars. *Sci. Hortic.* **2017**, *226*, 353–360. [CrossRef]
19. Rouphael, Y.; Giordano, M.; Cardarelli, M.; Cozzolino, E.; Mori, M.; Kyriacou, M.; Bonini, P.; Colla, G. Plant and seaweed-based extracts increase yield but differentially modulate nutritional quality of greenhouse spinach through biostimulant action. *Agronomy* **2018**, *8*, 126. [CrossRef]
20. Sestili, F.; Rouphael, Y.; Cardarelli, M.; Pucci, A.; Bonini, P.; Canaguier, R.; Colla, G. Protein hydrolysate stimulates growth in tomato coupled with N-dependent gene expression involved in N assimilation. *Front. Plant Sci.* **2018**, *9*, 1233. [CrossRef]
21. Carillo, P.; Colla, G.; El-Nakhel, C.; Bonini, P.; D'Amelia, L.; Dell'Aversana, E.; Pannico, A.; Giordano, M.; Sifola, M.I.; Kyriacou, M.C.; et al. Biostimulant application with a tropical plant extract enhances *Corchorus olitorius* adaptation to sub-optimal nutrient regimens by improving physiological parameters. *Agronomy* **2019**, *9*, 249. [CrossRef]
22. Carillo, P.; Colla, G.; Fusco, G.M.; Dell'Aversana, E.; El-Nakhel, C.; Giordano, M.; Pannico, A.; Cozzolino, E.; Mori, M.; Reynaud, H.; et al. Morphological and physiological responses induced by protein hydrolysate-based biostimulant and nitrogen rates in greenhouse spinach. *Agronomy* **2019**, *9*, 450. [CrossRef]
23. Caruso, G.; Stoleru, V.; De Pascale, S.; Cozzolino, E.; Pannico, A.; Giordano, M.; Teliban, G.; Cuciniello, A.; Rouphael, Y. Production, leaf quality and antioxidants of perennial wall rocket as affected by crop cycle and mulching type. *Agronomy* **2019**, *9*, 194. [CrossRef]
24. Caruso, G.; De Pascale, S.; Cozzolino, E.; Cuciniello, A.; Cenvinzo, V.; Bonini, P.; Colla, G.; Rouphael, Y. Yield and nutritional quality of Vesuvian Piennolo tomato PDO as affected by farming system and biostimulant application. *Agronomy* **2019**, *9*, 505. [CrossRef]
25. Schiavon, M.; Ertani, A.; Nardi, S. Effects of an alfalfa protein hydrolysate on the gene expression and activity of enzymes of the tricarboxylic acid (TCA) cycle and nitrogen metabolism in *Zea mays* L. *J. Agric. Food Chem.* **2008**, *56*, 11800–11808. [CrossRef] [PubMed]
26. Matsumiya, Y.; Kubo, M. Soybean peptide: Novel plant growth promoting peptide from soybean. In *Soybean and Nutrition*; El-Shemy, H., Ed.; InTech Europe Publisher: Rijeka, Croatia, 2011; pp. 215–230. [CrossRef]
27. Colla, G.; Rouphael, Y.; Canaguier, R.; Svecova, E.; Cardarelli, M. Biostimulant action of a plant-derived protein hydrolysate produced through enzymatic hydrolysis. *Front. Plant Sci.* **2014**, *5*, 448. [CrossRef] [PubMed]
28. Colla, G.; Rouphael, Y.; Lucini, L.; Canaguier, R.; Stefanoni, W.; Fiorillo, A.; Cardarelli, M. Protein hydrolysate-based biostimulants: Origin, biological activity and application methods. *Acta Hortic.* **2016**, *1148*, 27–34. [CrossRef]
29. Battacharyya, D.; Babgohari, M.Z.; Rathor, P.; Prithiviraj, B. Seaweed extracts as biostimulants in horticulture. *Sci. Hortic.* **2015**, *196*, 39–48. [CrossRef]

30. Sady, W.; Smoleń, S. *The Influence of Pentakeep V and Nitrogen Fertilization on the Yield and Biological Quality of Carrot and Spinach Crop*; Final Report for Cosmo Oil Co., Ltd.; Cosmo Oil Co., Ltd.: Tokio, Japan, 2007.
31. Smoleń, S.; Sady, W. The influence of nitrogen fertilization and Pentakeep V application on contents of nitrates in carrot. *Acta Hortic. Regiotec.* **2009**, *12*, 221–223.
32. Smoleń, S.; Sady, W.; Wierzbińska, J. The effect of plant biostimulation with 'Pentakeep V' and nitrogen fertilization on yield, nitrogen metabolism and quality of spinach. *Acta Sci. Pol. Hortorum Cultus* **2010**, *9*, 25–36.
33. Paul, K.; Sorrentino, M.; Lucini, L.; Rouphael, Y.; Cardarelli, M.; Bonini, P.; Reynaud, H.; Canaguier, R.; Trtílek, M.; Panzarová, K.; et al. Understanding the biostimulant action of vegetal-derived protein hydrolysates by high-throughput plant phenotyping and metabolomics: A case study on tomato. *Front. Plant Sci.* **2019**, *10*, 47. [CrossRef] [PubMed]
34. Paul, K.; Sorrentino, M.; Lucini, L.; Rouphael, Y.; Cardarelli, M.; Bonini, P.; Miras Moreno, M.B.; Reynaud, H.; Canaguier, R.; Trtílek, M.; et al. A combined phenotypic and metabolomic approach for elucidating the biostimulant action of a plant-derived protein hydrolysate on tomato grown under limited water availability. *Front. Plant Sci.* **2019**, *10*, 493. [CrossRef] [PubMed]
35. Re, R.; Pellegrini, N.; Proteggente, A.; Pannala, A.; Yang, M.; Rice-Evans, C. Antioxidant activity applying an improved abts radical cation decolorization assay. *Free Rad. Biol. Med.* **1999**, *26*, 1231–1237. [CrossRef]
36. Fogliano, V.; Verde, V.; Randazzo, G.; Ritieni, A. Method for measuring antioxidant activity and its application to monitoring the antioxidant capacity of wines. *J. Agric. Food Chem.* **1999**, *47*, 1035–1040. [CrossRef] [PubMed]
37. Lichtenthaler, H.K.; Buschmann, C. *Chlorophylls and Carote-Noids–Measurement and Characterisation by UV-VIS.–Cur-rent Protocols in Food Analytical Chemistry (CPFA), (Sup-plement 1), F4.3.1–F4.3.8*; John Wiley: New York, NY, USA, 2001. [CrossRef]
38. Sah, R.N. Nitrate-nitrogen determination: A critical review. *Commun. Soil Sci. Plant Anal.* **1994**, *25*, 2841–2869. [CrossRef]
39. Kampfenkel, K.; Van Montagu, M.; Inzé, D. Extraction and determination of ascorbate and dehydroascorbate from plant tissue. *Anal. Biochem.* **1995**, *225*, 165–167. [CrossRef]
40. Colla, G.; Cardarelli, M.; Bonini, P.; Rouphael, Y. Foliar applications of protein hydrolysate, plant and seaweed extracts increase yield but differentially modulate fruit quality of greenhouse tomato. *HortScience* **2017**, *52*, 1214–1220. [CrossRef]
41. Trevisan, S.; Manoli, A.; Quaggiotti, S.A. Novel biostimulant, belonging to protein hydrolysates, mitigates abiotic stress effects on maize seedlings grown in hydroponics. *Agronomy* **2019**, *9*, 28. [CrossRef]
42. Szczepanek, M.; Siwik-Ziomek, A. P and K accumulation by rapeseed as affected by biostimulant under different NPK and S fertilization doses. *Agronomy* **2019**, *9*, 477. [CrossRef]
43. Kyriacou, M.C.; Rouphael, Y.; Colla, G.; Zrenner, R.M.; Schwarz, D. Vegetable grafting: The implications of a growing agronomic imperative for vegetable fruit quality and nutritive value. *Front. Plant Sci.* **2017**, *8*, 741. [CrossRef]
44. Jannin, L.; Arkoun, M.; Etienne, P.; Laîné, P.; Goux, D.; Garnica, M.; Fuentes, M.; Francisco, S.S.; Baigorri, R.; Cruz, F. *Brassica napus* growth is promoted by *Ascophyllum nodosum* (L.) Le Jol. seaweed extract: Microarray analysis and physiological characterization of N, C, and S metabolisms. *J. Plant Growth Regul.* **2013**, *32*, 31–52. [CrossRef]
45. Ertani, A.; Schiavon, M.; Nardi, S. Transcriptome-wide identification of differentially expressed genes in *Solanum Lycopersicon* L. in response to an alfalfa-protein hydrolysate using microarrays. *Front. Plant Sci.* **2017**, *8*, 1159. [CrossRef] [PubMed]
46. Tsouvaltzis, P.; Koukounaras, A.; Siomos, S.A. Application of amino acids improves lettuce crop uniformity and inhibits nitrate accumulation induced by the supplemental inorganic nitrogen fertilization. *Int. J. Agric. Biol.* **2014**, *16*, 951–955. [CrossRef]
47. Colla, G.; Kim, H.J.; Kyriacou, M.C.; Rouphael, Y. Nitrate in fruits and vegetables. *Sci. Hortic.* **2018**, *237*, 231–238. [CrossRef]
48. Ali, M.; Cheng, Z.H.; Hayat, S.; Ahmad, H.; Ghani, M.I.; LIU, T. Foliar spraying of aqueous garlic bulb extract stimulates growth and antioxidant enzyme activity in eggplant (*Solanum melongena* L.). *J. Integr. Agric.* **2019**, *18*, 1001–1013. [CrossRef]

49. Tadros, M.J.; Omari, H.J.; Turk, M.A. The morphological, physiological and biochemical responses of sweet corn to foliar application of amino acids biostimulants sprayed at three growth stages. *Aust. J. Crop Sci.* **2019**, *13*, 412–417. [CrossRef]
50. Khanam, U.K.S.; Oba, S.; Yanase, E.; Murakami, Y. Phenolic acids, flavonoids and total antioxidant capacity of selected leafy vegetables. *J. Fun. Foods* **2012**, *4*, 979–987. [CrossRef]
51. Kyriacou, M.C.; Rouphael, Y.; Di Gioia, F.; Kyratzis, A.; Serio, F.; Renna, M.; De Pascale, S.; Santamaria, P. Micro-scale food production and the rise of microgreens. *Trends Food Sci. Technol.* **2016**, *57*, 103–115. [CrossRef]
52. Kyriacou, M.C.; El-Nakhel, C.; Graziani, G.; Pannico, A.; Soteriou, G.A.; Giordano, M.; Ritieni, A.; De Pascale, S.; Rouphael, Y. Functional quality in novel food sources: Genotypic variation in the nutritive and phytochemical composition of thirteen microgreens species. *Food Chem.* **2019**, *277*, 107–118. [CrossRef]
53. Kyriacou, M.C.; Rouphael, Y. Towards a new definition of quality for fresh fruits and vegetables. *Sci. Hortic.* **2018**, *234*, 463–469. [CrossRef]
54. Rouphael, Y.; Kyriacou, M.C. Enhancing quality of fresh vegetables through salinity eustress and biofortification applications facilitated by soilless cultivation. *Front. Plant Sci.* **2018**, *9*, 1254. [CrossRef] [PubMed]
55. Wang, Z.H.; Li, S.X.; Malhi, S. Effects of fertilization and other agronomic measures on nutritional quality of crops. *J. Sci. Food Agric.* **2008**, *88*, 7–23. [CrossRef]
56. Vasantharaja, R.; Stanley Abraham, L.; Inbakandan, D.; Thirugnanasambandam, R.; Senthilvelan, T.; Ayesha Jabeen, S.K.; Prakash, P. Influence of seaweed extracts on growth, phytochemical contents and antioxidant capacity of cowpea (*Vigna unguiculata* L. Walp). *Biocatal. Agric. Biotechnol.* **2019**, *17*, 589–594. [CrossRef]

agronomy

MDPI

Article

Seed Coating with Thyme Essential Oil or *Paraburkholderia phytofirmans* PsJN Strain: Conferring Septoria Leaf Blotch Resistance and Promotion of Yield and Grain Isotopic Composition in Wheat

Maissa Ben-Jabeur [1,*], Zayneb Kthiri [1], Kalthoum Harbaoui [2], Karima Belguesmi [2], Maria Dolores Serret [3], Jose Luis Araus [3,*] and Walid Hamada [1,*]

1 Laboratory of Genetics and plant breeding, National Institute of Agronomy (INAT), 43, Av Charles Nicolle, Tunis 1082, Tunisia; zayneb.kthiri@gmail.com

2 Regional Field Crops Research Center (CRRGC), Béja 9000, Tunisia; harbaoui_kalthoum@yahoo.fr (K.H.); k.belguesmi@yahoo.com (K.B.)

3 Integrative Crop Ecophysiology Group, Plant Physiology Section, Faculty of Biology, University of Barcelona, 08028 Barcelona, Spain; dserret@ub.edu

* Correspondence: maissa.benjabeur@hotmail.com (M.B.J.); jaraus@ub.edu (J.L.A.); W_hamada@yahoo.com (W.H.); Tel.: +216-53-708-857 (M.B.J.); +34-934-021-465 (J.L.A.); +216-52-928-483 (W.H.); Fax: +216-71799391 (M.B.-J. & W.H.); +34-934-112-842 (J.L.A.)

Received: 13 July 2019; Accepted: 18 September 2019; Published: 26 September 2019

Abstract: Septoria leaf blotch (SLB) is considered one of the most devastating diseases affecting global wheat production. Biostimulant application is among the modern approaches in plant protection to overcome the impact of SLB's fungicide resistance. In this manner, the effect of coating seeds with thyme essential oil or *Paraburkholderia phytofirmans* PsJN strain on SLB severity and yield components (spikes/m^2, straw yield (SY), grain yield (GY) and thousand kernel weight (TKW)) were assessed under field conditions for 3 years. The effect on physiological traits and nitrogen and carbon isotope composition ($\delta^{15}N_{grain}$, $\delta^{13}C_{grain}$) and nitrogen and carbon content (N_{grain}, C_{grain}) of grains was assessed in one year of study. The increasing SLB severity decreased all yield components, increased $\delta^{15}N_{grain}$ and C_{grain} content and slightly decreased $\delta^{13}C_{grain}$ as the resulting effect of *Zymoseptoria tritici* inducing stomatal opening and leaf necrosis. Across the years, both treatments alleviated the SLB adverse impact by reducing SLB severity, increasing spikes/m^2, SY, GY and TKW. Both treatments ameliorated grain quality by increasing C_{grain} content and decreasing $\delta^{13}C_{grain}$ and $\delta^{15}N_{grain}$. The difference between the performance of thyme oil or PsJN strain in terms of intensity and stability is discussed and considered to be linked to the different triggered systemic resistance and the associated amount of costs deriving from resource allocation towards defense processes.

Keywords: Septoria; wheat; *Paraburkholderia phytofirmans*; thyme essential oil; isotope

1. Introduction

Globally, wheat leads all crops in terms of cultivated area and continues to be the most important food grain source for humans [1]. The high consumption of hard (or durum) wheat in some countries is associated with a decrease in wheat production resulting from ongoing climate change causing a rise of drought stress and the emergence of more aggressive pathogens [2], which leads to above-average imports to meet needs for consumption. Septoria leaf blotch (SLB), caused by the hemibiotroph *Zymoseptoria tritici*, constitutes one of the major constraints affecting durum wheat global production resulting in yield losses [3] and shriveled grains, which is undesirable for industries as they result in low flour extraction rates in milling and provide poor quality for feeding livestock [4]. Since the introduction of fungicides in the 1980s, chemical control is currently one of the main approaches used to manage SLB [3,4]. However, fungicide resistance and its associated environmental impact is now a widespread problem [5].

Biostimulants are considered as products modifying biochemical and physiological processes in plants, neutralizing the adverse impact of weather conditions and reducing the occurrence of diseases by stimulating plant growth, strengthening plant defenses and improving nutrition efficiency leading to sustainable crop yield [6]. In this context, this study's interest focused towards assessing the effect of the biostimulants thyme oil and *Paraburkholderia phytofirmans* PsJN strain against SLB severity via the seed coating technique. Our previous experiments revealed that seed coating with both agents induced seed priming associated with increased germination, the emergence of seedlings, shoot and root development, and a decreased root to shoot ratio [7]. Moreover, coating seeds with either thyme oil or *P. phytofirmans* revealed great potential in controlling SLB under controlled conditions [8]. Thyme oil and PsJN strain differed in their mode of action. Thyme oil induced systemic programmed cell death (PCD) with higher frequency of formed papillae, high peroxidases activity and H_2O_2 amount, and low catalase and phenolic compounds, indicating systemic acquired resistance (SAR), and the necrotic area was reduced to 30% with reduced pycnidial density to 1.8%. While PsJN strain encountered hyphae and condensate for biofilm formation, the induced local PCD with less frequency of formed papillae, low peroxidases activity and H_2O_2 amount, and low catalase and phenolic compounds, indicated induced systemic resistance (ISR), and the necrotic area was reduced to 10% with reduced pycnidial density to 9.4%. Despite the potential of biostimulants in achieving disease control under controlled conditions, their performance under field conditions could be less imposing. Hence, the effect of thyme essential oil and PsJN strain under field conditions on SLB severity, yield components and carbon and nitrogen stable isotope composition in durum wheat grains are examined.

2. Materials and Methods

2.1. Plant Material

A Tunisian variety of durum wheat (*Triticum turgidum* L. subsp. *Durum* (Desf) Husn.); 'Karim', known for its sensitivity to SLB, was used.

2.2. Seed Coating Treatment

Just before sowing, the seeds were coated with either thyme essential oil or *Paraburkholderia phytofirmans* PsJN strain. Thyme essential oil was extracted by hydro distillation from dried aerial parts of *Thymbra capitata* (L.) Cav. (chemotype carvacrol, voucher specimen D 1186-3), and harvested during the flowering stage from the plain of Kef (Tunisia, 36°23′ N, 8°79′ E). The obtained essential oil was distributed into 1 mL-amber-glass vials and stored at 4 °C for subsequent use. The chemical composition of the oil was investigated and carvacrol was identified as the major compound according to Ben Jabeur et al. [9]. The concentration of thyme oil was adjusted to 5 ppm before use with adding 0.5% of dimethyl sulfoxide (DMSO) as a solubilizing agent to assure the homogenous application of the essential oil. The bacterial inoculum of *P.phytofirmans* PsJN strain (provided by Pr. Ait Barka, University of Reims, France) was produced by transferring one colony to 20 mL of King's B liquid medium, incubated at 27 °C at 150 rpm for 48 h. The bacteria were collected by centrifugation at 8000 rpm for 5 min and washed and the concentration was adjusted to 10^8 CFU.mL^{-1} before use with phosphate-buffered saline (PBS) (10 mM, pH 6.5). The coating product Agicote Rouge T17 (AEGILOPS Applications, Val de Reuil, France), specific for cereal seeds, containing propane-1,2-diol (5–10%), polyethylene glycol mono(tristyrylphenyl)ether (5–10%), and 1,2-benzisothiasol3(2H)-one (0.0357%), was used [10]. The coating technique consists of preparing the appropriate volume of the coating solution mixture based on the quantity of seeds required for each experimental plot. Each 10 g of wheat seeds required 40 µL of the coating product Agicote Rouge T17 and 400 µL of either thyme oil (5 ppm) or PsJN inoculum (10^8 CFU.mL^{-1}), (400 µL of water was used as a control). Then, the coating mixture was applied progressively to wheat seeds in a continuous rotation, using a portable rotating drum apparatus (SUNCOO, Atlanta, GA, USA) with a speed of 2800 rpm, at an ambient temperature (20 ± 2 °C) until complete adhesion and absorption, to assure the homogeneous distribution of the coating mixture among the seeds. The final concentration of products per seed was 10^{-5} µL of coated thyme oil/seed and 210^4 CFU of coated PsJN strain/seed. Prior to the evaluation of the effect of coating seeds with thyme oil, the effect of the coating product was evaluated in the laboratory. The positive or negative effects of the coating product on seed germination and seedling growth were not detected and its inertness was assured.

2.3. Experimental Design for Field Trials

The experiments were conducted at the experimental station in Oued-Beja (CRGC), located in the sub-humid bioclimatic zone of Tunisia, for three years; 2015–2016 and 2017 under rainfed conditions (Table 1). The soil type of the experimental area is mostly clay loam with pH 7.2 (Table 2). A complete random block design with three replicates was used. The plots size was 1 × 3 m spaced by 1.5 m. Each plot consisted of 6 rows; with a row spacing of 0.15 m. The sowing was carried out in the first week of December at a sowing density of 350 seeds /m^2. The plants were inoculated with 10^7 spores/ml of *Z. tritici* twice. After full emergence of the third leaf and at stem elongation, a CO_2-pressurized knapsack sprayer was used. Nitrogen (ammonium nitrate) was applied at 25 kg N/ha at sowing and at the stem elongation stage.

Table 1. The climatic conditions (temperature, precipitation, humidity) of the three years in the experimental station of Oued Beja.

Climatic Factors	Precipitation (mm)			T Min (°C)			T Max (°C)			Humidity (%)		
year	2015	2016	2017	2015	2016	2017	2015	2016	2017	2015	2016	2017
October	59.2	77.5	32.0	15.03	17.5	14.76	28.92	27.10	28.85	73.5	75.4	76.5
November	39.2	108.8	60.0	10.6	14.06	9.58	24.0	20.14	21.56	72.9	86.6	84.3
December	105.6	21.4	40.8	6.89	11.44	8.04	16.28	17.83	17.16	86.2	90.0	92.8
January	136.2	65	119.2	5.18	5.12	3.46	15.81	17.06	13.31	83.5	88.7	81.1
February	189.0	39.2	96.4	5.20	6.35	4.78	13.69	17.78	17.80	87.1	86.2	76.9
March	77.3	115.6	25.6	7.48	6.52	6.37	17.75	18.63	20.72	83.4	86.1	71.7
April	5.2	23.4	42.4	4.58	6.15	7.52	23.68	24.57	22.71	72.0	78.4	69.4
May	25.0	40.4	23.4	12.9	9.57	11.44	29.07	27.86	29.71	65.9	70.7	56.3
Sum/Average *	636.7	491.3	439.8	8.48	9.58	8.24	21.15	21.37	21.47	78.06	82.76	76.18

* Sums for precipitation; average values for the rest.

Table 2. Soil's physicochemical characteristics of Oued Beja station.

pH	7.2				
Soil type	Vertosol (texture: Clay loam)				
	Composition of Soil				
Depth	Clay (%)	Loam (%)	Sand (%)	Mineral N (ppm)	Total N (%)
0–20	67.5	22.5	10	859	0.17
20–40	65	23.7	11.3	934.7	0.16

2.4. Effect of Seed Coating with PsJN Strain and Thyme Oil on Plant Physiology, Disease Control and Yield Components

At anthesis, five leaves within each plot were selected for nondestructive measurements of leaf chlorophyll content, using a portable meter (SPAD 502 plus, Minolta, UK), and stomatal conductance of the flag leaf with a leaf porometer (Decagon, Pullman, Washington, USA). In addition, the following measurements were performed for each plot at the canopy level: The canopy normalized difference vegetation index (NDVI), with a spectroradiometer (GreenSeeker@Trimble, Westminster, Colorado, USA), canopy temperature using an infrared thermometer (Fluke, Everett, Washington, USA). For disease scoring, 15 plants were sampled from each plot, all leaves were taken for assessing the vertical disease progress and estimated for severity according to Eyal et al. [11]. Since the difference in vertical disease progress upon the samples was not observed, the diseases assessment was conducted at the leaf numbered flag leaf-3, the highest leaf showing symptoms. The leaves were scanned, and the images were analyzed using ImageJ software (the National Institute of Mental Health, Bethesda, MD, USA). The extent of the necrotic area was determined, according to Stewart and McDonald [12]. Briefly, the background was removed from each image and the total leaf area and green leaf area in the pixel was calculated using color thresholding in the red-green-blue (RGB) color space as formulated: Septoria severity (%) = (total leaf area-green leaf area)/total leaf area × 100. At harvest, 1 m^2 of each plot was hand harvested, and then straw yield (SY, Mg ha^{-1}), number of spikes/m^2, thousand kernels weight (TKW, g) and grain yield (GY, Mg ha^{-1}) were measured.

2.5. Effect of Seed Coating with PsJN Strain and Thyme Oil on Total Nitrogen and Carbon Content and Stable Carbon and Nitrogen Isotope Composition

The total N and C content and the stable nitrogen isotope signature in the dry matter of the mature grains sampled from each plot of the third field trial (2017) were analyzed at the Scientific Facilities of the University of Barcelona. Approximately 1mg of each sample and reference materials were weighed into tin capsules and measured with an elemental analyzer (Flash1112EA; Thermo Finnigan, Bremen, Germany) coupled with an isotope ratio mass spectrometer (Delta CIRMS, Thermo Finnigan, Bremen, Germany) operating in continuous flow mode in order to determine the total C and N content and the stable carbon ($^{13}C/^{12}C$) and nitrogen ($^{15}N/^{14}N$) isotopes' ratios. The ratios were expressed in δ notation [13], as δ13C = ($^{13}C/^{12}C$) sample/ ($^{13}C/^{12}C$) standard −1, where sample refers to the plant material and standard to Pee Dee Belemnite (PDB) calcium carbonate, and as $\delta^{15}N$ = ($^{15}N/^{14}N$) sample/ ($^{15}N/^{14}N$) standard −1, where sample refers to plant material and standard refers to N_2 in air.

2.6. Statistical Analysis

The effects of the treatments and years and their interaction on SLB severity and yield components were determined through a two-factor (treatment × year) analysis of variance (ANOVA) with RStudio 1.1.463 (R Foundation for Statistical Computing, Vienna, Austria). The effects of the treatments on physiological traits, yield components and grain stable isotopic compositions were determined through a one-factor ANOVA (treatment). The least significant difference (LSD) test was used to assess the differences between the treatment means. The clustered Pearson correlation matrices were generated in the RStudio environment using the mean values of all traits to study the relationships between all parameters analyzed within each treatment. The data of the non-inoculated control and inoculated control were correlated (Figure 1, IC) assessing for relationship between traits in wheat-*Z. tritici* interaction. The data of the inoculated control and plants treated with PsJN strain were correlated (Figure 1, CB), and the data of the inoculated control and plants treated with thyme oil were correlated (Figure 1, CT) for extracting the potential mode of action of each treatment in conferring disease resistance and yield improvement.

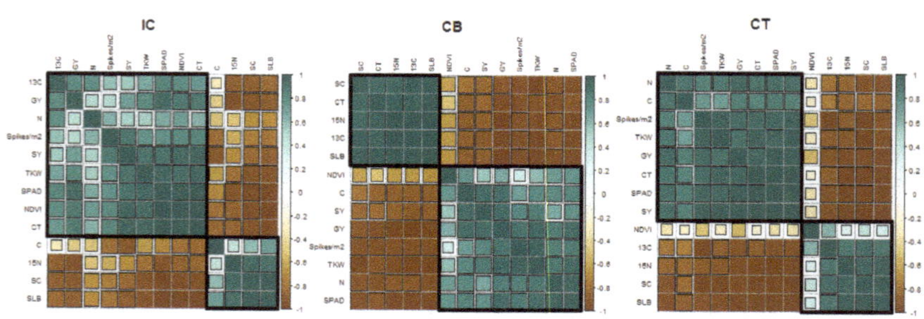

Figure 1. A correlation matrix for physiological traits, yield components and grain stable isotope composition (2017 year of study). Treatments; IC: inoculated control, CB: coated with PsJN strain,

CT: coated with thyme oil. Traits; 13 C: $\delta^{13}C_{grain}$, 15N: $\delta^{15}N_{grain}$, C: C_{grain}, N: N_{grain}, GY: Grain yield, SY: Straw yield, TKW: Thousand kernel weight, CT: Canopy temperature, SLB: SLB severity. The darker, bigger blue squares indicate a stronger positive correlation. The darker, bigger brown squares indicate stronger negative correlation.

3. Results

3.1. Climatic Features and Sources of Variances of 3 Years of Study

The data in Table 1 show that the experimental season 2016 is the season most favoring SLB compared to the other seasons tested. It was characterized by a higher amount of annual precipitations, lower maximal temperatures and high humidity. By contrast, the experimental season 2017, was characterized by drier weather due to a lower amount of precipitation, a higher maximal temperature and lower humidity. In fact, the analysis of variance revealed a highly significant ($p < 0.001$) effect for SLB severity (%), straw yield (SY) and grain yield (GY), and thousand kernel weight (TKW) was also significantly ($p < 0.01$) affected between the years. The effect of treatment (T) and the interaction year x treatment (Y × T) was highly significant ($p < 0.001$) for all four traits (Table 3).

3.2. Effect of SLB Severity on Wheat Yield Components in Control Plants

SLB was spotted in the non-inoculated control. Nevertheless, SLB severity was less compared to the inoculated control (Table 3). Therefore, a comparison between the inoculated control and non-inoculated control revealed that field artificial inoculation of wheat with *Z.tritici* increased SLB severity over naturally occurring levels, facilitating the study of the effect of treatment on wheat yield under infested conditions. Furthermore, SLB severity varied according to the variability in climatic conditions between the years. The highest severity occurred at the driest season 2017. SLB decreased significantly all yield components of the cultivar 'Karim' specifically and compared with the control. The grain yield reduced by 0.2, 0.3, and 0.5 Mg ha^{-1} in 2015, 2016, and 2017 respectively.

3.3. Effect of Seed Coating Treatment on SLB Severity and Yield Components

Both treatments showed a great potential in controlling SLB under field conditions (Table 3). The plants coated with thyme oil reduced SLB severity by 22%, 25.5%, and 53.2% in 2015, 2016, and 2017 respectively compared to the inoculated control. The plants coated with PsJN strain reduced SLB severity by 30%, 24%, and 48.3% in 2015, 2016, and 2017 respectively compared to the inoculated control. In season 2015, when water availability was high, PsJN strain was more efficient than thyme oil in reducing SLB severity. In seasons 2016 and 2017, when water availability decreased, thyme oil was more efficient than PsJN strain in reducing Septoria severity. In fact, a significant treatment by year interaction was observed for SLB.

The treatment with PsJN strain increased all yield components in the 3 seasons, not only with regard to the inoculated control but also compared with the non-inoculate control (Table 3), and the increased intensity varied among the years, most likely due to environmental factors. Contrastingly, thyme oil increased TKW compared to the inoculated control and decreased it compared to the non-inoculated control in all seasons. Furthermore, thyme oil had different effects on GY and SY among the 3 years. In 2015, in which the rainfall was more abundant in the vegetative growth stage (December–February) than the grain filling stage (April), thyme oil increased SY and decreased GY. In 2016, in which rainfall was limited in the vegetative growth stage (December-February) and abundant at the heading and anthesis (March), thyme oil decreased SY and increased GY. In 2017, in which rainfall was abundant in both vegetative growth stage (December–February) and grain filling stage (April), thyme oil increased both SY and GY.

Table 3. Effect of treatments on SLB severity and yield components of durum wheat evaluated in three-year-study.

	SLB Severity (%)				SY (Mg ha^{-1})				GY (Mg ha^{-1})				TKW (g)			
Treatment	NIC	IC	CB	CT	NIC	IC	CB	CT	NIC	IC	CB	CT	NIC	IC	CB	CT
Year																
2015	36.0^{cd}	50.0^{ab}	20.0^{ef}	28.0^{de}	8.7^{c}	6.0^{ef}	13.3^{a}	13.9^{a}	1.8^{ef}	1.6^{fg}	2.0^{de}	1.6^{g}	49.8^{cd}	37.4^{f}	59.1^{a}	48.7^{d}
2016	15.0^{fg}	34.5^{cd}	10.6^{fg}	9.0^{fg}	8.1^{cd}	7.1^{de}	11.4^{b}	6.6^{ef}	2.4^{c}	2.1^{d}	3.0^{a}	2.5^{bc}	51.3^{cd}	48.7^{d}	57.8^{a}	50.2^{cd}
2017	43.5^{bc}	59.1^{a}	10.8^{fg}	5.9^{g}	4.5^{gh}	3.7^{h}	5.7^{fg}	6.1^{ef}	1.7^{fg}	1.2^{h}	2.4^{c}	2.7^{ab}	53.0^{bc}	44.4^{e}	55.6^{ab}	53.2^{bc}
LSD																
2015	16.46	18.85	10.54	11.35	1.47	0.60	0.90	0.98	0.06	0.07	0.02	0.05	6.50	2.46	1.64	2.10
2016	6.66	18.62	2.83	2.10	0.09	0.44	1.21	0.55	0.04	0.02	0.26	0.11	0.20	0.25	3.01	0.19
2017	10.04	25.83	6.80	2.95	0.30	0.20	0.95	0.08	0.12	0.14	0.15	0.31	0.40	2.66	0.66	1.40
ANOVA																
Treatment (T)			45.348 ***				58.35 ***				55.38 ***				43.929 ***	
Year (Y)			16.629 ***				146.48 ***				80.97 ***				5.766 **	
Interaction (T × Y)			4.879 ***				20.60 ***				18.47 ***				6.371 ***	

The F values are shown, and the symbols indicate statistical significance (**, $p < 0.01$; ***, $p < 0.001$), values with different superscript letters are significantly different classes according to the LSD test ($p \leq 0.05$).LSD: least significant difference; SLB: Septoria leaf blotch; SY: Straw yield; GY: Grain yield; TKW: Thousand Kernels weigh; NIC: non-inoculated control; IC: inoculated control; CB: coated with PsJN strain; CT: coated with thyme oil.

3.4. Effect of Seed CoatingTreatment and SLB Severity on Physiological Traits, Yield Components and Grain Isotopic Composition

3.4.1. Effect of SLB in Control Plants

On the control plants inoculated with *Z.tritici*, during vegetative growth, the green leaf area was reduced compared with the other treatments (Figure 2), as shown by the reduction in the canopy vegetation index NDVI, and the decrease in leaf chlorophyll content (SPAD), while stomatal conductance increased and the carbon isotope composition ($\delta^{13}C$) of the grains slightly decreased. At harvest, SLB severity caused a reduction in GY and biomass as well as in the yield components spikes/m^2 and TKW and altered the grain composition by increasing C_{grain} content and $\delta^{15}N_{grain}$ (Table 4). SLB had no effect on N_{grain} content. The behavior of *Z.tritici*, the effect of SLB on the wheat physiological state, and the impact on yield components and grain composition was confirmed by the negative correlation between traits in cluster 1: SPAD, NDVI, spikes/m^2, GY, SY, TKW, canopy temperature, $\delta^{13}C_{grain}$ and the traits in cluster 2: SLB severity, stomatal conductance, C_{grain} content, $\delta^{15}N_{grain}$ (Figure 1, IC).

3.4.2. Effect of Seed Coating with PsJN Strain

Disease resistance was observed and characterized by a higher green leaf area (Figure 2) and SPAD values, and lower SLB severity and stomatal conductance compared to the inoculated control (Table 4). The plant growth promoting effect of coating seeds with PsJN strain was remarkably observed from (i) an increase in SPAD, and NDVI in the vegetative growth phase and increase in SY, GY, TKW at harvest (Table 4), and (ii) the positive correlation among SY, GY, SPAD, spikes/m^2, TKW, NDVI (Figure 1, CB, cluster 1). Concerning grain composition, the coating with PsJN strain and C_{grain} content was positively correlated to SY, GY, SPAD, spikes/m^2, TKW, NDVI (Figure 1, CB, cluster 1), and decreased $\delta^{15}N_{grain}$ and $\delta^{13}C_{grain}$, which is most likely related to a lower canopy temperature, stomatal conductance, and SLB severity compared to the inoculated control (Figure 1, CB, cluster 2). No effect was observed on N_{grain} content.

3.4.3. Effect of Seed Coating with Thyme Oil

Disease resistance was observed and characterized by a higher green leaf area (Figure 2) and SPAD values, lower SLB severity, and a lower stomatal conductance, resulting in a higher canopy temperature compared to the inoculated control. Coating seeds with thyme oil increased GY, SY, spikes/m^2 and TKW compared to the inoculated control (Table 4). Concerning grain composition, thyme oil increased C_{grain} content which was positively correlated to GY, SY, spikes/m^2, TKW, canopy temperature and SPAD. The effect of thyme oil on decreasing $\delta^{13}C_{grain}$ and $\delta^{15}N_{grain}$ content is most likely related to an increase in stomatal conductance mediated by a lower SLB severity (Figure 1, CT, cluster 2) and NDVI was the less correlated trait. No effect was observed on N_{grain} content.

Table 4. Effect of treatments on physiological traits, yield components and grain stable isotope composition in the 2017 year.

2017	SLB Severity (%)	SPAD	NDVI	Canopy Temperature (°C)	Stomatal Conductance ($mmol.m^{-2}.s^{-1}$)	$Spikes.m^{-2}$	Straw Yield ($Mg.\ ha^{-1}$)	Grain Yield ($Mg.\ ha^{-1}$)	Thousand Kernel Weigh (g)	N_{grain} (%, g DW)	Isotopic Composition $\delta^{15}N_{grain}$ (‰)	C_{grain} (%, g DW)	Isotopic Composition $\delta^{13}C_{grain}$ (‰)
NIC	43.56^{b}	49.46^{b}	0.78^{a}	19.70^{a}	157.50^{d}	176.33^{bc}	4.50^{b}	1.75^{b}	53.06^{a}	1.79^{b}	0.77^{b}	35.87^{b}	−24.62^{a}
IC	59.19^{a}	43.20^{c}	0.75^{b}	18.00^{b}	282.36^{a}	154.00^{c}	3.78^{b}	1.21^{c}	44.44^{b}	1.71^{b}	1.34^{a}	40.81^{a}	−24.75^{b}
CB	10.82^{c}	53.20^{a}	0.76^{b}	16.99^{c}	234.53^{b}	232.00^{a}	5.77^{a}	2.48^{a}	55.69^{a}	1.97^{a}	0.05^{c}	44.22^{a}	−25.31^{d}
CT	5.94^{d}	48.63^{b}	0.74^{b}	19.43^{a}	204.80^{c}	193.66^{b}	6.18^{a}	2.77^{a}	53.20^{a}	1.96^{a}	0.62^{b}	42.49^{a}	−24.99^{c}
LSD													
NIC	1.20	0.37	0.05	0.36	12.82	14.57	0.30	0.12	1.40	0.04	0.15	4.81	0.05
IC	0.46	0.36	0.00	0.17	10.26	10.14	0.20	0.14	2.66	0.03	0.14	0.77	0.03
CB	0.58	1.99	0.01	0.15	3.29	24.97	0.95	0.15	0.66	0.07	0.01	1.40	0.02
CT	0.22	0.30	0.01	0.15	4.55	9.81	0.08	0.31	0.40	0.06	0.05	0.27	0.09
ANOVA Treatment	3872.00 ***	46.89 ***	10.03 **	95.03 ***	109.30 ***	12.58 **	14.07 **	37.08 ***	30.10 ***	16.11 ***	69.74 ***	6.039 *	79.45 ***

The F values are shown, and the symbols indicate statistical significance (*, $p < 0.05$; **, $p < 0.01$; ***, $p < 0.001$), values with different superscript letters are significantly different classes according to the LSD test ($p \leq 0.05$). LSD: Least significant difference; NIC: non-inoculated control; IC: inoculated control; CB: coated with PsJN strain; CT: coated with thyme oil.

Figure 2. Pearson's correlation matrix of seed coating treatment on SLB symptoms in 3 representative leaves of wheat (2017 year of study). NIC: non-inoculated control, IC: inoculated control, CB: coated with PsJN strain, CT: coated with thyme oil.

4. Discussion

4.1. Effect of Climate on Variability of SLB Severity, Yield Components among the 3 Years

Despite the less favoring conditions for disease development in the dry season 2017, SLB severity was the highest. In fact, one of the fundamental concepts in plant pathology illustrates that plant disease occurrence requires a three-way interaction of a susceptible host, a virulent pathogen and an environment suitable for disease development, which is referred to as the disease triangle [14].The drought and temperature stresses, associated with climatic change as well as anthropogenic air pollutants as is the case of elevated O_3 levels, have the potential to: (i) Accelerate tissue necrosis favoring infection by necrotrophic pathogens, drawing nutrients from dead host tissues; (ii) reduce the major plant defense processes against pathogens due to reduced photosynthate production and the activation of the ABA-responsive signaling pathway [15,16]. SLB significantly decreased straw yield, grain yield and the yield components of the cultivar 'Karim' specifically and compared with the control in the three years of study, which agrees with the sensitive attitude of this cultivar reported [17].

4.2. Effect of SLB on PhysiologicalTraits, Yield Components, and Stable Isotopic Composition

SLB was spotted in the non-inoculated control due to the natural aerial epidemics in the experimental station zone considered as a hot spot for SLB [17]. In season 2017, the green status of plants (SPAD and NDVI) decreased with the increasing SLB severity as expected since symptoms of SLB involve chlorotic and necrotic lesions in leaves, thus reducing the green leaf area. Furthermore, SLB caused a decrease in canopy temperature (CT) and an increase in stomatal conductance (SC). This constitutes a part of *Z.tritici* hemibiotrophic behavior causing early malfunction of stomatal regulation through the stimulation of a stomatal opening leading to an increase in the transpiration rate and energy dissipation, and the subsequent decline of canopy temperature [18]. All these metabolic modifications provoked by SLB are thought to contribute to the decreasing grain yield, straw yield, number of spikes/m^2, and the decreasing grain quality through the modification of TKW, $\delta^{15}N_{grain}$, C_{grain} and $\delta^{13}C_{grain}$.

Carbon content in grains is derived from photosynthetic fixation occurring during grain filling, from diffusion of CO_2 from the air into the leaves (and the non-laminar parts) through stomata and carboxylation by Rubisco, and from earlier-assimilated carbon remobilized from vegetative organs [19]. Through these enzymatic and physical processes, C3 plants discriminate against ^{13}C in favor of ^{12}C leading to lower $\delta^{13}C/\delta^{12}C$ ratio [20]. The values of the $\delta^{13}C/\delta^{12}C$ ratio in C3 plants have been shown to vary depending on the balance between CO_2 diffusive supply (stomatal conductance) and the enzymatic demand for CO_2 (net photosynthetic assimilation), which defines the intercellular versus atmospheric ratio of CO_2 (Ci/Ca) in the photosynthetic organ [19–21]. In this context, multiple mechanisms could be involved in the alteration of carbon metabolism by SLB, decreasing $\delta^{13}C_{grain}$ content and increasing C_{grain} content: (i) The induced stomatal opening by SLB results in an increase of CO_2 supply to carboxylation sites; (ii) during the long latent biotrophic period, and referred as the symptomless growth phase, the pathogen suppresses the plant defense response which consumes the carbon skeleton components resulting in an increase in the carbon reserve [22]; (iii) during the necrotrophic phase, the pathogen releases the early suppressed plant defense resulting in the accumulation of ABA responsible for increasing the carbohydrate content in leaves and for enhancing their remobilization to grains [22,23]; (iv) in the necrotrophic phase, the pathogen causes a decrease in the photosynthetic capacity associated with less chlorophyll resulting in an increase in the Ci/Ca ratio, therefore decreasing the $\delta^{13}C$ [24]. On the other hand, the nitrogen content in grains is derived from direct nitrogen assimilation from roots during grain filling and from remobilization of earlier-assimilated nitrogen from vegetative organs to developing grains [25]. The natural variation of the stable nitrogen isotopes $^{15}N/^{14}N$ assessed through the nitrogen isotope composition ($\delta^{15}N$) is linked to nitrogen sources used by the plant (NH_4^+ uptake will induce ^{15}N enrichment compared to NO_3^-), to the activity of enzymes involved in the assimilation of ammonium (glutamine synthetase, GS) or nitrate (nitrate reductase, NR), to the nature of compounds resulting from nitrogen fractionation. Proteins are generally ^{15}N enriched compared to chlorophyll, lipids, amino sugars and alkaloids [26], and to volatilization, translocation, or nitrogen recycling in the plant [25]. SLB, decreasing $\delta^{15}N_{grain}$ ($^{15}N/^{14}N$) and not influencing total N_{grain} content at the same time, suggests that SLB both increased the isotopic fraction ^{15}N and decreased the isotopic fraction ^{14}N. In this context, multiple mechanisms could be involved in the decrease of the isotopic fraction ^{14}N by SLB: (i) During the long latent biotrophic period, pathogens successfully acquire primary and secondary nitrogen sources available in the living tissues by enzymatic digestion of host cell walls, by invading neighboring cells, or by inducing nutrient leakage from the surrounding tissues [27] resulting in decreased ^{14}N leaf storage in the vegetative growth stage; (ii) at the metabolic level, *Z.tritici* causes a decrease in N assimilation and remobilization via reducing the activity of the enzymes NR, GS and GDH starting from the first phase of infection leading to decreased $^{14}N_{leaf}$ and a resulting decrease in $^{14}N_{grain}$ [28]; (iii) SLB causing chlorotic and necrotic lesions induce N retention in the diseased plant parts, thus decreasing N remobilization to grain resulting in decreased $^{14}N_{grain}$ [29]; (iv) stomatal-opening induced by *Z.tritici* can cause an increase in N compounds volatilization resulting in decreased ^{14}N leaf storage, thus a decrease in later $^{14}N_{grain}$ content [26]. Moreover, the mechanism involved in the increase of the isotopic fraction $^{15}N_{grain}$ tends to be the effect of SLB on increasing grain protein (^{15}N enriched) content as a consequence of the loss of photosynthetic leaf area and, therefore, of carbohydrate availability to the developing grain [26,30].

4.3. *Effect of PsJN Strain*

Coating seeds with PsJN strain showed a great potential for controlling SLB under field conditions in the three years of study and tends to be the most stable treatment by increasing all yield components (GY, SY, spikes/$m^2$and TKW) despite the different climatic conditions. Disease resistance was associated to the alleviation of the plant damage induced by *Z.tritici* behavior characterized by less stomata openings and enhanced chlorophyll pigmentation observed in the 2017 year of study. This could be referred to the bacterial direct effect in altering the fungal development, and the indirect effect in triggering induced systemic resistance (ISR) within the plant tissues and promoting shoot and root growth [8]. The increase in photosynthesis (SPAD) and yield components is thought to be related to the effect of PsJN strain on: (i) Inducing seed priming resulting in metabolic changes that involve phenolic compound accumulation and growth promotion of root and shoot parts starting from the seedling emergence stage [7]; (ii) decreasing the plant ethylene level by decreasing ACC levels in plants via the bacterial 1-aminocyclopropane-1-carboxylate (ACC) deaminase activity resulting in a delay of senescence and prolonged photosynthetic activity of green tissue [31]; (iii) producing the growth regulator indole 3-acetic acid (IAA) that stimulates the development of the root system, thereby increasing nutrient absorption [32].

More specifically, in a way to understand the effect of the interaction PsJN strain-*Z.tritici* on carbon and nitrogen metabolism, the total carbon content (C_{grain}) and fractionation ($\delta^{13}C$, $\delta^{15}N$) were analyzed in the grains. The effect of PsJN strain on decreasing $\delta^{13}C_{grain}$ and increasing C_{grain} content compared with the inoculated and the non-inoculate controls suggests that this effect is mostly related to its potential in improving the plant water status due to the enhanced root development conferring a higher amount of captured water [7,24]. The effect of PsJN strain on decreasing $\delta^{15}N_{grain}$ compared to the inoculated control and simultaneously not influencing total N_{grain} content suggests that PsJN strain both increases the isotopic fraction $^{14}N_{grain}$ and decrease the isotopic fraction $^{15}N_{grain}$ and could be interpreted as: (i) The enhanced N uptake and assimilation during vegetative growth and remobilization during grain filling leading to increased $^{14}N_{grain}$ [25]; (ii) the enhanced photosynthesis and water status leading to nitrogen fractionation into chlorophyll, lipids, amino sugars rather than proteins in the vegetative growth resulting in decreased $\delta^{15}N_{grain}$ [25]; (iii) and/or as the consequence of the alleviation of SLB's adverse effects.

4.4. Effect of Thyme Oil

Coating seeds with thyme oil showed a great potential in controlling SLB under field conditions in the three years of study and seems to be more efficient in controlling SLB compared to PsJN strain according to SLB severity values. The thyme oil effect on yield components tends to be dependent on climatic conditions since the latter had different effects on GY and SY among the 3 years. In 2015, in which rainfall was more abundant in the vegetative growth stage (December–February) than the grain filling stage (April), thyme oil increased SY and decreased GY. In 2016, in which rainfall was limited in the vegetative growth stage (December-February) and abundant at the heading and anthesis (March), thyme oil decreased SY and increased GY. In 2017, in which rainfall was abundant in both vegetative growth stage (December–February) and grain filling stage (April), thyme oil increased both SY and GY. This suggests that thyme oil increases the growth rate of the assimilatory organ dependent on water availability. Thyme oil seems to be ineffective in promoting grain yield when there is an interaction disease x water deficit at the grain filling stage. This is thought to be the side effect of the activation of the systemic acquired resistance SAR [8], which induces the energy allocation towards defense related mechanisms and limits energy availability towards drought-tolerance mechanisms when water deficit occurs at the grain filling stage. According to the 2017 one year of study, disease resistance was branded by the absence of the plant damage induced by *Z.tritici* behavior, resistance was characterized by less stomata opening and the absence of chlorophyll deterioration which is most likely due to thyme oil's direct effect via hampering the fungal development and indirect effect via inducing SAR within plant tissues [8].The thyme oil effect behind enhanced GY, SY, spikes/m^2 and TKW of wheat is thought to be related to both: (i) The elicitor effect inducing seed priming resulting in the metabolic changes that involve peroxidase, phenolic compounds accumulation and the growth promotion of root and shoot parts starting from seedling emergence stage [7]; (ii) the alleviation of SLB's adverse effect. Concerning grain composition, the effect of thyme oil on increasing C_{grain} content and decreasing $\delta^{13}C_{grain}$ suggests that this effect is mostly related to thyme oil's potential in improving the plant water status due to the enhanced root elongation conferring a higher water uptake [7,24]. The thyme oil effect on decreasing $\delta^{15}N_{grain}$ compared to the inoculated control and simultaneously, not influencing the total N_{grain} content suggests that thyme oil both increases the isotopic fraction $^{14}N_{grain}$ and decreases the isotopic fraction $^{15}N_{grain}$ and could be explained by: (i) The enhanced N uptake during vegetative growth as a consequence of the thyme oil priming effect on inducing intracellular acidification of plant cells [7] was found to increase N uptake [33], leading to increased ^{14}N [25]; (ii)the enhanced water status leading to nitrogen fractionation into lipids, amino sugars rather than proteins in the vegetative growth resulting in decreased $\delta^{15}N_{grain}$ [20]; (iii) and/or as the consequence of the alleviation of SLB's adverse effect.

4.5. Comparison between Treatments and Insight to Cost/Gain Balance

The effect of PsJN strain and thyme oil differed among the three years of study. Concerning their effect on crop protection against SLB, in season 2015, when water availability was high, PsJN strain was more efficient than thyme oil in reducing SLB severity. Contrastingly, in seasons 2016 and 2017, when water availability decreased, thyme oil was more efficient than PsJN strain in reducing SLB severity. It is suggested that this difference is most likely due to their different induced type of resistance. Thyme oil triggers systemic acquired resistance causing the systemic stomatal closure [8], thus preserving water content and, by the way, decreasing the drought side effects. However, PsJN strain triggers induced systemic resistance (ISR) causing local stomatal closure only in the presence of a pathogen [8], thus maintaining the normal water dissipation rate. By this way, the energy needed for SLB resistance is expected to decrease due to energy allocation towards drought-tolerance mechanisms when a water deficit occurs, as in the years 2016 and 2017.

The better impact of PsJN strain on yield components and grain composition compared to thyme oil is suggested to be related also to the distinct defence mechanisms and can be explained by the selective cost–benefit scenario of inducible defences [22]. Thyme oil is considered to trigger constitutive defence and PsJN strain is considered to trigger induced defence [8]. The plant defence is a costly business, requiring energy and resources that would otherwise be used for growth and development [22,34]. In this context, the constitutive resistance triggered by thyme oil, where the activation occurs before the onset of the disease, is considered to be a costly advantage causing higher allocation of resources. While, the induced resistance triggered by PsJN strain, where defences are only activated following pathogen attack and only at the site of infection, is considered a less pricey advantage compared to constitutive resistance [22,34].

5. Conclusions

This study revealed that economic losses in durum wheat due to increased SLB severity can result from losses in straw yield, grain yield and grain quality. Coating seeds with either thyme oil or PsJN strain showed potential in counteracting the deleterious effects of SLB and the promotion of straw yield, and grain yield and quality. The data showed that the impact of thyme oil and PsJN differed in terms of intensity and stability. Further, it is considered to be linked to the different growth promoting effects and the different triggered systemic resistance and associated amount of costs deriving from resource allocation towards defense processes. This cost-benefit of induced resistance in the variety 'Karim' of durum wheat gives insight into the worth of studying the effects of PsJN strain or thyme oil in other varieties of wheat in order to seek better interaction which minimizes the costly effect of biostimulants.

Author Contributions: K.H. and K.B. managed and directed the field trials at the Regional Field Crops Research Center (CRRGC) at Béja, Tunisia. M.B.-J. and Z.K. conducted collection of samples and nondestructive field measurements. M.B.-J. analyzed the samples and wrote the paper under the supervision of J.L.A., M.D.S. and W.H.

Acknowledgments: This work was financially supported by the Ministry of Research and Higher Education of Tunisia [grant number LR02AGR02], ARIMNet, and the Spanish Ministry of Economy and Competitiveness (MEC) [grant number AGL2016-76527- R]. J.L.A. acknowledges the support of ICREA Academia, Government of Catalonia, Spain.

Conflicts of Interest: The authors declare that they have no conflicts of interest.

References

1. Curtis, B.C. Wheat in the World. In *Bread Wheat Improvement and Production*; Curtis, B.C., Rajaram, S., Macpherson, H.G., Eds.; Plant Prod Prot Series; FAO: Roma, Italy, 2002; Volume 30, pp. 1–18.
2. McDonald, B.A.; Mundt, C.C. How knowledge of pathogen population biology informs management of Septoria tritici blotch. *Phytopathology* **2016**, *106*, 948–955. [CrossRef] [PubMed]
3. Jørgensen, L.N.; Jahn, M.; Clark, B.; Antichi, D.; Góral, T.; Schepers, H.; Lucas, P.; Gouache, D.; Hornok, L. *Report on Best Control Practices of Diseases in Winter Wheat in 8 Countries in the EU*; ENDURE Project Report DR1.2; Endure: Aarhus, Denmark, 2008.
4. Matzen, N.; Jørgensen, J.R.; Holst, N.; Jørgensen, L.N. Grain quality in wheat—Impact of disease management. *Eur. J. Agron.* **2019**, *103*, 152–164. [CrossRef]
5. Goodwin, S.B.; Mbarek, S.B.; Dhillon, B.; Wittenberg, A.H.; Crane, C.F.; Hane, J.K.; Foster, A.J.; Van der Lee, T.A.J.; Grimwood, J.; Aerts, A.; et al. Finished genome of the fungal wheat pathogen *Mycosphaerella graminicola* reveals dispensome structure, chromosome plasticity, and stealth pathogenesis. *PLoS Genet.* **2011**, *7*, e1002070. [CrossRef] [PubMed]
6. Matysiak, K.; Miziniak, W.; Kaczmare, S.; Kierzek, R. Herbicides with natural and synthetic biostimulants in spring wheat. *Ciênc. Rural* **2018**, *48*, e20180405. [CrossRef]
7. Ben-Jabeur, M.; Djebali, N.; Kthiri, Z.; Ait-Barka, E.; Jacquard, C.; Hamada, W. Seed coating with *Paraburkholderia phytofirmans* PsJN strain or thyme essential oil enhances durum wheat seedling growth and modulates peroxidase activity and phenolics content. *Acta Physiol. Plant.* Under review.
8. Ben-Jabeur, M.; Kthiri, Z.; Somai-Jammali, L.; Ait-Barka, E.; Jacquard, C.; Hamada, W. Triggering Systemic Resistance by seed coating with thyme oil and *Paraburkholderia phytofirmans* PsJN strain against Septoria leaf blotch of durum wheat. *J. Plant Dis. Prot.*. Under review.
9. Francl, L.J. The Disease Triangle: A plant pathological paradigm revisited. *Plant Health Instr.* **2001**. [CrossRef]

10. Gouache, D.; Bensdoun, A.; Brun, F.; Oage, C.; Makowski, D.; Wallach, D. Modelling climate change impact on Septoria tritici Blotch (STB) in France: Accounting for climate model and disease model uncertainty. *Agric. Meteorol.* **2013**, *170*, 242–252. [CrossRef]
11. Eastburn, D.M.; McElrone, A.J.; Bilgin, D.D. Influence of atmospheric and climatic change on plant-pathogen interactions. *Plant Pathol.* **2011**, *60*, 54–69. [CrossRef]
12. Ferjaoui, S.; MBarek, S.; Bahri, B.; Slimane, R.; Hamza, S. Identification of resistance sources to septoria tritici blotch in old Tunisian durum wheat germplasm applied for the analysis of the *Zymoseptoria tritici*-durum wheat interaction. *J. Plant Pathol.* **2015**, *97*, 471–481.
13. Oerke, E.C.; Steiner, U. Potential of Digital Thermography for Disease Control. In *Precision Crop Protection—The Challenge and Use of Heterogeneity*; Oerke, E.C., Gerhards, R., Menz, G., Sikora, R., Eds.; Springer: Dordrecht, The Netherlands, 2010; Volume 5, pp. 167–182.
14. Ghashghaie, J.; Tcherkez, G. Isotope Ratio Mass Spectrometry Technique to Follow Plant Metabolism: Principles and Applications of 12C/13C Isotopes. In *Advances in Botanical Research: Metabolomics Coming of Age with its Technological Diversity*; Rolin, D., Ed.; Academic Press Elsevier: London, UK, 2013; Volume 67, pp. 377–405.
15. Farquhar, G.D.; O'Leary, M.H.; Berry, J.A. On the relationship between carbon isotope discrimination and the intercellular carbon dioxide concentration in leaves. *Aust. J. Plant Physiol.* **1982**, *9*, 121–137. [CrossRef]
16. Farquhar, G.D.; Ehleringer, J.R.; Hubick, K.T. Carbon isotope discrimination and photosynthesis. *Annu. Rev. Plant Physiol. Plant Mol. Biol.* **1989**, *40*, 503–537. [CrossRef]
17. Heil, M.; Baldwin, I.T. Fitness costs of induced resistance: Emerging experimental support for a slippery concept. *Trends Plant Sci.* **2002**, *7*, 61–67. [CrossRef]
18. Travaglia, C.; Cohen, A.C.; Reinoso, H.; Castillo, C.; Bottini, R. Exogenous abscisic acid increases carbohydrate accumulation and redistribution to the grains in wheat grown under field conditions of soil water restriction. *J. Plant Growth Regul.* **2007**, *26*, 285–289. [CrossRef]
19. Araus, J.L.; Amaro, T.; Zuhair, Y.; Nachit, M.M. Effect of leaf structure and water status on carbon isotope discrimination in field-grown durum wheat. *Plant Cell Environ.* **1997**, *20*, 1484–1494. [CrossRef]
20. Sanchez-Bragado, R.; Serret, M.D.; Araus, J.L. The nitrogen contribution of different plant parts to wheat grains: Exploring genotype, water, and nitrogen effects. *Front. Plant Sci.* **2017**, *7*, 19–86. [CrossRef] [PubMed]
21. Marshall, J.D.; Brooks, J.R.; Lajtha, K. Sources of variation in the stable isotopic composition of plants. In *Stable Isotopes in Ecology and Environmental Science*, 2nd ed.; Michener, R., Lajtha, K., Eds.; Wiley-Blackwell: Malden, MA, USA, 2007; Volume 2, pp. 22–60.
22. Divon, H.H.; Fluhr, R. Nutrition acquisition strategies during fungal infection of plants. *FEMS Microbiol. Lett.* **2007**, *266*, 65–74. [CrossRef]
23. Ben Slimane, R. *Effets de la Septoriose Foliaire sur la Sénescence et les Flux D'azote Pendant le Remplissage des Grains Chez le blé Tendre*; Agronomie, AgroParisTech: Paris, França, 2010.
24. Garry, G.; Tivoli, B.; Jeuffroy, M.H.; Citharel, J. Effects of *Ascochyta* blight by *Mycosphaerella pinodes* on the translocation of carbohydrates and nitrogenous compounds from the leaf and hull to the seed of dried-pea. *Plant Pathol.* **1996**, *45*, 769–777. [CrossRef]
25. McKendry, A.L.; Henke, G.E.; Finney, P.L. Effects of Septoria leaf blotch on soft red winter wheat milling and baking quality. *Cereal Chem.* **1995**, *72*, 142–146.
26. Glick, B.R. Bacteria with ACC deaminase can promote plant growth and help to feed the world. *Microbiol. Res.* **2014**, *169*, 30–39. [CrossRef]
27. Hirel, B.; Tétu, T.; Lea, P.J.; Dubois, F. Improving nitrogen use efficiency in crops for a sustainable agriculture. *Sustainability* **2011**, *3*, 1452–1485. [CrossRef]
28. Vessey, J.K.; Henry, L.T.; Chaillou, S.; Raper, C.D. Root-zone acidity affects relative uptake of nitrate and ammonium from mixed nitrogen sources. *J. Plant Nutr.* **1990**, *13*, 95–116. [CrossRef] [PubMed]
29. Boots, M.; Best, A. The evolution of constitutive and induced defences to infectious disease. *Proc. R. Soc. B* **2018**, *285*, 20180658. [CrossRef] [PubMed]
30. Ben Jabeur, M.; Ghabri, E.; Myriam, M.; Hamada, W. Thyme essential oil as a defense inducer of tomato against gray mold and Fusarium wilt. *Plant Phys. Biochem.* **2015**, *94*, 35–40. [CrossRef] [PubMed]
31. Eyal, Z.; Scharen, A.L.; Prescott, J.M.; van Ginkel, M. *The Septoria Diseases of Wheat: Concepts and Methods of Disease Management*; CIMMYT: El Batan, Mexico, 1987.

32. Stewart, E.L.; McDonald, B.A. Measuring quantitative virulence in the wheat pathogen *Zymoseptoria tritici* using high-throughput automated image analysis. *Phytopathology* **2014**, *104*, 985–992. [CrossRef] [PubMed]
33. Coplen, T.B. Explanatory Glossary of Terms Used in Expression of Relative Isotope Ratios and Gas Ratios. IUPAC Provisional Recommendations. Inorganic Chemistry Division, Comission on Isotopic Abundances and Atomic Weights. Available online: http://old.iupac.org/reports/provisional/abstract08/coplen_prs.pdf (accessed on 25 November 2017).
34. *Fiche de Données de Sécurité: Agicote Rouge Cereals*; Quick-FDS [18207-41570-12577-010076]; Aegilops Applications: Val de Reuil, France, 5 December 2017; Available online: http://www.utdf.coop/alfresco/service/api/node/content/workspace/SpacesStore/ec008e4e-1c48-4401-9a42-6df5cffcd4cb/101098.pdf?guest=true (accessed on 16 February 2018).

agronomy

Article

Biostimulants for Plant Growth Promotion and Sustainable Management of Phytoparasitic Nematodes in Vegetable Crops

Trifone D'Addabbo [1,*], Sebastiano Laquale [2], Michele Perniola [3] and Vincenzo Candido [2]

[1] Institute for Sustainable Plant Protection, CNR, Via Giovanni Amendola 122/d, 70126 Bari, Italy
[2] School of Agricultural, Forest, Food and Environmental Sciences–UNIBAS, Viale dell'Ateneo Lucano 10, 85100 Potenza, Italy; sebastiano.laquale@unibas.it (S.L.); vincenzo.candido@unibas.it (V.C.)
[3] Department of European and Mediterranean Cultures, UNIBAS, Via Lanera 20, 75100 Matera, Italy; michele.perniola@unibas.it
* Correspondence: trifone.daddabbo@ipsp.cnr.it; Tel.: +39-080-592-9233

Received: 28 August 2019; Accepted: 4 October 2019; Published: 7 October 2019

Abstract: The parasitism of root-knot nematodes, *Meloidogyne* spp., can cause heavy yield losses to vegetable crops. Plant biostimulants are often reported for a side-suppressive effect on these pests and many commercial products are increasingly included in sustainable nematode control strategies. Source materials of most biostimulants derived from plant or seaweed raw materials were documented for a reliable suppression of root-knot nematode species, whereas the suppressiveness of microbial biostimulants was found largely variable, as related to the crop and to environmental factors. Chitosan-based biostimulants were also stated for a variable phytonematode suppression, though clearly demonstrated only by a few number of studies. In a preliminary experimental case study, four commercial biostimulants based on quillay extract (QE), sesame oil (SO), seaweeds (SE), or neem seed cake (NC) were comparatively investigated for their effects against the root-knot nematode *M. incognita* on potted tomato. Soil treatments with all the four biostimulants resulted in a significant reduction of nematode eggs and galls on tomato roots, though NC and SO were significantly more suppressive than QE or SE. In addition, almost all biostimulant treatments also resulted in a significant improvement of tomato growth compared to the non-treated control. These preliminary results seem to confirm the literature data and clearly indicate the potential role of biostimulants for a safe nematode management both in organic and integrated crop systems.

Keywords: biostimulants; phytoparasitic nematodes; suppressiveness; sustainable management

1. Introduction

Phytoparasitic nematodes are among the most harmful pests of vegetable crops, responsible for an annual yield loss amounting to 9–15% of the world crop yield [1]. Most of these losses are due to root-knot nematode species, *Meloidogyne* spp., causing poor plant growth and reduced crop yield and quality and reducing plant resistance to other biotic and abiotic stresses [2]. Traditionally, control of these pests relied on soil treatments with synthetic nematicides, but the increasing demand for a higher crop safety to the environment and humans has led to a progressive dismission of these products, giving a strong impulse to the search and the implementation of control strategies based on natural mechanisms, such as the use of plant biostimulants [3].

Plant biostimulants derived from natural materials have been receiving a growing interest by researchers, farmers, and industrial companies, as considered an effective tool for improving crop productivity [4]. The previous unclear and misunderstanding legislation frame led to include among the biostimulants a large variety of products with different activities, such as growth enhancers, plant strengtheners or conditioners, resistance elicitors, as well to registration procedures variable among countries or even within the same country [5]. The uncertain legislative frame resulted in the immission in the market of a large variety of biostimulants stated for a suppressiveness on phytoparasitic nematodes, because of their content of raw materials (plants, seaweeds, microorganisms, and more) widely demonstrated for an activity against phytonematode species [6–8]. However, the recent EU Regulation 2019/1009 [9] has restricted the definition of fertilizing products and biostimulants and, therefore, many of these borderline products are destined to be classified as phytochemicals, dealing with more complex and expensive registration procedures.

Because of the increasing technical and economic relevance of these products, the aim of this study is to provide a review of the main groups of nematode-suppressive plant biostimulants actually available in the market and to indicate their potential for an effective but safe nematode management by a preliminary experimental case study on the root knot nematode *M. incognita* Kofoid et White (Chitw.) on tomato.

2. The State-of-the-Art

2.1. The Market Supply

A survey of the Italian market in 2018 revealed the presence of almost 40 different commercial plant biostimulants/strengtheners declaring a side activity on phytoparasitic nematodes on their labels (Table 1). More than 50% of these commercial products were based on plant raw materials, such as extracts, seed oils or green and seed biomasses, whereas another 25% was represented by seaweed derivatives. There was only one chitosan-based formulate, whereas the remaining others were microbial formulations. Only four products were clearly described as nematotoxic and the activity of other nine formulates was related to nematode repellence, disorientation, or antifeeding effects, whereas the remaining products were generically described as enhancers of plant resistance or of unfavorable soil conditions.

Table 1. Commercial biostimulants reporting an activity against phytoparasitic nematodes available in the Italian market at December 2018.

Commercial Name	Formulation [1]	Raw Materials	Activity [2]
Aegis ™	P	Micorrhizal fungi	1, 4, 5
Alg-a-Mic ™	L	Seaweed extract	1, 4, 5, 7
Algafit ™	L	Seaweed extract	1, 4
Ascogreen ™	L	Seaweed extract	1, 4
Biofence ™	P	Brassica meal	1, 3, 5, 6
Biofence 10 ™	P	Brassica meal	1, 3, 5, 6
Biofence FL ™	L	Brassica extract	1, 2, 4, 6
Bioki ™	p	Neem oil	1, 3, 7
Cogisin ™	L	Plant extracts	1, 2, 4, 7
Ecoessen NP ™	P	Bone meal, neem cake	1, 3, 6
Ekoprop Nemax ™	P	Mycorrhizal fungi	1, 2, 4
Ergo Bio ™	L	Humic and fulvic acids	1, 3, 4, 5, 8
Ergon ™	L	Seaweed extract	1, 4
Fertineem ™	L	Neem oil	1, 4
Force 4 ™	L	Seaweed extract	1, 2, 4, 5
Hunter ™	L	Plant extracts	1, 4
Ilsaneem ™	P	Neem cake	1, 2, 3, 7
Kendal Nem ™	L	Plant extracts	1, 2, 4, 6
Keos Guardian ™	L	Chitosan	1, 5
Micofort ™	P	Micorrhizal fungi	1, 2, 4, 5
Micosat F ™	P	Micorrhizal fungi	1, 4, 5
Micosat Jolly ™	P	Micorrhizal fungi	1, 4, 5
Mychodeep ™	P	Micorrhizal fungi	1, 2, 4

Table 1. *Cont.*

Commercial Name	Formulation [1]	Raw Materials	Activity [2]
Neem Soil ™	**P**	**Neem cake**	**1, 3, 4, 6**
Neem Care FL ™	L	Plant extracts	1, 2, 4
Nema 300 WW ™	L	Plant oils	1, 2, 4
Nemaforce ™	L	Humic and fulvic acids, plant extracts	1, 2, 4, 5, 7
Nematec ™	**L**	**Seaweed extract**	**1, 2, 4**
Nematiller ™	L	Plant extracts	1, 2, 4
Nematon EC ™	L	Sesame oil	1, 4
NeMax ™	**L**	**Sesame oil**	**1, 4**
Nutrich ™	P	Neem and pongamia cake	1, 2, 3, 4
Propoli oleoso ™	L	Propolis oil	1, 4
Rigenera Active	L	Seaweed macerate, plant extracts	1, 2, 4
Sesamin EC ™	L	Sesame oil	2, 4, 5
Tagete ™	L	Tagetes extract	1, 2, 4, 8
Tequil Multi ™	**L**	**Quillay and yucca extracts**	**1, 2, 4, 8**
Tyson ™	L	Propolis oil	1, 4, 8
Xedaneem ™	P	Neem cake	1, 6

[1] L = liquid; D = dry meals, P = pellets, G = granules; [2] 1 = biostimulant 2 = rooting; 3 = fertilizing; 4 = plant defense enhancement; 5 = increase of soil beneficial microflora; 6 = creation of a nematode-unfavorable environment; 7 = repellence, antifeeding, disorientation; 8 = toxicity. Products applied in the case study are reported in bold.

2.2. The Literature Review

Plant-derived biostimulants previously documented for an activity on phytonematodes were mostly liquid formulations of extracts and oils or, at a less instance, granular or powder seed meal or cake derivatives. A large number of plant biostimulants based on sesame seed oil [10], quillay water extract [11,12], or meals from biomasses or seeds of *Brassicaceae* plants and neem [13–15] were previously demonstrated for a suppressive activity on root-knot nematode populations on field and greenhouse tomato.

Seaweed extracts were found to cause an almost complete mortality of root-knot nematode juveniles and eggs in in vitro studies [16,17], as well as formulations of the extracts from seaweed species *Ascophyllum nodosum* L. and *Ecklonia maxima* Osbeck were reported for an effective control of root-knot nematodes also in soil experiments on tomato [18–20]. In addition to extract derivatives, a strong suppression of *Meloidogyne* spp. infestations on fruit or vegetable crops was described also for soil amendments with biomasses of seaweeds *Uva lactuca* L. and *Spatoglossus schroederi* Agardh (Kützing), may be due to their high content of phenolics and other bioactive compounds [21,22]. In addition to Meloidogyne species, suppressive activity of seaweed products was also detected on nematode parasites economically relevant to tropical or subtropical vegetable crops, such as *Helicotylenchus indicus* Siddiqui, *Belonolaimus longicaudatus* Rau, or *Radopholus similis* Cobb (Thorne) [23–26].

Literature studies are available also on the suppressive activity of chitosan and/or its derivatives, both alone or combined with other suppressive materials (agricultural wastes, plant compounds, biocontrol agents), either on root-knot nematodes [27–30] and other phytoparasitic species i.e., the soybean cyst nematode *Heterodera glycines* Ichinoe and the pinewood parasite *Bursaphelenchus xylophilus* (Steiner et Buhrer) Nickle [31–33].

Most of the microbial biostimulants reported as active on phytoparasitic nematodes were formulations of arbuscular mycorrhizal fungi [34,35]. Suppressiveness to root-knot nematodes of these products, either alone or combined with other microorganisms or plant extracts, was documented both in field and greenhouse [36–39]. Moreover, their activity was demonstrated also on other phytonematode parasites, such as *Nacobbus aberrans* Thorne et Allen or *Helicotylenchus multicinctus* (Cobb) Golden on field banana and greenhouse tomato, respectively [40,41]. In addition to mycorrhizal fungi, formulations of other fungal or bacterial biocontrol agents (*Trichoderma* spp., *Bacillus* spp.) or nitrogen fixers (*Azospirillum* spp., *Azotobacter* spp.) were also reported for controlling *M. incognita* in glasshouse tomato and field sunflower [42–44], or improving crop tolerance to the cyst nematode *Heterodera schachtii* Schmidt and more generically to soil phytoparasitic nematophauna [45,46].

3. An Experimental Case Study

3.1. Materials and Methods

A sandy soil (64.4% sand,18.7% silt, 16.9% clay, 0.8% organic matter, pH 7.5; 18.2% soil average humidity, 23.5% field capacity, 12.9% wilting point), artificially infested with the root-knot nematode *M. incognita* (8 eggs and juveniles mL^{-1} soil) was poured into 2.5 L clay pots. Soil was then treated with three commercial liquid biostimulants derived from quillay (*Quillaja saponaria* Molina) extract (Tequil Multi®, Fertenia) (QE), sesame (*Sesamum indicum* L.) oil (NeMax®, Sumitomo Chemical) (SO) or brown algae (*Laminaria* spp.) extract (AgriPrime Nematec®, BioAtlantis) (SE), and a granular formulation of neem (*Azadirachta indica* Juss) cake (Neem Soil®, Serbios) (NC). QE, SO and SE were applied at transplant and 15 and 30 days later at amounts corresponding to 60, 10, and 2 L ha^{-1}, respectively, whereas NC was incorporated to the soil at a 1000 kg ha^{-1} rate two weeks before transplanting. The same treatments were also provided to pots containing non-infested soil. Soil treated with the nematicide Oxamyl (OX), applied at a 10 L ha^{-1} field rate 3 days before tomato transplant and 15 days later, and non-treated soil, both infested (NT) and non-infested (NI) by *M. incognita*, were used as controls. One-month-old tomato seedlings (cv. Harvester) were transplanted in each pot, providing five replicates for each treatment in comparison.

The pots were arranged in a randomized block design in a plastic greenhouse at 25 °C, where they were maintained for 75 days, receiving a regular irrigation but no additional pesticide or fertilizer treatment. At the end of their permanence in the greenhouse, plants were uprooted and weight of green and root biomass was recorded for each plant. Root gall formation was estimated according to a 0–10 scale [47] and nematode multiplication on tomato roots was determined by extracting eggs and juveniles by the Hussey and Barker's method [48]. Data were statistically analyzed by ANOVA and treatment means were compared by the Fisher's Least Significant Difference Test at $P \leq 0.05$, using PlotIT 3.2 (Scientific Programming Enterprises, Haslett, MI) software.

3.2. Results

The number of nematode eggs and juveniles on tomato roots were always significantly lower in the soil treated with the four biostimulants or OX than in NT soil (Figure 1A). Moreover, the multiplication of *M. incognita* in pots treated with NC or SO was not statistically different from OX and significantly lower than the treatments with QE and SE. Finally, QE resulted to be significantly more suppressive than SE.

Treatments with the four biostimulants and OX also resulted in a significantly lower number of root galls in comparison with NT (Figures 1B and 2). As for nematode eggs and juveniles, the formation of galls in soil treated with NC and SO was statistically lower than QE and SE, though only NC was significantly not different from OX. No statistical difference occurred between the number of galls from QE and SE.

Tomato plant biomass in soil infested by *M. incognita*, either non-treated and treated with the biostimulants or OX, was always significantly lower than NI (Figure 3A). Green biomass from plants in soil treated with QE was significantly larger compared to all the other treatments and NT. Adversely, weight of green biomass from pots treated with the other three formulates was not significantly different from NT and statistically lower than OX.

Weight of the tomato roots from all the treatments but NC was significantly higher than the NT (Figure 3B). Moreover, QE resulted in a root biomass significantly heavier than the other three biostimulants and OX and not different from NI. Finally, SE resulted in a root growth statistically not different from OX but higher compared to NC and SO.

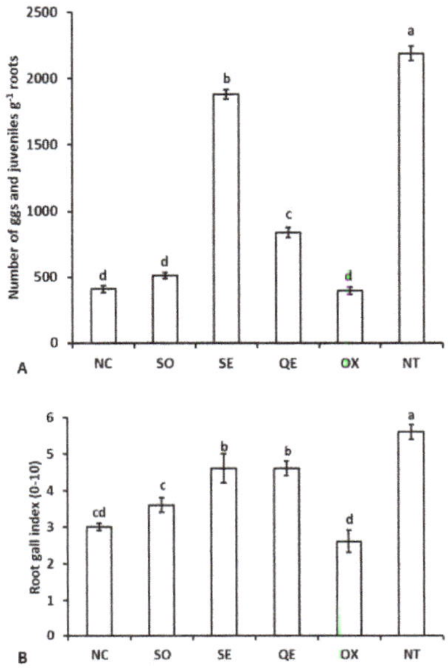

Figure 1. Multiplication of the root-knot nematode *Meloidogyne incognita* (**A**) and gall formation (**B**) on the roots of tomato cv. Harvester in soil non-treated (NT) or treated with commercial biostimulants based on neem cake (NC), sesame oil (SO), seaweed extract (SE), and quillay extract (QE) or with nematicide Oxamyl (OX). Bars tagged with the same letters are not statistically different ($P \leq 0.05$) according to the Least Significant Difference's Test.

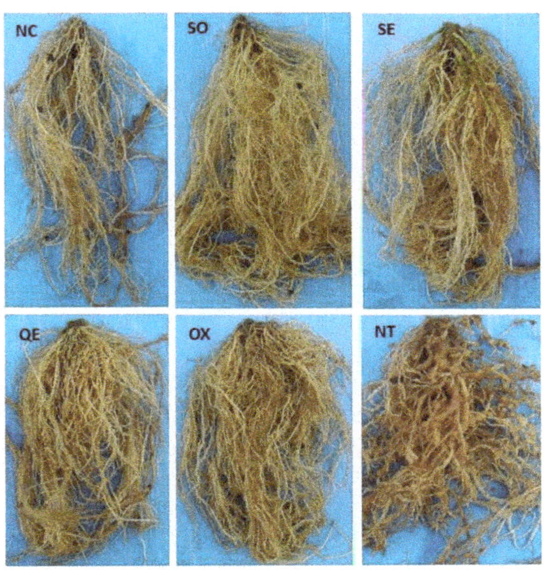

Figure 2. Roots of tomato plants cv. Harvester from soil treated with commercial biostimulants based on neem cake (NC), sesame oil (SE), seaweed extract (SE) and quillay extract (QE) or with nematicide Oxamyl (OX) and from non-treated soil (NT).

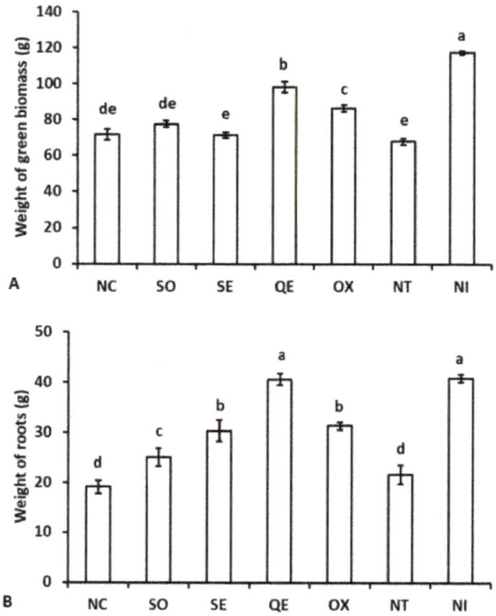

Figure 3. Weight of green biomass (**A**) and roots (**B**) of tomato plants cv. Harvester in soil non-treated (NT) or treated with commercial biostimulants based on neem cake (NC), sesame oil (SO), seaweed extract (SE), and quillay extract (QE) or with nematicide Oxamyl (OX). Bars tagged with the same letters are not statistically different ($P \leq 0.05$) according to the Least Significant Difference's Test.

4. Discussion

The experimental case study indicated that biostimulants can also provide a satisfactory nematode suppression, as confirming previous findings from literature studies. However, these results aim to be only indicative of the potential use of biostimulants in nematode management and need to be validated by future trials in field conditions, as well as different combinations of biostimulants should be also tested to verify a potential synergism among different products.

The mechanisms of biostimulants suppressiveness to nematodes are only partially known or simply hypothesized. Seaweed activity on phytoparasitic nematodes was generally attributed to their content of secondary metabolites, such as steroids, triterpenoids, alkaloids, and phenols, known for a nematicidal activity or as plant resistance elicitors [49,50]. Analogously, the suppressiveness to phytonematode populations of plant-based biostimulants is mainly related to nematotoxic metabolites both preformed in raw plant material (saponins, fatty acids, alkaloids and more) or released during the plant materials degradation in soil [51,52]. Induction of a systemic plant resistance to nematode penetration has been also documented for some active compounds of plant-derived biostimulants, such as neem azadiractin or chestnut (*Castanea sativa* Mill.) tannins [53,54]. Nematode suppression by microbial biostimulants was generally attributed to the induction of crop defense responses to nematode invasion [55,56]. Additional or alternative mechanisms, such as a competition for nutrients and space or the synthesis of nematicidal microbial metabolites have been also suggested [57–59]. The nematicidal effectiveness of chitosan products was generally attributed to the induction of a local or systemic plant resistance [60], though an enhancement of nematode-suppressive rhizospheric bacteria and fungi has been also hypothesized [36,40].

In our study, only QE was confirmed for a biostimulant effect on tomato growth, as limited only to the root biomass for SO and SE or nil for NC. The growth effect of QE can be attributed to the high content of triterpenic saponins, widely acknowledged for significant plant growth regulating properties [61], in *Q. saponaria* extracts.

Chemical composition of plant-based biostimulants can change according to a range of environmental and agronomic factors [62], as well as the nematode suppressiveness of microbial formulations may vary according to microbial strains, crop species/varieties, and environmental conditions [63]. Variable effects on soil phytonematode populations were also documented for chitosan products, as strictly dependent on the molecular weight of raw materials [32,64]. The unstable composition is a serious constraint to the full implementation of biostimulants in nematode management strategies, as leading to a fluctuating activity in field and, consequently, to a difficult certification of nematicidal performances and registration of commercial products [51]. A preliminary standardization of source raw materials and manufacturing processes should ensure constant suppressive performances and a successful market presence to the future commercial plant biostimulants addressed to nematode management. Moreover, preliminary toxicological screenings should be provided for any new biostimulant, as to exclude the presence of compounds with an unknown toxicological profile or the persistence of human pathogens in materials of animal origin [51].

In conclusion, plant biostimulants can also play a relevant role in the future nematode management strategies, as providing an acceptable nematode suppression in addition to their main activity of plant growth and ensuring a full safety to the other biotic soil components. It may be reasonably expected that the Regulation 2019/1009 [9] will lead to the disappearance of products with a direct toxicity to nematodes activity, because of the high costs of their registration as pesticides, as limiting the market to the products working through plant resistance improvement. A stand-alone application of these products can be reasonable only in organic crop systems, where few nematode control tools are available, or in short-cycle crops where the short pre-harvest intervals do not allow the use of most synthetic nematicides. However, a combination with other chemical or nonchemical control tools can justify the application of these products also in conventional crop systems. Benefit–cost ratio of treatments with the kind of products analyzed in this work should be always evaluated before their application as nematode suppressants, because of the high market price of these products which limit their use preferably to high value crops.

Author Contributions: Conceptualization, T.D., M.P., V.C.; data curation, S.L.; formal analysis, T.D., V.C.; investigation, S.L.; methodology, T.D., S.L.; software, S.L.; supervision, M.P.; validation, V.C.; visualization, S.L.; writing—original draft, T.D.; writing—review and editing, T.D., V.C.

Acknowledgments: The authors acknowledge the technical assistance of Fabio Catalano for the arrangement of greenhouse experiment and lab work.

Conflicts of Interest: The authors declare that the submitted work was carried out in the absence of any personal, professional, or financial relationships that could potentially be construed as a conflict of interest.

References

1. Nicol, J.M.; Turner, S.J.; Coyne, D.L.; Den Nijs, L.; Hockland, S.; Maafi, Z.T. Current nematode threats to world agriculture. In *Genomics and Molecular Genetics of Plant-Nematode Interactions*; Jones, J., Gheysen, G., Fenoll, C., Eds.; Springer: Dordrecht, NL, USA, 2011; pp. 21–43.
2. Sikora, R.A.; Fernandez, E. Nematode parasites of vegetables. In *Plant Parasitic Nematodes in Subtropical and Tropical Agriculture*; Luc, M., Sikora, R.A., Bridge, J., Eds.; CABI: Wallingford, UK, 2005; pp. 319–392.
3. Le Mire, G.; Nguyen, M.L.; Fassotte, B.; du Jardin, P.; Verheggen, F.; Delaplace, P.; Jijakli, H. Implementing plant biostimulants and biocontrol strategies in the agroecological management of cultivated ecosystems, A review. *Biotechnol. Agron. Soc. Environ.* **2016**, *20*, 299–313.
4. Brown, P.; Saa, S. Biostimulants in agriculture. *Front. Plant Sci.* **2015**, *6*, 671. [CrossRef] [PubMed]
5. Basak, A. Biostimulators. Definitions, classification and legislation. In *Biostimulators in Modern Agriculture: General Aspects*; Gawrońska, H., Ed.; Wie Jutra: Warsaw, Poland, 2008; pp. 7–17.
6. Askary, T.H.; Martinelli, P.R.P. *Biocontrol Agents of Phytonematodes*; CAB International: Wallingford, UK, 2015; p. 470.

7. D'Addabbo, T.; Laquale, S.; Lovelli, S.; Candido, V.; Avato, P. Biocide plants as a sustainable tool for the control of pests and pathogens in vegetable cropping systems. *Ital. J. Agron.* **2014**, *9*, 137–145. [CrossRef]
8. Oka, Y. Mechanisms of nematode suppression by organic soil amendments—A review. *Appl. Soil Ecol.* **2010**, *44*, 101–115. [CrossRef]
9. European Union. Regulation 2019/1009, 2019. (EU) of the European Parliament and of the Council of 5 June 2019 Laying Down Rules on the Making Available on the Market of EU Fertilising Products and Amending Regulations (EC) No 1069/2009 and (EC) No 1107/2009 and Repealing Regulation (EC) No 2003/2003. OJ L 170, 25.6.2019. pp. 1–114. Available online: http://data.europa.eu/eli/reg/2019/1009/oj (accessed on 6 August 2019).
10. D'Addabbo, T.; Radicci, V.; Lucarelli, G.; Carella, A.; Bernad, D.; Martin, E. Effectiveness of a formulation from *Pedaliaceae* plants (Nematon® EC) for the control of the root-knot nematode *Meloidogyne incognita* on greenhouse tomato. *Acta Hortic.* **2011**, *914*, 233–236. [CrossRef]
11. Giannakou, I.O. Efficacy of a formulated product containing *Quillaja saponaria* plant extracts for the control of root-knot nematodes. *Eur. J. Plant Pathol.* **2011**, *130*, 587–596. [CrossRef]
12. D'Addabbo, T.; Curto, G.; Greco, P.; Di Silvestro, D.; Coiro, M.I.; Lamberti, F.; Ferrari, V.; Santi, R.; Carella, A. Preliminary trials with extracts of *Quillaja saponaria* Molina for the control of root-knot nematodes. *Nematol. Medit.* **2005**, *33*, 29–34.
13. Curto, G.; Dallavalle, E.; Matteo, R.; Lazzeri, L. Biofumigant effect of new defatted seed meals against the southern root-knot nematode, *Meloidogyne incognita*. *Ann. Appl. Biol.* **2016**, *169*, 1–10. [CrossRef]
14. Rizvi, R.; Singh, G.; Safiuddin; Ali Ansari, R.; Ali Tiyagi, S.; Mahmood, I. Sustainable management of root-knot disease of tomato by neem cake and *Glomus fasciculatum*. *Cogent Food Agric.* **2015**, *1*, 1008859. [CrossRef]
15. Abbasi, P.A.; Riga, E.; Conn, K.L.; Lazarovits, G. Effect of neem cake soil amendment on reduction of damping-off severity and population densities of plant-parasitic nematodes and soilborne plant pathogens. *Can. J. Plant Pathol.* **2005**, *27*, 38–45. [CrossRef]
16. Khan, S.A.; Abid, M.; Hussain, F. Nematicidal activity of seaweeds against *Meloidogyne javanica*. *Pak. J. Nematol.* **2015**, *33*, 195–203.
17. Rizvi, M.A.; Shameel, M. In vitro nematicidal activities of seaweed extracts from Karachi coast. *Pak. J. Bot.* **2006**, *38*, 1245–1248.
18. Whapham, C.A.; Jenkins, T.; Blunden, G.; Hankins, S. The role of seaweed extracts, *Ascophyllum nodosum*, in the reduction in fecundity of *Meloidogyne javanica*. *Fundam. Appl. Nematol.* **1994**, *17*, 181–183.
19. Wu, Y.; Jenkins, T.; Blunden, G.; Whapham, C.; Hankins, S.D. The role of betaines in alkaline extracts of *Ascophyllum nodosum* in the reduction of *Meloidogyne javanica* and *M. incognita*. *Fundam. Appl. Nematol.* **1997**, *20*, 99–102.
20. Featonby-Smjth, B.C.; Van Staden, J. The effect of seaweed concentrate on the growth of tomato plants in nematode infested soil. *Sci. Hortic.* **1983**, *20*, 137–146. [CrossRef]
21. El-Ansary, M.S.M.; Hamouda, R.A. Biocontrol of root-knot nematode infected banana plants by some marine algae. *Russ. J. Mar. Biol.* **2014**, *40*, 140–146. [CrossRef]
22. Paracer, S.; Armin, C.; Tarjan, A.C.; Hodgson, L.M. Effective use of marine algal products in the management of plant parasitic nematodes. *J. Nematol.* **1987**, *19*, 194–200.
23. Naqvi, B.S.; Khan, A.; Shaikh, D.; Shaikh, M.R. Nematicidal properties of selected marine algae from Karachi coast. *J. Islam. Acad. Sci.* **1992**, *5*, 171–172.
24. De Waele, D.; McDonald, A.H.; De Waele, E. Influence of seaweed concentrate on the reproduction of *Pratylenchus zeae* (Nematoda) on maize. *Nematologica* **1988**, *34*, 71–77.
25. Morgan, K.T.; Tarjan, A.C. Management of sting nematode on centipedegrass with kelp extracts. *Proc. Flor. State Hortic. Soc.* **1980**, *93*, 97–99.
26. Tarjan, A.C. Kelp derivatives for nematode-infected citrus trees. *J. Nematol.* **1977**, *9*, 287.
27. Radwan, M.A.; Farrag, S.A.; Abu-Elamayem, M.M.; Ahmed, N.S. Extraction, characterization, and nematicidal activity of chitin and chitosan derived from shrimp shell wastes. *Biol. Fert. Soils* **2012**, *48*, 463–468. [CrossRef]
28. Escudero, N.; Lopez-Moya, F.; Ghahremani, Z.; Zavala-Gonzalez, E.A.; Alaguero-Cordovilla, A.; Ros-Ibañez, C.; Lacasa, A.; Sorribas, F.J.; Lopez-Llorca, L.V. Chitosan increases tomato root colonization by *Pochonia chlamydosporia* and their combination reduces root-knot nematode damage. *Front. Plant Sci.* **2017**, *8*, 1415. [CrossRef] [PubMed]

29. Asif, M.; Ahmad, F.; Tariq, M.; Khan, A.; Ansari, T.; Khan, F.; Siddiqui, A.M. Potential of chitosan alone and in combination with agricultural wastes against the root-knot nematode, *Meloidogyne incognita* infesting eggplant. *J. Plant Prot. Res.* **2017**, *57*, 288–295. [CrossRef]
30. Mota, L.C.; dos Santos, M.A. Chitin and chitosan on *Meloidogyne javanica* management and on chitinase activity in tomato plants. *Trop. Plant Pathol.* **2016**, *41*, 84–90. [CrossRef]
31. Liang, W.; Yu, A.; Wang, G.; Zheng, F.; Jia, J.; Xu, H. Chitosan-based nanoparticles of avermectin to control pine wood nematodes. *Int. J. Biol. Macromol.* **2018**, *112*, 258–263. [CrossRef]
32. Mwaheb, M.A.; Hussain, M.; Tian, J.; Zhang, X.; Hamid, M.I.; El-Kassim, N.A.; Hassan, G.M.; Xiang, M.; Liu, X. Synergetic suppression of soybean cyst nematodes by chitosan and *Hirsutella minnesotensis* via the assembly of the soybean rhizosphere microbial communities. *Biol. Contr.* **2017**, *115*, 86–94. [CrossRef]
33. Nunes da Silva, M.; Cardoso, A.R.; Ferreira, D.; Brito, M.; Pintado, M.E.; Vasconcelos, M.W. Chitosan as a biocontrol agent against the pinewood nematode (*Bursaphelenchus xylophilus*). *For. Pathol.* **2014**, *44*, 420–423. [CrossRef]
34. Candido, V.; Campanelli, G.; D'Addabbo, T.; Castronuovo, D.; Perniola, M.; Camele, I. Growth and yield promoting effect of artificial mycorrhization on field tomato at different irrigation regimes. *Sci. Hort* **2015**, *187*, 35–43. [CrossRef]
35. Colella, T.; Candido, V.; Campanelli, G.; Camele, I.; Battaglia, D. Effect of irrigation regimes and artificial mycorrhization on insect pest infestations and yield in tomato crop. *Phytoparasitica* **2014**, *42*, 235–246. [CrossRef]
36. Flor-Peregrín, E.; Verdejo-Lucas, S.; Talavera, M. Combined use of plant extracts and arbuscular mycorrhizal fungi to reduce root-knot nematode damage in tomato. *Biol. Agric. Hortic.* **2017**, *33*, 115–124. [CrossRef]
37. Flor-Peregrín, E.; Azcón, R.; Martos, V.; Verdejo-Lucas, S.; Talavera, M. Effects of dual inoculation of mycorrhiza and endophytic, rhizospheric or parasitic bacteria on the root-knot nematode disease of tomato. *Biocontr. Sci. Technol.* **2014**, *24*, 1122–1136. [CrossRef]
38. Affokpon, A.; Coyne, D.L.; Lawouin, L.; Tossou, C.; Agbèdè, R.D.; Coosemans, J. Effectiveness of native West African arbuscular mycorrhizal fungi in protecting vegetable crops against root-knot nematodes. *Biol. Fertil. Soils* **2011**, *47*, 207–217. [CrossRef]
39. Zhang, L.; Zhang, J.; Christie, P.; Li, X. Pre-inoculation with arbuscular mycorrhizal fungi suppresses root knot nematode (*Meloidogyne incognita*) on cucumber (*Cucumis sativus*). *Biol. Fertil. Soils* **2008**, *45*, 205. [CrossRef]
40. Marro, N.; Lax, P.; Cabello, M.; Doucet, M.E.; Becerra, A.G. Use of the arbuscular mycorrhizal fungus *Glomus intraradices* as biological control agent of the nematode *Nacobbus aberrans* parasitizing tomato. *Braz. Arch. Biol. Technol.* **2014**, *57*, 668–674. [CrossRef]
41. Jonathan, E.I.; Cannayane, I.; Samiyappan, R. Field application of biocontrol agents for the management of spiral nematode, *Helicotylenchus multicinctus*, in banana. *Nematol. Mediterr.* **2004**, *32*, 169–173.
42. Radwan, M.A.; Farrag, S.A.A.; Abu-Elamayem, M.M.; Ahmed, N.S. Biological control of the root-knot nematode, *Meloidogyne incognita* on tomato using bioproducts of microbial origin. *Appl. Soil Ecol.* **2012**, *56*, 58–62. [CrossRef]
43. El-Haddad, M.E.; Mustafa, M.I.; Selim, S.M.; El-Tayeb, T.S.; Mahgoob, A.E.; Aziz, N.H. The nematicidal effect of some bacterial biofertilizers on *Meloidogyne incognita* in sandy soil. *Braz. J. Microbiol.* **2011**, *42*, 105–113. [CrossRef]
44. Ismail, A.E.; Hasabo, S.A. Evaluation of some new Egyptian commercial biofertilizers, plant nutrients and a biocide against *Meloidogyne incognita* root knot nematode infecting sunflower. *Pak. J. Nematol.* **2000**, *18*, 39–49.
45. Biliavska, L.O.; Tsygankova, V.A.; Kozyritska, V.E.; Iutynska, G.O.; Andrusevich, Y.V.; Babich, O.A. Application of new microbial plant resistance/plant growth protection inducers for increasing Chinese cabbage plant tolerance against parasitic nematode *Heterodera schachtii* Schmidt. *Int. J. Res. Biosci.* **2016**, *5*, 64–82.
46. Khan, Z.; Tiyagi, S.A.; Mahmood, I.; Rizvi, R. Effects of N fertilization, organic matter, and biofertilizers on the growth and yield of chilli in relation to management of plant-parasitic nematodes. *Turk. J. Bot.* **2012**, *36*, 73–78.
47. Bridge, J.; Page, S.L.J. Estimation of root-knot nematode infestation levels on roots using a rating chart. *Int. J. Pest Manag.* **1980**, *26*, 296–298. [CrossRef]

48. Hussey, P.S.; Barker, K.R. A comparison of methods of collecting inocula of Meloidogyne spp., including a new technique. *Plant Dis. Rep.* **1973**, *57*, 1025–1028.
49. Chitwood, J.D. Phytochemical based strategies for nematode control. *Ann. Rev. Phytopathol.* **2002**, *40*, 21–49. [CrossRef]
50. Verkleij, F.N. Seaweed extracts in agriculture and horticulture: A review. *Biol. Agric. Hort.* **1992**, *8*, 309–324. [CrossRef]
51. Yakhin, O.I.; Lubyanov, A.A.; Yakhin, I.A.; Brown, P.H. Biostimulants in plant science: A global perspective. *Front. Plant Sci.* **2017**, *7*, 2049. [CrossRef]
52. Oliveira, R.D.; Dhingra, O.D.; Lima, A.O.; Jham, G.N.; Berhow, M.A.; Holloway, R.K.; Vaughn, S.F. Glucosinolate content and nematicidal activity of Brazilian wild mustard tissues against *Meloidogyne incognita* in tomato. *Plant Soil* **2011**, *341*, 155–164. [CrossRef]
53. Pretali, L.; Bernardo, L.; Butterfield, T.S.; Trevisan, M.; Lucini, L. Botanical and biological pesticides elicit a similar induced systemic response in tomato (*Solanum lycopersicum*) secondary metabolism. *Phytochemistry* **2016**, *130*, 56–63. [CrossRef]
54. Bargiacchi, E.; Miele, S.; Romani, A.; Campo, M. Biostimulant activity of hydrolyzable tannins from sweet chestnut (*Castanea sativa* mill.). *Acta Hortic.* **2013**, *1009*, 111–116. [CrossRef]
55. Sofo, A.; Nuzzaci, M.; Vitti, A.; Tataranni, G.; Scopa, A. Control of biotic and abiotic stresses in cultivated plants by the use of biostimulant microorganisms. In *Improvement of Crops in the Era of Climatic Changes*; Ahmad, P., Wani, M.R., Azooz, M.M., PhanTran, L.S., Eds.; Springer Science+Business Media: New York, NY, USA, 2014; pp. 107–117.
56. Vos, C.; Schouteden, N.; Van Tuinen, D.; Chatagnier, O.; Elsen, A.; De Waele, D.; Gianinazzi-Pearson, V. Mycorrhiza-induced resistance against the root–knot nematode *Meloidogyne incognita* involves priming of defense gene responses in tomato. *Soil Biol. Biochem.* **2013**, *60*, 45–54. [CrossRef]
57. Mahanty, T.; Bhattacharjee, S.; Goswami, M.; Bhattacharyya, P.; Das, B.; Ghosh, A.; Tribedi, P. Biofertilizers: A potential approach for sustainable agriculture development. *Env. Sci. Poll. Res.* **2017**, *24*, 3315–3335. [CrossRef] [PubMed]
58. Schouteden, N.; De Waele, D.; Panis, B.; Vos, C.M. Arbuscular mycorrhizal fungi for the biocontrol of plant-parasitic nematodes: A review of the mechanisms involved. *Front. Microbiol.* **2015**, *6*, 1280. [CrossRef] [PubMed]
59. Youssef, M.M.; Eissa, M.F. Biofertilizers and their role in management of plant parasitic nematodes, A review. *J. Biotechnol. Pharm. Res.* **2014**, *5*, 1–6.
60. Vasyukova, N.I.; Zinov'Eva, S.V.; Il'Inskaya, L.I.; Perekhod, E.A.; Chalenko, G.I.; Gerasimova, N.G.; Il'ina, A.V.; Varlamov, V.P.; Ozeretskovskaya, O.L. Modulation of plant resistance to diseases by water-soluble chitosan. *Appl. Biochem. Microbiol.* **2001**, *37*, 103–109. [CrossRef]
61. Reichert, C.L.; Salminen, H.; Weiss, J. Quillaja Saponin Characteristics and Functional Properties. *Ann. Rev. Food Sci. Technol.* **2019**, *10*, 43–73. [CrossRef]
62. Sharma, S.H.S.; Lyons, G.; McRoberts, C.; McCall, D.; Carmichael, E.; Andrews, F.; McCormack, R. Brown seaweed species from Strangford Lough: Compositional analyses of seaweed species and biostimulant formulations by rapid instrumental methods. *J. Appl. Phycol.* **2012**, *24*, 1141–1157.
63. Dong, L.Q.; Zhang, K.Q. Microbial control of plant-parasitic nematodes: A five-party interaction. *Plant Soil* **2006**, *288*, 31–45. [CrossRef]
64. Khalil, M.S.; Badawy, M.E. Nematicidal activity of a biopolymer chitosan at different molecular weights against root-knot nematode, *Meloidogyne incognita*. *Plant Prot. Sci.* **2012**, *48*, 170–178. [CrossRef]

agronomy

MDPI

Article

Chemical Composition of Winter Rape Seeds Depending on the Biostimulators Used

Marek Gugała [1], Anna Sikorska [2,*], Krystyna Zarzecka [1], Pavol Findura [3,4] and Urszula Malaga-Toboła [4]

1. Faculty of Agrobioengineering and Animal Husbandry, University of Natural Sciences and Humanities in Siedlce, ul. Prusa 14, 08-110 Siedlce, Poland; gugala@uph.edu.pl (M.G.); krystyna.zarzecka@uph.edu.pl (K.Z.)
2. Department of Agriculture, Vocational State School of Ignacy Mościcki in Ciechanów, ul. Narutowicza 9, 06-400 Ciechanów, Poland
3. Department of Biosystems Engineering; Slovak University of Agriculture Nitra, Tr. A. Hlinku 2, 949-76 Nitra, Slovak Republic; pavol.findura@uniag.sk
4. Faculty of Production and Power Engineering; University of Agriculture in Krakow, ul. Balicka 116B, 30-149 Krakow, Poland; Urszula.Malaga-Tobola@ur.krakow.pl

* Correspondence: anna.sikorska@puzim.edu.pl; Tel.: +48(23)672-30-75

Received: 21 October 2019; Accepted: 4 November 2019; Published: 5 November 2019

Abstract: Plant growth regulators may reduce the negative effect of environmental stress factors and can contribute to increasing the quality and quantity of the yield. The aim of the research was to determine the effect of biostimulators on the quality of seeds of three winter rape morphotypes. Three varieties of winter rape were used: Poznaniak (population variety), PX104 (hybrid variety restored with a semi-dwarf growth type) and Konkret (hybrid variety restored with a traditional growth type). The varieties were exposed to three treatments: the biostimulator Tytanit®, the biostimulator Asahi®SL and the biostimulator Silvit®, and the control with no biostimulators. Seeds were analysed for content of crude fat, total fat and crude fibres. The biostimulators reduced total protein content (on average from 0.8 to 1.75 g·kg^{-1} of d.m.) and increased the concentration of crude fat (on average from 0.71 to 1.93 g·kg^{-1} of d.m.) and crude fibre (on average from 0.15 to 0.84 g·kg^{-1} of d.m.) compared to the control. PX104 had the highest content of crude fat and total fat protein, and the lowest in crude fibre. The smallest protein content was found in seeds of the long-stem hybrid Konkret, while crude fat was lowest in the population form (Poznaniak), and crude fibre was lowest in long-stem hybrid (Konkret).

Keywords: anti-nutritional substances; fat; fibre; morphotype; protein

1. Introduction

Rapeseed (*Brassica napus* L. var. oleifera) is one of the most important oil-protein crops grown in the world. One of the many factors with a negative effect on the quantity and quality of rapeseed crops include unfavourable soil conditions and drought-related stress. Strong stress leads to damaged cell structures and disturbances in metabolism and as a result, can lead to photosynthesis and plant and metabolism disruption [1]. Rouphael and Colla [2] reports that plant biostimulators are products obtained from various organic or inorganic substances or microorganisms that improve plant growth, productivity and reduce the negative effects of environmental stress. Many authors [3–8] have shown that regulators of plant growth and development reduce the negative impact of abiotic stress factors. Petrozza et al. [9] showed that when a plant experiences stress, the biostimulator strengthens its stress tolerance mechanism.

Colla and Rouphael [10] and Rouphael et al. [11] emphasize that the use of biostimulators is increasingly becoming one of the basic elements of agricultural technology in many crop species around the world. Calvo et al. [12] forecast that the global market for biostimulants in consumption will increase by 14% per year.

According to El-Boray et al. [13], Przybysz et al. [14], Kocira et al. [15] and Zulfiqar et al. [16], preparations stimulating plant growth can be based on extracts of marine algae, free amino acids, humic compounds, effective microorganisms or phenolic compounds. Their use in plant cultivation has a positive effect on photosynthesis, regulation of water management and increasing the content of organic and inorganic compounds, which in turn, has a positive effect on the size and quality of the crop.

Grabowska et al. [17] and Kolomaznik et al. [18] stated that the effectiveness of biostimulators depends on many factors, including the correct selection of preparations, their dose, concentration and methods of application, as well as plant species and varieties and environmental factors.

The study assumes the hypothesis that the use of biostimulators may have a positive effect on the chemical composition of winter rapeseeds.

Due to few studies being available on the beneficial effects of growth bioregulators on the quality characteristics of winter rapeseed, and the wide interest in agricultural practice, research was undertaken to determine the effect of biostimulators on the chemical composition of three winter rapeseed varieties.

2. Materials and Methods

2.1. Arrangement of the Experiment and Research Location

The field experiment was carried out in 2013–2016 in three different fields at the Agricultural Experimental Station—Zawady (52°03′N; 22°33′E), belonging to the University of Natural Sciences and Humanities in Siedlce. The experiment was established in a random split-plot system in three repetitions (total number of plots $3 \times 4 = 12$, repeated in 3 successive crop rotations from 2013–2016). The surface of one plot was 21 m^{-2}. The examined factors were:

I—three varieties of winter rape: Poznaniak (population variety), PX104 (hybrid variety restored with a semi-dwarf growth type) and Konkret (hybrid variety restored with a traditional growth type).

II—four types of biostimulators:

1. control—no biostimulators;

2. biostimulator Tytanit® (active substance—titanium), applied in three doses of 0.20 dm^3 ha^{-1} in the autumn (2 October 2013, 6 October 2014, 5 October 2015) at the 4–8 leaf stage (BBCH 14–18) according to the rating of Biologische Bundesantalt, Bundessortenamt and Chemische Industrie (BBCH), in the spring (26 March 2014, 23 March 2015, 21 March 2016) after the onset of growth (BBCH 21–36), and at the budding stage—early flowering (30 April 2014, 29 April 2015,4 May 2016) (BBCH 50–61) [19];

3. biostimulator Asahi®SL (active substances: sodium orto nitrophenol, sodium para nitrophenol, sodium 5-nitroguaiacolate), applied in three doses of 0.60 $dm^3 \cdot ha^{-1}$ in the autumn (25 September 2013, 29 September 2014, 27 September 2015) at the stage of 3–5 leaves (BBCH 13–15), in the spring (26 March 2014, 23 March 2015, 21 March 2016) after the plants resumed growth (BBCH 28–30), and two weeks following the second application (10 April 2014, 7 April 2015, 4 April 2016);

4. biostimulator Silvit® (active substances: active silicon, potassium oxide, boron, zinc), applied in three doses of 0.20 $dm^3 \cdot ha^{-1}$, three weeks after emergence (2 October 2013, 6 October 2014, 5 October 2015) (BBCH 12–14), in spring (26 March 2014, 23 March 2015, 21 March 2016) after plants resumed growth (BBCH 28–30), and two weeks after the second application (10 April 2014, 7 April 2015, 4 April 2016).

The studies were carried out on soil classified according to WBR FAO (2014) [20] as the Haplic Luvisols group—sandy, belonging to a very good rye soil complex of the IVb botanical class. In the years of the experiment, the soil reaction (pH) ranged from 5.68 to 5.75. The soil was characterised by a low total nitrogen content (average from 0.80 to 0.90 g·kg^{-1}), phosphorus content (average from 0.33 to 0.55 g·kg^{-1}), potassium content (average from 0.61 to 0.67 g·kg^{-1}) calcium content (average from 0.82 to 0.85 g·kg^{-1}), magnesium content (average from 0.38 to 0.46 g·kg^{-1}) and sulphur content (average from 0.11 to 0.15 g·kg^{-1}). It has a low abundance in assimilable forms of phosphorus (average from 75.0 to 80. g·kg^{-1}) and an average assimilability of potassium (from 200.0 to 205.0 g·kg^{-1}) and magnesium (average from 59.0 to 61.0 g·kg^{-1}).

The phosphorus and potassium fertilization at the dose of 40.0 kg P·ha^{-1} and 110.0 kg K·ha^{-1} with the first dose of 40.0 kg N·ha^{-1} was used before sowing. Fertilization was used in the form of Lubofos for Rape at the dose of 600.0 kg, i.e., 21.0 kg N·ha^{-1}, 26.4 kg P·ha^{-1}, 92.1 kg K·ha^{-1}, 34.8 kg S·ha^{-1}, 1 2 kg B·ha^{-1}. Fertilization rates were supplemented by 55.9 kg·ha^{-1} of ammonium nitrate (19.0 kg N·ha^{-1}), 29.6 kg·ha^{-1} of triple superphosphate (13.6 kg P·ha^{-1}) and 29 kg·ha^{-1} of potassium salt (17.9 kg K·ha^{-1}). The second nitrogen dose of 100.0 kg·ha^{-1} was applied in spring before vegetation using ammonium nitrate at the dose of 255.5 kg·ha^{-1} and ammonium sulphate at the dose of 62.5 kg·ha^{-1}. The third dose of nitrogen 60.0 kg·ha^{-1} was applied at the beginning of budding using ammonium nitrate at the dose of 176.5 kg·ha^{-1}.

The three crops of rapeseed were harvested on 11 July 2014, 17 July 2015 and 14 July 2016, respectively.

2.2. Chemical Analysis of Seeds

The tests samples of winter rape seeds were analyzed for:

Crude fat (g·kg^{-1}of d.m.)—with the Soxhlet method, which extracted the fat with petroleum ether in a Soxhlet apparatus and determides its quantity by weight, total protein (g·kg^{-1} of d.m.) [21].

Total protein (g·kg^{-1}of d.m.)—with the Kjeldahl method where protein nitrogen was converted to ammonium sulphate with concentrated sulphuric acid in the presence of a catalyst, the solution was alkalised, distilled and titrated with hydrochloric acid-ammonia bound with boric acid, the conversion factor Nx6.25 was used, crude fibre (g·kg^{-1} of d.m.) [22].

Crude fibres (g·kg^{-1}of d.m.)—with the Wenden method consisting of the quantitative determination of organic substances insoluble during cooking in an acid solution.

2.3. Statistical Analysis

Research results were statistically analysed by ANOVA. The results of the study were statistically analysed using the analysis of variance. The significance of the sources of variation was tested by the Fischer-Snedecor "F" test, and the assessment of significance of differences at the significance level $p < 0.05$ between the compared averages used Tukey's multiple intervals. Statistical calculations were made based on our own algorithm written in Excel [23].

2.4. Weather Conditions

Climatic data from 2013–2016 was obtained from the Hydrological and Meteorological Station in Siedlce. During the years of conducting the experiment, varied weather conditions prevailed (Table 1). In the second growing season, the largest annual rainfall was recorded (average of 599.2 mm) and the smallest mean annual air temperature (average of 8.8 °C). In this period, the annual amount of rainfall was 171.7 mm higher compared to the long-term period. The last year of tests was the warmest and most dry. The annual rainfall was 43.8 mm lower than the average for the long-term period, and the average air temperature was higher by 1.3 °C compared to the average from 1996–2010. Based on the calculated Sielianinov hydrothermal coefficient, the first and last year of the study were optimal, while the growing season 2014–2015 was rather wet (K = 1.71).

Table 1. Characteristics of weather conditions in the years 2013–2016 (Poland).

Months	Rainfalls (mm)				Air Temperature (°C)			
	Multiyear Sum	Monthly Sum			Multiyear Mean	Monthly Mean		
	1996–2010	2013–2014	2014–2015	2015–2016	1996–2010	2013–2014	2014–2015	2015–2016
VIII	59.9	15.0	105.7	11.9	18.5	18.8	18.1	21.0
IX	42.3	94.3	26.3	47.1	13.5	11.7	14.1	14.5
X	24.2	32.8	3.0	37.0	7.9	9.3	8.5	6.5
XI	20.2	34.7	32.5	42.2	4.0	5.1	3.4	4.7
XII	18.6	15.4	90.4	16.5	−0.1	1.2	0.1	3.7
I	19.0	28.6	51.4	10.9	−3.2	−4.5	0.6	−4.5
II	16.0	34.0	0.7	29.0	−2.3	0.7	0.7	2.5
III	18.3	29.6	53.1	33.5	2.4	5.8	4.6	3.5
IV	33.6	45.0	30.0	28.7	8.0	9.8	8.2	9.1
V	58.3	92.7	100.2	54.8	13.5	13.5	12.3	15.1
VI	59.6	55.4	43.3	36.9	17.0	15.4	16.5	18.4
VII	57.5	10.0	62.6	35.2	19.7	20.8	18.7	19.1
VIII–VII	427.5	487.5	599.2	383.7	8.2	9.0	8.8	9.5

Sielianinovs hydrothermic coefficients *

	2013–2014	2014–2015	2015–2016
VIII	0.31	1.87	0.20
IX	2.63	0.66	1.20
X	1.01	0.22	2.15
III	1.48	4.63	3.49
IV	1.41	1.35	1.07
V	2.33	2.91	1.47
VI	1.23	0.84	0.72
VII	0.16	1.20	0.64
VIII–VII	1.32	1.71	1.37

* Index value [24]: extremely dry $k \le 0.4$, very dry $0.4 < k \le 0.7$, dry $0.7 < k \le 1.0$, rather dry $1.0 < k \le 1.3$, optimal $1.3 < k \le 1.6$, rather humid $1.6 < k \le 2.0$, humid $2.0 < k \le 2.5$, very humid $2.5 < k \le 3.0$, extremely humid $k > 3.0$.

3. Results and Discussion

3.1. The Content of Total Protein Depending on the Types of Biostimulators Used

Our own research showed that biostimulators significantly affected the reduction of total protein content in rapeseeds (Table 2). The smallest concentration was recorded on object 4, sprayed with the Silvit biostimulator. This value was lower on average by 1.75 g·kg^{-1} of d.m. compared to the control variant. Different results were obtained by Gugała et al. [25], who did not find a significant effect of the biostimulators Tytanit, Asahi SL or Silvit for the value of this feature. Similarly, Jarecki and Bobrecka-Jamro [26,27], Kozak et al. [28] and Matysiak et al. [29,30] did not prove the effect of bioregulators and foliar fertilizers containing micro- and macro-elements for the value of this feature. While Jankowski et al. [31], after a double foliar application with boron, increased the protein content in seeds by an average of 8.8 g·kg^{-1} of d.m. compared to the control object. In regards to seed protein, the present study's research showed the interaction of the types of biostimulators used in relation to the protein content in the seeds of the rapeseed morphotype varieties studied, which indicated the individual response of the rapeseed varieties to the biopreparations used (Table 2). The lowest protein was in the treatment of Konkret with Silvit and PX104 with Silvit, and in Pozniak with Tytanit. In all morphotypes, the highest protein content was recorded on the object where no natural growth stimulants were used. In the cultivar with the traditional growth type, the lowest protein content was found after the application of the Tytanit biostimulator, while in the other varieities it was after the use of the Silvit biostimulator. Equal protein content was found in the restored hybrids of the semi-dwarf type (PX104) after the application of the Asahi SL and Tytanit preparations. A similar tendency was observed in the restored hybrid with the traditional growth type.

The content of total seed protein was dependant on the genetic factor (Table 2). In our own research, the content of protein in the seeds of the studied winter rape varieties averaged from 361.37 to 373.42 g·kg^{-1} of d.m. The highest concentration was found in the semi-dwarf hybrid PX104, while in the long-stem hybrid (Konkret), it was lower on average by 12.05 g·kg^{-1} of d.m. Different results were obtained by Gugała et al. [25], who received the highest value of this feature in a hybrid with a traditional type of growth and the lowest in a semi-dwarf hybrid. Ratajczak et al. [32] did not show significant differences between heterosis morphotypes with a traditional and semi-dwarf type of growth or in the population Califorium variety.

3.2. The Content of Crude Fat Depending on the Types of Biostimulators Used

The bioregulators used in the experiment significantly influenced the increase of crude fat in winter rapeseeds (Table 2). The greatest value of this feature was noted after the use of the Asahi SL biostimulator, it was significantly smaller on the objects where Tytanit and Silvit were applied. The beneficial effect of the Asahi SL biostimulator on the fat content in seeds was also confirmed by Spychaj-Fabisiak et al. [33] and Gugała et al. [25]. Similarly, Kováčik et al. [34] confirmed that a two-fold application of the Tytanit biostimulator affected the increase of the fat content in rapeseeds compared to the control object. The lack of effect of biostimulators on the fat content in seeds has been demonstrated by Matysiak et al. [29,30]. The authors observed only a slight tendency to increase the value of this feature even by 3.9% in relation to the control object. Similarly, Szczepanek et al. [35] noted a small effect of stimulating plant preparations on this feature. Jankowski et al. [31] after using a boron-containing foliar preparation, found a significant increase in the content of crude fat only after its two applications in the BBCH50 and BBCH55 phases. Jarecki and Bobrecka-Jamro [25,26] did not prove the effect of foliar preparations containing micro- and macro-elements on the value of this feature.

The impact of the types of biostimulators used on the crude fat content in rapeseed depended on the genetic factor (as shown in Table 2). The lowest fat content in all tested cultivars was recorded on the control object. The population cultivar had the highest fat content after using the Asahi SL biostimulator, but after the application of all biopreparations in this cultivar, the differences in protein crude fat content were not statistically significant. The seeds of the restored hybrid with the traditional growth type were characterized by the highest content of crude fat after the application of Asahi SL, and under the influence of the other biostimulators, they were the same as on the control object. A similar tendency was observed in the semi-dwarf hybrid, with differences in the value of this trait on the objects with the Tytanit and Silvit biostimulator were not statistically significant.

The content of crude fat depending on the genetic factor is shown in Table 2. Our own research proved that the highest content of fat was a characteristic of the PX104 (restored hybrid with a semi-dwarf type of growth), it was significantly smaller by 17.66 g·kg^{-1} of d.m. in the long-stem hybrid (Konkret), while the smallest on average by 20.57 g·kg^{-1} was in the population form (Poznaniak). Different results of studies were obtained by Gugała et al. [25] who showed that the highest value of this feature was characteristic for a restored hybrid with a traditional type of growth, while the smallest the population (Monolit).

3.3. The Content of Crude Fibre Depending on the Types of Biostimulators Used

Natural plant preparations influenced the increase of the crude fibre content in winter oilseed rapeseeds on average from 0.15 to 0.84 g·kg^{-1} of d.m. (Table 2). The highest value of this feature was noted on object 3 with the Asahi SL biostimulator. Different results were obtained by Gugała et al. [25]. In this study, the biostimulants did not significantly alter the crude fibre content in seeds of the rapeseed cultivars (Table 2).

The content of crude fibre depending on the genetic factor is shown in Table 2. Our own studies indicate that the highest content of crude fibre was observed in the seeds of the PX104 variety, while the smallest was in the long-stem morphotype (Konkret). Different results were obtained by Gugała et al. [25], who did not find any statistical differences in the value of this feature between the studied morphotypes.

3.4. Chemical Composition Depending on Weather Conditions

The chemical composition of seeds depending on climatic conditions in the study years is shown in Table 2. In our own research, the highest content of total protein, fat and crude fibre was obtained in seeds collected in the second year of research, in which the total precipitation was 41.9 mm higher in May, and the average monthly temperature was smaller by 0.5 °C from the average multi-year. Similar results were obtained by Chmura et al. [36] and Gugała et al. [25]. According to the authors, during the period from the end of flowering to the technical maturity stage of high protein content, temperatures of 16.2 °C were maintained on average, regardless of the sum of rainfall. Mączyńska et al. [37] recorded a higher concentration of fat in colder years with a greater sum of precipitation, while it was lower in warm years. In our own studies, differences in the content of total protein and crude fat in the growing season of 2013–2014 and 2015–2016 were statistically insignificant, while the lowest content of crude fibre was found in seeds collected in the first year of research.

Table 2. Chemical composition of winter oilseed rapeseeds depending on factors of experience.

Culticars	Years			Types of Biostimulators Used				Mean
				Objects				
				1.	2.	3.	4.	
	2013–2014	2014–2015	2015–2016	Control Variant	Tytanit®	Asahi®SL	Silvit®	
Total protein [g·kg^{-1} d.m]								
Poznaniak	370.92	372.80	370.78	372.18	370.71	371.83	371.27	371.50a
Konkret	357.84	367.64	358.62	362.79	361.26	361.33	360.09	361.37b
PX104	373.13	374.33	372.81	374.11	373.62	373.49	372.46	373.42c
Mean	367.29a	371.59b	367.40a	369.69a	368.53ab	368.89bc	367.94d	
$LSD_{0.05}$ for: years—0.68; cultivars—0.68; types of biostimulators used—0.52; interaction: years x cultivars—1.18; cultivars x types of biostimulators used—0.82								
Crude fat [g·kg^{-1} d.m]								
Poznaniak	468.98	470.43	469.43	468.88	469.71	470.10	469.77	469.61a
Konkret	472.33	472.93	472.30	472.00	472.41	473.14	472.53	472.52b
PX104	489.78	490.70	490.06	488.78	489.64	492.19	490.10	490.18c
Mean	477.03a	478.02b	477.26a	476.55a	477.26be	478.48d	477.47ce	
$LSD_{0.05}$ for: years—0.35; cultivars—0.35; types of biostimulators used—0.52; interaction: years x cultivars—n.s.; cultivars x types of biostimulators used—0.5.								
Crude fiber [g·kg^{-1} d.m]								
Poznaniak	89.31	90.19	89.73	89.40	89.70	90.32	89.56	89.74a
Konkret	85.28	86.11	85.58	85.41	85.70	86.02	85.49	85.66b
PX104	69.48	70.44	70.09	69.61	70.00	70.58	69.82	70.00c
Mean	81.36a	82.25b	81.80c	81.47a	81.80b	82.31c	81.62d	
$LSD_{0.05}$ for: years—0.2; cultivars—0.2; types of biostimulators used—0.17; interaction: years x cultivars—n.s.; cultivars x types of biostimulators used—n.s.								

n.s.—non-significant differences. Different letters above the bars denote significant differences $p \leq 0.05$.

4. Conclusions

In summary, the applied biostimulators had an effect on reducing the total protein content (on average from 0.8 to 1.75 $g \cdot kg^{-1}$ of d.m.) and increasing the concentration of crude fat (on average from 0.71 to 1.93 $g \cdot kg^{-1}$ of d.m.) and crude fibre (on average from 0.15 to 0.84 $g \cdot kg^{-1}$ of d.m.) compared to the control object. The best quality of seeds was characteristic for the semi-dwarf PX104 variety. The smallest protein content was found in seeds of the long-stem hybrid Konkret, while crude fat was lowest in the population form (Poznaniak), and crude fibre was lowest in long-stem hybrid (Konkret). Diverse climatic conditions prevailing in the years of conducting the experiment influenced the chemical composition of rapeseeds. The highest content of total protein, crude fat and fibre were obtained in the second year of studies.

Author Contributions: Methodology, K.Z. and M.G.; software, M.G.; writing—original draft preparation, A.S.; review and editing—P.F. and U.M.-T.

Funding: The results of the research carried out under the research theme No. 363/S/13 were financed from the science grant granted by the Ministry of Science and Higher Education".

Conflicts of Interest: The authors declare no conflict of interest.

References

1. Kałużewicz, A.; Krzesiński, W.; Spiżewski, T.; Zaworska, A. Effect of biostimulants on several physiological characteristics and chlorophyll content in broccoli under drought stress and re-watering. *Not. Bot. Horti Agrobo.* **2017**, *45*, 197–202. [CrossRef]
2. Rouphael, Y.; Colla, G. Synergistic biostimulatory action: Designing the next generation of plant biostimulants for sustainable agriculture. *Front. Plant Sci.* **2018**, *9*, 1655. [CrossRef] [PubMed]
3. Wierzbowska, J.; Sienkiewicz, S.; Bowszys, T. Effect of growth regulators on the mineral balance in spring triticale. *J. Elem.* **2010**, *15*, 745–756. [CrossRef]
4. Marjańska-Cichoń, B.; Sapieha-Waszkiewicz, A. The effect of preparations Asahi SL and Tytanit on the growth and yielding of strawberry variety salut. *Prog. Plant Prot.* **2010**, *50*, 383–388. (In Polish)
5. Stadnik, M.J.; de Freitas, M.B. Algal polysaccharides as source of plant resistance inducers. *Trop. Plant Pathol.* **2014**, *39*, 111–118. [CrossRef]
6. Panda, D.; Pramanik, K.B.; Naya, R. Use of sea weed extracts as plant growth regulators for sustainable agriculture. *Int. J. Stress Manag.* **2012**, *3*, 404–411.
7. Sharma, S.H.S.; Fleming, C.; Selby, C.H.; Rao, J.R.; Trevor, M. Plant biostimulants: A review on the processing of macroalgae and use of extracts for crop management to reduce abiotic and biotic stresses. *J. Appl. Phycol.* **2014**, *26*, 465–490. [CrossRef]
8. Yakhin, O.I.; Lubyanov, A.A.; Yakhin, I.; Brown, P.H. Biostimulants in plant science: A global perspective. *Front. Plant Sci.* **2017**, *7*, 2049. [CrossRef]
9. Petrozza, A.; Santaniello, A.; Summerer, S.; di Tommaso, G.; di Tommaso, D.; Paparelli, E.; Piaggesi, A.; Perata, P.; Cellini, F. Physiological responses to Megafol® treatments in tomato plants under drought stress: A phenomic and molecular approach. *Sci. Hortic.* **2014**, *174*, 185–192. [CrossRef]
10. Colla, G.; Rouphael, Y. Biostimulants in horticulture. *Sci. Hortic.* **2015**, *196*, 1–2. [CrossRef]
11. Rouphael, Y.; de Micco, V.; Arena, C.; Raimondi, G.; Colla, G.; de Pascale, S. Effect of *Ecklonia maxima* seaweed extract on yield, mineral composition, gas exchange and leaf anatomy of zucchini squash grown under saline conditions. *J. Appl. Phycol.* **2017**, *29*, 459–470. [CrossRef]
12. Calvo, P.; Nelson, L.; Kloepper, J. Agricultural uses of plant biostimulants. *Plant Soil* **2014**, *383*, 3–4. [CrossRef]
13. El-Boray, M.S.; Mostafa, M.F.M.; Salem, S.E.; El-Sawwah, O.A.O. Improving yield and fruit quality of washington navel orange using foliar applications of some natural biostimulants. *Int. J. Plant Prod.* **2015**, *6*, 1317–1332. [CrossRef]
14. Przybysz, A.; Gawronska, H.; Gajc-Wolska, J. Biological mode of action of a nitrophenolates-based biostimulant: Case study. *Front. Plant Sci.* **2014**, *5*, 713. [CrossRef]

15. Kocira, A.; Świeca, M.; Kocira, S.; Złotek, U.; Jakubczyk, A. Enhancement of yield, nutritional and nutraceutical properties of two common bean cultivars following the application of seaweed extract (*Ecklonia maxima*). *Saudi J. Biol. Sci.* **2018**, *25*, 563–571. [CrossRef]
16. Zulfiqar, F.; Casadesús, A.; Brockman, H.; Munné-Bosch, S. An overview of plant-based natural biostimulants for sustainable horticulture with a particular focus on moringa leaf extracts. *Plant Sci.* 2019. [CrossRef]
17. Grabowska, A.; Kunicki, E.; Sekara, A.; Kalisz, A.; Wojciechowska, R. The effect of cultivar and biostimulant treatment on the carrot yield and its quality. *Veg. Crops Res. Bull.* **2012**, *77*, 37–48. [CrossRef]
18. Kolomaznik, K.; Pecha, J.; Friebrova, V.; Janacova, D.; Vasek, V. Diffusion of biostimulators into plant tissues. *Heat Mass Transf.* **2012**, *48*, 1505–1512. [CrossRef]
19. Meier, U.; Bleiholder, H.; Buhr, L.; Feller, C.; Hack, H.; Heß, M. The BBCH system to coding the phenological growth stages of plants-history and publications. *J. Kult.* **2009**, *61*, 41–52.
20. World Reference Base for Soil Resources. International soil classification system for naming soils and creating legends for soil. In *World Soil Resources Reports 106*; Field Experiment; Food and Agriculture Organization: Rome, Italy, 2014; Available online: http://www.fao.org (accessed on 1 November 2019).
21. PN-ISO 6492:2005, Feed—Determination of fat content. Available online: http://sklep.pkn.pl/pn-iso-6492-2005p.html (accessed on 1 November 2019).
22. PN-EN ISO 5983-2:2009, Feed—Determination of nitrogen content and calculation of total protein content. Available online: https://infostore.saiglobal.com/en-au/Standards/PN-EN-ISO-5983-2-2009-924549_SAIG_PKN_PKN_2182095/ (accessed on 1 November 2019).
23. Trętowski, J.; Wójcik, R. *Methodology of Agricultural Experiments*; Wyższa Szkoła Rolniczo-Pedagogiczna: Siedlce, Poland, 1991; pp. 331–334. (In Polish)
24. Skowera, B. Changes of hydrothermal conditions in the Polish area (1971–2010). *Fragm. Agron.* **2014**, *31*, 74–87. (In Polish)
25. Gugała, M.; Sikorska, A.; Zarzecka, K.; Kapela, K.; Mystkowska, M. The effect of biostimulators on the content of crude oil and total protein in winter oilseed rape (*Brassica napus* L.) seeds. *Acta Agric. Scand. Sect. B Soil Plant Sci.* **2019**, *69*, 121–125. [CrossRef]
26. Jarecki, W.; Bobrecka-Jamro, D. Reaction of winter rape to the extra feeding of its leaves. *Ann. UMCS Agric.* **2008**, *63*, 86–96. [CrossRef]
27. Jarecki, W.; Bobrecka-Jamro, D. The reaction of the Huzar cultivar of spring oilseed rape to the density of sowing and to leaf feeding with urea. *Rośliny Oleiste* **2011**, *32*, 117–126. (In Polish)
28. Kozak, M.; Malarz, W.; Bzowy-Wójtowicz, A.; Białkowska, A. Influence of biostimulators on yield of winter rapeseed. In Proceedings of the Prosperující Olejniny, Prague, Czech Republic, 9–10 December 2010; pp. 67–69.
29. Matysiak, K.; Adamczewski, K.; Kaczmarek, S. Response of some crops cultivated in Great Poland to application of Asahi SL. *Prog. Plant Prot.* **2011**, *51*, 1849–1857. (In Polish)
30. Matysiak, K.; Kaczmarek, S.; Kierzek, R. Effect of algae *Ecklonia maxima* (Kelpak SL) on winter oilseed rape. *Rośliny Oleiste* **2012**, *33*, 81–88. (In Polish) [CrossRef]
31. Jankowski, K.J.; Sokólski, M.; Dubis, B.; Krzebietke, S.; Żarczyński, P.; Hulanicki, P.; Hulanicki, P.S. Yield and quality of winter oilseed rape (*Brassica napus* L.) seeds in response to foliar application of boron. *Agric. Food Sci.* **2016**, *25*, 164–176. [CrossRef]
32. Ratajczak, K.; Sulewska, H.; Szymańska, G. New winter oilseed rape varieties—Seed quality and morphological traits depending on sowing date and rate. *J. Plant Prod. Sci.* **2017**, *20*, 262–272. [CrossRef]
33. Spychaj-Fabisiak, E.; Murawska, B.; Pacholczyk, Ł. Values of quality traits of oilseed rape seeds depending on the fertilization and plant density. *J. Elem.* **2011**, *16*, 115–124.
34. Kováčik, P.; Šimanský, V.; Wierzbowska, J.; Renčo, M. Impact of foliar application of biostimulator Mg-Titanit on formation of winter oilseed rape phytomass and its titanium content. *J. Elem.* **2016**, *21*, 1235–1251. [CrossRef]
35. Szczepanek, M.; Wilczewski, E.; Grzybowski, K. Response of winter oilseed rape (*Brassica napus* L.) on soil applied humus preparation and foliar potassium fertilizer. *Acta Sci. Pol. Ser. Agric.* **2016**, *15*, 85–94.

36. Chmura, K.; Dzieżyc, H.; Piotrowski, M. Influence of meteorological conditions on the fat and protein content in seed of winter oilseed rape. *Acta Agrophys.* **2016**, *23*, 163–173. (In Polish)
37. Mączyńska, A.; Krzyzińska, B.; Banachowska, J. Effect of fungicides applied at the flowering stage of winter oilseed rape on the yield quantity and quality of seeds. *Prog. Plant Prot.* **2015**, *55*, 231–236. (In Polish)

agronomy

MDPI

Article

Investigating the Impact of Biostimulants on the Row Crops Corn and Soybean Using High-Efficiency Phenotyping and Next Generation Sequencing

Nunzio Briglia [1], Angelo Petrozza [2], Frank A. Hoeberichts [3,†], Nathalie Verhoef [3] and Giovanni Povero [4,*]

1 Dipartimento delle Culture Europee e del Mediterraneo: Architettura, Ambiente, Patrimoni Culturali (DiCEM), Università degli Studi della Basilicata, via Lanera 20, 75100 Matera, Italy; nunzio.briglia@unibas.it
2 ALSIA Centro Ricerche Metapontum Agrobios, s.s. Jonica 106, km 448, 2, 75010 Metaponto, Italy; apetrozza@agrobios.it
3 NSure BV, Binnenhaven 5, 6709 PD Wageningen, The Netherlands; frank.hoeberichts@nsure.nl (F.A.H.); nathalie.verhoef@nsure.nl (N.V.)
4 Valagro SpA, Zona Industriale–Via Cagliari 1, 66041 Atessa, Chieti, Italy
* Correspondence: g.povero@valagro.com; Tel.: +39-36-6583-4540
† Current Affiliation: Keygene N.V., P.O. Box 216, 6700 AE Wageningen, The Netherlands.

Received: 18 September 2019; Accepted: 12 November 2019; Published: 16 November 2019

Abstract: Row crops represent the most important crops in terms of global cultivated area. Such crops include soybean, corn, wheat, rice, rapeseed, sunflower, and cotton. Row crops agriculture is generally an intensive system of farming used to obtain high yields by employing elevated quantities of organic and mineral fertilizers. Considering this, and the decrease in area of arable land, it becomes crucial to ensure high yield and quality using alternative strategies, such as the use of plant biostimulants. These compounds are increasingly recognized as sustainable solution to optimize nutrient uptake, crop yield, quality, and tolerance to abiotic stresses. In this work, by means of high-throughput plant phenotyping, we evaluated the effectiveness of a set of three new foliar biostimulant prototypes (coded as 52096, 52097, 52113) applied on corn and soybean at application rates 2.5 and 5 mL/L (corresponding to 1 and 2 L/ha respectively). This allowed us to select the most effective prototype (52097, commercial name "YieldOn®") in increasing digital biovolume (DB) and greener area (GGA) either in soybean (both application rates) or corn (rate 5 mL/L) and decreasing Stress Index (SI) in soybean (both application rates). Molecular mechanism of action of selected prototype 52097 was subsequently characterized through Next Generation Sequencing (NGS). In corn, genes involved in hormone (cytokinin and auxin) metabolism/catabolism, maltose biosynthesis, sugar transport and phloem loading were upregulated after application of prototype 52097. In soybean, genes involved in nitrogen metabolism, metal ion transport (mainly zinc and iron), sulfate reduction, and amino acid biosynthesis were induced. The proposed approach supports the integration of multiple omics to open new perspectives in the discovery, evaluation, and development of innovative and sustainable solutions to meet the increasing needs of row-crops agriculture.

Keywords: biostimulants; corn; imaging; industrial crops; maize; next generation sequencing; phenomics; plant phenotyping; row crops; soybean

1. Introduction

The increase in global population and the uncertainty produced by climate change represent big challenges for current and future agriculture [1,2]. Agricultural activity should ensure crop production systems that can tolerate increasingly adverse environmental conditions, such as drought, flooding, and other stressful events. At the same time, it should provide adequate yields to guarantee an economic return for farmers, and high-quality produce to satisfy the demands of consumers [3]. With a decreasing acreage of arable land and the limits of genetic potential of primary crops, to reach such objective it becomes necessary to increase crop yield, producing "more with less" [4–6] and to avoid overexploitation of natural resources, such as soil and water [7]. According to this, many research projects are supporting to design energy-efficient and eco-friendly cultivation systems, which are less dependent on the use of external inputs (e.g., fertilizers) [8,9].

To achieve these goals, the use of plant biostimulants (PBS) appears to be one of the most promising strategies [10]. According to the European Biostimulant Industry Council (EBIC, 2019) [11], plant biostimulants "contain substance(s) and/or microorganisms, whose function when applied to plants or the rhizosphere is to stimulate natural processes to enhance/benefit nutrient uptake, nutrient efficiency, tolerance to abiotic stress, and crop quality". The PBS formulations are generally proprietary compositions based on micro and macro-algae, plant extracts, hormone-like compounds, complex organic materials, amino acids or humic acids. Extensive reviews have recently discussed the discovery and the characterization of the activity of PBS derived from seaweeds, especially *Ascophyllum nodosum* [12–17]. In addition, several studies on the beneficial effects of natural PBS on plant growth, production and fruit quality in various crops have been recently published [18–21]. Physiological aspects in relation to the supply of PBS, like increased root and shoot growth, tolerance to abiotic stress, plant water uptake, and reduction of transplant shock, have also been reported [22–26]. Moreover, application of specific PBS may reduce fertilizer use and nutrient solution concentrations in hydroponic systems [27]. The development of PBS can therefore be used for the modulation of some plant physiological processes such as growth stimulation, stress mitigation, leading to increase yield and nutritional value of edible organs [16–18,28].

Considering the row crops sector, effective PBS are needed. Row/industrial crops such as soybean, corn, wheat, rice, rapeseed, sunflower, and cotton represent the most important crops in terms of global cultivated area [29]. It should be pointed out that row crops agriculture is generally based on an intensive farming system aimed at obtaining high yields by the use of high external inputs including organic and mineral fertilizers [9]. This is inconsistent with a vision of sustainable eco-compatible agricultural activity. Consequently, the use of PBS represents a sustainable strategy to contribute to ensure high yield and quality of product in this sector.

Recently, it was proposed to use transcriptomics together with plant phenomics to screen PBS and characterize their influence on plant physiology including the mechanisms activated by specific formulations [16,24]. Through transcriptomics, it is possible to identify possible modes of action of different substances and in turn predicting their role as biostimulants [30]. In addition to the transcriptomic profiling via microarrays, the novel technology Next Generation Sequencing (NGS) has been recently proposed as a tool to monitor the impact of PBS on the transcriptome of non-model plants, making it feasible to perform genomics in agricultural crops [31,32].

Using phenomics, it is possible to study the effect of PBS on plant biomass accumulation and the performances of the photosynthetic apparatus based on multi-spectrum, high-throughput image analysis to detect morphometric and specific physiological parameters (e.g., "Digital" Biovolume) [33,34]. This represents a step forward compared to "classical" in vitro and in vivo bioassays based on manual determination of simple physiological and morphological traits, evaluating nutrient uptake and growth stimulation through destructive quantification of root and shoot biomass. Such measurements result in a partial evaluation of PBS effects, without giving a real explanation of the mechanisms by which certain PBS exerts their effect(s). Among the different bioassays, the root growth inhibition of cress and the chicory hypocotyl growth are the most frequently used tests [35].

On the other hand, plant phenomics, based on multi-spectrum analysis of reflected or re-emitted light from the plant crown, stem and leaves provides a series of information related to plant structure and function, for example, plant water and nutritional status, pathogen infection, as well as on the plant's ability to intercept light. The use of high-throughput imaging analysis system allows to successfully integrate the experiments involving many variables, a large number of samples, and multiple comparisons [36]. Moreover, the high-throughput image analysis system is a non-invasive method that has the potential to determine the plant phenotypic response to experimental variable(s) (e.g., abiotic stress conditions), throughout the growing (or part of it) of experimental crops [24,25,37].

Hence, very recent papers showed that the use of a "multi-omics" approach, in particular metabolomics and plant phenotyping, represents an effective tool to examine plant performances under different experimental conditions [24,34]. The application of such integrated approach could offer a better explanation of the mechanism(s) of action of different PBS molecules or compounds on crops. This can be obtained by the identification of several biomarkers of PBS action as reported in Paul et al. (2019b) and Ugena et al. (2018) [34,36].

The aim of this study was the selection and characterization of a novel biostimulant formulation conceived to increase the yield of row crops. To achieve such objective, using a phenomic approach we investigated—on corn and soybean—the effect and physiological mechanism(s) of action of three different foliar biostimulant formulations/prototypes. This allowed the selection of the most effective one, subsequently characterized at transcriptomic level to understand its molecular action. This study, based on the integration of phenomic and genomic tools, opens new perspectives to release effective formulations for row-crops agriculture.

2. Materials and Methods

2.1. Plant Material and Growing Conditions

Experiments were performed on corn (*Zea mays* L., hybrid P0423, Pioneer) from November to December 2016, and soybean (*Glycine max* L. Merr.), from May to June 2017.

Plants were grown in a greenhouse under natural light conditions at the Plant Phenomics Platform, ALSIA-Metapontum Agrobios Research Center, Italy (N 40°23′ E 16°47′). Temperatures, humidity (RH%), and radiation (PAR μmol m^{-2} s^{-1}) are reported in Supplemental Table S1.

Both species were sowed directly into white pots (16 cm diameter, 20 cm height), containing 3.5 L of substrate consisting of a 50:50 mixture of peat and river sand. The day before sowing, 20 units of nitrogen, 40 units of phosphate, and 20 units of potassium oxide were added to the substrate mixture. Both soybean and corn plants were irrigated daily with 100 mL water, for 12 days. Afterwards, water was increased to 200 mL until the end of experiment to ensure an adequate water supply.

Three different biostimulant prototypes based on different combinations of seaweed and plant extracts formulated with selected micronutrients such as Mn, Zn, Mo (prototypes coded as 52096, 52097, 52113; proprietary composition Valagro SpA) were sprayed (using a portable atomizer sprayer) on both species at the third true leaf stage. Two rates were applied: 2.5 (lower dosage) and 5 mL/L (higher dosage) (corresponding to 1 and 2 L/ha respectively) during the experiment on soybean, while 5 mL/L was the unique rate applied during the experiment on corn. Untreated control plants were sprayed with distilled water. The experimental setup was composed by 5 biological replicates (plants) for each experimental condition using a completely randomized experimental design.

For NGS and qRT-PCR analysis, samples were collected just before (t = 0) and after the foliar application of PBS 52097 (5 ml/L) at 8, 24, 48, and 168 h. Both corn and soybean 3 leaves were removed from the plants and immediately submerged in liquid nitrogen. For each experimental condition (treatment and time-point), three biological replicates were collected from different plants at the third fully expanded leaf stage. Each biological replicate consisted of three entire leaves (central position) collected from three individual plants and pooled.

Plants used for sampling leaves were excluded from subsequent imaging acquisition or additional leaf sampling. All samples were collected at around 9:00 a.m.

2.2. Non-Destructive Measurements

The morphological and physiological characterization of the plants was carried out, non-destructively, by plant imaging. Images of plants were acquired, throughout the experiment, with the plant phenotyping platform Scanalyzer 3D (LemnaTec GmbH). Detailed information on the platform is reported in Briglia et al. (2019) [38]. Briefly, it is composed of 2 imaging chambers visible light (RGB) and fluorescence (FLUO), respectively. FLUO images, recorded into the fluorescence imaging chamber, were used to evaluate the photosynthetic performance through the "Stress Index" (see below). For each imaging chamber three images per plant were taken, one from top view of the plant and two from side view (0° and 90°).

2.2.1. Digital Biovolume Assessment

Plant growth was assessed through the digital biovolume (DB) [39] as follow:

$$\sum pixel\ sideview\ 0^{\circ} + \sum pixel\ sideview\ 90^{\circ} + \log_{10}\left(\sum pixel \frac{topview}{3}\right) \quad (1)$$

where pixel sideview 0°, 90° and top view are the plant pixel areas from all sides and top view images.

2.2.2. Color Classification

During the experiment the resulting RGB images were then analyzed by categorizing the pixels according to their color.

After the color segmentation process, that allow to separate the plant from the background, the RGB images were converted to HSI color space (Hue, Saturation, Intensity) and then the hue histogram was calculated. According to Casadesús et al. (2007) [40] the relative greener area (GGA) of each image was calculated as the sum of frequencies of the histogram classes included in the hue angle ranging from 80° to 180°. The GGA were used to evaluate the health status of the plant via colour classification (e.g., green: healthy and active leaf surface; yellow: degree of the plant senescence) [40].

2.2.3. Stress Index

The performance of photosynthetic system is not constant and depends mainly on the health and stress condition of a plant. When a plant is placed under stress, more fluorescent light of higher energy is released and this change in the pixel distribution can be measured using the fluorescence imaging chamber.

The Stress Index was calculated according to Petrozza et al. (2014) [25] as $(F_x - F_y)/(F_x + F_y)$ where F is the number of pixels in the x, y color classes, under the assumption that any impairments of the photosynthesis result in a change of pixel number at the x and y color class [25].

The x, y color classes were determined experimentally, by examining the hue histogram. Values of photosynthetic Stress Index vary from +1, poor photosynthetic efficiency, to −1, greater photosynthetic efficiency and should be considered only as a relative level when compared to other plants in the same experiment.

2.3. RNA-Seq Analysis

Single samples from leaves from untreated control plants (UTC) and from those treated with PBS 52097 (24 h after application) were used for RNA-sequencing. For each sample, total RNA was isolated using a CTAB-based protocol as described by Chang et al. (1993) [41]. RNA-seq libraries were prepared according to the so-called "dUTP method" to generate mRNA-seq libraries [42,43]. In short, mRNA was purified from 4 µg total RNA using oligo-dT beads, fragmented, and converted to cDNA. Libraries were subsequently made using the Illumina mRNA-Seq Sample Preparation Kit according to the manufacturer's instructions. An amount of 4 pmol of each library was sequenced by BaseClear B.V. (The Netherlands) using the Illumina HiSeq2500 system, with a read length of 50 nucleotides.

Single-end sequencing reads were filtered using the Illumina Casava pipeline version 1.8.3 and Illumina Chastity filtering. Additional filtering on the remaining reads was performed using the FASTQC quality control tool version 0.10.0. For RNA-seq analysis, sequence reads were mapped (per sample) to the reference using CLC Bio Genomics Workbench software (version 5.1.5). As a reference for corn, the publicly available B73 reference sequence (AGPv3.22, downloaded from the ZmGDB genome browser) consisting of 63,241 sequences was used. For soybean, the Glycine max_275_Wm82.a2.v1 primary transcripts [44], consisting of 88,647 sequences were used as a reference. To determine gene expression levels and differential gene expression, RPKM values (read counts corrected for library size and transcript length) were calculated using the CLC Bio software. Differentially expressed genes (DEGs) were selected by calculating the ratio of the RPKM value of treated samples over the RPKM value of untreated samples. Only genes with at least 50 reads and an RPKM value of over 5 in at least one sample were considered.

To functionally categorize the corn and soybean gene sequences, gene ontology (GO) terms were assigned to each assembled contig using Blast2GO software (version 3.1). GO terms provide a controlled vocabulary to describe the functions of genes across species. Blast2GO is an automated tool for the assignment of GO terms based on sequence similarity [45]. Statistical assessment of GO term enrichment in groups of DEGs were done using Fisher's Exact Test in combination with false discovery rate (FDR) correction for multiple testing.

2.4. qRT-PCR Analysis

qRT-PCR analysis was performed on single samples collected 8, 24, 48, and 168 h after PBS(s) application. Total RNA was isolated as described above. First-strand cDNA was prepared using 80 ng total RNA and qScript™ cDNA Supermix (Quanta Biosciences). Two and a half μL of 1:5 diluted first-strand cDNA was used as a template in the subsequent PCR, performed on a Bio-Rad CFX using 5 pmol of both primers (sequence of primers reported in Supplemental Table S2) and PerfeCta SYBR Green SuperMix UNG (Quanta Biosciences) in a final volume of 12.5 μL per reaction, according to the manufacturer's instructions. All transcript levels were normalized using a eukaryotic translation initiation factor gene (corn) or actin gene (soybean) as a control.

2.5. Statistical Analysis of Data

The statistical analysis was performed using R software (3.3.2 version; R foundation for Statistical Computing, Vienna, Austria). Phenotyping results were analyzed using one-way analysis of variance (ANOVA), and the means were compared with Duncan's New Multiple Range Test ($p < 0.05$).

3. Results and Discussion

3.1. Phenomic Parameters

3.1.1. Digital Biovolume

Plant development was assessed through the DB which is a morphometric measurement previously employed in high-throughput (HTP) studies to monitor the influence of abiotic stresses, mainly drought, on plant growth [34,38].

The DB of soybean plants was significantly improved after 10 days from the application of each of the three prototypes tested both at lower (2.5 mL/L) and higher (5 ml/L) concentration, in comparison with UTC (Figure 1A,B). However, the most consistent results (higher DB increase) were obtained using formulation 52097 at the lower dosage (2.5 mL/L; Figure 1A) reaching—at the end of experiment—the mean DB value of 49.58 k units (+72% compared to UTC plants). Such improvement in DB was clearly observed after 10 days from prototype application and maintained during time until the end of DB measurements in soybean, which was 26 days after treatment (Figure 1A).

Parallel measurements on corn confirmed that the 52097 PBS prototype exerted the higher increasing effect on DB in comparison with the other treatments and UTC plants (Figure 1C). In this case, beside an early plant response to the treatment observed already 7 days after treatment, the plants treated with 52097 maintained a greater DB throughout the experiment. No statistically significant differences between the 52113 and control plants were noted. At day 10 after treatment the plants treated with the 52097 reached a DB level of 27.92 k units, 81% higher than that of the UTC plants.

For both soybean and corn, plants treated with prototype 52097 showed a constant and consistent increase in DB accumulation compared to plants treated with 52096 or 52113.

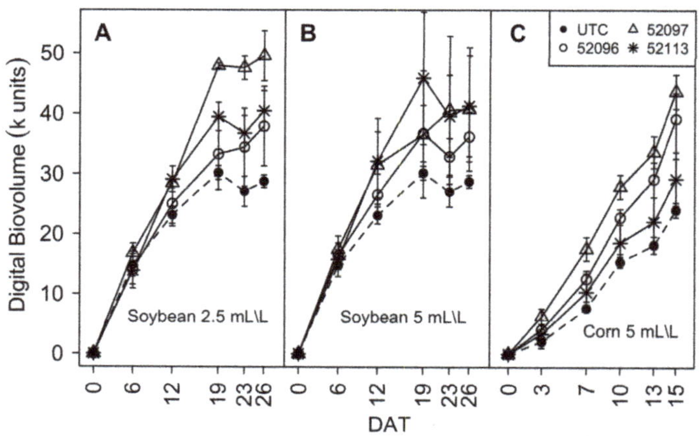

Figure 1. Mean values ± SE (n = 5) of digital biovolume (DB) measured on (**A**) soybean plants treated at lower dosage, (**B**) soybean plants treated at higher dosage, and (**C**) corn plants treated at higher dosage. Dashed lines and filled black circles (●) identify the untreated control plants (UTC).

3.1.2. Greener Area

The value of greener area (GGA) in soybean showed a general decrease during time (Figure 2A,B), as expected, due to the progression of phenological phases, which lead to senescence and yellowing. This was observed in untreated soybean plants, but also in treated with prototypes 52096 and 52113. Interestingly, soybean plants treated with prototype 52097 at both rates showed a consistent persistence of optimal (ranging from 0.7 to 0.8) GGA values during time (Figure 2A,B), even during the latest phenomic measurements (19–26 DAT).

For the first 12 days after treatment (DAT), no statistically significant differences were observed between the treatments, since all plants showed the same mean GGA value around 0.8 (Figure 2A,B). Starting from 14 days after treatment the first yellowing/sign of senescence were recorded on the untreated control plants and the plants treated with 52113 and 52096 prototypes. At the end of the experiment it was possible to see how, at both rates tested, the application of 52097 on soybean allowed a higher level of GGA, in particular around 0.77 (2.5 ml/L dosage) and 0.75 (5 mL/L dosage). On the other hand, control plants and plants treated with 52113 and 52096 prototypes reached values between 0.40 and 0.28 respectively (Figure 2A,B). It can be concluded that only prototype 52097 was able to preserve and improve GGA in comparison with the other experimental conditions. This can be attributed to a positive effect of prototype 52097 on the "stay-green" condition, that is known to allow plants to maintain high photosynthetic activity [46], and gain benefits on biomass accumulation, as confirmed by the data previously shown on DB (Figure 1).

The positive results on GGA observed after application of prototype 52097 were visibly clear by looking at the set of pictures taken during the cycle, by mean of the visible camera (Figure 2C).

Considering the same test on corn, all the formulations exerted a slight increase in GGA in comparison with untreated control (UTC) plants (Figure 2D), statistically significant at 7, 10, and 13 days after treatment with prototypes 52096 and 52097.

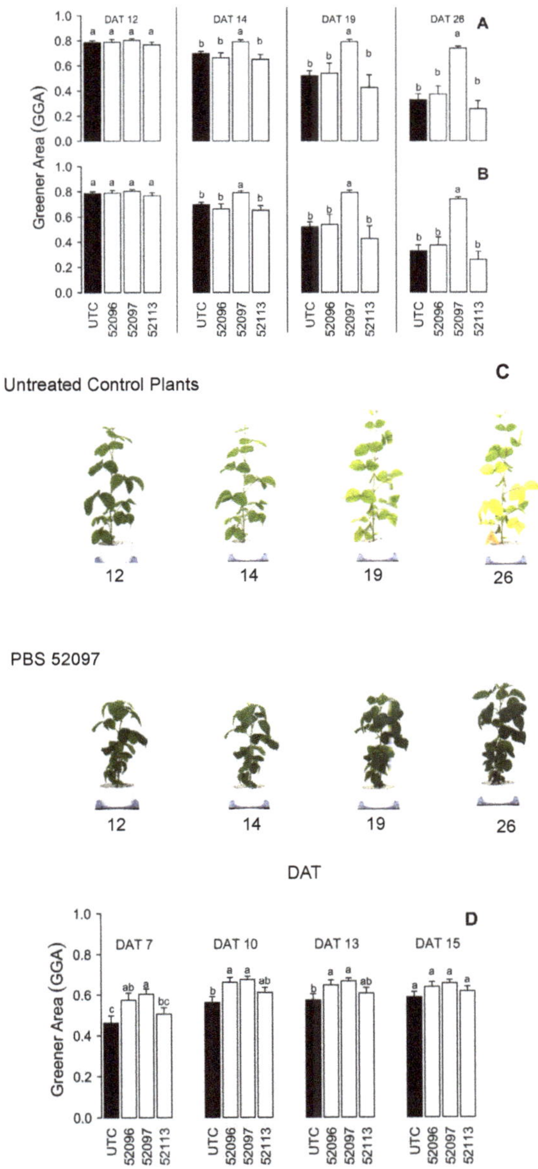

Figure 2. Mean values ± SE (n = 5) of greener area (GGA) measured on (**A**) soybean plants treated with biostimulants at lower dosage, (**B**) soybean plants treated with biostimulants at higher dosage at 12, 14, 19, and 26 days after treatment (DAT). (**C**) Acquired RGB images of representative UTC (top) and 52097-treated (bottom) soybean plants, showing the effect of PBS application on color and growth during the trial (from T0 to 26 days after treatment). (**D**) GGA measurements taken on corn plants at 7, 10, 13, and 15 days after treatment (DAT). Solid black bars identify the untreated control plants (UTC).

3.1.3. Stress Index

The measurement of Stress Index of treated and untreated soybean plants did not show statistically significant differences during the first four data acquisitions time (Figure 3). As expected, due to the plant cycle progression and senescence, during the last three data acquisitions a higher Stress Index—ranging from 0.4 to 0.7—was observed for UTC. Treating plants with 52096 or 52113 did not affect the Stress Index. Interestingly, soybean plants treated with 52097 showed a lower level of Stress Index than the other treatments, with values stable around 0.2 (Figure 3).

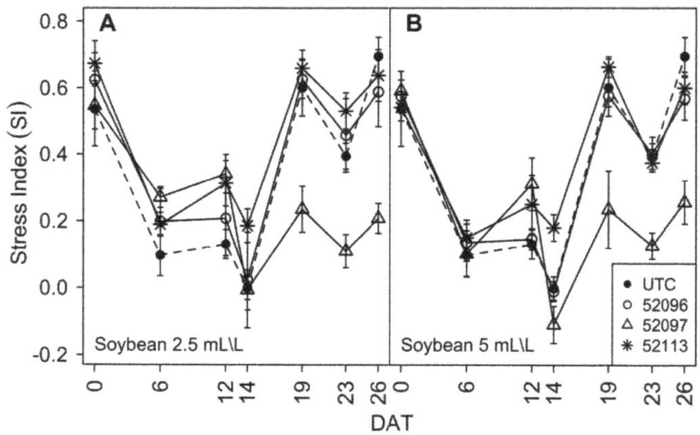

Figure 3. Mean values ± SE ($n = 5$) of Stress Index (SI) measured on (**A**) soybean plants treated at lower dosage, (**B**) soybean plants treated at higher dosage. Dashed lines and filled black circles (●) identify the untreated control plants (UTC).

This was not observed on corn plants, where both UTC and treated plants showed a Stress Index value ranging from 0.2 to 0.4 throughout the experiment (Supplemental Table S3).

3.2. Molecular Analyses

Based on the obtained results, although some differences were observed between soybean and corn, we selected compound 52097 as the best candidate for further NGS analyses. Leaf tissue from untreated soybean and corn plants was compared to its 52097-treated counterpart by RNA-seq analysis. Per sample, over 25 million single reads were generated and mapped to the relevant reference transcriptome (see Material and Methods). For corn, around 77% of the available sequencing reads could be mapped to this assembly, for soybean this was 89%. By comparing 52097-treated samples to untreated controls, differentially expressed genes (DEGs) were identified. Naturally, the number of DEGs depended on the fold-change threshold applied (Table 1). In general, the number of DEGs was higher in corn than in soybean. Lists of the 20 most upregulated genes for both crops are provided in Supplemental Table S4 (soybean) and Table S5 (corn).

Table 1. Number of differentially expressed genes (up and down-regulated) in corn and soybean 24 h after application of formulation 52097 when compared to mock-treated control plants.

FC	>2	>3	>4	>6	>9	
up	877	331	173	69	32	
down	1672	593	315	142	67	Maize
total	2549	924	488	211	99	
up	278	65	22	6	2	
down	321	59	17	1	1	Soybean
total	599	124	39	7	3	

3.3. Functional Annotation Using Gene Ontology

Of the 63,241 sequences present in the reference transcriptome from corn, 49,426 sequences (78%) could be functionally annotated using GO, meaning that one or more biological processes, molecular functions, or cellular localizations could be linked to these sequences based on sequence homology. For soybean, from the 88,647 genes present in the soybean transcriptome, 70,776 sequences (80%) could be functionally annotated using GO.

For corn, all DEGs up- or downregulated more than 3-fold were used for enrichment analysis. Several biological processes, including nitrogen assimilation, maltose biosynthesis, and cytokinin metabolism were enriched among the 331 upregulated genes from corn (Figure 4A). Analysis of the 593 downregulated corn transcripts resulted in 55 enriched GO-terms (biological process). By filtering out the most reduced GO-terms, that is, removing parent terms of already present statistically significant child GO terms, a list of 13 significantly enriched biological processes remained (Figure 4B). These terms included divalent metal ion transport, response to carbohydrate, phenylalanine degradation, and flavonoid biosynthesis.

For soybean, first, all DEGs up- or downregulated more than 3-fold were used for enrichment analysis. Analysis of the 65 upregulated soybean transcripts resulted in 25 enriched GO-terms (biological process), including metal ion transport, sulfate reduction, asparagine biosynthesis, and serine metabolism (not shown). However, there were no significantly enriched GO terms among the 59 downregulated soybean transcripts. For this reason, it was decided to include all soybean contigs with a FC greater than 2. This resulted in 16 significantly enriched (reduced) GO terms for the 278 upregulated contigs (Figure 4C), and 13 for the 321 downregulated contigs, including auxin-activated signaling, sulfur amino acid metabolism, iron transport, and sulfur compound biosynthesis (Figure 4D).

From both crops, two individual DEGs were selected using the results from the GO enrichment analysis. For corn, these were a cytokinin dehydrogenase (CKX; Figure 5A) and a glutamine synthetase (GS; Figure 5B). CKX catabolizes the plant hormone cytokinin and plays an important role in cytokinin regulated processes [47]. GS is required for nitrogen assimilation and allocation within the plant, and for nitrogen remobilization in both source and sink tissues [48]. GS is important for ammonium assimilation in roots, during senescence, and during photorespiration. Several studies have indicated that GS plays an essential role in plant development and yield formation in cereals. For example, in corn, leaf-localized GS are of specific importance for the development of the cob with respect to kernel number and kernel size [49]. A putative GS gene is induced by treatment with product 52097 (Figure 5B).

In addition, 3 more genes from corn were selected on the basis of their functionality: An SPX domain-containing protein (named after SYG1/Pho81/XPR1 proteins; Figure 5C), a NRT1/PTR FAMILY (NPF) protein (Figure 5D, and a polyol/monosaccharide transporter (PMT; Figure 5E). It is well described that plant growth and development are highly dependent on the availability of inorganic phosphate (Pi). Among the many proteins involved in the plant response to Pi starvation, proteins containing the SPX domain are key players involved in the maintenance of internal levels of Pi. Indeed, SPX genes have been reported to be induced upon Pi starvation in roots and shoots, and proteins harboring the SPX domain have been shown to be involved in P use efficiency [50]. Members of the plant NITRATE TRANSPORTER 1/PEPTIDE TRANSPORTER (NRT1/PTR) family display protein sequence homology with peptide transporters in animals. In comparison to their animal and bacterial counterparts, the plant NRT1/PTR family proteins transport a wide variety of substrates: nitrate, peptides, amino acids, dicarboxylates, glucosinolates, IAA, and ABA [51]. The transcript identified here shows the highest similarity to the first identified member of the NRT1/PTR family: NRT1.1. NRT1.1 is an Arabidopsis nitrate transporter that also functions as a nitrate sensor and can transport auxins. As such, it links nutrient and hormone signaling [51]. PMTs are proteins capable of transporting a range of sugar alcohols and monosaccharides including glucose, fructose, sorbitol, mannitol, xylitol, xylose, and galactose [52]. PMTs are believed to be involved in phloem loading [52]. Hence, the induction of a PMT gene identified in this study could point towards increased phloem loading.

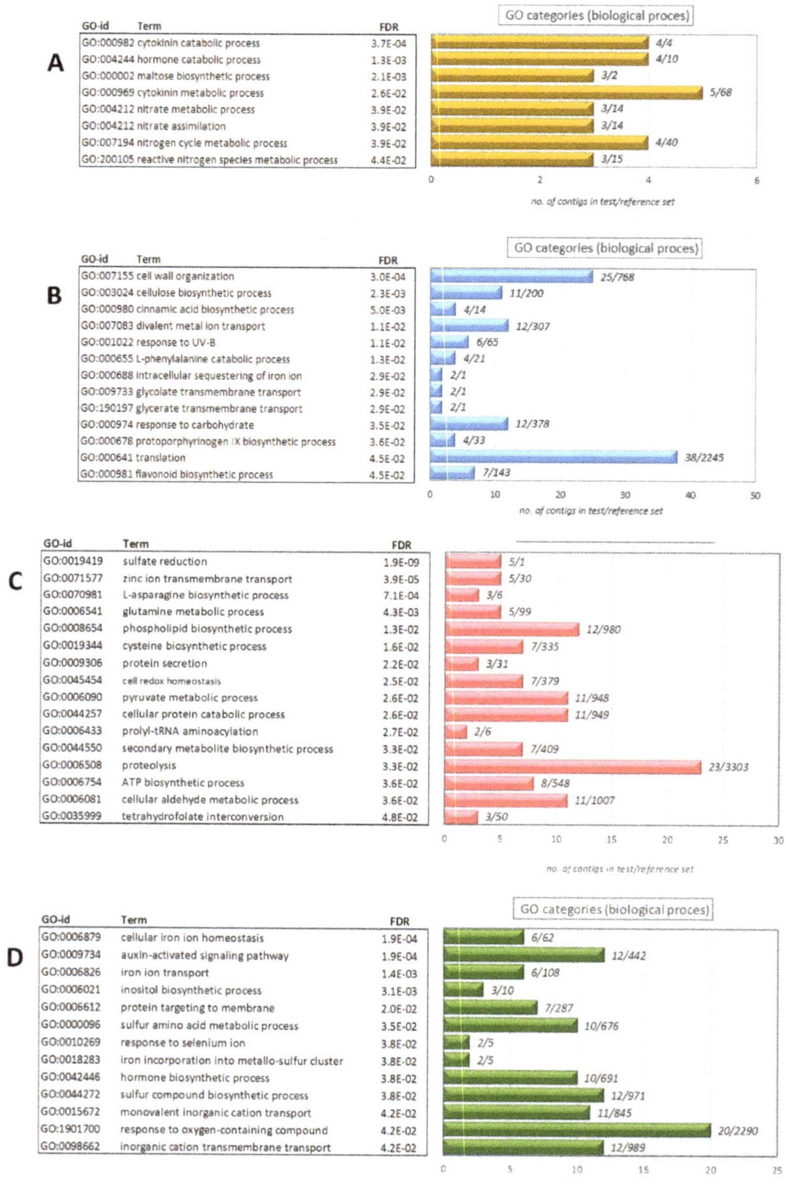

Figure 4. Gene ontology (GO) term enrichment analysis of the differentially expressed contigs from prototype 52097-treated plants (24 h after application; false discovery rate (FDR) < 0.05). (**A**) Upregulated GO terms in corn. (**B**) Down-regulated GO terms in corn. (**C**) Upregulated GO terms in soybean. (**D**) Down-regulated GO terms in soybean. The absolute number of contigs in the test set is represented by the bars in the graphs on the right (numbers of contigs in the test/reference sets are reproduced next to each bar).

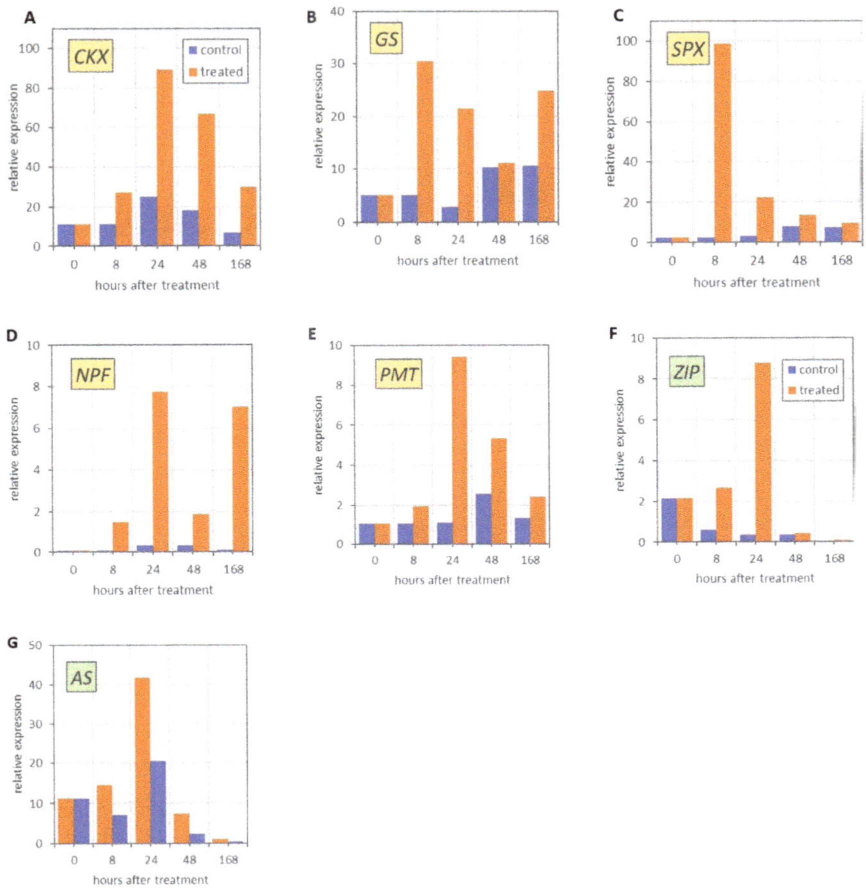

Figure 5. Gene expression of selected genes as determined by qRT-PCR. Samples were collected before (T0), and 8, 24, 48, and 168 h after treatment with formulation 52097. Expression of genes (**A**) *CKX (cytokinin dehydrogenase)*, (**B**) *GS (glutamine synthase)*, (**C**) *SPX domain-containing protein*, (**D**) *NPF* family protein (*NRT1/PTR*), and (**E**) *PMT* (*polyol/monosaccharide transporter*) was analysed in corn, while (**F**) *ZIP* (ZRT, IRT-like transporter) and (**G**) *AS* (*asparagine synthetase*) expression was assessed in soybean samples. Relative expression levels are in ddCt.

For soybean, we selected a ZIP (ZRT, IRT-like protein; Figure 5F) transporter and an asparagine synthetase (AS; Figure 5G). ZIP transporters are important during uptake and transport of zinc and iron and other divalent metal cations [53]. AS, like GS, functions in nitrogen metabolism [54].

For all these genes, expression was determined in leaf samples collected on several timepoints after application of prototype 52097. It was observed that data obtained by qRT-PCR corroborated the NGS results. Twenty-four hours after application, the differences in gene expression between treated and untreated leaves were very comparable. The additional timepoints showed that GS and SPX expression already peaked 8 h after application (Figure 5B,C), whereas the other 5 genes reached their maximum expression after 24 h (Figure 5A,D–G).

4. Conclusions

This study highlights the use of high-throughput/efficiency plant phenotyping (phenomics) together with Next Generation Sequencing to investigate the effectiveness and mechanism of action of new biostimulant formulations. Such formulations were specifically conceived as foliar applications to increase yield of different row/industrial crops, such as corn and soybean.

Phenomic-based measurements of digital biovolume, Greener Area, and Stress index allowed us to select 52,097 (commercial name "YieldOn®") as the most effective prototype among the ones tested. Subsequently, through NGS a deep characterization of the molecular mechanisms by which the biostimulant under investigation exerts its positive effect was performed. This analysis explained the mechanism of action of the biostimulant under investigation, which in corn upregulated specific processes like nitrogen and phosphate assimilation and metabolism, maltose biosynthesis, sugar transport and phloem loading, hormone (cytokinin) metabolism. In soybean nitrogen metabolism, metal ion transport (mainly zinc and iron), sulfate reduction, and amino acid biosynthesis were upregulated.

In conclusion, the results showed in this work support the integration of multiple "omics" as robust and objective tools in the discovery, evaluation, and development of innovative, sustainable, and targeted solutions to meet the emerging needs of row-crops agriculture.

Supplementary Materials: The following are available online at http://www.mdpi.com/2073-4395/9/11/761/s1, Table S1: Temperatures, humidity (RH%), and radiation (PAR µmol m^{-2} s^{-1}) parameters measured during the trial. Table S2: Sequence of primers used for qRT-PCR analysis; Table S3: Stress Index measurements in corn; Table S4: lists of the 20 most upregulated genes for soybean; Table S5: lists of the 20 most upregulated genes for corn.

Author Contributions: N.B. Conceptualization, main writing and structuring of the paper, High efficiency phenotyping Data Curation and Formal analysis; A.P. Methodology, High efficiency phenotyping Data Curation; F.A.H. Writing, Methodology, Next Generation Sequencing Data Curation and Formal Analysis; N.V. Methodology, Next Generation Sequencing Data Curation and G.P. Conceptualization, writing, review, editing and supervision.

Funding: The research conducted was funded by Valagro SpA, which provided biostimulant prototypes that were blindly identified by a numbered code to avoid any conflict of interest in the data production. The prototypes were named only at the end of the experiments.

Acknowledgments: The authors wish to thank Prem Warrior and Giuseppe Montanaro for their critical review of the manuscript. YieldOn® is a trademark registered by Valagro in Italy and other countries.

Conflicts of Interest: The authors declare that the research was conducted in the absence of any commercial or financial relationships that could be construed as a potential conflict of interest.

References

1. Takeda, S.; Matsuoka, M. Genetic approaches to crop improvement: Responding to environmental and population changes. *Nat. Rev. Genet.* **2008**, *9*, 444–457. [CrossRef] [PubMed]
2. Kang, Y.; Khan, S.; Ma, X. Climate change impacts on crop yield, crop water productivity and food security—A review. *Prog. Nat. Sci.* **2009**, *19*, 1665–1674. [CrossRef]
3. Eckardt, N.A.; Cominelli, E.; Galbiati, M.; Tonelli, C. The future of science: Food and water for life (Meeting Report). *Plant Cell* **2009**, *21*, 368–372. [CrossRef] [PubMed]
4. Food and Agriculture Organization of the United Nations [FAO]. Towards the Future we Want. End Hunger and Make the Transition to Sustainable Agricultural and Food Systems. Rome: FAO at Rio+20. 2012. Available online: http://www.fao.org/docrep/015/an894e/an894e00.pdf (accessed on 15 November 2019).
5. Carvalho, S.; Vasconcelos, M.W. Producing more with less: Strategies and novel technologies for plant-based food biofortification. *Food Res. Int.* **2013**, *54*, 961–971. [CrossRef]
6. International Food Policy Research Institute [IFPRI]. Producing More with Less? 2014. Available online: http://www.ifpri.org/blog/producing-more-less (accessed on 15 November 2019).
7. Montanaro, G.; Xiloyannis, C.; Nuzzo, V.; Dichio, B. Orchard management, soil organic carbon and ecosystem services in Mediterranean fruit tree crops. *Sci. Hort.* **2017**, *217*, 92–101. [CrossRef]
8. Pearson, C.J. Regenerative, Semiclosed Systems: A Priority for Twenty-First-Century Agriculture. *BioScience* **2007**, *57*, 409–418. [CrossRef]
9. Kremen, C.; Iles, A.; Bacon, C. Diversified farming systems: An agroecological, systems-based alternative to modern industrial agriculture. *Environ. Stud. Sci.* **2012**, *17*, 44. [CrossRef]
10. du Jardin, P. Plant biostimulants: Definition, concept, main categories and regulation. *Sci. Hort.* **2015**, *196*, 3–14. [CrossRef]

11. European Biostimulant Industry Council (EBIC). 2018. Available online: http://www.biostimulants.eu/ (accessed on 15 November 2019).
12. Khan, W.; Rayirath, U.P.; Subramanian, S.; Jithesh, M.N.; Rayorath, P.; Hodges, D.M. Seaweed extracts as biostimulants of plant growth and development. *J. Plant Growth Regul.* **2009**, *28*, 386–399. [CrossRef]
13. Craigie, J.S. Seaweed extract stimuli in plant science and agriculture. *J. Appl. Phycol.* **2011**, *23*, 371–393. [CrossRef]
14. Calvo, P.; Nelson, L.; Kloepper, J.W. Agricultural uses of plant biostimulants. *Plant Soil* **2014**, *383*, 3–41. [CrossRef]
15. Sharma, H.S.S.; Fleming, C.; Selby, C.; Rao, J.R.; Martin, T. Plant biostimulants: A review on the processing of macroalgae and use of extracts for crop management to reduce abiotic and biotic stresses. *J. Appl. Phycol.* **2014**, *26*, 465–490. [CrossRef]
16. Povero, G.; Mejia, J.F.; Di Tommaso, D.; Piaggesi, A.; Warrior, P. A Systematic Approach to Discover and Characterize Natural Plant Biostimulants. *Front. Plant Sci.* **2016**, *7*. [CrossRef] [PubMed]
17. Yakhin, O.I.; Lubyanov, A.A.; Yakhin, I.A.; Brown, P.H. Biostimulants in Plant Science: A Global Perspective. *Front. Plant Sci.* **2017**, *7*. [CrossRef] [PubMed]
18. Parađiković, N.; Vinković, T.; Vinković Vrček, I.; Žuntar, I.; Bojić, M.; Medić-Šarić, M. Effect of natural biostimulants on yield and nutritional quality: An example of sweet yellow pepper (*Capsicum annuum* L.) plants. *J. Sci. Food Agric.* **2011**, *91*, 2146–2152. [CrossRef] [PubMed]
19. Parađiković, N.; Vinković, T.; Vinković Vrček, I.; Tkalec, M. Natural biostimulants reduce the incidence of BER in sweet yellow pepper plants (*Capsicum annuum* L.). *Agric. Food Sci.* **2013**, *22*, 307–317. [CrossRef]
20. Bulgari, R.; Morgutti, S.; Cocetta, G.; Negrini, N.; Farris, S.; Calcante, A.; Spinardi, A.; Ferrari, E.; Mignani, I.; Oberti, R.; et al. Evaluation of Borage Extracts As Potential Biostimulant Using a Phenomic, Agronomic, Physiological, and Biochemical Approach. *Front. Plant Sci.* **2017**, *8*, 935. [CrossRef]
21. Saa, S.; Olivos-Del Rio, A.; Castro, S.; Brown, P.H. Foliar application of microbial and plant based biostimulants increases growth and potassium uptake in almond (*Prunus dulcis* [Mill.] D.A. Webb). *Front. Plant Sci.* **2015**, *6*. [CrossRef]
22. Adani, F.; Genevini, P.; Zaccheo, P.; Zocchi, G. The effect of commercial humic acid on tomato plant growth and mineral nutrition. *J. Plant. Nutr.* **1998**, *21*, 561–575. [CrossRef]
23. Parrado, J.; Bautista, J.; Romero, E.J.; García-Martínez, A.M.; Friaza, V.; Tejada, M. Production of a carob enzymatic extract: Potential use as a biofertilizer. *Bioresour. Technol.* **2008**, *99*, 2312–2318. [CrossRef]
24. Paul, K.; Sorrentino, M.; Lucini, L.; Rouphael, Y.; Cardarelli, M.; Bonini, P.; Reynaud, H.; Canaguier, R.; Trtílek, M.; Panzarová, K.; et al. Understanding the Biostimulant Action of Vegetal-Derived Protein Hydrolysates by High-Throughput Plant Phenotyping and Metabolomics: A Case Study on Tomato. *Front. Plant Sci.* **2019**, *10*. [CrossRef] [PubMed]
25. Petrozza, A.; Santaniello, A.; Summerer, S.; Di Tommaso, G.; Di Tommaso, D.; Paparelli, E.; Piaggesi, A.; Perata, P.; Cellini, F. Physiological responses to Megafol® treatments in tomato plants under drought stress: A phenomic and molecular approach. *Sci. Hort.* **2014**, *174*, 185–192. [CrossRef]
26. Alam, M.Z.; Braun, G.; Norrie, J.; Hodges, D.M. *Ascophyllum* extract application can promote plant growth and root yield in carrot associated with increased root-zone soil microbial activity. *Can. J. Plant Sci.* **2014**, *94*, 337–348. [CrossRef]
27. Vernieri, P.; Borghesi, E.; Tognoni, F.; Serra, G.; Ferrante, A.; Piagessi, A. Use of biostimulants for reducing nutrient solution concentration in floating system. *Acta Hortic.* **2006**, *718*, 477–484. [CrossRef]
28. Danzi, D.; Briglia, N.; Petrozza, A.; Summerer, S.; Povero, G.; Stivaletta, A.; Cellini, F.; Pignone, D.; De Paola, D.; Janni, M. Can high throughput phenotyping help food security in the Mediterranean area? *Front. Plant Sci.* **2019**. [CrossRef]
29. Leff, B.; Ramankutty, N.; Foley, J.A. Geographic distribution of major crops across the world. *Glob. Biogeochem. Cycles* **2004**. [CrossRef]
30. Santaniello, A.; Giorgi, F.M.; Di Tommaso, D.; Di Tommaso, G.; Piaggesi, A.; Perata, P. Genomic approaches to unveil the physiological pathways activated in *Arabidopsis* treated with plant-derived raw extracts. *Acta Hortic.* **2013**, *1009*, 161–174. [CrossRef]
31. Hoeberichts, A.; Povero, G.; Ibañez, M.; Strijker, A.; Pezzolato, D.; Mills, R.; Piaggesi, A. Next Generation Sequencing to characterise the breaking of bud dormancy using a natural biostimulant in kiwifruit (*Actinidia deliciosa*). *Sci. Hort.* **2017**, *225*, 252–263. [CrossRef]

32. Trevisan, S.; Manoli, A.; Ravazzolo, L.; Franceschi, C.; Quaggiotti, S. mRNA-Sequencing Analysis Reveals Transcriptional Changes in Root of Corn Seedlings Treated with Two Increasing Concentrations of a New Biostimulant. *J. Agric. Food Chem.* **2017**, *65*, 9956–9969. [CrossRef]
33. Furbank, R.T.; Tester, M. Phenomics-technologies to relieve the phenotyping bottleneck. *Trends Plant Sci.* **2011**, *16*, 635–644. [CrossRef]
34. Paul, K.; Sorrentino, M.; Lucini, L.; Rouphael, Y.; Cardarelli, M.; Bonini, P.; Miras Moreno, M.B.; Reynaud, H.; Canaguier, R.; Trtílek, M.; et al. A Combined Phenotypic and Metabolomic Approach for Elucidating the Biostimulant Action of a Plant-Derived Protein Hydrolysate on Tomato Grown Under Limited Water Availability. *Front. Plant Sci.* **2019**, *10*. [CrossRef]
35. Summerer, S.; Petrozza, A.; Cellini, F. High throughput plant phenotyping: A new and objective method to detect and analyse the biostimulant properties of different products. *Acta Hortic.* **2013**, *1009*, 143–148. [CrossRef]
36. Ugena, L.; Hýlová, A.; Podlešáková, K.; Humplík, J.F.; Doležal, K.; De Diego, N.; Spíchal, L. Characterization of Biostimulant Mode of Action Using Novel Multi-Trait High-Throughput Screening of Arabidopsis Germination and Rosette Growth. *Front. Plant Sci.* **2018**, *9*. [CrossRef] [PubMed]
37. Rouphael, Y.; Spíchal, L.; Panzarová, K.; Casa, R.; Colla, G. High-Throughput Plant Phenotyping for Developing Novel Biostimulants: From Lab to Field or From Field to Lab? *Front. Plant Sci.* **2018**, *9*. [CrossRef] [PubMed]
38. Briglia, N.; Montanaro, G.; Petrozza, A.; Summerer, S.; Cellini, F.; Nuzzo, V. Drought phenotyping in Vitis vinifera using RGB and NIR imaging. *Sci. Hort.* **2019**, *256*, 108555. [CrossRef]
39. Eberius, M.; Lima-Guerra, J. High-throughput plant phenotyping-data acquisition, transformation, and analysis. In *Bioinformatics*; Edwards, D., Stajich, J., Hansen, D., Eds.; Springer: New York, NY, USA, 2013; pp. 259–278. [CrossRef]
40. Casadesús, J.; Kaya, Y.; Bort, J.; Nachit, M.M.; Araus, J.L.; Amor, S.; Ferrazzano, G.; Maalouf, F.; Maccaferri, M.; Martos, V. Using vegetation indices derived from conventional digital cameras as selection criteria for wheat breeding in water-limited environments. *Ann. Appl. Biol.* **2007**, *150*, 227–236. [CrossRef]
41. Chang, S.; Puryear, J.; Cairney, J. A simple and efficient method to isolate RNA from pine trees. *Plant Mol. Biol. Rep.* **1993**, *11*, 114–117. [CrossRef]
42. Parkhomchuk, D.; Borodina, T.; Amstislavskiy, V.; Banaru, M.; Hallen, L.; Krobitsch, S.; Lehrach, H.; Soldatov, A. Transcriptome analysis by strand-specific sequencing of complementary DNA. *Nucl. Acids Res.* **2009**, *37*, 123. [CrossRef] [PubMed]
43. Levin, J.Z.; Yassour, M.; Adiconis, X.; Nusbaum, C.; Thompson, D.A.; Friedman, N.; Gnirke, A.; Regev, A. Comprehensive comparative analysis of strand-specific RNA sequencing methods. *Nat. Methods* **2010**, *7*, 709–715. [CrossRef]
44. Schmutz, J.; Cannon, S.B.; Schlueter, J. Genome sequence of the palaeopolyploid soybean. *Nature* **2010**, *463*, 178–183. [CrossRef]
45. Conesa, A.; Götz, S.; García-Gómez, J.M.; Terol, J.; Talón, M.; Robles, M. Blast2GO: A universal tool for annotation, visualization and analysis in functional genomics research. *Bioinformatics* **2005**, *21*, 3674–3676. [CrossRef] [PubMed]
46. Thomas, H.; Smart, C.M. Crops that stay green. *Ann. Appl. Biol.* **1993**, *123*, 193–219. [CrossRef]
47. Jameson, P.E.; Song, J. Cytokinin: A key driver of seed yield. *J. Exp. Bot.* **2016**, *67*, 593–606. [CrossRef] [PubMed]
48. Bernard, S.M.; Habash, D.Z. The importance of cytosolic glutamine synthetase in nitrogen assimilation and recycling. *New Phytol.* **2009**, *182*, 608–620. [CrossRef]
49. Thomsen, H.C.; Eriksson, D.; Møller, I.S.; Schjoerring, J.K. Cytosolic glutamine synthetase: A target for improvement of crop nitrogen use efficiency? *Trends Plant Sci.* **2014**, *19*, 656–663. [CrossRef]
50. Secco, D.; Wang, C.; Arpat, B.A.; Wang, Z.; Poirier, Y.; Tyerman, S.D. The emerging importance of the SPX domain-containing proteins in phosphate homeostasis. *New Phytol.* **2012**, *193*, 842–851. [CrossRef]
51. Léran, S.; Varala, K.; Boyer, J.C.; Chiurazzi, M.; Crawford, N.; Daniel-Vedele, F. A unified nomenclature of NITRATE TRANSPORTER 1/PEPTIDE TRANSPORTER family members in plants. *Trends Plant Sci.* **2014**, *19*, 5–9. [CrossRef]
52. Slewinski, T. Diverse transporters and their homologs in vascular plants: A physiological perspective. *Mol. Plant.* **2011**, *4*, 641–662. [CrossRef]

53. Hall, J.L.; Williams, L.E. Transition metal transporters in plants. *J. Exp. Bot.* **2003**, *54*, 2601–2613. [CrossRef]
54. Gaufichon, L.; Reisdorf-Cren, M.; Rothstein, S.J.; Chardon, F.; Suzuki, A. Biological functions of asparagine synthetase in plants. *Plant Sci.* **2010**, *179*, 141–153. [CrossRef]

agronomy

MDPI

Article

Bacillus subtilis CBR05 for Tomato (*Solanum lycopersicum*) Fruits in South Korea as a Novel Plant Probiotic Bacterium (PPB): Implications from Total Phenolics, Flavonoids, and Carotenoids Content for Fruit Quality

Murugesan Chandrasekaran [1], Se Chul Chun [2], Jae Wook Oh [3,*], Manivannan Paramasivan [4], Ramesh Kumar Saini [5] and Jesudoss Joseph Sahayarayan [6]

1 Department of Food Science and Biotechnology, Sejong University, Gwangjin-gu, Seoul 05006, Korea; chandrubdubio@gmail.com
2 Department of Environmental Health Science, Konkuk University, Gwangjin-gu, Seoul 05029, Korea; scchun@konkuk.ac.kr
3 Department of Stem Cell and Regenerative Biotechnology, Konkuk University, Gwangjin-gu, Seoul 05029, Korea
4 Department of Microbiology, Bharathidasan University, Tiruchirappalli 620024, India; manivannan3065@gmail.com
5 Department of Crop Science, Konkuk University, Gwangjin-gu, Seoul 05029, Korea; saini_1997@yahoo.com
6 Department of Bioinformatics, Alagappa University, Karaikudi 630 003, India; jjsrbioinformatics2016@gmail.com
* Correspondence: ohjw@konkuk.ac.kr

Received: 7 September 2019; Accepted: 29 November 2019; Published: 3 December 2019

Abstract: Plant growth-promoting rhizobacteria (PGPR) are naturally occurring soil biota which benefit plants by improving plant productivity and immunity. The aim of the present work was to evaluate the effect of the inoculation of PGPR strain, *Bacillus subtilis* CBR05 on the quality of tomato fruits produced under greenhouse conditions. Results were compared with mock-inoculated control and market sample. We found a significant increase in total phenol and flavonoid contents of tomato fruits in PGPR strain *B. subtilis* CBR05 inoculated plants compared to those of mock-inoculated control and market sample. Moreover, *B. subtilis* CBR05 inoculation stimulated antioxidant activities and levels of carotenoid (β carotene and lycopene) content in plants. In addition, the inoculation of the strain *B. subtilis* CBR05 produced the highest content of lycopene (21.08 μg/g FW) in tomato fruits as compared to mock-inoculated plants. Our results show that the PGPR strain *B. subtilis* CBR05 is a versatile soil bacterium that enhances tomato production by elevating antioxidant activities and carotenoid (β carotene and lycopene) levels in fruit.

Keywords: *Bacillus subtilis*; tomato; antioxidant activity; carotenoids; probiotics; PGPR

1. Introduction

Tomato (*Solanum lycopersicum*) is regarded as the second most vegetable crop next to potato in the agricultural implications of human consumption. According to agricultural statistics, tomatoes along with sweet corn and snap beans constitute 93% of crop production and processing strategies (Agricultural Statistics, United States Department of Agriculture (USDA), 2016). The positive benefits of tomato consumption have been rigorously proved against a variety of diseases like chronic degenerative diseases, owing to the escalated content of significant phytochemicals with potent health benefits, like the carotenoids (β-carotene and lycopene), the glycoalkaloids (dehydrotomatine and α-tomatine), ascorbic acid, tocopherols, and many phenolic and flavonoid compounds [1–3]. Tomato also contributes as a major dietary ingredient for Vitamin A and C which implies increased per individual consumption in the United States and many Western countries [2,4]. Fruit ripening in tomato comprises a cascade of events on biochemical, physiological, and structural perspectives involving the influence of secondary metabolites that confer flavor, aroma, texture, and appearance of the tomato [5,6]. Accumulation of large quantities of pigments, especially lycopene and β carotene, inside the plastoglobules of chromoplast provides a visual indication that the fruit is mature and suitable for consumption [4].

Agronomic practices are recognized as a vital factor in determining the nutritional quality of tomato crops [7,8]. The nutrient contents in tomato fruits depend on the environment in which they grow [9,10]. Nowadays, the use of crop modeling to identify effective farmer strategies to counteract adverse future climatic conditions has become a standard in climate change impact assessments [11–13]. Over the past few years, a variety of methods have been proposed to comprehensively assess fruit quality and its relationship with water, including principal component analysis (PCA), analytic hierarchy process (AHP), gray relational analysis (GRA), and technique for order preference by similarity to ideal solution (TOPSIS) [14–17]. The challenge of producing fresh fruits and vegetables is increasing both yield and quality to satisfy consumers as the environment changes in ways that are deleterious to crop species [18]. The quality of agricultural products is affected by many pre- and postharvest factors [9]. The utilization of biofertilizers that mitigate these adverse environmental effects has become a feasible and beneficial production practice. Plant growth-promoting rhizobacteria (PGPR) may be considered as preharvest biotic factors that mitigate adverse environmental effects and promote improved crop yield and quality [19,20].

Among various PGPR approaches, *Bacillus* species are considered as likely candidates due to their broad-spectrum antagonistic activity against phytopathogens, production of long-lived and stress-tolerant spores, secondary metabolites, lytic enzymes, resistance to adverse environments, and plant growth promotion [19,21–23]. *Bacillus subtilis* plays a significant role in improving plant growth and tolerance to both biotic and abiotic stresses. PGPR strains also act as bio-stimulants of phytohormones and peptide synthesis [20,24], but studies of the PGPR strain, *B. subtilis* CBR05 on tomato have not yet appeared. Preharvest factors that directly affect crop yield and quality can be summarized into biological factors comprising pathological, entomological, and animal issues, which was found to be nullified upon increased usage of PGPR. The dire need for assessment of tomato and tomato-based products is given significant attention concerning nutrition and quality relying on the nature of the variety, maturity at harvest, effective transport, and storage [25]. Characterization of the carotenoids, mainly β-carotene and lycopene during storage and various ripening stages, shows drastic developments in sustainable yield and quality parameters of tomato [26]. This information adds to our understanding of temporal differentiation of nutritionally significant phytochemicals during ripening of tomato fruits. The objective of this study was to evaluate the effects of *B. subtilis CBR05* on the quality of tomato fruits under greenhouse conditions.

2. Material and Methods

2.1. Chemicals and Reagents

Authentic standards of carotenoid, all-*E*-lutein, was purchased from Cayman Chemical Company, Michigan, USA. All-*E*-β-carotene were purchased from Sigma-Aldrich, St. Louis, MO, USA. Only the HPLC grade of organic solvents was employed in carotenoid extraction (Daejung, Siheung-*si*, Korea).

2.2. Bacterial Strain and Culture Conditions

B. subtilis CBR05 isolated from our lab maintained on tryptic soy agar (TSA) plates. For long-time storage, bacterial cultures were maintained in tryptic soy broth at −80 °C. For experimental purposes, the cultures were transferred to TSA (MBCell, Seoul, Korea) and incubated at 30 °C for 24 h. The inoculum mixture of the strain was prepared by culturing in nutrient broth and incubating at 28 °C with constant shaking at 130 rpm. The bacterial cells were centrifuged at 10,000 g for 10 min at 4 °C. The cell precipitate was resuspended in 10 mM $MgCl_2$ and the cell concentration of bacterial suspensions of *B. subtilis* CBR05 was adjusted to 10^8 colony-forming unit (CFU)/mL (OD_{600} = 1.0) for further studies.

2.3. Plant and Growth Conditions

Tomato seeds (Korean cultivar, Kwangbok) utilized for this study were obtained from a Korean seed resource center, Seoul, South Korea. They were surface sterilized in sodium hypochlorite, rinsed several times with distilled water, and planted onto pots containing sterilized growth media (Peat moss with perlite in a ratio of 3:1). Two sets of plants (three plants per set) were maintained, one without PGPR (mock-inoculated control) inoculation and the other with PGPR strain, *B. subtilis* CBR05 inoculum. All treatments were placed randomly in the greenhouse and replicated 3 times. Plant growth-promoting *B. subtilis* CBR05 were applied under sterile conditions to the base of the plants close to the roots to ensure better colonization. Plants were maintained under greenhouse conditions at a temperature of 25 °C with watering carried out every alternate day, to make a better availability for nutrition and plant growth promotion.

2.4. Antioxidant Assays

2.4.1. DPPH Assay

1,1-diphenyl-2-picrylhydrazyl (DPPH) assay was performed to assess the fractions exhibiting scavenging property of free radicals in vitro [27]. Then, 0.2 mM solution of DPPH in ethanol was added to the fraction of aliquot at concentrations (100 µg/mL). The mixture was allowed to stand for 30 min and the absorbance was measured at 517 nm using a UV-Visible spectrophotometer. The percent scavenging activity was determined and Trolox was used as the standard.

2.4.2. ABTS (2,2′-azino-bis (3-ethylbenzothiazoline-6-sulfonic acid) Assay

A Trolox equivalent antioxidant capacity (TEAC)/ABTS assay was conducted based on the method of Ramos et al. [28]. The ABTS solution (7 mM) was oxidized with potassium peroxodisulfate (2.45 mM) for 16–18h at room temperature. The ABTS solution was diluted with solvents. An aliquot (100 µg/mL) was mixed with diluted ABTS solution and the absorbance was read at 734 nm. Trolox and ascorbic acid were used as reference standards.

2.5. Determination of Total Phenolic Contents

The Folin–Ciocalteu method was used to estimate various concentrations concerned with the total phenolic content. The extracts were dissolved in absolute methanol and later 200 µL of the extract was mixed with 800 µL of 1 N Folin–Ciocalteu reagent (1: 10). After 5 min at room temperature, 3 mL of sodium carbonate (15%) was added to the extracts. Following incubation for 30 min at room temperature, the absorbance was read at 765 nm using a UV-spectrophotometer. A standard curve for gallic acid equivalents (GAE) (milligrams per gram of extract (mg GAE/ge)) was utilized to evaluate the concentration of total phenolic compounds. Analyses were performed in triplicate per each extract.

2.6. Determination of Total Flavonoid Contents

Screening the content of flavonoids was done by a modified protocol as reported previously [29]. The extracts were dissolved in absolute methanol. In a 15 mL conical tube, 1 mL of a sample was mixed with 0.3 mL of 5% sodium nitrite, followed by incubation for 5 min. After incubation, 0.3 mL of aluminum chloride (10%) and 2 mL of sodium hydroxide (1 mol/L) were added to the reaction mixture, and the absorbance was read at 496 nm with a UV-spectrophotometer, using catechin as the standard. Quercetin equivalents (QE) present per g of extract (mg QE/ge) was used to quantify the expression levels.

2.7. Extraction and Quantification of Carotenoids

Carotenoids were extracted in triplicates and quantified according to previously established protocol with minor modifications [30,31]. All the preparations were performed in low light conditions to avoid the degradation of carotenoids. Three independent biological samples were extracted separately. Briefly, one whole tomato fruit was finely chopped and mixed thoroughly. Five grams of chopped fruits (exact to 0.001 g) from each treatment were separately transferred into test tubes containing 20 mL of acetone and 0.1% butylated hydroxytoluene (BHT: *w/v*). The samples were homogenized with a mechanical homogenizer and centrifuged at 5000× *g* (5 min at 4 °C temperature). The supernatant was recovered and pelleted samples were repeatedly extracted until the pallets became colorless. Supernatants from all extractions were pooled and vacuum-dried in a rotary evaporator (Temperature < 35 °C) (Büchi RE 111, Switzerland).

2.8. HPLC Analysis

The extract was recovered with 10 mL of methylene chloride (CH_2Cl_2) containing 0.1% BHT and transferred to an amber color HPLC vial for HPLC analysis. The chromatographic separation was achieved using an Agilent Model 1100 HPLC instrument (Agilent Technologies Canada Inc., Mississauga, ON, Canada) equipped with a degasser, autosampler, dual pump, and diode array detector (DAD). Samples were scanned (200–800 nm) with 0.05 min (1 s) response time, 8.0 mm slit width, and a detection wavelength of 450 (for most of the carotenoids) and 470 nm (for lycopene). The bandwidth was ±16 nm for all detection wavelengths. Similarly, 600 nm was used as a reference wavelength with ±50 nm bandwidth in all detections. The column used was a YMC, C30 carotenoid column, 250 × 4.6 mm, 5 µm (YMC, Wilmington, NC, USA), and the chromatographic data were recorded with ChemStation LC 3D software. The column thermostat was maintained at 25 °C temperature. Then, 20 µL of standards and samples were injected with an autosampler. The solvent system consisted of Methanol: methyl tertiary butyl ether (MTBE): water (81:15:4) (Mobile phase A) and MTBE: Methanol (91:9) (Mobile phase B). The gradient elution was 0%–100% B in 90 min, and 5-min post-run at a flow rate of 1 mL/min.

2.9. Statistics

All of the experiments were conducted in triplicate and results were tabulated as the Mean ± standard deviation (SD). Statistical significance of the data was determined using one-way analysis of variance (one-way ANOVA) followed by Fisher LSD (Least Significant Difference) test. Data analyses were performed using Sigmastat v8.02 (Systat Software Inc., San Jose, CA, USA). A p value of ≤ 0.05 was considered significant.

3. Results

In the present investigation, the effects of the PGPR strain, *B. subtilis* CBR05 inoculation on the maintenance of carotenoids, total phenolics, flavonoid contents, and antioxidant properties were evaluated. The results revealed that the PGPR strain, *B. subtilis* CBR05 has the capacity to improve the plant growth and change some of the tomato fruit quality characteristics under greenhouse conditions. Tomato plants treated with PGPR strain, *B. subtilis* CBR05 showed significantly higher biomass compared with mock-inoculated controls. Significant increases in root length and dry weight, over mock-inoculated controls, were achieved in green-house conditions (data not shown). In addition, antioxidant activities of three tomato fruits (mock-inoculated, *B. subtilis* CBR05 inoculated, and market fruit) were determined using both DPPH and ABTS radical scavenging method. The extract of fruits from plants inoculated with *B. subtilis* CBR05 strain was the most active against DPPH and ABTS radical and that of fruits from the market, the differences being significant when compared to mock-inoculated control (Figure 1).

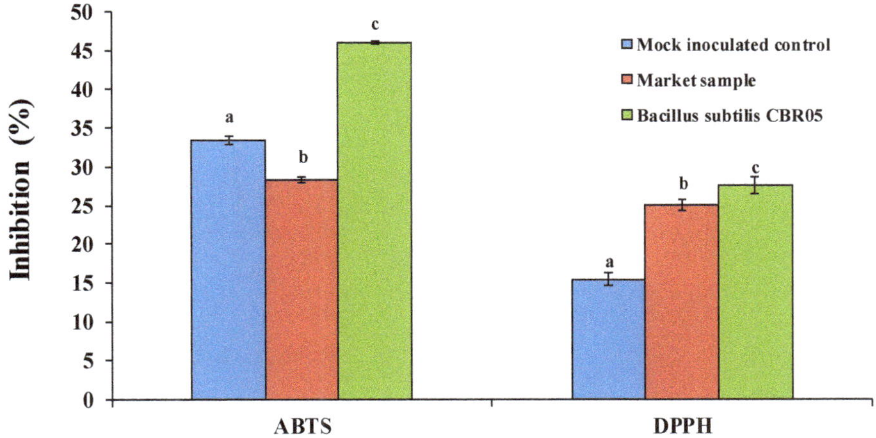

Figure 1. Antioxidant assays ((2,2′-azino-bis (3-ethylbenzothiazoline-6-sulfonic acid) (ABTS) and 1,1-diphenyl-2-picrylhydrazyl (DPPH)). Values are mean ± standard deviation from three replicates. Bars followed by the same letter(s) are not significantly different ($p \leq 0.05$).

The total phenolic content of the fruit extracts is shown in Figure 2. Among the fruits, the PGPR strain, *B. subtilis* CBR05 inoculated plants had the highest phenolic content followed by market fruits. Moreover, the flavonoid contents were also elevated in PGPR treated tomato fruits as compared to those of both mock-inoculated control and market fruits (Figure 2). We did not find any significant differences between mock-inoculated control and market fruit. The results of this study show that *B. subtilis* CBR05 is an effective probiotic agent for the promotion of tomato fruit quality.

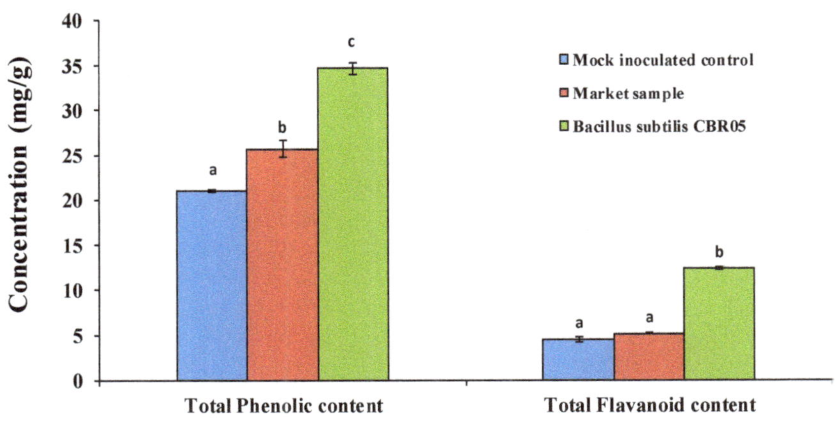

Figure 2. Total phenol and flavonoids contents. Values are mean ± standard deviation from three replicates. Bars followed by the same letter(s) are not significantly different ($p \leq 0.05$).

The carotenoids, such as lycopene and β-carotene, were extracted and quantified (Table 1). The contents and composition of carotenoids were found to differ among control, *B. subtilis* CBR05 inoculated, and market fruits (Figure 3). Carotenoid levels (lycopene) in tomato fruits in plants treated with PGPR strain, *B. subtilis* CBR05 were higher than both the control and market fruits (Table 1; Figure 3).

Table 1. Contents of carotenoids in tomato fruits.

Treatment	Carotenoids	Content (μg/g FW)
Mock inoculated control	All-*E*-β-Carotene	5.63 ± 0.28 [a]
	All-*E*-Lycopene	7.48 ± 1.96 [a]
Market tomato	All-*E*-β-Carotene	4.65 ± 0.96 [a]
	All-*E*-Lycopene	10.51 ± 3.34 [a]
Bacillus subtilis CBR05	All-*E*-β-Carotene	3.53 ± 0.12 [a]
	All-*E*-Lycopene	21.08 ± 0.32 [b]

In each experiment, means followed by different superscript letters are significantly different among the treatment groups ($p \leq 0.05$). Values are mean ± standard deviation from three replicates.

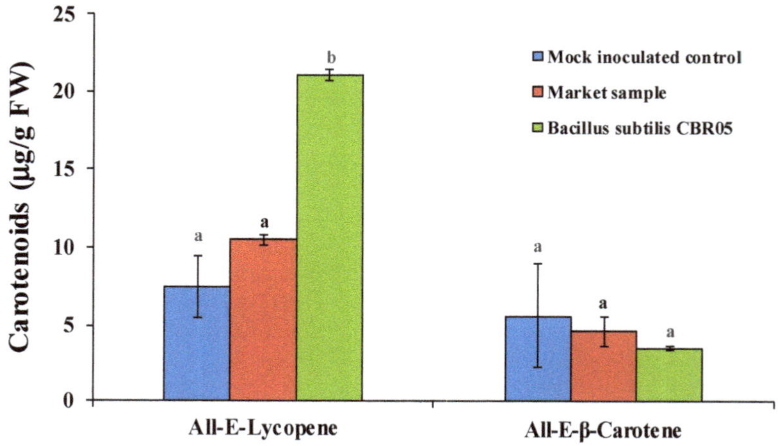

Figure 3. Carotenoids' contents (lycopene and β-carotene). Values are mean ± standard deviation from three replicates. Bars followed by the same letter(s) are not significantly different ($p \leq 0.05$).

The present investigation extends their results by comparing the contents of carotenoids from the fruits using HPLC–DAD (Figure 4). Using this methodology, all-E- β -carotene and all-E- lycopene were identified as the major carotenoids in tomato fruits, based on retention time with standards and by comparing the peak spectra recorded with a DAD during the analysis. The chromatograms (470 nm) and the peak spectra of major identified peaks were shown in Figure 4. The other minor carotenoids were not quantified due to the unavailability of standard compounds. We did not find any significant differences in β-carotene. However, in the present study, we have recorded a significantly higher amount of lycopene (All-E-lycopene) in fruits during ripening in PGPR strain, *B. subtilis* CBR05 inoculated plants than those of both mock-inoculated control and market fruits. These results validate the productive roles of the *B. subtilis* CBR05 in enhancing the nutritional potential of tomato fruits.

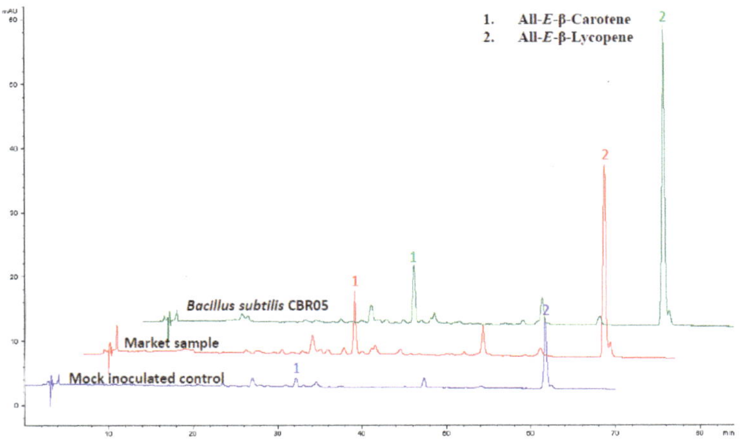

Figure 4. HPLC chromatogram of carotenoids.

4. Discussion

The use of PGPR is increasing in agriculture and may offer an attractive alternative to synthetic chemicals and fertilizers. Plant growth-promoting microorganisms are efficient microbial competitors that can promote plant growth by producing phytohormones and/or by increasing available nutrients through production of secondary metabolites or act as biocontrol agents to protect plants from infection by phytopathogens [19,21–24]. There have been many reports on PGPR and their effective roles [19–24]. Insufficient experimental work has been reported to speculate on the mechanisms of PGPR effects on fruit quality. In the present study, PGPR strain, *B. subtilis* CBR05 isolated from rice were used as inoculants for tomato plants grown under greenhouse conditions. Our results showed that PGPR inoculations significantly increased the total biomass and root length compared to those in the control. *B. subtilis* CBR05 appears to impart plant growth promotion effects that are distinct from other commercial biocontrol agents. Tomato fruits are a good source of antioxidant compounds that can reduce harmful oxidation reactions in the human body, thus preventing various diseases associated with free radical oxidation, such as cardiovascular and neurological disorders and cancer [1–4].

Antioxidant activity has been widely used to test the ability of plant extracts to act as free radical scavengers [32]. In the present investigation, *B. subtilis* CBR05 had a net positive effect on the antioxidant activity measured by the DPPH and ABTS scavenging capacity, which seems to indicate that the bacteria acted as a regulator of the synthesis of antioxidant compounds in the plant (Figure 1). Strong scavenging of ion radical was exhibited by the inoculated tomato fruits, thus showing that *B. subtilis* CBR05 inoculation increased the radical scavenging capacity of tomato fruits. In a previous study, *B. licheniformis* inoculated plants had increased antioxidant profiles in tomato plants under greenhouse conditions [32]. Similar results concerned with enhanced fruit quality and marketable grade have been reported for other crops under the influence of PGPR [33] but this report is the first of its kind to specifically address PGPR strain, *B. subtilis* CBR05 in improvement of fruit quality. Moreover, PGPR enhances fruit characteristics based on the mediation of increased availability of nutrients to plants like phosphorous and iron, enhancing the nutritional status of the plants in the rhizosphere [33–35].

Phenolic contents perform an essential role in plant resistance and defense against phytopathogens, which are closely linked with reactive oxygen species (ROS). This study reveals that among the selected tomato fruits, PGPR strain, *B. subtilis* CBR05 inoculated tomato fruits had the highest amount of phenolics (Figure 2). Some phenolic compounds may prevent oxidative damage in vivo and thus protect against the development of the disease such as cardiac disease and cancer [1,2]. This might be considered as useful for health purposes. In addition, inoculation of *B. subtilis* significantly increased flavonoid content compared to those of both control and market fruits. Similarly, increases in total flavonoids content by *B. licheniformis* have been reported for tomato fruits [32]. In our previous studies, we also found that *B. subtilis* CBR05 inoculation enhanced the accumulation of peroxidase and polyphenol oxidase enzymes, which are involved in the metabolism of phenols and flavonoids [22,23]. Hence, this shows that *B. subtilis* colonization induces resistance against biotic and abiotic stress agents. The results of antioxidant assays also revealed that tomatoes are a rich source of antioxidants, thus their habitual consumption can potentially help to combat the oxidative stress.

Regulation of carotenoid biosynthesis and high-accumulation lycopene during tomato fruit development is widely studied [36–38]. In the present study, we found a significantly higher amount of lycopene (All-E-lycopene) in *B. subtilis* CBR05 inoculated tomato fruits (Figures 3 and 4). Lycopene possesses the highest antioxidant potential among the carotenoids and several other antioxidants found in fruits and vegetables [39]. Thus, the addition of PGPR enhances lycopene content in tomato fruits and can potentially contribute to antioxidant levels of diets. This potent antioxidant activity of lycopene protects from a variety of ROS and reactive nitrogen species (RNS), thus helping in preventing chronic diseases in humans [31,36]. Similar to the lycopene contents, the DPPH and ABTS antioxidant activity of *B. subtilis* CBR05 inoculated tomato fruits was much higher than both the control and market fruits. Earlier, we also reported that defense-related enzymes in tomato after treatment with *B. subtilis* CBR05 efficiently combated *X. campestris* pv, *vesicotoria* [40] and induction of defense-related enzymes like superoxide dismutase, catalase, peroxidase, and polyphenol oxidase assessment revealed the up-regulation of glucanase and phenyl ammonia lyase indicating induced systemic resistance (ISR) in tomato. The earlier results established that the antioxidant capabilities of tomatoes are naturally present. Further, it was specifically proved that *B. subtilis* CBR05 mechanism of disease resistance against *X. campestris* pv, *vesicotoria* was confirmed for the involvement of the de novo pathway involved in Vitamin B6 biosynthesis [41]. In addition, known bacterial elicitors of ISR are microbial associated molecular patterns triggering immunity. When this fails, microbial effector-triggered immunity is induced and leads to programmed cell death. It increases the plant's systemic resistance to subsequent pathogen challenge by PGPR. Moreover, plant probiotic bacterium (PPB) could be used to reduce the use of chemicals (fertilizers, pesticides) in agriculture. This could lead to improved quality at reduced costs and could provide the basis for more sustainable agriculture [42]. Organic agriculture has been widely promoted and adopted to establish better sustainability in food production and crop protection. *Micromonospora* has been regarded as a PPB due to rhizobia helper bacteria properties in *Medicago sativa* L. [43]. Similarly, *Phyllobacterium* and *B. licheniformis* also show promising benefits for increasing vitamin C content across various functional foods and have been considered as PPB, devoid of economic loss [32,44].

In the present investigation, we extracted and quantified phenols, flavonoids, and carotenoids, in the mock-inoculated control, *B. subtilis* CBR05 inoculated plants and market fruits. Among them, PGPR strain, *B. subtilis* CBR05 inoculated tomato fruits were found to have the richest source of lycopene, total phenolics, and flavonoids contents. Additionally, the PGPR strain, *B. subtilis* CBR05 inoculated tomato fruits showed potent antioxidant activities. The significantly higher lycopene content and radical-quenching activity of PGPR strain, *B. subtilis* CBR05 inoculation confer tomato fruits with more nutritional value, and their consumption can minimize oxidative stress-mediated chronic diseases. Thus, biofertilizers based on PGPR may be a viable alternative to improve the nutraceutical quality of greenhouse-produced tomato fruits.

Author Contributions: Conceptualization, M.C., and S.C.C.; methodology, M.C., R.K.S., and S.C.C.; formal analysis, M.C., R.K.S., M.P., and S.C.C.; writing—Original draft preparation, M.C., M.P., J.J.S., and J.W.O.; writing and editing, M.C., J.J.S., M.P., and J.W.O.

Funding: This research received no external funding.

Conflicts of Interest: The authors declare no conflict of interest.

References

1. Choi, S.H.; Kim, D.S.; Kozukue, N.; Kim, H.J.; Nishitani, Y.; Mizuno, M.; Levin, C.E.; Friedman, M. Protein, free amino acid, phenolic, β-carotene, and lycopene content, and antioxidative and cancer cell inhibitory effects of 12 greenhouse-grown commercial cherry tomato varieties. *J. Food Compos. Anal.* **2014**, *34*, 115–127. [CrossRef]
2. Franceschi, S.; Bidoli, E.; La Vecchia, C.; Talamini, R.; D'Avanzo, B.; Negri, E. Tomatoes and risk of digestive-tract cancers. *Int. J. Cancer* **1994**, *59*, 181–184. [CrossRef] [PubMed]
3. Saini, R.K.; Zamany, A.J.; Keum, Y.S. Ripening improves the content of carotenoid, alpha-tocopherol, and polyunsaturated fatty acids in tomato (*Solanum lycopersicum* L.) fruits. *3 Biotech* **2017**, *7*, 43. [CrossRef] [PubMed]
4. Klee, H.J.; Giovannoni, J.J. Genetics and control of tomato fruit ripening and quality attributes. *Annu. Rev. Genet.* **2011**, *45*, 41–59. [CrossRef]
5. Alexander, L.; Grierson, D. Ethylene biosynthesis and action in tomato: A model for climacteric fruit ripening. *J. Exp. Bot.* **2002**, *53*, 2039–2055. [CrossRef]
6. Konozy, E.H.E.; Rogniaux, H.; Causse, M.; Faurobert, M. Proteomic analysis of tomato (*Solanum lycopersicum*) secretome. *J. Plant Res.* **2012**, *126*, 251–266. [CrossRef]
7. Barrett, D.M.; Weakley, C.; Diaz, J.V.; Watnik, M. Qualitative and nutritional differences in processing tomatoes grown under commercial organic and conventional production system. *J. Food Sci.* **2007**, *72*, 441–451. [CrossRef]
8. Erba, D.; Casiraghi, M.C.; Ribas-Agustí, A.; Cáceres, R.; Marfà, O.; Castellari, M. Nutritional value of tomatoes (*Solanum lycopersicum* L.) grown in greenhouse by different agronomic techniques. *J. Food Compos. Anal.* **2013**, *31*, 245–251. [CrossRef]
9. Dumas, Y.; Dadomo, M.; Di Lucca, G.; Grolier, P. Effects of environmental factors and agricultural techniques on antioxidant content of tomatoes. *J. Sci. Food Agric.* **2003**, *83*, 369–382. [CrossRef]
10. Wang, X.; Xing, Y. Evaluation of the effects of irrigation and fertilization on tomato fruit yield and quality: A principal component analysis. *Sci. Rep.* **2017**, *7*, 350. [CrossRef] [PubMed]
11. Beveridge, L.; Whitfield, S.; Challinor, A. Crop modelling: Towards locally relevantand climate-informed adaptation. *Clim. Chang.* **2018**, *147*, 475–489. [CrossRef]
12. Challinor, A.J.; Müller, C.; Asseng, S.; Deva, C.; Nicklin, K.J.; Wallach, D.; Vanuytrecht, E.; Whitfield, S.; Ramirez-Villegas, J.; Koehler, A.K. Improving the use of crop models for risk assessment and climate change adaptation. *Agric. Syst.* **2018**, *159*, 296–306. [CrossRef] [PubMed]
13. Giuliani, M.M.; Gatta, G.; Cappelli, G.; Gagliardi, A.; Donatelli, M.; Fanchini, D.; De Nart, D.; Mongiano, G.; Bregaglio, S. Identifying the most promising agronomic adaptation strategies for the tomato growing systems in Southern Italy via simulation modeling. *Eur. J. Agron.* **2019**, *111*, 125937. [CrossRef]
14. Wang, F.; Kang, S.; Du, T.; Li, F.; Qiu, R. Determination of comprehensive quality index for tomato and its response to different irrigation treatments. *Agric. Water Manag.* **2011**, *98*, 1228–1238. [CrossRef]
15. Shi, S.; Chen, D.; Fu, J.; Liu, L.; Wei, Y.; Shu, B.; Li, W.; Liang, Q.; Liu, C. Comprehensive evaluation of fruit quality traits in longan 'Fengliduo' × 'Dawuyuan' sexual progenies. *Sci. Hortic.* **2015**, *192*, 54–59. [CrossRef]
16. Wang, C.; Gu, F.; Chen, J.; Yang, H.; Jiang, J.; Du, T.; Zhang, J. Assessing the response of yield and comprehensive fruit quality of tomato grown in greenhouse to deficit irrigation and nitrogen application strategies. *Agric. Water Manag.* **2015**, *161*, 9–19. [CrossRef]
17. Luo, H.; Li, F. Tomato yield, quality and water use efficiency under different drip fertigation strategies. *Sci. Hortic.* **2018**, *235*, 181–188. [CrossRef]
18. Mader, P.; Fliessbach, A.; Dubois, D.; Gunst, L.; Fried, P.; Niggli, U. Soil fertility and biodiversity in organic farming. *Science* **2002**, *296*, 1694–1697. [CrossRef] [PubMed]
19. Vessey, J.K. Plant growth promoting rhizobacteria as biofertilizers. *Plant Soil* **2003**, *255*, 571–586. [CrossRef]
20. Mena-Violante, H.G.; Olalde-Portugal, V. Alteration of tomato fruit quality by root inoculation with plant growth-promoting rhizobacteria (PGPR): *Bacillus subtilis* BEB-13bs. *Sci. Hortic.* **2007**, *113*, 103–106. [CrossRef]
21. Zehnder, G.W.; Murphy, I.F.; Sikora, E.J.; Kloepper, J.W. Application to rhizobacteria for induced resistance. *Eur. J. Plant Pathol.* **2001**, *107*, 39–50. [CrossRef]
22. Chandrasekaran, M.; Chun, S.C. Induction of defence-related enzymes in tomato (*Solanum lycopersicum*) plants treated with *Bacillus subtilis* CBR05 against *Xanthomonas campestris* pv. *vesicatoria*. *Biocontrol Sci. Technol.* **2016**, *26*, 1366–1378. [CrossRef]

23. Chandrasekaran, M.; Chun, S.C. Expression of PR-protein genes and induction of defense-related enzymes by *Bacillus subtilis* CBR05 in tomato (*Solanum lycopersicum*) plants challenged with *Erwinia carotovora* subsp. *carotovora*. *Biosci. Biotechnol. Biochem.* **2016**, *80*, 2277–2283. [CrossRef] [PubMed]
24. Glick, B.R.; Penrose, D.M.; Li, J. A model for the lowering of plant ethylene concentrations by plant growth promoting bacteria. *J. Theor. Biol.* **1998**, *190*, 63–68. [CrossRef] [PubMed]
25. Verheul, M.; Maessen, H.; Grimstad, O.S. Optimizing a year-round cultivation system of tomato under artificial light. *Acta Hortic.* **2012**, *956*, 389–394. [CrossRef]
26. Moco, S.; Capanoglu, E.; Tikunov, Y.; Bino, R.J.; Boyacioglu, D.; Hall, R.D.; Vervoort, J.; De Vos, R.C. Tissue specialization at the metabolite level is perceived during the development of tomato fruit. *J. Exp. Bot.* **2007**, *58*, 4131–4146. [CrossRef]
27. Blois, M.S. Antioxidant determinations by the use of a stable free radical. *Nature* **1958**, *181*, 1199–1200. [CrossRef]
28. Ramos, A.; Rivero, R.; Victoria, M.C.; Visozo, A.; Piloto, J.; Garcia, A. Assessment of mutagenicity in *Parthenium hysterophorus* L. *J. Ethnopharmacol.* **2001**, *77*, 25–30. [CrossRef]
29. Chen, L.; Xin, X.; Yuan, Q.; Su, D.; Liu, W. Phytochemical properties and antioxidant capacities of various colored berries. *J. Sci. Food Agric.* **2014**, *94*, 180–188. [CrossRef]
30. Rodriguez-Amaya, D.B. *A Guide to Carotenoid Analysis in Foods*; ILsi Press: Washington, DC, USA, 2001.
31. Saini, R.K.; Nile, S.H.; Park, S.W. Carotenoids from fruits and vegetables: Chemistry, analysis, occurrence, bioavailability and biological activities. *Food Res. Int.* **2015**, *76*, 735–750. [CrossRef]
32. Ochoa-Velasco, C.E.; Valadez-Blanco, R.; Salas-Coronado, R.; Sustaita-Rivera, F.; Hernández-Carlos, B.; García-Ortega, S.; Santos-Sánchez, N.F. Effect of nitrogen fertilization and *Bacillus licheniformis* biofertilizer addition on the antioxidants compounds and antioxidant activity of greenhouse cultivated tomato fruits (*Solanum lycopersicum* L. var. Sheva). *Sci. Hortic.* **2016**, *201*, 338–345. [CrossRef]
33. Baset Mia, M.A.; Shamsuddin, Z.H. Nitrogen fixation and transportation by rhizobacteria: A scenario of rice and banana. *Int. J. Bot.* **2010**, *6*, 235–242.
34. Bar-Ness, E.; Chen, Y.; Hadar, Y.; Marchner, H.; Romheld, V. Siderophores of Pseudomonas putida as an iron source for dicot and monocot plants. *Plant Soil* **1991**, *130*, 231–241. [CrossRef]
35. Lucas, G.J.A.; Probanza, A.; Ramos, B.; Palomino, M.R.; Gutierrez Mañero, F.J. Effect of inoculation of *Bacillus licheniformis* on tomato and pepper. *Agronomie* **2004**, *24*, 169–176.
36. Martí, R.; Roselló, S.; Cebolla-Cornejo, J. Tomato as a source of carotenoids and polyphenols targeted to cancer prevention. *Cancer* **2016**, *8*, 58. [CrossRef]
37. Ronen, G.; Cohen, M.; Zamir, D.; Hirschberg, J. Regulation of carotenoid biosynthesis during tomato fruit development: Expression of the gene for lycopene epsilon-cyclase is downregulated during ripening and is elevated in the mutant Delta. *Plant J. Cell Mol. Biol.* **1999**, *17*, 341–351. [CrossRef]
38. Su, L.; Diretto, G.; Purgatto, E.; Danoun, S.; Zouine, M.; Li, Z.; Roustan, J.P.; Bouzayen, M.; Giuliano, G.; Chervin, C. Carotenoid accumulation during tomato fruit ripening is modulated by the auxin–ethylene balance. *BMC Plant Biol.* **2015**, *15*, 114. [CrossRef]
39. Zhang, Y.; Liu, Y.; Lv, Q. DFT study on the quenching mechanism of singlet oxygen by lycopene. *RSC Adv.* **2016**, *6*, 98498–98505. [CrossRef]
40. Chandrasekaran, M.; Belachew, S.T.; Yoon, E.; Chun, S.C. Expression of β-1,3-glucanase (GLU) and phenylalanine ammonia-lyase (PAL) genes and their enzymes in tomato plants induced after treatment with *Bacillus subtilis* CBR05 against *Xanthomonas campestris* pv. *vesicatoria*. *J. Gen. Plant Pathol.* **2016**, *83*, 7–13. [CrossRef]
41. Chandrasekaran, M.; Manivannan, P.; Chun, S.C. *Bacillus subtilis* CBR05 induces Vitamin B6 biosynthesis in tomato through the *de novo* pathway in contributing disease resistance against *Xanthomonas campestris* pv. *vesicatoria*. *Sci. Rep.* **2019**, *9*, 6495. [CrossRef]
42. Jiménez-Gómez, A.; Flores-Félix, J.D.; García-Fraile, P.; Mateos, P.F.; Menéndez, E.; Velázquez, E.; Rivas, R. Probiotic activities of rhizobium laguerreae on growth and quality of spinach. *Sci. Rep.* **2018**, *8*, 295. [CrossRef]

43. Martínez-Hidalgo, P.; García, J.M.; Pozo, M.J. Induced systemic resistance against *Botrytis cinerea* by *Micromonospora* strainsisolated from root nodules. *Front. Microbiol.* **2015**, *6*, 922. [CrossRef] [PubMed]
44. Flores-Felix, J.D.; Silva, L.R.; Rivera, L.P.; Marcos-Garcia, M.; Garcia-Fraile, P.; Martinez-Molina, E.; Mateos, P.F.; Velazquez, E.; Andrade, P.; Rivas, R. Plants probiotics as a tool to produce highly functional fruits: The case of *Phyllobacterium* and vitamin C in strawberries. *PLoS ONE* **2015**, *10*, e0122281. [CrossRef]

agronomy

MDPI

Article

Effects of Water Stress and Modern Biostimulants on Growth and Quality Characteristics of Mint

Hosam O. Elansary [1,2,*], Eman A. Mahmoud [3], Diaa O. El-Ansary [4] and Mohamed A. Mattar [5,6,*]

1 Plant Production Department, College of Food and Agriculture Sciences, King Saud University, Riyadh 11451, Saudi Arabia
2 Floriculture, Ornamental Horticulture, and Garden Design Department, Faculty of Agriculture (El-Shatby), Alexandria University, Alexandria 21545, Egypt
3 Department of Food Industries, Damietta University, Damietta 3417, Egypt; emanmail2005@du.edu.eg
4 Precision Agriculture Laboratory, Department of Pomology, Faculty of Agriculture (El-Shatby), Alexandria University, Alexandria 21545, Egypt; diaa.elansary@alexu.edu.eg
5 Department of Agricultural Engineering, College of Food and Agriculture Sciences, King Saud University, Riyadh 11451, Saudi Arabia
6 Agricultural Engineering Research Institute (AEnRI), Agricultural Research Centre, Giza 12618, Egypt
* Correspondence: helansary@ksu.edu.sa (H.O.E.); mmattar@ksu.edu.sa (M.A.M.); Tel.: +966-581216322 (H.O.E.); +966-11-4676024 (M.A.M.); Fax: +966-11-4678502 (M.A.M.)

Received: 5 November 2019; Accepted: 15 December 2019; Published: 18 December 2019

Abstract: Natural biostimulants combine different elicitors that may influence economic properties of herbal crops, such as mint. Mint (*Mentha longifolia* L.) plants were subjected to three water levels based on container substrate capacity (CSC; 100% CSC, 70% CSC, and 50% CSC) and/or applications of four biostimulants (CRADLE™, Mobilizer™, Nanozim De'Lite™ [ND], and Nanozim NXT™ [NN]). ND and NN exhibited higher vegetative growth and root dry weight than the control (without biostimulants) and other treatments. NN produced the highest fresh and dry mint yields under all water levels. Irrigation water-use efficiency (IWUE) of NN was highest (2.78 kg m^{-3}) with 70% CSC, whereas the control produced the lowest IWUE (1.85 kg m^{-3}) with 100% CSC. Biostimulants boosted physiological and metabolic responses, including gas exchange, leaf water potential, relative water content, and proline accumulation of stressed plants. NN treatment with 70% CSC had the highest essential oil (EO) ratio (3.35%). Under 70% and 50% CSC with NN treatment, the proportion of 1,8-cineol increased and that of pulegone decreased in EOs. Increased antioxidant activities, reduced H_2O_2 levels, and increased catalase and superoxide dismutase activities were observed. Applications of ND and NN during water stress conditions increased economic and medicinal properties of mint EOs with applications in the agricultural and pharmaceutical industries.

Keywords: *Mentha longifolia*; biostimulants; *Ascophyllum nodosum*; humic acid; antioxidants

1. Introduction

Mint plants have a long history as traditional medicinal plants [1]. *Mentha longifolia* L. belongs to the family Lamiaceae and naturally occurs in Egypt, Saudi Arabia, and most Arabian countries. The fresh/dried plants are mainly used as an herbal medicine for the treatment of indigestion, menstrual pain, coughs, asthma, fever, and headaches [2,3]. The fresh leaves are used in soft drinks and as garnishes for salads in some countries. The essential oil (EO) is used in the pharmaceutical, cosmetic, and food industries [4]. The EOs exhibit strong antimicrobial activity against several microorganisms [3].

Water stress is one of the major limiting factors for agriculture and food safety worldwide [5]. This stress causes reduced vegetative growth and great losses to farmers. Different studies have focused on the effects of water stress on the growth parameters and EO yield. Zade et al. [6] reported that water stress decreased peppermint plant fresh and dry weight, leaf number, plant height, and root dry weight but nonetheless increased EOs compared to that of normal irrigation in greenhouse and field experiments. Figueroa-Pérez et al. [7] showed that water stress decreased fresh and dry weights of peppermint but increased composition of plant secondary metabolites and antioxidant capacity. Ekren et al. [8] reported that plant height and yield of purple basil were negatively affected by water stress, whereas the EO content increased and irrigation water-use efficiencies were not significantly affected. Shormin et al. [9] showed that the harmful effects of water stress on Japanese mint yield could not be compensated by high nitrogen quantities. Farahani et al. [10] also reported the highest content of EO in balm occurred at 60% field capacity (FC). However, in other studies, Khorasaninejad et al. [11] showed that water stress had negative effects on some growth parameters and EO content of peppermint plants. Razmjoo et al. [12] found that this stress reduced some growth parameters and EO content of chamomile.

Several approaches have been applied to control water stress, such as the use of biostimulants. Modern biostimulants have been produced to increase the productivity and the quality of horticultural crops and help the plants tolerate stress conditions. Some of these biostimulants are mixtures of seaweed extracts, humic acid, and macro and micro elements, whereas other products contain mixtures of mycorrhiza and seaweed extracts, as well as other micro elements. Seaweed extracts work as elicitors for plant secondary metabolites, including EOs and may increase the pharmaceutical properties against microorganisms [13,14]. However, the effects of the mixtures of seaweed extracts and other elicitors, such as humic acid and specific minerals have not been investigated for mint plants. Further, water stress may cause significant changes in the EO composition and these changes might cause parallel changes in the antimicrobial properties of respective EOs.

In this investigation, our goal was to determine the effects of water stress and commercial biostimulants on the growth, physiology, secondary metabolites, and antioxidant activities of mint (*Mentha longifolia* L.). We propose that these natural biostimulants modulate growth, EO ratio, and EO constitutes, leading to enhanced bioactivity of mint plants. These effects indeed have the potential to have future agricultural industry applications.

2. Materials and Methods

2.1. Plant Material

Uniformly rooted cuttings of mint (*Mentha longifolia* L.) were brought from nurseries of the Alexandria University farm in February 2018 and 2019 (as two successive growing seasons). The species was identified and vouchered by Hosam Elansary in the Faculty of Agriculture, Alexandria University. The sandy soil (75.5%, 13.2%, and 11.3% of sand, silt and clay, respectively) samples were air dried and sieved with a 2 mm mesh. The soil had an FC of 20.5%, wilting point of 9.6%, electrical conductivity of 0.36 mS cm^{-1}, organic matter of 1.4%, pH of 6.2, total nitrogen of 0.085%, total phosphorus of 0.05%, and total sulfur of 0.03%. After proper soil preparation, the plants were grown in 2.1 L plastic pots containing the natural sandy soil supplemented with Crystalon® (65 kg N ha^{-1} as urea, 40 kg P_2O_5 ha^{-1} as triple superphosphate, 34 kg K_2O ha^{-1} as potassium sulfate, and 2 g L^{-1} media) in the greenhouse. The temperature inside the greenhouse ranged between 15.0 °C (night) and 27.3 °C (day) and the relative humidity ranged between 67% and 72% during the growing period. The photosynthetic active radiation was approximately 1000 μmol m^{-2} s^{-1} at noon. Daily watering by drip irrigation was applied to reach the full pot substrate FC. Pots were irrigated equally for 30 days after transplantation (DAT). Mint plants were harvested at 90 DAT. Container substrate capacity (CSC) is the maximum amount of water that can be retained by the substrate after the discharge because of gravity [15]. Before planting, the gravimetric method was used to determine CSC or FC by watering the plants to saturate the soil then the pots were left to drain for 60 min and the volume of drained water was quantified and the difference between the supplied and drained water volumes were considered the volumetric water retained by the soil (i.e., *CSC*). The amount of water applied (*AWA*) to compensate for the soil water deficit to reach the FC is calculated as follows:

$$AWA = (CSC - \theta_v)\ D\ A \tag{1}$$

where θ_v is soil water content at the irrigation event, D is the soil depth, and A is the surface area of the pot.

2.2. Treatments

The plants were subjected to three watering levels of CSC (100%, 70%, and 50%) after 30 DAT and/or single biostimulant of four commercial biostimulants, namely, CRADLE™, Mobilizer™, Nanozim NXT™, and Nanozim De'Lite™ (Biostadt, Mumbai, India). CRADLE (CR) powder is a mycorrhizal biofertilizer developed by InGene Organics, India and was used at g L^{-1} growing soil. Mobilizer (Mob) is a granular mycorrhizal biofertilizer mixed with kelp seaweed extract (*Macrocystis pyrifera*), humic acid, and amino acids and was applied at g L^{-1} growing soil. Nanozim De'Lite (ND) is a granular formulation of 25% ($w\ w^{-1}$) seaweed (*Ascophyllum nodosum*), 25% (w/w) carbohydrates, 2% ($w\ w^{-1}$) amino acid, and 1% (w/w) potassium (K_2O), and was used at 1 g L^{-1} with irrigation water. Nanozim NXT (NN) is a liquid mixture of 15% ($w\ w^{-1}$) seaweed (*Ascophyllum nodosum*), 5% (w/w) humic acid, 1% ($w\ w^{-1}$) potassium (K_2O), 0.01% ($w\ w^{-1}$) phosphorus (P_2O_5), 0.05% (w/w) alginic acid, 0.05% ($w\ w^{-1}$) hydrolyzed protein, and several micronutrients and was applied at 1.5 mL L^{-1} of irrigation water. The doses of the biostimulants and method of applications matched the manufacturer recommendations and untreated plants with biostimulants were considered the controls. Plants were grouped into three blocks containing 10 replicates per treatment [3 water levels (100%, 70%, and 50% CSC) × (4 biostimulants +1 control "without biostimulant") = 15 treatments] and totaling 450 plants (150 plants/block × 3 blocks) in a completely randomized design.

2.3. Measurements

2.3.1. Morphological and Physiological

Following 9 weeks of treatments, several morphological measurements were determined including leaf number ($plant^{-1}$), leaf area ($cm^2\ plant^{-1}$), plant heights (cm), plant fresh weight (g), plant dry weight (g), and root dry weight (g). Irrigation water-use efficiency (IWUE, kg m^{-3}) was calculated by dividing the fresh weight of the plant (kg) by the total *AWA* (m^3) to each treatment during the growing period [16]. A digital area meter was used to determine the leaf area. The dry weights were determined following drying at 35 °C in an oven until reaching a constant weight.

Gas exchange measurements were performed on fully expanded leaves, under clear, sunny conditions using a portable photosynthesis system analyzer (ADC BioScientific, LCi, Bioscientific, Ltd., Hoddesdon, UK) and included photosynthetic rate (A), transpiration rate (E), and stomatal conductance (gs). Leaf midday water potential and midday relative water content were calculated at the end of the experiments at noon following the methods of Elansary et al. [17]. Leaf proline composition was also determined following the methods of Elansary et al. [18].

2.3.2. Essential Oil and Gas Chromatography/Mass Spectrometry (GC/MS)

The EOs were obtained by hydro-distillation of dried leaves for 1 h in Clevenger type glass equipment in the Department of Plant Production, King Saud University. The EO ratio was determined per treatment and the EOs were maintained dry by subjecting samples to anhydrous sodium sulfate, then stored at 4 °C. A Thermo Scientific, Trace GC Ultra was used coupled with a mass spectrometer (ISQ). A TG-1MS column (narrow bore, length 30 m × 0.32 mm ID, 0.25 μm film thickness) was used and the carrier gas was helium. The machine was programed with a starting temperature of 45 °C, then a gradual increase was made to 165 °C (4 °C min^{-1}), followed by an increase to 280 °C (15 °C min^{-1}), and ending with holding time of 15 min. A 2 μL sample of each EO was injected at 250 °C on a splitless mode flow (1 mL min^{-1}) for splitless time (3 min) followed by another split flow (10 mL min^{-1}). The FID was also accomplished in the same column and program. A homologous series of *n*-alkanes (C_{10}–C_{36}) was used to identify the compounds by the retention time and indices were coupled with a mass spectral search program (NIST Ver. 2.0) and WILEY libraries. Selected references from the literature were also used for comparison purposes [1,13].

2.3.3. Antioxidant Potential

Lipid peroxidation levels expressed as thiobarbituric acid reactive substances (TBARS), catalase (CAT) activity, H_2O_2, and superoxide dismutase (SOD) activities were quantified in frozen petal tissues [19].

2.4. Statistical Analyses

Data obtained from both years (2018 and 2019) were subjected to analyses of variance (ANOVA) with a completely randomized design to determine the significance of differences among treatments in SPSS Version 22 software. Standard errors (SE) were calculated from the data presenting for the mean of 20 replicates from both years. The least significant differences (LSD) method at $p \leq 0.05$ was used to compare all means at each watering level [20].

3. Results

3.1. Morphological Responses

Plants subjected to different water levels of *CSC* and/or biostimulants showed different morphological responses as shown in Table 1. The plants were subjected to three watering levels of *CSC* (100%, 70%, and 50%) and/or single biostimulants (CR, Mob, ND, and NN). Under 100% *CSC*, NN-treated plants showed the highest leaf number and leaf area and was followed by ND, Mob, and CR. ND and NN treatments at 70% *CSC* and 50% CSC showed the highest increase in leaf number and leaf area as compared to those of other treatments, as well as those of the control. The tallest plants and highest root dry weight were found treatments with NN. CR and Mob showed no significant differences under 50% *CSC*. On the other hand, when water stress increased (i.e., 70% *CSC* and 50% *CSC*), the values for leaf number (20.19% and 45.77%), leaf area (24% and 52.40%), plant height (17.30% and 41.35%), and root dry weight (15.70% and 49.59%) decreased significantly. Plant fresh and dry weights increased significantly in biostimulant-treated plants compared to those of the control, and the highest increase was for plants subjected to NN followed by ND under different water levels (Table 2). NN-treated plants with 70% *CSC* water level had the highest IWUE (average of 2.78 kg m^{-3}) in both growth seasons, which was statistically different from that of all other treatments.

3.2. Physiological and Metabolic Performance

Figure 1a–c shows the effects of water stress and biostimulants on gas exchange, namely, A, E, and gs of mint plants during the growing seasons in 2018 and 2019 (shown as averages). Under 100% *CSC*, the A of mint plants showed a ($p < 0.05$) significant increase in plants subjected to NN (8.35 μmol CO_2 m^{-2} s^{-1}) compared to that of other treatments (Figure 1a). The reduction in irrigation water to 70% *CSC* and 50% *CSC* caused significantly reduced A values in control plants. However, biostimulant treatment increased the values of A, wherein NN-treated plants under 70% *CSC* and 50% *CSC* had the highest values (7.4 and 4.9 μmol CO_2 m^{-2} s^{-1}, respectively), followed by ND-treated plants (7.3 and 4.6 μmol CO_2 m^{-2} s^{-1}, respectively).

Table 1. Average morphological responses for mint plants grown under greenhouse conditions during the growth season in 2018 and 2019, subjected to three water levels, 100%, 70%, and 50% *CSC*, and four biostimulants, CR, Mob, ND, and NN, as well as a control (without biostimulants).

Variable	Leaf Number (plant^{-1})						Leaf Area (cm^2 plant^{-1})					
Water Levels/Biosti.	Control	CR	Mob	ND	NN	Means	Control	CR	Mob	ND	NN	Means
100% CSC	57.00 ^{e}	62.10 ^{d}	67.20 ^{c}	72.30 ^{b}	77.20 ^{a}	67.16 ^{A}	228.00 ^{e}	247.10 ^{d}	268.80 ^{c}	288.20 ^{b}	307.80 ^{a}	267.98 ^{A}
70% CSC	47.00 ^{e}	51.20 ^{d}	53.10 ^{c}	56.80 ^{b}	59.90 ^{a}	53.60 ^{B}	178.60 ^{e}	193.56 ^{d}	200.78 ^{c}	216.80 ^{b}	228.62 ^{a}	203.67 ^{B}
50% CSC	29.00 ^{d}	34.80 ^{c}	35.90 ^{c}	39.30 ^{b}	43.10 ^{a}	36.42 ^{C}	101.50 ^{d}	121.20 ^{c}	125.65 ^{c}	137.55 ^{b}	151.85 ^{a}	127.55 ^{C}
Means	44.33 ^{E}	49.37 ^{D}	52.07 ^{C}	56.13 ^{B}	60.07 ^{A}		169.37 ^{E}	187.29 ^{D}	198.41 ^{C}	214.18 ^{B}	229.42 ^{A}	
Water levels Biosti. Water levels × Biosti.	*** ($LSD_{0.05}$ = 0.877) *** ($LSD_{0.05}$ = 1.132) *** ($LSD_{0.05}$ = 1.961)						*** ($LSD_{0.05}$ = 2.502) *** ($LSD_{0.05}$ = 3.230) *** ($LSD_{0.05}$ = 5.594)					

Variable	Plant Height (cm)						Root Dry Weight (g)					
Water Levels/Biosti.	Control	CR	Mob	ND	NN	Means	Control	CR	Mob	ND	NN	Means
100% CSC	27.60 ^{e}	29.10 ^{d}	30.20 ^{c}	32.10 ^{b}	35.30 ^{a}	30.86 ^{A}	1.05 ^{e}	1.13 ^{d}	1.19 ^{c}	1.29 ^{b}	1.38 ^{a}	1.21 ^{A}
70% CSC	23.10 ^{e}	24.40 ^{d}	25.70 ^{c}	26.90 ^{b}	27.50 ^{a}	25.52 ^{B}	0.82 ^{d}	1.01 ^{c}	1.04 ^{c}	1.12 ^{b}	1.21 ^{a}	1.02 ^{B}
50% CSC	16.20 ^{d}	17.30 ^{c}	17.80 ^{c}	18.90 ^{b}	20.30 ^{a}	18.10 ^{C}	0.56 ^{d}	0.59 ^{c}	0.60 ^{c}	0.64 ^{b}	0.67 ^{a}	0.61 ^{C}
Means	22.30 ^{E}	23.60 ^{D}	24.57 ^{C}	25.97 ^{B}	27.70 ^{A}		0.81 ^{E}	0.91 ^{D}	0.94 ^{C}	1.02 ^{B}	1.06 ^{A}	
Water levels Biosti. Water levels × Biosti.	*** ($LSD_{0.05}$ = 0.282) *** ($LSD_{0.05}$ = 0.365) *** ($LSD_{0.05}$ = 0.631)						*** ($LSD_{0.05}$ = 0.010) *** ($LSD_{0.05}$ = 0.013) *** ($LSD_{0.05}$ = 0.022)					

Data represent the mean calculated from n = 100, 60, or 20 for each water level, biostimulants or their interaction, respectively, in two growth seasons. *** ($p \leq 0.001$). Means differing in lowercase letters (a–e) within each row indicate significant differences at $p \leq 0.05$. Means differing in uppercase letters (A–E) within each mean row and each mean column indicate significant differences according to the least significant difference (LSD) test at $p \leq 0.05$. CSC (container substrate capacity), Control (without biostimulants), CR (CRADLE™), Mob (Mobilizer™), ND (Nanozim De'Lite™), NN (Nanozim NXT™), and Biosti. (biostimulant).

Table 2. Average plant fresh and dry weights and irrigation water-use efficiency (IWUE) for mint plants grown under greenhouse conditions during the growth season in 2018 and 2019, subjected to three water levels, 100%, 70%, and 50% *CSC*, and four biostimulants, CR, Mob, ND, and NN, as well as a control (without biostimulants).

Variable	Water Applied (mm)	Plant Fresh Weight (g)					
Water Levels/Biosti.		Control	CR	Mob	ND	NN	Means
100% CSC	35.91	7.38 ^{d}	8.01 ^{c}	8.48 ^{b}	8.47 ^{b}	9.24 ^{a}	8.32 ^{A}
70% CSC	27.00	6.02 ^{d}	7.53 ^{c}	7.39 ^{c}	7.81 ^{b}	8.34 ^{a}	7.42 ^{B}
50% CSC	22.14	4.69 ^{b}	4.71 ^{b}	4.83 ^{b}	5.18 ^{a}	5.31 ^{a}	4.94 ^{C}
Means		6.03 ^{E}	6.75 ^{D}	6.90 ^{C}	7.15 ^{B}	7.63 ^{A}	
Water levels Biosti. Water levels ×Biosti.		*** ($LSD_{0.05}$ = 0.086) *** ($LSD_{0.05}$ = 0.111) *** ($LSD_{0.05}$ = 0.193)					

Variable	Plant Dry Weight (g)						IWUE (kg m^{-3})					
Water Levels/Biosti.	Control	CR	Mob	ND	NN	Means	Control	CR	Mob	ND	NN	Means
100% CSC	6.64 ^{e}	7.05 ^{d}	7.38 ^{c}	7.79 ^{b}	8.41 ^{a}	7.45 ^{A}	1.85 ^{d}	2.01 ^{c}	2.13 ^{b}	2.12 ^{b}	2.32 ^{a}	2.08 ^{B}
70% CSC	5.12 ^{d}	6.48 ^{c}	6.50 ^{c}	6.87 ^{b}	7.42 ^{a}	6.48 ^{B}	2.01 ^{d}	2.51 ^{c}	2.46 ^{c}	2.60 ^{b}	2.78 ^{a}	2.47 ^{A}
50% CSC	3.75 ^{d}	3.96 ^{c}	4.01 ^{c}	4.25 ^{b}	4.46 ^{a}	4.09 ^{C}	1.91 ^{b}	1.92 ^{b}	1.96 ^{b}	2.11 ^{a}	2.16 ^{a}	2.01 ^{C}
Means	5.17 ^{E}	5.83 ^{D}	5.96 ^{C}	6.31 ^{B}	6.76 ^{A}		1.92 ^{E}	2.15 ^{D}	2.18 ^{C}	2.28 ^{B}	2.42 ^{A}	
Water levels Biosti. Water levels ×Biosti.	*** ($LSD_{0.05}$ = 0.075) *** ($LSD_{0.05}$ = 0.097) *** ($LSD_{0.05}$ = 0.168)						*** ($LSD_{0.05}$ = 0.028) *** ($LSD_{0.05}$ = 0.036) *** ($LSD_{0.05}$ = 0.062)					

Data represent the mean calculated from n = 100, 60, or 20 for each water level, biostimulants or their interaction, respectively, in two growth seasons. *** ($p \leq 0.001$). Means differing in lowercase letters (a–e) within each row indicate significant differences at $p \leq 0.05$. Means differing in uppercase letters (A–E) within each mean row and each mean column indicate significant differences according to the least significant difference (LSD) test at $p \leq 0.05$. CSC (container substrate capacity), Control (without biostimulants), CR (CRADLE™), Mob (Mobilizer™), ND (Nanozim De'Lite™), NN (Nanozim NXT™), and Biosti. (biostimulant).

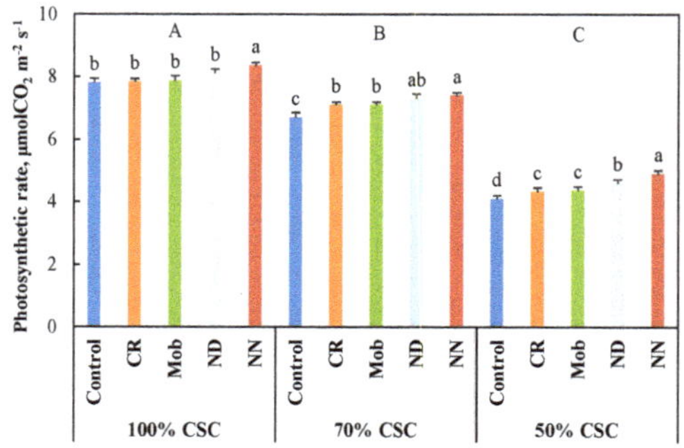

(a)

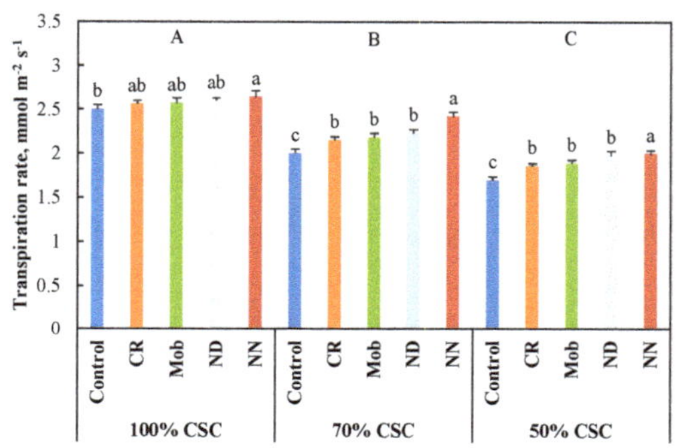

(b)

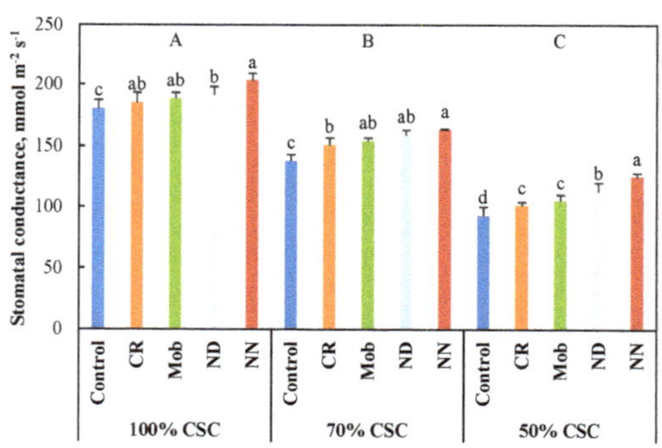

(c)

Figure 1. *Cont.*

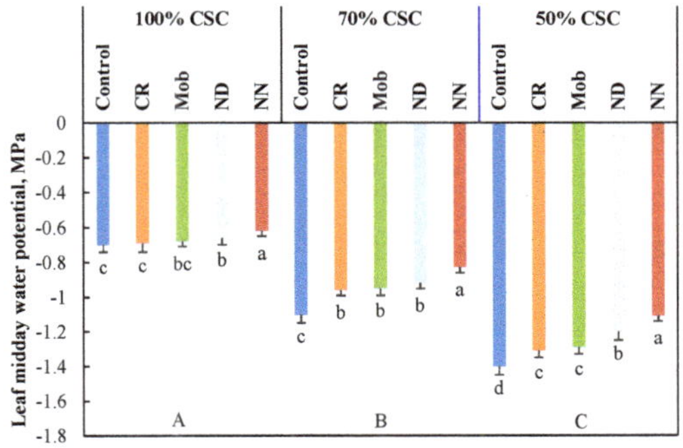

(d)

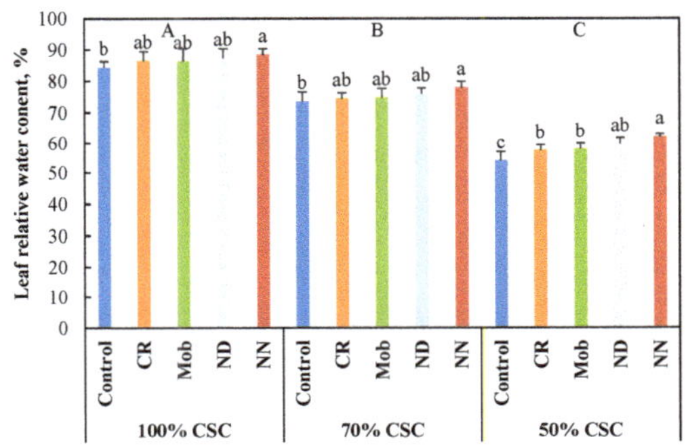

(e)

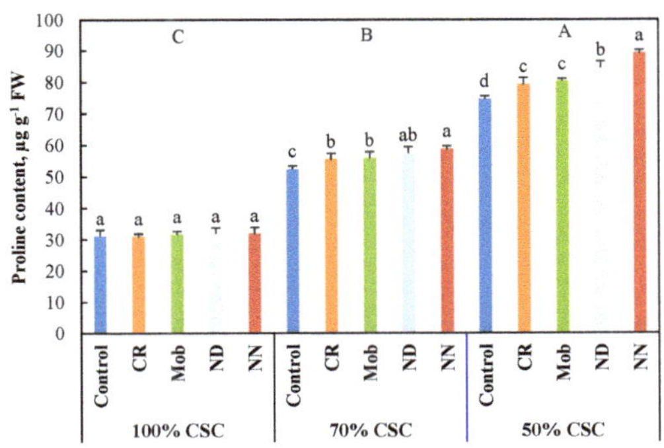

(f)

Figure 1. Photosynthetic rate (**a**), transpiration rate (**b**), stomatal conductance (**c**), leaf midday water potential (**d**), leaf relative water content (**e**), and proline content (**f**) responses of mint plants, average of two growing seasons, as affected by three water levels, 100%, 70%, and 50% *CSC* (container substrate capacity), and four biostimulants, CR (CRADLE™), Mob (Mobilizer™), ND (Nanozim De'Lite™), and NN (Nanozim NXT™), and a Control (without biostimulants). FW (fresh weight). Different capital letters on top indicate significant differences between water levels at $p \leq 0.05$. Different letters on top of columns indicate significant differences between biostimulants across water levels at $p \leq 0.05$. Bars indicate the means ± SE of the mean ($n = 20$).

The values of E were higher in plants subjected to biostimulants compared to that of the control (Figure 1b). Treating plants with CR, Mob, and ND showed no significant differences in E under different *CSC* treatments. However, the E in NN-treated plants was significantly ($p < 0.05$) increased by 5.6%, 21%, and 17.6%, respectively, compared with that of the control plants under 100%, 70%, and 50% *CSC*.

The values of gs increased in plants treated with NN (203, 163, and 125 mmol m^{-2} s^{-1}, respectively) compared to that of other biostimulant treatments under 100%, 70%, and 50% *CSC* (Figure 1c). Statistically significant differences in gs under 50% *CSC* were found only between the ND treatment and both CR and Mob treatments, but significant effects on gs to the same treatments were not observed for 100% and 70% *CSC*. Irrespective of the biostimulant treatments, there were significant ($p < 0.05$) differences between the water levels treatments, where the 100% *CSC* was superior.

Leaf midday water potential increased in plants subjected to different biostimulants under 70% *CSC* (average of −0.95 MPa) and 50% *CSC* (average of −1.26 MPa) compared to that of 100% *CSC* (average of −0.67 MPa), as shown in Figure 1d. The NN treatment had the lowest water potential under different water level treatments. The leaf relative water content increased significantly ($p < 0.05$) in plants subjected to NN under different water levels (Figure 1e), where NN-treated plants with 100% *CSC* had the highest value of 88.3%. The proline content (Figure 1f) increased in plants subjected stress conditions only (70% and 50% *CSC*) and the increases were higher in plants subjected to biostimulants than that of the control (without biostimulants) treatment. The NN-treated plants with 50% CSC had the highest proline content value of 89.1 μg g^{-1} fresh weight.

3.3. EO Ratio and Constitutes

The EO ratio was increased in response to biostimulant treatments as shown in Figure 2. Control treatments had the lowest EO ratio of 2.7%, 2.8%, and 2.1% fresh weight, respectively, under 100%, 70%, and 50% *CSC*. In NN-treated plants, the average EO ratio was increased by 21.57%, followed by that of ND (15.81%), in relation to that of the control plants under water level treatments. There was a significant difference in the EO ratio ($p < 0.05$) of mint plants between different biostimulant treatments within each water level, except between CR and Mob ($p > 0.05$) under water stress conditions of 70% and 50% *CSC*. Irrespective of the biostimulant treatments, the EO ratio under 70% *CSC* was not significantly higher than that of 100% *CSC*.

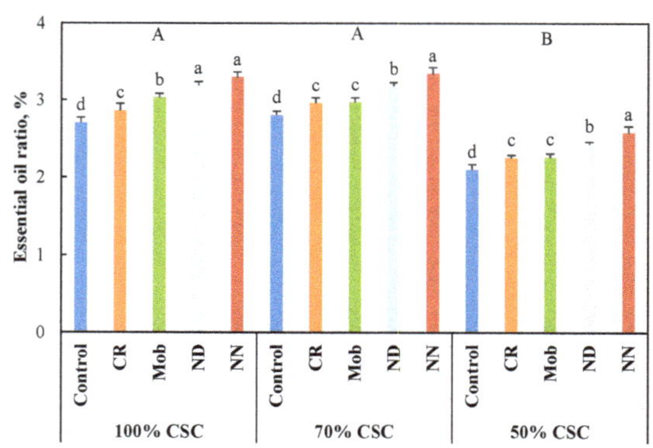

Figure 2. Essential oil ratio of mint plants, average of two growing seasons, as affected by three water levels: 100, 70, and 50% *CSC* (container substrate capacity), and four biostimulants: CR (CRADLE™), Mob (Mobilizer™), ND (Nanozim De'Lite™), and NN (Nanozim NXT™), in addition to Control (without biostimulants). Different capital letters on top are significant differences between water levels at $p \leq 0.05$. Different letter on top columns indicate significant differences between biostimulants across water levels at $p \leq 0.05$. Bars give the means ± SE of the mean ($n = 20$).

Major EO constitutes in all treatments were 1-menthone, isopulegone, pulegone, α-pinene, 1,8-cineol, and α-terpineol ratios as shown in Tables 3 and 4. The 1-menthone, isopulegone and pulegone were significantly ($p < 0.001$) reduced in plants subjected to water stress and biostimulant treatments (Table 3). Irrespective of the biostimulant treatments, 70% and 50% *CSC* plants exhibited decreased 1-menthone by 5.3% and 9.6%, respectively, compared to that of 100% *CSC* plants. Likewise, isopulegone for these plants was decreased by 5.3% and 8.6% and pulegone by 6.9% and 8.5%, respectively. Irrespective of the water level treatments, ND and NN treatments significantly reduced 1-menthone, isopulegone, and pulegone, by 23.8% and 33.3%; 8.1% and 18.6%; 11.1% and 18% on average, respectively, compared to that of the control treatment. However, the application of biostimulants significantly ($p < 0.001$) increased the α-pinene, 1,8-cineol, and α-terpineol ratios under different water stress conditions (Table 4). These ratios were significantly ($p < 0.001$) different between 100%, 70%, and 50% *CSC* plants, where 100% CSC showed the lowest values. The NN-treated plants yielded the highest ratios for α-pinene (4.3%), 1,8-cineol (36.1%), and α-terpineol (3.4%) at 50% *CSC*.

3.4. Antioxidant Activities

There was a significantly ($p < 0.05$) reduced accumulation of lipid peroxidation and H_2O_2 in plants subjected to different biostimulants, as shown in Figure 3. The NN-treated plants yielded the lowest values of lipid peroxidation (57, 46, and 27 μmol TBARS g^{-1} fresh weight, respectively) and H_2O_2 (2.6, 4.6, and 5.9 μmol g^{-1} fresh weight, respectively) under 100%, 70%, and 50% *CSC*. However, control plants had the highest values of lipid peroxidation (64, 54, and 37 μmol TBARS g^{-1} fresh weight, respectively) and H_2O_2 (2.9, 5.2, and 7.2 μmol g^{-1} fresh weight, respectively). It was observed that there were no significant differences between the control and CR treatments in lipid peroxidation and H_2O_2 under different water levels, except for the 70% *CSC* for lipid peroxidation. In 50% *CSC*, there were only significant differences between the Mob and ND treatments. On the contrary, there were significant ($p < 0.05$) increases in the activities of CAT and SOD of leaf extracts of biostimulant-treated plants under normal and water stress conditions (Figure 3). The highest CAT and SOD activity values were found in NN treatments (increasing 25.5% and 40.3%, respectively), followed by that of ND-treated plants (increasing 17.9% and 26.8%, respectively) comparing with those of the control treatment, which had the lowest values. The CAT activity was significantly ($p < 0.05$) increased by 30% and 58.2% on average, respectively, in plants subjected to water stress (70% and 50% *CSC*) compared to that of the normal (100% *CSC*) condition, whereas SOD activity was increased by 79.3% and 123.7% on average, respectively.

Table 3. Average 1-menthone, isopulegone, and pulegone for mint plants grown under greenhouse conditions during the growth season in 2018 and 2019, subjected to three water levels, 100%, 70%, and 50% *CSC*, and four biostimulants, CR, Mob, ND, and NN, as well as a control (without biostimulants).

Variable	**1-Menthone (%)**						**Isopulegone (%)**						**Pulegone (%)**					
Water Levels/Biosti.	**Control**	**CR**	**Mob**	**ND**	**NN**	**Means**	**Control**	**CR**	**Mob**	**ND**	**NN**	**Means**	**Control**	**CR**	**Mob**	**ND**	**NN**	**Means**
100% CSC	2.20^{a}	2.10^{c}	1.90^{c}	1.70^{d}	1.50^{e}	1.88^{A}	5.60^{a}	$5.50^{a,b}$	5.40^{b}	5.20^{c}	4.90^{d}	5.32^{A}	56.46^{a}	54.27^{b}	53.51^{b}	50.32^{c}	46.93^{d}	52.30^{A}
70% CSC	2.10^{a}	2.00^{b}	1.80^{c}	1.60^{d}	1.40^{e}	1.78^{B}	5.40^{a}	5.30^{b}	5.20^{c}	5.00^{d}	4.30^{e}	5.04^{B}	53.34^{a}	50.12^{b}	49.68^{b}	47.08^{c}	43.22^{d}	48.69^{B}
50% CSC	2.00^{a}	2.00^{a}	1.70^{b}	1.50^{c}	1.30^{d}	1.70^{C}	5.30^{a}	5.10^{b}	5.00^{b}	4.80^{c}	4.10^{d}	4.86^{C}	52.13^{a}	49.52^{b}	48.31^{c}	46.52^{d}	42.71^{e}	47.84^{C}
Means	2.10^{A}	2.04^{B}	1.80^{C}	1.60^{D}	1.40^{E}		5.44^{A}	5.30^{B}	5.20^{C}	5.00^{D}	4.43^{E}		53.98^{A}	51.30^{B}	50.50^{C}	47.98^{D}	44.29^{E}	
Water levels			*** ($LSD_{0.05}$ = 0.016)						*** ($LSD_{0.05}$ = 0.046)						*** ($LSD_{0.05}$ = 0.545)			
Biosti.			*** ($LSD_{0.05}$ = 0.021)						*** ($LSD_{0.05}$ = 0.059)						*** ($LSD_{0.05}$ = 0.703)			
Water levels × Biosti.			*** ($LSD_{0.05}$ = 0.036)						*** ($LSD_{0.05}$ = 0.102)						ns			

Data represent the mean calculated from n = 100, 60, or 20 for each water level, biostimulants or their interaction, respectively, in two growth seasons. ^{ns} (not significant) and *** ($p \leq 0.001$). Means differing in lowercase letters (a–e) within each row indicate significant differences at $p \leq 0.05$. Means differing in uppercase letters (A–E) within each mean row and each mean column indicate significant differences according to the least significant difference (LSD) test at $p \leq 0.05$. CSC (container substrate capacity), Control (without biostimulants), CR (CRADLE™), Mob (Mobilizer™), ND (Nanozim De'Lite™), NN (Nanozim NXT™), and Biosti. (biostimulant).

Table 4. Average α-pinene, 1,8-cineol, and α- terpineol ratios of mint plants grown under greenhouse conditions during the growth season in 2018 and 2019, subjected to three water levels, 100%, 70%, and 50% *CSC*, and four biostimulants, CR, Mob, ND, and NN, as well as a control (without biostimulants).

Variable	**α-Pinene (%)**						**1,8-Cineol (%)**						**α- Terpineol (%)**					
Water Levels/Biosti.	**Control**	**CR**	**Mob**	**ND**	**NN**	**Means**	**Control**	**CR**	**Mob**	**ND**	**NN**	**Means**	**Control**	**CR**	**Mob**	**ND**	**NN**	**Means**
100% CSC	0.90^{e}	1.80^{d}	3.40^{c}	3.60^{b}	4.00^{a}	2.74^{C}	24.28^{c}	25.09^{c}	25.20^{c}	27.31^{b}	29.11^{a}	26.20^{A}	1.70^{e}	1.90^{d}	2.30^{c}	2.50^{b}	2.70^{a}	2.22^{C}
70% CSC	1.50^{e}	2.50^{d}	3.80^{b}	3.70^{c}	4.20^{a}	3.14^{B}	25.62^{c}	26.62^{c}	26.70^{c}	28.61^{b}	35.71^{a}	28.65^{B}	1.90^{e}	2.10^{d}	2.40^{c}	2.70^{b}	3.10^{a}	2.44^{B}
50% CSC	2.10^{e}	3.10^{d}	3.70^{c}	4.10^{b}	4.30^{a}	3.46^{A}	26.82^{c}	27.79^{c}	27.92^{c}	29.82^{b}	36.11^{a}	29.70^{C}	2.10^{e}	2.30^{d}	2.50^{c}	3.00^{b}	3.40^{a}	2.66^{A}
Means	1.50^{E}	2.47^{D}	3.63^{C}	3.80^{B}	4.17^{A}		25.57^{D}	26.50^{C}	26.61^{C}	28.58^{B}	33.64^{A}		1.90^{E}	2.10^{D}	2.40^{C}	2.73^{B}	3.07^{A}	
Water levels			*** ($LSD_{0.05}$ = 0.047)						*** ($LSD_{0.05}$ = 0.544)						*** ($LSD_{0.05}$ = 0.018)			
Biosti.			*** ($LSD_{0.05}$ = 0.061)						*** ($LSD_{0.05}$ = 0.702)						*** ($LSD_{0.05}$ = 0.024)			
Water levels × Biosti.			*** ($LSD_{0.05}$ = 0.105)						*** ($LSD_{0.05}$ = 1.216)						*** ($LSD_{0.05}$ = 0.041)			

Data represent the mean calculated from n = 100, 60, or 20 for each water level, biostimulants or their interaction, respectively, in two growth seasons. *** ($p \leq 0.001$). Means differing in lowercase letters (a–e) within each row indicate significant differences at $p \leq 0.05$. Means differing in uppercase letters (A–E) within each mean row and each mean column indicate significant differences according to the least significant difference (LSD) test at $p \leq 0.05$. CSC (container substrate capacity), Control (without biostimulants), CR (CRADLE™), Mob (Mobilizer™), ND (Nanozim De'Lite™), NN (Nanozim NXT™), and Biosti. (biostimulant).

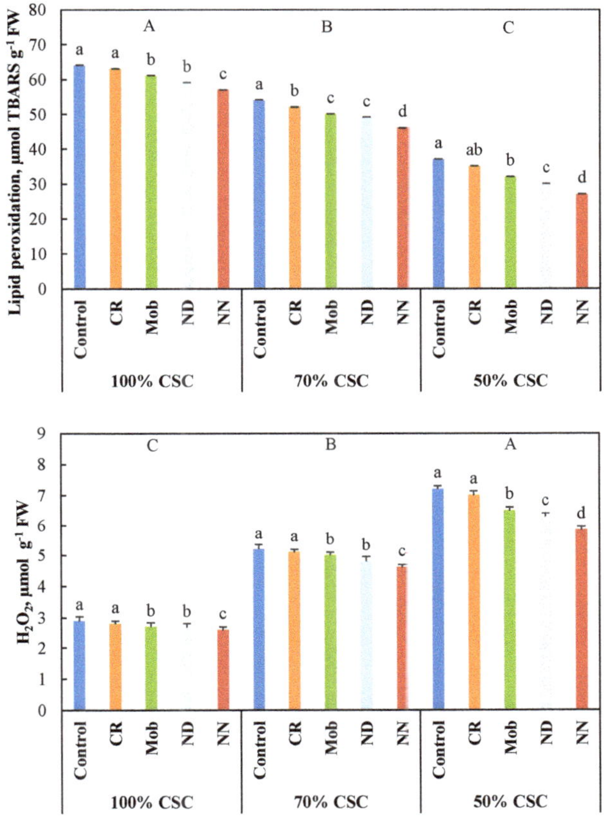

Figure 3. *Cont.*

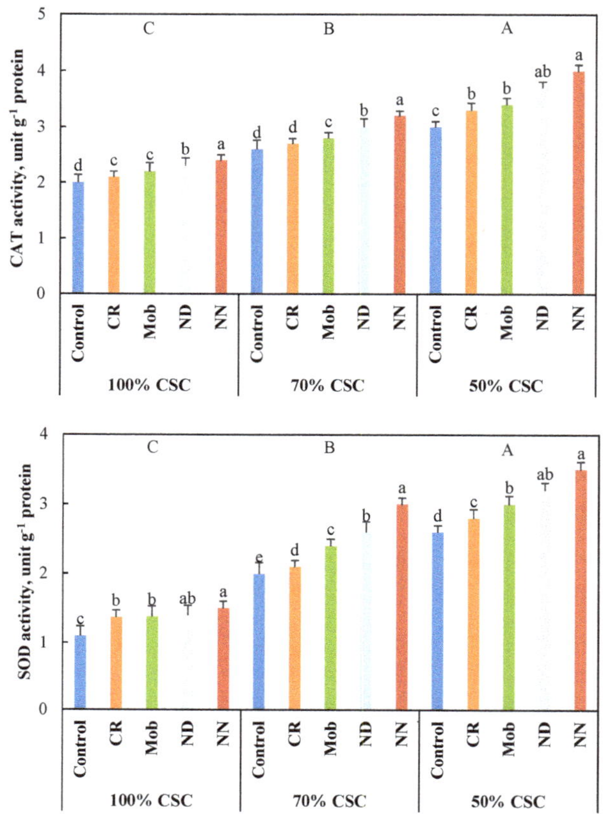

Figure 3. Lipid peroxidation, H_2O_2, catalase (CAT), and superoxide dismutase (SOD) activities of mint plants, average of two growing seasons, as affected by three water levels, 100%, 70%, and 50% *CSC* (container substrate capacity), and four biostimulants, CR (CRADLE™), Mob (Mobilizer™), ND (Nanozim De'Lite™), and NN (Nanozim NXT™), and a control (without biostimulants). FW (fresh weight). Different capital letters on top indicate significant differences between water levels at $p \leq 0.05$. Different letters on top columns indicate significant differences between biostimulants across water levels at $p \leq 0.05$. Bars provide the means ± SE of the mean ($n = 20$).

4. Discussion

Water stress is one of the major limiting factors of the growth and productivity of plants worldwide [21]. The amount of irrigation water applied influenced the biomass and EO yields of mint. The fresh and dry weights of mint were decreased with the irrigation water stress because of vegetative growth (i.e., leaf number and plant height), which decreased under water deficit conditions. Reduction in growth parameters as a consequence of drought has also been described in peppermint [6,7,11], Japanese mint [9], purple basil [8], balm [10], and chamomile [12]. The irrigation water level of 50% *CSC* had a negative effect on EO yield of mint. This is in agreement with earlier findings in peppermint [11] and chamomile [12], and in contrary to the results found in the previous studies from Ekren et al. [8] in purple basil and Farahani et al. [10] in balm.

The application of biostimulants under water stress conditions (70% and 50% *CSC*) showed enhanced growth by means of increased leaf number, plant height, root dry weight, fresh and dry weights, and IWUE. These morphological improvements are mainly attributed to the composition of these biostimulants. The most active biostimulant in this study was NN, which is composed of a unique mixture of important compounds: seaweed extract (15%), humic acid (5%), macro (potassium 1% and phosphorus) and micro elements, alginic acid, and hydrolyzed protein, as described in the materials and methods. The major constitutes of the NN biostimulant were seaweed extracts (*Ascophyllum nodosum*) and humic acid. The application of *Ascophyllum nodosum* extracts as plant biostimulants has been reported in several studies [19,22]. Humic acid may increase the leaf area, stem diameter, plant dry weight in different plants [23,24] and may ameliorate stress conditions in tomatoes [25]. Potassium, phosphorus, and microelements play critical roles in the growth and morphology of most plants [26,27]. However, the mixture was superior in the ameliorating effects against water stress in mint plants compared to other commercial biostimulants. The second biostimulant showing relatively high morphological performance was ND, which is mainly composed of *Ascophyllum nodosum* extracts (25%), carbohydrates (25%), (*w*/*w*) amino acid (2%) and potassium (1%). ND showed slightly lower morphological promoting effects than that of NN. CR and Mob are mainly composed of mycorrhizal biofertilizer. However, Mob contains additional components. including seaweed extract (*Macrocystis pyrifera*), humic acid, and amino acids, which may explain the slight increased vegetative performance of Mob compared to that of CR. Furthermore, *Macrocystis pyrifera* has been reported to have stimulatory effects on plant growth [28].

Gas exchange parameters (*gs*, *E*, and *A*) are important indicators of the physiological performance of plants under stress conditions. The increase in *gs* under stress conditions in response to external factors is strongly related to enhanced gas exchange through the stomata [29]. The increased gas exchange is normally reflected as enhanced transpiration and photosynthesis rates in the leaves [30]. There were increases in the gas exchange in plants treated with different biostimulants under water stress conditions, which indicated that these biostimulants acted as effective stress ameliorants. Leaf water potential and relative water content reductions might be associated with stress conditions [31,32]. They increased in this study in plants subjected to different biostimulants, indicating enhanced metabolic performance of treated plants. Furthermore, the increased proline composition in biostimulant-treated plants reflected enhanced stress tolerance as found in previous studies using other external elicitors [22,33].

The main constitutes of the EO were pulegone and 1,8-cineol. A previous study on the same species from Egypt reported comparable composition of both compounds [1]. There were fluctuations in the main constitutes of EO, as well as specific compounds, including pulegone, isopulegone, 1-menthone, 1,8-cineol, α-pinene, and α-terpineol. Secondary metabolites, such as cineole are usually associated with terpenes [34] and this explains the parallel increase in 1,8-cineol, α-terpineol, and pinene. 1-menthone and isopulegone are metabolites of pulegone. The pulegone is not favored in the EO composition of mint plants because of its carcinogenic effects at high doses [1], whereas, cineol is a favored compound in EOs because of its medicinal applications and pharmaceutical potential [35,36]. The application of NN showed the highest increase in 1,8-cineol and related terpenes ratios and lowest compositions of pulegone compared to that of the control and other biostimulant treatments. This result suggests that NN application may have future applications in medicinal plants, such as mints. The use of the NN biostimulant is a novel approach for enhancing the chemical composition of the EOs of mint plants by reducing hazardous compounds and increasing useful ones as found in this study.

In this study, the application of seaweed extract-based biostimulants mixed with humic acid and/or macro elements represented a novel tool for the enhancement of the medicinal properties of major medicinal plants, such as mints. The achievement of enhanced antioxidant activities of the EOs of mint might be of great importance for agricultural and related pharmaceutical industries. The oil of mints is routinely used in perfume and cosmetic preparations, as well as in the food industries, such as in chocolate and soft drinks. The development of new EO compositions with increased antioxidant properties will increase the additive value of the medicinal crop and will assist in reducing dependence on synthetic antioxidants to control ROS accumulation.

5. Conclusions

This study revealed an association between the application of specific biostimulants and the increase/decrease of the main EO composition (cineol and pulegone) of mint plants. The application of this finding is related to the agricultural, medicinal, and pharmaceutical industries. There were increases in the morphological characteristics, physiological performance, and EO ratio of biostimulant-treated plants. The morphological and physiological enhancements indicated increased tolerance to water stress. Further, biostimulant-treated plants showed higher antioxidant activities, reduced accumulation of H_2O_2, and increased CAT and SOD activities, which indicated an antioxidant stress tolerance activation mechanism in treated plants. The application of biostimulants to mint plants increased the quantity and quality of produced EOs and enhanced the medicinal properties, as well as that of the traditional medicinal crop. ND and NN are recommended under water stress conditions in mint.

Author Contributions: H.O.E., conceptualization, funding acquisition, investigation, supervision, project administration, methodology, writing—original draft; M.A.M., formal analysis, investigation, data curation, methodology, project administration, writing—review and editing; E.A.M., supervision, visualization; D.O.E.-A., supervision, visualization. All authors have read and agreed to the published version of the manuscript.

Funding: This work was financially supported by the Deanship of Scientific Research at King Saud University through research group number RG-1440-022.

Acknowledgments: The authors extend their appreciation to the Deanship of Scientific Research at King Saud University for their financial support of the present study.

Conflicts of Interest: The authors declare no conflict of interest.

References

1. Elansary, H.O.; Ashmawy, N.A. Essential oils of mint between benefits and hazards. *J. Essent. Oil-Bear. Plants* **2013**, *16*, 429–438. [CrossRef]
2. Stringaro, A.; Vavala, E.; Colone, M.; Pepi, F.; Mignogna, G.; Garzoli, S.; Cecchetti, S.; Ragno, R.; Angiolella, L. Effects of *Mentha suaveolens* essential oil alone or in combination with other drugs in candida albicans. *Evid. Based Complement. Alternat. Med.* **2014**, *2014*, 9. [CrossRef] [PubMed]
3. Khoury, M.; Stien, D.; Eparvier, V.; Ouaini, N.; El Beyrouthy, M. Report on the medicinal use of eleven Lamiaceae species in Lebanon and rationalization of their antimicrobial potential by examination of the chemical composition and antimicrobial activity of their essential oils. *Evid. Based Complement. Altern. Med.* **2016**, *2016*, 17. [CrossRef] [PubMed]
4. Sarkic, A.; Stappen, I. Essential oils and their single compounds in cosmetics—A critical review. *Cosmetics* **2018**, *5*, 11. [CrossRef]
5. Mancosu, N.; Snyder, R.L.; Kyriakakis, G.; Spano, D. Water scarcity and future challenges for food production. *Water* **2015**, *7*, 975–992. [CrossRef]
6. Zade, N.S.E.; Sadeghi, A.; Moradi, P. Streptomyces strains alleviate water stress and increase peppermint (*Mentha piperita*) yield and essential oils. *Plant Soil* **2019**, *434*, 441–452. [CrossRef]
7. Figueroa-Pérez, M.G.; Rocha-Guzmán, N.E.; Pérez-Ramírez, I.F.; Mercado-Silva, E.; Reynoso-Camacho, R. Metabolite profile, antioxidant capacity, and inhibition of digestive enzymes in infusions of peppermint (*Mentha piperita*) grown under drought stress. *J. Agric. Food Chem.* **2014**, *62*, 12027–12033. [CrossRef]
8. Ekren, S.; Sönmez, Ç.; Özçakal, E.; Kurttaş, Y.S.K.; Bayram, E.; Gürgülü, H. The effect of different irrigation water levels on yield and quality characteristics of purple basil (*Ocimum basilicum* L.). *Agric. Water Manag.* **2012**, *109*, 155–161. [CrossRef]
9. Shormin, T.; Khan, M.A.H.; Alamgir, M. Response of different levels of nitrogen fertilizer and water stress on the growth and yield of Japanese mint (*Mentha arvensis* L.). *Bangladesh J. Sci. Ind. Res.* **2009**, *44*, 137–145. [CrossRef]
10. Farahani, H.A.; Valadabadi, S.A.; Daneshian, J.; Khalvati, M.A. Evaluation changing of essential oil of balm (*Melissa officinalis* L.) under water deficit stress conditions. *J. Med. Plants Res.* **2009**, *3*, 329–333.
11. Khorasaninejad, S.; Mousavi, A.; Soltanloo, H.; Hemmati, K.; Khalighi, A. The effect of drought stress on growth parameters, essential oil yield and constituent of Peppermint (*Mentha piperita* L.). *J. Med. Plant Res.* **2011**, *5*, 5360–5365.

12. Razmjoo, K.; Heydarizadeh, P.; Sabzalian, M.R. Effect of salinity and drought stresses on growth parameters and essential oil content of *Matricaria chamomila*. *Int. J. Agric. Biol.* **2008**, *10*, 451–454.
13. Elansary, H.O.; Yessoufou, K.; Shokralla, S.; Mahmoud, E.A.; Skalicka-Woźniak, K. Enhancing mint and basil oil composition and antibacterial activity using seaweed extracts. *Ind. Crops Prod.* **2016**, *92*, 50–56. [CrossRef]
14. El-Esawi, M.A.; Elansary, H.O.; El-Shanhorey, N.A.; Abdel-Hamid, A.M.E.; Ali, H.M.; Elshikh, M.S. Salicylic acid-regulated antioxidant mechanisms and gene expression enhance rosemary performance under saline conditions. *Front. Physiol.* **2017**, *8*, 716. [CrossRef]
15. Fonteno, W.; Hardin, C.; Brewster, J. *Procedures for Determining Physical Properties of Horticultural Substrates Using the NCSU Porometer*; Horticultural Substrates Laboratory, North Carolina State University: Raleigh, NC, USA, 1995.
16. Howell, T.A. Challenges in increasing water use efficiency in irrigated agriculture. In Proceedings of the International Symposium on Water and Land Management for Sustainable Irrigated Agriculture, University of C¸ Ukurova, Adana, Turkey, 4–8 April 2006; pp. 53–63.
17. Elansary, H.O.; Salem, M.Z.M. Morphological and physiological responses and drought resistance enhancement of ornamental shrubs by trinexapac-ethyl application. *Sci. Hortic.* **2015**, *189*, 1–11. [CrossRef]
18. Elansary, H.O.; Yessoufou, K. Growth regulators and mowing heights enhance the morphological and physiological performance of seaspray turfgrass during drought conditions. *Acta Physiol. Plant.* **2015**, *37*, 232. [CrossRef]
19. Elansary, H.O.; Yessoufou, K.; Abdel-Hamid, A.M.E.; El-Esawi, M.A.; Ali, H.M.; Elshikh, M.S. Seaweed extracts enhance salam turfgrass performance during prolonged irrigation intervals and saline shock. *Front. Plant Sci.* **2017**, *8*, 830. [CrossRef]
20. Snedecor, G.W.; Cochran, W.G. *Statistical Methods, New York*, 8th ed.; The Iowa State University Press Publishing: Ames, IA, USA, 1989; p. 503.
21. Osakabe, Y.; Osakabe, K.; Shinozaki, K.; Tran, L.P. Response of plants to water stress. *Front. Plant Sci.* **2014**, *5*, 86. [CrossRef]
22. Al-Ghamdi, A.A.; Elansary, H.O. Synergetic effects of 5-aminolevulinic acid and Ascophyllum nodosum seaweed extracts on Asparagus phenolics and stress related genes under saline irrigation. *Plant Physiol. Biochem.* **2018**, *129*, 273–284. [CrossRef]
23. Melo, R.O.d.; Oliveira, H.P.d.; Silveira, K.C.; Baldotto, L.E.B.; Baldotto, M.A. Initial performance of maize in response to humic acids and plant growth-promoting bacteria. *Revista Ceres* **2018**, *65*, 271–277. [CrossRef]
24. Schoebitz, M.; López, M.D.; Serrí, H.; Martínez, O.; Zagal, E. Combined application of microbial consortium and humic substances to improve the growth performance of blueberry seedlings. *J. Soil Sci. Plant Nutr.* **2016**, *16*, 1010–1023. [CrossRef]
25. Abdellatif, I.M.Y.; Abdel-Ati, Y.Y.; Abdel-Mageed, Y.T.; Hassan, M.A.M. Effect of humic acid on growth and productivity of tomato plants under heat stress. *J. Hortic. Res.* **2017**, *25*, 59–66. [CrossRef]
26. Razaq, M.; Zhang, P.; Shen, H.; Salahuddin. Influence of nitrogen and phosphorous on the growth and root morphology of Acer mono. *PLoS ONE* **2017**, *12*. [CrossRef] [PubMed]
27. Wang, M.; Zheng, Q.; Shen, Q.; Guo, S. The critical role of potassium in plant stress response. *Int. J. Mol. Sci.* **2013**, *14*, 7370–7390. [CrossRef]
28. Godlewska, K.; Michalak, I.; Tuhy, Ł.; Chojnacka, K. Plant growth biostimulants based on different methods of seaweed extraction with water. *Biomed Res. Int* **2016**, *2016*, 5973760. [CrossRef]
29. Ekinci, M.; Yildirim, E.; Sahin, U. Changes in gas exchange capacity and selected physiological properties of squash seedlings (*Cucurbita pepo* L.) under well-watered and drought stress conditions AU—Ors, Selda. *Arch. Agron. Soil Sci.* **2016**, *62*, 1700–1710.
30. Kusumi, K.; Hirotsuka, S.; Kumamaru, T.; Iba, K. Increased leaf photosynthesis caused by elevated stomatal conductance in a rice mutant deficient in SLAC1, a guard cell anion channel protein. *J. Exp. Bot.* **2012**, *63*, 5635–5644. [CrossRef]
31. Argyrokastritis, I.G.; Papastylianou, P.T.; Alexandris, S. Leaf water potential and crop water stress index variation for full and deficit irrigated cotton in mediterranean conditions. *Agric. Agric. Sci. Procedia* **2015**, *4*, 463–470. [CrossRef]
32. Elansary, H.O.; El-Ansary, D.O. Trinexapac-ethyl application enhanced physiological performance of *Prunus persica* (L.) Batsch under drought conditions. *Bangladesh J. Bot.* **2017**, *46*, 1367–1373.

33. Elansary, H.O. Basil morphological and physiological performance under trinexapac-ethyl foliar sprays and prolonged irrigation intervals. *Acta Physiol. Plant.* **2015**, *37*, 92. [CrossRef]
34. Myburg, A.A.; Grattapaglia, D.; Tuskan, G.A.; Hellsten, U.; Hayes, R.D.; Grimwood, J.; Jenkins, J.; Lindquist, E.; Tice, H.; Bauer, D.; et al. The genome of Eucalyptus grandis. *Nature* **2014**, *510*, 356. [CrossRef] [PubMed]
35. Kim, K.Y.; Seo, H.J.; Min, S.S.; Park, M.; Seol, G.H. The effect of 1,8-cineole inhalation on preoperative anxiety: A randomized clinical trial. *J. Evid. Based Complement. Altern. Med.* **2014**, *2014*, 7. [CrossRef] [PubMed]
36. Abu-Darwish, M.S.; Cabral, C.; Ferreira, I.V.; Gonçalves, M.J.; Cavaleiro, C.; Cruz, M.T.; Al-bdour, T.H.; Salgueiro, L. Essential oil of common sage (*Salvia officinalis* L.) from Jordan: Assessment of safety in mammalian cells and its antifungal and anti-inflammatory potential. *J. Biomed Res. Int.* **2013**, *2013*, 9.

agronomy

MDPI

Review

Arbuscular Mycorrhizal Fungi and Associated Microbiota as Plant Biostimulants: Research Strategies for the Selection of the Best Performing Inocula

Luca Giovannini [1,†], Michela Palla [1,†], Monica Agnolucci [1], Luciano Avio [1], Cristiana Sbrana [2], Alessandra Turrini [1] and Manuela Giovannetti [1,*]

[1] Department of Agriculture, Food and Environment, University of Pisa, Via del Borghetto 80, 56124 Pisa, Italy; luca.giovannini@phd.unipi.it (L.G.); michela.palla@for.unipi.it (M.P.); monica.agnolucci@unipi.it (M.A.); luciano.avio@unipi.it (L.A.); alessandra.turrini@unipi.it (A.T.)

[2] CNR, Institute of Agricultural Biology and Biotechnology, UOS Pisa, Via Moruzzi 1, 56124 Pisa, Italy; sbrana@ibba.cnr.it

* Correspondence: manuela.giovannetti@unipi.it; Tel.: +39-050-221-6643

† These Authors share the first Authorship.

Received: 18 December 2019; Accepted: 9 January 2020; Published: 11 January 2020

Abstract: Arbuscular mycorrhizal fungi (AMF) are beneficial soil microorganisms establishing mutualistic symbioses with the roots of the most important food crops and playing key roles in the maintenance of long-term soil fertility and health. The great inter- and intra-specific AMF diversity can be fully exploited by selecting AMF inocula on the basis of their colonization ability and efficiency, which are affected by fungal and plant genotypes and diverse environmental variables. The multiple services provided by AMF are the result of the synergistic activities of the bacterial communities living in the mycorrhizosphere, encompassing nitrogen fixation, P solubilization, and the production of phytohormones, siderophores, and antibiotics. The tripartite association among host plants, mycorrhizal symbionts, and associated bacteria show beneficial emerging properties which could be efficiently exploited in sustainable agriculture. Further in-depth studies, both in microcosms and in the field, performed on different AMF species and isolates, should evaluate their colonization ability, efficiency, and resilience. Transcriptomic studies can reveal the expression levels of nutrient transporter genes in fungal absorbing hyphae in the presence of selected bacterial strains. Eventually, newly designed multifunctional microbial consortia can be utilized as biofertilizers and biostimulants in sustainable and innovative production systems.

Keywords: arbuscular mycorrhizal symbiosis; mycorrhizosphere; AMF associated bacteria; plant growth-promoting bacteria; biofertilizers; phosphate-solubilizing bacteria; siderophore production

1. Introduction

In the next decades, the major challenge for agriculture will be the adoption of a new paradigm, sustainable intensification, to meet human needs for the production of enough food at a global scale while maintaining environmental quality and reducing the input of chemical fertilizers and pesticides [1]. These objectives may be pursued by giving more attention to beneficial soil microorganisms that play key roles in the maintenance of long-term soil fertility and health, the reduction of chemical inputs in agriculture, the promotion of plant nutrition, and the production of safe and high-quality food [2]. Among them, arbuscular mycorrhizal (AM) fungi (AMF) represent a key functional group, positively affecting plant growth, nutrition, and health. AMF are obligately biotrophic organisms that establish mutualistic symbioses with the roots of all major land plant taxa, including the most important food crops such as cereals, pulses, fruit trees, vegetables, medicinal plants, and other economically relevant species such as sunflower, cotton, sugarcane, tobacco, coffee, tea, cocoa, rubber, and cassava [3]. Within food crops, the only exceptions are represented by genera and species belonging to Brassicaceae and Chenopodiaceae, which are non-mycorrhizal plants.

In exchange for plant photosynthates, AMF facilitate the uptake and transfer of mineral nutrients, such as phosphorus (P), nitrogen (N), sulfur, potassium, calcium, copper and zinc, from the soil to their host plants by means of the extraradical mycelium (ERM) extending from colonized roots into the soil [3]. Such a fungal structure represents one of the critical elements of the AM symbiosis, as the flow of nutrients translocated to the root cells of host plants is highly dependent on its structure, extent, and interconnectedness. ERM functions as an efficient absorbing system, given the high surface-to-volume ratio of the mycelium, which is able to uptake soil nutrients beyond the depletion zone around roots, and the presence of nutrient transporter genes in the hyphae [4]. Besides plant nutrition improvement, AMF facilitate the completion of biogeochemical cycles, increase plant tolerance to biotic and abiotic stresses, carbon sequestration and soil aggregation [5], and the content of health-promoting phytochemicals [6,7] (Figure 1).

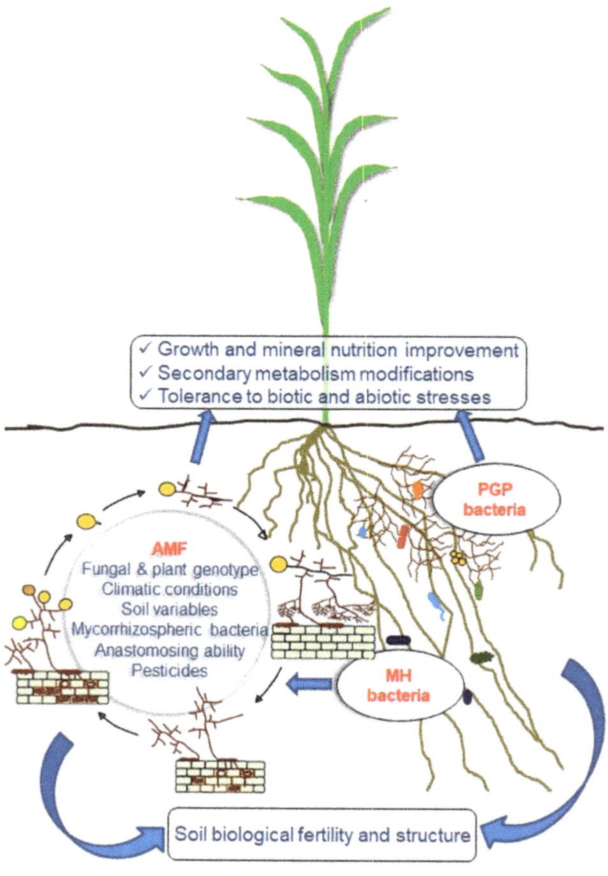

Figure 1. Schematic drawing representing the impacts of arbuscular mycorrhizal fungi (AMF) and beneficial bacteria on plant performance and soil fertility. On the left: a visual representation of the AMF life cycle and factors affecting the different AMF developmental stages; on the right: mycorrhizal helper (MH) and plant growth promoting (PGP) bacteria synergistically interacting with AMF.

Several studies showed that the multiple services provided by AMF are the result of the synergistic activity of diverse bacterial communities living in the mycorrhizosphere, strictly associated with their spores and extraradical mycelium and playing diverse plant growth-promoting (PGP) roles, from nitrogen fixation and P solubilization and mineralization to the production of indole acetic acid (IAA), siderophores, and antibiotics [8,9]. Such microbiota was identified not only by culture-independent methods but also by culture-dependent approaches, which allowed their functional characterization, aimed at detecting the best performing bacterial strains, to be used in combination with selected AMF as biofertilizers and biostimulants in innovative and sustainable food production systems [10].

The aim of this review is to provide an overview of the recent developments which contributed to disclose the biostimulant properties of AMF and their associated bacteria and to propose the best research strategies for the selection of functional isolates and consortia to be utilized as high-quality inocula in sustainable agriculture.

2. Arbuscular Mycorrhizal Fungi

AMF belong to the phylum Glomeromycota, encompassing ten out of eleven families: Acaulosporaceae, Ambisporaceae, Archaesporacea, Claroidoglomeraceae, Diversisporaceae, Gigasporaceae, Glomeraceae, Pacisporaceae, Paraglomeraceae, and Sacculosporaceae (http://www.amf-phylogeny.com/, accessed on 7 January 2020). Given their status of obligate biotrophs, the AMF life cycle cannot be completed in the absence of host plants. It starts with an asymbiotic phase, during which spores germinate in response to physical factors such as moisture, temperature and pH, producing hyphae with a limited lifespan [11]. In the presence of root exudates from host plants, a differential hyphal morphogenesis occurs, with germling hyphae reorienting the direction of elongation and initiating a differential branching pattern [12–14]: this pre-symbiotic phase is followed by physical contact between AMF hyphae and host roots, with the differentiation of appressoria, which give rise to hyphae growing intercellularly within the root cortex, eventually penetrating in root cells and producing highly branched hyphal tree-like structures similar to haustoria, the arbuscules. Arbuscules are the key structures of mycorrhizal symbioses, as at their level nutrient exchanges between the two partners take place: AMF obtain carbon (up to 20% of plant photosynthates) and lipids from the host plant and release mineral nutrients absorbed and translocated by ERM [15–17]. Two types of root colonization have been detected: *Arum*-type and *Paris*-type [3]. In the *Arum*-type, the AM symbiont spreads intercellularly between cortical root cells, forming terminal arbuscules on intracellular hyphal branches [18]. In the *Paris*-type, the fungus grows directly from cell to cell within the cortex and forms intracellular hyphal coils and intercalary arbuscules along the coils. However, most of the data available on AMF derive from studies carried out on the *Arum*-type mycorrhizal symbioses, which are widely distributed in natural and agricultural ecosystems. Beyond arbuscules, several AMF species produce intraradical vesicles, which are spore-like storage structures containing lipids. After receiving host carbon, the fungal symbiont is able to grow extraradically, colonize the surrounding soil, absorb mineral nutrients to be transferred to the host plant, interact with rhizosphere and soil microorganisms, colonize the roots of other plant living nearby (even belonging to species, genera and families different from their host), and also translocate mineral nutrients from one host to another [19,20]. The life cycle is closed by the formation of asexual spores by ERM, functional to the maintenance of a high mycorrhizal potential of the soil and, consequently, of soil biological fertility (Figure 1).

3. AMF Functional Diversity: Colonization Ability and Efficiency

So far, 323 AMF species have been described (http://www.amf-phylogeny.com/amphylo_species.html, accessed 3 December 2019), though only a few species have been investigated for their functional diversity, in order to detect and select the best isolates to be used in agriculture. As a consequence, most of the available commercial inocula are prepared with *Rhizoglomus irregulare* (syn. *Rhizophagus irregularis*, formerly *Glomus intraradices*) and *Funneliformis mosseae* (formerly *Glomus mosseae*), that are generalist symbionts, widespread all over the world in almost all soils and climatic zones [3]. In order to exploit the great inter- and intra-specific diversity, the general criteria to be applied when selecting the most efficient AMF isolates are outlined here.

The two fundamental fungal characteristics to be taken into account are colonization ability, which refers to fungal capacity of a rapid and extensive root colonization, and efficiency, represented by fungal symbiotic performance, in terms of plant growth and nutrition.

3.1. Colonization Ability

A high root colonization ability is the essential prerequisite for any AMF isolate to be designed for agricultural utilization, as it should be able to compete with highly competitive native AMF. AMF colonization ability does not depend only on fungal genotype, but also on soil characteristics and plant genotype, which may influence the different steps of mycorrhizal establishment, from spore germination to appressorium formation and intraradical growth.

The first variable affecting the competitive ability of an AMF strain is represented by spore dormancy, which may be relieved by storage at 5–10 °C for 5–6 weeks; nevertheless, it is extremely important to know which AMF isolates produce dormant spores when selecting strains for inoculation. As an example, different species of the genera *Glomus*, *Funneliformis*, and *Acaulospora* show spore dormancy, while species such as *Gigaspora gigantea* and *Gigaspora margarita* are able to germinate as early as one day after incubation [21]. It is unfortunate that only a few works have investigated this critical element, which should be further studied not only at the species but, most importantly, at the isolate level, as the producers of commercial inocula often reproduce their own strains.

A key fungal characteristic directly linked to AMF establishment and persistence in the field is represented by spore germination, which is affected by different factors such as soil pH and nutrient content, temperature, soil bacteria, and pesticides. Poor information is available on soil variables, suggesting that the different AMF strains show optimum germination when cultivated in environments with characteristics similar to those from which they were originally isolated. Thus, for example, *Acaulospora laevis*, predominant in low pH soils, germinates well at pH 4–5, while *Dentiscutata heterogama* (formerly *Gigaspora heterogama*), isolated from warm climates, germinates best at 34 °C [21], although nine AMF, isolated and maintained in tropical areas, showed very different germination rates, ranging from 8% to 78%, when cultured in the same environmental conditions [22]. It has long been known that spore germination can be stimulated by soil microorganisms, from Actinobacteria to Pseudomonads, although the most relevant role is played by bacteria living in intimate association with AMF, often located on and within spore wall layers (mycorrhizospheric bacteria) [9]. Actually, many bacterial taxa able to degrade biopolymers were recently detected in spore homogenates by culture-independent methods, suggesting a possible chitinolytic activity on chitin of spore walls that could enhance spore germination [23,24]. It is interesting to note that a recent molecular work reported the ability of six AMF isolates to recruit different bacterial communities on their spores, belonging to Actinomycetales, Bacillales, Burkholderiales, Pseudomonadales, and Rhizobiales, possibly exerting an activity on spore germination [25]. As to pesticides, their effects on spore germination are different depending on the target organisms. Several fungicides, like copper hydroxide and mancozeb, were able to inhibit spore germination of *F. mosseae* in vivo, while flutolanil, azoxystrobin, fenpropimorph, and fenhexamid inhibited germination of *R. irregulare* spores in vitro [26,27]. On the other hand, other fungicides, such as fosetil Al, metalaxyl and different herbicides, seem to exert no activity on spore germination even if the results obtained on the same substance in different investigations were often contradictory [28].

After germination, another important variable affecting the competitive ability of AMF towards native fungi is represented by the ability of germlings to produce an extensive and interconnected hyphal network, which is essential for increasing the chance of coming into contact with a host root. Germling growth may be affected by the same environmental variables quoted above, but depends largely on fungal genotype as it can range from 0.25 up to 104 and 544 mm of hyphal length per germling in the same experimental in vitro conditions [11]. It is important to underline that the possibility to contact host roots and to establish the symbiosis is greatly extended by the ability of germling hyphae to become interconnected through hyphal fusions (anastomoses): this capacity represents a fundamental survival strategy for AMF germlings, which can plug into compatible extraradical networks, gaining immediate access to plant-derived carbon [29]. Anastomosis formation is highly related to the fungal genotype, as species belonging to the families of Glomeraceae and Acaulosporaceae show a high frequency of hyphal fusions, while members of the family Gigasporaceae do not form fusions interconnecting different hyphae [30,31]. The length, viability, and interconnectedness of germling hyphae are affected by various pesticides: for example, fungicides containing benomyl and fenhexamid, even at doses below the recommended field rate, inhibited hyphal growth of *F. mosseae*, affected mycelial viability, and induced abnormal hyphal branching, while the herbicide glufosinate ammonium decreased mycelial growth and viability, and also the anastomosis rate [32,33].

When AMF germlings come into contact with a host root, a differential hyphal morphogenesis is induced, characterized by an increase in hyphal branching, functional to the production of appressoria on the root surface [12,34]. Appressoria are swollen, multinucleate structures formed as early as 36 h after the contact between germlings and roots [35], and represent the signs of fungal recognition of the host plant. A prompt production of a large number of appressoria, which is requisite for a rapid root colonization, characterizes the most infective AMF, as it makes them highly competitive with native symbionts. Several works investigated this AMF functional trait: an old, but not obsolete work, reported that the number of appressoria may range from 2.6 to 21.1 and from 4.6 to 10.7 per mm of root length in field-grown strawberry and apple, respectively [36], while more recent works found 10.2–80.5 appressoria per plant in parsley and aubergine inoculated in microcosms with *F. mosseae* [37,38]. The same fungus showed variable results depending on host plants: for example, it produced 3.6 appressoria per mm of root in *Medicago truncatula*, 9.7 in *Prunus cerasifera*, and 1.26 in *Trifolium pratense* [39–41]. On the other hand, *G. margarita* produced only 0.01 appressoria per mm of root when inoculated on *Allium cepa* [42]. The dynamics of appressoria formation was monitored in a time-course experiment, showing that the first structures were produced after 36, 48, and 60 h, depending on the fungal genotype [35].

Appressoria produce intraradical hyphae able to establish the mycorrhizal symbiosis by rapidly spreading in the apoplastic space between root cortical cells, although the levels of root colonization greatly vary among AMF and plant genotypes. While such variability among different AMF species have been assessed in countless experiments aimed at evaluating fungal performance in terms of plant growth, the susceptibility of different plant genotypes to mycorrhizal colonization has been investigated only in recent works, reporting large differences among 11 sunflower cultivars (range 8.6–78.7%) and 108 durum wheat varieties (range 10–44%) [43,44].

3.2. Efficiency

The efficiency of the different AM fungal isolates is generally interpreted as their ability to increase plant growth and nutrient uptake, and evaluated by considering the relevant fungal variables such as ERM development, extent, interconnectedness, viability, and rate of nutrient uptake and translocation, that are directly linked to the occurrence of fungal transporter genes in the absorptive extraradical hyphae [4].

ERM length density, assessed after destructive extraction from the soil, showed a large variability among AMF species, ranging from 1.1–6.9 to 3–5 and 10 m/g soil in *Acaulospora laevis*, *F. caledonius* (formerly *Glomus caledonium*), and *Scutellospora calospora*, respectively [11]. Recent works have reported higher hyphal lengths (up to 22 m/g soil) produced by *R. irregulare* isolate BEG 87 [45]. It is worth mentioning the ERM growth rate, which was 738–1067 and 3.1–3.8 mm/day in bidimensional and tri-dimensional experimental systems, respectively [20,46].

ERM structure and interconnectedness have been investigated by nondestructive tests, which provided both qualitative [47,48] and quantitative data. For example, ERM produced by members of the family Glomeraceae, widely distributed in agricultural soils, is highly interconnected by means of anastomoses between contacting hyphae (67–77% in *F. mosseae*), reaching the value of

100–410 anastomoses per gram of soil [20,31,49]. On the contrary, hyphae of members of the families Gigasporaceae, Ambisporaceae, and Paraglomeraceae are not able to fuse after contact, in vivo [50]. Nevertheless, within the Glomeraceae family, self-incompatible interactions between contacting hyphae may occur, with frequencies ranging from 5% to 32% [29,50,51]. Further extensive studies addressed such a clue, revealing major differences among three glomalean AMF: in particular, when grown in symbiosis with five different plant species, *F. mosseae* and *R. irregulare* ERM showed anastomosis frequency of 26–48% and 36–54%, respectively, while *F. coronatus* never exceeded 7.7% [52]; length and density affect AMF symbiotic performance, positively correlating with plant growth responses and nutrient levels [53]. Specifically, AMF isolates showing a high anastomosing ability are able to tolerate soil disturbance, such as tillage, by producing large mycorrhizal networks capable of re-establish interconnections after disruption [54–57]. ERM length and structure may be affected by pesticides, as reported by a recent work performed using a whole-plant experimental system, i.e., in *F. mosseae*, ERM length and density decreased in the presence of the herbicides dicamba and glufosinolate and the fungicides benomyl and fenhexamid, while ERM length and density increased in the presence of two mycorrhizospheric bacteria, *Ensifer meliloti* (formerly *Sinorhizobium meliloti*) and *Enterobacter ludwigii* [58]. Such recent novel data stress the need for further studies to evaluate the impact of agrochemicals and biocontrol agents on ERM structure and activity in a large number of AMF taxa in order to detect the most resilient isolates able to maintain a high mycorrhizal inoculum potential in soil.

Beyond the mentioned phenotypic parameters, viability, which is the most important factor affecting ERM functionality in soil, has been poorly investigated. A few studies reported that metabolic activity occurred in 63–96%, 96–100%, and 100% of extraradical hyphae in *R. irregulare*, *F. mosseae*, and *Rhizoglomus clarum* (formerly *Glomus clarum*), respectively [46,59,60]. A recent study posed the interesting question of whether ERM could survive and maintain colonization ability after plant harvest, thus representing a source of inoculum for the successive crops. The authors, utilizing an in vivo whole-plant experimental system and two worldwide distributed glomalean AMF, *F. mosseae* and *R. irregulare*, revealed that ERM viability and functionality are uncoupled from the host plant lifespan, as, after shoot removal, its growth from detached roots was comparable with that from intact plants and continuous for at least 150 days [61]. Accordingly, ERM represents a long-term survival structure able to maintain mycorrhizal potential and biological fertility in agricultural soils.

AMF efficiency is highly correlated with the rate of P translocation to the host plant: alas, only scanty information is available, showing that in *F. mosseae*, P fluxes in hyphae were 3.4×10^{-8} mol cm^{-2} s^{-1} [62]. However, as the transfer of nutrients flowing in the extraradical hyphae can occur exclusively through appressoria, which are the unique structures connecting soil-based to root-based mycelium, a high number of appressoria produced on the root surface is a key factor affecting not only AMF colonization ability but also their efficiency.

Studies on the occurrence of nutrient transporter genes in AMF extraradical hyphae have mostly been performed in vitro, using transgenic root organ cultures and few species, i.e., *R. irregulare* and *R. intraradices*. The results showed that a number of nutrient transporter genes (ammonium, phosphorus, zinc) are differentially regulated, depending on the availability of various mineral or organic compounds [4,63]. However, as transformed roots show an altered hormonal balance and sugar acquisition, possibly affecting the physiology of the mycorrhizal symbiosis, diverse whole-plant experimental systems were devised, encompassing other AMF species, *F. mosseae*, *F. coronatus*, and *G. margarita* [64]. Further extensive investigations focusing on nutrient transporters gene expression in extraradical mycorrhizal mycelium produced by a large number of AMF isolates are needed in order to achieve a deeper knowledge of differences in AMF efficiency and to select the best performing symbionts to be used as inocula, if also meeting the other quality characteristics concerning colonization ability and efficiency.

4. AMF Efficiency in the Enhancement of Plant Health-Promoting Compounds

In the light of the new findings on plant secondary metabolism being modulated by AMF, the concept of efficiency should be expanded to take into consideration the production of health-promoting compounds, a theme of the highest concern not only to scientists but also to consumers and producers as phytochemicals may reduce oxidative damages, prevent chronic and heart diseases, and decrease the risk of mortality from cancer [65–67]. The levels of such compounds, mainly represented by carotenoids, glucosinolates, polyphenols, including flavonoids, isoflavones and anthocyanins, are affected by different variables such as plant genotype, agronomic techniques, soil characteristics, and also by mycorrhizal symbioses [6].

For example, sweet basil (*Ocimum basilicum*) inoculated with *Glomus* spp. increased the production of rosmarinic and caffeic acids, and of essential oils [68,69], while *R. intraradices* affected the gene expression of key enzymes involved in basil rosmarinic acid biosynthetic pathway [70]. *Echinacea purpurea* inoculated with *R. irregulare* and *G. margarita* showed higher concentrations of caffeic acid derivatives, alkylamides, and terpenes [71], while *R. irregulare* inoculated on *Stevia rebaudiana* enhanced its content of the health-promoting compound steviol glycoside [72]. Interestingly, diverse AMF isolates differentially affected the production of specific phytochemicals; for example, the levels of thymol derivatives in the roots of *Inula ensifolia* were more enhanced by *R. clarus* than by *R. irregulare* [73], while in basil leaves the production of camphor and alfa-terpineol were enhanced by *Gigaspora rosea* but not by *G. margarita,* which decreased the total content of essential oils, in particular that of eucalyptol, linalool, and eugenol [68].

Despite the good results obtained by utilizing medicinal plants and herbs, only a few food crops have been investigated for their levels of health-promoting compounds upon mycorrhizal inoculation, i.e., lettuce, onion, tomato, maize, artichoke, strawberry, pepper, and sweet potato [7]. Most experimental works utilized either AMF inocula composed of a mixture of species, obtained from commercial producers or single species inocula, often represented by *R. irregulare* or *F. mosseae*. Also, molecular studies focused on the assessment of the levels of transcripts encoding the enzymes of the pathways relevant to the production of health-promoting secondary metabolites mainly utilized the same two species [7]. This has impaired the evaluation of the efficiency of different AMF, aimed at selecting the best performing symbionts in the production of beneficial phytochemicals. Accordingly, in the years to come, in-depth investigations should fully exploit the wide physiological and genetic diversity of AMF, testing the highest possible range of diverse species, isolates, and lineages within isolates. In addition, transcriptomic studies would allow the identification of AMF strains differentially expressing genes relevant to the biosynthesis of nutraceutical compounds in food plants.

5. Mycorrhizospheric Bacteria and Their Functional Significance

It has long been known that AMF colonization ability and efficiency may be mediated by a third partner of the symbiosis, the diverse and abundant bacterial communities living in the mycorrhizosphere, i.e., associated with mycorrhizal roots, spores, sporocarps, and extraradical hyphae [74]. Later, by ultrastructural studies, bacteria were detected in spore wall layers, within the peridial hyphae surrounding spores [75,76], and inside the cytoplasm [77–80]. Culture-dependent approaches allowed the isolation of many different bacterial taxa from the mycorrhizosphere of *Glomus versiforme*, *R. clarus*, *G. margarita*, *F. mosseae*, and *R. irregulare* [81–84]. A recent work isolated from *Rhizoglomus irregulare* (formerly *R. intraradices*) spores as many as 374 bacterial strains [85]. Culture-independent methods provided an in-depth description of the different bacterial taxa associated with spores: for example, PCR denaturing gradient gel electrophoresis (PCR-DGGE) identified the bacterial communities associated with *F. geosporus*, *Septoglomus constrictum*, and *G. margarita* spores [23,24], and those strongly associated with the spores of six AMF isolates, three belonging to *F. mosseae*, one to *F. coronatus*, and two to *R. irregulare*—the 48 relevant sequences were affiliated with Actinomycetales, Bacillales, Burkholderiales, Pseudomonadales, Rhizobiales, and Mollicutes-related endobacteria [25].

The mycorrhizospheric microbiota showed different functional activities, ranging from the role of "mycorrhiza helper" (MH) [86] to that of "plant growth promoters" (PGP) (Figure 1). MH bacteria may increase spore germination and mycorrhizal symbiosis establishment: for example, *Streptomyces* spp., *Pseudomonas* sp., and *Corynebacterium* sp. improved the germination of *F. mosseae*, *G. versiforme*, and *G. margarita* spores [81,87–89]. The enhancement of spore germination was ascribed to Actinobacteria, a group of bacteria frequently associated with AMF spores, able to hydrolyze chitin, the main component of spore walls [23,25,76,90]. Other MH bacteria, such as *Klebsiella pneumoniae*, *Trichoderma* sp., and *Paenibacillus validus*, increased germlings hyphal growth [91–93], while one bacterial strain belonging to Oxalobacteriaceae enhanced not only spore germination and germling growth but also root colonization [94]. In addition, the development of AMF extraradical mycelium (ERM) may be promoted by strains of *Paenibacillus rhizosphaerae*, *Azospirillum* sp., *Rhizobium etli*, *Pseudomonas* spp., *Burkolderia cepacia*, and *E. meliloti* [45,95–98] (Figure 1).

PGP bacteria show multifunctional activities, encompassing nitrogen fixation, P solubilization and mineralization, the production of indole acetic acid (IAA), siderophores, and antibiotics while supplying fundamental nutrients and growth factors [8,9]. Such activities represent key characteristics to be taken into account when selecting the best AMF and bacterial combinations for the production of inocula for agricultural use. For example, as P is rapidly immobilized in the soil, forming insoluble compounds with aluminium/iron and with calcium in acid and alkaline soil and thus becoming unavailable to plants, P-solubilizing bacteria may work in synergy with AMF to increase P availability and plant P uptake. Indeed, P-mobilizing bacteria, such as *Streptomyces* spp., *Leifsonia* sp., *Bacillus pumilus*, *Lisinobacillus fusiformis*, and *E. meliloti*, isolated from AMF spores of *R. irregulare*, showed synergistic action with AMF, promoting the mineralization of soil phytate and facilitating P uptake by mycorrhizal plants [45,99]. Similarly, the isolation from the mycorrhizosphere of bacterial strains possessing the *nifH* gene amplicon suggested a possible role in plant acquisition of nitrogen [85]. On the other hand, some PGP bacteria are able to produce IAA, a phytohormone of the auxin class, which plays a key role in the regulation of plant growth, increasing plant cell division and root formation, thus affecting water and nutrient uptake [100–102]. Accordingly, IAA producing bacteria isolated from *R. irregulare* and *F. mosseae*, such as *E. meliloti* and *Paenibacillus favisporus*, enhanced the growth of AMF extraradical hyphae, the fungal structure fundamental for absorbing and translocating P from the soil to plant roots [45,95]. An important role in the promotion of plant growth is played by mycorrhizospheric bacteria able to protect plants against soil-borne pathogens, either by directly producing antibiotics or indirectly producing siderophores, high-affinity iron-chelating compounds which mediate iron acquisition by pathogenic microorganisms [85,103–105]. Moreover, the facilitation of plant iron acquisition by siderophores-producing bacteria represents an additional benefit, as iron is an essential element in key biochemical processes like photosynthesis and respiration [106,107]. Interestingly, many of the bacteria isolated from AMF spores showed multiple PGP activities, i.e., 17 actinobacterial strains were able to produce siderophores and IAA to mineralize phytate and solubilize inorganic phosphate, and ten putative N-fixers to produce siderophore and solubilize P [85]. A recent work confirmed such data, reporting the occurrence of diverse bacterial functional taxa in a commercial AMF inoculum: 14 isolates showed the best combination of PGP traits, such as the production of IAA and siderophores, while 6 of them were also able to solubilize P, i.e., *Bacillus megaterium*, *Streptomyces* sp., and *Enterobacter* spp. [108]. These strains, both as single- and multi-strain inocula, deserve further in-depth studies in order to evaluate their efficiency as biofertilizers and biostimulants, able to boost plant growth, nutrition and health in sustainable food production systems (Figure 1). New remarkable findings showed that several members of the mycorrhizospheric microbiota may establish a more intimate relationship with their host plants as root endophytes [109,110]: considering that they can reach 10^5–10^7 CFU per g of root [111,112], their possible beneficial effects should be further investigated in the years to come.

6. Conclusions and Perspectives for Future Studies

The multiple beneficial activities of AMF and their associated bacteria discussed so far highlight the complex networks of interactions taking place in the mycorrhizosphere, functional to plant growth, nutrition, and health. The tripartite association among host plants, fungal symbionts and their associated bacteria shows beneficial emerging properties that could be efficiently exploited in sustainable food production. Although much is known on a very small number of AMF species, often studied singly in sterile conditions, very little is known about the high physiological and genetic inter- and intra-specific diversity of AMF and their associated microbiota. Further in-depth studies should be performed on different AMF species and isolates, and on their associated bacteria, both singly and in various combinations, in order to evaluate their colonization ability and efficiency when inoculated with a number of plant hosts. The studies carried out in microcosms should be followed by investigations in the field to assess the ability of the selected AMF and bacteria to compete with native microorganisms and to maintain their beneficial activities. Once detected as the best performing inocula, they could be differentiated by assessing their resilience against diverse environmental conditions, from soil types to drought, salt, biotic stresses, and pesticides. Transcriptomic studies could reveal the expression levels of nutrient transporter genes in fungal absorbing hyphae in the presence of selected efficient bacterial strains, possibly leading to the detection of the best synergistic combinations of AMF and associated bacterial communities, enhancing nutrient availability and plant performance. At the same time, transcriptomics could increase knowledge on the differential expression of genes encoding enzymes relevant to the biosynthesis of nutraceutical compounds in food plants. Eventually, newly designed multifunctional microbial consortia could be commercially reproduced and utilized as biofertilizers and biostimulants in sustainable and innovative production systems.

Author Contributions: M.A., A.T. and M.G. conceived the topic of the paper and wrote the original draft. L.G., M.P., L.A. and C.S. participated in the preparation and review of the manuscript. L.G. and M.P. provided editing assistance. All authors have read and agreed to the published version of the manuscript.

Funding: This work was funded by UNIVERSITY OF PISA, ITALY, grant: Fondi di Ateneo.

Conflicts of Interest: The authors declare no conflict of interest.

References

1. Food and Agriculture Organization. *A Policymaker's Guide to the Sustainable Intensification of Smallholder Crop Production*; FAO: Rome, Italy, 2011; Available online: http://www.fao.org/3/a-i2215e.pdf (accessed on 7 January 2020).
2. Philippot, L.; Raaijmakers, J.M.; Lemanceau, P.; Van Der Putten, W.H. Going back to the roots: The microbial ecology of the rhizosphere. *Nat. Rev. Microbiol.* **2013**, *11*, 789–799. [CrossRef]
3. Smith, S.E.; Read, D.J. *Mycorrhizal Symbiosis*, 3rd ed.; Academic Press: London, UK, 2008.
4. Casieri, L.; Ait Lahmidi, N.; Doidy, J.; Veneault-Fourrey, C.; Migeon, A.; Bonneau, L.; Courty, P.E.; Garcia, K.; Charbonnier, M.; Delteil, A.; et al. Biotrophic transportome in mutualistic plant-fungal interactions. *Mycorrhiza* **2013**, *23*, 597–625. [CrossRef]
5. Gianinazzi, S.; Gollotte, A.; Binet, M.N.; van Tuinen, D.; Redecker, D.; Wipf, D. Agroecology the key role of arbuscular mycorrhizas in ecosystem services. *Mycorrhiza* **2010**, *20*, 519–530. [CrossRef]
6. Sbrana, C.; Avio, L.; Giovannetti, M. Beneficial mycorrhizal symbionts affecting the production of health-promoting phytochemicals. *Electrophoresis* **2014**, *35*, 1535–1546. [CrossRef]
7. Avio, L.; Turrini, A.; Giovannetti, M.; Sbrana, C. Designing the ideotype mycorrhizal symbionts for the production of healthy food. *Front. Plant Sci.* **2018**, *9*, 1089. [CrossRef]
8. Barea, J.M.; Azcón, R.; Azcón-Aguilar, C. Mycorrhizosphere interactions to improve plant fitness and soil quality. *Antonie Leeuwenhoek* **2002**, *81*, 343–351. [CrossRef] [PubMed]
9. Rouphael, Y.; Franken, P.; Schneider, C.; Schwarz, D.; Giovannetti, M.; Agnolucci, M.; De Pascale, S.; Bonini, P.; Colla, G. Arbuscular mycorrhizal fungi act as biostimulants in horticultural crops. *Sci. Hortic.* **2015**, *196*, 91–108. [CrossRef]

10. Turrini, A.; Avio, L.; Giovannetti, M.; Agnolucci, M. Functional complementarity of arbuscular mycorrhizal fungi and associated microbiota: The challenge of translational research. *Front. Plant Sci.* **2018**, *9*, 1407. [CrossRef] [PubMed]
11. Giovannetti, M.; Avio, L. Biotechnology of arbuscular mycorrhizas. In *Applied Mycology and Biotechnology: Agriculture and Food Production*; Khachatourians, G.G., Arora, D.K., Eds.; Elsevier: Amsterdam, The Netherlands, 2002; Volume 2, pp. 275–310.
12. Giovannetti, M.; Sbrana, C.; Avio, L.; Citernesi, A.S.; Logi, C. Differential hyphal morphogenesis in arbuscular mycorrhizal fungi during pre-infection stages. *New Phytol.* **1993**, *125*, 587–593. [CrossRef]
13. Akiyama, K.; Matsuzaki, K.I.; Hayashi, H. Plant sesquiterpenes induce hyphal branching in arbuscular mycorrhizal fungi. *Nature* **2005**, *435*, 824–827. [CrossRef]
14. Sbrana, C.; Giovannetti, M. Chemotropism in the arbuscular mycorrhizal fungus *Glomus mosseae*. *Mycorrhiza* **2005**, *15*, 539–545. [CrossRef]
15. Kiers, E.T.; Duhamel, M.; Beesetty, Y.; Mensah, J.A.; Franken, O.; Verbruggen, E.; Fellbaum, C.R.; Kowalchuk, G.A.; Hart, M.M.; Bago, A.; et al. Reciprocal rewards stabilize cooperation in the mycorrhizal symbiosis. *Science* **2011**, *333*, 880–882. [CrossRef]
16. Jiang, Y.; Wang, W.; Xie, Q.; Liu, N.; Liu, L.; Wang, D.; Zhang, X.; Yang, C.; Chen, X.; Tang, D.; et al. Plants transfer lipids to sustain colonization by mutualistic mycorrhizal and parasitic fungi. *Science* **2017**, *356*, 1172–1175. [CrossRef]
17. Luginbuehl, L.H.; Menard, G.N.; Kurup, S.; Van Erp, H.; Radhakrishnan, G.V.; Breakspear, A.; Oldroyd, G.E.D.; Eastmond, P.J. Fatty acids in arbuscular mycorrhizal fungi are synthesized by the host plant. *Science* **2017**, *356*, 1175–1178. [CrossRef]
18. Smith, F.A.; Smith, S.E. Tansley Review No. 96 structural diversity in (vesicular)–arbuscular mycorrhizal symbioses. *New Phytol.* **1997**, *137*, 373–388. [CrossRef]
19. Giovannetti, M.; Sbrana, C.; Avio, L.; Strani, P. Patterns of below-ground plant interconnections established by means of arbuscular mycorrhizal networks. *New Phytol.* **2004**, *164*, 175–181. [CrossRef]
20. Mikkelsen, B.L.; Rosendahl, S.; Jakobsen, I. Underground resource allocation between individual networks of mycorrhizal fungi. *New Phytol.* **2008**, *180*, 890–898. [CrossRef]
21. Giovannetti, M.; Avio, L.; Sbrana, C. Fungal spore germination and pre-symbiotic mycelial growth–physiological and genetic aspects. In *Arbuscular Mycorrhizas: Physiology and Function*; Koltai, H., Kapulnik, Y., Eds.; Springer: Dordrecht, The Netherlands, 2010; pp. 3–32. [CrossRef]
22. de Novais, C.B.; Sbrana, C.; Júnior, O.J.S.; Siqueira, J.O.; Giovannetti, M. Vegetative compatibility and anastomosis formation within and among individual germlings of tropical isolates of arbuscular mycorrhizal fungi (Glomeromycota). *Mycorrhiza* **2013**, *23*, 325–331. [CrossRef]
23. Roesti, D.; Ineichen, K.; Braissant, O.; Redecker, D.; Wiemken, A.; Aragno, M. Bacteria associated with spores of the arbuscular mycorrhizal fungi *Glomus geosporum* and *Glomus constrictum*. *Appl. Environ. Microbiol.* **2005**, *71*, 6673–6679. [CrossRef]
24. Long, L.; Zhu, H.; Yao, Q.; Ai, Y. Analysis of bacterial communities associated with spores of *Gigaspora margarita* and *Gigaspora rosea*. *Plant Soil* **2008**, *310*, 1–9. [CrossRef]
25. Agnolucci, M.; Battini, F.; Cristani, C.; Giovannetti, M. Diverse bacterial communities are recruited on spores of different arbuscular mycorrhizal fungal isolates. *Biol. Fertil. Soils* **2015**, *51*, 379–389. [CrossRef]
26. Giovannetti, M.; Turrini, A.; Strani, P.; Sbrana, C.; Avio, L.; Pietrangeli, B. Mycorrhizal fungi in ecotoxicological studies: Soil impact of fungicides, insecticides and herbicides. *Prev. Today* **2006**, *2*, 47–61.
27. Buysens, C.; Dupré de Boulois, H.; Declerck, S. Do fungicides used to control *Rhizoctonia solani* impact the non-target arbuscular mycorrhizal fungus *Rhizophagus irregularis*? *Mycorrhiza* **2015**, *25*, 277–288. [CrossRef] [PubMed]
28. Hage-Ahmed, K.; Rosner, K.; Steinkellner, S. Arbuscular mycorrhizal fungi and their response to pesticides. *Pest Manag. Sci.* **2019**, *75*, 583–590. [CrossRef] [PubMed]
29. Sbrana, C.; Fortuna, P.; Giovannetti, M. Plugging into the network: Belowground connections between germlings and extraradical mycelium of arbuscular mycorrhizal fungi. *Mycologia* **2011**, *103*, 307–316. [CrossRef] [PubMed]
30. Giovannetti, M.; Azzolini, D.; Citernesi, A.S. Anastomosis formation and nuclear and protoplasmic exchange in arbuscular mycorrhizal fungi. *Appl. Environ. Microbiol.* **1999**, *65*, 5571–5575. [CrossRef]

31. Voets, L.; De La Providencia, I.E.; Declerck, S. Glomeraceae and Gigasporaceae differ in their ability to form hyphal networks. *New Phytol.* **2006**, *172*, 185–188. [CrossRef]
32. Chiocchio, V.; Venedikian, N.; Martinez, A.E.; Ana Menendez, A.M.; Ocampo, J.A.; Godeas, A. Effect of the fungicide benomyl on spore germination and hyphal length of the arbuscular mycorrhizal fungus *Glomus mosseae*. *Int. Microbiol.* **2000**, *3*, 173–175.
33. de Novais, C.B.; Giovannetti, M.; de Faria, S.M.; Sbrana, C. Two herbicides, two fungicides and spore-associated bacteria affect *Funneliformis mosseae* extraradical mycelium structural traits and viability. *Mycorrhiza* **2019**, *29*, 341–349. [CrossRef]
34. Buee, M.; Rossignol, M.; Jauneau, A.; Ranjeva, R.; Bécard, G. The pre-symbiotic growth of arbuscular mycorrhizal fungi is induced by a branching factor partially purified from plant root exudates. *Mol. Plant Microbe Interact.* **2000**, *13*, 693–698. [CrossRef]
35. Giovannetti, M.; Citernesi, A.S. Time-course of appressorium formation on host plants by arbuscular mycorrhizal fungi. *Mycol. Res.* **1993**, *97*, 1140–1142. [CrossRef]
36. Mosse, B. The regular germination of resting spores and some observations on the growth requirements of an *Endogone* sp. causing vesicular-arbuscular mycorrhiza. *Trans. Br. Mycol. Soc.* **1959**, *42*, 273–286. [CrossRef]
37. Breuninger, M.; Requena, N. Recognition events in AM symbiosis: Analysis of fungal gene expression at the early appressorium stage. *Fungal Genet. Biol.* **2004**, *41*, 794–804. [CrossRef] [PubMed]
38. Turrini, A.; Sbrana, C.; Pitto, L.; Ruffini Castiglione, M.; Giorgetti, L.; Briganti, R.; Bracci, T.; Evangelista, M.; Nuti, M.P.; Giovannetti, M. The antifungal Dm-AMP1 protein from *Dahlia merckii* expressed in *Solanum melongena* is released in root exudates and differentially affects pathogenic fungi and mycorrhizal symbiosis. *New Phytol.* **2004**, *163*, 393–403. [CrossRef]
39. Morley, C.D.; Mosse, B. Abnormal vesicular-arbuscular mycorrhizal infections in white clover induced by lupin. *Trans. Br. Mycol. Soc.* **1976**, *67*, 510–513. [CrossRef]
40. Fortuna, P.; Citernesi, S.; Morini, S.; Giovannetti, M.; Loreti, F. Infectivity and effectiveness of different species of arbuscular mycorrhizal fungi in micropropagated plants of Mr S 2/5 plum rootstock. *Agronomie* **1992**, *12*, 825–829. [CrossRef]
41. Staehelin, C.; Charon, C.; Boller, T.; Crespi, M.; Kondorosi, Á. *Medicago truncatula* plants overexpressing the early nodulin gene enod40 exhibit accelerated mycorrhizal colonization and enhanced formation of arbuscules. *Proc. Natl. Acad. Sci. USA* **2001**, *98*, 15366–15371. [CrossRef]
42. Tawaraya, K.; Hashimoto, K.; Wagatsuma, T. Effect of root exudate fractions from P-deficient and P-sufficient onion plants on root colonisation by the arbuscular mycorrhizal fungus *Gigaspora margarita*. *Mycorrhiza* **1998**, *8*, 67–70. [CrossRef]
43. Turrini, A.; Giordani, T.; Avio, L.; Natali, L.; Giovannetti, M.; Cavallini, A. Large variation in mycorrhizal colonization among wild accessions, cultivars, and inbreds of sunflower (*Helianthus annuus* L.). *Euphytica* **2016**, *207*, 331–342. [CrossRef]
44. De Vita, P.; Avio, L.; Sbrana, C.; Laidò, G.; Marone, D.; Mastrangelo, A.M.; Cattivelli, L.; Giovannetti, M. Genetic markers associated to arbuscular mycorrhizal colonization in durum wheat. *Sci. Rep.* **2018**, *8*, 10612. [CrossRef]
45. Battini, F.; Grønlund, M.; Agnolucci, M.; Giovannetti, M.; Jakobsen, I. Facilitation of phosphorus uptake in maize plants by mycorrhizosphere bacteria. *Sci. Rep.* **2017**, *7*, 4686. [CrossRef] [PubMed]
46. Giovannetti, M.; Fortuna, P.; Citernesi, A.S.; Morini, S.; Nuti, M.P. The occurrence of anastomosis formation and nuclear exchange in intact arbuscular mycorrhizal networks. *New Phytol.* **2001**, *151*, 717–724. [CrossRef]
47. Friese, C.F.; Allen, M.F. The spread of VA mycorrhizal fungal hyphae in the soil: Inoculum types and external hyphal architecture. *Mycologia* **1991**, *83*, 409–418. [CrossRef]
48. Bago, B.; Azcón-Aguilar, C.; Goulet, A.; Piché, Y. Branched absorbing structures (BAS): A feature of the extraradical mycelium of symbiotic arbuscular mycorrhizal fungi. *New Phytol.* **1998**, *139*, 375–388. [CrossRef]
49. Giovannetti, M.; Avio, L.; Sbrana, C. Functional significance of anastomosis in arbuscular mycorrhizal networks. In *Mycorrhizal Networks*; Horton, T., Ed.; Springer: Dordrecht, The Netherlands, 2015; Volume 224, pp. 41–67.
50. Purin, S.; Morton, J.B. In situ analysis of anastomosis in representative genera of arbuscular mycorrhizal fungi. *Mycorrhiza* **2011**, *21*, 505–514. [CrossRef]
51. Purin, S.; Morton, J.B. Anastomosis behavior differs between asymbiotic and symbiotic hyphae of *Rhizophagus clarus*. *Mycologia* **2013**, *105*, 589–602. [CrossRef]

52. Pepe, A.; Giovannetti, M.; Sbrana, C. Different levels of hyphal self incompatibility modulate interconnectedness of mycorrhizal networks in three arbuscular mycorrhizal fungi within the Glomeraceae. *Mycorrhiza* **2016**, *26*, 325–332. [CrossRef]
53. Avio, L.; Pellegrino, E.; Bonari, E.; Giovannetti, M. Functional diversity of arbuscular mycorrhizal fungal isolates in relation to extraradical mycelial networks. *New Phytol.* **2006**, *172*, 347–357. [CrossRef]
54. Helgason, T.; Daniell, T.J.; Husband, R.; Fitter, A.H.; Young, J.P.W. Ploughing up the wood-wide web? *Nature* **1998**, *394*, 431. [CrossRef]
55. Simard, S.W.; Durrell, D.M. Mycorrhizal networks: A review of their extent, function and importance. *Can. J. Bot.* **2004**, *82*, 1140–1165. [CrossRef]
56. Avio, L.; Castaldini, M.; Fabiani, A.; Bedini, S.; Sbrana, C.; Turrini, A.; Giovannetti, M. Impact of nitrogen fertilization and soil tillage on arbuscular mycorrhizal fungal communities in a Mediterranean agroecosystem. *Soil Biol. Biochem.* **2013**, *67*, 285–294. [CrossRef]
57. Njeru, E.M.; Avio, L.; Sbrana, C.; Turrini, A.; Bocci, G.; Bàrberi, P.; Giovannetti, M. First evidence for a major cover crop effect on arbuscular mycorrhizal fungi and organic maize growth. *Agron. Sustain. Dev.* **2014**, *34*, 841–848. [CrossRef]
58. de Novais, C.B.; Avio, L.; Giovannetti, M.; de Faria, S.M.; Siqueira, J.O.; Sbrana, C. Interconnectedness, length and viability of arbuscular mycorrhizal mycelium as affected by selected herbicides and fungicides. *Appl. Soil Ecol.* **2019**, *143*, 144–152. [CrossRef]
59. Schubert, A.; Marzachi, C.; Mazzitelli, M.; Cravero, M.C.; Bonfante-Fasolo, P. Development of total and viable extraradical mycelium in the vesicular–arbuscular mycorrhizal fungus *Glomus clarum* Nicol. & Schenck. *New Phytol.* **1987**, *107*, 183–190. [CrossRef]
60. Hamel, C.; Fyles, H.; Smith, D.L. Measurement of development of endomycorrhizal mycelium using three different vital stains. *New Phytol.* **1990**, *115*, 297–302. [CrossRef]
61. Pepe, A.; Giovannetti, M.; Sbrana, C. Lifespan and functionality of mycorrhizal fungal mycelium are uncoupled from host plant lifespan. *Sci. Rep.* **2018**, *8*, 10235. [CrossRef]
62. Sanders, F.E.; Tinker, P.B. Phosphate flow into mycorrhizal roots. *Pestic. Sci.* **1973**, *4*, 385–395. [CrossRef]
63. Ferrol, N.; Azcón-Aguilar, C.; Pérez-Tienda, J. Arbuscular mycorrhizas as key players in sustainable plant phosphorus acquisition: An overview on the mechanisms involved. *Plant Sci.* **2018**, *280*, 441–447. [CrossRef]
64. Pepe, A.; Sbrana, C.; Ferrol, N.; Giovannetti, M. An in vivo whole-plant experimental system for the analysis of gene expression in extraradical mycorrhizal mycelium. *Mycorrhiza* **2017**, *7*, 659–668. [CrossRef]
65. Johnson, I.T. Glucosinolates: Bioavailability and importance to health. *Int. J. Vitam. Nutr. Res.* **2002**, *71*, 26–31. [CrossRef]
66. Duthie, S.J. Plant polyphenols in cancer and heart disease: Implications as nutritional antioxidants. *Nutr. Res. Rev.* **2000**, *13*, 79–106. [CrossRef] [PubMed]
67. Lund, E. Non-nutritive bioactive constituents of plants: Dietary sources and health benefits of glucosinolates. *Int. J. Vitam. Nutr. Res.* **2003**, *73*, 135–143. [CrossRef] [PubMed]
68. Copetta, A.; Lingua, G.; Berta, G. Effects of three AM fungi on growth, distribution of glandular hairs, and essential oil production in *Ocimum basilicum* L. var. Genovese. *Mycorrhiza* **2006**, *16*, 485–494. [CrossRef]
69. Toussaint, J.P.; Kraml, M.; Nell, M.; Smith, S.E.; Smith, F.A.; Steinkellner, S.; Schmiderer, C.; Vierheilig, H.; Novak, J. Effect of *Glomus mosseae* on concentrations of rosmarinic and caffeic acids and essential oil compounds in basil inoculated with *Fusarium oxysporum* f. sp. *basilici*. *Plant Pathol.* **2008**, *57*, 1109–1116. [CrossRef]
70. Battini, F.; Bernardi, R.; Turrini, A.; Agnolucci, M.; Giovannetti, M. *Rhizophagus intraradices* or its associated bacteria affect gene expression of key enzymes involved in the rosmarinic acid biosynthetic pathway of basil. *Mycorrhiza* **2016**, *26*, 699–707. [CrossRef] [PubMed]
71. Gualandi, R.J., Jr. Fungal Endophytes Enhance Growth and Production of Natural Products in *Echinacea Purpurea* (Moench.). Master's Thesis, University of Tennessee, Knoxville, TN, USA, 2010. Available online: https://trace.tennessee.edu/utk_gradthes/713 (accessed on 7 January 2020).
72. Tavarini, S.; Passera, B.; Martini, A.; Avio, L.; Sbrana, C.; Giovannetti, M.; Angelini, L.G. Plant growth, steviol glycosides and nutrient uptake as affected by arbuscular mycorrhizal fungi and phosphorous fertilization in *Stevia rebaudiana* Bert. *Ind. Crops Prod.* **2018**, *111*, 899–907. [CrossRef]
73. Zubek, S.; Stojakowska, A.; Anielska, T.; Turnau, K. Arbuscular mycorrhizal fungi alter thymol derivative contents of *Inula ensifolia* L. *Mycorrhiza* **2010**, *20*, 497–504. [CrossRef]

74. Rambelli, A. The rhizosphere of mycorrhizae. In *Ectomycorrhizae: Their Ecology and Physiology*; Marks, G.C., Kozlowski, T.T., Eds.; Academic Press: New York, NY, USA, 1973; pp. 299–343.
75. Walley, F.L.; Germida, J.J. Failure to decontaminate *Glomus clarum* NT4 spores is due to spore wall-associated bacteria. *Mycorrhiza* **1996**, *6*, 43–49. [CrossRef]
76. Filippi, C.; Bagnoli, G.; Citernesi, A.S.; Giovannetti, M. Ultrastructural spatial distribution of bacteria associated with sporocarps of *Glomus mosseae*. *Symbiosis* **1998**, *24*, 1–12.
77. Mosse, B. Honey-coloured, sessile *Endogone* spores: II. Changes in fine structure during spore development. *Arch. Mikrobiol.* **1970**, *74*, 129–145. [CrossRef]
78. MacDonald, R.M.; Chandler, M.R. Bacterium-like organelles in vesicular-arbuscular mycorrizal fungus *Glomus caledonium*. *New Phytol.* **1981**, *89*, 241–246. [CrossRef]
79. MacDonald, R.M.; Chandler, M.R.; Mosse, B. The occurrence of bacterium-like organelles in vesicular–arbuscular mycorrizal fungi. *New Phytol.* **1982**, *90*, 659–663. [CrossRef]
80. Bianciotto, V.; Bandi, C.D.; Minerdi, M.; Sironi, H.; Tichy, V.; Bonfante, P. An obligately endosymbiotic mycorrhizal fungus itself harbors obligately intracellular bacteria. *Appl. Environ. Microbiol.* **1996**, *62*, 3005–3010. [CrossRef]
81. Mayo, K.; Davis, R.E.; Motta, J. Stimulation of germination of spores of *Glomus versiforme* by spore-associated bacteria. *Mycologia* **1986**, *78*, 426–431. [CrossRef]
82. Xavier, L.J.C.; Germida, J.J. Bacteria associated with *Glomus clarum* spores influence mycorrhizal activity. *Soil Biol. Biochem.* **2003**, *35*, 471–478. [CrossRef]
83. Bharadwaj, D.P.; Lundquist, P.O.; Persson, P.; Alström, S. Evidence for specificity of cultivable bacteria associated with arbuscular mycorrhizal fungal spores. *FEMS Microbiol. Ecol.* **2008**, *65*, 310–322. [CrossRef]
84. Cruz, A.F.; Horii, S.; Ochiai, S.; Yasuda, A.; Ishii, T. Isolation and analysis of bacteria associated with spores of *Gigaspora margarita*. *J. Appl. Microbiol.* **2008**, *104*, 1711–1717. [CrossRef]
85. Battini, F.; Cristani, C.; Giovannetti, M.; Agnolucci, M. Multifunctionality and diversity of culturable bacterial communities strictly associated with spores of the plant beneficial symbiont *Rhizophagus intraradices*. *Microbiol. Res.* **2016**, *183*, 68–79. [CrossRef]
86. Frey-Klett, P.; Garbaye, J.A.; Tarkka, M. The mycorrhiza helper bacteria revisited. *New Phytol.* **2007**, *176*, 22–36. [CrossRef]
87. Mugnier, J.; Mosse, B. Spore germination and viability of a vesicular arbuscular mycorrhizal fungus, *Glomus mosseae*. *Trans. Br. Mycol. Soc.* **1987**, *88*, 411–413. [CrossRef]
88. Tylka, G.L.; Hussey, R.S.; Roncadori, R.W. Axenic germination of vesicular–arbuscular mycorrhizal fungi: Effects of selected *Streptomyces* species. *Phytopathology* **1991**, *81*, 754–759. [CrossRef]
89. Carpenter-Boggs, L.; Loynachan, T.E.; Stahl, P.D. Spore germination of *Gigaspora margarita* stimulated by volatiles of soil-isolated actinomycetes. *Soil Biol. Biochem.* **1995**, *27*, 1445–1451. [CrossRef]
90. Ames, R.N.; Mihara, K.L.; Bayne, H.G. Chitin-decomposing actynomycetes associated with a vesicular–arbuscular mycorrhizal fungus from a calcareous soil. *New Phytol.* **1989**, *111*, 67–71. [CrossRef]
91. Will, M.E.; Sylvia, D.M. Interaction of rhizosphere bacteria, fertilizer, and vesicular-arbuscular mycorrhizal fungi with sea oats. *Appl. Environ. Microbiol.* **1990**, *56*, 2073–2079. [CrossRef]
92. Calvet, C.; Barea, J.M.; Pera, J. In vitro interactions between the vesicular-arbuscular mycorrhizal fungus *Glomus mosseae* and some saprophytic fungi isolated from organic substrates. *Soil Biol. Biochem.* **1992**, *24*, 775–780. [CrossRef]
93. Hildebrandt, U.; Ouziad, F.; Marner, F.-J.J.; Bothe, H. The bacterium *Paenibacillus validus* stimulates growth of the arbuscular mycorrhizal fungus *Glomus intraradices* up to the formation of fertile spores. *FEMS Microbiol. Lett.* **2006**, *254*, 258–267. [CrossRef]
94. Pivato, B.; Offre, P.; Marchelli, S.; Barbonaglia, B.; Mougel, C.; Lemanceau, P. Bacterial effects on arbuscular mycorrhizal fungi and mycorrhiza development as influenced by the bacteria, fungi, and host plant. *Mycorrhiza* **2009**, *19*, 81–90. [CrossRef]
95. Bidondo, L.F.; Silvani, V.; Colombo, R.; Pérgola, M.; Bompadre, J.; Godeas, A. Pre-symbiotic and symbiotic interactions between *Glomus intraradices* and two *Paenibacillus* species isolated from AM propagules. In vitro and in vivo assays with soybean (AG043RG) as plant host. *Soil Biol. Biochem.* **2011**, *43*, 1866–1872. [CrossRef]
96. Ordoñez, Y.M.; Fernandez, B.R.; Lara, L.S.; Rodriguez, A.; Uribe-Vélez, D.; Sanders, I.R. Bacteria with phosphate solubilizing capacity alter mycorrhizal fungal growth both inside and outside the root and in the presence of native microbial communities. *PLoS ONE* **2016**, *11*, e0154438. [CrossRef]

97. Ravnskov, S.; Jakobsen, I. Effects of *Pseudomonas fluorescens* DF57 on growth and P uptake of two arbuscular mycorrhizal fungi in symbiosis with cucumber. *Mycorrhiza* **1999**, *8*, 329–334. [CrossRef]
98. Ravnskov, S.; Larsen, J.; Jakobsen, I. Phosphorus uptake of an arbuscular mycorrhizal fungus is not effected by the biocontrol bacterium *Burkholderia cepacia*. *Soil Biol. Biochem.* **2002**, *34*, 1875–1881. [CrossRef]
99. Zhang, L.; Fan, J.; Ding, X.; He, X.; Zhang, F.; Feng, G. Hyphosphere interactions between an arbuscular mycorrhizal fungus and a phosphate solubilizing bacterium promote phytate mineralization in soil. *Soil Biol. Biochem.* **2014**, *74*, 177–183. [CrossRef]
100. Khalid, A.; Arshad, M.; Kahir, Z.A. Screening plant growth promoting rhizobacteria for improving growth and yield of wheat. *J. Appl. Microbiol.* **2004**, *96*, 473–480. [CrossRef] [PubMed]
101. Aloni, R.; Aloni, E.; Langhans, M.; Ullrich, C.I. Role of cytokinin and auxin in shaping root architecture: Regulating vascular differentiation, lateral root initiation, root apical dominance and root gravitropism. *Ann. Bot.* **2006**, *97*, 883–893. [CrossRef] [PubMed]
102. Duca, D.; Lorv, J.; Patten, C.L.; Rose, D.; Glick, B.R. Indole-3- acetic acid in plant–microbe interactions. *Antonie Leeuwenhoek* **2014**, *106*, 85–125. [CrossRef] [PubMed]
103. Budi, S.W.; van Tuinen, D.; Martinotti, G.; Gianinazzi, S. Isolation from *Sorghum bicolor* mycorrhizosphere of a bacterium compatible with arbuscular mycorrhiza development and antagonistic towards soil-borne fungal pathogens. *Appl. Environ. Microbiol.* **1999**, *65*, 5148–5150. [CrossRef] [PubMed]
104. Li, B.; Ravnskov, S.; Xie, G.; Larsen, J. Biocontrol of *Pythium* damping-off in cucumber by arbuscular mycorrhiza-associated bacteria from the genus *Paenibacillus*. *Biocontrol* **2007**, *52*, 863–875. [CrossRef]
105. Bharadwaj, D.P.; Lundquist, P.O.; Alström, S. Arbuscular mycorrhizal fungal spore-associated bacteria affect mycorrhizal colonization, plant growth and potato pathogens. *Soil Biol. Biochem.* **2008**, *40*, 2494–2501. [CrossRef]
106. Kobayashi, T.; Nishizawa, N.K. Iron uptake, translocation, and regulation in higher plants. *Ann. Rev. Plant Biol.* **2012**, *63*, 131–152. [CrossRef]
107. Robin, A.; Vansuyt, G.; Hinsinger, P.; Meyer, J.M.; Briat, J.F.; Lemanceau, P. Iron dynamics in the rhizosphere: Consequences for plant health and nutrition. *Adv. Agron.* **2008**, *99*, 183–225. [CrossRef]
108. Agnolucci, M.; Avio, L.; Pepe, A.; Turrini, A.; Cristani, C.; Bonini, P.; Cirino, V.; Colosimo, F.; Ruzzi, M.; Giovannetti, M. Bacteria associated with a commercial mycorrhizal inoculum: Community composition and multifunctional activity as assessed by Illumina sequencing and culture-dependent tools. *Front. Plant Sci.* **2018**, *9*, 1956. [CrossRef] [PubMed]
109. Le Cocq, K.; Gurr, S.J.; Hirsch, P.R.; Mauchline, T.H. Exploitation of endophytes for sustainable agricultural intensification. *Mol. Plant Pathol.* **2017**, *18*, 469–473. [CrossRef]
110. Agnolucci, M.; Palla, M.; Cristani, C.; Cavallo, N.; Giovannetti, M.; De Angelis, M.; Gobbetti, M.; Minervini, F. Beneficial Plant Microorganisms Affect the Endophytic Bacterial Communities of Durum Wheat Roots as Detected by Different Molecular Approaches. *Front. Microbiol.* **2019**, *10*, 2500. [CrossRef] [PubMed]
111. Compant, S.; Clément, C.; Sessitsch, A. Plant growth-promoting bacteria in the rhizo- and endosphere of plants: Their role, colonization, mechanisms involved and prospects for utilization. *Soil Biol. Biochem.* **2010**, *42*, 669–678. [CrossRef]
112. Hardoim, P.R.; van Overbeek, P.S.; Berg, G.; Pirttilä, A.M.; Compant, S.; Campisano, A.; Döring, M.; Sessitsch, A. The hidden world within plants: Ecological and evolutionary considerations for defining functioning of microbial endophytes. *Microbiol. Mol. Biol. Rev.* **2015**, *79*, 293–320. [CrossRef] [PubMed]

agronomy

MDPI

Article

The Influence of Bio-Stimulants and Foliar Fertilizers on Yield, Plant Features, and the Level of Soil Biochemical Activity in White Lupine (*Lupinus albus* L.) Cultivation

Alicja Niewiadomska [1], Hanna Sulewska [2], Agnieszka Wolna-Maruwka [1], Karolina Ratajczak [2,*], Zyta Waraczewska [1] and Anna Budka [3]

1 Department of General and Environmental Microbiology, Poznań University of Life Sciences, ul. Szydłowska 50, 60-656 Poznań, Poland; alicja.niewiadomska@up.poznan.pl (A.N.); amaruwka@up.poznan.pl (A.W.-M.); zyta.waraczewska@wp.pl (Z.W.)

2 Department of Agronomy, Poznan University of Life Sciences, ul. Dojazd 11, 60-632 Poznań, Poland; hanna.sulewska@up.poznan.pl

3 Department of Mathematical and Statistical Methods, Poznań University of Life Sciences, ul. Wojska Polskiego 28, 60-637 Poznań, Poland; anna.budka@up.poznan.pl

* Correspondence: karolina.ratajczak@up.poznan.pl

Received: 17 December 2019; Accepted: 17 January 2020; Published: 20 January 2020

Abstract: The aim of this study is to assess the effect of two biostimulators (Titanit, Rooter) and six foliar fertilizers (Optysil, Metalosate Potassium, Bolero Bo, ADOB 2.0 Zn IDHA, ADOB B, ADOB 2.0 Mo) on white lupine. In addition, we evaluated the enzymatic activity of dehydrogenase, acid, and alkaline phosphatases, catalase, the level of biological nitrogen fixation, yield, plant biometric, chlorophyll fluorescence and chlorophyll content. A field experiment was conducted between 2016 and 2018 at the Gorzyń Experimental and Educational Station, Poznań University of Life Sciences in Poland. The best effects in plant yield were obtained after the application of Optysil or ADOB Zn IDHA. The three years results of dehydrogenase (DHA), alkaline phosphatase (PAL), and the biological index of soil fertility (BIF), show that the bio-stimulants and most of the foliar fertilizers used did not always stimulate the activity of these enzymes and index in the white lupine crops, as compared with the control treatment. Analysis of the results of the acid phosphatase activity (PAC) shows that during the entire white lupine growing season the foliar fertilizers and bio-stimulants decreased the activity of this enzyme. This effect was not observed when the Metalosate potassium foliar fertilizer was applied. The field analyses of biological nitrogen fixation showed that the fertilizers and bio-stimulants significantly stimulated nitrogenase activity under the white lupine plantation. The best effects in plant yield were obtained after application Optysil or ADOB Zn IDHA.

Keywords: soil enzymatic activity; biological index fertility; nitrogenase activity; microelements fertilization (Ti, Si, B, Mo, Zn)

1. Introduction

The degradation of the soil environment, excessive use of chemicals, depletion of natural resources, as well as the decreasing biodiversity instigated the European Union to make a decision about the need for integrated crop cultivation and protection [1]. Since 2014 the recommendations concerning integrated protection and cultivation have been in force in Poland. At present we can see the transitional phase between conventional and sustainable agriculture. In order to meet the assumptions of sustainable agriculture it is necessary to diversify the crop structure and minimize the excessive share of cereals. It is also necessary to use integrated methods of agricultural production, so it might be particularly important to restore legume plantations [2].

The significance of legumes in sustainable agriculture is increasing because they improve the physicochemical properties of soil, increase the content of organic matter by leaving large quantities of crop residues, and reduce the need to apply nitrogen fertilizers. White lupine (*Lupinus albus* L.) is one of the most important crops in this group of plants in Poland. It has been the longest known crop species of the *Lupinus* genus. Because of its very high content of protein and fat, especially in seeds, it has been used for human nutrition for thousands of years, despite its high content of bitter alkaloids [3]. It was only in 1930 that low-alkaloid forms were obtained. Because of the introduction of new varieties, the cultivation of white lupine with low alkaloid content became popular in Poland. Between 2005 and 2015 the area of cultivation of large-seeded legumes increased almost four times so that in 2015 they covered an area of 407,000 ha [4].

Lupine species have the largest share in this group of crops. On the other hand, the area of plantations with small-seeded legumes, such as clover and alfalfa, did not fluctuate much in that decade and in 2015 they covered an area of 93,000 ha [5].

Legumes are characterized by the ability to coexist with the nitrogen-fixing diazotrophic bacteria (*Rhizobium*). In order to increase the protein content in plants, which depends on the system developed by the plant and rhizobia, it is necessary to find agents improving the efficiency of this symbiosis.

Scientists are more and more interested in bio-stimulants, which are defined as materials containing one or more active substances and/or microorganisms. They improve the uptake of nutrients by plants, their tolerance to abiotic and biotic stress, and the quality of crops [6]. Bio-stimulants also increase the activity of rhizosphere microorganisms and soil enzymes, as well as they stimulate hormone production and photosynthesis [7]. They also promote the overall plant growth, including increased biomass and crop yields [8]. In the group of synthetic bio-stimulants, there are preparations containing growth regulators, phenolic compounds, inorganic salts, and beneficial nutrients [9,10], which naturally occur in plants in trace amounts (e.g., titanium and silicon). They act mainly by the stimulation of numerous physiological processes, which has a positive effect on plant yield and crop quality. Nutrients assimilable by plants, reduces the impact of stress, which affects the growth and development of plants. They regulate the uptake of macro- and microelements, alleviate the negative effects of periodic water shortage, high salinity, as well as activates the natural immune mechanisms of plants. They also strengthen cell walls and reduce the susceptibility of plants to mechanical damage [11]. Microelements regulate biochemical processes occurring in plants, being part of most enzymes or acting as their activators, therefore their deficit may lead to the inhibition of specific enzymatic reactions, which in turn leads to disorders of many biochemical and physiological processes, adversely affecting the growth and plant development [12,13]. There are many fertilizers that are enriched with amino acids, organic compounds, or surfactants. For example, potassium in fertilizer is in the form of very small molecules complexed with a unique set of natural amino acids. In turn, boron in the fertilizer is in the form of sodium pentaborate decahydrate, and the addition of sorbitol ensures rapid uptake of the fertilizer through the leaves of fertilized plants and high efficiency of the fertilizer. Zinc in modern fertilizers is chelated with the biodegradable IDHA chelating agent, because of which it also gains a form that is very well absorbed by plants. This fertilizer increases the plants' resistance to drought and diseases and increases the germination of seeds. It is produced in the form of microgranules, based on modern microgranulation technology. The manufacturer of molybdenum fertilizer has developed a liquid formula of the fertilizer additionally enriched with biodegradable tensides, which decreases the surface tension of the working liquid and increases the efficiency of covering the leaf blade during spraying increases [14].

Essential plant nutrients are mainly applied to soil and plant foliage in order to achieve maximum economic yields. Soil application is more common and most effective for nutrients that are required in high quantities. However, under certain circumstances, foliar fertilization is more economic and effective. Because of the intensified cultivation foliar fertilization has become an indispensable agrotechnical procedure. Plants exhibit the highest demand for potassium and nitrogen (more than 200 kg in terms of the yield per 1 ha), and the lowest demand for zinc, boron, copper, and molybdenum. Plants need only a few grams of molybdenum in terms of the yield per hectare. This means that foliar fertilization is particularly recommended and effective when it is necessary to supply micronutrients to crops [15].

Each agrotechnical treatment, i.e., the use of fertilizers or bio-stimulants, may cause changes in the soil environment. There have been numerous studies showing various effects of these treatments on the count of selected groups of microorganisms and the amount of soil enzymes they secrete [16].

Measurement of the activity of soil enzymes provides information about the quality of soil. This procedure is important as it indicates the biochemical activity of soil. Enzymes are thought to be good and sensitive indicators because they quickly react to changes in soil caused by natural and anthropogenic factors. Apart from that, it is easy to measure their activity, which affects the main microbiological reactions involving the cycles of nutrients in soil. Studies also showed that agrotechnical procedures influence the enzymatic activity more than other biochemical parameters [17].

The aim of this study is to assess the effect of selected bio-stimulants (Tytanit, Rooter) and foliar fertilizers (Optysil, Metalosate potassium, Bolero Bo, ADOB 2.0 Zn IDHA, ADOB B, ADOB 2.0 Mo) on the yield and plant features, activity of dehydrogenase, acid and alkaline phosphatases, and catalase, as well as the level of biological nitrogen fixation based on the activity of nitrogenase in a white lupine plantation.

2. Material and Methods

2.1. Experimental Design

A field experiment was conducted between 2016 and 2018 at the Gorzyń Experimental and Educational Station, Poznań University of Life Sciences. The GPS coordinates of the experiment are as follows: N-52.56589, E-015.90556, 65 m AMSL. Each year one-factor experiment was conducted as randomized block design in four replications with the following nine factor levels: 1. control treatment—plants not treated with preparations; 2. Tytanit; 3. Optysil; 4. Metalosate Potassium; 5. Rooter; 6. Bolero Mo; 7. ADOB Zn IDHA; 8. ADOB B; 9. ADOB 2.0 Mo. Each fertilizer was applied in a timely manner, according to the manufacturer's recommendations (Table 1).

An experiment was conducted on white lupine (*Lupinus albus* L.) of the Butan cultivar. The lupine seeds were inoculated with the effective strain of *Bradyrhizobium lupinus* root nodule bacteria directly before sowing by using nitragina. Nitragina is a single-component graft, containing a specific bacterial strain for a specific legume plant, in which peat is a carrier. The Butan cultivar can be grown all over Poland, this variety is insensitive to delayed sowing; its growing period is 2–14 days shorter than that of traditional varieties and it is less susceptible to diseases caused by *Fusarium* fungi. The cultivar is more valuable as a feed and it has high content of protein (32–37%) and fat (10–12%), while the content of alkaloids is about 30–40% lower.

The seeds were sown (4 April 2016, 4 April 2017 and 7 April 2018) on plots with an area of 21 m^2, with a distance of rows of 15 cm, and sowing density of 75 seeds per 1 m^2.

According to the FAO/WRB classification [18], the soil in the experimental plots is a typical lessive soil formed from light loamy sands, deposited in a shallow layer on light loam (*Haplic Luvisols*) (Table 2). The soil texture was determined by means of a sieve (sand fraction) for the silt and clay fraction [19].

The agrotechnical and cultivation treatments were carried out in accordance with the principles of good agricultural and experimental practice for this species [20]. In the autumn before winter plowing, basic macronutrients were supplied to the soil in the form of multi-component fertilizer Polifoska 4 in the amount of 350 kg ha^{-1} (N—4%, P—12%, K—32%). Before sowing, urea in the amount of 30 kg ha^{-1} was used.

Table 1. The terms and doses of bio-stimulants and fertilizers applied in the experiment.

Bio-Stimulants/Foliar Fertilizers		Term and Dose of Bio-Stimulant	Fertilizer Characteristics
Bio-stimulants	Tytanit	I: leaf and shoot development (BBCH 13–29)—0.3 dm^3 ha^{-1} II: inflorescence development (BBCH 51–59)—0.3 dm^3 ha^{-1} III: beginning of pod development (BBCH 71)—0.3 dm^3 ha^{-1}	Liquid, mineral stimulant containing titanium (Ti). It increases the yield volume and development of plants, improves yield quality parameters and increases plants' natural resistance to stress factors. Composition: 8.5 g Ti $(dm^3)^{-1}$
	Rooter	BBCH 13–14—1 dm^3 ha^{-1}	Bio-stimulant—it stimulates the growth of the root system, accelerates regeneration and improves the uptake of soil minerals. Composition: P_2O_5 13.0%; K_2O 5.0%
Foliar fertilizers	Optysil	I: leaf and shoot development (BBCH 15–29)—0.5 dm^3 ha^{-1} II: inflorescence development (BBCH 51–55)—0.5 dm^3 ha^{-1} III: beginning of pod development (BBCH 71–73)—0.5 dm^3 ha^{-1}	Liquid, silicon antistressor stimulating the growth and development of plants, activating their natural immune systems and increasing tolerance to unfavorable cultivation conditions. Composition: 200 g SiO_2 $(dm^3)^{-1}$
	Metalosate Potassium	2–3 treatments every 10–14 days during intensive growth—3 dm^3 ha^{-1}	Liquid foliar fertilizer containing an easily absorbable form of potassium, which supplements potassium deficit in plants with amino acids. Composition: K_2O 24%
	Bolero Mo	Before florescence—1.5 dm^3 ha^{-1}	Liquid foliar fertilizer containing boron and molybdenum to supplement the deficit of these elements in plants. Composition: B 8.2%; Mo 0.8%
	ADOB 2.0 Zn IDHA	Before florescence—1 dm^3 ha^{-1}	Foliar fertilizer containing zinc (Zn) fully chelated by biodegradable chelating agent IDHA. Composition: Zn 100 g kg^{-1} (weight percentage content 10, chelated by IDHA)
	ADOB B	I: before florescence—2 dm^3 ha^{-1} II: after florescence on pods—1 dm^3 ha^{-1}	Liquid, highly concentrated foliar fertilizer containing boron that regulates auxin activity and participates in cell division. Composition: N 78 g kg^{-1}; B 150 g kg^{-1}
	ADOB 2.0 Mo	early stages of development—0.15 dm^3 ha^{-1}	Liquid, single-component fertilizer which increases the rate and efficiency of use of nitrogen by plants and improves interaction with iron. Composition: Mo 20%

Table 2. The texture of soil sampled at a depth of 0–25 cm and the soil chemical properties of the 3-year experiment.

	Percentage of Soil Fractions			Texture Class
Fraction [mm]	Sand 2–0.05	Silt 0.05–0.002	Clay <0.002	
	78	18	4	LS
Soil Chemical Properties				
pH in 1 mol KCl			6.0	
Phosphorus P ($mg \cdot kg^{-1}$)			70.1	
Potassium K ($mg \cdot kg^{-1}$)			99.3	
Magnesium Mg ($mg \cdot kg^{-1}$)			56.7	
Manganese Mn ($mg \cdot kg^{-1}$)			303.4	
Zinc Zn ($mg \cdot kg^{-1}$)			10.9	
Copper Cu ($mg \cdot kg^{-1}$)			2.6	
Iron Fe ($mg \cdot kg^{-1}$)			1525.2	
Boron B ($mg \cdot kg^{-1}$)			>20	
Organic carbon (%)			0.5	
Percent of caries			0.8	

LS—loamy sand.

The agrotechnical procedures were carried out in accordance with the rules adopted for the species used in the test. White lupine was sown in early April. The following products were used for weed control: Afalon Dispersive 450 EC (1.1 L ha^{-1}) in April, Basagran 480 SL (2.6 L ha^{-1}) and Betanal MaxPro 209 OD (1.25 L ha^{-1}) in May. Fusilade Forte 150 EC (1.0 L ha^{-1}) was additionally applied in June. The following products were sprayed to protect the plants from diseases: Gwarant 500 SC (2.0 L ha^{-1}) in May and Korazzo 250 SC (1.0 L ha^{-1}) in mid- and late June.

2.2. Weather Conditions

During the growing seasons in 2016 and 2017 the weather conditions were similar in terms of temperature and rainfall. During the growing season the highest average air temperature was noted in July both in 2016 (19.5 °C) and 2017 (18.9 °C), whereas the lowest temperature was noted in April, i.e., 8.7 °C in 2016 and 7.5 °C in 2017. However, the weather conditions in 2018 were different than in the previous years (Figure 1). The highest average temperature was noted in August (21.2 °C), whereas the lowest was noted in May (12.7 °C). As far as the average monthly temperature from April to September is concerned, 2018 was the warmest—it was 2.9 °C warmer than 2016 and 1.7 °C warmer than 2017. In 2016 there was drought only at the end of the growing season. Likewise, in 2017 there was no rainfall deficit. On the contrary, it was a wet year, especially from June to August. On the other hand, in 2018 rainfall was unevenly distributed and there were droughts that were particularly unfavorable for plants in May, June, and August.

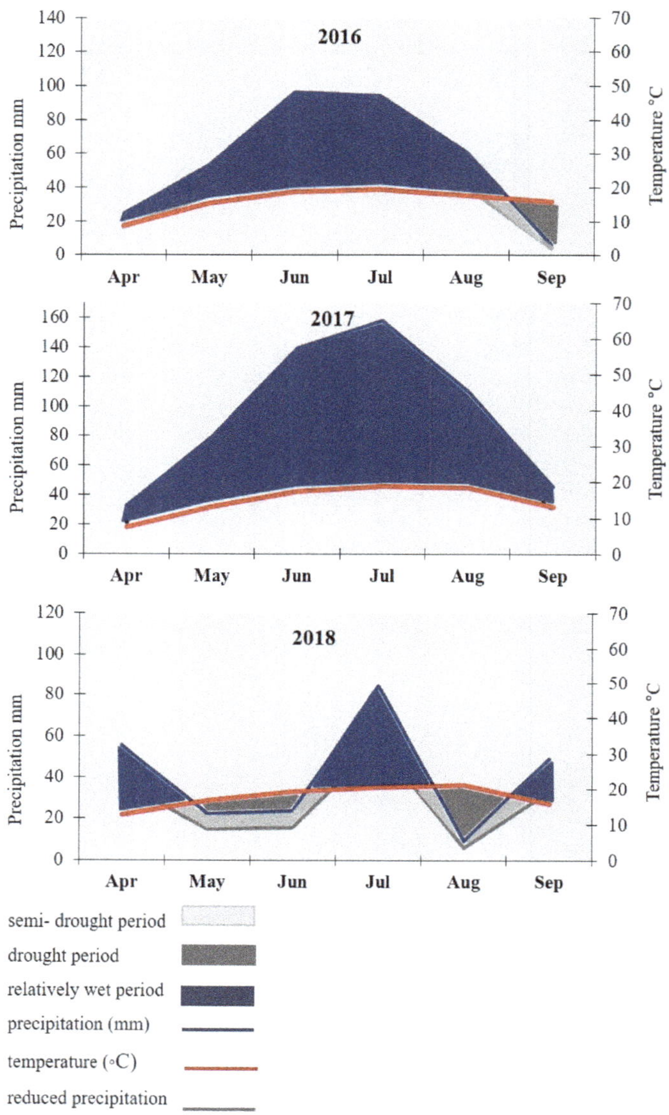

Figure 1. Climate graphs according to Walter [21] characterizing weather conditions in Gorzyń.

2.3. *Influence of Fertilizers on Nitrogenase Activity (Diazotrophy)*

Nitrogenase activity was estimated using the acetylene reduction assay (ARA) at the beginning of the plants' flowering [22]. For this purpose, five plants were randomly selected in plots, in a given experimental treatment and directly were placed tightly in sealed test vials (2000 mL) at the field, purified C_2H_4 was injected to obtain an acetylene concentration of 10% (*v*/*v*) in the gas phase (air). After an hour, 1 mL of the gas phase was taken from inside of the test vessels with a Hamilton gas-tight syringe and stored in small glass vials, which were sealed with rubber septa and aluminum seals. Ethylene concentration was determined using gas chromatograph CHROM 5 (Laboratorni Přistroje, Praha, Czech Republic, 1980). Nitrogenase activity was determined based on the quantity of acetylene reduced to ethylene and expressed in nmolC_2H_4 produced per plant per hour (nMC_2H_4 plant^{-1} h^{-1}). The results are the mean value of five replications from each measurement.

2.4. Plant Biometric Assessment

The plant height (from soil surface to the highest plant point) and number of pods per plant were measured. Shoot, root, and nodule dry mass were determined after drying for 2 days at 70 °C until reaching constant weight. All the biometric traits were measured on 10 randomly selected plants from each object and replication during plant vegetation and before harvest. The total one-sided area of leaf per unit ground surface area expressed by the leaf area index (LAI) and was measured at three randomly selected places of each plot at the BBCH stage 69 using a SunScan Canopy Analysis System type SSI. Lupine was harvested at one stage (BBCH 90–92) with a 1.35-m wide plot combine. The yield of clean seeds was determined in dt·ha^{-1}, given at a standardized (15%) water content and thousand seed weight was measured using a seed counter.

2.5. Chlorophyll Fluorescence and Chlorophyll Content Measurements

A fluorimeter (OS5p; Opti-Sciences, Inc., Hudson, NY, USA) was used to measure the efficiency of the photosynthetic apparatus. Prior to fluorescence measurements, the upper surface of three healthy leaves at the top of one plant from three randomly selected sites for each plot was covered with leaf clips for 30 min. Leaf fluorescence was then measured with a light pulse of 15,000 μmol m^{-2} s^{-1} at a wavelength of 660 nm The assessed parameter was maximum photosynthetic efficiency of PSII (F_v/F_m), which was calculated using the following formula: $F_v/F_m = (F_m - F_0)/F_m$, on the basis of the measured parameters: minimal fluorescence (F_0), maximal fluorescence (F_m), variable fluorescence (F_v) [23].Chlorophyll content meter (CCM-200plus; Opti-Sciences, USA) was used to estimate the chlorophyll content index (CCI) on the same leaves that were used for chlorophyll α fluorescence measurements. CCM-200plus measures the chlorophyll absorbance and calculates the chlorophyll content index, which is proportional to the concentration of chlorophyll in the sample.

2.6. Soil Sampling for Biochemical Analyzes

Soil samples collected from the arable layer (0–20 cm) were used as the research material for biochemical analyses. Each year they were collected at four terms: First term—at the plants' emergence (BBCH 5–10), Second term—at the plants' full growth (BBCH 35–40), third term—at the at the plants' florescence (BBCH 51–59), fourth term—after harvest.

Soil samples were taken from five places of each experimental plot, in four replications for each of the nine treatments of the experiment. In this way, at each analysis term we received 36 samples of soil, each of 1 kg.

2.7. Soil Enzymatic Activity

The analyses of soil enzymatic activity in individual treatments were based on the colorimetric method applied to measure the dehydrogenase activity (DHA), where 1% triphenyl tetrazolium chloride (TTC) was used as the substrate. The activity was measured after 24-h incubation at a temperature of 30 °C and a wavelength of 485 nm and it was expressed as μmol triphenyl formazane TPF 24 h^{-1} g^{-1}dm of soil [24].

Apart from that, the biochemical analyses of soil involved the determination of activities of acid (EC 3.1.3.2) phosphomonoesterases (PAC) and alkaline phosphomonoesterases (PAL) with the method developed by Tabatabai and Bremner [25]. The activities were determined with disodium *p*-nitrophenyl phosphate tetrahydrate used as a substrate after 1 h incubation at 37 °C and at a wavelength of 400 nm. The results were converted into μmol (p-nitrophenol) PNP h^{-1} g^{-1}dm of soil.

Catalase activity was measured by means of permanganometry, according to the method developed by Johnsons and Temple [26], where 0.3% H_2O_2 was the substrate. After 20-min incubation at room temperature (about 20 °C) 0.02 M $KMnO_4$ was titrated to a light pink colour and expressed as μmol H_2O_2 g^{-1} dm min^{-1}.

2.8. Biological Index of Fertility

The biological index of fertility (BIF) was calculated using the dehydrogenase activity (DHA) and catalase activity (CAT) according to the Stefanic method [27] using the following formula: (DHA + kCAT)/2, where k was the factor of proportionality which equaled 0.01.

2.9. Statistical Analyses

The dynamics of changes in the soil enzymatic activity was statistically analyzed. As there were no significant differences between the parameters in the research years, they were treated as replicates and the results were analyzed by two-way ANOVA using Statistica 12.0 software. The fertilization method and the term of analysis were the factors differentiating the traits under study to estimate the soil biochemical activity parameters. Homogeneous subsets of mean were identified via Duncan's test at a significance level of $p = 0.05$. Yield, biometric, physiological traits of plants, and nitrogenase activity were tested once a year for the experiment. Hence, one-way analysis of variance (ANOVA) was used with Duncan's confidence interval, which was applied at a significance level of $p = 0.05$. As there were no significant differences between the parameters in the experimental years, they were treated as replicas.

Principal component analysis (PCA) was used to visualize the multidimensional dependencies between the soil biochemical activity and the types of fertilization [28]. In order to show the existing regularities (correlations) between biometric and physiological parameters of plants in individual years of research, a Pearson correlation matrix was determined, which was illustrated using a heatmap. The colors indicate the correlation coefficient values (from darkest—value −1, to the lightest—value +1). Cluster analysis enables grouping of the studied physiological parameters of the plants in the experiment in such a way that the degree of correlation between parameters within one group was the highest and between groups the smallest [29]. The agglomeration Ward method (Ward Hierarchical Clustering) and the Euclidean distance were used to create a tree diagram.

3. Results

3.1. Yield, Biometric, and Physiological Traits of White Lupine Plants

The studied biostimulators/foliar fertilizers modified the yield and yield components of white lupine. The yield of white lupine seeds was low and ranged from 11.67 dt·ha^{-1} (ADOB B) to 13.88 dt·ha^{-1} (Optysil) and depended significantly on the bio-stimulants or foliar fertilizers that were applied (Table 3). After applying Optysil or ADOB Zn IDHA (13.63 dt·ha^{-1}), the yields were significantly higher when compared to the control plants by 1.82 and 1.57 dt·ha^{-1}, respectively.

Thousand seed weight (TSW) was significantly higher than the control plants when ADOB Zn IDHA (322.7 g) was applied. All tested preparations significantly stimulated the height of white lupine. The strongest stimulation was obtained by Metalosate potassium, which increased the height of white lupine (40.5 cm) by 6.2 cm when compared to the control. Apart from these fertilizers, in the group that most strongly stimulated this trait were: Optysil (39.8 cm), ADOB 2.0 Mo (38.9 cm), and Bolero Mo (38.6 cm).

ADOB Zn IDHA (318.4 pc·m^{-2}) and Tytanit (300.8 pc·m^{-2}) significantly increased the number of pods compared to the control, and the increase was 96.6 and 79 pc·m^{-2}, respectively.

Studies have also shown changes in nodulation and physiological parameters of the plant. Dry mass of root nodules was significantly stimulated after application of ADOB Zn IDHA (0.212 g) by 0.067 g when compared to the control treatment.

Chlorophyll fluorescence (F_v/F_m), showing the level of plant stress, was measured in the BBCH 69 (end of flowering) and BBCH 79 phases (75% of the pods reached typical length). At the end of flowering, the best plant condition, expressed by the F_v/F_m parameter, was obtained after the application of ADOB Zn IDHA (0.815) or Metalosate potassium (0.813) and both values significantly exceeded those obtained both with the control and all other treatments. In the assessment made at a later developmental phase, the tested biostimulators/foliar fertilizers did not significantly differentiate this parameter.

ADOB Zn IDHA application significantly stimulated the content of chlorophyll in leaves, expressed in CCI, which was 50.9 and exceeded the control by 17.4, as well as all other objects. In addition, significantly higher CCI values than in the control object were obtained after using Tytanit (46.7), ADOB B (42.9), or ADOB 2.0 Mo (40.7).

In turn, the significantly highest LAI value in the experiments was obtained after application of Rooter. The LAI value was 2.03 and exceeded the control by 0.62, for which the lowest LAI value was determined.

Table 3. The influence of the bio-stimulants and fertilizers on yield, biometric, and physiological traits of white lupine.

Objects	Seed Yield, dt·ha^{-1}	TSW, g	Height, cm	Number of Pods, pc.·m^{-2}	Plant Dry Mass, g	Root Nodules Dry Mass, g	F_v/F_m BBCH 69	CCI	LAI
1	12.06 bc	302.4 bc	34.3 e	221.8 bc	5.05	0.145 bc	0.784 cd	33.5 d	1.41 g
2	11.82 bc	295.2 c	37.8 bcd	300.8 a	5.72	0.147 bc	0.796 b	46.7 b	1.73 de
3	13.88 a	301.7 bc	39.8 ab	250.5 b	6.46	0.160 bc	0.792 bc	23.1 f	1.81 c
4	11.96 bc	305.8 bc	40.5 a	189.1 c	5.28	0.170 b	0.813 a	24.2 f	1.70 de
5	13.18 ab	313.9 ab	37.1 cd	235.0 bc	5.36	0.142 bc	0.776 d	25.6 f	2.03 a
6	12.76 abc	306.1 bc	38.6 abc	250.0 b	5.14	0.129 c	0.774 d	30.2 e	1.88 b
7	13.63 a	322.7 a	36.4 d	318.4 a	6.17	0.212 a	0.815 a	50.9 a	1.61 f
8	11.67 c	310.4 b	37.9 bcd	273.4 ab	5.12	0.169 bc	0.779 d	42.9 c	1.68 e
9	11.94 bc	309.0 b	38.9 abc	240.7 b	5.24	0.170 bc	0.797 b	40.7 c	1.76 cd
p-value	0.001	0.000	0.000	0.000	0.236	0.001	0.000	0.000	0.000

1. control—no bio-stimulants or foliar fertilizers applied to the plants; 2. plant + Tytanit; 3. plant + Optysil; 4. plant + Metalosate potassium; 5. plant + Rooter; 6. plant + Bolero Mo; 7. plant + ADOB Zn IDHA; 8. plant + ADOB B; 9. plant + ADOB 2.0 Mo; lack of homogeneous groups means no significant differences at the level of $p < 0.05$, a, b, c, d, e, f, g-homogeneous groups (Duncan's test. $p < 0.05$); TSW-thousand seed weight, F_v/F_m—maximum photosynthetic efficiency of PSII, CCI—chlorophyll content index, LAI—leaf area index.

The results of the experiment showed that foliar fertilizers and bio-stimulants affected the enzymatic activity of the soil and the biological index of fertility (BIF), as well as the nitrogenase activity in the white lupine plantation. The two-way analysis of variance showed that the foliar fertilization/bio-stimulants did not have a significant influence on the enzymatic activity and the soil biological index of fertility (BIF). Only the term of the test (development phase, based on BBCH scale) had a highly significant influence on the enzymatic activity and the biological index of fertility (BIF) of the soil (Table 4). One-way analysis of variance showed that foliar fertilization/bio-stimulants had a significant influence on nitrogenase activity.

Table 4. The test *F* statistics and the significance levels of the two-way analysis of variance for the soil bioactivity. The traits under analysis were affected by two factors, i.e., foliar fertilization and the term of the test.

Parameter	Fertilization	Development Phase	Interaction
	White Lupine *Butan*		
Dehydrogenase	13.393 ^{ns}	159.989 ***	41.123 ^{ns}
Alkaline phosphatase	7.036 ^{ns}	51.672 ***	5.37 ^{ns}
Acid phosphatase	14.907 ^{ns}	116.200 ***	10.116 ^{ns}
Catalase	192.47 ^{ns}	1558.42 ***	121.42 ^{ns}
BIF	2.90 ^{ns}	131.96 ***	2.71 ^{ns}
Nitrogenase	14.08 ***	-	-

F test statistics and significance levels of two-way analysis of variance for activity of enzymes associated with herbicides and terms research fixed factors *** $p = 0.001$, ns—no signification.

3.2. *Biological Fixation of Nitrogen under Lupine Plantation*

The field analyses of the biological fixation of nitrogen showed that the fertilizers and bio-stimulants significantly stimulated the nitrogenase activity in the white lupine plantation (Figure 2). During the three years in all the experimental treatments nitrogenase exhibited higher activity than in the control plot and differences were statistically significant. The highest nitrogenase activity was noted after the application of the ADOB B and ADOB Zn IDHA. The activity of the enzyme was respectively six and four times higher than in the control plot. Apart from the control treatment, the lowest biological fixation of nitrogen was noted after the application of Metalosate potassium.

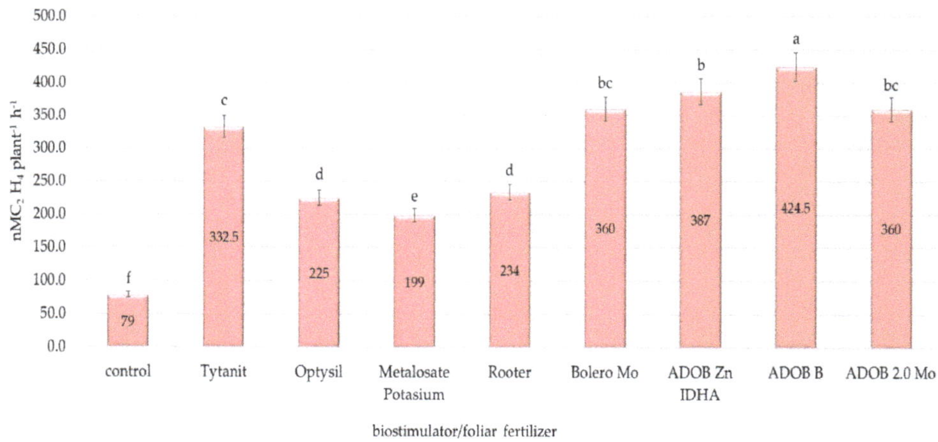

Figure 2. The influence of the bio-stimulants and fertilizers on the level of biological fixation of nitrogen. Abbreviation: means values ± standard errors; a, b, c, d, e, f—homogenous groups according to Duncan's test at level $p = 0.05$.

The heat map presents correlations between all biometric and physiological characteristics of white lupine plants studied (Figure 3). Based on this visualization, relatively higher correlations were found between some features, including: PN (number of pods, pc.·m^{-2}), TSW (thousand seed weight), H (height plant), PDM (plant dry mass), Y (seed yield), and PDM, Y, LAI (leaf area index) and Fv/Fm^1 (maximum photosynthetic efficiency of PSII BBCH–69). In turn, BNF (biological nitrogen fixation) and RNDM (root nodules dry mass) are negatively correlated with LAI, Y, PDM, H, TSW, PN, and Fv/Fm^2 (maximum photosynthetic efficiency of PSII BBCH–78). Additionally, based on cluster analysis, groups of related biometric and physiological traits of plants were determined. Three groups have been designated. The first group that is the most distinct from the others contains: RNDM, BNF, CCI, and Fv/Fm^1. The other two groups are: LAI, Y, PDM and H, TSW, PN, Fv/Fm^2.

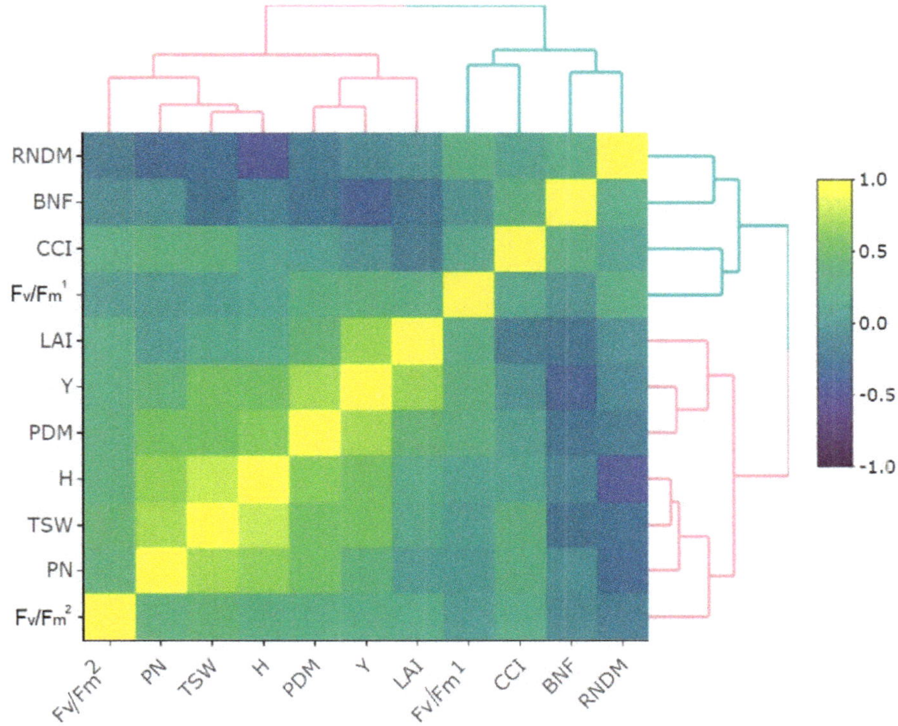

Figure 3. Correlations between all biometric and physiological characteristics of white lupine plants Abbreviation: RNDM—root nodules dry mass, g—BNF—biological nitrogen fixation, CCI—chlorophyll content index, Fv/Fm^1—maximum photosynthetic efficiency of PSII BBCH–69, LAI—leaf area index Y—seed yield, PDM—plant dry mass, g, H—height plant, TSW—thousand seed weight, PN—number of pods, pc.·m^{-2}; Fv/Fm^2—maximum photosynthetic efficiency of PSII BBCH–78.

3.3. Analysis of Soil Biochemical Activity

Only the ADOB 2.0 Mo and Metalosate potassium foliar fertilizers stimulated the dehydrogenase activity throughout the growing season, as compared with the control treatment. After the application of the bio-stimulants the level of the enzyme activity was similar to the activity in the control treatment. However, when the Optysil and ADOB B were applied, the activity decreased but not statistically significantly. The experiment also showed that the peak of the dehydrogenase activity significant occurred at the third term of analyses, when the plants began flowering (BBCH 51–59). The results of the analysis of the dehydrogenase activity in the soil under the white lupine plantation are shown in Figure 4.

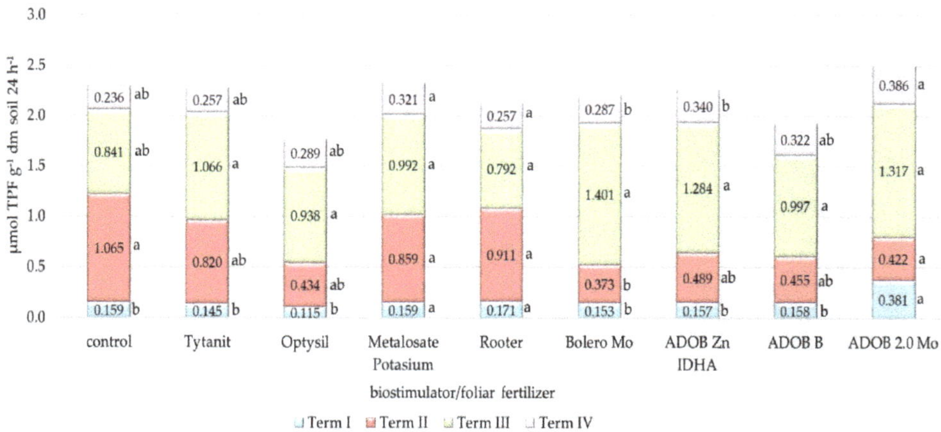

Figure 4. The influence of the bio-stimulants and fertilizers on the dehydrogenase activity. Abbreviation: a, b—homogenous groups according to Duncan's test at level $p = 0.05$; I term—at the plants' emergence (BBCH 5–10), II term—at the plants' full growth (BBCH 35–40), III term—at the at the plants' florescence (BBCH 51–59), IV term—after harvest.

The analysis of the results of the acid phosphatase activity (PAC) shows that during the entire white lupine growing season the foliar fertilizers and bio-stimulants decreased the activity of this enzyme, as compared with the control treatment (Figure 5). This effect was not observed when the Metalosate potassium foliar fertilizer was applied. During the second term of analyses, shortly before flowering, the acid phosphatase activity in all the experimental treatments was higher than in the control treatment. It was very high after the application of the Bolero Mo (0.170 µmol PNP h^{-1} kg^{-1}dm of soil) and ADOB 2.0 Mo (0.171 µmol PNP h^{-1} g^{-1}dm of soil).

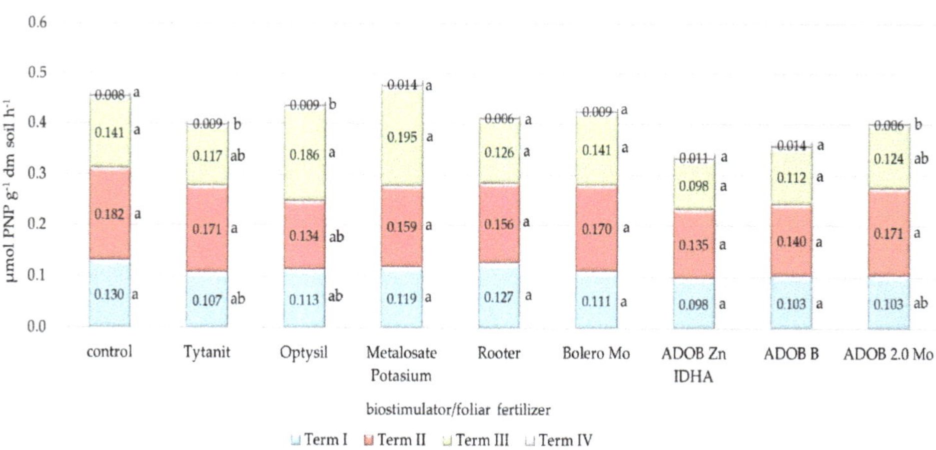

Figure 5. The influence of the bio-stimulants and fertilizers on the acid phosphatase level. Abbreviation: a, b—homogenous groups according to Duncan's test at level $p = 0.05$; I term—at the plants' emergence (BBCH 5–10), II term—at the plants' full growth (BBCH 35–40), III term—at the at the plants' florescence (BBCH 51–59), IV term—after harvest.

The bio-stimulants and most of the foliar fertilizers did not increase the alkaline phosphatase (PAL) activity in the white lupine plantation, as compared with the control treatment (Figure 6). The ADOB 2.0 Mo and Bolero Mo stimulated the activity of this enzyme, which respectively increased by

14% and 5%, as compared with the control treatment. The enzyme exhibited statistically significantly increased activity shortly before they began flowering (II term).

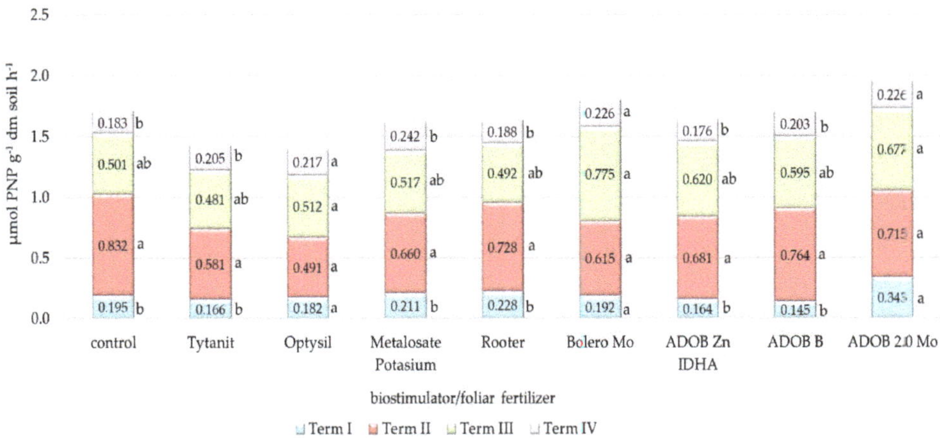

Figure 6. The influence of the bio-stimulants and fertilizers on the alkaline phosphatase level. Abbreviation: a, b—homogenous groups according to Duncan's test at level $p = 0.05$; I term—at the plants' emergence (BBCH 5–10), II term—at the plants' full growth (BBCH 35–40), III term—at the at the plants' florescence (BBCH 51–59), IV term—after harvest.

All the preparations stimulated the catalase activity, as compared with the control treatment (Figure 7), but not significantly. The enzyme significantly exhibited high activity, i.e., when the plants started flowering (III term) in all the experimental treatments. The catalase activity ranged from 98.510 µmol $H_2O_2g^{-1}$ dm min^{-1} after the application of the Tytanit to 135.819 µmol $H_2O_2g^{-1}$ dm min^{-1} after the application of the Bolero Mo.

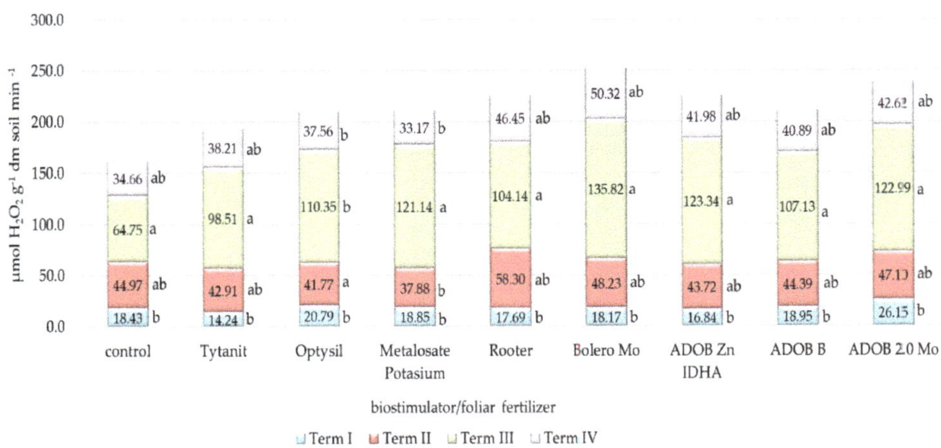

Figure 7. The influence of the bio-stimulants and fertilizers on the catalase activity. Abbreviation: a, b—homogenous groups according to Duncan's test at level $p = 0.05$; I term—at the plants' emergence (BBCH 5–10), II term—at the plants' full growth (BBCH 35–40), III term—at the at the plants' florescence (BBCH 51–59), IV term—after harvest.

The biological index of fertility (BIF), which was calculated on the basis of the dehydrogenase and catalase activity, was not always higher after the application of the bio-stimulants and foliar fertilizers

(Figure 8). The highest value of this indicator was noted after the application of the Optysil and the lowest after ADOB Zn IDHA. The BIF was significantly high at the beginning of flowering, as it ranged from 5.17 after the application of ADOB Zn IDHA to 12.34 after the application of ADOB 2.0 Mo. The indicator was also high after the application of the Bolero Mo and Optysil.

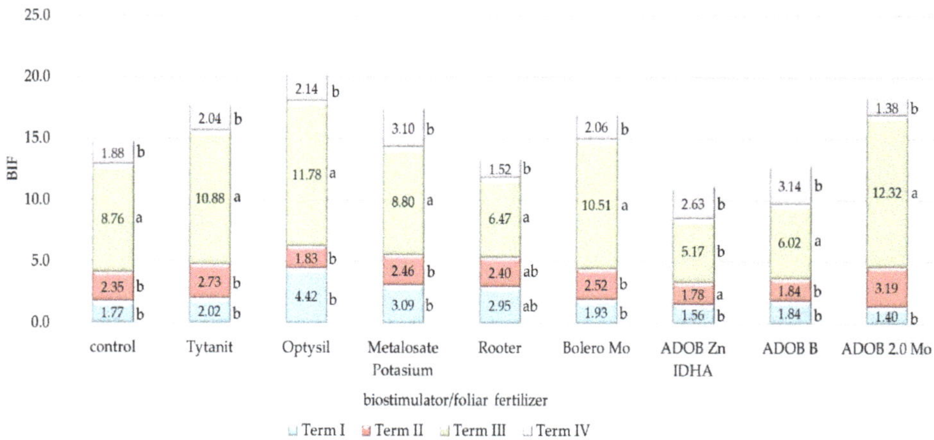

Figure 8. The influence of the bio-stimulants and fertilizers on the BIF. Abbreviation: a, b—homogenous groups according to Duncan's test at level $p = 0.05$; I term—at the plants' emergence (BBCH 5–10), II term—at the plants' full growth (BBCH 35–40), III term—at the at the plants' florescence (BBCH 51–59), IV term—after harvest.

Principal component analysis (PCA) was used to show how the foliar fertilizers and bio-stimulants affected the white lupine plantation. The first two principal components accounted for over 89.2% of the total variation (Figure 9). The parameters of the soil biochemical activity in 2018 differed significantly from 2016 to 2017. This effect may have been caused by the weather conditions (Figure 1). In 2018 the season was the warmest of all the research years. The average temperature difference between 2018 and the previous years was 2.9 °C in August and 1.7 °C in May. As the thermal conditions were very similar in 2016 and 2017, the PCA showed similar dependencies for these two years. In 2016 the fertilizer preparations and bio-stimulants significantly affected the catalytic activity of acid phosphatase (PAC) at all the terms of analyses. This dependency was not observed for the other parameters of soil biochemical activity.

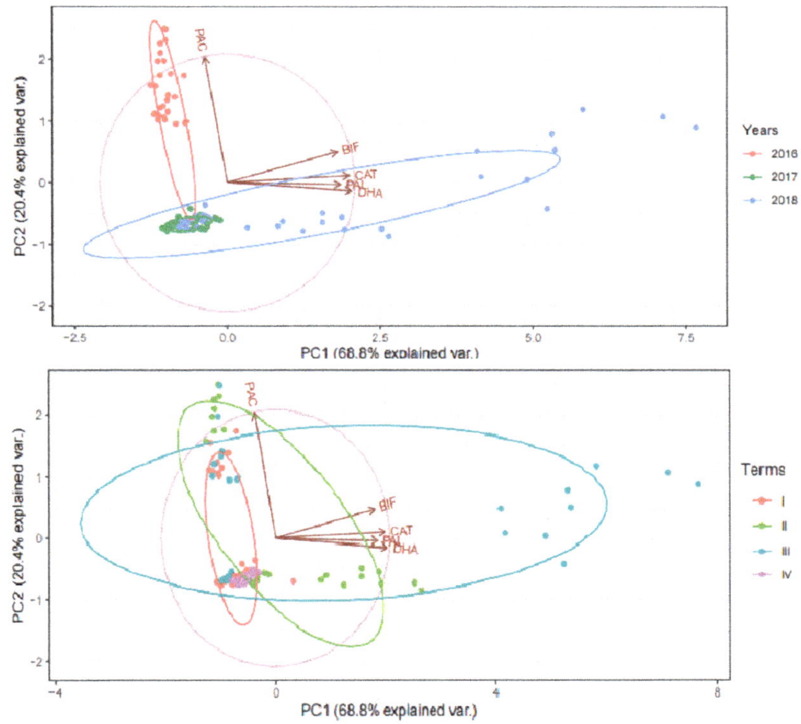

Figure 9. The dependence between the soil enzymatic activity and all treatments with fertilizers and bio-stimulants at the terms of analyses. Abbreviation: I term—at the plants' emergence (BBCH 5–10), II term—at the plants' full growth (BBCH 35–40), III term—at the at the plants' florescence (BBCH 51–59), IV term—after harvest. BIF—index of fertility, CAT—catalase activity, PAC—acid phosphomonoesterases, PAL—alkaline phosphomonoesterases, DHA—dehydrogenase activity.

In 2017 the application of the fertilizers did not cause significant differences in the activity of the soil enzymes or the biological index of soil fertility. In dry 2018 the preparations did not significantly affect the catalytic activity of the test parameters only during plants' emergence (I term). However, at the plants' full growth (II term), the foliar fertilizers and bio-stimulants strongly influenced the catalytic activity of catalase (CAT), dehydrogenase (DHA), alkaline phosphatase (PAL), and the biological index of soil fertility (BIF). Apart from that, the principal component analysis showed that in 2018 the indicators of soil biochemical activity were affected most strongly by foliar fertilizers and bio-stimulants the flowering of the plants (III term) and after the harvest (IV term).

4. Discussion

4.1. Yield, Biometric, and Physiological Traits

Silicon, iron, manganese, boron, copper, molybdenum, and zinc are the basic micronutrients. The silicon content in most plants is comparable to the content of calcium, magnesium, and phosphorus. Many studies have shown the positive effects of silicon on plants, their development, yield, and sensitivity to biotic and abiotic stress [30]. In many tests, silicon has been shown to significantly influence the regulation of nutrient uptake such as: calcium, magnesium, and phosphorus. In other studies [31], silicon fertilization increased the yield of sugar beet roots by 13.7–15.9%, as well as the yield of many other species [11], especially in the form of spraying plants under stress conditions. According to Fageria and Baligar [32] and Duffy [33] Zn is the microelement most limiting crop yield.

Zinc is taken up in small amounts and it participates in all major functions of the plant, increases nitrogen uptake, and activates CO_2 binding in later stages. Hence, it is necessary in plant nutrition and its importance in plant production is growing [13]. Similarly, Kaya et al. [34] obtained the highest common bean plants (*Phaseolus vulgaris* L.), with the largest number of pods and seeds per plant after application of a foliar mixture of zinc.

The preparations used in our study also stimulated the tested biometric parameters of the plants. Plant height was stimulated the most after application of Metalosate potassium (by 8.5%) when compared to the control treatments. In turn, the number of nodules was most strongly stimulated by ADOB Zn IDHA (by 68.4%) and LAI by Rooter (by 69.5%). Other fertilizers containing boron, molybdenum, silicon, and titanite also increased the parameters indicated above. These results are consistent with the results of Raj and Raj [12] regarding the beneficial effects of Zn on plant efficiency, physiological parameters, plant height, and nodulation formation. Our results are also consistent with field studies of Omer et al. [35], in which the treatments of molybdenum application did not modify any of the studied lentil characteristics, except for the height of the plant. Also Rahman et al. [36] showed that the use of molybdenum in its deficiency in soil, stimulates the formation of nodules. Of the physiological traits studied, chlorophyll fluorescence (Fv/Fm) was most strongly stimulated by ADOB Zn IDHA (by 3.9%) and Metalosate potassium (by 3.7%). In turn, the CCI index was most strongly stimulated by ADOB Zn IDHA, whose application resulted in an increase of this parameter by 51.9% when compared to the control treatment. The results of research on *Vigna sinensis* [37] and on *Celosia* [38] showed that Zn spraying on plants caused a significant increase in chlorophyll content. In a study conducted by Artyszak et al. [39], foliar fertilization with silicon increased LAI and effective quantum efficiency of PSII—ΦPSII, as well as positively affected the growth and development of many plant species [40,41].

4.2. Biological Fixation of Nitrogen

The bio-stimulants and foliar fertilizers which improved the biological fixation of nitrogen in the white lupine plantation contained important macro- or microelements. Scientific reports suggest that some elements are particularly significant to the nitrogen fixation process.

Mineral nutrients may influence N_2 fixation in legumes at various stages of the symbiotic process: infection and nodule development, nodule function, and host plant growth. For healthy and vigorous growth, intact plants need to take up relatively large amounts of some inorganic elements: ions of nitrogen (N), potassium (K), calcium (Ca), phosphorus (P) and sulphur (S), and small quantities of other elements: iron (Fe), nickel (Ni), chlorine (Cl), manganese (Mn), zinc (Zn), boron (B), copper (Cu), and molybdenum (Mo). Molybdenum and iron are especially important because they are components of the nitrogenase complex in rhizobia which is required for nitrogen fixation. They are components of nitrogenase—the bacterial enzyme that enables the diazotrophy process. The nitrogenase protein consists of two subunits: the larger one containing the FeMo cofactor and the smaller one containing iron alone [42]. Plants growing on acidic, moist, and poorly buffered soils do not have sufficient supply of molybdenum. When molybdenum is applied in a field to the leaves of legumes, the nitrogen fixation of these plants is more efficient, and the mass of their root nodules and the yield of seeds increase [43,44]. The use of ADOB 2.0 Mo with high molybdenum content in our experiment confirmed this fact. There are small amounts of boron in plants, but this micronutrient plays an important role in various physiological processes. It affects the separation of plant tissues and it is necessary for the optimal growth of plants. Boron-deficient plants have less bacteria of the *Rhizobium* genus and fewer infection threads [44]. The significant increase in the level of biological fixation of nitrogen may have been caused by the application of the foliar fertilizer containing boron (ADOB B). Our research also proved that zinc supplied with the ADOB Zn IDHA foliar fertilizer significantly increased the nitrogenase activity. Although plants absorb moderate amounts of zinc, this element has significant influence on bacteria of the *Rhizobium* genus. The research by Wani et al. [45] showed that higher concentrations of this element in soil stimulate bacteria of the *Rhizobium* genus to produce phytohormones (including indoleacetic acid), which promote the growth of plants by increasing the number of root nodules, their dry mass, and the content of leghemoglobin in the nodules.

Many researchers have studied the role of phosphorus in symbiotic systems. Phosphorus plays a crucial role in the nitrogen fixation process [46,47]. The Rooter bio-stimulant, which contained phosphorus and potassium, stimulated this process considerably. Phosphorus participates in a wide range of molecular and biochemical processes. Apart from that, some phosphate bonds are carriers of the energy used in cells. The presence of phosphorus in soil affects the plant's ability to produce root nodules, especially the weight and the number of nodules [48], which translates into the level of nitrogen fixation.

When the supply of phosphorus is insufficient, plants often suffer from nitrogen deficiency. Sulphur and potassium are less important for symbiotic systems than the aforementioned elements. Nevertheless, potassium ions are very desirable in saline soils because they function as an osmolyte. In view of the fact that nearly half of irrigated soils around the world are considered saline, the addition of potassium helps to maintain the bacteria-plant system [48,49].

4.3. Biochemical Activity

The activities of soil enzymes are considered sensitive indicators of important microbial reactions involved in nutrient cycles and they respond to changes in the soil caused by natural or anthropogenic factors. In this regard, soil enzyme activities are often used to evaluate the impact of human activity on soil life [50].

Soil enzymes are a group of catalysts that significantly affect the ecological properties of the pedosphere. These are both extracellular enzymes and the ones that are present in microorganisms (both in proliferating cells and in endospores). Enzymes control the course of all chemical reactions in microbial cells, e.g., the synthesis of proteins, nucleic acids, and carbohydrates [51]. Soil enzymes are involved in the decomposition of organic substances released into the soil during the plant's growth as well as the formation and decomposition of humus in the soil. They release and transfer minerals to plants. In spite of the dynamic nature of the microbiological and biochemical properties, soil enzymes are accurate and significant determinants of soil fertility, and they are important indicators of changes taking place in the soil [52,53].

Dehydrogenases (DHA) are enzymes belonging to the group of oxidoreductases. They are responsible for catalyzing the oxidation of organic compounds. Active dehydrogenases are present only inside living cells and they indicate the presence of physiologically active microorganisms. Dehydrogenases are commonly found in the pedosphere, where they are involved in the decomposition of organic compounds. Measurement of the dehydrogenase activity in soil shows the intensity of respiratory metabolism of soil microorganisms, mainly actinobacteria and bacteria.

Our research showed that only some foliar fertilizers (ADOB 2.0 Mo and Metalosate Potassium) stimulated the dehydrogenase activity in the white lupine plantation, however, the results were not significant.

Dehydrogenase exhibited high activity at the beginning of the plants' flowering phase (BBCH 51–59). It may have been caused by an increased secretion from the root system during that period [54,55]. In consequence, the count of microorganisms increased [56].

Also macro- and microelements applied in the form of foliar fertilizers and biostimulators could affect dehydrogenase activity. Bielińska et al. [57] observed that fertilizing preparations with nitrogen, phosphorus, and potassium increased the content of these enzymes in the soil. There was a similar effect observed in our study after the application of the Metalosate potassium foliar fertilizer. There were analogous results of experiments on similar bio-conditioners conducted by [58] and [53]. According to Bilen et al. [59], boron improves the dehydrogenase activity. Taran et al. [60] showed that molybdenum stimulated the production of these enzymes by the root nodules of legumes. They also observed that the content of titanium might be positively correlated with the soil biochemistry.

The results of the experiment showed that both the bio-stimulants (Tytanit and Rooter) and foliar fertilizers positively affected the acid phosphatase activity, which was lower than in the control treatment. The Metalosate potassium foliar fertilizer did not cause this effect. This shows that the preparations used in our experiment positively influenced the plants' ability to absorb phosphorus. It is necessary to remember that phosphorus-deficient plants are characterized by increased secretion of acid phosphatase through the root system into the soil. Ciereszko et al. [61] found that the deficit of this macroelement stimulated plants' secretion of acid phosphatases. Lemanowicz et al. [62] and Niewiadomska et al. [56] also suggest these relationships in their studies on the effect of the PRP SOL fertilizer containing phosphorus, potassium, zinc, boron, and molybdenum on the lupine plantation. They observed a decrease in the catalytic activity of this enzyme because of the activation of the compounds that were inaccessible to plants. Bielińska and Mocek-Płóciniak [63] made similar observations. Wang et al. [64] also found that these enzymes exhibited higher activity in the experimental treatment without phosphorus fertilization.

The alkaline phosphatase activity increased significantly only after the application of the ADOB 2.0 Mo and Bolero Mo foliar fertilizers. This effect may have been caused by the increased activity of soil microorganisms, which were stimulated by organic phosphorus compounds secreted into the soil by white lupine plants. Waldrip et al. [65] proved that the content of organic forms of phosphorus was correlated with the activity of alkaline phosphatases in the soil.

All the preparations used in the experiment significantly stimulated catalase activity. As early as 1963, Koter [66] found that the catalase activity increased when plants were fertilized with boron. Hu and Zhu [67] observed that the catalase and dehydrogenase activity increased when plants were fertilized with silicon. Such elements as copper and zinc are essential constituents of physiological processes in all living organisms, including microorganisms. Some soils suffer from zinc deficits, which is why they are enriched with fertilizers containing this element to satisfy the nutritional requirements of crops and improved soil activity [68].

The results of the enzymatic analyses of the dehydrogenase and catalase activities enabled the calculation of the biological index of soil fertility (BIF). The treatments with the Optysil and ADOBE 2.0 Mo preparations had influence on the BIF values, as compared with the control sample. The use of the Optysil preparation resulted in particularly high values in the soil samples collected at the beginning of the flowering phase. The BIF value resulted from the significant influence of these fertilizers on the activity of catalase and dehydrogenase. Siwik-Ziomek and Szczepanek [69] indicated that mineral fertilization, which increases the yield of crops, indirectly affects root secretion, and thus increases the biochemical activity of soil at specific phases of plants' development.

5. Conclusions

When non-root fertilization is applied to plants, they take up all necessary elements chiefly through their leaves as well as the stalk and the whole aerial system. A strong stimulating effect on the yield of white lupine plants in comparison with the control object was obtained after the application of silicon (Optysil) or chelated zinc (ADOB Zn IDHA). The use of zinc in foliar fertilizers (ADOB Zn IDHA) in comparison with control treatment stimulated most of the tested features/parameters: TSW, number of pods per 1 m^2, root nodules dry matter, photochemical efficiency of PSII (F_v/F_m), and chlorophyll content (CCI). However, it is noteworthy that this way of "feeding" cannot substitute soil fertilization. It can only be used to quickly supply necessary nutrients to plants at difficult phases so as not to slow down their growth. The bio-stimulants and foliar fertilizers used in our study improved some of the biochemical parameters of soil activity and the nitrogen fixation process in the white lupine plantation. This effect may have been caused by the higher rate of penetration and better uptake of nutrients applied to the plants' leaves. Although macro- and micronutrients differ in their penetration rates, this process can be accelerated up to about a dozen times by non-root fertilization. The downside of foliar fertilization is the fact that only a limited amount of fertilizer can be supplied to plants in this way. Therefore, this method is particularly effective when plants need to be provided with the elements they need in smaller amounts, e.g., iron, boron, and molybdenum. Not only is foliar fertilization a more efficient method of supplying micronutrients, but it is also safer for the environment and the plants. The search for methods that improve the yield and biochemical parameters of the soil environment is in agreement with the sustainable agriculture policy.

Author Contributions: A.N., H.S., A.W.-M., K.R., and Z.W. conceived and designed the experiments; A.N., H.S., A.W.-M., K.R., and Z.W. performed the field experiments and analyzed the data; A.B. statistical analysis; A.N., Z.W., H.S. wrote the paper; A.N., Z.W. revised the manuscript. All authors have read and agreed to the published version of the manuscript.

Funding: This research was funded by the Ministry of Agriculture and Rural Development of Poland, (it was not a grant. And The APC was funded by framework of Ministry of Science and Higher Education programme Project No. 005/RID/2018/19.

Acknowledgments: This research was financed by the Ministry of Agriculture and Rural Development of the Republic of Poland in the framework of the Multi-annual Program "Increasing the use of domestic feed protein for the production of high quality animal products in conditions of balanced development" implemented at the Department of Agronomy of the University of Life Sciences in Poznan. The publication was co-financed within the framework of Ministry of Science and Higher Education programme as, Regional Initiative Excellence" in years 2019-2022, Project No. 005/RID/2018/19.

Conflicts of Interest: The authors declare no conflict of interest.

References

1. Directive 2009/128/EC of the European Parliament and of the Council of 21 October 2009 Establishing a Framework for Community Action to Achieve the Sustainable Use of Pesticides:1-16. Available online: https://ec.europa.eu/food/plant/pesticides/sustainable_use_pesticides_en (accessed on 9 October 2019).
2. Małecka-Jankowiak, I.; Blecharczyk, A.; Sawinska, Z.; Waniorek, W. Effect of lupins and pea on the subsequent winter wheat yield depending of tillage and nitrogen fertilization. *Fragm. Agron.* **2018**, *35*, 67–79.
3. Prusinski, J. White lupin (*Lupinus albus* L.)—Nutritional and health values in human nutrition—A review. *Czech J. Food Sci.* **2017**, *35*, 95–105.
4. GUS. *Agriculture in 2015 Year*; GUS: Warsaw, Poland, 2016; p. 177.
5. Dobosz, R.; Krawczyk, R. Comparison of the development ability of the northern root-knot nematode *Meloidogyne hapla* Chitwood, 1949 on selected species and varieties of Fabaceae Lindl. *Prog. Plant Prot.* **2017**, *57*, 159–163.
6. Calvo, P.; Nelson, L.; Kloepper, J.W. Agricultural uses of plant biostimulants. *Plant Soil* **2014**, *383*, 3–41. [CrossRef]
7. Fageria, N.K.; Filho, M.B.; Moreira, A.; Guimarães, C.M. Foliar fertilization of crop plants. *J. Plant Nutr.* **2009**, *32*, 1044–1064. [CrossRef]
8. Kocira, S.; Szparaga, A.; Kocira, A.; Czerwińska, E.; Wójtowicz, A.; Bronowicka-Mielniczuk, U.; Koszel, M.; Findura, P. Modeling biometric traits, yield and nutritional and antioxidant properties of seeds of three soybean cultivars through the application of biostimulant containing seaweed and amino acids. *Front. Plant. Sci.* **2018**, *9*, 388. [CrossRef]
9. Przybysz, A.; Gawronska, H.; Gajc-Wolska, J. Biological mode of action of a nitrophenolates-based biostimulant: Case study. *Front. Plant Sci.* **2014**, *5*, 713. [CrossRef]
10. Du Jardin, P. Plant biostimulants: Definition, concept, main categories and regulation. *Sci. Hortic.* **2015**, *196*, 3–14. [CrossRef]
11. Artyszak, A. Effect of silicon fertilization on crop yield quantity and quality—A literature review in Europe. *Plants* **2018**, *7*, 54. [CrossRef]
12. Raj, A.B.; Raj, S.K. Zinc and boron nutrition in pulses: A review. *J. App. Nat. Sci.* **2019**, *11*, 673–679. [CrossRef]
13. Fageria, N.K. *The Use of Nutrients in Crop Plants*; CRC Press: Boca Raton, FL, USA, 2009.
14. Sulewska, H.; Ratajczak, K. Chemical composition of selected preparations supporting plant development and evaluation of their activity in soybean cultivation. *Chem. Ind.* **2017**, *96*, 1352–1355.
15. Karaca, A.; Cetin, S.C.; Turgay, O.C.; Kizilkaya, R. Soil enzymes as indication of soil quality. In *Soil Enzymology*; Springer: Berlin/Heidelberg, Germany, 2010; pp. 119–148.

16. Nannipieri, P.; Kandeler, E.; Ruggiero, P. Enzyme activities and microbiological and biochemical processes in soil. In *Enzymes in the Environment. Activity, Ecology and Applications*; Marcel Dekker Inc.: New York, NY, USA, 2002; pp. 1–33.
17. Saviozzi, A.; Levi-Minzi, R.; Cardelli, R.; Riffaldi, R. A comparison of soil quality in adjacent cultivated, forest and native grassland soils. *Plant Soil* **2001**, *233*, 251–259. [CrossRef]
18. IUSS Working Group WRB. *World Reference Base for Soil Resources 2006, First Update 2007*; World Soil Resources Reports No.103; FAO: Rome, Italy, 2007.
19. Van Reeuwijk, L.P. *Procedures for Soil Analysis*, 6th ed.; ISRIC: Wageningen, The Netherlands, 2002.
20. Krawczyk, R.; Mrówczyński, M. (Eds.) *Metodyka Integrowanej Ochrony Łubinu Wąskolistnego, Żółtego i Białego dla Producentów*; Instytut Ochrony Roślin-Państwowy Instytut Badawczy: Poznan, Poland, 2012.
21. Walter, H. *Vegetation Zones and Climate*; PWRiL: Warsaw, Poland, 1976; pp. 26–31.
22. Sawicka, A. *The Ecological Aspects of Dinitrogen Fixation*; Dissertations Scientific: 134; University of Agriculture in Poznan: Poznan, Poland, 1983.
23. Kalaji, M.H.; Guo, P. Chlorophyll fluorescence: A useful tool in barley plant breeding programs. In emphPhotochemistry Research Progress; Sánchez, A., Gutierrez, S.J., Eds.; Nova Science Publishers: Hauppauge, NY, USA, 2008; pp. 447–471.
24. Thalmann, A. Zur Methodik der Bestimmung der Dehydrogenase aktivität im Boden mittels triphenytetrazoliumchlorid (TTC). *Landwirtsch Forsch.* **1968**, *21*, 249–258.
25. Tabatabai, M.A.; Bremner, J.M. Use of p-nitrophenyl phosphate for assay of soil phosphatase activity. *Soil Biol. Biochem.* **1969**, *1*, 301–307. [CrossRef]
26. Johnson, J.L.; Temple, K.L. Some variables affecting the measurement of "catalase activity" in soil 1. *Soil Sci. Soc. Am.* **1964**, *28*, 207–209. [CrossRef]
27. Stefanic, F.; Ellade, G.; Chirnageanu, J. Researches concerning a biological index of soil fertility. In *Proceeding of the 5th Symposium of Soil Biology*; Nemes, M.P., Kiss, S., Papacostea, P., Stefanic, C., Rusan, M., Eds.; Romanian National Society of Soil Science: Bucharest, Romania, 1984; pp. 35–45.
28. Abdi, H.; Lynne, J.W. Principal component analysis. *Wiley Interdiscip. Rev. Comput. Stat.* **2010**, *2*, 433–459. [CrossRef]
29. Murtagh, F.; Legendre, P. Ward's hierarchical clustering method: Clustering criterion and agglomerative algorithm. *arXiv* **2011**, arXiv:1111.6285.
30. Sacała, E. Role of silicon in plant resistance to water stress. *J. Elementol.* **2009**, *14*, 619–630. [CrossRef]
31. Artyszak, A.; Gozdowski, D.; Kuchcińska, K. The effect of silicon foliar fertilization in sugar beet—*Beta vulgaris* (L.) ssp. vulgaris conv. crassa (Alef.) prov. altissima (Döll). *Turk. J. Field Crop.* **2015**, *20*, 115–119. [CrossRef]
32. Fageria, N.K.; Baligar, V.C. Growth components and zinc recovery efficiency of upland rice genotypes. *Pesqui. Agropecu. Bras.* **2005**, *40*, 1211–1215. [CrossRef]
33. Duffy, B. Zinc and plant disease. In *Mineral Nutrition and Plant Disease*; Datnoff, L.E., Elmer, W.H., Huber, D.M., Eds.; APS Press: St. Paul, MN, USA, 2007; pp. 155–175.
34. Kaya, M.U.; Atak, M.E.; Khawar, K.M.; Ciftci, C.Y.; Ozcan, S. Effect of pre-sowing seed treatment with zinc and foliar spray of humic acids on yield of common bean (*Phaseolus vulgaris* L.). *Int. J. Agric. Biol.* **2005**, *7*, 875–878.
35. Omer, F.A.; Abbas, D.N.; Khalaf, A.S. Effect of molybdenum and potassium application on nodulation, growth and yield of lentil (*Lens culinaris* MEDIC). *Pak. J. Bot.* **2016**, *48*, 2255–2259.
36. Rahman, M.M.; Bhuiyan, M.M.H.; Sutradhar, G.N.C.; Rahman, M.M.; Paul, A.K. Effect of phosphorus, molybdenum and *Rhizobium* inoculation on yield and yield attributes of mung bean. *Int. J. Sustain. Crop Prod.* **2008**, *3*, 26–33.
37. Hassanein, M.S.; Shalaby, M.A.F.; Rashad, E.M. Improving growth and yield of some faba bean cultivars by using some plant growth promoters in newly cultivated land. *Ann. Agric. Sci.* **2000**, *38*, 2141–2155.
38. Tobbal, Y.F.M. Physiological Studies on the Effect of Some Nutrients and Growth Regulators on Plant Growth and Metabolism. Ph.D. Thesis, Faculty of Science Al-Azhar University, Cairo, Egypt, 2006.
39. Artyszak, A.; Gozdowski, D.; Kucińska, K. Effect of foliar fertilization with silicon on the chosen physiological features and yield of sugar beet. *Fragm. Agron.* **2016**, *33*, 7–14.
40. Guntzer, F.; Keller, C.; Meunier, J.D. Benefits of plant silicon for crops: A review. *Agron. Sustain. Dev.* **2012**, *32*, 201–213. [CrossRef]

41. Marschner, H. *Mineral Nutrition of Higher Plants*, 2nd ed.; Academic Press: London, UK, 2002.
42. Weisany, W.; Raei, Y.; Allahverdipoor, K.H. Role of some of mineral nutrients in biological nitrogen fixation. *Bull. Environ. Pharm. Life Sci.* **2013**, *2*, 77–84.
43. Vieira, R.F.; Cardoso, E.J.B.N.; Vieira, C.; Cassini, S.T.A. Foliar application of molybdenum in common beans. I. Nitrogenase and reductase activities in a soil of high fertility. *J. Plant Nutr.* **1998**, *21*, 169–180. [CrossRef]
44. Brown, P.H.; Bellaloui, N.; Wimmer, M.A.; Bassi, E.S.; Ruiz, J.; Hu, H.; Pfeffer, H.; Dannel, F.; Römheld, V. Boron in plant biology. *Plant Biol.* **2002**, *4*, 205–223. [CrossRef]
45. Wani, P.A.; Khan, S.; Zaidi, A. Impact of zinc-tolerant plant growth-promoting rhizobacteria on lentil grown in zinc-amended soil. *Agron. Sustain. Dev.* **2008**, *28*, 449–455. [CrossRef]
46. Bucher, M. Functional biology of plant phosphate uptake at root and mycorrhiza interfaces. *New Phytol.* **2007**, *173*, 11–26. [CrossRef] [PubMed]
47. Bonilla, I.; Bolaños, L. Mineral nutrition for legume-rhizobia symbiosis: B, Ca, N, P, S, K, Fe, Mo, Co, and Ni. A review. *Sustain. Agric. Rev.* **2009**, *1*, 253–274.
48. Abusuwar, A.O.; Omer, E.A. Effect of intercropping, phosphorus fertilization and rhizobium inoculation on the growth and nodulation of some leguminous and cereal forages. *Agric. Biol. J. N. Am.* **2011**, *2*, 109–124. [CrossRef]
49. Zahran, H.H. Rhizobium-legume symbiosis and nitrogen fixation under severe conditions and in arid climate. *Microbiol. Mol. Biol. Rev.* **1999**, *63*, 968–989. [CrossRef] [PubMed]
50. Karaca, A.; Cetin, S.C.; Turgay, O.C.; Kizilkaya, R. Effects of heavy metals on soil enzyme activities. In *Soil Heavy Metals*; Springer: Berlin/Heidelberg, Germany, 2010; pp. 237–262.
51. Das, S.K.; Varma, A. Role of enzymes in maintaining soil health. In *Soil Enzymology, Soil Biology*; Springer-Verlag: Berlin/Heidelberg, Germany, 2011; pp. 25–42.
52. Bielińska, E.; Mocek-Płóciniak, A. Impact of the tillage system on the soil enzymatic activity. *Arch Environ. Prot.* **2012**, *38*, 75–82. [CrossRef]
53. Panasiewicz, K.; Niewiadomska, A.; Sulewska, H.; Wolna-Maruwka, A.; Borowiak, K.; Budka, A.; Ratajczak, K. The effect of sewage sludge and BAF inoculant on plant condition and yield as well as biochemical and microbial activity of soil in willow (*Salix viminalis* L.) culture as an energy crop. *Peer J.* **2019**, *7*, e6434. [CrossRef]
54. De-la-Peña, C.; Badri, D.V.; Lei, Z.; Watson, B.S.; Brăndao, M.M.; Silva-Filho, M.C.; Sumner, L.W.; Vivanco, J.M. Root secretion of defense-related proteins is development-dependent and correlated with flowering time. *J. Biol. Chem.* **2010**, *285*, 30654–30665. [CrossRef]
55. Badri, D.V.; De-la-Peña, C.; Lei, Z.; Manter, D.K.; Chaparro, J.M.; Guimarães, R.L.; Sumner, L.W.; Vivanco, J.M. Root secreted metabolites and proteins are involved in the early events of plant-plant recognition prior to competition. *PLoS ONE* **2012**, *7*, e46640. [CrossRef]
56. Niewiadomska, A.; Sulewska, H.; Wolna-Maruwka, A.; Ratajczak, K.; Głuchowska, K.; Waraczewska, Z.; Budka, A. An assessment of the influence of co-inoculation with endophytic bacteria and rhizobia, and the influence of PRP SOL and PRP EBV fertilizers on the microbial parameters of soil and nitrogenase activity in yellow lupine (*Lupinus luteus* L.) cultivation. *Polish J. Environ. Stud.* **2018**, *27*, 2687–2702. [CrossRef]
57. Bielińska, E.J.; Futa, B.; Bik-Mołodzińska, M.; Szewczuk, C.; Sugier, D. The impact of fertilizing agents on the enzymatic activity of soils. *J. Res. Appl. Agric. Eng.* **2013**, *58*, 15–19.
58. Swędrzyńska, D.; Zielewicz, W.; Przybył, P.; Starzyk, J. Effect of bio soil conditioner on microbial state and enzymatic activity of soil under *Lolium perenne* cultivation. *Grassl. Sci.* **2013**, *16*, 111–128.
59. Bilen, S.; Bilen, M.; Bardhan, S. The effects of boron management on soil microbial population and enzyme activities. *Afri. J. Biotech.* **2011**, *10*, 5311–5319.
60. Taran, N.Y.; Gonchar, O.M.; Lopatko, K.G.; Batsmanova, L.M.; Patyka, M.V.; Volkogon, M.V. The effect of colloidal solution of molybdenum nanoparticles on the microbial composition in rhizosphere of *Cicer arietinum* L. *Nanoscale Res. Lett.* **2014**, *9*, 289–297. [CrossRef] [PubMed]
61. Ciereszko, I.; Szczygła, A.; Żebrowska, E. Phosphate deficiency affects acid phosphatase activity and growth of two wheat varieties. *J. Plant Nutr.* **2011**, *34*, 815–829. [CrossRef]
62. Lemanowicz, J.; Siwik-Ziomek, A.; Koper, J. Effect of spring barley nitrogen fertilization on the changes in the content of phosphorus and the activity of alkaline and acid phosphatase in soil. *Ecol. Chem. Eng. A* **2012**, *19*, 1497–1507.

63. Bielińska, E.; Mocek-Płóciniak, A. *Phosphatases in the Soil Environment*; UP Pub.: Poznan, Poland, 2009; ISBN 978-83-7160-554-3.
64. Wang, J.B.; Chen, Z.H.; Chen, L.J.; Zhu, A.N.; Wu, Z.J. Surface soil phosphorus and phosphatase activities affected by tillage and crop residue input amounts. *Plant Soil Environ.* **2011**, *57*, 251–257. [CrossRef]
65. Waldrip, H.M.; He, Z.; Erich, M.S. Effects of poultry manure amendment on phosphorus uptake by ryegrass, soil phosphorus fractions and phosphatase activity. *Biol. Fert. Soils* **2011**, *47*, 407–418. [CrossRef]
66. Koter, M. Research on the impact of boron on the growth and development of plants. *Soil Sci. Ann.* **1963**, *13*, 185–211.
67. Hu, X.J.; Zhu, Z.J. Effect of silicon on resistance of powdery mildew and the activities of antioxidative enzymes in leaf apoplast of cucumber. *Acta Agric. Zhejiang* **2008**, *20*, 67–71.
68. Wyszkowska, J.; Borowik, A.; Kucharski, M.A.; Kucharski, J. Effect of cadmium, copper and zinc on plants, soil microorganisms and soil enzymes. *J. Elementol.* **2013**, *18*, 2587–2602. [CrossRef]
69. Siwik-Ziomek, A.; Szczepanek, M. Soil extracellular enzyme activities and uptake of N by oilseed rape depending on fertilization and seaweed biostimulant application. *Agron. J.* **2019**, *9*, 480. [CrossRef]

agronomy

MDPI

Article

Biostimulant Seed Coating Treatments to Improve Cover Crop Germination and Seedling Growth

Yi Qiu [1,2,†], Masoume Amirkhani [1,*,†], Hilary Mayton [1], Zhi Chen [2] and Alan G. Taylor [1]

1 Cornell AgriTech, Horticulture Section, Cornell University, SIPS, Geneva, New York, NY 14850, USA; qyi0508@emails.imau.edu.cn (Y.Q.); hsm1@cornell.edu (H.M.); agt1@cornell.edu (A.G.T.)
2 College of Mechanical and Electrical Engineering, Inner Mongolia Agricultural University, Hohhot, Inner Mongolia 010018, China; sgchenzhi@imau.edu.cn
* Correspondence: ma862@cornell.edu
† Co-first authors.

Received: 28 December 2019; Accepted: 19 January 2020; Published: 22 January 2020

Abstract: Biostimulant seed coating formulations were investigated in laboratory experiments for their potential to increase maximum germination, germination rate, germination uniformity, and seedling growth of red clover (*Trifolium pratense* L.) and perennial ryegrass (*Lolium perenne* L.) seeds. Red clover and perennial ryegrass seeds were coated with different combinations of soy flour, diatomaceous earth, micronized vermicompost, and concentrated vermicompost extract. Coated and non-coated seeds of red clover and perennial ryegrass were evaluated for germination and growth after 7 and 10 days, respectively. Red clover seed was maintained at a constant 20 °C with a 16/8 h photoperiod, whereas for perennial ryegrass seed, the germinator was maintained at 15/25 °C, with the same photoperiod as red clover. Coated treatments significantly improved germination rate and uniformity with no reduction in total germination, compared to the non-treated controls in red clover. In contrast, for perennial ryegrass, the total germination percentage of all coated seeds was reduced and displayed a delayed germination rate, compared with the non-treated controls. Shoot length, seedling vigor index, and dry weight of seedlings of coated seed treatments of both crops were significantly higher when compared to controls for both species. In addition to growth metrics, specific surface mechanical properties related to seed coating quality of seeds of both species were evaluated. Increasing the proportion of soy flour as a seed treatment binder in the coating blend increased the integrity and compressive strength of coated seeds, and the time for coatings to disintegrate. These data show that seed coating technologies incorporating nutritional materials and biostimulants can enhance seedling growth and have the potential to facilitate the establishment of cover crops in agriculture and land reclamation.

Keywords: seed coating; cover crop; vermicompost; biostimulant; growth enhancement

---

1. Introduction

Exponential growth in human global population, from 1.7 billion in 1900 to approximately 7.6 billion in 2019, has led to the over use and degradation of agricultural landscapes, including grasslands used for grazing, forage, and food production [1,2]. The rapid growth of populations in pastoral areas, including Inner Mongolia, China, have caused intensive overutilization of grasslands. Approximately, 40% of land area in China is classified as grassland and accounts for 13% of the world's grassland [2,3]. Overgrazing and conversion of grassland to cropland has led to declines in overall agricultural productivity due to increased soil erosion, degraded soil structure, and reduced soil fertility. Recently, China implemented vegetation restoration programs to improve biodiversity in agriculture environments, soil health, and productivity, and to reduce erosion and desertification [2]. Legumes and ryegrasses are widely used as cover crops to reduce desertification and restore productivity on degraded grasslands [4] and are commonly used for land reclamation and restoration of abandoned mine land [5,6]. Perennial ryegrass (*Lolium perenne* L.) is a cool season grass native to southern Europe, the Middle East, Central Asia, and North Africa [7]. Ryegrass is often used to stabilize soils for erosion control and is frequently seeded with red (*Trifolium pratense* L.) or white (*Trifolium repens* L.) clovers for increased productivity in grazing and to provide nitrogen and aid in weed suppression [8]. Red clover is particularly tolerant to drought conditions and helps to improve soil structure due to its large, fast-growing (more than 60 cm/year) tap root [8]. The benefits of cover cropping in both organic and conventionally managed systems are well documented [9].

Cover crops increase soil organic matter, soil structure, nutrient retention and cycling, and reduce soil erosion [8]. However, under drought conditions, and in areas with poor soils such as arid degraded grasslands, germination and subsequent growth of cover crops are inadequate, and sowing is often unsuccessful. Seed enhancements, which can include seed priming, coating, and conditioning are frequently used to improve seed delivery during planting, and to increase seed germination, stand uniformity, seedling growth, and suppress disease [10]. Seed priming increased germination rate and overall seedling emergence in a study investigating perennial ryegrass for fall seeding under cool temperatures and improved wheat stand establishment under marginal soil conditions [11,12]. Seed treatments with fungicides and fertilizers enhanced stand establishment of perennial ryegrass in field experiments in New Zealand [13,14]. Seed enhancements via seed coating can also provide micro and macronutrients or biostimulant materials to increase germination, seedling vigor, and stand establishment [15]. Seed coating technology has been used as a promising and effective approach for enhancing establishment and yield of different grass and forage species (*Lolium perenne, Trifolium pratense, Elymus dahuricus*) in semi-arid areas of China such as Inner Mongolia [16–18].

Biostimulants are materials that can augment plant growth when applied to plants and seeds, but are not classified as fertilizers, pesticides, or soil amendments [19]. Commonly applied biostimulants include microbial inoculants, beneficial bacteria and fungi, nitrogen containing compounds, biopolymers, and plant extracts [19]. Research and use of biostimulants in agriculture has increased in recent years in an effort to reduce reliance on less sustainable conventional pesticides and fertilizers, which are often overused, in agricultural cropping systems [15,19–22]. Seed treatments require even smaller amounts of active ingredients per hectare than foliar applied treatments, primarily due to the reduced surface area treated, and increase germination and plant growth when compared to non-treated seed [23].

Modern seed coating technology utilizes different approaches depending on the shape and size of the seed and the type and amount of materials added to seeds [10,24–27]. Currently, seed pelleting, film coating, and seed encrusting are the most common coating/treatment procedures used in the seed industry to enhance plant and seedling performance. While seed pelleting, often employed to develop more uniform seeds for mechanical planting, can increase seed weight from 200 to > 5000%, film coating or encrusting utilizes much smaller quantities of materials resulting in a build-up in seed weight of between 0.5–10% and 20–200%, respectively [24]. The physical properties and thickness of the seed treatment/coating are critical factors that influence seed germination and seedling vigor. A thick hard seed coating can reduce, delay, or cause abnormal germination or may even be toxic, while a minimal, fragile seed coating can break or disintegrate before planting or not have a high enough dosage of an active ingredient to be effective. Therefore, specialized seed coating formulations must be developed and evaluated in order to be utilized effectively for any given plant species and agronomic purpose.

The specific objectives of this research were to explore plant-derived bio-based biostimulant seed coatings to enhance germination and growth of two cover crop species, red clover and perennial ryegrass, as an approach for seeding cover crops for grassland restoration. Previous research on seed coatings of broccoli and tomato with soy flour and compost materials showed promising results related to maximum germination, germination uniformity, and seedling vigor [15,25–27].

2. Materials and Methods

2.1. Seed and Coating Materials

Two species of cover crops were selected to evaluate biostimulant seed coatings in this study. Red clover 'VSN-variety not stated' seed was obtained from King's AgriSeeds, Inc., Lancaster, PA, USA, and 'Tetraprime' perennial ryegrass seed was provided by SeedWay, LLC, Penn Yan, NY, USA. The red clover and perennial ryegrass seeds were coated with different combinations of soy flour (SF), diatomaceous earth (DE), micronized vermicompost (MVC-2 and 3), and concentrated vermicompost extract (CVE) to identify the most stable and effective coating formulations (Table 1). Specific treatments and ratios of coating materials evaluated were (SF:DE 30:70, 40:60, 50:50 and 60:40, SF:MVC-2 (30:70), SF:MVC-3 (30:70), SF:DE:CVE (30:70)). A mechanical Ro-Tap shaker (Ro-Tap Testing Sieve Shaker No. 1506; The W.S. Tyler Co., Cleveland, OH) was utilized to sieve the SF to obtain a particle size smaller than (<75 μ), as previous studies have shown that smaller particle size results in more even distribution of the SF coating on seeds [15]. Seed coating biostimulant materials used in this research were previously analyzed by the Cornell Soil Health Nutrient Analysis Laboratories and recently reported in [27].

Table 1. Materials used as seed coating biostimulant treatment formulations in this study.

Coating Materials	Abbreviation	Source
Soy Flour	SF	Archer Daniels Midland Co., Decatur, IL, USA
Diatomaceous Earth	DE	Perma-Guard, Inc., Albuquerque, NM, USA
Concentrated Vermicompost Extract (liquid)	CVE	Worm Power, Avon, NY, USA (concentrated by Caloris Engineering, Easton, MD, USA)
Micronized Vermicompost	MVC-2	Worm Power, Avon, NY, USA
Micronized Vermicompost	MVC-3	TerraVesco, Sonoma Valley Worm Farm, Sonoma, CA, USA

2.2. Seed Coating

A 15-cm diameter, R-6 (Universal Coating Systems, Independence, OR, USA) laboratory-scale rotary pan coater was used to coat seeds in all experiments (Figure 1). Each seed coating treatment consisted of two components: dry powder and liquid. SF and DE, SF and MVC-2, and SF and MVC-3 were applied as dry powder to the seed surface with distilled water. For the SF, DE, and CVE treatments, water was replaced with liquid compost extract. For each treatment, the powder and liquid were applied to the surface of the seeds in incremental amounts as they rotated in the R6 to achieve uniform results. To clean the residual dust of each coating batch and avoid cross contamination of treatments, the R-6 pan was cleaned with a sponge and hot water-liquid soap solution. This was followed by cleaning with disinfectant wipes (three times), and R-6 with high pressure air flow was applied around the pan and cylinder to ensure completely drying. In this study, coating combinations of SF:DE and SF:MVC for both crops were applied on separate days.

Twenty-five grams of seed were used for red clover (1000 seed weight = 2.5 g) treatments, and 15 g of seed were used for the perennial ryegrass (1000 seed weight = 1.5 g). Therefore, we had an equal amount of treated seeds (~10,000 seeds) for each treatment of each crop. The total build up percentage was approximately 30% for clover and 70% for perennial ryegrass (Figure 2). Variation in the percent build up needed to achieve uniform coverage reflects the need for specifically developed seed coatings for each seed type and species. The size and shape of seeds influences uniformity of treatments on the seeds. After coating, the seeds were dried at room temperature for 24 h (h) until completely dry [15,28]. To improve observation of the seed coating uniformity of the SF:DE and SF:DE:CVE blend, a red dye (Pro-Ized red colorant, Bayer Corp, Research Triangle Park, NC, USA) was added to the binder (1.0 mL dye per 10 mL binder). Due to the dark brown color of the MVC materials, no dye was used for SF and MVC combinations; the natural color showed coating quality and uniformity of application.

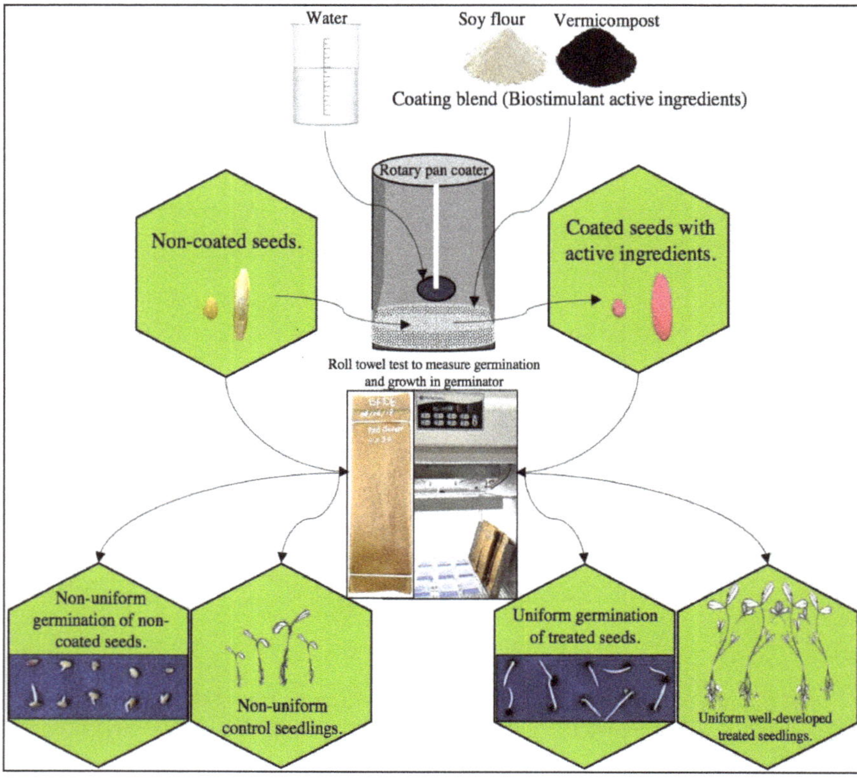

Figure 1. Figure of seed coating methodology used as an approach for application of biostimulant compounds for sustainable agriculture.

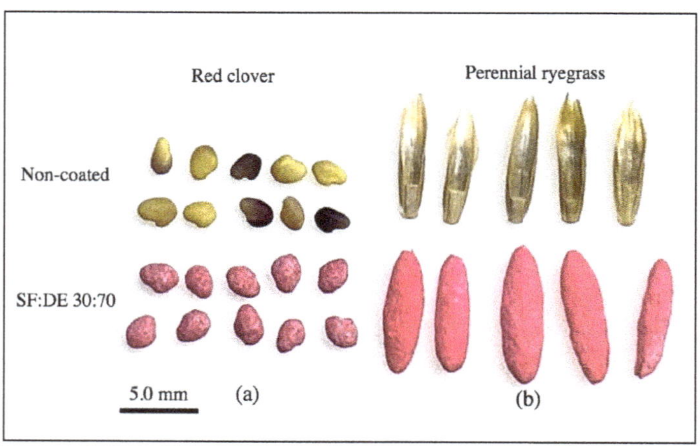

Figure 2. Non-coated and coated (SF:DE 30:70) red clover seeds with 30% build up (**a**) and non-coated and coated (SF:DE 30:70) perennial ryegrass with 70% build up (**b**). To improve observation of seed coating uniformity, a red dye was added to the binder.

2.3. Seed Coat Physical Properties

2.3.1. Seed Coating Integrity Test

The strength of the coating is an important quality as it relates to germination and potential for damage during handling, transportation, and planting. The surface material of coated seeds must have good mechanical properties to ensure that they do not crack or disintegrate before sowing. A Ro-Tap sieve shaker (The W.S. Tyler Co., Cleveland, OH, USA) was used to test the integrity of the coated seeds [15]. Four replications of 1.5 g of coated red clover seeds and four replications of 1.5 g of perennial ryegrass seeds from each coating formulation (treatments listed in Table 2) were tested to assure reliability and reproducibility. Samples were weighed and shaken for 2 min using a standard Ro-tap test shaker with U.S. Standard Testing Sieve No. 25 (0.71 mm opening) and a solid catch pan. Each sample was weighed again, and the percentage of coating loss was calculated according to the weight before and after the Ro-tap procedure. The weight of coating material, which passed through a No. 25 sieve was reported as weight loss (WL %) of material.

Table 2. Results of seed coating physical property testing, weight loss (WL, %), disintegration time (DT, min), compressive strength (Force N), time to break (TB) seed coating measured in s, relaxation time (RT) after the seed coating was fractured measured in s for coated seeds of red clover and perennial ryegrass.

Crop	Treatment	WL (%)	DT (min)	Force (N)	TB (s)	RT (s)
Red clover	SF:DE 30:70	1.5 a *	58 a	16.2 a	4.4 a	0.3 a
	SF:DE 40:60	1.2 ab	75 b	19.2 b	5.1 b	0.36 b
	SF:DE 50:50	0.6 bc	100 c	20.6 b	5.6 c	0.48 c
	SF:DE 60:40	0.4 c	103 c	23.9 c	5.5 c	0.5 c
	SF:MVC-2	1.2 ab	78 b	16.8 a	4.4 a	0.38 b
	SF:MVC-3	1.1 ab	83 b	16.5 a	4.7 a	0.32 a
	SF:DE:CVE	1.0 ab	80 b	17.5 b	4.9 a	0.31 a
Perennial ryegrass	SF:DE 30:70	1.4 A *	40 A	15.3 A	5.7 A	0.2 A
	SF:DE 40:60	1.2 AB	58 B	17.9 B	6.0 B	0.38 B
	SF:DE 50:50	0.5 B	90 C	20.9 C	6.4 C	0.51 C
	SF:MVC-2	1.3 B	60 B	15.6 A	5.4 A	0.24 A
	SF:MVC-3	0.9 BC	55 B	16.1 A	5.5 A	0.22 A
	SF:DE:CVE	0.9 BC	60 B	15.8 A	5.8 A	0.31 B

* Different letters within each column for each crop indicate significant differences using a Least Significant Difference (LSD) test at a significance level of $p < 0.05$. Lower case letters represent significant differences in red clover treatments and upper case letters denote perennial ryegrass treatment differences.

2.3.2. Mechanical Property Test

A texture analyzer (TA-XTplusC, Texture Technologies Corp., Hamilton, MA) was used to test the compressive strength of coated seeds. The TA-XTplusC is a precision instrument used to measure the surface mechanical properties of coated seeds and the compressive strength of a single seed. The arm of the texture analyzer containing a weighing sensor moves in a downward motion to compress the coated seed placed on the base of the analyzer and then returns to its original position. Data are assessed as the compressive strength (Force N) and time to breakage (TB, measured in seconds (s)) required to fracture the seed coating. The relaxation time (RT), which is the time required to completely rupture the seed coating was also measured [29,30]. After the seed coating was completely broken, the force (N) increases until the seed embryo was crushed (Figure 3). Texture analyzer software (Exponent Connect, version 7.0.2.0, S. Hamilton, MA, USA, 2018) was used to record force for TB and RT [26]. Ten coated seeds were randomly selected from batches of different formulations (SF:DE = 30:70, 40:60, 50:50 and 60:40, SF:MVC-2 (30:70), SF:MVC-3 (30:70), SF:DE:CVE (30:70)) to test their surface compressive strength for both red clover and perennial ryegrass coated seeds.

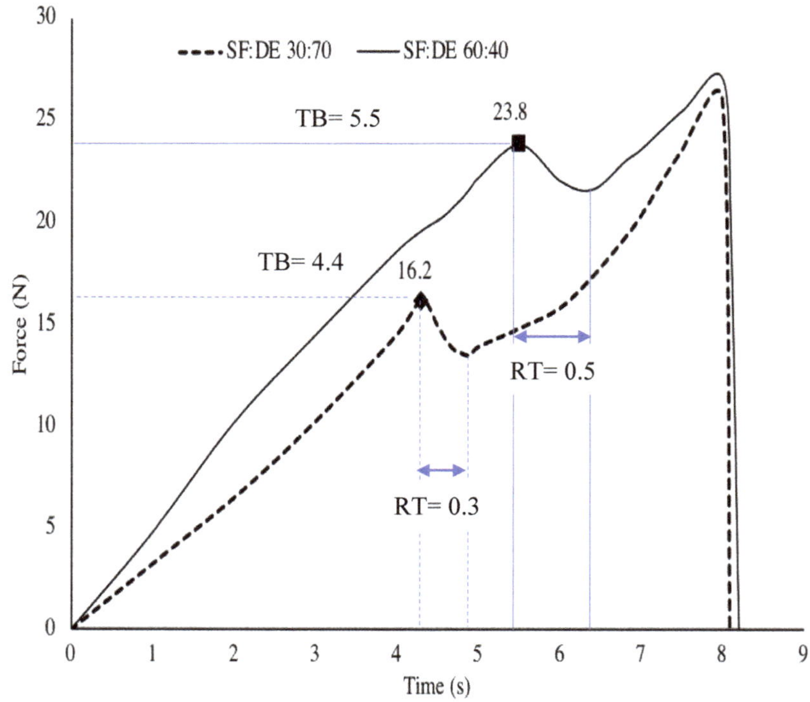

Figure 3. The values of peak load force required to break the seed coat of a single seed from two different seed coating blends of soy flour (SF) and diatomaceous earth (DE) tested at room temperature (SF:DE 30:70 and 60:40) for red clover are 16.2 and 23.8 N, respectively. The maximum force value (N) is a measure of coating strength and shows the maximum force needed to break the seed coat. Time to break seed coating (TB) and Relaxation Time (RT) after seed coat fracture until the coat is completely broken, both measured in s, are shown for a single seed. Force, TB, and RT data shown in Table 2 are the means of 10 seeds (Equipment: TA-XTplusC, Texture Technologies Corp., Hamilton, MA, USA, Software: Exponent Connect, version 7.0.2.0).

2.3.3. Seed Coating Hydration Test

The wet strength of a seed coating is largely dependent on the adhesion of the components after immersion in water. In theory, the slower the decomposition rate of coated seeds in water is, the more likely it is to delay germination. The hydration test was used to investigate the integrity of the coating materials when immersed in water. Hydration tests were conducted to determine the potential for seed coatings to prevent or delay germination. Four replicates of 100 coated seeds with different proportions of soy flour and diatomaceous earth, SF:DE (30:70, 40:60, 50:50 and 60:40), and the soy flour micronized vermicompost treatments SF: MVC-2, SF: MVC-3 were placed in 5 mL of distilled water to determine disintegration time. Disintegration time was measured in minutes.

2.4. Seed Germination and Seedling Growth Measurements

Four replicates of 50 treated and non-coated control seeds were placed on two 30 cm × 45 cm moistened germination paper towels (Anchor Paper Company, St. Paul, MN, USA); then, an additional moistened standard germination paper towel was placed on top of the seeds. The towels were rolled and positioned in a germinator (Percival germinator, Model I-36LL, Perry, IA, USA). For perennial ryegrass seed, the germinator was maintained at 15/25 °C, with a 16/8 h photoperiod [31]; red clover, was maintained at a constant 20 °C with the same photoperiod [31]. Radical emergence (>2 mm) was used to determine successful seed germination. The number of newly germinated seeds for both red clover and perennial ryegrass was recorded every 24 h. For perennial ryegrass, the total germination percentage (Gmax %) was recorded after 10 days. The Gmax % for red clover was recorded after seven days. Gmax %, the number of germinated seeds and germination uniformity (GU), (GU = time required for 90% germination subtracted by time required for 10% germination) were calculated [32] for each treatment. In addition, germination rate (T50, the time in h to reach 50% total germination) was calculated according to the equation developed by Coolbear et al. [33].

Root and shoot measurements (cm) were conducted in separate roll towel experiments (under the same growing conditions described above for seed germination) for each treatment using four replicates of 50 seeds. The seed vigor index (SVI) was equal to Gmax % multiplied by seedling length (combined root and shoot lengths) divided by 100 [34]. Seedlings were measured a week after full emergence for both crops. After measuring shoot and root lengths, all seedlings from each treatment were dried in an oven at 80 °C for 48 h to obtain the dry weight data.

2.5. Statistical Analysis

In all experiments, normality tests were conducted prior to ANOVA and all data passed the normal distribution test at a significance level of 0.05. Analysis of variance (ANOVA) ($\alpha = 0.05$) and Fisher's least significant difference test for seed coating physical property and Dunnett test for germination and seedling growth data were performed on each of the significant variables measured by Minitab Express [35]. All Gmax % (Tables 2–4) and WL % (Table 2) data were arcsine transformed for analysis. Data for Gmax % and WL% are presented as non-transformed means (Tables 2–4). Pearson correlation was conducted for coating physical properties data using Minitab Express (Table 5).

Table 3. Germination and growth metrics of soy flour formulations as measured by total germination (Gmax %), germination rate (T50) measured in hours (h), germination uniformity (GU) measured in hours (h), shoot and root length (cm), and Seedling Vigor Index (SVI = Gmax % × Seedling length) of different coating formulas of SF:DE for red clover and perennial ryegrass.

Crop	Treatment	Gmax (%)	T50 (h)	GU (h)	Shoot (cm)	Root (cm)	SVI
Red clover	Control	95 b *	35 a	37 a	3.6 b	2.4 b	5.7 b
	SF:DE 30:70	98 a	27 b	27 b	4.1 a	3.0 a	7.0 a
	SF:DE 40:60	99 a	29 b	25 b	4.3 a	2.9 a	7.1 a
	SF:DE 50:50	96 b	30 b	27 b	4.1 a	2.9 a	6.7 a
	SF:DE 60:40	96 b	34 a	28 b	3.9 a	2.7 a	6.3 b
Perennial ryegrass	Control	85 A *	75 B	39 B	6.5 B	5.5 B	10.2 B
	SF:DE 30:70	83 A	79 A	40 B	7.6 A	6.2 A	11.5 A
	SF:DE 40:60	83 A	80 A	42 B	7.4 A	6.3 A	11.4 A
	SF:DE 50:50	80 B	83 A	47 A	7.2 A	5.9 A	10.5 A

* Different letters within each column for each crop indicate significant differences using a Dunnett test at a significance level of $p < 0.05$. Lower case letters represent significant differences between each red clover seed coating treatment compared with the control and upper case letters denote each of perennial ryegrass treatment differences compared with the control.

Table 4. Germination and growth metrics of soy flour/vermicompost formulations as measured by total germination (Gmax %), germination rate (T50) measured in hours (h), germination uniformity (GU) measured in hours (h), shoot and root length (cm), seedling dry weight (DW) recorded in grams (g), and Seedling Vigor Index (SVI = Gmax % × Seedling length) from evaluation of different coating materials for red clover and perennial ryegrass. The proportion of all coating materials is 30:70 (30% SF and 70% of DE or MVC).

Crop	Treatment	Gmax (%)	T50 (h)	GU (h)	Shoot (cm)	Root (cm)	DW (g)	SVI
Red clover	Control	94 b *	36 a	35 a	3.7 b	2.5 b	0.05 b	6.0 b
	SF:DE	98 a	26 b	27 b	4.2 a	2.8 a	0.07 a	6.9 a
	SF:MVC-2	99 a	26 b	24 b	4.4 a	2.9 a	0.07 a	7.3 a
	SF:MVC-3	99 a	25 b	25 b	4.7 a	3.0 a	0.08 a	7.6 a
	SF:DE:CVE	98 a	27 b	25 b	4.5 a	3.2 a	0.07 a	7.5 a
Perennial ryegrass	Control	87 A *	76 A	41 B	6.5 B	5.9 B	0.10 B	10.8 B
	SF:DE	85 A	78 A	43 B	7.9 A	6.5 A	0.13 A	12.3 A
	SF:MVC-2	86 A	77 A	40 B	8.1 A	6.8 A	0.14 A	12.8 A
	SF:MVC-3	82 B	81 A	48 A	8.3 A	6.6 A	0.13 A	12.2 A
	SF:DE:CVE	85 A	77 A	43 B	8.4 A	6.6 A	0.15 A	12.8 A

* Different letters within each column for each crop indicate significant differences using a Dunnett test at a significance level of $p < 0.05$. Lower case letters represent significant differences between each red clover seed coating treatment compared with the control and upper case letters denote each of perennial ryegrass treatment differences compared with the control.

Table 5. Correlation coefficients between disintegration time (DT, min), weight loss (WL, %), and compressive strength (Force, N) from seed coating applications of soy flour and diatomaceous earth on red clover and perennial ryegrass seeds.

Crop		WL (%)	Force (N)
Red clover	DT (min)	−0.99 ***	+0.92 **
	WL (%)	-	−0.94 **
Perennial ryegrass	DT (min)	−0.99 ***	+0.99 ***
	WL (%)	-	−0.96 ***

, * Significant at $p < 0.001$, 0.0001, respectively.

3. Results and Discussion

3.1. Seed Coating Physical Properties

The integrity and physical properties of coated and pelleted seeds are critical for overall performance. The production of dust can lead to health and environmental risks; therefore, it is important to quantitatively analyze the potential for breakage and weight loss that may occur during transportation and handling. In contrast, a seed coating that is too hard or impermeable to water may hinder germination. In this study, physical properties of the various seed coating formulations were tested by employing three different tests, including mechanical, texture, and hydration analyses.

Experimental results from the seed coating mechanical Ro-tap and Texture analysis (TA-XTplusC) tests are presented in Table 2 and Figure 3. For both crops, increasing the proportion of soy flour in the coating blend increased the compressive strength (Force N) of coated seeds. The time required to break the seed coating (TB), measured in s and relaxation time (RT) after breaking seed coating (Table 2) increased as soy flour proportion increased (Table 2). For example, for red clover, the average force (N) increased from 16.2 to 23.9 N as the soy flour content increased from 30% to 60%, which is an increase of approximately 48%. As soy flour content increased, the TB of coated seeds gradually increased from 4.4 to 5.5 s for red clover and 5.7 to 6.4 s for perennial ryegrass (Table 2). Although the same seed coating blend were used for both crops, the TB ranges were different most likely due to the difference in build-up percentage. There was a 1.1 s delay in breakage time for red clover when the content of soy flour increased from 30% to 60%, and 0.7 s delay for perennial ryegrass when the content of soy flour increased from 30% to 50%. The force value to break down the coating for perennial ryegrass significantly increased by approximately 37% as soy flour increased from 30% (15.3) to 50% (20.9). Additionally, the weight loss of coated seeds from the Ro-tap test gradually decreased from 1.5% to 0.4% for red clover and 1.4% to 0.5% for perennial ryegrass (Table 2) as the proportion of soy flour in the seed coating increased. Interestingly, the mechanical properties of each seed formulation were not significantly different, even though the shape and surface properties of the seeds of both species differed.

There was no significant difference in both crops in terms of weight loss and disintegration time of coated seeds in water (Table 2) of soy flour and vermicompost. For red clover, when seed coatings with soy flour and vermicompost (30:70) (SF:MVC-2, SF:MVC-3 and SF:DE:CVE) were compared with SF:DE 30:70, the force (N) to break the seed coating increased by 0.6, 0.4, and 1.3 N, respectively and time to break (TB) the seed coating were non-significant, indicating that soy flour could serve as a binder for both types of materials used in this study (DE and vermicompost). There was no significant difference observed on force and TB data among treatments for perennial ryegrass (Table 2). Relaxation time after breaking red clover seed coating (RT) of SF:MVC-2, SF:MVC-3, and SF:DE:CVE increased slightly by 0.08, 0.02, and 0.01s, and the RT of perennial ryegrass increased by 0.04, 0.02, and 0.11s compared with that of SF:DE 30:70. In conclusion, the surface mechanical strength of different seed coating formulations with soy flour and vermicompost blends were non-significant but slightly higher than that of soy flour and diatomaceous earth.

The hydration test measures the time required to dissolve the coating materials. The proportion of soy flour in the seed coating blends had a significant effect on disintegration time (DT) (Table 2). A higher proportion of soy flour in the coating blend for red clover increased the DT from 58 to 105 min. This pattern was also observed for perennial ryegrass, as the proportion of soy flour ratio increased from 30% to 50%, the DT significantly increased from 40 to 90 min (Table 2). There were no significant differences in WL or DT for micronized vermicompost and compost extract (SF:MVC-2, SF:MVC-3 and SF:DE:CVE) seed treatments on either crop species (Table 2).

According to the American Seed Trade Association (ASTA), the key to a successful seed treatment is high physical integrity with low dust production. Determination of mechanical integrity of coated seeds is an important step in order to meet environmental safety standards [36]. Amirkhani et al. [15,27] tested the mechanical properties of several different broccoli coated seeds with Ro-tap and Texture Analyzer methods. The weight loss percentage that Amirkhani et al. [27] reported were slightly higher than the data collected from the red clover and perennial ryegrass seed treatments evaluated in this study for the same seed coating formulations. Total peak load force to break the seed coats for broccoli seeds were slightly lower than the value that was observed in this study for the SF:DE combination. Overall, the mechanical integrity of the coated red clover and perennial ryegrass seeds were more stable for the same seed coating formulation treatments, compared to broccoli. This difference might be because of the size and shape of clover and perennial ryegrass seeds contrasted to the broccoli seeds. Accinelli et al. [37] also attributed differences in seed dust emission for seed coating treatments of maize (*Zea mays* L.) and canola (*Brassica napus* L.) to seed physical characteristics. The mechanical integrity data observed for the different formulations in this study are in accordance with European Standards (Italy and France) and meet the benchmarks for safety of dust production of coated seeds [38].

3.2. Germination and Seedling Growth of Soy Flour and Diatomaceous Earth Seed Coating

In seed coatings with diatomaceous earth, soy flour served as the biostimulant component of the formulations. The results presented in Table 3 show that all coated treatments of red clover seeds significantly improved T50 and GU with no reduction in Gmax % compared to the non-treated control (Table 3), except for the SF:DE 60:40 treatment. Although soy flour proportions higher than 40% resulted in stronger and more durable seed coating mechanical properties, it had a negative effect on germination parameters (Gmax % and T50). For example, seeds treated with 30% and 40% soy flour had 98 and 99% Gmax % and T50 of 27 and 29 h, respectively, However, increasing the soy flour to

60% resulted in maximum germination of 96% and delayed the T50 to 34 h (Table 3). The negative effect in germination was attributed to the hard mechanical barrier of the seed coating with high soy flour content.

In contrast, for perennial ryegrass seeds, the Gmax % of all coated seeds was slightly reduced and showed delayed germination rates compared with the non-treated control. Control seeds of perennial ryegrass had the greatest Gmax % (85%) and significantly faster T50 (75 h) compared to all coating formulations. Application of 50% soy flour to the seed coating (SF:DE 50:50) reduced the Gmax % to 80% and T50 by 8 h and GU, compared with the non-treated control seeds. Due to the high percentage of coating build up (70%), the delay and a slight reduction in Gmax % was not unexpected. Several studies have indicated that the seed coating can act as a mechanical barrier for water absorbance and radical emergence [10,15].

Shoot and root length and seedling vigor index (SVI) are important indicators that determine whether the treated seeds promote seedling growth. Shoot and root length and seedling vigor index of treated seeds were significantly higher than those of non-treated control seeds for both crops (Table 3). The lowest application of soy flour (SF:DE 30:70) to the red clover seeds resulted in 4.1 cm shoot length, 3.0 cm root length, and 7.0 SVI, respectively, which were 14, 25, and 23 % higher than those of non-treated seeds. The same application rate of soy flour (SF:DE 30:70) to perennial ryegrass seeds improved the shoot growth by 17% and increased both root length and SVI values by approximately 13% compared to the control seeds. Amirkhani et al. [15,25–27] reported similar results for seed coatings that combined soy flour with diatomaceous earth. In their research, the seed coating blends had significant and positive effects on the above and below ground growth parameters of broccoli, tomato, radish, and hemp. They hypothesized that since soy flour is a plant-based protein and a rich source of several amino acids, it may have led to the increase in plant shoot and root growth and dry matter content and influenced uptake of nitrogen.

3.3. Germination and Seedling Growth of Soy Flour and Vermicompost Seed Coating

In addition to soy flour and diatomaceous earth, co-application of soy flour and vermicompost as rich sources of nutritional materials were tested as seed coatings and their effect on germination and seedling growth were recorded for red clover and perennial ryegrass. Shoot and root length, dry weight, and seedling vigor index of seedlings of all coated seed treatments were significantly higher compared to the non-treated controls (Table 4).

All treated red clover seeds germinated significantly faster (approximately 10 h) and had higher Gmax % (Table 4 and Figure 4) than non-coated seeds. They also germinated more uniformly than the non-treated control seeds. Gmax % was $\geq$ 98% for all treated seeds, which was significantly higher than control with 94% Gmax %. Red clover data showed that the shoot and root length and seedling vigor index of treated seeds were significantly higher than the non-treated control seeds. For example, compared with non-treated seeds control, the shoot length of SF:DE, SF:MVC-2, SF:MVC-3, and SF:DE:CVE increased by 14%, 19%, 27%, and 22%, respectively. Moreover, the root length of treated seeds increased by 12%, 16%, 20%, and 28%, respectively. All treatments showed a 40 to 60% increase in seedling dry weight (DW) compared with the control. The seedling vigor indexes (SVI) were 15%, 22%, 27%, and 25% higher than control, respectively (Table 4).

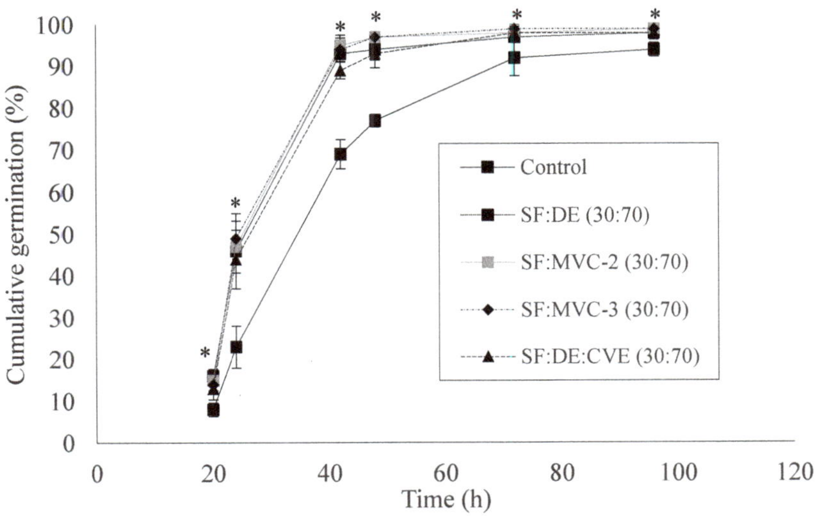

Figure 4. Cumulative germination percentage of red clover non-treated control seeds versus biostimulant coated seeds. * Significant at $p \leq 0.05$.

For perennial ryegrass, application of soy flour (SF:DE) and co-application with vermicomposts (SF:MVC-2 and 3, and SF:DE:CVE) increased shoot length by 22%, 25%, 28%, and 29% and root length by 10%, 15%, 12%, and 12%, respectively, compared to the non-treated control. The highest root length was observed in the SF:MVC-2 treatment. All treatments had higher DW than the control. Additionally, the highest SVI was observed in the SF:DE:CVE treatment, which was approximately 40% higher than the SVI of non-treated control (Table 4).

Statistical analysis (Pearson's correlation) of seed coating formulations and germination showed significant negative correlations between seed coating WL (%) and DT (min) (r = −0.99 ***). There was also a significant positive correlation between DT (min) and force (N) (r = +0.92 **). A significant negative correlation between WL (%) and force (N) (r = −0.94 **) from SF:DE coating formulations of red clover evaluated was observed (Table 5). For perennial ryegrass, the significant correlation coefficient between WL (%) and DT (min) was r = −0.99 *** and the correlation between WL (%) and force (N) was r = −0.96 ***. Lastly, for perennial ryegrass, the correlation between DT (min) and force (N) was r = +0.99 *** (Table 5). These data indicate that a higher proportion of SF in the seed coating formulation resulted in harder coatings but had only a slight impact on the Gmax %. For red clover, increasing the soy flour from 30% to 60% in seed coating formula reduced the Gmax % by 2%; however, T50 was significantly delayed by 7 h (Table 3). Similarly, for perennial ryegrass, increasing soy flour from 30% to 50% in the seed coating resulted in a 3% reduction in Gmax % and a minor delay on T50 (4 h).

In the present study, the seedling growth data for both cover crops evaluated indicate that seed coating can be an efficient and effective delivery method for application of nutritional biostimulant materials at the time of sowing for rangeland and grassland restoration. Several previous studies showed that applications of plant-based proteins and vermicompost can improve biometric growth parameters, related to production and yield of horticultural, field, and cover crops. Karlsons et al. [39] showed that a 10% addition of vermicompost in pure sand significantly increased fresh and dry weight of winter rye shoots by 578% and 265%, respectively. Tognetti et al. [40] found that application of vermicompost to degraded volcanic soil (to the extent of 20 and 40 g/kg soil) sown with ryegrass (*L. perene*) significantly increased ryegrass yields compared to the control due to the large nutrient concentrations and high microbial populations, when mixed with the soil. The positive effect of vermicompost on plant growth in this study agrees with the results of Alwaneen [41] on alfalfa and Amirkhani et al. [27] on broccoli. Amirkhani et al. [27] found that dairy manure-based vermicompost can supply essential nutrients to plants to enhance growth as well as increase the organic matter contents of soil for higher crop production. Moreover, in the recent decade, several researchers have been working on treating plants with biostimulants to stimulate crop productivity and increase stress tolerance under dynamic abiotic stresses [42–44]. The cover crop seeds treated with biostimulants in combination with other bio-effectors, such as superabsorbent polymers to investigate the response of these plants to drought, can be an area of future studies.

4. Conclusions

Seed coating technology can be an effective strategy to maximize early stand establishment of cover crops. Biostimulants applied as seed coatings have the potential to effectively promote seedling growth, and early stand establishment of red clover and perennial ryegrass. In this study, biostimulant seed coatings promoted the seedling growth of red clover and perennial ryegrass seeds and accelerated the germination of red clover compared to the non-treated control seeds. More rapid germination could aid in establishment under arid conditions and in areas with poor soils. However, further studies are needed to determine if vermicompost and plant-based proteins can be developed for economical commercial applications as seed treatments. The seed industry commonly includes fertilizers and *Rhizobium*, nitrogen fixing bacterial inoculants for red clover seeding. Additional research will be needed to determine if the biostimulant materials used in these experiments are compatible with seed inoculants. The use of biostimulants in combination with vermicompost and other biofertilizers as seed coatings may offer a great opportunity to increase plant production and long-term sustainability in agricultural landscapes.

Author Contributions: Y.Q., M.A., and H.M. contributed equally in writing the manuscript. Z.C. was involved in reviewing the manuscript. A.G.T. assisted with conceptualization of the overall experiments, funding, writing, and reviewing the manuscript. All authors have read and agreed to the published version of the manuscript.

Funding: This material is based upon work that is supported by the National Institute of Food & Agriculture, US Department of Agriculture, Multi-state Project W-3168, under accession #1007938. Y.Q. was sponsored by China Scholarship Council.

Acknowledgments: The authors appreciate the technical assistance of Michael Loos in laboratory experiments and Zhen Wang for assistance with this project.

Conflicts of Interest: The authors declare that the research was conducted without any financial relationships that could be construed as a potential conflict of interest.

References

1. Roser, M.; Ritchie, H.; Ortiz-Ospina, E. *World Population Growth*; Our World in Data; Oxford University-Global Change Data Lab (GCDL), 2019; Available online: https://ourworldindata.org/world-population-growth (accessed on 20 January 2020).
2. Mu, S.; Zhou, S.; Chen, Y.; Li, J.; Ju, W.; Odeh, I.O.A. Assessing the impact of restoration-induced land conversion and management alternatives on net primary productivity in Inner Mongolian grassland, China. *Glob. Planet. Chang.* **2013**, *108*, 29–41. [CrossRef]
3. Ni, J. Carbon storage in grasslands of China. *J. Arid Environ.* **2002**, *50*, 205–218. [CrossRef]
4. Qiu, Y.; Chen, Z.; Hou, Z.F.; Song, T.; Mi, L.K.; Shao, Z.W. Numerical simulation and experiment on improving pelleted coating of forage grass seeds by vibration force field. *Trans. CSAE* **2017**, *33*, 86–93.
5. Sheoran, V.; Sheoran, A.S.; Poonia, P. Soil reclamation of abandoned mine land by revegetation: A review. *Int. J. Soil Sediment Water* **2010**, *3*, 13.
6. Skousen, J.; Zipper, C.E. Reclamation guidelines for surface mined land in South-Western Virginia. Chapter: Revegetation species and practices. *Va. Coop. Ext. Publ.* **1996**, *460*, 120.
7. Balfourier, F.; Imbert, C.; Charmet, G. Evidence for phylogeographic structure in lolium species related to the spread of agriculture in Europe. A cpDNA study. *Theor. Appl. Genet.* **2000**, *101*, 131–138. [CrossRef]

8. McKenna, P.; Cannon, N.; Conway, J.; Dooley, J.; Davies, W.P. Red clover (*Trifolium pratense*) in conservation agriculture: A compelling case for increased adoption. *Int. J. Agric. Sustain.* **2018**, *16*, 342–366. [CrossRef]
9. Daryanto, S.; Fu, B.; Wang, L.; Jacinthe, P.A.; Zhao, W. Quantitative synthesis on the ecosystem services of cover crops. *Earth-Sci. Rev.* **2018**, *185*, 357–373. [CrossRef]
10. Taylor, A.G.; Allen, P.S.; Bennett, M.A.; Bradford, K.J.; Burris, J.S.; Misra, M.K. Seed enhancements. *Seed Sci. Res.* **1998**, *8*, 245–256. [CrossRef]
11. Snapp, S.; Price, R.; Morton, M. Seed priming of winter annual cover crops improves germination and emergence. *Agron. J.* **2008**, *100*, 1506–1510. [CrossRef]
12. Harris, D.; Raghuwanshi, B.S.; Gangwar, J.S.; Singh, S.C.; Joshi, K.; Rashid, A.; Hollington, P.A. Evaluation by farmers of on-farm seed priming in wheat in India, Nepal and Pakistan. *Exp. Agric.* **2001**, *37*, 403–415. [CrossRef]
13. Vartha, E.W.; Clifford, P.T.P. Effects of seed coating on establishment and survival of grasses, surface-sown on tussock grasslands. *N. Z. J. Exp. Agric.* **1973**, *1*, 39–43. [CrossRef]
14. Falloon, R.E.; Fletcher, R.H. Increased herbage production from perennial ryegrass following fungicide seed treatment. *N. Z. J. Agric. Res.* **1983**, *26*, 1–5. [CrossRef]
15. Amirkhani, M.; Netravali, A.N.; Huang, W.; Taylor, A.G. Investigation of soy protein-based biostimulant seed coating for broccoli seedling and plant growth enhancement. *HortScience* **2016**, *51*, 1121–1126. [CrossRef]
16. Ou, C.M.; Mao, P.S. Progress in research and application of coating technology for grasses. *Seed* **2019**, *11*, 63–67. (In Chinese)
17. Li, C.Y.; Zhang, F.; Liu, C.H.; Yu, F.S.; Li, Y.C. Screening of coating materials of the forage seeds. *J. Northeast Agric. Univ.* **2013**, *4*, 94–100, (In Chinese with English abstract).
18. He, Z.X.; Mao, P.S.; Sun, Y.; Li, M. Review of grass seed coating technology. *Acta Agrestia Sin.* **2016**, *2*, 270–277, (In Chinese with English abstract).
19. Rouphael, Y.; Colla, G. Synergistic biostimulatory action: Designing the next generation of plant biostimulants for sustainable agriculture. *Front. Plant Sci.* **2018**, *9*, 1655. [CrossRef]
20. Colla, G.; Hoagland, L.; Ruzzi, M.; Cardarelli, M.; Bonini, P.; Canaguier, R.; Rouphael, Y. Biostimulant action of protein hydrolysates: Unraveling their effects on plant physiology and microbiome. *Front. Plant Sci.* **2017**, *8*, 2202. [CrossRef]
21. Wilson, H.T.; Amirkhani, M.; Taylor, A.G. Evaluation of gelatin as a biostimulant seed treatment to improve plant performance. *Front. Plant Sci.* **2018**, *9*, 1006. [CrossRef]
22. Calvo, P.; Nelson, L.; Kloepper, J.W. Agricultural uses of plant biostimulants. *Plant Soil* **2014**, *383*, 3–41. [CrossRef]
23. Schmitt, A.; Koch, E.; Stephan, D.; Kromphardt, C.; Jahn, M.; Krauthausen, H.J. Evaluation of non-chemical seed treatment methods for the control of Phoma valerianellae on lamb's lettuce seeds. *J. Plant Dis. Prot.* **2009**, *116*, 200–207. [CrossRef]
24. Taylor, A.G. Seed treatments. In *Encyclopedia of Applied Plant Sciences*; Thomas, B., Murphy, D.J., Murray, B.G., Eds.; Elsevier Acad. Press: Oxford, UK, 2003; pp. 1291–1298.
25. Amirkhani, M.; Netravali, A.N.; Taylor, A.G. Improving seedling growth uniformity and seed vigor index by using plant-based protein seed coating in tomato and broccoli NYSAES research symposium. In Proceedings of the NYSAES Research Symposium 2017, Geneva, NY, USA, 23 June 2017; p. 6.
26. Amirkhani, M.; Mayton, H.S.; Loos, M.T.; Netravali, A.N.; Taylor, A.G. Biostimulant effect of a plant-based protein applied as a seed coating on selected crops. In Proceedings of the 3rd Annual Cornell AgriTech Research Symposium, Geneva, NY, USA, 29 June 2018; p. 4.
27. Amirkhani, M.; Mayton, H.S.; Netravali, A.N.; Taylor, A.G. A seed coating delivery system for bio-based biostimulants to enhance plant growth. *Sustainability* **2019**, *11*, 5304. [CrossRef]
28. Talley, B. *Status and New Developments in Seed Coatings*; Summit Seed Coating: Caldwell, ID, USA, 2011; Available online: https://uknowledge.uky.edu/ky_alfalfa/2012/Session/6/ (accessed on 28 August 2019).
29. Cespi, M.; Bonacucina, G.; Misici-Falzi, M.; Golzi, R.; Boltri, L.; Palmieri, G.F. Stress relaxation test for the characterization of the viscoelasticity of pellets. *Eur. J. Pharm. Biopharm.* **2007**, *67*, 476–484. [CrossRef] [PubMed]
30. Texture Analysis. Texture: Measure and Analyze Properties. Disintegration. Available online: https://www.stablemicrosystems.com/MeasureDisintegration.html (accessed on 15 August 2019).
31. AOSA. Rules for testing seeds. Principles and procedures. *Assoc. Off. Seed Anal.* **2014**, *1*, 6–25.

32. Egli, D.B.; Hamman, B.; Rucker, M. Seed vigor and uniformity of seedling emergence in soybean. *Seed Technol.* **2010**, *32*, 87–95.
33. Coolbear, P.; Francis, A.; Grierson, D. The effect of low temperature pre-sowing treatment on the germination performance and membrane integrity of artificially aged tomato seeds. *J. Exp. Bot.* **1984**, *35*, 1609–1617. [CrossRef]
34. Abdul-Baki, A.A.; Anderson, J.D. Vigor determination in soybean seed by multiple criteria. *Crop Sci.* **1973**, *13*, 630–633. [CrossRef]
35. Minitab (Version Express for PC). 2019. Available online: http://www.minitab.com/en-us/products/express (accessed on 28 August 2019).
36. American Seed Trade Association (ASTA). The Guide to Seed Treatment Stewardship. For Applicators. Available online: www.seed-treatment-guide.com (accessed on 5 December 2019).
37. Accinelli, C.; Abbas, H.K.; Little, N.S.; Kotowicz, J.K.; Mencarelli, M.; Shier, W.T. A liquid bioplastic formulation for film coating of agronomic seeds. *Crop Prot.* **2016**, *89*, 123–128. [CrossRef]
38. Foqué, D.; Devarrewaere, W.; Verboven, P.; Nuyttens, D. Physical and chemical characteristics of abraded seed coating particles. *Asp. Appl. Biol.* **2014**, *122*, 85–94.
39. Karlsons, A.; Osvalde, A.; Una Andersone-Ozola, U.; Ievinsh, G. Vermicompost from municipal sewage sludge affects growth and mineral nutrition of winter rye (*Secale cereale*) plants. *J. Plant Nutr.* **2015**, *39*, 765–780. [CrossRef]
40. Tognetti, C.; Laos, F.; Mazzarino, M.J.; Hernández, M.T. Composting vs. vermicomposting: A comparison of end product quality. *Compost Sci. Util.* **2005**, *13*, 6–13. [CrossRef]
41. Alwaneen, W.S. Effect of cow manure vermicompost on some growth parameters of alfalfa and *Vinca rosa* plants. *Asian J. Plant Sci.* **2016**, *15*, 81–85. [CrossRef]
42. Fleming, T.R.; Fleming, C.C.; Levy, C.C.B.; Carlos, R.; Franck, H.; Nolasco, J.B.; Liu, F.Q. Biostimulants enhance growth and drought tolerance in Arabidopsis Thaliana and exhibit chemical priming action. *Ann. Appl. Biol.* **2019**, *174*, 153–165. [CrossRef]
43. Rouphael, Y.; Cardarelli, M.; Bonini, P.; Colla, G. Synergistic action of a microbial-based biostimulant and a plant derived-protein hydrolysate enhances lettuce tolerance to alkalinity and salinity. *Front. Plant Sci.* **2017**, *8*, 131. [CrossRef]
44. Van Oosten, M.J.; Pepe, O.; De Pascale, S.; Silletti, S.; Maggio, A. The role of biostimulants and bioeffectors as alleviators of abiotic stress in crop plants. *Chem. Biol. Technol. Agric.* **2017**, *4*, 5. [CrossRef]

agronomy

MDPI

Article

Effects of Different Microbial Inocula on Tomato Tolerance to Water Deficit

Giuseppe Mannino [1,†], Luca Nerva [2,3,†], Takoua Gritli [4], Mara Novero [1], Valentina Fiorilli [1], Mnasri Bacem [4], Cinzia Margherita Bertea [1], Erica Lumini [3], Walter Chitarra [2,3] and Raffaella Balestrini [3,*]

1 Life Sciences and Systems Biology Department, University of Torino, 10125 Torino (TO), Italy; giuseppe.mannino@unito.it (G.M.); mara.novero@unito.it (M.N.); valentina.fiorilli@unito.it (V.F.); cinzia.bertea@unito.it (C.M.B.)

2 Council for Agricultural Research and Economics - Research Centre for Viticulture and Enology CREA-VE, 31015 Conegliano (TV), Italy; luca.nerva@crea.gov.it (L.N.); walter.chitarra@crea.gov.it (W.C.)

3 National Research Council, Institute for Sustainable Plant Protection, 10125 Torino (TO), Italy; erica.lumini@ipsp.cnr.it

4 Laboratory of Legumes, Centre of Biotechnology of Borj-Cedria, BP 901, Hammam-lif 2050, Tunisia; takouagritli1@gmail.com (T.G.); mnbacemm@yahoo.com (M.B.)

* Correspondence: raffaella.balestrini@ipsp.cnr.it; Tel.: +39-011-650-2927

† These authors equally contributed as first authors.

Received: 28 December 2019; Accepted: 22 January 2020; Published: 24 January 2020

Abstract: Several recent reports have highlighted some of the mechanisms involved in the enhanced tolerance to abiotic stresses induced by root-associated microorganisms, although additional efforts are still required to exploit and optimize these strategies. Particularly, arbuscular mycorrhizal fungi (AMF) play an important role as "bio-fertilizing microorganisms", establishing mutualistic symbioses with the roots of most crops. In this work, different microbial inocula (a single AMF species, a combination of three different AMF species, a combination of two plant growth-promoting bacteria (PGPB) strains and a more complex commercial inoculum) have been used to inoculate tomato plants (cv San Marzano nano), in order to verify their effects on the tolerance to a water deficit condition in pots, through the evaluation of biochemical stress markers and hormonal profiles (ABA and IAA). Results showed differences among tomato responses to water limitation depending on microbial inocula, confirming the importance to characterize the optimal plant/microorganism genotype combination(s) to maximize plant performance and tolerance. These findings open new perspectives for a better exploitation of these microorganisms.

Keywords: tomato; AM fungi; PGPB; water deficit; biostimulant

1. Introduction

Environmental stresses are becoming a serious threat and productivity is declining at an unprecedented level [1]. Water deficit, salinity, extreme temperatures, flooding, nutritional limitations, pest and pathogen attacks are key threats to plant growth and crop productivity and constitute major constraints to actual agriculture [2]. The extent of agricultural soil affected by water stress and exposed to a loss of fertility is predicted to progressively increase due to climate change [3]. Conventional agriculture's dependence on chemical fertilizers and pesticides has encouraged the thriving of industries producing these products that are not only hazardous for human consumption but also exert negative effects on the environment [1]. Biofertilizers, and biostimulants, could help to solve the problem of feeding an increasing global population at a time where agriculture faces several environmental stresses [1]. A number of recent reports have highlighted some of the mechanisms involved in the enhanced tolerance to abiotic stresses induced by root-associated microorganisms [4–6]; however, additional efforts are still required to exploit the useful aspects of the different root-associated microorganisms and to optimize these strategies, supporting their application to current agricultural practices [1]. Arbuscular mycorrhizal (AM) fungi play an important role as "bio-fertilizing microorganisms" as they establish mutualistic symbioses with the roots of most crops [7,8]. These symbiotic fungi colonize plant roots and enhance the uptake of water and nutrients by the host plants, while, they receive the carbon compounds. These fungi are considered essential elements for plant nutrition, mainly in low-nutrient conditions, as their hyphae can extend for many meters in the ground, helping the plants to acquire mineral soil nutrients. Since AM fungi play an instrumental role in protecting the plants against abiotic stresses such as nutrient deficiency, extreme temperatures [9] and drought [10–14], they can benefit their hosts in both wild and agricultural environments [15]. Consequently, AM fungi are thought to have a great impact in natural environments [16–18], as in managed conditions in agriculture, horticulture and forestry [8]. Although there is no hard symbiont specificity in AM interactions, the efficiency of these associations depends on the interacting partner genotypes and the environmental conditions [19,20]. Recent findings suggest a certain degree of functional specialization in AM interactions [8]. Some plant/fungus genotype combinations are more efficient than others in terms of nutrition or stress tolerance/resistance [13,14,20–25]. Despite the low host specificity of AM under controlled conditions, the presence of several symbionts might result in the most effective mutualistic combination [8]. Berruti and colleagues [26] demonstrated that the AM fungal communities originating from cells containing the arbuscules, which represent the functional structures in AM symbiosis, and the whole root samples of *Camellia* plants (grown in natural soil) differed remarkably. These results suggested that not all the AM fungal isolates present in soil could form a functional symbiosis. Symbiotic fungi, however, are only part of the soil and root-associated microbiota. Plant growth-promoting bacteria (PGPB), genera like *Bacillus*, *Azospirillum* or *Pseudomonas*, also exert beneficial effects on plant metabolism and primes tolerance mechanisms against biotic and abiotic stresses [6]. Interestingly, it was recently demonstrated that grapevine roots differently respond to a pure AM inoculum with respect to a mixed inoculum containing different microbial isolates/strains [27]. Here, we have used different microbial inocula on the commercial tomato cv San Marzano nano to verify the impact on the tolerance to a water deficit condition. One of the most important challenges in this research area is to dissect the actual mechanism of mode of action for different strains/isolates to evaluate their efficacy, alone or in combination, towards its use in sustainable agriculture.

2. Materials and Methods

2.1. Inoculation of Tomato Plants and Growth Conditions

Tomato (*Solanum lycopersicum* 'San Marzano nano') seeds were surface sterilized in sodium hypochlorite for 20 min, washed five times in sterile water, and germinated on wet filter paper. For this pot experiment, pots (10 cm × 10 cm × 12 cm) with a volume of 1 L containing substrate (sterilized quartz sand) were arranged on a growth chamber in a randomized block design including five treatments: (i) non-inoculated control (CTRL); (ii) AM fungi mono fungal inoculum (Myc_Rhizo); (iii) AM fungi multi fungal inoculum (MULTISTRAIN); (iv) PGPB (LC3.5 + 5.2) and (v) a mixed commercial inoculum containing both bacteria and fungi (Commercial MIX). Each treatment was replicated ten times (10 pots), each pot contained one seedling.

Tomatoes were inoculated with AM fungi at planting time by placing the inoculum in the planting hole and in contact with the roots, as follows: 10 plants were inoculated with 15 g/pot of a mono fungal inoculum (Myc_Rhizo) based on *Rhizoglomus irregulare* BEG140 and 10 plants with 15g/pot of multi fungal inoculum (MULTISTRAIN) constituted by *Claroideoglomus claroideum* BEG96, *Funneliformis caledonium* BEG97 and *F. geosporum* BEG199; both the pure AM fungi inocula were provided by Symbiom Ltd., (Lanškroun, Czech Republic). The commercial MIX Opera-Rizon (MsBiotech, Larino, Italy) was used to inoculate other 10 tomato plants. This formulate, as reported in the product label, consists of AM fungal species (*Glomus* spp. 0.001%) and rhizospheric bacteria (1 × 10^2 Colony Forming Unit CFU). In addition, two PGPB (LC3.5 and LC5.2) were selected based on their high levels production of auxin: 46 μg/mL and 24 μg/mL for LC3.5 and LC5.2 respectively (Gritli and Bacem unpublished results). Both bacterial strains were isolated from roots of *Lathyrus cicera* in the northern of Tunisia. They were prepared as following: pure colonies of PGPB strains were multiplied separately in Luria Bertani (LB) broth by incubating them in a shaker for 72 and 24 h respectively at 27 °C. The optical density was adjusted to 1 (at 660 nm for PGPR. One mL of the bacterial suspension (10^9 CFU/mL) of the two PGPRs (LC3.5 + LC5.2) was inoculated to 10 pots, where 15 g/pot of the carrier material (without AM fungi) was applied.

Ten plants were left as non-inoculated control plants: in these pots 15 g of carrier material (without AM fungi) was applied. Plants were grown in controlled conditions, with a temperature of 23 °C/21 °C day/night, 16/8-h light/dark photoperiod, and relative humidity 65%. From transplanting to the beginning of the water deficit experiment (after about 6–7 weeks), all the plants were watered twice per week with tap water and, once per week, with a modified Long-Ashton nutrient solution [28] containing 3.2 μM inorganic phosphate.

Out of 50 plants, 25 (five plants for each treatment) were used as controls (irrigated or non-stressed, NS) and maintained in a well-watered state (at container capacity). The remaining 25 plants (five plants for each treatment) were subjected to a water stress (WS) treatment. To this aim, about 6 weeks after fungal inoculation irrigation was withheld and the experiment was stopped when the first plants reached a stress level (measured by infrared gas analyzer, IRGA, ADC-LCPro+ system; Analytical Development Company Ltd, Hoddesdon, UK).

2.2. Miniprep Bacterial Strains DNA Isolation and 16S SSU rRNA Amplification

The two PGPB strains were subjected to molecular characterization by means of amplification with conventional PCR using DNAs isolated from bacterial strains LC3.5 and LC5.2 as a template. An almost complete small subunit (SSU) bacterial ribosomal RNA gene (16S) was amplified with bacterial universal primers 27F-1492R that amplified a fragment of about 1465 bp [29]. The PCR were carried out in a final volume of 25 μL containing 10 μL of Platinum Hot Start PCR Master Mix (2X), 0.5 μL of each primer (10 μM), template DNA (1 μL) and 13 μL of PCR-grade water. Bacterial PCR amplification was performed using a T3000 thermal cycler (Biometra, Göttingen, Germany) with the following profile: initial denaturation for 5 min at 95 °C; 35 cycles of denaturation (60 s at 94 °C), annealing (60 s at 58 °C) and extension (60 s at 72 °C) and a further 7 min at 72 °C. All the PCR products were checked using 1.5% (*w/v*) agarose gel stained with ethidium bromide (Merck KGaA, Darmstadt, Germany). The two PCR products replicates for each strain were pooled and purified using Wizard SV Gel and a PCR Clean-Up System kit (Promega, Madison, WI, USA). Purified PCR products were sequenced, using either the universal primer 27F or 1492R, by LMU sequencing services (Munich, Germany). The two sequences were deposited at NCBI (accession # MN879506 and MN879507 for LC3.5 and LC5.2, respectively).

2.3. Eco-Physiological Parameters

Measurements of transpiration rate (E), stomatal conductance (g_s) and net photosynthetic rate (A_N) were performed on adult, non-senescing leaves at the same physiological age (in the middle part of the plant, considering the third to fourth leaf from the shoot apex). Intrinsic water use efficiency (iWUE) was calculated as the ratio between A_N and g_s. Measurements were taken with an IRGA instrument. During measurements, light intensity in the leaf chamber was set at 1200 μ mol photons m^{-2} s^{-1}, temperature was 25 °C, and the concentration of CO_2 was maintained between 450 and 470 ppm. Measurements were taken between 10:00. and 13:00. The chlorophyll content index (CCI) was determined at the end of the experiment (about 9 DAT) using the portable chlorophyll meter SPAD 502 (CCM-200; Opti-Sciences, Hudson, NH, USA).

2.4. Assessment of Symbiosis Development

At the end of the experiments, plants were harvested, and plant height and fresh weight (not shown) were recorded. A part of the root apparatus of at least three plants (up to five depending on the treatment) was stained with 0.1% Cotton Blue in lactic acid. For each plant, sixty randomly chosen 1-cm-long root segments were mounted on slides and fungal colonization was quantified according with the Trouvelot system [30] using MYCOCALC software, while the remaining root systems were stored at −80 °C until further analyses.

2.5. Preparation of Extracts and Biochemical Parameter Evaluation

After the measurement of plant morphological parameters, leaf and root samples were dried by lyophilization, then, grounded and homogenized using a mortar and pestle. About 30 mg of the obtained dried powders were extracted with 90% (*v/v*) methanol using a 1:50 (*w/v*) ratio. Samples were mixed by vortexing for 5 min, and sonicated for 15 min at 8 °C. Following a centrifugation step (10 min at 8000*g*, 4 °C), the supernatants were filtered using a filter tips, and directly used for chemical determinations.

2.5.1. Determination of Total Chlorophyll Content (TCC)

The leaf extracts were employed for the determination of the total chlorophyll content (TCC), according to Lichtenthaler and Buschmann [31]. Briefly, 1 mL of appropriate diluted sample was subjected to spectrophotometric measurements at 665.2 nm and 652.4 nm. TCC, expressed as μg per g of dry weight (d.wt), was calculated for each sample using the following equation:

$$TCC = \frac{[(16.82\ x\ Abs_{665.2} - 9.28\ x\ Abs_{652.4}) + (36.92\ x\ Abs_{652.4} - 16.54\ x\ Abs_{665.2})]\ x\ V_{extr}\ x\ DF}{WH}. \quad (1)$$

V_{extr} = volume, expressed as mL, used for the extraction process; *DF* = dilution factor and *WH* = weight of each sample expressed in grams.

2.5.2. Determination of Proline Concentration (TpC)

The proline concentration (TpC) was determined according to Carillo and Gibon [32]. Briefly, 500 μL of undiluted samples were incubated with 1 mL of the reaction mix containing 1% (*w/v*) ninhydrin solubilized in 60% (*v/v*) acetic acid and 20% (*v/v*) ethanol. The mixture was incubated at 95 °C for 20 min in the dark, and then centrifuged at 10.000 rpm for 1 min at room temperature in a table microfuge. The absorbance was then measured at 520 nm. Quantification was performed using an external calibration curve prepared using a pure standard of proline, whose concentration ranged from 0.01 to 1.00 mmol.

2.5.3. Determination of Total Polyphenol Content (TPC)

The total polyphenol content (TPC) was evaluated both in leaf and root extracts following the method of Ainsworth and Gillespie [33]. Briefly, each sample was appropriately diluted in 90% (*v/v*) methanol and then 930 μL were incubated with 30 μL of Folin–Ciocalteu reagent and 40 μL of 20% (*w/v*) sodium carbonate (Na_2CO_3). The samples were then incubated for 1 min at 80 °C and for 20 min at room temperature in the dark. Then the absorbance was monitored at 725 nm. An external calibration curve using gallic acid (GA), ranged between 50 and 400 mmol, was employed to quantify TPC in the samples. The results were expressed as μmol of gallic acid equivalent (GAE) per g of dry weight (d.wt).

2.6. Determination of Abscisic Acid (ABA) and Indole-3-Acetic Acid (IAA) Content

About 500 mg of homogenized leaf and root samples were freeze-dried and transferred with 1 mL of methanol:water (8:2 *v/v*) acidified with 0.1% (*v/v*) of acetic acid in an ultrasonic bath for 1 h. Samples were centrifuged for 10 min at 4 °C and 15,000 rpm, and the supernatant was analyzed by high-performance liquid chromatography (HPLC, Agilent, Waldbronn, Germany). Original standards of abscisic acid (ABA, purity ≥ 98.5%, Merck KGaA, Darmstadt, Germany) and indole acetic acid (IAA, purity ≥ 99%, Merck KGaA, Darmstadt, Germany) were used for the identification by comparing retention time and UV spectra. The quantification was made by external calibration method. The HPLC apparatus was an Agilent 1220 Infinity LC system (Agilent R, Waldbronn, Germany) model G4290B equipped with gradient pump, auto-sampler and column oven set at 30 °C. A 170 Diode Array Detector (Gilson, Middleton, WI, USA) set at 265 nm was used as detector. A Nucleodur C18 analytical column (250 mm × 4.6 mm i.d., 5 μm, Macherey Nagel) was used. The mobile phases consisted in water acidified with formic acid 0.1% (A) and acetonitrile (B), at a flow rate of 0.500 mL min^{-1} in gradient mode, 0–6 min: 30% of B, 6–16 min: from 30% to 100% B and 16–21 min: 100% B; 20 μL was injected for each sample.

2.7. Statistical Analysis

All measurements are the average of five different biological replicates. Each biological replicate was analyzed three times in each experiment. The content of chlorophylls (TCC), proline (TpC), polyphenols (TPC), ABA and IAA were reported both as relative content (Figures 1–6) and as absolute content (Supplementary Tables S1–S4). The relative content was calculated comparing the content observed in inoculated plants (treated with MULTISTRAIN, Myc_Rhizo, LC3.5 + 5.2 or the Commercial MIX) with the content of unstressed and/or untreated plants (control plants). In both cases, data are expressed as mean values ± standard deviation (SD). Significant differences were evaluated by performing one-way ANOVA followed by Tukey's HSD test ($p \leq 0.05$) or *t*-test ($p \leq 0.05$) using SPSS ver. 24 software.

3. Results and Discussion

The effect of several microbial inocula on tomato tolerance to a water deficit condition was verified. The beneficial effects of root-associated microbes (i.e., AM fungi and PGPB) on plant growth and performance under water limitation have already been reported for several plant species [5,34], including the tomato genotype considered in the present research [13,14,35]. These previous works already showed a different plant response to a water deficit condition depending on the AM fungal species associated to plant roots. An untargeted metabolomic analysis in tomato roots colonized by three AM fungi of different genera showed that some responses to drought and salt stress were common to all AM fungi tested, while others were specifically related to single isolates [25]. Here, several microbial inocula (a single AM fungal species, a combination of three different AM fungal species, a combination of two PGPB strains and a more complex commercial inoculum) were tested for the effect on tomato tolerance to water limitation. Both the bacterial strains used in this work (LC3.5 and LC5.2) showed sequence identity with *Bacillus* spp. In detail, a sequence identity with *Bacillus subtilis* (first hit: MN704441.1, query cover 100%, *e*-value 0.0, identity 99.78) and *B. megaterium* (first hit: MK791705.1, query cover 100%, *e*-value 0.0, identity 98.64%) was found for LC3.5 and LC5.2, respectively. The AM fungal colonization using several formulates was also evaluated, showing some relevant differences among the two AM fungal inocula (Myc_Rhizo and MULTISTRAIN), while the presence of AM fungal structures were not observed in the roots inoculated with the mixed inoculum (Commercial MIX; Table 1). A very low colonization by AM fungi was already observed using a commercial mixed inoculum on grapevine rootstocks [27]. In detail, a different grapevine root transcriptome profile was observed after inoculation with a pure AM inoculum (*Funneliformis mosseae*) and the mixed one, although this last elicited an important transcriptional regulation probably due to the predominantly presence of PGPB.

Table 1. AM fungal colonization using three different inocula. F%, Frequency of mycorrhization in root system; M%, Intensity of mycorrhizal colonization in the root system; a%, arbuscule abundance in mycorrhizal parts of root fragments; A%, arbuscule abundance in the whole root system. Values are expressed as a mean ± SD ($n = 3$). Data were subjected to statistical analysis using SYSTAT 10 software, applying the nonparametric Kruskal-Wallis test adopting a probability level of $p < 0.05$. Data followed by different superscript letters indicate significant statistical differences among samples.

	F%	M%	a%	A%
Myc_Rhizo NS	33.33 ± 10.27 ^{a}	2.69 ± 1.16 ^{a}	68.09 ± 10.84 ^{a}	1.75 ± 0.46 ^{a}
Myc_Rhizo WS	23.67 ± 12.21 ^{a}	2.59 ± 2.38 ^{a,b}	68.53 ± 26.69 ^{a}	1.4 ± 0.88 ^{a,b}
MULTISTRAIN NS	28.02 ± 8.02 ^{a}	4.23 ± 2.62 ^{b}	41.13 ± 27.38 ^{a}	1.39 ± 1.33 ^{b}
MULTISTRAIN WS	35.96 ± 5.35 ^{a}	4.28 ± 2.74 ^{b}	52.3 ± 29.09 ^{a}	2.24 ± 2.11 ^{b}
Commercial MIX NS	0 ± 0 ^{b}	0 ± 0 ^{c}	0 ± 0 ^{b}	0 ± 0 ^{c}
Commercial MIX WS	0 ± 0 ^{b}	0 ± 0 ^{c}	0 ± 0 ^{b}	0 ± 0 ^{c}

3.1. Impact of Treatments and Water Stress on Eco-Physiological Parameters

Eco-physiological parameters were recorded at the end of the experiment, considering both gas exchanges and CCI (Table 2 and Table S1). In general, assimilation rates (A) decreased under water deficit condition with the lowest values in the treatment with bacteria (LC3.5 + 5.2 and commercial mix), while no effect was observed on *e*-values. Brilli et al. [34] already found that the tomato inoculation with a PGPR (*Pseudomonas chlororaphis*) did not affect the physiological parameters. In addition, the highest "A" values under water stress for the AM-inoculated plants are in agreement with the data from Chitarra et al. [13] and Volpe et al. [14] on the same tomato genotype, although a difference between two AM fungal species was observed. As expected, stomatal conductance (g_s) decreased under water limitation mainly in inoculated plants (Figure 2). This is not surprising since a different timing in reaching a stress level has been already reported from AM-colonized and non-colonized tomato plants [13]. Similarly, although not statistically significantly different, a decrease in g_s was observed in the presence of bacteria inoculation [34]. On the contrary, an increasing trend in iWUE was observed in WS inoculated plants, in agreement with previous works [13,14,34]. Regarding CCI, a general decrease was observed in WS plants. Interestingly, a different impact of the several inocula was observed (e.g., MULTISTRAIN vs. Myc_Rhizo) both in NS and WS plants, confirming species-specificity in affecting physiological traits. Concerning plant height, no significant results have been obtained among treatments and stress conditions. Taken together our results highlighted that symbiotic fungi (i.e., AM fungi) differently affect plant traits important for the tolerance to stressful conditions with respect to root-associated bacteria.

Table 2. Eco-physiological and biometric parameters. Photosynthetic rate (A), transpiration rate (E), stomatal conductance (gs), intrinsic water use efficiency (iWUE), stem height and SPAD in non-stressed (NS) and water stress (WS) leaves no treated (control) or treated with the different inocula (MULTISTRAIN, Myc_Rhizo, LC3.5 + 5.2 or the commercial mix). Values are represented as mean ± SD. For each column, within the same series (NS or WS) different lowercase letters indicate significant differences at $p \leq 0.05$ as measured by an ANOVA-one way followed by a Tukey's multiple range test. Letter "a" denotes the highest value. Among the WS series, the symbols "*" ($p \leq 0.05$), "**" ($p \leq 0.005$) and "***" ($p \leq 0.001$) indicate significant differences between NS and WS leaves (Student's *t*-test).

		A	E	gs	iWUE (A/gs)	Stem Height	CCI
		µmol CO_2 m^{-2} s^{-1}	mmol m^{-2} s^{-1}	mol m^{-2} s^{-1}	µmol CO_2 mol^{-1} H_2O	cm	
NS	**Control**	2.73 ± 0.27 ^{a}	0.86 ± 0.16 ^{a}	0.14 ± 0.04 ^{a}	22.49 ± 3.83 ^{a}	23.56 ± 2.03 ^{a}	31.83 ± 2.29 ^{b}
	Multistrain	3.17 ± 0.24 ^{a}	0.74 ± 0.19 ^{a}	0.12 ± 0.02 ^{a}	27.04 ± 6.99 ^{a}	16.92 ± 1.99 ^{a}	40.71 ± 0.88 ^{a}
	Myc_Rhizo	2.79 ± 0.15 ^{a}	0.65 ± 0.06 ^{a}	0.11 ± 0.01 ^{a}	27.91 ± 1.55 ^{a}	16.92 ± 4.19 ^{a}	31.47 ± 1.59 ^{b}
	LC3.5 + 5.2	3.42 ± 0.38 ^{a}	1.32 ± 0.67 ^{a}	0.14 ± 0.05 ^{a}	26.56 ± 11.53 ^{a}	21.53 ± 2.23 ^{a}	39.71 ± 2.81 ^{a}
	Commercial MIX	3.43 ± 1.24 ^{a}	0.76 ± 0.24 ^{a}	0.11 ± 0.04 ^{a}	32.03 ± 11.34 ^{a}	24.03 ± 3.43 ^{a}	39.91 ± 3.41 ^{a}
WS	**Control**	1.28 ± 0.11 ^{ab,***}	0.83 ± 0.25 ^{a}	0.13 ± 0.05 ^{a}	11.08 ± 5.25 ^{a,*}	18.72 ± 2.39 ^{a,***}	23.85 ± 2.95 ^{b,***}
	Multistrain	1.96 ± 0.59 ^{a,***}	0.66 ± 0.32 ^{a}	0.09 ± 0.06 ^{a}	24.09 ± 11.82 ^{a}	18.02 ± 2.27 ^{a}	36.22 ± 2.28 ^{a,**}
	Myc_Rhizo	1.36 ± 0.2 ^{a,b,***}	0.62 ± 0.32 ^{a}	0.04 ± 0.01 ^{a,***}	21.36 ± 16.67 ^{a}	17.56 ± 1.41 ^{a}	26.57 ± 2.8 ^{b,**}
	LC3.5 + 5.2	0.49 ± 0.35 ^{b,***}	1.15 ± 0.51 ^{a}	0.05 ± 0.02 ^{a,**}	13.62 ± 10.23 ^{a}	17.46 ± 2.24 ^{a,*}	33.34 ± 2.48 ^{a,***}
	Commercial MIX	0.53 ± 0.25 ^{b,**}	0.45 ± 0.12 ^{a}	0.04 ± 0.01 ^{a,*}	22.33 ± 4.24 ^{a}	26.93 ± 8.41 ^{a}	38.39 ± 0.49 ^{a}

3.2. Effect of Water Deficit on Total Chlorophyll Content (TCC), Total Proline Content (TpC) and Total Polyphenol Content (TPC)

Water stress implicates morphological, biochemical and molecular changes [36], and may affect plant growth during different developmental stages [37]. As a first point, we evaluated how a water stress (WS) condition could change some biochemical parameters of tomato plants grown in the absence of specific treatments. The biochemical profile of stressed plants was compared to that of unstressed ones grown in well-watered (WW) conditions (Figure 1; Tables S2 and S3). In our experimental conditions, the exposure of tomato plants to WS negatively influenced the TCC and TPC, while the TpC was positively affected. The major effect of drought in plants is correlated with the decrease of photosynthetic processes, leading not only to the reduction of leaf expansion, but also to fruit production [38]. The main reason for a decrease in photosynthesis is due to changes in photosynthetic pigment levels that are part of the photosynthetic apparatus [39]. The decrease in TCC (0.67 ± 0.08), observed here, was already reported in tomato plants subjected to WS [40,41]. Other abiotic stresses such as heat [42] and salt stress [43,44] also affected the amount of these molecules.

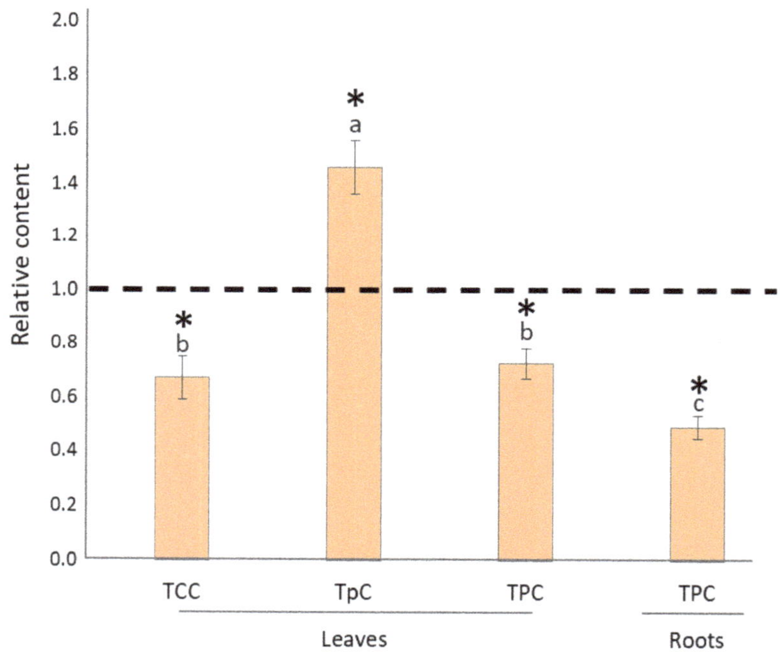

Figure 1. Effects of water-stress (WS) on the tomato total content of chlorophylls (TCC), proline (TpC) and polyphenols (TPC). Data for each quantification are expressed as relative content, comparing the measurements obtained for water-stressed plants with those of unstressed plants. The dotted line indicates the basal expression of NS. Absolute quantification of each parameter is also reported in Supplementary Table S2. Bars with different lowercase letters indicate significantly different values at $p \leq 0.05$ as measured by a one way-ANOVA followed by a Tukey's HSD post hoc test (see Supplementary Table S4 for additional information). The symbol "*" indicates significant differences ($p \leq 0.05$) between untreated-water stressed and untreated-non stressed plants, as measured by a t-test.

On the other hand, the role of polyphenols, which represent the soluble antioxidant defenses of the plant, in stressed samples has been widely discussed and contrasting. Although several studies already reported the increase of polyphenols to contrast the oxidative damage generated after the exposure to different abiotic stresses [45], in other cases a substantial decrease of these molecules was observed [46,47]. This response is probably due to the loss of the plant capability to synthesize *ex novo* the soluble antioxidant defenses. In our experiment, the TPC, evaluated on both leaves and roots, decreased after the exposure to WS. Moreover, a stronger effect was observed in leaves compared to roots (0.72 ± 0.06 and 0.49 ± 0.04, respectively).

Finally, in order to respond to unbalanced water repartition, plants generally accumulate compatible solutes with the aim to raise osmotic pressure and thereby to maintain both turgor and driving gradient for water uptake [32,48]. Among these solutes, proline plays a key role in these processes. The accumulation of proline in leaves can be considered as a strong indicator of abiotic stresses such as drought, salt and heath stresses. In accordance with our results (1.45 ± 0.09), an increase of proline in leaves of stressed plants was previously reported, not only in tomato but also in other plants [32,48–51].

3.3. MULTISTRAIN, Myc_Rhizo, LC3.5 + 5.2 and the Commercial MIX are Able to Recover the Biochemical Parameters in Water Stressed Plants

In order to check if the treatments with the different microbial inocula were able to restore the correct plant homeostasis, tomato plants were inoculated with four different inocula. To allow the successfully establishing of the relationship between roots and employed microorganisms, plants were grown in well-watered conditions for a period of about six weeks before to start with water limitation. Figure 2 shows the change in TCC, TpC and TPC values of treated-WS plants compared to untreated-WS plants (dotted-line). All the treatments promoted a recovery of the TCC and TPC amount in WS plants, suggesting beneficial properties of the formulations, and a decrease of water stress in treated plants. However, significant differences ($p \leq 0.05$) among the four treatments were found. In particular, Myc_Rhizo (Figure 2B) was the most effective in increasing TCC in leaves (1.86 ± 0.06), followed by the commercial mix (1.37 ± 0.05; Figure 2D). The highest recovery in term of TPC in the leaves was recorded in WS plants treated with Myc_Rhizo (3.04 ± 0.05; Figure 2B), while the best recovery of TPC in roots was observed with MULTISTRAIN (1.62 ± 0.08; Figure 2A) followed by Myc_Rhizo (1.44 ± 0.04; Figure 2B).

Finally, TpC was also affected by the different treatments, with Myc_Rhizo and MULTISTRAIN (Figure 2A,B) that showed again the highest decrease (0.42 ± 0.05 and 0.65 ± 0.03, respectively). Moreover, a very strong and negative correlation was found between TCC/TpC ($\rho = -0.89$) and leaf-TPC/TpC ($\rho = -0.96$), as revealed by Pearson analysis (Table S4). On the other hand, no correlation was found between root-TPC and TpC ($\rho = -0.15$).

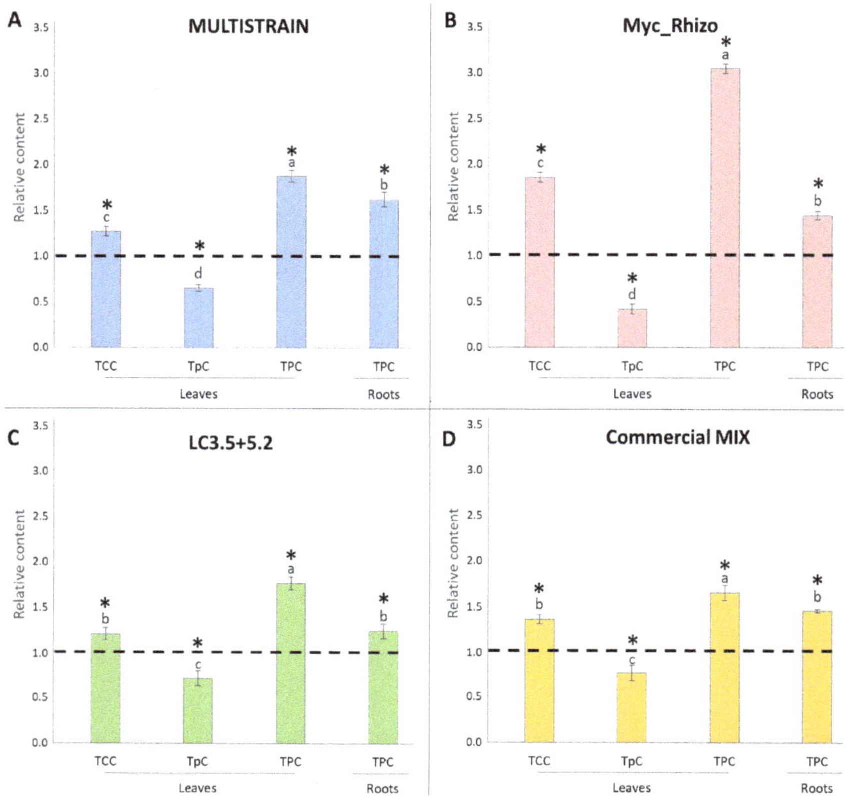

Figure 2. Effect of the treatment with MULTISTRAIN (**A**), Myc_Rhizo (**B**), LC3.5 + 5.2 (**C**) and the commercial mix (**D**) on the total content of chlorophylls (TCC), proline (TpC) and polyphenols (TPC) evaluated on water-stressed plants. Data for each quantification are expressed as relative content, comparing the measures obtained by treated-and untreated-water stress plants. The dotted line indicates the basal level of untreated water stressed plants. Absolute quantification of each parameter is also reported in Supplementary Table S2. Bars with different lowercase letters indicate significantly different values at $p \leq 0.05$ as measured by a one way-ANOVA followed by a Tukey's HSD post hoc test (see Supplementary Table S4 for additional information). The symbol "*" indicates significant differences ($p \leq 0.05$) between treated-WS and untreated-WS plants as measured by a *t*-test.

3.4. MULTISTRAIN, Myc_Rhizo, LC3.5 + 5.2 and the Commercial MIX Affect Total Chlorophyll, Polyphenol and Proline Content in Absence of Stress

In order to evaluate the performance of different formulations without a water stress condition, NS plants treated with MULTISTRAIN, Myc_Rhizo, LC3.5 + 5.2 or of the commercial mix were analyzed. Figure 3 shows the relative content of treated-non stressed plants in comparison to untreated-non stressed plants (dotted-line). As a general trend, the treatments with MULTISTRAIN (Figure 3A) and LC3.5 + 5.2 (Figure 3C) did not change the content of the analyzed biochemical parameters, with the exception of TpC in non-stressed plants treated with LC3.5 + 5.2. A more evident effect was instead observed in non-stressed plants inoculated with Myc_Rhizo or with the commercial mix (Figure 3B,D). In these cases, TpC and TPC in the leaves statistically ($p < 0.05$) increased with respect to untreated non-stressed plants (1.20 ± 0.18 and 1.20 ± 0.09 for Myc_Rhizo and commercial mix, respectively). On the other hand, TPC decreased in roots (Figure 3B,D). The slightly significant changes in the biochemical parameters could be associated to the functional traits of the considered microorganisms that led to a priming status, also in the absence of stress, as previously reported ([6] and reference therein). Physiological, transcriptional and metabolic changes stimulated by the colonization of soil root-associated microorganisms can prime plants for enhanced defense ahead of abiotic and biotic stress occurrence [52]. Evidence of a possible priming of the plant defensive system induced by AM-inoculation was recently suggested in *Arundo donax* [53], where a significant increase in proline accumulation in AM-colonized roots was reported. Brilli et al. [34] suggested that *Pseudomonas chlororaphis* acted as a 'priming stimulus' triggering in inoculated tomatoes enhanced tolerance to water stress. Interestingly, a simultaneous increase in the activity of superoxide dismutase (SOD) and catalase (CAT), and in proline accumulation was observed in tomato leaves from inoculated plants, independently by the stress level (well-watered or water stressed plants). However, the contribution of the root-associated microorganisms in plant adaptation to environmental stress factors needs to be still extensively evaluated, particularly in natural conditions, where a complex soil microbiota is present, and upon multiple stresses.

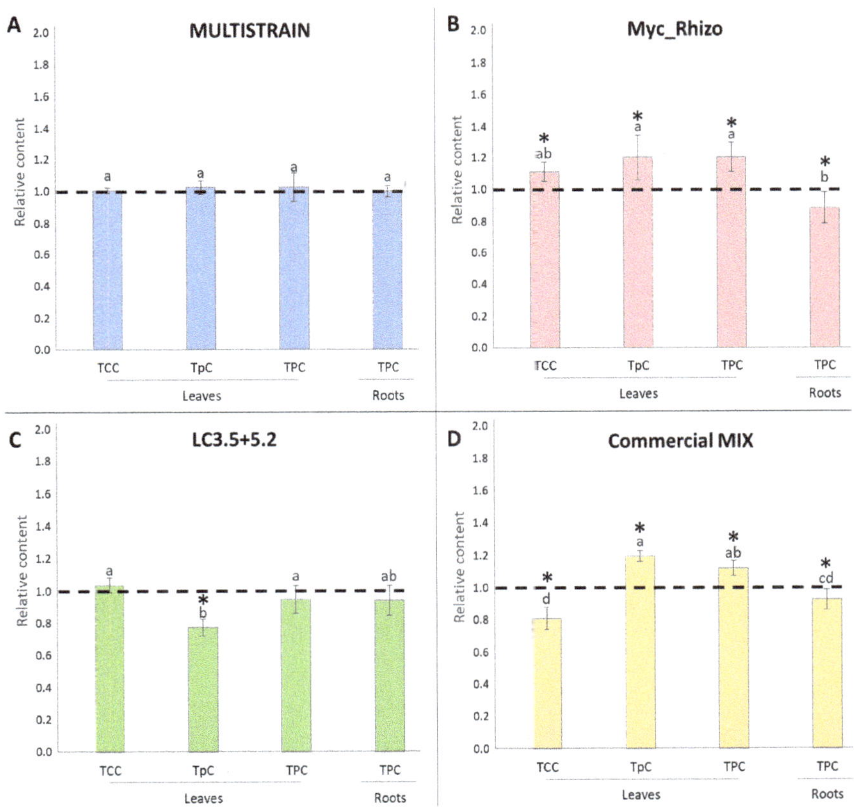

Figure 3. Effect of the treatment with MULTISTRAIN (**A**), Myc_Rhizo (**B**), LC3.5 + 5.2 (**C**) and the commercial mix (**D**) on the total content of chlorophylls (TCC), proline (TpC) and polyphenols (TPC) evaluated on unstressed plants. Data for each quantification are expressed as relative content, comparing the measurements obtained by treated- and un-treated-non stressed plants. The dotted line indicates the basal level of untreated-non stressed plants. Absolute quantification of each parameter is also reported in Supplementary Table S2. Bars with different lowercase letters indicate significantly different values at $p \leq 0.05$ as measured by a one way-ANOVA followed by a Tukey's HSD post hoc test (see Supplementary Table S4 for additional information). The symbol "*" indicates significant differences ($p \leq 0.05$) between treated-NS and untreated-NS plants, as measured by a *t*-test.

3.5. Effects of the Different Formulations Applied on ABA and IAA Content

Regardless of the water regime conditions the pattern of ABA and IAA content were strongly affected by the applied consortia. The plant hormone ABA is a chemical signal produced in leaves and roots, largely studied because of its pivotal roles in stomata movement and molecular-mediated responses under water stress [54]. In general, ABA content was less affected in roots of treated plants with respect to the controls in both WS and NS conditions (Figure 4).

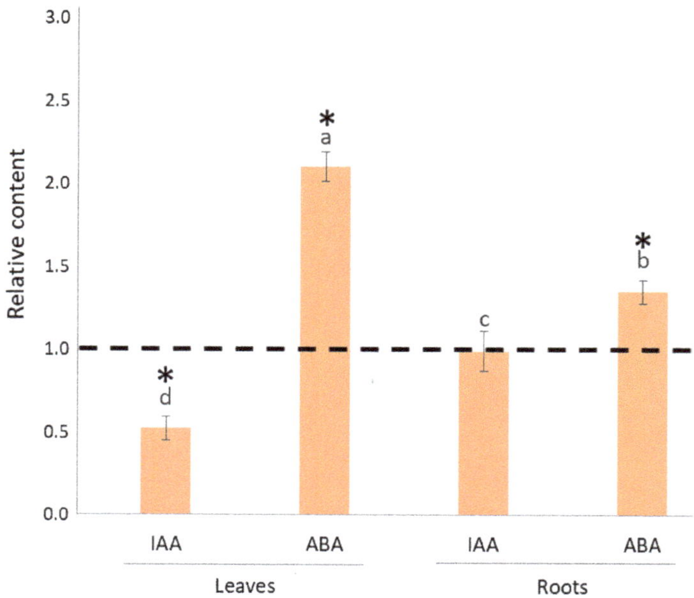

Figure 4. Effects of water-stress (WS) on the tomato content of indole acetic acid (IAA) and abscisic acid (ABA) evaluated both on leaves and roots. Data for each quantification are expressed as relative content, comparing the measurements obtained for water-stressed plants with those of unstressed plants. The dotted line indicates the basal expression of non-stressed plants. Absolute quantification of each parameter is also reported in Supplementary Table S2. Bars with different lowercase letters indicate significantly different values at $p \leq 0.05$ as measured by a one way-ANOVA followed by a Tukey's HSD post hoc test (see Supplementary Table S4 for additional information). The symbol "*" indicates significant differences ($p \leq 0.05$) between untreated-water stress and untreated-non stressed plants, as measured by a *t*-test.

Under the NS condition, the ABA content in roots was significantly higher ($p \leq 0.05$) in the treated plants when compared to their controls, suggesting an ABA-primed status induced by the microorganisms added in the substrates (Figure 5). In NS leaves, MULTISTRAIN, Myc_Rhizo and LC3.5 + 5.2 showed significantly higher levels of ABA with respect to the controls. As expected, under WS conditions, ABA content was generally higher with respect to NS and only in roots of Myc_Rhizo and leaves of the commercial mix was significantly higher with respect to their controls ($p \leq 0.05$), pointing out a microbial-mediated role in WS sensing and in turn ABA synthesis on inoculated plants (Figure 6) [4,34].

In almost all conditions tested, under NS conditions, IAA content showed an opposite trend for ABA) confirming their negative correlation as previously reported by Saeedipour and Moradi [55], with the exception of LC3.5 + 5.2. Interestingly, under WS conditions, all the treatments showed significantly higher IAA levels in leaves whilst lower levels were observed in roots with respect to their controls ($p \leq 0.05$; Figure 6).

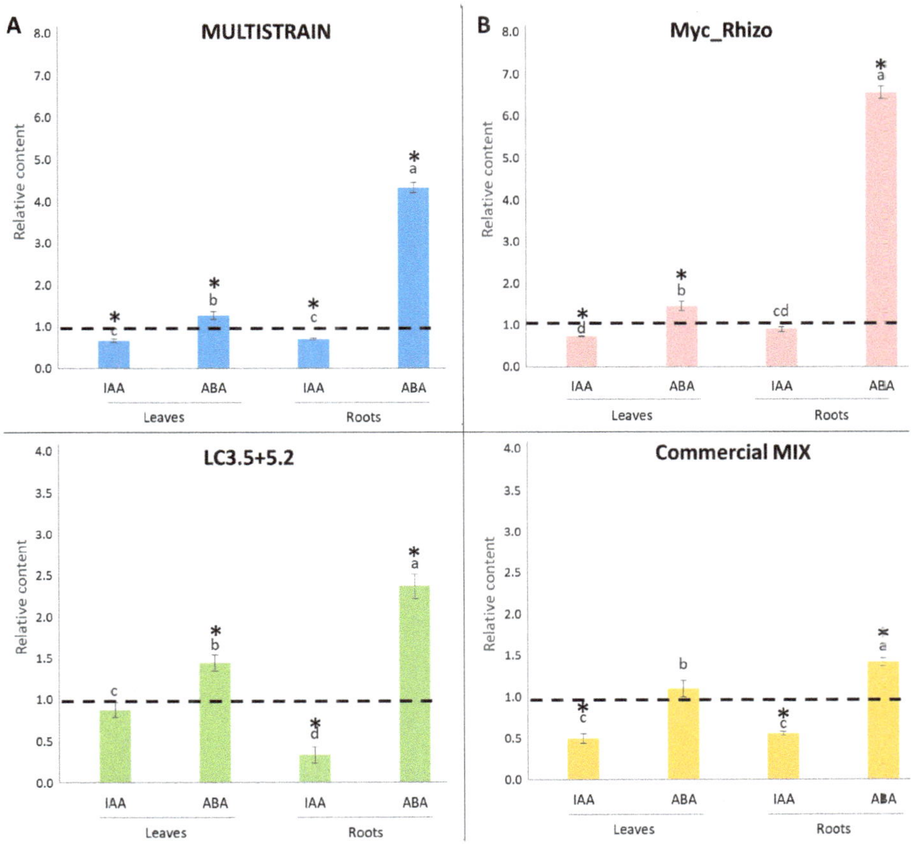

Figure 5. Effect of the treatment with MULTISTRAIN (**A**), Myc_Rhizo (**B**), LC3.5 + 5.2 (**C**) and the commercial mix (**D**) on the content of indole acetic acid (IAA) and abscisic acid (ABA) evaluated both on leaves and roots of unstressed plants. Data for each quantification are expressed as relative content, comparing the measurements obtained by treated- and untreated-non stressed plants. The dotted line indicates the basal expression of untreated-non stressed plants. Absolute quantification of each parameter is also reported in Supplementary Table S2. Bars with different lowercase letters indicate significantly different values at $p \leq 0.05$ as measured by a one way-ANOVA followed by a Tukey's HSD post hoc test (see Supplementary Table S4 for additional information). The symbol "*" indicates significant differences ($p \leq 0.05$) between treated- and untreated-non stressed plants, as measured by a *t*-test.

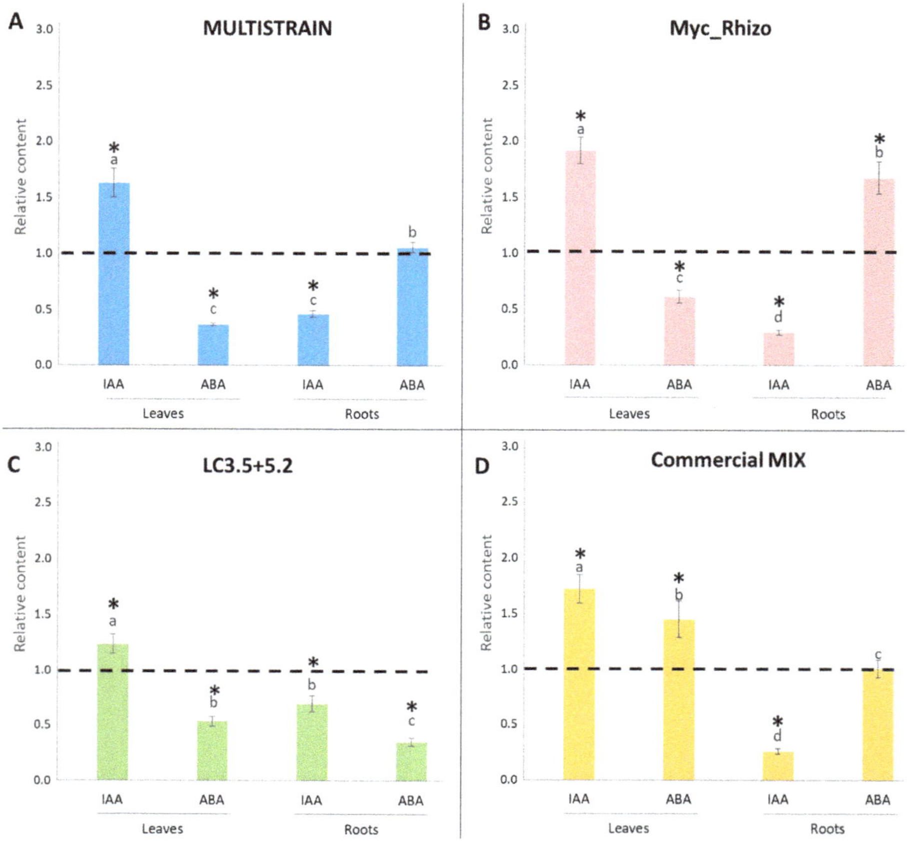

Figure 6. Effect of the treatment with MULTISTRAIN (**A**), Myc_Rhizo (**B**), LC3.5 + 5.2 (**C**) and the commercial mix (**D**) on the content of indole acetic acid (IAA) and abscisic acid (ABA) evaluated both on leaves and roots of water-stressed plants. Data for each quantification are expressed as relative content, comparing the measurements obtained by treated- and untreated-water stressed plants. The dotted line indicates the basal expression of untreated water stressed plants. Absolute quantification of each parameter is also reported in Supplementary Table S2. Bars with different lowercase letters indicate significantly different values at $p \leq 0.05$ as measured by a one way-ANOVA followed by a Tukey's HSD post hoc test (see Supplementary Table S4 for additional information). The symbol "*" indicates significant differences ($p \leq 0.05$) between treated- and untreated-non stressed plants, as measured by a *t*-test.

4. Conclusions

In conclusion, our results confirmed the fact that several microbial inocula have different impacts on the tomato's response under a water stress condition. Although aspects related to the persistence of the inocula at the end of the experiment were not considered, our results showed that the biochemical response of tomato to a stressful factor changed depending on the applied consortia of root-associated microorganisms. The latter were also able to induce a different effect on physiological traits. Moreover, the importance of symbiotic fungi, i.e., the AM fungi, in inducing a primed status and, in turn, a tolerance to water deficit was highlighted, reinforcing the consolidated evidence of the positive role played by these "biostimulants". However, many factors can affect the success of inoculation and persistence of inoculated microorganisms in soil, including compatibility with the target environment, the degree of spatial competition with other soil organisms in the target niche and the timing of inoculation. For this reason, further efforts should be done, mainly for bacteria species, to implement methods for monitoring and characterizing the degree of root/rhizosphere colonization of the microbial inoculants.

Supplementary Materials: The following are available online at http://www.mdpi.com/2073-4395/10/2/170/s1. Supplementary Table S1. Statistical analysis of absolute values of photosynthetic rate (A), transpiration rate (E), stomatal conductance (g_s), intrinsic water use efficiency (iWUE), stem eight and CCI. Supplementary Table S2. Absolute determination of Total Content of Chlorophylls (TCC), Proline (TpC), Polyphenols (TPC), Indole Acetic Acid (IAA) and Abscisic Acid (ABA). Supplementary Table S3. Statistical analysis of absolute determination of Total Content of Chlorophylls (TCC), Proline (TpC), Polyphenols (TPC), Indole Acetic Acid (IAA) and Abscisic Acid (ABA). Table S4. Statistical analysis of the relative content of Total Content of Chlorophylls (TCC), Proline (TpC), Polyphenols (TPC), Indole Acetic Acid (IAA) and Abscisic Acid (ABA).

Author Contributions: All authors have read and agree to the published version of the manuscript. Conceptualization, R.B., E.L. and W.C.; investigation, G.M., L.N., T.G., C.M.B., E.L., M.N., W.C. and R.B.; formal analysis, G.M., L.N., W.C.; resources, T.G., M.B.; writing—original draft preparation, G.M., R.B., E.L., W.C.; writing—review and editing, G.M., L.N., C.M.B., M.N., V.F., W.C. and R.B.; supervision, R.B. All authors have read and agreed to the published version of the manuscript.

Funding: This research received no external funding.

Acknowledgments: The authors thank Maria Teresa Della Beffa for the help in plant preparation and maintenance, and Miroslav Vosatka and Aleš Látr for providing the two AM fungal inocula produced by Symbiom.

Conflicts of Interest: The authors declare no conflict of interest.

References

1. Bhardwaj, D.; Ansari, M.W.; Sahoo, R.K.K.; Tuteja, N. Biofertilizers function as key player in sustainable agriculture by improving soil fertility, plant tolerance and crop productivity. *Microbiol. Cell Fact.* **2014**, *13*, 66. [CrossRef]
2. Abewoy, D. Review on impacts of climate change on vegetable production and its management practices. *Adv. Crop Sci. Tech.* **2017**, *6*, 330. [CrossRef]
3. Pachauri, R.K.; Allen, M.R.; Barros, V.R.; Broome, J.; Cramer, W.; Christ, R; Church, J.A.; Clarke, L.; Dahe, Q.; Dasgupta, P.; et al. *Climate Change 2014: Synthesis Report. Contribution of Working Groups I, II and III to the Fifth Assessment Report of the Intergovernmental Panel on Climate Change*; IPCC: Geneva, Switzerland, 2014; 151p.
4. Balestrini, R.R.; Chitarra, W.W.; Antoniou, C.C.; Ruocco, M.M.; Fotopoulos, V. Improvement of plant performance under water deficit with the employment of biological and chemical priming agents. *J. Agric. Sci.* **2018**, *156*, 680–688. [CrossRef]
5. Balestrini, R.R.; Chitarra, W.W.; Fotopoulos, V.V.; Ruocco, M.; Lukac, M. *Potential Role of Beneficial Soil Microorganisms in Plant Tolerance to Abiotic Stress*; Lukac, M., Gamboni, M., Grenni, P., Eds.; Soil biological communities and ecosystem resilience; Sustainability in Plant and Crop Protection Series; Springer Cham: Bern, Switzerland, 2017; pp. 191–207.
6. Alagna, F.F.; Balestrini, R.R.; Chitarra, W.W.; Marsico, A.D.D.; Nerva, L. *Getting ready with the priming: Innovative weapons against biotic and abiotic crop enemies in a global changing scenario*; Hossain, M.A., Liu, F., Burritt, D.J., Fujita, M., Huang, B., Eds.; Priming-Mediated Stress and Cross-Stress Tolerance in Crop Plants; Elsevier: London, UK, 2020; pp. 35–56.
7. Bucher, M.M.; Hause, B.B.; Krajinski, F.F.; Küster, H. Through the doors of perception to function in arbuscular mycorrhizal symbioses. *New Phytol.* **2014**, *204*, 833–840. [CrossRef] [PubMed]
8. Chen, M.M.; Arato, M.M.; Borghi, L.L.; Nouri, E.; Reinhardt, D. Beneficial services of arbuscular mycorrhizal fungi—From ecology to application. *Front Plant Sci.* **2018**, *4*, 1270. [CrossRef] [PubMed]
9. Bunn, R.; Lekberg, Y.; Zabinski, C. Arbuscular mycorrhizal fungi ameliorate temperature stress in thermophilic plants. *Ecology* **2009**, *90*, 1378–1388. [CrossRef]
10. Augé, R.M. Water relations, drought and vesicular-arbuscular mycorrhizal symbiosis. *Mycorrhiza* **2001**, *11*, 3–42. [CrossRef]
11. Ruíz-Sánchez, M.; Armada, E.; Muñoz, Y.; García de Salamone, I.E.; Aroca, R.; Ruíz-Lozano, J.M.; Azcón, R. Azospirillum and arbuscular mycorrhizal colonization enhance rice growth and physiological traits under well-watered and drought conditions. *J. Plant Physiol.* **2011**, *168*, 1031–1037. [CrossRef]

12. Rapparini, F.; Penuelas, J. Mycorrhizal fungi to alleviate drought stress on plant growth. In *Use of Microbes for the Alleviation of Soil Stresses*; Miransari, M., Ed.; Springer: New York, NY, USA, 2014.
13. Chitarra, W.; Pagliarani, C.; Maserti, B.; Lumini, E.; Siciliano, I.; Cascone, P.; Shubert, A.; Gambino, G.; Balestrini, R.; Guerrieri, E. Insights on the impact of arbuscular mycorrhizal symbiosis on tomato tolerance to water stress. *Plant Physiol.* **2016**, *171*, 1009–1023. [CrossRef]
14. Volpe, V.; Chitarra, W.; Cascone, P.; Volpe, M.G.; Bartolini, P.; Moneti, G.; Pieraccini, G.; Di Serio, C.; Maserti, B.; Guerrieri, E.; et al. The association with two different arbuscular mycorrhizal fungi differently affects water stress tolerance in tomato. *Front Plant Sci.* **2018**, *9*, 1480. [CrossRef] [PubMed]
15. Wu, Q.S. *Arbuscular Mycorrhizas and Stress Tolerance of Plants*; Springer: Singapore, 2017.
16. Read, D.J. Towards ecological relevance—Progress and pitfalls in the path towards an understanding of mycorrhizal functions in nature. In *Mycorrhizal Ecology*; Springer: Berlin, Germany, 2002; pp. 3–29.
17. Smith, S.E.; Read, D. *Introduction, in Mycorrhizal Symbiosis*, 3rd ed.; Academic Press: London, UK, 2008; pp. 1–9.
18. Van der Heijden, M.G.A.; Martin, F.M.; Selosse, M.A.; Sanders, I.R. Mycorrhizal ecology and evolution: The past, the present and the future. *New Phytol.* **2015**, *205*, 1406–1423. [CrossRef] [PubMed]
19. Walder, F.; Niemann, H.; Natarajan, M.; Lehmann, M.F.; Boller, T.; Wiemken, A. Mycorrhizal networks: Common goods of plants shared under unequal terms of trade. *Plant Physiol.* **2012**, *59*, 789–797. [CrossRef] [PubMed]
20. Smith, F.A.; Smith, S.E. How harmonious are arbuscular mycorrhizal symbioses? Inconsistent concepts reflect different mindsets as well as results. *New Phytol.* **2015**, *205*, 1381–1384. [CrossRef] [PubMed]
21. Feddermann, N.; Finlay, R.; Boller, T.; Elfstrand, M. Functional diversity in arbuscular mycorrhiza—The role of gene expression, phosphorous nutrition and symbiotic efficiency. *Fungal Ecol.* **2010**, *3*, 1–8. [CrossRef]
22. Jung, S.C.; Martinez-Medina, A.; Lopez-Raez, J.A.; Pozo, M.J. Mycorrhiza-induced resistance and priming of plant defenses. *J. Chem. Ecol.* **2012**, *38*, 651–664. [CrossRef]
23. Mensah, J.A.; Koch, A.M.; Antunes, P.M.; Kiers, E.T.; Hart, M.; Bücking, H. High functional diversity within species of arbuscular mycorrhizal fungi is associated with differences in phosphate and nitrogen uptake and fungal phosphate metabolism. *Mycorrhiza* **2015**, *25*, 533–546. [CrossRef]
24. Rivero, J.; Gamir, J.; Aroca, R.; Pozo, M.J.; Flors, V. Metabolic transition in mycorrhizal tomato roots. *Front. Microbiol.* **2015**, *6*, 598. [CrossRef]
25. Rivero, J.; Álvarez, D.; Flors, V.; Azcón-Aguilar, C.; Pozo, M.J. Root metabolic plasticity underlies functional diversity in mycorrhiza-enhanced stress tolerance in tomato. *New Phytol.* **2018**, *220*, 1322–1336. [CrossRef]
26. Berruti, A.; Borriello, R.; Lumini, E.; Scariot, V.; Bianciotto, V.; Balestrini, R. Application of laser microdissection to identify the mycorrhizal fungi that establish arbuscules inside root cells. *Front. Plant Sci.* **2013**, *4*, 135. [CrossRef]
27. Balestrini, R.; Salvioli, A.; Dal Molin, A.; Novero, M.; Gabelli, G.; Paparelli, E.; Marroni, F.; Bonfante, P. Impact of an arbuscular mycorrhizal fungus versus a mixed microbial inoculum on the transcriptome reprogramming of grapevine roots. *Mycorrhiza* **2018**, *27*, 417–430. [CrossRef] [PubMed]
28. Hewitt, E.J. *Sand and Water Culture Methods Used in the Study of Plant Nutrition*, 2nd ed.; Commonwealth Agricultural Bureau: London, UK, 1966.
29. Lane, D.J. 16S/23S rRNA sequencing. In *Nucleic Acid Techniques in Bacterial Systematics*; Stackebrandt, E., Goodfellow, M., Eds.; John Wiley & Sons: New York, NY, USA, 1991; pp. 115–175.
30. Trouvelot, A.; Kough, J.L.; Gianinazzi-Pearson, V. Mesure du taux de mycorhization VA d'un système radiculaire. Recherche de méthodes d'estimation ayant une signification fonctionnelle. In *Physiological and Genetical Aspects of Mycorrhizae*; Gianinazzi-Pearson, V., Gianinazzi, S., Eds.; INRA Press: Paris, France; pp. 217–221.
31. Lichtenthaler, H.K.; Buschmann, C. Chlorophylls and carotenoids: Measurement and characterization by UV-VIS spectroscopy. *Curr. Protoc. Food Anal. Chem.* **2001**, *1*, F4.3.1–F4.3.8. [CrossRef]
32. Carillo, P.; Gibon, Y. Protocol: Extraction and determination of proline. *PrometheusWiki* 2011.
33. Ainsworth, E.A.; Gillespie, K.M. Estimation of total phenolic content and other oxidation substrates in plant tissues using Folin–Ciocalteu reagent. *Nat. Protoc.* **2007**, *2*, 875. [CrossRef] [PubMed]

34. Brilli, F.; Pollastri, S.; Raio, A.; Baraldi, R.; Neri, L.; Bartolini, P.; Podda, A.; Loreto, F.; Maserti, B.E.; Balestrini, R. Root colonization by Pseudomonas chlororaphis primes tomato (*Lycopersicum esculentum*) plants for enhanced tolerance to water stress. *J. Plant Physiol.* **2019**, *232*, 82–93. [CrossRef] [PubMed]
35. Balestrini, R.; Rosso, L.C.; Veronico, P.; Melillo, M.T.; De Luca, F.; Fanelli, E.; Colagiero, M.; Salvioli di Fossalunga, A.; Ciancio, A.; Pentimone, I. Transcriptomic responses to water deficit and nematode infection in mycorrhizal tomato roots. *Front Microbiol.* **2019**, *10*, 1807. [CrossRef]
36. Farooq, M.; Wahid, A.; Kobayashi, N.; Fujita, D.; Basra, S.M.A. Plant drought stress: Effects, mechanisms and management. In *Sustainable Agriculture*; Springer: Dordrecht, The Netherlands, 2009; pp. 153–188.
37. Ji, K.; Wang, Y.; Sun, W.; Lou, Q.; Mei, H.; Shen, S.; Chen, H. Drought-responsive mechanisms in rice genotypes with contrasting drought tolerance during reproductive stage. *J. Plant Physiol.* **2012**, *169*, 336–344. [CrossRef]
38. Lakso, A.N. The effects of water stress on physiological processes in fruit crops. In Proceedings of the I International Symposium on Water Relations in Fruit Crops. *Acta Hortic.* **1985**, *171*, 275–290. [CrossRef]
39. Dekov, I.; Tsonev, T.; Yordanov, I. Effects of water stress and high-temperature stress on the structure and activity of photosynthetic apparatus of *Zea mays* and *Helianthus annuus*. *Photosynthetica* **2000**, *38*, 361–366. [CrossRef]
40. Castrillo, M.; Calcagno, A.M. Effects of water stress and re watering on ribulose-1, 5-bis-phosphate carboxylase activity, chlorophyll and protein contents in two cultivars of tomato. *J. Hortic. Sci.* **1989**, *64*, 717–724. [CrossRef]
41. Zgallaï, H.; Steppe, K.; Lemeur, R. Effects of different levels of water stress on leaf water potential, stomatal resistance, protein and chlorophyll content and certain anti-oxidative enzymes in tomato plants. *J. Integr. Plant Biol.* **2006**, *48*, 679–685. [CrossRef]
42. Murkowski, A. Heat stress and spermidine: Effect on chlorophyll fluorescence in tomato plants. *Biol. Plant.* **2001**, *44*, 53–57. [CrossRef]
43. Pandey, P.; Irulappan, V.; Bagavathiannan, M.V.; Senthil-Kumar, M. Impact of combined abiotic and biotic stresses on plant growth and avenues for crop improvement by exploiting physio-morphological traits. *Front. Plant Sci.* **2017**, *8*, 537. [CrossRef] [PubMed]
44. Zribi, L.; Fatma, G.; Fatma, R.; Salwa, R.; Hassan, N.; Néjib, R.M. Application of chlorophyll fluorescence for the diagnosis of salt stress in tomato "*Solanum lycopersicum* (variety Rio Grande)". *Sci. Hortic.* **2009**, *120*, 367–372. [CrossRef]
45. Akula, R.; Ravishankar, G.A. Influence of abiotic stress signals on secondary metabolites in plants. *Plant Signal. Behav.* **2011**, *6*, 1720–1731. [CrossRef]
46. Griesser, M.; Weingart, G.; Schoedl-Hummel, K.; Neumann, N.; Becker, M.; Varmuza, K.; Liebner, F.; Schuhmacher, R.; Forneck, A. Severe drought stress is affecting selected primary metabolites, polyphenols, and volatile metabolites in grapevine leaves (*Vitis vinifera* cv. Pinot noir). *Plant Physiol. Biochem.* **2015**, *88*, 17–26. [CrossRef]
47. Cheruiyot, E.K.; Mumera, L.M.; Ng'etich, W.K.; Hassanali, A.; Wachira, F. Polyphenols as potential indicators for drought tolerance in tea (Camellia sinensis L.). *Biosci. Biotechnol. Biochem* **2007**, 707310505.
48. Errabii, T.; Gandonou, C.B.; Essalmani, H.; Abrini, J.; Idaomar, M.; Skali-Senhaji, N. Growth, proline and ion accumulation in sugarcane callus cultures under drought-induced osmotic stress and its subsequent relief. *African J. Biotechnol.* **2006**, *5*, 16.
49. Maralian, H.; Ebadi, A.; Haji-Eghrari, B. Influence of water deficit stress on wheat grain yield and proline accumulation rate. *African J. Agric. Res.* **2010**, *5*, 286–289.
50. Summart, J.; Thanonkeo, P.; Panichajakul, S.; Prathepha, P.; McManus, M.T. Effect of salt stress on growth, inorganic ion and proline accumulation in Thai aromatic rice, Khao Dawk Mali 105, callus culture. *African J. Biotechnol.* **2010**, *9*, 2.
51. Kamran, M.; Shahbaz, M.; Ashraf, M.; Akram, N.A. Alleviation of drought-induced adverse effects in spring wheat (*Triticum aestivum* L.) using proline as a pre-sowing seed treatment. *Pak. J. Bot* **2009**, *41*, 621–632.
52. Mauch-Mani, B.; Baccelli, I.; Luna, E.; Flors, V. Defense priming: An adaptive part of induced resistance. *Ann. Rev. Plant Biol.* **2017**, *68*, 485–512. [CrossRef] [PubMed]
53. Pollastri, S.; Savvides, A.; Pesando, M.; Lumini, E.; Volpe, M.G.; Ozudogru, E.A.; Faccio, A.; De Cunzo, F.; Michelozzi, M.; Lambardi, M.; et al. Impact of two arbuscular mycorrhizal fungi on *Arundo donax* L. response to salt stress. *Planta* **2018**, *247*, 573–585. [CrossRef] [PubMed]

54. Chitarra, W.; Balestrini, R.; Vitali, M.; Pagliarani, C.; Perrone, I.; Schubert, A.; Lovisolo, C. Gene expression in vessel-associated cells upon xylem embolism repair in *Vitis vinifera* L. petioles. *Planta* **2014**, *239*, 887–899. [CrossRef] [PubMed]
55. Saeedipour, S.; Moradi, F. Relationship of endogenous ABA and IAA to accumulation of grain protein and starch in two winter wheat cultivars under post-anthesis water deficit. *J. Agric. Sci.* **2012**, *4*, 147. [CrossRef]

agronomy

MDPI

Article

Biostimulants Application Alleviates Water Stress Effects on Yield and Chemical Composition of Greenhouse Green Bean (*Phaseolus vulgaris* L.)

Spyridon A. Petropoulos [1,*], Ângela Fernandes [2], Sofia Plexida [1], Antonios Chrysargyris [3], Nikos Tzortzakis [3], João C. M. Barreira [2], Lillian Barros [2] and Isabel C. F. R. Ferreira [2,*]

1 Department of Agriculture, Crop Production and Rural Environment, University of Thessaly, 38446 N. Ionia, Magnissia, Greece; sofiaplexida@uth.gr

2 Centro de Investigação de Montanha (CIMO), Instituto Politécnico de Bragança, Campus de Santa Apolónia, 5300-253 Bragança, Portugal; afeitor@ipb.pt (Â.F.); jbarreira@ipb.pt (J.C.M.B.); lillian@ipb.pt (L.B.)

3 Department of Agricultural Sciences, Biotechnology and Food Science, Cyprus University of Technology, 3036, Lemesos, Cyprus; a.chrysargyris@cut.ac.cy (A.C.); nikolaos.tzortzakis@cut.ac.cy (N.T.)

* Correspondence: spetropoulos@uth.gr (S.A.P.); iferreira@ipb.pt (I.C.F.R.F.); Tel.: +0030-2421-093-196 (S.A.P.); +351-273303219 (I.C.F.R.F.).

Received: 23 December 2019; Accepted: 25 January 2020; Published: 27 January 2020

Abstract: The increasing scarcity of water demands proper water management practices to ensure crop sustainability. In this study, the effect of drought stress and biostimulants application on the yield and chemical composition of green pods and seeds of common bean (*Phaseolus vulgaris* L.) was evaluated. For this purpose, four commercially available biostimulant products, namely Nomoren (G), EKOprop (EK), Veramin Ca (V), and Twin-Antistress (TW), were tested under two irrigation regimes: normal irrigation (W+) and water-holding (W-) conditions. The highest increase (20.8%) of pods total yield was observed in EKW+ treatment due to the formation of more pods of bigger size compared to control treatment (CW+). In addition, the highest yield under drought stress conditions was recorded for the GW- treatment (5691 ± 139 kg/ha). Regarding the effects of biostimulants on the protein and ash content of pods, the application of VW+ treatment (first harvest of pods; 201 ± 1 and 79 ± 1 g/kg dw for proteins and ash content, respectively) and GW+ (second harvest of pods; 207.1 ± 0.1 and 68.4 ± 0.5 g/kg dw for proteins and ash content, respectively) showed the best results. For seeds, the application of GW+ treatment resulted in the highest content for fat, protein, and ash content (52.7 ± 0.1, 337 ± 1, 56 ± 1 g/kg dw) and energetic value (5474 ± 3 kcal/kg dw). γ-tocopherol was the main detected tocopherol in pods and seeds, and it was significantly increased by the application of TWW- (first harvest of pods; 6410 ± 40 μg/kg dw), VW- (second harvest of pods; 3500 ± 20 μg/kg dw), and VW+ (seeds; 39.8 ± 0.1 g/kg dw) treatments. EKW- treatment resulted in the lowest oxalic acid content for both pod harvests (26.3 ± 0.1 g/kg dw and 22.7 ± 0.2 g/kg dw for the first and second harvest of pods, respectively) when compared with the rest of the treatments where biostimulants were applied, although in all the cases, the oxalic acid content was considerably low. Fructose and sucrose were the main sugars detected in pods and seeds, respectively, while the highest content was recorded for the TWW- (first harvest of pods) and GW- (second harvest of pods and seeds) treatments. The main detected fatty acids in pods and seeds were α-linolenic, linoleic, and palmitic acid, with a variable effect of the tested treatments being observed. In conclusion, the application of biostimulants could be considered as an eco-friendly and sustainable means to increase the pod yield and the quality of common bean green pods and seeds under normal irrigation conditions. Promising results were also recorded regarding the alleviation of negative effects of drought stress, especially for the application of arbuscular mycorrhizal fungi (AMF; G treatment), which increased the total yield of green pods. Moreover, the nutritional value and chemical composition of pods and seeds was positively affected by biostimulants application, although a product specific effect was recorded depending on the irrigation regime and harvesting time (pods and seeds).

Keywords: arbuscular mycorrhizal fungi; biofertilizers; common bean; *Glomus* spp.; organic acids; pod quality; seaweed extracts; seed quality; tocopherols; total sugars

1. Introduction

The increasing concerns for food security in a rapidly growing world population has rendered the necessary intensification of agricultural production for the achievement of higher crop yield and total production. Protected vegetables cultivation is the most intensified cropping system and requires high amounts of fertilizers and pesticides [1,2]. However, despite the fact that higher fertilizer rates result in increased total yield, this practice is not always favorable when the quality of the final product is also considered. On the contrary, it is very common for excessive fertilization to stimulate vegetative growth and increase susceptibility to pathogens, which may result in increased product losses, as well as high nutrient losses due to leaching [3].

In addition, the increasing scarcity of water availability for human activities and irrigation in particular is a worldwide phenomenon and demands appropriate water management practices to ensure crop sustainability and economic activities related to water, especially in semi-arid and arid regions [4]. The use of biostimulants can diminish effects of environmental abiotic stress factors such as water stress, improve soil water-holding capacity and root conformation, and increase root growth with beneficial effects on nutrient and water use efficiency and yield; hence, the past decade has witnessed tremendous growth in the use of biostimulants in the farming sector [5–7]. The use of biostimulants containing arbuscular mycorrhizal fungi (AMF), saprophytic fungi, or algae extracts is considered an environmental friendly technique for the alleviation of adverse impact of osmotic stress, by increasing wáter and the nutrient uptake of crops and tolerance to biotic and abiotic stress [8–10].

Plant biostimulants usually consist of amino acids and peptide mixtures [11]. They also contain a wide number of bioactive compounds that are able to improve various physiological processes that stimulate plant growth and increase nutrient use efficiency without adverse effects on crop yield and final product quality, while at the same time reduce chemical fertilizers inputs [5,12]. However, the effect of biostimulants may differ from species to species, while it greatly depends on environmental factors during and after application, as well as on the dose and time of application [13,14]. For example, the application of saprophytic fungi (*Trichoderma harzianum* ALL-42) was associated with increased shoot biomass production and the number of lateral shoots in *Phaseolus vulgaris* plants due to the beneficial effects of root colonization by fungi on plant root growth [15]. On the other hand, seaweed extracts (*Ascophyllum nodosum*) increased the plant growth and overall yield of leafy vegetables such as spinach and lettuce [16–18], while in bean plants, the application of extracts enhanced root growth and plant development, especially when water stress conditions were imposed [19]. The biostimulatory activity of symbiotic bacteria such as *Bacillus* sp. is mostly associated with adaptation mechanisms for improved water retention though alterations in plant cell wall composition and hormones production (e.g., indole-3-acetic acid (IAA)) [6]. Therefore, environmental friendly methods such as applying biostimulants for stimulating early growth in vegetable crops and ensuring high yields are innovative agricultural practices that have to be further investigated in order to improve our understanding of their functions and the involved mechanisms of action [20].

Common bean (*Phaseolus vulgaris* L.) is a drought-sensitive vegetable crop, and water stress may have a detrimental effect on crop yield [21] and the chemical composition of pods and seeds [22,23]. So far, there is limited literature regarding the use of biostimulants on common bean plants, whereas various studies have tested the effects of biostimulants on other legume species, especially under drought stress conditions. In particular, the application of *Pseudomonas aeruginosa* GGRJ21 strain on mung bean (*Vigna radiata* (L.) R. Wilczek) under greenhouse and field conditions up-regulated the expression of drought stress-responsive genes, which resulted in better plant growth and development under water stress conditions [24]. Foliar application of amino acids on faba beans (*Vicia faba* L.)

subjected to salt stress showed significant ameliorative effects that were mainly associated with the use of amino acids as carbon and nitrogen pools, which further increased photosynthetic apparatus efficiency [25]. Kumar et al. [26] reported the synergistic effects of *Pseudomonas putida* and *Bacillus amyloliquefaciens* on chickpea plants subjected to water stress through the evaluation of several traits, including the activation of plant defense and soil enzymes and plant growth parameters. Moreover, the inoculation of common bean plants with *Azospirillum brasilense* altered root morphology, which allowed plants to overcome water stress without increasing plant biomass compared to non-inoculated plants [27]. In another study, Klimek-Kopyra et al. [28] suggested that biostimulants application on seeds of seven winter pea cultivars (*Pisum sativum* L.) may increase frost tolerance through the increased germination percentage and growth rate of seedlings, although a varied response depending on biostimulant x cultivar combination was observed. In contrast, Galvão et al. [19] suggested that the application of *Bacillus amyloliquefaciens BV 03* and/or the combination of *B amyloliquefaciens BV 03* with *A. nodosum* extracts did not alleviate water deficit effects on common bean plants. According to Dourado-Neto et al. [29] the use of hormones with biostimulant activity (combination of kinetin, indole butyric acid, and gibberellic acid) on common bean plants through seed treatment, sowing, or foliar spraying may increase the number of grains per pod and grains yield. Moreover, licorice root extracts may have a beneficial role on mitigating the negative effect of salt stress on *P. vulgaris* plants' growth and yield, as well as on the total soluble carbohydrates, soluble sugars, and nutrients content [30]. The combined application of salicylic acid and *Moringa oleifera* leaves extracts has been also reported to mitigate salinity stress effects on common bean plants through the improvement of green pods and seeds yield and the physicochemical characteristics of pods and seeds [31]. Other examples of biostimulants use on common beans crop include the application of *Lolium perenne* foliage extracts as potent cell defense elicitors [32] and the positive effect of aqueous extracts of moringa leaves and garlic cloves on the yield and chemical composition of snap beans [33,34].

Most of the studies regarding the mitigating effects of biostimulants to abiotic stressors refer to high salinity or nutrient deficiency stress. The main goal of this study was to record the effects of natural biostimulants on the yield, nutritional value, and chemical composition under drought conditions. For this purpose, a drought-sensitive species, namely the common bean (*P. vulgaris*), was selected and grown in a protected environment under water stress conditions, and the use of commercially available biostimulants products was evaluated as an environmentally friendly and sustainable method for increasing the yield and quality of end-products through the improvement of the chemical composition of the final products (pods and/or seeds) without compromising yield.

2. Materials and Methods

2.1. Plant Material and Growing Conditions

The experiment was carried out during the growing period of summer–autumn 2017. Sowing took place on 11 August 2017 and seeds of bean (*Phaseolus vulgaris* L.) were sown directly in soil within the unheated plastic greenhouse at the experimental farm of the University of Thessaly, Greece. Seeds were sown in double rows with a spacing of 50 cm between the rows, and the plant density was 2 plants/m^2 (20,000 plants/ha), while each treatment consisted of 6 plants and was replicated three times (180 plants in total). The soil at 0–30 cm depth was clay (26% sand, 32% silt, and 42% clay); pH: 8.0 (1:1 soil/H2O); organic matter content: 3.1%; CaCO3: 10.8%; available P (Olsen method): 70.9 mg/kg; total N (Kjeldahl method): 1.8 g/kg; exchangeable K2O (ammonium acetate method): 195 mg/kg; electrical conductivity (ECe): 0.95 dS/m. The growth conditions throughout the experimental period are presented in Figure 1. Two factors were applied in a split-plot factorial design, namely water stress and biostimulants. Biostimulants treatments included: (1) Control (C: no biostimulants added), (2) Nomoren (G; Anthis S.A., Greece) (3) Twin-Antistress (TW; Microspore Hellas–Sacom Hellas, Greece), (4) Veramin Ca (V; Microspore Hellas–Sacom Hellas, Greece), and (5) EKOprop (EK; Anthesis S.A., Greece). Regarding the detailed composition of each product, Nomoren contains 20% arbuscular mycorrhizal fungi (AMF) (*Glomus* spp.). Twin-Antistress contains natural microorganisms based on *Bacillus subtilis*, as well as yeast and *Ascophyllum nodosum* extracts, as well as N (organic): 1%, organic carbon: 10%, and organic matter (<50 kDa): 30%. Veramin Ca contains an amino acid complex of vegetable origin with *Aloe vera* extract, and CaO: 15.6%. EKOprop contains a mixture of arbuscular mycorrhizal fungi (*Glomus* spp: 1%), rhizospere symbiotic bacteria (*Bacillus* spp., *Streptomyces* spp., *Pseudomonas* spp.,: 1.6 × 109 CFU/g in total), and saprophytic fungi (*Trichoderma* spp.: 5 × 105 CFU/g) (Table 1).

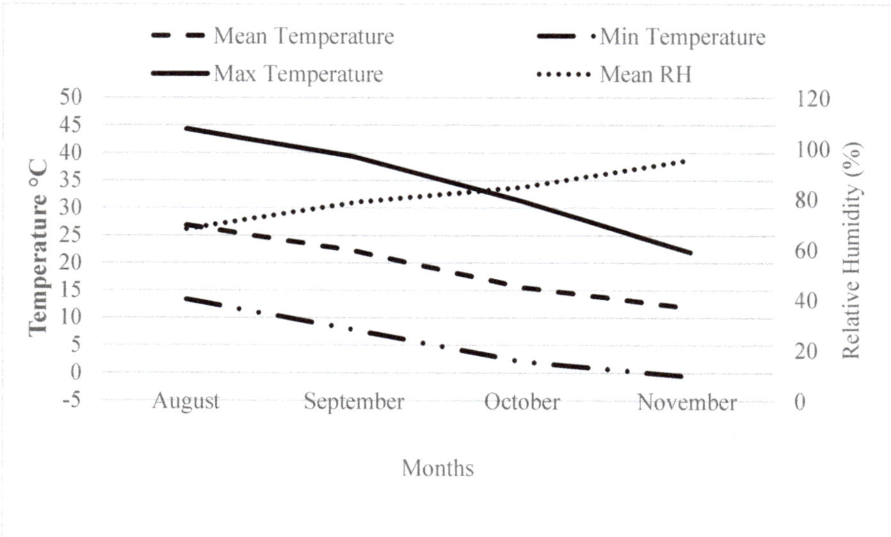

Figure 1. Environmental conditions (mean, max, and min temperature and mean relative humidity (RH)) throughout the experimental period.

Water stress treatments were previously described by the authors and scheduled with the use of tensiometers (Irrometer-Moisture Indicator, Irrometer, Riverside, CA) including: (a) normally irrigated plants (W+) where irrigation was applied approximately twice a week and when tensiometer readings were between 10% and 15%, and (b) water-stressed plants (W-) where water holding was applied with irrigation being implemented approximately once a week and when tensiometer readings were between 40% and 50% [35,36]. Tensiometer readings are percent levels that correspond to soil moisture content ranging from field capacity (0%) to dry soil (100%). Irrigation was applied through a drip irrigation system with one dripper per plant and a water flow rate of 4.0 L/h for each dripper. The total amount of irrigation water was 350 m3/ha (17.5 L per plant) for normally irrigated plants and 210 m^3/ha (10.5 L per plant) for water-stressed plants. Biostimulants were applied according to the directions for use of each product at 10, 20 and 30 days after sowing (DAS) as following: (G) was applied with irrigation water at 5 L/ha for each dose; (TW) was applied with irrigation water 5 L/ha for each dose; (V) was applied with foliar spraying at 500 g/100 L H2O for each dose; and (EK) was applied with irrigation water at 1 kg/ha for each dose. Water holding started after the second application of biostimulants (21 DAS). The harvest of green pods took place at marketable maturity at 60 DAS (first harvest) and 70 DAS (second harvest), while seeds were collected from fully mature green pods at 103 DAS. All harvests were carried out on the same plants. After harvest, the fresh and dry weight of pods, as well the fresh and dry weight of seeds, number of seeds per pod, and 100 seeds weight were recorded. The number of seeds per pod and the weight of 100 seeds was evaluated from 10 pods for each plot (30 pods per treatment). Batch samples of pods and seeds were put in deep-freezing conditions, lyophilized, ground with a mortar and pestle, and stored at freezing conditions (−80 °C) until further analyses.

Table 1. Detailed composition and application guides for the tested biostimulants.

Product	Composition	Application Method	Dose
Nomoren (G)	20% of arbuscular mycorrhizal fungi (AMF) (*Glomus* spp.)	Irrigation water	5 L/ha
Twin-Antistress (TW)	Natural microorganisms based on *Bacillus subtilis*, and yeast and *Ascophyllum nodosum* extracts, as well as N (organic): 1%, organic carbon: 10%, and organic matter (<50 kDa): 30%	Irrigation water	5 L/ha
Veramin Ca (V)	Amino acid complex of vegetable origin with *Aloe vera* extract, and CaO: 15.6%	Foliar spraying	500 g/100 L H_2O
EKOprop (EK)	Mixture of arbuscular mycorrhizal fungi (*Glomus* spp: 1%), rhizospere symbiotic bacteria (*Bacillus* spp., *Streptomyces* spp., *Pseudomonas* spp.; 1.6×10^9 CFU/g in total), and saprophytic fungi (*Trichoderma* spp.: 5×10^5 CFU/g)	Irrigation water	1 kg/ha

2.2. *Chemical Analyses*

2.2.1. Nutritional Value

Sample were analyzed in terms of nutritional compounds (moisture, fat, ash, proteins, and carbohydrates) according to the Association of Analytical Communities (AOAC) methods [37]. Moisture was determined by pods and seeds drying at 105 ± 5 °C until constant weight. Crude protein was evaluated by the macro-Kjeldahl method (N × 6.25) using an automatic distillation and titration unit (model Pro-Nitro-A, JP Selecta, Barcelona, Spain), ash content was determined by incineration at 550 ± 15 °C, and the crude fat was determined by extraction with petroleum ether using a Soxhlet apparatus (Behr Labor Technik, Dusseldorf, Germany). Total carbohydrates (g/kg dw) were determined by difference according to the equation: 1000 – (g moisture + g fat + g ash + g proteins), and energy (kcal/kg dw) was determined according to the equation: 4 × (g proteins + g carbohydrates) + 9 × (g fat).

2.2.2. Minerals Composition

Mineral composition analysis was performed in forced-air dried (at 72 °C) and ground to powder pods and seeds, after dry ashing and extraction with 2 N HCl according to the method described by Chrysargyris et al. [38]. Atomic absorption spectrophotometry (PG Instruments AA500FG, Leicestershire, UK) was used for Ca, Mg, Mn, Zn, and Cu content determination, while flame photometry (Lasany Model 1832, Lasany International, Haryana, India) was used for and Na and K content determination. Nitrogen and phosphorus content were determined by Kjeldahl (Digest Automat K-439 and Distillation Kjelflex K-360, BÜCHI, Flawil, Switzerland) and spectrophotometry methods (Multiskan GO, Thermo Fisher Scientific, Waltham, Massachusetts, USA), respectively. The determination of minerals composition was performed only in pods of the second harvest and seeds. Pods of the first harvest were not evaluated due to insufficient amounts of samples for specific treatments, which did not allow a complete set of data. Results are expressed on a dry weight basis.

2.2.3. Tocopherols

Tocopherols were determined in the lyophilized samples by HPLC fluorescence, following a procedure previously described using tocol (Matreya, Pleasant Gap, Pensylvania, USA) as internal standard [39]. Tocopherols standards (α-, β-, γ-, and δ-isoforms, Sigma-Aldirch, St. Louis, MO, USA) were used for compounds identification, and quantification was assessed by the internal standard method. Results were obtained using the Clarity 2.4 software (DataApex, Prague, Czech Republic) and expressed in μg/kg dw and mg/kg dw for pods and seeds, respectively.

2.2.4. Organic Acids

Organic acids were determined in the lyophilized sample and determined by a high-performance liquid chromatography system equipped with a diode array detector (HPLC-DAD), following a procedure previously described [40]. Compounds were identified and quantified by comparison of the retention time, spectra, and peak area recorded at 245 nm and 215 nm (for ascorbic acid and remaining acids, respectively), with those obtained from commercial standards (oxalic, malic, fumaric, and ascorbic acids, Sigma-Aldrich, St. Louis, MO, USA). The results were recorded and processed using LabSolutions Multi LC-PDA software (Shimadzu Corporation, Kyoto, Japan) and were expressed in g/kg dw and mg/kg dw for pods and seeds, respectively.

2.2.5. Free Sugars

Free sugars were determined by HPLC coupled to a refractive index (RI) detector (Knauer, Smartline system 1000, Berlin, Germany) using the internal standard (IS; melezitose). The lyophilized sample was extracted using a methodology previously described [40]. Compounds were identified by comparison with standards (Sigma-Aldrich, St. Louis, MO, USA), and quantification was performed by the IS method. Results were processed using the Clarity 2.4 software (DataApex, Prague, Czech Republic) and expressed in g per kg dw.

2.2.6. Fatty Acids

Fatty acids profile was characterized after a transesterification procedure and according to the method previously described [41]. The analysis was carried out with gas-liquid chromatography with flame ionization detection (GC-FID; DANI1000, Contone, Switzerland). Fatty acids identification and quantification (Clarity DataApex 4.0 Software, Prague, Czech Republic) were performed by comparing the relative retention times of fatty acid methyl ester (FAME) peaks from samples with standards (reference standard mixture 47885-U, Sigma, St. Louis, MO, USA). Results were expressed in the percentage of each fatty acid.

2.3. Statistical Analysis

2.3.1. Experimental Layout and Statistical Treatment

The experimental design was laid out in a split plot arrangement with each main plot consisting of water stress treatments (W+ or W-), while fully randomized sub-plots comprised the biostimulants treatments. Each subplot contained 6 plants and each main plot contained 30 plants. Pod yield components were evaluated in 18 plants for each treatment (n = 18), whereas for seed yield, 30 randomly selected pods from each treatment were measured. In order to constitute a representative and adequate sample of the tested treatments, batches of several samples of pods and seeds were taken at random from each plot in order to obtain three different samples. Then, these batches were powdered to obtain homogenous samples. For each methodology, three extractions were carried out, and the analyses were performed in triplicate. Statistical analysis was conducted with the aid of Statgraphics 5.1.plus (Statpoint Technologies, Inc., Warrenton, VA, USA). Data were evaluated by a two-way analysis of variance (two-way ANOVA), and significant interactions of the tested factors (water regime and biostimulant treatment) were observed. Therefore, all the means for each pod harvest and seeds were compared separately by using the Tukey's honestly significant difference (HSD) test (p = 0.05).

2.3.2. Linear Discriminant Analysis (LDA)

Linear discriminant analysis (LDA) was applied to evaluate the overall effects of different biostimulants, independently of water level, in each phenological stage (first and second harvest of pods and seeds). The stepwise technique, considering the Wilk's Λ test with the usual probabilities of F

(3.84 to enter and 2.71 to be removed) for variable selection, was employed. With this procedure, it was aimed to estimate the association between the single categorical dependent variables (biostimulant treatments: C, G, TW, V, EK) and the quantitative independent variables (analyzed parameters: proximate composition, organic acids, tocopherols, sugars, fatty acids). In all cases, a leaving-one-out cross-validation procedure was carried out to assess the model performance.

3. Results and Discussion

3.1. Yield and Growth Parameters

The yield and growth characteristics of pods and seeds are presented in Table 2. Yield was positively affected by the application of EK treatment in normally irrigated plants (EKW+) where higher yields compared to the control and the rest of biostimulant treatments were observed (5284 ± 120 kg/ha, 3701 ± 88 kg/ha, and 8985 ± 196 kg/ha, for the first harvest, second harvest, and total yield, respectively). This increase in pod yield was attributed to the higher number of pods harvested from both harvests in plants treated with the specific biostimulants, while the mean pod weight was also the highest in EKW+ treatment only for aggregated results. On the contrary, the application of V treatment in water-stressed plants (VW-) resulted in the lowest yields for the first and second harvest and consequently in the total yield of green pods (2213 ± 90 kg/ha, 1749 ± 59 kg/ha, and 3962 ± 147 kg/ha for the first harvest, second harvest, and total yield, respectively). Additionally, the number of pods per plant and consequently total yield was higher in normally irrigated plants comparing to water-stressed ones, while GW- and TWW- treatments were the most effective at alleviating the negative effects of stress conditions. Similar results have been reported by Aimo et al. [42], who suggested that the *Crocus sativus* yield was increased after AMF application due to the higher number of flowers. This was also the case in our study, where the products containing AMF (G) or a mixture of AMF, saprophytic fungi, and rizosphere bacteria (EK) resulted in higher yields in water-stressed and normally irrigated plants, respectively. According to German et al., the inoculation of common bean plants with *Azospirillum brasilense* increased tap root length as well as the proportion of long and thin roots at the early growth stages, which are critical for plant adaptation to water stress conditions [27]. Moreover, Weber et al. reported an increase of fruit setting and total yield in strawberry plants as the result of *Ascophyllum nodosum* extracts application [43]. In the present study, the application of products containing *A. nodosum* extracts had a positive effect on total yield under water stress conditions (TWW-) when compared to non-biostimulant treated plants (CW-), although the product containing AMFs (GW-) was shown to be more effective. According to Arthur et al. [44], biostimulants such as seaweed extracts may contain plant hormones (cytokins and auxins) that induce flower formation. Seaweed extracts (*Ascophyllum nodosum*) have been also reported to have a positive effect on the plant growth of lettuce, carrot, and strawberry through increased hormone activity and K uptake [18,45,46]. Moreover, the increased yield for plants treated with biostimulants is associated with improved plant tolerance to abiotic stress conditions, which according to Battacharyya et al. [47] could be attributed to various protective mechanisms such as the regulation of related genes, the accumulation of osmolytes, the improvement in water-use efficiency, and other effects on the plant rhizosphere. Moreover, Ahmad et al. [10] reported that the inoculation of Indian mustard plants with *Trichoderma harzianum* alleviated osmotic stress effects through the activation of plant antioxidant mechanisms.

Table 2. Yield and growth characteristics of bean plants in relation to water stress and biostimulants application (mean ± SD).

Treatment	1st Harvest			2nd Harvest			Total				
	Number of Pods/Plant	Mean Pod Weight (g)	Yield (kg/ha)	Number of Pods/Plant	Mean Pod Weight (g)	Yield (kg/ha)	Number of Pods/Plant	Mean Pod Weight (g)	Yield (kg/ha)	100 Seeds Weight (g)	Seeds per Pot
CW+ [Y]	24 ± 1c	9 ± 1b	4296 ± 105d	17 ± 4c	9 ± 1b	3138 ± 43c	41 ± 4d	9 ± 1c	7434 ± 120d	101 ± 2f	5.4 ± 0.5a
VW+	26 ± 2b	9.7 ± 0.6a	4933 ± 154b	18 ± 1b	9 ± 1b	3318 ± 89b	44 ± 4b	10 ± 1b	8311 ± 233b	112 ± 3d	4.8 ± 0.9b
EKW+	27 ± 1a	9.7 ± 0.8a	5284 ± 120a	18.6 ± 0.8a	10 ± 1a	3701 ± 88a	46 ± 3a	10 ± 2a	8985 ± 196a	108 ± 2e	5.0 ± 0.8b
GW+	26 ± 2b	9 ± 1b	4621 ± 85.0c	16.4 ± 0.8c	10 ± 2a	3318 ± 81b	42 ± 2c	9 ± 1c	7939 ± 160c	116 ± 2b	5.5 ± 0.9a
TWW+	24 ± 2c	8 ± 1d	3799 ± 109e	17.5 ± 0.9c	10 ± 2a	3357 ± 96b	42 ± 3c	9 ± 1c	7156 ± 187e	116 ± 2b	5.5 ± 0.9a
CW-	14 ± 3f	8.8 ± 0.9c	2456 ± 75h	13 ± 3f	9 ± 2b	2435 ± 77e	27 ± 2h	9 ± 1c	4891 ± 137h	114 ± 3bc	5.0 ± 0.9b
VW-	14 ± 2f	8 ± 1d	2213 ± 90i	12.0 ± 0.6g	7.3 ± 0.8d	1749 ± 59f	26 ± 2i	7.8 ± 0.8f	3962 ± 147j	115 ± 3b	4.4 ± 0.5c
EKW-	16.7 ± 0.8e	8 ± 1d	2515 ± 50h	14.1 ± 0.9e	8 ± 2c	2147 ± 62c	31 ± 2g	7.6 ± 0.7g	4662 ± 80i	122 ± 2a	5.0 ± 0.6b
GW-	20.2 ± 0.9d	8 ± 1d	3172 ± 79f	15 ± 1d	8 ± 2c	2519 ± 81de	36 ± 2e	8 ± 1e	5691 ± 139f	113 ± 3cd	5.1 ± 0.5b
TWW-	16.7 ± 0.8e	8.2 ± 0.8d	2748 ± 64g	15.4 ± 0.8d	8 ± 2c	2603 ± 71d	32 ± 2f	8.3 ± 0.9d	5351 ± 127g	114 ± 2bc	5.0 ± 0.7b

[Y] W+: indicates normal irrigation regime; W-: indicates water-holding irrigation regime; C: Control; V: Veramin Ca; EK: EKOprop; G: Nomoren; TW: Twin-Antistress. Means in the same column followed by different Latin letters are significantly different according to Tukey's honestly significant difference (HSD) test ($p = 0.05$).

Regarding the weight of 100 seeds, the highest value was observed for the EK treatment under water stress conditions (EKW-), while normally irrigated plants with no added biostimulants (CW+) presented the lowest values (122 ± 2 and 101 ± 2 for EKW- and CW+ treatments, respectively) (Table 2). Moreover, the weight of 100 seeds and number of seeds per pod were higher and lower, respectively, for water-stressed plants regardless of biostimulants treatment when compared to the control treatment of normally irrigated plants, indicating that water stress may affect the fertilization process and consequently the number of seeds per pod. The beneficial use of plant-growth promoting rhizobacteria under water stress conditions has been also reported by Sarma and Saikia [24], who suggested that the inoculation of mung bean plants with *Pseudomonas aeruginosa* strains alleviated water stress effects through the scavenging of oxidative enzymes. Moreover, Korir et al. [48] reported a synergistic effect of plant growth promoting bacteria with common bean rhizobia that enhanced plant growth and development, while Farouk and Abdul Qados [49] suggested that folic acid application increased the seed yield and chemical composition of pea plants (*Pisum sativum*).

3.2. Nutritional Value

The nutritional value of the pods and seeds is presented in Table 3. The application of biostimulants did not have a beneficial effect on the moisture content of pods in the first harvest when compared with the control treatment for either normally irrigated (CW+) or water-stressed plants (CW-). Similar trends were observed under prolonged water stress (second harvest), while for normally irrigated plants, the VW+ treatment resulted in the highest moisture content values. For both harvests, the lowest values were recorded for pods harvested from water-stressed plants treated with the V treatment (VW-). These findings could be probably attributed to a functional allocation equilibrium where under biostimulant treatments, plants allocated resources and biomass in fruit; thus, a reduction in moisture content (or similarly an increased dry matter content) was observed [50]. Moreover, plants under stress tend to accumulate minerals and metabolites as a means to maintain high water potential [5]. However, considering that these trends were observed both in water-stressed and normally irrigated plants highlights the need for further investigation. On the other hand, the moisture content of seeds was beneficially affected by the various biostimulants in plants grown under water stress conditions, especially for G treatment (GW-), where the highest values were observed. Regarding the rest of the nutritional value parameters, a varied effect of biostimulants and irrigation treatments was observed in terms of fat, protein, ash, and carbohydrates content and the energetic value of pods and seeds. In particular and for the first harvest of pods, protein, ash content, and energetic value were beneficially affected under the normal irrigation regime, while carbohydrates content was the highest for water-stressed plants with no added biostimulants. Similarly, a beneficial effect of GW+ treatment on protein and ash content was observed for the pods of the second harvest, whereas fat and carbohydrates content were the highest for water-stressed plants that received no biostimulants (CW-) or the G treatment (GW-). The energetic value was the highest for normally irrigated plants that received no biostimulants. Seeds' nutritional value was beneficially affected by the application of G treatment under normal irrigation conditions (GW+) when proteins, fat, and ash content were considered, whereas EK and G treatments increased energetic value under the same irrigation treatments (5468 ± 6 kcal/kg dw and 5474 ± 3 kcal/kg dw, respectively). The highest carbohydrates content values were recorded for normally irrigated plants where no biostimulant or the V treatment was applied (CW+ and VW+, respectively), as well as for water-stressed plants that received the EK treatment (EKW-). Significant differences were also observed between normally irrigated and water-stressed plants in a biostimulant treatment-specific manner, although no direct comparisons between the corresponding treatments were performed due to the presence of significant interactions among the tested factors. Nevertheless, it is worth highlighting the beneficial effects of biostimulants on the protein content of green pods under water stress conditions compared to the corresponding control treatment (CW-). The effect of biostimulants on plant metabolism and the quality of end-products has been previously reported by Colla et al. [51], while Przygocka-Cyna and Grzebisz [52] have associated the use of biostimulants with the improvement in plant nutrient uptake and therefore with the better nutritional value of the end products. Moreover, according to Elsheikh et al., the inoculation of faba bean (*Vicia faba*) plants with arbuscular mycorrhiza increased the protein content in the seeds, regardless of irrigation conditions, suggesting the improved nutritional status of plants as the main reason for this increase [53]. Similarly, Farouk and Abdul Qados [49] suggested that folic acid and hydrogen peroxide application may improve the nutritional value of pea seeds through the increase of protein and carbohydrates content, while Elsheikh and Mohamedzein reported that the inoculation of groundnut with *Glomus* sp. and *Bradyrhizobium* sp. increased the protein content of seeds. Regarding the biostimulatory activity of seaweed extracts, Kocira et al. [54] and Castellanos-Barriga et al. [55] reported contrasting effects of *Ecklonia maxima* and *Ulva lactuca* extracts on the nutritional value of common bean (*P. vulgaris*) and mung bean (*Vigna radiata*) seeds, respectively. These differences in the literature reports highlight the variable biostimulatory effects of seaweed extracts, which contain a wide range of compounds associated with improved plant nutrient uptake, phytohormone-like activities, tolerance to abiotic stressors, and the modulatory effects of plant metabolism and physiology [47].

Table 3. Nutritional (g/kg dw) and energetic value (kcal/kg dw) of the studied pods and seeds of beans in relation to irrigation regime (mean ± SD).

			1st Harvest of Pods			
Treatment	**Moisture (%)**	**Fat**	**Proteins**	**Ash**	**Carbohydrates**	**Energy**
CW+ ¥	96.4 ± 0.7a	48.2 ± 0.8e	165 ± 6h	59.6 ± 0.4h	728 ± 3b	4630 ± 20f
VW+	94.2 ± 0.4c	45.6 ± 0.6g	201 ± 1a	79 ± 1a	674 ± 2i	4821 ± 3a
EKW+	95.0 ± 0.6b	48 ± 2e	197 ± 4b	73.7 ± 0.2b	681 ± 1h	4790 ± 20b
GW+	92.5 ± 0.7de	53.7 ± 0.2b	174.5 ± 0.1e	71.6 ± 0.5c	700 ± 1f	4658 ± 1e
TWW+	93 ± 1d	50.1 ± 0.5c	193 ± 3c	70 ± 3d	688 ± 1g	4760 ± 10c
CW-	93 ± 2d	49.4 ± 0.8cd	150 ± 02i	66 ± 2f	735 ± 2a	4553 ± 8h
VW-	88 ± 1g	48.7 ± 0.8de	181.5 ± 0.5d	67 ± 1e	703 ± 2e	4713 ± 1d
EKW-	93 ± 1d	45.5 ± 0.3g	169.5 ± 0.4g	61 ± 1g	724 ± 1c	4665 ± 1e
GW-	89.4 ± 0.8f	47 ± 3f	171.6 ± 0.3f	66 ± 2f	715 ± 1d	4671 ± 9e
TWW-	92 ± 1e	67 ± 1a	173.2 ± 0.1e	69.5 ± 0.1d	690 ± 1g	4599 ± 3g
			2nd Harvest of pods			
CW+	91 ± 1b	27.1 ± 0.2e	204.6 ± 0.3b	49.6 ± 0.7cd	719 ± 1f	4915 ± 2a
VW+	92 ± 2a	32 ± 2d	174 ± 1e	56.9 ± 0.8b	737 ± 1e	4739 ± 2f
EKW+	91.1 ± 0.5b	37.1 ± 0.2b	195.9 ± 0.3c	51 ± 1c	716 ± 1f	4831 ± 2d
GW+	90.7 ± 0.4c	34 ± 2c	207.1 ± 0.1a	68.4 ± 0.5a	690 ± 1g	4898 ± 5b
TWW+	90 ± 2d	27.3 ± 0.2e	175 ± 1e	49 ± 2d	748 ± 1d	4768 ± 2e
CW-	91.1 ± 0.5b	43 ± 2a	150.2 ± 0.1h	49 ± 1d	758 ± 1c	4580 ± 6g
VW-	87.6 ± 0.8f	32 ± 2d	174.3 ± 0.2e	52 ± 2c	741 ± 3e	4742 ± 4f
EKW-	89 ± 1e	21.4 ± 0.7f	189.2 ± 0.4d	45 ± 1e	745 ± 1d	4861 ± 1c
GW-	90 ± 2d	18 ± 1h	164.0 ± 0.4g	49 ± 2d	769 ± 1a	4747 ± 5f
TWW-	90 ± 1d	19.8 ± 0.8g	171 ± 1f	45 ± 2e	764 ± 1b	4776 ± 2e
			Seeds			
CW+	69.8 ± 0.3a	40.0 ± 0.1c	305 ± 6e	51.4 ± 0.5cd	604 ± 4a	5360 ± 20g
VW+	66.7 ± 0.7d	37 ± 1d	310.2 ± 0.2d	53 ± 2b	600 ± 2a	5403 ± 5f
EKW+	67.7 ± 0.9b	33.2 ± 0.1g	320 ± 2b	53 ± 2b	593 ± 3c	5468 ± 6ab
GW+	67 ± 2c	52.7 ± 0.1a	337 ± 1a	56 ± 1a	554 ± 1e	5474 ± 3a
TWW+	66.3 ± 0.7e	45 ± 3b	317.6 ± 0.1c	49.7 ± 0.1f	588 ± 2d	5410 ± 7e
CW-	62 ± 1g	40.2 ± 0.1c	322.7 ± 0.1b	51 ± 1de	587 ± 1d	5453 ± 1c
VW-	65.8 ± 0.9f	35 ± 2e	320.2 ± 0.1b	51.7 ± 0.3c	593 ± 2c	5461 ± 6b
EKW-	67.3 ± 0.6c	34 ± 1f	317 ± 1c	48.1 ± 0.8g	601 ± 1a	5448 ± 6c
GW-	69.8 ± 0.5a	34 ± 1f	317.1 ± 0.4c	50.9 ± 0.4e	598 ± 1b	5450 ± 2c
TWW-	66.2 ± 0.4e	35.1 ± 0.6e	316.5 ± 0.4c	50 ± 1f	598 ± 1b	5442 ± 3d

¥ W+: indicates normal irrigation regime; W-: indicates water-holding irrigation regime; C: Control; V: Veramin Ca; EK: EKOprop; G: Nomoren; TW: Twin-Antistress. Means in the same column and the same harvest (first and second harvest of pods and seeds) followed by different Latin letters are significantly different according to Tukey's honestly significant difference (HSD) test ($p = 0.05$).

3.3. Mineral Composition

The mineral composition of pods and seeds is presented in Table 4. The combination of biostimulants application and irrigation regime had a varied effect on mineral content of pods and seeds with no specific trends being observed. In particular, the application of V treatment under water stress conditions (VW-) increased the nitrogen content of pods without being significantly different from normally irrigated control plants (CW+). Similarly, the highest values of nitrogen content in seeds were recorded for GW- and TWW-. Positive effects were observed for K content in the pods of normally irrigated plants that received G treatment (GW+), whereas no significant differences were observed in the K content of seeds for the tested treatments. EKW+ and GW+ resulted in the highest values for P content in pods and seeds, respectively, whereas contrasting effects of the irrigation regime × TW treatment combination were observed on Ca content where TWW- and TWW+ increased its content in pods and seed, respectively. The Na content of pods was the highest when EK treatment was applied regardless of the irrigation regime, while similar results were observed in seeds for the control and V treatment. The Mg content of pods and seeds increased when no biostimulants or the V treatment were applied on water-stressed plants, respectively. Moreover, the Cu content of pods was beneficially affected by the EK treatment, regardless of the irrigation regime, whereas for seeds, the highest values were recorded for the TWW+ treatment. Regarding the Zn content, the highest values in pods and seeds were observed for the EKW+ and VW- treatments, respectively. Finally, the Mn content of pods increased when V and G treatments were applied, regardless of the irrigation regime, whereas for seeds, the application of biostimulants had a negative effect on Mn content when compared to control treatments (CW- and CW+). The impact of biostimulants on the nutrient content of agricultural products could be attributed to the fact that they usually contain various minerals in their composition [5]. Mineral uptake from plants may help in maintaining high stomatal conductance and leaf water potential; therefore, biostimulants may improve the nutritional status of plants and induce tolerance to abiotic stress factors such as drought stress [56]. In addition, according to Chrysargyris et al. [18], the application of *Ascophyllum nodosum* seaweed extracts alleviated the negative effects of K deficiency on lettuce plants, while the beneficial effect of AMFs as biostimulants has been associated with higher P uptake from plants [57]. This was the case in our study under normal irrigation conditions where the P content of pods was the highest for the EK treatment. The results from the study of Colla et al. [58] confirm the beneficial effect of biostimulant application on the Ca content of tomato fruit for the seaweed extracts treatment, which was also observed in our study for TW treatment, regardless of the irrigation regime. Moreover, it has been reported in several studies that the inoculation of plants with mixtures of bacteria has better results in nutrient mobilization and uptake compared to inoculation with a single bacterium [26], which was also the case in our study where EK treatment increased the P, Na, Cu, and Zn content in the pods of normally irrigated plants.

Table 4. Mineral composition of the studied pods (second harvest) and seeds of beans in relation to the irrigation regime, expressed on a dry weight basis (mean ± SD).

Treatment	N (g/kg)	K (g/kg)	P (g/kg)	Na (g/kg)	Ca (g/kg)	Mg (mg/kg)	Cu (mg/kg)	Zn (mg/kg)	Mn (mg/kg)
					Pods				
CW+ [¥]	26.9 ± 0.3a	17.7 ± 0.3c	6.3 ± 0.4b	0.05 ± 0.03c	6.6 ± 1.6e	104 ± 5h	138 ± 17c	24 ± 1d	4.1 ± 0.4b
VW+	24.6 ± 0.2b	18.7 ± 0.4b	4.2 ± 0.4e	0.07 ± 0.06c	4.3 ± 0.9h	235 ± 17c	78 ± 18e	29 ± 2c	4.5 ± 0.7a
EKW+	24.6 ± 0.3b	19.1 ± 0.3b	7.3 ± 0.2a	0.16 ± 0.02a	5.4 ± 0.7g	131 ± 21g	205 ± 18a	38 ± 4a	3.8 ± 0.4c
GW+	23.0 ± 0.5cd	20.0 ± 0.4a	6.2 ± 0.4b	0.07 ± 0.03c	6.1 ± 1f	219 ± 46d	170 ± 15b	21 ± 1e	4.5 ± 0.1a
TWW+	23.2 ± 0.8c	19.0 ± 0.5b	4.2 ± 0.4e	0.17 ± 0.03a	10.7 ± 3.2c	314 ± 6b	162 ± 19b	33 ± 3b	4.0 ± 0.5bc
CW-	22.4 ± 0.3e	16.7 ± 0.3d	5.2 ± 0.7d	0.03 ± 0.01d	4 ± 1.3h	341 ± 16a	98 ± 18d	29 ± 2c	3.8 ± 0.2c
VW-	26.5 ± 0.5a	15.1 ± 0.2f	6.0 ± 0.2c	0.14 ± 0.02b	11.5 ± 3.5b	171 ± 12e	46 ± 16f	28 ± 2c	4.2 ± 0.8ab
EKW-	23.2 ± 0.2c	17.9 ± 0.6c	5.5 ± 0.1d	0.16 ± 0.08a	8.5 ± 2.6d	147 ± 8f	200 ± 27a	28 ± 3c	3.8 ± 0.5c
GW-	21.5 ± 0.1f	17.1 ± 0.8d	5.3 ± 0.7d	0.07 ± 0.01c	10.2 ± 2.3c	311 ± 14b	169 ± 21b	21.8 ± 0.3e	4.4 ± 0.3a
TWW-	22.8 ± 0.5d	16.1 ± 0.3e	5.9 ± 0.2c	0.06 ± 0.01c	13.2 ± 4.7a	311 ± 35b	145 ± 12c	20 ± 3e	4.1 ± 0.2b
					Seeds				
CW+	42.9 ± 0.3b	1.6 ± 0.2a	13.3 ± 1.4b	0.18 ± 0.08a	11.1 ± 2.3e	7 ± 1f	158 ± 45c	13 ± 2g	5 ± 1a
VW+	41.3 ± 0.4d	1.6 ± 0.3a	9.4 ± 1.8g	0.16 ± 0.05a	6.8 ± 1.7f	11 ± 3cd	99 ± 38e	21 ± 1b	0.5 ± 0.2g
EKW+	41.9 ± 0.8c	1.6 ± 0.2a	12.7 ± 2.3c	0.05 ± 0.01b	10.9 ± 1.4e	6 ± 3fg	193 ± 11b	13 ± 2g	4.4 ± 0.9cd
GW+	41.9 ± 0.1c	1.7 ± 0.3a	14.0 ± 6.6a	0.05 ± 0.01b	12.6 ± 2.5c	5 ± 4g	165 ± 15c	18 ± 3cd	3.5 ± 0.4e
TWW+	41.9 ± 0.1c	1.6 ± 0.3a	8.0 ± 0.9h	0.03 ± 0.01b	15.2 ± 1.5a	13 ± 1b	199 ± 39a	17 ± 3de	4.6 ± 0.4bc
CW-	39.5 ± 0.3e	1.6 ± 0.4a	10.0 ± 2.5f	0.15 ± 0.04a	10.9 ± 2.4e	10.0 ± 0.6de	120 ± 50d	13 ± 1g	5.1 ± 0.3a
VW-	42.8 ± 0.5b	1.6 ± 0.1a	7.2 ± 0.5i	0.15 ± 0.06a	3.7 ± 1.3g	15 ± 3a	90 ± 8f	22.6 ± 0.3a	2 ± 1f
EKW-	42.2 ± 0.8c	1.6 ± 0.9a	11.9 ± 2.6d	0.03 ± 0.01b	11.7 ± 1.7d	2.9 ± 0.3h	190 ± 49b	16 ± 2ef	4.2 ± 0.3d
GW-	44.7 ± 0.2a	1.6 ± 0.6a	11.0 ± 1.2e	0.04 ± 0.01b	14.1 ± 1.3b	9.2 ± 0.6e	168 ± 30c	19.2 ± 0.9c	4.2 ± 0.4d
TWW-	44.7 ± 0.2a	1.6 ± 0.1a	7.5 ± 1.2hi	0.04 ± 0.01b	12.3 ± 3.7c	12 ± 2bc	160 ± 43c	15.1 ± 0.2f	4.8 ± 0.5b

[¥] W+: indicates normal irrigation regime; W-: indicates water-holding irrigation regime; C: Control; V: Veramin Ca; EK: EKOprop; G: Nomoren; TW: Twin-Antistress. Means in the same column and the same plant part (pods and seeds) followed by different Latin letters are significantly different according to Tukey's honestly significant difference (HSD) test ($p = 0.05$).

3.4. *Tocopherols*

The main detected tocopherols in pods were γ-tocopherol, followed by α-tocopherol, while seeds contained mainly γ-tocopherol and less amounts of δ- and α-tocopherol (Table 5). The application of TW treatment resulted in a significant increase of γ-tocopherol (104% and 18.3% for the TWW+ and TWW- treatments, respectively) and total tocopherols (82.3% and 19.4% for the TWW+ and TWW- treatments, respectively) in the pods of the first harvest compared to the control treatments (CW+ and CW-). In contrast, the application of G treatment had a negative effect on the γ-tocopherol content in pods of the first harvest under water stress conditions (reduced by 45.7%), while EK treatment resulted in the lowest content of α-tocopherol for the same irrigation regime (1800 ± 40 μg/kg dw). In the second harvest, the highest values for γ- and total tocopherols were recorded for the VW- treatment (3500 ± 20 μg/kg dw and 5250 ± 10 μg/kg dw, respectively), while pods from the control treatment (CW-) contained the highest amounts of α-tocopherol (1810 ± 20 μg/kg dw). In addition, drought stress increased the individual and total tocopherols content for all the biostimulant treatments, except for G treatment, where normally irrigated plants had a higher content of tocopherols compared to water-stressed ones. The observed increase of tocopherols in pods under prolonged water stress conditions could be attributed to the induction of self-defense mechanisms by biostimulants application and the production of antioxidant compounds such as tocopherols [59]. However, the variable effects of the tested biostimulants indicate a diverse plants x biostimulant interaction, as well as the induction of different mechanisms in each combination depending on the biostimulant composition and the severity of stress [60]. Therefore, although in the first harvest TWW- treatment induced tocopherols biosynthesis as a non-enzymatic antioxidant mechanism, under prolonged stress conditions, VW- (second harvest) and VW+ (seeds) treatments were beneficial to tocopherols content. Regarding seeds, the presence of α- and γ-tocopherol has been previously reported in common bean seeds by Kan et al. [61]. In our study, γ-, δ-, and the total tocopherols content was the highest under normal irrigation conditions and for those plants that did not receive biostimulants or the V treatment was applied. In contrast, EK treatment had a negative effect on tocopherols content under water stress conditions. According to the literature, Ca and amino acids supplementation (as in the case of V treatment in our study) may induce the biosynthesis of non-enzymatic antioxidants such as tocopherols and increase tolerance against drought stress [62–65].

Table 5. Composition in tocopherols of the studied pods (μg/kg dw) and seeds (mg/kg dw) of beans in relation to the irrigation regime (mean ± SD).

Treatment	α-Tocopherol	γ-Tocopherol	Total Tocopherols
		1st Harvest of Pods	
CW+ [¥]	770 ± 20f	2970 ± 20i	3740 ± 10i
VW+	880 ± 40c	4880 ± 30f	5760 ± 10f
EKW+	930 ± 30b	5230 ± 50d	6160 ± 20d
GW+	306 ± 3g	4020 ± 60g	4330 ± 50g
TWW+	760 ± 20f	6060 ± 90b	6820 ± 70b
CW-	798 ± 3e	5416 ± 1c	6214 ± 3c
VW-	810 ± 30d	4990 ± 50e	5800 ± 90e
EKW-	278 ± 3i	3910 ± 30h	4190 ± 20h
GW-	290 ± 20h	2940 ± 60j	3240 ± 80j
TWW-	1010 ± 10a	6410 ± 40a	7420 ± 30a

Table 5. *Cont.*

Treatment	α-Tocopherol	γ-Tocopherol		Total Tocopherols
		1st Harvest of Pods		
		2nd Harvest of pods		
CW+	288 ± 6g	1560 ± 60i		1840 ± 60i
VW+	236 ± 6h	1670 ± 10h		1900 ± 10h
EKW+	266 ± 3g	1530 ± 30j		1800 ± 40j
GW+	508 ± 6e	3280 ± 20b		3790 ± 30d
TWW+	160 ± 10i	1940 ± 40g		2100 ± 30g
CW-	1810 ± 20a	2820 ± 30d		4630 ± 50b
VW-	1750 ± 20b	3500 ± 20a		5250 ± 10a
EKW-	720 ± 20d	2440 ± 20e		3160 ± 10e
GW-	326 ± 8f	2000 ± 40f		2320 ± 50f
TWW-	1210 ± 10c	3090 ± 20c		4300 ± 40c
		Seeds		
	α-Tocopherol	γ-Tocopherol	δ-Tocopherol	Total Tocopherols
CW+	0.96 ± 0.01a	39.1 ± 0.1b	2.48 ± 0.03a	42.5 ± 0.1a
VW+	0.52 ± 0.01c	39.8 ± 0.1a	2.28 ± 0.01b	42.7 ± 0.1a
EKW+	0.50 ± 0.01c	38.5 ± 0.1c	1.95 ± 0.02f	40.9 ± 0.1c
GW+	0.52 ± 0.01c	36.7 ± 0.1d	2.21 ± 0.03c	39.5 ± 0.1d
TWW+	0.53 ± 0.02c	38.4 ± 0.1c	2.01 ± 0.03e	41.0 ± 0.1c
CW-	0.46 ± 0.01d	36.1 ± 0.1e	2.16 ± 0.05d	38.7 ± 0.1e
VW-	0.57 ± 0.01b	35.4 ± 0.1f	1.91 ± 0.02f	38.0 ± 0.1f
EKW-	0.96 ± 0.05a	32.0 ± 0.1h	1.48 ± 0.02g	34.5 ± 0.1h
GW-	0.59 ± 0.02b	39.0 ± 0.1b	1.91 ± 0.02f	41.5 ± 0.1b
TWW-	0.45 ± 0.02d	34.7 ± 0.1g	2.14 ± 0.08d	37.3 ± 0.1g

[¥] W+: indicates normal irrigation regime; W-: indicates water-holding irrigation regime; C: Control; V: Veramin Ca; EK: EKOprop; G: Nomoren; TW: Twin-Antistress. Means in the same column and the same harvest (1st and 2nd pod harvests and seeds) followed by different Latin letters are significantly different according to Tukey's honestly significant difference (HSD) test ($p = 0.05$).

3.5. Organic Acids

The main detected organic acids in pods were malic and oxalic acid, while ascorbic acid was detected in less amounts in specific treatments of the first harvest (Table 6). On the other hand, malic and oxalic acid were the main organic acids detected in seeds, followed by ascorbic acid and traces of fumaric acid. The application of the VW+ treatment resulted in the highest content of oxalic and malic acid and total organic acids in pods of the first harvest (26.3 ± 0.1 g/kg dw, 23.1 ± 0.1 g/kg/dw, and 49.4 ± 0.1 g/kg/dw, respectively). Similarly, in the second harvest, the highest content of oxalic and malic acid was recorded for the GW+ and VW- treatments (22.7 ± 0.2 g/kg/dw and 24.6 ± 0.2 g/kg/dw, respectively), while total organic acids content was most abundant in the GW+ treatment (46.9 ± 0.1 g/kg/dw). Regarding seeds, VW- treatment resulted in the highest content of oxalic and malic acid, and total organic acids (752 ± 1 mg/kg/dw, 1440 ± 40 mg/kg/dw, and 2780 ± 30mg/kg/dw, respectively), whereas the highest amounts of ascorbic acid were detected in GW+ treatment (715 ± 4 mg/kg dw). Considering the antinutritional properties of oxalic acid, it is worth mentioning that EKW- treatment resulted in the lowest content for both pod harvests when compared with the rest of the treatments where biostimulants were applied, although in all the cases, the oxalic acid content was considerably low. On the other hand, in the case of seeds, the application of EKW+ and TWW+ treatments significantly reduced the oxalic acid content compared to the control and the rest of the biostimulant treatments. Although there are reports in the literature that suggest that organic acids increase under stress conditions, according to Zushi and Matsuzoe [66], this increase could be attributed only to a concentration effect due to the increase in dry matter content under stress conditions. According to other studies, the composition of biostimulants may significantly affect the organic acids composition, especially those biostimulant products that contain microorganisms such as Twin-Antistress and EKOprop in our study [36,67]. However, although the total organic acids content of pods harvested from water-stressed plants was in general higher in biostimulant-treated plants compared to the control treatment (CW-), the application of the EKW- treatment resulted in a significant reduction of organic acids content. This finding is reflected to the reduced total pod yield for this treatment (see Table 2), suggesting a non-effective alleviating mechanism against water stress related to biostimulant product composition. Moreover, the effect of V treatment on the oxalic acid content of seeds under water stress conditions (VW-) could be attributed to Ca addition, which is associated with calcium oxalate formation for the removal of excessive calcium or oxalic acid [67].

3.6. Free Sugars

The sugars composition of pods and seeds is presented in Table 6. The main detected sugar in the pods of both harvests was fructose, followed by glucose and sucrose, whereas in seeds, only sucrose was detected. Similarly with our study, Kan et al. [61] detected sucrose as the main sugar in common bean seeds, while they also detected the presence of glucose; this difference could be attributed to the different harvesting stages (fully dried seeds comparing to fully developed green seeds in our study), which may affect hydrolysis and the transformation of sugars after harvest [68]. In the first harvest, the highest content of individual and total sugars were recorded in TWW- (fructose: 198 ± 1 g/kg dw), CW- (glucose: 135 ± 5 g/kg dw; total sugars: 333 ± 7 g/kg dw), and VW- (sucrose: 7.6 ± 1 g/kg dw) treatments. Under prolonged water stress (second harvest), the application of G treatment (GW-) resulted in the highest content of fructose, glucose, and total sugars (232 ± 7 g/kg dw, 140 ± 7 g/kg dw, and 380 ± 10 g/kg dw, respectively), while the sucrose content was the highest for the TWW- treatment (15.3 ± 0.8 g/kg/dw). Similarly, the highest content of sucrose in seeds was recorded for the GW- treatment (22.9 ± 0.7 g/kg dw). The low levels of sucrose in pods could be attributed to the inhibitory activity of hexose sugars (fructose and glucose) to sucrose synthase activity [68]. Moreover, for most of the tested biostimulants and control treatments, the total and individual sugars content was higher in water-stressed plants than normally irrigated plants, especially for G treatment—that resulted in the highest total sugars content, which could be associated with osmoprotective effects against water stress [69]. Considering the involvement of soluble sugars in plant defense mechanisms as well as in the regulation of stress and growth-related genes, the findings of our study suggest an efficient defense mechanism against water stress for the AMF-containing biostimulant product (G treatment), as already justified by the increased pods yield under water stress conditions for the same treatment (see Table 2). According to the literature, inoculation with AMF is associated with increased soluble sugars content in *Ipomea batatas* and *Vigna subterranea* under drought stress, since sugars may serve as organic carbon pools to be used for photosynthates and biomass production [70]. Apart from the osmoregulatory role of sugars in plant defense mechanisms against stress, sucrose content is also related with secondary metabolites biosynthesis, which may also contribute to the overall non-enzymatic tolerance of plants under stress [71].

Table 6. Composition in organic acids and sugars of the studied pods and seeds of beans in relation to the irrigation regime (mean ± SD).

	Oxalic Acid (g/kg dw)	Malic Acid (g/kg dw)	Ascorbic Acid (g/kg dw)	Total Organic Acids (g/kg dw)	Fructose (g/kg dw)	Glucose (g/kg dw)	Sucrose (g/kg dw)	Total Sugars (g/kg dw)
Treatment				1st Harvest of pods				
CW+ ¥	14.0 ± 0.1h	18.4 ± 0.2d	tr	32.5 ± 0.1h	151 ± 4f	84.5 ± 0.1g	5.8 ± 0.4b	241 ± 4h
VW+	26.3 ± 0.1a	23.1 ± 0.1a	tr	49.4 ± 0.1a	168 ± 5e	90 ± 2f	nd	257 ± 7f
EKW+	21.8 ± 0.2c	20.4 ± 0.5c	0.10 ± 0.01b	42.4 ± 0.4b	194 ± 3b	107 ± 3c	nd	301 ± 6c
GW+	19.3 ± 0.1e	21.7 ± 0.4b	tr	41.0 ± 0.3d	173 ± 0d	85 ± 3g	4.7 ± 0.3d	263 ± 4e
TWW+	23.2 ± 0.1b	18.1 ± 0.2de	0.5 ± 0.1a	41.8 ± 0.3c	173 ± 1d	101 ± 2d	3.5 ± 0.1f	277 ± 2d
CW-	18.5 ± 0.1f	16.5 ± 0.2f	0.50 ± 0.03a	35.5 ± 0.1f	194 ± 2b	135 ± 5a	4.3 ± 0.1e	333 ± 7a
VW-	17.0 ± 0.1g	20.6 ± 0.4c	0.10 ± 0.01b	37.7 ± 0.4e	153 ± 8f	84 ± 6g	7.6 ± 0.1a	250 ± 10g
EKW-	16.4 ± 0.1g	17.9 ± 0.3e	tr	34.3 ± 0.3g	171 ± 4d	106 ± 3c	3.1 ± 0.3g	280 ± 8d
GW-	20.9 ± 0.1d	16.7 ± 0.1f	tr	37.7 ± 0.2e	184 ± 2c	94 ± 6e	3.1 ± 0.3g	281 ± 8d
TWW-	21.2 ± 0.1cd	20.4 ± 0.3c	0.50 ± 0.01a	42.1 ± 0.4bc	198 ± 1a	110 ± 3b	5.2 ± 0.4c	314 ± 3b
				2nd Harvest of pods				
CW+ ¥	10.9 ± 0.1d	15.0 ± 0.4h	tr	25.9 ± 0.4h	170 ± 2e	82 ± 4g	3.5 ± 0.4e	256 ± 6h
VW+	10.7 ± 0.1e	17.8 ± 0.1d	tr	28.5 ± 0.1f	180 ± 2d	91 ± 2f	2.4 ± 0.2f	273 ± 4f
EKW+	10.5 ± 0.1f	17.4 ± 0.1e	tr	28.0 ± 0.1g	181 ± 4d	101 ± 3d	4.3 ± 0.2d	287 ± 6d
GW+	22.7 ± 0.2a	24.1 ± 0.3b	tr	46.9 ± 0.1a	195 ± 1b	105 ± 1c	1.76 ± 0.01g	301 ± 1c
TWW+	8.9 ± 0.2h	16.4 ± 0.3g	tr	25.3 ± 0.4j	170 ± 1e	106 ± 3c	3.5 ± 0.7e	280 ± 3e
CW-	10.9 ± 0.2d	20.6 ± 0.3c	tr	31.5 ± 0.1d	180 ± 5d	96 ± 2e	14 ± 2b	290 ± 9d
VW-	9.9 ± 0.2g	24.6 ± 0.2a	tr	34.5 ± 0.4c	162 ± 6f	90 ± 3f	13.1 ± 0.5c	265 ± 9g
EKW-	8.3 ± 0.1i	17.1 ± 0.2f	tr	25.5 ± 0.3i	191 ± 3c	110 ± 10b	13.3 ± 0.9c	318 ± 14b
GW-	16.3 ± 0.1b	20.4 ± 0.2c	tr	36.8 ± 0.2b	232 ± 7a	140 ± 7a	4.2 ± 0.4d	380 ± 10a
TWW-	13.2 ± 0.1c	17.3 ± 0.3e	tr	30.6 ± 0.4e	163 ± 4f	77 ± 3h	15.3 ± 0.8a	256 ± 8h

Table 6. *Cont.*

	Seeds						
	Oxalic acid (mg/kg dw)	Malic acid (mg/kg dw)	Ascorbic acid (mg/kg dw)	Fumaric acid (mg/kg dw)	Total organic acids (mg/kg dw)	Sucrose (g/kg dw)	Total sugars (g/kg dw)
CW+ ¥	224 ± 3g	1170 ± 20c	657 ± 6b	tr	2050 ± 30d	19.5 ± 0.3c	19.5 ± 0.3c
VW+	293 ± 1f	1383 ± 7b	612 ± 3c	tr	2290 ± 10b	19.3 ± 0.9c	19.3 ± 0.9c
EKW+	185 ± 8h	970 ± 30d	524 ± 2e	tr	1680 ± 40f	22.3 ± 0.5b	22.3 ± 0.5b
GW+	600 ± 10b	910 ± 60e	715 ± 4a	tr	2220 ± 60c	16.6 ± 0.5f	16.6 ± 0.5f
TWW+	185 ± 3h	316 ± 3h	488 ± 9g	tr	990 ± 20i	18.9 ± 0.4d	18.9 ± 0.4d
CW-	530 ± 10d	tr	530 ± 10e	tr	1040 ± 40h	19.2 ± 0.6c	19.2 ± 0.6c
VW-	752 ± 1a	1440 ± 40a	580 ± 10d	tr	2780 ± 30a	15.5 ± 0.7g	15.5 ± 0.7g
EKW-	579 ± 8c	690 ± 30f	515 ± 7f	tr	1780 ± 40e	17.1 ± 0.5e	17.1 ± 0.5e
GW-	530 ± 10d	tr	396 ± 3i	tr	964 ± 1j	22.9 ± 0.7a	22.9 ± 0.7a
TWW-	440 ± 10e	490 ± 20g	449 ± 9h	tr	1380 ± 20g	15 ± 1h	15 ± 1h

¥ W+: indicates normal irrigation regime; W-: indicates water-holding irrigation regime; C: Control; V: Veramin Ca; EK: EKOprop; G: Nomoren; TW: Twin-Antistress. Means in the same column and the same harvest (1st and 2nd pod harvests and seeds) followed by different Latin letters are significantly different according to Tukey's honestly significant difference (HSD) test ($p = 0.05$). Tr: traces.

3.7. Fatty Acids

The main fatty acids composition is presented in Table 7. Seventeen individual fatty acids were detected in pods and seeds regardless of the irrigation treatment and harvest (data not shown). Pods were abundant in α-linolenic (C18:3n3), linoleic (C18:2n6c), and palmitic acid (C16:0) followed by stearic (C18:0), oleic (C18:1n9c), behenic (C22:0), and lignoceric acid (C24:0), which were detected in lower amounts. Similarly, in seeds, the most abundant fatty acids were α-linolenic, linoleic, and palmitic acid, followed by stearic and oleic acid. In the first harvest of pods, GW- and EKW- treatments had a beneficial effect on palmitic and linoleic acid content in water-stressed plants (25.6 ± 0.3% and 35.02 ± 0.01%, respectively), whereas in normally irrigated plants, TWW+ treatment resulted in the highest content of α-linolenic acid (42.76 ± 0.06%). In the second harvest, fatty acids composition showed a varied response, with control and TW treatment resulting in the highest content of linoleic and palmitic acid for normally irrigated plants (43.3 ± 0.1% and 28.90 ± 0.06, respectively), whereas α-linolenic acid content was the highest for the GW- treatment (34.33 ± 0.09%). For seeds, the highest amounts of α-linolenic, linoleic and palmitic acid were recorded in the treatments of GW-, CW+, and VW+ (59.08 ± 0.02%, 29.54 ± 0.02%, and 11.86 ± 0.03%, respectively). Polyunsaturated fatty acids (PUFA) were the most abundant fatty acids class, followed by saturated (SFA) and monounsaturated fatty acids (MUFA) in both seeds and pods due to the high amounts of α-linolenic and linoleic acids. Overall, the ratios of PUFA/SFA and n-6/n-3 fatty acids were higher than 0.45 and lower than 4.0 for all the tested treatments, respectively, which according to Petropoulos et al. [72] is indicative for the good nutritional value of a food product. Moreover, the increase of PUFAs under water stress conditions comparing to the control treatment (CW-) for all the biostimulant treatments except for GW- (first harvest) and EKW- (second harvest) treatments indicates the stimulation of plant antioxidant mechanisms which effectively quenched the reactive oxygen species (ROS) that appear after stress initiation and induce lipid peroxidation and decrease fatty acids content [10,73]. Fatty acids may also serve as organic carbon pools to be used for photosynthates and biomass production [70]. Therefore, considering that inoculation with AMFs and bacteria induces synergistic effects between plants and symbionts that may improve plant nutrient and water uptake, this could be the reason for the increased content of fatty acids.

Table 7. The main fatty acids composition (%) of the studied pods and seeds of common bean (mean ± SD).

	CW+ [¥]	VW+	EKW+	GW+	TWW+	CW-	VW-	EKW-	GW-	TWW-
					1st Harvest of pods					
C16:0	21.72 ± 0.04d	20.87 ± 0.07f	19.35 ± 0.05i	24.1 ± 0.1b	18.4 ± 0.1j	23.55 ± 0.06c	19.9 ± 0.1h	20.26 ± 0.05g	25.6 ± 0.3a	21.18 ± 0.09e
C18:0	5.30 ± 0.01c	4.87 ± 0.01d	3.95 ± 0.01g	9.81 ± 0.02a	3.65 ± 0.01h	5.35 ± 0.03c	4.23 ± 0.02f	4.27 ± 0.01f	5.59 ± 0.03b	4.48 ± 0.01e
C18:1n9c	1.75 ± 0.02d	1.34 ± 0.01g	1.51 ± 0.01f	2.08 ± 0.01b	1.60 ± 0.01e	1.79 ± 0.01d	1.72 ± 0.01d	2.18 ± 0.01a	1.93 ± 0.01c	1.53 ± 0.01f
C18:2n6c	27.60 ± 0.03f	28.69 ± 0.04d	28.65 ± 0.01d	26.98 ± 0.01g	29.94 ± 0.03b	28.52 ± 0.05d	28.60 ± 0.01d	32.02 ± 0.01a	27.74 ± 0.08e	29.33 ± 0.04c
C18:3n3	38.50 ± 0.01f	39.53 ± 0.04d	42.21 ± 0.04b	32.62 ± 0.01j	42.76 ± 0.06a	35.51 ± 0.02h	41.36 ± 0.06c	36.83 ± 0.05g	33.1 ± 0.1i	38.7 ± 0.1e
C22:0	1.27 ± 0.03b	0.96 ± 0.09d	0.82 ± 0.01f	1.06 ± 0.06c	0.77 ± 0.01g	1.47 ± 0.05a	0.86 ± 0.03e	0.88 ± 0.05e	1.46 ± 0.05a	0.99 ± 0.02d
C24:0	1.09 ± 0.02f	1.29 ± 0.02c	1.15 ± 0.01e	1.03 ± 0.02g	1.04 ± 0.01g	1.32 ± 0.04c	1.24 ± 0.02d	1.12 ± 0.02ef	1.56 ± 0.02a	1.39 ± 0.01b
SFA	31.06 ± 0.03d	29.33 ± 0.02e	26.80 ± 0.07h	37.64 ± 0.03a	25.02 ± 0.07h	33.35 ± 0.08c	27.62 ± 0.06g	27.87 ± 0.06f	35.9 ± 0.2b	29.42 ± 0.06e
MUFA	2.75 ± 0.06c	2.30 ± 0.05f	2.18 ± 0.02g	2.62 ± 0.01d	2.17 ± 0.01g	2.43 ± 0.01e	2.26 ± 0.02f	3.12 ± 0.01a	2.99 ± 0.02b	2.25 ± 0.01f
PUFA	66.19 ± 0.03f	68.37 ± 0.07e	71.02 ± 0.05b	59.74 ± 0.02i	72.81 ± 0.08a	64.22 ± 0.09g	70.12 ± 0.07c	69.01 ± 0.07d	61.1 ± 0.2h	68.33 ± 0.06e
					2nd Harvest of pods					
C16:0	18.8 ± 0.2g	26.81 ± 0.08b	22.4 ± 0.1d	21.4 ± 0.1e	28.90 ± 0.06a	23.6 ± 0.1c	17.09 ± 0.06i	23.6 ± 0.1c	19.1 ± 0.2f	17.7 ± 0.1h
C18:0	4.38 ± 0.03f	9.68 ± 0.02a	4.81 ± 0.01d	4.00 ± 0.01g	6.78 ± 0.01b	5.35 ± 0.02c	3.69 ± 0.01h	5.30 ± 0.03c	4.68 ± 0.03e	4.38 ± 0.01f
C18:1n9c	1.48 ± 0.01h	1.71 ± 0.01f	1.75 ± 0.01f	1.46 ± 0.01h	2.29 ± 0.01b	1.68 ± 0.01g	1.95 ± 0.01e	2.13 ± 0.01d	2.21 ± 0.02c	2.82 ± 0.01a
C18:2n6c	28.46 ± 0.02d	24.80 ± 0.01g	28.46 ± 0.04d	28.89 ± 0.05c	26.04 ± 0.02f	28.39 ± 0.01d	32.46 ± 0.02b	28.03 ± 0.04e	34.33 ± 0.09a	28.41 ± 0.01d
C18:3n3	43.3 ± 0.1a	31.42 ± 0.01h	35.0 ± 0.1g	40.02 ± 0.07d	29.67 ± 0.01i	36.12 ± 0.03e	40.43 ± 0.07c	35.08 ± 0.03g	35.63 ± 0.07f	42.78 ± 0.02b
C22:0	0.64 ± 0.02h	1.1 ± 0.1c	1.7 ± 0.1a	1.04 ± 0.04d	0.927 ± 0.005f	0.95 ± 0.02e	1.18 ± 0.06b	1.68 ± 0.06a	0.46 ± 0.01i	0.720 ± 0.008g
C24:0	0.901 ± 0.008f	1.32 ± 0.01a	1.03 ± 0.01e	1.17 ± 0.02d	1.27 ± 0.05b	1.22 ± 0.03c	0.91 ± 0.02f	1.20 ± 0.04cd	0.930 ± 0.006f	0.84 ± 0.01g
SFA	26.1 ± 0.1h	40.95 ± 0.03a	31.75 ± 0.02e	29.1 ± 0.1f	40.23 ± 0.09b	32.94 ± 0.05d	24.3 ± 0.1j	33.6 ± 0.1c	26.8 ± 0.2g	25.50 ± 0.07i
MUFA	1.93 ± 0.01h	2.63 ± 0.04e	4.6 ± 0.2a	1.78 ± 0.01i	3.90 ± 0.07b	2.37 ± 0.03g	2.52 ± 0.01f	2.99 ± 0.02d	3.08 ± 0.02c	3.06 ± 0.02c
PUFA	71.9 ± 0.1b	56.42 ± 0.01i	63.6 ± 0.1g	69.1 ± 0.1e	55.87 ± 0.02j	64.68 ± 0.02f	73.1 ± 0.1a	63.43 ± 0.09h	70.2 ± 0.2d	71.44 ± 0.05c
					Seeds					
C16:0	11.60 ± 0.04b	11.86 ± 0.03a	11.16 ± 0.06c	10.46 ± 0.04i	10.73 ± 0.05g	10.93 ± 0.04d	10.58 ± 0.01h	10.83 ± 0.05f	10.89 ± 0.01e	10.38 ± 0.02j
C18:0	2.61 ± 0.01b	2.69 ± 0.01a	2.19 ± 0.01h	2.25 ± 0.01g	2.10 ± 0.01i	2.54 ± 0.01c	2.38 ± 0.01e	2.43 ± 0.01d	2.33 ± 0.01f	2.27 ± 0.01g
C18:1n9c	2.35 ± 0.01b	1.80 ± 0.01d	1.95 ± 0.02c	1.54 ± 0.01h	1.58 ± 0.02g	2.56 ± 0.01a	1.70 ± 0.01e	1.63 ± 0.01f	1.46 ± 0.01i	1.79 ± 0.01d
C18:2n6c	29.54 ± 0.02a	25.28 ± 0.01e	27.36 ± 0.03b	26.00 ± 0.01c	24.89 ± 0.03h	24.97 ± 0.01f	25.84 ± 0.01d	24.31 ± 0.03j	24.52 ± 0.01i	24.93 ± 0.02g
C18:3n3	51.48 ± 0.03i	55.96 ± 0.02g	55.54 ± 0.02h	57.75 ± 0.05d	58.41 ± 0.05c	56.88 ± 0.01f	57.46 ± 0.04e	58.72 ± 0.03b	59.08 ± 0.02a	58.77 ± 0.03b
SFA	16.15 ± 0.06b	16.66 ± 0.01a	14.63 ± 0.05f	14.32 ± 0.03h	14.45 ± 0.01g	15.30 ± 0.02c	14.66 ± 0.03f	15.07 ± 0.07d	14.78 ± 0.01e	14.15 ± 0.02i
MUFA	2.72 ± 0.01a	2.02 ± 0.02e	2.39 ± 0.03b	1.84 ± 0.01f	2.17 ± 0.06c	2.77 ± 0.01a	1.97 ± 0.02e	1.82 ± 0.01f	1.59 ± 0.01g	2.08 ± 0.02d
PUFA	81.13 ± 0.05i	81.32 ± 0.03h	82.98 ± 0.02f	83.84 ± 0.05a	83.39 ± 0.07d	81.93 ± 0.01g	83.37 ± 0.05d	83.11 ± 0.06e	83.63 ± 0.01c	83.77 ± 0.04b

C16:0 palmitic acid, C18:0 stearic acid, C18:1n9c oleic acid, C18:2n6c linoleic acid, C18:3n3 c, C22:0 behenic acid, C24:0 lignoceric acid; SFA: saturated fatty acids; MUFA: monounsaturated fatty acids; PUFA: polyunsaturated fatty acids; n6/n3: ratio of omega-6/omega-3 fatty acids. [¥] W+: indicates normal irrigation regime; W-: indicates water-holding irrigation regime; C: Control; V: Veramin Ca; EK: EKOprop; G: Nomoren; TW: Twin-Antistress. Means in the same row followed by different Latin letters are significantly different according to Tukey's honestly significant difference (HSD) test ($p = 0.05$).

3.8. LDA Analysis

3.8.1. First Harvest

In the first harvest, the linear discriminant analysis (LDA) selected PUFA, C17:0, ascorbic acid, glucose, α-tocopherol, C16:1, C14:0, C16:0, carbohydrates, and C22:1 as variables with discriminant ability, which is equivalent to say that these were the parameters showing the most profound changes in result of using different biostimulants (Figure 2). Function 1 separated primarily samples treated with G, which was placed in the farthest position in the negative side of the axis; in turn, the biostimulants with the most similar effect according to this function (which was the most important, as it included 85.1% of the observed variance) were V and TW. In turn, Function 2 separated mainly C samples, which was mostly due to their higher contents in carbohydrates and glucose (in this case, specifically in W-samples). The most noticeable effect of Function 3 was the individualization of markers corresponding to EK samples, which was mostly due to the levels of PUFA, C16:0, and C14:0.

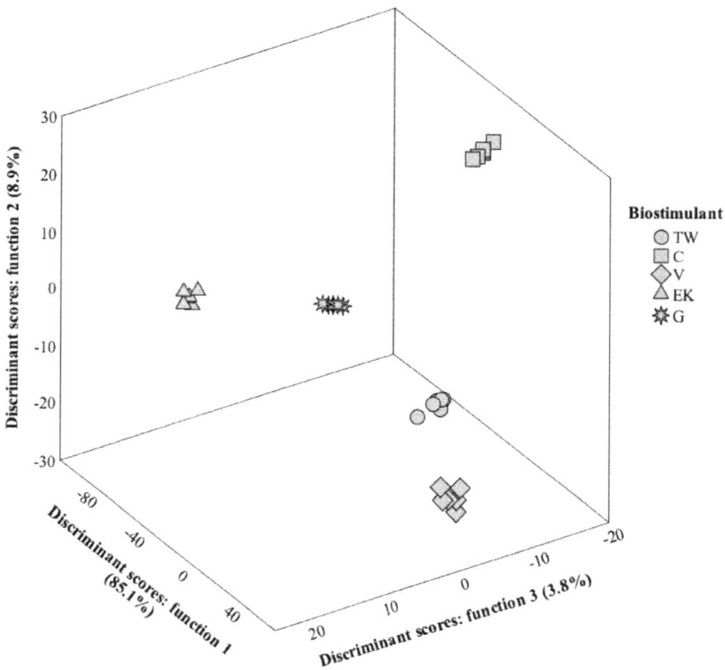

Figure 2. Canonical discriminant functions coefficients defined from the evaluated parameters plotted to show the effect of biostimulants treatments on *Phaseolus vulgaris* green pods of the first harvest under different irrigation regimes (normal irrigation and water stress).

3.8.2. Second Harvest

Concerning the second harvest, the LDA selected C22:0, fructose, organic acids, C21:0, proteins, C16:1, C24:0, C20:1, C18:0, C18:1n9c, MUFA, and sugars as the variables with the highest differences as a result of using different biostimulants (highest discriminant ability) (Figure 3). Function 1 was especially effective in separating samples treated with TW or EK, which was mostly due to their higher C18:1n9c contents (regardless of the irrigation treatment). On the other hand, Function 2 separated samples treated with G, which was placed in the farthest position in comparison to control samples. Considering all the functions together, it was possible to conclude that the biostimulant treatment V was the one that induced the least differences in comparison to the control.

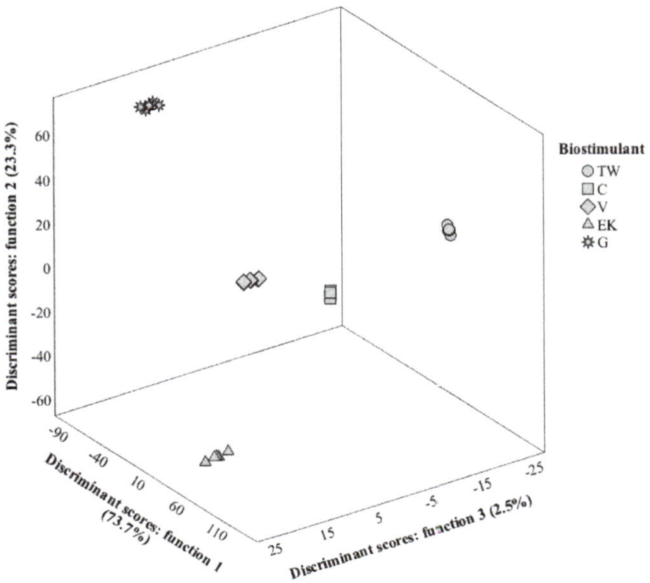

Figure 3. Canonical discriminant functions coefficients defined from the evaluated parameters plotted to show the effect of biostimulants treatments on *Phaseolus vulgaris* green pods of the second harvest under different irrigation regimes (normal irrigation and water stress).

3.8.3. Seeds

The effects of the tested biostimulants on seeds were also more pronounced for fatty acids, as indicated by the variables classified as being discriminant: C20:1, C20:0, sucrose, C22:1, lipids, C17:0, C15:0, organic acids, PUFA, C16:0, C18:2n6c, and C16:1 (Figure 4). According to Function 1, all biostimulants had similar effects (markers are almost vertically aligned), while untreated samples (C) were completely individualized (negative side of the axis); among the selected variables, the one showing the highest correlation with this function was C20:1, which showed higher percentages in C samples, independently of water level. Function 2, in turn, was mostly correlated to C16:1 and sucrose, contributing mainly to separate samples treated with EK, while Function 3 was more highly correlated with lipids content, contributing to separate samples treated with G.

In all the former LDAs, the classification performance was 100% accurate both for originally grouped cases as well as for cross-validated ones.

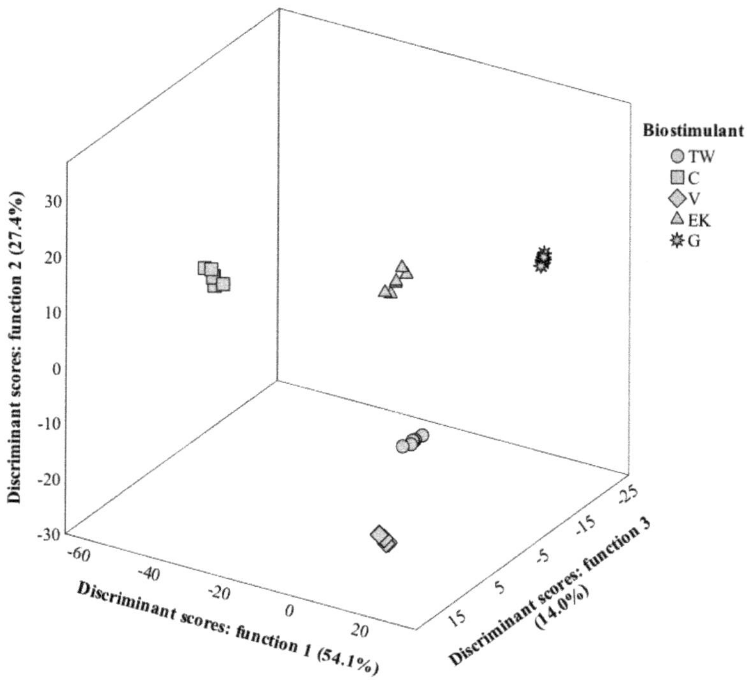

Figure 4. Canonical discriminant functions coefficients defined from the evaluated parameters plotted to show the effect of biostimulants treatments on *Phaseolus vulgaris* seeds under different irrigation regimes (normal irrigation and water stress).

4. Conclusions

The results of the present study showed a varied effect of biostimulants and water treatments on pod yield and the quality of common bean green pods and seeds, while significant differences were also observed between normally irrigated and water-stressed plants in a biostimulants treatment-specific manner. Promising results were also recorded regarding the alleviation of negative effects of drought stress where the application of arbuscular mycorrhizal fungi (AMF; G treatment) increased the crop yield of green beans. Moreover, the nutritional value and chemical composition of pods and seeds was positively affected by biostimulants application, although a product specific effect was recorded depending on the irrigation regime and harvesting time (pods and/or seeds). In conclusion, the application of biostimulants could be considered as an eco-friendly and sustainable tool to increase the pod yield and quality of common bean green pods and seeds under normal irrigation and/or drought stress conditions. Considering that the tested biostimulants contain beneficial microorganisms such as AMF, symbiotic rizosphere bacteria, and saprophytic fungi, its application not only could benefit crops but it could also improve soil properties and preserve soil quality. However, future research is needed to investigate in depth the mechanisms of action of biostimulant product, the application dose efficiency, as well as the most effective application regime and the possible effect of genotype x biostimulants interactions.

Author Contributions: S.A.P. conceived and designed the research, administered and supervised the project, carried out the cultivation, wrote the original draft, and reviewed and edited the final manuscript; A.F. performed chemical analyses, data curation, and methodology; S.P. carried out the cultivation and prepared the original draft; A.C. performed chemical analyses, data curation, and methodology; N.T. performed chemical analyses, data curation, prepared the original draft, and edited the final manuscript; J.C.M.B. performed the LDA analysis of the data and the interpretation of the statistical analysis results; L.B. performed chemical analyses, data curation, and methodology, wrote the original draft, and reviewed and edited the final manuscript; I.C.F.R.F. obtained funding, administered and supervised the project, and reviewed and edited the final manuscript. All authors have read and agreed to the published version of the manuscript.

Funding: The authors are grateful to the Foundation for Science and Technology (FCT, Portugal) and FEDER under Programme PT2020 for financial support to CIMO (UID/AGR/00690/2013), A. Fernandes and L. Barros contract.

Conflicts of Interest: The authors declare no conflict of interest.

References

1. Huang, Z.A.; Jiang, D.A.; Yang, Y.; Sun, J.W.; Jin, S.H. Effects of nitrogen deficiency on gas exchange, chlorophyll fluorescence, and antioxidant enzymes in leaves of rice plants. *Photosynthetica* **2004**, *42*, 357–364. [CrossRef]
2. Zhao, D.; Reddy, K.R.; Kakani, V.G.; Reddy, V.R. Nitrogen deficiency effects on plant growth, leaf photosynthesis, and hyperspectral reflectance properties of sorghum. *Eur. J. Agron.* **2005**, *22*, 391–403. [CrossRef]
3. Liebman, M.; Davis, A.S. Integration of soil, crop and weed management in low-external-input farming systems. *Weed Res.* **2000**, *40*, 27–47. [CrossRef]
4. Postel, S.L. Entering an era of water scarcity: The challenges ahead. *Ecol. Appl.* **2000**, *10*, 941–948. [CrossRef]
5. Bulgari, R.; Cocetta, G.; Trivellini, A.; Vernieri, P.; Ferrante, A. Biostimulants and crop responses: A review. *Biol. Agric. Hortic.* **2015**, *31*, 1–17. [CrossRef]
6. Van Oosten, M.J.; Pepe, O.; De Pascale, S.; Silletti, S.; Maggio, A. The role of biostimulants and bioeffectors as alleviators of abiotic stress in crop plants. *Chem. Biol. Technol. Agric.* **2017**, *4*, 1–12. [CrossRef]
7. Ziosi, V.; Zandoli, R.; Di Nardo, A.; Biondi, S.; Antognoni, F.; Calandriello, F. Biological activity of different botanical extracts as evaluated by means of an array of in vitro and in vivo bioassays. *Acta Hortic.* **2012**, *1009*, 61–66. [CrossRef]
8. Abdel Latef, A.A.H.; Chaoxing, H. Does inoculation with *Glomus mosseae* improve salt tolerance in pepper plants? *J. Plant Growth Regul.* **2014**, *33*, 644–653. [CrossRef]
9. Abu Seif, Y.I.; El-Miniawy, S.E.D.M.; Abu El-Azm, N.A.I.; Hegazi, A.Z. Response of snap bean growth and seed yield to seed size, plant density and foliar application with algae extract. *Ann. Agric. Sci.* **2016**, *61*, 187–199. [CrossRef]
10. Ahmad, P.; Hashem, A.; Abd-Allah, E.F.; Alqarawi, A.A.; John, R.; Egamberdieva, D.; Gucel, S. Role of Trichoderma harzianum in mitigating NaCl stress in Indian mustard (*Brassica juncea* L) through antioxidative defense system. *Front. Plant Sci.* **2015**, *6*, 1–15. [CrossRef]
11. Halpern, M.; Bar-Tal, A.; Ofek, M.; Minz, D.; Muller, T.; Yermiyahu, U. The use of biostimulants for enhancing nutrient uptake. *Adv. Agron.* **2015**, *130*, 141–174.
12. Koleška, I.; Hasanagić, D.; Todorović, V.; Murtić, S.; Klokić, I.; Paradiković, N.; Kukavica, B. Biostimulant prevents yield loss and reduces oxidative damage in tomato plants grown on reduced NPK nutrition. *J. Plant. Interact.* **2017**, *12*, 209–218. [CrossRef]
13. Rouphael, Y.; Franken, P.; Schneider, C.; Schwarz, D.; Giovannetti, M.; Agnolucci, M.; De Pascale, S.; Bonini, P.; Colla, G. Arbuscular mycorrhizal fungi act as biostimulants in horticultural crops. *Sci. Hortic. (Amsterdam)* **2015**, *196*, 91–108. [CrossRef]
14. Kunicki, E.; Grabowska, A.; Sękara, A.; Wojciechowska, R. The effect of cultivar type, time of cultivation, and biostimulant treatment on the yield of spinach (*Spinacia oleracea* L.). *Folia Hortic.* **2010**, *22*, 9–13. [CrossRef]
15. Pereira, J.L.; Queiroz, R.M.L.; Charneau, S.O.; Felix, C.R.; Ricart, C.A.O.; Lopes Da Silva, F.; Steindorff, A.S.; Ulhoa, C.J.; Noronha, E.F. Analysis of *Phaseolus vulgaris* response to its association with *Trichoderma harzianum* (ALL-42) in the presence or absence of the phytopathogenic fungi Rhizoctonia solani and Fusarium solani. *PLoS ONE* **2014**, *9*, 1–23. [CrossRef]
16. Rouphael, Y.; Giordano, M.; Cardarelli, M.; Cozzolino, E.; Mori, M.; Kyriacou, M.C.; Bonini, P.; Colla, G. Plant-and seaweed-based extracts increase yield but differentially modulate nutritional quality of greenhouse spinach through biostimulant action. *Agronomy* **2018**, *8*, 126. [CrossRef]
17. Xu, C.; Leskovar, D.I. Effects of *A. nodosum* seaweed extracts on spinach growth, physiology and nutrition value under drought stress. *Sci. Hortic. (Amsterdam)* **2015**, *183*, 39–47. [CrossRef]

18. Chrysargyris, A.; Xylia, P.; Anastasiou, M.; Pantelides, I.; Tzortzakis, N. Effects of *Ascophyllum nodosum* seaweed extracts on lettuce growth, physiology and fresh-cut salad storage under potassium deficiency. *J. Sci. Food Agric.* **2018**, *98*, 5861–5872. [CrossRef]
19. Galvão, Í.M.; dos Santos, O.F.; de Souza, M.L.C.; de Jesus Guimarães, J.; Kühn, I.E.; Broetto, F. Biostimulants action in common bean crop submitted to water deficit. *Agric. Water Manag.* **2019**, *225*, 105762. [CrossRef]
20. De Pascale, S.; Rouphael, Y.; Colla, G. Plant biostimulants: Innovative tool for enhancing plant nutrition in organic farming. *Eur. J. Hortic. Sci.* **2018**, *82*, 277–285. [CrossRef]
21. Karimzadeh Soureshjani, H.; Nezami, A.; Kafi, M.; Tadayon, M. Responses of two common bean (*Phaseolus vulgaris* L.) genotypes to deficit irrigation. *Agric. Water Manag.* **2019**, *213*, 270–279. [CrossRef]
22. Hummel, M.; Hallahan, B.F.; Brychkova, G.; Ramirez-Villegas, J.; Guwela, V.; Chataika, B.; Curley, E.; McKeown, P.C.; Morrison, L.; Talsma, E.F.; et al. Reduction in nutritional quality and growing area suitability of common bean under climate change induced drought stress in Africa. *Sci. Rep.* **2018**, *8*, 1–11. [CrossRef] [PubMed]
23. Arruda, I.M.; Moda-Cirino, V.; Koltun, A.; dos Santos, O.J.A.P.; Moreira, R.S.; Moreira, A.F.P.; Gonçalves, L.S.A. Physiological, biochemical and morphoagronomic characterization of drought-tolerant and drought-sensitive bean genotypes under water stress. *Physiol. Mol. Biol. Plants* **2018**, *24*, 1059–1067. [CrossRef] [PubMed]
24. Sarma, R.K.; Saikia, R. Alleviation of drought stress in mung bean by strain *Pseudomonas aeruginosa* GGRJ21. *Plant. Soil* **2014**, *377*, 111–126. [CrossRef]
25. Sadak, M.S.; Abdelhamid, M.T.; Schmidhalter, U. Effect of foliar application of aminoacids on plant yield and some physiological parameters in bean plants irrigated with seawater. *Acta Biol. Colomb.* **2015**, *20*, 141–152.
26. Kumar, M.; Mishra, S.; Dixit, V.; Kumar, M.; Agarwal, L.; Chauhan, P.S.; Nautiyal, C.S. Synergistic effect of *Pseudomonas putida* and *Bacillus amyloliquefaciens* ameliorates drought stress in chickpea (*Cicer arietinum* L.). *Plant. Signal. Behav.* **2016**, *11*, 1–9. [CrossRef]
27. German, M.A.; Burdman, S.; Okon, Y.; Kigel, J. Effects of *Azospirillum brasilense* on root morphology of common bean (*Phaseolus vulgaris* L.) under different water regimes. *Biol. Fertil. Soils* **2000**, *32*, 259–264. [CrossRef]
28. Klimek-Kopyra, A.; Kliszcz, A.; Ślizowska, A.; Kot, D. Application of biostimulants influences shoot and root characteristics of seedlings of winter pea (Pisum sativum L.). *Acta Agrobot.* **2019**, *72*, 1–8. [CrossRef]
29. Dourado-Neto, D.; Dario, G.J.A.; Barbieri, A.P.P.; Martin, T.N. Biostimulant action on agronomic efficiency of corn and common beans. *Biosci. J.* **2014**, *30*, 371–379.
30. Rady, M.M.; Desoky, E.S.M.; Elrys, A.S.; Boghdady, M.S. Can licorice root extract be used as an effective natural biostimulant for salt-stressed common bean plants? *S. Afr. J. Bot.* **2019**, *121*, 294–305. [CrossRef]
31. Rady, M.M.; Mohamed, G.F. Modulation of salt stress effects on the growth, physio-chemical attributes and yields of *Phaseolus vulgaris* L. plants by the combined application of salicylic acid and *Moringa oleifera* leaf extract. *Sci. Hortic. (Amsterdam)* **2015**, *193*, 105–113. [CrossRef]
32. Selby, C.; Carmichael, E.; Sharma, H.S.S. Bio-refining of perennial ryegrass (*Lolium perenne*): Evaluation of aqueous extracts for plant defence elicitor activity using French bean cell suspension cultures. *Chem. Biol. Technol. Agric.* **2016**, *3*, 1–7. [CrossRef]
33. Elzaawely, A.A.; Ahmed, M.E.; Maswada, H.F.; Xuan, T.D. Enhancing growth, yield, biochemical, and hormonal contents of snap bean (*Phaseolus vulgaris* L.) sprayed with moringa leaf extract. *Arch. Agron. Soil Sci.* **2017**, *63*, 687–699. [CrossRef]
34. Elzaawely, A.A.; Ahmed, M.E.; Maswada, H.F.; Al-Araby, A.A.; Xuan, T.D. Growth traits, physiological parameters and hormonal status of snap bean (*Phaseolus vulgaris* L.) sprayed with garlic cloves extract. *Arch. Agron. Soil Sci.* **2018**, *64*, 1068–1082. [CrossRef]
35. Petropoulos, S.A.; Taofiq, O.; Fernandes, Â.; Tzortzakis, N.; Ciric, A.; Sokovic, M.; Barros, L.; Ferreira, I.C. Bioactive properties of greenhouse-cultivated green beans (*Phaseolus vulgaris* L.) under biostimulants and water-stress effect. *J. Sci. Food Agric.* **2019**, *99*, 6049–6059. [CrossRef] [PubMed]
36. Pereira, C.; Dias, M.I.; Petropoulos, S.A.; Plexida, S.; Chrysargyris, A.; Tzortzakis, N.; Calhelha, R.C.; Ivanov, M.; Stojković, D.; Soković, M.; et al. The Effects of Biostimulants, Biofertilizers and Water-Stress on Nutritional Value and Chemical Composition of Two Spinach Genotypes (*Spinacia oleracea* L.). *Molecules* **2019**, *24*, 4494. [CrossRef]
37. AOAC International. *Official Methods of Analysis of AOAC International*; Horwitz, W., Latimer, G., Eds.; AOAC International: Gaithersburg, MD, USA, 2016.

38. Chrysargyris, A.; Nikolaidou, E.; Stamatakis, A.; Tzortzakis, N. Vegetative, physiological, nutritional and antioxidant behavior of spearmint (*Mentha spicata* L.) in response to different nitrogen supply in hydroponics. *J. Appl. Res. Med. Aromat. Plants* **2017**, *6*, 52–61. [CrossRef]
39. Pereira, C.; Barros, L.; Carvalho, A.M.; Ferreira, I.C.F.R. Use of UFLC-PDA for the analysis of organic acids in thirty-five species of food and medicinal plants. *Food Anal. Methods* **2013**, *6*, 1337–1344. [CrossRef]
40. Barros, L.; Pereira, E.; Calhelha, R.C.; Dueñas, M.; Carvalho, A.M.; Santos-Buelga, C.; Ferreira, I.C.F.R. Bioactivity and chemical characterization in hydrophilic and lipophilic compounds of *Chenopodium ambrosioides* L. *J. Funct. Foods* **2013**, *5*, 1732–1740. [CrossRef]
41. Barros, L.; Pereira, C.; Ferreira, I.C.F.R. Optimized analysis of organic acids in edible mushrooms from Portugal by Ultra-Fast Liquid Chromatography and Photodiode Array Detection. *Food Anal. Methods* **2013**, *6*, 309–316. [CrossRef]
42. Aimo, S.; Gosetti, F.; D'Agostino, G.; Gamalero, E.; Gianotti, V.; Bottaro, M.; Gennaro, M.C.; Berta, G. Use of arbuscular mycorrhizal fungi and beneficial soil bacteria to improve yield and quality of saffron (*Crocus sativus* L.). *Acta Hortic.* **2010**, 159–164. [CrossRef]
43. Weber, N.; Schmitzer, V.; Jakopic, J.; Stampar, F. First fruit in season: Seaweed extract and silicon advance organic strawberry (*Fragaria×ananassa* Duch.) fruit formation and yield. *Sci. Hortic. (Amsterdam)* **2018**, *242*, 103–109. [CrossRef]
44. Arthur, G.D.; Stirk, W.A.; Van Staden, J. Effect of a seaweed concentrate on the growth and yield of three varieties of *Capsicum annuum*. *S. Afr. J. Bot.* **2003**, *69*, 207–211. [CrossRef]
45. Alam, M.Z.; Braun, G.; Norrie, J.; Hodges, D.M. Effect of *Ascophyllum* extract application on plant growth, fruit yield and soil microbial communities of strawberry. *Can. J. Plant. Sci.* **2013**, *93*, 23–36. [CrossRef]
46. Alam, M.Z.; Braun, G.; Norrie, J.; Mark Hodges, D. *Ascophyllum* extract application can promote plant growth and root yield in carrot associated with increased root-zone soil microbial activity. *Can. J. Plant. Sci.* **2014**, *94*, 337–348. [CrossRef]
47. Battacharyya, D.; Babgohari, M.Z.; Rathor, P.; Prithiviraj, B. Seaweed extracts as biostimulants in horticulture. *Sci. Hortic. (Amsterdam)* **2015**, *196*, 39–48. [CrossRef]
48. Korir, H.; Mungai, N.W.; Thuita, M.; Hamba, Y.; Masso, C. Co-inoculation effect of rhizobia and plant growth promoting rhizobacteria on common bean growth in a low phosphorus soil. *Front. Plant. Sci.* **2017**, *08*. 1–10. [CrossRef]
49. Farouk, S.; Abdul Qados, A.M.S. Enhancing seed quality and productivity as well as physio-anatomical responses of pea plants by folic acid and/or hydrogen peroxide application. *Sci. Hortic. (Amsterdam)* **2018**, *240*, 29–37. [CrossRef]
50. Poorter, H.; Nagel, O. The role of biomass allocation in the growth response of plants to different levels of light, CO2, nutrients and water: A quantitative review. *Aust. J. Plant. Physiol.* **2000**, *27*, 595–607.
51. Colla, G.; Nardi, S.; Cardarelli, M.; Ertani, A.; Lucini, L.; Canaguier, R.; Rouphael, Y. Protein hydrolysates as biostimulants in horticulture. *Sci. Hortic. (Amsterdam)* **2015**, *196*, 28–38. [CrossRef]
52. Przygocka-Cyna, K.; Grzebisz, W. Effect of biofertilizers on nutrient uptake by vegetables grown in a short cropping sequence. *J. Elem.* **2018**, *23*, 807–823. [CrossRef]
53. Elsheikh, E.A.E.; El Tinay, A.H.; Fadul, I.A. Effect of nutritional status of faba bean on proximate composition, anti-nutritional factors and in vitro protein digestibility (IVPD). *Food Chem.* **1999**, *67*, 379–383. [CrossRef]
54. Kocira, A.; Świeca, M.; Kocira, S.; Złotek, U.; Jakubczyk, A. Enhancement of yield, nutritional and nutraceutical properties of two common bean cultivars following the application of seaweed extract (*Ecklonia maxima*). *Saudi, J. Biol. Sci.* **2018**, *25*, 563–571. [CrossRef] [PubMed]
55. Castellanos-Barriga, L.G.; Santacruz-Ruvalcaba, F.; Hernández-Carmona, G.; Ramírez-Briones, E.; Hernández-Herrera, R.M. Effect of seaweed liquid extracts from *Ulva lactuca* on seedling growth of mung bean (*Vigna radiata*). *J. Appl. Phycol.* **2017**, *29*, 2479–2488. [CrossRef]
56. Mancuso, S.; Azzarello, E.; Mugnai, S.; Briand, X. Marine bioactive substances (IPA extract) improve foliar ion uptake and water stress tolerance in potted *Vitis vinifera* plants. *Adv. Hortic. Sci.* **2006**, *20*, 156–161.
57. Gosling, P.; Mead, A.; Proctor, M.; Hammond, J.P.; Bending, G.D. Contrasting arbuscular mycorrhizal communities colonizing different host plants show a similar response to a soil phosphorus concentration gradient. *New Phytol.* **2013**, *198*, 546–556. [CrossRef]

58. Colla, G.; Cardarelli, M.; Bonini, P.; Rouphael, Y. Foliar applications of protein hydrolysate, plant and seaweed extracts increase yield but differentially modulate fruit quality of greenhouse tomato. *HortScience* **2017**, *52*, 1214–1220. [CrossRef]
59. Huang, D.; Ou, B.; Rior, R.L. The chemistry behind antioxidant capacity assays. *J. Agric. Food Chem.* **2005**, *53*, 1841–1856. [CrossRef]
60. Kellős, T.; Tímár, I.; Szilágyi, V.; Szalai, G.; Galiba, G.; Kocsy, G. Stress hormones and abiotic stresses have different effects on antioxidants in maize lines with different sensitivity. *Plant. Biol.* **2008**, *10*, 563–572. [CrossRef]
61. Kan, L.; Nie, S.; Hu, J.; Wang, S.; Bai, Z.; Wang, J.; Zhou, Y.; Jiang, J.; Zeng, Q.; Song, K. Comparative study on the chemical composition, anthocyanins, tocopherols and carotenoids of selected legumes. *Food Chem.* **2018**, *260*, 317–326. [CrossRef]
62. Elkelish, A.A.; Alnusaire, T.S.; Soliman, M.H.; Gowayed, S.; Senousy, H.H.; Fahad, S. Calcium availability regulates antioxidant system, physio-biochemical activities and alleviates salinity stress mediated oxidative damage in soybean seedlings. *J. Appl. Bot. Food Qual.* **2019**, *92*, 258–266.
63. El-Bassiouny, H.M.S.; El-Monem, A.A.A.; Abdallah, M.M.-S.; Soliman, K.M. Role of arbuscular mycorrhiza, α-tocopherol and nicotinamide on the nitrogen containing compounds and adaptation of sunflower plant to Water stress. *Biosci. Res.* **2018**, *15*, 2068–2088.
64. Laxa, M.; Liebthal, M.; Telman, W.; Chibani, K.; Dietz, K.J. The role of the plant antioxidant system in drought tolerance. *Antioxidants* **2019**, *8*. [CrossRef] [PubMed]
65. Casadesús, A.; Polo, J.; Munné-Bosch, S. Hormonal effects of an enzymatically hydrolyzed animal protein-based biostimulant (pepton) in water-stressed tomato plants. *Front. Plant. Sci.* **2019**, *10*, 1–11. [CrossRef] [PubMed]
66. Zushi, K.; Matsuzoe, N.; Kitano, M. Developmental and tissue-specific changes in oxidative parameters and antioxidant systems in tomato fruits grown under salt stress. *Sci. Hortic. (Amsterdam)* **2009**, *122*, 362–368. [CrossRef]
67. Palmieri, F.; Estoppey, A.; House, G.L.; Lohberger, A.; Bindschedler, S.; Chain, P.S.G.; Junier, P. Chapter Two - Oxalic acid, a molecule at the crossroads of bacterial-fungal interactions. *Adv. Appl. Microbiol.* **2019**, *106*, 49–77.
68. Weber, H.; Borisjuk, L.; Wobus, U. Sugar import and metabolism during seed development. *Trends Plant. Sci.* **1997**, *2*, 169–174. [CrossRef]
69. Rathinasabapathi, B. Metabolic engineering for stress tolerance: Installing osmoprotectant synthesis pathways. *Ann. Bot.* **2000**, *86*, 709–716. [CrossRef]
70. Begum, N.; Qin, C.; Ahanger, M.A.; Raza, S.; Khan, M.I.; Ashraf, M.; Ahmed, N.; Zhang, L. Role of Arbuscular Mycorrhizal Fungi in Plant Growth Regulation: Implications in Abiotic Stress Tolerance. *Front. Plant. Sci.* **2019**, *10*, 1–15. [CrossRef]
71. Rosa, M.; Prado, C.; Podazza, G.; Interdonato, R.; González, J.A.; Hilal, M.; Prado, F.E. Soluble sugars-metabolism, sensing and abiotic stress. A complex network in the life of plants. *Plant. Signal. Behav.* **2009**, *4*, 388–393. [CrossRef]
72. Petropoulos, S.; Ntatsi, G.; Levizou, E.; Barros, L.; Ferreira, I. Nutritional profile and chemical composition of *Cichorium spinosum* ecotypes. *LWT - Food Sci. Technol.* **2016**, *73*, 95–101. [CrossRef]
73. López-Bucio, J.; Pelagio-Flores, R.; Herrera-Estrella, A. Trichoderma as biostimulant: Exploiting the multilevel properties of a plant beneficial fungus. *Sci. Hortic. (Amsterdam)* **2015**, *196*, 109–123. [CrossRef]

agronomy

MDPI

Article

Changes in Biochemistry and Yield in Response to Biostimulants Applied in Bean (*Phaseolus vulgaris* L.)

Anna Kocira [1,*], Joanna Lamorska [1], Rafał Kornas [1], Natalia Nowosad [1], Marzena Tomaszewska [1], Danuta Leszczyńska [2], Katarzyna Kozłowicz [3] and Sylwester Tabor [4]

1 Institute of Agricultural Sciences, State School of Higher Education in Chełm, Pocztowa 54, 22-100 Chełm, Poland; j.lamorska@gmail.com (J.L.); rkornas@pwsz.chelm.pl (R.K.); nnowosad@pwsz.chelm.pl (N.N.); mtomaszewska@pwsz.chelm.pl (M.T.)

2 Institute of Soil Science and Plant Cultivation, State Research Institute, Czartoryskich 8, 24-100 Puławy, Poland; leszcz@iung.pulawy.pl

3 Department of Biological Bases of Food and Feed Technologies, University of Life Sciences in Lublin, Głęboka 28, 20-612 Lublin, Poland; katarzyna.kozlowicz@up.lublin.pl

4 Department of Production Engineering, Logistics and Applied Computer Science, Agricultural University Kraków, Balicka 116B, 30-149 Kraków, Poland; sylwester.tabor@urk.edu.pl

* Correspondence: akocira@pwsz.chelm.pl; Tel.: +48-82-565-8895

Received: 22 December 2019; Accepted: 28 January 2020; Published: 31 January 2020

Abstract: Biostimulants are preparations that favorably impact the growth, development, and yield of plants. The research objective was to examine the effect of the frequency of use of Kelpak, Terra Sorb Complex and Fylloton biostimulants on improving the yield and nutritional properties of beans. Been seeds (variety Oczko) were sown in the first week of May in 2015, 2016, and 2017. During the growing season, Fylloton (1%), Terra Sorb Complex (0.5%), and Kelpak (1%) biostimulants were applied by single (BBCH 12-13) and double spraying of plants (BBCH 12-13, BBCH 61). All variants of treatment with biostimulants were compared with the control. Single application of Kelpak had a positive effect on increasing the number of pods. The double application of Kelpak increased the number and yield of seeds and protein contents. Double application of Fylloton increased the number of seeds, and application of Terra Sorb Complex increased the protein content in the beans. Application of all biostimulants increased the flavonoid content. Biostimulants containing seaweed (Kelpak–*Ecklonia maxima* extract) or amino-acid extracts (Fylloton–*Ascophyllum nodosum* extract and amino acids or Terra Sorb Complex–amino acids) increased the seed yield, while improving its quality by increasing the content of protein, polyphenols, and flavonoids. It was found that the double application of Kelpak biostimulant stimulated the yield and quality of beans to a greater extent.

Keywords: bean; biostimulants; amino acids; seaweed extract; yield; protein; phenols; flavonoids

---

1. Introduction

Agricultural production seeks technological solutions to improve the quality of yields. Therefore, biostimulants are increasingly popular as preparations that favorably impact the growth, development, and yield of plants [1,2], and they are safe for humans and environmentally friendly at the same time. Du Jardin [3] defines biostimulants as “any substance or microorganism applied to plants with the aim of enhancing nutrition efficiency, abiotic stress tolerance, and/or crop quality traits, regardless of its nutrient content”. In addition, in order to develop legal provisions regarding the registration of biostimulants based on their specificity of operation, the European Biostimulants Industry Council (EBIC) was created. Currently, however, their registration is based on legal provisions on fertilizers and pesticides, and, for some of them, there is a marketing gap in many European Union (EU) member states [3–6].

Depending on the origin, there are natural or synthetic biostimulants. The former are obtained from biological material, and the latter are structurally similar and functionally identical to biological material [7]. The group of natural biostimulants includes preparations based on free amino acids, humic compounds, seaweed or fruit extracts, chitin and its derivative, chitosan, or microbial inoculants (free-living bacteria, fungi, and arbuscular mycorrhizal fungi) [8–10]. Of this group, biostimulants containing seaweed extract and protein hydrolysates are the most important category of substances that stimulate plant growth and development [11,12].

Biostimulants affect the metabolic processes occurring in the plant, stimulating the synthesis or activity of phytohormones, facilitating the growth of the root system, and improving the uptake, translocation, and utilization of nutrients, which determines the quality of the obtained yield [3,8–12]. Moreover, biostimulants increase plant resistance to abiotic stress factors such as drought, frost, salinity, and environmental contamination with heavy metals, which is probably caused by changes in the enzymatic activity of antioxidant compounds and their increased synthesis [8,13].

In bean cultivation, the most commonly used are extracts of brown algae, e.g., of the species *Ascophyllum nodosum*, *Laminaria* spp., *Ecklonia maxima*, *Sargassum* spp., and *Fucus* spp. [14–17]. The positive effect of seaweed extracts on plant growth, development, and yield is attributed to the presence of phytohormones and low-molecular-weight compounds [18,19]. Some authors suggest that the polysaccharides and polyphenols present in the extract are also important, since they are allelochemicals which increase plant resistance to stress conditions [20–22]. Generally, organic and mineral compounds occur in seaweed extracts, the content of which depends on the algae species, their harvest date, and the applied extraction process [8]. The most important bioactive ingredients include proteins, enzymes, and amino acids (glycoproteins, metalloproteins, exogenous amino acids such as aspartic acid, glutamic acid, alanine), phytohormones (auxins, cytokinins, gibberellins, abscisic acid, polyamines, betaine, ethylene, brassinosteroids, brasinolide, castasterone), polyphenols (florotanins, ecol, floroglucin), phytalexins, vitamins (C, B_2, B_{12}, D_3, E, K, niacin, panthotenic and folic acid), oligosaccharides, polysaccharides (agar, hyaluronic acid, alginic acid and its salts, carrageenans, fucans, mannitol, sorbitol, laminarin), macro- and microelements (Mg, Cu, Fe, Br, Zn, I, Mn), and essential unsaturated fatty acids (arachidonic, eicosapentenoic, γ-linolenic). Although vitamin A is not present in seaweed extracts, the presence of its precursors, i.e., carotene and possibly fucoxanthin was detected [14,23–31].

The positive effect of seaweed extracts on plants is visible in the stimulation of phytohormone synthesis, the uptake and translocation of nutrients, and soil conditioning, which is done by improving water–air conditions and the activity of beneficial soil microorganisms [32,33]. It was demonstrated that, even in low concentrations, seaweed extracts induce a number of physiological processes in the plant, contributing to their better growth, flowering, yield size, and quality, and improving the nutritional and storage quality of crop plants. In addition, the use of seaweed extracts increases the tolerance of plants to unfavorable growing conditions, for example, salinity, drought, or extreme temperatures [12,34,35]. The use of seaweed extract has a positive effect on plant growth and the size and quality of the obtained yield of tomato, eggplant, pepper, lettuce, beans, soybean, and wheat [13,25,36–46]. Undoubtedly, the beneficial effects of biostimulants on plants, under both optimal and stressful growing conditions, can be associated with stimulation of enzymatic activity related to carbon and nitrogen metabolism, the Krebs cycle, and glycolysis. Treatment of plants with these preparations may induce activity similar to that of phytohormones (auxin, gibberellin), which in effect improves their nutrition by modifying the structure of the root system [8,9,47].

The biostimulant Kelpak (Kelp Products Ltd.) is based on an extract from *Ecklonia maxima* (Osbeck) Papenfuss and contains auxins (11 mg·dm^{-3}), cytokinins (0.031 mg·dm^{-3}), alginates (1.5 g·L^{-1}), amino acids (total 441.3 mg·100 g^{-1}), mannitol (2261 mg·L^{-1}), neutral sugars (1.08 g·L^{-1}), and small amounts of macroelements (N 0.09%, P 90.7 mg·kg^{-1}, K 7163.3 mg·kg^{-1}, Ca 190.4 mg·kg^{-1}, Mg 337.2 mg·kg^{-1}, Na 1623.7 mg·kg^{-1}) and microelements (mean composition: Mn 17.3 mg·kg^{-1}, Fe 40.7 mg·kg^{-1}, Cu 13.5 mg·kg^{-1}, Zn 17.0 mg·kg^{-1}, B 33.0 mg·kg^{-1}) [14,48]. The very high auxin-to-cytokinin ratio is responsible for stimulating the growth and development of the root system, which in turn contributes to better uptake and translocation of macro- and microelements, and it is associated with a significant increase in crops [48].

Biostimulants based on protein products and protein hydrolysates consist of a mixture of peptides, animal or vegetal amino acids, and single amino acids. Amino acids are a building material for proteins, but they are also precursors of phytohormones. They are involved in the synthesis of, e.g., vitamins, enzymes, terpenes, amines, purines, pyrimidines, and alkaloids [49,50]. They also play an important role in the process of pollination and fruit formation [51]. The application of exogenous amino acids, which are active in metabolic signaling (glutamate, histidine, proline, glycine, betaine), induces plant defense mechanisms by increasing their resistance to abiotic stress factors [8,52]. Due to the presence of specific peptides and precursors of phytohormone biosynthesis (tryptophan, which is the main precursor of IAA (Indole-3-acetic acid) biosynthesis and bioactive peptides), protein hydrolysates affect the hormonal balance of plants, which is related to stimulating plant growth [53]. The use of these biopreparations positively impacts the quality of agricultural produce, increasing the content of carotenoids, flavonoids, polyphenols, and ascorbic acid [54–57], and reducing the amount of undesirable compounds, e.g., nitrates [56].

Terra Sorb Complex (Bioiberica, S.A.U.) is a biostimulant containing 20% free vegetal amino acids and 5.5% total N (including 5% organic N), 0.8% MgO, 1.5% B, 1% Fe, 0.1% Mn, 0.001% Mo, 0.1% Zn, and 25% organic matter.

The beneficial effect of biostimulants with seaweed extracts or amino acids on plant growth and development, and on the quantity and quality of the yield, regardless of its developmental stage, was confirmed in numerous studies [13,34,35,39–41,57–62]. An interesting solution is to combine these two components in a single preparation, as in the case of the Fylloton biostimulant (Biolchim Poland), which contains the extract of *Ascophyllum nodosum* (L.) Le Jolis, as well as vegetal amino acids. The composition of this preparation includes *Ascophyllum nodosum* extract, amino acid complexes of vegetal origin 37.5%, organic nitrogen 6%, organic carbon of biological origin 11%, and organic substance 35%.

Bean is an economically important legume that is sensitive to low temperatures in the early stages of its development and flowering. The use of biostimulants that positively affect the metabolic processes occurring in the plant, especially in the time of climate change, which causes stress factors for this sensitive plant, can be one of the elements contributing to the improvement in the quantity and quality of bean yield. Plant response to the biostimulant often depends on the variety, as demonstrated in earlier studies [15,39,59,60]. There are also no reports regarding the reaction of two-colored coat seed of bean to treatment with biostimulants. In view of the above and based on the importance being given to the improving crop yields, the research objective was to investigate the effect of the use of Kelpak, Terra Sorb Complex, and Fylloton biostimulants on improving the yield and nutritional properties of common bean (*Phaseolus vulgaris* L.) variety Oczko.

2. Materials and Methods

2.1. Plant Materials and Growth Conditions

The research material came from field studies carried out in the year 2015–2017 in Perespa (50°66′ north (N); 23°63′ east (E)), Poland, on common beans (*Phaseolus vulgaris* L.), variety Oczko. The experiment was established in a random block system, in four replications, on an area of 10 m^2. The experiment was established on an alkaline (pH in 1 M KCl-7.4) soil of the brown rendzina subtype. Soil fertility level was as follows: phosphorus medium (12.6–14.2 mg P_2O_5 in 100 g of soil), potassium medium (15.3–17.1 mg K_2O in 100 g of soil), and magnesium medium (6.2–6.8 mg Mg in 100 g of soil). In each year of research, the forecrop for common beans was winter wheat. Tillage for bean was carried out in accordance with good agricultural practice [63]. Pre-winter plowing was performed in the first week of November. In the spring, soil treatment combined with mineral fertilization was performed. Mineral fertilizers were in the following doses: 30 kg N·ha^{-1}, 60 kg P_2O_5·ha^{-1}, and 120 kg K_2O·ha^{-1}.

Mineral fertilization was used at a constant level throughout all experimental combinations. Beans of common bean, variety Oczko (with a red and white bean coat), were sown with a mechanical precision bean drill in the first week of May (2 May in 2015, 2016, and 2017) at a depth of 3–4 cm, in rows 45 cm apart, using 30 plants per 1 m^2. Biostimulants Terra Sorb Complex, Kelpak, and Fylloton were used during the growing season, according to the experiment design (Table 1), and the obtained results were compared with the control, in which pure water was used for double spraying the plants. In individual years of research, biostimulants were used in time frames dependent on the development phase of plants, as shown in Table 1. Plants single sprayed with the biostimulant in BBCH 12-13, in the second period (plant stage of BBCH 61), were sprayed with pure water.

Table 1. Overview of biostimulant application in bean variety Oczko cultivation.

Biostimulant	Concentration	Number of Sprays and Plant Developmental Stages (BBCH)	Volume of Working Solution/ Working Pressure	Date		
				2015	2016	2017
Fylloton (F)	1%	Single spraying: BBCH 12–13 (SS)	300 L·ha^{-1}/0.30 MPa	June 5	June 7	June 9
		Double spraying: BBCH 12–13, BBCH 61 (DS)		June 5, June 20	June 7, June 23	June 9, June 26
Terra Sorb Complex (TS)	0.5%	Single spraying: BBCH 12–13 (SS)	300 L·ha^{-1}/0.30 MPa	June 5	June 7	June 9
		Double spraying: BBCH 12–13, BBCH 61 (DS)		June 5, June 20	June 7, June 23	June 9, June 26
Kelpak (K)	1%	Single spraying: BBCH 12–13 (SS)	300 L·ha^{-1}/0.30 MPa	June 5	June 7	June 9
		Double spraying: BBCH 12–13, BBCH 61 (DS)		June 5, June 20	June 7, June 23	June 9, June 26

Abbreviations: single spraying BBCH 12-13—single spraying at the 2–3-leaf stage; double spraying BBCH 12-13, BBCH 61—double spraying first at the 2–3-leaf stage and second at the beginning of bean blooming.

The plants were sprayed using a GARLAND FUM 12B backpack sprayer. The Lechler LU 120–03 atomizer was used, at a working pressure of 0.30 MPa, using 300 L of the working liquid per 1 ha. All variants of treatment with biostimulants were compared with the control, where plants were treated with the same volume of water (no biostimulant was applied). No pesticides were used in the cultivation, as pathogens, pests and weeds did not exceed the damage threshold. Plants were weeded manually. The average temperature and rainfall during the bean growing season are shown in Table 2. The weather station (W200P, Vector Instruments Ltd., Rhyl, UK) was located in the experimental field, in which the experiment was carried out, at 210 m above sea level.

After harvesting plants in the third week of August (22 August 2015; 27 August 2016; 24 August 2017), 20 plants were randomly selected from each plot, and the number of pods, number of seeds, seed yield, and weight of one thousand seeds was determined. The beans obtained from each plot were dried, ground in a laboratory mill, and sieved with a 0.310-mm sieve. The flours were stored at −20 °C and used for further chemical analysis.

Table 2. Conditions during the growing seasons in bean variety Oczko cultivation in 2015–2017.

Month	Year						Average from 2002–2015	
	2015		2016		2017			
	T (°C) Average (min/max)	Rainfall (mm)	T (°C) Average (min/max)	Rainfall (mm)	T (°C) Average (min/max)	Rainfall (mm)	T (°C)	Rainfall (mm)
IV	8.2 (−1.7/24.3)	30.1	9.2 (−1.2/22.6)	68.4	7.7 (−1.6/23.3)	37.2	8.6	41.9
V	12.7 (1.5/24.9)	108.6	13.8 (2.6/26.7)	61.3	13.7 (−1.4/26.9)	100.0	12.6	64.1
VI	17.4 (6.6/30.5)	14.1	18.1 (4.2/31.5)	97.1	18.3 (5.7/30.2)	38.6	17.8	68.3
VII	19.6 (8.4/33.4)	59.2	19.5 (8.8/31.2)	107.6	18.5 (5.3/32.9)	61.1	18.8	79.4
VIII	21.6 (5.6/35.5)	23.4	18.2 (7.1/30.7)	95.3	19.5 (4.3/34.4)	25.5	19.5	71.5
IX	15.1 (4.2/34.5)	137.6	15.2 (1.6/28.7)	41.2	13.2 (−0.3/27.3)	100.4	14.0	69.6
Average/Total	15.8	373.0	17.1	470.9	15.2	362.8	15.2	394.8

Abbreviation: T—temperature.

2.2. Determination of Polyphenols

A ground sample of bean seeds of 0.25 g was weighed, to which 4 cm^3 of extraction solution (acetone:water:hydrochloric acid 70:29:1) was added. The solutions were shaken for 1 h. Then, 100 μL of distilled water and 0.4 mL of Folina–Ciocalteu reagent were added to 100 μL of extract, and, after 10 min, 2 mL of 10% Na_2O_3 solution was added. After 30 min, the absorbance of methanol was measured at $\lambda = 725$ nm. The polyphenol content was calculated in mg·100 g^{-1}, from the gallic acid calibration curve (1 mg·mL^{-1}).

2.3. Determination of Flavonoids

A ground sample of bean seeds of 0.25 g was weighed, to which 4 cm^3 of extraction solution (acetic acid:methanol 1:19) was added. The solutions were shaken for 1 h. Then, 0.1 mL of a 2% $AlCl_3{\cdot}5H_2O$ methanolic solution was added to 1 mL of the extract, together with 1.4 mL of CH_3COOH methanolic solution (1:19). The sample was then incubated at 20 °C for 30 min. Absorbance was measured at $\lambda = 425$ nm against methanol. The flavonoid content was calculated in mg·100 g^{-1}, from the calibration curve for quercetin (0.2 mg·mL^{-1}).

2.4. Determination of Proteins

The protein content of bean extracts was determined using the Bradford reagent, according to the method of Redmile-Gordon et al. [57] with modifications. The Bradford reagent (150 μL) was applied on a microplate and 50 μL of assay or standard protein (BSA, bovine serum albumin) was added. The samples were shaken at room temperature for 15 min. Absorbance at 595 nm was measured using an Epoch Microplate Spectrophotometer (BioTek-USA). The resulting protein was expressed in mg·g^{-1} of dry weight (DW).

2.5. Statistical Analysis

The statistical analysis was performed using the Statistica 10PL program by StatSoft®. The normal distribution of variables was tested using the Shapiro–Wilk test. The one-way (for 2015, 2016, and 2017) and the two-way (for average 2015–2017) analysis of variance was used. The significance of the mean was determined using the Tukey test, at a significance level of $p < 0.05$.

3. Results

Based on the two-way ANOVA analysis, the effect of the number of applications and biostimulant treatment on seed yield, number of seeds, and phenol content in bean seeds was found (Table 3). The effect of biostimulant treatment on the number of pods, the weight of one thousand seeds, and the content of proteins and flavonoids in bean seeds were demonstrated. The interaction of biostimulant treatment with its number of applications regarding the impact on the number of seeds, as well as the content of protein and phenols in bean seeds, was found.

Table 3. Two-way ANOVA of number of applications and treatment of bean variety Oczko (average 2015–2017).

Effect	Sum of Squares	Degrees of Freedom	Mean Squares	F Ratio	*p*-Values
		Seed yield			
Intercept	5,439,549	1	5,439,549	23,099.18	0.000000
Number of applications	3181	1	3181	13.51	0.000697
Treatment	34,486	3	11,495	48.82	0.000000
Number of applications × treatment	1510	3	503	2.14	0.110568
Error	9419	40	235		
		1000 seed weight			
Intercept	10,599,621	1	10,599,621	44,665.77	0.000000
Number of applications	314	1	314	1.32	0.257138
Treatment	11,114	3	3705	15.61	0.000001
Number of applications × treatment	281	3	94	0.40	0.757166
Error	9492	40	237		
		Number of pods			
Intercept	2,570,576	1	2,570,576	10,667.40	0.000000
Number of applications	290	1	290	1.20	0.279127
Treatment	15,025	3	5008	20.78	0.000000
Number of applications × treatment	1506	3	502	2.08	0.117750
Error	9639	40	241		
		Number of seeds			
Intercept	24,743,716	1	24,743,716	33,040.26	0.000000
Number of applications	22,838	1	22,838	30.50	0.000002
Treatment	209,494	3	69,831	93.25	0.000000
Number of applications × treatment	12,377	3	4126	5.51	0.002905
Error	29,956	40	749		
		Protein			
Intercept	20,792.66	1	20,792.66	88,921.99	0.000000
Number of applications	0.39	1	0.39	1.65	0.206357
Treatment	31.28	3	10.43	44.59	0.000000
Number of applications × treatment	2.36	3	0.79	3.37	0.027788
Error	9.35	40	0.23		
		Total phenols			
Intercept	731.7033	1	731.7033	3095.779	0.000000
Number of applications	1.3763	1	1.3763	5.823	0.020490
Treatment	6.8737	3	2.2912	9.694	0.000062
Number of applications × treatment	2.5409	3	0.8470	3.583	0.021918
Error	9.4542	40	0.2364		
		Total flavonoids			
Intercept	0.159506	1	0.159506	7312.597	0.000000
Number of applications	0.000001	1	0.000001	0.024	0.877973
Treatment	0.006708	3	0.002236	102.506	0.000000
Number of applications × treatment	0.000174	3	0.000058	2.663	0.060961
Error	0.000873	40	0.000022		

Number of applications (1 or 2); treatment (Fylloton; Terra Sorb Complex, Kelpak, control).

Treating plants with the biostimulants had a positive effect on increasing the number of pods (Table 4). In 2015, no significant effect of the biostimulants on the studied trait was identified; however, a tendency to increase the number of pods was observed after foliar application of biostimulants, especially Kelpak. On the other hand, in 2016, after treating the plants with biostimulants based on seaweed extract, i.e., both after a single application of Kelpak (*Ecklonia maxima* extract) and after a single or double application of Fylloton (*Ascophyllum nodosum* extract, amino acids), a significant increase in this trait was identified (increases by 36%, 28%, and 33%, respectively, as compared to the control). Treatment of plants with Kelpak in 2017 significantly increased the number of pods, regardless of the number of applications (increase by 18% for a single application, and by 19% for a double application, as compared to the control). A synthesis of the three years of research (2015–2017) confirmed that a single spraying of plants with the Kelpak biostimulant in the BBCH 12-13 phase significantly increased the number of pods (by 26%), as compared to the control.

Table 4. Effect of Fylloton, Terra Sorb Complex, and Kelpak biostimulants treatment on number of pods and seeds of bean variety Oczko.

Parameters	Biostimulant Treatment	Season			Average 2015–2017
		2015	2016	2017	
Number of pods (per m^2)	F_1	224 ± 7.8 n.s.	240 ± 8.5 a	231 ± 11.3 ab	232 ± 4.0 b
	F_2	234 ± 9.9 n.s.	249 ± 8.5 a	245 ± 4.9 ab	243 ± 7.8 ab
	TS_1	223 ± 29.7 n.s.	207± 7.8 bc	249 ± 7.8 ab	226 ± 4.7 b
	TS_2	248 ± 12.0 n.s.	234 ± 9.9 ab	254 ± 9.8 ab	245 ± 2.6 ab
	K_1	257 ± 10.6 n.s.	254 ± 7.8 a	258 ± 12.0 a	256 ± 3.1 a
	K_2	252 ±18.4 n.s.	225 ± 6.4 ab	260 ± 6.3 a	245 ± 6.1 ab
	C	203 ±17.7 n.s.	187 ± 9.2 c	219 ± 9.2 b	203 ± 5.9 c
Number of seeds (per m^2)	F_1	776 ± 9.2 a	649 ± 9.2 d	686 ± 24.0 bc	703 ± 1.9 c
	F_2	809 ±17.0 a	774 ± 17.7 b	799 ± 23.3 a	794 ± 3.8 a
	TS_1	713 ±19.1 b	720 ± 4.9 c	693 ± 21.9 bc	708 ± 2.6 c
	TS_2	762 ±6.4 ab	752 ± 6.4 bc	751 ± 21.8 ab	755 ± 3.1 b
	K_1	764 ± 22.6 ab	777 ± 9.9 b	754 ± 20.5 ab	765 ± 2.6 b
	K_2	793 ± 15.6 a	835 ± 5.7 a	780 ± 25.5 a	803 ± 5.2 a
	C	602 ± 14.8 c	611 ± 19.1 d	613 ± 5.7 c	608 ± 9.4 d

Abbreviations: F_1, single spraying of Fylloton; F_2, double spraying of Fylloton; TS_1, single spraying of Terra Sorb Complex; TS_2, double spraying of Terra Sorb Complex; K_1, single spraying of Kelpak; K_2, double spraying of Kelpak; C, control; n.s., not significant. Means in the columns, concerning the selected traits, followed by different small letters are significantly different at $p < 0.05$.

Analysis of variance showed that, regardless of the number of applications, the treatment of plants with Fylloton significantly increased the number of seeds in 2015, as did the double spraying with Kelpak (increases by 29%, 34%, and 32% respectively, as compared to the control) (Table 4). In the second year of research, the best results were obtained after double spraying with biostimulant based on the extract of *Ecklonia maxima*, with a significant increase of 37%, as compared to the control. In 2017, it was found that spraying plants with biostimulants containing seaweed extracts, i.e., Fylloton and Kelpak, significantly increased the number of seeds, by 30% and 27%, respectively, as compared to the control. The average of three years of research demonstrated that an increase of this trait, by 31% and 32%, respectively, as compared to the control, was obtained after a double foliar application of Fylloton and Kelpak.

Double application of Kelpak, in 2015 and 2016, had the most beneficial effect on seed yield, increasing this trait by 22% and 38%, respectively, as compared to the control (Figure 1). In 2017, the best effects in increasing the bean crop were obtained after double treatment of plants with Fylloton and Kelpak biostimulants, when a 25% increase of this trait was obtained, as compared to the control. A synthesis of the three years of research demonstrated that double spraying plants with Kelpak biostimulant was most beneficial for increasing the bean yield (increase by 28% as compared to the control).

Figure 1. Effect of Fylloton, Terra Sorb Complex, and Kelpak biostimulant treatment on seed yield of bean variety Oczko. Abbreviations: F_1, single spraying of Fylloton; F_2, double spraying of Fylloton; TS_1, single spraying of Terra Sorb Complex; TS_2, double spraying of Terra Sorb Complex; K_1, single spraying of Kelpak; K_2, double spraying of Kelpak; C, control. Means over the study years followed by different small letters are significantly different at $p < 0.05$.

Analysis of variance showed that the use of biostimulants in bean cultivation resulted in a reduction of the weight of one thousand seeds (Figure 2). In the first year of research, a significant increase in the weight of one thousand seeds was obtained in the control, by 11%–15%, as compared to the combination with Fylloton or Terra Sorb Complex. However, in 2016 there were no significant differences in the weight of one thousand seeds between the combinations that included the biostimulants and the control. Plants in the control plot in 2017 were characterized by a higher weight of one thousand seeds by 8%, as compared to the double application of the Terra Sorb Complex. In turn, the average of three years of research confirmed that the highest weight of one thousand seeds (an increase of 9%, as compared to the double use of Fylloton) was obtained in the control.

Foliar application of biostimulants increased the protein content in the seeds (Table 5). In 2015 and 2017, the best effects were obtained after applying Terra Sorb Complex as a single or double spraying of plants, which increased this trait by 12% and 13% (in 2015), and by 9% and 10% (in 2017), respectively, as compared to the control. Double treatment of plants with the Terra Sorb Complex biostimulant in 2016 had the most beneficial effect on increasing the protein content by 13%, as compared to the control. A synthesis of the three years of research showed that, regardless of the number of applications, the use of an amino acid-based biostimulant significantly increased the protein content, as did the double application of the biostimulant containing the of the *Ecklonia maxima* extract (11%, 12%, and 10% increases, respectively, as compared to the control).

Figure 2. Effect of Fylloton, Terra Sorb Complex, and Kelpak biostimulant treatment on the weight of one thousand seeds of bean variety Oczko. Abbreviations: F_1, single spraying of Fylloton; F_2, double spraying of Fylloton; TS_1, single spraying of Terra Sorb Complex; TS_2, double spraying of Terra Sorb Complex; K_1, single spraying of Kelpak; K_2, double spraying of Kelpak; C, control. Means over the study years followed by different small letters are significantly different at $p < 0.05$.

Table 5. Effect of Fylloton, Terra Sorb Complex, and Kelpak biostimulant treatment on protein, phenol, and flavonoid content of bean seeds.

Parameters	Biostimulant Treatment	Season			Average 2015–2017
		2015	2016	2017	
Protein (% DM)	F_1	21.19 ± 0.23 bc	20.59 ± 0.29 bc	21.30 ± 0.14 ab	21.03 ± 0.38 a
	F_2	20.82 ± 0.10 c	20.19 ± 0.14 c	21.01 ± 0.10 ab	20.67 ± 0.60 ab
	TS_1	21.81 ± 0.02 a	21.20 ± 0.05 ab	22.04 ± 0.02 a	21.68 ± 0.43 a
	TS_2	21.99 ± 0.04 a	21.38 ± 0.10 a	22.27 ± 0.18 a	21.88 ± 0.46 a
	K_1	20.74 ± 0.27 c	20.05 ± 0.27 c	21.14 ± 0.43 ab	20.64 ± 0.55 ab
	K_2	21.79 ± 0.08 ab	21.15 ± 0.16 ab	21.61 ± 0.59 ab	21.52 ± 0.33 a
	C	19.53 ± 0.15 d	18.94 ± 0.17 d	20.18 ± 0.58 b	19.55 ± 0.62 b
Total phenols ($mg \cdot g^{-1}$ DM)	F_1	3.76 ± 0.0 b	4.02 ± 0.02 c	4.13 ± 0.03 c	3.97 ± 0.19 n.s.
	F_2	5.96 ± 0.05 a	3.56 ± 0.01 d	4.11 ± 0.02 c	4.54 ± 1.29 n.s.
	TS_1	4.01 ±0.01 b	4.43 ± 0.01 b	4.36 ± 0.03 b	4.26 ± 0.22 n.s.
	TS_2	4.01 ± 0.03 b	4.48 ± 0.12 ab	3.74 ± 0.06 d	4.07 ± 0.37 n.s.
	K_1	2.82 ± 0.22 d	3.89 ± 0.09 c	3.54 ±0.08 e	3.42 ± 0.54 n.s.
	K_2	3.83 ± 0.01 b	4.70 ± 0.04 a	4.63 ± 0.04 a	4.39 ± 0.48 n.s.
	C	3.27 ± 0.08 c	3.30 ± 0.05 e	3.29 ± 0.01 f	3.29 ± 0.02 n.s.
Total flavonoids ($mg \cdot g^{-1}$ DM)	F_1	0.061± 0.001 bc	0.075 ± 0.001 a	0.078 ± 0.001 a	0.071 ± 0.009 a
	F_2	0.058 ± 0.001 c	0.066 ± 0.001 ab	0.068 ± 0.001 b	0.064 ± 0.005 a
	TS_1	0.059 ± 0.001 c	0.060 ± 0.008 b	0.069 ± 0.001 b	0.063 ± 0.006 a
	TS_2	0.063 ± 0.001 ab	0.061 ± 0.001 ab	0.065 ± 0.001 b	0.063 ± 0.002 a
	K_1	0.064 ± 0.002 ab	0.062 ± 0.003 ab	0.066 ± 0.001 b	0.064 ± 0.002 a
	K_2	0.067 ± 0.001 a	0.073 ± 0.001 ab	0.075 ± 0.001 a	0.072 ± 0.004 a
	C	0.036 ± 0.001 d	0.038 ± 0.001 c	0.041 ± 0.001 c	0.038 ± 0.003 b

Abbreviation: F_1, single spraying of Fylloton; F_2, double spraying of Fylloton; TS_1, single spraying of Terra Sorb Complex; TS_2, double spraying of Terra Sorb Complex; K_1, single spraying of Kelpak; K_2, double spraying of Kelpak; C, control; n.s., not significant; DM, dry matter. Means in the columns, concerning the selected traits, followed by different small letters are significantly different at $p < 0.05$.

Double application of biostimulants containing seaweed extracts most favorably impacted the content of phenols in the seeds (Table 5). In 2015, an increase in the phenolic compound content in beans was noted after a double application of Fylloton (an increase of this trait by 82%, as compared to the control). However, after a single application of Kelpak, this trait was found to be reduced, as compared to the control. In turn, in 2016 and 2017, the best effects in increasing the phenol content were observed after a double application of Kelpak (increases by 42% and 41%, respectively, as compared to the control). No significant differences were found for this trait in the average for the years 2015–2017, but only a tendency to increase after a double application of biostimulants containing seaweed.

The application of preparations containing seaweed had the most beneficial effect on increasing the flavonoid content in the seeds (Table 5). In the first year of research, this characteristic was increased by 87%, as compared to the control, after a double application of Kelpak. In 2016, a single application of Fylloton had the most beneficial effect on the increase of the flavonoid content (by 98%, as compared to the control). On the other hand, in 2017, a significant increase in the flavonoid content was noted both with a single application of Fylloton and double of Kelpak (increases by 90% and 82%, respectively, as compared to the control). The average of three years of research showed that the application of all biostimulants significantly increased the flavonoid content, by 64%–88%, as compared to the control.

4. Discussion

The results of our research show a positive effect of natural biostimulants on bean yield, as well as on its quality. A more beneficial effect in modifying the yield components (number of pods and seeds, seed yield) was obtained after using biostimulants based on seaweed extract, especially upon double application of the *Ecklonia maxima* (Kelpak) extract. Only in the case of the weight of one thousand seeds was a reduction of this trait observed as a result of using biostimulants. The use of other biostimulants, i.e., Fylloton, which contains an *Ascophyllum nodosum* extract and amino acids, as well as Terra Sorb Complex containing amino acids, also had a positive effect on the yield components. In turn, the conditions favorable for setting pods and seeds, and increasing the weight of seeds were the most beneficial in 2016, when the double use of Kelpak resulted in the largest increase in these characteristics compared to control. This year, there were also favorable temperature and humidity conditions conducive to flowering, and setting pods and seeds. The significant increase in protein content was most positively influenced by weather conditions in 2015 and 2016, when the highest value of this feature was obtained after double application of Terra Sorb Complex. The application of biostimulants based on marine algae had a positive effect on the flavonoid content in all years of research. In 2015, however, a significant increase in polyphenols was found after double application of Fylloton.

Numerous studies conducted on arable crops confirmed the beneficial effect of seaweed extracts on increasing yield components [11,12,47,64–66]. Previous research on other bean and soybean varieties confirmed the stimulating effect of the *Ecklonia maxima* or *Ascophyllum nodosum* extracts on the number of pods and seeds, and the weight of beans [39–41,67]. The lack of effect of Kelpak on the one-thousand-seed weight of bean was also confirmed by previous studies on common bean [66]. Use of extracts from *Kappaphycus alvarezii* and *A. nodosum* increased the number of pods, seeds, and yield in soybean [68,69], even under stress conditions (reduced NPK fertilization) [70]. Bean plants reacted favorably to the foliar application of *Caulerpa racemosa* and *A. nodosum* extract by increasing the number of pods and seeds, one thousand weight of bean, and seed yield in common beans [71,72], mung beans [73], and broad beans [74].

The foliar application of amino acids positively affects the yield of many plants, even growing under stress [11,75,76]. In previous studies, the studied bean responded positively to foliar application of the Terra Sorb Complex; however, the yielding effect depended on the variety, concentration, and number of applications of the biostimulant, as well as climatic conditions prevailing in a given study year [59]. The Aura variety (with white seeds) responded more favorably to a single application of a 0.5% concentration, and the Toska variety (with red seeds) responded more favorably to a single application of a 0.3% concentration of this biostimulant. The plants increased the number of pods and seeds, as well as the seed yield; however, no effect of this biostimulant was found for the weight of one thousand seeds. Foliar application of biostimulants containing amino acids increased the number of pods and seeds, weight of one thousand seeds, and yield in seeds of beans [71,77], peas [78], and broad beans [79,80].

The positive effect of seaweed extracts on plant growth and development and, as a result, on the increase in yield is undoubtedly associated with the presence of phytohormones, especially cytokinins [81]. Together with auxins, cytokinins regulate many physiological processes, including those affecting plant growth and development [82–84]. Aremu et al. [85] and Masondo et al. [86] also observed a positive effect of Kelpak on increasing the content of cytokinins. In addition, Kulkarni et al. [87] found an increase in the content of *cis*-zeatin, dihydrozeatin, and isopentenyladenine after using the *Ecklonia maxima* extract (Kelpak). The many active substances and compounds included in Kelpak suggest that it is not only cytokinin that is responsible for the growth and development of plants, but also probably the cross-reactions of these compounds with other bioactive molecules included in biostimulants that are based on seaweed extracts [70].

Thanks to the content of endogenous auxins, seaweed extracts have a positive effect on the growth and development of the root system [88]. This improves the uptake of water and nutrients and, in effect, stimulates the growth and development of plants, contributing to the improvement of yield quantity and quality [81]. The application of biostimulants based on seaweed extracts also has a positive effect on plant growth and development due to the content of gibberellins (GA1, GA3, GA4, GA5, GA6, GA7, GA13) [14], which affect seed germination, stem elongation, leaf expansion, and flower and seed development [89–91], as well as of gibberellin-like substances, e.g., terpenoids and tocopherol [92,93]. In the seaweed extract (Kelpak), the presence of brassinosteroids, brassinolide and castasterone [10], was identified. As phytohormones, brassinosteroids promote cell division and elongation, stimulate stem and root growth, and initiate flowering and flower development, as well as fruit development and increases in seed yield. Under stress conditions, they protect plants against abiotic and biotic stress [94,95].

The positive effect of amino acid-based preparations on plant growth and development probably results from the fact that, at the molecular level, they stimulate the plant's defense response to biotic and abiotic stress factors [96]. The amino acids contained in them are easily absorbed by plants. They participate in the synthesis of a number of organic compounds and affect the uptake of macro- and microelements [97]. Garcia et al. [98] showed that foliar application of amino acids and peptides together with nutrients increases the content of potassium, calcium, magnesium, iron, copper, and zinc in leaves, affecting their nutritional condition and promoting improved growth and development of plants. Applied on a leaf, preparations of this type exhibit phytohormone-like effects, comparable to that of auxin and gibberellin [54]. They also contribute to increasing the content of phytohormones (gibberellins, cytokinins, auxins) [76]. In addition, the use of protein hydrolysates in plant cultivation has a beneficial effect on the uptake of water and nutrients, resulting in increasing crop yielding thanks to the increased microbial and enzymatic activity of the soil, improved mobility and solubility of microelements (iron, zinc, manganese, copper), modified structure of the root system (its length, compaction, and number of lateral roots), or the increased synthesis of nitrate and glutamine reductase, as well as the activity of iron reductase [11,54,99–102]. Numerous reports confirmed that protein hydrolysates, such as auxins and gibberellins, have hormone-like effects, stimulating root and shoot growth. This, in turn, has a positive effect on crop productivity [11,53,54,100,102–106].

In our research, the use of biostimulants based on amino acids and seaweed extracts had a positive effect on the nutritional value of beans by increasing their protein content. Application of seaweed extract and amino acids had a positive effect on increasing the protein content in bean, pea, and faba bean seeds [59,71,74,78,80]. This was confirmed by Rouphael [62]; in their research, the increase in protein content in spinach plants was obtained after using biostimulants containing an extract of *Ecklonia maxima* and *Ascophyllum nodosum* and legume-derived protein hydrolysate. The protein content of legumes leaves was also determined by seaweed extracts. Numerous authors found an increase in this trait after the application of *Ulva rigida, Fucus spiralis, Hypnea musciformis,* and

Colpomenia sinuosa extracts in bean leaves [16,107,108]. However, the use of biostimulants is not always beneficial for the protein content in beans [15,109]. Schubert and Mengel [110] demonstrated that amino-acid uptake is an important mechanism for recovering carbon and nitrogen that was lost in the rhizosphere. Ertani et al. [100] found that the stimulation of nitrogen assimilation is responsible for accelerating the growth and metabolism of nitrogen in plants treated with protein hydrolysates. This is due to an increase in the activity of two key enzymes, nitrate reductase and glutamine synthetase, thereby contributing to increasing protein firmness.

After the application of biostimulants containing seaweed extracts, researchers observed an increase in the phenolic content. Ertani [56] showed that the treatment of plants with biostimulants stimulates numerous metabolic pathways in plants. The pathways are also associated with the synthesis of secondary metabolites, including phenolic compounds, which play an important role in protecting plants against stress factors. In turn, a frequent indicator of plant resistance to biotic factors is the content of phenolic compounds, which are precursors to more complex phenolic structures, such as flavonoids and lignins [60,111]. The presence of bioactive compounds in biostimulants, including phytohormones, amino acids, protein, and phenols, is responsible for the physiological response of plants treated with these preparations [56,112,113]. Eckol (phenolic compound) found in seaweed extracts affects the phenylpropanoid pathway in the biochemical synthesis of phenolic acids [114]. Aremu et al. [85,115] showed that the timing of eckol plants increases the content of phenolic compounds, such as *p*-hydroxybenzoic and ferulic acids. In turn, the Kelpak application increases the content of caffeic acid, ferulic acid [85,116], protocatechuic acid, *p*-hydroxybenzoic acid, gentisic acid, *p*-coumaric acid, and *trans*-cinnamic acid in *Eucomis autumnalis* [85]; however, the content of phenolic compounds depended on the biostimulant concentration. In turn, Rouphael et al. [62] found an increased content of phenolic compounds after the treatment of plants with biostimulants containing alginians, fucoidans, and laminarins that affect endogenic hormonal homeostasis [117]. Treatment of plants with eckol also had a positive effect on the flavonoid content, increasing the amount of kempferol in plants several times. However, foliar application of Kelpak increased the content of kaempferol in tubers and entire plants, as well as taxifolin in leaves of *Eucomis autumnalis* [85,115]. Increasing the content of bioactive compounds in plants treated with biostimulants is associated with a mechanism that includes the stimulation of the chalcone isomerase enzyme, involved in the biosynthesis of flavanone precursors [117].

Paul et al. [118] found that tomato plants treated with protein hydrolysates were characterized by a higher content of, e.g., low-molecular-weight phenolic compounds, phytohormones (polyamines), hydroxy-carotenoids, poly-hydroxy fatty acids, and membrane lipids (glycoand phospholipids). They suggest that the metabolic changes caused by treating plants with protein hydrolysates can be correlated with a relatively small number of processes that converge toward the ROS-related (reactive oxygen species-related) plant signaling network. Increasing the content of secondary metabolites, such as phenols and carotenoids, which play a key role in protecting plants against oxidative stress [62,102,119], suggests fine-tuning of ROS signaling in plants after the application of protein hydrolysates [118]. In addition, the use of animal protein hydrolysates had a positive effect on the content of protein, phenols, and flavonoids in bananas [120], and vegetal protein hydrolysates stimulated an increase in the content of phenolic compounds and anthocyanins in grape [55].

So far, there are few reports confirming the beneficial effect of a preparation consisting of a combination of seaweed extract and amino acids. In our research, combining the Fylloton biostimulant, which induces the extract's effect, with *Ascophyllum nodosum* and amino acids increased the yield components, particularly the number of seeds after its double application. Moreover, previous studies, conducted on three soybean varieties, confirmed that this preparation has a positive effect on the number of pods and seeds, as well as seed yield [40]. The application of Fylloton in the cultivation of winter oilseed rape positively influenced the increase in the number of pods, the yield of seeds, and the weight of one thousand seeds, especially after applying the biostimulant together with the Perfektmikro micronutrient fertilizer containing EDTA-chelated (ethylenediaminetetraacetic acid-chelated) manganese, copper, iron, and zinc, as well as molybdenum, boron, and nitrogen [121].

5. Conclusions

All studied biostimulants had a positive effect on quantity and quality of bean yield. Double application of Kelpak biostimulant (*Ecklonia maxima* extract) stimulated morphological features and seed yield, as well as the content of polyphenols and flavonoids in seeds to a greater extent. In turn, the biostimulant containing amino acids (Terra Sorb Complex) significantly increased the protein content in beans. In contrast, Fylloton containing *Ascophyllum nodosum* extract and amino acids had also a more favorable effect on the number of seeds. In 2015 and 2017, biostimulants containing seaweed extracts had the most beneficial effect on bean yield and its quality. On the other hand, in 2016, treatment of plants with Kelpak had a more beneficial effect on the studied features.

Author Contributions: All authors read and agreed to the published version of the manuscript. A.K. and J.L. conceptualized and designed the research; R.K. and M.T. performed the field experiments; J.L, N.N., and K.K. analyzed the plant material; S.T. carried out statistical analysis of the data; A.K., R.K., and D.L. wrote the paper; M.T. and S.T. revised the manuscript.

Funding: This research received no external funding.

Conflicts of Interest: The authors declare no conflicts of interest.

References

1. Caruso, G.; De Pascale, S.; Cozzolino, E.; Cuciniello, A.; Cenvinzo, V.; Bonini, P.; Colla, G.; Rouphael, Y Yield and nutritional quality of Vesuvian Piennolo tomato PDO as affected by farming system and biostimulant application. *Agronomy* **2019**, *9*, 505. [CrossRef]
2. Caruso, G.; De Pascale, S.; Cozzolino, E.; Giordano, M.; El-Nakhel, C.; Cuciniello, A.; Cenvinzo, V.; Colla, G.; Rouphael, Y. Protein hydrolysate or plant extract-based biostimulants enhanced yield and quality performances of greenhouse perennial wall rocket grown in different seasons. *Plants* **2019**, *8*, 208. [CrossRef] [PubMed]
3. Du Jardin, P. Plant biostimulants: Definition, concept, main categories and regulation. *Sci. Hortic.* **2015**, *196*, 3–14. [CrossRef]
4. Matyjaszczyk, E. The introduction of biostimulants on the Polish market. The present situation and legal requirements. *Przem. Chem.* **2015**, *10*, 1841–1844. [CrossRef]
5. Matyjaszczyk, E. Problems of implementing compulsory integrated pest management. *Pest. Manag. Sci.* **2019**, *75*, 2063–2067. [CrossRef]
6. Traon, D.; Amat, L.; Zotz, F.; Du Jardin, P. *A Legal Framework for Plant Biostimulants and Agronomic Fertiliser Additives in the EU. Report for the European Commission Enterprise & Industry Directorate—General*; Publications Office of the European Union: Luxembourg, 2014.
7. Matyjaszczyk, E. "Biorationals" in integrated pest management strategies. *J. Plant Dis. Prot.* **2018**, *125*, 523–527. [CrossRef]
8. Calvo, P.; Nelson, L.; Kloepper, J.W. Agricultural uses of plant biostimulants. *Plant Soil* **2014**, *383*, 3–41. [CrossRef]
9. Rouphael, Y.; Cardarelli, M.; Bonini, P.; Colla, G. Synergistic action of a microbial-based biostimulant and a plant derived-protein hydrolysate enhances lettuce tolerance to alkalinity and salinity. *Front. Plant Sci.* **2017**, *8*, 131. [CrossRef]
10. Colla, G.; Rouphael, Y. Biostimulants in horticulture. *Sci. Hortic.* **2015**, *196*, 1–2. [CrossRef]
11. Colla, G.; Hoagland, L.; Ruzzi, M.; Cardarelli, M.; Bonini, P.; Canaguier, R.; Rouphael, Y. Biostimulant action of protein hydrolysates: unraveling their effects on plant physiology and microbiome. *Front. Plant Sci.* **2017**, *8*, 2202. [CrossRef]
12. Battacharyya, D.; Babgohari, M.Z.; Rathor, P.; Prithiviraj, B. Seaweed extracts as biostimulants in horticulture. *Sci. Hortic.* **2015**, *196*, 39–48. [CrossRef]
13. Basak, A. Biostimulators. Definitions, classification and legislation. In *Biostimulators in Modern Agriculture, General Aspects*; Gawrońska, H., Ed.; Editorial House Wieś Jutra: Warszawa, Poland, 2008; pp. 7–17.

14. Stirk, W.A.; Tarkowská, D.; Turecová, V.; Strnad, M.; van Staden, J. Abscisic acid, gibberellins and brassinosteroids in Kelpak®, a commercial seaweed extract made from *Ecklonia maxima*. *J. Appl. Phycol.* **2014**, *26*, 561–567. [CrossRef]
15. Kocira, A.; Świeca, M.; Kocira, S.; Złotek, U.; Jakubczyk, A. Enhancement of yield, nutritional and nutraceutical properties of two common bean cultivars following the application of seaweed extract (*Ecklonia maxima*). *Saudi J. Biol. Sci.* **2018**, *25*, 563–571. [CrossRef]
16. Latique, S.; Chernane, H.; Mansori, M.; El Kaoua, M. Seaweed liquid fertilizer effect on physiological and biochemical parameters of bean plant (*Phaesolus vulgaris* variety Paulista) under hydroponic system. *Eu. Sci. J.* **2013**, *9*, 174–191.
17. Sharma, S.H.S.; Lyons, G.; McRoberts, C.; McCall, D.; Carmichael, E.; Andrews, F.; Swan, R.; McCormack, R.; Mellon, R. Biostimulant activity of brown seaweed species from Strangford Lough: compositional analyses of polysaccharides and bioassay of extracts using mung bean (*Vigno mungo* L.) and pak choi (*Brassica rapa chinensis* L.). *J. Appl. Phycol.* **2012**, *24*, 1081–1091. [CrossRef]
18. Stirk, W.A.; Van Standen, J. Flow of cytokinins through the environment. *Plant. Growth Regul.* **2010**, *62*, 101–116. [CrossRef]
19. Tarakhovskaya, E.R.; Maslov, Y.I.; Shishova, M.F. Phytohormones in algae. *Russ. J. Plant. Physiol.* **2007**, *54*, 163–170. [CrossRef]
20. Klarzynski, O.; Descamps, V.; Plesse, B.; Yvin, J.C.; Fritig, B. Sulfated fucan oligosaccharides elicit defense responses in tobacco and local and systemic resistance against Tobacco Mosaic Virus. *Mol. Plant-Microbe Interact.* **2003**, *16*, 115–122. [CrossRef]
21. Zhang, Q.; Zhang, J.; Shen, J.; Silva, A.; Dennis, D.A. A simple 96-well microplate method for estimation of total polyphenol content in seaweeds. *J. Appl. Phycol.* **2006**, *18*, 445–450. [CrossRef]
22. Rioux, L.E.; Turgeon, S.L.; Beaulieu, M. Characterization of polysaccharides extracted from brown seaweeds. *Carbohydr. Polym.* **2007**, *69*, 530–537. [CrossRef]
23. Stirk, W.A.; Novak, O.; Hrandecka, V.; Pencik, A.; Rolcik, J.; Strnad, M.; Van Staden, J. Endogenous cytokinins, auxins and abscisic acid in *Ulva fasciata* (Chlorophyta) and *Dictyota humifusa* (Phaeophyta): towards understanding their biosynthesis and homoeostasis. *Eur. J. Phycol.* **2009**, *44*, 231–240. [CrossRef]
24. Yokoya, N.S.; Stirk, W.A.; Van Staden, J.; Novak, O.; Tureckova, V.; Pencik, A.; Strnad, M. Endogenous cytokinins, auxins, and abscisic acid in red algae from Brazil. *J. Phycol.* **2010**, *46*, 1198–1205. [CrossRef]
25. Matysiak, K.; Kaczmarek, S.; Leszczyńska, D. Influence of liquid seaweed extract of *Ecklonia maxima* on winter wheat cv Tonacja. *J. Res. Appl. Agric. Engin.* **2012**, *57*, 44–47.
26. Papenfus, H.B.; Stirk, W.A.; Finnie, J.F.; Van Staden, J. Seasonal variation in the polyamines of *Ecklonia maxima*. *Botanica Marina* **2012**, *55*, 539–546. [CrossRef]
27. Rengasamy, K.R.R.; Kulkarni, M.G.; Stirk, W.A.; Van Staden, J. Eckol—a new plant growth stimulant from the brown seaweed *Ecklonia maxima*. *J. Appl. Phycol.* **2015**, *27*, 581–587. [CrossRef]
28. Rengasamy, K.R.R.; Kulkarni, M.G.; Stirk, W.A.; Van Staden, J. Eckol improves growth, enzyme activities, and secondary metabolite content in maize (*Zea mays* cv. Border King). *J. Plant Growth Regul.* **2015**, *34*, 410–416. [CrossRef]
29. Panda, D.; Pramanik, K.; Nayak, B.R. Use of sea weed extracts as plant growth regulators for sustainable agriculture. *Int. J. Bioresour. Stress Manag.* **2012**, *3*, 404–411.
30. Crough, I.J.; Van Staden, J. Effect of seaweed concentrate on the establishment and yield of greenhouse tomato plants. *J. Appl. Phycol.* **1992**, *4*, 291–296. [CrossRef]
31. Stirk, W.A.; Bálint, P.; Tarkowská, D.; Strnad, M.; van Staden, J.; Ördög, V. Endogenous brassinosteroids in microalgae exposed to salt and low temperature stress, *Eur. J. Phycol.* **2018**. [CrossRef]
32. Ecoforce, A. Extracto de Algas para Agricultura Ecológica—Fertilizantes Ecoforce. Available online: http://fertilizantesecoforce.es/es/blog/index/list/cat/agricultura-ecologica/?p=19 (accessed on 1 December 2019).
33. Van Oosten, M.J.; Pepe, O.; De Pascale, S.; Silletti, S.; Maggio, A. The role of biostimulants and bioeffectors as alleviators of abiotic stress in crop plants. *Technol. Agric.* **2017**, *4*, 5. [CrossRef]
34. Sosnowski, J.; Malinowska, E.; Jankowski, K.; Redzik, P. Morpho-chemical diversity in *Festuca pratensis* and *Lolium perenne* depending on concentrations of *Ecklonia maxima* extract. *Appl. Ecol. Env. Res.* **2016**, *14*, 369–379. [CrossRef]
35. Drobek, M.; Frąc, M.; Cybulska, J. Plant biostimulants: importance of the quality and yield of horticultural crops and the improvement of plant tolerance to abiotic stress—A review. *Agronomy* **2019**, *9*, 335. [CrossRef]

36. Ali, N.; Farrell, A.; Ramsubhag, A.; Jayaraman, J. The effect of *Ascophyllum nodosum* extract on the growth, yield and fruit quality of tomato grown under tropical conditions. *J. Appl. Phycol.* **2016**, *28*, 1353–1362. [CrossRef]
37. Mikiciuk, M.; Dobromilska, R. Assessment of yield and physiological indices of small sized tomato cv. 'Bianka F1' under the influence of biostimulators of marine algae origin. *Acta Sci. Pol., Hortorum Cultus* **2014**, *13*, 31–41.
38. Chanthini, K.-P.; Stanley-Raja, V.; Thanigaivel, A.; Karthi, S.; Palanikani, R.; Shyam Sundar, N.; Sivanesh, H.; Soranam, R.; Senthil-Nathan, S. Sustainable agronomic strategies for enhancing the yield and nutritional quality of wild tomato, *Solanum lycopersicum* (l) var *cerasiforme* Mill. *Agronomy* **2019**, *9*, 311. [CrossRef]
39. Kocira, S.; Kocira, A.; Kornas, R.; Koszel, M.; Szmigielski, M.; Krajewska, M.; Szparaga, A.; Krzysiak, Z. Effect of seaweed extract on yield and protein content of two common bean (*Phaseolus vulgaris* L.) cultivars. *Legume Res.* **2017**, LR-383. [CrossRef]
40. Kocira, S.; Szparaga, A.; Kocira, A.; Czerwińska, E.; Wójtowicz, A.; Bronowicka-Mielniczuk, U.; Koszel, M.; Findura, P. Modelling biometric traits, yield and nutritional and antioxidant properties of seeds of three soybean cultivars through the application of biostimulant containing seaweed and amino acids. *Front. Plant Sci.* **2018**, *9*. [CrossRef]
41. Kocira, S.; Szparaga, A.; Kuboń, M.; Czerwińska, E.; Piskier, T. Morphological and biochemical responses of *Glycine max* (L.) Merr. to the use of seaweed extract. *Agronomy* **2019**, *9*, 93. [CrossRef]
42. Szczepanek, M.; Siwik-Ziomek, A. P and K Accumulation by rapeseed as affected by biostimulant under different NPK and S fertilization doses. *Agronomy* **2019**, *9*, 477. [CrossRef]
43. Abd El-Gawad, H.G.; Osman, H.S. Effect of exogenous application of boric acid and seaweed extract on growth, biochemical content and yield of eggplant. *J. Hortic. Sci. Ornam. Plants* **2014**, *6*, 133–143. [CrossRef]
44. Manna, D.; Sarkar, A.; Maity, T.K. Impact of Biozyme on growth, yield and quality of chilli (*Capsicum annuum* L.). *J. Crop Weed* **2012**, *8*, 40–43.
45. Kumar, N.A.; Vanlalzarzova, B.; Sridhar, S.; Baluswami, M. Effect of liquid seaweed fertilizer of *Sargassum wightii* Grev. on the growth and biochemical content of green gram (*Vigna radiata* (L.) R. Wilczek). *Recent Res. Sci. Technol.* **2012**, *4*, 40–45.
46. Di Mola, I.; Cozzolino, E.; Ottaiano, L.; Giordano, M.; Rouphael, Y.; Colla, G.; Mori, M. Effect of vegetal- and seaweed extract-based biostimulants on agronomical and leaf quality traits of plastic tunnel-grown baby lettuce under four regimes of nitrogen fertilization. *Agronomy* **2019**, *9*, 571. [CrossRef]
47. Colla, G.; Cardarelli, M.; Bonini, P.; Rouphael, Y. Foliar applications of protein hydrolysate, plant and seaweed extracts increase yield but differentially modulate fruit quality of greenhouse tomato. *HortScience* **2017**, *52*, 1214–1220. [CrossRef]
48. Lötze, E.; Hoffman, E.W. Nutrient composition and content of various biological active compounds of three south African-based commercial seaweed biostimulants. *J. Appl. Phycol.* **2016**, *28*, 1379–1386. [CrossRef]
49. Moreira Sisalema, J.M. Comportamiento Agronómico del Cultivo de Soya (*Glycine max* L.), a la Aplicación de Tres Extractos de Algas Marinas, en la Zona de Puebloviejo. Available online: http://dspace.utb.edu.ec/bitstream/49000/5048/1/TE-UTB-FACIAG-ING%20AGROP-000018.pdf (accessed on 1 December 2019).
50. Marschner, H. *Mineral Nutrition of Higher Plants*; Academic Press: London, UK, 1995.
51. Pratelli, R.; Pilot, G. Altered amino acid metabolism in glutamine dumper1 plants. *Plant. Signal. Behav.* **2007**, *2*, 182–184. [CrossRef]
52. Stitt, M.; Müller, C.; Matt, P.; Gibon, Y.; Carillo, P.; Morcuende, R.; Scheible, W.R.; Krapp, A. Steps towards an integrated view of nitrogen metabolism. *J. Exp. Bot.* **2002**, *53*, 959–970. [CrossRef]
53. Colla, G.; Nardi, S.; Cardarelli, M.; Ertani, A.; Lucini, L.; Canaguiere, R.; Rouphael, J. Protein hydrolysates as biostimulants in horticulture. *Sci. Hortic.* **2015**, *196*, 28–38. [CrossRef]
54. Colla, G.; Rouphael, Y.; Canaguier, R.; Svecova, E.; Cadarelli, M. Biostimulant action of a plant-derived protein hydrolysate produced through enzymatic hydrolysis. *Front. Plant Sci.* **2014**, *5*, 448. [CrossRef]
55. Parrado, J.; Escudero-Gilete, M.L.; Frriaza, V.; Garcia-Martinez, A.; González-Miret, M.L.; Bautista, J.D.; Heredia, F.J. Enzymatic vegetable extract with bioactive components: influence of fertilizer on the colour and anthocyanins of red grapes. *J. Sci. Food Agric.* **2007**, *87*, 2310–2318. [CrossRef]
56. Ertani, A.; Pizzeghello, D.; Francioso, O.; Sambo, P.; Sanchez-Cortes, S.; Nardi, S. *Capsicum chinensis* L. growth and nutraceutical properties are enhanced by biostimulants in a long-term period: Chemical and metabolomic approaches. *Front. Plant Sci.* **2014**, *5*, 375. [CrossRef] [PubMed]

57. Di Mola, I.; Ottaiano, L.; Cozzolino, E.; Senatore, M.; Giordano, M.; El-Nakhel, C.; Sacco, A.; Rouphael, Y.; Colla, G.; Mori, M. Plant-based biostimulants influence the agronomical, physiological, and qualitative responses of baby rocket leaves under diverse nitrogen conditions. *Plants* **2019**, *8*, 522. [CrossRef] [PubMed]
58. Liu, X.Q.; Ko, K.Y.; Kim, S.H.; Lee, K.S. Effect of amino acid fertilization on nitrate assimilation of leafy radish and soil chemical properties in high nitrate soil. *Commun. Soil Sci. Plant Anal.* **2008**, *39*, 269–281. [CrossRef]
59. Kocira, S.; Kocira, A.; Szmigielski, M.; Piecak, A.; Sagan, A.; Malaga-Tobała, U. Effect of an amino acids-containing biostimulator on common bean crop. *Przem. Chem.* **2015**, *94*, 1732–1736. [CrossRef]
60. Kocira, S.; Szparaga, A.; Kocira, A.; Czerwińska, E.; Depo, K.; Erlichowska, B.; Deszcz, E. Effect of applying a biostimulant containing seaweed and amino acids on the content of fiber fractions in three soybean cultivars. *Legume Res.* **2018**, LR-412. [CrossRef]
61. Rouphael, Y.; De Micco, V.; Arena, C.; Raimondi, G.; Colla, G.; De Pascale, S. Effect of *Ecklonia maxima* seaweed extract on yield, mineral composition, gas exchange and leaf anatomy of zucchini squash grown under saline conditions. *J. Appl. Phycol.* **2017**, *29*, 459–470. [CrossRef]
62. Rouphael, Y.; Giordano, M.; Cardarelli, M.; Cozzolino, E.; Mori, M.; Kyriacou, M.C.; Bonini, P.; Colla, G. Plant and seaweed-based extracts increase yield but differentially modulate nutritional quality of greenhouse spinach through biostimulant action. *Agronomy* **2018**, *8*, 126. [CrossRef]
63. Duer, I.; Fotyma, M.; Madej, A. *Kodeks Dobrej Praktyki Rolniczej*, 3rd ed.; Fundacja Programów Pomocy dla Rolnictwa: Warszawa, Poland, 2004. (In Polish)
64. Craigie, J.S. Seaweed extract stimuli in plant science and agriculture. *J. Appl. Phycol.* **2011**, *23*, 371–393. [CrossRef]
65. Sharma, H.S.S.; Fleming, C.; Selby, C.; Rao, J.R.; Martin, T. Plant biostimulants: A review on the processing of microalgae and use of extracts for crop management to reduce abiotic and biotic stresses. *J. Appl. Phycol.* **2014**, *26*, 465–490. [CrossRef]
66. Bulgari, R.; Cocetta, G.; Trivellini, A.; Vernieri, P.; Ferrante, A. Biostimulants and crop responses: A review. *Biol. Agric. Hortic.* **2015**, *31*, 1–17. [CrossRef]
67. Kocira, A.; Kornas, R.; Kocira, S. Effect assessment of Kelpak on the bean yield (*Phaseolus vulgaris* L.). *J. Cent. Eur. Agric.* **2013**, *14*, 545–554. [CrossRef]
68. Rathore, S.S.; Chaudhary, D.R.; Boricha, G.N.; Gosh, A.; Bhatt, B.P.; Zodape, S.T.; Patolia, J.S. Effect of seaweed extract on the growth, yield and nutrient uptake of soybean (*Glycine max*) under rainfed conditions. *S. Afr. J. Bot.* **2009**, *75*, 351–355. [CrossRef]
69. Karthikeyan, K.; Shanmugam, M. Development of a protocol for the application of commercial bio-stimulant manufactured from *Kappaphycus alvarezii* in selected vegetable crops. *J. Exp. Biol. Agric. Sci.* **2016**, *4*, 92–102.
70. Tandon, S.; Dubey, A. Effects of Biozyme (*Ascophyllum nodosum*) biostimulant on growth and development of soybean [*Glycine max* (L.) Merill]. *Commun. Soil Sci. Plant Anal.* **2015**, *46*, 845–858. [CrossRef]
71. Zewail, R.M.Y. Effect of seaweed extract and amino acids on growth and productivity and some biocostituents of common bean (*Phaseolus vulgaris* L.) plants. *J. Plant. Production Mansoura Univ.* **2014**, *5*, 1441–1453. [CrossRef]
72. Abo-Sedera, F.A.; Shams, A.S.; Mohamed, M.H.M.; Hamoda, A.H.M. Effect of organic fertilizer and foliar spray with some safety compounds on growth and productivity of snap bean. *Ann. Agric. Sci. Moshtohor* **2016**, *54*, 105–118.
73. Sujatha, K.; Vijayalakshmi, V. Foliar application of *Caulerpa racemosa* seaweed extract as bio-stimulant for enhancement of growth and yield of blackgram (*Vigna mungo* L.). *IJOART* **2013**, *2*, 216–230.
74. Jasim, A.H.; Obaid, A.S. Effect of foliar fertilizers spray, boron and their interaction on broad bean (*Vicia faba* L.) yield. *Sci. Pap. B Hortic.* **2014**, *58*, 271–276.
75. Delchev, G.; Zhelyazkova, T.; Stoyanova, A. Stability valuation of some mixtures between foliar fertilizers and combined herbicides for the grain yield of durum wheat. *Turk. J. Agric. Natur. Sci., Special Issue* **2014**, *1*, 1128–1133.
76. El-Gamal, I.S.; Abd El-Aal, M.M.M.; El-Desouky, S.A.; Khedr, Z.M.; Abo Shady, K.A. Effect of some growth substances on growth, chemical compositions and root yield productivity of sugar beet (*Beta vulgaris* L.) plant. *Middle East. J. Agric. Res.* **2016**, *5*, 171–185.
77. Abdel-Mawgoud, A.M.R.; El-Bassiouny, A.M.; Ghoname, A.; Abou-Hussein, S.D. Foliar application of amino acids and micronutrients enhance performance of green bean crop under newly reclaimed land conditions. *Aust. J. Basic & Appl. Sci.* **2011**, *5*, 51–55.

78. Shafeek, M.R.; Hafez, M.M.; Mahmoud, A.R.; Ali, A.R. Comparative effect on N-fixing bacterial with foliar application of amino acid mixed on growth and yield of pea plants (*Pisum sativum* L.). *Middle East. J. Appl. Sci.* **2014**, *4*, 755–761.
79. El-Ghamry, A.M.; Abd El-Hai, K.M.; Ghoneem, K.M. Amino and humic acids promote growth, yield and disease resistance of faba bean cultivated in clayey soil. *Aust. J. Basic Appl. Sci.* **2009**, *3*, 731–739.
80. Sadak, M.S.H.; Abdelhamid, M.T.; Schmidhalter, U. Effect of foliar application of aminoacids on plant yield and some physiological parameters in bean plants irrigated with seawater. *Acta Biol. Colomb.* **2015**, *20*, 141–152.
81. Khan, W.; Rayirath, U.P.; Subramanian, S.; Jithesh, M.N.; Rayorath, P.; Hodges, D.M.; Critchley, A.T.; Craigie, J.S.; Norrie, J.; Prithiviraj, B. Seaweed extracts as biostimulants of plant growth and development. *J. Plant Growth Regul.* **2009**, *28*, 386–399. [CrossRef]
82. Santner, A.; Calderon-Villalobos, L.I.A.; Estelle, M. Plant hormones are versatile chemical regulators of plant growth. *Nat. Chem. Biol.* **2009**, *5*, 301–307. [CrossRef] [PubMed]
83. Depuydt, S.; Hardtke, C.S. Hormone signalling crosstalk in plant growth regulation. *Curr. Biol.* **2011**, *21*, 365–373. [CrossRef]
84. Sakakibara, H. Cytokinins: Activity, biosynthesis, and translocation. *Ann. Rev. Plant Biol.* **2006**, *57*, 431–449. [CrossRef]
85. Aremu, A.O.; Plačková, L.; Grúz, J.; Biba, O.; Novák, O.; Stirk, W.A.; Doležal, K.; Van Staden, J. Seaweed-derived biostimulant (Kelpak) influences endogenous cytokinins and bioactive compounds in hydroponically grown *Eucomis autumnalis*. *J. Plant Growth Regul.* **2016**, *35*, 151–162. [CrossRef]
86. Masondo, N.A.; Aremu, A.O.; Kulkarni, M.G.; Petřík, I.; Plačková, L.; Šubrtová, M.; Novák, O.; Grúz, J.; Doležal, K.; Strnad, M.; et al. Elucidating the role of Kelpak® on the growth, phytohormone composition, and phenolic acids in macronutrient-stressed *Ceratotheca triloba*. *J. Appl. Phycol.* **2019**, *31*, 2687–2697. [CrossRef]
87. Kulkarni, M.G.; Rengasamy, K.R.R.; Pendota, S.C.; Grúz, J.; Plačková, L.; Novák, O.; Doležal, K.; Van Staden, J. Bioactive molecules derived from smoke and seaweed *Ecklonia maxima* showing phytohormone-like activity in *Spinacia oleracea* L. *New Biotechnol.* **2019**, *48*, 83–89. [CrossRef] [PubMed]
88. Crouch, I.J.; Smith, M.T.; Van Staden, J.; Lewis, M.J.; Hoad, G.V. Identification of auxins in a commercial seaweed concentrate. *J. Plant Physiol.* **1992**, *139*, 590–594. [CrossRef]
89. Jennings, R.C. Gibberellins as endogenous growth regulators in green and brown algae. *Planta* **1968**, *80*, 34–42. [CrossRef]
90. Yamaguchi, S. Gibberellin metabolism and its regulation. *Annu. Rev. Plant Biol.* **2008**, *59*, 225–251. [CrossRef]
91. Tanimoto, E. Gibberellins. In *Plant Roots—The Hidden Half*; Waisel, Y., Eshel, A., Kafkafi, U., Eds.; Marcel Dekker: New York, NY, USA, 2002; pp. 405–416.
92. Gopala, R.P. Gibberellin-like behaviour of α-tocopherol in green gram *Vigna radiata*. *Geobios* **1984**, *11*, 21–25.
93. Jensen, A. Tocopherol content of seaweed and seaweed meal. I. Analytical methods and distribution of tocopherols in benthic algae. *J. Sci. Food Agric.* **1969**, *20*, 449–453.
94. Bajguz, A.; Hayat, S. Effects of brassinosteroids on the plant responses to environmental stresses. *Plant Physiol. Biochem.* **2009**, *47*, 1–8. [CrossRef]
95. Divi, U.K.; Krishna, P. Overexpression of the brassinosteroid biosynthetic gene AtDWF4 in *Arabidopsis* seeds overcomes abscisic acid-induced inhibition of germination and increases cold tolerance in transgenic seedlings. *J. Plant Growth Regul.* **2010**, *29*, 385–393. [CrossRef]
96. Cambri, D.; Filippini, L.; Apone, F.; Arciello, S.; Colucci, G.; Portoso, D. Effect of Aminoplant® on expression of selected genes in *Arabidopsis thaliana* L. plants. In *Biostimulators in Modern Agriculture: General Aspects*; Gawrońska, H., Ed.; Wieś Jutra: Warszawa, Poland, 2008; pp. 77–82.
97. Maini, P. The experience of the first biostimulant, based on amino acids and peptides: A short retrospective review on the laboratory researches and the practical results. *Fertilitas Agrorum* **2000**, *1*, 29–43.
98. Garcia, A.L.; Madrid, R.; Gimeno, V.; Rodriguez-Ortega, W.M.; Nicolas, N.; Garcia-Sanchez, F. The effects of amino acids fertilization incorporated to the nutrient solution on mineral composition and growth in tomato seedlings. *Span. J. Agric. Res.* **2011**, *9*, 852–861. [CrossRef]
99. Cerdán, M.; Sánchez-Sánchez, A.; Oliver, M.; Juárez, M.; Sánchez-Andreu, J.J. Effect of foliar and root applications of amino acids on iron uptake by tomato plants. *Acta Hortic.* **2009**, *830*, 481–488. [CrossRef]

100. Ertani, A.; Cavani, L.; Pizzeghello, D.; Brandellero, E.; Altissimo, A.; Ciavatta, C.; Nardi, S. Biostimulant activities of two protein hydrolysates on the growth and nitrogen metabolism in maize seedlings. *J. Plant Nutr. Soil Sci.* **2009**, *172*, 237–244. [CrossRef]
101. García-Martínez, A.M.; Díaz, A.; Tejada, M.; Bautista, J.; Rodríguez, B.; María, C.S.; Revilla, E.; Parrado, J. Enzymatic production of an organic soil biostimulant from wheat condensed distiller solubles: effects on soil biochemistry and biodiversity. *Process. Biochem.* **2010**, *45*, 1127–1133. [CrossRef]
102. Lucini, L.; Rouphael, Y.; Cardarelli, M.; Canaguier, R.; Kumar, P.; Colla, G. The effect of a plant-derived protein hydrolysate on metabolic profiling and crop performance of lettuce grown under saline conditions. *Sci. Hortic.* **2015**, *182*, 124–133. [CrossRef]
103. Matsumiya, Y.; Kubo, M. Soybean peptide: Novel plant growth promoting peptide from soybean. In *Soybean and Nutrition*; El-Shemy, H., Ed.; InTech, 2011. Available online: http://www.intechopen.com/books/soybean-and-nutrition/soybean-peptide-novel-plant-growth-promoting-peptide-from-soybean (accessed on 1 December 2019).
104. Schiavon, M.; Ertani, A.; Nardi, S. Effects of an alfalfa protein hydrolysate on the gene expression and activity of enzymes of TCA cycle and N metabolism in *Zea mays* L. *J. Agric. Food Chem.* **2008**, *56*, 11800–11808. [CrossRef]
105. Halpern, M.; Bar-Tal, A.; Ofek, M.; Minz, D.; Muller, T.; Yermiyahu, U. The use of biostimulants for enhancing nutrient uptake. *Adv. Agron.* **2015**, *130*, 141–174. [CrossRef]
106. Nardi, S.; Pizzeghello, D.; Schiavon, M.; Ertani, A. Plant biostimulants: physiological responses induced by protein hydrolyzed-based products and humic substances in plant metabolism. *Sci. Agric.* **2016**, *73*, 18–23. [CrossRef]
107. Selvam, G.G.; Sivakumar, K. Micromorphological study of *Vigna mungo* L. using seaweed liquid fertilizer from *Hypnea musciformis* (Wulf.) Lamouroux. *Indian J. Mar. Sci.* **2016**, *45*, 1199–1207.
108. John, P.P.J.; Yuvaraj, P. Effect of Seaweed Liquid Fertilizer of *Colpomenia sinuosa* (Mert. ex Roth) Derbes & Solier (Brown Seaweed) on *Vigna radiata* (L.) R. Wilczek. in Koothankuzhi, Tirunelveli district, Tamil Nadu, India. *Int. J. Pure Appl. Biosci.* **2014**, *2*, 177–184.
109. Gugała, M.; Sikorska, A.; Zarzecka, K.; Findura, P.; Malaga–Toboła, U. Chemical composition of winter rape seeds depending on the biostimulators used. *Agronomy* **2019**, *9*, 716. [CrossRef]
110. Schubert, S.; Mengel, K. Important factors in nutrient availability: root morphology and physiology. *Z. Pflanzenernähr. Bodenkd.* **1989**, *152*, 169–174. [CrossRef]
111. Czewińska, E.; Szparaga, A. Antibacterial and antifungal activity of plant extracts. *Rocz. Ochr. Sr.* **2015**, *17*, 209–229.
112. Jindo, K.; Martim, S.A.; Navarro, E.C.; Pérez-Alfocea, F.; Hernandez, T.; Garcia, C.; Aguiar, N.O.; Canellas, L.P. Root growth promoting by humic acids from composted and non-composted urban organic wastes. *Plant Soil* **2012**, *353*, 209–220. [CrossRef]
113. Ertani, A.; Schiavon, M.; Muscolo, A.; Nardi, S. Alfalfa plant-derived biostimulant stimulate short-term growth of salt stressed *Zea mays* L. plants. *Plant. Soil* **2013**, *364*, 145–158. [CrossRef]
114. Cheynier, V.; Comte, G.; Davies, K.M.; Lattanzio, V.; Martens, S. Plant phenolics: Recent advances on their biosynthesis, genetics, and ecophysiology. *Plant. Physiol. Biochem.* **2013**, *72*, 1–20. [CrossRef]
115. Aremu, A.O.; Masondo, N.A.; Rengasamy, K.R.R.; Amoo, S.O.; Grúz, J.; Biba, O.; Šubrtová, M.; Pencık, A.; Novák, O.; Doležal, K.; et al. Physiological role of phenolic biostimulants isolated from brown seaweed *Ecklonia maxima* on plant growth and development. *Planta* **2015**, *241*, 1313–1324. [CrossRef]
116. Świeca, M.; Gawlik-Dziki, U.; Kowalczyk, D.; Złotek, U. Impact of germination time and type of illumination on the antioxidant compounds and antioxidant capacity of *Lens culinaris* sprouts. *Sci. Hortic.* **2012**, *140*, 87–95. [CrossRef]
117. Fan, D.; Hodges, D.M.; Critchley, A.T.; Prithiviraj, B. A commercial extract of brown macroalga (*Ascophyllum nodosum*) affects yield and the nutritional quality of spinach in vitro. *Commun. Soil Sci. Plant Anal.* **2013**, *44*, 1873–1884. [CrossRef]
118. Paul, K.; Sorrentino, M.; Lucini, L.; Rouphael, Y.; Cardarelli, M.; Bonini, P.; Reynaud, H.; Canaguier, R.; Trtílek, M.; Panzarová, K.; et al. Understanding the biostimulant action of vegetal-derived protein hydrolysates by high-throughput plant phenotyping and metabolomics: a case study on tomato. *Front. Plant Sci.* **2019**, *10*, 47. [CrossRef]

119. Shalaby, S.; Horwitz, B.A. Plant phenolic compounds and oxidative stress: integrated signals in fungal–plant interactions. *Curr. Genet.* **2015**, *61*, 347–357. [CrossRef]
120. Gurav, R.G.; Jadhav, J.P. A novel source of biofertilizer from feather biomass for banana cultivation. *Environ. Sci. Pollut. Res. Int.* **2013**, *20*, 4532–4539. [CrossRef] [PubMed]
121. Wenda-Piesik, A.; Kazek, M.; Ropińska, P. Impact of amino acid biostimulation and microelements fertilization in foliar application on productivity of winter oilseed rape. *Fragm. Agron.* **2017**, *34*, 119–129.

agronomy

MDPI

Article

Appraisal of Combined Applications of *Trichoderma virens* and a Biopolymer-Based Biostimulant on Lettuce Agronomical, Physiological, and Qualitative Properties under Variable N Regimes

Youssef Rouphael [1,*], Petronia Carillo [2], Giuseppe Colla [3,*], Nunzio Fiorentino [1], Leo Sabatino [4], Christophe El-Nakhel [1], Maria Giordano [1], Antonio Pannico [1], Valerio Cirillo [1], Edris Shabani [1,5], Eugenio Cozzolino [6], Nadia Lombardi [1], Mauro Napolitano [1] and Sheridan L. Woo [7,8,9]

1 Department of Agricultural Sciences, University of Naples Federico II, 80055 Portici, Italy; nunzio.fiorentino@unina.it (N.F.); Nakhel_Christophe@hotmail.com (C.E.-N.); maria.giordano@unina.it (M.G.); antonio.pannico@unina.it (A.P.); valerio.cirillo@unina.it (V.C.); edris.shabani@gmail.com (E.S.); nadia.lombardi@unina.it (N.L.); mauro.napolitano@gmail.com (M.N.)
2 Department of Environmental, Biological and Pharmaceutical Sciences and Technologies, University of Campania "Luigi Vanvitelli", 81100 Caserta, Italy; petronia.carillo@unicampania.it
3 Department of Agriculture and Forest Sciences, University of Tuscia, 01100 Viterbo, Italy
4 Dipartimento Scienze Agrarie, Alimentari e Forestali, Università di Palermo, Viale delle Scienze 4, 90128 Palermo, Italy; leo.sabatino@unipa.it
5 Faculty of Agriculture, Shahid Chamran University of Ahvaz, Ahvaz, Iran
6 Council for Agricultural Research and Economics (CREA)—Research Center for Cereal and Industrial Crops, 81100 Caserta, Italy; eugenio.cozzolino@crea.gov.it
7 Department of Pharmacy, University of Naples Federico II, 80131 Naples (NA), Italy; woo@unina.it
8 Task Force on Microbiome Studies, University of Naples Federico II, 80055 Portici, Italy
9 National Research Council, Institute for Sustainable Plant Protection, 80055 Portici, Italy
* Correspondence: youssef.rouphael@unina.it (Y.R.); giucolla@unitus.it (G.C.)

Received: 7 January 2020; Accepted: 28 January 2020; Published: 1 February 2020

Abstract: The current research elucidated the agronomical, physiological, qualitative characteristics and mineral composition of lettuce (*Lactuca sativa* L. var. *longifolia*) after treatments with a beneficial fungus *Trichoderma virens* (TG41) alone or in combination with a vegetal biopolymer-based biostimulant (VBP; 'Quik-link'). The experiment consisted of lettuce plants grown in three N conditions: sub-optimal (0N kg ha^{-1}), optimal (70N kg ha^{-1}), and supra-optimal (140N kg ha^{-1}) N levels. Lettuce grown under 0N fertilization showed a significant increase in fresh yield when inoculated with TG41 alone (45%) and a greater increase with TG41 + VBP biostimulant (67%). At 48 days after transplanting, both the TG41 alone or TG41+VBP biostimulant induced higher values of CO_2 assimilation in comparison to the control. The mineral concentrations in leaf tissues were greater by 10% for K and 12% for Mg with the TG41+VBP treatments compared to the untreated lettuce. The lettuce plants receiving either TG41 alone or TG41+VBP biostimulants had a significantly lower nitrate content than any of the untreated controls. In non-fertilized conditions, plants treated with TG41+VBP biostimulants produced lettuce of higher premium quality as indicated by the higher antioxidant activity, total ascorbic acid (+61%–91%), total phenols (+14%) and lower nitrate content when compared to the untreated lettuce.

Keywords: microbial biostimulant; non-microbial biostimulant; Lactuca sativa L. var. *longifolia*; mineral profile; physiological mechanism; photosynthesis; nitrate; functional quality

1. Introduction

Rapid growth in the world population will determine an increase in global food demand that is expected to double by 2050 [1]. The intensification in agricultural production appears to be the only useful strategy to meet the rapidly growing food demand in the future, although this imposes stress to the agroecosystem [1], presents serious problems to the ecosystem and health [2–4], since it requires high-input resource cropping systems (such as greenhouse horticulture), that are not ecologically sustainable [5]. In actual fact, greenhouse farming systems use the highest amount of synthetic nitrogen (N) fertilizers per unit area of cultivated produce than any other cropping system [6–9].

Nitrogen-containing compounds are typically applied as chemical fertilizers in agriculture [10,11]. Nitrogen overuse and/or the imbalance between N and other nutrients, such as phosphorus, increases N losses while reducing nitrogen use efficiency (NUE) by the plant, which affects yield and, consequently, profit margins for farmers [12]. Moreover, the accumulation of excess nitrate in edible plant parts can be reduced to the nitrite form, which can cause diseases, such as methemoglobinemia, to which children are particularly at risk [13,14]. However, to date, the efforts to reduce N fertilizer use while at the same time attempting to increase NUE have been proven ineffective. This can be attributed to the inability of crop plants to adapt to low N availability conditions, which limit the activation of the physiological processes necessary for increasing crop production [7,15].

Recently, promising strategies that could aid a shift from N-intensive agriculture to a more eco-friendly approach that reduces the use of N fertilizers while simultaneously increasing NUE and yields, proposes the integrated use of non-chemical plant biostimulants (PBs) in cropping systems [16–20]. PBs are products able to enhance plant growth and development that include several substances with bioactive properties (seaweed and plant extracts, humic and fulvic acids, protein hydrolysates, and silicon), as well as some plant growth-promoting microorganisms (mycorrhizal fungi and plant growth-promoting rhizobacteria) [21–24]. Other plant beneficial microbes include fungi, such as *Trichoderma,* that have multiple plant beneficial capabilities, such as pathogen/pest control, increased nutrient uptake, stimulation of photosynthesis, and carbohydrate metabolism processes, that positively influence crop productivity and quality [25–29]. Several *Trichoderma* spp. are registered as microbial biological control agents in Plant Protection Products commercialized for the control of a broad-spectrum plant diseases [27,30]. Biocontrol mechanisms include direct antagonism with the production of secondary metabolites (i.e., hydrolytic enzymes, antibiotics), competition, and induced plant resistance [26–28,31–33]. Furthermore, many species, among *T. harzianum, T. virens, T. asperellum,* and *T. atroviride*, also act as plant biostimulants, able to enhance nutrient uptake and plant growth, or conferring plant tolerance to abiotic stress [25,34–38]. The direct and indirect benefits to the plants depend upon *Trichoderma*-plant molecular crosstalk, and exchange of diverse chemicals and small peptides that stimulate various plant responses [39,40]. These include the fungi metabolites, proven to have auxin and ethylene-like activity, that induce a reorganization of gene expression patterns in shoots and roots with significant changes in the plant metabolic machinery and a consequent improvement in plant resilience and yield [26,40–42]. These released compounds specifically modify plant root architecture, increasing root length, density and branching, and nutrient uptake (P, Fe, Mn, and Zn), in addition to acting as mediators in the plant microbiome for communication, warning signals, and pest management [25–27,43,44]. Recently, experiments conducted by Fiorentino and co-workers [25] on lettuce and rocket, grown under three different N fertilization rates and inoculated with two *Trichoderma* strains, demonstrated that, in particular, one strain of *T. virens* G41 (ex-*Gliocladium virens* GV41) was able to enhance NUE in lettuce, also favoring the uptake of native N present in the soil. Specifically, the benefits of inoculating plants with this *Trichoderma* strain were more evident when cultivation was performed under sub-optimal N conditions [25].

Another prominent category of PBs that has demonstrated beneficial effects on root stimulation, similar to those exerted by *Trichoderma,* is that of vegetal biopolymer-based products (VBP) that contain lateral root promoting peptides (LRPP) and lignosulphonates. In particular, the lignosulfonates obtained from sulfite pulping processes during cellulose extraction from wood are used in a variety of industries, but they have also been used as fertilizers in crops [45]. They have proven auxin and gibberellin-like activities, probably due to the biological action of phenol metabolites able to interact with plant phytohormones and enzymes affecting carbon–nitrogen metabolism [45]. Lucini et al. [46] indicated that when the vegetal-based biopolymer was applied as a drench to melon, it altered the plant hormone profile by inducing an increase in ABA intermediates, brassinosteroids, and cytokinins in a dose-dependent manner. This mechanism stimulated root growth and consequently resulted in a 'nutrient acquisition response' improving resources use efficiency (RUE), thus enhancing plant biomass production and resistance to transplant stress. In addition, the authors reported that brassinosteroids may play a key role both in root system architecture changes as well as in shoot interference with hormone signaling and secondary metabolites, such as phenolic acids and carotenoids, plus the modulation of photosynthesis.

Romaine lettuce requires varying levels of N during the 65–75-day production cycle that depends upon the plant growth and development stages, plus N availability in the rhizosphere. N availability affects the morphological and physiological plant attributes [47] that influence the marketability of the leafy produce (i.e., leaf size) and consumer perception (i.e., visual green color). From this perspective, depending on the farming conditions, growing season, and genotypes, the combined application of *Trichoderma* and vegetal-biopolymer biostimulants could be particularly useful to enhance lettuce production due to their abilities to increase NUE, favoring nutrient uptake and utilization efficiency. Furthermore, the appropriate incorporation of N in the plant is important since the nitrate content in vegetable products must be within the limits established by the market according to EU regulation no. 1258/2011, whereby the levels should not exceed 3000–5000 mg kg^{-1} fw.

In a recent opinion, Rouphael and Colla [23], indicated that the scientific community and private companies should focus on exploiting the potential synergistic biostimulatory action of microbes with non-microbial PBs combinations to design and develop second-generation plant products (biostimulant 2.0) with specific targeted biostimulant actions. A few experimental investigations have demonstrated the beneficial effects on crop performance of combining microbial inoculants (i.e., *Rhizophagus intraradices* or plant growth promoting bacteria or *R. irregulare* and *T. atroviride*) with humic acids [48,49] or protein hydrolysates [50]. Previous indications by Fiorentino et al. [25] suggested that the nutrient content of leafy horticulture crops could vary according to cultivation in diverse fertilizer conditions and in the presence/absence of a fungal inoculant. However, to date, nothing is known about the effects of *Trichoderma* alone or in combination with a vegetal biopolymer based-biostimulant on the agronomical, physiological and qualitative responses of an important leafy vegetables, such as Romaine lettuce (*Lactuca sativa* L. var. *longifolia*). This study will investigate the effect of a beneficial microbe (*T. virens* TG41) when used alone or in combination with a VBP biostimulant ('Quik-link'), under supra-optimal, optimal, and suboptimal N regimes, on Romaine lettuce production and marketability characteristics. This study will increase understanding of the processes involving these two different types of plant biostimulants and the effects on plant N acquisition response, for which the comprehension is pivotal to increasing NUE, as well as attempting to decrease N environmental inputs and reduce risks to consumer health.

2. Materials and Methods

2.1. Experimental Setup, Design, and Crop Management

An experiment was performed on lettuce (*Lactuca sativa* L. var. *longifolia* cv. 'Romana Bionda Lentissima a Montare'—Esasem, Casaleone, Verona) from November 4, 2015 to January 19, 2016, in a protected greenhouse structure (unheated) at the Department of Agricultural Sciences, University of Naples Federico II located at Portici, Italy. The soil was classified as a sandy loam texture (73% sand, 19% silt, 8% clay), with a pH of 7.0, electrical conductivity of 0.5 dS m^{-1}, an organic matter of 1.25% (w/w) and a total N of 1.1 g kg^{-1}. The NO_3-N, $NH_4{}^+$-N, available P, and exchangeable K were 95, 7, 35, and 950 mg kg^{-1}, respectively.

A split-plot design with three replicates (randomized blocks) was adopted with fertilization (3 levels) as the main factor and biostimulant applications (3 levels) as the sub-factor. The three N fertilization levels were suboptimal (0 kg ha^{-1}; 0N), optimal (70 kg N ha^{-1}; 70N) and supra-optimal (140 kg N ha^{-1}; 140N), while the three biostimulant applications were non-inoculated control, inoculated *Trichoderma virens* G41 (TG41), and *T. virens* + vegetal biopolymer-based biostimulant (TG41 + VBP). The cultivated area of each experimental plot (27 experimental plots in total) was 3.5 m^2. Lettuce were transplanted on November 4th (at the 3 true-leaf stage) in double rows with a plant density of 14 plants per square meter. A biodegradable black mulch film (15 μm thick MaterBi®, Novamont, Novara, Italy) was used and maintained throughout the entire greenhouse experiment.

N total amount was applied as ammonium nitrate (NH_4NO_3 34%) into two identical doses, at 6 and 27 days after transplanting (DAT) by fertigation using a drip irrigation system with in-line emitters (flow rate: 3.3 L h^{-1}; distances: 35 cm). Foliar pests, such as cutworms, were controlled with two applications of Decis Evo (active ingredient 25 g L^{-1} of deltamethrin—Bayer Crop Science, Milano, Italy) at the rate of 0.4 L ha^{-1}, whereas a copper-based fungicide (Cupravit 35 WG containing 350 g kg^{-1} of copper as copper oxychloride—Bayer Crop Science, Milano) was sprayed twice at the rate of 2.5 kg ha^{-1} to control downy mildew caused by *Bremia lactucae* Regel.

2.2. Fungal and Vegetal Biostimulants

A spore suspension of *T. virens* strain G41 (final concentration 1×10^7spores mL^{-1}; TG41) was used to inoculate the lettuce seedlings at time of transplant by using a root dip method (with submergence for 10 min); then a repeated inoculation was conducted at 18 DAT by watering 25 mL of the inoculum $plant^{-1}$. The vegetal biopolymer-based (VBP) biostimulant ('Quik-link®', Italpollina, Rivoli Veronese, Italy) was used in the current experiment. The product has a density of 1.21 kg L^{-1}, a pH (1:5) of 4.7, an electrical conductivity; EC (1:5) of 20 mS cm^{-1}, 25 g kg^{-1} of organic N as peptides and free amino acids, 160 g kg^{-1} of organic C, lignosulphonates, and micronutrients, such as iron, manganese, zinc, copper, and molybdenum, in the following concentrations 10.0, 7.0, 3.0, 1.0, and 0.2 g kg^{-1}, respectively [46]. Peptides and free amino acids were obtained through enzymatic hydrolysis of a vegetal source of proteins, as reported by Carillo et al. [7]. The peptides in the product have a high biological activity being signaling molecules (e.g., lateral root promoting peptides—LRPP). The commercial product was applied at the base of each plant (100 mL, containing 6 L ha^{-1} of 'Quik-link') at transplant, plus 17 (stage BBCH41-head beginning to form) and 45 DAT (stage BBCH45%–50% of the expected head size).

2.3. Fungal Colony Forming Units in Soil Rhizosphere and Trichoderma-VBP Compatibility

Soil samples were collected from the plant rhizosphere at the time of harvest. The number of fungal colonies forming units was determined, as indicated in Fiorentino et al. [25]. Briefly, a 1% (w/v) soil suspension was prepared in water, in serial dilutions, then 100 μL aliquots of each sample were spread on the surface of 90 mm culture plates containing Rose Bengal-Chloramphenicol agar (HiMedia Pvt. Ltd., Mumbai–India) supplemented with 0.1% (v/v) Igepal (Sigma–Aldrich, Milano, Italy), and incubated for 3–7 days at 25 °C. The emerging fungal colonies were counted daily.

In vitro tests were performed with varying doses of the VBP biostimulant and the *Trichoderma* inoculum, including the doses used for the field treatments to determine if the 'Quik-link' product inhibited the germination and growth of the fungus.

2.4. Fresh and Dry Yield, SPAD index and CIE (lab) Measurements

At harvest (76 DAT), the lettuce fresh yield was assessed in sampling areas of 2 m^2 from the center of the 27 experimental plots. The shoot dry biomass was determined (after oven drying at 80 °C for 72 h). The dried leaf tissues were conserved for mineral analysis. At 45 and 75 DAT, the soil plant analysis development (SPAD) index (i.e., non-destructive measurement of chlorophyll content) was measured on undamaged and expanded lettuce leaves using a portable SPAD-502 chlorophyll meter (Konica-Minolta, Tokyo, Japan). Twelve measurements were conducted on four randomly picked lettuce plants per experimental plot, then averaged to a single SPAD value for each replicate [51]. Subsequently, on the same date, measurements were performed using a Minolta CR-300 Chroma Meter (Minolta Camera Co. Ltd., Osaka, Japan) to evaluate the *Commission Internationale de L'Eclairage* (CIE) color space parameters for L* (lightness) and chroma coordinates: a* (−a* greenness) and b* (+b* yellowness). In each experimental plot, 10 healthy leaves were measured and averaged to represent a single color value [52].

2.5. Net CO_2 Assimilation Rate and Stomatal Resistance Measurements

At 33, 40, and 48 DAT, measurements of leaf gas exchange were carried out within 2 h across solar noon on the youngest fully expanded lettuce leaves, using nine replicates for each treatment. Measurements of net CO_2 assimilation rate (A_{co2}; μmol CO_2 m^{-2} s^{-1}) and stomatal resistance (r_s; m^2 s^{-1} mol^{-1}) were recorded using a portable gas exchange analyzer (LCA-4; ADC BioScientific Ltd., UK). Photosynthetically active radiation, relative humidity, and carbon dioxide concentration (PAR, 850 ± 100, 1000 ± 100, and 600 ± 100 μmol m^{-2} s^{-1}, RH 60 ± 5, 55 ± 5, and 60 ± 5%, and 400 ± 5, 410 ± 5, and 400 ± 5 ppm, at 33, 40, and 48 DAT, respectively) were set at ambient value, and the airflow rate was 400 mL s^{-1}.

2.6. Mineral Composition Analysis

Plant material was dried and pulverized using a cutting–grinder head (IKA, MF10.1, Staufen, Germany), then the powder was extracted in Milli-Q water (Merck Millipore, Darmstadt, Germany) for 10 min at 80 °C in a thermostatic bath (ShakeTemp SW22, Julabo, Seelbach, Germany) and centrifuged at 6000 rpm for 10 min as indicated in Rouphael et al. [50]. A Dionex ICS-3000 system (Sunnyvale, CA, USA) equipped with suppressed conductivity detection was used to determine the ion content of the samples. The ion separation of the samples was carried out with two different ion-exchange columns: An IonPac CS12A column (250 × 4 mm) was used for the cation separation eluted with 20 mM methanesulfonic acid (flow rate 1 mL min^{-1}), and an IonPac AS11-HC column (250 × 4 mm) was used for the anion separation eluted with a potassium hydroxide gradient (flow rate 1.5 mL min^{-1}). Nitrogen (total N) concentration in leaf tissue was determined according to the Kjeldahl method [53].

2.7. Antioxidant Capacity, Total Phenols, and Total Ascorbic Acid Analysis

Lipophilic and hydrophilic antioxidant capacity and total phenols were determined on freeze-dried tissue samples, whereas the total ascorbic acid was assessed on fresh material and measured using a spectrophotometer (Hach DR 2000, Hach Co., Loveland, CO, USA) according to the protocols of Re et al. [54], Fogliano et al. [55], Singleton et al. [56], and Kampfenkel et al. [57], respectively. Solution absorbances were assessed at 505, 734, 525, and 765 nm for the lipophilic and hydrophilic antioxidant fractions, total polyphenols, and total ascorbic acid, respectively.

2.8. Data Elaboration, Statistical Analysis, Principal Component Analysis, and Heat Map

The statistical analyses were all carried out using the software IBM SPSS Statistics 21. All data were subjected to two-way analysis of variance, and mean values were separated according to Duncan test with $p < 0.05$. Principal component analysis (PCA) was performed on the whole morphological and physiological data set, and the eigen values, total variance of the first three principal components (PCs) as well as the loading scores and plots were determined [58–60]. A heat map summarizing the agronomical, physiological, and qualitative responses of lettuce to plant biostimulant applications and N fertilization levels was also generated using the https://biit.cs.ut.ee/clustvis/ online program package with Euclidean distance as the similarity measure and hierarchical clustering with complete linkage [6].

3. Results

3.1. Fungal Concentration in the Soil

The total number of fungal colonies (including *Trichoderma*) recovered from soil rhizosphere in the nine treatments ranged between 2.0×10^5 and 6.5×10^5 colony forming units (CFU) g^{-1} of soil and was significantly ($p < 0.05$) influenced by the interaction of the two tested factors: N fertilization level (N) and VBP biostimulant application. In particular, results indicated that the highest fungal CFU was observed in soils from lettuce plants inoculated with TG41 under suboptimal 0N conditions (6.5×10^5 CFU g^{-1} of soil), in comparison to any of non-inoculated plants under suboptimal, optimal, or supra-optimal N conditions (average 2.5×10^5 CFU g^{-1} of soil), whereas the treatments with TG41 (at 70N and 140N) or TG41 +VBP biostimulant (at 0N and 70N) exhibited intermediate values (average 3.9×10^5 CFU g^{-1} of soil) (data not shown). Moreover, the in vitro tests performed with the beneficial microbe (TG41) and non-microbial VBP biostimulant at the dose applied in the greenhouse experiment did not demonstrate any inhibition of the germination and growth of the fungi concentration (69.2 CFU in the absence and 68.5 CFU in the presence of the 'Quik-link-product), suggesting compatibility between the two biostimulants.

3.2. Growth Responses, SPAD Index and Leaf Colorimetry

A significant ($p < 0.01$) interaction between N fertilization level and biostimulant application was observed on fresh yield and dry biomass. For instance, the use of the TG41-based biostimulant alone or in combination with the VBP biostimulant positively affected both fresh and dry yield of lettuce plants under both sub-optimal (0 kg ha^{-1}) and optimal (70 kg ha^{-1}) N conditions, but the beneficial effect was not apparent in the over N fertilization condition (140 kg ha^{-1}) (Figure 1). Lettuce grown in the absence of N fertilization demonstrated a highly significant increase in fresh yield of 67% when inoculated with the combined TG41+VBP biostimulants. Instead, a more moderate increase of 45% was observed over the untreated 0N condition with the inoculation of *T. virens* G41 alone. Moreover, under optimal N fertilization (70 kg ha^{-1}), only lettuce plants inoculated with TG41 alone exhibited significantly higher fresh yields. Treatments with TG41 alone or TG41+VBP increased marketable dry yield by 16% when compared to the untreated control, but no significant differences were noted between the two different biostimulant inoculations (Figure 1). No effects on lettuce yield were observed with either of the biostimulants at the supra-optimal 140N fertilization.

The SPAD index in *Lactuca sativa* L. var. *longifolia*, as an indication of chlorophyll content, was significantly affected by N fertilization levels (at 75 DAT) and by biostimulant applications (at 45 and 75 DAT), with effects in the N × T interaction (Table 1). At 75 DAT, the highest SPAD index values were recorded with TG41 + VBP biostimulant combination (Table 1). The visual appearance, particularly the greenness of leaf color, is a primary parameter used by the consumer in product preference and selection choice [61]. In general, neither the N fertilization level nor biostimulant application had a significant effect on the leaf greenness (−a* values) in lettuce (Table 1). Overall, the N application levels resulted in a greater lightness in the color of the lettuce leaves, with the lowest L* values recorded in the 140 kg N ha^{-1} treatment, which also corresponded to a decrease in the chroma coordinate (b*; Table 1).

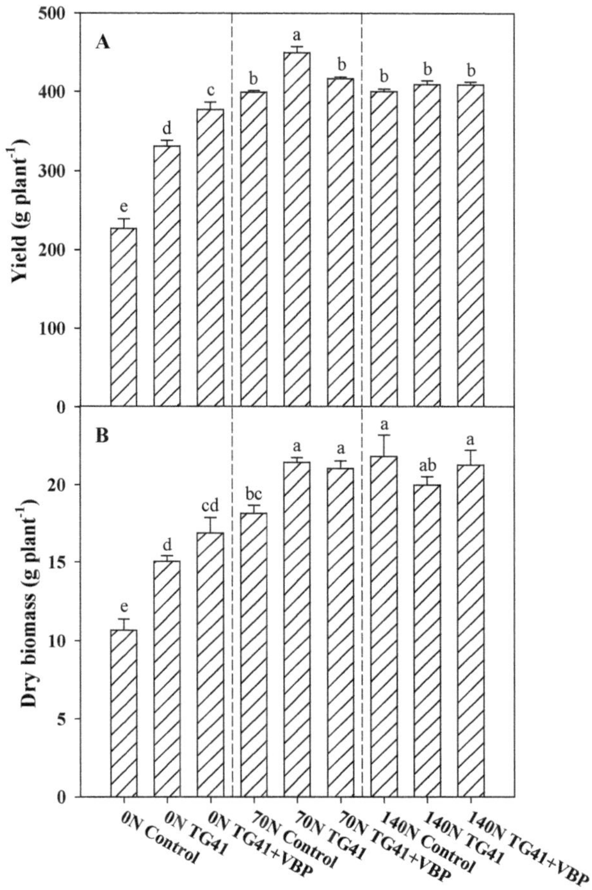

Figure 1. Fresh yield (**A**) and dry biomass (**B**) of Romaine lettuce grown in greenhouse in relation to N fertilization level (0N = 0 kg ha^{-1}, 70N = 70 kg ha^{-1}, 140N = 140 kg ha^{-1}) and biostimulant application (Untreated = Control, TG41 = *T. virens* G41, and TG41+VBP = vegetal biopolymer-based biostimulant). Mean values with the same letter were not different, according to Duncan's test ($p < 0.05$).

Table 1. Effects on Romaine lettuce soil plant analysis development (SPAD) index and *Commission Internationale de L'Eclairage* (CIE) color space parameters for: L* (lightness) and chroma coordinates: a* (−a* greenness) and b* (+b* yellowness) in relation to N fertilization level (0N = 0 kg ha^{-1}, 70N = 70 kg ha^{-1}, 140N = 140 kg ha^{-1}) and biostimulant application (Untreated=Control, TG41=*T. virens* G41, and TG41+VBP=vegetal biopolymer-based biostimulant) during the cultivation cycle.

Treatments	SPAD Index		L	a*	b*
	45 DAT	75 DAT			
Nitrogen rate (N)	NS	***	**	NS	*
Biostimulant (B)	*	**	NS	NS	NS
N × B	NS	NS	NS	NS	*
Nitrogen rate (kg ha^{-1})					
0	38.27	37.92 b	42.81 a	−16.61	24.53 a
70	38.94	39.35 a	42.60 a	−16.37	24.20 ab
140	38.47	39.24 a	41.54 b	−16.23	23.62 b
Biostimulant					
Control	37.54 b	38.39 b	41.94	−16.33	23.86
TG41	39.50 a	38.56 b	42.65	−16.49	24.37
TG41+VBP	38.63 a	39.50 a	42.36	−16.38	24.13
N × B					
0N Control	37.23	38.00	42.08	−16.25	23.50 c
0N TG41	39.30	37.45	43.08	−16.93	25.18 a
0N TG41+VBP	38.27	38.30	43.28	−16.65	24.93 ab
70N Control	37.50	38.69	42.95	−16.53	24.55 abc
70N TG41	40.60	39.15	42.98	−16.53	24.53 abc
70N TG41+VBP	38.73	40.15	41.88	−16.05	23.53 bc
140N Control	37.90	38.48	40.80	−16.20	23.53 bc
140N TG41	38.60	39.23	41.90	−16.03	23.40 c
140N TG41+VBP	38.90	40.14	41.93	−16.45	23.93 abc

* $p < 0.05$; ** $p < 0.01$; *** $p < 0.001$; NS, not significant. Mean values with the same letter in each column were not different according to Duncan's test ($p < 0.05$). DAT: days after transplanting.

3.3. Leaf Gas Exchange: Net CO_2 Assimilation Rate and Stomatal Resistance

The physiological parameters, in particular, the net CO_2 assimilation rate (A_{CO2}) and stomatal resistance (r_s) in the Romaine lettuce plants throughout the cultivation cycle in the greenhouse, were evaluated as a function of N fertilization level and biostimulant application as displayed in Table 2. The A_{CO2} was significantly affected by the biostimulant treatments for all measured data, and to a lesser degree, by the N fertilization level (only at 48 DAT). Irrespective of the N fertilization level (N × B interaction= ns) at 33 and 40 DAT, both the TG41 alone or in combination with the VBP-based biostimulant induced higher values of A_{CO2} in comparison to the control treatment that was not significantly different between the two biostimulant treatments. At 48 DAT, the A_{CO2} increased in the following order with the applications: TG41+VBP > TG41 > control (Table 2). On the other hand, augmenting the N fertilization level resulted in a linear increase in A_{CO2} from 0 to 140 kg ha^{-1} but only at 48 DAT (Table 2).

Contrary to A_{CO2}, the r_s was not affected neither by N fertilization level nor by biostimulant application at 33 and 48 DAT, while at 40 DAT, the r_s was only influenced by the two biostimulant applications (Table 2). Particularly, on this date, the r_s was significantly lower on average by 26% when lettuce plants were inoculated with *Trichoderma* alone or in combination with the commercial product 'Quik-link' (Table 2).

3.4. Mineral Composition in Leaf Tissue

The results regarding the mineral profile in Romaine lettuce leaves are presented in Table 3. For all the macronutrients and sodium analyzed, no significant differences were observed in the N fertilization level and biostimulant application interaction. In particular, neither N fertilization rate nor biostimulant treatment had a significant effect on Ca and Na concentrations in lettuce leaves (average 7.0 and 1.4 g kg^{-1} dry weight, respectively; Table 3). The concentrations of N and P in leaf tissues were significantly affected by N fertilization rate. Concentrations of N and P increased as the N fertilization level increased, with the highest values recorded at 140 kg ha^{-1} for N and at 70 and 140 kg ha^{-1} for P (Table 3).

The effects of TG41 and TG41+VBP biostimulant, when averaged over all N fertilization rates, affected the K and Mg concentrations in leaf tissues which were higher by 10% and 12%, respectively, than in untreated lettuce plants, but with no significant difference noted between the two biostimulant treatments (Table 3).

Table 2. Net CO_2 assimilation rate and stomatal resistance of greenhouse Romaine lettuce plants measured during the production cycle in relation to N fertilization level (0N = 0 kg ha^{-1}, 70N = 70 kg ha^{-1}, 140N = 140 kg ha^{-1}) and biostimulant application (Untreated=Control, TG41=*T. virens* G41, and TG41+VBP=vegetal biopolymer-based biostimulant).

Treatments	Net CO_2 Assimilation Rate (µmol CO_2 m^{-2} s^{-1})			Stomatal Resistance (m^2 s^1 mol^{-1})		
	33 DAT	40 DAT	48 DAT	33 DAT	40 DAT	48 DAT
Nitrogen rate (N)	NS	NS	***	NS	NS	NS
Biostimulant (B)	***	***	***	NS	*	NS
N × B	NS	NS	NS	NS	NS	NS
Nitrogen rate (kg ha^{-1})						
0	14.51	21.26	13.20 c	3.32	4.06	4.35
70	15.16	20.93	14.12 b	3.94	4.40	3.80
140	14.61	20.45	15.78 a	2.96	4.52	3.36
Biostimulant						
Control	11.74 b	18.49 b	12.48 c	4.59	5.25 a	4.27
TG41	15.25 a	21.60 a	14.76 b	2.77	4.03 b	3.63
TG41+BVP	17.34 a	22.55 a	15.86 a	3.02	3.70 b	3.61
N × B						
0N Control	10.92	18.59	10.93	4.01	5.05	4.83
0N TG41	15.08	22.12	13.23	2.64	3.37	4.20
0N TG41+VBP	17.52	23.08	15.44	3.29	3.77	4.01
70N Control	12.44	18.73	12.06	5.80	4.91	3.98
70N TG41	15.17	21.67	14.54	3.11	4.49	3.61
70N TG41+VBP	17.86	22.38	15.76	2.93	3.79	3.82
140N Control	11.94	18.15	14.44	3.67	5.79	4.01
140N TG41	15.60	21.01	16.52	2.46	4.23	3.08
140N TG41+VBP	16.29	22.20	16.37	2.74	3.53	3.01

* $p < 0.05$; *** $p < 0.001$; NS, not significant. Mean values with the same letter in each column were not different according to Duncan's test ($p < 0.05$). DAT: days after transplanting.

Table 3. Total nitrogen (N), phosphorus (P), potassium (K), calcium (Ca), magnesium (Mg), and sodium (Na) concentrations of greenhouse Romaine lettuce plants at time of harvest (75 DAT) in relation to N fertilization level (0N = 0 kg ha^{-1}, 70N = 70 kg ha^{-1}, 140N = 140 kg ha^{-1}) and biostimulant application (Untreated=Control, TG41=*T. virens* G41, and TG41+VBP=vegetal biopolymer-based biostimulant).

Treatments	N	P	K	Ca	Mg	Na
	(mg g^{-1} dw)	(mg g^{-1} dw)	(mg g^{-1} dw)	(mg g^{-1} dw)	(mg g^{-1} dw)	(mg g^{-1} dw)
Nitrogen rate (N)	**	***	NS	NS	NS	NS
Biostimulant (B)	NS	NS	**	NS	*	NS
N × B	NS	NS	NS	NS	NS	NS
Nitrogen rate (kg ha^{-1})						
0	36.90 b	2.02 b	55.02	6.83	3.63	1.52
70	37.63 b	2.58 a	53.71	7.53	3.69	1.42
140	39.04 a	2.93 a	53.40	6.62	3.42	1.35
Biostimulant						
Control	37.43	2.48	50.65 b	6.47	3.31 b	1.54
TG41	37.57	2.59	55.90 a	7.06	3.70 a	1.34
TG41+VBP	38.58	2.46	55.57 a	7.46	3.73 a	1.41
N × B						
0N Control	37.03	1.98	51.36	6.39	3.46	1.60
0N TG41	35.70	2.03	56.17	6.27	3.55	1.52
0N TG41+VBP	37.98	2.05	57.53	7.84	3.90	1.43
70N Control	37.60	2.45	52.59	7.18	3.46	1.44
70N TG41	37.38	2.55	54.09	7.88	3.92	1.38
70N TG41+VBP	37.93	2.75	54.45	7.53	3.69	1.45
140N Control	37.65	3.00	48.01	5.85	3.01	1.58
140N TG41	39.63	3.20	57.45	7.02	3.62	1.13
140N TG41+VBP	39.85	2.58	54.75	7.00	3.62	1.35

* $p < 0.05$; ** $p < 0.01$; *** $p < 0.001$; NS, not significant. Mean values with the same letter were not different according to the Duncan's test ($p < 0.05$). DAT: days after transplanting.

3.5. Nitrate, Antioxidant Capacity, and Bioactive Content

The registered nitrate content among the 9 experimental conditions (890–1496 mg kg^{-1} fresh weight) was within the limits imposed by the European Regulation No. 1258/2011 for the commercialization of fresh lettuce (3000–5000 mg kg^{-1} on a fresh weight basis). In our study, nitrate content was affected by both N fertilization level and biostimulant application, without significant effects in the N×B interaction (Table 4). As expected, our results demonstrated that increasing N fertilization from 0 to 140 kg ha^{-1} elicited a significant increase in nitrate content compared to non-fertilized plants, whereas lettuce plants cultivated under optimal N fertilization (70 kg ha^{-1}) exhibited intermediate values (Table 4). Interestingly, the nitrate content was significantly lowered in lettuce plants receiving treatments of either TG41 alone and the combined TG41+VBP biostimulants (not significant between them) compared to the untreated control (Table 4).

The hydrophilic and lipophilic antioxidant fractions of greenhouse lettuce ranged from 1.44 to 1.61 mmol ascorbic acid eq. 100 g^{-1} dw and from 2.69 to 4.62 mmol trolox 100 g^{-1} dw, respectively. Neither N fertilization level nor the biostimulant application had a significant effect on the hydrophilic antioxidant activity. Moreover, significant effects were noted on lipophilic antioxidant activity (LAA) with both N and biostimulant treatments, but not the N×B interaction. Irrespective of N fertilization treatments, the application of TG41+VBP demonstrated a significant increase in LAA (+13%) compared to the treatment of TG41 alone and the non-inoculated control (Table 4). Moreover, antioxidant molecules, in particular, total phenols and total ascorbic acid, were significantly influenced by either tested factors of N fertilization and biostimulant application. When averaged over the nitrogen treatments, the lettuce plants cultivated under supra-optimal conditions (i.e., 140 kg ha^{-1}) were characterized by low-quality bioactive compounds in terms of both total phenols and total ascorbic acid (Table 4). Interestingly, the biostimulants-treated plants with TG41 alone and particularly in the combination of TG41+VBP, produced a major amplification of total phenols (+14%) and total ascorbic acid (+61%–91%) in comparison to untreated lettuce plants (Table 4).

Table 4. Nitrate content, hydrophilic (HAA), and lipophilic (LAA) antioxidants activities, total phenols and total ascorbic acid (TAA) content of greenhouse Romaine lettuce at time of harvest in relation to N fertilization level (0N = 0 kg ha^{-1}, 70N = 70 kg ha^{-1}, 140N = 140 kg ha^{-1}) and biostimulant application (Untreated = Control, TG41 = *T. virens* G41, and TG41+VBP = vegetal biopolymer-based biostimulant).

Treatments	Nitrate	HAA	LAA	Phenols	TAA
	(mg kg^{-1} fw)	(mmol eq. ascorbic acid $100g^{-1}$ dw)	(mmol eq. trolox $100g^{-1}$ dw)	(mg eq. gallic acid g^{-1} dw)	(mg $100g^{-1}$ fw)
Nitrogen rate (N)	*	NS	***	***	***
Biostimulant (B)	**	NS	***	*	***
N × B	NS	NS	NS	NS	***
Nitrogen rate (kg ha^{-1})					
0	1019.09 b	1.54	4.06 b	55.94 a	22.66 a
70	1119.46 ab	1.56	2.84 c	54.44 a	13.81 b
140	1319.38 a	1.47	4.88 a	47.38 b	11.20 c
Biostimulant					
Control	1356.88 a	1.56	3.66 b	49.39 b	10.53 c
TG41	1052.41 b	1.47	3.85 b	52.28 ab	16.97 b
TG41+VBP	1048.64 b	1.53	4.26 a	56.10 a	20.17 a
N × B					
0N Control	1152.55	1.56	3.74	52.84	13.88 de
0N TG41	890.63	1.47	3.99	56.07	21.73 b
0N TG41+VBP	1014.10	1.58	4.44	58.91	32.36 a
70N Control	1422.03	1.63	2.69	53.09	9.54 ef
70N TG41	1063.18	1.61	2.95	54.06	19.27 bc
70N TG41+VBP	873.18	1.44	2.88	56.18	12.61 def
140N Control	1496.08	1.51	4.56	42.23	8.15 f
140N TG41	1203.43	1.35	4.62	46.71	9.91 ef
140N TG41+VBP	1258.65	1.55	5.45	53.19	15.54 cd

* $p < 0.05$; ** $p < 0.01$; *** $p < 0.001$; NS, not significant. Mean values with the same letter in each column were not different according to the Duncan's test ($p < 0.05$). DAT: days after transplanting.

3.6. Heat Map Analysis of all Measured Plant Parameters

An aggregated data heat-map analysis of the measured agronomic and physiological parameters was conducted to produce a visual comparison of the effects determined by the tested treatment factors on the Romaine lettuce plants. In Figure 2, the analysis revealed two dendrograms: on the top (Dendrogram 1), a classification that corresponded principally to the biostimulant applications, and on the left (Dendrogram 2), the parameters that influenced this distribution. Dendrogram 1 revealed two main groups: on the left, the cluster corresponded to controls for each of the three N levels that were all untreated with the biostimulant conditions; then on the right of the heat map, two clusters that contained the other six treatments, consisting of a mix of the N levels receiving the biostimulant applications (Figure 2).

In particular, in the left cluster of Dendrogram 1, the 140N Control was well separated from the other two controls (0N and 70N) due to the higher r_s at 40 and 48 DAT; nitrate, Na, P, and dry biomass (in the first/highest cluster of Dendrogram 2), as well as the lower values for the parameters in the second cluster, mainly for the parameters of L*, total phenols, Mg, and K content. On the right side of Dendrogram 1, two clusters were identified, the first on the left included treatments 70N TG41+VBP biostimulant, separated from the 140N level with the biostimulants TG41 or TG41+VBP, that showed in particular lower Na, r_s at 40 and 48 DAT, hydrophilic (HAA), b* and total phenols parameters, but higher LAA, a* value, P and N content, SPAD index and A_{CO2} at 33 and 48 DAT. The grouping on the right included 70N TG41, 0N TG41, and 0N TG41+VBP treatments. Within this cluster, the 0N treatments with the biostimulants were clearly separated from the 70 N TG41 by higher HAA, leaf number (LN), and lower LAA in this latter treatment. Instead, the two 0N levels receiving the biostimulants were distinguished by the parameter groupings found in Dendrogram 2, whereby 0N TG41 could be attributed to the lower values for the parameters found in the third cluster (mainly due to a*, SPAD Index, N), as well as the lower r_s 40 DAT and nitrate, but higher b*; whereas 0N TG41+VBP biostimulant could principally be identified by the all the higher parameters found in the second cluster—specifically total ascorbic acid (TAA). Interestingly, the first cluster in Dendrogram 2 clearly demonstrated the differential effects of the biostimulant treatments (untreated ones had high parameters for all N levels), while the second cluster clearly revealed the consequence of supra-optimal N levels (all parameter values were low), and the outcome of the combined biostimulants in the low N level condition (all parameter values were high), comparatively to the *Trichoderma* alone (i.e., TG41) at 0N.

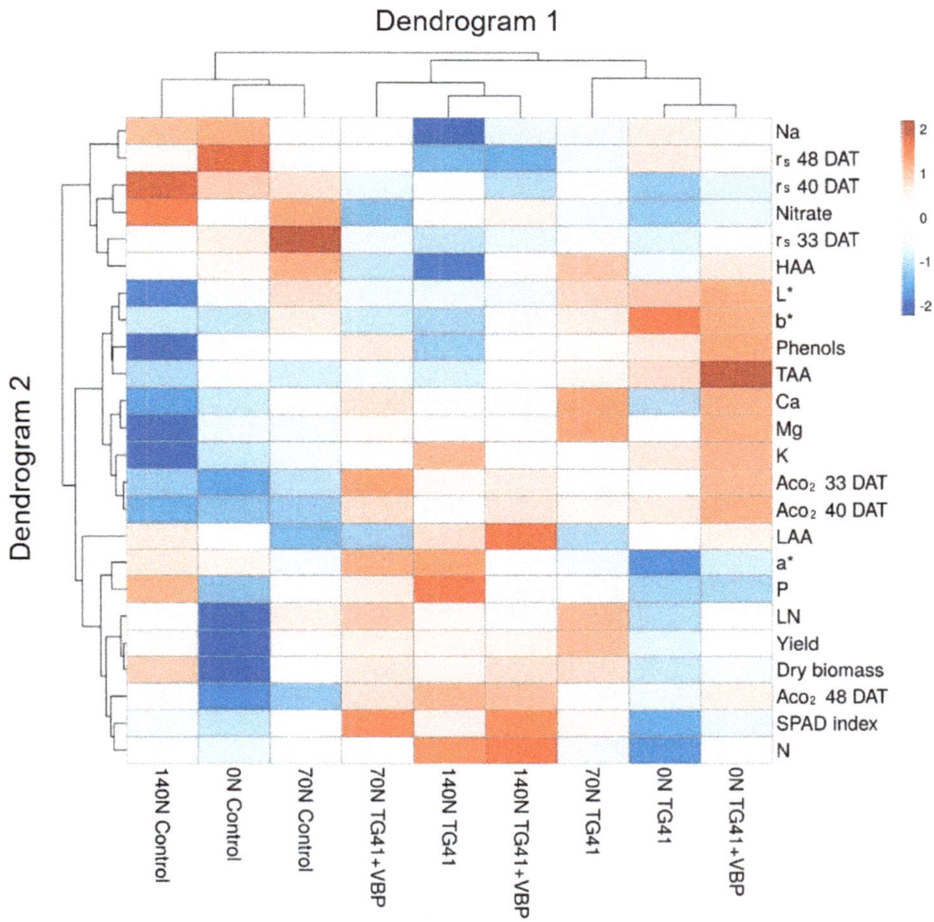

Figure 2. Cluster heat map analysis summarizing greenhouse lettuce plant responses to a factorial experiment with three N fertilization levels (0N = 0 kg ha^{-1}, 70N = 70 kg ha^{-1}, 140N = 140 kg ha^{-1}) and biostimulant application (Untreated=Control, TG41=*T. virens* G41, and TG41+VBP=vegetal biopolymer-based biostimulant). Control plants were not treated with TG41 and/or VBP. The figure was generated using the https://biit.cs.ut.ee/clustvis/ online program package with Euclidean distance as the similarity measure and hierarchical clustering with complete linkage. A_{CO2}: net CO_2 assimilation rate; r_s: stomatal resistance; HAA: hydrophilic antioxidant activity; LAA: lipophilic antioxidant activity; TAA: total ascorbic acid; SPAD: soil plant analysis development; DAT: days after transplanting.

3.7. Principal Component Analysis of all Measured Plant Parameters

Principal component analysis was carried out on the whole experimental data set, and the loading plot and scores are reported in Figure 3. The analysis indicated that the variables in the first three principal components (PCs) were highly correlated, with eigen values greater than 1. thus explaining for 80.4% of the total variance, with PC1, PC2, and PC3 accounting for 36.1%, 32.0%, and 12.4%, respectively. The variable distribution along PC1 was clearly attributed to the biostimulant treatments, while N fertilization levels contributed to that on PC2 (Figure 3). TG41 and TG41+VBP biostimulant treated plants were distributed in the positive quadrants of PC1 except for 0N TG41, while all control treatments (untreated lettuce plants) were distributed in the negative side of PC1. In particular, 0N TG41+VBP biostimulant and 70N TG41 were in the upper right quadrant, while

70N TG41+VBP biostimulant, 140N TG41+VBP biostimulant, and 140N TG41 were in the lower right quadrant. Moreover, in PC2, 0N TG41 was positioned in the positive side of the upper left quadrant, with 0N and 70N untreated control treatments, while the 140N control was in the lower left negative quadrant (Figure 3). PC1 was positively correlated to A_{CO2} at 33, 40, and 48 DAT, K, Mg, and Ca content, yield (fresh weight), dry biomass, and SPAD index. PC1 was also negatively correlated with r_s at 33, 40, 48 DAT, and also with nitrate content. PC2 was positively correlated with L* and b* colorimetric parameters, total phenols, and TAA, while it was negatively correlated to P content and a* colorimetric parameter. In addition, the treatments with 70N TG41+VBP biostimulant and 140N TG41 produced lettuce with a higher yield, leaf number, SPAD index, and A_{CO2} at 48 DAT. Interestingly, the non-fertilized 0N lettuce plants treated with TG41+VBP produced lettuce with higher premium quality (higher total phenols and TAA and lower nitrate content) (Figure 3). Finally, the upper and lower left quadrant depicted the three non-treated control treatments with the lowest quality characteristics (high Na and nitrate content; Figure 3).

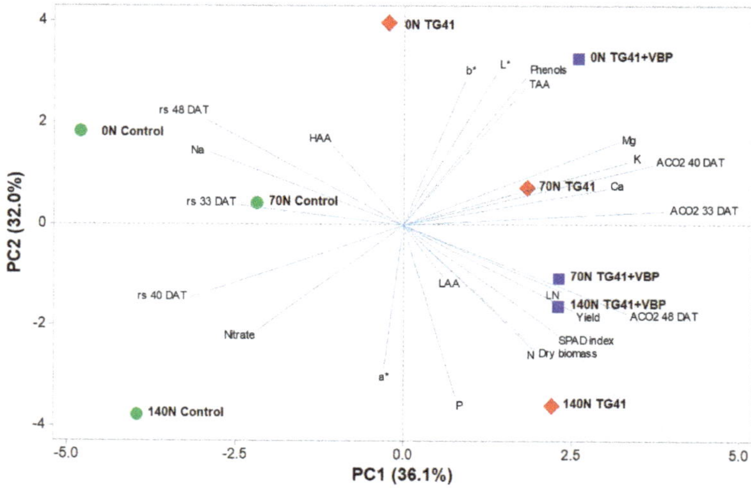

Figure 3. Principal component loading plot and scores of principal component analysis of all morpho-physiological and qualitative parameters analyzed in Romaine lettuce plants submitted to a factorial experiment with three N fertilization levels (0N = 0 kg ha^{-1}, 70N = 70 kg ha^{-1}, 140N = 140 kg ha^{-1}) and biostimulant application (Untreated = Control, TG41 = *T. virens* G41, and TG41+VBP = vegetal biopolymer-based biostimulant). A_{CO2}: net CO_2 assimilation rate; r_s: stomatal resistance; HAA: hydrophilic antioxidant activity; LAA: lipophilic antioxidant activity; TAA: total ascorbic acid; SPAD: soil plant analysis development; DAT: days after transplanting.

4. Discussion

Our findings indicated that the suboptimal fertilizer condition (0 kg N ha^{-1}) sharply reduced yield, dry biomass, and A_{CO2}, particularly at 48 DAT, whereas r_s and sodium content increased. In fact, at 0 and 70 kg ha^{-1}, the lower leaf N availability may affect photosynthetic performance and rate due to N remobilization from photosynthetic enzymes and pigments [62]. The decreased SPAD index, which is significantly correlated to chlorophyll concentration as indicated by absorbance measurements [63], corresponded to the decrease in photosynthetic capacity, and an increase in the sensitivity to photo-inhibition [64]. However, the application of TG41 alone, but especially in combination with the VBP, to lettuce grown in sub-optimal N induced significant changes in morphology and physiology, as noted with increased yield and dry biomass. Therefore, under low-input conditions (0 kg N ha^{-1}), the combination of the microbial inoculant with the biopolymer-based biostimulant exhibited an important synergistic effect, thus confirming the beneficial effects on crop productivity as previously reported by several authors [48–50]. Both PB treatments enhanced photosynthetic activity, SPAD index, and leaf nutritional status, as reflected by higher K and Mg and lower Na concentrations, that indicate a more efficient accumulation and translocation of assimilates to photosynthetic sinks that improve crop performance, but are not associated to the external N fertilization level applied [50]. Under optimal N conditions (i.e., recommended rate of 70 kg ha^{-1}), the treatment with TG41 alone had the best effect on fresh yield, combined with high Mg and antioxidant contents, as well as low nitrate and Na. In this N regime, the addition of the VBP-based biostimulant to the fungal inoculant did not improve the morpho-physiological parameters, nor the mineral profile in the leaves. As mentioned above, growth under suboptimal N conditions increased leaf cell susceptibility to light-induced oxidative damages, a condition that plants are not capable of overcoming. However, the application of the combined microbial and VBP PBs induced a strong production of TAA, phenols, and probably glutathione, a metabolite that works cooperatively with ascorbic acid to generate antioxidant effects that safely detoxify accumulated reactive oxygen species (ROS), thus protecting the plant and increasing the photosynthetic rate [6,7].

The application of 140 kg N ha^{-1} to lettuce was an excess condition that determined a plateau in yield and dry biomass but not in the N content, although there was an increase in nitrate and Na, as well as r_s at 40 DAT. This demonstrated that plants supplied with high levels of N were not able to assimilate and reduce all the nitrate supplied, risking negative consequences by the accumulation of these compounds in the vacuoles. This was also reported by Di Mola et al. [65] in rocket plants and by Wang et al. [66] in leafy vegetables, whereby optimal and particularly supra-optimal N treatments were not always characterized by the best quality traits in the produce, but on the contrary, resulted in damage to the commercial, nutritional, and functional quality traits. These effects were similar to those noted in our lettuce plants under supra-optimal N conditions, i.e., low macronutrients and total ascorbic acid, high nitrate and sodium content. The application of both PBs under supra-optimal N level (e.g., 140 kg ha^{-1}) significantly enhanced the N content and SPAD index while reducing nitrate content without affecting the CO_2 assimilation rate and the accumulation of beneficial nutrients. The strong increase in the SPAD index at 140 kg ha^{-1} in plants inoculated with TG41 or TG41+VBP biostimulants was also observed at 70 kg ha^{-1}, suggesting that the biostimulants were able to increase the number and efficiency of photosynthesis systems and light-harvesting complexes (LHC), that allowed plants to "fine-tune" photosynthesis in the fluctuating spectral quality and light intensity conditions, thus avoiding ROS formation and photo-oxidation. This also allowed a higher use efficiency of nitrate, as confirmed by the lower concentration of this ion in leaf tissues when compared to the untreated control because of a more efficient reduction and assimilation processes [6,7].

Our results correspond to previous findings on the plant growth-promoting effect of fungi inoculants containing *Trichoderma* [25,26,29,30,33,38,67]. The presumed mechanisms behind the beneficial morpho-physiological effects on lettuce plants by TG41 could be due to the release of signaling molecules with auxin and ethylene-like activity [28], in particular, bioactive volatile compounds [43], which increased nutrient bioavailability to the plant, that improved their uptake, translocation, and accumulation within the plant [35]. In addition, it has also been demonstrated that *Trichoderma* in the rhizosphere stimulates root growth and reshapes its architecture, morphological changes which are pivotal for improving nutrient uptake, in particular, nitrate, Ca, Mg, and K [29,30,35,41]. The synergistic action of TG41 with the VBP biostimulant is of particular interest because it resulted in the production of premium quality lettuce traits, as is clearly exhibited by the PCA. The vegetal-biopolymer biostimulant action was probably due to the presence of phenol metabolites with auxin and gibberellin-like activities, that interacted with phytohormones and enzymes stimulating the activity of carbon–nitrogen metabolism and plant development [45,46]. Another putative mechanism behind the stimulation of plant growth and yield in response to VBP drench application could involve the increased presence of bioactive molecules, such as signaling peptides (LRPP) and lignosulphonates, which are typical compounds present in VBP [46]. A previous study reported that lignosulphonate treatments can improve N uptake and assimilation in plants through the stimulation of glutamate synthase and glutamine synthetase, as well as by triggering photosynthetic activity through the stimulation of both rubisco enzyme activity, thus improving plant performance [45]. The improved NUE in lettuce treated with PBs enhanced not only the chlorophyll content (as represented by the increased SPAD index) but also the synthesis of antioxidant metabolites that were capable of re-activating photosynthetic activity that under sub-optimal N conditions without PBs, would be severely compromised. Finally, the synergistic beneficial effect on root system architecture, as previously shown by Colla et al. [38,68], determined a '*nutrient acquisition response*' improving resource use efficiency (RUE) that enhanced plant biomass production and the quality of the produce.

5. Conclusions

Our study on the leafy vegetable crop Romaine lettuce confirmed that inoculations with *Trichoderma* TG41 under optimal N conditions (70 kg ha^{-1}) were able to improve the leaf nutritional status as indicated with the higher potassium and magnesium content and lower sodium content, plus providing the best yield performance of all tested conditions in terms of plant fresh and dry weight. Interestingly, the combined biostimulant applications of *Trichoderma* with the vegetal biopolymer-based product, in suboptimal fertilizer conditions of low N availability (0N kg ha^{-1}), was more effective than the treatment of the microbial inoculant alone not only in improving yield but also in producing a premium quality marketable lettuce with higher lipophilic antioxidant activity and total ascorbic acid content. Together these biostimulants positively influenced plant morpho-physiological processes that improved the assimilation of nitrate and macronutrients and stimulated root system architecture reshaping, thus permitting increased bioabsorption or '*nutrient acquisition response*'. Moreover, the assimilatory pathways were stimulated, for which nitrate was used to synthesize chlorophyll (increased SPAD index) and the antioxidant metabolites, which, in turn, re-activated the CO_2 assimilation activity normally decreased under sub-optimal N conditions. Therefore, the combination of microbial and non-microbial plant biostimulants represents a promising, efficient, and sustainable strategy for improving yield and quality of horticultural crops, such as lettuce, as well as improving cultivation in N compromised fields or low fertilizer input scenarios.

Author Contributions: Conceptualization, Y.R., G.C., and S.L.W.; methodology, Y.R., G.C., N.F., and S.L.W.; software, L.S., C.E.-N., A.P., M.G., and V.C.; validation, L.S., C.E.-N., A.P., M.G., E.S., N.L., M.N., and V.C.; formal analysis, A.P. and V.C.; data curation, C.E.-N., A.P., E.C., M.G., E.S., N.L., M.N., and V.C.; writing—original draft preparation, P.C. and Y.R.; writing—review and editing, Y.R., P.C., G.C., N.F., and S.L.W.; visualization, Y.R., P.C., G.C., and S.L.W.; supervision, Y.R., P.C., G.C., and S.L.W.; project administration, Y.R. and S.L.W.; funding acquisition, G.C. and S.L.W. All authors have read and agreed to the published version of the manuscript.

Funding: This work was supported by the following projects: MIURPON [grant number Linfa 03PE_00026_1; grant number Marea 03PE_00106]; POR FESR CAMPANIA 2014/2020- O.S. 1.1 [grant number Bioagro 559]; European Union Horizon 2020 Research and Innovation Program, ECOSTACK [grant agreement no. 773554]; ProBio IZSM (Portici), n. D01 6309, 14/12/2016.

Acknowledgments: The authors would like to thank Antonio Pignataro and Mauro Senatore for their technical assistance in the greenhouse.

Conflicts of Interest: The authors declare no conflict of interest.

References

1. FAO. *Global Agriculture towards 2050, High-Level Expert Forum, How to Feed the World 2050*; Food and Agriculture Organization of the United Nations: Rome, Italy, 2009.
2. D'Antuono, L.F.; Rossini, F. Yield potential and ecophysiological traits of the Altamurano linseed (*Linum usitatissimum* L.), a landrace of southern Italy. *Genet. Resour. Crop Evol.* **2006**, *53*, 65–75. [CrossRef]
3. Rossini, F.; Loreti, P.; Provenzano, M.E.; De Santis, D.; Ruggeri, R. Agronomic performance and beer quality assessment of twenty hop cultivars grown in central Italy. *Ital. J. Agron.* **2016**, *11*, 180–187. [CrossRef]
4. Lassaletta, L.; Billen, G.; Garnier, J.; Bouwman, L.; Velazquez, E.; Mueller, N.D.; Gerber, J.S. Nitrogen use in the global food system: past trends and future trajectories of agronomic performance, pollution, trade, and dietary demand. *Environ. Res. Lett.* **2016**, *11*, 095007. [CrossRef]

5. FAO. *The Future of Food and Agriculture–Alternative Pathways to 2050*; Food and Agriculture Organization of the United Nations: Rome, Italy, 2018.
6. Carillo, P.; Colla, G.; El-Nakhel, C.; Bonini, P.; D'Amelia, L.; Dell'Aversana, E.; Pannico, A.; Giordano, M.; Sifola, M.I.; Kyriacou, M.C.; et al. Biostimulant application with a tropical plant extract enhances *Corchorus olitorius* adaptation to sub-optimal nutrient regimens by improving physiological parameters. *Agronomy* **2019**, *9*, 249. [CrossRef]
7. Carillo, P.; Colla, G.; Fusco, G.M.; Dell'Aversana, E.; El-Nakhel, C.; Giordano, M.; Pannico, A.; Cozzolino, E.; Mori, M.; Reynaud, H.; et al. Morphological and physiological responses induced by protein hydrolysate-based biostimulant and nitrogen rates in greenhouse spinach. *Agronomy* **2019**, *9*, 450. [CrossRef]
8. Rouphael, Y.; Colla, G. The influence of drip irrigation or subirrigation on zucchini squash grown in closed-loop substrate culture with high and low nutrient solution concentrations. *HortScience* **2009**, *44*, 306–311. [CrossRef]
9. Gupta, S.K.; Gupta, A.B.; Gupta, R. Pathophysiology of nitrate toxicity in humans in view of the changing trends of the global nitrogen cycle with special reference to India. In *The Indian Nitrogen Assessment*; Abrol, Y.P., Adhya, T.K., Aneja, V.P., Eds.; Elsevier: Amsterdam, The Netherland, 2017; pp. 459–468. [CrossRef]
10. Raun, W.R.; Johnson, G.V. Improving nitrogen use efficiency for cereal production. *Agron. J.* **1999**, *91*, 357–363. [CrossRef]
11. Lassaletta, L.; Billen, G.; Grizzetti, B.; Anglade, J.; Garnier, J. 50 year trends in nitrogen use efficiency of world cropping systems: the relationship between yield and nitrogen input to cropland. *Environ. Res. Lett.* **2014**, *9*, 105011. [CrossRef]
12. Ertani, A.; Cavani, L.; Pizzeghello, D.; Brandellero, E.; Altissimo, A.; Ciavatta, C.; Nardi, S. Biostimulant activity of two protein hydrolyzates in the growth and nitrogen metabolism of maize seedlings. *J. Plant Nutr. Soil Sci.* **2009**, *172*, 237–244. [CrossRef]
13. Colla, G.; Kim, H.J.; Kyriacou, M.; Rouphael, Y. Nitrate in fruits and vegetables. *Sci. Hortic.* **2018**, *237*, 221–238. [CrossRef]
14. Kyriacou, M.C.; Rouphael, Y. Towards a new definition of quality for fresh fruits and vegetables. *Sci. Hortic.* **2018**, *234*, 463–469. [CrossRef]
15. Masclaux-Daubresse, C.; Daniel-Vedele, F.; Dechorgnat, J.; Chardon, F.; Gaufichon, L.; Suzuki, A. Nitrogen uptake, assimilation and remobilization in plants: Challenges for sustainable and productive agriculture. *Ann. Bot.* **2010**, *105*, 1141–1157. [CrossRef] [PubMed]
16. Colla, G.; Rouphael, Y. Biostimulants in horticulture. *Sci. Hortic.* **2015**, *196*, 1–2. [CrossRef]
17. Rouphael, Y.; Spíchal, L.; Panzarová, K.; Casa, R.; Colla, G. High-throughput plant phenotyping for developing novel biostimulants: From lab to field or from field to lab? *Front. Plant Sci.* **2018**, *9*, 1197. [CrossRef]
18. Rouphael, Y.; Kyriacou, M.C.; Petropoulos, S.A.; De Pascale, S.; Colla, G. Improving vegetable quality in controlled environments. *Sci. Hortic.* **2018**, *234*, 275–289. [CrossRef]
19. Du Jardin, P. Plant biostimulants: Definition, concept, main categories and regulation. *Sci. Hortic.* **2015**, *196*, 3–14. [CrossRef]
20. EU. Regulation of the European Parliament and of the council laying down rules on the making available on the market of EU fertilising products and amending Regulations (EC) No 1069/2009 and (EC) No 1107/2009 and repealing Regulation (EC) No 2003/2003. Available online: https://eur-lex.europa.eu/legal-content/EN/TXT/?uri=OJ:L:2019:170:TOC (accessed on 31 December 2019).
21. Canellas, L.P.; Olivares, F.L.; Aguiar, N.O.; Jones, D.L.; Nebbioso, A.; Mazzei, P.; Piccolo, A. Humic and fulvic acids as biostimulants in horticulture. *Sci. Hortic.* **2015**, *196*, 15–27. [CrossRef]
22. Ruzzi, M.; Aroca, R. Plant growth-promoting rhizobacteria act as biostimulants in horticulture. *Sci. Hortic.* **2015**, *196*, 124–134. [CrossRef]
23. Rouphael, Y.; Colla, G. Synergistic Biostimulatory action: designing the next generation of plant biostimulants for sustainable agriculture. *Front. Plant Sci.* **2018**, *9*, 1655. [CrossRef]
24. Colla, G.; Hoagland, L.; Ruzzi, M.; Cardarelli, M.; Bonini, P.; Canaguier, R.; Rouphael, Y. Biostimulant action of protein hydrolysates: unraveling their effects on plant physiology and microbiome. *Front. Plant Sci.* **2017**, *8*. [CrossRef]

25. Fiorentino, N.; Ventorino, V.; Woo, S.L.; Pepe, O.; De Rosa, A.; Gioia, L.; Romano, I.; Lombardi, N.; Napolitano, M.; Colla, G.; et al. *Trichoderma*-based biostimulants modulate rhizosphere microbial populations and improve N uptake efficiency, yield, and nutritional quality of leafy vegetables. *Front. Plant Sci.* **2018**, *9*. [CrossRef] [PubMed]
26. López-Bucio, J.; Pelagio-Flores, R.; Herrera-Estrella, A. *Trichoderma* as biostimulant: exploiting the multilevel properties of a plant beneficial fungus. *Sci. Hortic.* **2015**, *196*, 109–123. [CrossRef]
27. Woo, S.; Ruocco, M.; Vinale, F.; Nigro, M.; Marra, R.; Lombardi, N.; Pascale, A.; Lanzuise, S.; Manganiello, G.; Lorito, M. *Trichoderma*-based products and their widespread use in agriculture. *Open Mycol. J.* **2014**, *8*. [CrossRef]
28. Vinale, F.; Sivasithamparam, K.; Ghisalberti, E.L.; Marra, R.; Woo, S.L.; Lorito, M. *Trichoderma*–plant–pathogen interactions. *Soil Biol. Biochem.* **2008**, *40*, 1–10. [CrossRef]
29. Stewart, A.; Hill, R. Applications of *Trichoderma* in plant growth promotion. In *Biotechnology and Biology of Trichoderma*; Gupta, V.K., Schmoll, M., Herrera-Estrella, A., Upadhyay, R.S., Druzhinina, I., Tuohy, M.G., Eds.; Elsevier: Amsterdam, The Netherland, 2014; Chapter 31; pp. 415–428. [CrossRef]
30. Harman, G.E. Overview of new insights into mechanisms and uses of *Trichoderma* based products. *Phytopathology* **2004**, *94*, 138.
31. Hermosa, R.; Viterbo, A.; Chet, I.; Monte, E. Plant-beneficial effects of *Trichoderma* and of its genes. *Microbiol.* **2012**, *158*, 17–25. [CrossRef] [PubMed]
32. Harman, G.E.; Howell, C.R.; Viterbo, A.; Chet, I.; Lorito, M. *Trichoderma* species–opportunistic, avirulent plant symbionts. *Nat. Rev. Microbiol.* **2004**, *2*, 43–56. [CrossRef]
33. Lorito, M.; Woo, S.L. *Trichoderma*: A multi-purpose tool for integrated pest management. In *Principles of Plant-Microbe Interactions*; Lugtenberg, B., Ed.; Springer International Publishing: Manhattan, NY, USA, 2015; pp. 345–353. [CrossRef]
34. Baker, R. *Trichoderma* spp. as plant-growth stimulants. *CRC Crit. Rev. Biotechnol.* **1988**, *7*, 97–106. [CrossRef]
35. Altomare, C.; Norvell, W.A.; Bjorkman, T.; Harman, G.E. Solubilization of phosphates and micronutrients by the plant-growth-promoting and biocontrol fungus *Trichoderma harzianum Rifai* 1295-22. *Appl Environ Microbiol.* **1999**, *65*, 2926–2933. [CrossRef]
36. De Santiago, A.; Quintero, J.M.; Avilés, M.; Delgado, A. Effect of *Trichoderma asperellum* strain T34 on iron, copper, manganese, and zinc uptake by wheat grown on a calcareous medium. *Plant Soil* **2011**, *342*, 97–104. [CrossRef]
37. Mastouri, F.; Björkman, T.; Harman, G.E. *Trichoderma harzianum* enhances antioxidant defense of tomato seedlings and resistance to water deficit. *Mol. Plant-Microbe Interact.* **2012**, *25*, 1264–1271. [CrossRef] [PubMed]
38. Colla, G.; Rouphael, Y.; Di Mattia, E.; El-Nakhel, C.; Cardarelli, M. Co-inoculation of *Glomus intraradices* and *Trichoderma atroviride* acts as a biostimulant to promote growth, yield and nutrient uptake of vegetable crops. *J. Sci. Food Agric.* **2015**, *95*, 1706–1715. [CrossRef] [PubMed]
39. Woo, S.L.; Scala, F.; Ruocco, M.; Lorito, M. The molecular biology of the interactions between *Trichoderma* spp., phytopathogenic fungi, and plants. *Phytopathology* **2006**, *96*, 181–185. [CrossRef] [PubMed]
40. Manganiello, G.; Sacco, A.; Ercolano, M.R.; Vinale, F.; Lanzuise, S.; Pascale, A.; Napolitano, M.; Lombardi, N.; Lorito, M.; Woo, S.L. Modulation of tomato response to *Rhizoctonia solani* by *Trichoderma harzianum* and its secondary metabolite harzianic acid. *Front. Microbiol.* **2018**, *9*, 1966. [CrossRef] [PubMed]
41. Contreras-Cornejo, H.A.; Macías-Rodríguez, L.; Cortés-Penagos, C.; López-Bucio, J. *Trichoderma virens*, a plant beneficial fungus, enhances biomass production and promotes lateral root growth through an auxin-dependent mechanism in *Arabidopsis*. *Plant Physiol.* **2009**, *149*, 1579–1592. [CrossRef] [PubMed]
42. Prisa, D.; Sarrocco, S.; Vannacci, G. Microrganismi biostimolanti: *Trichoderma* sp. in substrati innovativi e alternativi alla torba per piante ornamentali. *Italus Hort.* **2014**, 17–28.
43. Salwan, R.; Kumari, N.; Sharma, V. Bioactive Volatile Metabolites of *Trichoderma*: An overview. In *Secondary Metabolites Plant Growth Promoting Rhizomicroorganisms*; Singh, H., Keswani, C., Reddy, M., Sansinenea, E., García-Estrada, C., Eds.; Springer: Singapore, 2019; pp. 87–111. [CrossRef]
44. Coppola, M.; Diretto, G.; Digilio, M.C.; Woo, S.L.; Giuliano, G.; Molisso, D.; Pennacchio, F.; Lorito, M.; Rao, R. Transcriptome and metabolome reprogramming in tomato plants by *Trichoderma harzianum* strain T22 primes and enhances defense responses against aphids. *Front. Physiol.* **2019**, *10*. [CrossRef]

45. Ertani, A.; Francioso, O.; Tugnoli, V.; Righi, V.; Nardi, S. Effect of commercial lignosulfonate-humate on *Zea mays* L. metabolism. *J. Agric. Food Chem.* **2011**, *59*, 11940–11948. [CrossRef]
46. Lucini, L.; Rouphael, Y.; Cardarelli, M.; Bonini, P.; Baffi, C.; Colla, G. A vegetal biopolymer-based biostimulant promoted root growth in melon while triggering brassinosteroids and stress-related compounds. *Front. Plant Sci.* **2018**, *9*. [CrossRef]
47. Broadley, M.R.; Escobar-Gutièrrez, A.J.; Burns, A.; Burns, G. What are the effects of nitrogen deficiency on growth components of lettuce? *New Phytol.* **2000**, *147*, 519–526. [CrossRef]
48. Bettoni, M.M.; Mogor, A.F.; Pauletti, V.; Goicoechea, N. Growth and metabolism of onion seedlings as affected by the application of humic substances, mycorrhizal inoculation and elevated CO_2. *Sci. Hortic.* **2014**, *180*, 227–235. [CrossRef]
49. Nikbakht, A.; Pessarakli, M.; Daneshvar-Hakimi-Maibodi, N.; Kafi, M. Perennial ryegrass growth responses to mycorrhizal infection and humic acid treatments. *Agron. J.* **2014**, *106*, 585–595. [CrossRef]
50. Rouphael, Y.; Cardarelli, M.; Bonini, P.; Colla, G. Synergistic action of a microbial-based biostimulant and a plant derived-protein hydrolysate enhances lettuce tolerance to alkalinity and salinity. *Front. Plant Sci.* **2017**, *8*, 131. [CrossRef]
51. Kumar, P.; Lucini, L.; Rouphael, Y.; Cardarelli, C.; Kalunke, R.M.; Colla, G. Insight into the role of grafting and arbuscular mycorrhiza on cadmium stress tolerance in tomato. *Front. Plant Sci.* **2015**, *6*, 477. [CrossRef]
52. Colonna, E.; Rouphael, Y.; Barbieri, G.; De Pascale, S. Nutritional quality of leafy vegetables harvested at two light intensities. *Food Chem.* **2016**, *199*, 702–710. [CrossRef]
53. Bremner, J.M. Total nitrogen. In *Methods of Soil Analysis*; Black, C.A., Evans, D.D., White, D.D., Ensminger, E., Clark, F.E., Eds.; American Society of Agronomy: Madison, WI, USA, 1965; pp. 149–1178. [CrossRef]
54. Re, R.; Pellegrini, N.; Proteggente, A.; Pannala, A.; Yang, M.; Rice-Evans, C. Antioxidant activity applying an improved Abts radical cation decolorization assay. *Free Rad. Biol. Med.* **1999**, *26*, 1231–1237. [CrossRef]
55. Fogliano, V.; Verde, V.; Randazzo, G.; Ritieni, A. Method for measuring antioxidant activity and its application to monitoring the antioxidant capacity of wines. *J. Agric. Food Chem.* **1999**, *47*, 1035–1040. [CrossRef]
56. Singleton, V.L.; Orthofer, R.; Lamuela-Raventos, R.M. Analysis of total phenols and other oxidation substrates and antioxidants by means of Folin-Ciocalteu reagent. Oxidants Antioxidants Part A. *Meth. Enzymol.* **1999**, *299*, 152–178. [CrossRef]
57. Kampfenkel, K.; Van Montagu, M.; Inzé, D. Extraction and determination of ascorbate and dehydroascorbate from plant tissue. *Anal. Biochem.* **1995**, *225*, 165–167. [CrossRef]
58. Rouphael, Y.; Kyriacou, M.C.; Carillo, P.; Pizzolongo, F.; Romano, R.; Sifola, M.I. Chemical eustress elicits tailored responses and enhances the functional quality of novel food *Perilla Frutescens*. *Molecules* **2019**, *24*, 185. [CrossRef]
59. El-Nakhel, C.; Petropoulos, S.A.; Pannico, A.; Kyriacou, M.C.; Giordano, M.; Colla, G.; Troise, A.D.; Vitaglione, P.; De Pascale, S.; Rouphael, Y. The bioactive profile of lettuce produced in a closed soilless system as configured by combinatorial effects of genotype and macrocation supply composition. *Food Chem.* **2020**, in press. [CrossRef] [PubMed]
60. El-Nakhel, C.; Pannico, A.; Kyriacou, M.C.; Giordano, M.; De Pascale, S.; Rouphael, Y. Macronutrient deprivation eustress elicits differential secondary metabolites in red and green-pigmented butterhead lettuce grown in closed soilless system. *J. Sci. Food Agric.* **2019**, *99*, 6962–6972. [CrossRef] [PubMed]
61. Kyriacou, M.C.; Rouphael, Y.; Colla, G.; Zrenner, R.M.; Schwarz, D. Vegetable grafting: The implications of a growing agronomic imperative for vegetable fruit quality and nutritive value. *Front. Plant Sci.* **2017**, *8*, 741. [CrossRef] [PubMed]
62. Bassi, D.; Menossi, M.; Mattiello, L. Nitrogen supply influences photosynthesis establishment along the sugarcane leaf. *Sci. Rep.* **2018**, *8*, 2327. [CrossRef]
63. Van Oosten, M.J.; Dell'Aversana, E.; Ruggiero, A.; Cirillo, V.; Gibon, Y.; Woodrow, P.; Maggio, A.; Carillo, P. Omeprazole treatment enhances nitrogen use efficiency through increased nitrogen uptake and assimilation in corn. *Front. Plant Sci.* **2019**, *10*, 1–17. [CrossRef]
64. Kim, E.H.; Li, X.P.; Razeghifard, R.; Anderson, J.M.; Niyogi, K.K.; Pogson, B.J.; Chow, W.S. The multiple roles of light-harvesting chlorophyll a/b-protein complexes define structure and optimize function of Arabidopsis chloroplasts: A study using two chlorophyll b-less mutants. *Biochim. Biophys. Acta* **2009**, *1787*, 973–984. [CrossRef]

65. Di Mola, I.; Ottaiano, L.; Cozzolino, E.; Senatore, M.; Giordano, M.; El-Nakhel, C.; Sacco, A.; Rouphael, Y.; Colla, G.; Mori, M. Plant-based biostimulants influence the agronomical, physiological, and qualitative responses of baby rocket leaves under diverse nitrogen conditions. *Plants* **2019**, *8*, 522. [CrossRef]
66. Wang, Z.H.; Li, S.X.; Malhi, S. Effects of fertilization and other agronomic measures on nutritional quality of crops. *J. Sci. Food Agric.* **2008**, *88*, 7–323. [CrossRef]
67. Hermosa, R.; Rubio, B.; Cardoza, R.E.; Nicolas, C.; Monte, E.; Gutièrrez, S. The contribution of *Trichoderma* to balancing the costs of plant growth and defense. *Int. Microbio.* **2013**, 69–80. [CrossRef]
68. Colla, G.; Rouphael, Y.; Canaguier, R.; Svecova, E.; Cardarelli, M. Biostimulant action of a plant-derived protein hydrolysateproduced through enzymatic hydrolysis. *Front. Plant Sci.* **2014**, 5. [CrossRef]

agronomy

MDPI

Article

A Composite Bioinoculant Based on the Combined Application of Beneficial Bacteria and Fungi

Henrietta Allaga [1], Bettina Bóka [1], Péter Poór [2], Viktor Dávid Nagy [1], Attila Szűcs [1], Istvánné Stankovics [3], Miklós Takó [1], László Manczinger [1], Csaba Vágvölgyi [1], László Kredics [1,*] and Péter Körmöczi [4]

1. Department of Microbiology; Faculty of Science and Informatics; University of Szeged; Közép fasor 52, 6726 Szeged, Hungary; henrietta.allaga@gmail.com (H.A.); boka.tina@gmail.com (B.B.); viktor.david.nagy@gmail.com (V.D.N.); a.szunyi@windowslive.com (A.S.); tako78@bio.u-szeged.hu (M.T.); manczing@bio.u-szeged.hu (L.M.); csaba@bio.u-szeged.hu (C.V.)
2. Department of Plant Biology, Faculty of Science and Informatics, University of Szeged, Közép fasor 52, 6726 Szeged, Hungary; poorpeti@bio.u-szeged.hu
3. Felső-Bácskai Agrolabor Ltd., Rákóczi Ferenc u. 19., 6430 Bácsalmás, Hungary; agrolabor@agrolabor.hu
4. Department of Field Crops Research, National Agricultural Research and Innovation Centre, Alsó kikötő sor 9, 6726 Szeged, Hungary; kormoczipeti@gmail.com

* Correspondence: kredics@bio.u-szeged.hu; Tel.: +36-62-544-516

Received: 20 December 2019; Accepted: 31 January 2020; Published: 3 February 2020

Abstract: A composite soil bioinoculant containing beneficial bacteria and fungi was developed for biocontrol of plant pathogens, phosphorous mobilization, stem degradation, humification, and nitrogen fixation. A *Trichoderma asperellum* isolate with outstanding in vitro antagonistic abilities toward a series of plant pathogenic fungi was included as a potential biocontrol component. The selected strain was also shown to promote growth and increase photosynthetic activity of tomato plants. For phosphorous mobilization and stem degradation, a *Trichoderma atrobrunneum* strain was selected, which produced cellulose-degrading enzymes even in the absence of stem residues, while this ability increased 10–15-fold in the presence of ground maize stem. The strain was also shown to produce large amounts of enzymes liberating organically bound phosphorous, as well as cellulase and xylanase activities in solid-state fermentation on various plant residues. A *Streptomyces albus* strain with excellent peroxidase-producing abilities was selected as a potential humus-producing component, while an *Azotobacter vinelandii* strain with the potential to provide excess nitrogen for crops was included for nitrogen fixation. The assembled soil bioinoculant had positive effect on the uptake of certain important macro- and microelements (potassium, sodium, and manganese) from the soil by field-grown tomato plants. The applied screening strategy proved to be applicable for the assembly of a composite soil bioinoculant with notable application potentials.

Keywords: biocontrol; plant growth promotion; soil inoculant; *Trichoderma*; *Azotobacter*; *Streptomyces*

1. Introduction

Chemical pesticides and fertilizers are applied world-wide in agricultural production. Pesticides are used for the prevention and control of plant pests and diseases in order to reduce or eliminate yield losses and maintain product quality. However, there are serious concerns regarding the risks resulting from occupational exposure to them, as well as from environmental pollution leading to the presence of their residues in the food-chain and drinking water [1]. Chemical fertilizers are used to supply plants with necessary elements (primarily phosphorous and nitrogen), thereby improving crop productivity; however, their application is resulting in pollution with phosphates and nitrates. The agricultural run-off of phosphates deriving from fertilizers contributes to the eutrophication of fresh water bodies and also presents a serious threat to the biodiversity in terrestrial ecosystems [2], while the increased run-off of nitrogen fertilizers results in nitrate pollution of surface and groundwater [3]. Therefore, the need for alternative, environment-friendly, microbial soil treatment strategies with favorable effects on crop plants is emerging all over the world. Microbial abilities of biocontrol, plant growth promotion, stem degradation, phosphorous solubilization, humification, and nitrogen-fixation can be exploited for the development of microbial soil inoculants to be applied in sustainable agricultural production.

One of the main challenges in the agricultural use of beneficial microorganisms as plant growth promoters and/or biocontrol agents (BCAs) is their frequently inconsistent field performance [4,5], which may be due to a series of biotic and abiotic factors. Among the abiotic factors, physicochemical properties of the rhizosphere such as pH, temperature, water activity, and the chemical composition of the soil are varying in space and time, which substantially influences the performance of biocontrol and plant growth promoting microorganisms. Particular agents may exert different activities under different soil environmental conditions. Inconsistent field performance has long been identified as the major impediment to the wide-scale commercialization of beneficial microorganisms for agricultural applications [6]. A possible strategy to counteract inconsistencies due to varying environmental conditions is the development of consortial soil inoculants consisting of multiple beneficial organisms. The combination of efficient plant growth promoting microorganisms and BCAs may result in an increased consistency of field performance during different periods of the growing season, thereby enabling a more predictable increase in crop yields [7].

The aim of this study was to assemble a consortial soil bioinoculant based on the combined application of beneficial bacteria and fungi with the potential of increasing pathogen control, plant growth and crop yield, stem residue degradation, phosphorous mobilization, humification, and nitrogen fixation in treated agricultural soils.

2. Materials and Methods

2.1. Examined Strains

The microbial strains involved in this study derived from the Szeged Microbiology Collection, Szeged, Hungary (SZMC). The *Trichoderma* strains included in the study (*Trichoderma asperellum* SZMC 20866, and SZMC 20786; *Trichoderma harzianum* species complex (THSC) members SZMC 20761, SZMC 20762, and SZMC 20869; *Trichoderma atroviride* SZMC 20780 and SZMC 20781; *Trichoderma virens* SZMC 20779; *Trichoderma gamsii* SZMC 20783; and *Trichoderma hamatum* SZMC 20784) were isolated from Hungarian agricultural soil samples and initially identified by sequence analysis of the internal transcribed spacer (ITS) region [8]. However, as ITS sequence analysis is not able to discriminate between species belonging to THSC, the species level identity of strains SZMC 20761, SZMC 20762, and SZMC 20869 was determined during this study by the sequence analysis of a fragment of the *tef1* alpha gene [9] as *Trichoderma guizhouense, T. guizhouense* and *Trichoderma atrobrunneum* (GenBank accession numbers MN750371, MN750372, MN750373), respectively. Fungal isolates were maintained on yeast extract-glucose medium (5 g L^{-1} glucose, 5 g L^{-1} KH_2PO_4, 1 g L^{-1} yeast extract, 20 g L^{-1} agar).

Streptomyces sp. isolates (SZMC 0282, SZMC 0232, 00001, 00002, 00004, 00005, 00006, 00007, 00008, 00009, 00010, 00012, 00013, 00014, 00015, 00017, 00019, 00020, 00021, 00022, 00023, 00024, 00025, 00026, 00027, 00028, 00029, 00030, 00031, 00032, 00033, 00034, 00035, 00036, 00037, 00038, 00039, 00040, 00041, 00042, 00043, 00044, 00045, 00046, and 00047) and *Azotobacter vinelandii* SZMC 22195 were derived from soil samples. *Streptomyces microflavus* DSM 40561 was derived from the DSMZ (Deutsche Sammlung von Mikroorganismen und Zellkulturen) strain collection and was included as control in the peroxidase-producing assays. Bacterial strains were maintained on glucose-yeast extract-malt extract *Streptomyces* medium (GYM-STR: 4 g L^{-1} glucose, 4 g L^{-1} yeast extract, 10 g L^{-1} malt extract, 2 g L^{-1} $CaCO_3$, and 20 g L^{-1} agar).

2.2. Determination of Biocontrol Index (BCI) Values

Dual confrontation tests were performed in vitro in Petri dishes (90 mm in diameter) according to the method described by Szekeres et al. [10]. During the experiments, 11 different strains of plant pathogenic fungi were confronted with the *Trichoderma* strains, including 3 *Armillaria* species (*Armillaria mellea* SZMC 23638, *Armillaria ostoyae* SZMC 23080, and *Armillaria gallica* SZMC23076), 4 *Fusarium solani* species complex (FSSC) isolates (SZMC 11057F, SZMC 6241J, SZMC 11067F, and SZMC 11070F), as well as *Phoma cucurbitacearum* (SZMC 16088), *Alternaria alternata* (SZMC 16085), *Botrytis cinerea* (SZMC 6244J), and *Rhizoctonia solani* (SZMC 6252J). The tests were carried out in three replicates on malt extract agar (MEA) medium (10 g L^{-1} glucose, 2.5 g L^{-1} yeast extract, 20 mL L^{-1} 20% malt extract, 20 g L^{-1} agar). After the incubation period, digital photos were taken with a Nikon Coolpix P7700 camera (Nikon, Tokyo, Japan) from each Petri plate, and the area visibly covered by the *Trichoderma* strain as well as the area covered by *Trichoderma* and the pathogen together were calculated for each plate with the aid of the Image J software (http://imagej.nih.gov/ij). BCI values were calculated with Excel 2010 (Microsoft, Redmond, WA, USA) according to the formula: BCI = (area of *Trichoderma* colony/total area occupied by the colonies of both *Trichoderma* and the plant pathogenic fungus) × 100 [10].

2.3. Liquid and Solid-State Fermentations

For the investigation of cellulase and phosphatase activities, culturing was performed in 100 mL flasks containing 20 mL liquid minimal (10 g L^{-1} glucose, 5 g L^{-1} $(NH_4)_2SO_4$, 5 g L^{-1} KH_2PO_4, 0.1 g L^{-1} $MgSO_4 \times 7H_2O$) or maize stem medium (2 g L^{-1} dried maize stem ground with a coffee grinder (Bosch, Gerlingen, Germany), 1 g L^{-1} $NaNO_3$). The minimal medium was inoculated with *Trichoderma* conidia to a concentration of 2×10^5 mL^{-1}, while the maize stem medium was inoculated with the total amount of 5-days-old fungal mycelium pre-grown in 20 mL liquid minimal medium, filtered with the aid of filter paper and vacuum pump and washed with sterile distilled water. After 5 days of incubation at 25 °C in an IKA KS 4000 IC Control shaker (ProfiLab24, Berlin, Germany) at 180 rpm, samples were filtered, and the culture filtrates were used for extracellular enzyme activity measurements.

Extracellular enzyme activities of *T. atrobrunneum* SZMC 20869 from THSC and the industrially important *Trichoderma reesei* strain QM9414 (SZMC 22616) were also compared in solid-state fermentation (SSF) experiments using maize, wheat, sunflower, and canola stem residues as substrates. One gram amounts of ground plant residues were placed into 50 mL Erlenmeyers flask and moisturized with 5 mL distilled water. After sterilization, the substrates were inoculated with 2×10^5 *Trichoderma* conidia. On the 8th day of fermentation, extractions were performed by adding 20 mL distilled water to the cultures and incubating for 3 h at 4 °C. The fluid phases were filtered through sterile gauze sheets into 15 mL centrifuge tubes and centrifuged at 4600 rpm for 10 min, 2 times by transferring the fluid phase to a new centrifuge tube. The 8× dilution of the fluid samples in distilled water were used for extracellular enzyme activity measurements.

2.4. Enzyme Activity Measurements

Cellobiohydrolase, β-glucosidase, β-xylosidase, and acidic phosphatase enzyme activities were measured with the chromogenic substrates p-nitrophenyl-β-D-cellobioside, p-nitrophenyl-β-D-glucopyranoside, p-nitrophenyl-β-D-xylopyranoside, and p-nitrophenyl-phosphate (Sigma-Aldrich, Budapest, Hungary), respectively. Enzyme reactions were carried out in three replicates in the wells of 96-well microtiter plates (Sarstedt, Nümbrecht, Germany) by mixing 100 μL culture filtrate or SSF extract with 100 μL substrate solution (1 mg mL^{-1} in distilled water). After 1 h of incubation at room temperature, enzyme reactions were stopped with 10% (w/v) Na_2CO_3 and the optical densities measured at 405 nm with a Spectrostar Nano microplate reader (BMG Labtech, Ortenberg, Germany).

Peroxidase assays of bacteria were carried out in liquid *Streptomyces* induction medium (STR-IND) 6 g L^{-1} yeast extract, 8 g L^{-1} xylan, 0.1 g L^{-1} $(NH_4)_2SO_4$, 0.3 g L^{-1} NaCl, 0.1 g L^{-1} $MgSO_4$, 0.02 g

L^{-1} $CaCO_3$, 0.6 mL L^{-1} TE (0.1 g L^{-1} $FeSO_4$, 0.002 g L^{-1} $MnSO_4 \times 7H_2O$, 0.09 g L^{-1} $ZnSO_4 \times 7H_2O$) inoculated with the examined *Streptomyces* strains. After 7 days of incubation (28 °C, 150 rpm), the samples were centrifuged at 7000 rpm for 10 min. The reaction mixture contained 0.2 mL phosphate buffer (100 mM, pH 7.2), 0.2 mL 2,4-dichlorophenoxy acetic acid (25 mM), 0.2 mL 4-aminoantipyrine (4AAP, 16 mM), 0.2 mL ferment broth, and 0.2 mL hydrogen peroxide (50 mM). The reaction mixtures were put into a 53 °C thermostat for 1 min, which was followed by the measurement of the optical density at 510 nm with a Spectrostar Nano microplate reader (BMG Labtech, Ortenberg, Germany). All measurements were carried out in three replicates.

In order to investigate the peroxidase enzyme production, dye decolorization assays were also performed. For this purpose, we used STR-IND medium supplemented with 20 g L^{-1} agar and Remazol Brilliant Blue (RBB), Methyl Orange (MO), or Neutral Red (NR) dyes at a concentration of 1 g L^{-1}. The plates were inoculated with *Streptomyces* isolates with the aid of inoculation loop onto the middle of the Petri plates. Color changes were observed around the colonies after 1 week of incubation.

2.5. Growth Assay in Nitrogen-Free Medium

The growth kinetics of *A. vinelandii* strain SZMC 22195 were tested in nitrogen source-free liquid medium (5 g L^{-1} glucose, 5 g L^{-1} mannitol, 0.1 g L^{-1} $CaCl_2 \times 2H_2O$, 0.1 g L^{-1} $MgSO_4 \times 7H_2O$, 0.005 g L^{-1} $Na_2MoO_4 \times 2H_2O$, 0.9 g L^{-1} K_2HPO_4, 0.1 g L^{-1} KH_2PO_4, 0.01 g L^{-1} $FeSO_4 \times 7H_2O$, pH 7.3). During an incubation period of 1 week (30 °C, 120 rpm), the optical densities of the liquid cultures were measured on days 1, 2, 4, and 7 at 620 nm with a Spectrostar Nano microplate reader.

2.6. Plant Material for Growth Chamber Experiments

Seeds of tomato (*Solanum lycopersicum* Mill. L. cvar. Ailsa craig) were germinated at 26 °C for 3 days in the dark, and the seedlings were subsequently transferred to 6 × 6 cm pots filled with vermiculite (Terracult GmbH, Siegburg, Germany) for 6 weeks. Plants were irrigated every third day with nutrient solution containing 2 mM $Ca(NO_3)_2$, 1 mM $MgSO_4$, 0.5 mM KCl, 0.5 mM KH_2PO_4, and 0.5 mM Na_2HPO_4, pH 6.0. The concentrations of micronutrients were 0.001 mM $MnSO_4$, 0.005 mM ZnSO4, 0.0001 mM (NH4)6Mo7O24, 0.01 mM H_3BO_4, and 0.02 mM Fe(III)-EDTA. The plants were grown in a controlled environment under 300 µmol m^{-2} s^{-1} light intensity (emitted F36W/GRO lamps, Feilo Sylvania, Erlangen, Germany), with 12/12-h light/dark period, day/night temperatures of 24/22 °C, and relative humidity of 55–60%. Plants were treated with 20 µL of *Trichoderma* suspension (1×10^6 conidia mL^{-1}) after the 3-days-long germination. Samples for measurements were prepared in each replicate from the second, fully expanded young leaves of tomato plants. After harvest, the plant height and root length as well as the biomass production were recorded in 5 replicates.

2.7. Measurement of Stomatal Conductance, CO_2 Assimilation, and Total Soluble Sugar Content

Stomatal conductance and CO_2 assimilation were measured in 3 replicate samples by a portable photosynthesis system (LI-6400, LI-COR Inc., Lincoln, NE, USA), as described by Poór et al. [11]. Data were recorded after 15 min light adaptation on 300 µmol m^{-2} s^{-1} light intensity and under constant conditions (25 °C, 65 ± 10% relative humidity, and controlled CO_2 supply of 400 µmol mol^{-1}).

Total sugar contents were determined according to Dubois et al. [12]. One gram of leaf samples was homogenized in 10 mL distilled water and incubated in a 90 °C water bath for 45 min. Samples were centrifuged (12,000× *g* for 15 min, at 4 °C), and 40 µL of the supernatant was mixed with 400 µL of 1.8% phenol and 2 mL of concentrated sulfuric acid. The absorbance was measured from 5 replicate samples by a spectrophotometer at 490 nm.

2.8. Chlorophyll a Fluorescence Measurements

Chlorophyll a fluorescence was detected in 3 replicate samples with the portable photosynthesis system (LI-6400, LI-COR Inc., Lincoln, NE, USA) described above [11]. Leaves were dark-adapted for 15 min before the measurement of the minimal fluorescence (F0) using weak measuring light. The maximal fluorescence (Fm) was measured by applying a pulse (800 ms) of saturating light (12,000 μmol m^{-2} s^{-1}). The leaves were then illuminated continuously with actinic light (300 μmol m^{-2} s^{-1}). After 20 min, the light-adapted steady-state fluorescence (Fs) was recorded and the maximum fluorescence level (Fm′) in the light-adapted state was determined with saturating pulses. The actinic light was next turned off and the minimum fluorescence level in the light-adapted state (F0′) was determined by illuminating the leaf with 3-s far-red light (5 μmol m^{-2} s^{-1}). The following chlorophyll fluorescence parameters were calculated: the maximal quantum yield of PSII photochemistry, Fv/Fm = (Fm − F0/Fm); the actual quantum yield of PSII electron transport in the light adapted-state, ΦPSII = (Fm′ − Fs)/Fm′ [13]; the photochemical quenching coefficient, qP = (Fm′ − Fs)/(Fm′ − F0′) [14]; and the non-photochemical quenching NPQ = Fm/Fm′ − 1 [15].

2.9. Pigment Analysis

For pigment analysis, a two-step extraction was applied. Fifty milligrams of leaf samples were homogenized in ice-cold 100% (v/v) acetone (1 mL) and extracted for 24 h. Samples were centrifuged (12,000× *g* for 15 min at 4 °C). The pellet was extracted again with 80% (v/v) acetone (1 mL) for 24 h. After spinning down (12,000× *g*, 15 min, 4 °C), the supernatants were collected. The pigment composition was measured in 5 replicate samples as according to Lichtenthaler and Wellburn [16].

2.10. Field Experiment

A field study was performed in tomato culture (*Solanum lycopersicum* cvar. ACE-55) on sandy loam soil according to the yarn number of Arany (K_A), which is a humus-rich soil with good nutrient availability and water management. The GPS coordinates of the examined area are 46°05′01.05′′ N, 19°26′28.83′′ E. Three-week-old tomato seedlings were planted on 11 May 2019. A total of 220 seedlings were planted in the field with 40 cm row distance and 40 cm plant distance, with 20 seedlings in a row. At the beginning of the experiment, a bioinoculant preparation consisting of a mixture of two *Trichoderma* strains and two bacteria was prepared. The concentration of the bioinoculant was adjusted to 10^6 conidia mL^{-1} for both selected *Trichoderma* components, and 10^8 cells mL^{-1} for both selected bacteria. The application of the soil inoculant was performed after 100× dilution at a concentration of 100 mL L^{-1}, while the control area was not treated. Three control and eight treated rows were examined. No organic manure or chemical fertilizer was applied to the area during the soil preparation. Changes in the contents of soil macro- and microelements were measured three times during cultivation (I: 22 June, II: 6 July, and III: 3 August 2019) from rhizosphere samples taken from a depth of 15 cm. The total numbers of the control and treated plants were 60 and 160, respectively. During the experiment, the same plant protection measures were applied both in the control and treated area: Cuproxat FW (5 mL L^{-1}), Topaz (0.5 mL L^{-1}), Mospilan (200 mg L^{-1}), Wuxal (2 mL L^{-1}), Humusz (1 mL L^{-1}), as well as calcium (5 mL L^{-1}) and magnesium (5 mL L^{-1}) in the form of foliar fertilizer were used 3 times during the cultivation (20 May, 8 June, and 13 July 2019). Control and treated plants were examined separately.

2.11. Soil Examination Methods

Soil sampling and analysis were carried out according to the test methods prescribed by the Hungarian Standard [17,18]. The soil tests were carried out by the Felső-Bácskai Agrolabor Ltd., Bácsalmás, Hungary. Carbonate content was determined with a Scheibler's calcimetre (Bovimex, Székesfehérvár, Hungary). The total salt content was measured by the electric conductivity using a HI98311 conductivity meter (HANNA Instruments, Szeged, Hungary). These methods were based on the Hungarian Standard (MSZ) MSZ-08-0206-2:1978. The soil texture was determined by the yarn number of Arany (MSZ-08-0205:1978). The macro- and microelement content and the humus content were measured by a Lambda 25 UV/VIS Spectrophotometer (PerkinElmer, Waltham, MA, USA) according to the Hungarian standards MSZ 20135:1999 and MSZ-21470-52:1983, respectively. N-content was measured by the Kjeldahl method, K_2O with flame emission spectrophotometry (FES), while Mg and Ca content with flame atomic absorption spectroscopy (FAAS) after acidic (H_2SO_4-$HClO_4$)

digestion [19]. Samples were prepared with microwave digestion for microelement analysis (Cu, Mn, Zn, and Fe content) measured with FAAS [17,19,20].

2.12. Statistical Analysis

Data presented as average values resulted from at least 3 independent experiments. Statistical analyses were carried out for the measurement data with Sigma plot v11.0 software (Systat Software Inc., Erkrath, Germany) using Student's t test, and the differences were considered significant if $p < 0.05$ (*), $p < 0.01$ (**), or $p < 0.001$ (***). The statistical analyses of the soil results and crop yield data were performed with the GraphPad Prism v8.3.0 software (GraphPad Software Inc., San Diego, CA, USA) applying two-way ANOVA, and the results were considered significant if $p < 0.05$ (*), $p = 0.02$ (**), $p = 0.001$ (***), and $p < 0.0001$ (****).

3. Results

3.1. Selection of the Components for the Soil Inoculant

The potential biocontrol component of the soil inoculant was selected based on the results of dual confrontation tests between *Trichoderma* strains and plant pathogenic fungi. As shown in Table 1 and Figure 1, the two examined *T. asperellum* strains were the most effective against many of the tested pathogens. The BCI values of *T. asperellum* SZMC 20786 were the highest against FSSC SZMC 11067F and SZMC 11070F and *Alternaria alternata* 16085, while against *Rizoctonia solani* and *Armillaria gallica* they even reached 100, which means that the *Trichoderma* could completely overgrow and inhibit these plant pathogens. Except for *T. atrobrunneum* SZMC 20869 and *T. virens* SZMC 20779, all other *Trichoderma* strains could completely inhibit the *R. solani* strain SZMC 6252J. Based on the results, strain *T. asperellum* SZMC 20786 was selected as the potential biocontrol component of the soil inoculant.

Eight *Trichoderma* strains were included in the screening for cellulose-degrading and phosphatase-producing abilities (Figure 2). High levels of β-glucosidase and cellobiohydrolase enzyme activities could be measured in the case of three *Trichoderma* strains, which included two isolates of THSC (*T. guizhouense* SZMC 20761, *T. atrobrunneum* SZMC 20869) and one isolate of *T. hamatum* (SZMC 20784). The cellulolytic activities of SZMC 20869 were inducible with maize stem powder (Figure 2A,B), while the other two strains possessed high enzyme activity values only in liquid minimal medium. Only low phosphatase enzyme activities could be detected for the examined *Trichoderma* strains, except for the above mentioned *T. atrobrunneum* SZMC 20869 strain (Figure 2C), for which increased enzyme activity levels could be observed in liquid medium supplemented with maize stem powder. Based on the results, strain *T. atrobrunneum* SZMC 20869 from THSC was selected as the potential stem-degrading and phosphate-mobilizing component of the soil inoculant.

In the dye decolorization assays performed with *Streptomyces* isolates, color changes could be observed only when RBB was applied. Only the isolates *S. albus* SZMC 0232 and SZMC 0282 and *S. microflavus* DSM 40561 gave positive reactions with the RBB dye. The results of peroxidase assays are shown in Figure 3, indicating that only the isolates *S. albus* SZMC 0232 and SZMC 0282 showed an increased peroxidase production. The activities of these two strains were about twice as high as those of the control strain *S. microflavus* DSM 40561. Based on the results, strain *S. albus* SZMC 0282 was selected as the potential humus-producing component of the soil inoculant.

Table 1. Biocontrol index values of the examined *Trichoderma* strains against plant pathogenic fungi.

	Plant Pathogens										
Examined *Trichoderma* Strains	FSSC SZMC 11057F	FSSC SZMC 6241J	FSSC SZMC 11067F	FSSC SZMC 11070F	*Armillaria mellea* MUCL 31056	*Armillaria ostoyae* SZMC 23080	*Armillaria gallica* SZMC 23076	*Alternaria alternata* SZMC 16085	*Phoma cucurbitacearum* SZMC 16088	*Botrytis cinerea* SZMC 6244J	*Rhizoctonia solani* SZMC 6252J
T. asperellum SZMC 20866	79.77 ± 1.22	83.71 ± 1.35	71.38 ± 3.04	75.61 ± 5.35	33.91 ± 6.67	100.00 ± 0.00	100.00 ± 0.00	64.32 ± 0.53	81.36 ± 1.26	100.00 ± 0.00	100.00 ± 0.00
T. asperellum SZMC 20786	77.76 ± 0.96	83.50 ± 6.97	88.84 ± 5.48	85.76 ± 2.12	47.68 ± 5.22	92.33 ± 3.51	100.00 ± 0.00	65.77 ± 1.42	80.63 ± 0.50	100.00 ± 0.00	100.00 ± 0.00
T. atroviride SZMC 20780	62.41 ± 2.90	63.39 ± 5.10	37.63 ± 2.15	51.19 ± 0.96	93.76 ±2.89	100.00 ± 0.00	100.00 ± 0.00	51.43 ± 0.42	59.21 ± 1.37	100.00 ± 0.00	100.00 ± 0.00
T. atroviride SZMC 20781	63.91 ± 0.76	71.32 ± 4.40	39.93 ± 2.14	48.39 ± 2.53	97.43 ± 3.07	98.02 ± 2.00	100.00 ± 0.00	54.16 ± 0.27	56.82 ± 1.20	100.00 ± 0.00	100.00 ± 0.00
T. gamsii SZMC 20783	60.82 ± 1.42	67.42 ± 0.79	39.97 ± 1.74	59.36 ± 1.34	82.21 ± 9.56	100.00 ± 0.00	100.00 ± 0.00	49.22 ± 2.06	60.14 ± 0.75	50.70 ± 7.26	100.00 ± 0.00
T. hamatum SZMC 20784	68.22 ± 0.73	64.10 ± 0.67	47.88 ± 1.75	62.52 ± 0.63	96.71 ± 3.41	97.00 ± 0.00	92.75 ± 2.95	61.28 ± 1.38	67.96 ± 1.19	46.82 ± 1.06	100.00 ± 0.00
T. guizhouense SZMC 20761	63.02 ± 5.40	58.81 ± 1.72	63.39 ± 1.69	64.92 ± 0.87	96.70 ± 3.23	99.93 ± 0.11	100.00 ± 0.00	60.77 ± 1.06	61.19 ± 1.35	44.20 ± 4.21	100.00 ± 0.00
T. atrobrunneum SZMC 20869	42.19 ± 2.53	47.46 ± 2.08	37.83 ± 0.17	38.40 ± 0.95	54.38 ± 1.98	100.00 ± 0.00	88.43 ± 4.27	18.02 ± 1.02	14.92 ± 1.89	42.26 ± 4.55	74.87 ± 7.34
T. guizhouense SZMC 20762	63.69 ± 1.23	58.47 ± 5.30	47.91 ± 0.37	59.77 ± 1.79	97.42 ± 2.4	100.00 ± 0.00	100.00 ± 0.00	59.33 ± 0.53	57.84 ± 1.95	49.43 ± 3.03	100.00 ± 0.00
T. virens SZMC 20779	40.66 ± 1.51	37.20 ± 1.38	53.43 ± 0.62	49.10 ± 2.91	53.48 ± 2.76	100.00 ± 0.00	86.19 ± 5.77	25.07 ± 2.33	35.59 ± 1.91	45.32 ± 8.37	62.37 ± 5.36

FSSC: *Fusarium solani* species complex; MUCL: Belgian Coordinated Collections of Microorganisms/MUCL Agro-food & Environmental Fungal Collection; SZMC: Szeged Microbiology Collection.

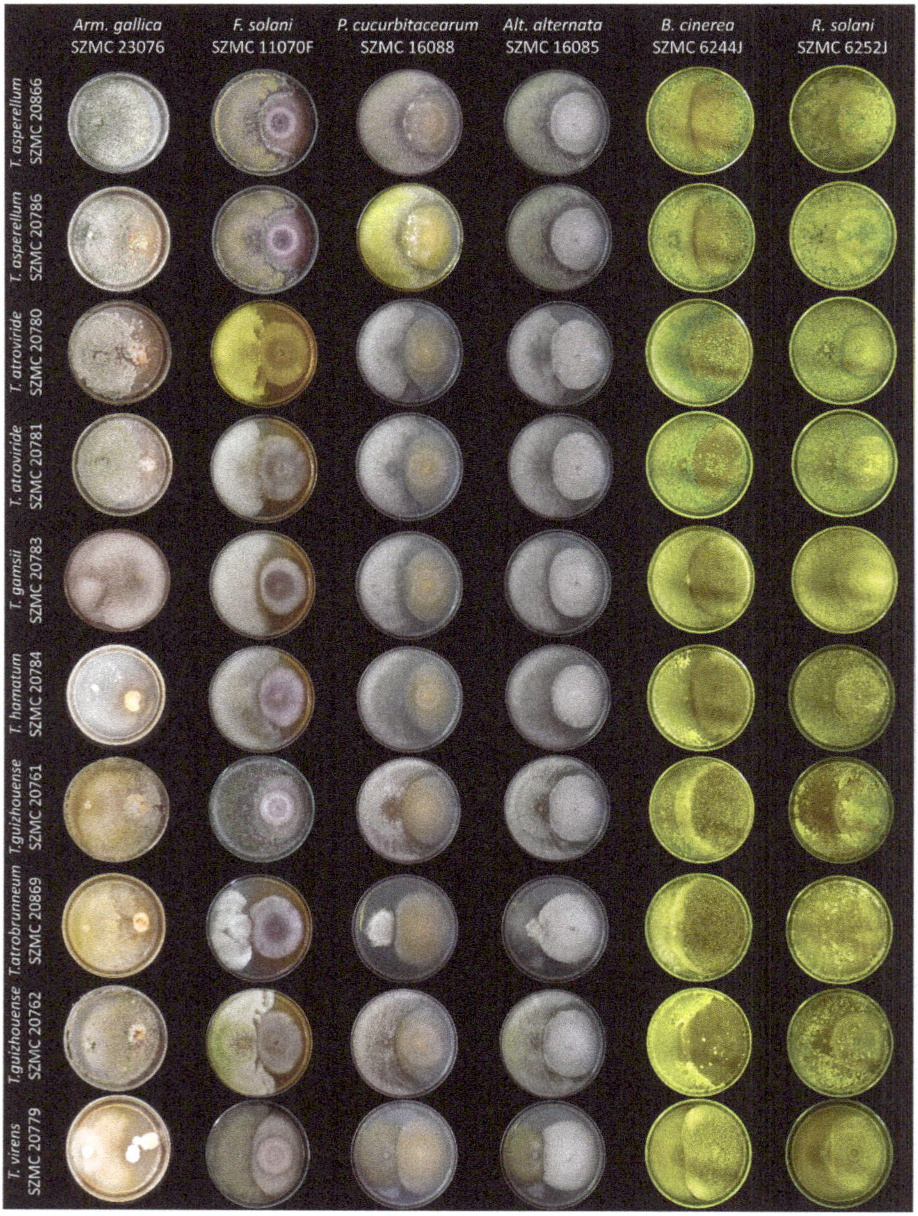

Figure 1. In vitro antagonism of *Trichoderma* strains against different plant pathogenic fungi examined in dual confrontation tests.

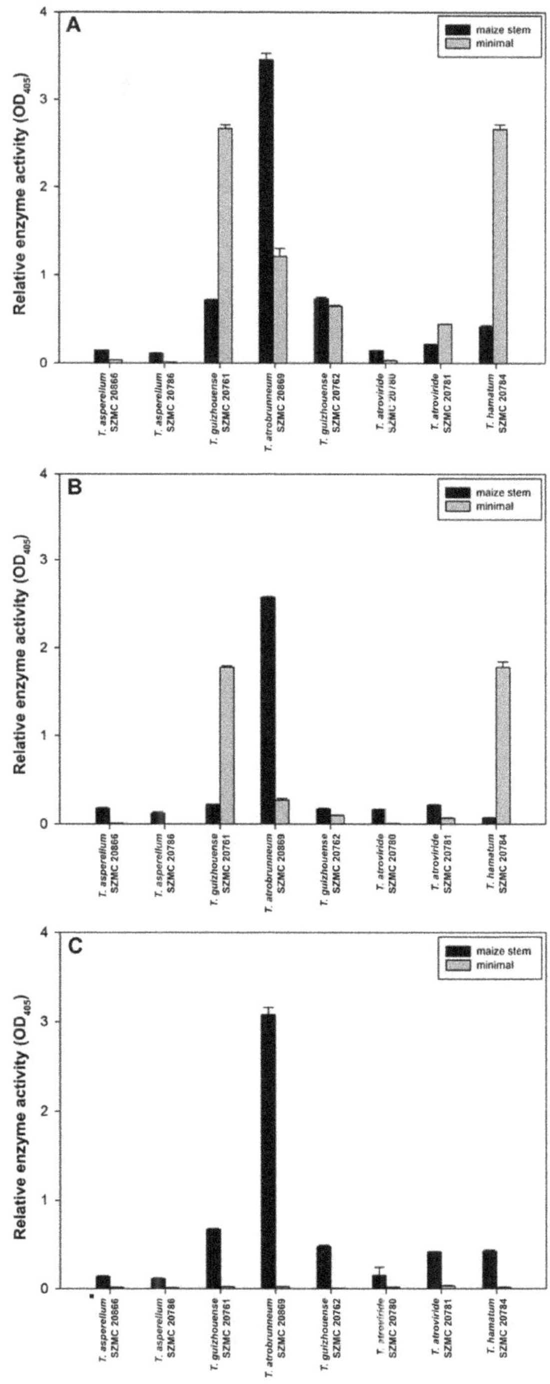

Figure 2. Extracellular enzyme activities of *Trichoderma* strains in liquid minimal and maize stem medium (mean ± SE, n = 3). (**A**): β-glucosidase, (**B**): cellobiohydrolase, (**C**): phosphatase.

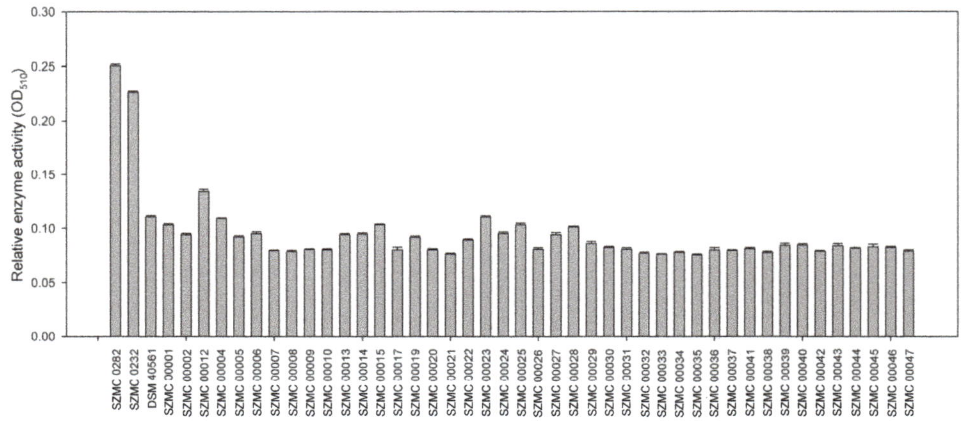

Figure 3. Peroxidase activities of *Streptomyces* isolates (mean ± SE, *n* = 3).

As the growth kinetics of strain *A. vinelandii* SZMC 22195 revealed that it reached the concentration of 10^8 cells ml^{-1} after 4 days of incubation in nitrogen-free liquid medium and the cell concentration increased further until day 7 (Figure 4), this strain was selected as the nitrogen-fixing component of the soil inoculant.

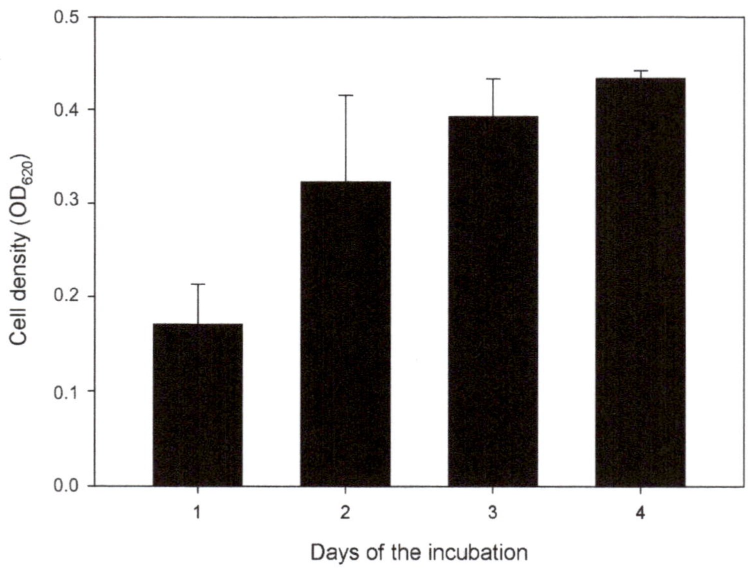

Figure 4. Growth kinetics of *Azotobacter vinelandii* SZMC 22195 in nitrogen-free liquid medium (mean ± SE, *n* = 3).

3.2. *Influence of* T. asperellum *Strain SZMC 20786 on the Shoot and Root Growth and Photosynthetic Activity of Tomato Plants*

In the case of the *T. asperellum* strain SZMC 20786 selected for the composite bioinoculant, direct plant growth promotion and effects on photosynthetic activity were examined on tomato plants. Significant increases in the fresh weight of the roots and shoots could be recorded in comparison to the control plants (Figure 5). The results deriving from the measurements of stomatal conductance and CO_2 assimilation are shown in Figure 6. Treatment with strain SZMC 20786 resulted in a non-significant increase in stomatal conductance (Figure 6A), coupled with a significant increase in both the CO_2 assimilation (Figure 6B) and the total sugar content (Figure 6C). indicating an increased photosynthetic activity in the plants treated with *T. asperellum*.

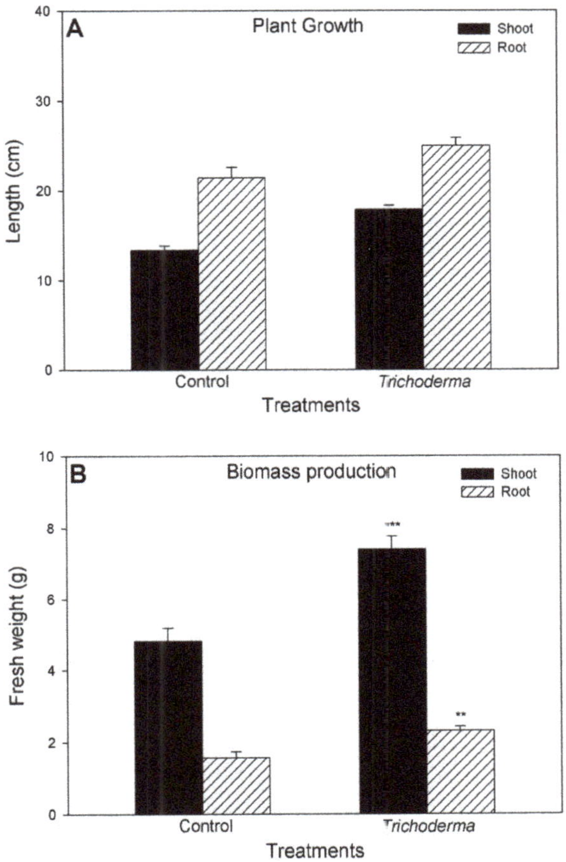

Figure 5. Effect of *T. asperellum* SZMC 20786 treatment on the growth parameters of tomato plants. (**A**) Root and shoot growth (mean ± SE, $n = 10$), (**B**) root and shoot biomass (mean ± SE, $n = 10$). Treated samples marked with asterisks are significantly different from the untreated control at $p \leq 0.01$ (**), or $p \leq 0.001$ (***).

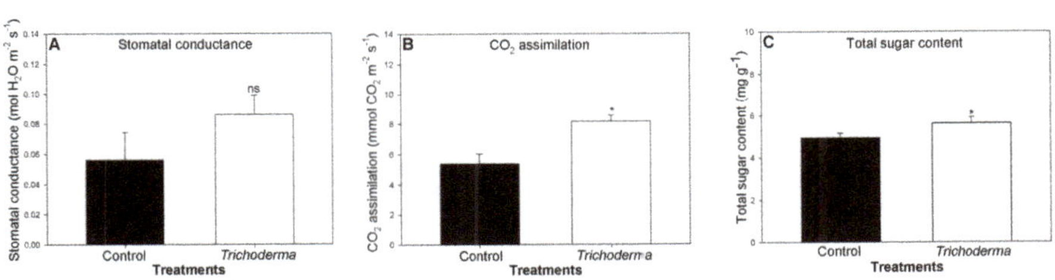

Figure 6. Changes in stomatal conductance (**A**), CO_2 assimilation (**B**), and total sugar content (**C**) in the leaves of tomato plants 6 weeks after treatment with *T. asperellum* SZMC 20786 (mean ± SE, $n = 5$). Treated samples marked with asterisks are significantly different from the untreated control at $p \leq 0.05$ (*), ns: not significant.

This is also supported by the changes of the chlorophyll a fluorescence induction parameters: the maximal quantum yield of PSII photochemistry (Fv/Fm), the actual quantum yield of PSII electron transport in the light-adapted state (ΦPSII), the photochemical quenching coefficient (qP), and the non-photochemical quenching (NPQ) were increased in plants treated with *T. asperellum* strain SZMC 20786; however, a significant change could be measured only for qP (Figure 7). The *T. asperellum* treatment also resulted in non-significant increases in the levels of chlorophyll a + b and carotenoids (Figure 8).

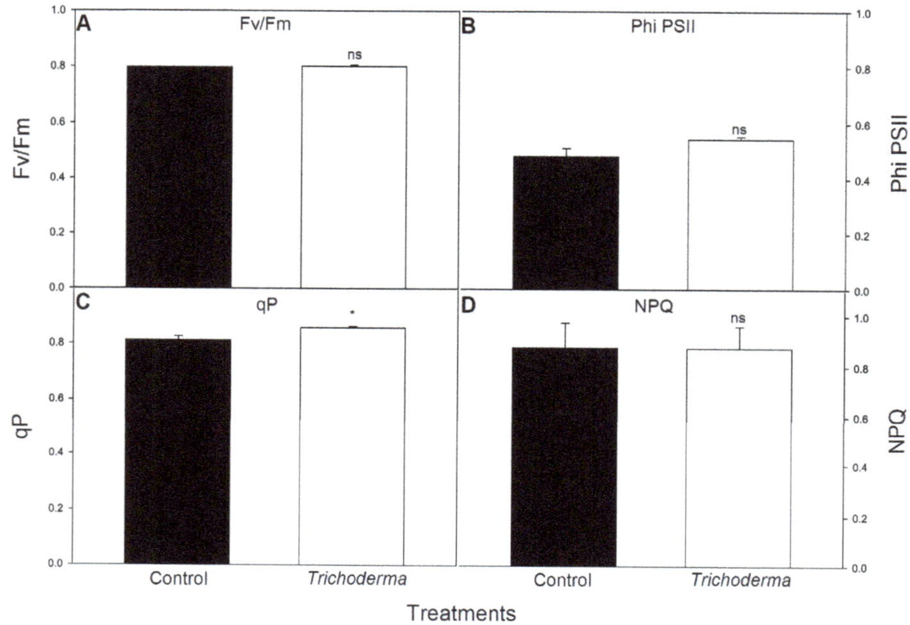

Figure 7. Changes of chlorophyll *a* fluorescence induction parameters ((**A**): Fv/Fm; (**B**): ΦPSII; (**C**): qP; (**D**): NPQ) in the leaves of tomato plants 6 weeks after treatment with *T. asperellum* SZMC 20786 (mean ± SE, $n = 5$). Treated samples marked with asterisks are significantly different from the untreated control at $p \leq 0.05$ (*), ns: not significant.

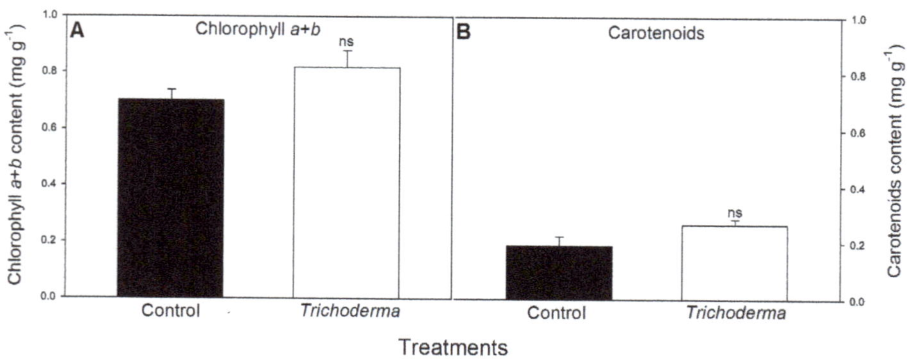

Figure 8. Changes in the content of chlorophyll a + b (**A**) and carotenoids (**B**) in the leaves of tomato plants 6 weeks after treatment with *T. asperellum* SZMC 20786 (mean ± SE, $n = 5$). ns: not significant.

3.3. Solid State Fermentation of Plant Stem Residues with *T. atrobrunneum* in Comparison with *T. reesei*

The extracellular enzyme activities of *T. atrobrunneum* SZMC 20869 (THSC) selected for the composite bioinoculant were compared with those of the industrially important, hypercellulolytic *T. reesei* strain QM9414 (SZMC 22616) in SSF experiments performed on the residues of 4 different crop plants (wheat, maize, sunflower, and canola) as substrates (Figure 9). Strain SZMC 20869 was able to produce β-glucosidase, cellobiohydrolase, β-xylosidase, and phosphatase activities on all four examined stem residues. Although the industrial strain of *T. reesei* produced larger amounts of β-glucosidase, cellobiohydrolase, and β-xylosidase—the three examined plants' cell-wall-degrading enzymes (PCWDEs)—on maize stem and canola stem residues, and was also a better producer of the two cellulolytic enzyme activities on sunflower stem residues than *T. atrobrunneum* SZMC 20869; the selected *T. atrobrunneum* strain was more efficient in production of all examined PCWDEs on wheat straw as the substrate. Regarding phosphatase activities, strain SZMC 20869 had been proved to be better than *T. reesei* on sunflower stem residues and equal to it on maize stem residues.

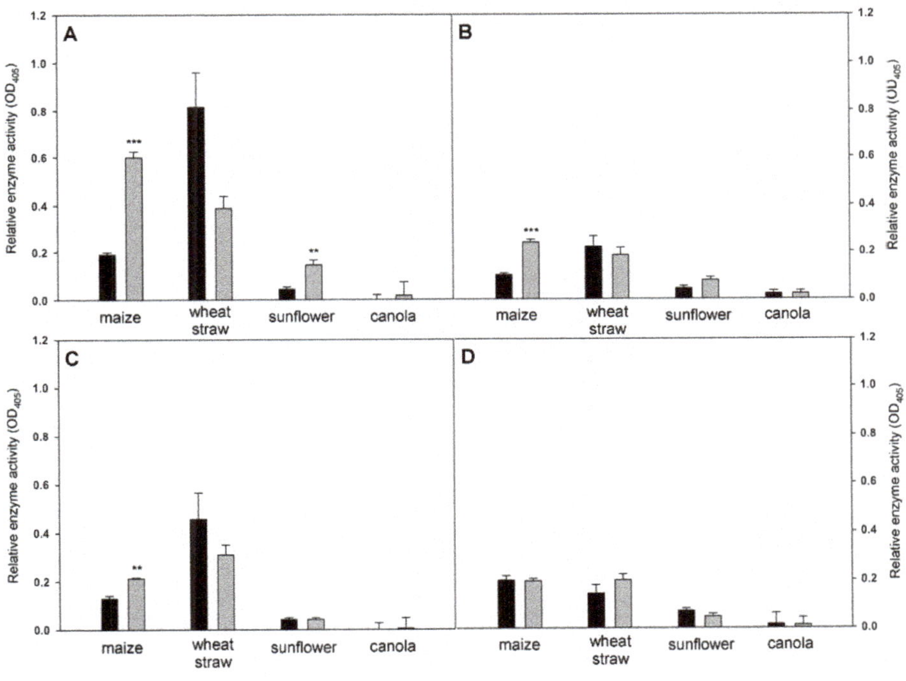

Figure 9. Extracellular enzyme activities of *T. atrobrunneum* SZMC 20869 and *Trichoderma reesei* SZMC 22616 after 8 days of solid-state fermentation on stem residues of wheat, maize, sunflower, and canola plants as substrates. (**A**) β-glucosidase, (**B**) cellobiohydrolase, (**C**) β-xylosidase, and (**D**) phosphatase (mean ± SE, $n = 3$). ■: *T. atrobrunneum*; ■: *T. reesei*. Columns of *T. reesei* marked with asterisks are significantly different at $p \leq 0.01$ (**), or $p \leq 0.001$ (***).

3.4. Field Experiment with the Combination of the Selected Bioinoculant Strains in Tomato Culture

Soil initial chemical and physical characteristics of the experimental area were as follows: pH: 7.6 (KCl), K_A: 35, carbonate: 10.6 m/m%, humus content: 1.84 m/m%, P_2O_5: 2423 mg/kg, K_2O: 669 mg/kg, NOX-N: 11 mg/kg; SO_4-S: 39 mg/kg, Cu: 4.4 mg/kg, Zn: 7.6 mg/kg, Mn: 13 mg/kg, Mg: 233 mg/kg, and Na: 99 mg/kg.

Changes in the soil macroelement, microelement, and humus content were monitored three times during the cultivation (Table 2). The initial potassium, sodium, and manganese contents of the total area determined from soil samples taken on 11th May were 669 mg kg^{-1}, 99 mg kg^{-1}, and 13 mg kg^{-1}, respectively. According to the statistical analysis, the K_2O, Na, and Mn content of the soil changed significantly in the treated area compared to the control area. This could be the result of a more efficient uptake of K_2O, Na, and Mn in the case of the treated plants; however, it is also possible that the elements P, Na, and Mn could have been sequestered in the microbial biomass. The average tomato fruit crop yields per row were 22,470 g and 22,810 g for the control and treated rows, respectively; however, the increase in the crop size in the case of the treated rows did not prove to be significant. The average total green mass of control and treated rows were 11,613 g and 11,032 g respectively, with no significant difference.

Table 2. Changes in the macro- and microelement content of the soil during the cultivation period of tomato.

Parameter	Measure	Sampling Time	Mean ± SD	
			Control Rows	Treated Rows
Humus	m/m %	I.	1.89 ± 0.04	1.98 ± 0.11
W		II. III.	1.57 ± 0.09 1.90 ± 0.10	1.90 ± 0.33 1.93 ± 0.19
P_2O_5	mg kg^{-1}	I.	2478.33 ± 182.57	1895.88 ± 125.12
		II. III.	1677.67 ± 69.04 1694.00 ± 86.41	1546.88 ± 135.31 1145.25 ± 368.56
K_2O	mg kg^{-1}	I.	702.67 ± 23.70	558.00 ± 76.90 *
		II. III.	701.67 ± 26.66 821.33 ± 54.63	451.75 ± 64.14 *** 480.13 ± 85.53 ****
SO_4-S	mg kg^{-1}	I.	16.33 ± 0.47	17.00 ± 1.22
		II. III.	15.00 ± 0.82 7.33 ± 0.94	16.88 ± 2.32 7.25 ± 2.11
NO_x-N	mg kg^{-1}	I.	3.93 ± 0.54	4.68 ± 0.96
		II. III.	4.67 ± 1.96 5.17 ± 0.66	4.64 ± 1.25 5.24 ± 1.96
Na	mg kg^{-1}	I.	49.33 ± 0.94	45.00 ± 4.00
		II. III.	45.00 ± 12.75 46.33 ± 4.11	25.88 ± 12.69 * 33.88 ± 5.37
Mg	mg kg^{-1}	I.	209.67 ± 2.62	220.75 ± 2.59
		II. III.	194.33 ± 8.22 229.00 ± 11.22	194.75 ± 12.16 213.38 ± 25.70
Mn	mg kg^{-1}	I.	14.33 ± 2.05	10.75 ± 0.66
		II. III.	19.33 ± 1.25 19.67 ± 1.25	17.63 ± 1.58 15.88 ± 3.18 *
Zn	mg kg^{-1}	I.	6.20 ± 0.78	5.75 ± 0.72
		II. III.	5.70 ± 0.37 4.73 ± 0.40	6.44 ± 0.74 5.11 ± 1.16
Cu	mg kg^{-1}	I.	6.23 ± 1.11	4.36 ± 0.44
		II. III.	8.20 ± 0.33 9.20 ± 1.55	7.76 ± 0.44 9.15 ± 3.00

Soils samplings were performed on I: 22 June, II: 6 July, and III: 3 August 2019. Data marked with asterisks are significantly different from the untreated control at $p < 0.05$ (*), $p = 0.001$ (***), $p < 0.001$ (****).

4. Discussion

The soil inoculant developed during this study contains two *Trichoderma* strains (*T. asperellum* and *T. atrobrunneum*) and two bacteria (*A. vinelandii* and *S. albus*) with potentially synergistic beneficial effects.

Members of the genus *Trichoderma* are geographically widespread filamentous ascomycetes from Hypocreales [21], which have long been known as agriculturally important, beneficial fungi with antagonistic abilities toward plant pathogenic fungi. *Trichoderma* antagonism is based on a series of different mechanisms including the competition for space and nutrients, antibiosis, mycoparasitism [22], plant growth promotion [23], enhancement of plant resistance to diseases [24,25], and relieving abiotic stress in plants [26]. These properties make many representatives of the genus *Trichoderma* (e.g., the THSC, *T. asperellum*, *T. atroviride*, or *T. virens*) to potential ingredients of soil inoculant and biocontrol preparations. However, when the practical application of a *Trichoderma* strain is planned, an exact, sequence-based, species-level identification is important to prevent the spread of species known as the causal agents of the green mold disease in mushroom cultivation [27–29] or of opportunistic infections in immunocompromised humans [30].

Strain *T. asperellum* SZMC 20786 was selected as a component of the composite soil bioinoculant due to its good in vitro antagonistic performance against different plant pathogenic fungi and its abilities to promote the growth of tomato plants and increase their photosynthetic activities. According to the literature, one of the mostly studied strains of "*T. asperellum*" for plant growth promotion is strain T203 [31], which, however, was recently reidentified as *T. asperelloides* [32]; thus, the number of studies about the plant growth promoting activities of *T. asperellum* sensu stricto is restricted. Qi and Zhao [33] demonstrated the plant growth promoting activities of *T. asperellum* srain Q1 on cucumber plants, and the positive effects were detected even when the plants were subjected to salt stress. In our study, *T. asperellum* strain SZMC 20786 showed positive effects on the CO_2 assimilation, total sugar content, and the photochemical quenching coefficient of tomato leaves. Similar results were obtained by Doni et al. [34], who found plant growth promotion as well as increased stomatal conductance and CO_2 assimilation in rice plants treated with *Trichoderma* sp. isolates. Other studies reported about the positive effects on the photosynthetic pigments exerted by *T. harzianum* strains on tomato [35] and wheat plants [36], as well as by *T. hamatum* on mungbean [37].

Another beneficial trait of many *Trichoderma* strains is their efficient ability to produce PCWDEs including cellulases and xylanases, which can be exploited both in the biotechnological industry and in the agriculture for the degradation of cellulose and xylan-containing materials, e.g., stem residues [38,39]. In accordance with these, another *Trichoderma* component, a *T. atrobrunneum* isolate possessing good cellulase, xylanase, and phosphatase enzyme production capabilities has been included in the assembled soil inoculant.

The species *A. vinelandii* involves Gram-negative, aerobic, free-living soil-inhabiting gamma-proteobacteria from the *Pseudomonadaceae* family. This species is capable of direct nitrogen fixation from the atmosphere by three distinct nitrogenase systems under fully aerobic conditions, thereby providing plant roots with bioavailable nitrogen source [40]. This aerobic bacterium possesses various protection mechanisms for nitrogenase against oxygen, which include alginate formation [41]. Furthermore, phytohormone and siderophore synthesis as well as phosphate solubilization are also among the abilities of *Azotobacter* species. These properties were suggested to be directly involved in their plant growth promotion effect [42]. Considering the above facts, an *A. vinelandii* strain has also been included in the soil inoculant.

As peroxidases were shown to play an important role in the humification properties of *Streptomyces* species [43], the potential humus-producing component of the bioinoculant was selected from *Streptomyces* isolates with a peroxidase assay, which revealed a *S. albus* strain as the best peroxidase-producing isolate. The species *S. albus* is among the geographically most widely distributed members of the genus *Streptomyces*; it could be isolated from various habitats including sea sediments, sponges, and insects [44]. This species was shown to be able to biosynthesize heterologously diverse and important natural products and was suggested to encode important natural product gene clusters [44]. Based on the genome sequence of *S. albus*, the secretion of a series of degradative enzymes (including amylases, chitinases, glucanases, proteinases/peptidases, and a cellulase) with supposed roles in breaking down heterogeneous alternative food sources in soil could be predicted [45]. The feather-degrading abilities of *S. albus* could be exploited during the development of an eco-friendly biofertilizer feather compost [46].

Several publications are available in the literature about the combined application of beneficial microorganisms for plant growth promotion and biological control. There are studies about the co-application of multiple bacteria including the combination of *Bacillus* and *Pseudomonas* for growth promotion and biological control of soil-borne diseases in pepper and tomato [47], as well as to increase rice yields [48], or the application of a mixture of fluorescent pseudomonads to suppress take-all disease of wheat [49]. There are also examples of co-application of fungi and bacteria, for instance the combination of *T. koningii* with fluorescent pseudomonads for the control of take-all disease of wheat [50], or the combination of *T. harzianum* with an *Alcaligenes* strain for the reduction of the incidence of rot disease caused by *Phytophthora capsici* in black pepper [51]. Other microbial combinations have been applied to promote the growth of tomato [52], to control tobacco diseases [53], or to improve the salinity tolerance of *Vicia faba* [54]. The combination of microorganisms could also increase the dry matter yield and nutrient uptake by wheat grown in a sandy soil [55].

5. Conclusions

The selection of the components for the composite soil inoculant assembled in this study was driven by the idea to combine various crop protective and plant growth promoting traits (biocontrol against plant pathogenic fungi, phosphorous mobilization, stem degradation, humification, and nitrogen fixation) of different microorganisms into a preparation, which may also have the potential to exert an increased consistency of field performance under various environmental conditions. The screening strategy performed during this study proved to be applicable for the assembly of a promising composite soil bioinoculant with notable application potentials.

Author Contributions: H.A., P.K., C.V., L.M. and L.K. contributed to the design and implementation of the research and evaluation of the results, and participated in the preparation of the manuscript; P.K. and L.K. performed the in vitro antagonism and liquid fermentation experiments; B.B. and L.M. contributed to the selection of the bacterial components of the bioinoculant; P.K. and P.P. designed and performed the experiments on tomato plants; V.D.N. and M.T. contributed to the solid-state fermentation experiments; H.A. performed and evaluated the field experiment; I.S. performed the measurements of the soil samples taken; and A.S. performed data curation and statistical analyses. The Tables and Figures were prepared by H.A., P.K., B.B., P.P., V.D.N. and C.V. All authors have read and agreed to the published version of the manuscript.

Funding: This study was supported by the Hungarian Government and the European Union within the frames of the Széchenyi 2020 Programme (GINOP-2.3.2-15-2016-00052) and the Hungary-Serbia IPA Cross-border Co-operation Programme (PLANTSVITA project, HUSRB/1602/41/0031). L.K. and M.T. are grantees of the János Bolyai and Bolyai Plus Research Scholarships.

Acknowledgments: The authors thank Sarolta Szabó and Gábor Tarnai (BioeGO Ltd.) for their support during the performance of the field experiment.

Conflicts of Interest: The authors declare no conflict of interests.

References

1. Damalas, C.A.; Eleftherohorinos, I.G. Pesticide exposure, safety issues, and risk assessment indicators. *Int. J. Environ. Res. Public Health* **2011**, *8*, 1402–1419. [CrossRef] [PubMed]
2. Ceulemans, T.; Stevens, C.J.; Duchateau, L.; Jacquemyn, H.; Gowing, D.J.; Merckx, R.; Wallace, H.; van Rooijen, N.; Goethem, T.; Bobbink, R.; et al. Soil phosphorus constrains biodiversity across European grasslands. *Glob. Chang. Biol.* **2014**, *20*, 3814–3822. [CrossRef] [PubMed]
3. Sekhon, G.S. Fertilizer-N use efficiency and nitrate pollution of groundwater in developing countries. *J. Contam. Hydrol.* **1995**, *20*, 167–184.
4. Meyer, S.L.F.; Roberts, D.P. Combinations of biocontrol agents for management of plant-parasitic nematodes and soilborne plant-pathogenic fungi. *J. Nematol.* **2002**, *34*, 1. [PubMed]

5. Syed Ab Rahman, S.F.; Singh, E.; Pieterse, C.M.J.; Schenk, P.M. Emerging microbial biocontrol strategies for plant pathogens. *Plant Sci.* **2018**, *267*, 102–111. [CrossRef] [PubMed]
6. Weller, D.M.; Thomashow, L.S.; Cook, R.J. Biological control of soil-borne pathogens of wheat: Benefits, risks and current challenges. In *Biological Control: Benefits and Risks*; Hokkanen, H.M.T., Lynch, J.M., Eds.; Cambridge University Press: Cambridge, UK, 1995; pp. 149–160.
7. Körmöczi, P.; Danilovic, G.; Manczinger, L.; Jovanovic, L.; Pankovic, D.; Vágvölgyi, C.; Kredics, L. Species composition of *Trichoderma* isolates from the rhizosphere of vegetables grown in Hungarian soils. *Fresen. Environ. Bull.* **2013**, *22*, 1736–1741.
8. Javaid, A. Effects of biofertilizers combined with different soil amendments on potted rice plants. *Chilean J. Agric. Res.* **2011**, *71*, 157–163. [CrossRef]
9. Chen, L.; Bóka, B.; Kedves, O.; Nagy, V.D.; Szűcs, A.; Champramary, S.; Roszik, R.; Patocskai, Z.; Münsterkötter, M.; Huynh, T.; et al. Towards the biological control of devastating forest pathogens from the genus *Armillaria*. *Forests* **2019**, *10*, 1013. [CrossRef]
10. Szekeres, A.; Leitgeb, B.; Kredics, L.; Manczinger, L.; Vágvölgyi, C. A novel, image analysis-based method for the evaluation of in vitro antagonism. *J. Microbiol. Meth.* **2006**, *65*, 619–622. [CrossRef]
11. Poór, P.; Gémes, K.; Horváth, F.; Szepesi, Á.; Simon, M.L.; Tari, I. Salicylic acid treatment via the rooting medium interferes with stomatal response, CO_2 fixation rate and carbohydrate metabolism in tomato, and decreases harmful effects of subsequent salt stress. *Plant Biol.* **2011**, *13*, 105–114.
12. Dubois, M.; Gibbs, K.A.; Hamilton, J.K.; Roberts, D.A.; Smith, F. Colorimetric methods for the determination of sugars and related substances. *Anal. Chem.* **1956**, *28*, 350–352. [CrossRef]
13. Genty, B.; Briantais, J.M.; Baker, N.R. The relationship between the quantum yield of photosynthetic electron transport and quenching of chlorophyll fluorescence. *Biochim. Biophys. Acta* **1989**, *990*, 87–92. [CrossRef]
14. Bilger, W.; Schreiber, U. Energy-dependent quenching of dark level chlorophyll fluorescence in intact leaves. *Photosynth. Res.* **1986**, *10*, 303–308. [CrossRef] [PubMed]
15. Bilger, W.; Björkman, O. Role of the xanthophyll cycle in photoprotection elucidated by measurements of light-induced absorbance changes, fluorescence and photosynthesis in *Hedera canariensis*. *Photosynth. Res.* **1990**, *25*, 173–185. [CrossRef] [PubMed]
16. Lichtenthaler, H.K.; Wellburn, A.R. Determinations of total carotenoids and chlorophylls a and b of leaf extracts in different solvents. *Biochem. Soc. Trans.* **1983**, *11*, 591–592. [CrossRef]
17. Ábrahám, B.É.; Őri, N.; Szabó, S.; Blaskó, L.; Zsigrai, G.Y. Effects of whitening process on the quality of different grain sorghum hybrids. *Eur. J. Plant Sci. Biotechnol.* **2012**, *6*, 90–93.
18. Szolnoki, Z.; Farsang, A.; Puskás, I. Cumulative impacts of human activities on urban garden soils: Origin and accumulation of metals. *Environ. Pollut.* **2013**, *177*, 106–115. [CrossRef]
19. Stefanovits, P.; Filep, G.; Füleky, G. *Talajtan*; Mezőgazda Kiadó: Budapest, Hungary, 1999; pp. 470p. (In Hungarian)
20. Junsomboon, J.; Jakmunee, J. Determination of potassium, sodium and total alkalies in Portland cement, fly ash, admixtures, and water of concrete by a simple flow injection flame photometric system. *J. Autom. Meth. Manag. Chem.* **2011**, *2011*, 742656. [CrossRef]
21. Kredics, L.; Hatvani, L.; Naeimi, S.; Körmöczi, P.; Manczinger, L.; Druzhinina, I.S. Biodiversity of the genus *Hypocrea/Trichoderma* in different habitats. In *Biotechnology and Biology of Trichoderma*; Gupta, V.K., Schmoll, M., Herrera-Estrella, A., Upadhyay, R.S., Druzhinina, I., Tuohy, M., Eds.; Elsevier Science B.V.: Amsterdam, The Netherlands, 2014; pp. 3–24.
22. Olmedo Monfil, V.; Casas-Flores, S. Molecular mechanisms of biocontrol in *Trichoderma* spp. and their applications in agriculture. In *Biotechnology and Biology of Trichoderma*; Gupta, V.K., Schmoll, M., Herrera-Estrella, A., Upadhyay, R.S., Druzhinina, I., Tuohy, M., Eds.; Elsevier Science B.V.: Amsterdam, The Netherlands, 2014; pp. 429–454.
23. Stewart, A.; Hill, R. Applications of *Trichoderma* in plant growth promotion. In *Biotechnology and Biology of Trichoderma*; Gupta, V.K., Schmoll, M., Herrera-Estrella, A., Upadhyay, R.S., Druzhinina, I., Tuohy, M., Eds.; Elsevier Science B.V.: Amsterdam, The Netherlands, 2014; pp. 415–428.
24. Saldajeno, M.G.B.; Naznin, H.A.; Elsharkawy, M.M.; Shimizu, M.; Hyakumachi, M. Enhanced resistance of plants to disease using *Trichoderma* spp. In *Biotechnology and Biology of Trichoderma*; Gupta, V.K., Schmoll, M., Herrera-Estrella, A., Upadhyay, R.S., Druzhinina, I., Tuohy, M., Eds.; Elsevier Science B.V.: Amsterdam, The Netherlands, 2014; pp. 477–494.

25. Contreras-Cornejo, H.A.; Macías-Rodríguez, L.; López-Bucio, J.S.; López-Bucio, J. Enhanced plant immunity using *Trichoderma*. In *Biotechnology and Biology of Trichoderma*; Gupta, V.K., Schmoll, M., Herrera-Estrella, A., Upadhyay, R.S., Druzhinina, I., Tuohy, M., Eds.; Elsevier Science B.V.: Amsterdam, The Netherlands, 2014; pp. 495–504.
26. Zaidi, N.W.; Dar, M.H.; Singh, S.; Singh, U.S. Trichoderma species as abiotic stress relievers in plants. In *Biotechnology and Biology of Trichoderma*; Gupta, V.K., Schmoll, M., Herrera-Estrella, A., Upadhyay, R.S., Druzhinina, I., Tuohy, M., Eds.; Elsevier Science B.V.: Amsterdam, The Netherlands, 2014; pp. 515–526.
27. Hatvani, L.; Kocsubé, S.; Manczinger, L.; Antal, Z.; Szekeres, A.; Druzhinina, I.S.; Komoń-Zelazowska, M.; Kubicek, C.P.; Nagy, A.; Vágvölgyi, C.; et al. The green mould disease global threat to the cultivation of oyster mushroom (*Pleurotus ostreatus*): A review. In *Science and Cultivation of Edible and Medicinal Fungi: Mushroom Science XVII: Proceedings of the 17th Congress of the International Society for Mushroom Science*; Gruening, M., Ed.; International Society for Mushroom Science: Cape Town, South Africa, 2008; pp. 485–495.
28. Kredics, L.; García Jimenez, L.; Naeimi, S.; Czifra, D.; Urbán, P.; Manczinger, L.; Vágvölgyi, C.; Hatvani, L. A challenge to mushroom growers: The green mould disease of cultivated champignons. In *Current Research, Technology and Education Topics in Applied Microbiology and Microbial Biotechnology*; Méndez-Vilas, A., Ed.; FORMATEX: Badajoz, Spain, 2010; pp. 295–305.
29. Błaszczyk, L.; Siwulski, M.; Sobieralski, K.; Fruzynska-Jozwiak, D. Diversity of *Trichoderma* spp. causing *Pleurotus* green mould diseases in Central Europe. *Folia Microbiol.* **2013**, *58*, 325–333. [CrossRef]
30. Hatvani, L.; Manczinger, L.; Vágvölgyi, C.; Kredics, L. Trichoderma as a human pathogen. In *Trichoderma-Biology and Applications*; Mukherjee, P.K., Horwitz, B.A., Singh, U.S., Mukherjee, M., Schmoll, M., Eds.; CAB International: Wallingford, UK, 2013; pp. 292–313.
31. Viterbo, A.; Landau, U.; Kim, S.; Chernin, L.; Chet, I. Characterization of ACC deaminase from the biocontrol and plant growth-promoting agent *Trichoderma asperellum* T203. *FEMS Microbiol. Lett.* **2010**, *305*, 42–48. [CrossRef]
32. Brotman, Y.; Lisec, J.; Méret, M.; Chet, I.; Willmitzer, L.; Viterbo, A. Transcript and metabolite analysis of the *Trichoderma*-induced systemic resistance response to *Pseudomonas syringae* in *Arabidopsis thaliana*. *Microbiology* **2012**, *158*, 139–146. [CrossRef] [PubMed]
33. Qi, W.; Zhao, L. Study of the siderophore-producing *Trichoderma asperellum* Q1 on cucumber growth promotion under salt stress. *J. Basic Microbiol.* **2013**, *53*, 355–364. [CrossRef] [PubMed]
34. Doni, F.; Isahak, A.; Zain, C.R.C.M.; Yusoff, W.M.W. Physiological and growth response of rice plants (*Oryza sativa* L.) to *Trichoderma* spp. inoculants. *AMB Exp.* **2014**, *4*, 45. [CrossRef]
35. Azarmi, R.; Hajieghrari, B.; Giglou, A. Effect of *Trichoderma* isolates on tomato seedling growth response and nutrient uptake. *Afr. J. Biotechnol.* **2011**, *10*, 5850–5855.
36. Al-Ezerjawi, N.H.; Kadhim, J.H. Effect of two isolates of *Trichoderma harzianum* on total nitrogen, chlorophyll a and b contents and yield of wheat (*Triticum aestivum.* L.) class Eba'a-95. *Int. J. Sci. Res.* **2014**, *3*, 1078–1083.
37. Badar, R.; Qureshi, S.A. Comparative effect of *Trichoderma hamatum* and host-specific *Rhizobium* species on growth of *Vigna mungo*. *J. Appl. Pharm. Sci.* **2012**, *2*, 128–132. [CrossRef]
38. Do Vale, L.H.F.; Filho, E.X.F.; Miller, R.N.G.; Ricart, C.A.O.; De Sousa, M.V. Cellulase systems in Trichoderma: An overview. In *Biotechnology and Biology of Trichoderma*; Gupta, V.K., Schmoll, M., Herrera-Estrella, A., Upadhyay, R.S., Druzhinina, I., Tuohy, M., Eds.; Elsevier Science B.V.: Amsterdam, The Netherlands, 2014; pp. 229–244.
39. Błaszczyk, L.; Siwulski, M.; Sobieralski, K.; Lisiecka, J.; Jędryczka, M. *Trichoderma* spp.—Application and prospects for use in organic farming and industry. *J. Plant Prot. Res.* **2014**, *54*, 309–317. [CrossRef]
40. Nosrati, R.; Owlia, P.; Saderi, H.; Olamaee, M.; Rasooli, I.; Akhavian, T.A. Correlation between nitrogen fixation rate and alginate productivity of an indigenous *Azotobacter vinelandii* from Iran. *Iran J. Microbiol.* **2012**, *4*, 153–159.
41. Daiz-Barrera, A.; Soto, E. Biotechnological uses of *Azotobacter vinelandii*: Current state, limits and prospects. *Afr. J. Biotechnol.* **2010**, *9*, 5240–5250.
42. Rubio, E.J.; Montecchia, M.S.; Tosi, M.; Cassán, F.D.; Perticari, A.; Correa, O.S. Genotypic characterization of Azotobacteria isolated from Argentinean soils and plant-growth-promoting traits of selected strains with prospects for biofertilizer production. *Sci. World J.* **2013**, *2013*, 519603. [CrossRef]
43. Cozzolino, A.; Piccolo, A. Polymerization of dissolved humic substances catalyzed by peroxidase. Effects of pH and humic composition. *Org. Geochem.* **2002**, *33*, 281–294. [CrossRef]

44. Seipke, R.F. Strain-level diversity of secondary metabolism in *Streptomyces albus*. *PLoS ONE* **2015**, *10*, e0116457. [CrossRef]
45. Zaburannyi, N.; Rabyk, M.; Ostash, B.; Fedorenko, V.; Luzhetskyy, A. Insights into naturally minimised *Streptomyces albus* J1074 genome. *BMC Genomics* **2014**, *15*, 97. [CrossRef]
46. Nayaka, S.; Vidyasagar, G.M. Development of eco-friendly bio-fertilizer using feather compost. *Ann. Plant Sci.* **2013**, *2*, 238–244.
47. Domenech, J.; Reddy, M.S.; Kloepper, J.W.; Ramos, B.; Gutierrez-Mañero, J. Combined application of the biological product LS213 with *Bacillus*, *Pseudomonas* or *Chryseobacterium* for growth promotion and biological control of soil-borne diseases in pepper and tomato. *BioControl* **2006**, *51*, 245–258. [CrossRef]
48. Nguyen, H.T.; Deaker, R.; Kennedy, I.R.; Roughley, R.J. The positive yield response of field-grown rice to inoculation with a multi-strain biofertiliser in the Hanoi area, Vietnam. *Symbiosis* **2003**, *35*, 231–245.
49. Pierson, E.A.; Weller, D.M. Use of mixtures of fluorescent pseudomonads to suppress take-all and improve the growth of wheat. *Phytopathology* **1994**, *84*, 940–947. [CrossRef]
50. Duffy, B.K.; Simon, A.; Weller, D.M. Combination of *Trichoderma koningii* with fluorescent pseudomonads for control of take-all on wheat. *Phytopathology* **1996**, *86*, 188–194. [CrossRef]
51. Anith, K.N.; Manomohandas, T.P. Combined application of *Trichoderma harzianum* and *Alcaligenes* sp. strain AMB 8 for controlling nursery rot disease of black pepper. *Indian Phytopathol.* **2001**, *54*, 335–339.
52. El-Katatny, H.M. Enzyme production and nitrogen fixation by free, immobilized and coimmobilized inoculants of *Trichoderma harzianum* and *Azospirillum brasilense* and their possible role in growth promotion of tomato. *Food Technol. Biotechnol.* **2010**, *48*, 161–174.
53. Maketon, M.; Apisitsantikul, J.; Siriraweekul, C. Greenhouse evaluation of *Bacillus subtilis* AP-01 and *Trichoderma harzianum* AP-001 in controlling tobacco diseases. *Braz. J. Microbiol.* **2008**, *39*, 296–300. [CrossRef] [PubMed]
54. Rabie, G.H.; Almadini, A.M. Role of bioinoculants in development of salt-tolerance of *Vicia faba* plants under salinity stress. *Afr. J. Biotechnol.* **2005**, *4*, 210–222.
55. Singh, S.; Kapoor, K.K. Inoculation with phosphate-solubilizing microorganisms and a vesicular-arbuscular mycorrhizal fungus improves dry matter yield and nutrient uptake by wheat grown in a sandy soil. *Biol. Fertil. Soils* **1999**, *28*, 139–144. [CrossRef]

agronomy

MDPI

Article

Chemical Traits of Fermented Alfalfa Brown Juice: Its Implications on Physiological, Biochemical, Anatomical, and Growth Parameters of Celosia

Nóra Bákonyi [1], Szilvia Kisvarga [2], Döme Barna [1], Ibolya O. Tóth [1], Hassan El-Ramady [1,3], Neama Abdalla [1,4], Szilvia Kovács [1], Margaréta Rozbach [5], Csaba Fehér [5], Nevien Elhawat [1,6], Tarek Alshaal [1,3,*] and Miklós Gábor Fári [1]

1 Department of Agricultural Botany, Plant Physiology and Biotechnology (MEK), Debrecen University, Böszörményi Street 138, 4032 Debrecen, Hungary; nbakonyi@agr.unideb.hu (N.B.); barnadoeme@gmail.com (D.B.); olahne@agr.unideb.hu (I.O.T.); ramady2000@gmail.com (H.E.-R.); neama_ncr@yahoo.com (N.A.); szkovacs@agr.unideb.hu (S.K.); nevienelhawat@gmail.com (N.E.); miklos0810@gmail.com (M.G.F.)

2 Research Institute for Fruitgrowing and Ornamentals, National Agricultural Research and Innovation Center (NARIC) Szent-Györgyi Street 4., 2100 Gödöllő, Hungary; fullerina@gmail.com

3 Soil and Water Department, Faculty of Agriculture, Kafrelsheikh University, Kafr El-Sheikh 33516, Egypt

4 Plant Biotechnology Department, Genetic Engineering Division, National Research Center, 33-El-Behouth Street, Cairo 12622, Egypt

5 Department of Applied Biotechnology and Food Science (BME-ABÉT), Szent Gellért Square 4, 1111 Budapest, Hungary; rozbachgreti@gmail.com (M.R.); fehercsaba87@gmail.com (C.F.)

6 Department of Biological and Environmental Sciences, Faculty of Home Economic, Al-Azhar University, Sperpay Number 4 Rd, Tanta, 31732, Egypt

* Correspondence: alshaaltarek@gmail.com

Received: 3 December 2019; Accepted: 3 February 2020; Published: 7 February 2020

Abstract: Brown juice is a byproduct of fractionated green biomass during leaf protein isolation. It represents approximately 45%–50% of the total pressed fresh biomass. Disposal of brown juice is a serious issue in leaf protein production due to its high biological oxygen demand and carbohydrates content. The current study aimed to find a possible potential use of brown juice. Therefore, chemical and biochemical properties of brown juice—derived from alfalfa green biomass—were determined before and after fermentation by lactic acid bacteria. Additionally, the growth stimulation potential of fermented brown juice on plumed cockscomb (*Celosia argantea* var. plumose 'Arrabona') plants were tested. Celosia seedlings were sprayed at different rates of fermented brown juice (i.e., 0.5%, 1%, 2.5%, 5%, and 10%) and tap water was applied as control. The results revealed that lactic acid bacteria successfully enhanced the stabilization of brown juice via reducing sugars content and increasing organic acids content. After fermentation, contents of glucose monomers were 15 times lower; while concentrations of lactic and acetic acids increased by 7- and 10-fold, respectively. This caused a reduction in the pH of fermented brown juice by 13.9%. Treating Celosia plants at lower rates of fermented brown juice (up to 1.0%) significantly induced their growth dynamics and antioxidant capacity. Higher values of vegetative parameters were measured in treated plants compared to control. The brown juice treatments caused significant changes in histological parameters as well. The activity of catalase and peroxidase increased in plants that received fermented brown juice especially at low rates. Moreover, an increase in water-soluble protein and phenol was measured in different tissues of plants sprayed with fermented brown juice. Malondialdehyde content was lowered in treated plants compared to control. Fermented brown juice at high rates slightly reduced the amount of photosynthetic pigments; however, this reduction was not reported for low rates of fermented brown juice. These results surely illustrate the potential use of fermented alfalfa brown juice as a growth stimulator for crops particularly at rates below 2.5%.

Keywords: deproteinized leaf juice; fermentation; lactic acid bacteria; plant nutrition; antioxidant capacity; ornamental plants

1. Introduction

Due to the continuous growth in the global population (7.2 billion) and malnutrition, the global demand for the protein will increase in the next years [1]. The lack of protein supply has existed as a health problem for many years and is considered as one of the main types of malnutrition in developing and developed countries [2]. Over the next decades, a dramatic increase in the global protein demand is expected and overall protein consumption is predicted to nearly double by 2050. These rapid changes will create serious and accelerated pressure on land and water resources and their scarcity [3]. To meet the increased protein demand there are several approaches to introduce novel protein sources or alternatives [4,5]. The extraction of proteins from forage crops such as alfalfa, clover or grass is a potential process for the production of leaf protein concentrates (LPC), which can be utilized as feed or food but also hydrolyzed into amino acids for the cosmetics or pharma industries [6]. Alfalfa or lucerne plant is well known as the king of forage. It is a perennial flowering plant belonging to the legume family Fabaceae. This plant has several advantages including high-quality leaf protein (50%–60%), strong adaptability, high nutritional value, good taste, wide distribution, and stable productivity [7]. It can also yield crude protein 2-, 3-, and 4-fold higher than peas, soybean, and wheat, respectively [8]. Therefore, alfalfa nowadays is considered as the most promising crop for LPC. Isolation of leaf protein in form of LPC aims to extract solid or insoluble proteins (i.e., the protein of mitochondria, chloroplasts, nucleoprotein, and cell wall) and soluble proteins (i.e., the soluble fraction of mitochondrial proteins, chloroplast matrix, and cytoplasm proteins). Therefore, the thermal treatment of green juice obtained by pressing the fresh biomass is needed to coagulate these types of proteins. During coagulation of leaf protein, a brown liquid byproduct is produced, and it is referred to as "brown juice". One kilogram of fresh alfalfa biomass can produce up to 500 g of brown juice [9]. These large amounts of brown juice are rich in protein and phenols as well as micronutrients. Plant phenolic compounds are known to be able to modulate important physiological routes like signal transduction and transcriptional regulation. Phenolic compounds in brown juice associated with auxin bioregulators [10] prove that the disposal of these amounts of valuable brown juice is a waste. Disposal is high in its costs and will waste the nutritional value of this byproduct, which would be easily adaptable to the circular economy concept; a technology that generates no further waste by utilizing all the produced renewable resources [11–13]. The main product, the leaf protein produced by coagulation, is widely studied [14]; however, the brown liquid, also known as whey or brown juice [15], has limited literature especially in the case of the plant nutrition aspect. Brown juice is mentioned in some articles as DPJ (Deproteinized Plant Juice) [16] or deproteinized leaf extracts or leaf juice, deproteinized whey [17] as a byproduct of plant protein-producing technologies. DPJ can be applied for several purposes; for instance, as a fertilizer for plants, milk for calves, excellent fodder for cattle and rabbits, medium for microbial growth, and also for in vitro rhizogenesis [18–21]. The dry matter and protein content of brown juice range from 13% to 15% and 16%–20%, respectively, whereas the cellulose content is 25%–30% [18]. The alfalfa brown juice has a dry matter content of 4%–8% which is influenced by the species, varieties, weather conditions, phenophase, methods of harvest, and processing.

Several microorganisms like lactic acid bacteria (LAB) are useful, having advantageous features, and can be found in a range of locations from soil and natural water, to the surface of plants up to the human intestinal tract [22]. These microorganisms have been applied for decades in the fermentation processes of raw materials because of their beneficial effects. It has been validated that ferments containing lactic acid bacteria (or other PGPB—plant growth-promoting bacteria) (isolated from different sources) have plant growth-promoting properties [23]. Lactic acid bacteria containing ferments were proven to be effective biofertilizers, biocontrol agents, and biostimulants because they promote plant health, growth, and resilience as they improve nutrient availability [24], however, the functional roles of these bacteria in the phytomicrobiome have not been discovered yet [25]. *Celosia* genus is native to tropical America and Africa. *Celosia argentea* is a food crop in West Africa as well as a medicinal plant in China and India with considerable pharmacological properties [26]. Among 13 green leafy vegetables, *Celosia argentea* was one of the few that had exceptionally high iron (13.5 mg 100 g^{-1}), calcium (188 mg 100 g^{-1}), sodium (240.6 mg 100 g^{-1}), ascorbic acid (26 mg 100 g^{-1}), and β-carotene (4.42 mg 100 g^{-1}) content. The edible portion of *Celosia argentea* was found to be 55 g 100 g^{-1} fresh weight which was one of the highest, while its moisture and protein content was found to be 87.6 and 3.2 g 100 g^{-1}, respectively [27]. Plumosa Group of *Celosia argentea* is an attractive ornamental plant characterized by a wide range in size and color of flowers. Plumosa cultivars can grow from dwarf to tall. The inflorescence of narrow pyramidal, plume-like, is consistent with tiny, vivaciously colored (e.g., orange, red, purple, yellow) flowers.

This research aimed to enhance the stability of stored brown juice through fermentation by lactic acid bacteria; assess physiochemical traits of alfalfa brown juice before and after fermentation; determine whether the different fermented brown juice concentrations have any impact on the formation of the stem's anatomy; and evaluate the potential of fermented brown juice as a growth stimulator using *Celosia argantea* var. *plumosa* as a model plant.

2. Materials and Methods

2.1. Brown Juice Production and Its Characteristics

2.1.1. Source of Alfalfa Biomass

A field experiment of alfalfa (*Medicago sativa* L. var. Hunor-40) was carried out during 2017 and 2018, under the GINOP (2.2.1-15-2017-00051) project labeled Proteomill [28], at the experimental farm of Tedej Zrt., Hajdúnánás, Hungary. The seeds were sown on chernozem soil at the rate of 25 kg ha^{-1}. All recommended agronomic practices such as irrigation, weed control, and fertilization were done. The alfalfa fresh biomass was used as a source for brown juice. The first cut of alfalfa plants was carried out in the middle of May 2018 directly before the flowering stage since at this time protein in alfalfa biomass is at its highest content. Plants were harvested early morning and directly transferred in special boxes to the laboratory to avoid the degradation of protein by protease enzyme.

2.1.2. Extraction of Brown Juice

Alfalfa fresh biomass was fractionated into the fiber, leaf protein concentrate (LPC), and deproteinized plant juice (DPJ, brown juice) as follows: fresh biomass was pressed and pulped mechanically using Angel Juicer (5500, Angel Ltd., Praha, Czech Republic) into fiber and green juice fractions. Later, the green juice was thermally treated at 80 °C in order to coagulate mainly the chloroplastic and cytoplasmic proteins. After thermal coagulation, the mixture was left at room temperature for approximately 10 min, then the coagulant was separated from brown juice using moistened 100% natural unbleached cotton cloth filter (pore size = 10 microns).

2.1.3. Fermentation of Brown Juice

Fermentation of brown juice was necessary to increase the stability of brown juice and its storage period because fresh brown juice rapidly spoils due to high sugar and protein content. After cooling, the brown juice was transferred into a 20-L container and inoculated with AdiSil LG-100 Perfect (Fides Agro, Šardice, Czech Republic) containing heterofermentative lactic acid bacterial cultures (10^{11} CFU g^{-1}, *Pediococcus acidilactici, Lactobacillus paracasei, Lactobacillus plantarum*) at the rate of 0.01 g L^{-1}. The inoculated samples were kept at 35 °C for 48 h.

2.1.4. Determination of Lactic Acid Bacteria

The qualitative measurement of lactic acid bacteria in the fermented brown juice was determined at the end of the fermentation process by methylene blue test [29]. Briefly, 1 mL methylene blue reagent was added to 10 mL fermented brown juice, and then the samples were incubated at 37 °C for 48 h. The time needed for the disappearing of blue color is an indication of lactic acid bacteria density in the solution.

2.1.5. Chemical Properties of Brown Juice

The pH of brown juice was measured by pH-meter (Mettler Toledo S20 Seven Easy, Switzerland). Electrical conductivity (EC) was determined using EC-meter (Thermo Scientific, Orion Model 209A$^+$ type, Germany). Degree Brix was recorded manually by a refractometer (RBR32-ATC, Germany). The content of macro- and micro-elements in brown juice before and after fermentation was measured using HNO_3-H_2O_2 wet digestion method as described by Kovács et al. [30]. Briefly, 1 g lyophilized brown juice was weighed into a Kjeldahl digestion tube, then 10 mL HNO_3 (99%, VWR International, USA) was added. The mixture was placed on the heater at 100 °C for 45 min; after cooling 5 mL H_2O_2 (30%, Sigma-Aldrich, St. Louis, MO, USA) was added for complete oxidation of organic materials and samples were kept on the heater for additional 45 min at 120 °C. After cooling the sample volume was brought to 50 mL using distilled water and then filtered using MN 640 W filter paper. The elemental content of brown juice was measured by ICP-OES spectrometer (Perkin Elmer made OPTIMA 3300 DV, Pittsboro, NC, USA).

Total phenol content in brown juice was determined spectrophotometerically using Ultrospec spectrophotometer (2100 pro, Amersham BioSciences, Amersham, United Kingdom) as previously described by Boór and Bélafiné Bakó [31]. Determination of total N content was carried out by Kjeldahl method [32] (Sparks et al., 1996). Concentrations of glucose and organic acids were determined by HPLC using BioRad (Hercules, CA, USA) Aminex HPX-87H (300 × 7.8 mm) column at 65 °C, and a refractive index detector. The eluent was 5 mmol L^{-1} H_2SO_4 at a flow rate of 0.5 mL min^{-1}. The injection volume was 40 µL. Concentrations of fructose, xylose, and arabinose were determined by HPLC using Phenomenex (Torrance, CA, USA) Rezex RPM-Monosaccharide Pb^{+2} (300 × 7.8 mm) column at 80 °C, and a refractive index detector. The eluent was ultrapure (milli-Q) water at a flow rate of 0.5 mL min^{-1}. The injection volume was 40 µL. Total sugars include monomer sugars and sugar oligomers solubilized. Monomer sugar concentrations were determined by HPLC after a sample preparation of 5 min boiling followed by centrifugation (5000 rpm) to eliminate residual proteins. To determine the oligomer sugar content of the samples, weak acid hydrolysis was performed. The samples were mixed with 8 w/w % H_2SO_4 at a volume ratio of 1:1 and treated at 120 °C in the autoclave for 15 min to decompose sugar oligomers into monomers, which were determined by HPLC.

2.2. *Celosia Experiment*

This experiment was carried out to assess the potential use of brown juice as a plant growth stimulator. In the present study, Celosia (*Celosia argantea* var. *plumosa* 'Arrabona') was used as a model plant for examining physiological, biochemical, and anatomical responses to fermented brown juice in our department and the National Agricultural Research and Innovation Center (NARIC, Budapest, Hungary). Celosia seeds were obtained from NARIC.

2.2.1. Experimental Design

A greenhouse pot experiment was carried out at the NARIC. The experimental layout was the Randomized Complete Block design (RCB) with 15 replicates. A polyethylene pot (7 × 7 × 8 cm) was filled with potting soil for horticultural crops (Klassman-Deilmann TS 3 FINE type, Geeste, Germany). The physical and chemical properties of potting soil are structure fine, pH (H_2O) 6, N 140 mg L^{-1}, P (P_2O_5) 100 mg L^{-1}, K (K_2O) 180 mg L^{-1}, Mg 100 mg L^{-1}, S 150 mg L^{-1}. Seeds of Celosia were sown in nursery substrate on 16th July 2018 and 4 days later germinated seeds were fertilized using different rates of brown juice. After two weeks, identical and healthy seedlings were transferred to the pots. Fermented brown juice was applied as a foliar application at rates of 0.5%, 1.0%, 2.5%, 5%, and 10%. The final application volume was 250 mL and equally shared among all replicates of the same treatment. The control plants were sprayed with tap water. Brown juice was applied once a week from starting the experiment on 16th July until 14th August, then we applied brown juice twice a week (Tuesdays and Fridays) until the end of the experiment on 11th September. At the end of the experiment, the following vegetative parameters were measured: root and stem length, root and stem volume, root and stem fresh and dry mass, and the number of leaves.

2.2.2. Determination of Water-Soluble Protein and Antioxidant Enzymes

Water-soluble protein fraction of lyophilized root, stem, and leaf tissues was determined using Coomassie Brilliant Blue G-250 according to Bradford [33] in triplicate with bovine serum albumin as standard. Briefly, 20 mg plant tissue was ground into homogenate in the mortar with quartz sand, then transferred into a volumetric flask, and then suspended in 100 mL distilled water to extract water-soluble protein fraction. The solution was centrifuged at 3000 rpm for 5 min. The supernatant was used for the assay of water-soluble protein content using UV-160A spectrophotometer (Shimadzu, Japan) at 595 nm. Peroxidase activity was determined in lyophilized roots, stems, and leaves of Celosia plants according to Roxas et al. [34]. Briefly, 100 mg plant tissue was macerated in 4 mL of phosphate buffer 0.01 M (pH 6.0). The homogenate was centrifuged at 13,000 rpm for 10 min to collect the supernatant. The supernatant was used to measure peroxidase activity using UV-160A spectrophotometer (Shimadzu, Japan) at 460 nm for 1 min. The unit of peroxidase activity was defined with the increase of one unit of absorbance per mL^{-1} min^{-1} g^{-1} of dry matter. Catalase (CAT) activity in lyophilized Celosia leaves was measured by following the decomposition of hydrogen peroxide at 240 nm according to Woodbury et al. [35]. The reaction included 0.2 mL supernatant, 1.5 mL phosphate buffer (pH 7.8, 0.2 M), and 1 mL distilled water. The colorimetric determination of CAT was conducted by the model UV-160A spectrophotometer (Shimadzu, Japan) at 240 nm. The biochemical reaction was initiated by adding 0.3 mL 0.1 M H_2O_2. The activity of CAT was expressed as μmol H_2O_2 consumed/mg protein/min.

2.2.3. Malondialdehyde Measurement

The malondialdehyde (MDA) content was determined from roots, stems, and leaves of Celosia plants by the method of Zhang and Huang [36]. Briefly, 100 mg lyophilized sample was homogenized in 1 mL 0.1% (*w*/*v*) TCA solution using cold mortar and pestle. The homogenates were centrifuged at 10,000× *g* for 10 min. Then, 4 mL of 0.5% thiobarbituric acid (TBA) in 20% TCA solution was added into 1 mL of supernatant and incubated at 96 °C for 30 min. The tubes were cooled by transferring into an ice bath. The absorbance of the supernatant was recorded at 532 nm. The standard curve was generated from MDA standard. The concentration of MDA of samples was calculated from the absorbance knowing calibration curve.

2.2.4. Photosynthetic Pigment

The photosynthetic pigment content of Celosia leaves was measured spectrophotomertically based on methods described by Porra et al. [37]. For the sample preparation, the leaf disc was cut and the chlorophyll content was extracted by N'N dimethyl-formamide overnight. The absorbance was measured by spectrophotometer (Amersham Biosciences Ultrospec 2100 Pro UV/Visible) on 663 and 645 nm wavelengths and from these data, the chl a, b, a + b, and a/b ratio were calculated.

2.2.5. Histology

We used three specimens per treatment for the stem's histological examination. Each plant was cut into smaller pieces and the third internodes (from beneath) fixed separately in a mixture of glycerin:alcohol:water (1:1:1) for a week. Then, several cross-sections were prepared using blades, after clarification, they were stained with Toluidin-blue. All analyses were performed under a light microscope (Zeiss Axioscope 2+; Zeiss International, Oberkochen, Ostalbkreis, Germany) with a compatible camera, and the Scope Photo software (Scopetek, München, Germany) was used for processing the images. For the measurement, we used at least 15 different cross-sections per internodes. The measured parameters were thick at the epidermis, primary cortex, pith, primary and secondary vascular tissue.

2.3. Statistical Analysis

Before the ANOVA test, Levene's Test for Equality of Variances was performed. The Levene's test for different variables at the six treatments of brown juice (i.e., 0%, 0.5%, 1%, 2.5%, 5%, and 10%) was negative, $p < 0.05$, and then the variances showed homogeneity. Results of the experiments were subjected to one-way (for fresh and dry weight, chlorophyll pigments, protein, MDA, POD, and catalase) and two-way (for root and shoot lengths, root and shoot volumes, and number of leaves) ANOVA by 'SigmaPlot 12.0' software and the means were compared by Duncan's Multiple Range Test [38] at $p < 0.05$.

3. Results

3.1. Characteristics of Brown Juice

3.1.1. Chemical Traits of Brown Juice

The fermentation of brown juice significantly changed its chemical properties (Table 1). Inoculation of fresh brown juice by lactic acid bacteria under anaerobic conditions caused a 13.9% reduction in pH. The degree Brix slightly increased after fermentation as it changed from 7.03 to 7.20. Total phenolic content dropped down after fermentation by almost 33.4%. Moreover, EC of fermented brown juice was 25.2% lower than fresh brown juice. Additionally, the density of brown color, that brown juice has, was reduced as its absorbance at 430 nm was diminished by 35.9%.

Table 1. Physiochemical characteristics of alfalfa brown juice before and after fermentation using lactobacillus.

Parameter	Before	After
pH	4.54 ± 0.03	3.91 ± 0.05
Brix † (%)	7.03 ± 0.02	7.20 ± 0.01
Total phenolic content (μg mL^{-1})	36.5 ± 1.19	24.26 ± 0.55
Electrical conductivity (dS m^{-1})	11.13 ± 0.11	8.47 ± 0.06
Color-absorbance (at 430 nm)	0.594 ± 0.006	0.381 ± 0.004
Lactic acid bacteria (CFU × 10^8 per mL)	11.33 ± 4.04	8.00 ± 4.36
Sugars content (g L^{-1})		
Glucose monomer *H*	21.19 ± 0.64	1.33 ± 0.03
Glucose oligomer *H*	2.80 ± 0.58	Nd ‡
Xylose monomer *Pb*	12.0 ± 0.06	nd
Xylose oligomer *Pb*	1.90 ± 0.03	0.60 ± 0.02
Arabinose monomer *Pb*	nd	0.10 ± 0.01
Arabinose oligomer *Pb*	1.50 ± 1.16	0.64 ± 0.01
Fructose monomer *Pb*	3.70 ± 0.02	0.89 ± 0.01
Fructose oligomer *Pb*	nd	nd
Acids content (g L^{-1})		
Acetic acid *H*	1.5 ± 0.02	10.4 ± 0.03
Lactic acid *H*	5.0 ± 0.25	50.1 ± 0.68
Propionic acid *H*	nd	1.2 ± 0.02

Notes: † Degree Brix = water-soluble sugar content (one degree Brix means 1 g of sucrose in 100 mL aqueous solution); ‡ nd = not detected; sample size ($n = 6$); *H*-samples run on Aminex HPX 87 H column; *Pb*-samples run on Aminex HPX 87 Pb column.

3.1.2. Contents of Sugars and Organic Acids in the Brown Juice

Furthermore, the effect of lactic acid bacteria was not only reflected in the chemical characteristics of brown juice but also was noticed in sugars content. Interestingly, contents of monomer and oligomer forms of glucose, xylose, arabinose, and fructose reduced after fermentation, except arabinose monomer which was below the detected limit in fresh brown juice and became 0.1 g L^{-1} after fermentation; also, no fructose oligomer was detected either in fresh or fermented brown juice samples (Table 1). The highest decrease was found for glucose monomer as it lowered by 16 times in fermented brown juice compared to fresh brown juice. Fructose monomer, also, was four times lower in fermented brown juice, while arabinose oligomer recorded a decrease of 57.3% (Table 1). In contrast to sugars content, organic acids such as acetic, lactic, and propionic acids were considerably increased after fermentation by lactic acid bacteria. The content of lactic acid was 10-fold higher in fermented brown juice, as the highest recorded increase for any measured organic acid, while acetic acid content changed by seven times higher. Propionic acid content was below the detected limit in fresh brown juice; however, after fermentation it increased, recording 1.2 g L^{-1} (Table 1).

3.1.3. Macro- and Microelements Content of Brown Juice

Content of macro-and microelements of brown juice meaningfully changed due to fermentation by lactic acid bacteria (Table 2). Fermentation of brown juice resulted in a substantial reduction in the concentration of N, P, K, and S by 11%, 32%, 38%, and 21%, respectively. Otherwise, the contents of other elements displayed in Table 2 were found to be considerably higher after treating brown juice with lactic acid bacteria under anaerobic conditions. Interestingly, concentrations of Ca, Mg, Mo, Sr, and Ba were increased by 55%, 63%, 36%, 54%, and 109%, respectively. Furthermore, Na, Mn, Fe, Zn, B, and Al contents were 14.5-, 2.0-, 11.0-, 2.5-, 1.5-, and 5.7-fold higher, respectively, in fermented brown juice than fresh brown juice. No Cu was detected in brown juice either fresh or fermented.

Table 2. Content of macro- and microelements (mg L^{-1}) in alfalfa brown juice before and after fermentation.

Elements	Before	After
N	18.24 ± 0.66 †	16.19 ± 0.01
P	286 ± 30	238 ± 5.98
K	6090 ± 571	5276 ± 153
Ca	1270 ± 70	2326 ± 66.58
Mg	379 ± 16	739 ± 24.83
Na	31.03 ± 8.70	452 ± 15.02
S	352 ± 22	425 ± 6.57
Mn	1.61 ± 0.13	4.66 ± 0.09
Mo	0.29 ± 0.12	0.45 ± 0.01
Fe	2.04 ± 0.45	32.20 ± 0.86
Cu	0.06 ± 0.05	nd ‡
Zn	2.03 ± 0.19	8.59 ± 0.21
Sr	5.46 ± 0.24	8.80 ± 0.23
B	3.69 ± 0.33	11.61 ± 0.33
Al	0.24 ± 0.27	1.67 ± 0.13
Ba	0.39 ± 0.02	0.94 ± 0.02

Notes: † Standard deviation; ‡ not detected; sample size (n = 6).

3.2. *Fermented Brown Juice as A Growth Stimulator*

The possible utilization of fermented brown juice as a growth stimulator was evaluated. Celosia seedlings were treated with different doses of fermented brown juice through foliar application.

3.2.1. Growth Dynamic of Celosia

Spraying of Celosia seedlings with fermented brown juice significantly induced the development of stems (Figure 1). The application of brown juice at low concentrations had better effects on plant growth than higher concentrations. Spraying Celosia plants with 0.5% of fermented brown juice resulted in the tallest stem (26.0 cm); however, higher concentrations drastically diminished stem length. For instance, at the rate of 10% fermented brown juice stem length was 15.6 cm (Figure 1A). The root system of Celosia plants responded to fermented brown juice differently to the shoot part. All rates of brown juice resulted in very similar lengths of root systems except the rate of 10% which caused a significant reduction in root systems (16.9 cm). However, the tallest root system was found in control plants sprayed with tap water (Figure 1A).

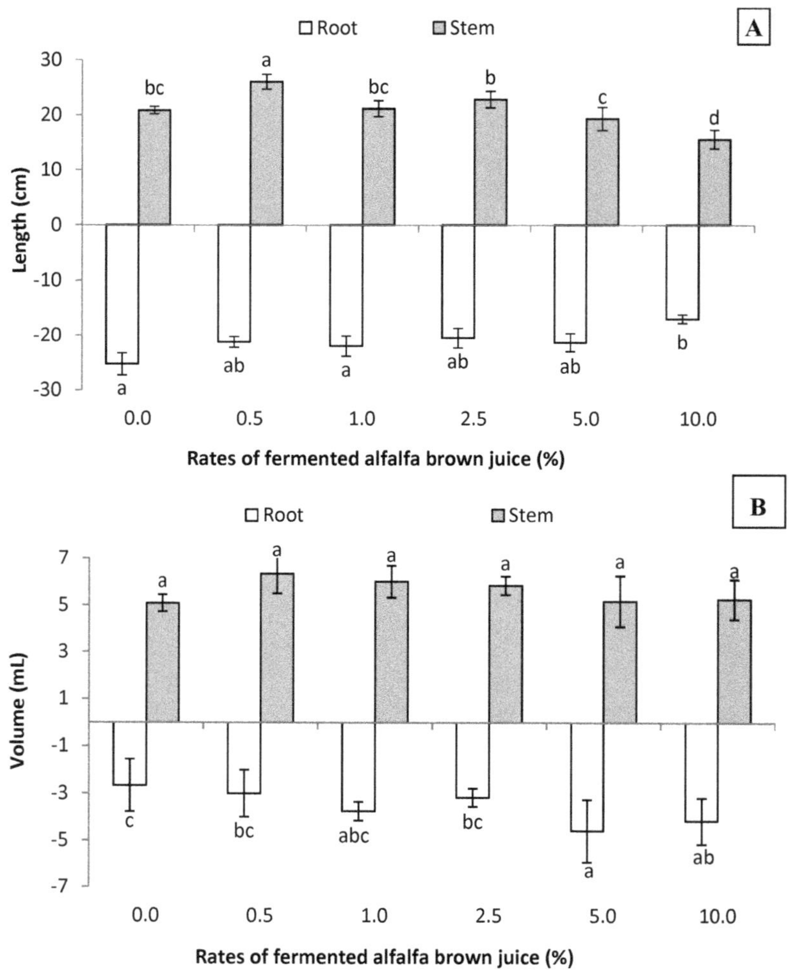

Figure 1. Length (**A**) and volume (**B**) of root and stem systems of Celosia plants fertilized at different rates of fermented alfalfa brown juice applied as a foliar application. Sample size ($n = 6$). Different letters above the same columns show significant differences at the level of $p < 0.05$.

Although, length of shoot and root systems is considered as a good indicator for plant growth and its response to the newly added fertilizers and/or stimulators, alone it does not precisely describe the real status of plant health. Therefore, to have a comprehensive description of the shoot and root systems, their volumes should be also measured. This is very essential particularly to describe the root system and its architecture as shoot parts respond to growth conditions in a proportional way. Concerning stem volume, similar findings as for its length were reported. At lower rate of fermented brown juice (0.5%) the highest volume of stem (6.0 cm^3) was measured while increasing the rate of fermented brown juice gradually and significantly declined the stem volume and lowest volume (2.3 cm^3) was measured for plants sprayed at 10% fermented brown juice (Figure 1B). Results of root volume presented in Figure 1B displayed that although control plants had the tallest root length, its volume was the lowest among all the treatments. This means that control plants had long roots but unbranched ones with few lateral roots. All treated Celosia plants with fermented brown juice showed higher root volume compared to control plants. The highest root volume was noticed at plants sprayed with 5% of fermented brown juice. Additionally, results show that higher rates of fermented brown juice, i.e., 5% and 10% resulted in higher measured root volumes (Figure 1B).

Fresh mass of different Celosia tissues (roots, stems, and leaves) significantly responded to spraying the plants with different rates of fermented brown juice as shown in Figure 2. Fresh mass of roots, stems, and leaves of all plant parts was higher for plants treated with fermented brown juice compared to control plants sprayed with tap water. The highest fresh mass of roots, stems, and leaves was 4.30, 8.87, and 8.49 g plant^{-1}, respectively, that measured at rates of 2.5%, 0.5%, and 5% fermented brown juice, respectively (Figure 2A). Control plants displayed the lowest dry mass of roots, stems, and the number of leaves 0.17, 0.31, and 0.55 g plant^{-1}, respectively; while sprayed plants with 2.5% fermented brown juice showed the highest determined dry mass 0.44, 0.64, and 0.86 g plant^{-1}, for roots, stems, and leaves, respectively (Figure 2B). All rates of fermented brown juice, except 10%, significantly increased the number of leaves per plant (Figure 2C). Applying fermented brown juice at the rate of 10% significantly decreased the number of leaves not only compared to other fermented brown juice rates but also control plants. The highest number of leaves per plant was 18 and was counted for plants treated with 1% fermented brown juice. However, the differences between treatments of 0.5%, 1%, 2.5%, and 5% of fermented brown juice were not significant.

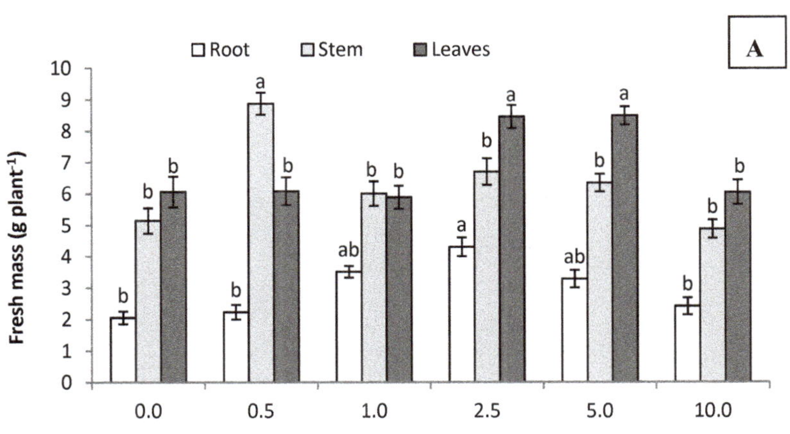

Figure 2. *Cont.*

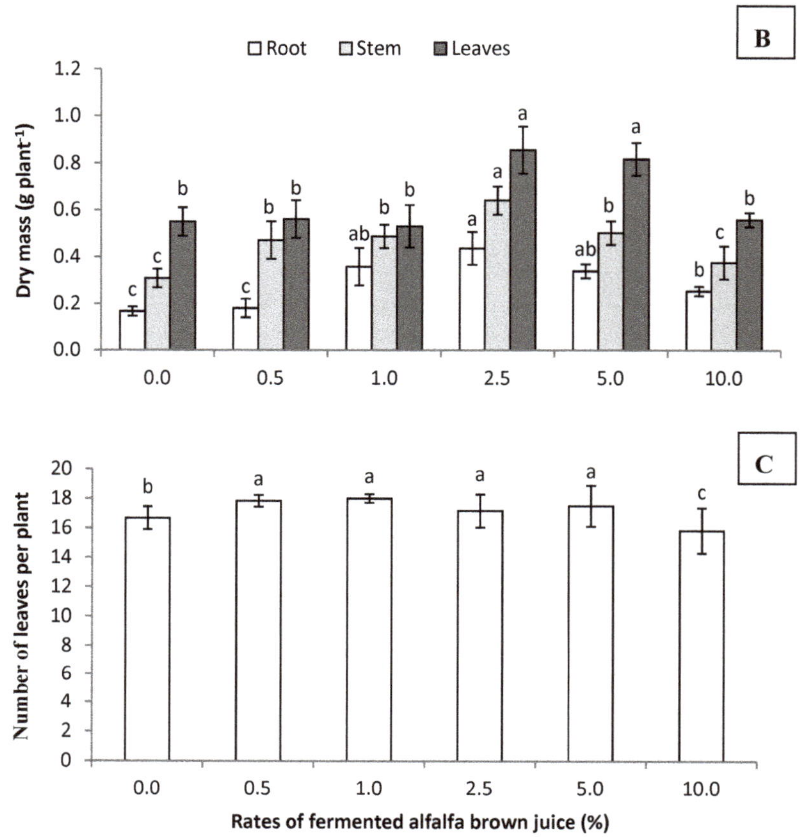

Figure 2. Fresh (**A**) and dry (**B**) masses and the number of leaves (**C**) of different plant tissues (roots, stems, and leaves) of Celosia plants sprayed at different rates of fermented alfalfa brown juice. Sample size ($n = 6$). Different letters above the same columns show significant differences at the level of $p < 0.05$.

3.2.2. Antioxidant Capacity of Celosia Plants Treated with Fermented Brown Juice

Spraying Celosia plants with fermented brown juice significantly induced the activity of catalase (CAT) enzyme in the leaves (Figure 3A). All treated plants had higher activities of CAT enzyme compared to control plants (sprayed with tap water). However, increasing the rate of applied brown juice gradually reduced the CAT activity up to 5%, but this reduction was still higher than the control. Treated Celosia plants at the rate of 10% achieved the highest CAT activity among all treatments (0.290 µmoL H_2O_2 consumed mg^{-1} protein min^{-1}).

Different Celosia plant tissues (i.e., root, stem, and leaf) showed a significant response of peroxidase enzyme activity (POD) to added fermented brown juice (Figure 3B). Higher rates of fermented brown juice above 1.0% resulted in higher POD activity in the root system than both lower rates and control plants. The root POD activity in treatments of 2.5%, 5%, and 10% of fermented brown juice was higher than lower rates and the control; however, no statistically significant differences were calculated among these treatments. Interestingly, applying fermented brown juice at the rate of 1% resulted in the lowest determined activity of POD in the root system among all treatments including the control plants. The activity of POD in the stem tissue of Celosia plants was totally in contrast to POD activity in the root system (Figure 3B). The high rates of fermented brown juice above 1% showed lower POD activity in the stem than low rates (i.e., 0.5% and 1%) and control plants. The lowest POD activity in stems was noticed when plants were sprayed at 10% fermented brown juice, while the highest measured POD activity in the stem was found for plants that received 0.5% fermented brown juice (Figure 3B). Except for treatments of 2.5% and 5% fermented brown juice, all other treatments including control plants showed similar POD activity in leaf tissue without significant differences. The highest leaf POD activity was measured in the leaves of treated plants with 2.5% fermented brown juice, while at the rate of 5% fermented brown juice the lowest leaf POD activity was determined (Figure 3B).

Malondialdehyde (MDA) content in different tissues of Celosia plants was measured as a marker for the degree of lipid peroxidation of unsaturated fatty acids due to oxidative stress. In the root system of Celosia plants, the highest measured value of MDA content was denoted in control plants. All treated plants with fermented brown juice had lower root MDA content than control plants. However, the response of treated plants with fermented brown juice hesitated as no clear trend was seen. The lowest applied rate 1% fermented brown juice showed the lowest root MDA content, while the highest root MDA content was measured in the root system of plants sprayed with 2.5% fermented brown juice (Figure 3C). In contrast to the root system, stem MDA content was found to increase as the rate of fermented brown juice increased up to 5% then reduced at the rate 10% recording the lowest MDA content in stem tissue among all treated plants with fermented brown juice. However, the lowest MDA content in the stem was displayed in control plants. Leaves of control plants showed higher MDA content than plants that received different rates of fermented brown juice. No significant differences were found in the MDA content of leaves of plants sprayed at the rates of 1%, 2.5%, and 5% fermented brown juice (Figure 3C). However, the lowest leaf MDA content was determined in the leaves of plants treated with 0.5% fermented brown juice.

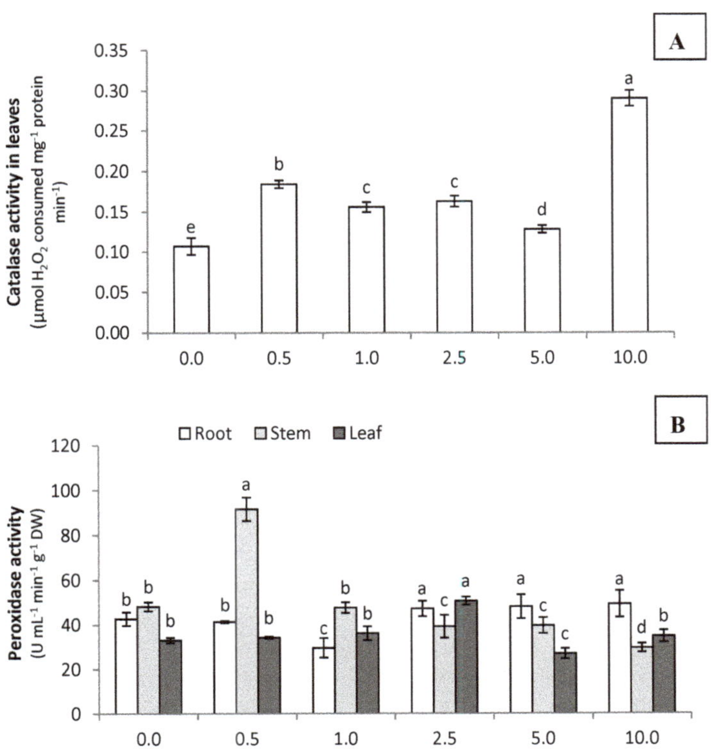

Figure 3. *Cont.*

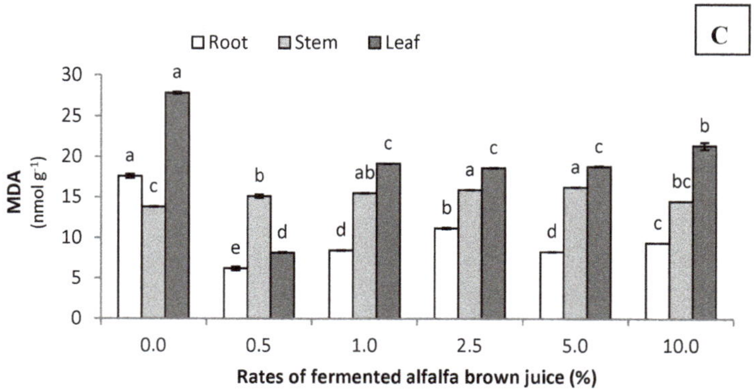

Figure 3. The activity of catalase (**A**) and peroxidase (**B**) and malondialdehyde content (**C**) in different plant tissues (roots, stems, and leaves) of Celosia plants sprayed at different rates of fermented alfalfa brown juice. Sample size ($n = 6$). Different letters above the same columns show significant differences at the level of $p < 0.05$.

3.2.3. Phenolic, Protein, and Photosynthetic Pigments Contents

The results of water-soluble phenol content are depicted in Figure 4A. Different plant tissues of Celosia plants possessed different water-soluble phenol contents as the root system showed the lowest content, while the highest water-soluble phenol content was measured in leaves. The addition of fermented brown juice as a foliar application to Celosia plants significantly affected the water-soluble phenol content in the root system. The highest water-soluble phenol content (46.7 $\mu g\ g^{-1}$) was measured in the root system of plants that received 0.5% fermented brown juice, while, when plants were allowed to grow in the presence of 10% fermented brown juice, the water-soluble phenol content was 12.8 $\mu g\ g^{-1}$ (Figure 4A). The root system of the control plant displayed 18.6 $\mu g\ g^{-1}$ water-soluble phenol content. In stem tissues, water-soluble phenol content in plants treated with fermented brown juice showed lower water-soluble phenol content than control plants. Increasing the rate of fermented brown juice up to 2.5% gradually increased the content of water-soluble phenol in stem tissues, then a linear increase was recorded when rates of fermented brown juice were increased up to 10%. The highest stem water-soluble phenol content (56.8 $\mu g\ g^{-1}$) was measured for control plants (Figure 4A). Except for treatment of 0.5% fermented brown juice, all fermented brown juice rates showed higher water-soluble phenol content in leaf tissues. The highest water-soluble phenol content ($\mu g\ g^{-1}$) was measured in leaves of plants that received 1% fermented brown juice; then a gradual decrease was noticed with increasing the rate of fermented brown juice up to 10% (Figure 4A).

The content of water-soluble protein was higher in leaf tissue followed by the root system, while the lowest content was denoted in stem tissue. Significantly, the application of fermented brown juice improved water-soluble protein content in the root system. Treatments of 1%, 2.5%, and 10% fermented brown juice displayed higher water-soluble protein content than control and treatments of 0.5% and 5% (Figure 4B). The lowest water-soluble protein content (2.18 $mg\ g^{-1}$) was measured in the root system of plants treated with 0.5% of fermented brown juice. Similar results were found in stem tissue of Celosia plants, where all treated plants with fermented brown juice had higher water-soluble protein content than control except treatment of 0.5% fermented brown juice. Although the content of water-soluble protein in leaves was higher than measured in the root system, the trend in which roots and leaves responded to spraying with fermented brown juice was almost the same. Leaf water-soluble protein contents in plants of treatments of 0.5% and 5% were the lowest among all treatments including the control. Other fermented brown juice rates enhanced the water-soluble protein content in leaf tissue over the control plants (Figure 4B).

Significant differences were noticed in a few cases among treatments for chlorophyll pigments content (Figure 4C). Content of *chl a* was reduced gradually with increasing the rate of applied fermented brown juice. Application of fermented brown juice at low rates (i.e., 0.5%) significantly improves the *chl a* content recording the highest value among all other treatments but was similar to control plants. On the other hand, the *chl b* content was found to respond negatively to increasing the rate of applied fermented brown juice as a gradual significant reduction was noticed. Content of total *chl a + b* displayed a similar tendency as it slightly decreased with increasing the rate of fermented brown juice. The low rate of fermented brown juice showed a slightly higher content than the control, but this increase was not significant (Figure 4C). Except treatment of 0.5% fermented brown juice, carotenoids content in all treatments including the control showed similar values as no significant differences were statistically measured. The lowest carotenoids content was determined in leaves of plants that received 0.5% fermented brown juice (Figure 4C).

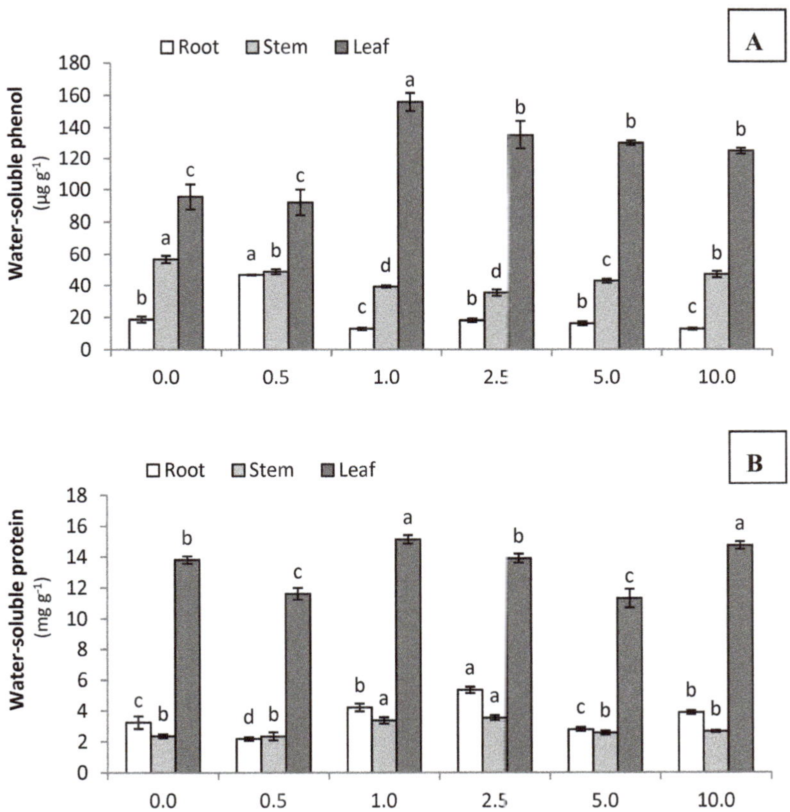

Figure 4. *Cont.*

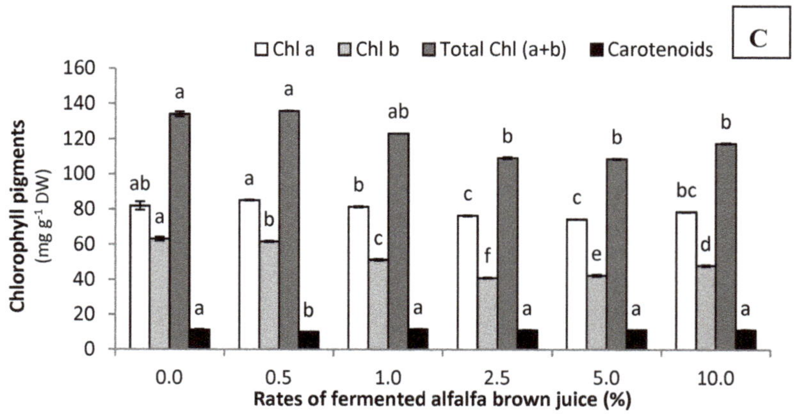

Figure 4. Water-soluble phenol (**A**), water-soluble protein (**B**), and chlorophyll pigments contents (**C**) in different plant tissues of Celosia plants sprayed at different rates of fermented alfalfa brown juice. Sample size ($n = 6$). Different letters above the same columns show significant differences at the level of $p < 0.05$.

3.2.4. Anatomical Features of Celosia Stem after Brown Juice Application

Regarding the cross-sections, 10–15 cm from the apex were analyzed, and the tissue structure was representative of an older Celosia's stem anatomy, with successive cambia [39,40]. Stems were covered by the epidermis (single row); beneath its primer cortex containing angular collenchyma (four to six cells thick) was visible. In the pith primary vascular bundles were located surrounded by a cylinder of anomalous cambium. Secondary and primary vascular tissues were separated by the conjunctive tissue [41]. Both the conjunctive tissue and the inner part of the pith were composed of parenchymatous cells (Figure 5).

There is no fundamental difference in the tissue structure in connection with the treatment, but there are significant differences in the thickness of the tissues, which support the differences that are visible to the naked eye too (e.g., thicker, stiffer stem). All levels of concentration caused a reduction in the thickness of the epidermis, while it was the 1% treatment that caused a reduction to a greater extent. The thickness of the primary cortex reinforced with angular collenchyma was decreased by most treatments, except the 0.5% and 10% treatments, where statistically verified thickening was observed. The proportion of pith involved vascular tissues increased for all treatments. The more concentrated brown juice treatments resulted in significantly thicker primary vascular tissues, except the 1% and the 10% treatments. Growing of the secondary vascular tissue was the highest at the 0.5% treatment, where the new successive cambium formed almost entirely closed xylem and phloem, significantly contributing to the strength of the stem (Table 3).

Table 3. Impact of different concentrations of fermented brown juice on stem tissue of *Celosia argentea* var. *plumosa* (μm) (mean ± SD, $n = 45$).

	Epidermis	Cortex	Pith	Primary Vascular Bundle	Secondary Vascular Tissue
Cont.	31.44 ± 5.28 a	356.14 ± 57.69 ab	1873.30 ± 295.29 b	236.47 ± 79.43 b	221.47 ± 51.79 b
0.5%	30.81 ± 6.72 a	374.10 ± 99.50 ab	1903.79 ± 187.39 b	271.65 ± 73.59 ab	295.59 ±76.55 a
1%	25.83 ± 3.59 b	322.42 ± 61.76 bc	2033.30 ± 205.45 b	242.23 ± 41.22 b	212.56 ± 51.56 b
2.5%	28.50 ± 4.42 ab	261.69 ± 19.64 d	2011.11 ± 198.88 b	303.98 ± 88.94 a	230.18 ± 73.69 b
5%	31.43 ± 3.81 a	339.46 ± 68.14 b	2227.77 ± 310.33 a	348.30 ±122.47 a	256.93 ± 66.78 a
10%	28.69 ± 3.49 ab	392.07 ± 91.05 a	1915.50 ± 209.11 b	310.12 ± 102.47 a	274.80 ± 89.62 a

Notes: Different letters in each column indicate statistically significant differences ($p < 0.05$).

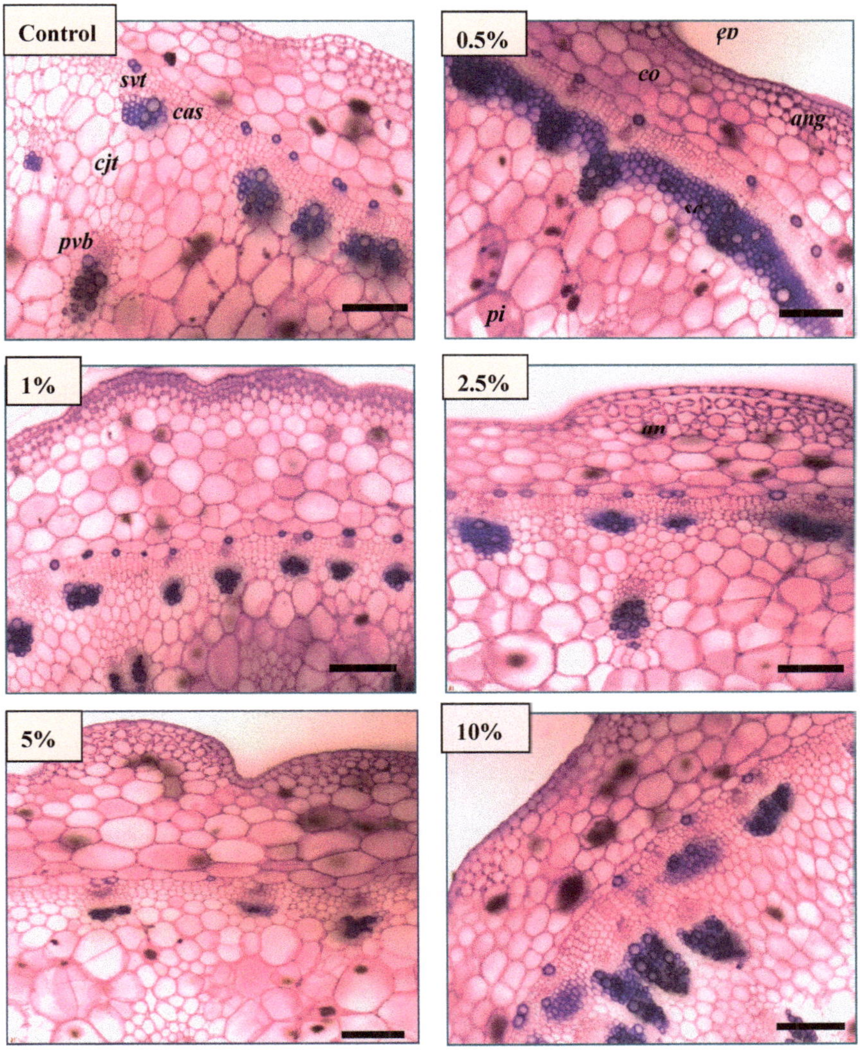

Figure 5. Anatomical sections of *Celosia argentea* var. *plumosa* stem. ***ep*** epidermis, ***co*** cortex, ***ang*** angular collenchyma, ***pi*** pith, ***sc*** successive cylinder (***svt*** secondary vascular tissue), ***ca*** cambium, ***cjt*** conjunctive tissue, ***pvb*** primary vascular bundle after spraying Celosia plants with different rates of fermented brown juice (i.e., control, 0.5%, 1%, 2.5%, 5%, and 10%). *Scale bar is* 200 μm.

To sum up, it can be stated that the brown juice treatments (applied as foliar) influence the proportions of the Celosia stem's tissue. As a result of the treatments, the thinning of the epidermis and the intense growth of the vascular tissues (especially the secondary vascular tissue) can be projected. The growth rate of secondary tissues within the pith is the highest at 0.5% treatment.

4. Discussion

Recently, isolation of protein from plant green leaves has gained increasing attention as an attempt to bridge the gap between protein production and demand due to the dramatic increase in population and the increase in living standards. Brown juice (referred to as deproteinized plant juice or DPJ as well) is a byproduct generated during the coagulation of soluble protein in green juice through thermal treatment. Brown juice has gained less attention than LPC and press cake. It represents nearly 50% of pressed and pulped fresh biomass [9]. Therefore, these huge amounts could be an obstacle facing the acceleration of this approach and its acceptance by both politicians and the public. Disposal of alfalfa brown juice is a serious issue in LPC production due to its high biological oxygen demand (BOD) and carbohydrates content [42]. Due to its richness in free amino acids, peptides, soluble sugars, vitamins, and many macro- and microelements, it can be directed towards animal feeding and production of many chemicals [43]. Additionally, it can be used as a fertilizer, growth stimulator and/or growth medium for microorganisms [42]. Although some pieces of literature have been reporting the possible utilization of brown juice as a ruminant feed [44], few studies have been focusing on brown juice as a fertilizer [42,45].

During our recent experiments on LPC production from alfalfa biomass, it became clear that the storage of the brown juice at room conditions leads to fast spoiling. Therefore, we had to store it below 4 °C. This may be due to its high carbohydrate content, which represents a suitable environment for bacteria to grow [42]. Therefore, converting sugars into organic acids and subsequent decrease in the pH of brown juice through fermentation using lactic acid bacteria seemed to be an ideal solution since fermented brown juice is stable and this facilitates its handling.

Lactic acid bacteria have long been known for their role in the fermentation of carbohydrates. Consequently, it has wide applications in medicine and food processing. Nowadays, lactic acid bacteria have been found to play an important role in agriculture, bioenergy production, and bioremediation of the environment [46]. In the present study we, firstly, aimed to stabilize alfalfa brown juice through reducing its water-soluble sugars content and pH using lactic acid bacteria. Accordingly, sugars content in brown juice was reduced after fermentation, because lactic acid bacteria use sugars as energy and carbon sources [47]. As shown in Table 1, most of the sugars in brown juice were found to be below the quantification limits after fermentation indicating that lactic acid bacteria consumed them. Comparable results have been previously presented by many researchers [48,49]. On the other hand, organic acids (e.g., lactic, acetic, and propionic acids) were increased in fermented brown juice causing a subsequent reduction in pH (Table 1). Novik, et al. [46] reported lactic acid as the main acid produced after fermentation of water-soluble sugars such as glucose and fructose either monomer or oligomer by lactic acid bacteria. Similar findings were cited by Bautista-Trujillo et al. [48], who observed a decrease in pH of maize silage after inoculation by lactic acid bacteria due to the increase in the production of organic acids, mainly lactic and acetic acids. Moreover, they reported an increase of 46.3% in lactic acid content. However, lactic acid content was found to increase by 8-fold after fermentation compared to unfermented brown juice (Table 1). This high increase in lactic acid content may be attributed to the initial low pH of fresh brown juice (4.54), which helps to hydrolyze the oligo- and polysaccharides, therefore they become available for lactic acid bacteria [48]. Additionally, another possible reason for high lactic acid content could be attributed to the high Mn content in brown juice. Cheng et al. [49] stated that applying Mn to Jerusalem artichoke juice enhanced the lactic acid production by lactic acid bacteria up to 12 g L^{-1}. Moreover, Dimitrovski et al. [50] stated that the fermentation of Jerusalem artichoke tuber juice by lactic acid bacteria reduced its pH from 6.5 to 4.7 after 30 h. In the current study, at the end of the fermentation process, the pH of brown juice was 3.91. Lactic acid bacteria significantly reduced the absorbance of brown juice by 35.9%. This result was supported by that previously cited by Kwaw et al. [51]. They studied the effect of different strains of lactic acid bacteria on colorimetric properties of mulberry juice, reporting a 6.9% reduction in the color. They referred this reduction to the increase in content of the monomeric anthocyanin. Although an increase in total phenolic content has been previously reported for fermented mulberry juice [51] and pomegranate juice [52], our results displayed a decrease of 33.4% after inoculation of brown juice by lactic acid bacteria. Except N, P, K, and S other macro- and microelements were higher in fermented alfalfa brown juice (Table 2). The reduction in concentrations of N, P, K, and S could be attributed to the fact that they are essential elements for the growth of lactic acid bacteria. However, similar findings were described by Kim [53], who stated that fermented kale juice had higher elemental composition than unfermented juice. Moreover, he cited significant differences between kale juice fermented by different lactic acid bacterial strains. The increase in the concentration of microelements, in particular, may be due to the increase of brown juice acidity. Although, the concentration of macronutrients (e.g., N, P, and K) was reduced after fermentation, the content of macro- and microelements is still high, and this makes the fermented brown juice a potential growth stimulator. On the whole, these results are supported by earlier findings of Ream et al. [45]. They reported that alfalfa brown juice contains a relatively high content of N and K, in addition to small amounts of P, Ca, Mg, and other microelements.

In the present study, fermented alfalfa brown juice as a growth stimulator was evaluated using the Celosia plant as a model. Brown juice was applied at different rates by foliar application. The foliar application of fermented brown juice showed a significant potential on the development of Celosia seedlings in comparison with control. Noticeably, increasing the application rate of brown juice sprayed on Celosia seedlings led to a considerable reduction in shoot parts, particularly the stem length. From a horticultural point of view, this seems to be a good result since the target is the flower not the vegetative growth of Celosia. Application of brown juice at low rates such as 0.5% and 1.0% enhanced the growth and resulted in high values of stem length, the volume of stem and root, fresh masses of stem and root, and number of leaves. Shorter but more branched root systems were observed when Celosia seedlings were sprayed with brown juice (Figure 2). This phenomenon is supported by data of length and the volume of roots (Figure 1). The beneficial effect of alfalfa brown juice could be attributed to its high content of macro- (i.e., N, P, K, Ca, and Mg) and microelements (i.e., S, Mn, Fe, Cu, Zn, and Mo); all in phyto-available forms (Table 2). Similarly, Ream et al. [45] observed that using brown juice as a fertilizer added at an annual rate of 1.25 cm induced the growth and yield of alfalfa, corn, and bromegrass; while, at the higher rate (2.5 cm) a reduction in yield and plant damage were noticed in all crops. However, they referred to the damage in plant growth caused by high rates of brown juice to unknown reasons; moreover, they considered it a not serious problem since the added amount of brown juice can be controlled. These results are supported by findings of Reddy et al. [42], who earlier stated that the application of alfalfa brown juice at low rates enhanced germination and growth of cowpea, mung bean, and groundnut; while high rates inhibited the germination and reduced the plant growth. They reported that alfalfa brown juice can be used as a fertilizer if it would be added at a lower level than 10%. Additionally, they could not give a reason for such damaging effects of high rates of brown juice, except what previously was mentioned by Pirie [54], who stated that alfalfa brown juice contains some phytotoxic organic compounds.

In our experiment, the reduction in plant growth of treated Celosia plants at high rates of fermented brown juice can be explained by high EC value and low pH of brown juice solutions. Increasing the rates of brown juice caused a gradual increase in EC and a decrease in pH as shown in Table 1. At treatment of 0.5% brown juice, EC (dS m^{-1}) and pH were 0.12 and 4.21, respectively; while, at the highest applied rate 10% they were 1.99 and 4.38, respectively. Low pH is not favorable for the development of plants; it reduces photosynthesis due to a reduction in stomatal conductance [55]. This might explain why we found diminished growth of Celosia plants treated at high rates of brown juice. However, in the current study, the lowest pH was 4.38 when Celosia plants received 10% fermented brown juice. Long et al. [56] had earlier reported a reduction in citrus growth below pH 4, while higher pH did not inhibit the growth and seedlings reached their maximum growth at pH 5. The reduction in growth may be attributed to H^+-toxicity which damages leaves. Absorption of nutrients applied as foliar application depends on the pH of the solution. Extreme pH (below 2 and above 12) was reported to burn the leaves. Moreover, some elements prefer different pH values for their optimum absorption by plant leaves.

Antioxidant enzymes such as CAT and POD are among the most important antioxidant enzymes which play a vital role in scavenging reactive oxygen species generated in cells due to different biotic and abiotic stresses [57]. Thus, enhancing the activity of these enzymes is considered an important step in improving the plants' tolerance to different kinds of stress [58,59]. The results abstracted from this research showed that the application of alfalfa brown juice after fermentation by lactic acid bacteria significantly increased the activity of CAT and POD in different Celosia tissues. However, the low rates of brown juice seemed to be more effective than higher ones, as a reduction in the activities was noticed. These results are confirmed by results of MDA, as treated plants with fermented brown juice had lower MDA content than control plants (untreated plants) regardless of the type of plant tissues. These results demonstrate that fermented brown juice can potentially be exploited as a growth stimulator, particularly at low rates. Besides, fermented brown juice had a significant effect on water-soluble phenol and protein contents, as they were higher in treated plants in comparison to control ones. The high rates of brown juice were found to reduce the photosynthetic pigment content. This could be attributed to low pH at high rates of fermented brown juice. This result was in accordance with that cited by Solati et al. [60], who reported a decrease in chlorophyll content due to low pH.

Brown juice could be very useful as a soil fertilizer/conditioner particularly in alkaline soils and/or sandy soil due to its rich composition in macro- and microelements and sugars. These could induce the microbial growth in soil increasing soil fertility. Additionally, sugars play an important role in soil stabilization through maintaining soil aggregation, which subsequently leads to better water holding capacity [42,61]. While delivering deeper insights into the possible use of alfalfa brown juice as a growth stimulator and trying to precisely determine the most effective rate and application method, there are many issues that should be addressed in the future [62]. Results, undoubtedly, suggest that brown juice tolerance can be plant species dependent; therefore, more studies on different plant species at different rates of brown juice are crucially needed. Additionally, phytotoxicity of brown juice should be the focus of future studies.

Table 4. pH and electrical conductivity (EC, dS m^{-1}) values of alfalfa brown juice solutions before and after fermentation at the beginning of the experiment.

Rates of Brown Juice (%)	pH		EC	
	Before	*After*	*Before*	*After*
0.5	4.65 ± 0.02 †	4.21 ± 0.07	0.15 ± 0.01	0.12 ± 0.03
1.0	4.67 ± 0.02	4.16 ± 0.00	0.28 ± 0.01	0.54 ± 0.58
2.5	4.68 ± 0.01	4.16 ± 0.01	0.67 ± 0.07	0.46 ± 0.02
5.0	4.72 ± 0.01	4.18 ± 0.00	1.20 ± 0.04	0.81 ± 0.05
10.0	4.72 ± 0.01	4.19 ± 0.00	2.26 ± 0.04	1.44 ± 0.12

Notes: † Standard deviation.

5. Conclusions

The present study highlights the possible use of alfalfa brown juice as a growth stimulator. Brown juice is a serious problem in LPC production, where its disposal represents a threat to the environment due to its high content of water-soluble sugars as well as macro- and microelements. Fermentation of brown juice using lactic acid bacteria significantly improved its nutritional value and stability, because these bacteria—as our data showed—produce a significant amount of organic acids i.e., lactic, acetic, and propionic acids through their metabolism making the nutrients more available and the pH of row material (brown juice) lower, thereby stabilizing it. Most water-soluble sugars were under the detectable level after fermentation as the bacteria used them as carbon source. Moreover, the concentration of nutrients increased—showing the effect of bacteria for nutrient availability—after fermentation except N, P, K, and S showed a slight decrease. In this study, treating *Celosia argentea*, a valuable ornamental species with significant food and medical uses, with low rates of fermented brown juice through foliar application significantly improved the growth, as all of the vegetative parameters such as stem and root length, shoot and root volume, fresh mass of stem and root, and the number of leaves increased. The brown juice treatments in low (0.5%) concentration caused positive changes in histological parameters, in the growth rate of secondary tissues. Additionally, fermented brown juice showed a considerable impact on the antioxidant capacity of Celosia plants, as CAT and

POD activities increased while MDA content decreased. Moreover, both water-soluble phenol and protein were found to increase in treated plants with fermented brown juice compared to the control showing the beneficial effect of lactic acid fermentation and chemical properties of brown juice. These results conclude and state the potential use of fermented alfalfa brown juice as a sustainable growth stimulator for crops with a particular interest in horticultural crops. Our data regarding the chemical and microbiological properties of brown juice and the effects (listed plant responses) it triggered confirm the scientific investigations where plant growth-promoting properties of lactic acid bacteria contribute greatly to the maintenance of the health of plants (also strengthening disease resistance) and by consuming these plants, they are also beneficial in human digestive processes. It should be noted that the sample size number was modest in our study that is why it was difficult to draw strong far-going conclusions, however preliminary conclusions support the fact that phytomicrobiome engineering can be a promising strategy for sustainable agriculture, but the data available is limited to understand properly these complex symbiotic relationships. Therefore, examination of fermented alfalfa brown juice's effect on physiological, biochemical, and anatomical parameters of other horticultural and agricultural crops is in progress.

Author Contributions: Conceptualization, N.B., S.K. and M.G.F.; methodology, D.B., N.E. and I.O.T.; software, D.B.; validation, D.B. and I.O.T.; formal analysis, N.A. and S.K.; investigation, M.R. and C.F.; resources, H.E.-R. data curation, T.A.; writing—original draft preparation, N.B. and D.B.; writing—review and editing, N.B., T.A., H.E.-R. and D.B.; visualization, S.K.; supervision, M.G.F.; project administration, N.B.; funding acquisition, N.B. All authors have read and agreed to the published version of the manuscript.

Funding: This research received no external funding.

Acknowledgments: The research was financed by the Higher Education Institutional Excellence Programme (NKFIH-1150-6/2019) of the Ministry of Innovation and Technology in Hungary, within the framework of the Biotechnology thematic programme of the University of Debrecen."; "Complex Rural Economic and Sustainable Development, Elaboration of its Service Networks in the Carpathian Basin (Project ID: EFOP-3.6.2-16-2017-00001, Hungary)" research project. Also, support was given by Tempus Public Foundation (TPF), Hungary and Széchenyi 2020 under the GINOP-2.2.1-15-2017-00051.

Conflicts of Interest: The authors declare no conflict of interest.

References

1. Tripathi, A.D.; Mishra, R.; Maurya, K.K.; Singh, R.B.; Wilson, D.W. Estimates for world population and global food availability for global health. In *The Role of Functional Food Security in Global Health*; Academic Press: Cambridge, MA, USA.
2. Alexandratos, N.; Bruinsma, J. *World Agriculture towards 2030/2050: The 2012 Revision*; ESA Working Paper No. 12-03; FAO: Rome, Italy, 2012.
3. Henchion, M.; Hayes, M.; Mullen, A.M.; Fenelon, M.; Tiwari, B. Future Protein Supply and Demand: Strategies and Factors Influencing a Sustainable Equilibrium. *Foods* **2017**, *6*, 53. [CrossRef] [PubMed]
4. Boer, J.; Aiking, H. Strategies towards healthy and sustainable protein consumption: A transition framework at the levels of diets, dishes, and dish ingredients. *Food Qual. Prefer.* **2019**, *73*, 171–181. [CrossRef]
5. Tenorio, A.T.; Kyriakopoulou, K.E.; Suarez-Garcia, E.; van den Berg, C.; van der Goot, A.J. Understanding differences in protein fractionation from conventional crops, and herbaceous and aquatic biomass—Consequences for industrial use. *Trends Food Sci. Technol.* **2018**, *71*, 235–245. [CrossRef]
6. Kromus, S.; Wachter, B.; Koschuh, W.; Mandl, M.; Krotscheck, C.; Narodoslawsky, M. The green biorefinery Austria-development of an integrated system for green biomass utilization. *Chem. Biochem. Eng. Q.* **2004**, *18*, 7–12.
7. Zhang, W.; Grimi, N.; Jaffrin, M.Y.; Ding, L.; Tang, B. A short review on the research progress in alfalfa leaf protein separation technology. *J. Chem. Technol. Biotechnol.* **2017**, *2*, 2894–2900.
8. Tenorio, A.T.; Gieteling, J.; de Jong, G.A.H.; Boom, R.M.; van der Goot, A.J. Recovery of protein from green leaves: Overview of crucial steps for utilisation. *Food Chem.* **2016**, *203*, 402–408. [CrossRef]
9. Manwatkar, V.G.; Gogle, D.P. The Effect of Deproteinised Juice (DPJ) on Seed Germination and Seedling Growth of Different Plants (by Paper Towel Method). *Int. J. Life Sci.* **2014**, *2*, 65–68.

10. Cheyniera, V.; Comte, G.; Davies, K.M.; Lattanzio, V.; Martens, S. Plant phenolics: Recent advances on their biosynthesis, genetics, and ecophysiology. *Plant. Physiol. Biochem.* **2013**, *72*, 1–20. [CrossRef]
11. Salah, E.H. *Sustainability of Agricultural and Rural Waste Management*, 1st ed.; Cradle-to-Cradle for Sustainable Development eBook; Academic Press: New York, NY, USA, 2007; 424p, ISBN 9780080550145.
12. Popp, J.; Lakner, Z.; Harangi-Rakos, M.; Fari, M. The effect of bioenergy expansion: Food, energy, and environment. *Renew. Sustain. Energy Rev.* **2014**, *32*, 559–578. [CrossRef]
13. Obi, F.O.; Ugwuishiwu, B.O.; Nwakaire, J.N. Agricultural waste concept, generation, utilization and management. *Niger. J. Technol.* **2016**, *35*, 957–964. [CrossRef]
14. Grela, E.R.; Pietrzak, K. Production technology, chemical composition and use of alfalfa protein-xanthophyll concentrate as dietary supplement. *J. Food Process. Technol.* **2014**, *5*, 10.
15. Pirie, N.W. *Leaf Protein and its By-products in Human and Animal Nutrition*; Cambridge University Press: Cambridge, UK, 1987; p. 209.
16. Iliyas, S. Effect of different flours on α -amylase production. *Int. J. Bioassays* **2014**, *3*, 3011–3015.
17. Jadhav, R.K.; Chhaya, G. Influence of deproteinised foliage fluid liquid biofertilizer on nitrate reductase activity of Eleusine coracana plants. *Curr. Bot.* **2018**, *9*, 33–36. [CrossRef]
18. Zanin, V. A New Nutritional Idea for Man: Lucerne Leaf Concentrate. APEF, Association Pour la Promotion des Extraits Foliaires en Nutrition. 1998. Available online: http://www.nutrition-luzerne.org/anglais/pdf/EtudeZaninenglish.pdf (accessed on 29 October 2019).
19. Thomsen, M.H.; Bech, D.; Kiel, P. Manufacturing of Stabilised Brown Juice for L-lysine production—From University Lab Scale over Pilot Scale to Industrial Production. *Chem. Biochem. Eng. Q.* **2003**, *18*, 37–46.
20. Shende, G.C.; Gogle, D.P. To study the effect of various concentration of deproteinized leaf juice (DPJ) of selected plants on growth of Aspergillus niger. *Int. J. Life Sci.* **2016**, *6*, 186–188.
21. Jadhav, R.K.; Kadam, R.; Joshi, D. Physiology of Gramineae Crops by Deproteinised Foliage Extract and its Influence on Photosynthetic Chemistry. *J. Phytochem.* **2019**, *110*, 149–163.
22. Hegyi, F. Examination of Lactobacilli by Advanced Colorimetric Method. In Hungarian. Laktobacilluszok Vizsgálata továbbfejlesztett Kolorimetriás Módszerrel. Ph.D. Thesis, Corvinus University, Budapest, Hungary, 2014; pp. 8–93.
23. Lamont, J.R.; Wilkins, O.; Bywater-Ekegard, M.; Smith, D.L. From yogurt to yield: Potential applications of lactic acid bacteria in plant production. *Soil Biol. Biochem.* **2017**, *111*, 1–9. [CrossRef]
24. Smith, D.L.; Subramanian, S.; Lamont, J.R.; Bywater-Ekeg€ard, M. Signaling in the phytomicrobiome: Breadth and potential. *Front. Plant Sci.* **2015**, *6*, 709. [CrossRef]
25. Smith, D.L.; Praslickova, D.; Ilangumaran, G. Inter-organismal signaling and management of the phytomicrobiome. *Front. Plant. Sci.* **2015**, *6*, 722. [CrossRef]
26. Tang, Y.; Xin, H.-L.; Guo, M.-L. Review on research of the phytochemistry and pharmacological activities of Celosia argentea. *Rev. Bras. Farmacogn.* **2016**, *26*, 787–796. [CrossRef]
27. Gupta, S.; Lakshmi, A.J.; Manjunath, M.N.; Prakash, J. Analysis of nutrient and antinutrient content of underutilized green leafy vegetables. *LWT Food Sci. Technol.* **2005**, *38*, 339–345. [CrossRef]
28. Fári, M.G.; Domokos-Szabolcsy, É. Method for Producing Plant Protein Coagulum. Hun Patent WO/2019/150144, 8 August 2019.
29. Gibbs, P.A. Novel uses for lactic acid fermentation in food preservation. *J. Appl. Bacteriol.* **1987**, *63*, 51S–58S. [CrossRef]
30. Kovács, B.; Győri, Z.; Prokisch, J.; Loch, J.; Dániel, P. A study of plant sample preparation and inductively coupled plasma emission spectrometry parameters. *Commun. Soil Sci. Plant Anal.* **1996**, *27*, 1177–1198. [CrossRef]
31. Boór, A.; Bélafiné Bakó, K. Determination of antioxidant content in sloe (*Prunus spinosa* L.) and dogwood (*Cornus mas* L.) fruits. In Proceedings of the Conference of Chemical Engineering '2010, Singapore, 26–28 February 2010; pp. 55–58. (in Hungarian).
32. Sparks, D.L.; Page, A.L.; Helmke, P.A.; Loppert, R.H.; Soltanpour, P.N.; Tabatabai, M.A.; Johnston, C.T.; Summner, M.E. *Methods of Soil Analysis: Chemical Methods*; Part 3; ASA and SSSA: Madison, WI, USA, 1996.
33. Bradford, M. A rapid and sensitive method for the quantitation of microgram quantities of protein utilizing the principle of protein-dye binding. *Anal. Biochem.* **1979**, *72*, 248–254. [CrossRef]

34. Roxas, V.P.; Smith, R.K.; Allen, E.R.; Allen, R.D. Overexpression of glutathione S-transferase/glutathione peroxidase enhances the growth of transgenic tobacco seedlings during stress. *Nat. Biotechnol.* **1997**, *15*, 988–991. [CrossRef] [PubMed]
35. Woodbury, W.; Spencer, A.K.; Stahmann, M.A. An improved procedure using ferricyanide for detecting catalase isozymes. *Anal. Biochem.* **1971**, *44*, 301–305. [CrossRef]
36. Zhang, Z.; Huang, R. Analysis of Malondialdehyde, Chlorophyll Proline, Soluble Sugar, and Glutathione Content in Arabidopsis seedling. *Bio-Protocol* **2013**, *3*, 1–8. [CrossRef]
37. Porra, R.J.; Thompson, W.A.; Kriedemann, P.E. Determination of accurate extinction coefficients and simultaneous equations for assaying chlorophylls a and b extracted with four different solvents: Verification of the concentration of chlorophyll standards by atomic absorption spectroscopy. *Biochim. Biophys. Acta* **1989**, *975*, 384–394. [CrossRef]
38. Duncan, D.B. Multiple range and multiple F-tests. *Biometric* **1955**, *11*, 1–42. [CrossRef]
39. Balfour, E. Anomalous secondary thickening in Chenopodiaceae, Nyctaginaceae and Amaranthaceae. *Phytomorphology* **1965**, *15*, 111–122.
40. Myśkow, E.; Gola, E.; Tulik, M. Continuity of procambium and anomalous cambium during formation of successive cambia in Celosia argentea. *J. Plant. Growth Regul.* **2019**, *38*, 1458–1466. [CrossRef]
41. Carlquist, S. Successive cambia revisited: Ontogeny, histology, diversity, and functional significance. *J. Torrey Bot. Soc.* **2007**, *134*, 301–332. [CrossRef]
42. Reddy, G.U.; Deshmukh, V.R.; Joshi, R.N.; Kayama, R. Utilization of alfalfa (*Mediacago sativa* L.) whey as a fertilizer in irrigation. *J. Jpn. Grassl. Sci.* **1987**, *33*, 32–37.
43. Monteros, J.M.; Bouton, H.J. The future of alfalfa and forage crops. In Proceedings of the Western Alfalfa & Forage Conference, Reno, Nevada, 2–4 December 2009.
44. Martin, N.P. Alfalfa: Forage crop of the future. In *Proceedings of the 28th Kentucky Alfalfa Conference*; Lacefield, G., Forsythe, C., Eds.; CaveCity Convention Center: Cave, KY, USA, 2008; pp. 17–26.
45. Ream, H.W.; Smith, D.; Walgenbach, R. Effects of deproteinized alfalfa juice applied to alfalfa-bromegrass, bromegrass, and corn. *Agron. J.* **1977**, *69*, 685–689. [CrossRef]
46. Novik, G.; Olga Meerovskaya, O.; Savich, V. Waste Degradation and Utilization by Lactic Acid Bacteria: Use of Lactic Acid Bacteria in Production of Food Additives, Bioenergy and Biogas. In *Food Additives*; Karunaratne, D.N., Pamunuwa, G., Eds.; IntechOpen: Rijeka, Croatia, 2017. [CrossRef]
47. Zubaidah, E.; Akhadiana, W. Comparative study of inulin extracts from dahlia, yam, and gembili tubers as prebiotic. *Food Nutr. Sci.* **2013**, *4*, 8–12. [CrossRef]
48. Bautista-Trujillo, G.U.; Cobos, M.A.; Ventura-Canseco, L.M.C.; Ayora-Talavera, T.; Abud-Archila, M.; Oliva-Llaven, M.A.; Dendooven, L.; Gutierrez-Miceli, F.A. Effect of Sugarcane Molasses and Whey on Silage Quality of Maize. *Asian J. Crop Sci.* **2009**, *1*, 34–39.
49. Cheng, X.; Dong, Y.; Su, P.; Xiao, X. Improvement of the fermentative activity of lactic acid bacteria starter culture by the addition of Mn. *Appl. Biochem. Biotechnol.* **2014**, *174*, 1752–1760. [CrossRef]
50. Dimitrovski, D.; Velickova, E.; Dimitrovska, M.; Langerholc, T.; Winkelhausen, E. Synbiotic functional drink from Jerusalem artichoke juice fermented by probiotic Lactobacillus plantarum PCS26. *J. Food Sci. Technol.* **2016**, *53*, 766–774. [CrossRef]
51. Kwaw, E.; Ma, Y.; Tchabo, W.; Apaliya, M.T.; Wu, M.; Sackey, A.S.; Xiao, L.; Tahir, H.E. Effect of lactobacillus strains on phenolic profile, color attributes and antioxidant activities of lactic-acid fermented mulberry juice. *Food Chem.* **2018**, *1*, 148–154. [CrossRef]
52. Valero-Cases, E.; Nuncio-Jáuregui, N.; Frutos, M.J. Influence of fermentation with different lactic acid bacteria and in vitro digestion on the biotransformation of phenolic compounds in fermented pomegranate juices. *J. Agric. Food Chem.* **2017**, *65*, 6488–6496. [CrossRef]
53. Kim, S.Y. Production of fermented kale juices with Lactobacillus strains and nutritional composition. *Prev. Nutr. Food Sci.* **2017**, *22*, 231–236. [PubMed]
54. Pirie, N.W. *Leaf Protein and Other Aspects of Fodder Fractionation*; Cambridge University Press: London, UK, 1978; pp. 118–123.
55. Anugoolprasert, O.; Kinoshita, S.; Naito, H.; Shimizu, M.; Ehara, H. Effect of low pH on the growth, physiological characteristics and nutrient absorption of sago palm in a hydroponic system. *Plant Prod. Sci.* **2012**, *15*, 125–131. [CrossRef]

56. Long, A.; Zhang, J.; Yang, L.T.; Ye, X.; Lai, N.W.; Tan, L.L.; Lin, D.; Chen, L.S. Effects of low pH on photosynthesis, related physiological parameters, and nutrient profiles of citrus. *Front. Plant Sci.* **2017**, *8*, 185. [CrossRef] [PubMed]
57. Anjum, N.A. Book review: Oxidative damage to plants-antioxidant networks and signaling. *Front. Plant. Sci.* **2015**, *6*, 452. [CrossRef]
58. Szollosi, R. Superoxide dismutase (SOD) and abiotic stress tolerance in plants: An overview. In *Oxidative Damage to Plants*; Academic Press: New York, NY, USA, 2014; pp. 89–129.
59. Santamaría-Fernández, M.; Karkov Ytting, N.; Lübeck, M. Influence of the development stage of perennial forage crops for the recovery yields of extractable proteins using lactic acid fermentation. *J. Clean. Prod.* **2019**, *218*, 1055–1064. [CrossRef]
60. Solati, Z.; Jørgensen, U.; Eriksen, J.; Søegaard, K. Dry matter yield, chemical composition and estimated extractable protein of legume and grass species during the spring growth. *J. Sci. Food Agric.* **2017**, *97*, 3958–3966. [CrossRef] [PubMed]
61. Kovács, Z.; Fári, M.G.; Kaszás, L.; Koroknai, J.Á.; Domokosné-Szabolcsy, É. Obtention of functional leaf protein concentrate (F-LPC) based on Ereky-process: Historical survey and preliminary results. *New Biotechnol.* **2018**, *44*, s100–s101. [CrossRef]
62. Kamm, B.; Hille, C.; Schönicke, P. Green biorefinery demonstration plant in Havelland (Germany). *Biofuels Bioprod. Biorefin.* **2010**, *6*, 253–262. [CrossRef]

agronomy

MDPI

Article

Nitrogen Use and Uptake Efficiency and Crop Performance of Baby Spinach (*Spinacia oleracea* L.) and Lamb's Lettuce (*Valerianella locusta* L.) Grown under Variable Sub-Optimal N Regimes Combined with Plant-Based Biostimulant Application

Ida Di Mola [1,*], Eugenio Cozzolino [2], Lucia Ottaiano [1], Sabrina Nocerino [1], Youssef Rouphael [1], Giuseppe Colla [3], Christophe El-Nakhel [1] and Mauro Mori [1]

[1] Department of Agricultural Sciences, University of Naples Federico II, 80055 Portici, Italy; lucyottaiano@gmail.com (L.O.); sabrinanocerino@libero.it (S.N.); youssef.rouphael@unina.it (Y.R.); nakhel_christophe@hotmail.com (C.E.-N.); mori@unina.it (M.M.)

[2] Council for Agricultural Research and Economics (CREA)–Research Center for Cereal and Industrial Crops, 81100 Caserta, Italy; eugenio.cozzolino@crea.gov.it

[3] Department of Agriculture and Forest Sciences, University of Tuscia, 01100 Viterbo, Italy; giucolla@unitus.it

* Correspondence: ida.dimola@unina.it

Received: 14 January 2020; Accepted: 12 February 2020; Published: 15 February 2020

Abstract: An optimized nitrogen (N) fertilization may have a positive effect on leafy vegetables by increasing growth, yield and nutrient content of plants. Nevertheless, crop performance must be coupled with an increase in Nitrogen Use Efficiency (NUE) in order to limit external N inputs and to avoid N surpluses associated with environmental and health problems. The aim of the current study was to assess the effects of a legume-derived plant hydrolysates (LDPH; Trainer®) and N fertilization levels (0, 2.25 and 4.5 g N m^{-2} for spinach and 0, 2.5 and 5.0 g N m^{-2} for lamb's lettuce; N0%, N50%, N100%, respectively) on agronomical, biochemical, qualitative responses and NUE of these two important greenhouse leafy vegetables. Spinach and lamb's lettuce were sprayed four times during the growing period (at a concentration of 4 mL L^{-1} of LDPH). In baby spinach, the LDPH application elicited a significant increase at the three levels of N fertilization: +16.8%, +14.2%, and 39.4% at 0, 2.25 and 4.5 g N m^{-2}, respectively. Interestingly, in lamb's lettuce, the N50% plants treated with LDPH reached similar values of marketable yield in comparison to treated and non-treated plants under N100% conditions. The presumed mechanism involved in the enhancement of yield response in the two leafy greens could be associated to a better activity of the photosystem II (higher SPAD index), biochemical (higher content of chlorophyll a, b and total) and leaf nitrate status. The foliar application of LDPH produced a major fortification in lipophilic and hydrophilic antioxidant activities (+11.6 and 6.3% for spinach and lamb's lettuce, respectively). The biostimulant application also improved N-use efficiency and N-uptake efficiency compared to untreated plants: +17.8% and +18.8%, and +50% and +73.3%, for spinach and lamb's lettuce, respectively.

Keywords: N fertilization; nitrogen use efficiency; antioxidant activity; leaf quality; protein hydrolysate; *Spinacia oleracea* L.; sustainable agriculture; *Valerianella locusta* L.

1. Introduction

Chemical fertilizers, especially nitrogen (N), are basically the main input for boosting yield and concomitantly one of the most expensive inputs in terms of economics and environment. Many crops require high amounts of this element to maximize yield [1], but N fertilization requires a particular care because it is involved in many environmental and health risks [2]. The main environmental impacts of N can be summarized in the contamination of surface and groundwater resources and greenhouse gases emissions [3,4]. The effects on human health strongly depend on the accumulation of nitrate in edible plant tissue; when nitrate is reduced to nitrite in human body it can cause methemoglobinemia, which is dangerous to children [5–7]. Moreover, nitrite can also react with several chemical compounds (amines and amides), producing N-nitrous compounds, known as probably carcinogenic to humans [7–9].

On the other hand, it is certainly necessary to adapt the correct management of N fertilization through a balanced application of the elements in order to reach the right dose, nevertheless by choosing the convenient chemical form and application time. Moreover, another possible perspective is to raise the nitrogen use efficiency (NUE) that is linked to the capacity of plants to uptake nutrients, nevertheless to their systems of transport, storage and mobilization and to the N loss into the environment [10]. NUE is expressed as the harvestable yield per the amount of available N in the soil or per N supply [11–13].

In recent years, the approach to improve NUE, passed through biotechnology and plant breeding strategies, but currently it is necessary to evaluate alternative means, which are environmentally friendly, such as the use of plant biostimulants. These products can be used to complement fertilizers in order to reduce the inputs and increase the NUE [14]. They act in several ways: on plant growth, physiology, carbon and nitrogen metabolism, productivity, product quality and tolerance to abiotic stress [14]. Moreover, some studies found that plant biostimulants, particularly commercial legume-derived proteins, have a great potential to reduce nitrate accumulation in the leaves of some green leafy vegetables [15]. It is of a major result because these crops have the genetic predisposition to greatly accumulate nitrate in their leaves [16]. It is known that the different crops ability of nitrate accumulation can depend on different localization and activity of nitrate reductase (NR) [17,18], but also on unbalanced relationship between nitrate uptake and NR activity, as well as the different capacity of uptake, translocation and accumulation of plants [16]. Moreover, this behavior is worsened by specific environmental conditions, where nitrate accumulation increases at low solar radiation [19–21].

In addition, the cultivation in protected environment causes a similar effect, because the plastic film cover reduces the solar radiation transmission. Likewise, the photoperiod and growing period affect nitrate accumulation; in fact, both conditions are matched to conditions of low solar radiation.

Green leafy vegetables play a key part in the economical market of many countries, both in the Mediterranean area and Nord-Europe, because they are widely used in ready-to-eat salads. In addition to typical leafy greens such as lettuce and rocket, also spinach and lamb's lettuce are largely spreading. Italy is a leading country in the European production of green leafy vegetables destined for the ready-to-eat market, with more than 150 kilotons harvested per year in protected conditions [22,23]. Among these crops, spinach is the less-efficient in terms of N uptake and use [24], requiring high rates of fertilization to grow well and reach higher leaf quality (dark green leaves) [25].Instead, lamb's lettuce is still under-studied, and its behavior regarding NUE under different N regimes is unknown.

Previous studies regarding vegetable crops including leafy greens have documented that the application of plant biostimulants triggers several molecular and physiological processes, accompanied by improvement in growth, yield, quality, NUE and tolerance to abiotic stress [22,26–37]. The capacity of biostimulants to improve NUE is the utmost reason for which they are spreading in the market, considering their economic and environmental motives [38]. However, relatively few researches regarding biostimulants effects on plants grown under sub-optimal N conditions are available [33,35,39–41], especially about green leafy vegetables. The reduction of N inputs in leafy vegetables is very important, both for containing the phenomenon of nitrate accumulation in leaves and for reducing the economic and environmental impacts of fertilization. Di Mola et al. [42] reported that the foliar application of different biostimulants (in particular seaweed extract and legume-derived hydrolysate protein) on greenhouse baby lettuce boosted plant growth, mainly in sub-optimal N fertilization. Furthermore, in baby rocket cultivated under greenhouse conditions, Di Mola et al. [43]

found that the application of plant-based biostimulants boosted the marketable yield at low N levels compared to the control.

The aim of this study was to assess the effect of foliar application of legume-derived protein hydrolysates on N demand and uptake efficiency of two important leafy greens. Therefore, two experiments were carried out for evaluating the possible beneficial effects of a plant-derived protein hydrolysates applied on greenhouse spinach and lamb's lettuce grown under variables N conditions, in order to depict its influence on NUE, yield and leaf quality.

2. Materials and Methods

2.1. Experimental Setting, Leafy Vegetables Tested and Cultural Practices

Two consecutive experiments were carried out in a plastic tunnel during winter 2018/2019 and spring 2019 seasons at the experimental site "Gussone Park" of the Department of Agricultural Sciences (40°48.870′ N; 14°20.821′ E; 70 m a.s.l.), University of Naples Federico II, Italy. The two tested crops were cultivated in large pots (diameter 0.70 m and height 0.60 m) filled with sandy soil, with the following physical and chemical proprieties: pH 7.4, 2.5% organic matter, 0.9 g kg^{-1} total N (Kjeldhal method), 252.6 mg kg^{-1} P_2O_5 and 490.9 mg kg^{-1} of K_2O.

For the first experiment (Winter 2018/19), baby spinach (*Spinacia oleracea* L. cv. Platypus RZ F1, Rijk Zwaan, Bologna, Italy), a widely spread cultivar in Italy with dark green leaves, was sown on January 17th (1000 seeds per square meter) and harvested on March 12th. While for the second experiment (Spring 2019), lamb's lettuce (*Valerianella locusta* L. cv. Princess HM CLAUSE, Torino, Italy) was sown on March 26th (1200 seeds per square meter)—this cultivar is characterized by deep green leaves and a high adaptability to different growing seasons—and harvested in five different dates from May 10th till the 25th, upon reaching the marketable size according the different treatments. The germination time was 8 and 10 days after sowing and the plant densities after germination were 900 and 1100, for spinach and lamb's lettuce, respectively. For both crops, there were no differences between the treatments in terms of plant density.

Considering the chemical composition of soil, no phosphorus or potassium was given to either crop; while N was added as ammonium nitrate (34%) in a single application 27 and 20 days after the sowing, for spinach and lamb's lettuce, respectively. Water was not a limiting factor; the crop evapotranspiration was calculated with the Hargreaves method and the deficit was fully restored by sprinkler irrigation.

2.2. Experimental Design, Nitrogen Fertilization and Biostimulant Application

A factorial combination of three nitrogen fertilization levels and two biostimulant applications (treated and non-treated control) distributed in a randomized complete-block design were adopted for both experiments. Each treatment was replicated three times accounting a total of 18 pots (3 N levels × 2 biostimulant applications × 3 replicates).

The optimal nitrogen dose was calculated based on the balance method that considers all inputs and outputs. For the first experiment (spinach) N levels were: optimal dose (N100%) −4.5 g m^{-2}, sub-optimal dose (N50%) −2.25 g m^{-2} and no fertilization (N0%). While for the second experiment (lamb's lettuce) N levels were: optimal dose (N100%) −5.0 g m^{-2}, sub-optimal dose (N50%) −2.5 g m^{-2} and no fertilization (N0%).

The plant-based biostimulant used for both green leafy vegetables was a legume-derived protein hydrolysates, promoted as Trainer ® by Italpollina S.p.A. The legume-derived PH biostimulant obtained through enzymatic hydrolysis contains 75% of free amino acids and peptides, 22% of carbohydrates and 3% of mineral nutrients. The detailed aminogram of the product along with the phenolics, flavonoids, and elemental composition were reported in detail by Rouphael et al. [22]. For both crops, the treated plants were sprayed four times at 21, 27, 33 and 39 days after sowing, at a concentration of 4 mL L^{-1}. Untreated control spinach and lamb's lettuce plants were only sprayed with water. Each pot was sprayed with a solution volume of 38.5 mL (=1000 L ha^{-1}) corresponding to a biostimulant application rate of 0.000154 mL per pot (=4 L of biostimulant per ha).

2.3. Marketable Yield and Sampling

In both experiments, the whole area of all the pots at harvest was cut and leaves were weighed in order to measure the marketable fresh yield. In addition, a representative sub-sample of each replicate was dried in a forced air oven at 70 °C and then weighed in order to determine dry weight and then to calculate leaf dry matter content and subsequently used for N content determination (total N and nitrate) by chemical analysis. For qualitative analysis, fresh samples were also collected from each replicate and conserved at −80 °C.

2.4. Nitrogen Determination, N-use Efficiency and Uptake Efficiency

The Kjeldahl method [44] was used to determine the concentration of N in dried leaves samples that were mineralized with sulfuric acid, while nitrate content was determined using the Foss FIAstar 5000 continuous flow Analyzer (FOSS analytical Denmark).

Nitrogen use efficiency (NUE) was calculated by dividing yield by N application dose plus the available N in the soil and expressed as ton per kg. In addition, N uptake efficiency was determined as the ratio between N content in the leaves and N application dose and it was expressed as kg kg^{-1}.

2.5. Leaf Quality: Antioxidant Activity and Compounds, Chlorophyll Content and SPAD Index

Lipophilic (LAA) and hydrophilic (HAA) antioxidant activities were determined using the protocols of Re et al. [45] and Fogliano et al. [46], respectively. The two extract fractions, lipophilic and hydrophilic, were measured by the means of a Hach DR 2000 spectrophotometer at 734 and 505 nm, respectively.

The Kampfenkel et al. [47] method was used to determine ascorbic acid spectrophotometrically. A wavelength of 525 nm was set in order to measure the absorbance of the extract. Total phenols were also assessed spectrophotometrically, and the absorbance solution was detected at 765 nm, based on the Singleton et al. method [48].

Leaves chlorophyll content was measured spectrophotometrically: the first step was the extraction of fresh material by ammoniacal acetone as described by Wellburn [49], then the absorbance of solutions was measured at 662 and 647 for chlorophyll a and b, respectively.

The soil plant analysis development (SPAD) index was measured at harvest, on 15 leaves by replicate, using a portable SPAD-502 chlorophyll meter.

2.6. Statistical Processing

In both experiments, a two-way ANOVA was conducted using the SPSS 21 software package. Duncan's Multiple Range Test (DMRT; significance level 0.05) was adopted for mean comparisons on each of the independent measured variables.

3. Results

3.1. Marketable Yield and SPAD Index

The effects of both tested factors (N fertilization rates and biostimulant application) on marketable fresh yield and SPAD index were reported in Figure 1A,B and Figure 2A,B, where the relevant F and P values and the degrees of freedom are reported in Table 1.

Table 1. Analysis of variance of marketable fresh yield and SPAD index of spinach and lamb's lettuce (Figure 1A,B and Figure 2A,B).

	Spinach		Lamb's Lettuce	
	Yield	**SPAD Index**	**Yield**	**SPAD Index**
Nitrogen × Biostimulant				
f value	25.198	9.580	6.195	7.554
Degrees of freedom	17	17	17	17
p value	0.001	0.01	0.05	0.01

In particular, the marketable yield of baby spinach was positively influenced by N fertilization, but it was further boosted by biostimulant application (Figure 1A). The LDPH application elicited a significant increase at all the levels of N: +16.8%, +14.2%, and 39.4% at 0, 2.25 and 4.5 g N per square meter, respectively.

As with baby spinach, the marketable yield of lamb's lettuce increased with higher N dose and it was positively affected by biostimulant foliar application (Figure 1B). However, no significant difference was recorded between LDPH-treated and non-treated control plants at the higher N fertilization level (N100%). Interestingly, the N50% plants treated with LDPH reached significantly similar values of marketable yield in comparison to treated and non-treated plants under N100% conditions.

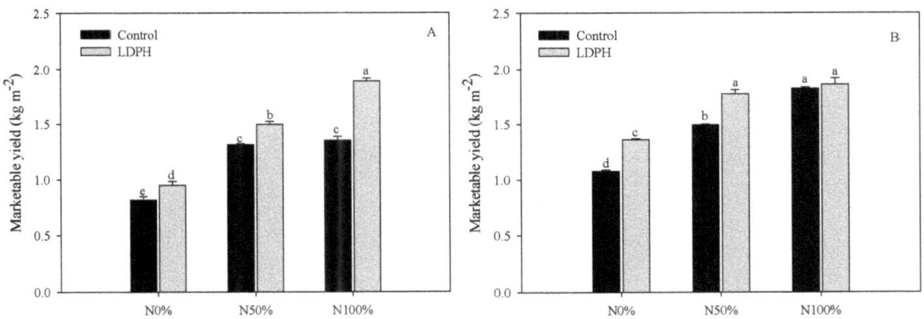

Figure 1. Marketable yield of baby spinach (**A**) and lamb's lettuce plants (**B**) as affected by nitrogen (N) fertilization levels (0, 2.25 and 4.5 g N m^{-2} and 0, 2.5 and 5.0 g N m^{-2}; N0%, N50%, N100%, respectively) and biostimulant application (non-treated control and LDPH: Legume-derived protein hydrolysates). Different letters indicate significant differences according to the DMR test ($p < 0.05$). Vertical bars indicate ± standard error of means.

The SPAD index statistically increased with the higher availability of N and also with the foliar application of LDPH in both baby spinach (Figure 2A) and lamb's lettuce (Figure 2B). The average increase of the SPAD index of fertilized and sprayed spinach plants was 8.6% compared to fertilized unsprayed plants. At 0 g N per square meter, the SPAD index of spinach plants treated with LDPH was +7% compared to untreated N0% plants. Finally, for lamb's lettuce the SPAD index increases due to biostimulant application were less marked: +5.2% and +2.9% for fertilized (N50% and N100% plants) and non-fertilized plants (N0%).

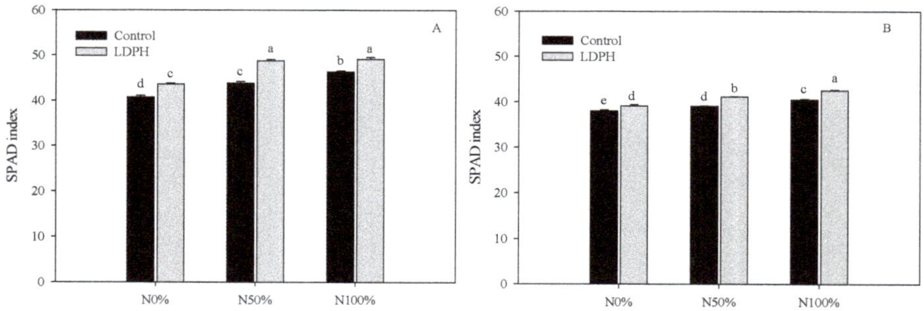

Figure 2. The SPAD index of baby spinach (**A**) and lamb's lettuce plants (**B**) as affected by nitrogen (N) fertilization levels (0, 2.25 and 4.5 g N m^{-2} and 0, 2.5 and 5.0 g N m^{-2}; N0%, N50%, N100%, respectively) and biostimulant application (non-treated control and LDPH: Legume-derived protein hydrolysates). Different letters indicate significant differences according to the DMR test ($p < 0.05$). Vertical bars indicate ± standard error of means.

3.2. N-Use and Uptake Efficiency

The results regarding the two efficiency parameters: N use efficiency and N uptake efficiency in baby spinach and lamb's lettuce are presented in Table 2. In both leafy vegetables, significant effects were noted on N use efficiency with both N and biostimulant treatments, but not the N × B interaction, whereas N uptake efficiency was only affected by foliar biostimulant application (Table 2).

Table 2. Nitrogen use and uptake efficiency of baby spinach and lamb's lettuce plants as affected by nitrogen (N) fertilization levels (0, 2.25 and 4.5 g N m^{-2} and 0, 2.5 and 5.0 g N m^{-2}; N0%, N50%, N100%, respectively) and biostimulant applications (control and LDPH: Legume-derived protein hydrolysates).

Treatments	Spinach		Lamb's Lettuce	
	N-Use Efficiency	N-Uptake Efficiency	N-Use Efficiency	N-Uptake Efficiency
	(t kg^{-1})	(kg kg^{-1})	(t kg^{-1})	(kg kg^{-1})
Fertilization				
N0%	0.35 a (0.29–0.40)	0.14 (0.29–0.45)	0.49 a (0.48–0.56)	0.20 (0.15–0.26)
N50%	0.31 ab (0.25–0.36)	0.17 (0.25–0.36)	0.33 b (0.25–0.34)	0.21 (0.12–0.22)
N100%	0.25 b (0.19–0.30)	0.14 (0.19–0.30)	0.25 c (0.17–0.26)	0.20 (0.12–0.23)
Biostimulant				
Control	0.28 b (0.23–0.31)	0.12 b (0.09–0.14)	0.32 b (0.28–0.35)	0.15 b (0.07–0.16)
LDPH	0.33 a (0.29–0.37)	0.18 a (0.15–0.21)	0.38 a (0.34–0.41)	0.26 a (0.21–0.30)
Significance				
Fertilization (F)	*	NS	**	NS
Biostimulant (B)	*	**	*	**
F × B	NS	NS	NS	NS

NS, *, ** Non-significant or significant at $p \leq 0.05$ and 0.01. Different letters within each column indicate significant differences according to Duncan's test ($p \leq 0.05$). The numbers in parenthesis are the data of 90% confidence interval.

When averaged over the N treatments, the baby spinach plants sprayed with the plant-based biostimulant showed a 17.8% and 50.0% increase compared to untreated plants, for N-use efficiency and N-uptake efficiency, respectively (Table 2). Moreover, irrespective of biostimulant application, the N0% and N50% plants had the highest values of NUE (Table 2). The trends of the two efficiency parameters in lamb's lettuce were similar to those of spinach but were always higher. In particular, the NUEs of unfertilized plants were significantly higher than in N50% and N100% plants around +69% and +59.5%, respectively (Table 2). When averaged over the N treatments, the foliar application of LDPH improved N-use efficiency and N-uptake efficiency compared to untreated plants, by 18.8% and 73.3% respectively (Table 2).

3.3. Total Chlorophyll, Chlorophyll a and b and Nitrate content

In spinach, the N fertilization levels statistically affected the content of total chlorophyll, chlorophyll a and b, as well as nitrate content in leaves. This latter was the only parameter also affected by the biostimulant application (Table 3). Particularly, chlorophyll (a, b, and total) content increased when N dose was raised. The two treatments N50% and N100% were not significantly different, but N100% was significantly higher than N0% (+10%, +37.3%, and +20.3% respectively).

Table 3. Chlorophyll a and b, total chlorophyll and nitrate content of baby spinach plants as affected by nitrogen (N) fertilization levels (0, 2.25 and 4.5 g N m^{-2}; N0%, N50%, N100%, respectively) and biostimulant application (control and LDPH: Legume-derived protein hydrolysates).

Treatments	Chlorophyll a ($mg\ g^{-1}$ fw)	Chlorophyll b ($mg\ g^{-1}$ fw)	Total Chlorophyll ($mg\ g^{-1}$ fw)	Nitrate ($mg\ kg^{-1}$ fw)
Fertilization				
N0%	0.905 b (0.846–0.965)	0.547 b (0.418–0.675)	1.452 b (1.268–1.639)	84.9 c (−209.1–379.0)
N50%	0.976 ab (0.917–1.035)	0.716 ab (0.587–0.844)	1.692 ab (1.508–1.876)	2932.8 b (20638.7–3226.8)
N100%	1.015 a (0.955–1.074)	0.786 a (0.657–0.914)	1.801 a (1.616–1.984)	3867.5 a (3573.4–4161.6)
Biostimulant				
Control	0.957 (0.908–1.005)	0.681 (0.576–0.786)	1.637 (1.487-1.787)	476.5 b (236.3–716.6)
LDPH	0.974 (0.925–1.022)	0.685 (0.58–0.790)	1.659 (1.509–1.809)	4113.7 a (3873.6–4353.8)
Significance				
Fertilization (F)	*	*	*	**
Biostimulants (B)	NS	NS	NS	**
F × B	NS	NS	NS	NS

NS, *, ** Non-significant or significant at $p \leq 0.05$ and 0.01. Different letters within each column indicate significant differences according to Duncan's test ($p \leq 0.05$). The numbers in parenthesis are the data of 90% confidence interval.

As expected, our results demonstrated that increasing N fertilization from 0 to 5.0 g m^{-2} elicited a significant linear increase in nitrate content compared to non-fertilized plants. Particularly, at N100%, the nitrate content in LDPH-treated plants exceeded the limits imposed by the European Regulation No. 1258/2011 for the commercialization of fresh spinach (3500 mg kg^{-1} on fresh weight basis) as determined by the cultivation practices, growing conditions and latitude (Table 3).

In lamb's lettuce, all parameters were affected by both factors, but not by their interaction (Table 4). The chlorophyll a, b and total content increased with increasing N level; N100% had the highest values and was statistically different from the other two treatments: +19%, +24.7% and +21% over the mean value of N0% and N100%, for chlorophyll a, b and total chlorophyll, respectively. Moreover, the increases due to biostimulant applications were 26.6%, 44.0% and 32.3% for chlorophyll a, b and total chlorophyll, respectively.

Table 4. Chlorophyll a and b, total chlorophyll and nitrate content of lamb's lettuce plants as affected by nitrogen (N) fertilization levels (0, 2.5 and 5.0 g N m^{-2}; N0%, N50%, N100%, respectively) and biostimulant applications (control and LDPH: Legume-derived protein hydrolysates).

Treatments	Chlorophyll a ($mg\ g^{-1}$ fw)	Chlorophyll b ($mg\ g^{-1}$ fw)	Total Chlorophyll ($mg\ g^{-1}$ fw)	Nitrate ($mg\ kg^{-1}$ fw)
Fertilization				
N0%	0.673 b (0.621–0.726)	0.347 c (0.312–0.372)	1.020 c (0.954–1.067)	102.8 b (−925.1–1130.7)
N50%	0.722 b (0.672–0.777)	0.397 b (0.373–0.429)	1.120 b (1.075–1.189)	3191.8 a (2163.8–4219.7)
N100%	0.831 a (0.780–0.885)	0.464 a (0.430–0.498)	1.295 a (1.228–1.362)	3210.0 a (2182.0–4237.9)
Biostimulant				
Control	0.655 b (0.613–0.698)	0.330 b (0.302–0.359)	0.985 b (0.931–1.040)	562.4 b (−315.4–1363.1)
LDPH	0.829 a (0.788–0.874)	0.475 a (0446–0.504)	1.304 a (1.251–1.361)	3774.0 a (2973.2–4651.8)
Significance				
Fertilization (F)	**	**	**	**
Biostimulants (B)	**	**	**	**
F × B	NS	NS	NS	NS

NS, ** Non-significant or significant at $p \leq 0.05$ and 0.01. Different letters within each column indicate significant differences according to Duncan's test ($p \leq 0.05$). The numbers in parenthesis are the data of 90% confidence interval.

Furthermore, in lamb's lettuce nitrate content in leaves increased when nitrogen dose was raised, but without significant differences between N50% and N100%, and it was higher in the plants sprayed with biostimulants compared to untreated plants. For this crop, the European Community has not fixed any threshold, but if we consider the limit imposed for fresh spinach, only the value of biostimulant-sprayed plants overcame it.

3.4. *Leaf Quality: Antioxidant Activity and Compounds*

In spinach, LAA and the content of total phenols and ascorbic acid (AsA) were significantly affected by N fertilization, while the biostimulant application influenced only LAA. HAA was neither affected by N fertilization treatments nor by biostimulant application (Table 5). Irrespective of biostimulant application, LAA, total phenols, and AsA were significantly higher in N0% plants in comparison to N100% plants, around 3.9%, 29.8%, and 41.8% respectively. Interestingly, when averaged over N treatments, the foliar application of LDPH boosted LAA compared to untreated plants by 11.6% (Table 5).

In lamb's lettuce, all the measured leaf quality parameters (LAA, HAA, total phenols, and AsA) were significantly affected by N fertilization levels, while only HAA was affected by the biostimulant application (Table 6). Regarding LAA, total phenols, and AsA, the trends were similar to those observed for spinach; where the values of N0% plants were higher (+8.3%, +23.3%, and +26.9%, respectively) compared to N100% plants. Instead, HAA had an opposite trend: it was higher in fertilized plants (+18.5% compared to unfertilized plants) and it was also higher in the plants sprayed with biostimulant (+6.3%).

Table 5. Lipophilic (LAA) and hydrophilic (HAA) antioxidant activity, total phenols and ascorbic acid (AsA) of baby spinach plants as affected by nitrogen (N) fertilization levels (0, 2.25 and 4.5 g N m^{-2}; N0%, N50%, N100%, respectively) and biostimulant applications (untreated control and LDPH: Legume-derived protein hydrolysates).

Treatments	LAA (mM Trolox eq. $100g^{-1}$ dw)	HAA (mM AA eq. $100g^{-1}$ dw)	Total Phenols (mg Gallic Acid eq. g^{-1} dw)	AsA ($mg\ g^{-1}$ fw)
Fertilization				
N0%	22.65 a (22.16–23.13)	8.08 (7.38–8.77)	3.22 a (2.97–3.46)	33.49 a (31.46–35.51)
N50%	22.02 ab (21.54–22.50)	8.11 (7.41–8.80)	2.88 ab (2.63–3.12)	27.45 b (25.42–29.46)
N100%	21.80 b (21.32–22.28)	8.15 (7.44–8.84)	2.48 b (2.23–2.72)	23.62 c (21.56–25.60)
Biostimulant				
Control	20.95 b (20.55–21.34)	8.05 (7.48–8.62)	2.36 (2.69–3.09)	28.55 (16.86–40.23)
LDPH	23.37 a (22.97–23.76)	8.17 (7.60–8.74)	2.37 (2.62–3.03)	27.82 (16.13–39.51)
Significance				
Fertilization (F)	*	NS	**	**
Biostimulants (B)	**	NS	NS	NS
F × B	NS	NS	NS	NS

NS, *, ** Non-significant or significant at $p \leq 0.05$ and 0.01. Different letters within each column indicate significant differences according to Duncan's test ($p \leq 0.05$). The numbers in parenthesis are the data of 90% confidence interval.

Table 6. Lipophilic (LAA) and hydrophilic (HAA) antioxidant activity, total phenols and ascorbic acid (AsA) of lamb's lettuce plants as affected by nitrogen (N) fertilization levels (0, 2.5 and 5.0 g N m^{-2}; N0%, N50%, N100%, respectively) and biostimulant applications (control and LDPH: Legume-derived protein hydrolysates).

Treatments	LAA (mM Trolox eq. $100g^{-1}$ dw)	HAA (mM AA eq. $100g^{-1}$ dw)	Total phenols (mg Gallic Acid eq. g^{-1} dw)	AsA ($mg\ g^{-1}$ fw)
Fertilization				
N0%	30.08 a (28.91–31.25)	6.26 b (5.920–6.60)	10.16 a (9.520–10.80)	63.04 a (56.543–69.54)
N50%	28.49 ab (27.31–29.66)	7.18 a (6.840–7.52)	8.55 b (7.913–9.19)	53.41 b (46.911–59.91)
N100%	27.77 b (26.60–28.94)	7.65 a (7.311–7.99)	7.62 c (6.974–8.25)	49.68 b (43.177–56.17)
Biostimulant				
Control	29.40 (28.44–30.36)	6.82 b (6.537–7.09)	8.96 (8.439–9.48)	56.85 (51.537–62.15)
LDPH	28.16 (27.20–29.11)	7.25 a (6.968–7.52)	8.59 (8.067–9.11)	53.91 (48.603–59.21)
Significance				
Fertilization (F)	*	**	**	*
Biostimulants (B)	NS	*	NS	NS
F × B	NS	NS	NS	NS

NS, *, ** Non-significant or significant at $p \leq 0.05$ and 0.01. Different letters within each column indicate significant differences according to Duncan's test ($p \leq 0.05$). The numbers in parenthesis are the data of 90% confidence interval.

4. Discussion

In order to increase the supply of food produced on the available arable land—since the global population will reach 10 billion by 2050—growers must boost the yield of their produce, through the massive use of technical means, in particular N fertilization. Nowadays, it is impossible to adopt an agriculture that is not sustainable and environmentally friendly. Therefore, the objective of boosting crop productivity must occur through the reduction of N fertilizers, but also through the improvement of nitrogen use efficiency (NUE), that assures reasonable yield and a profit margin for farmers [50].

Several researches have highlighted that plant-based biostimulants have a triggering effect on growth and yield, but they are also capable of improving the NUE in consideration of both economic and environmental motives [38,51]. The plant-based biostimulant used in this test was Trainer®, a legume-derived protein hydrolysate (containing free amino acids and signaling molecules such as small soluble peptides), for which previous researches have already demonstrated its ability to boost crops' resources use efficiency (RUE) [15,52]—especially N uptake and assimilation [39]—as well as productivity [6,32] and quality [53,54]. Our results highlighted the ability of LDPH to enhance yield of both baby spinach and lamb's lettuce (+24.6% and +13.5% for plant sprayed with Trainer® compared to control plants, respectively), which is in line with Carillo et al.'s [35] findings on spinach, and Di Mola et al. [42,43] on other two important leafy greens (lettuce and baby leaf rocket) cultivated under variable N regimes. The positive effects of the foliar application of LDPH, irrespective of the N fertilization treatments, were more pronounced in spinach than in lamb's lettuce, demonstrating a species-specific response [15,55], especially that the same commercial plant-based biostimulant was used. The different responses between the two leafy vegetables species could be attributed to the different leaf permeability and cuticle morphology as well as the stomatal aperture and thus the efficacy of the plant biostimulant [38]. Therefore, our results highlight, that further study is warranted to assess the physiological and molecular mechanisms behind the biostimulant action and to investigate the specificity of species dependent responses in impacting leaf characteristics and consequently interacting with the different bioactive compounds of plant biostimulants. Interestingly, in our study the marketable fresh yield of LDPH-treated spinach and lamb's lettuce grown under N50% was similar to those grown under N100% (especially the non-treated plants). A number of biochemical and physiological aspects may have contributed to this result, including (i) a higher chlorophyll content (a, b and total) and SPAD index in biostimulant-treated than in non-treated plants, and (ii) improved leaf status in terms of nitrate content, triggering a more efficient translocation of assimilates to potential photosynthetic sinks, thus boosting plant growth and yield [35,42,43]. Moreover, several authors attributed the stimulation action and the increased N assimilation in response to LDPH application to multiple mechanisms of action involving (i) the hormones-like activities (i.e., auxin and giberrellins-like activities), (ii) the increase in the activity of the key enzymes glutamine synthetase and nitrate reductase, and (iii) the upregulation of specific genes responsible in N assimilation and pigment synthesis [27,33,56–58].

Similar to the effect N fertilization on agronomic performance, our findings highlighted the higher NUE of baby spinach and lamb's lettuce, even without N fertilization. The current results are in agreement with the findings of several researches such as Abdelraouf [59], Canali et al., 2011 [60], and Zhang et al. [61], which in spinach observed a linear decrease in NUE when N dose increased. Moreover, our findings about N uptake are in line to the results of Canali et al. [60], which observed that this parameter was not affected by variable nitrogen regimes.

Interestingly, our findings also indicated that foliar application of LDPH can be considered an efficient tool to reduce N additional inputs to the cropping system, hence cutting down the production costs for farmers and N surpluses into the environment [62]. Mainly because the LDPH-treated baby spinach and lamb's lettuce plants exhibited both higher NUE and higher N-uptake efficiency, irrespective of the N fertilization levels. The positive effect of foliar application of LDPH on the two N efficiency parameters can be attributed to the improvement of root architecture (i.e., more vigorous root apparatus) which is related to an overall increase in nutrient accessibility caused by its power to boost the capacity of absorption, translocation and assimilation of macro and micro minerals, especially when N is limiting plant growth [27,56,63]. This phenomenon associated to the PH-induced remolding of root advocating N uptake and translocation was described by Colla et al. [64], as "nutrient acquisition response". The stimulation of root system architecture—in particular the increase in root hair density and length—was observed previously by several authors on a wide range of agronomic and horticultural species such as corn, sunflower, tomato, eggplant, lettuce and *Brassica* genus [27,57,64,65].

Although the application of fertilizers (nitrogen, phosphorus and potassium) generally increases the crop yield; alternatively, the excessive application of synthetic fertilizers—especially N—can result in undesirable nutritional quality changes such as a decrease in some bioactive compounds (phenols and vitamin C) and soluble sugars [66]. This was the case in the current study, whereby baby spinach and lamb's lettuce cultivated under N100% negatively modulated the synthesis and accumulation of antioxidant molecules such as total phenols and ascorbic acid along with low antioxidant activity. Similar trends were reported recently by Di Mola et al. [42,43] on baby lettuce and rocket grown under optimal and supra-optimal N regimes.

Concerning the effect of LDPH application on the quality of the two tested leafy greens, some findings demonstrated that the application of protein hydrolysates-based biostimulant was able to modify plant primary and secondary metabolism [15,55], leading to the synthesis and accumulation of phytochemicals with health-promoting properties. This was the case in the current greenhouse experiment, since baby spinach and lamb's lettuce plants treated with the commercial protein hydrolysates Trainer® positively modulated both the lypophilic and hydrophilic antioxidant capacity, which are considered important traits in evaluating the quality of food including leafy vegetables [23]. However, the foliar LDPH application did not affect the concentration of total phenols and ascorbic acid in both leafy vegetables. A variable effect of three commercial plant biostimulants containing mainly free amino acids (Aminovert, Megafol and Veramin) was also observed on the chemical composition, phenolic profile and bioactive properties of two greenhouse spinach cultivars [67]. Therefore, future research should focus on designing the ideotype plant biostimulants and identifying the best species × biostimulant × fertilization (N) combination(s) for the production of healthy and nutrient-dense leafy vegetables.

5. Conclusions

Sustainable agriculture is the greatest challenge of our century, and plant-based biostimulants represent an efficient and concrete possibility to reach this objective by maintaining high production and improving the NUE of leafy greens with several economic, nutritional and environmental benefits. The positive effects of the LDPH biostimulant were manifested in terms of marketable fresh yield in baby spinach, irrespective of N fertilization treatments and at low N rates (N0% and N50%) in lamb's lettuce. Such benefits were likely derived from the signaling molecules (such as small peptides) as a result of augmented leaf nitrate content, SPAD index and pigments synthesis. These stimulation actions of the LDPH application were more pronounced under sub-optimal (N0% and N50%) than under optimal (N100%) N regimes. Interestingly, foliar LDPH application in both tested leafy vegetables boosted NUE and N uptake efficiency, which is fundamental for both economic and environmental reasons. Our results also demonstrated that the foliar application of LDPH can promote the antioxidant capacity which is important for the human diet and may constitute an added value for both growers and consumers. Overall, our findings suggest that the application of protein hydrolysates can be a sustainable practice in intensive greenhouse cropping systems to enhance crop productivity and NUE under both optimal and sub-optimal (low-input conditions) N regimes.

Author Contributions: Conceptualization, M.M. methodology, I.D.M., S.N. and E.C. software, L.O., I.D.M. and C.E.-N. validation, M.M., I.D.M., E.C. and L.O. formal analysis, C.E.-N. and L.O. investigation, I.D.M. resources, M.M. data curation, I.D.M. and L.O. writing—original draft preparation, I.D.M. writing-review and editing, Y.R.,

C.E.-N. and G.C. visualization, L.O., I.D.M. and C.E.-N. supervision, Y.R., G.C. and M.M. project administration, M.M. funding acquisition, M.M. All authors have read and agreed to the published version of the manuscript.

Funding: This research received no external funding.

Conflicts of Interest: The authors declare no conflict of interest.

References

1. Ruiz, J.M.; Rivero, R.M.; Cervilla, L.M.; Castellano, R.; Romero, L. Grafting to improve nitrogen-use efficiency traits in tobacco plants. *J. Sci. Food Agric.* **2006**, *86*, 1014–1021. [CrossRef]
2. Gupta, S.K.; Gupta, A.B.; Gupta, R. Pathophysiology of nitrate toxicity in humans in view of the changing trends of the global nitrogen cycle with special reference to India. In *The Indian Nitrogen Assessment*; Abrol, Y.P., Adhya, T.K., Aneja, V.P., Eds.; Elsevier: Amsterdam, The Netherlands, 2017; pp. 459–468.
3. Lassaletta, L.; Billen, G.; Grizzetti, B.; Anglade, J.; Garnier, J. 50 year trends in nitrogen use efficiency of world cropping systems: The relationship between yield and nitrogen input to cropland. *Environ. Res. Lett.* **2014**, *9*, 105011. [CrossRef]
4. Sutton, M.A.; Bleeker, A.; Howard, C.M.; Bekunda, M.; Grizzetti, B.; de Vries, W.; van Grinsven, H.J.M.; Abrol, Y.P.; Adhya, T.K.; Billen, G.; et al. Our Nutrient World: The challenge to produce more food and energy with less pollution. *Cent. Ecol. Hydrol.* **2013**, *8*, 95–108.
5. Aires, A.; Carvalho, R.; Rosa, E.A.S.; Saavedra, M.J. Effects of agriculture production systems on nitrate and nitrite accumulation on baby-leaf salads. *Food Sci. Nutr.* **2013**, *1*, 3–7. [CrossRef]
6. Kyriacou, M.C.; Rouphael, Y. Towards a new definition of quality for fresh fruits and vegetables. *Sci. Hortic.* **2018**, *234*, 463–469. [CrossRef]
7. Colla, G.; Kim, H.J.; Kyriacou, M.C.; Rouphael, Y. Nitrate in fruits and vegetables. *Sci. Hortic.* **2018**, *237*, 231–238. [CrossRef]
8. Santamaria, P. Nitrate in vegetables: Toxicity, content, intake and EC regulation. *Sci. Food Agric.* **2006**, *86*, 10–17. [CrossRef]
9. Fallovo, C.; Rouphael, Y.; Rea, E.; Battistelli, A.; Colla, G. Nutrient solution concentration and growing season affect yield and quality of *Lactuca sativa* L. var. acephala in floating raft culture. *J. Sci. Food Agric.* **2009**, *89*, 1682–1689. [CrossRef]
10. Hawkesford, M.; Kopriva, S.; De Kok, L. Nutrient use efficiency in plants–Concepts and approaches. In *Plant Ecophysiol*; Springer International Publishing: Basel, Switzerland, 2014. [CrossRef]
11. Sisson, V.A.; Rufty, T.W.; Williamson, R.E. Nitrogen-use efficiency among flue-cured tobacco genotypes. *Crop Sci.* **1991**, *31*, 1615–1620. [CrossRef]
12. Moll, R.H.; Kamprath, E.J.; Jackson, W.A. Analysis and interpretation of factors which contribute to efficiency of nitrogen utilization. *Agron. J.* **1982**, *74*, 562–564. [CrossRef]
13. Colla, G.; Rouphael, Y.; Mirabelli, C.; Cardarelli, M. Nitrogen-use efficiency traits of mini-watermelon in response to grafting and nitrogen-fertilization doses. *J. Plant Nutr. Soil Sci.* **2011**, *174*, 933–941. [CrossRef]
14. Calvo, P.; Nelson, L.; Kloepper, J.W. Agricultural uses of plant biostimulants. *Plant Soil* **2014**, *383*, 3–41. [CrossRef]
15. Colla, G.; Hoagland, L.; Ruzzi, M.; Cardarelli, M.; Bonini, P.; Canaguier, R.; Rouphael, Y. Biostimulant action of protein hydrolysates: Unraveling their effects on plant physiology and microbiome. *Front. Plant Sci.* **2017**, *8*, 2202. [CrossRef]
16. Maynard, D.N.; Baker, A.V.; Minotti, P.L.; Peck, N.H. Nitrate accumulation in vegetables. *Adv. Agron.* **1976**, *28*, 71–118. [CrossRef]
17. Blom-Zandstra, M.; Eenink, A.H. Nitrate content and reduction in different genotypes of lettuce. *J. Am. Soc. Hortic. Sci.* **1986**, *111*, 908–911.
18. Pate, J.S. Uptake, assimilation and transport of nitrogen compounds by plants. *Soil Biol. Biochem.* **1973**, *5*, 109–119. [CrossRef]
19. Blom-Zandstra, M. Nitrate accumulation in vegetables and its relationship to quality. *Ann. Appl. Biol.* **1989**, *115*, 553–561. [CrossRef]
20. Steingröver, E.; Ratering, P.; Siesling, J. Daily changes in uptake, reduction and storage of nitrate in spinach grown at low light intensity. *Physiol. Plant.* **1986**, *66*, 550–556. [CrossRef]

21. Steingröver, E.; Siesling, J.; Ratering, P. Effect on one nigth with "low light" on uptake, reduction and storage of nitrate in spinach. *Physiol. Plant.* **1986**, *66*, 557–562. [CrossRef]
22. Rouphael, Y.; Giordano, M.; Cardarelli, M.; Cozzolino, E.; Mori, M.; Kyriacou, M.; Bonini, P.; Colla, G. Plant and seaweed-based extracts increase yield but differentially modulate nutritional quality of greenhouse spinach through biostimulant action. *Agronomy* **2018**, *8*, 126. [CrossRef]
23. Colonna, E.; Rouphael, Y.; Barbieri, G.; De Pascale, S. Nutritional quality of leafy vegetables harvested at two light intensities. *Food Chem.* **2016**, *199*, 702–710. [CrossRef]
24. Biemond, H.; Vos, J.; Struik, P.C. Effects of nitrogen on accumulation and partitioning of dry matter and nitrogen of vegetables. 3. Spinach. *Neth. J. Agric. Sci. Wagening. J. Sci.* **1996**, *44*, 227–239.
25. Smolders, E.; Buysse, J.; Merckx, R. Growth analysis of soil-grown spinach plants at different N-regimes. *Plant Soil* **1993**, *154*, 73–80. [CrossRef]
26. Rouphael, Y.; De Micco, V.; Arena, C.; Raimondi, G.; Colla, G.; De Pascale, S. Effect of Ecklonia maxima seaweed extract on yield, mineral composition, gas exchange and leaf anatomy of zucchini squash grown under saline conditions. *J. Appl. Phycol.* **2017**, *29*, 459–470. [CrossRef]
27. Ertani, A.; Cavani, L.; Pizzeghello, D.; Brandellero, E.; Altissimo, A.; Ciavatta, C.; Nardi, S. Biostimulant activity of two protein hydrolyzates in the growth and nitrogen metabolism of maize seedlings. *J. Plant Nutr. Soil Sci.* **2009**, *172*, 237–244. [CrossRef]
28. Ertani, A.; Schiavon, M.; Muscolo, A.; Nardi, S. Alfalfa plant-derived biostimulant stimulate short-term growth of salt stressed *Zea mays* L. plants. *Plant Soil* **2013**, *64*, 145–158. [CrossRef]
29. Botta, A. Enhancing plant tolerance to temperature stress with amino acids: An approach to their mode of action. *Acta Hortic.* **2013**, *1009*, 29–35. [CrossRef]
30. Lucini, L.; Rouphael, Y.; Cardarelli, M.; Canaguier, R.; Kumar, P.; Colla, G. The effect of a plant-derived biostimulant on metabolic profiling and crop performance of lettuce grown under saline conditions. *Sci. Hortic.* **2015**, *182*, 124–133. [CrossRef]
31. Lucini, L.; Rouphael, Y.; Cardarelli, M.; Bonini, P.; Baffi, C.; Colla, G. A vegetal biopolymer-based biostimulant promoted root growth in melon while triggering brassinosteroids and stress-related compounds. *Front. Plant Sci.* **2018**, *9*, 472. [CrossRef]
32. Rouphael, Y.; Colla, G.; Giordano, M.; El-Nakhel, C.; Kyriacou, M.C.; De Pascale, S. Foliar applications of a legume-derived protein hydrolysate elicit dose dependent increases of growth, leaf mineral composition, yield and fruit quality in two greenhouse tomato cultivars. *Sci. Hortic.* **2017**, *226*, 353–360. [CrossRef]
33. Sestili, F.; Rouphael, Y.; Cardarelli, M.; Pucci, A.; Bonini, P.; Canaguier, R.; Colla, G. Protein hydrolysate stimulates growth in tomato coupled with N-dependent gene expression involved in N assimilation. *Front. Plant Sci.* **2018**, *9*, 1233. [CrossRef] [PubMed]
34. Carillo, P.; Colla, G.; El-Nakhel, C.; Bonini, P.; D'Amelia, L.; Dell'Aversana, E.; Pannico, A.; Giordano, M.; Sifola, M.I.; Kyriacou, M.C.; et al. Biostimulant application with a tropical plant extract enhances Corchorus olitorius adaptation to sub-optimal nutrient regimens by improving physiological parameters. *Agronomy* **2019**, *9*, 249. [CrossRef]
35. Carillo, P.; Colla, G.; Fusco, G.M.; Dell'Aversana, E.; El-Nakhel, C.; Giordano, M.; Pannico, A.; Cozzolino, E.; Mori, M.; Reynaud, H.; et al. Morphological and physiological responses induced by protein hydrolysate-based biostimulant and nitrogen rates in greenhouse spinach. *Agronomy* **2019**, *9*, 450. [CrossRef]
36. Caruso, G.; Stoleru, V.; De Pascale, S.; Cozzolino, E.; Pannico, A.; Giordano, M.; Teliban, G.; Cuciniello, A.; Rouphael, Y. Production, leaf quality and antioxidants of perennial wall rocket as affected by crop cycle and mulching type. *Agronomy* **2019**, *9*, 194. [CrossRef]
37. Caruso, G.; De Pascale, S.; Cozzolino, E.; Cuciniello, A.; Cenvinzo, V.; Bonini, P.; Colla, G.; Rouphael, Y. Yield and nutritional quality of Vesuvian Piennolo tomato PDO as affected by farming system and biostimulant application. *Agronomy* **2019**, *9*, 505. [CrossRef]
38. Rouphael, Y.; Colla, G. Synergistic biostimulatory action: Designing the next generation of plant biostimulants for sustainable agriculture. *Front. Plant Sci.* **2018**, *9*, 1655. [CrossRef]
39. Sady, W.; Smole'n, S. *The Influence of Pentakeep V and Nitrogen Fertilization on the Yield and Biological Quality of Carrot and Spinach Crop*; Final Report for Cosmo Oil Co., Ltd.; Cosmo Oil Co., Ltd.: Tokyo, Japan, 2007.
40. Smole'n, S.; Sady, W.; Wierzbi'nska, J. The effect of plant biostimulation with 'Pentakeep V' and nitrogen fertilization on yield, nitrogen metabolism and quality of spinach. *Acta Sci. Pol. Hortorum Cultus* **2010**, *9*, 25–36. [CrossRef]

41. Smole'n, S.; Sady, W. The influence of nitrogen fertilization and Pentakeep V application on contents of nitrates in carrot. *Acta Hortic. Regiotect.* **2009**, *12*, 221–223.
42. Di Mola, I.; Cozzolino, E.; Ottaiano, L.; Giordano, M.; Rouphael, Y.; Colla, G.; Mori, M. Effect of vegetal- and seaweed extract-based biostimulants on agronomical and leaf quality traits of plastic tunnel-grown baby lettuce under four regimes of nitrogen fertilization. *Agronomy* **2019**, *9*, 571. [CrossRef]
43. Di Mola, I.; Ottaiano, L.; Cozzolino, E.; Senatore, M.; Giordano, M.; El-Nakhel, C.; Sacco, A.; Rouphael, Y.; Colla, G.; Mori, M. Plant-Based Biostimulants Influence the Agronomical, Physiological, and Qualitative Responses of Baby Rocket Leaves under Diverse Nitrogen Conditions. *Plants* **2019**, *8*, 522. [CrossRef]
44. Bremner, J.M. Total nitrogen. In *Methods of Soil Analysis. Agronomy Monograph*; Part 2; Black, C.A., Evans, D.D., White, I.L., Ensminger, L.E., Clark, F.E., Eds.; American Society of Agronomy: Madison, WI, USA, 1965; Volume 9, pp. 1149–1178.
45. Re, R.; Pellegrini, N.; Proteggente, A.; Pannala, A.; Yang, M.; Rice-Evans, C. Antioxidant activity applying an improved ABTS radical cation decolorization assay. *Free Radic. Biol. Med.* **1999**, *26*, 1231–1237. [CrossRef]
46. Fogliano, V.; Verde, V.; Randazzo, G.; Ritieni, A. Method for measuring antioxidant activity and its application to monitoring the antioxidant capacity of wines. *J. Agric. Food Chem.* **1999**, *47*, 1035–1040. [CrossRef] [PubMed]
47. Kampfenkel, K.; Van Montagu, M.; Inzé, D. Extraction and determination of ascorbate and dehydroascorbate from plant tissue. *Anal. Biochem.* **1995**, *225*, 165–167. [CrossRef] [PubMed]
48. Singleton, V.L.; Orthofer, R.; Lamuela-Raventós, R.M. Analysis of total phenols and other oxidation substrates and antioxidants by means of Folin-Ciocalteu reagent. In *Methods in Enzymology*; Academic Press: Cambridge, MA, USA, 1999; Volume 299, pp. 152–178. [CrossRef]
49. Wellburn, A.R. The spectral determination of chlorophylls a and b, as well as total carotenoids, using various solvents with spectrophotometers of different resolution. *J. Plant Physiol.* **1994**, *144*, 307–313. [CrossRef]
50. Hirel, B.; Tétu, T.; Lea, P.J.; Dubois, F. Improving nitrogen use efficiency in crops for sustainable agriculture. *Sustainability* **2011**, *9*, 1452–1485. [CrossRef]
51. Fiorentino, N.; Ventorino, V.; Woo, S.L.; Pepe, O.; De Rosa, A.; Gioia, L.; Romano, I.; Lombardi, N.; Napolitano, M.; Colla, G.; et al. Trichoderma-based biostimulants modulate rhizosphere microbial populations and improve N uptake e_ciency, yield and nutritional quality of leafy vegetables. *Front. Plant Sci.* **2018**, *9*, 743. [CrossRef]
52. Rouphael, Y.; Kyriacou, M.C.; Petropoulos, S.A.; De Pascale, S.; Colla, G. Improving vegetable quality in controlled environments. *Sci. Hortic.* **2018**, *234*, 275–289. [CrossRef]
53. Colla, G.; Cardarelli, M.; Bonini, P.; Rouphael, Y. Foliar applications of protein hydrolysate, plant and seaweed extracts increase yield but differentially modulate fruit quality of greenhouse tomato. *HortScience* **2017**, *52*, 1214. [CrossRef]
54. Ertani, A.; Pizzeghello, D.; Francioso, O.; Sambo, P.; Sanchez-Cortes, S.; Nardi, S. *Capsicum chinensis* L. growth and nutraceutical properties are enhanced by biostimulants in a long-term period: Chemical and metabolomic approaches. *Front. Plant Sci.* **2014**, *5*, 375.
55. Colla, G.; Nardi, S.; Cardarelli, M.; Ertani, A.; Lucini, L.; Canaguier, R.; Youssef, R. Protein hydrolysates as biostimulants in horticulture. *Sci. Hortic.* **2015**, *196*, 28–38. [CrossRef]
56. Schiavon, M.; Ertani, A.; Nardi, S. Effects of an alfalfa protein hydrolysate on the gene expression and activity of enzymes of the tricarboxylic acid (TCA) cycle and nitrogen metabolism in *Zea mays* L. *J. Agric. Food Chem.* **2008**, *56*, 11800–11808. [CrossRef] [PubMed]
57. Matsumiya, Y.; Kubo, M. Soybean peptide: Novel plant growth promoting peptide from soybean. In *Soybean and Nutrition*; El-Shemy, H., Ed.; In Tech Europe Publisher: Rijeka, Croatia, 2011; pp. 215–230.
58. Colla, G.; Svecova, E.; Rouphael, Y.; Cardarelli, M.; Reynaud, H.; Canaguier, R.; Planques, B. Effectiveness of a plant—Derived protein hydrolysate to improve crop performances under different growing conditions. *Acta Hortic.* **2013**, *1009*, 175–179. [CrossRef]
59. Abdelraouf, E.A.A. The Effects of Nitrogen Fertilization on Yield and Quality of Spinach Grown in High Tunnels. *Alex. Sci. Exch. J.* **2016**, *37*, 488–496.
60. Canali, S.; Montemurro, F.; Tittarelli, F.; Masetti, O. Is it possible to reduce nitrogen fertilization in processing spinach? *J. Plant Nutr.* **2011**, *34*, 534–546. [CrossRef]

61. Zhang, J.; Yue, Y.; Sha, Z.; Kirumba, G.; Zhang, Y.; Bei, Z.; Cao, L. Spinach-irrigating and fertilizing for optimum quality, quantity, and economy. *Acta Agric. Scand. Sect. B Soil Plant Sci.* **2014**, *64*, 590–598. [CrossRef]
62. Vernieri, P.; Borghesi, E.; Tognoni, F.; Serra, G.; Ferrante, A.; Piagessi, A. Use of biostimulants for reducing nutrient solution concentration in floating system. *ISHM Acta Hortic.* **2006**, *718*, 477–484. [CrossRef]
63. Lea, P.J.; Sodek, L.; Parry, M.A.J.; Shewry, P.R.; Halford, N.G. Asparagine in plants. *Ann. Appl. Biol.* **2007**, *150*, 1–26. [CrossRef]
64. Colla, G.; Rouphael, Y.; Canaguier, R.; Svecova, E.; Cardarelli, M. Biostimulant action of a plant-derived protein hydrolysate produced through enzymatic hydrolysis. *Front. Plant Sci.* **2014**, *5*, 448. [CrossRef]
65. Ugolini, L.; Cinti, S.; Righetti, L.; Stefan, A.; Matteo, R.; D'Avino, L.; Lazzeri, L. Production of an enzymatic protein hydrolyzate from defatted sunflower seed meal for potential application as a plant biostimulant. *Ind. Crops Prod.* **2015**, *75*, 15–23. [CrossRef]
66. Wang, Z.H.; Li, S.X.; Malhi, S. Effects of fertilization and other agronomic measures on nutritional quality of crops. *J. Sci. Food Agric.* **2008**, *88*, 7–23. [CrossRef]
67. Pereira, C.; Dias, M.I.; Petropoulos, S.A.; Plexida, S.; Chrysargyris, A.; Tzortzakis, N.; Calhelha, R.C.; Ivanov, M.; Stojković, D.; Soković, M.; et al. The Effects of Biostimulants, Biofertilizers and Water-Stress on Nutritional Value and Chemical Composition of Two Spinach Genotypes (*Spinacia oleracea* L.). *Molecules* **2019**, *24*, 4494. [CrossRef] [PubMed]

agronomy

MDPI

Brief Report

Exploratory Study on the Foliar Incorporation and Stability of Isotopically Labeled Amino Acids Applied to Turfgrass

Rachel M. McCoy [1,2], George W. Meyer [1,2], David Rhodes [1], George C. Murray [3], Thomas G. Sors [4] and Joshua R. Widhalm [1,2,*]

1 Department of Horticulture and Landscape Architecture, Purdue University, West Lafayette, IN 47907, USA; mccoy26@purdue.edu (R.M.M.); meyer335@purdue.edu (G.W.M.); davidrhodes124@gmail.com (D.R.)
2 Center for Plant Biology, Purdue University, West Lafayette, IN 47907, USA
3 EnP Investments, LLC, 2001 West Main Street, Mendota, IL 61342, USA; george@foliarpak.com
4 Purdue Institute of Inflammation, Immunology and Infectious Disease, Purdue University, West Lafayette, IN 47907, USA; tsors@purdue.edu
* Correspondence: jwidhalm@purdue.edu; Tel.: +1-765-496-3891

Received: 13 February 2020; Accepted: 3 March 2020; Published: 5 March 2020

Abstract: There is increasing interest in the use of amino acid-based biostimulant products due to their reported abilities to improve a number of quality characteristics in a variety of specialty crops. However, when it comes to the foliar application of amino acids to turfgrass, there are still many basic questions about their uptake forms and incorporation into cellular metabolism. In this study, we shed light on the fate of amino acids exogenously applied to turfgrass foliage through a series of time-course, isotopic-labeling studies in creeping bentgrass (*Agrostis stolonifera* L.) leaves. Using both ^{15}N-labeled and ^{15}N,^{13}C double-labeled L-glutamate applied exogenously to creeping bentgrass foliage, we measured the uptake of glutamate and its integration into γ-aminobutyric acid (GABA) and L-proline, two amino acids with known roles in plant stress adaptation. Our results demonstrate that glutamate is rapidly absorbed into creeping bentgrass foliage and that it is utilized to produce GABA and proline. Based on the labeling patterns observed in the endogenous pools of glutamate/glutamine, GABA, and the proline from applied glutamate-[$^{13}C_5{}^{15}N_1$], we can further conclude that glutamate is predominantly taken up intact and that mineralization into other forms of nitrogen is a minor fate. Taken together, the collective findings of this study provide evidence that amino acids exogenously applied to turfgrass foliage can be rapidly absorbed, and serve as stable sources of precursor molecules to be integrated into the metabolism of the plant.

Keywords: biostimulants; amino acids; isotopic labeling; turfgrass

1. Introduction

The use of biostimulants to promote quality traits in specialty crops has gone up over the last decade [1]. With an estimated annual growth rate of more than 10% each year, the projected market for biostimulants is estimated to be at $4.9 billion by 2025 [2]. Biostimulants is a broad term referring to extracts, lysates, purified natural compounds or microorganisms that are applied to crops in small amounts to enhance aspects like health, resiliency, and/or vigor [1,3] but whose primary role is not to fertilize or protect against pathogens [4].

Amino acids and small peptide-based biostimulants have received increased attention for their positive effects on plant performance [5]. Whilst externally applied amino acids are poorly taken up by roots because of competition with soil microbes, foliar application has the potential to improve availability due to reduced competition [6]. As a result, amino acids are emerging in many foliarly applied products marketed to golf course superintendents and sports turf managers with claims of enhanced growth, greening, and increased resistance to stress. Despite the substantial sales of such products from a variety of companies in the turfgrass market, there have been limited studies on the uptake by and the fate of amino acids in turfgrass foliage. Using ^{15}N-labeled glycine, L-glutamate, and L-proline, it was previously demonstrated that the nitrogen from these applied amino acids was absorbed into creeping bentgrass (*Agrostis stolonifera* L.) foliage to similar degrees as other nitrogen fertilizer forms [7]. Assuming that no mineralization into other transportable forms of nitrogen occurred on the leaf's surface, this suggests that amino acids can be directly taken up through bentgrass foliage. This now raises questions about the metabolic fate of exogenously applied amino acids once inside the plant. The objective of this study was to determine the uptake form, stability, and incorporation of amino acids exogenously applied into metabolism in turfgrass foliage. To accomplish this, we conducted a series of exploratory tracer studies using ^{15}N- and ^{15}N,^{13}C-labeled glutamate, applied exogenously to bentgrass foliage, and we measured their integration into endogenous amino acid pools and derived metabolites.

2. Materials and Methods

2.1. Plant Growth Conditions, General Experimental Procedures, and Reagents

Turfgrass used in this experiment was PennTrio Bentgrass (Tee-2-Green Corporation, Hubbard, OR, USA) which is a creeping bentgrass (*Agrostis stolonifera* L.) mix that contains equal parts Penncross, Penneagle, and Pennlinks. Turfgrass was grown in a controlled environment in 8″ pots at 23–24 °C with an average humidity of 45% under daylight spectrum fluorescent lighting with 12-h days. Plants were watered weekly and fertilized once at germination with a fertilizer containing 12% nitrogen, 6% phosphorus (P_2O_5), 6% potassium (K_2O), and micronutrients boron, copper, iron, manganese, and zinc. Unlabeled amino acid standards were purchased from Sigma-Aldrich (St. Louis, MO, USA). Stable isotopes were purchased from Cambridge Isotopes (Tewksbury, MA, USA). All other reagents were purchased from Fisher Scientific (Pittsburgh, PA, USA). For gas chromatography-mass spectrometry (GC-MS) experiments, an Agilent 7890B GC (Agilent Technologies, Santa Clara, CA, USA) connected with an Agilent 5966A mass spectrometer (Agilent Technologies, Santa Clara, CA, USA) were used. All analyses were done using Agilent Chemstation software.

2.2. Stable Isotope Labeling of Turfgrass

Labeling was conducted by spraying the foliage of the potted plants with a mixture of each stable-isotopically labeled amino acid in water at a rate of 804 L per hectare, at the concentrations indicated below. The first trial used 10 mM glutamate-[$^{15}N_1$] (Cambridge Isotopes) with sampling at 0, 1, 4, 8, 24, and 48 h post application. In the second trial, 4 mM glutamate-[$^{13}C_5$-$^{15}N_1$] was applied and sampling occurred at 0, 0.25, 0.5, 1, 4, 8, 24, 48, and 72 h post application. At each timepoint, the aboveground tissue was cut and rinsed to remove any residue and then the leaves were transferred directly into methanol to quench metabolism, and stored at 4 °C until extraction.

2.3. Extraction and Quantification of Amino Acids

Amino acids were extracted according to a protocol adapted from Rhodes et al. [8] using a ration of 10 mL methanol for every 500 mg of creeping bentgrass leaves. Extracts were spiked with 25 μL of 10 mM α-aminobutyrate, vortexed well, and then incubated in the dark at 4 °C for 2 d to extract metabolites. Next, for every 10 mL methanol, 5 mL chloroform and 6 mL water were added and incubated for 1 h at room temperature to allow phase separation. The aqueous phase was collected and evaporated to dryness under N_2 gas using a Techne sample concentrator. The dried aqueous phase was resuspended in 1 mL of water and applied to a Dowex-50-H^+ 200 mesh column. The column was washed with 7 mL water, and amino acids were eluted from the column with 6 mL of 6 M NH_4OH and dried. Amino acids were derivatized for GC-MS analysis, as described previously [9], with 1 μL of each derivatized sample being analyzed by GC-MS on an Agilent 19091s-433 HP-5MS capillary column (30 m × 0.25 mm; film thickness 0.25 μm) as described previously [8]. Labeling percentage was calculated by dividing the intensity of the shifted molecular ion by the sum of the shifted and unshifted ion and corrected for natural isotope abundance. See Supplementary Tables S1 and S2 for masses analyzed for each labeled and unlabeled amino acid.

3. Results and Discussion

3.1. Nitrogen from Foliar Applied Glutamate is Incorporated into Proline and γ-aminobutyric acid (GABA)

To investigate whether amino acids are absorbed by turfgrass leaves and incorporated into cellular metabolism, we measured time course labeling in the endogenous pools of glutamate and some major glutamate-derived amino acids from glutamate-[$^{15}N_1$] applied to the foliage of creeping bentgrass. In addition to serving as the precursor for the synthesis of chlorophylls and proteins, glutamate functions as a hub metabolite in plant amino acid metabolism (Figure 1). Glutamate is a substrate for producing L-glutamine from ammonia; it serves as the primary α-amino donor for aminotransferases involved in synthesizing multiple amino acids, and its carbon skeleton and amino group are directly incorporated into L-arginine, L-proline, and γ-aminobutyric acid (GABA) [10]. The accumulation of GABA, a non-proteinogenic amino acid found ubiquitously in plants, functions in adaptive responses to mitigate plant stress, including defense against drought and insect herbivory [11]. The overproduction of proline was also demonstrated to be a metabolic response involved in plant stress tolerance. Proline functions as an osmolyte to maintain cell turgor, stabilizes membranes to prevent electrolyte leakage, and helps prevent oxidative bursts by lowering the concentrations of reactive oxygen species [12]. Therefore, because enhanced resiliency to environmental stresses underlies one of the major purported benefits of amino acid-based biostimulant products, we focused on labeling in GABA and proline from glutamate-[$^{15}N_1$] exogenously applied to creeping bentgrass foliage.

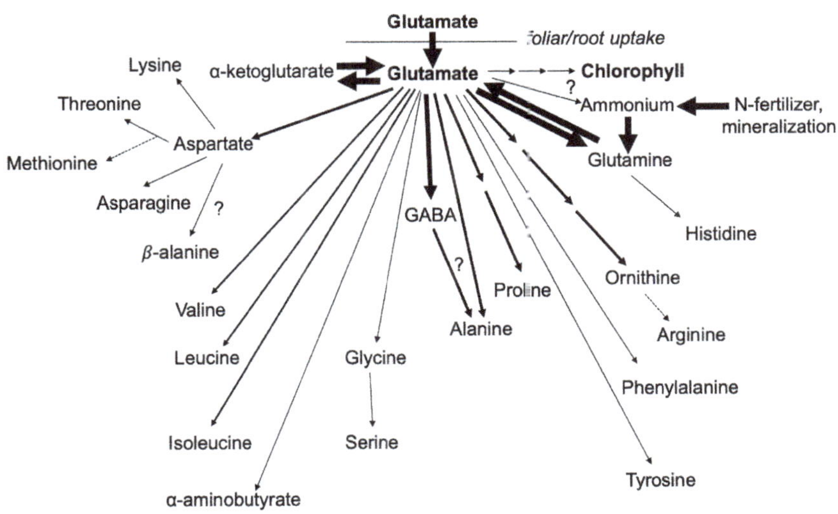

Figure 1. Glutamate occupies a central position in plant amino acid metabolism. The arrows indicate the multiple fates for the carbon backbone and/or amino group of glutamate in plant cells. The arrow thickness approximately correlates with relative flux toward each metabolite. The arrows labeled with a question mark (aspartate to β-alanine (aspartate decarboxylase), GABA to alanine (GABA: pyruvate aminotransferase), and glutamate to ammonium (glutamate dehydrogenase)) denote metabolic fates that are controversial.

To ensure that the endogenous precursor pool was labeled highly enough to detect possible labeling in GABA and proline, we first examined labeling in glutamate by looking at the glutamate/glutamine pool. Note that in the current sample preparation protocol, glutamine is converted into glutamate during derivatization, so the two amino acids are quantified together by GC-MS as glutamate. Within 1 h of foliar application with glutamate-[$^{15}N_1$], the glutamate/glutamine pool was labeled by 60% and remained constant over the 48-h experiment (Figure 2). The pool of GABA, which is formed via the irreversible decarboxylation of glutamate in plant cytoplasm by glutamate decarboxylase (GDC) [10], was labeled by 29% within 1 h of glutamate-[$^{15}N_1$] application, increased to over 40% labeled 4 h post application, and then remained relatively constantly labeled for the duration of the experiment (Figure 2). The rapid incorporation of glutamate into GABA is consistent with the observation that the expression of the gene encoding GDC in rice roots increased nearly 10-fold in response to exogenous application of glutamate [13]. Labeling in proline, whose biosynthesis from glutamate can take place in chloroplasts or cytoplasm [14], was in comparison expectedly delayed (Figure 2). The proline pool was labeled by 12% 4 h after application with glutamate-[$^{15}N_1$], increased to 23% labeled by 8 h, and then remained constant until 48 h. Taken together with the fact that glutamate must be present in cytoplasm to produce GABA and in the cytoplasm or chloroplast to synthesize proline, these data are consistent with not only glutamate-[$^{15}N_1$] being absorbed into the foliage of creeping bentgrass, but also with it being taken up by cells where it can be utilized to produce metabolites with well-established roles in plant stress adaptation.

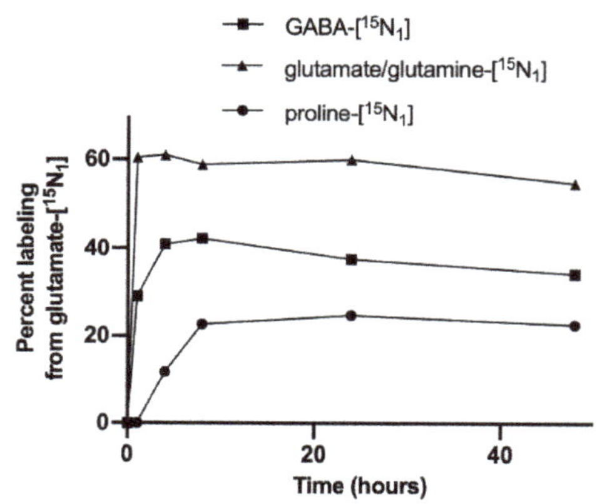

Figure 2. Time course of percent labeling of glutamate/glutamine-[$^{15}N_1$], GABA-[$^{15}N_1$], and proline-[$^{15}N_1$], from 10 mM glutamate-[$^{15}N_1$] applied to the foliage of creeping bentgrass (*Agrostis stolonifera* L.).

3.2. The Carbon Skeleton from Foliar Applied Glutamate is also Incorporated into Proline and GABA

In a previous labeling study by Stiegler et al. [7], it was found that the uptake of nitrogen from glycine, glutamate, and proline into creeping bentgrass foliage is equal to or less than that of nitrogen from urea. Thus, it is possible that glutamate-[$^{15}N_1$] applied to creeping bentgrass foliage in the current study was mineralized on the leaf surface and that ammonia-[^{15}N] was absorbed and then re-assimilated into glutamine/glutamate (Figure 1) before being used to synthesize GABA-[$^{15}N_1$] and proline-[$^{15}N_1$] (Figure 2). To definitively determine whether glutamate-[$^{15}N_1$] was taken up intact or mineralized before absorption, we performed the same time course labeling experiment with glutamate-[$^{13}C_5{}^{15}N_1$]. By using double-labeled glutamate, in which the nitrogen and all carbon atoms are labeled, it is possible to differentiate between the uptake of mineralized ammonia-[^{15}N] and the intact amino acid.

Similarly to what was observed with glutamate-[$^{15}N_1$] (Figure 2), applied glutamate-[$^{13}C_5{}^{15}N_1$] rapidly labeled the glutamate/glutamine pool (Figure 3A). The predominant form detected was the fully intact form, glutamate/glutamine-[$^{13}C_5{}^{15}N_1$], which represented approximately 55% of the total pool and remained relatively constant for the duration of the experiment. The second most abundantly labeled form detected was glutamate/glutamine-[$^{13}C_5$]. It was found to represent approximately 10% of the total pool and then attenuated to nearly 0% by 24 h after application. This form would a priori derive from the metabolism of glutamate-[$^{13}C_5{}^{15}N_1$] to α-ketoglutarate-[$^{13}C_5$] that is transaminated back to glutamate-[$^{13}C_5$] with an unlabeled nitrogen. The least abundant form detected was glutamate/glutamine-[$^{15}N_1$], representing less than 3% of the total pool by 1 h post application and rapidly decreasing thereafter. This form likely results from the labeled nitrogen of absorbed glutamate-[$^{13}C_5{}^{15}N_1$] being used to transaminate an unlabeled α-ketoglutarate to produce glutamate-[$^{15}N_1$]. This form could also originate if applied glutamate-[$^{13}C_5{}^{15}N_1$] was mineralized on the leaf surface to produce ammonia-[^{15}N] that was absorbed and then re-assimilated back into amino acid metabolism to produce glutamate-[$^{15}N_1$] (Figure 1). Regardless of how it was formed, because glutamate-[$^{15}N_1$] accounted for such a small fraction of the total glutamate/glutamine pool compared to the ^{13}C-labeled forms, this suggests that the intact amino acid was the predominant form absorbed by turfgrass foliage.

Next, we examined whether the glutamate-[$^{13}C_5{}^{15}N_1$] applied to creeping bentgrass foliage labeled GABA and proline like what was observed with glutamate-[$^{15}N_1$] (Figure 2). Peak labeling in GABA occurred 1 h after application with glutamate-[$^{13}C_5{}^{15}N_1$], though labeling was already detectable at 15 min (Figure 3B). Unlike the first experiment (Figure 2), there was a decrease in labeled GABA pools following the initial peak (Figure 3B). This likely reflects the fact that less glutamate-[$^{13}C_5{}^{15}N_1$] was administered. Previous work in rice by Kan et al. [13] showed that the expression of the gene encoding GDC displays a sensitive dosage-dependent induction in response to glutamate. The subsequent decline and increase in GABA labeling is likely related to the incorporation of GABA into the GABA shunt, a bypass pathway in which the GABA produced in the cytoplasm is imported into the mitochondria, where it is converted to succinate that can enter the tricarboxylic acid (TCA) cycle. The GABA shunt is the major source of succinate in foliage during the day (reviewed in Michaeli and Fromm, 2015 [15]).

The most abundantly labeled form of GABA detected was GABA-[$^{13}C_4{}^{15}N_1$], which represented approximately 9% of the total pool. This isotopic form likely originated from decarboxylation of glutamate-[$^{13}C_5{}^{15}N_1$], the predominant labeled form found in the glutamate pool (Figure 3A). The other isotopic forms of GABA detected after 1 h, GABA-[$^{13}C_4$] and GABA-[$^{15}N_1$], represented 7.5% and 3.3% of the total pool, respectively (Figure 3B). Because GABA-[$^{15}N_1$] is a priori synthesized from glutamate-[$^{15}N_1$], the observation that GABA-[$^{15}N_1$] was the least abundant labeled form present is consistent with glutamate-[$^{15}N_1$] being the minor form in the glutamate pool (Figure 3A). Along the same lines, proline-[$^{13}C_5{}^{15}N_1$] and proline-[$^{13}C_5$] were more abundant than proline-[$^{15}N_1$]; however, like in the previous experiment (Figure 2), labeling was delayed, peaking at 24 h post application with glutamate-[$^{13}C_5{}^{15}N_1$] (Figure 3C). Thus, in all cases, the double-labeled ^{13}C and ^{15}N isotopes and the single-labeled ^{13}C isotopes of glutamate, GABA, and proline were more abundant than the single-labeled ^{15}N isotopic forms (Figure 3A–C). These data imply that intact amino acids are taken up by turfgrass foliage rather than being mineralized to other transportable forms of nitrogen. The data also indicate that once inside the plant, exogenously applied amino acids are imported into cells where they can be rapidly and directly incorporated into metabolism.

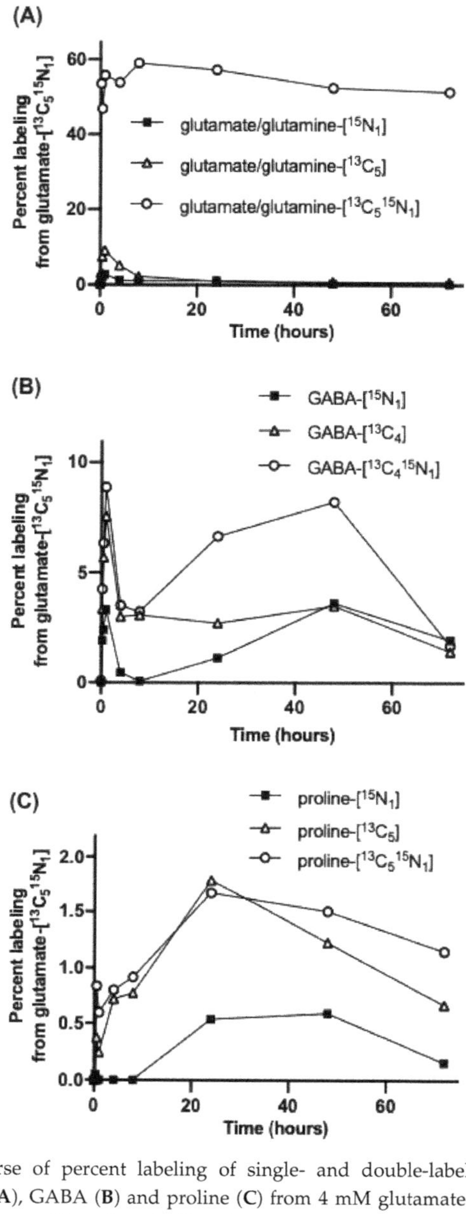

Figure 3. Time course of percent labeling of single- and double-labeled isotopic forms of glutamate/glutamine (**A**), GABA (**B**) and proline (**C**) from 4 mM glutamate-[$^{13}C_5{}^{15}N_1$] applied to foliage of creeping bentgrass (*Agrostis stolonifera* L.).

4. Conclusions

In this exploratory study, we investigated questions about the uptake forms and the incorporation of exogenously applied amino acids on turfgrass foliage. Through time course labeling studies with glutamate-[$^{15}N_1$] and glutamate-[$^{13}C_5{}^{15}N_1$], we demonstrated that glutamate is rapidly absorbed intact into creeping bentgrass leaves and directly utilized as a precursor to synthesize GABA and proline, two well-studied glutamate-derived metabolites with roles in plant stress adaptation. Our results also provide evidence that the mineralization of glutamate into other nitrogen forms is likely a minor fate of the amino acids applied to the foliage, though future work measuring the formation and foliar uptake of other nitrogen forms should be performed to independently investigate this question. Furthermore, the labeling in the endogenous pools of glutamate/glutamine remained stable for 72 h, the latest point measured in this study. Taken together, the collective findings of our work suggest that amino acids applied to turfgrass foliage, like those in some specialty turf care products, can be rapidly absorbed and serve as stable sources of precursor molecules to be integrated into the metabolism of the plant.

Supplementary Materials: The following are available online at http://www.mdpi.com/2073-4395/10/3/358/s1, Table S1: Fragments of each labeled and unlabeled amino acid from glutamate-[$^{15}N_1$] as analyzed by GC-MS; Table S2: Fragments of each labeled and unlabeled amino acid from glutamate-[$^{13}C_5{}^{15}N_1$] as analyzed by GC-MS.

Author Contributions: Conceptualization, D.R., T.G.S, G.C.M., and J.R.W.; methodology, D.R., T.G.S, G.C.M., and J.R.W.; formal analysis, R.M.M., G.W.M., D.R., and J.R.W.; investigation, R.M.M., G.W.M., and D.R.; resources, G.C.M and J.R.W.; writing—original draft preparation, R.M.M., G.W.M., D.R., G.C.M., and J.R.W.; writing—review and editing, R.M.M., G.W.M., D.R., G.C.M., T.G.S., and J.R.W.; visualization, R.M.M., G.W.M., D.R., and J.R.W.; supervision, D.R., G.C.M., and J.R.W.; project administration, D.R., G.C.M., T.G.S., and J.R.W.; funding acquisition, D.R., G.C.M., and J.R.W. All authors have read and agreed to the published version of the manuscript.

Funding: This research was funded in part by EnP Investments, LLC (Mendota, IL USA), start-up funds from Purdue University to J.R.W., and by the USDA National Institute of Food and Agriculture Hatch Project number 177845.

Acknowledgments: We thank Kyle Ladenburger of EnP Investments, LLC for his assistance in performing amino acid applications and collecting tissue.

Conflicts of Interest: G.C.M. is the president of EnP Investments, LLC (manufacturer of the Foliar-Pak brand, Mendota, IL USA) where he is also the chief formulator and inventor. He was involved in the design of the study and collection of the samples, and decision to publish, but was not involved in the analyses or interpretation of data. The remaining authors declare no competing financial or non-financial interests.

References

1. Rouphael, Y.; Colla, G. Synergistic biostimulatory action: Designing the next generation of plant biostimulants for sustainable agriculture. *Front. Plant Sci.* **2018**, *871*, 1–7. [CrossRef] [PubMed]
2. Markets & Markets Biostimulants Market by Active Ingredient (Humic Substances, Amino Acids, Seaweed Extracts, Microbial Amendments), Crop Type (Fruits & Vegetables, Cereals, Turf & Ornamentals), Application Method, Form, and Region – Global Forecast to 2025. Available online: https://www.marketsandmarkets.com/Market-Reports/biostimulant-market-1081.html?gclid=EAIaIQobChMI08H2466A6AIVUdHeCh0yBg7xEAAYASAAEgKOuPD_BwE (accessed on 4 March 2020).
3. Zhang, X.; Schmidt, R.E. Hormone-containing products' impact on antioxidant status of tall fescue and creeping bentgrass subjected to drought. *Crop Sci.* **2000**, *40*, 1344–1349. [CrossRef]
4. Du Jardin, P. Plant biostimulants: Definition, concept, main categories and regulation. *Sci. Hortic.* **2015**, *196*, 3–14. [CrossRef]
5. Colla, G.; Hoagland, L.; Ruzzi, M.; Cardarelli, M.; Bonini, P.; Canaguier, R.; Rouphael, Y. Biostimulant action of protein hydrolysates: Unraveling their effects on plant physiology and microbiome. *Front. Plant Sci* **2017**, *8*, 1–14. [CrossRef] [PubMed]
6. Colla, G.; Nardi, S.; Cardarelli, M.; Ertani, A.; Lucini, L.; Canaguier, R.; Rouphael, Y. Protein hydrolysates as biostimulants in horticulture. *Sci. Hortic.* **2015**, *196*, 28–38. [CrossRef]
7. Stiegler, J.C.; Richardson, M.D.; Karcher, D.E.; Roberts, T.L.; Norman, R.J. Foliar Absorption of Various Inorganic and Organic Nitrogen Sources by Creeping Bentgrass. *Crop Sci.* **2013**, *53*, 1148. [CrossRef]
8. Rhodes, D.; Hogan, A.L.; Deal, L.; Jamieson, G.C.; Haworth, P. Amino acid metabolism of *Lemna minor* L.: II. responses to chlorsulfuron. *Plant Physiol.* **1987**, *84*, 775–780. [CrossRef] [PubMed]
9. McCoy, R.M.; Utturkar, S.M.; Crook, J.W.; Thimmapuram, J.; Widhalm, J.R. The origin and biosynthesis of the naphthalenoid moiety of juglone in black walnut. *Hortic. Res.* **2018**, *5*, 1–11. [CrossRef] [PubMed]
10. Forde, B.G.; Lea, P.J. Glutamate in plants: Metabolism, regulation, and signalling. *J. Exp. Bot.* **2007**, *58*, 2339–2358. [CrossRef] [PubMed]
11. Bown, A.W.; Shelp, B.J. Plant GABA: Not Just a Metabolite. *Trends Plant Sci.* **2016**, *21*, 811–813. [CrossRef] [PubMed]

12. Hayat, S.; Hayat, Q.; Alyemeni, M.N.; Wani, A.S.; Pichtel, J.; Ahmad, A. Role of proline under changing environments: A review. *Plant Signal. Behav.* **2012**, *7*, 1456–1466. [CrossRef] [PubMed]
13. Kan, C.C.; Chung, T.Y.; Wu, H.Y.; Juo, Y.A.; Hsieh, M.H. Exogenous glutamate rapidly induces the expression of genes involved in metabolism and defense responses in rice roots. *BMC Genomics* **2017**, *18*, 1–17. [CrossRef] [PubMed]
14. Szepesi, Á.; Szollosi, R. Mechanism of Proline Biosynthesis and Role of Proline Metabolism Enzymes Under Environmental Stress in Plants. In *Plant Metabolites and Regulation Under Environmental Stress*; Academic Press: Cambridge, MA, USA, 2018; pp. 337–353.
15. Michaeli, S.; Fromm, H. Closing the loop on the GABA shunt in plants: Are GABA metabolism and signaling entwined? *Front. Plant Sci.* **2015**, *6*, 1–7. [CrossRef] [PubMed]

agronomy

MDPI

Article

The Use of a Plant-Based Biostimulant Improves Plant Performances and Fruit Quality in Tomato Plants Grown at Elevated Temperatures

Silvana Francesca [1], Carmen Arena [2], Bruno Hay Mele [3], Carlo Schettini [4], Patrizia Ambrosino [5], Amalia Barone [1] and Maria Manuela Rigano [1,*]

[1] Department of Agricultural Sciences, University of Naples Federico II, Portici, 80055 Naples, Italy; silvana.francesca@unina.it (S.F.); ambarone@unina.it (A.B.)

[2] Department of Biology, University of Naples "Federico II", via Cinthia 4, 80126 Napoli, Italy; carena@unina.it

[3] Stazione Zoologica Anton Dohrn, Villa Comunale, 80121 Napoli, Italy; bruno.haymele@szn.it

[4] ALMA SEGES Soc. Coop., Eboli, 84025 Salerno, Italy; carlo@almaseges.com

[5] Agriges S.r.l., San Salvatore Telesino, 82035 Benevento, Italy; patrizia.ambrosino@agriges.com

* Correspondence: mrigano@unina.it; Tel.: +39-0812532125

Received: 21 February 2020; Accepted: 3 March 2020; Published: 6 March 2020

Abstract: Abiotic stresses can cause a substantial decline in fruit quality due to negative impacts on plant growth, physiology and reproduction. The objective of this study was to verify if the use of a biostimulant based on plant and yeast extracts, rich in amino acids and that contains microelements (boron, zinc and manganese) can ensure good crop yield and quality in tomato plants grown at elevated temperatures (up to 42 °C). We investigated physiological responses of four different tomato landraces that were cultivated under plastic tunnel and treated with the biostimulant CycoFlow. The application of the biostimulant stimulated growth (plants up to 48.5% taller) and number of fruits (up to 105.3%). In plants treated with the biostimulant, antioxidants contents were higher compared to non-treated plants, both in leaves and in fruits. In particular, the content of ascorbic acid increased after treatments with CycoFlow. For almost all the traits studied, the effect of the biostimulant depended on the genotype it was applied on. Altogether, the use of the biostimulant on tomato plants led to better plant performances at elevated temperatures, that could be attributed also to a stronger antioxidant defence system, and to a better fruit nutritional quality.

Keywords: antioxidants; biostimulant; tomato; fruit quality; abiotic stress

1. Introduction

Tomato (*Solanum lycopersicum* L.) is one of the most consumed vegetables worldwide also owing to the development of products such as soups, juices, purees, and sauces [1]. Tomato is an essential component of the Mediterranean diet and of other traditional diets. However, heat can negatively affect vegetative and reproductive growth phases in tomato resulting in up to 70% harvest losses [2,3]. Indeed, in tomato, when temperatures exceed 35 °C different physiological functions result adversely affected including seed germination, seedling and vegetative growth, flowering and fruit set and ripening [3]. High temperature stress leads also to inhibition of chlorophyll biosynthesis and of photosystem II activity [4]. Indeed, photosynthesis is one of the processes most affected by elevated temperatures [5].

Considering the importance of this crop, the development of new management practices to enhance tolerance to abiotic stresses, including heat stress, could contribute to global food production. The use of biostimulants is proposed as an innovative solution to address the novel challenge to improve the sustainability of agricultural systems and reduce the use of chemical fertilizers [6,7]. The most accepted and complete definition of a biostimulant is the one from Du Jardin that defines a plant biostimulant as "any substance or microorganism that applied to plants, regardless of its nutrients content, is able to enhance nutrition efficiency and also abiotic stress tolerance and quality traits" [8]. Du Jardin allocated the biostimulants into eight classes: humic substances, complex organic materials, beneficial chemical elements, inorganic salts, seaweed extracts, chitin and chitosan derivatives, anti-transpirant and free amino acids and considered other N-containing substances with microorganism a potential ninth category. The mechanisms activated in plants by the different biostimulants are still not known as they can act directly on plant metabolism and physiology or indirectly on soil conditions [9]. The effects of biostimulants compounds include stimulation of enzyme activities of glycolysis, Krebs cycle, nitrate assimilation, and of hormonal activities [10]. It has been also demonstrated that biostimulants application is able to enhance tolerance to different abiotic stresses, such as drought [11,12], salinity [7,13,14], and thermal stresses [15]. For example, it has been demonstrated that applications of algal extracts are able to promote tolerance to drought, salinity, and heat, while extracts rich in amino acids can help increasing tolerance to thermal stresses [16,17]. Lettuce plants (*Lactuca sativa*) treated with a mixture derived from enzymatic hydrolysis of proteins and subjected to cold showed higher fresh weights and better stomatal conductance compared to non-treated plants [18]. In another work, perennial ryegrass (*Lolium perenne* L.) treated with hydrolyzed amino acids had improved photosynthetic efficiency compared to non-treated plants at high temperatures (36 °C) [15]. In general, the application of amino acids was found to exert positive effects on plant growth due to their use for the biosynthesis of a large number of non-protein nitrogenous compounds (pigments, vitamins, coenzymes, purine, and pyrimidine bases). Therefore, amino acids applications could directly influence the physiological activity in plant growth and yield also under abiotic stress [19]. Protein hydrolysates can also improve soil respiration, microbial biomass and activity and impact on plant nutrition by forming complexes and chelates between amino acids and soil nutrients [20].

To improve the tolerance to high temperatures the use of biostimulants has been previously investigated, even if it is presently unclear to what extent these compounds are able to improve the physiological performances of tomato plants under elevated temperatures [7]. We hypothesize that the use of an amino acid-based biostimulant could stimulate natural processes to enhance plant performances also at elevated temperatures. Indeed, the use of protein hydrolysates could directly stimulate carbon and nitrogen metabolism and indirectly enhance nutrient availability, nutrient uptake and nutrient use-efficiency in plants [21]. To verify this hypothesis, we used a novel plant-based biostimulant named CycoFlow and we performed physiological and biochemical analyses on four different tomato landraces grown at elevated temperatures and treated or not with this biostimulant. We reasoned that treatments with CycoFlow could facilitate stress adaptation because of its putative cytokinin-like action and its high concentration of glycine betaine known to mitigate the effect of heat stress [7,22]. Considering climate changes and the expected rise of temperatures in the next few years, to understand the contribution of biostimulants to ensure good plant performances at high temperatures may become increasingly important.

2. Materials and Methods

2.1. Plant Growth, Experimental Design, and Treatments

One-month-old tomato seedlings (landraces E17, E36, E107, PDVIT, described in Table 1) were transplanted in May 2018 under walk-in plastic tunnel (22 × 8 m^2) in Battipaglia in the Campania Region in Southern Italy (40°57′68″N 14°95′97″E). The tunnel was covered in polyethylene sheet and was open on both sides. Microclimatic conditions and temperatures were not regulated but were recorded during the growing season. All four genotypes have an indeterminate growth habit. The genotype E17 is characterized by large fruits (200–500 g), the genotype E107 is characterized by medium-sized fruits (70–100 g) and the E36 and the PDVIT genotypes are characterized by small cherry fruits (Table 1). Only the mature fruits of the E107 genotype are yellow while the fruits of the other genotypes are red. Tomato plants were grown following the standard cultural practices of the area. The experimental design consisted of a completely randomized design with three replicates *per* treatment and ten plant *per* each biological replication. There were two different groups: one control, which did not receive any biostimulant, and one that was treated with the biostimulant. The biostimulant CycoFlow (Agriges, Benevento, Italy) was produced by mixing sugar cane molasses with yeast extract obtained by autolysis of previously grown *Saccharomyces cerevisiae* yeasts. It is rich in high free amino acids, peptides, nucleotides, B-vitamins, trace elements, and other growth factors. Its chemical composition contains total nitrogen of 4.5% and organic carbon of 19.5%. The aminogram of the Biostimulant Cyco Flow is reported in Supplementary Table S1. The product contains also Boron (0.2%), Manganese (1%) and Zinc (1.2%). The biostimulant has a pH of 5.0, a density of 1200 kg/m^3 and an EC value of 15.0 dS/m. The Biostimulant, in liquid formulation, was initially applied directly to the soil (400 mL *per* plant) at the moment of transplanting, and thereafter every 15 days, until the end of the cultivation cycle for a total of four total applications. CycoFlow was applied by fertigation at a final concentration of 3 g/L. The control and the treatment groups received the same amount of water. No fertilizer has been applied. During the whole growing period climatic data (Figure S1) were recorded using the weather station VantagePro2 from Davis Instrument Corp. At the end of the cultivation cycle, plants were harvested and separated into leaves, stems, roots and fully ripe fruits. Plant height, numbers of leaves *per* plant, fresh weight of biomass, total number of fruits, weight of fruit and final yield were recorded. Dry biomass (in grams) was determined by drying plant tissues to constant weight in a forced-air-oven at 80 °C for 72 hours. Measurements were done on three randomly selected plants *per* each biological replication *per* genotypes for each treatment.

Table 1. Details of the tomato genotypes used in this study.

Genotype	Origin	Common Accession	Fruit Size	Fruit Color
E17	Italy	Pantano Romanesco	Big (200–250 g)	red
E36	Italy	Riccia San Vito	Small (25–30 g)	red
E107	Spain	E-L-19	Medium (70–100 g)	yellow
PDVIT	Italy	Caramella	Small (10–15 g)	red

2.2. Pollen Viability

Pollen viability was analyzed using five flowers *per* plant sampled from three different plants *per* replicate. In the laboratory, pollen grains were spread on microscope slides. One droplet of DAB solution (SIGMA) was added on each pollen sample; slides were gently warmed with a gas lighter and mounted with a cover slip [23]. Scoring was made using an LEITZ Laborlux12 microscope.

2.3. Ascorbic Acid Quantification

Reduced ascorbic acid (AsA) and total ascorbic acid (AsA + dehydroascorbate − DHA) measurements were carried out by using a colorimetric method [24] with modifications reported by Rigano et al. [25,26]. Briefly, 500 mg of frozen powder from tomato fruits or leaves were extracted with 300 µL of ice cold 6% trichloroacetic acid (TCA) and the mixture was then incubated for 15 min on ice and centrifuged at 14,000 rpm for 20 min. For reduced AsA evaluation, to 20 µL of supernatant were added 20 µL of 0.4 M phosphate buffer (pH 7.4), 10 µL of double distilled (dd) H_2O and 80 µL of color reagent solution. This solution was prepared by mixing solution A (31% (*w*/*v*) H_3PO_4, 4.6% (*w*/*v*) TCA and 0.6% (*w*/*v*) $FeCl_3$) with solution B (4% (*w*/*v*) 2,2′-Dipyridyl). For total AsA, to 20 µL of sample, 20 µL of 5 mM dithiotreitol in 0.4 M phosphate buffer (pH 7.4) were added and the mixture was incubated for 20 min at 37 °C. Ten microliters of N-ethyl maleimide (NEM; 0.5% (*w*/*v*) in water) were added and left for 1 min at room temperature. Eighty microliters of color reagent were added as previously described for reduced AsA. Both the final mixtures were incubated at 37 °C for 40 min and measured at 525 nm by using a Nano Photometer TM (Implen, Munich, Germany). Three separated biological replicates for each sample and three technical assays for each biological repetition were measured. The concentration was expressed in mg/100 g of fresh weight (FW).

2.4. Total Carotenoids and Chlorophylls Content

The evaluation of total carotenoids and chlorophylls was carried out according to the method reported by Wellburn [27] and by Zouari et al. [28] as modified by Rigano et al. [2]. To obtain the lipophilic extract, 0.25 grams of sample were extracted with 24 mL of acetone/hexane (40/60, *v/v*). The mixture was centrifuged at 15,000 rpm for 5 min at 4 °C. Supernatants were collected and stored at −20 °C until analyses. For carotenoids and chlorophylls a and b levels determination, absorbance of lipophilic extracts was read at 470, 663, and 645 nm, respectively. For lycopene and β-carotene levels absorbance was read at 505 and 453 nm, respectively. Results were converted into mg/100 g FW. Three separated biological replicates for each sample and three technical assays for each biological repetition were measured.

2.5. Antioxidant Activity Determination

Hydrophilic antioxidant activity (HAA) was evaluated in the water-soluble fraction, obtained by adding to 2 g of frozen powder 25 mL of 80% methanol, using the ferric reducing/antioxidant power (FRAP) method [29] with slight modifications. The FRAP assay was carried out by adding in a vial 2.5 mL of acetate buffer at pH 3.6, 0.25 mL of TPTZ solution (10 mM) in 40 mM HCl, 0.25 mL of $FeCl_3 \cdot 6H_2O$ solution (12 mM), and 150 μL of methanolic extract. The mixture was incubated for 30 min in the dark, and then readings of the colored products (ferrous tripyridyltriazine complex) were taken at 593 nm using a spectrophotometer. Results were expressed as micromoles of Trolox equivalents (TE) per 100 g FW. Lipophilic antioxidant activity (LAA) determination was carried out according to the 2,2′-azinobis-(3-ethylbenzothiazoline-6-sulfonic acid) (ABTS) method, using the lipophilic extract obtained as described in the previous paragraph [30]. The ABTS assay was based on the reduction of the ABTS•+ radical action by the antioxidants present in the sample. A solution constituted by 7.4 mM ABTS•+ (5 mL) mixed with 140 mM $K_2S_2O_8$ (88 μL) was prepared and stabilized for 12 h. This mixture was then diluted by mixing ABTS•+ solution with ethanol (1:88) to obtain an absorbance of 0.70 ± 0.10 unit at 734 nm using a spectrophotometer. Methanolic extracts (100 μL) were allowed to react with 1 mL of diluted ABTS•+ solution for 2.5 min, and then the absorbance was taken at 734 nm using a spectrophotometer. All biological replicates of samples were analyzed in triplicate. Results were expressed as micromoles of TE per 100 g FW.

2.6. Fluorescence Emission Measurements

Fluorescence emission measurements were performed on five replicates *per* each treatment, coming from five different plants. A portable FluorPen FP100max fluorometer, equipped with a light sensor (Photon System Instruments, Brno, Czech) was used for measurements, following the procedure reported in Figlioli et al. [31]. The ground fluorescence signal, F_o, was induced on 40′ dark adapted leaves, by a blue LED internal light of about 1–2 μmol m^{-2} s^{-1}. The maximal fluorescence level in the dark, F_m, was induced by a 1s saturating light pulse of 3000 μmol m^{-2} s^{-1}. The maximum quantum efficiency of PSII photochemistry, F_v/F_m, was calculated as $(F_m - F_o)/F_m$, according to Kitajima and Butler [32].

2.7. Leaf Functional Traits Determination

Fully expanded leaves, without apparent damages, were collected to determine the functional leaf traits following Arena et al. [33]. Leaf area (LA) was measured by the program Image J 1.45 (Image Analysis Software) and expressed in *per* square centimeter, specific leaf area (SLA) was measured as the ratio of leaf area to leaf dry mass and expressed as square centimeter *per* gram dry weight (DW). For dry mass determination, leaves were dried at 70 °C for 48 h. Leaf dry matter content (LDMC) was measured as the oven-dry mass of a leaf divided by its water-saturated fresh mass and expressed as gram *per* gram of water saturated leaf mass (WSLM). Relative water content in leaves (RWC) was calculated by dividing the amount of water in the fresh leaf tissue by the water in the leaf tissue after rehydration multiplied by 100 [34].

2.8. Statistical Analysis

Data were subjected to analysis of variance using a two-way ANOVA. To separate means within each parameter, the Tukey-HSD's test was performed. Differences at $p < 0.05$ were considered to be significant. ANOVA was performed by using SPSS (Statistical Package for Social Sciences) Package 6, version 23.0. To explore the overall data, we used the R environment for statistical computing and graphics R Core Team (2018). We first selected variables of interest for each genotype, treatment and plant part (4 × 2 × 2) then calculated the arithmetic mean (n = 3), and finally used the scale function to center the data around the mean and scale it using the standard deviation. The transformed data were visualized using a heatmap (heatmap function). To aid interpretation of the data, we also performed an SVD-based Principal Component Analysis over the multivariate matrix (function prcomp in base R) after normalization.

3. Results

3.1. Phenotypic and Physiological Analyses

In this study four different tomato genotypes were transplanted under a plastic walk-in tunnel with a delay of one month compared to the usual transplanting period (tomato plants in the South of Italy are usually transplanted in April), thus imposing a high-temperature condition during flowering and fruit setting. Indeed, the maximum temperature of 32 °C during the day, which represents a critical threshold in the sensitive stages of reproductive development, was frequently exceeded in this trial [3] (Figure S1). The four different tomato landraces were treated with a plant-based biostimulant named CycoFlow. According to ANOVA analyses, the treatment with the biostimulant increased the height of genotypes E107 and PDVIT by 48.5% and 30.1%, respectively (Supplementary Table S2). Generally, the number of leaves was lower in the biostimulant treated group compared to the control, independently of the genotype it was applied to (no significant interaction G X T). For the fresh biomass parameter, in PDVIT the treatment with CycoFlow increased the above ground fresh biomass by 68.4% (Figure 1a). Genotypes E17 and E36 showed, instead, lower values in treated plants compared to non-treated ones (−53.8% and −21.1%, respectively). A slightly higher pollen viability was also observed in the genotypes treated with the biostimulant compared to the respective controls (Figure 1b). In particular, in the genotype E107 the treatment with the biostimulant increased pollen viability by 125%. Generally, the treatment with the biostimulant increased the number of fruits, independently of the genotype (no significant interaction G X T). In particular, the treatment with the biostimulant increased the number of fruits in the genotype PDVIT by 105.3% (Figure 1c). The medium fruit weight was significantly affected only by the factor genotype (Supplementary Table S2). Generally, the final yield (kg *per* plant) showed a tendency to be higher in all the samples from the treated genotypes, even though these differences were not significant (Figure 1d). Interestingly, the yield was significantly affected only by the factor treatment (Supplementary Table S2).

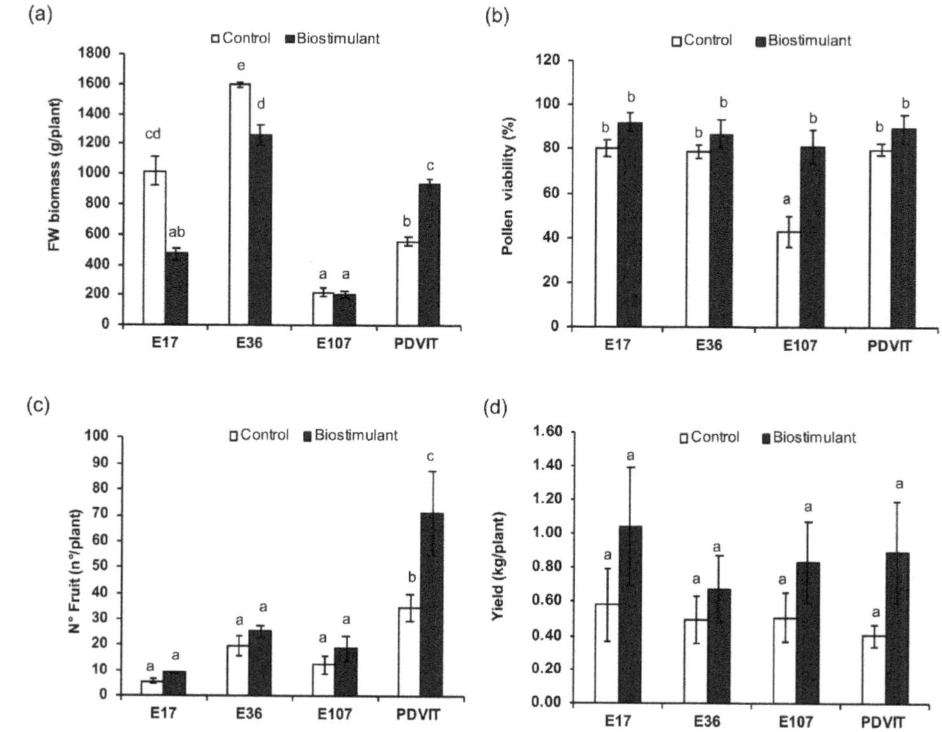

Figure 1. Effect of CycoFlow on: (**a**) Fresh weight (FW) biomass, (**b**) pollen viability, (**c**) fruit number, and (**d**) final yield in four tomato genotypes. Values are mean ± SE. Different letters indicate significant differences based on Tukey-HSD test ($p \leq 0.05$).

The treatment with the biostimulant CycoFlow also increased the maximal PSII photochemical efficiency (F_v/F_m) in the E107 and PDVIT genotypes (Figure 2). The monitoring of leaf functional traits evidenced that biostimulant application did not affect these traits significantly (Supplementary Table S3).

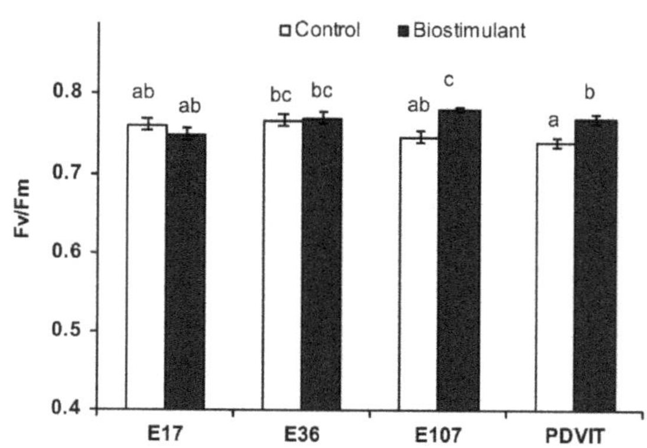

Figure 2. Maximal photochemical efficiency (F_v/F_m) in leaves of four tomato genotypes. Data are mean ± SE (n = 5). Different letters indicate significant differences based on Tukey-HSD test ($p \leq 0.05$).

3.2. Leaf and Fruit Antioxidant Content

The main interaction effects of the biostimulant Cyco Flow on the content of antioxidants in leaves from treated and non-treated plants is reported in Table 2.

Table 2. Analyses of variance and mean comparison for reduced and total ascorbic acid (AsA), total phenols, carotenoids, chlorophylls a and b and total lipophilic and hydrophilic antioxidant activities (LAA and HAA, respectively) in leaves of different tomato cultivars treated with the biostimulant CycoFlow applied by fertirrigation four times. Means ± SD within rows and columns followed by the different letter are significantly different based on Tukey-HSD test ($p \leq 0.05$).

		E17	E36	E107	PDVIT	SIGNIFICANCE	
Reduced AsA (mg/100 g FW)	control	6 ± 0.43 a	7.95 ± 1.33 a	10.83 ± 1 ab	18.20 ± 0.91 bc	G	**
	treated	20.05 ± 3.30 c	20.12 ± 1.42 c	17.41 ± 1.91 bc	19.34 ± 1.33 c	T	***
						GXT	**
Total AsA (mg/100 g FW)	control	16.79 ± 0.73 ab	14.45 ± 0.51 a	24.52 ± 2.03 bc	21.28 ± 0.86 bc	G	***
	treated	21.15 ± 0.90 bc	24.40 ± 2.55 cd	20.27 ± 0.83 cd	26.85 ± 0.69 d	T	***
						GXT	***
Phenols (mg/100 g FW)	control	43.38 ± 0.98 e	26.91 ± 1.19 a	35.14 ± 0.48 c	35.30 ± 0.56 c	G	***
	treated	25.33 ± 1.20 a	25.58 ± 0.27 a	31.57 ± 0.52 b	39.57 ± 0.54 d	T	***
						GXT	***
Carotenoids (mg/100 g FW)	control	23.91 ± 1.06 ab	26.06 ± 0.53 abc	23.80 ± 0.75 a	28.73 ± 0.23 de	G	***
	treated	23.78 ± 0.48 a	30.17 ± 0.24 e	28.10 ± 0.47 cde	26.42 ± 0.46 bcd	T	***
						GXT	***
Chl a (mg/100 g FW)	control	108.78 ± 3.05 a	113.30 ± 4.3 6 ab	128.22 ± 5.34 bc	140.30 ± 4.25 c	G	***
	treated	110.13 ± 1.37 a	137.08 ± 2.07 c	138.61 ± 3.32 c	130.20 ± 2.80 bc	T	**
						GXT	***
Chl b (mg/100 g FW)	control	38.65 ± 3.96 a	37.45 ± 2.12 a	45.84 ± 3.67 ab	55.75 ± 3.74 b	G	***
	treated	37.29 ± 2.73 a	55.41 ± 2.11 b	59.47 ± 2.69 b	45.85 ± 5.72 ab	T	**
						GXT	***
LAA (mg/100 g FW)	control	18.88 ± 0.14 a	18.75 ± 0.07 a	18.86 ± 0.04 a	18.62 ± 0.05 b	G	***
	treated	18.98 ± 0.04 a	19.07 ± 0.21 a	19.90 ± 0.08 b	19.75 ± 0.10 a	T	***
						GXT	***
HAA (mg/100 g FW)	control	828.58 ± 140.08 a	493.19 ± 220.27 bc	599.85 ± 118.33 ab	434.30 ± 88.34 cd	G	***
	treated	255.57 ± 91.31 d	390.49 ± 25.34 bc	510.33 ± 53.59 ab	438.26 ± 125.38 bc	T	***
						GXT	***

G = genotype; T = treatment; * = $p \leq 0.05$; ** = $p \leq 0.01$; *** = $p \leq 0.001$.

For the hydrophilic antioxidants, the treatment with the biostimulant increased the content of reduced AsA in the genotypes E17 and E36 and of total AsA in the leaves of the genotypes E36 and PDVIT. In particular, in the genotype E107 a 60.8% higher content of total AsA was registered in leaves treated with the biostimulant. As for the content of phenolic compounds, two genotypes (E17 and E107) showed lower contents of total phenols in the leaf after treatment with the biostimulant. In particular, in the E17 genotype a 41.6% decrease in the treated compared to the non- treated samples was demonstrated. Only in the PDVIT genotype the treatment with the biostimulant increased phenols content. It has been reported that phenolics compounds are the most important contributors to HAA [35]. Accordingly, in the leaves of the treated plants, HAA was lower in E17 compared to the respective non-treated control. For the lipophilic antioxidants, the treatment with the biostimulant increased the content of carotenoids in the genotypes E36 and E107 and the content of chlorophylls a and b only in the genotype E36. Particularly, the E36 genotype showed a 15.8% higher content of carotenoids in the treated leaves compared to the non-treated one, and 17.35% and 48% higher levels of chlorophyll a and b, respectively. The treatment with the biostimulant also increased total lipophilic antioxidant activities in E107 and surprisingly also in PDVIT, suggesting that other compounds outside of carotenoids contributed to this parameter.

In Table 3 is reported the content of hydrophilic antioxidants determined in red ripe fruit from genotypes treated or non-treated with the biostimulant CycoFlow. In general, the content of hydrophilic antioxidants in the fruits was higher in almost all the genotypes treated with biostimulants compared to the non-treated ones. The treatment with the biostimulant increased the content of reduced AsA independently of the genotype it was applied on (not significant interaction G X T). The content of reduced AsA was 28.7%–58.7% higher in fruits from treated genotypes compared to non-treated genotypes. Moreover, a content 112.8% higher of total AsA was registered in fruits from PDVIT treated with the biostimulant compared to the respective non-treated control. Contrary to what seen in the leaf, the content of total phenols in berries of treated E17 and E36 genotypes was higher compared to the non-treated control. In particular, in the E17 genotype 72.8% higher values were registered. Moreover, a significantly higher antioxidant activity HAA was demonstrated in fruits from E36 plants treated with CycoFlow, according to ANOVA analyses. Assessing the content of lipophilic antioxidants, the treatment with the biostimulant had no effects on the content of carotenoids and chlorophylls but only on the total lipophilic antioxidant activity, as reported in Supplementary Table S4. In particular, LAA was higher in fruits from the treated genotypes E17, E36, and E107.

Table 3. Analyses of variance and mean comparison for reduced and total ascorbic acid (AsA), total phenols, hydrophilic antioxidant activities (HAA) in fruits of different tomato cultivars treated with the biostimulant CycoFlow applied by fertirrigation four times. Means ± SD within rows and columns followed by the different letter are significantly different based on Tukey-HSD test ($p \leq 0.05$).

		E17	E36	E107	PDVIT	SIGNIFICANCE	
Reduced AsA (mg/100 g FW)	**control**	33.31 ± 2.99 a	39.56 ± 2.30 ab	47.14 ± 1.66 bc	50.36 ± 1.84 bc	G	***
	treated	47.36 ± 1.60 bc	59.87 ± 4.34 cd	74.79 ± 3.25 e	64.83 ± 2.34 de	T	***
						G X T	ns
Total AsA (mg/100 g FW)	**control**	61.97 ± 0.57 ab	78.03 ± 3.29 bc	85.40 ± 3.75 c	52.87 ± 4.24 a	G	**
	treated	79.34 ± 4.44 bc	87.15 ± 2.35 c	93.36 ± 6.19 cd	112.53 ± 4.08 d	T	***
						G X T	***
Phenols (mg/100 g FW)	**control**	9.62 ± 0.46 a	13.70 ± 0.68 b	16.17 ± 0.58 c	22.35 ± 0.37 e	G	***
	treated	16.62 ± 0.46 c	18.92 ± 0.76 d	16.88 ± 0.44 c	22.55 ± 0.19 e	T	***
						G X T	***
HAA (mg/100 g FW)	**control**	129.28 ± 33.95 a	189.22 ± 49.66 b	179.38 ± 20.62 bc	309.06 ± 39.51 d	G	***
	treated	151.57 ± 8.71 c	304.38 ± 30.92 c	212.47 ± 7.08 c	333.03 ± 46.91 d	T	***
						G X T	***

G = genotype; T = treatment; * = $p \leq 0.05$; ** = $p \leq 0.01$; *** = $p \leq 0.001$; ns = not significant.

3.3. Heat Map Analysis

A heat map providing the morphological, biochemical, and physiological changes in leaves and fruits of four different tomato genotypes in response to the addition of one biostimulant is displayed in Figure 3. With regard to leaves, the heat-map identified two main clusters which divided the analyzed samples differently (Figure 3, panel a). The first cluster separated the control genotypes E107 and E17 from the other genotypes and respective treated samples, the second cluster associated the treated genotypes E107, E17, and PDVIT in a sub-group and control PDVIT and E36 genotypes in another sub-group (Figure 3). Our data indicate that biostimulant application was the main clustering factor for E107, E17, and PDVIT genotypes, on the basis of differences in some leaf traits, F_v/F_m, phenols, yield and HAA, suggesting that the biostimulant utilization produces significant effect on many metabolites. The heat map built on tomato fruits clearly separated the treated PDVIT genotype from all others, in particular for number of fruits and reduced AsA (Figure 3, panel b), indicating this genotype as the most responsive to biostimulant application for fruit characteristics. A remarkable separation was also evident for control E107 and E36 compared to treated genotypes, grouped in two sub-clusters on the basis of pigments (chlorophylls and carotenoids) and LAA. A PCA analyses was also performed (Supplementary Figure S2). The PCA output further showed an evident separation between the treated and the non- treated genotypes.

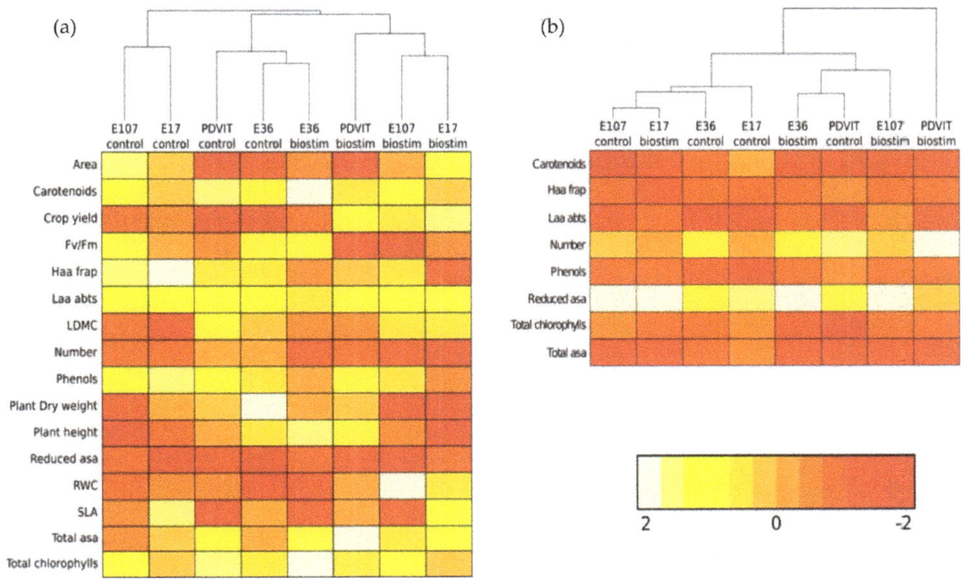

Figure 3. Cluster heat map analysis summarizing the behavior of the different tomato genotypes E36, E17, E107, PDVIT treated or non-treated with the biostimulant CycoFlow in leaf (panel **a**) and in fruit (panel **b**). The heat map was generated using the R environment for statistical computing and graphics (https://www.R-project.org/online) program package with Euclidean distance as the similarity measure and hierarchical clustering with complete linkage.

4. Discussion

In this paper four different tomato landraces were grown at elevated temperatures under a plastic walk-in tunnel and were treated or not with a plant-based biostimulant named CycoFlow. The higher height demonstrated in the majority of the tomato plants treated with CycoFlow compared to non-treated plants is in agreement with previous studies on different plant species and biostimulants [36–40]. Probably, the presence of signaling molecules in the biostimulant, such as free amino acids, promoted endogenous phytohormonal biosynthesis thus stimulating growth and also fruit setting [41]. Indeed, several authors demonstrated that the application of plant-based biostimulants exhibited cytokinin-like activity promoting cell division [42]. Moreover, cytokinins mitigate stresses induced by free radicals by direct scavenging and also by preventing ROS formation inhibiting xanthine oxidation [39]. Also, the treatment with CycoFlow overall increased the number of fruits, as previously demonstrated also in tomatoes treated with other biostimulants [10,36,39–41]. For example, Rouphael et al. [41] demonstrated that application of a protein hydrolysate in tomato increased in one cultivar the fruit mean weight and in another cultivar the number of fruits. In this study, in the genotype E107, the higher number of fruits observed was also linked to a higher pollen vitality observed after CycoFlow treatment. This result could be due to a combination of multiple effects. While the cytokinin-like activity could have favored cell division, the high level of proline present in the biostimulant, an amino acid whose natural content in the flower organs is ten times higher than that in the leaves, may have played an important role [31]. Indeed, it is known that also the amino acid proline promotes the translocation of nutrients towards developing flowers (sink) [43]. The positive effects of biostimulants based on amino acid on growth and yield is also due to the fact that the amino acids present in plant-based biostimulants stimulate plant defenses, participate in the synthesis of organic compounds (such as amines, purines, pyrimidines, vitamins) and affect the uptake of macro and micronutrients [37]. The CycoFlow effects observed in this study on yield and yield components are even more remarkable considering the elevated temperatures (up to 43 °C) reached under the plastic walk-in tunnel in Battipaglia. Indeed, this temperature normally impairs fertilization and reduces pollen viability [10]. It can be hypothesized that the presence of glycine betaine in the CycoFlow may have enhanced the tolerance of tomato plants to elevated temperatures. Indeed, it has been previously demonstrated that during tomato germination glycine betaine applied exogenously improved tolerance to high temperatures and enhanced the expression of heat shock genes [44]. At elevated temperatures, the glycine betaine compound may have also a crucial role in the repair of photodamaged PSII, in maintaining the activity of Rubisco and in alleviating the inhibition of gas exchanges [22]. Accordingly, a higher maximal photochemical efficiency was observed in the genotypes E107 and PDVIT treated with the biostimulant. These results are consistent with other papers, which demonstrated that applications of plant- and animal-based biostimulants are able to enhance photosynthetic rates and ensure a higher carbon assimilation efficiency [45,46]. For example, under drought stress conditions, Arabidopsis plants treated with an *Ascophyllum nodosum*-extract maintained a better photosynthetic performance compared to non-treated plants during the dehydration period, showing a higher capacity to dissipate thermally the excess of energy in the PSII reaction centers [47]. These results were linked to the fact that pre-treatments with the Ascophyllum-extracts induced partial stomatal closures and also modifications of the expression levels of genes involved in ABA-responsive and antioxidant system pathways [47]. Accordingly, our data indicate that biostimulant treatment induced the activation of the antioxidant defense system, as demonstrated by the higher content of reduced and total AsA in treated leaves. Although the precise reasons for these increases are not explained, it is known that biostimulants components, including glycine betaine, can promote the activity of specific enzymes involved in antioxidant homeostasis [22,41,48]. The ability to maintain an optimal chlorophyll content during heat stress is another key heat tolerance trait in tomato [49]. Interestingly, herein we observed higher contents of carotenoids and chlorophylls in two genotypes (E36 and E107) treated with the biostimulant compared to the non-treated samples. The higher chlorophylls content detected in these genotypes could be linked to limited chlorophyll degradation and leaf senescence [9]. In particular, this could be the case for the genotype E107 that demonstrated a higher maximal photochemical efficiency after treatment with the biostimulant.

The biostimulant-mediated effects on photosynthesis and secondary metabolism could also enhance fruit quality [10]. Indeed, one interesting finding of this study is the positive effect of the biostimulant CycoFlow on the quality of the tomato fruits. In general, the content of hydrophilic antioxidants in the fruits, including AsA, was higher in almost all the genotypes treated with biostimulants compared to the non-treated ones. Higher content of reduced AsA was observed in all the genotypes and of total AsA in the genotypes E17 and PDVIT. This result confirms data previously obtained in other studies that demonstrated an increase in AsA content in tomato, in kiwi fruits and in peppers after the application of plant-based biostimulants [36,41]. Contrary to what seen in the leaf, the content of total phenols in berries of treated E17 and E36 genotypes was higher compared to the non-treated control. Moreover, a significantly higher antioxidant activity HAA was demonstrated in fruits from E36 plants treated with CycoFlow. These results are in agreement with results previously obtained in other crops (soybean seeds, common bean, tomato, corn), even if the reported effects depended on the type of biostimulants, their concentrations and the number of applications [37]. Assessing the content of lipophilic antioxidants, the treatment with the biostimulant had no effects on the content of carotenoids and chlorophylls but only on the total lipophilic antioxidant activity. Similar results were obtained by Chehade et al. [36] in tomato. On the contrary, Rouphael et al. [41] demonstrated that in tomato foliar applications of a legume-derived protein hydrolysate had an effect also on lycopene content. Also, Colla et al. [10] demonstrated that foliar applications of protein hydrolysate, plant and seaweed extract affected lycopene content in greenhouse tomato. In the future, foliar application of CycoFlow will be also tested in order to verify if the results obtained in this study are also linked to the used application regimen.

Altogether, the genotypic factors remain decisive in the response obtained in the different tomato lines to the biostimulant. Indeed, for almost all the traits considered the effect of the biostimulant depended on the cultivar it was applied to, as seen by the interaction between the effect of the biostimulant and cultivars in most of the studied parameters. These variations can be explained by the differences in the genetic background between the different cultivars that were used in this study [33]. Indeed, the four genotypes here tested differed in terms of fruit shape and size and also in terms of fruit color (e.g., fruit of E107 is yellow). The geographical origin is also different with the E107 genotype coming from Spain and the other coming from Italy. These further highlight the fact that one biostimulant should be tested on a certain number of cultivars in order to assess its mechanisms of action.

5. Conclusions

In this paper we investigated the effects of the application of one plant-based biostimulant named CycoFlow on the nutritional quality and yield of tomatoes grown in walk-in tunnel under elevated temperatures. The application of the CycoFlow biostimulant had a clear effect on plant growth and final crop quality. Indeed, CycoFlow application had a significant effect on the content of hydrophilic antioxidants in both tomato leaves and fruits. In particular, the content of AsA increased after treatments with CycoFlow. Herein, the biostimulant application improved plant performances and fruit quality mostly in the genotypes E107 and PDVIT. In particular, in the genotype PDVIT application with CycoFlow determined a higher plant height, a higher number of fruits, a higher pollen vitality, a higher photochemical efficiency, a higher accumulation of AsA and a higher antioxidant activity. Additional studies are now planned in order to investigate if different applications regimen, such as foliar application, can also influence the observed effects.

Supplementary Materials: The following are available online at http://www.mdpi.com/2073-4395/10/3/363/s1, Table S1: Amino acid composition expressed in g/100 g of the biostimulant CycoFlow, Table S2: Analyses of variance and mean comparison for height, number of leaves, fresh weight (FW) and dry weight (DW) biomass, number of fruits, medium fruit weight, yields, and pollen viability (%) per plants of different tomato cultivars treated with the biostimulant CycoFlow applied by fertirrigation four times. Means ± SD within rows and columns followed by the different letter are significantly different based on Tukey-HSD test ($p \leq 0.05$). Table S3: Analyses of variance and mean comparison for maximal PSII photochemical efficiency (F_v/F_m), leaf area (LA), specific leaf area (SLA), leaf dry matter content (LDMC) and relative water content (RWC) *per* plants of different tomato cultivars treated with the biostimulant CycoFlow applied by fertirrigation four times. Means ± SD within rows and columns followed by the different letter are significantly different based on based on Tukey-HSD test ($p \leq 0.05$). Table S4: Analyses of variance and mean comparison for total lipophilic antioxidant activities (LAA), carotenoids, chlorophylls a and b (Chl A and Chl B, respectively) content in fruit of different tomato cultivars treated with the biostimulant CycoFlow applied by fertirrigation four times. Means ± SD within rows and columns followed by the different letter are significantly different based on Tukey-HSD test ($p \leq 0.05$). Figure S1: Maximum temperatures recorded in the experimental field located in Battipaglia during the day from May to August 2018. Figure S2: Principal component analysis (PCA) of phenotypic and physiological traits in tomato plants treated or not with the biostimulant CycoFlow. The treated genotypes are indicated by the letter T after the name.

Author Contributions: Conceptualization: S.F., A.B. and M.M.R.; Data Curation: S.F., M.M.R., C.A., B.H.M.; Funding acquisition: A.B.; Writing—original draft: S.F., M.M.R., C.A., A.B.; Writing—review and editing: S.F., M.M.R., C.A., A.B., P.A., C.S. All authors have read and agreed to the published version of the manuscript.

Funding: The authors have received funding from the European Union's Horizon 2020 research and innovation program through the TomGEM project under grant agreement No 679796. The APC was also funded by the TomGEM project under grant agreement No. 679796.

Conflicts of Interest: The authors declare no conflict of interest.

References

1. Del Giudice, R.; Petruk, G.; Raiola, A.; Barone, A.; Monti, D.M.; Rigano, M.M. Carotenoids in Fresh and Processed Tomato (*Solanum lycopersicum*) Fruits Protect Cells from Oxidative Stress Injury. *J. Sci. Food Agric.* **2016**, *97*, 1616–1623. [CrossRef] [PubMed]

2. Rigano, M.M.; Arena, C.; Di Matteo, A.; Sellitto, S.; Frusciante, L.; Barone, A. Eco-physiological response to water stress of drought-tolerant and drought-sensitive tomato genotypes. *Plant Biosyst.* **2016**, *150*, 682–691. [CrossRef]
3. Ruggieri, V.; Calafiore, R.; Schettini, C.; Rigano, M.M.; Olivieri, F.; Frusciante, L.; Barone, A. Exploiting genetic and genomic resources to enhance heat-tolerance in tomatoes. *Agronomy* **2019**, *9*, 22. [CrossRef]
4. Wen, J.; Jiang, F.; Weng, Y.; Sun, M.; Shi, X.; Zhou, Y.; Yu, L.; Wu, Z. Identification of heat-tolerance QTLs and high-temperature stress-responsive genes through conventional QTL mapping, QTL-seq and RNA-seq in tomato. *BMC Plant Biol.* **2019**, *19*, 398. [CrossRef] [PubMed]
5. Szymanska, R.; Slesak, I.; Orzechowska, A.; Kruk, J. Physiological and biochemical responses to high light and temperature stress in plants. *Environ. Exp. Bot.* **2017**, *139*, 165–177. [CrossRef]
6. Povero, G.; Mejia, J.F.; Di Tommaso, D.; Pioggesi, A.; Warrior, P. A systematic approach to discover and characterize natural plant biostimulant. *Front. Plant Sci.* **2016**, *7*, 435. [CrossRef]
7. Di Stasio, E.; Van Oosten, M.J.; Silletti, S.; Raimondi, G.; Dell'Aversana, E.; Carillo, P.; Maggio, A. *Ascophyllum nodosum*- based algal extracts act as enhancers of growth, fruit quality, and adaptation to stress in salinized tomato plants. *J. Appl. Physiol.* **2018**, *30*, 2675–2686. [CrossRef]
8. Du Jardin, P. Plant biostimulant: Definition, concept, main categories and regulation. *Sci. Hortic.* **2015**, *196*, 3–14. [CrossRef]
9. Di Mola, I.; Ottaiano, L.; Cozzolino, E.; Senatore, M.; Giordano, M.; El-Nakhel, C.; Sacco, A.; Youssef, R.; Colla, G.; Mori, M. Plant-Based Biostimulants Influence the Agronomical, Physiological, and Qualitative Responses of Baby Rocket Leaves under Diverse Nitrogen Conditions. *Plants* **2019**, *8*, 522. [CrossRef]
10. Colla, G.; Cardarelli, M.; Bonini, P.; Rouphael, Y. Foliar Application of Protein Hydrolysate, Plant and Seaweed Extracts Increase Yield but Differentially Modulate Fruit Quality of Greenhouse Tomato. *Hortic. Sci.* **2017**, *52*, 1214–1220. [CrossRef]
11. Feitosa de Vasconcelos, A.C.; Zhang, X.; Ervin, E.H.; De Castro Kiehl, J. Enzymatic antioxidant responses to biostimulants in maize and soybean subjected to drought. *Sci. Agric.* **2009**, *66*, 395–402. [CrossRef]
12. Petrozza, A.; Santaniello, A.; Summerer, S.; Di Tommaso, G.; Di Tommaso, D.; Paparelli, E.; Piaggesi, A.; Perata, P.; Cellini, F. Physiological responses to Megafol®treatments in tomato plants under drought stress: A phenomic and molecular approch. *Sci. Hortic.* **2014**, *174*, 185–192. [CrossRef]
13. Ertani, A.; Schiavon, M.; Muscolo, A.; Nardi, S. Alfalfa plant-dericed biostimulant stimulate short-term growth of salt stressed *Zea mays* L. plants. *Plant Soil* **2013**, *364*, 145–158. [CrossRef]
14. Lucini, L.; Rouphael, Y.; Cardarelli, M.; Canaguier, R.; Kumar, P.; Colla, G. The effect of a plant-derived biostimulant on metabolic profiling and crop performance of lettuce grown under saline conditions. *Sci. Hortic.* **2015**, *182*, 124–133. [CrossRef]
15. Botta, A. Enhancing plant tolerance to temperature stress with amino acids: An approach to their mode of action. *Acta Hortic.* **2013**, *1009*, 29–35. [CrossRef]
16. Verkleij, F.N. Seaweed extracts in agriculture and horticulture: A review. *Biol. Agric. Hortic.* **1992**, *8*, 309–324. [CrossRef]
17. Battacharyya, D.; Babgohari, M.Z.; Rathor, P.; Prithiviraj, B. Seaweed extracts as biostimulants in horticulture. *Sci. Hortic.* **2015**, *30*, 39–48. [CrossRef]
18. Van Oosten, M.J.; Pepe, O.; De Pascale, S.; Silletti, S.; Maggio, A. The role of biostimulants and bioeffectors as alleviators of abiotic stress in crop plants. *Chem. Biol. Technol. Agric.* **2017**, *4*, 5. [CrossRef]
19. Hammad, S.A.R.; Ali, A.M.O. Physiological and biochemical studies on drought tolerance of wheat plants by application of amino acids and yeast extract. *Ann. Agric. Sci.* **2014**, *59*, 133–145. [CrossRef]
20. De Pascale, S.; Rouphel, Y.; Colla, G. Plant biostimulants: Innovative tool for enhancing plant nutrition in organic farming. *Eur. J. Hortic. Sci.* **2017**, *82*, 277–285. [CrossRef]
21. Paul, K.; Sorrentino, M.; Lucini, L.; Rouphael, Y.; Cardarelli, M.; Bonini, P.; Moreno, M.B.M.; Reynaud, H.; Canaguier, R.; Trtilek, M.; et al. A combined phenotypic and metabolic approach for elucidating the biostimulant action of a plant-derived protein hydrolysate on tomato grown under limited water availability. *Front. Plant Sci.* **2019**, *10*, 493. [CrossRef] [PubMed]
22. Sorwong, A.; Sakhonwasee, S. Foliar Application of Glycine Betaine Mitigates the Effect of Heat Stress in Three Marigold (Tageteserecta) Cultivars. *Hortic. J.* **2015**, *84*, 161–171. [CrossRef]
23. Dafni, A. *Pollination Ecology: A Practical Approach*; Oxford University Press: Oxford, UK, 1992; p. xiv +250.

24. Stevens, R.; Buret, M.; Garchery, C.; Carretero, Y.; Causse, M. Technique for rapid small-scale analysis of vitamin C levels in fruit and application to a tomato mutant collection. *J. Agric. Food Chem.* **2006**, *54*, 6159–6165. [CrossRef] [PubMed]
25. Rigano, M.M.; Raiola, A.; Tenore, G.C.; Monti, D.M.; Del Giudice, R.; Frusciante, L.; Barone, A. Quantitative trait loci pyramiding can improve the nutritional potential of tomato (*Solanum lycopersicum*) fruits. *J. Agric. Food Chem.* **2014**, *62*, 11519–11527. [CrossRef] [PubMed]
26. Rigano, M.M.; Lionetti, V.; Raiola, A.; Bellincampi, D.; Barone, A. Pectic enzymes as potential enhancers of ascorbic acid production through the D-galacturonate pathway in Solanaceae. *Plant Sci.* **2018**, *266*, 55–63. [CrossRef] [PubMed]
27. Wellburn, A.R. The spectral determination of chlorophylls a and b, as well as total carotenoids, using various solvents with spectrophotometers of different resolution. *J. Plant Physiol.* **1994**, *144*, 307–313. [CrossRef]
28. Zouari, I.; Salvioli, A.; Chialva, M.; Novero, M.; Miozzi, L.; Tenore, G.C.; Bagnaresi, P.; Bonfante, P. From root to fruit: RNA-Seq analysis shows that arbuscular mycorrhizal symbiosis may affect tomato fruit metabolism. *BMC Genom.* **2014**, *15*, 221. [CrossRef]
29. Benzie, I.F.F.; Strain, J.J. The ferric reducing ability of plasma (FRAP) as a measure of "antioxidant power": The FRAP assay. *Anal Biochem.* **1996**, *239*, 70–76. [CrossRef]
30. Miller, J.N.; Rice-Evans, C.A. Factors influencing the antioxidant activity determined by the ABTS•+ radical cation assay. *Free Radic. Res.* **1997**, *26*, 195–199. [CrossRef]
31. Figlioli, F.; Sorrentino, M.C.; Memoli, V.; Arena, C.; Maisto, G.; Giordano, S.; Capozzi, F.; Spagnuolo, V. Overall plant responses to Cd and Pb metal stress in maize: Growth pattern, ultrastructure, and photosynthetic activity. *Environ. Sci. Pollut. Res. ESPR* **2019**, *26*, 1781–1790. [CrossRef]
32. Kitajima, M.; Butler, W.L. Quenching of chlorophyll fluorescence and primary photochemistry in chloroplasts by dibromothymoquinone. *Biochim. Biophys. Acta-Bioenerg.* **1975**, *376*, 105–115. [CrossRef]
33. Arena, C.; De Micco, V.; De Maio, A.; Mistretta, C.; Aronne, G.; Vitale, L. Winter and summer leaves of *Cistus incanus*: Differences in leaf morpho-functional traits, photosynthetic energy partitioning and poly-ADP-ribose polymerase (PARP) activity. *Botany* **2013**, *91*, 805–813. [CrossRef]
34. Hossain, M.I.; Khatun, A.; Talukder, M.S.A.; Dewan, M.M.R.; Uddin, M.S. Effect of drought o physiology and yield contributing characters of sunflower. *Bangladesh J. Agric Res.* **2010**, *35*, 113–124. [CrossRef]
35. Paradikovic, N.; Vinkovic, T.; VinkovicVrcek, I.; Zuntar, I.; Bojic, M.; Medic-Saric, M. Effect of natural biostimulants on yield and nutritional quality: An example of sweet yellow pepper (*Capsicum annuum* L.) plants. *J. Sci. Food Agric.* **2011**, *91*, 2146–2152. [CrossRef]
36. Chehade, L.A.; Al Chami, Z.; De Pascali, S.A.; Cavoski, I.; Fanizzi, F.P. Biostimulants from food processing by-products: Agronomic, quality and metabolic impacts on organic tomato (*Solanum lycopersicum* L.) *J. Sci. Food Agric.* **2017**, *98*, 1426–1436. [CrossRef]
37. Kocira, S.; Szparaga, A.; Kocira, A.; Czerwinska, E.; Wojtowicz, A.; Bronowicka-Mielniczuk, U.; Koszel, M.; Findura, P. Modeling biometric traits, yield and nutritional and antioxidant properties of seeds of three soybean cultivars through the application of biostimulant containing seaweed and amino acids. *Front. Plant Sci.* **2018**, *9*, 388. [CrossRef]
38. Cristiano, G.; Pallozzi, E.; Conversa, G.; Tufarelli, V.; De Lucia, B. Effect of an Animal-Derived Biostimulant on the Growth and Physiological Parameters of Potted Snapdragon (*Antirrhinum majus* L.). *Front. Plant Sci.* **2018**, *9*, 861. [CrossRef]
39. Polo, J.; Mata, P. Evaluation of a biostimulant (pepton) based in enzymatic hydrolyzed animal protein in comparison to seaweed extracts on root development, vegetative growth, flowering, and yield of gold cherry tomatoes grown under low stress ambient field conditions. *Front. Plant Sci.* **2018**, *8*, 2261. [CrossRef]
40. Hamasaki, T.; Kitagawa, T.; Yasuhara, T. Efficacy of yeast cell wall extract, a Byproduct of beer brewing, in Tomato (*Solanum lycopersicum*) culture. *IPCBEE* **2014**, *76*, 21–25. [CrossRef]
41. Rouphael, Y.; Colla, G.; Giordano, M.; El-Nakhel, C.; Kyriacou, M.C.; De Pascale, S. Foliar applications of a legume-derived protein hydrolysate elicit dose-dependent increases of growth, leaf mineral composition, yield and fruit quality in two greenhouse tomato cultivars. *Sci. Hortic.* **2017**, *226*, 353–360. [CrossRef]
42. Matsuo, S.; Kikuchi, K.; Fukuda, M.; Honda, I.; Imanishi, S. Roles and regulation of cytokinins in tomato fruit development. *J. Exp. Bot.* **2012**, *63*, 5569–5579. [CrossRef] [PubMed]

43. Sato, S.; Kamiyama, M.; Iwara, T.; Makita, N.; Furukawa, H.; Ikeda, H. Moderate increase of mean daily temperature adversely affects fruit set of *Lycopersicum esculentum* by distrupting specific physiological processes in male reproductive development. *Ann. Bot.* **2006**, *97*, 731–738. [CrossRef] [PubMed]
44. Li, S.; Li, F.; Wang, J.; Zhang, W.; Meng, Q.; Chen, T.H.H.; Murata, N.; Yang, X. Glycinebetaine enhances the tolerance of tomato plants to high temperature during germination of seeds and growth of seedlings. *Plant Cell Environ.* **2011**, *34*, 1931–1943. [CrossRef] [PubMed]
45. Cristiano, G.; Pallozzi, E.; Conversa, G.; Tufarelli, V.; De Lucia, B. Effects of an animal-derived biostimulant on the growth and physiological parameters of potted snapdragon (*Antirrhinum majus* L.). *Front. Plant Sci.* **2017**, *9*, 861. [CrossRef]
46. Xu, J.; Wolters-Arts, M.; Mariani, C.; Huber, H.; Rieu, I. Heat stress affects vegetative and reproductive performance and trait correlations in tomato (*Solanum lycopersicum*). *Euphytica* **2017**, *213*, 156. [CrossRef]
47. Santaniello, A.; Scartazza, A.; Gresta, F.; Loreti, E.; Biasone, A.; Di Tommaso, D.; Piaggesi, A.; Perata, P. *Ascophyllum nodosum* Seeweed Extract Alleviates Drought Stress in Arabidopsis by Affecting Photosynthetic Performance and Related Gene Expression. *Front. Plant Sci.* **2017**, *8*, 1362. [CrossRef]
48. Ertani, A.; Pizzeghello, D.; Francioso, O.; Sambo, P.; Sanchez-Cortes, S.; Nardi, S. *Capsicum chinensis* L. growth and nutraceutical properties are enhanced by biostimulants in a long-term period: Chemical and metabolomic approaches. *Front. Plant Sci.* **2014**, *5*, 375. [CrossRef]
49. Parrado, J.; Bautista, J.; Romero, E.J.; Garcìa-Martìnez, A.M.; Friaza, V.; Tejada, M. Production of a carob enzymatic extract: Potential use as a biofertilizer. *Bioresour. Technol.* **2008**, *99*, 2312–2318. [CrossRef]

agronomy

MDPI

Article

Changes in Assimilation Area and Chlorophyll Content of Very Early Potato (*Solanum tuberosum* L.) Cultivars as Influenced by Biostimulants

Wanda Wadas [1,*] and Tomasz Dziugieł [2]

1 Institute of Agriculture and Horticulture, Faculty of Agrobioengineering and Animal Husbandry, Siedlce University of Natural Sciences and Humanities, B. Prusa 14, 08-110 Siedlce, Poland

2 Main Inspectorate of Plant Health and Seed Inspection, Voivodeship Plant Health and Seed Inspection Service, Żółkiewskiego 17, 05-075 Warszawa-Wesoła, Poland; tomasz_dziugiel@wp.pl

* Correspondence: wanda.wadas@uph.edu.pl

Received: 7 February 2020; Accepted: 10 March 2020; Published: 12 March 2020

Abstract: This paper analyses the effects of foliar application of the seaweed extracts Bio-algeen S90 (*Ascophyllum nodosum*) and Kelpak SL (*Ecklonia maxima*), as well as the humic and fulvic acids ini HumiPlant (leonardite extract), on the assimilation area and chlorophyll content of very early potato cultivars ('Denar', 'Lord', Miłek'). The field experiment was carried out in central-eastern Poland over three growing seasons, using Luvisol. The biostimulants were applied according to the manufacturers' recommendations. The use of biostimulants resulted in enlargement of the assimilation area, but had no effect on the specific leaf area (SLA) or chlorophyll content (Soil Plant Analysis Development (SPAD) value). The assimilation area was larger, on average, by 0.0505 m^2 and leaf area index (LAI) was higher by 0.30 compared with the plants from the control group without a biostimulant. The SLA and SPAD depend on the cultivar and weather conditions, or nitrogen and magnesium content in soil, to a greater extent. The biostimulants enhanced abiotic stress tolerance and increased marketable tuber yield (diameter above 30 mm) 75 days after planting (the end of June), on average by 2.15 $t \cdot ha^{-1}$. Bio-algeen S90 and Keplak SL produced better results in a warm and very wet growing season, whereas HumiPlant produced better results in a year with lower air temperature and with drought periods during potato growth. No correlations were found between the tuber yield and assimilation area or between the tuber yield and SPAD value, although a significant negative correlation was found between the tuber yield and SLA.

Keywords: seaweed extract; humic acids; leaf area index (LAI); specific leaf area (SLA); Soil Plant Analysis Development (SPAD) index; tuber yield

1. Introduction

In recent years, the growth and productivity of crop plants have been greatly influenced by abiotic stresses. Periods of high temperature and drought are becoming more frequent in regions with extensively crop production, such as Central Europe, South-Central Asia, south-eastern South America and the south-eastern United States [1]. Under climate change conditions, biostimulants play an important role in sustainable crop production. These natural products (seaweed extracts, humic substances, hydrolysed proteins, and amino acids containing products or microorganism) contain a bioactive substance which enhances nutrition efficiency, abiotic stress tolerance, and/or crop quality traits, regardless of its nutrients content [2–5]. In recent years, the use of seaweed extracts and humic substances as plant growth stimulants has been increasing. Seaweed extracts and humic acids can promote plant growth, enhance abiotic stress tolerance as well as increase nutrient use efficiency [6–10].

Many plant growth-stimulating compounds (auxins, cytokinins, gibberellins, betaines, polysaccharides, polyamines, abscisic acids, brassinosteroids, and minerals) have been identified from seaweed. The chemical composition of seaweed extracts depends on the algae species and on the method of extraction. Brown algae (*Phaeophyta*) are most commonly used for the manufacture of extracts used as biostimulants of plant growth, including *Ascophyllum nodosum* and *Ecklonia maxima* [7,8,11]. An increase in leaf area and chlorophyll content are common plant responses to seaweed extract treatment. Cytokinins present in the seaweed extracts stimulate cell division, resulting in enlarged leaf area, and also stimulate chlorophyll biosynthesis, whereas betaines slow chlorophyll degradation and delay leaf senescence [7,8]. *Ascophyllum nodosum* extracts applied on foliage or to soil caused an increase in the leaf chlorophyll content of French bean, tomato, barley, maize, wheat, pepper, and strawberry [8,11,12]. A one-year study carried out in Iraq showed an increase in chlorophyll content in potato following the application of brown seaweed *Sargassum* extracts [13]. Foliar application of seaweed extracts *Ascophyllum nodosum* and *Ecklonia maxima* increased potato yield [14–16]. Biostimulants based on seaweed extracts improved plant growth and yield of wheat, barley, maize, potato, tomato, pepper, onion, and carrot [7,8,10,11].

The biological activity of humic substances depends on their source, chemical structure, and concentration. Humic substances may influence both respiration and photosynthesis. One of the effects of humic substances applied to growing plants was an increase in chlorophyll content, which can affect photosynthesis [17]. Leonardite is the most common commercial source of humic substances. Leonardite humic acids stimulate melon and soybean growth and chlorophyll synthesis [6]. A one-year study carried out in Iraq showed an increase in chlorophyll content in potato following the application of humic and fulvic acids in HumiMax [13]. A one-year study carried out in Egypt showed that the application of humic acid under water stress conditions enhanced the leaf chlorophyll content of very early potato cultivars [18]. Application of humic substances originating from leonardite increased potato yield and nutrient uptake [19]. In most experiments, foliar or soil application of humic and fulvic acids increased potato yield [13,20,21], but one study showed no clear effect of humic and fulvic acids on the potato yield [22]. Humic and fulvic acids improved plant growth and yield quality of wheat, maize, tomato, pepper and cucumber [2,22–24]. The effect of humic acids on plant growth depends of their source and concentration, and on the date and method (foliar or soil) of application, as well as the plant species and environmental conditions [9,17].

There is a relationship between leaf chlorophyll content and Soil Plant Analysis Development (SPAD) index [25]. Leaf SPAD values is related to nutrient plant status, especially nitrogen [26,27]. There was a relationship found between SPAD value and potato yield. A higher SPAD does not always guarantee a higher potato yield [28–31]. Plant-based biostimulants increased SPAD index and marketable yield of tomato and rocket [32–34].

To date, few studies have been focused on the effect of seaweed extract and humic acid application in early crop potato culture. The aim of the study was to determine the effect of foliar application of brown seaweed extracts and humic acids on the asssimilation area and chlorophyll content of very early potato cultivars. In the current study, it was hypothesised that seaweed extracts and humic acids could contribute to increasing assimilation area and chlorophyll content and, as a result, increase the early crop potato yield. The assumption was also made that the response to the application of these biostimulants depends on the cultivar and environmental conditions.

2. Materials and Methods

2.1. Experimental Site and Season

The study was carried out in central-eastern Poland (52°03′N, 22°33′E), over three growing season 2012–2014, on Luvisol with a low total nitrogen content, a high content of available phosphorus, a medium-to-high content of potassium and a low-to-medium content of magnesium, with an acidic-to-slightly-acid reaction. Spring triticale was grown as a potato forecrop. Farmyard manure was applied in autumn, at rate of 25 t·ha^{-1}, and mineral fertilizers were applied at rates of 80 kg N (ammonium nitrate), 35 kg P (superphosphate) and 100 kg K (potassium sulphate) per hectare in spring. Potato cultivation was carried out according to common agronomical practice.

The thermal and moisture conditions during the potato growth period were different (Table 1). The mean air temperatures were above or similar to the long-term average. In 2012, total precipitation was similar and, in 2013 and 2014, above the long-term average, although it was unevenly distributed during the potato growth period. The most favourable hydrothermal conditions for early crop potato culture were in the warm and moderately wet growing season of 2012. The next year, 2013 was warm and with heavy rainfall, whereas 2014 was cool with heavy rainfall after plant emergence and a drought in the period of tuber growth.

Table 1. Hydrothermal conditions during potato growing period.

Month	Temperature (°C)				Rainfall (mm)				Hydrothermal Index		
	2012	2013	2014	Many year	2012	2013	2014	Many year	2012	2013	2014
April	8.9	7.4	9.8	8.3	29.9	36.0	45.0	41.2	1.1	1.6	1.5
May	14.6	15.3	13.5	12.2	53.4	105.9	92.7	53.0	1.2	2.2	2.2
June	16.3	18.0	15.4	16.8	76.2	98.8	55.4	63.8	1.5	1.8	1.2

Hydrothermal index value: up to 0.4 extremely dry; 0.41–0.7 very dry; 0.71–1.0 dry; 1.01–1.3 rather dry; 1.31–1.6 optimal; 1.61–2 rather humid; 2.01–2.5 humid; 2.51–3 very humid; >3 extremely humid [35].

2.2. Plant Material and Experimental Design

The field experiment was established in a split-plot design with three replications. The experimental factors were: (1) plant biostimulant; and (2) cultivar. The potato plants were treated with three biostimulants: Bio-algeen S90 and Keplak SL containing seaweed extracts, and HumiPlant based on humic and fulvic acids. Bio-algeen S90 is an extract from *Ascophyllum nodosum* which contains amino acids, vitamins, alginic acids and other active components of seaweeds, as well as macronutrients (N, P, K, Ca, Mg) and micronutrients (B, Fe, Cu, Mn, Zn, Se, Co). Kelpak SL is an extract from *Ecklonia maxima* containing auxin (11 mg·dm^{-3}) and cytokinin (0.031 mg·dm^{-3}). HumiPlant is an extract from leonardite which contains humic acid (12%) and fulvic acid (6%) as well as macronutrients (K, Ca, Mg, S) and micronutrients (Fe, Mn, B, Mo, Zn, Cu). The biostimulants were applied according to the manufacturers' recommendations: Bio-algeen S90–2 dm^3·ha^{-1} at the beginning of leaf development stage (BBCH 10–11) and 2 dm^3·ha^{-1} two weeks after the first treatment, Kelpak SL–2 dm^3·ha^{-1} at the leaf development stage (BBCH 14–16) and 2 dm^3·ha^{-1} two weeks after the first treatment, HumiPlant–2 dm^3·ha^{-1} at the leaf development stage (BBCH 14–16) and 2 dm^3·ha^{-1} one week after the first treatment. Potato plants sprayed with water were used as a control without a biostimulant.

The most popular very early potato cultivars (Denar, Lord and Miłek) in the research area were grown. In successive years, 6-weeks pre-sprouted seed potatoes were planted on April 12, April 18 and April 7 with a row spacing of 0.25 m and 0.675 m between rows. The plots were six rows wide and 4 m long (96 plants per plot). Potatoes were harvested 75 days after planting (the end of June).

2.3. Determination of Assimilation Area, Chlorophyll Content and Tuber Yield

At the tuber formation stage (BBCH 41–43), the assimilation area, leaf area index (LAI), specific leaf area (SLA), and chlorophyll content (SPAD value) were determined. The measurements were made on four successive randomized plants per plot. The assimilation area was measured by the weight method [36]. SLA was calculated as the ratio of assimilation area/weight of leaves [37].

The chlorophyll content was estimated with non-destructive methods using a portable SPAD-502 chlorophyll meter (Minolta, Osaka, Japan). The measurements were made on the youngest fully expanded leaf, i.e., the fourth or fifth leaf from the top.

The total and marketable tuber yield were determined. The marketable tuber yield constituted tubers with a transverse diameter above 30 mm, excluding cracked and deformed tubers. The marketable tuber yield was determined on the basis of the total tuber yield of ten successive plants per plot using a hand calibrator with a square hole.

2.4. Statistical Analysis

The results of the study were analysed statistically with an analysis of variance (ANOVA) for the split-pot design. The significance of differences between the compared averages was verified using Tukey's test at the significance level $p \leq 0.05$.

3. Results

3.1. Assimilation Area

The effect of biostimulants on the assimilation area depended on the weather conditions during potato growth (Table 2). In the year with the highest air temperature and heavy rainfall after plant emergence (2013), the greatest enlargement of the assimilation area was caused by Kelpak SL, whereas in the year with the lowest air temperature and heavy rainfall after plant emergence (2014), the greatest enlargement of assimilation area was caused by Bio-algeen S90. The assimilation areas were larger, on average, by 0.0624 m^2 (11.5%) and 0.0941 m^2 (10%) respectively, and the leaf area index (LAI) was higher by 0.37 and 0.56 compared with the plants from the control group without a biostimulant. Regardless of the biostimulant applied, the assimilation area was largest in the year with the highest air temperature and moderate rainfall at the end of May (Table 3).

Table 2. Assimilation area in relation to plant biostimulant, potato growing season and cultivar.

Plant Biostimulant	Years			Cultivar		
	2012	2013	2014	Denar	Lord	Miłek
			Assimilation leaf area (m^2)			
Without biostimulant	0.7131 b	0.5411 b	0.9438 b	0.7214 a	0.6425 b	0.8341 a
Bio-algeen S90	0.7847 a	0.5746 ab	1.0379 a	0.8120 a	0.8008 a	0.7845 a
Kelpak SL	0.7964 a	0.6035 a	0.9434 b	0.7871 a	0.7636 ab	0.7926 a
HumiPlant	0.7963 a	0.5947 ab	0.9170 b	0.7592 a	0.6903 b	0.8584 a
			LAI ($m^2 \cdot m^{-1}$)			
Without biostimulant	4.22 b	3.21 b	5.59 b	4.27 a	3.81 b	4.49 a
Bio-algeen S90	4.65 a	3.41 ab	6.15 a	4.81 a	4.75 a	4.65 a
Kelpak SL	4.72 a	3.58 a	5.59 b	4.66 a	4.43 ab	4.70 a
HumiPlant	4.72 a	3.52 ab	5.43 b	4.46 a	4.09 b	5.09 a
			SLA ($m^2 \cdot kg^{-1}$)			
Without biostimulant	2.87 b	3.63 a	3.33 a	3.37 a	3.26 a	3.20 a
Bio-algeen S90	3.20 a	3.53 a	3.41 a	3.38 a	3.36 a	3.41 a
Kelpak SL	3.07 ab	3.65 a	3.32 a	3.37 a	3.33 a	3.34 a
HumiPlant	3.12 a	3.54 a	3.32 a	3.21 a	3.31 a	3.46 a

Means within columns followed by the same letters do not differ significantly at $p \leq 0.05$.

The potato cultivars tested showed different responses to the biostimulants applied (Table 2). The type of biostimulant had a greatest effect on the assimilation area of the 'Lord' cultivar. The greatest enlargement of the assimilation area of 'Lord' was caused by Bio-algen S90. Following the application of this biostimulant, the assimilation area of 'Lord' was larger, on average, by 0.1583 m^2 (24.5%) and the LAI value was higher by 0.94 compared with the plants from the control without biostimulant. The differences were highest in the year with a low air temperature and heavy rainfall after the plant emergence (2014). Despite the biostimulant applied, the assimilation area was higher for 'Miłek' than for 'Denar' and 'Lord' (Table 3).

Table 3. Assimilation area in relation to potato growing season and cultivar.

Year and Cultivar	Weight of Leaves (kg)	Assimilation Leaf Area (m^2)	LAI ($m^2 \cdot m^{-1}$)	SLA ($m^2 \cdot kg^{-1}$)
		Year		
2012	0.254 b	0.7726 b	4.58 b	3.06 c
2013	0.169 c	0.5785 c	3.43 c	3.59 a
2014	0.287 a	0.9605 a	5.69 a	3.35 b
		Cultivar		
Denar	0.235 ab	0.7699 b	4.56 b	3.33 a
Lord	0.222 b	0.7243 c	4.29 c	3.31 a
Miłek	0.251 a	0.8174 a	4.84 a	3.35 a

Means within columns followed by the same letters do not differ significantly at $p \leq 0.05$.

Only in a warm and moderately wet growing season (2012), following application of Bio-algeen S90 and HumiPlant, was the specific leaf area (SLA) higher, on average, by 0.29 $m^2 \cdot kg^{-1}$ compared with the plants from the control group without biostimulant (Table 2). With the use of Kelpak SL, the difference was smaller and not statistically confirmed. The SLA depended to a greater extent on the weather conditions during potato growth. Irrespective of the treatment (with or without biostimulant), the SLA was highest in the year with the highest air temperature and heavy rainfall after plant emergence (Table 3). The type of biostimulant and cultivar interaction effect on SLA was not statistically confirmed (Table 2). Regardless of the treatment, the SLA values of the potato tested cultivars were similar (Table 3).

3.2. Chlorophyll Content (SPAD Value)

The biostimulants used in the experiment had no significant effect on the chlorophyll content in leaves (Figure 1). The SPAD value depended to a greater extent on the cultivar and weather or soil conditions during potato growth. Irrespective of the treatment (with or without biostimulant), the SPAD values were higher for 'Denar' and 'Lord' than 'Miłek'. The SPAD was highest in the warm and wet growing season (2013) and, at the same time, the highest content of total nitrogen and available magnesium in soil (Figure 2).

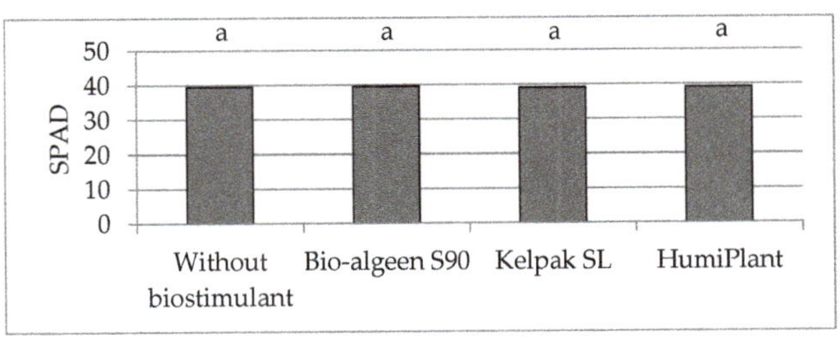

Figure 1. Effect of plant biostimulants on chlorophyll content (Soil Plant Analysis Development (SPAD) value); average of the three year tests on three cultivars. Means followed by the same letters do not differ significantly at $p \leq 0.05$.

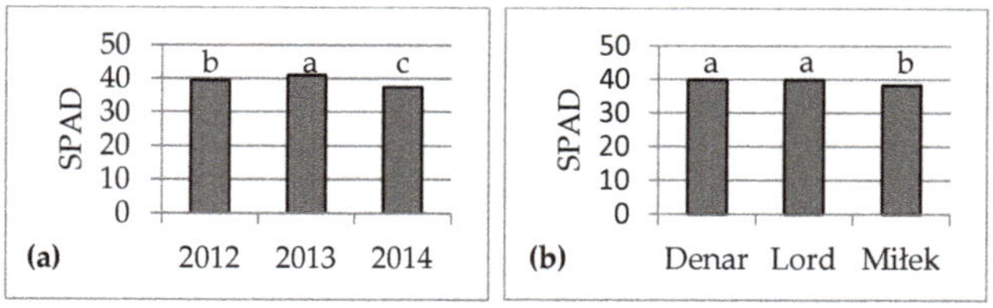

Figure 2. Chlorophyll content (SPAD value) in relation to potato growing season (**a**) and cultivar (**b**). Means followed by the same letters do not differ significantly at $p \leq 0.05$.

3.3. Relationship between Tuber Yield, Assimilation Area and Chlorophyll Content (SPAD Value)

The biostimulants used in the experiment had no effect on the weight of leaves [38], but caused enlargement of the assimilation area (Table 4). Over the three years of the study, the assimilation area was larger, on average, by 0.0505 m^2 (7%) and the LAI was higher by 0.30 compared with the plants from the control group without a biostimulant. The biostimulants had no significant effect on the SLA and SPAD (Figure 1).

Table 4. Effect of plant biostimulants on assimilation area; average of the three year tests on three cultivars.

Plant biostimulant	Weight of Leaves (kg)	Assimilation Leaf Area (m^2)	LAI ($m^2 \cdot m^{-1}$)	SLA ($m^2 \cdot kg^{-1}$)
Without biostimulant	0.234 a	0.7327 b	4.34 b	3.28 a
Bio-algeen S90	0.243 a	0.7991 a	4.74 a	3.38 a
Kelpak SL	0.268 a	0.7811 a	4.63 a	3.345 a
HumiPLant	0.230 a	0.7693 a	4.56 a	3.33 a

Means within columns followed by the same letters do not differ significantly at $p \leq 0.05$.

The biostimulants used in the experiment had a significant effect on the tuber yield [38]. The yield-increasing effects of biostimulants were comparable (Table 5). In the three years of the study, the total tuber yield was higher, on average, by 2.64 $t \cdot ha^{-1}$ (7.7%) and marketable tuber yield (diameter above 30 mm) by 2.15 $t \cdot ha^{-1}$ (6.5%). The yield-increasing effect of biostimulants depended on weather conditions during the potato growing season. Bio-algeen S90 and Kelpak SL caused the highest increase in tuber yield in the warm and very wet growing season (2013), and HumiPlant in the year with a low air temperature and a drought in the period of tuber growth (2014).

The tuber yield was not significantly correlated with the weight and assimilation leaf area or LAI (Table 6). A significant negative correlation was found between the marketable tuber yield and SLA. No significant correlation was found between the marketable tuber yield and SPAD value.

Table 5. Tuber yield in relation to plant biostimulant, potato growing season and cultivar.

Plant Biostimulant	Years			Cultivar			Mean
	2012	2013	2014	Denar	Lord	Miłek	
Total tuber yield ($t \cdot ha^{-1}$)							
Without biostimulant	40.26 a	31.46 b	31.38 b	33.54 a	34.49 a	35.07 a	34.37 b
Bio-algeen S90	40.45 a	36.88 a	33.19 ab	36.97 a	37.66 a	35.89 a	36.84 a
Kelpak SL	41.43 a	36.30 a	33.40 ab	35.63 a	37.64 a	37.89 a	37.04 a
HumiPlant	42.29 a	33.85 b	35.27 a	36.76 a	37.99 a	36.66 a	37.14 a
Marketable tuber yield ($t \cdot ha^{-1}$)							
Without biostimulant	39.49 a	29.75 b	29.34 b	31.84 a	33.10 a	33.64 a	32.86 b
Bio-algeen S90	38.96 a	34.64 a	30.62 b	34.49 a	35.47 a	34.26 a	34.74 ab
Kelpak SL	40.22 a	34.12 a	30.62 b	33.50 a	35.46 a	36.00 a	34.99 a
HumiPlant	41.43 a	31.61 ab	32.83 a	34.95 a	36.21 a	34.71 a	35.29 a

Means within columns followed by the same letters do not differ significantly at $p \leq 0.05$.

Table 6. Correlation coefficient between tuber yield and assimilation area and SPAD.

Plant Growth Characteristics	Total Tuber Yield	Marketable Tuber Yield
Weight of leaves	+0.1496	+0.1356
Assimilation leaf area	−0.0199	−0.0488
LAI	−0.0206	−0.0494
SLA	−0.5537*	−0.5767*
SPAD	+0.1886	+0.1894

* significant at $p \leq 0.05$.

3.4. *Effect of Experimental Factors on Assimilation Area, Chlorophyll Content and Tuber Yield*

The effect of the experimental factors and their interactions on potato assimilation area and chlorophyll content (SPAD value) are presented in Table 7.

Table 7. Effect of experimental factors on assimilation area, chlorophyll content (SPAD value) and tuber yield.

Experimental Factors	Weight of Leaves	Assimilation Leaf Area	LAI	SLA	SPAD	Total Tuber Yield	Marketable Tuber Yield
Year (Y)	**	**	**	**	**	**	**
Biostimulant (B)	ns	**	**	ns	ns	*	*
Y × B	ns	**	**	*	ns	*	*
Cultivar ©	*	**	**	ns	**	ns	ns
Y × C	ns	**	**	ns	ns	**	**
B × C	*	**	**	ns	ns	ns	ns
Y × B × C	ns	*	*	**	ns	ns	ns

* significant at $p \leq 0.05$, ** significant at $p \leq 0.01$, ns—non-significant.

4. Discussion

In sustainable crop production, biostimulants play an important role in improving plant growth and crop quality. Assimilation area and chlorophyll content are important parameters of assessment plant growth. The biostimulants used in the experiment caused enlargement of assimilation area, but had no effect on the chlorophyll content (SPAD value) in leaves of very early potato cultivars. SPAD value depended on the cultivar and weather or soil conditions to a greater extent. The effect of foliar application of seaweed extracts on potato assimilation area was comparable to humic and fulvic acids. In the three years of the study, following biostimulant application, the average leaf area index (LAI) was 4.64, being higher by 0.30 compared to the average for the untreated control group. Potato cultivars showed different responses to the applied biostimulants. Studies have shown the highest light absorption efficiency values at the LAI value of 3, which corresponded to maximum ground cover. If potato LAI exceeds 3, the intercepted photosynthetically active radiation value changes very little [39,40]. According to Howlader and Hoque [41], irrespective of potato cultivars, LAI increased progressively over time, reaching a peak at 60 days after planting and thereafter declining. The rate of assimilation area expansion showed the interaction between genotype and environment and varied by year [42], which was confirmed in the present study. The effect of seaweed extracts on potato assimilation area depended on the weather conditions after plant emergence. In the year with the highest air temperature and heavy rainfall after plant emergence, the assimilation area was larger after the application of Kelpak SL (*Ecklonia maxima*), whereas in the year with the lowest air temperature and with heavy rainfall after plant emergence, the assimilation area was larger after the application of Bio-algeen S90 (*Ascophyllum nodosum*). Potato plants are very sensitive to heat stress. In general, heat stress increases plant height, reduces leaf size, increases leaf chlorophyll content, and severely reduces tuber mass [43]. Kelpak SL contains auxins and cytokinins in a ratio of 350/1. Exogenous auxin plays an important role in plant stress resistance. The action of auxin depends on its concentration, the light conditions and carbohydrate content in the plant [44]. Exogenous cytokinins also play an important role in plant adaptation to environmental stresses [45]. Cytokinins present in the seaweed extracts stimulate cell division, resulting in enlarged leaf area [7,8], which was confirmed in the present study.

The leaf area index describes the growth of lowland fields, whereas the growth of individual plants is characterized by the specific leaf area (SLA). Biostimulants caused enlargement of the assimilation area, but had no effect on the SLA. The SLA for potato depends on the cultivar and growth stage, and temperature [42], which was confirmed in the present study. Early foliar expansion of potato is associated with a strong increase in SLA [41].

Foliar or soil application of *Ascophyllum nodosum* extracts caused an increase in the chlorophyll content of some agriculture (barley, wheat, maize) and horticulture (French bean, tomato, pepper, strawberry) plants [8,11,12], which was not confirmed in the present study. A study carried out in Egypt showed that the application of humic acid under water stress conditions enhanced the chlorophyll content of very early potato 'Spunta' grown on sandy soil [18], which was not confirmed in the present study with very early potato cultivars grown on loamy soil (Luvisol). A one-year study carried out in Iraq showed that foliar application of humic and fulvic acids caused an increase in the chlorophyll content of medium-early potato cultivar [13]. The effect of humic acids depends on their source and concentration, and on the date and method of application, as well as the plant species and cultivar [9]. The increase in chlorophyll alone does not necessarily result in higher yields [17,26].

The biostimulants used in the experiment enhanced tolerance to abiotic stress and improved crop quality. In the three years of the study, the marketable tuber yield (diameter above 30 mm) was higher, on average, by 2.15 $t \cdot ha^{-1}$. Bio-algeen S90 and Keplak SL containing seaweed extracts produced better results in a warm and very wet growing season, whereas HumiPlant based on humic and fulvic acids produced better results in a year with lower air temperature and with drought periods during potato growth.

A correlation between the tuber yield and assimilation area was not found. Li et al. [46] found a significant positive correlation between LAI and tuber yield, which suggests that the enlargement of leaf area could enhance the export of photosynthetic products and cause an increase in tuber yield. According to Ascione et al. [47], the tuber growth rate is only slightly correlated with LAI, and still less so with SLA, which was not confirmed in the present study. A significant negative correlation was found between the total and marketable (diameter above 30 mm) tuber yield and SLA.

No correlation was found between the tuber yield of three very early potato cultivars and SPAD value measured on the fourth or fifth leaf from the top at the tuber formation stage (BBCH 41-43), which suggest that the biostimulants used in the experiment had no effect on the plant nitrogen status. Bărăscu et al. [30] found a significant negative correlation between SPAD measured on the fourth and fifth leaves from the top and the tuber weight of two mid-early potato cultivars, which could have been associated with oxidative stress [29]. SPAD index as an indicator of crop nitrogen status may be used for the prediction of the potato yield, however a higher SPAD does not always guarantee a higher tuber yield [26,28,31]. SPAD value is a useful indicator for selecting the high yield cultivars in the early period, however, no single threshold leaf SPAD value can be used for all potato cultivars. The SPAD value can predict the level of tuber yield if the value is calibrated for a particular potato cultivar [28,31]. Establishing threshold SPAD value is quite difficult due to the influence of climate and technical factors. SPAD values can be affected by leaf age and position, as well as, time of the day [26,27]. As a rule SPAD measurements are carried out on the third-fifth leaf from the top. Recently it was demonstrated that there is a significance difference in SPAD values between the upper and lower leaves among potato cultivars. It was shown that cultivar affects the SPAD values of the fourth and eighth leaf, but does not affect SPAD value of the fourth-eighth leaves and the difference between SPAD of the fourth and eighth leaf. Therefore the SPAD values of the fourth-eighth leaves could be applied as a general index of nitrogen status across different potato cultivars [27].

5. Conclusions

In conclusion, the foliar application of seaweed extracts *Ascophyllum nodosum* (Bio-algeen S90) and *Ecklonia maxima* (Kelpak SL), as well as humic and fulvic acids from leonardite (HumiPlant), resulted in enlargement of the assimilation area of very early potato cultivars, but had no effect on the SLA or chlorophyll content (SPAD value). The assimilation area was larger, on average, by 0.0505 m^2 (7%), and LAI was higher by 0.30 compared with the plants from the control group without a biostimulant. The SLA and SPAD depend on the cultivar and weather conditions, or nitrogen and magnesium content, in soil to a greater extent. These biostimulants enhanced abiotic stress tolerance and increased marketable tuber yield (diameter above 30 mm) 75 days after planting (the end of June), on average, by 2.15 $t \cdot ha^{-1}$. Bio-algeen S90 and Keplak SL containing seaweed extracts produced better results in a warm and very wet growing season, whereas HumiPlant based on humic and fulvic acids produced better results in a year with lower air temperature and with drought periods during potato growth. No correlation was found between the tuber yield and assimilation area or between the tuber yield and SPAD value, although a significant negative correlation was found between the tuber yield and SLA.

Author Contributions: Conceptualization, W.W. and T.D.; methodology and formal analysis, W.W.; investigation, W.W. and T.D.; writing—original draft preparation, W.W. and T.D.; writing—review and editing, W.W. All authors have read and agreed to the published version of the manuscript.

Funding: This research was financed from the science grant granted by the Polish Ministry of Science and Higher Education, research theme number 218/05/S.

Conflicts of Interest: The authors declare no conflict of interest.

References

1. Carrão, H.; Naumann, G.; Barbosa, P. Mapping global patterns of drought risk: An empirical framework based on sub-national estimates of hazard, exposure and vulnerability. *Glob. Environ. Chang.* **2016**, *39*, 108–124. [CrossRef]
2. Calvo, P.; Nelson, L.; Kloepper, J.W. Agricultural uses of plant biostimulans. *Plant Soil* **2014**, *383*, 3–41. [CrossRef]
3. du Jardin, P. Plant biostimulants: Definition, concept. main categories and regulation. *Sci. Hort.* **2015**, *196*, 3–14. [CrossRef]
4. Van Oosten, M.J.; Pepe, O.; De Pascale, S.; Silletti, S.; Maggio, A. The role of biostimulants and bioeffectors as alleviators of abiotic stress in crop plants. *Chem. Biol. Technol. Agric.* **2017**, *4*, 1–12. [CrossRef]
5. Bulgari, R.; Franzoni, G.; Ferrante, A. Biostimulants application in horticultural crops under abiotic stress conditions. *Agronomy* **2019**, *9*, 306. [CrossRef]

6. Chen, Y.; Clapp, C.E.; Magen, H. Mechanism of plant growth stimulation by humic substances: The role of organo-iron complex. *Soil Sci. Plant Nutr.* **2004**, *50*, 1089–1095. [CrossRef]
7. Khan, W.; Rayirath, U.P.; Subramanian, S.; Jithesh, M.N.; Rayorath, P.; Hodges, D.M.; Critchley, A.T.; Craige, J.S.; Norrie, J.; Prithiviraj, B. Seaweed extracts as biostimulants of plant growth and development. *J. Plant Growth Regul.* **2009**, *28*, 386–399. [CrossRef]
8. Sharma, H.S.S.; Fleming, C.; Selby, C.; Rao, J.R.; Martin, T. Plant biostimulants: A review on the processing of macroalgae and use of extracts for crop management to reduce abiotic and biotic stresses. *J. Appl. Phycol.* **2014**, *26*, 465–490. [CrossRef]
9. Canellas, L.P.; Olivares, F.L.; Aguiar, N.O.; Jones, D.L.; Nebbioso, N.; Mazzei, P.; Piccolo, A. Humic and fulvic acids as biostimulants in horticulture. *Sci. Hort.* **2015**, *196*, 15–27. [CrossRef]
10. Begum, M.; Bordoloi, B.C.; Singha, D.D.; Ojha, N.J. Role of seaweed extract on growth, yield and quality of some agricultural crops: A review. *Agric. Rev.* **2018**, *39*, 321–326. [CrossRef]
11. Battacharyya, D.; Babgohari, M.Z.; Rathor, P.; Prithiviraj, B. Seaweed extracts as biostimulants in horticulture. *Sci. Hort.* **2015**, *196*, 39–48. [CrossRef]
12. Shukla, P.S.; Manti, E.G.; Adil, M.; Bajpai, S.; Critchley, A.T.; Prithiviraj, B. *Ascophyllum nodosum*—Based biostimulants: Sustainable applications in agriculture for the stimulation of plant growth, stress tolerance, and disease management. *Front. Plant Sci.* **2019**, *10*, 655. [CrossRef] [PubMed]
13. Sarhan, T.Z. Effect of humic acid and seaweed extracts on growth and yield of potato plant (*Solanum tuberosum* L.) Desiree cv. *Mesop. J. Agric.* **2011**, *39*, 19–25. [CrossRef]
14. Yanbo, X. Effect of bio-stimulant Kelpak on potato yield. *Chin. Potato J.* **2008**, *22*, 134–136.
15. Haider, M.W.; Ayyub, C.M.; Pervez, M.A.; Asad, H.U.; Manan, A.; Raza, S.A.; Ashraf, I. Impact of foliar application of seaweed extract on growth, yield and quality of potato (*Solanum tuberosum* L.). *Soil Environ.* **2012**, *31*, 157–162.
16. Wierzbowska, J.; Cwalina-Ambroziak, B.; Głosek-Sobieraj, M.; Sienkiewicz, S. Effect of biostimulators on yield and selected chemical properties of potato. *J. Elem.* **2015**, *20*, 575–768. [CrossRef]
17. Nardi, S.; Pizzeghello, D.; Muscolo, A.; Vianello, A. Physiological effects of humic substances on higher plants. *Soil Biol. Biochem.* **2002**, *34*, 1527–1536. [CrossRef]
18. Selim, E.M.; Shedeed, S.I.; Asaad, F.F.; El-Neklawy, A.S. Interactive effect of humic acid and water stress on chlorophyll and mineral nutrient contents of potato plants. *J. App. Sci. Res.* **2012**, *8*, 531–537.
19. Verlinden, G.; Pycke, J.; Mertens, J.; Debersaques, F.; Verheyen, K.; Baert, G.; Bries, J.; Haesaert, G. Application of humic substances results in consistent increase in crop yield and nutrient uptake. *J. Plant Nutr.* **2009**, *32*, 1407–1426. [CrossRef]
20. Alenazi, M.; Wahb-Allah, M.A.; Abdel-Razzak, H.S.; Ibrahim, A.A.; Alsadon, A. Water regimes and humic acid application influences potato growth, yield, tuber quality and water use efficiency. *Am. J. Potato Res.* **2016**, *93*, 463–473. [CrossRef]
21. Zhang, Y.; Zhou, Q.; Li, Y.; Xu, C.; He, S.; Huang, Y.; Yang, H.; Wang, M.; Liu, L.; Tang, M. Effect of fulvic acid on agronomic traits and yield of autumn potato. *Agric. Sci. Technol.* **2017**, *18*, 1448–1451.
22. Suh, H.Y.; Yoo, K.S.; Suh, S.G. Effect of foliar application of fulvic acid on plant growth and fruit quality of tomato (*Lycoporsicon esculentum* L.). *Hort. Environ. Biotechnol.* **2014**, *55*, 455–461. [CrossRef]
23. Karakurt, Y.; Unlu, H.; Unlu, H.; Padem, H. The influence of foliar and soli fertilization of humic acid on yield and quality of pepper. *Acta Agric. Scand. B-PS* **2009**, *59*, 233–237. [CrossRef]
24. Ozdamar Unlu, H.; Unlu, H.; Karakurt, Y.; Padem, H. Change in fruit yield and quality in response to foliar and soil humic acid application in cucumber. *Sci. Res. Essays* **2011**, *6*, 2800–2803.
25. Udding, J.; Gelang-Alfredsson, G.; Piikkki, K. Evaluating the relationship between leaf chlorophyll concentration and SPAD-502 chlorophyll meter readings. *Photosynth. Res.* **2007**, *91*, 37–46. [CrossRef]
26. Gianquinto, G.; Goffart, J.P.; Olivier, M.; Guarda, G.; Colauzzi, M.; Dalla Costa, L.; Delle Vedove, G.; Vos, J.; Mackerron, D.K.L. The use of hand-held chlorophyll meters as a tool to assess the nitrogen status and guide nitrogen fertilization of potato crop. *Potato Res.* **2004**, *47*, 35–80. [CrossRef]
27. Li, R.; Chen, J.; Qin, Y.; Fan, M. Possibility of using a SPAD chlorophyll meter to establish a normalized threshold index of nitrogen status in different potato cultivars. *J. Plant Nutr.* **2019**, *42*, 834–841. [CrossRef]
28. Su, Y.; Guo, H.; Chen, Y. Relationship between SPAD readings chlorophyll contents and yield of potato (*Solanum tuberosum* L.). *Southwest China J. Agric. Sci.* **2007**, *4*, 690–693.

29. Raminez, D.A.; Yactayo, W.; Gutiérrez, R.; Mares, V.; De Mendiburu, F.; Posadas, A.; Quiroz, R. Chlorophyll concentration in leaves is an indicator of potato tuber yield in water-shortage conditions. *Sci. Hort.* **2014**, *168*, 202–209. [CrossRef]
30. Bărăscu, N.; Duda, M.M.; Olteanu, G. Study of dynamics SPAD and NDVI values of potato plants according to the differentiated fertilization. *Bull. UASVM Agric.* **2016**, *73*, 5–14. [CrossRef]
31. Dvořák, P.; Král, M. Effect of organic mulching on soil water potential and SPAD values as factors on yield of potatoes (*Solanum tuberosum* L.). *Tarim Bilim. Derg.* **2019**, *25*, 147–154. [CrossRef]
32. Caruso, G.; De Pascale, S.; Cozzolino, E.; Cuciniello, A.; Cenvinzo, V.; Bonini, P.; Colla, G.; Rouphael, Y. Yield and nutrition quality of vesuvian piennolo tomato PDO as affected by farming system and biostimulant application. *Agronomy* **2019**, *9*, 505. [CrossRef]
33. Dima, S.-O.; Neamtu, C.; Desliu-Avram, M.; Ghiurea, M.; Capra, L.; Radu, E.; Stoica, R.; Faraon, V.-A.; Zamfiropol-Cristea, V.; Constantinescu-Aruxandei, D.; et al. Plant biostimulant effects of baker's yeast vinasse and seleniu on tomatoes through foliar fertilization. *Agronomy* **2020**, *10*, 133. [CrossRef]
34. Di Mola, I.; Ottaiano, L.; Cozzolino, E.; Senatore, M.; Giordano, M.; El-Nakhel, C.; Sacco, A.; Rouphael, Y.; Colla, G.; Mori, M. Plant-based biostimulants influence the agronomical, physiological, and qualitative responses of baby rocket leaves under diverse nitrogen conditions. *Plants* **2019**, *8*, 522. [CrossRef] [PubMed]
35. Skowera, B. Changes of hydrothermal conditions in the Polish area (1971–2010). *Fragm. Agron.* **2014**, *31*, 74–87. (In Polish)
36. Wadas, W.; Kalinowski, K. Effect of titanium on assimilation leaf area and chlorophyll content of very early-maturing potato cultivars. *Acta Sci. Pol. Agric.* **2017**, *16*, 87–98.
37. Pietkiewicz, S. An indicator-based analysis of plant growth. *Wiad. Bot.* **1985**, *29*, 29–42. (In Polish)
38. Wadas, W.; Dziugieł, T. Growth and marketable potato (*Solanum tuberosum* L) tuber yield in response to foliar application of seaweed extract and humic acids. *Appl. Ecol. Environ. Res.* **2019**, *6*, 13219–13230.
39. de la Casa, A.; Ovando, G.; Bressanini, L.; Martinez, J.; Rodriguez, A. Radiation use efficiency in potato crop estimated from the foliage coverage. *Agriscientia* **2011**, *28*, 21–30. [CrossRef]
40. Camargo, D.C.; Montoya, F.; Moreno, M.A.; Ortega, J.F.; Córcoles, J.I. Impact of water deficit on light interception, radiation use efficiency and leaf area index in a potato crop (*Solanum tuberosum* L.). *J. Agric. Sci.* **2016**, *154*, 662–673. [CrossRef]
41. Howlader, O.; Hoque, M.A. Growth analysis and yield performance of four potato (*Solanum tuberosum* L.) varieties. *Bangladesh J. Agric. Res.* **2018**, *43*, 267–280. [CrossRef]
42. Van Delden, A.; Pecio, A.; Haverkort, A.J. Temperature response of early foliar expansion of potato and wheat. *Ann. Bot.* **2000**, *86*, 355–369. [CrossRef]
43. Tang, R.; Niu, S.; Zhang, G.; Chen, G.; Haroon, M.; Yang, Q.; Rajora, O.P.; Li, X.Q. Physiological and growth responses of potato cultivars to heat stress. *Botany* **2018**, *96*, 897–912. [CrossRef]
44. Kolachevskaya, O.O.; Lomin, S.N.; Arkhipov, D.V.; Romanov, G.A. Auxin in potato: Molecular aspects and emerging roles in tuber formation and stress resistance. *Plant Cell Rep.* **2019**, *38*, 681–698. [CrossRef] [PubMed]
45. Ha, S.; Vankova, R.; Yamaguchi-Shinozaki, K.; Shinozaki, K.; Tran, L.-S.P. Cytokinins: Metabolism and function in plant adaptation to environmental stresses. *Trends Plant Sci.* **2012**, *17*, 172–179. [CrossRef] [PubMed]
46. Li, J.; Zhong, Y.; Guo, H. Correlation of leaf area with yield and quality in potato. *Chin. Potato J.* **2013**, *27*, 34–37.
47. Ascione, S.; Ruggiero, C.; Vitale, C. Contribution of growth components on relative, plant, crop and tuber growth rate of nine potato cultivars in southern Italy. *Int. J. Sci.* **2013**, *2*, 1–11.

agronomy

MDPI

Article

Effect of *Pterocladia capillacea* Seaweed Extracts on Growth Parameters and Biochemical Constituents of Jew's Mallow

Mohamed Ashour [1], Ahmed A. El-Shafei [2,3,*], Hanan M. Khairy [4], Doaa Y. Abd-Elkader [5], Mohamed A. Mattar [2,6,*], Abed Alataway [3,*] and Shimaa M. Hassan [5]

1 Invertebrate Aquaculture Lab., Aquaculture Division, National institute of Oceanography and Fisheries (NIOF), Alexandria Branch, Alexandria 21576, Egypt; microalgae_egypt@yahoo.com
2 Department of Agricultural Engineering, College of Food and Agriculture Sciences, King Saud University, Riyadh 11451, Saudi Arabia
3 Prince Sultan Bin Abdulaziz International Prize for Water Chair, Prince Sultan Institute for Environmental, Water and Desert Research, King Saud University, Riyadh 11451, Saudi Arabia
4 Hydrobiology Lab., Marine Environmental Division, National Institute of Oceanography and Fisheries, Alexandria 21545, Egypt; hanan_khairy@yahoo.com
5 Department of Vegetable, Faculty of Agriculture (El-Shatby), Alexandria University, Alexandria 21545, Egypt; doaa_004@yahoo.com (D.Y.A.-E.); shaymaa.hassan@alexu.edu.eg (S.M.H.)
6 Agricultural Engineering Research Institute (AEnRI), Agricultural Research Centre, Giza 12618, Egypt
* Correspondence: aelshafei1bn.c@ksu.edu.sa (A.A.E.-S.); mmattar@ksu.edu.sa or mohamedmattar@gmail.com (M.A.M.); aalataway@ksu.edu.sa (A.A.); Tel.: +966-11-4678504 (A.A.E.-S.); +966-11-4676024 (M.A.M.); +96-601-146-75576 (A.A.); Fax: +966-11-4678502 (M.A.M.)

Received: 13 February 2020; Accepted: 15 March 2020; Published: 19 March 2020

Abstract: We performed field experiments to evaluate the influence of two extraction treatments, seaweed (*Pterocladia capillacea* S.G. Gmelin) water extraction (WE) and ultrasound-assisted water extraction (USWE) at three concentrations (5%, 10%, and 15%), as well as control NPK traditional mineral fertilizer on the growth, yield, minerals, and antioxidants of Jew's Mallow (*Corchorus olitorius* L.) during the two seasons of 2016 and 2017 in Egypt. Plant height, number of leaves, and fresh weight of WE10 treatment were the highest ($p < 0.05$) as 59.67 cm, 10.67 and 2.41 kg m^{-2} in 2016, respectively, and 57.33 cm, 11.00 and 2.32 kg m^{-2} in 2017, respectively. WE10 and USWE5 treatments produced the highest dry matter (17.07%) in 2016 and (16.97%) in 2017, respectively. WE10 plants had an increased water productivity of 41.2% relative to control plants in both seasons. The highest chlorophyll 'a' was recorded after the WE10 treatment in 2016 and 2017 (17.79 μg g^{-1} and 17.84 μg g^{-1}, respectively). The highest levels of total antioxidant capacity, total phenolics, and total flavonoids were also recorded after the WE10 treatment. Application of WE10 boosted growth, yield, minerals, and antioxidants of Jew's Mallow. The CROPWAT model was used to estimate the evapotranspiration, irrigation water requirements, and yield response to irrigation scheduling. Our data showed a yield reduction in the initial growth stage if a limited amount of water was provided. Therefore, irrigation water should be provided during the most important stages of crop development with the choice of effective irrigation practices to avoid water losses, as this helps to maximize yield.

Keywords: seaweed extract; ultrasound-assisted water; foliar spray; *Pterocladia capillacea*; bio-fertilizer; growth parameters; antioxidants; Jew's Mallow; CROPWAT model

1. Introduction

Vegetables and their products contain non-enzymatic antioxidants and micro-nutrients that stabilize free radicals and in turn, increase the capacity of the plant to fight against pathogens that may affect humans and animals [1–4]. For example, the antioxidant compounds in vegetables and fruits could help prevent oxidative stress, diabetes, neurodegenerative disorders, cardiovascular disease, and cancer [5,6]. Jute (*Corchorus olitorius* L.), or Jew's mallow, belongs to the *Tiliaceae* family. *C. olitorius* thought to have originated from South China, from where it was introduced to India and Pakistan. However, a wild variety has been discovered in many areas in India, China, Australia, and Africa, particularly in Southeastern Nigeria. Jute leafy vegetable is commonly used in the preparation of soup [7]. The young shoot tips can be consumed raw or cooked and contain elevated concentrations of protein and vitamin C [8,9]. Jute is generally suggested for pregnant and nursing women because it is thought to be rich in iron [10].

Chemical fertilizers have been used in large quantities to compensate the nutrients deficiency in the soil. It has been observed that this use affects soil, plants, and human health. Their potential carcinogenicity and toxicity have been demonstrated, particularly after the reduction of nitrate to nitrite, or just reacting with amines and/or amides in the formation of N-nitroso compounds, N-nitrosamines, and other nitrogen compounds with high levels of nitrate [11]. Screening of native algal species must be considered to achieve a successful commercial and biotechnological potential of native algal species [12]. Seaweeds are the most promising plants from marine ecosystems and are used as a source of food and medicine. The coast of Egypt has a wide range of wild seaweed available throughout the year, even the Mediterranean coast [13] or the Red Sea coast [14]. Along the Egyptian Mediterranean coast, especially near Alexandria, red algae (*Pterocladia capillacea*; Rhodophyta) are the most dominant native seaweeds. Khairy and El-Shafay [13] studied the seasonal variations (spring, summer and autumn 2010) of biochemical composition of *P. capillacea* collected from Abu Qir Bay, Mediterranean Coast of Alexandria, Egypt. In 2010 spring season, *P. capillacea* achieved the highest significant protein (23.72%) and lipid (2.71%), while in 2010 summer season, *P. capillacea* achieved the highest significant carbohydrate (50.96%), ash (15.81%), and moisture (10.19%). Total fatty acids (248–515 μg/g), total saturated fatty acids (189–360 μg/g), total mono-unsaturated fatty acids (29–77 μg/g), total poly-unsaturated fatty acids (30–78 μg/g), total amino acids (2836–3924 μg/g), total essential amino acids (1136–1445 μg/g), and total non-essential amino acids (1700–2445 μg/g) of seasonally collected *P. capillacea* species were observed. Moreover, Khairy and El-Sheikh [15] observed the mineral composition and antioxidant activities of *P. capillacea* species collected seasonally (spring, Summer and autumn 2010) from Abu-Qir Bay, Mediterranean coast of Alexandria, and they concluded that this species is a rich in carotenoids, phenolic compounds, DPPH free radicals and minerals, therefore, this species can be used as potential source of health food in human diets and may be of use to food industry. In general, marine algae are a rich source of protein, lipids, carbohydrates, polysaccharides, minerals, antioxidants, and other bioactive compounds that can serve in multiple biological activities related to different industries [15,16].

Seaweed extracts can be used as fertilizer for flowering plants, vegetables, and grain crops [17–19]. Furthermore, they have been marketed as fertilizer additives, which are better than other fertilizers [20,21]. Using such extracts (bio-fertilizers) in cultivation many protect the soil and improve crop quality. Therefore, applying them to seeds or adding them to the soil stimulates plant growth [22]. Liquid extracts obtained from seaweeds have gained popularity as foliar sprays for many crops. These extracts contain cytokines, growth promoting hormones, elements, vitamins, and amino acids [23,24]. Some unknown bioactive component in seaweed acts to illicit the plant's own production of plant hormones through internal metabolic pathways [25].

Booth [20] reported that the efficacy of seaweeds as extracts was due to the presence of several metabolites and trace elements. The green seaweed *Enteromorpha* has a high potential for commercial exploitation because of its abundant and varied chemical composition, quality, and concentration of basic nutrients [26]. *Enteromorpha* sp. contains 28 times more calcium than spinach, 26 times more than nopal, and 13 times more than quelite [27]. *Ulva lactuca* and *Enteromorpha intestinalis* are used as seaweed liquid extract for many crops [21,28]. Rama Rao [29] reported good yields of *Zizyphus rugosa* fruits, when leaves sprayed with seaweed liquid extracts obtained from *Sargassum*. Seaweed extracts are now available commercially as Maxicrop (Sea-Born), Algifert (Marinure), Goemar GA14, Kelpak 66, Seaspray, Seasol, Cytex, and Seacrop. It has been reported that seaweed extracts are better than other extracts [21,23].

Traditional methods employed for extracting bioactive compounds are time consuming and have low extraction efficiencies. To overcome these disadvantages, novel technologies for extraction of bioactive compounds from marine algae have been investigated including the use of microwaves [30], enzymes [31], and super-critical fluids [32]. Recently, ultrasonic technologies have been used to enhance the extraction efficiencies of bioactive compounds (total phenolics, fucose, and uronic acid) from brown seaweed *Ascophyllum nodosum* [33–35] and starch from microalgae *Chlamydomonas fasciata* Ettl NIES-437 [36]. Moreover, ultrasonic assisted extraction was utilized in various industrial fields including phenolic compounds from citrus peel [37], lycopene from tomatoes [38], and anthocyanins from raspberries [39]. Ultrasound-assisted extraction is a simple and employed to improve extraction of bioactive compounds from seaweed [34–36,40].

Full irrigation should be practiced to maximize water productivity [41]. Weekly skipping of irrigation during seed filling may substantially reduce seed yield and water productivity. Skipping during seed germination may be a viable option when water is scarce and land is not limiting. Economic evaluation will provide guidance to policy makers at basin scales for formulating improved and efficient water management plans under all varying weather conditions. CROPWAT is a software for irrigation planning and management [42,43]. Its main functions are: To calculate reference evapotranspiration (ET_o), crop water requirements, and crop irrigation requirements, which may be used to develop irrigation schedules under multiple management conditions and water supply schemes, to estimate rain-fed production and drought effects, and to evaluate the efficiency of irrigation practices. The CROPWAT model has been validated in previous studies for estimating the dynamics main components of soil water balance [44–47]. We undertook this study to investigate the effect of *P. capillacea* seaweed liquid extracts on the growth, yield, minerals, and antioxidants of Jew's Mallow (*C. olitorius* L.).

2. Materials and Methods

2.1. Seaweed

2.1.1. Sampling

P. capillacea seaweed was collected in spring 2016 from the submerged rocky site near Boughaz El-Maadya, Abu-Qir Bay, Alexandria (31.3000° N and 30.1667° E) in Egypt. Harvested samples were transferred to the Microalgae and Invertebrates Aquaculture Laboratories, National Institute of Oceanography and Fisheries (NIIOF), Alexandria, Egypt. Epiphytes were removed from samples, and the seaweed samples were cleaned, washed, and air-dried in shadow. Dried seaweed samples were powdered and stored at room temperature in plastic bag for further analysis and utilization.

2.1.2. Biochemical Composition

Protein, lipid, carbohydrates, and ash of identified seaweed *P. capillacea*, collected in spring 2016, were determined. Total proteins were extracted according to Rauch [48] and determined according to Hartree [49]. Total carbohydrates were extracted according to Myklestad and Haug [50] and determined according to Dubois et al. [51]. Total lipid was calculated according to Bligh and Dyer [52]. Fatty acids and amino acids were extracted and estimated as described by El-Shenody et al. [14].

2.1.3. Seaweed Liquid Extracts Preparation

In this study, seaweed crude liquid extracts of *P. capillacea* were prepared using two extraction methods: three treatments using water extract (WE) and the other three treatments using

Ultrasound-Assisted Water Extraction (USWE), as shown in Figure 1. For WE, 100 g seaweed powder was soaked in 1 L distilled water in a 60 °C water bath for 60 min (extraction phase I). The residual filtrate was filtered and soaked in 1 L distilled water (1:10, w/v) in a 60 °C water bath for 60 min (extraction phase II) and this process was repeated a third time (extraction phase III). Each extraction phase was filtered through Whatman No. 3 filter paper and the supernatants of the three phases (I, II, and III) were combined to the final WE volume of 3 L and stored at −20 °C. For USWE, the three extraction phases were prepared as described above for WE, but after each phase the mixture was subjected to ultrasonication. The USWE extraction was performed at 60 °C for 5 min and 99% amplitude of 20 kHz; these conditions were adjusted and stable for all three extraction phases. After three USWE extraction phases, the supernatants were combined to a final volume of 3 L, and stored at −20 °C. The final combined supernatants, both WE and USWE, were considered to be a 100% crude extract that was utilized as seaweed foliar spray.

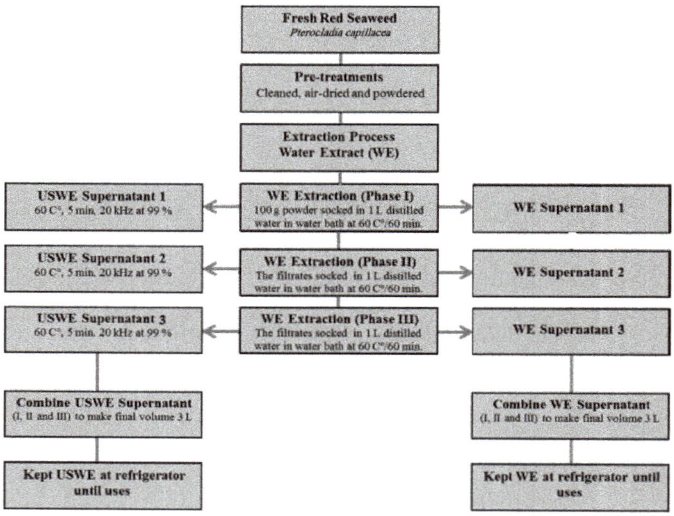

Figure 1. Procedures for water extraction (WE) and ultrasound-assisted water extraction (USWE) of pre-treated *Pterocladia capillacea*.

2.2. Experimental Design

The field experiment with Jew's Mallow (*C. olitorius* cv. Balady) was conducted for two successive growing seasons (2016–2017) at Abeis Experimental Farm, Alexandria University, Alexandria (31.2001° N and 29.9187° E) in Egypt. Before sowing, soil samples were collected (0–30 cm depth) to determine physical and chemical properties following Page [53] (Table 1). Climatic data, such as maximum and minimum air temperature (T_{max} and T_{min}), relative humidity (RH), wind speed (u_2), and rainfall (P), were collected at a meteorological station near the experimental field location (Figure 2) to calculate daily ET_o using the Penman–Monteith FAO-56 equation [54]. Evapotranspiration was estimated during growth using crop coefficient (Kc) values [54] multiplied by ET_o. The experimental area of 220.5 m^2 was divided into three replicate blocks separated by 2-m buffer zones. Each block consisted of seven plots including one traditional fertilizer and six seaweed extract treatments. Each plot covered an area of 10.5 m^2 (3 × 3.5 m). A randomized complete block design (RCBD) was used. Commercial seeds were sown on March 20, 2016 and March 22, 2017 at the rate of 28 kg ha^{-1} [55]. The site was irrigated five times during the first 20 days after sowing to allow germination and establishment before the application of extract treatments. After that, irrigation was carried out every six to seven days for all treatments. The first dose (0.5 m^3 ha^{-1}) of growth fertilizer or seaweed extract was applied 10 days after sowing (DAS), the second one (0.75 m^3 ha^{-1}) was applied 18 DAS, and the third dose (1 m^3 ha^{-1}) was adding 26 DAS. Harvesting included two cuttings at 45 and 70 DAS.

Table 1. Soil physical and chemical properties.

Soil Physical Properties					
Season	**2016**	**2017**	**Season**	**2016**	**2017**
Sand (%)	43.3	42.8	Saturation moisture (m^3 m^{-3})	0.49	0.52
Silt (%)	25.5	23.5	Field capacity (m^3 m^{-3})	0.40	0.41
Clay (%)	31.2	33.7	Wilting point (m^3 m^{-3})	0.17	0.17
Soil texture	Clay loam	Clay loam	Total available moisture (m m^{-1})	0.22	0.24
Bulk density (g cm^{-3})	1.48	1.3	Infiltration rate (mm h^{-1})	3.44	3.20
Soil chemical properties					
pH	8.45	8.88	Total Nitrogen (%)	0.19	0.15
E.C. (dS m^{-1})	3.01	3.0	Phosphorus (ppm)	0. 41	0.44
Soluble cations (meq L^{-1})			**Soluble anions (meq L^{-1})**		
Ca^+	2.08	1.97	CO_3^{--}	0.0	0.0
Mg^{++}	1.98	1.88	HCO_3^-	1.43	1.28
Na^+	2.47	2.39	Cl^-	2.05	1.95
K^+	0.40	0.37	SO_2^{--}	3.46	3.37

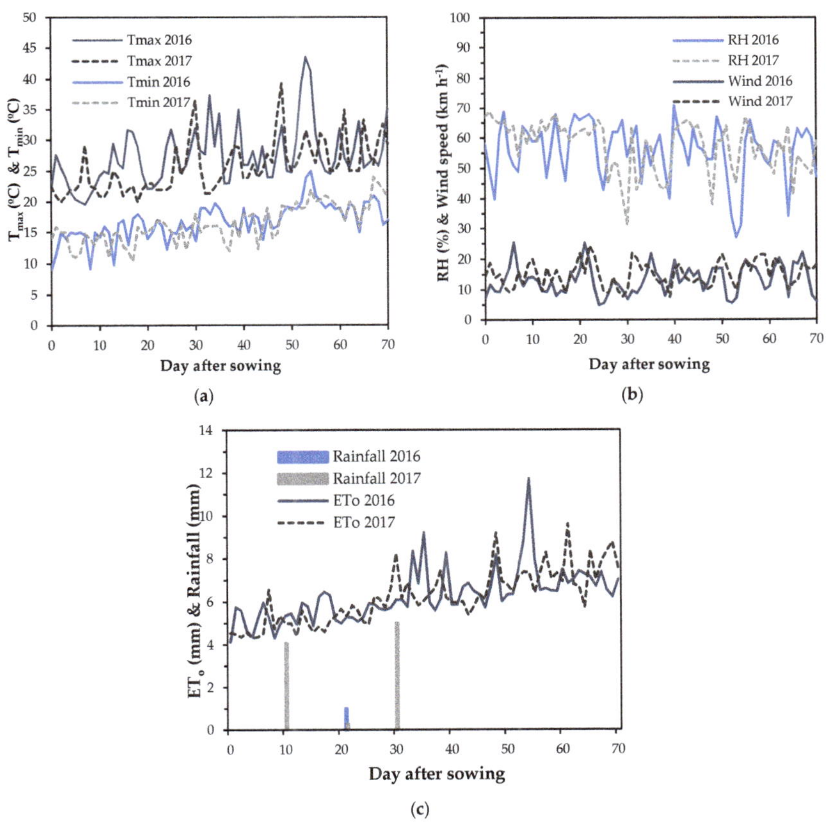

Figure 2. Daily climate parameters in the 2016 and 2017 experimental periods during the Jew's Mallow growing season. (**a**) daily maximum and minimum air temperature (T_{max} and T_{min}, °C), (**b**) relative humidity (RH, %) and wind speed (km h^{-1}), and (**c**) reference evapotranspiration (ET_o, mm) and rainfall (mm).

2.3. Treatments

The following seven treatments were used: mineral NPK fertilizer (control); water extracted seaweed at 5%, 10%, and 15% (WE5, WE10, and WE15); and ultrasound-assisted water extraction seaweed at 5%, 10%, and 15% (USWE5, USWE10, and USWE15).

NPK fertilization was carried out according to the recommendations for commercial production of Jew's Mallow plant. The NPK treatment dose consisted of ammonium nitrate NH_4NO_3 (33%N) at the rate of 300 kg ha^{-1}, calcium superphosphate ($Ca(H_2PO_4)_2.H_2O$ (15% P_2O_5); 525 kg ha^{-1}); and potassium sulphate (K_2SO_4 (48% K_2O); 125 kg ha^{-1}). Nitrogen fertilizer was applied thrice at 7, 15, and 21 DAS. Phosphorus fertilizer was mixed during soil preparation. Potassium fertilizer was applied at 15 DAS.

2.4. Measurements

2.4.1. Agronomic and Physiological

Plants were harvested (cut) twice, at 45 DAS and 70 DAS, from the center of each plot (treatment) per season to determine leaf and stem fresh weight (kg m^{-2}). Five plants were randomly chosen from each plot to measure plant height and; number of leaves. Ratios between dry leaf weight and fresh leaf weight were determined after drying 70 °C in a forced-air oven until reaching a constant weight. Water productivity (WP, kg m^{-3}) was used to evaluate treatments, calculated by dividing total fresh weight (kg m^{-2}) at harvest by the amount of water applied (supplemental irrigation plus rainfall, m^3 m^{-2}) to the crop. Chlorophyll 'a', Chlorophyll 'b', and total carotene (μg g^{-1}) as described by Dere, et al. [56].

2.4.2. Nutrient Contents

Plant nutrient content (N, P, and K) was analyzed and expressed as percentage on leaf dry weight basis. Total N and P contents were determined calorimetrically using a spectrophotometer at 662 and 650 nm, following the methods of Evenhuis [57]. K was quantified by atomic absorption spectrometry as described by Cottenie, et al. [58].

2.4.3. Antioxidant Activities

Antioxidant activities of crude extracts of WE, USWE and Jew's Mallow were observed. Free radical scavenging activity against DPPH (2,2-diphenyl-1-picrylhydrazy) was determined as described by Suresh Kumar, et al. [59]. The total antioxidant content (TAC; mg g^{-1}) was determined with a Phosphomolybdate assay using ascorbic acid as the standard [60]. The total phenolic content (TPC; mg g^{-1}) was determined by using the Folin–Ciocalteu method as modified by Suresh Kumar, et al. [59]. Total flavonoid content (TVC; μg g^{-1}) was determined according to the method of Chang, et al. [61] with Quercetin as the standard.

2.5. CROPWAT Model

Before Jew's Mallow cultivation, the water application depth and irrigation timing intervals were calculated using the CLIMWAT 2.0 and CROPWAT models. CLIMWAT 2.0 is climatic software [62] presenting the monthly agro-climatic data of over 5000 stations worldwide, including the Alexandria-Nouzha agroclimatic station, which was the nearest to the experimental site (4 km). The CROPWAT model was used for calculation of crop water requirements and the development of irrigation schedules using the option to irrigate at critical depletion and refill soil to field capacity. Irrigation times and the amounts were estimated based on the efficiency of the basin irrigation system and applied for both growth seasons (Figure 3). At the end of each season, the CROPWAT model with the options of user defined application depth and irrigation at user defined intervals were used to evaluate the irrigation schedule. The input data for the CROPWAT version 8.0 model [63] required the following data:

- The daily climatic (T_{max}, T_{min}, RH, daylight hours, and u_2) and P data for the seasons of 2016 and 2017 were accessed from the Meteorological Data of Central Laboratory for Agricultural Climate (Figure 1).
- A cropping pattern consisting of the crop type, planting date, growing stage (20, 20, 25, and 8 days for initial, development, mid-season, and late-season stages, respectively), *Kc* (0.7 for initial, 1.15 for mid-, and 0.95 for late-season stage) and critical depletion fraction; *P* (0.3 for initial and development, 0.45 for mid-season stages, and 0.5 for late season stage), rooting depth; Z_r (0.18 m for initial stage and 0.5 m for maximum (mid- and late-season)), and yield response factor; k_y (0.8 for initial, 0.4 for development, 1.2 for mid-, and 1 for late-season). The crop values were assumed as data for a small vegetable according to Allen, et al. [54].
- Soil type: Total available soil moisture, maximum infiltration rate and initial soil moisture depletion were obtained from measured data (Table 1).

The output of CROPWAT model consists of daily root zone depletion ($D_{r,i}$, Equation (1)), deep percolation (DP_i), actual water use by crop $(ET_c)_{actual}$, efficiency of the irrigation schedule (*EIS*, Equation (2)), deficiency of the irrigation schedule (*DIS*, Equation (3)) and yield reduction (Y_R, Equation (4)) were collected and analyzed using the following equation:

$$D_{r,i} = D_{r,i-1} + (ET_{c,i})_{actual} - P_i - I_i + RO_i + DP_i \quad (1)$$

where $D_{r,i}$, and $D_{r,i-1}$ are at days *i* and *i*−1, P_i is total rainfall over day *i*, I_i is net irrigation on day *i*, RO_i is water loss by runoff from the soil surface on day *i*, in our study the *RO* is equal to zero, and DP_i is water loss by deep percolation on day *i*.

$$EIS = \frac{\sum(I_i - DP_i)}{\sum I_i} \times 100 \quad (2)$$

$$DIS = \frac{Sesaonal\ (ET_c)_{\text{potential}} - Sesaonal\ (ET_c)_{\text{actual}}}{Sesaonal\ (ET_c)_{\text{potential}}} \times 100 \quad (3)$$

$$Y_R = K_y\left(1 - \frac{(ET_c)_{\text{actual}}}{(ET_c)_{\text{potential}}}\right) \quad (4)$$

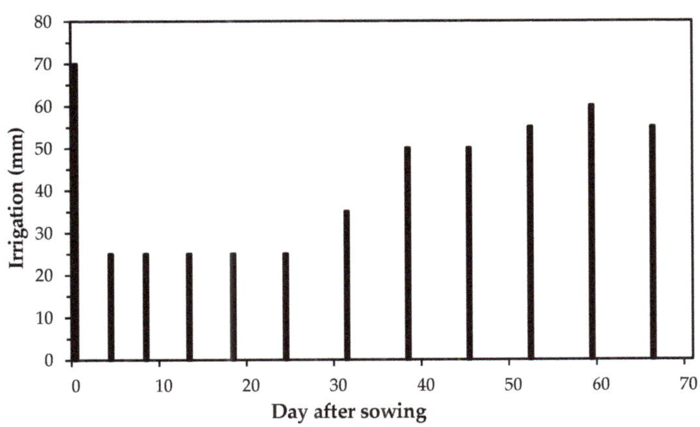

Figure 3. Amounts of irrigation water (mm) applied during the Jew's Mallow growing season.

2.6. Statistical Analyses

Analysis of variance (ANOVA) with RCBD was performed on data obtained from both growing seasons (2016 and 2017) using the IBM SPSS Version 23 software to determine the significance of differences among treatments. Standard errors (SE) were presented for the mean of data from both growing seasons. Differences among means of replicates were measured using Duncan method at $p \leq 0.05$ [64].

3. Results

3.1. Biochemical Composition Seaweed P. Capillacea

Nutritional compositions of the red seaweed species *P. capillacea* collected during spring season of 2016 were investigated. Lipid, protein, carbohydrate, and ash percentages, based on dry weight, were 2.46%, 18.47%, 51.36%, and 13.71%, respectively. Moreover, total fatty acids, total saturated fatty acids, total mono-unsaturated fatty acids, total poly-unsaturated fatty acids, total amino acids, total essential amino acids, and total non-essential amino acids were 247.6, 188.6, 29.1, 29.9, 2836, 1136.3, and 1700 $\mu g\ g^{-1}$, respectively. The antioxidant activities result of WE and USWE crude extracts observed that no significant differences ($p < 0.05$) were found in total antioxidant content (22.48 and 22.33 mg g^{-1}) and total phenolic content (17.79 and 16.85 mg g^{-1}) in WE and USWE, respectively, while USWE achieved a significant difference ($p < 0.05$) in total flavonoid content (45.68 $\mu g\ g^{-1}$) and total carotene (2.03 $\mu g\ g^{-1}$) in comparing to WE (34.77 $\mu g\ g^{-1}$ and 1.29 $\mu g\ g^{-1}$, respectively).

3.2. Agronomic Traits

Table 2 shows the significant differences in in plant height ($p < 0.05$). WE10 treatment produced the tallest plants, followed by WE5 in both seasons. In WE10 and WE5 treated plants, height increased by 39.8% and 28.1%, respectively, compared with mineral fertilizer treated plants (control treatment) in 2016 while plant height increased by 28.3% and 23.2% in 2015. The USWE10 treatment had the smallest effect on height in both years, with increases of 4.6% and 0.7% in 2016 and 2017, respectively, relative to control. Statistically significant differences ($p < 0.05$) were found between treatments for leaf number, where WE5 and WE10 treatments had the highest values in both seasons.

Table 2. Effects of water extract (WE) and ultrasound-assisted Water Extraction (USWE) on growth characteristics of Jew's Mallow during 2016 and 2017 growing seasons.

Treatment	Plant Height (cm)		Leaf Number ($Plant^{-1}$)		Fresh Weight ($kg\ m^{-2}$)		Dry Matter (%)	
	2016	2017	2016	2017	2016	2017	2016	2017
Control	42. 7 ± 0.5 ^{d}	44. 7 ± 0.6 ^{d}	9.3 ± 1.1 ^{ab}	9.0 ± 1.0 ^{b}	1.64 ± 0.1 ^{d}	1.65 ± 0.1 ^{e}	13.9 ± 1.1 ^{c}	14.1 ± 1.6 ^{b}
WE5	54.7 ± 2.0 ^{b}	55.0 ± 2.3 ^{ab}	10.7 ±0.5 ^{a}	11.0 ± 0.2 ^{a}	2.01 ± 0.1 ^{c}	2.07 ± 0.1 ^{bc}	16.4 ± 0.5 ^{ab}	16.3 ± 0.4 ^{a}
WE10	59. 7 ± 1.5 ^{a}	57.3 ± 3.2 ^{a}	10.6 ±1.1 ^{a}	11.0 ± 1.0 ^{a}	2.41 ± 0.1 ^{a}	2.32 ± 0.1 ^{a}	17.1 ± 1.0 ^{a}	16.6 ± 0.4 ^{a}
WE15	52.0 ± 2.0 ^{c}	52.5 ± 2.2 ^{abc}	9.0 ±0.1 ^{b}	9.0 ± 0.1 ^{b}	1.91 ± 0.1 ^{c}	1.84 ± 0.1 ^{de}	16.7 ± 0.9 ^{ab}	16.2 ± 0.4
USWE5	44. 7 ± 0.5 ^{d}	47.7 ± 1.8 ^{cd}	9.3 ±0.6 ^{ab}	9.3 ± 0.5 ^{b}	2.09 ± 0.0 ^{bc}	2.15 ± 0.2 ^{ab}	16.1 ± 1.0 ^{ab}	16.9 ± 1.1 ^{a}
USWE10	44.6 ± 1.5 ^{d}	45.0 ± 2.2 ^{d}	9.0 ±0.2 ^{b}	9.0 ± 0.1 ^{b}	2.22 ± 0.1 ^{ab}	2.16 ± 0.1 ^{ab}	16.4 ± 0.2 ^{ab}	16.6 ± 0.9 ^{a}
USWE15	50.0 ± 1.0 ^{c}	49.9 ± 0.4 ^{bcd}	9.7 ±0.6 ^{ab}	9.6 ± 0.6 ^{b}	1.88 ± 0.2 ^{c}	1.90 ± 0.1 ^{cd}	14.9 ± 1.5 ^{bc}	15.1 ± 1.5 ^{b}

Control: NPK fertilization; WE5, WE10, and WE15: water extracted seaweed at 5%, 10%, and 15%, respectively; USWE5, USWE10, and USWE15: ultrasound-assisted water extraction seaweed at 5%, 10%, and 15%, respectively. Data are means ± SE. Different superscript letters in each column indicate significant differences ($p \leq 0.05$).

There was a significant difference ($p < 0.05$) in fresh weight and dry matter between the treatments in 2016 and 2017. Fresh weight was the highest in WE10 treated plants in both seasons, followed by the USWE10 treatment. For WE10 and USWE10 treatments, the fresh weight increased by 47% and 35.4% in 2016, respectively; then 40.6% and 30.9% in 2017, compared to control treatment. Across all treatments, WE15 and USWE15 treatments reduced the fresh weight in both seasons. There were significant differences ($p < 0.05$) between bio- and mineral fertilizer-treated plants dry matter in 2016 and 2017 (Table 2). The highest dry matter value was recorded in plants that received the WE10 and USWE5 treatments, which was 23% and 19.9%higher, respectively, than control in 2016.

3.3. Water Productivity

Water requirement varied from 3.8 to 9.1 and 3.5 to 8.5 mm day^{-1} from the early stage to the peak demand period (mid-season) for 2016 and 2017, respectively. Water productivity (WP) values determined for treatments in 2016 and 2017 are shown in Figure 4. In both seasons, there were significant differences ($p < 0.05$) between WP values. The highest WP values were recorded with the WE10 treatment 41.2% higher than control. Among extract treatments, USWE15 and WE15 had the lowest WP in 2016 and 2017, which was lower by 23% and 20.7% than WE10, respectively.

3.4. Physiological Traits

Water extraction treatments had the highest content of chlorophyll 'a' in both seasons (average, 17.49 µg g^{-1}), while the lowest chlorophyll 'a' content was observed with the control treatment (average, 9.4 µg g^{-1}, Table 3). WE10 and WE15 treated plants showed significant increases in chlorophyll 'b' content compared to control treatment in both seasons. The chlorophyll 'b' content for the WE10 and WE15 treated plants (average, 13 µg g^{-1} and 13.3 µg g^{-1}, respectively) was two-fold higher than the content in the control treatment in 2016 and 2017. Conversely, USWE5 and USWE10 application resulted in the lowest chlorophyll 'b' content in both growing seasons, 16.8% and 26.8% lower than control, respectively. The lowest carotene content was achieved by control treatment in 2016 and 2017 (2.9 and 2.8 µg g^{-1}, respectively; Table 3). The highest carotene content in 2016 and 2017 was measured with the USWE10 treatment (71.2% and 72% higher than control, respectively), followed by the USWE15 treatment (53.7% and 54.3% higher than control, respectively).

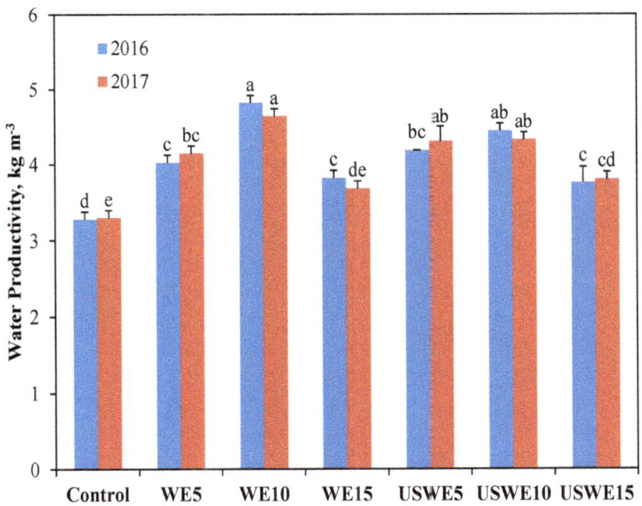

Figure 4. Water productivity (kg m^{-3}) of Jew's Mallow as a function of water extract (WE) and ultrasound-assisted water extract (USWE) treatments in the 2016 and 2017 growing seasons. Different letters (a, b, etc.) above bars indicate a significant difference among treatments in each season.

Table 3. Effects of water extract (WE) and Ultrasound-Assisted Water Extraction (USWE) on chlorophyll 'a', chlorophyll 'b', and carotene concentrations in Jew's Mallow during the 2016 and 2017 growing seasons.

Treatment	Chlorophyll 'a' ($\mu g\ g^{-1}$)		Chlorophyll 'b' ($\mu g\ g^{-1}$)		Carotene ($\mu g\ g^{-1}$)	
	2016	2017	2016	2017	2016	2017
Control	9.4 ± 1.9 ^{d}	9.4 ± 1.9 ^{d}	6.5 ± 0.2 ^{bc}	6.5 ± 0.2 ^{bc}	2.9 ± 0.2 ^{d}	2.8 ± 0.2 ^{c}
WE5	17.0 ± 0.7 ^{a}	17.1 ± 0.7 ^{a}	6.2 ± 0.0 ^{bc}	6.2 ± 0.0 ^{c}	3.1 ± 0.3 ^{cd}	3.1 ± 0.3 ^{cd}
WE10	17.8 ± 0.2 ^{a}	17.8 ± 0.3 ^{a}	13.0 ± 1.2 ^{a}	12.9 ± 1.1^{a}	3.0 ± 0.2 ^{d}	3.1 ± 0.2 ^{cd}
WE15	17.7 ± 1.7 ^{a}	17.5 ± 1.6 ^{a}	13.3 ± 0.8 ^{a}	13.3 ± 0.8 ^{a}	3.1 ± 0.2 ^{cd}	3.0 ± 0.2 ^{cd}
USWE5	12.3 ± 0.1 ^{c}	12.4 ± 0.0 ^{c}	5.4 ± 2.3 ^{c}	5.4 ± 2.3 ^{c}	3.9 ± 0.3 ^{bc}	3.8 ± 0.2 ^{bc}
USWE10	10.3 ± 0.7 ^{cd}	10.3 ± 0.7 ^{cd}	4.8 ± 0.6 ^{c}	4.7 ± 0.7 ^{c}	4.9 ± 0.8 ^{a}	4.9 ± 0.8 ^{a}
USWE15	14. 7 ± 1.7 ^{b}	14.8 ± 1.7 ^{b}	8.5 ± 1.9 ^{b}	8.5 ± 1.8 ^{b}	4.4 ± 0.6 ^{ab}	4.4 ± 0.6 ^{ab}

Control: NPK fertilization; WE5, WE10, and WE15: water extracted seaweed at 5%, 10%, and 15%, respectively; USWE5, USWE10, and USWE15: ultrasound-assisted water extraction seaweed at 5%, 10%, and 15%, respectively. Data are means ± SE. Different superscript letters in each column indicate significant differences ($p \leq 0.05$).

3.5. N, P, and K

Nutrient content (i.e., N, P, and K) of Jew's Mallow plant treated with different seaweed extracts in comparison to control treatment are presented in Table 4. Control treatment had the highest N content in 2016 and 2017 (1.78% and 1.71%, respectively), while WE10 had the lowest N content (1.20% and 1.33%), respectively. USWE10 treatment had the highest P content in 2016 and 2017 (0.74% and 0.77%, respectively). The highest K content (1.90%) was achieved by WE15 treatment in both seasons.

Table 4. Effects of water extract (WE) and ultrasound-assisted Water Extraction (USWE) on N, P, and K content in Jew's Mallow during the 2016 and 2017 growing seasons.

Treatment	N (%)		P (%)		K (%)	
	2016	2017	2016	2017	2016	2017
Control	1.78 ± 0.20 ^{a}	1.71 ± 0.25 ^{a}	0.67 ±0.06 ^{ab}	0.70 ±0.07 ^{ab}	1.40 ±0.01 ^{c}	1.40 ±0.02 ^{b}
WE5	1.33 ± 0.13 ^{b}	1.40 ±0.04 ^{b}	0.64 ±0.01 ^{b}	0.66 ±0.04 ^{b}	1.80 ±0. 10 ^{b}	1.90 ±0.04 ^{a}
WE10	1.20 ±0.10 ^{b}	1.33 ±0.12 ^{b}	0.70 ±0.06 ^{ab}	0.77 ±0.01 ^{a}	1.80 ±0.02 ^{b}	1.80 ±0.01 ^{a}
WE15	1.41 ±0.04 ^{b}	1.49 ±0.23 ^{ab}	0.71 ±0.05 ^{ab}	0.73 ±0.06 ^{ab}	1.90 ±0.03 ^{a}	1.90 ±0.01 ^{a}
USWE5	1.39 ±0.10 ^{b}	1.37 ±0.02 ^{b}	0.66 ±0.06 ^{ab}	0.69 ±0.07 ^{ab}	1.80 ±0.04 ^{b}	1.80 ±0.02 ^{a}
USWE10	1.24 ±0.12 ^{b}	1.32 ±0.11 ^{b}	0.74 ±0.03 ^{a}	0.77 ±0.02 ^{a}	1.20 ±0. 10 ^{d}	1.30 ±0.03 ^{b}
USWE15	1.41 ±0.04 ^{b}	1.39 ±0.06 ^{b}	0.68 ±0.03 ^{ab}	0.74 ±0.03 ^{ab}	1.20 ±0. 11 ^{d}	1.30 ±0.02 ^{b}

Control: NPK fertilization; WE5, WE10, and WE15: water extracted seaweed at 5%, 10%, and 15%, respectively; USWE5, USWE10, and USWE15: ultrasound-assisted water extraction seaweed at 5%, 10%, and 15%, respectively. Data are means ± SE. Different superscript letters in each column indicate significant differences ($p \leq 0.05$).

3.6. Antioxidant Activity

The highest DPPH percentage was achieved by WE10 in 2016 and 2017 (40.78% and 40.74%, respectively). The lowest DPPH percentage was recorded in USWE15 in 2016 and 2017 (8.75% and 8.74%, respectively; Figure 5). The highest TAC was recorded in WE10 in both seasons (43.97 and 44.22 mg g^{-1}, respectively), followed by the USWE10 treatment ((35.69 and 36.38 mg g^{-1}; Table 5). The lowest TAC was recorded with control (26.30 mg g^{-1}) in 2017 and USWE15 treatment (26.00 mg g^{-1}) in 2018. In both seasons, the highest significant TPC was obtained with WE10 treatment (116.28 and 115.81 mg g^{-1}, respectively), while the lowest TPC was obtained with the USWE15 treatment (49.62 and 49.61 mg g^{-1}, respectively). Although USWE5 had a higher TVC value, significant differences between extracts treatments were not observed, except for USWE15 treatment.

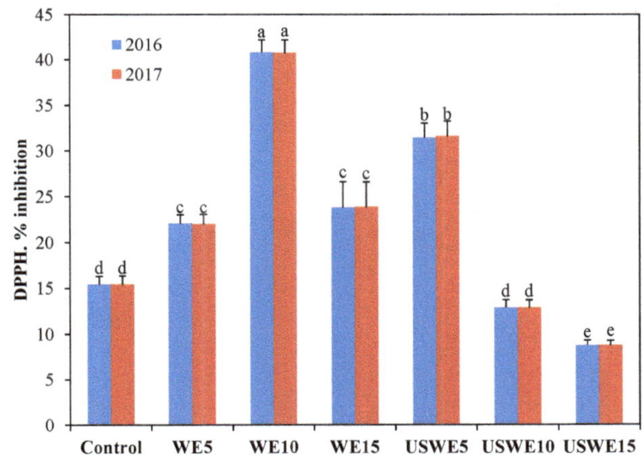

Figure 5. DPPH inhibition (% inhibition) of Jew's Mallow with water extract (WE) and ultrasound-assisted water extract (USWE) in the 2016 and 2017 growing seasons. Different letters (a, b, etc.) above bars indicate a significant difference among treatments in each season.

Table 5. Effects of water extract (WE) and ultrasound-assisted Water Extraction (USWE) on total antioxidant content (TAC), total phenolic content (TPC), and total flavonoid content (TVC) of Jew's Mallow during the 2016 and 2017 growing seasons.

Treatment	TAC (mg g^{-1})		TPC (mg g^{-1})		TVC (µg g^{-1})	
	2016	2017	2016	2017	2016	2017
Control	26.30 ± 3.43 ^{c}	26.22 ± 0.40 ^{c}	97.15 ± 11.78 ^{b}	98.15 ± 10.92 ^{b}	1168 ± 47.4 ^{ab}	1170 ± 49.2 ^{ab}
WE5	34.65 ± 3.36 ^{b}	32.69 ± 1.27 ^{b}	75.22 ± 2.19 ^{cd}	75.40 ± 2.43 ^{cd}	1193 ± 23.9 ^{a}	1194 ± 20.7 ^{a}
WE10	43.97 ± 1.04 ^{a}	44.22 ± 2.40 ^{a}	116.28 ± 6.59 ^{a}	115.81 ± 7.77 ^{a}	1208 ± 26.4 ^{a}	1204 ± 14.7 ^{a}
WE15	27.28 ± 1.37 ^{c}	28.13 ± 2.53 ^{c}	77.92 ± 11.12 ^{cd}	78.33 ± 10.70 ^{cd}	1191 ± 20.9 ^{a}	1187 ± 29.7 ^{a}
USWE5	35.35 ± 1.55 ^{b}	35.52 ± 1.32 ^{b}	66.56 ± 1.69 ^{d}	65.46 ± 1.47 ^{d}	1244 ± 3.1 ^{a}	1232 ± 7.3 ^{a}
USWE10	35.69 ± 2.41 ^{b}	36.38 ± 2.53 ^{b}	49.62 ± 6.54 ^{e}	49.61 ± 751 ^{e}	1205 ± 80.8 ^{a}	1209 ± 75.1 ^{a}
USWE15	26.78 ± 2.28 ^{c}	26.00 ± 2.54 ^{c}	80.20 ± 0.83 ^{c}	79.69 ± 0.56 ^{c}	1113 ± 31.1 ^{b}	1105 ± 30.1 ^{b}

Control: NPK fertilization; WE5, WE10, and WE15: water extracted seaweed at 5%, 10%, and 15%, respectively; USWE5, USWE10, and USWE15: ultrasound-assisted water extraction seaweed at 5%, 10%, and 15%, respectively. Data are means ± SE. Different superscript letters in each column indicate significant differences ($p \le 0.05$).

3.7. CROPWAT Model

Figure 6 shows the depletion curve before and after each irrigation event during the 2016 and 2017 growth seasons. The highest values of depletion were 52 and 48 mm in the mid-season for each year, respectively. The depletion values were between those of field capacity and readily available moisture except for the initial 20 days of both seasons. Thus, there was no water stress. In the initial stage, there was a maximum *DP* of 28 mm at the first irrigation event in both seasons and after that it decreased to 2.5 and 8.5 mm, respectively, in the 2016 and 2017 growing seasons. The effective rainfall means modeled for 2016 and 2017 were 9.4 and 1 mm, respectively leaving deficits of 387 and 401.7 mm to be made up from irrigation. Thus, effective rainfall showed an ineffective pattern across the growth stages. By applying the basin irrigation system, the application water efficiencies were 78% and 79%, respectively, in 2016 and 2017. Hence, the values of *EIS* were 95.1% and 98.6%, respectively, in 2016 and 2017. The values of $(ET_c)_{actual}$ were 393.2 and 399.8 mm, respectively in 2016 and 2017; the values of *DIS* were 0.8% and 0.7%. On the other hand, there were no yield reductions across the growth stages, with the maximum values of 3.7% and 3.1% in the initial stage, respectively in 2016 and 2017.

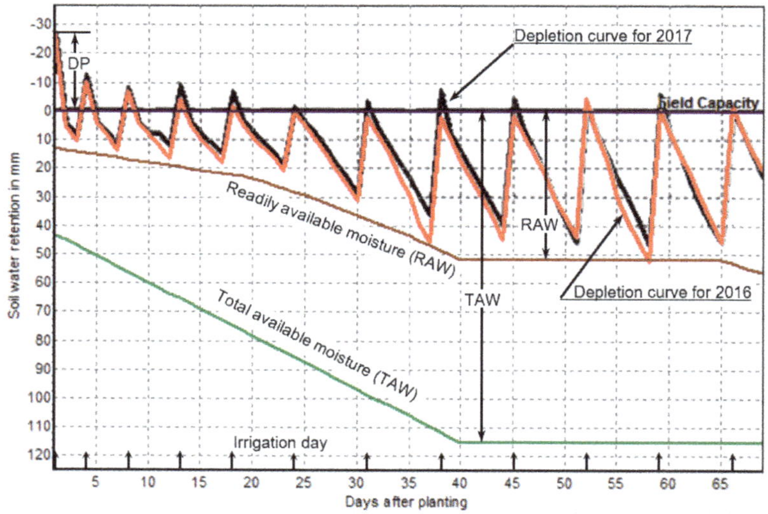

Figure 6. Soil water depletion during the 2016 and 2017 growing seasons.

4. Discussion

Marine algae are considered very important bioindicator for the marine ecosystem [13–15]. Many studies have reported that the constituents, diversity, and communities of marine algae are affected by variations in environmental parameters and nutrient limitation [65–70]. During the last few years, the attention on scientific and commercial interest to biotechnological applications of algae as a sustainable source and global commercial for aquaculture [71–75], biofuel [12,76], extracts [77,78], food supplement, pharmaceuticals, and cosmetics were increased [12].

In current study, data of biochemical composition (protein, lipid and carbohydrate) of *P. capillacea* showed that the large component is carbohydrate (51.36%), followed by protein (21.49%) and lipid (2.06%). The presented data may be act as an indicator for related bioactive secondary metabolites of *P. capillacea* liquid extract. However, our data is in the same line of the results presented by Khairy and El-Shafay [13] who found that, during spring season of 2010, the highest component is carbohydrate (50.49%), followed by protein (23.72%) and lipid (2.71%). Many authors reported that the biochemical constituents of marine algae are affected by variations in environmental conditions and nutrient availability [65–70].

The nutrient contents of seaweed *P. capillacea* used in current study were investigated previously by Khairy and El-Sheikh [14], at the same collected location of current study too, who observed that mineral were potassium (50.9 mg/100g), calcium (68.4 mg/100g), magnesium (22.1 mg/100g), cupper (0.5 mg/100g), ferrous (18.37 mg/100g), and zinc (0.19 mg/100g). In current study, although the applied seaweed extract is a rich source of nutrient, it not characterized as a nutrient fertilizer because of many consecrations like its constituent of bioactive compound which act as a plant growth promoting. Interestingly, the *P. capillacea* seaweed species is reported as a potential source for human healthy food because its constituent of bioactive compounds *like* protein, lipid, carbohydrate, fatty acids (saturated, mono-unsaturated and poly-unsaturated fatty acids), amino acids (essential and non-essential), carotenoids, phenolic compounds, and DPPH [13,14]. Hence, *P. capillacea*, collected from the same study location, is reported as a rich source of alkaloids, flavonoids, steroids, terpenoids, phlobatannins and many other phytochemicals and secondary metabolites [79].

Moreover, *P. capillacea* as a red alga is characterized as a rich source of different phytohormones [40,80,81]. It well known that some unknown bioactive component in seaweed acts to illicit the plant's own production of plant hormones through internal metabolic pathways [25].

Seaweeds and its extracts are becoming of increasing importance because of their bioactive compounds and their potential application in different industries. Liquid seaweed extract is commonly used as commercial agricultural biostimulants because of many considerations.

In current study, to enhance the efficiency of seaweed liquid extract, we evaluate two extraction methods; (1) water (WE); and (2) ultrasound-assisted water extraction (USWE). The effect of different seaweed extracts as a foliar spray on quantity (growth and yield) and quality (minerals and antioxidants activity) of Jew's Mallow (*C. olitorius* L.), comparing to NPK traditional fertilizers were observed. In general, Jew's Mallow (*C. olitorius* L.) treated with liquid seaweed extract (either WE or USWE) achieved the highest significant quantity (yield) and quality (antioxidant activity, P %, and K %), comparing to NPK traditional fertilizers, which only achieved the highest significant N %. Jew's Mallow (*C. olitorius* L.) treated with WE10 and USWE10 were achieved the highest significant yield (fresh weight), and P %. The highest significant Chlorophyll a and b; total antioxidant activity and total phenolic compounds were achieved by WE10, while the highest significant carotene and total flavonoid compounds were achieved by USWE10. In general, in the present study, it was observed that the seaweed liquid extract prepared from *P. capillacea* presented to Jew's Mallow gave better results in all aspects of growth to yield when compared to NPK traditional fertilizers. Using ultrasound-assisted water extraction (USWE) method was significantly improved the total flavonoid and carotene content in *P. capillacea* USWE crude extract, which is positively reflected on these compounds of Jew's Mallow (*C. olitorious* L.), when comparing to WE. Carotenes are indispensable to plants and act as precursors for the biosynthesis of phytohormones and strigolactones, improve the plant development and responses to unstable environmental, and serve as a source of pro-vitamin A [82].

In the present study, WE10-treated plants showed the best response in plant height and leaf number. Similarly, Stephenson [40] reported that seaweed liquid extract prepared from *Ascophyllum* and *Laminaria* accelerated maize growth. Blunden and Wildgoose [83] reported a marked increase in lateral root development in potato plants as a result of treatment with seaweed extract. Similar results were obtained with *Padina* biofertilizer, which induced maximum growth in *Cajanus cajan* [84]. Thirumaran, et al. [85] reported similar findings 20% seaweed liquid extract from brown algae *Rosenvingea intricate* had an increased growth of *Cyamopsis tetragonoloba*. Similarly, Whapham, et al. [86] observed that the application of seaweed *Ascophyllum nodosum* liquid extract increased the chlorophyll content in cucumber cotyledons, tomato, and guar plants [83].

Seaweed liquid extracts can be an effective way to some crop plants to increase both the nutrient content of the soil and crop yield. Hence, seaweeds play a vital role in agriculture, where irrational use of chemical fertilizer and pesticides is a cause of concern. Extensive regional trials with the product are needed to determine the environmental limitations of biological activity and to monitor the survival and dispersal of the inoculate [87]. Hence, use of modern agriculture in conjunction with traditional farming practices is the sustainable solution for the future. The expansion of nature source of other manures, seaweed extract application will be useful in enriching the production in the place of costly chemical fertilizer. The use of seaweed liquid extracts helps to avoid environmental pollution by high doses of chemical fertilizer. The beneficial effects of seaweed extract on terrestrial plants are improving the overall growth, yield and the ability to with stand adverse environmental conditions [88].

From the outputs of the CROPWAT model for 2016 and 2017 growing seasons, it appeared that additional irrigation was required to meet the daily crop water requirements as rainfall had minor effects or none. This high irrigation requirement may be attributed to the low rainfall during the growing seasons. Our data indicate that irrigation is crucial in the initial growth stage of Jew's Mallow due to high *DP* caused by basin irrigation system. To avoid yield reductions in Jew's Mallow cultivation, large quantities of water should be applied during the initial stage. In areas where water is a restricting factor in crop production, a well-designed irrigation schedule can improve water productivity when full irrigation is not plausible. However, a certain yield reduction should be expected due to the relationship between ET_c and yield of some crops [44,89–91].

5. Conclusions

Seaweeds are one of the most important marine resources for food, industrial raw materials, therapeutic and botanical applications. In current study, to enhance the efficiency of seaweed liquid extract, we evaluate two extraction methods; (1) water (WE); and (2) ultrasound-assisted water extraction (USWE). The effect of different seaweed extracts as a foliar spray on quantity (growth and yield) and quality (minerals and antioxidants activity) of Jew's Mallow *C. olitorious* L., comparing to NPK traditional fertilizers were observed. The present study observed that the seaweed liquid extract prepared from *P. capillacea* (either WE or USWE) presented to Jew's Mallow *C. olitorious* L. gave better results in all aspects of quantity and quality when compared to NPK traditional fertilizers. No significant differences of quantity (yield) of *C. olitorious* L. treated with WE10 and USWE10. Water extraction (WE) method improves the Chlorophyll 'a' and 'b'; total antioxidant activity and total phenolic compounds of Jew's Mallow *C. olitorious* L. While, using ultrasound-assisted water extraction (USWE) method improves the carotene and total flavonoid compounds of *P. capillacea* USWE crude extract which positively reflected on the contents of these compounds in Jew's Mallow *C. olitorious* L., when comparing to WE. Carotenes are indispensable to the plants and act as precursors for the biosynthesis of phytohormones and strigolactones, improve the plant development and responses to unstable environmental, and serve as a source of pro-vitamin A. Thus, USWE is an attractive novel technology enhancing the efficiency of seaweed liquid extract on Jew's Mallow. The CROPWAT model has shown that an adequate amount of water is vital, especially during the initial growth stage of Jew's Mallow, but also in other stages. Therefore, it is important to adopt efficient irrigation practices to maximize yields while reducing adverse effects on water resources.

Author Contributions: M.A., conceptualization, funding acquisition, investigation, methodology, writing—original draft, project administration; A.A.E.-S., formal analysis, investigation, data curation, methodology, writing—original draft, writing—review and editing; H.M.K., visualization, methodology, writing—original draft; D.Y.A.-E., supervision, visualization, methodology, writing—original draft; M.A.M., project administration, methodology, writing—review and editing, writing—original draft; A.A., Funding acquisition, project administration, resources, supervision, writing—original draft; S.M.H., methodology, investigation, supervision, project administration, writing—original draft. All authors have read and agree to the published version of the manuscript.

Funding: This work was supported by the Vice Deanship of Research Chairs at King Saud University.

Acknowledgments: This project was financially supported by the Vice Deanship of Research Chairs at King Saud University.

Conflicts of Interest: The authors declare no conflicts of interest.

References

1. Butnariu, M.; Butu, A. Chemical Composition of Vegetables and Their Products. In *Handbook of Food Chemistry*; Cheung, P.C.K., Mehta, B.M., Eds.; Springer: Berlin/Heidelberg, Germany, 2015; pp. 627–692.
2. Tang, L.; Hamid, Y.; Sahito, Z.A.; Gurajala, H.K.; He, Z.; Feng, Y.; Yang, X. Evaluation of variation in essential nutrients and hazardous materials in spinach (Spinacia oleracea L.) genotypes grown on contaminated soil for human consumption. *J. Food Compos. Anal.* **2019**, *79*, 95–106. [CrossRef]
3. Hassan, S.M.; Abd Alhafez, Z.A. Efficiency of plastic cover types on essential oil composition and vegetative growth of dill (*Anethum graveolens* L.). In Proceedings of the ISHS Acta horticulturae, VIII International Symposium on Light in Horticulture, East Lansing, MI, USA, 22 May 2016.
4. Oke, J.M.; Hamburger, M.O. Screening of some nigerian medicinal plants for antioxidant activity using 2, 2, diphenyl-picryl-hydrazyl radical. *Afr. J. Biomed. Res.* **2002**, *5*, 77–79.
5. Devasagayam, T.P.A.; Tilak, J.C.; Boloor, K.K.; Sane, K.S.; Ghaskadbi, S.S.; Lele, R.D. Free radicals and antioxidants in human health: Current status and future prospects. *JAPI* **2004**, *52*, 794–804. [PubMed]
6. Yoshikawa, T.; Toyokuni, S.; Yamamoto, Y.; Naito, Y. *Free Radicals in Chemistry, Biology and Medicine*; OICA International: London, UK, 2000.
7. Oboh, G.; Raddatz, H.; Henle, T. Characterization of the antioxidant properties of hydrophilic and lipophilic extracts of Jute (*Corchorus olitorius*) leaf. *Int. J. Food Sci. Nutr.* **2009**, *60*, 124–134. [CrossRef] [PubMed]

8. Shittu, T.A.; Ogunmoyela, O.A. Water blanching treatment and nutrient retention in some Nigerian green leafy vegetables. In Proceedings of the 25th Annual Conference of the Nigerian Institute of Food Science and Technology, Lagos, Nigeria, 5–9 November 2001; pp. 64–65.
9. Helaly, A.; Alkharpotly, A.E.-B.A.; Mady, E.; Craker, L.E. Characterization of Four Molokhia (*Corchorus olitorius*) Landraces by Morphology and Chemistry. *J. Med. Act. Plants* **2017**, *5*, 1–6.
10. Oyedele, D.J.; Asonugho, C.; Awotoye, O.O. Heavy metals in soil and accumulation by edible vegetables after phosphate fertilizer application. *Electron. J. Environ. Agric. Food Chem.* **2006**, *5*, 1446–1453.
11. Brkić, D.; Bošnir, J.; Bevardi, M.; Bošković, A.G.; Miloš, S.; Lasić, D.; Krivohlavek, A.; Racz, A.; Mojsović-Ćuić, A.; Trstenjak, N.U. Nitrate in leafy green vegetables and estimated intake. *Afr. J. Tradit. Complement. Altern. Med.* **2017**, *14*, 31–41.
12. Ashour, M.; Elshobary, M.E.; El-Shenody, R.; Kamil, A.-W.; Abomohra, A.E.-F. Evaluation of a native oleaginous marine microalga *Nannochloropsis oceanica* for dual use in biodiesel production and aquaculture feed. *Biomass Bioenergy* **2019**, *120*, 439–447. [CrossRef]
13. Khairy, H.M.; El-Shafay, S.M. Seasonal variations in the biochemical composition of some common seaweed species from the coast of Abu Qir Bay, Alexandria, Egypt. *Oceanologia* **2013**, *55*, 435–452. [CrossRef]
14. El-Shenody, R.A.; Ashour, M.; Ghobara, M.M.E. Evaluating the chemical composition and antioxidant activity of three Egyptian seaweeds: *Dictyota dichotoma, Turbinaria decurrens*, and *Laurencia obtusa*. *Braz. J. Food Technol.* **2019**, *22*, 623–630. [CrossRef]
15. Khairy, H.M.; El-Sheikh, M.A. Antioxidant activity and mineral composition of three Mediterranean common seaweeds from Abu-Qir Bay, Egypt. *Saudi J. Biol. Sci.* **2015**, *22*, 623–630. [CrossRef] [PubMed]
16. Elshobary, M.; El-Shenody, R.; Ashour, M.; Zabed, H.M.; Qi, X. Antimicrobial and antioxidant characterization of bioactive components from *Chlorococcum minutum*, a newly isolated green microalga. *Food Biosci.* **2020**, in press.
17. Abetz, P. Seaweed extracts: Have they a place in Australian agriculture or horticulture? *J. Aust. Inst. Agric. Sci.* **1980**, *46*, 23–29.
18. Ferreira, M.I.; Lourens, A.F. The efficacy of liquid seaweed extract on the yield of canola plants. *S. Afr. J. Plant Soil* **2002**, *19*, 159–161. [CrossRef]
19. Stirk, W.A.; Arthur, G.D.; Lourens, A.F.; Novák, O.; Strnad, M.; van Staden, J. Changes in cytokinin and auxin concentrations in seaweed concentrates when stored at an elevated temperature. *J. Appl. Phycol* **2004**, *16*, 31. [CrossRef]
20. Booth, E. The Manurial Value of Seaweed. *Bot. Mar.* **1965**, *8*, 138–143. [CrossRef]
21. Gandhiyappan, K.; Perumal, P. Growth promoting effect of seaweed liquid fertilizer (*Enteromorpha intestinalis*) on the sesame crop plant (*Sesamum indicum* L.). *Seaweed Resour. Util* **2001**, *23*, 23–25.
22. Blunden, G. Agricultural uses of seaweeds and seaweed extracts. In *Seaweed Resources in Europe: Uses and Potential*; Guiry, M.D., Blunden, G., Eds.; John Wiley and Sons: Hoboken, NJ, USA, 1991; pp. 65–81.
23. Ajmal, M.; Ali, H.I.; Saeed, R.; Akhtar, A.; Tahir, M.; Mehboob, M.Z.; Ayub, A. Biofertilizer as an alternative for chemical fertilizers. *J. Agric. Allied Sci.* **2018**, *7*, 1–7.
24. Bokil, K.K.; Mehta, V.C.; Datar, D.S. Seaweeds as manure: II. Pot culture manurial experiments on wheat [India]. *Phykos* **1974**, *13*, 1–5.
25. Craigie, J.S. Seaweed extract stimuli in plant science and agriculture. *J. Appl. Phycol.* **2011**, *23*, 371–393. [CrossRef]
26. Chapman, V.J.; Chapman, D.J. Seaweed as Animal Fodder, Manure and for Energy. In *Seaweeds and Their Uses*; Chapman, V.J., Chapman, D.J., Eds.; Springer: Berlin/Heidelberg, Germany, 1980; pp. 30–61. [CrossRef]
27. Aguilera-Morales, M.; Casas-Valdez, M.; Carrillo-Domínguez, S.; González-Acosta, B.; Pérez-Gil, F. Chemical composition and microbiological assays of marine algae *Enteromorpha* spp. as a potential food source. *J. Food Compos. Anal.* **2005**, *18*, 79–88. [CrossRef]
28. Arunkumar, K.; Puhazhendi, T.; Subramanian, S.; Thangaraj, N.; Rengaswamy, R. Alleviating effect of seaweed liquid fertilizer on water stressed black gram *Vigna mungo* (L.) Hepper. *Seaweed. Res. Utlin* **2002**, *24*, 151–158.
29. Rama Rao, K. Effect of seaweed liquid fertilizer on the yield of *Zizyphus rugosa* (Bores). In Proceedings of the International Symposium on Marine Algae, Bhavnagar, India, 9–12 January 1979; p. 27.

30. Wijesinghe, W.A.J.P.; Jeon, Y.-J. Enzyme-assistant extraction (EAE) of bioactive components: A useful approach for recovery of industrially important metabolites from seaweeds: A review. *Fitoterapia* **2012**, *83*, 6–12. [CrossRef] [PubMed]
31. Rodriguez-Jasso, R.M.; Mussatto, S.I.; Pastrana, L.; Aguilar, C.N.; Teixeira, J.A. Microwave-assisted extraction of sulfated polysaccharides (fucoidan) from brown seaweed. *Carbohydr. Polym.* **2011**, *86*, 1137–1144. [CrossRef]
32. Klejdus, L.; Lojková, L.; Vlcek, J. Hyphenated Solid Phase Extraction/Supercritical Fluid Extraction Methods for Extraction of Phenolic Compounds from Algae. *Curr. Anal. Chem.* **2014**, *10*, 86–98. [CrossRef]
33. Kadam, S.U.; Tiwari, B.K.; Smyth, T.J.; O'Donnell, C.P. Optimization of ultrasound assisted extraction of bioactive components from brown seaweed Ascophyllum nodosum using response surface methodology. *Ultrason. Sonochem.* **2014**, *23*, 308–316. [CrossRef]
34. Kadam, S.U.; Tiwari, B.K.; O'Donnell, C.P. Application of novel extraction technologies for bioactive from marine algae. *J. Agric. Food Chem.* **2013**, *61*, 4667–4675. [CrossRef]
35. Kadam, S.U.; Tiwarim, B.K.; Smyth, T.J.; O'Connell, S.; O'Donnell, C.P. Effect of Ultrasound Pretreatment on the Extraction Kinetics of Bioactives from Brown Seaweed (Ascophyllum nodosum). *J. Sep. Sci. Technol.* **2015**, *50*, 670–675. [CrossRef]
36. Asada, C.; Doi, K.; Sasaki, C.; Nakamura, Y. Efficient extraction of starch from microalgae using ultrasonic homogenizer and its conversion into ethanol by simultaneous saccharification and fermentation. *Nat. Resour.* **2012**, *3*, 175–179. [CrossRef]
37. Ma, Y.-Q.; Chen, J.-C.; Liu, D.; Ye, H.X.-Q. Simultaneous extraction of phenolic compounds of citrus peel extracts: Effect of ultrasound. *Ultrason. Sonochem.* **2009**, *16*, 57–62. [CrossRef]
38. Lianfu, Z.; Zelong, L. Optimization and comparison of ultrasound/microwave assisted extraction (UMAE) and ultrasonic assisted extraction (UAE) of lycopene from tomatoes. *Ultrason. Sonochem.* **2008**, *15*, 731–737. [CrossRef] [PubMed]
39. Chen, F.; Sun, Y.; Zhao, G.; Liao, X.; Hu, X.; Wu, J.; Wang, Z. Optimization of ltrasound-assisted extraction of anthocyanins in red raspberries and identification of anthocyanins in extract using high-performance liquid chromatography–mass spectrometry. *Ultrason. Sonochem.* **2007**, *14*, 767–778. [CrossRef] [PubMed]
40. Stephenson, W.A. Seaweeds in agriculture and horticulture. In *Ratequer, Perumai Valley*, 3rd ed.; Faber & Faber: London, UK, 1974; p. 241.
41. Adeboye, O.B.; Schultz, B.; Adekalu, K.O.; Prasad, K. Crop water productivity and economic evaluation of drip-irrigated soybeans (Glyxine max L. Merr.). *Agric. Food Secur.* **2015**, *4*, 10. [CrossRef]
42. Surendran, U.; Sushanth, C.M.; Joseph, E.J.; Al-Ansari, N.; Yaseen, Z.M. FAO CROPWAT Model-Based Irrigation Requirements for Coconut to Improve Crop and Water Productivity in Kerala, India. *Sustainability* **2019**, *11*, 5132. [CrossRef]
43. Clarke, D.; Smith, M.; El-Askari, K. CropWat for Windows: User Guide. Available online: http://www.fao.org/land-water/databases-and-software/cropwat/en/ (accessed on 1 March 2016).
44. Moseki, O.; Murray-Hudson, M.; Kashe, K. Crop water and irrigation requirements of Jatropha curcas L. in semi-arid conditions of Botswana: Applying the CROPWAT model. *Agric. Water Manag.* **2019**, *225*, 105754. [CrossRef]
45. Tsakmakis, I.D.; Zoidou, M.; Gikas, G.D.; Sylaios, G.K. Impact of Irrigation Technologies and Strategies on Cotton Water Footprint Using AquaCrop and CROPWAT Models. *Environ. Process.* **2018**, *5*, 181–199. [CrossRef]
46. Studer, C.; Spoehel, S. Potential and Actual Water Savings through Improved Irrigation Scheduling in Small-Scale Vegetable Production. *Agronomy* **2019**, *9*, 888. [CrossRef]
47. Flores-Velazquez, J.; Rojano, F.; Aguilar, E. ETo estimation to schedule irrigation in a green wall using Cropwat. In Proceedings of the 2019 ASABE Annual International Meeting, Boston, MA, USA, 7–10 July 2019.
48. Rauch, T. The estimation of microalgal protein content, its meaning the evolution of algal biomass, comparison method for extracting for protein. *Hydrobiologia* **1981**, *78*, 237. [CrossRef]
49. Hartree, E.F. Determination of protein: A modification of the Lowry method that gives a linear photometric response. *Anal. Riochem.* **1972**, *48*, 422–427. [CrossRef]
50. Myklestad, S.; Haug, A. Production of carbohydrates by the marine Chaetoceros affinis var. willei (Gran.) Hustedt. I. Effect of the concentration of nutrients in the culture medium. *J. Exp. Mar. Biol. Ecol.* **1972**, *9*, 125–136. [CrossRef]

51. Dubois, M.; Giles, K.A.; Hamilton, J.K.; Rebers, P.A.; Smith, F. Calorimetric method for determination of sugars and related substances. *Anal. Chem.* **1956**, *28*, 350–356. [CrossRef]
52. Bligh, E.G.; Dyer, W.J. A rapid method of total lipid extraction and purification. *Can. J. Biochem. Physiol.* **1959**, *37*, 911–917. [CrossRef]
53. Page, A.L. *Methods of Soil Analysis. Part 2. Chemical and Microbiological Properties*; American Society of Agronomy, Soil Science Society of America: Madison, WI, USA, 1982.
54. Allen, R.G.; Pereira, L.S.; Raes, D.; Smith, M. *Crop Evapotranspiration-Guidelines for Computing Crop Water Requirements-FAO Irrigation and Drainage Paper No. 56*; FAO—Food and Agriculture Organization of the United Nations: Rome, Italy, 1998.
55. Ghoneim, I.M.; El-araby, S.M. Effect of organic manure source and biofertilizer type on growth, productivity and chemical composition of Jew's Mallow (*Corchorus Olitorious* L.) plants. *J. Agric. Env. Sci. Alex. Univ. Egypt* **2003**, *2*, 88–105.
56. Dere, Ş.; Güneş, T.; Sivaci, R. Spectrophotometric determination of chlorophyll-A, B and total carotenoid contents of some algae species using different solvents. *Turkish J. Bot.* **1998**, *22*, 13–18.
57. Evenhuis, B. *Simplified Methods for Foliar Analysis Vol. Communication 70*; Department of Agriculture Research Product, Research Division, Koninklijk Institut voor den Tropen: Amsterdam, NL, USA, 1978.
58. Cottenie, A.; Verloo, M.; Kiekens, L.; Velghe, G.; Camerlynck, R. *Chemical Analysis of Plants and Soils, Laboratory of Analytical and Agrochemistry*; State University of Ghent: Ghent, Belgium, 1982.
59. Kumar, G.S.; Antony Muthu Prabhu, A.; Seethalashmi, P.G.; Bhuvanesh, N.; Kumaresan, S. Self-catalyzed syntheses, structural characterization, DPPH radical scavenging-, cytotoxicity-, and DFT studies of phenoxyaliphatic acids of 1,8-dioxo-octahydroxanthene derivatives. *J. Mol. Struct.* **2014**, *1059*, 51–60. [CrossRef]
60. Ahmad, M.Z.; Ali, M.; Mir, S.R. Anti-diabetic activity of *Ocimum sanctum* L. roots and isolation of new phytoconstituents using two-dimensional nuclear magnetic resonance spectroscopy. *J. Pharmacogn. Phytother.* **2012**, *4*, 75–85. [CrossRef]
61. Chang, C.-C.; Yang, M.-H.; Wen, H.-M.; Chern, J.-C. Estimation of total flavonoid content in propolis by two complementary colorimetric methods. *J. Food Drug Anal.* **2002**, *10*, 178–182.
62. Swennenhuis, J. *CROPWAT Version 8.0*; FAO—Food and Agriculture Organization of the United Nations: Rome, Italy, 2009.
63. Grieser, J. *Climwat 2.0 for CROPWAT*; FAO—Food and Agriculture Organization of the United Nations: Rome, Italy, 2006.
64. Snedecor, G.W.; Cochran, W.G. *Statistical Methods*, 8th ed.; Iowa State University Press: Ames, IA, USA, 1989.
65. Ashour, M.; Abo-Taleb, H.A.; Abou-Mahmoud, M.M.; El-Feky, M.M.M. Effect of the integration between plankton natural productivity and environmental assessment of irrigation water, El-Mahmoudia Canal, on aquaculture potential of *Oreochromis niloticus*. *Turk. J. Fish. Aquat. Sci.* **2018**, *18*, 1163–1175. [CrossRef]
66. Ashour, M.; Kamel, A. Enhance growth and biochemical composition of *Nannochloropsis oceanica* cultured under nutrient limitation using commercial agricultural fertilizers. *J. Mar. Sci. Res. Dev.* **2017**, *7*, 233. [CrossRef]
67. Nour, M.A.; Zaki, M.A.; Ashour, M.; Kamel, A.; Taha, S.M.; Omar, E. Effect of different *Nannochloropsis oculata* diets (cultured under different nitrogen and phosphorus regime) on growth and biochemical composition of rotifers *Brachionus plicatilis*. In Proceedings of the 12th International Congress on Microbial Interaction and Applications of Beneficial Microbes, Munich, Germany, 17–18 July 2017.
68. Omar, E.; Zaki, M.; Ashour, M.; Kamel, A.; Taha, S.M.; Nour, M.A. Effect of nutrient limitations on growth and biochemical composition of outdoor cultured *Chaetoceros calcitrans*. In Proceedings of the 12th International Congress on Microbial Interaction and Applications of Beneficial Microbes, Munich, Germany, 17–18 July 2017.
69. Nour, A.M.; Ashour, M.; Kamil, A.; Taha, S.M.; Zaki, M.A.; Omar, E. The use of artificial fertilizers for culture of marine microalgae: 1. Growth and biochemical constituents of *Nannochloropsis oculata*. World Congress and Expo on Applied Microbiology, Munich, Germany. *J. Microb. Biochem. Technol.* **2015**, *5*, 120.
70. Nour, A.M.; Ashour, M.; Kamil, A.; Taha, S.M.; Zaki, M.A.; Omar, E. The use of artificial fertilizers for culture of marine microalgae: 2. Growth and biochemical constituents of *Chaetoceros calcitrans*. World Congress and Expo on Applied Microbiology, Munich, Germany, 2015. *J. Microb. Biochem. Technol.* **2015**, *5*, 119.

71. Sharawy, Z.Z.; Ashour, M.; Abbas, E.; Ashry, O.; Helal, M.; Nazmi, H.; Kelany, M.; Kamel, A.; Hassaan, M.; Rossi, W.J.; et al. Effects of dietary marine microalgae, *Tetraselmis suecica* on production, gene expression, protein markers and bacterial count of Pacific white shrimp *Litopenaeus vannamei*. *Aquac. Res.* **2020**, in press.
72. Soliman, H.A.; Fatma, A.A.; AbouZeid, A.E.; Ashour, M. Optimum growth conditions of three isolated diatoms species; *Skeletonema costatum*, *Chaetoceros calcitrans* and *Detonula confervacea* and their utilization as feed for marine penaeid shrimp larvae. *EJAR* **2010**, *36*, 161–183.
73. Soliman, H.A.; Fatma, A.A.; AbouZeid, A.E.; Ashour, M. Population growth rate, fecundity, filtration and ingestion rate of marine rotifer *Brachionus plicatilis* fed with motile and immotile microalgae. *EJAR* **2008**, *34*, 426–439.
74. Heneash, A.; Ashour, M.; Matar, M. Effect of Un-live Microalgal diet, *Nannochloropsis oculata* and *Arthrospira* (*Spirulina*) platensis, comparing to yeast on population of rotifer, *Brachionus plicatilis*. *Mediterr. Aquac. J.* **2015**, *7*, 48–54. [CrossRef]
75. Abo-Taleb, H.; Ashour, M.; El-Shafei, A.; Alataway, A.; Maaty, M.M. Biodiversity of Calanoida Copepoda in Different Habitats of the North-Western Red Sea (Hurghada Shelf). *Water* **2020**, *12*, 656. [CrossRef]
76. Ashour, M. *Marine Microalgae: Aquaculture and Biodiesel Production*, 1st ed.; LAP Lambert Academic Publisher: Saarbrücken, Germany, 2015; pp. 1–240. ISBN 978-3659691683.
77. Ashour, M. Sustainable development and the future of marine natural products in Egypt: SMIAF-Prototype project. In Proceedings of the 4th International Conference and Exhibition on Natural Products Medicinal Plants & Marine Drugs, Rome, Italy, 11 June 2018.
78. Hassan, S.M.; Ashour, M.; Soliman, A.A.F. Anticancer activity, antioxidant activity, mineral contents, vegetative and yield of *Eruca sativa* using foliar application of autoclaved cellular extract of *Spirulina platensis* extract, comparing to N-P-K fertilizers. *J. plant Prod. Mansoura Univ.* **2017**, *8*, 529–536. [CrossRef]
79. Mohy El-Din, S.M.; El-Ahwany, A.M.D. Bioactivity and phytochemical constituents of marine red seaweeds (*Jania rubens*, *Corallina mediterranea* and *Pterocladia capillacea*. *J. Taibah Univ. Sci.* **2016**, *10*, 471–484. [CrossRef]
80. Tarakhovskaya, E.R.; Maslov, Y.I.; Shishova, M.F. Phytohormones in algae. *Russ J. Plant Physiol.* **2007**, *54*, 163. [CrossRef]
81. Nelson, W.R.; van Staden, J. The Effect of Seaweed Concentrate on Wheat culms. *J. Plant Physiol.* **1984**, *115*, 433–437. [CrossRef]
82. Sun, T.; Yuan, H.; Cao, H.; Yazdani, M.; Tadmor, Y.; Li, L. Carotenoid Metabolism in Plants: The Role of Plastids. *Mol. Plant.* **2018**, *11*, 58–74. [CrossRef] [PubMed]
83. Blunden, G.; Wildgoose, P.B. The effects of aqueous seaweed extract and kinetin on potato yields. *J. Sci. Food Agric.* **1977**, *28*, 121–125. [CrossRef]
84. Mohan, V.R.; Venkataraman Kumar, V.; Murugeswari, R.; Muthuswami, S. Effect of crude and commercial seaweed extracts on seed germination and seedling growth in *Cajanus cajan* L. *Phykos* **1994**, *33*, 47–51.
85. Thirumaran, G.; Arumugam, M.; Arumugam, R.; Anantharaman, P. Effect of seaweed liquid fertilizer on growth and pigment concentration of Cyamopsis tetrogonolaba (L) Taub. *Am. Eurasian J. Agron.* **2009**, *2*, 50–56.
86. Whapham, C.A.; Blunden, G.; Jenkins, T.; Hankins, S.D. Significance of betaines in the increased chlorophyll content of plants treated with seaweed extract. *J. Appl. Phycol.* **1993**, *5*, 231. [CrossRef]
87. Kannan, L.; Selvan, T. Effect of seaweed biofertilizer on green gram. In Proceedings of the International Symposium on Phycology, Madras, India, 25 May–15 June 1987; p. 19.
88. Subramanian, S.K. Studies on the Effect of Seaweed Liquid Fertilizer [SLF]. In Proceedings of the International Symposium on Phycology, Madras, India, 25 May–15 June 1987; p. 33.
89. Davidson, R.J. EEG Measures of Cerebral Asymmetry: Conceptual and Methodological Issues. *Int. J. Neurosci.* **1988**, *39*, 71–89. [CrossRef]
90. Bennett, D.R.; Harms, T.E. Crop yield and water requirement relationships for major irrigated crops in southern Alberta. *Can. Water Resour. J.* **2011**, *36*, 159–170. [CrossRef]
91. Fereres, E.; Garda-Vila, M. Irrigation management for efficient crop production. *Crop Sci.* **2019**. [CrossRef]

agronomy

MDPI

Article

Appraisal of Biodegradable Mulching Films and Vegetal-Derived Biostimulant Application as Eco-Sustainable Practices for Enhancing Lettuce Crop Performance and Nutritive Value

Eugenio Cozzolino [1,†], Maria Giordano [2,†], Nunzio Fiorentino [2], Christophe El-Nakhel [2], Antonio Pannico [2], Ida Di Mola [2], Mauro Mori [2], Marios C. Kyriacou [3], Giuseppe Colla [4] and Youssef Rouphael [2,*]

1 Council for Agricultural Research and Economics (CREA)—Research Center for Cereal and Industrial Crops, 81100 Caserta, Italy; eugenio.cozzolino@crea.gov.it

2 Department of Agricultural Sciences, University of Naples Federico II, 80055 Portici, Italy; maria.giordano@unina.it (M.G.); nunzio.fiorentino@unina.it (N.F.); nakhel_christophe@hotmail.com (C.E.N.); antonio.pannico@unina.it (A.P.); ida.dimola@unina.it (I.D.M.); mori@unina.it (M.M.)

3 Department of Vegetable Crops, Agricultural Research Institute, 1516 Nicosia, Cyprus; m.kyriacou@ari.gov.cy

4 Department of Agriculture and Forest Sciences, University of Tuscia, 01100 Viterbo, Italy; giucolla@unitus.it

* Correspondence: youssef.rouphael@unina.it

† Co-first authors (E.C. and M.G.).

Received: 6 March 2020; Accepted: 19 March 2020; Published: 20 March 2020

Abstract: Scientists, extensions specialists, and growers are seeking sustainable agricultural practices that are able to cope with these objectives in order to ensure global food security and minimize environmental damage. The use of mulching films and plant biostimulants in agriculture seems to be a valid solution for tackling these rising concerns. A greenhouse experiment was conducted in order to elucidate the morpho-physiological and nutritive characteristics of lettuce (*Lactuca sativa* L.) in response to foliar application of a tropical plant extract (PE) biostimulant and the use of plastic mulches. Two biodegradable mulch treatments (Mater-Bi® 1 and Mater-Bi® 2) were compared to black polyethylene (LDPE) and bare soil. Biodegradable mulch film Mater-Bi® 1 produced a comparable marketable fresh yield to the commercial standard polyethylene (LDPE), whereas Mater-Bi® 2 exhibited the highest crop productivity. When averaged over biostimulant application, lettuce plants grown with biodegradable film Mater-Bi® 2 exhibited superior quality traits in terms of K, Ca, total ascorbic acid, and carotenoids content. The combination of film mulching (LDPE, Mater-Bi® 1 or Mater-Bi® 2) with the tropical plant extract biostimulant exhibited a positive and significant synergistic effect (+30%) on yield. The PE-biostimulant induced higher values of SPAD index and total chlorophyll content when compared to untreated greenhouse lettuce. The mineral content of leaf tissues was greater by 10% and 17% (for P and Ca, respectively) when compared to the untreated lettuce (no PE application). Nitrate content was significantly reduced by 23% in greenhouse lettuce plants receiving PE as compared to the untreated control. The positive effect of Mater-Bi® 2 film on the ascorbic acid content has also been highlighted when combined with the biostimulant application, where a major amplification of total ascorbic acid (+168%) was recorded in comparison to the untreated lettuce. Overall, our work can assist leafy vegetables growers in adopting good agricultural practices, such as biodegradable plastic mulches and vegetal-derived biostimulants, to improve the sustainability of greenhouse production.

Keywords: eco-friendly practices; *Lactuca sativa* L.; total ascorbic acid; tropical plant extract; Mater-Bi®; nitrate; mineral composition; SPAD index; functional quality

1. Introduction

A widespread agricultural practice across the world consists of covering the soil around plants with plastic films. The introduction of this technique in agriculture dates back to the 1970s, and its success is still linked to multiple benefits. In fact, plastic films can: (i) increase soil temperature and keep it constant throughout the first 20–30 cm layer, so that plants' roots develop faster [1,2]; (ii) reduce soil evapotranspiration and preserve moisture; (iii) prevent soil erosion and excessive leaching of nutrients from plants' rhizosphere; and, (iv) improve the performance of plants in a quantitative and qualitative manner [1,3–5]. In addition, mulching films suppress weeds growth, protect crops against pests and various diseases, and reduce the use of pesticides and herbicides. Based on their color (black, clear or white), they absorb and/or reflect sunlight, differently varying soil temperature, thus affecting crop growth and productivity [5]. Plastic films are widely used for growing vegetables under both open-field and greenhouse conditions [6]. Moreover, these films are mainly made by low-density polyethylene (LDPE) [3], having a strong resistance and high durability, even though, like all petroleum products, they are non-compostable and non-biodegradable. The presence of LDPE residues in the soil beyond the duration of a crop cycle is associated to soil contamination with phthalate and phthalic acid esters due to thermal degradation [7]. Therefore, farmers must manually or mechanically collect from the field and recycle or dispose them to comply with the legislative directives of each country. Unfortunately, the frequent illegal burning of plastic mulches by farmers is becoming a common practice, with the aim of reducing production costs by avoiding disposal expenses, which results in a consequent emission of toxic and harmful substances for humans and the environment [1,3]. In such a way, plastic mulching films increase plastic wastes that are used in agriculture, such as pipes and fittings; agricultural packaging, such as bags, liners, and containers [3]. Therefore, there is an urgent need to use compostable and biodegradable materials in modern agriculture. Nowadays, research is projected towards the creation of films made of biopolymers, such as starch, polylactic acid, and cellulose. These materials are derived from renewable resources, such as corn, potato, and rice [1,6]. Their degradation is in compliance with the European laws and Italian ones (UNI 10785, 1999) on biodegradability (EN 13432, 2000). In fact, these materials are entirely degraded by soil microorganisms and they are mineralized in carbon dioxide and methane, water, and biomass, without the production of toxic substances. Any biodegradable material is designed to disappear within the soil in 5–6 months after the end of the crop [2].

Efficient management of natural resources, such as water and soil, is needed in a scenario where the world population is growing, and agriculture must meet an increasing food demand. On the other hand, the use of plant biostimulants in agriculture has been recognized during the last two decades as an efficient tool to boost yield under optimal and sub-optimal conditions, to improve quality as well as increase nutrient uptake and use efficiency of field and horticultural crops [8–11]. Under the new European Union Regulation 2019/1009, plant biostimulants are specified based on their agronomical effects on crops (i.e., claims), and they include humic substances, protein hydrolysates, algae and plant extracts, inorganic compounds (e.g., silicon), growth-promoting bacteria, and mycorrhizal fungi. Many recent studies on vegetal-based biostimulants have shown to increase the tolerance of crops to abiotic stress (extreme temperature, drought, and salinity), and improve the quality of the produce, in terms of organoleptic and nutraceutical characteristics [11]. They have also contributed to the reduction of unwanted substances content, such as nitrates and heavy metals, in crops [12]. Among these, plant extracts that mainly contain signaling molecules (i.e., small peptides and free amino acids) can influence both primary and secondary metabolism in plants, by stimulating glycolysis enzymes' activity, Krebs' cycle, and nitrates' assimilation [13,14]. Moreover, it has been shown that vegetal-derived plant biostimulants effects involve the size modifications of roots by increasing the length and the number of root hairs, as well as the intake of both macroelements and microelements, leading to better crop performance and the nutritive value of the final produce [8,13,14].

Lettuce (*Lactuca sativa* L.) belongs to the *Asteraceae* family and it is one of the most intensively produced leafy vegetables being widespread all over the world. It is valued for its organoleptic properties and is considered an important source for health-promoting metabolites (carotenoids, chlorophylls, macro and trace elements, phenolics, and vitamins), which are crucial in human nutrition [15,16]. Lettuce has a high water and low fat content, which makes it ideal for dietary plans [15]. Italy dedicates vast areas to lettuce production, and has a broad market, which places it as a European leader in this sector [10]. More importantly, production systems and agronomic practices are pre-harvest factors that can determine the quantitative and qualitative variations in lettuce bioactive compounds [17].

On the basis of the above-mentioned considerations, the aim of our work was to combine two eco-sustainable agricultural practices, such as the use of biodegradable films and plant-based biostimulant (tropical plant extract), and test their effect on the morpho-physiological performance, mineral composition, and nutritive value of greenhouse lettuce plants. The films used were two biodegradable mulching films, namely Mater-Bi® with different composition, which effect was compared with that of a polyethylene film and bare soil. The findings of the study will elucidate the biostimulant × mulch interaction to select the best combination (s) able to improve crop performance and nutritive value of this important leafy vegetable. We also believe that these results will be of great interest for horticulturists, extension specialists, and scientists.

2. Materials and Methods

2.1. Greenhouse Growth Conditions, Treatments and Experimental Design

The experimental test was implemented in a protected environment made of an unheated greenhouse, which was located at the experimental farm of the Department of Agriculture, University of Naples Federico II, Portici—Naples (lat. 40°49′ N; long 14°20′ E, 37 a.s.l). The main physical and chemical characteristics of the soil at the experimental site were: sandy loam texture (74% sand, 20% silt, 6% clay), electrical conductivity of 0.5 dS m^{-1}, neutral pH-7.0, total nitrogen (N) of 0.12%, and organic matter of 1.20% (w/w). The nitrate N, ammoniacal N, Olsen phosphorus, and exchangeable potassium were 105, 12, 40, and 936 mg kg^{-1}, respectively. The butterhead lettuce F1 hybrid SINTIA RZ (42–160; Rijk Zwaan, Der Lier, The Netherlands) was used in this test. This lettuce is very resistant to tip burn and bolting and it is characterized by bright green leaves. SINTIA RZ was selected as the most representative commercial cultivar that was used in Italy during the autumn and winter growing seasons under protected environment. On 16 September 2017 three mulching films (M) were installed, two black biodegradable films, namely Mater-Bi® PC 17 N1 (15 µm thick, commercial; Novamont S.p.A, Novara, Italy) and Mater-Bi® PC 17 N2 (15 µm thick, experimental; Novamont S.p.A, Novara, Italy), and one traditional black low-density polyethylene (LDPE) plastic film (50 µm thick, Idroland s.r.l., Bari, Italy). The compositions of the two biodegradable films are composed of thermoplastic starch and copolyester. The two Mater-Bi® mulching films differ in the presence of Masterbatch (PC 17 N2), a solid additive that is used for imparting color or other properties to plastics, with innovative characteristics to improve the color of mulches with low impact on the original polymer. The soil additive is a concentrated mixture of pigments that was made through a heating process and it includes a carrier resin (e.g., wax) that is cut into granules and then added to plastics.

The greenhouse consisted of a galvanized steel frame with plastic covering material, two non-automated side openings, and a mechanized roof opening. The total greenhouse surface corresponded to an area of 162 square meters (27 m × 6 m). The soil was prepared with low energy inputs consisting of a manual grubbing-up of weeds and then a shallow hoeing (20–25 cm) to allow for a leveling of the soil in a single pass. Water was not a limiting factor, the crop evapotranspiration was calculated with the Hargreaves method, and the water deficit was fully restored by using a drip irrigation system. The irrigation system consisted of a main polyethylene pipeline with a diameter of 32 mm with a low operating pressure of 2 atm, while a series of semi-compensating dripping wings (16 mm diameter and 10 cm interpolation) were laterally attached.

The lettuce seedlings were transplanted in the greenhouse on September 25th on raised furrows. On each furrow, the lettuce seedlings were arranged in double rows, at a plant density of 12.3 plants per m^2. The antiperonosporic protection was performed with Metalaxil seven days after transplantation in order to limit the development of fungal pathogens.

Figure 1 presents the trend of minimum, maximum, and mean daily air temperature inside the greenhouse during the cropping cycle. The soil temperature measurements (minimum, maximum, and mean temperature) were also recorded with microchip sensors (0.5 °C sensitivity) that were placed at 10 cm depth. All of the measurements were collected on a data logger (Davis Vantage Pro2, CA, USA). Nitrogen fertilization was applied by fertigation with ammonium nitrate at eight and 16 days after transplantation (DAT). Half of the plots were treated with Auxym® (Italpollina USA Inc., Anderson, IN, USA) product in order to assess the action of the biostimulant (B). This biostimulant is obtained from fermented tropical plant biomass. It contains phytohormones, amino acids, vitamins, phytochelatins, and enzymes. Auxym® contains as well micro and macroelements (g/kg): N 8.3, P 4.0, K 25.0, Ca 0.9, Mg 1.2, Fe 6.6, Mn 6.4, B 4.4, Zn 0.4, and Cu 0.2 [18]. The biostimulant was applied -at a concentration of 2 mL per liter of water- on plants by a sprayer shoulder pump and application took place five times at seven days' intervals starting 10 DAT (i.e., foliar application).

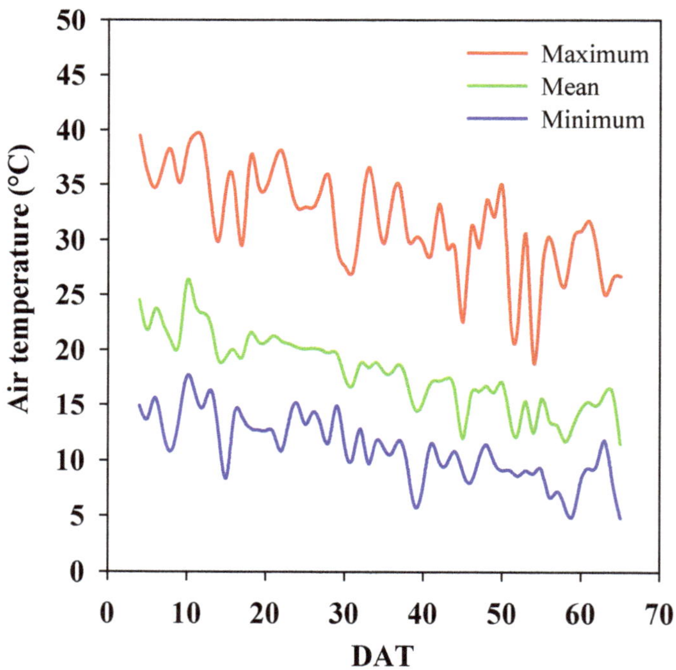

Figure 1. Daily maximum, mean, and minimum values of air temperature recorded inside the greenhouse during the growing period of lettuce.

The experimental scheme provided a two factors factorial combination that resulted in eight treatments in which the factors were mulching (M; three mulching films and bare soil) and biostimulant application (B; control treatment and foliar application of biostimulant). Each treatment was replicated three times and all of the treatments were organized in a randomized complete-block design, resulting in a total of 24 experimental plots.

2.2. Growth Analysis, Yield, Harvest and Quality Analysis Sampling

The harvest was manually carried out by cutting the plants at the crown area, just when commercial weight was attained. For each replicate, a total of 15 representative plants were collected. Each plant was first weighed as a whole (leaves and stem) in order to determine the total yield, while the commercial yield was estimated after separation and weighing of leaves. In both cases, yield was expressed in g plant^{-1}. Finally, the leaves were counted and the leaf area (cm^2 plant^{-1}) was determined using a LiCor 3100C leaf area meter (LI-COR Biosciences, Lincoln, NE, USA). Five fresh plants from each treatment were randomly sampled, and then stored at −80 °C until the determination of bioactive compounds content.

2.3. Soil Plant Analysis Development (SPAD) Index and Color Measurements

Fifteen SPAD index measurements were performed by a chlorophyll meter (Minolta SPAD-502, Tokyo, Japan) and averaged to a single value on five fully expanded lettuce leaves per replicate. Leaf color (space parameters L*, a* and b*) was recorded with a Minolta chroma meter (CR-300, Minolta Camera Co. Ltd., Tokyo, Japan), on the center of the upper leaf surface with special care to avoid the central vein.

2.4. Total Chlorophyll and Carotenoid Content Determination

On 1 g of fresh leaf samples, the total chlorophylls and carotenoids content was determined following the Lichtenhaler and Buschman [19] method. Fresh sample was extracted in pure acetone, for 15 min in the dark. Subsequently, the absorbance of the extracted solutions was measured at 662, 645, and 470 nm, while using a Hach DR 2000 spectrophotometer (Hach Co., Loveland, CO, USA).

2.5. Dry Matter, Nitrate and Macromineral Content Analysis

After the determination of fresh yield, the leaves were put to a ventilated stove at a temperature of 65 °C for 72 h until constant weight for dry weight determination. The dry mater content was expressed as percentage (%). Mineral analysis was carried out in 250 mg of dry ground leaves (IKA, MF 10.1, Staufen, Germany), which were sieved and diluted in 50 mL of ultrapure water (Milli-Q, Merck Millipore, Darmstadt, Germany). A syringe with a 0.45 µm pore filter (Phenomenex, Torrance, CA, USA) was used to inject each sample into an ion chromatography (ICS-3000, Dionex, Sunnyvale, CA, USA). For macrocations determination, an IonPac CG12A (4 × 250 mm) guard column and IonPac CS12A (4 × 250 mm) analytical column were used. While, for macroanions determination, an IonPac AG11-HC (4 × 50 mm) guard column and IonPac AS11-HC analytical column (4 × 250 mm) were used. All of the macrominerals were expressed on a dry weight (dw) basis (g kg^{-1}), while the nitrate content was expressed as mg kg^{-1} fw based on the respective leaf sample dry matter content.

2.6. Hydrophilic Antioxidant Activity Determination

In order to measure the hydrophilic antioxidant activity (HAA), 200 mg of lyophilized sample were extracted twice with distilled water, following the *N,N*-dimethyl-p-phenylenediamine (DMPD) method [20]. An aliquot of 20 µL of extract was combined with 2 mL of DMPD + solution. The bleaching of solution was proportional to the amount of antioxidant compounds concentration. The reduction in absorbance, as measured by UV Vis spectrophotometry at 505 nm, allows for determining the antioxidant activity. For this purpose, an ascorbate external standard calibration curve was used.

2.7. Total Phenols and Total Ascorbic Acid Content Determination

The total phenols content was assessed with the Folin–Ciocalteau procedure [21]. 250 mg of lyophilized sample were extracted with 10 mL of methanol/water (60:40 *v*/*v*). After an incubation of 90 min., the absorption was measured at 765 nm while employing a UV-Vis spectrophotometer. The results were calculated using an external gallic acid calibration curve (Sigma Aldrich Inc., St. Louis,

MO, USA). Total ascorbic acid content was measured according to the method of Kampfenkel et al. [22], and it was quantified by a spectrophotometer at 525 nm against an external ascorbate standard calibration curve.

2.8. Statistical Analysis

The normal distribution of the data was verified through the Shapiro–Wilk's and Kolmogorov–Smirnov's procedures. All of the data were subjected to Two-way ANOVA using SPSS 20 software package. For mulching factor, the treatment means were confronted utilizing Duncan's Multiple Range Test that was performed at $p \leq 0.05$, while, for the biostimulant effect, the means were compared using the *t*-test.

3. Results

3.1. Soil Temperature Trends

The minimum, maximum and mean soil temperatures under the three tested mulches were influenced by the composition of the utilized mulching material (Figure 2). The differences between the minimum and mean soil temperatures between LDPE and the two biodegradable mulches (Mater-Bi® 1 and Mater-Bi® 2) were notable during the first 15–20 days after transplanting, whereas the differences became narrowerer towards the end of the growing cycle (Figure 2). Concerning the maximum soil temperatures, the Mater-Bi® 1 film had similar maximum soil temperature values to LDPE and slightly higher ones than Mater-Bi® 2. However, the soil minimum, maximum, and mean temperatures trends that were recorded in bare soil were regularly lower than those reached among the three tested mulch materials (Figure 2). The soil temperature trends were similar under the three mulching films in all cases, since they had the highest values at the beginning of the crop cycle and underwent a gradual decrease afterwards, especially towards the end of the growing period. The average minimum soil temperature varied between 13.2–23.9 °C in LDPE, 12.7–22.2 °C in Mater-Bi® 1, 12.7–22.2 °C in Mater-Bi® 2, and 10.6–20.8 °C in bare soil. Finally, the average maximum soil temperature fluctuated between 15.1–30.0 °C in LDPE, 15.1–29.5 °C in Mater-Bi® 1, 14.7–28.3 °C in Mater-Bi® 2, and 12.0–25.2 °C in bare soil.

3.2. Yield and Biometric Parameters

The combination of LDPE or biodegradable mulching materials (Mater-Bi® 1 or Mater-Bi® 2) with the PE-based biostimulant positively affected the total and marketable yields of greenhouse lettuce when compared to the untreated plants, although the beneficial effect of biostimulant application was not apparent for bare soil treatment (Table 1). According to the average effect of the mulching films, a tendency to higher total yield values was recorded for Mater-Bi® 2 (319.5 g plant^{-1}), with a 22% increase as compared to bare soil (261.3 g plant^{-1}), even though no significant differences were recorded between the three mulching treatments. However, this trend became apparent for marketable yield with significantly higher values for Mater-Bi® 2 when compared to Mater-Bi® 1 or LDPE and especially to bare soil (Table 1). The positive effect of PE-treated lettuce plants that were cultivated under the three mulching materials (LDPE, Mater-Bi® 1, or Mater-Bi® 2) was mainly attributed to an increment in the total leaf area and not to an increase in the plant leaf number based on the M × B interaction (Table 1). Moreover, the effect of PE foliar application, when averaged over all mulching treatments, was shown to affect leaf number, which was higher by 10% in PE-treated than in untreated greenhouse lettuce plants. Finally, our findings demonstrated that lettuce plants that were grown under LDPE or Mater-Bi® 2 elicited a significant increment in the number of leaves confronted to the bare soil treatment, whereas the plants cultivated under Mater-Bi® 1 exhibited intermediate values (Table 1).

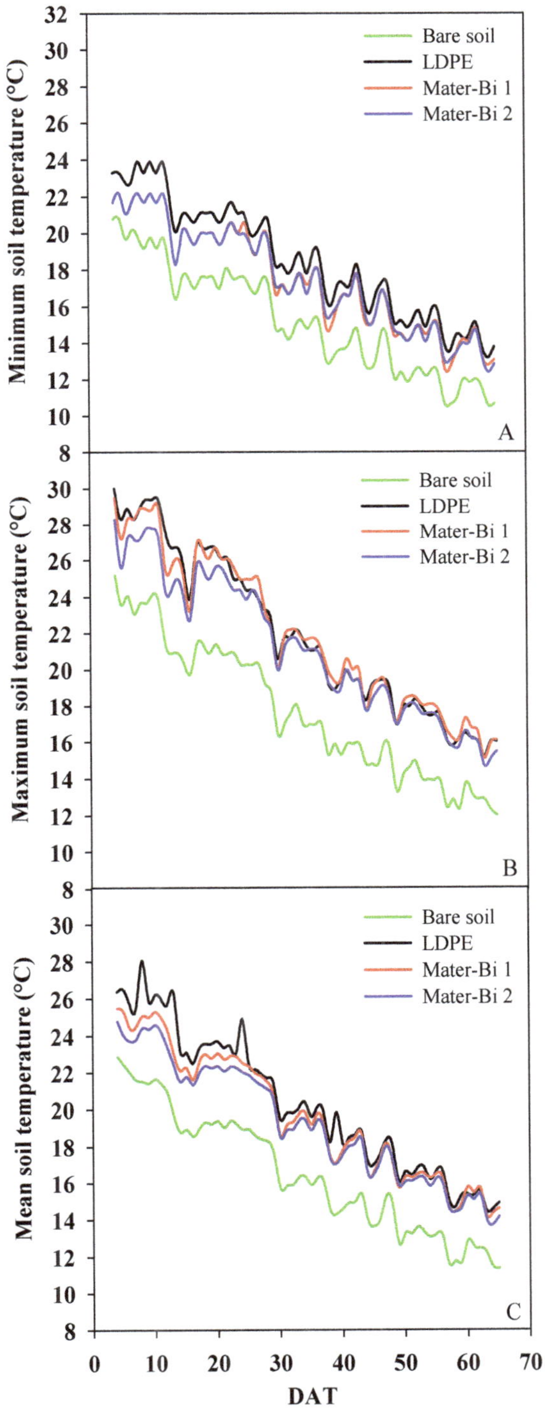

Figure 2. Daily minimum (**A**), maximum (**B**), and mean (**C**) values of soil temperature recorded at a depth of 10 cm in bare soil, LDPE, Mater-Bi® 1 and Mater-Bi® 2 mulches.

Table 1. Mean comparison and analysis of variance for total and marketable yield, leaf number, and total leaf area of untreated and biostimulant-treated greenhouse lettuce grown under low-density polyethylene (LDPE) mulch and biodegradable (Mater-Bi® 1 and Mater-Bi® 2) mulching materials in relation to bare soil.

Source of Variance	Total Yield (g plant^{-1})	Marketable Yield (g plant^{-1})	Leaf Number (no. plant^{-1})	Leaf Area (cm^2 plant^{-1})
Mulch (M)				
Bare soil	261.3 ± 7 b	243.9 ± 7 c	36.5 ± 0.4 c	3414 ± 58 c
LDPE	298.0 ± 18 a	274.3 ± 17 b	44.5 ± 1.2 a	3885 ± 193 a
Mater-Bi® 1	302.8 ± 21 a	274.7 ± 19 b	40.6 ± 1.4 b	3566 ± 168 b
Mater-Bi® 2	319.5 ± 17 a	296.5 ± 16 a	41.9 ± 1.0 a	3667 ± 168 ab
	***	***	***	*
Biostimulant (B)				
Control	266.4 ± 6	243.3 ± 5	38.9 ± 0.8	3375 ± 73
Tropical plant extract (PE)	324.5 ± 12	301.4 ± 10	42.8 ± 1.1	3891 ± 100
t-test	0.000	0.000	0.009	0.000
M × B				
Bare soil without biostimulant	260.6 ± 14 b	238.0 ± 13 b	35.5 ± 0.1	3433 ± 109 b
LDPE without biostimulant	263.9 ± 15 b	240.3 ± 12 b	42.7 ± 1.6	3545 ± 241 b
Mater-Bi® 1 without biostimulant	256.8 ± 5 b	232.9 ± 4 b	38.6 ± 0.5	3198 ± 47 b
Mater-Bi® 2 without biostimulant	284.8 ± 11 b	262.0 ± 8 b	38.9 ± 0.6	3323 ± 103 b
Bare soil + PE	262.0 ± 8 b	249.9 ± 6 b	37.4 ± 0.2	3395 ± 67 b
LDPE + PE	332.2 ± 13 a	308.4 ± 13 a	46.4 ± 1.0	4226 ± 114 a
Mater-Bi® 1 + PE	348.7 ± 9 a	316.5 ± 7 a	44.9 ± 0.3	3934 ± 57 a
Mater-Bi® 2 + PE	354.9 ± 4 a	331.0 ± 4 a	42.6 ± 0.8	4011 ± 112 a
	**	*	NS	*

NS, *, **, *** Non-significant or significant at $p \leq 0.05$, 0.01, and 0.001, respectively. Different letters in the same column indicate significant differences according to DMR test ($p = 0.05$). Means of biostimulant effect are compared according to Student's *t*-test ($p = 0.05$). All data are expressed as mean ± SE.

3.3. SPAD index, Chlorophyll Content and Colorimetric Indices

The non-destructive (SPAD index) and destructive measurement of chlorophylls content were significantly affected by mulching materials and biostimulant applications, with no significant effects from the M × B interaction (Table 2). The PE-based biostimulant provoked greater values of SPAD index and chlorophyll content (+6% and 30%, respectively) in comparison to the untreated control, irrespective of the mulching materials (Table 2). Moreover, when averaged over biostimulant applications, the total chlorophyll content was enhanced by 33% in mulched lettuce plants (avg. 51.3 mg 100 g^{-1} fw) when compared to bare soil (avg. 38.6 mg 100 g^{-1} fw), with no significant differences being observed among the three mulching materials (Table 2).

Concerning the Hunter color parameters, the ANOVA highlighted no significant M x B interaction for all of the examined color parameters (Table 2). In general, neither mulching nor biostimulant application had a significant effect on leaf yellowness (+b*; avg. 33.4) of greenhouse lettuce. Moreover, the use of LDPE as a mulching material resulted in greater lightness (i.e., lowest L* values) of greenhouse lettuce leaves (Table 2). Finally, the foliar application of PE-based biostimulant provoked greater values of brightness and greenness, in comparison to the untreated control, irrespective of the mulching materials (Table 2).

Table 2. Mean comparison and analysis of variance for Soil Plant Analysis Development (SPAD) index, total chlorophyll content, and Hunter color parameters L* (brightness), a* (−a* = green) and b* (+b* = yellow) of untreated and biostimulant-treated greenhouse lettuce grown under LDPE mulch and biodegradable (Mater-Bi® 1 and Mater-Bi® 2) mulching materials in relation to bare soil.

Source of Variance	SPAD Index	Total Chlorophyll (mg 100 g^{-1} fw)	L*	a*	b*
Mulch (M)					
Bare soil	27.3 ± 0.4 c	38.6 ± 0.0 b	54.4 ± 1.7 b	−19.4 ± 0.4	33.6 ± 0.4
LDPE	29.4 ± 0.4 b	50.3 ± 0.1 a	51.8 ± 2.1 c	−19.8 ± 0.4	33.3 ± 0.6
Mater-Bi® 1	30.3 ± 0.6 a	51.8 ± 0.1 a	60.2 ± 1.1 a	−18.6 ± 0.6	33.9 ± 0.9
Mater-Bi® 2	29.2 ± 0.4 b	51.7 ± 0.1 a	56.5 ± 2.0 b	−18.9 ± 2.0	32.8 ± 0.4
	***	*	***	NS	NS
Biostimulant (B)					
Control	28.3 ± 0.4	41.9 ± 0.0	52.1 ± 1.2	−19.9 ± 0.2	33.9 ± 0.3
Tropical plant extract (PE)	30.0 ± 0.5	54.4 ± 0.0	59.3 ± 0.9	−18.3 ± 0.3	32.9 ± 0.5
t-test	0.043	0.039	0.000	0.000	0.084
M × B					
Bare soil without biostimulant	26.7 ± 0.2	34.5 ± 0.0	50.6 ± 0.3	−20.4 ± 0.2	34.3 ± 0.4
LDPE without biostimulant	28.7 ± 0.9	39.3 ± 0.1	47.7 ± 0.5	−20.6 ± 0.3	33.6 ± 0.6
Mater-Bi® 1 without biostimulant	29.5 ± 0.7	40.8 ± 0.1	58.0 ± 1.0	−19.3 ± 0.3	34.8 ± 0.4
Mater-Bi® 2 without biostimulant	28.4 ± 0.1	52.9 ± 0.0	52.1 ± 1.2	−19.7 ± 0.5	33.2 ± 0.7
Bare soil + PE	27.9 ± 0.1	42.7 ± 0.1	58.1 ± 0.7	−18.5 ± 0.2	33.0 ± 0.5
LDPE + PE	30.0 ± 0.4	61.4 ± 0.1	56.0 ± 2.1	−19.0 ± 0.4	33.2 ± 1.1
Mater-Bi® 1 + PE	31.1 ± 0.6	62.9 ± 0.1	62.3 ± 0.5	−17.9 ± 1.0	32.9 ± 1.9
Mater-Bi® 2 + PE	30.0 ± 0.2	50.5 ± 0.1	60.9 ± 0.4	−18.0 ± 0.2	32.6 ± 0.3
	NS	NS	NS	NS	NS

NS, *, *** Non-significant or significant at $p \leq 0.05$ or 0.001, respectively. Different letters in the same column indicate significant differences according to DMR test ($p = 0.05$). Means of biostimulant effect are compared according to Student's *t*-test ($p = 0.05$). All data are expressed as mean ± SE.

3.4. Dry Matter Percentage and Leaf Mineral Profile

The leaf dry matter percentage and nitrate content were significantly influenced by M × B interaction (Table 3). The recorded leaf dry matter percentage across the eight experimental treatments ranged from 3.5 to 4.2%, with the lowest values being recorded in bare soil without biostimulant application (Table 3). The recorded nitrate content across the eight experimental treatments (836–2685 mg kg^{-1} fw) was within the limits set by the EU Commission Regulation No 1258/2011 for the commercialization of fresh lettuce (3000–5000 mg kg^{-1} fw). Our results also demonstrated that the presence of mulching materials, in particular, the use of Mater-Bi® 2, evoked a significant increment in nitrate content confronted to bare soil in both untreated and biostimulant-treated lettuce plants. Interestingly, the nitrate content was significantly reduced by 23% in greenhouse lettuce plants receiving foliar application with tropical plant extract (1566 mg kg^{-1} fw) confronted to the control (2037 mg kg^{-1} fw) (Table 3).

Table 3. Mean comparison and analysis of variance for leaf dry matter percentage and mineral composition of untreated and biostimulant-treated greenhouse lettuce grown under LDPE mulch and biodegradable (Mater-Bi® 1 and Mater-Bi® 2) mulching materials in relation to bare soil.

Source of Variance	Dry Matter (%)	Nitrate (mg kg^{-1} fw)	P (g kg^{-1} dw)	K (g kg^{-1} dw)	Ca (g kg^{-1} dw)	Mg (g kg^{-1} dw)	S (g kg^{-1} dw)	Na (g kg^{-1} dw)
Mulch (M)								
Bare soil	3.7 ± 0.1 c	955 ± 54 d	7.9 ± 0.3 b	90.2 ± 2.1 a	6.4 ± 0.3 b	3.5 ± 0.2	0.8 ± 0.1	2.2 ± 0.1 b
LDPE	4.2 ± 0.0 a	1898 ± 74 c	9.4 ± 0.4a	77.3 ± 2.7 b	4.8 ± 0.4 c	2.7 ± 0.2	1.0 ± 0.1	2.3 ± 0.2 b
Mater-Bi® 1	4.1 ± 0.1 ab	1984 ± 152 b	9.0 ± 0.2 a	79.6 ± 1.2 b	5.9 ± 0.4 bc	3.3 ± 0.1	1.1 ± 0.1	2.3 ± 0.1 b
Mater-Bi® 2	3.9 ± 0.1 bc	2368 ± 142 a	7.5 ± 0.3 b	88.7 ± 1.2 a	8.3 ± 0.6 a	4.0 ± 0.1	1.1 ± 0.1	2.7 ± 0.0 a
	***	***	***	***	***	NS	NS	*
Biostimulant (B)								
Control	3.9 ± 0.1	2037 ± 180	8.1 ± 0.3	82.1 ± 2.3	5.8 ± 0.5	3.3 ± 0.2	0.9 ± 0.1	2.5 ± 0.1
Tropical plant extract (PE)	4.0 ± 0.1	1566 ± 135	8.9 ± 0.3	85.9 ± 1.8	6.8 ± 0.4	3.5 ± 0.2	1.1 ± 0.1	2.2 ± 0.1
t-test	0.758	0.048	0.050	0.206	0.049	0.320	0.157	0.024
M × B								
Bare soil without biostimulant	3.5 ± 0.1 c	1075 ± 2 f	7.3 ± 0.3	88.6 ± 2.7	6.0 ± 0.6	3.4 ± 0.3	0.8 ± 0.1	2.5 ± 0.1
LDPE without biostimulant	4.2 ± 0.1 a	2063 ± 10 c	8.7 ± 0.5	72.8 ± 2.7	4.1 ± 0.0	2.5 ± 0.1	1.0 ± 0.1	2.5 ± 0.1
Mater-Bi® 1 without biostimulant	4.2 ± 0.1 a	2325 ± 8 b	8.9 ± 0.2	78.9 ± 2.4	5.7 ± 0.4	3.3 ± 0.2	1.0 ± 0.1	2.4 ± 0.2
Mater-Bi® 2 without biostimulant	3.9 ± 0.2 b	2685 ± 3 a	7.3 ± 0.2	88.0 ± 2.4	7.9 ± 1.2	3.9 ± 0.2	0.9 ± 0.1	2.7 ± 0.1
Bare soil + PE	3.9 ± 0.1 b	836 ± 9 g	8.5 ± 0.2	91.9 ± 3.6	6.9 ± 0.4	3.6 ± 0.2	0.9 ± 0.1	2.0 ± 0.1
LDPE + PE	4.2 ± 0.1 a	1733 ± 10 d	10.0 ± 0.3	81.8 ± 3.1	5.5 ± 0.8	2.9 ± 0.3	1.1 ± 0.1	1.9 ± 0.3
Mater-Bi® 1 + PE	3.9 ± 0.1 b	1643 ± 13 e	9.0 ± 0.4	80.4 ± 0.5	6.2 ± 0.7	3.3 ± 0.2	1.1 ± 0.1	2.2 ± 0.2
Mater-Bi® 2 + PE	3.8 ± 0.0 b	2051 ± 3 c	7.9 ± 0.5	89.4 ± 1.0	8.7 ± 0.4	4.2 ± 0.1	1.1 ± 0.1	2.7 ± 0.1
	*	***	NS	NS	NS	NS	NS	NS

NS, *, *** Non-significant or significant at $p \leq 0.05$ and 0.001, respectively. Different letters in the same column indicate significant differences according to DMR test ($p = 0.05$). Means of biostimulant effect are compared according to Student's *t*-test ($p = 0.05$). All data are expressed as mean ± SE.

Neither mulching materials nor PE-application had a significant influence on Mg and S concentrations in greenhouse lettuce leaves (avg. 3.4 and 1.0 g kg^{-1} dw, respectively). The concentrations of target macronutrients and sodium in leaf tissues were significantly affected by mulching materials, with the highest values of K, Ca, and Na being recorded in lettuce plants that were grown under Mater-Bi® 2 mulching material (Table 3). Interestingly, PE biostimulant treatment, as averaged over mulching materials (M × B interaction = ns), affected P, Ca, and Na leaf tissues concentrations, which were greater by 10% and 17% (for P and Ca, respectively) and lower by 12% (for Na) when compared to the untreated lettuce (Table 3).

3.5. Antioxidant Activity and Bioactive Compounds

The hydrophilic antioxidant fraction of lettuce ranged from 5.6 to 7.5 mmol ascorbate eq. 100 g^{-1} dw (Table 4). Regardless of mulching materials, the antioxidant capacity in lettuce that was treated with the commercial biostimulant Auxym® was significantly higher (+9%) as compared to the untreated control (Table 4). Neither mulching materials nor PE-application had a significant influence on total phenols content in lettuce leaves (avg. 3.4 mg gallic acid eq. 100 g^{-1} dw). Moreover, phytochemicals with antioxidant properties, such as total ascorbic acid and carotenoids, were affected by both the tested factors (mulching materials, biostimulant application, and their combination). When averaged over the biostimulant application, the use of the Mater-Bi® 2 film evoked a significant increase in the biosynthesis and the accumulation of carotenoids (Table 4). The positive effect of Mater-Bi® 2 film on total ascorbic acid content has also been highlighted in the interaction with the biostimulant, where a major increase of total ascorbic acid (+168%) was recorded in comparison to the untreated and PE-treated lettuce grown in bare soil (Table 4).

Table 4. Mean comparison and analysis of variance for hydrophilic antioxidant activity, total phenols, total ascorbic acid and carotenoid contents of untreated and biostimulant-treated greenhouse lettuce grown under LDPE mulch and biodegradable (Mater-Bi® 1 and Mater-Bi® 2) mulching materials in relation to bare soil.

Source of Variance	Hydrophilic Antioxidant Activity (mmol ascorbate eq. 100 g^{-1} dw)	Total Phenols (mg gallic acid eq. 100 g^{-1} dw)	Ascorbic Acid (mg 100 g^{-1} fw)	Carotenoids (mg g^{-1} fw)
Mulch (M)				
Bare soil	7.3 ± 0.1	3.9 ± 0.2	7.2 ± 0.6 b	18.4 ± 2.2 ab
LDPE	6.4 ± 0.3	3.1 ± 0.3	8.1 ± 1.1 b	13.1 ± 1.4 b
Mater-Bi® 1	6.8 ± 0.4	3.4 ± 0.3	7.7 ± 2.3 b	17.4 ± 1.4 ab
Mater-Bi® 2	7.2 ± 0.3	3.3 ± 0.1	13.7 ± 3.1 a	21.0 ± 2.1 a
	NS	NS	*	*
Biostimulant (B)				
Control	6.6 ± 0.2	3.3 ± 0.2	6.0 ± 0.7	16.0 ± 1.3
Tropical plant extract (PE)	7.2 ± 0.2	3.6 ± 0.2	12.3 ± 1.6	18.9 ± 1.6
t-test	0.036	0.241	0.002	0.158
M × B				
Bare soil without biostimulant	7.1 ± 0.2	3.7 ± 0.5	6.1 ± 0.3 c	14.3 ± 1.9
LDPE without biostimulant	5.6 ± 0.1	3.0 ± 0.3	7.1 ± 0.8 bc	12.7 ± 3.1
Mater-Bi® 1 without biostimulant	6.7 ± 0.3	3.3 ± 0.6	3.0 ± 0.4 c	16.2 ± 1.1
Mater-Bi® 2 without biostimulant	6.9 ± 0.2	3.2 ± 0.2	8.0 ± 1.4 bc	20.8 ± 1.5
Bare soil + PE	7.5 ± 0.1	4.1 ± 0.0	8.3 ± 0.6 bc	22.5 ± 2.0
LDPE + PE	7.0 ± 0.1	3.1 ± 0.6	9.1 ± 1.9 bc	13.4 ± 0.6
Mater-Bi® 1 + PE	6.9 ± 0.8	3.6 ± 0.1	12.4 ± 2.0 b	18.6 ± 2.6
Mater-Bi® 2 + PE	7.5 ± 0.4	3.5 ± 0.1	19.3 ± 3.7 a	21.2 ± 4.3
	NS	NS	*	NS

NS, * Non-significant or significant at $p \leq 0.05$, respectively. Different letters in the same column indicate significant differences according to DMR test ($p = 0.05$). Means of biostimulant effect are compared according to Student's *t*-test ($p = 0.05$). All data are expressed as mean ± SE.

4. Discussion

The use of biodegradable mulching films and plant-based biostimulants has revolutionized modern agriculture in the last two decades. Nevertheless, no scientific studies have assessed the combinatorial effect of these two agricultural practices on crop performance and nutritional value of an important greenhouse leafy vegetable, such as lettuce. Our findings indicated that biodegradable mulching film Mater-Bi® 1 produced comparable marketable fresh yield to the commercial standard polyethylene (LDPE), while Mater-Bi® 2 exhibited the highest crop productivity. It is well established that plastic mulching films increase soil temperature in comparison to bare ground. This was the case in the current experiment, since the soil minimum, maximum, and mean temperature trends that were recorded in bare soil were always lower by 2.3–3.3 °C, 3.5–4.2 °C, and 2.8–3.8 °C, respectively, than those that were observed among the three tested mulching materials. The differences in fresh yield could be also associated to differences in soil temperatures, when temperature is a limiting factor (autumn–winter growing season; [23]). The results that were recorded in this greenhouse experiment endorse the previous study, where the span of soil temperature under the different mulching materials had a pronounced effect on marketable lettuce yield [24–27]. Our findings concerning the beneficial effect of mulching versus bare soil were also reported in previous studies on open-field and greenhouse vegetables. For instance, melon plants had more fruits and higher fruit mean weight when grown with biodegradable films and LDPE, as compared to bare soil [2]. An increase in marketable yield in the presence of polyethylene and biodegradable (Mater-Bi®) films when compared to bare soil was also observed in pumpkin [24], tomato [1,4,25], strawberry [3,26], garlic chives [5], as well as lettuce [27]. The use of plastic films may have preserved soil moisture and prevented water evaporation and the excessive leaching of nutrients in the rhizosphere [1].

Interestingly, the combination of film mulching (LDPE, Mater-Bi® 1, or Mater-Bi® 2) with the tropical plant extract biostimulant exhibited a positive and important synergistic effect (+30%) on both total and marketable yield. Particularly, the higher marketable production that was observed in greenhouse lettuce plants that were grown under mulching films and treated with PE-biostimulant, was due to an increase in the leaf area and not to the number of leaves per plant. The increase in crop productivity and biometric parameters of lettuce plants grown under protected cultivation has been previously reported in several research studies testing the action of this tropical plant extract biostimulant on leafy and fruit vegetables, such as tomato, jute, wall rocket, and lettuce [18,28,29]. The biostimulant action of the commercial product Auxym® on PE-treated lettuce plants could be associated to the presence of signaling molecules, such as carbohydrates, vitamins, but especially free amino acids and soluble peptides [14,18,30]. The hormone-like activity of plant-derived peptides that are contained in Auxym® has been proposed in many scientific papers, where the foliar application of vegetal-based biostimulants elicited auxin- and gibberellin-like activities and, thus, boosted yield [31,32]. Since many other signaling peptides have been identified in plant cells controlling growth, development, and stress responses of plants [33], it is expected that more signaling-peptide based PE will be developed in the near future. Furthermore, some indirect effects of amino acids can be postulated. The amino acid L-tryptophan is a precursor of indole compounds (thus including auxins), while L-methionine is known as the precursor of ethylene [30]. Finally, these bioactive compounds that are present in the plant-based biostimulants can act on the primary metabolism, increasing the photosynthetic activity of the plants, and it can act as well on root growth, which might increase water and nutrient absorption efficiency, thus resulting in a yield increase [18]. This was the case in the current study, where plants that were grown under plastic mulching films and treated with tropical plant extract were characterized by better physiological and biochemical status. The greater SPAD index and chlorophyll content of lettuce leaves corroborated this, thus confirming the better photosynthetic efficiency that leads to better plant performance. Similar results on the stimulation of the physiological and biochemical status of biostimulant-treated plants were also previously observed in greenhouse tomato [18], spinach [34], lettuce [35], and jute [28].

The leaf appearance in peculiar color is among the visual characteristics of leafy vegetables that steadily govern consumer preference and selection choice [36]. Lettuce green color is directly dependent upon chlorophyll synthesis in leaf tissue. Plant extract-biostimulant application affected lettuce greenness color (−a*) to the extent it affected chlorophyll content, as observed earlier in a broad span of leafy greens, such as spinach, lamb's lettuce, and baby lettuce [29,37].

A negative aspect in the quality of leafy vegetables is, certainly, the high content of nitrates, as they are involved in the onset of different diseases [38]. Generally, vegetables that belong to the *Brassicaceae, Chenopodiaceae,* and *Asteraceae* families [18] may accumulate nitrates in their leaves. Significant genotypic variations in nitrate accumulation are shown for lettuce [39–41]. The nitrate concentration in plants is closely related to nitrate reductase activities [42]. This reality has prompted the European Commission to regulate the nitrate limits for lettuce. In our experiment, nitrate concentrations for plants cultivated with LDPE films, Mater Bi® 1, and Mater Bi® 2 films (1898–2368 mg kg^{-1} fw), were within the set limits for fresh lettuce according to Commission regulation (EU) No 1258/2011 (3000–5000 mg kg^{-1} fw) [43]. The PE-biostimulant application decreased nitrates concentration in lettuce leaves by 23% (avg. 1566 mg kg^{-1}), as compared to the control (avg. 2037 mg kg^{-1}). This positive effect could be linked to the presence of a high content of free amino acids in the biostimulant product, which, once absorbed by the plant, might exert the inhibition of the nitric ion transporters that are present in the root. On the other hand, the ability of the plant-based biostimulant Auxym® to reduce nitrates accumulation could be associated with the regulation of nitrogen metabolism in plants, which involves the activity of nitrate and nitrite reductase, glutamate synthase, as well as glutamine synthetase [14,18,44]. Various studies confirmed our results, such as that of Bulgari [45] performed on iceberg lettuce, which showed that a biostimulant of vegetal origin enriched with micro-elements (one), kept nitrate levels well under the limit required by the EC. Similar results were also obtained in spinach, on which the effect of amino acid-based biostimulant (Aminoplant) was evaluated [46]. Other studies on corn, soy, and wheat also showed that exogenous amino acids application can significantly reduce nitrate absorption [18].

Scientists recommend that people should consume fruits and vegetables daily, because they satisfy 11%, 35%, 7%, and 24% of the daily intake of P, K, Ca, and Mg, respectively [10]. These macronutrients help against certain diseases, such as blood pressure imbalances, hypertension (K), and osteoporosis (P, Ca, and Mg) [15]. For lettuce, several authors reported a potassium content between 48–72 mg g^{-1}, phosphorus 4–6 mg g^{-1}, magnesium 1.4–2.8 mg g^{-1}, and calcium 4–10 mg^{-1} on a dry weight basis [15,16,47]. In our work, the use of biodegradable films influenced the biofortification of macronutrinets in lettuce leaves. In particular, the use of polyethylene film Mater-Bi® 1 increased P content, confronted to the control, whereas lettuce plants that were grown under Mater-Bi® 2 exhibited higher values of K and Ca when compared to the bare soil treatment. Our results match with previous studies on the 'nutrient acquisition response' of plant-based biostimulant application on tomato [18], jute [28], and spinach [29]. In addition to the accumulation of macronutrients in leaf tissues of biostimulant-treated plants, the use of PE reduced sodium concentration in lettuce leaves by 12%, confronted to the control, which is in harmony with Carillo et al. [28] findings. This is a very important aspect, because Na causes hypertension and cardiovascular diseases [48].

Furthermore, lettuce is considered to be a good source of nutraceutical molecules, such as vitamin C and carotenoids [15]. These molecules represent the radical scavenging power that protects plants from the oxidative damage caused by free radicals. In our work, when averaged over biostimulant application, lettuce plants that were grown under Mater-Bi® 2 had the highest total ascorbic acid and carotenoids content. Similarly, Morra et al. [3] recorded a higher antioxidant activity, total polyphenols, and anthocyanins in two strawberry cultivars grown under biodegradable Mater-Bi film as compared to those cultivated with LDPE or in bare ground. These results are also confirmed for melon plants that are grown with biodegradable mulching films [27,49]. The Mater Bi® 2 behavior could be related to the fact that below this film there is a greater evaporation of the soil, which results in a lower accumulation of water in plants. Therefore, this mild condition of stress might trigger the plant to synthesize defensive molecules [27,49]. More compelling, these secondary metabolites are also crucial to human well-being [50,51].

In our work, the foliar application of PE on greenhouse lettuce also influenced antioxidant activity and health-promoting secondary metabolites, since the antioxidant potential increased by 10% when compared to the untreated control plants. The latter is a notable qualitative functional parameter in leafy vegetables, since it is correlated to the synergetic effect of low-molecular weight biologically active compounds, such as phenolic compounds and carotenoids [50]. Ertani et al. [52] showed an increase in antioxidant activity, lycopene, phenols, and ascorbic acid of *Capsicum chinense* L., in response to the application of plant extract based biostimulants. The synergistic action of Mater-Bi® 2 with tropical plant extract is of significant interest for scientists and nutritionists, because it resulted in the production of superior greenhouse lettuce leaves in terms of vitamin C content (+168% as compared to the control). In fact, as also shown by Carillo et al. [28], signaling compounds that are present in the tropical plant extract Auxym®, like glutamic and aspartic acids are involved in the stimulation of primary and secondary metabolism, thus, leading to a greater synthesis of antioxidant molecules, such as vitamin C [28].

5. Conclusions

In recent years, horticultural research has focused on improving farming practices in the framework of a more sustainable agricultural, including the use of biodegradable mulching films and vegetal-based plant biostimulants to improve the crop performance and nutritive quality of the produced commodities. Our greenhouse experiment on lettuce confirmed that the use of biodegradable plastic mulching materials, especially Mater-Bi® 2, could be considered as an alternative to LDPE and bare soil cultivation. This biodegradable mulching material increased marketable yield irrespective of the biostimulant application, due to many agronomic benefits, in particular, the better microclimate (minimum and maximum soil temperatures) in the rhizosphere. Our results also demonstrated, that lettuce plants grown under biodegradable film especially Mater-Bi® 2 exhibited superior quality traits in terms of K, Ca, total ascorbic acid, and carotenoids. Interestingly, the foliar application of PE-biostimulant in the presence of mulching materials was able to improve the total and marketable yield and biometric traits. The synergistic effect of mulching with plant-based biostimulant was linked to better physiological and biochemical status (higher SPAD index and chlorophyll content) and a higher nutrient acquisition response (higher P and Ca and lower Na content). The PE-biostimulant treated lettuce had a lower nitrate content and higher antioxidant scavenging capacity than the non-treated control, while the combination of Mater-Bi® 2 and PE-biostimulant resulted in the production of premium greenhouse lettuce leaves in terms of vitamin C content. The outcomes of the current study can encourage leafy vegetables producers to replace LDPE films with biodegradable ones in combination with plant-based biostimulants in order to attain high productivity and reach consumer expectations for high quality produce. In addition, the substitution of plastic mulching with biodegradable ones can significantly tackle the environmental issues that are related to the disposal of mulching materials at the end of the cropping cycles. The absence of dumping costs for farmers could likely offset the higher costs due to biodegradable mulching, favoring the application of biodegradable mulching materials on a wide scale.

Author Contributions: Conceptualization, E.C. and Y.R. methodology, E.C. software, C.E.-N., A.P., M.G., I.D.M., and M.M. formal analysis, C.E.-N., A.P., M.G., I.D.M., and M.M. investigation, E.C. resources, Y.R. data curation, C.E.-N., A.P., M.G., I.D.M., and M.M. writing—original draft preparation, M.G. writing-review and editing, M.G., N.F., C.E.-N., M.C.K., G.C. and Y.R. visualization, E.C. and Y.R. supervision, E.C., N.F. and Y.R. project administration, Y.R. funding acquisition, Y.R. All authors have read and agreed to the published version of the manuscript.

Funding: This research received no external funding.

Acknowledgments: The authors are grateful to Vincenzo Esposito for his technical assistance in the greenhouse experiment.

Conflicts of Interest: The authors declare no conflict of interest.

References

1. Moreno, M.M.; Moreno, A. Effect of different biodegradable and polyethylene mulches on soil properties and production in a tomato crop. *Sci. Hortic.* **2008**, *116*, 256–263. [CrossRef]
2. Filippi, F.; Magnani, G.; Guerrini, S.; Ranghino, F. Agronomic evaluation of green biodegradable mulch on melon crop. *Ital. J. Agron.* **2011**, *6*, e18. [CrossRef]
3. Morra, L.; Bilotto, M.; Cerrato, D.; Coppola, R.; Leone, V.; Mignoli, E.; Pasquariello, M.S.; Petriccione, M.; Cozzolino, E. The Mater-Bi® biodegradable film for strawberry (*Fragaria* x *ananassa Duch.*) mulching: Effects on fruit yield and quality. *Ital. J. Agron.* **2016**, *11*, 203–206. [CrossRef]
4. Cowan, J.S.; Miles, C.; Andrews, K.P.; Inglis, D.A. Biodegradable mulch performed comparably to polyethylene in high tunnel tomato (*Solanum lycopersicum* L.) production. *J. Sci. Food Agric.* **2013**, *94*, 1854–1864. [CrossRef] [PubMed]
5. Adamczewska-Sowinska, K.; Turczuk, J. Yielding and biological value of garlic chives (*Allium tuberosum Rottl.* ex *Spreng.*) depending to the type of mulch. *J. Elem.* **2016**, *21*, 7–19. [CrossRef]
6. Scaringelli, M.A.; Giannoccaro, G.M.; Lopolito, A. Adoption of biodegradable mulching films in agriculture: Is there a negative prejudice towards materials derived from organic wastes? *Ital. J. Agron.* **2016**, *11*, 92–99. [CrossRef]
7. Thompson, A.A.; Samuelson, M.B.; Ignatius Kadoma, I.; Erick Soto-Cantu, E.; Rhae Drijber, R.; Wortman, S.E. Degradation rate of bio-based agricultural mulch is influenced by mulch composition and biostimulant application. *J. Polym. Environ.* **2019**, *27*, 498–509. [CrossRef]
8. Du Jardin, P. Plant biostimulants: Definition, concept, main categories and regulation. *Sci. Hortic.* **2015**, *196*, 3–14. [CrossRef]
9. Colla, G.; Rouphael, Y. Biostimulants in horticulture. *Sci. Hortic.* **2015**, *196*, 1–2. [CrossRef]
10. Rouphael, Y.; Kyriacou, M.C.; Vitaglione, P.; Giordano, M.; Pannico, A.; Colantuono, A.; De Pascale, S. Genotypic variation in nutritional and antioxidant profile among iceberg lettuce cultivars. *Acta Sci. Pol. Hortorum Cultus* **2017**, *16*, 37–45. [CrossRef]
11. Rouphael, Y.; Colla, G. Editorial: Biostimulants in agriculture. *Front. Plant Sci.* **2020**, *11*, 40. [CrossRef]
12. Rouphael, Y.; Colla, G. Synergistic biostimulatory action: Designing the next generation of plant biostimulants for sustainable agriculture. *Front. Plant Sci.* **2018**, *9*. [CrossRef]
13. Battacharyya, D.; Babgohari, M.Z.; Rathor, P.; Prithiviraj, B. Seaweed extracts as biostimulants in horticulture. *Sci. Hortic.* **2015**, *196*, 39–48. [CrossRef]
14. Colla, G.; Nardi, S.; Cardarelli, M.; Ertani, A.; Lucini, L.; Canaguier, R.; Rouphael, Y. Protein hydrolysates as biostimulants in horticulture. *Sci. Hortic.* **2015**, *196*, 28–38. [CrossRef]
15. Kim, M.J.; Moon, Y.; Tou, J.C.; Mou, B.; Waterland, N.L. Nutritional value, bioactive compounds and health benefits of lettuce (*Lactuca sativa* L.). *J. Food Comp. Anal.* **2016**, *49*, 19–34. [CrossRef]
16. Baslam, M.; Idoia, G.; Nieves, G. The arbuscular mycorrhizal symbiosis can overcome reductions in yield and nutritional quality in greenhouse-lettuces cultivated at inappropriate growing seasons. *Sci. Hortic.* **2013**, *164*, 145–154. [CrossRef]
17. Rouphael, Y.; Cardarelli, M.; Bassal, A.; Leonardi, C.; Giufrida, F.; Colla, G. Vegetable quality as affected by genetic, agronomic and environmental factors. *J. Food Agric. Environ.* **2012**, *10*, 680–688.
18. Colla, G.; Cardarelli, M.; Bonini, P.; Rouphael, Y. Foliar applications of protein hydrolysate, plant and seaweed extracts increase yield but di_erentially modulate fruit quality of greenhouse tomato. *HortScience* **2017**, *52*, 1214. [CrossRef]
19. Lichtentahler, H.K.; Buschmann, C. Chlorophylls and carotenoids; Measurement and characterization by UV-VIS spectroscopy. *Curr. Protocols Food Anal. Chem.* **2001**, *1*. [CrossRef]
20. Fogliano, V.; Verde, V.; Randazzo, G.; Ritieni, A. Method for measuring antioxidant activity and its application to monitoring the antioxidant capacity of wines. *J. Agric. Food Chem.* **1999**, *47*, 1035–1040. [CrossRef]
21. Singleton, V.L.; Orthofer, R.; Lamuela-Raventós, R.M. Analysis of Total Phenols and Other Oxidation Substrates and Antioxidants by Means of Folin-Ciocalteu Reagent. In *Methods in Enzymology*; Academic Press: Cambridge, MA, USA, 1999; Volume 299, pp. 152–178.

22. Kampfenkel, K.; Van Montagu, M.; Inzé, D. Extraction and determination of ascorbate and dehydroascorbate from plant tissue. *Anal. Biochem.* **1995**, *225*, 165–167. [CrossRef] [PubMed]
23. Brown, J.E.; Goff, W.D.; Dangler, J.M.; Hogue, W.; West, M.S. Plastic mulch color inconsistently affects yield and earliness of tomato. *HortScience* **1992**, *27*, 1135. [CrossRef]
24. Miles, C.; Shuresh Ghimire, S.; Wszelaki, A.; Moore, J. Biodegradable Plastic Mulch in Organic Vegetable Production Systems. In Proceedings of the New England Fruit and Vegetable Conference, Manchester, NH, USA, 17 December 2015.
25. Cirujeda, A.; Aibar, J.; Anzalone, A.; Martín-Closas, L.; Meco, R.; Moreno, M.M.; Pardo, A.; Pelacho, A.M.; Rojo, F.; Royo-Esnal, A.; et al. Biodegradable mulch instead of polyethylene for weed control of processing tomato production. *Agron. Sustain. Dev.* **2012**, *2*, 889–897. [CrossRef]
26. Cozzolino, E.; Leone, V.; Lombardi, P.; Piro, F. Telo biodegradabile su fragola, buoni effetti su resa e qualità. *L'Informatore Agrario* **2011**, *29*, 46–48.
27. Cozzolino, E.; Leone, V.; Lombardi, P. Nuovi teli pacciamanti in prova su lattuga sotto terra. *L'Informatore Agrario* **2014**, *24*, 37.
28. Carillo, P.; Colla, G.; El-Nakhel, C.; Bonini, P.; D'Amelia, L.; Dell'Aversana, E.; Pannico, A.; Giordano, M.; Sifola, M.I.; Kyriacou, M.C.; et al. Biostimulant application with a tropical plant extract enhances *Corchorus olitorius* adaptation to sub-optimal nutrient regimens by improving physiological parameters. *Agronomy* **2019**, *9*, 249. [CrossRef]
29. Di Mola, I.; Cozzolino, E.; Ottaiano, L.; Giordano, M.; Rouphael, Y.; Colla, G.; Mori, M. Effect of vegetal- and seaweed extract-based biostimulants on agronomical and leaf quality traits of plastic tunnel-grown baby lettuce under four regimes of nitrogen fertilization. *Agronomy* **2019**, *9*, 571. [CrossRef]
30. Colla, G.; Hoagland, L.; Ruzzi, M.; Cardarelli, M.; Bonini, P.; Canaguier, R.; Rouphael, Y. Biostimulant action of protein hydrolysates: Unraveling their effects on plant physiology and microbiome. *Front. Plant Sci.* **2017**, *8*, 2202. [CrossRef]
31. Ertani, A.; Schiavon, M.; Muscolo, A.; Nardi, S. Alfalfa plant-derived biostimulant stimulate short-term growth of salt stressed *Zea mays* L. plants. *Plant Soil* **2013**, *64*, 145–158. [CrossRef]
32. Colla, G.; Rouphael, Y.; Canaguier, R.; Svecova, E.; Cardarelli, M. Biostimulant action of a plant-derived protein hydrolysate produced through enzymatic hydrolysis. *Front. Plant Sci.* **2014**, *5*, 448. [CrossRef]
33. Tavormina, P.; De Coninck, B.; Nikonorova, N.; De Smet, I.; Cammue, P.A.B. The plant peptidome: An expanding repertoire of structural features and biological functions. *Plant Cell* **2015**, *27*, 2095–2118. [CrossRef] [PubMed]
34. Rouphael, Y.; Giordano, M.; Cardarelli, M.; Cozzolino, E.; Mori, M.; Kyriacou, M.C.; Bonini, P.; Colla, G. Plant and seaweed-based extracts increase yield but differentially modulate nutritional quality of greenhouse spinach through biostimulant action. *Agronomy* **2018**, *8*, 126. [CrossRef]
35. Luziatelli, F.; Ficca, A.G.; Colla, G.; Baldassarre Švecová, E.; Ruzzi, M. Foliar application of vegetal-derived bioactive compounds stimulates the growth of beneficial bacteria and enhances microbiome biodiversity in lettuce. *Front. Plant Sci.* **2019**, *10*, 60. [CrossRef]
36. Kyriacou, M.C.; Rouphael, Y.; Colla, G.; Zrenner, R.M.; Schwarz, D. Vegetable grafting: The implications of a growing agronomic imperative for vegetable fruit quality and nutritive value. *Front. Plant Sci.* **2017**, *8*, 741. [CrossRef]
37. Di Mola, I.; Ottaiano, L.; Cozzolino, E.; Senatore, M.; Giordano, M.; El-Nakhel, C.; Sacco, A.; Rouphael, Y.; Colla, G.; Mori, M.. Plant-based biostimulants influence the agronomical, physiological, and qualitative responses of baby rocket leaves under diverse nitrogen conditions. *Plants* **2019**, *8*, 522. [CrossRef] [PubMed]
38. Santamaria, P. Nitrate in vegetables: Toxicity, content, intake and EC regulation. *J. Sci. Food Agric.* **2006**, *86*, 10–17. [CrossRef]
39. Burns, I.G.; Zhang, K.; Turner, M.K.; Meacham, M.; Al-Redhiman, K.; Lynn, J.; Broadley, M.R.; Hand, P.; Pink, D. Screening for genotype and environment effects on nitrate accumulation in 24 species of young lettuce. *J. Sci. Food Agric.* **2011**, *91*, 553–562. [CrossRef] [PubMed]
40. Burns, I.G.; Durnford, J.; Lynn, J.; McClement, S.; Hand, P.; Pink, D. The influence of genetic variation and nitrogen source on nitrate accumulation and iso-osmotic regulation by lettuce. *Plant Soil* **2012**, *352*, 321–339. [CrossRef]

41. López, A.; Javier, G.; Fenoll, J.; Pilar, H.; Pilar, F. Chemical composition and antioxidant capacity of lettuce: Comparative study of regular-sized (*Romaine*) and baby-sized (*Little Gem* and *Mini Romaine*) types. *J. Food Comp. Anal.* **2014**, *33*, 39–48. [CrossRef]
42. Pinto, E.; Fidalgo, F.; Teixeira, J.; Aguiarc, A.A.; Ferreira, I.M.P.L.V.O. Influence of the temporal and spatial variation of nitrate reductase, glutamine synthetase and soil composition in the N species content inlettuce (*Lactuca sativa*). *Plant Sci.* **2014**, *219–220*, 35–41. [CrossRef] [PubMed]
43. Commission Regulation (EU). Commission Regulation (EU) n° 1258/2011 of 2 of December 2011 amending Regulation (EC) n° 1881/2006 as regards maximum level for nitrates in foodstuffs. *Off. J. Eur. Union* **2011**.
44. Calvo, P.; Nelson, L.; Kloepper, J.W. Agricultural uses of plant biostimulants. *Plant Soil* **2014**, *383*, 3–41. [CrossRef]
45. Bulgari, R.; Cocetta, G.; Trivellin, A.; Martinetti, L.; Ferrante, A. Prodotti biostimolanti ed effetti sulle colture ortofloricole. *Acta Ital. Horts* **2015**, *15*, 55–63.
46. Kunicki, E.; Grabowska, A.; Sekara, A.; Wojciechowska, R. The effect of cultivar type, time of cultivation and biostimulant treatment on the yield of spinach (*Spinacia oleracea* L.). *Folia Hortic.* **2010**, *22*, 9–13. [CrossRef]
47. Giordano, M.; El-Nakhel, C.; Pannico, A.; Kyriacou, M.C.; Stazi, S.R.; De Pascale, S.; Rouphael, Y. Iron biofortification of red and green pigmented lettuce in closed soilless cultivation impacts crop performance and modulates mineral and bioactive composition. *Agronomy* **2019**, *9*, 290. [CrossRef]
48. Krummel, R.D. Nutricíon en la Hypertension. In *10th Nutricíon y Dietoterapia*; De Krause Mahan, L.K., Escott-Stump, S., Eds.; McGraw-Hill: Contadero, Mexico, 2000; pp. 649–664.
49. Cozzolino, E.; Bilotto, M.; Leone, V.; Zampella, L.; Petriccione, M.; Cerrato, D.; Morra, L. Produzione e qualità di melone retatosu pacciamatura in in Mater-Bi®. *Colt. Protette* **2015**, *6*, 66–71.
50. Khanam, U.K.; Oba, S.; Yanase, E.; Murakami, Y. Phenolic acids, flavonoids and total antioxidant capacity of selected leafy vegetables. *J. Func. Foods* **2012**, *4*, 979–987. [CrossRef]
51. Subhasree, B.; Baskar, R.; Keerthana, R.L.; Susan, R.L.; Rajasekaran, P. Evaluation of antioxidant potential in selected green leafy vegetables. *Food Chem.* **2009**, *115*, 1213–1220. [CrossRef]
52. Ertani, A.; Pizzeghello, D.; Francioso, O.; Sambo, P.; Sanchez-Cortes, S.; Nardi, S. *Capsicum chinensis* L. growth and nutraceutical properties are enhanced by biostimulants in a long-term period: Chemical and metabolomic approaches. *Front. Plant Sci.* **2014**, *5*, 375.

agronomy

MDPI

Article

Response of Soil Bacterial Community and Pepper Plant Growth to Application of *Bacillus thuringiensis* KNU-07

HyungWoo Jo [1,2], Setu Bazie Tagele [1], Huy Quang Pham [1], Min-Chul Kim [1], Seung-Dae Choi [1], Min-Ji Kim [1], Yeong-Jun Park [1], Jerald Conrad Ibal [1], Gun-Seok Park [1,3] and Jae-Ho Shin [1,4,*]

1 School of Applied Biosciences, Kyungpook National University, Daegu 41566, Korea; hyungwoo0501@gmail.com (H.J.); setubazie@gmail.com (S.B.T.); huypham@knu.ac.kr (H.Q.P.); skalscjf13@naver.com (M.-C.K.); csd506@knu.ac.kr (S.C.); tbd01188@knu.ac.kr (M.-J.K.); yjpark1091@knu.ac.kr (Y.-J.P.); jerald.ibal@gmail.com (J.C.I.); gspark@atogen.co.kr (G.-S.P.)

2 R&I Center COSMAX BTI Inc, Seongnam-si, Gyeonggi-do 13486, Korea

3 Atogen Co., Ltd., Yuseong-gu, Daejeon 34015 Korea

4 Department of Integrative Biology, Kyungpook National University, Daegu 41566, Korea

* Correspondence: jhshin@knu.ac.kr (J.H.-S.); Tel.: +82-53-950-5716; Fax: +82-53-953-7233

Received: 20 March 2020; Accepted: 9 April 2020; Published: 10 April 2020

Abstract: Many *Bacillus* species are among the plant growth-promoting rhizobacteria (PGPR) that promote the growth of many different plant species. This study aimed to investigate the effects of *Bacillus thuringiensis* KNU-07 on the growth of pepper plants and the soil microbiota. We also designed primers specific for the strain KNU-07 to monitor the population in pepper-cultivated soil. Accordingly, a strain-specific primer pair was designed using a database constructed from 16,160 complete bacterial genomes. We employed quantitative PCR (qPCR) to track the abundance of the strain KNU-07 introduced into pepper-cultivated soil using the strain-specific primers. Our study revealed that the strain was found to possess plant growth-promoting (PGP) activities, and it promoted the growth of pepper plants. The soil bacterial community structure due to the application of the PGPR strain was significantly changed after six weeks post-inoculation. In addition, based on qPCR analysis, the population of the introduced strain declined over time. In this study, application of a PGPR strain increased the growth of pepper plants and changed the soil bacterial community structure. The successful results of monitoring of a bacterial strain's population using a single strain-specific primer pair can provide important information about the quantification of bio-inoculants under non-sterile soil conditions.

Keywords: *Bacillus thuringiensis*; *Capsicum annuum*; PGPR; microbiome; strain-specific primer; tracking

1. Introduction

Plant growth-promoting rhizobacteria (PGPR), which are found in the vicinity of crop roots, increase the growth and health of the plants [1]. Considering the growing public concern about the use of chemical fertilizers, there is increasing high demand to use PGPR, such as *Bacillus* spp., *Pseudomonas* spp., *Streptomyces* spp., etc. [2–4]. *Bacillus* is one of the most important genera that provides plants with potent plant growth-promoting effects, and many *Bacillus* species have been successfully used for agricultural purposes as commercial bio-inoculants [5,6]. Some strains of *Bacillus thuringiensis* have been used as a PGPR to improve soil fertility and enhance crop growth [7–9]. In addition, although the effect of PGPR on the indigenous soil bacterial communities and their functional properties has been studied, there is very limited information on the effects of *B. thuringiensis* on bacterial communities in the soil [10,11]. The beneficial effects of *B. thuringiensis* on plants are due to direct and indirect mechanisms, including nitrogen fixation, siderophore production, plant nutrient solubilization, and plant growth hormone production [12–14]. However, the plant responses are often variable due to inconsistent performance of inoculants under field conditions [15].

The ability of inoculants to survive in the soil is an important factor for their ability to function under field conditions [15,16]. Hence, quantification of inoculants in the soil is helpful to determine the success of PGPR under field conditions [1,6,17]. However, measuring the spatiotemporal dynamics of PGPR in the environment remains challenging [18,19]. Various culture-dependent and culture-independent methods have been employed to track and quantify bio-inoculants in the soil [20,21]. However, culture-dependent methods are only successful under sterile conditions, and less than 1% of the soil microbial diversity is recovered with such methods [21,22]. On the other hand, culture-independent methods, such as reporter nucleic acid-based, gene-based, and immunological methods, are capable of detecting less abundant, slow-growing, and unculturable bacteria [8–10]. However, most culture-independent methods are incapable of monitoring population dynamics at a species level, making it difficult to determine the fate of some strains [23,24].

In this study, we investigated the effects of the PGPR strain KNU-07 (hereafter referred to as KNU-07) on the growth of pepper plants and soil bacterial community composition. More importantly, we developed a strain-specific primer pair and developed a qPCR protocol to track the quantity of *Bacillus thuringiensis* KNU-07 in pepper-cultivated soil in a greenhouse. KNU-07 increased the growth of pepper plants, and the application of KNU-07 significantly changed the soil bacterial community structure after six weeks. The established strain-specific primer was successful in quantifying and monitoring KNU-07 in non-sterile soil conditions.

2. Materials and Methods

2.1. Bacterial Strains and Growth Conditions

The genome of the rhizospheric bacterial strain used in this study, KNU-07, was assembled using PacBio RSII and comprised 6,152,737 bp. KNU-07 was cultured in Luria-Bertani (LB) broth and LB agar (Difco Laboratories, Sparks, MD) and incubated for 24 h at 30 °C. Bacterial strains used for in vitro PCR are indicated in Table S1.

2.2. Bioinformatics Approach for Designing a Strain-Specific Primer for KNU-07

2.2.1. Designing a Primer to Target a Unique Sequence of KNU-07

Strain-specific primers were designed using a Python script that was developed in house. The complete genome sequence of KNU-07 was cut into truncated fragments of 500 bp. BLASTN was used to search for each fragment in a custom database constructed by downloading 16,160 complete bacterial genomes from the NCBI Genome Browse section (updated on 27 December 2019). The BLASTN searches were performed with the following parameters: ungapped alignment (-ungapped), no filter query sequence with dust (-dust no), and apply filtering locations as soft masks (-soft_masking false). Unique fragments containing regions that had no overlap with any complete genome in the NCBI custom database were chosen for further analysis. A primer pair targeting the unique sequence was designed using the web-based tool Primer-BLAST (NCBI Primer-BLAST).

To validate the selected unique fragment, BLASTN was used as mentioned above. The primer pair was checked to ensure homology to the KNU-07 unique sequence region, and the specificity of the primer was validated by performing in silico PCR evaluation using ecoPCR software on the NCBI bacteria complete genome database [25]. The number of mismatches in the binding regions of the target sequence for either forward or reward primers were set to 0 to 2. The targeted PCR product size was set to a minimum of 50 bp and a maximum of 500 bp. For comparison, a universal primer pair, 27F/1492R, which amplifies a region of the 16S rRNA gene in prokaryotes, was used as a positive control (Table 1). Finally, an ecotaxstat script was used to summarize taxonomy information from the in silico PCR products.

Table 1. KNU-07-specific primers and universal primers used in this study.

Primer Name	Primer Sequence (5′→3′)	Reference
Strain-specific primers		
KNU07F	TGCTCTTTCTGGATTATTCCTTGAG	This study
KNU07R	CATCCTTTTGTAGAAGGTATTGCCA	This study
Universal primers		
27F	AGAGTTTGATCMTGGCTCAG	Lane, 1991
1492R	TACGGYTACCTTGTTACGACTT	Lane, 1991
515F	GTGCCAGCMGCCGCGG	Lane, 1991
907R	CCGTCAATTCMTTTRAGTTT	Lane, 1991

2.2.2. In Vitro Validation of the Strain-Specific Primer Pair

PCR was performed to verify whether the primer pair designed by in silico PCR analysis amplified the unique sequence of KNU-07. In addition, we conducted additional PCR assays to investigate whether the strain-specific primer pair amplified genomic DNA from other bacterial species and environmental samples in vitro (Tables S1 and S2). Each PCR reaction contained 10 ng of template DNA, 0.2 μL of each primer (10 μM), 5 μL of EmeraldAmp GT PCR Master Mixture (Takara Korea Biomedical Inc., Seoul, Korea), and sterile distilled water to a total volume of 20 μL. PCR amplifications were carried out using the following cyclic program: initial denaturation at 95 °C for 7 min, followed by 30 cycles of 30 s denaturation at 95 °C, 55 °C annealing for 30 s, 72 °C extension for 30 s, and a final extension at 72 °C for 5 min.

2.3. *In Vitro Plant Growth-Promoting (PGP) Traits Assay*

An in vitro assay was carried out to evaluate the effect of KNU-07 on plant growth potential. The indole acetic acid (IAA) production potential was evaluated following the method of Gordon and Weber [26]. The concentration of IAA was quantified based on a standard curve of pure IAA (Sigma-Aldrich, St. Louis, MO, USA). IAA identity and its purity were confirmed by gas chromatography–mass spectrometry with a SIM (6890N network GC system, and 5973 network mass selective detector; Agilent, CA, USA). In addition, the potential of the strain to exhibit urease activity [27], siderophore production [28], and phosphate solubilization [29] was determined.

2.4. *Greenhouse Experiment*

2.4.1. Plant Material and Bacterial Strain Preparation

Seeds of hot pepper (*Capsicum annuum* cv. CM334) were used in this experiment. KNU-07 was incubated at 30 °C for 24 h at 200 rpm. The pellet was collected after centrifugation, washed, and resuspended in sterile distilled water. The bacterial inoculum was adjusted using a sterile distilled water to concentrations of 7.8×10^6 cells mL^{-1} soil and 7.8×10^8 cells mL^{-1}.

2.4.2. In Vivo Assay

The effect of strain KNU-07 on the growth of pepper plants in pots under greenhouse conditions for two months was assessed. Pepper seeds were surface-sterilized with ethanol (70%) for 1 min and soaked in a disinfectant solution (Clorox, distilled water, and 0.05% Triton X-100 in a 3:2:2 ratio (v/v/v)) for 5 min and washed 7–10 times with sterile, deionized, distilled water. Pepper seeds were vernalized for 48 h in a refrigerator at 4 °C and germinated by placing the seeds on sterile, wet filter paper in a growth chamber for 7 days at 30 °C. The germinated seeds were then sown in plastic trays containing mixed soil. The mixed soil was composed of garden soil and Biosangto-Mix soil (Heung Nong Co.,

Ltd., Pyeongtaek, Republic of Korea) in a 1:9 ratio (v/v). The pepper seedlings were incubated in a growth chamber (25 °C, 65% relative humidity, and cycles of 16 h light and 8 h dark). After two weeks, uniform-sized pepper seedlings having shoots approximately 5 cm in height were each transplanted into a pot containing 300 g of mixed soil. To assess the effect of KNU-07, the soil of some pots was inoculated with 3.85 mL of KNU-07 at one of the following concentrations: 1.0×10^5 cells g^{-1} soil and 1.0×10^7 cells g^{-1} soil. Application of bio-inoculants at 1.0×10^5 cells g^{-1} soil is a very common practice in South Korea. Seedlings treated only with sterile distilled water served as the non-inoculated control. The experiment was replicated three times with five plants per replication. After 11 weeks of treatment, plant growth data, including plant shoot length, root length, and total biomass, were recorded.

2.4.3. DNA Extraction from Pure Cultures and Soil Samples

To analyze the soil bacterial community, the soil where *Capsicum annuum* cv. CM334 was growing in each pot was sampled weekly. Very small amounts of soil sample (less than one gram) were taken at five different sites in each pot and pooled into a composite sample per pot. Genomic DNA from the soil and the strain culture was extracted using a Power Soil DNA Isolation Kit (MoBio Laboratories, Carlsbad, CA, USA) according to the manufacturer's protocol. DNA concentrations were determined using a Qubit 2.0 Fluorometer (Thermo Fisher Scientific). For strain-specific PCR assays, KNU-07 genomic DNA was extracted using a Wizard Genomic DNA Purification Kit (Promega Corporation, Madison, WI, USA).

2.4.4. DNA Library Preparation and Amplicon Sequencing

The diversity of the soil bacterial community was assessed by amplifying and analyzing the V4-V5 hypervariable region of 16S rRNA gene using the universal primer pair 515F/ 907R (Table 1). The V4-V5 primer pair was tailored with Ion Torrent PGM adapter and barcode sequences, which are unique to each sample. The PCR reaction (50 μL) contained 1 ng template DNA, 1 μL of each primer, and 25 μL of EmeraldAmp GT PCR Master Mixture (Takara, Japan). The PCR conditions were as follows: initial denaturation at 95 °C for 7 min; 10 cycles of denaturation at 95 °C for 30 s, annealing at 55 °C for 30 s, and extension at 72 °C for 30 s; 30 cycles of denaturation at 95 °C for 30 s, annealing and extension at 72 °C for 45 s; and a final extension at 72 °C for 5 min.

Ion Torrent PGM sequencing technology and data analysis were used to sequence the amplified products. The quality of the amplified DNA library was assessed using an Agilent 2100 Bioanalyzer with a High-Sensitivity DNA (HS DNA) Kit (Agilent Technology, Santa Clara, CA, USA). The amplified DNA library was diluted to 6 pM to perform emulsion PCR with Ion Sphere Particles (ISPs) in the Ion OneTouch System II (Thermo Fisher Scientific), followed by enrichment for template-positive Ion Sphere Particles using Dynabeads MyOne Streptavidin C1 beads (Thermo Fisher Scientific, Waltham, MA, USA). Each sample was loaded on barcoded chips (Ion 316 Chip Kit v2 BC). Sequencing was carried out using the Ion Torrent PGM system and an Ion PGM Hi-Q Sequencing Kit (Thermo Fisher Scientific). The Torrent Suite Software, along with Ion Torrent PGM specific pipeline software, was employed to generate sequence reads, trim adapter sequences, filter, and exclude poor signal profile reads. Quality-filtered sequence reads were analyzed using a QIIME package (V1.9.1). Operational taxonomic units (OTUs) having 97% similarity were selected by an average neighbor algorithm and were identified using the sequence database of the National Center for Biotechnology Information (NCBI).

2.4.5. Continuous Tracking of KNU-07 Using qPCR

The abundance of inoculated KNU-07 in pepper-cultivated soil was monitored using qPCR with strain-specific primers (Table 1). The total bacteria in the soil were quantified using qPCR with the universal primer pair (27F/1492R) targeting a 16S rRNA gene (Table 1). Each PCR reaction mixture (10 μL) consisted of 10 ng of DNA from the soil sample, 0.3 μL of each KNU07 specific primer (10 μM), and MG 2X qPCR mix (SYBR green, MGmed). qPCR reactions were performed in triplicate under the following cycling conditions: 95 °C for 15 min, followed by 35 cycles of denaturing at 95 °C for 30 s, annealing at 55 °C for 30 s and extension at 72 °C for 30 s. Gel electrophoresis using a CFX Real-Time PCR Detection System (BioRad) was conducted to ensure the appropriate size of the amplified products.

A standard curve based on copy number was used to determine the abundance of KNU-07 and total bacteria in the soil. Briefly, a six-fold serial dilution of amplicons of KNU-07 unique sequence was prepared in triplicate. The copy number of each concentration was calculated based on the amplicon concentration and length. A regression equation was calculated based on the cycle threshold (Ct) value to the known amount of serially diluted copy number of the unique sequence. By using the standard curve, the abundance of KNU-07 was deduced and expressed as the number of genome equivalents. A genome equivalent corresponds to the number of KNU-07 cells. In addition, the abundance of the total bacteria was determined using qPCR with a universal primer pair that amplifies a conserved region of the 16S rRNA genes of multiple bacteria species.

2.5. Nucleotide Accession Numbers

The complete genome sequence of *B. thuringiensis* KNU-07 was deposited in GenBank under accession number CP016588. The unique DNA sequence of KNU-07, which was used to design the strain-specific primers, is located at the sequence position 1,904,488 bp to 1,904,728 bp. The NGS data of all raw sequence reads were deposited in the NCBI Short Read Archive (SRA) database under accession number SRP243872.

2.6. Statistical Analysis

The alpha diversity of KNU-07-treated and control samples was analyzed using taxonomic diversity indices, such as the Shannon index, Simpson's index, and the number of observed OTUs. The community diversity difference was analyzed based on principal coordinate analysis (PCoA) using Bray–Curtis distances in QIIME1. A dissimilarity analysis of Bray–Curtis based on permutational multivariate analysis of variance (PERMANOVA, ADONIS function) [30] and an analysis of similarity (ANOSIM function) [31] were conducted to determine the impact of KNU-07 application on the soil bacterial community composition. The abundance of predicted gene function of the soil bacterial community in each experimental sample was determined by the PICRUSt pipeline using an OTU table normalized to the 16S rRNA gene copy number [32]. The data of predicted function were analyzed using the STAMP software package [33]. All data of greenhouse experiments were arranged in a randomized design with at least three replications. Analysis of variance (ANOVA) was performed for plant growth parameters using SAS software version 9.4 [34], and treatment means were separated using post hoc Tukey significant difference (HSD) tests.

3. Results

3.1. In Silico and in Vitro PCR Verification of KNU-07-Specific Primer Pairs

The genome of KNU-07 was truncated into 500 bp fragments, and 10,687 fragments were found. Among these, 81 unique windows were identified, and one window (located at 1,904,488 bp to 1,904,728 bp) was selected for designing the primers. A primer targeting a unique sequence of KNU-07 was designed to have 25 bp (Table 1) using the Primer-BLAST tool on the NCBI web site.

For the in silico PCR analysis, complete genomes of 16,160 bacteria comprising 52 phyla, 173 orders, 1130 genera, and 3747 species were used. A universal primer pair, 27F/1492R, targeting the bacterial 16S rRNA gene matched perfectly with 76% of the species tested (no mismatch), 89% of species had one mismatch, and 92% of species had two mismatches. On the other hand, our strain-specific primer pair targeting the unique sequence of KNU-07 had a perfect match to only one genome, that of *B. thuringiensis* KNU-07 (Table 2). Even by increasing the number of mismatches, no bacterial species other than KNU-07 was found to match, indicating that the primer pair was highly specific to KNU-07.

Table 2. In silico PCR verification that strain-specific primers target sequences unique to KNU07.

Category	Primer	Taxonomic Level	Total Taxa	Number of Mismatches per Primer		
				Perfect Match	1 Mismatch	2 Mismatches
Bacterial 16S rRNA gene	27F/1492R	Phylum	52	26 (50%)	35 (67%)	41 (79%)
		Class	81	48(59%)	63 (78%)	70 (86%)
		Order	173	114 (66%)	140 (81%)	148 (86%)
		Family	367	262 (71%)	313 (85%)	323 (88%)
		Genus	1130	830 (73%)	993 (88%)	1019 (90%)
		Species	3747	2843 (76%)	3331 (89%)	3453 (92%)
KNU-07 unique region	KNU07F/KNU07R	Phylum	52	1*	1*	1*
		Class	81	1*	1*	1*
		Order	173	1*	1*	1*
		Family	367	1*	1*	1*
		Genus	1130	1*	1*	1*
		Species	3747	1*	1*	1*

* Perfect matches with *B. thuringiensis* KNU-07.

To verify whether our designed primer pair specifically detected KNU-07, an in vitro PCR assay was conducted using DNA samples from pure cultures of 28 bacterial strains. The results showed that the primer pair amplified the expected band size of 241 bp from strain KNU-07; however, no visible band was detected with any other bacterial strain, including *Bacillus* spp., other than KNU-07 (Figure S1). Our strain-specific primer can precisely detect and distinguish KNU-07 from other tested *Bacillus* species. Furthermore, the discrimination power of the KNU-07 strain-specific primer was verified using 28 diverse environmental DNA samples. The results confirmed that the strain-specific primer was able to successfully amplify KNU-07 with the expected band size of 241 bp, while no visible band was detected in any environmental sample (Figure S1), demonstrating that the strain-specific primer was selective in detecting KNU-07.

3.2. PGP Activity of KNU-07

KNU-07 was positive for in vitro PGP activities, including IAA production, siderophore production, phosphate solubilization, and urease activities (Figure S2). In addition, gas chromatography/mass spectrometry experiments revealed that the amount of IAA produced by KNU-07 with and without a tryptophan supplement was 4.886 and 0.167 $\mu g\ mL^{-1}$, respectively. The in vivo effects of KNU-07 inoculated at different concentrations on the growth of pepper plants was determined under non-sterile conditions. After 11 weeks of growth, a significant ($p < 0.05$) difference was found between bacterized and non-inoculated pepper seedlings (Figure 1). Plants treated with high concentrations of KNU-07 exhibited significant increases in root length (30.7%), shoot length (19.7%), and total dry biomass (30.7%) compared to non-inoculated control plants (Figure 1).

3.3. Response of the Soil Bacterial Community to KNU-07

The effect of KNU-07 on the composition of the soil bacterial community in pepper-cultivated soil was analyzed by the Ion Torrent PGM platform based on 16S rRNA gene amplicon sequences. In this study, 3146 observed OTUs, 27 phyla, and 408 genera were identified by a BLASTN search against the Green gene database (data not shown). The results revealed that the alpha diversity indices, the Shannon index, and the number of observed OTU increased similarly over time in inoculated and non-inoculated control samples (Figure 2A). The Simpson's index showed that diversity increased with each treatment, except when KNU-07 was applied at the highest concentration (1.0×10^7 cells g^{-1} soil) (Figure 2B). During the first two weeks, Simpson's index was low in soil treated with the highest concentration of KNU-07. However, after three weeks, Simpson's index increased over time (Figure 2C).

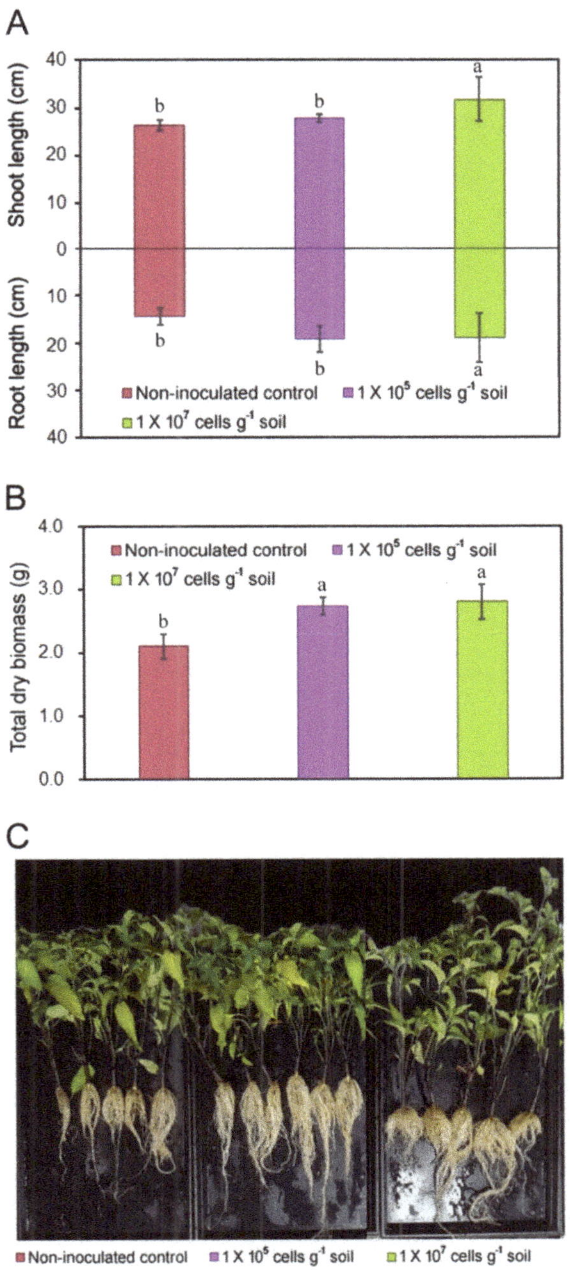

Figure 1. Effect of KNU-07 inoculation on the growth of pepper plants over 11 weeks post-inoculation in greenhouse conditions. The numerical value of (**A**) root length and shoot length and (**B**) total dry biomass. (**C**) Pictorial view of pepper plants inoculated with the indicated concentrations of *B. thuringiensis* KNU-07. Non-inoculated plants served as control. Mean values having different letters in each of the growth parameters are significantly different ($p \leq 0.05$).

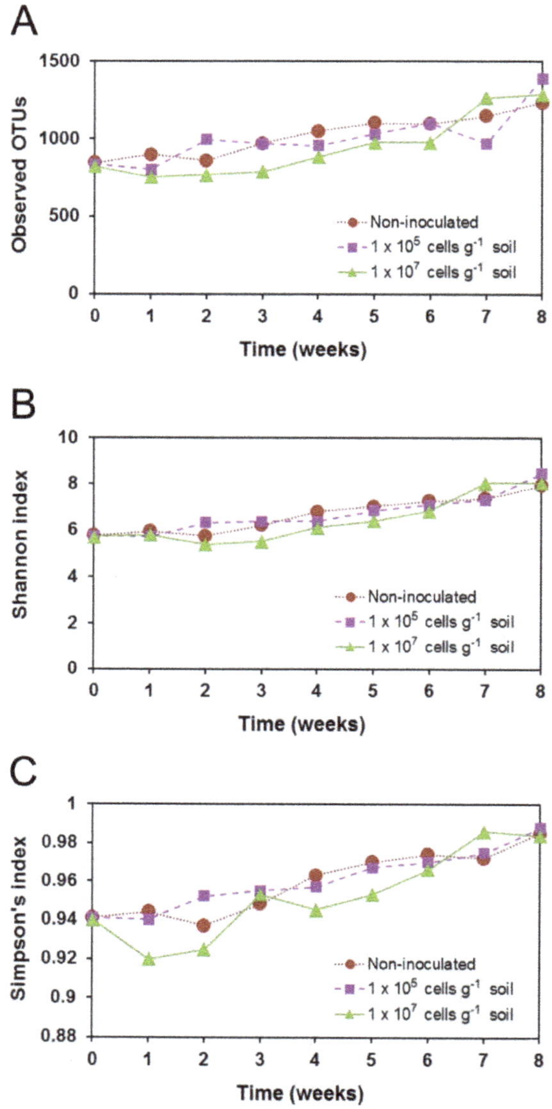

Figure 2. Taxonomic α-diversity analysis: (**A**) observed operational taxonomy units (OTUs); (**B**) Shannon index; (**C**) Simpson's index. Non-inoculated control was not inoculated with *B. thuringiensis* KNU-07.

Xanthomonadales and Saprospirales were the two most abundant orders in this study regardless of the KNU-07 application, and the abundance of these orders gradually decreased over time (Figure 3). In contrast, the abundance of orders Rhizobiales and Ellin329 increased over time in all samples, including controls. The abundance of Acidobacteriales decreased over time in all samples. The abundance of Bacillales, the order to which KNU-07 belongs, was comparatively high in KNU-07-inoculated soil during the first three weeks, but then it decreased (Figure 3). At genera level, the abundance of *Bacillus* spp. was comparatively high in soil inoculated with a high concentration of KNU-07 (1.0×10^7 cells g^{-1} soil) (Figure 4). At a higher concentration of KNU-07, although the abundance of *Bacillus* spp. was decreasing over time, the abundance of *Bacillus* spp. was still higher than the remaining treatments. The abundance of *Bacillus* spp. in the soil inoculated with a lower concentration of KNU-07 (1.0×10^5 cells g^{-1} soil) and non-inoculated control was comparatively higher in the last three weeks (Figure 4). However, it is important to note that the resolution power of NGS of 16S rRNA coding region is not strong enough to discriminate KNU-07 from indigenous *Bacillus* spp. Hence, we designed a strain-specific primer for KNU-07 to monitor the population dynamics of KNU-07 using qPCR with strain-specific primers.

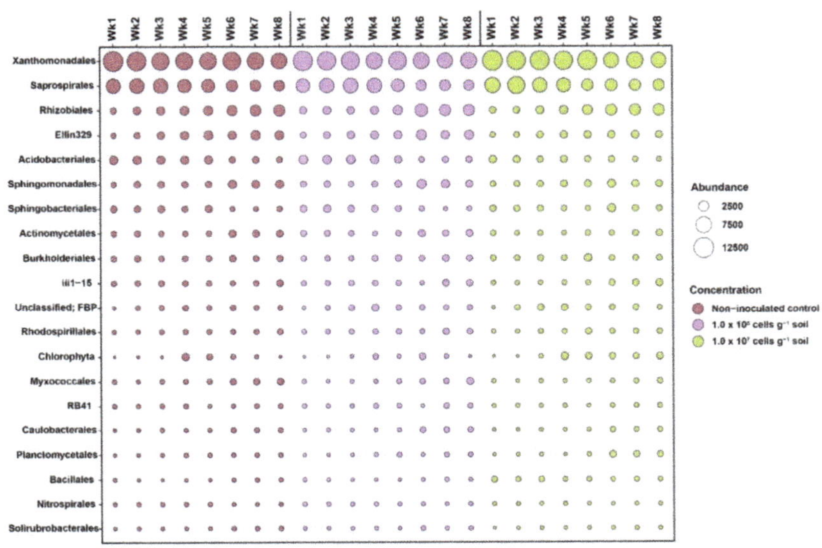

Figure 3. Bubble plot showing the abundance of the bacterial community at an order level based on the 16S rRNA gene in pepper-cultivated soil inoculated with strain KNU-07.

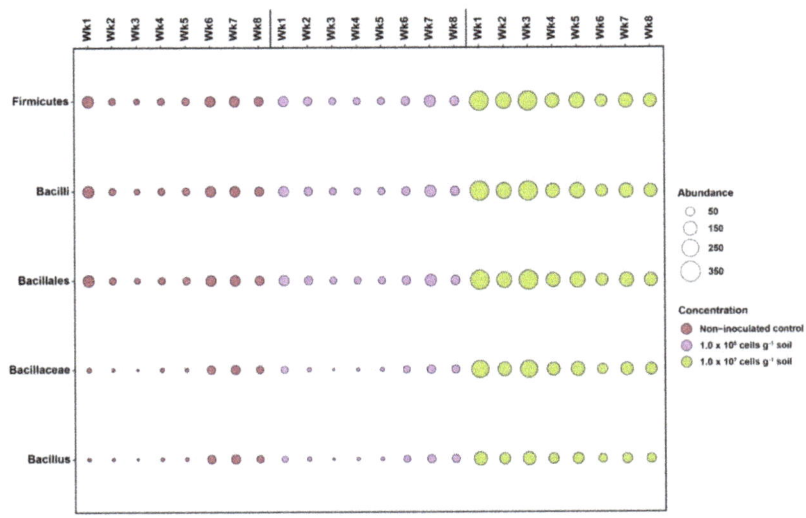

Figure 4. Bubble plot showing the abundance of bacterial taxa, where KNU-7 belongs, over eight weeks post-inoculation using 16S rRNA gene sequencing.

The results of the beta diversity analysis based on principal coordinate analysis (PCoA) at an OTU level revealed that the soil bacterial community compositions were separated over time in all treatments including control (Figure 5). More importantly, bacterial community compositions of the soil treated with a high concentration of KNU-07 were separated from non-inoculated control samples in the last three weeks (Figure 5). These test results were similar to non-parametric statistical analyses based on ADONIS and ANOSIM. The analysis confirmed that the beta diversity between KNU-07-bacterized and non-inoculated control samples was significantly ($p < 0.05$) separated six weeks post-inoculation (Table S3).

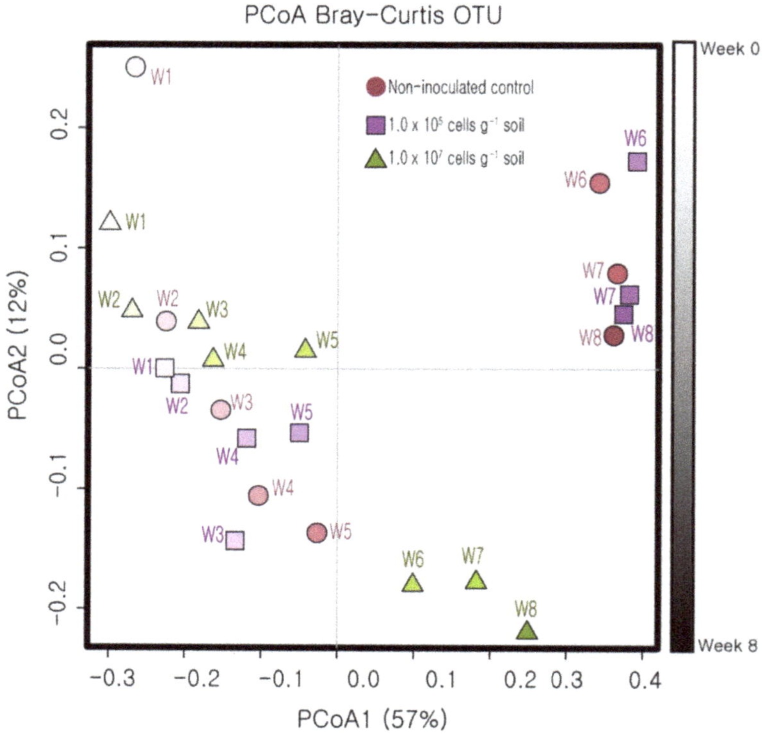

Figure 5. Principal coordinate analysis of 16S rRNA genes of total bacteria based on the Bray–Curtis similarity index at 97% identity (operational taxonomic unit level) for eight weeks (W1–W8). PCoA1 and PCoA2 explained 57% and 12% of the variance, respectively.

We employed the PICRUSt program to predict the function of the soil bacterial community based on 16S rRNA gene data (Figure 6). The PICRUSt functional analysis showed that pathways related to germination and sporulation were overrepresented before six weeks in samples that received an application of KNU-07 (1.0×10^7 cells g^{-1} soil) (Figure 6). After six weeks, the pathways that were positively impacted by the application of KNU-07 (1.0×10^7 cells g^{-1} soil) were energy metabolism and metabolism of cofactors and vitamins (Figure 6).

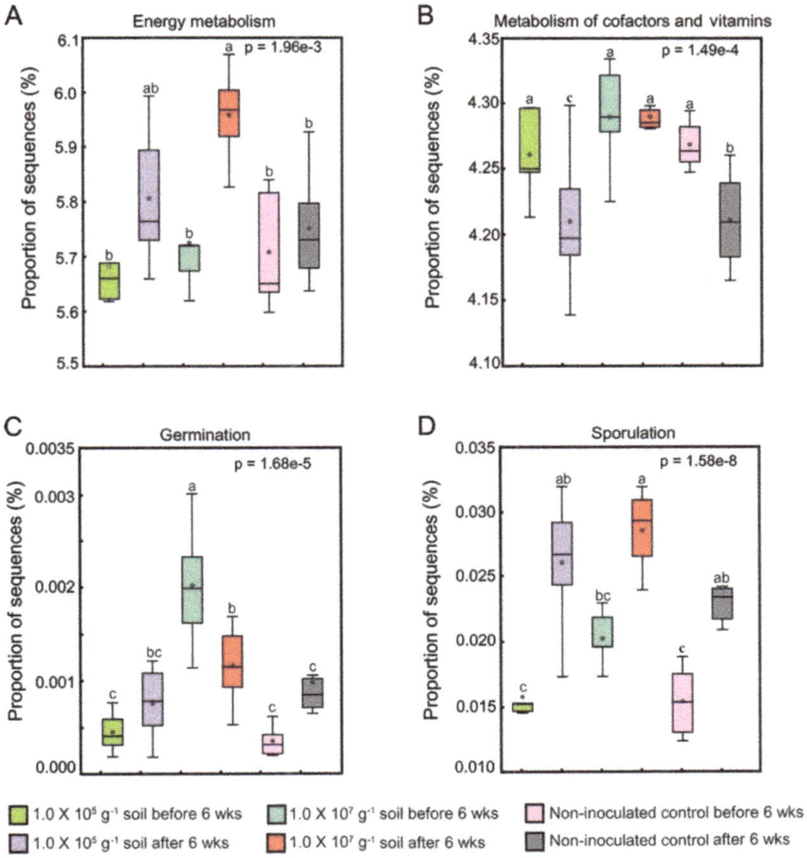

Figure 6. Predicted metabolic function from 16S rRNA gene sequences of soil bacterial community collected from KNU-07 bacterized and non-bacterized samples using PICRUSt and STAMP analysis before and after six weeks inoculation with the indicated concentrations of *B. thuringiensis* KNU-07. (**A**) Energy metabolism, (**B**) metabolism of cofactors and vitamins, (**C**) germination, and (**D**) sporulation. Non-inoculated control (Control). Non-inoculated plants served as controls. Mean values having different letters in each parameter are significantly different ($p \leq 0.05$).

3.4. Tracking of KNU-07 Population Using qPCR

The results of qPCR data showed soil treated with KNU-07 at higher concentrations had the highest abundance of KNU-07 throughout the experiments. As expected, KNU-07 cells were not detected in any non-inoculated control soil at any time (Figure 7). The abundance of KNU-07 decreased over time, regardless of the initial concentration of the KNU-07 inoculum (Figure 7). KNU-07 cells were detected long after inoculation (six weeks) from soils initially inoculated with a high concentration of KNU-07 (1.0×10^7 cells g^{-1} soil). However, KNU-07 cells were detected in soil initially inoculated with a low concentration of KNU-07 (1.0×10^5 cells g^{-1} soil) only within 3 weeks of inoculation (Figure 7). After eight weeks of inoculation, KNU-07 cells were not detected in any sample. We also investigated the total bacteria population using the 16S rRNA gene to determine whether there was a decrease in the total bacteria population, as was observed for KNU-07. The results of qPCR data showed that the abundance of total bacteria in all samples, including controls, increased slightly over time (Figure 7).

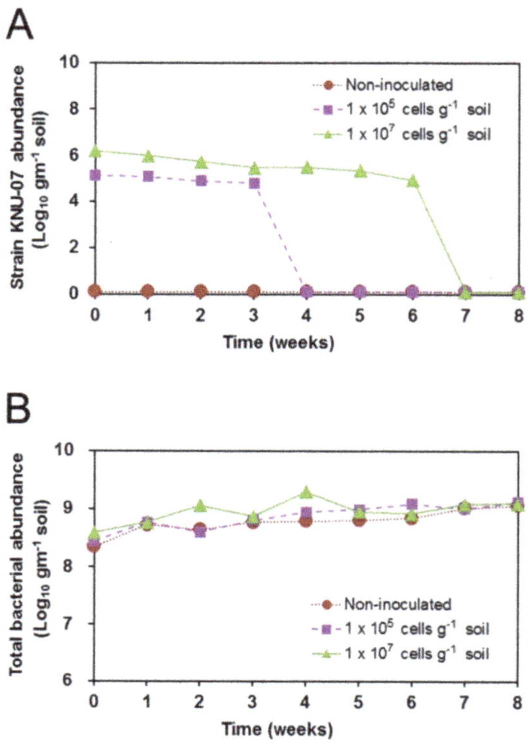

Figure 7. Abundance of (**A**) *B. thuringiensis* KNU-07 based on unique sequence copies and (**B**) total bacteria based on 16S rRNA gene copies in the soil over eight weeks post-inoculation using qPCR.

4. Discussion

In this study, we assessed the effect of KNU-07 on the growth of pepper plants and the soil bacterial community and designed a strain-specific primer pair to track the population dynamics of KNU-07 in soil using a qPCR-based method. Similar to the results of our in vitro assays, IAA production, siderophore production, phosphate solubilization, and urease activity by several strains of *B. thuringiensis* have been previously reported [12–14]. An increase in IAA production in the presence of L-tryptophan may attributed to the nature of the strain to utilize L-tryptophan as a physiological precursor [35]. Our in vivo assays also showed that KNU-07 promoted the growth of pepper plants after inoculation into the soil. Strains of *B. thuringiensis* have been used to promote the growth of plants, and our findings are consistent with these reports [36–38]. Previous studies reported that PGPR strains possessing siderophore production play a great role in helping plants to acquire iron for plant growth [13]. In addition, IAA production, phosphate solubilization, and urease activities play important roles in enhancing nutrient and water uptake and thereby enhance plant growth [12,14].

The change in soil bacterial community structure due to the presence of KNU-07 was less visible before six weeks post-inoculation. However, the community structure was separated after six weeks. Ke et al. [39] reported that the inoculation of soil with *Pseudomonas stutzeri* A1501 significantly changed the indigenous soil bacterial community structure after 2 months of inoculation, and our findings are consistent with this report. Similarly, Wang et al. [40] discussed the significant effect of bio-inoculants on soil microbial communities. In this study, the abundance of the Ellin 329 and Rhizobiales orders were higher in all samples. This may be due to the loss of Acidobacteriales [41,42]. There was also a change in soil microbial community structure over time. In our study, KNU-07-bacterized plants exhibited superior growth relative to control plants. Plant age has been reported to influence the dynamics of the soil microbiome [43,44], and our findings are consistent with these reports.

Predicting the function of the total bacterial community provides information about its interaction with the surrounding environment [45]. Hence, we employed PICRUSt to predict changes in the function of the soil microbiota due to KNU-07 inoculation. Several metabolic pathways that facilitate growth in plants were overrepresented following application KNU-07 (1.0×10^7 cells g^{-1} soil). He et al. [46] reported that rhizobacteria inoculation had beneficial effects on the function of the bacterial community. Predicted metabolic functions related to sporulation and germination were significantly affected during the first week after inoculation with KNU-07 at the highest concentration. The elevated abundance of predicted genes related to sporulation and germination might arise from the inoculated KNU-07, which belongs to the Bacillales order. Sporulation is a survival mechanism of *Bacillus* spp. in response to unfavorable environmental conditions [47]. More importantly, after six weeks post-inoculation, KNU-07 pathways related to energy metabolism and the metabolism of cofactors and vitamins were found to be overrepresented. This might give the pepper plants growing in inoculated soil better nutrition and plant growth [39,41,48].

Quantifying the abundance of a microbial inoculant in the soil is one of the best strategies for tracking [20,24]. Tracking helps to investigate the potential of inoculated microbes because PGPR is based on their persistence in the soil. Tracking bio-inoculants in the soil has been performed using different methods, including dilution plating and microscopy [49,50]. However, such methods can be laborious, time-consuming, and limited to sterile conditions [51]. Interestingly, a few recent studies proposed the possibility of tracking bacterial populations in field samples by using strain-specific primers in qPCR-based protocols [24,52]. To the best of our knowledge, this study is the first report of monitoring *B. thuringiensis* abundance in non-sterile soil using a single strain-specific primer pair in a qPCR-based method. The abundance of KNU-07 was relatively stable during the first two weeks post-inoculation and decreased over time regardless of the initial KNU-07 concentration. Coy et al. [53] reported that the population of *Bacillus sphaericus* drastically declined after six weeks post-inoculation. The bacterial population of antagonistic bacteria has also been reported to decline over time [54]. These decreases in the abundance of soil bio-inoculants might be attributed to physical and biological factors found in the soil environment [55]. Another factor that might cause a decrease in the abundance of KNU-07 may be microbial competition [53,56]. In this study, the amplicon sequence data of 16S rRNAs revealed that there was a gradual increase in the abundance of the total bacterial over time.

The design of a strain-specific primer pair and being able to track the strain in the soil by qPCR offers important information about the fate of PGPR under non-sterile soil conditions, which is an important step in registering a microbe as a PGPR product. Nevertheless, further studies are needed to identify ways to increase the survival of KNU-07 under different soil environmental conditions.

Supplementary Materials: The following are available online at http://www.mdpi.com/2073-4395/10/4/551/s1, Figure S1: Effectiveness of the single primer pair for specific detection of *B. thuringiensis* KNU-07. (A) Lane M: Doctor protein 1 kb plus ladder, lane 1: negative control, lane 2: KNU-07, lane 3–30: different bacterial strains samples (Table S1). (B) Lane M: Doctor protein 1 kb plus ladder, lane 1: negative control, lane 2: KNU-07, lane 3–30: soil samples isolated from different locations (Table S2). Primer pair KNU07F/ KNU07R without template KNU-07 DNA served as the negative control, Figure S2. Potential of some *Bacillus* spp. (1–8) and KNU-07 (9) for indole acetic acid production (A), siderophore activity (B), urease activity (C) and phosphatase activity (D). *1 = Bacillus licheniformis* KACC 10476, *2 = Bacillus megaterium* KACC 10482, *3 = Bacillus polymyxa* KACC 10485, *4 = Bacillus subtilis* KACC 10854, *5 = Bacillus pumilus* KACC 10917, *6 = Bacillus macerans* KACC 11233, *7 = Bacillus amyloliquefaciens* KACC 12067, *8 = Bacillus velezensis* KACC 14004. Table S1: Bacterial strains used in this study, Table S2: Sources of soil samples used for in vitro PCR assays, Table S3. Statistical analysis of bacterial community structure at an operational taxonomic unit level in the last three weeks.

Author Contributions: Conceptualization: H.J. and J.H.S.; methodology: H.J. and H.Q.P; software: H.Q.P; validation: H.J., S.B.T., M.-C.K. and S.C; formal analysis: H.J. and J.-H.S.; investigation: H.J., S.B.T. and J.-H.S.; resources: G.-S.P. and J.-H.S.; data curation: H.J., S.-D.C., M.-J.K. and Y.-J.P.; writing (original draft preparation): H.J., H.Q.P. and S.B.T.; writing (review and editing): S.B.T. and J.-H.S.; visualization: H.J., S.B.T., J.C.I. and J.-H.S.; supervision: J.H.S.; project administration: H.J. and J.-H.S.; funding acquisition: J.-H.S. All authors have read and agreed to the published version of the manuscript.

Acknowledgments: This work was supported by the Strategic Initiative for Microbiomes in Agriculture and Food (Grant number 918010-4), Ministry of Agriculture, Food and Rural Affairs, Republic of Korea; and the Cooperative Research Program for Agriculture Science and Technology Development (Project number PJ013383), Rural Development Administration, Republic of Korea.

Conflicts of Interest: Author H.J. and G.-S.P. were employed by the company COSMAX BTI Inc. and Atogen Co., Ltd., respectively. The remaining authors declare that the research was conducted in the absence of any commercial or financial relationships that could be construed as a potential conflict of interest.

References

1. Compant, S.; Clément, C.; Sessitsch, A. Plant growth-promoting bacteria in the rhizo- and endosphere of plants: Their role, colonization, mechanisms involved and prospects for utilization. *Soil Biol. Biochem.* **2010**, *42*, 669–678. [CrossRef]
2. Anusha, B.G.; Gopalakrishnan, S.; Naik, M.K.; Sharma, M. Evaluation of *Streptomyces* spp. and *Bacillus* spp. for biocontrol of *Fusarium* wilt in chickpea (*Cicer arietinum* L.). *Arch. Phytopathol. Plant Prot.* **2019**, *52*, 417–442. [CrossRef]
3. Gamez, R.; Cardinale, M.; Montes, M.; Ramirez, S.; Schnell, S.; Rodriguez, F. Screening, plant growth promotion and root colonization pattern of two rhizobacteria (*Pseudomonas fluorescens* Ps006 and *Bacillus amyloliquefaciens* Bs006) on banana cv. Williams (*Musa acuminata* Colla). *Microbiol. Res.* **2019**, *220*, 12–20. [CrossRef] [PubMed]
4. Mhatre, P.H.; Karthik, C.; Kadirvelu, K.; Divya, K.L.; Venkatasalam, E.P.; Srinivasan, S.; Ramkumar, G.; Saranya, C.; Shanmuganathan, R. Plant growth promoting rhizobacteria (PGPR): A potential alternative tool for nematodes bio-control. *Biocatal. Agric. Biotechnol.* **2019**, *17*, 119–128. [CrossRef]
5. Chowdhury, S.P.; Hartmann, A.; Gao, X.W.; Borriss, R. Biocontrol mechanism by root-associated *Bacillus amyloliquefaciens* FZB42—A review. *Front. Microbiol.* **2015**, *6*, 780. [CrossRef]
6. Shafi, J.; Tian, H.; Ji, M. *Bacillus* species as versatile weapons for plant pathogens: A review. *Biotechnol. Biotechnol. Equip.* **2017**, *31*, 446–459. [CrossRef]
7. Babu, A.G.; Kim, J.D.; Oh, B.T. Enhancement of heavy metal phytoremediation by *Alnus firma* with endophytic *Bacillus thuringiensis* GDB-1. *J. Hazard. Mater.* **2013**, *250*, 477–483. [CrossRef]
8. Azizoglu, U. *Bacillus thuringiensis* as a Biofertilizer and Biostimulator: A Mini-Review of the Little-Known Plant Growth-Promoting Properties of Bt. *Curr. Microbiol.* **2019**, *76*, 1379–1385. [CrossRef]
9. Singh, S.; Kumar, V.; Sidhu, G.K.; Datta, S.; Dhanjal, D.S.; Koul, B.; Janeja, H.S.; Singh, J. Plant growth promoting rhizobacteria from heavy metal contaminated soil promote growth attributes of *Pisum sativum* L. *Biocatal. Agric. Biotechnol.* **2019**, *17*, 665–671. [CrossRef]
10. Wang, M.; Chen, S.; Han, Y.; Chen, L.; Wang, D. Responses of soil aggregates and bacterial communities to soil-Pb immobilization induced by biofertilizer. *Chemosphere* **2019**, *220*, 828–836. [CrossRef]
11. You, C.; Zhang, C.; Kong, F.; Feng, C.; Wang, J. Comparison of the effects of biocontrol agent *Bacillus subtilis* and fungicide metalaxyl-mancozeb on bacterial communities in tobacco rhizospheric soil. *Ecol. Eng.* **2016**, *91*, 119–125. [CrossRef]
12. Wang, T.; Liu, M.Q.; Li, H.X. Inoculation of phosphate-solubilizing bacteria *Bacillus thuringiensis* B1 increases available phosphorus and growth of peanut in acidic soil. *Acta Agric. Scand. Sect. B Soil Plant Sci.* **2014**, *64*, 252–259.
13. Hollensteiner, J.; Wemheuer, F.; Harting, R.; Kolarzyk, A.M.; Diaz Valerio, S.M.; Poehlein, A.; Brzuszkiewicz, E.B.; Nesemann, K.; Braus-Stromeyer, S.A.; Braus, G.H.; et al. *Bacillus thuringiensis* and *Bacillus weihenstephanensis* inhibit the growth of phytopathogenic *Verticillium* species. *Front. Microbiol.* **2017**, *7*, 2171. [CrossRef] [PubMed]
14. Khan, N.; Bano, A.; Rahman, M.A.; Guo, J.; Kang, Z.; Babar, M.A. Comparative physiological and metabolic analysis reveals a complex mechanism involved in drought tolerance in chickpea (Cicer arietinum L.) induced by PGPR and PGRs. *Sci. Rep.* **2019**, *9*, 2097. [CrossRef]
15. Gange, A.C.; Gadhave, K.R. Plant growth-promoting rhizobacteria promote plant size inequality. *Sci. Rep.* **2018**, *8*, 13828. [CrossRef]

16. Goswami, D.; Thakker, J.N.; Dhandhukia, P.C. Portraying mechanics of plant growth promoting rhizobacteria (PGPR): A review. *Cogent Food Agric.* **2016**, *2*, 1127500. [CrossRef]
17. Cao, Y.; Pi, H.; Chandrangsu, P.; Li, Y.; Wang, Y.; Zhou, H.; Xiong, H.; Helmann, J.D.; Cai, Y. Antagonism of Two Plant-Growth Promoting *Bacillus velezensis* Isolates Against *Ralstonia solanacearum* and *Fusarium oxysporum*. *Sci. Rep.* **2018**, *8*, 4360. [CrossRef]
18. Mosimann, C.; Oberhänsli, T.; Ziegler, D.; Nassal, D.; Kandeler, E.; Boller, T.; Mäder, P.; Thonar, C. Tracing of two *Pseudomonas* strains in the root and rhizoplane of maize, as related to their plant growth-promoting effect in contrasting soils. *Front. Microbiol.* **2017**, *7*, 2150. [CrossRef]
19. Aufrecht, J.A.; Timm, C.M.; Bible, A.; Morrell-Falvey, J.L.; Pelletier, D.A.; Doktycz, M.J.; Retterer, S.T. Microfluidics: Quantifying the Spatiotemporal Dynamics of Plant Root Colonization by Beneficial Bacteria in a Microfluidic Habitat (Adv. Biosys. 6/2018). *Adv. Biosyst.* **2018**, *2*, 1800048. [CrossRef]
20. Rilling, J.I.; Acuña, J.J.; Nannipieri, P.; Cassan, F.; Maruyama, F.; Jorquera, M.A. Current opinion and perspectives on the methods for tracking and monitoring plant growth-promoting bacteria. *Soil Biol. Biochem.* **2019**, *130*, 205–219. [CrossRef]
21. Goulart, M.C.; Cueva-Yesquén, L.G.; Hidalgo Martinez, K.J.; Attili-Angelis, D.; Fantinatti-Garboggini, F. Comparison of specific endophytic bacterial communities in different developmental stages of *Passiflora incarnata* using culture-dependent and culture-independent analysis. *Microbiologyopen* **2019**, *8*, e896. [CrossRef] [PubMed]
22. Kraková, L.; Šoltys, K.; Otlewska, A.; Pietrzak, K.; Purkrtová, S.; Savická, D.; Puškárová, A.; Bučková, M.; Szemes, T.; Budiš, J.; et al. Comparison of methods for identification of microbial communities in book collections: Culture-dependent (sequencing and MALDI-TOF MS) and culture-independent (Illumina MiSeq). *Int. Biodeterior. Biodegrad.* **2018**, *131*, 51–59. [CrossRef]
23. Stets, M.I.; Alqueres, S.; Souza, E.M.; Pedrosa, F.O.; Schmid, M.; Hartmann, A.; Cruz, L.M. Quantification of *Azospirillum brasilense* FP2 bacteria in wheat roots by strain-specific quantitative PCR. *Appl. Environ. Microbiol.* **2015**, *81*, 6700–6709. [CrossRef]
24. Mendis, H.C.; Thomas, V.P.; Schwientek, P.; Salamzade, R.; Chien, J.T.; Waidyarathne, P.; Kloepper, J.; De La Fuente, L. Strain-specific quantification of root colonization by plant growth promoting rhizobacteria *Bacillus firmus* I-1582 and *Bacillus amyloliquefaciens* QST713 in non-sterile soil and field conditions. *PLoS ONE* **2018**, *13*, e0193119. [CrossRef]
25. Ficetola, G.F.; Coissac, E.; Zundel, S.; Riaz, T.; Shehzad, W.; Bessière, J.; Taberlet, P.; Pompanon, F. An in silico approach for the evaluation of DNA barcodes. *BMC Genom.* **2010**, *11*, 434. [CrossRef] [PubMed]
26. Gordon, S.A.; Weber, R.P. Colorimetric estimation of indoleacetic acid. *Plant Physiol.* **1951**, *26*, 192–195. [CrossRef]
27. Kurtzman, C.P.; Fell, J.W.; Boekhout, T.; Robert, V. Methods for isolation, phenotypic characterization and maintenance of yeasts. In *The Yeasts*; Elsevier: London, UK; Amsterdam, The Netherlands, 2011; ISBN 9780444521491.
28. Pérez-Miranda, S.; Cabirol, N.; George-Téllez, R.; Zamudio-Rivera, L.S.; Fernández, F.J. O-CAS, a fast and universal method for siderophore detection. *J. Microbiol. Methods* **2007**, *31*, 446–459. [CrossRef]
29. Kyaw, A.; Maung-U, K.; Toe, T. Determination of inorganic phosphate with molybdate and Triton X-100 without reduction. *Anal. Biochem.* **1985**, *145*, 230–234. [CrossRef]
30. Anderson, M.J. A new method for non-parametric multivariate analysis of variance. *Austral Ecol.* **2001**, *26*, 32–46.
31. CLARKE, K.R. Non-parametric multivariate analyses of changes in community structure. *Aust. J. Ecol.* **1993**, *18*, 117–143. [CrossRef]
32. Langille, M.G.I.; Zaneveld, J.; Caporaso, J.G.; McDonald, D.; Knights, D.; Reyes, J.A.; Clemente, J.C.; Burkepile, D.E.; Vega Thurber, R.L.; Knight, R.; et al. Predictive functional profiling of microbial communities using 16S rRNA marker gene sequences. *Nat. Biotechnol.* **2013**, *31*, 814–821. [CrossRef] [PubMed]
33. Parks, D.H.; Tyson, G.W.; Hugenholtz, P.; Beiko, R.G. STAMP: Statistical analysis of taxonomic and functional profiles. *Bioinformatics* **2014**, *30*, 3123–3124. [CrossRef] [PubMed]
34. SAS Institute. *SAS Version 9.4*; SAS Institute Inc.: Cary, NC, USA, 2012.

35. Spaepen, S.; Vanderleyden, J.; Remans, R. Indole-3-acetic acid in microbial and microorganism-plant signaling. *FEMS Microbiol. Rev.* **2007**, *31*, 425–448. [CrossRef] [PubMed]
36. Tagele, S.B.; Lee, H.G.; Kim, S.W.; Lee, Y.S. Phenazine and 1-undecene producing *Pseudomonas chlororaphis* subsp. *aurantiaca strain KNU17Pc1 for growth promotion and disease suppression in Korean maize cultivars. J. Microbiol. Biotechnol.* **2019**, *29*, 66–78.
37. Viljoen, J.J.F.; Labuschagne, N.; Fourie, H.; Sikora, R.A. Biological control of the root-knot nematode *Meloidogyne incognita* on tomatoes and carrots by plant growth-promoting rhizobacteria. *Trop. Plant Pathol.* **2019**, *44*, 284–291. [CrossRef]
38. Vidal-Quist, J.C.; Rogers, H.W.J.; Mahenthiralingam, E.; Berry, C. *Bacillus thuringiensis* colonises plant roots in a phylogeny-dependent manner. *FEMS Microbiol. Ecol.* **2013**, *86*, 474–489. [CrossRef]
39. Ke, X.; Feng, S.; Wang, J.; Lu, W.; Zhang, W.; Chen, M.; Lin, M. Effect of inoculation with nitrogen-fixing bacterium *Pseudomonas stutzeri* A1501 on maize plant growth and the microbiome indigenous to the rhizosphere. *Syst. Appl. Microbiol.* **2019**, *42*, 248–260. [CrossRef]
40. Wang, J.; Li, Q.; Xu, S.; Zhao, W.; Lei, Y.; Song, C.; Huang, Z. Traits-based integration of multi-species inoculants facilitates shifts of indigenous soil bacterial community. *Front. Microbiol.* **2018**, *9*, 1692. [CrossRef]
41. Bouskill, N.J.; Lim, H.C.; Borglin, S.; Salve, R.; Wood, T.E.; Silver, W.L.; Brodie, E.L. Pre-exposure to drought increases the resistance of tropical forest soil bacterial communities to extended drought. *ISME J.* **2013**, *7*, 384–394. [CrossRef]
42. Harbison, A.B.; Carson, M.A.; Lamit, L.J.; Basiliko, N.; Bräuer, S.L. A novel isolate and widespread abundance of the candidate alphaproteobacterial order (Ellin 329), in southern Appalachian peatlands. *FEMS Microbiol. Lett.* **2016**, *363*, fnw151. [CrossRef]
43. Van Overbeek, L.; Van Elsas, J.D. Effects of plant genotype and growth stage on the structure of bacterial communities associated with potato (*Solanum tuberosum* L.). *FEMS Microbiol. Ecol.* **2008**, *64*, 283–296. [CrossRef] [PubMed]
44. Degrune, F.; Theodorakopoulos, N.; Colinet, G.; Hiel, M.P.; Bodson, B.; Taminiau, B.; Daube, G.; Vandenbol, M.; Hartmann, M. Temporal dynamics of soil microbial communities below the seedbed under two contrasting tillage regimes. *Front. Microbiol.* **2017**, *8*, 1127. [CrossRef] [PubMed]
45. Han, L.; Wang, Z.; Li, N.; Wang, Y.; Feng, J.; Zhang, X. *Bacillus amyloliquefaciens* B1408 suppresses Fusarium wilt in cucumber by regulating the rhizosphere microbial community. *Appl. Soil Ecol.* **2019**, *136*, 55–66. [CrossRef]
46. He, Y.; Pantigoso, H.A.; Wu, Z.; Vivanco, J.M. Co-inoculation of *Bacillus* sp. and *Pseudomonas putida* at different development stages acts as a biostimulant to promote growth, yield and nutrient uptake of tomato. *J. Appl. Microbiol.* **2019**, *127*, 196–207. [CrossRef]
47. Omony, J.; de Jong, A.; Krawczyk, A.O.; Eijlander, R.T.; Kuipers, O.P. Dynamic sporulation gene co-expression networks for *Bacillus subtilis* 168 and the food-borne isolate *Bacillus amyloliquefaciens*: A transcriptomic model. *Microb. Genom.* **2018**, *4*, e000157. [CrossRef] [PubMed]
48. Mendes, L.W.; Kuramae, E.E.; Navarrete, A.A.; Van Veen, J.A.; Tsai, S.M. Taxonomical and functional microbial community selection in soybean rhizosphere. *ISME J.* **2014**, *8*, 1577–1587. [CrossRef]
49. Taute, K.M.; Gude, S.; Tans, S.J.; Shimizu, T.S. High-throughput 3D tracking of bacteria on a standard phase contrast microscope. *Nat. Commun.* **2015**, *6*, 1–9. [CrossRef]
50. Banik, A.; Mukhopadhaya, S.K.; Sahana, A.; Das, D.; Dangar, T.K. Fluorescence resonance energy transfer (FRET)-based technique for tracking of endophytic bacteria in rice roots. *Biol. Fertil. Soils* **2016**, *52*, 277–282. [CrossRef]
51. Segarra, G.; Puopolo, G.; Porcel-Rodríguez, E.; Giovannini, O.; Pertot, I. Monitoring *Lysobacter capsici* az78 using strain specific qpcr reveals the importance of the formulation for its survival in vineyards. *FEMS Microbiol. Lett.* **2016**, *363*, fnv243. [CrossRef]
52. Ricketts, C.; Pickler, L.; Maurer, J.; Ayyampalayam, S.; García, M.; Ferguson-Noel, N.M. Identification of strain-specific sequences that distinguish a *Mycoplasma gallisepticum* vaccine strain from field isolates. *J. Clin. Microbiol.* **2017**, *55*, 244–252. [CrossRef]
53. Coy, R.M.; Held, D.W.; Kloepper, J.W. Rhizobacterial colonization of bermudagrass by *Bacillus* spp. in a Marvyn loamy sand soil. *Appl. Soil Ecol.* **2019**, *141*, 10–17. [CrossRef]

54. Khabbaz, S.E.; Abbasi, P.A. Isolation, characterization, and formulation of antagonistic bacteria for the management of seedlings damping-off and root rot disease of cucumber. *Can. J. Microbiol.* **2014**, *60*, 25–33. [CrossRef] [PubMed]
55. Andreote, F.D.; De Araújo, W.L.; De Azevedo, J.L.; Van Elsas, J.D.; Da Rocha, U.N.; Van Overbeek, L.S. Endophytic colonization of potato (*Solanum tuberosum* L.) by a novel competent bacterial endophyte, *Pseudomonas putida* strain P9, and its effect on associated bacterial communities. *Appl. Environ. Microbiol.* **2009**, *75*, 3396–3406. [CrossRef] [PubMed]
56. Ishizawa, H.; Kuroda, M.; Inoue, K.; Inoue, D.; Morikawa, M.; Ike, M. Colonization and competition dynamics of plant growth-promoting/inhibiting Bacteria in the phytosphere of the Duckweed Lemna minor. *Microb. Ecol.* **2019**, *77*, 440–450. [CrossRef] [PubMed]

agronomy

MDPI

Article

Fermented Alfalfa Brown Juice Significantly Stimulates the Growth and Development of Sweet Basil (*Ocimum basilicum* L.) Plants

Szilvia Kisvarga [1], Döme Barna [2,*], Szilvia Kovács [2], Gábor Csatári [2], Ibolya O. Tóth [2], Miklós Gábor Fári [2], Péter Makleit [2], Szilvia Veres [2], Tarek Alshaal [2,3] and Nóra Bákonyi [2]

1 Research Institute for Fruit growing and Ornamentals, National Agricultural Research and Innovation Center (NARIC), Szent-Györgyi Street 4, 2100 Gödöllő, Hungary; fullerina@gmail.com

2 Department of Agricultural Botany, Plant Physiology and Biotechnology, Institute of Crop Sciences (MEK), Debrecen University, Böszörményi Street 138, 4032 Debrecen, Hungary; szkovacs@agr.unideb.hu (S.K.); gaborcsat96@gmail.com (G.C.); olahne@agr.unideb.hu (I.O.T.); fari@agr.unideb.hu (M.G.F.); pmakleit@agr.unideb.hu (P.M.); szveres@agr.unideb.hu (S.V.); alshaaltarek@gmail.com (T.A.); nbakonyi@agr.unideb.hu (N.B.)

3 Soil and Water Department, Faculty of Agriculture, Kafrelsheikh University, Kafr El-Sheikh 33516, Egypt

* Correspondence: barna.dome@agr.unideb.hu

Received: 17 April 2020; Accepted: 2 May 2020; Published: 7 May 2020

Abstract: Fertilization management is a key issue in plant nutrition to produce plants with good quality and quantity. Deproteinized leaf juice or brown juice (BJ) is a by-product during the isolation of leaf protein from biomass crops such as alfalfa. The idea of using BJ as a biostimulant fits well in the aspect of circular economy since BJ is currently a problematic issue of the leaf protein production approach. Fractionation of one-kilogram fresh biomass results in approximately 500 cm^3 BJ. Due to fast spoil of fresh BJ, if left on room temperature, it is found that fermentation of fresh BJ using lactic acid bacteria and reducing its pH increases its stability and storage on room temperature. In the present study, we examined the effect of fermented alfalfa BJ on vegetative, physiological, and anatomical properties of the versatile sweet basil (*Ocimum basilicum* L. 'Bíborfelhő') plants. Sweet basil seedlings were sprayed at different doses of fermented alfalfa BJ (i.e., 0.5%, 1.0%, and 2.5%) and tap water served as a control (0.0% BJ). The results revealed that foliar application of fermented alfalfa BJ significantly improved the biometrical features of sweet basil plants. Plants treated with fermented BJ showed significantly higher values of all the measured parameters compared to the control (0.0%), except for the number of leaves per plants where control plants (0.0%) had more leaves. However, the leaves of control plants (0.0%) were smaller than treated plants as data of leaf area showed. Fermented alfalfa BJ significantly increased the content of photosynthetic pigments (chl a and chl b), relative chlorophyll (SPAD value), lengths of stem and root, fresh masses of stem, root, and leaves, volumes of stem and root, and leaf area. Despite all rates of fermented BJ displayed higher values over control plants (0.0%), the rate of 0.5% was the best one supported by results. Application of fermented alfalfa BJ influenced the anatomical parameters as well. These findings demonstrate the possible use of fermented alfalfa BJ as a promising novel plant biostimulant.

Keywords: sweet basil; alfalfa brown juice; fermentation; biostimulation; chlorophyll pigments; histological changes

1. Introduction

The increase in the world's population leads to several issues that we already have to face, and we must find sustainable solutions for them to save our life on the planet. One of these concerns is the sustainable supply of high-quality protein. To solve the protein issue, several approaches were proposed to identify novel protein sources or alternatives [1–3]. Alfalfa (*Medicago sativa* L.) leaf protein concentrate (LPC) is a promising element either in human or animal diet as the human population of the Earth (7.2 billion) is growing rapidly causing a high demand for animal protein [4,5]. The isolation of leaf protein resulted in four products, i.e., green juice, fiber, leaf protein concentrate, and deproteinized juice or referred to as brown juice (BJ). The amount of produced BJ during the isolation of leaf protein and the utilization of this product is a huge obstacle for the wider recognition of LPC production. Fractionation of one-kilogram fresh biomass results in 450–550 mL BJ [6–8]. Therefore, disposal of BJ is a big challenge that the leaf protein isolate approach faces. Another reason making disposal of BJ risky is the high content of several bioactive components such as sugars, free amino acids, minerals and vitamins that BJ contains [7,9,10]. Therefore, finding an alternative use of BJ is urgent due to environmental concerns, besides maximizing the benefit from this waste that is very rich in several valuable compounds and nutrients. Instead of dumping it we could make a valuable product, thus making a step towards the circular economy concept [11,12]. The BJ contains about 40% carbohydrates (mainly monosaccharides, like glucose and fructose) and 3% nitrogen-based on dry mass [12]; additionally, numerous biologically active components like phenols, amino acids, macro- and microelements and biostimulators, etc. [7,11,13]. However, fresh BJ cannot be stored at room temperature, after a few days it gets spoiled at pH 5–6 due to its high sugar content. In our previous study, inoculation of BJ with lactic acid bacteria was not only found to increase BJ stability at room temperature but also substantially improved the nutritional characteristics of BJ [7] through converting sugars into organic acids decreasing the pH to almost 4.5. Therefore, BJ seems to be an ideal component in animal feeding programs as well as plant nutrition and soil stabilization. Several authors have previously suggested BJ as a plant fertilizer, fodder, and growth medium for microbes [10,14–16]. Fermented BJ can be applied as a very effective foliar biostimulant. We have observed remarkable effects of fermented alfalfa BJ on the growth dynamic of plumed cockscomb [7]. Additionally, lactic acid bacteria as a plant growth-promoting bacteria represents an additional benefit of fermentation of BJ as it promotes plant growth [12,17].

Sweet basil (*Ocimum basilicum* L.) is a well-known annual herb, member of the Lamiaceae family. It is one of the 150 species of the Ocimum genus, a very important medicinal, spice and fresh vegetable, culinary herb and industrial plant, cultivated for aromatic and medicinal use on large areas in many countries [18–23]. It is native to India, Asia and Africa, but grows in many regions of the world, including Italy, Thailand, Vietnam and Laos [24]. Sweet basil is popularly used in traditional Chinese medicine, because of its selected purified components and antiviral activity [25]. The latest scientific developments revealed the strong pharmacological action and nutritional aspects because of antioxidant content [25], additionally, it is a rich source of acylated and glycosylated anthocyanins being a valuable source for the food industry [26].

The aim of this study was to examine the impact of the fermented BJ on the biometrical, physiological and anatomical features of sweet basil plants.

2. Materials and Methods

2.1. Sources of Brown Juice and Plant Materials

The fresh alfalfa BJ was obtained from the Proteomill Green Protein Biorefinery Factory (Tedej Ltd., Hajdúnánás, Hungary). The fresh BJ was fermented using lactic acid bacteria to avoid its fast spoiling. The physicochemical properties of fresh and fermented BJ as well as the fermentation process were described by Bákonyi et al. [7]. The seeds of sweet basil (*Ocimum basilicum* 'Bíborfelhő') were obtained from the National Agricultural Research and Innovation Center (NARIC, Budapest, Hungary).

2.2. Experimental Study

A pot experiment under greenhouse conditions was carried out at the NARIC to assess the possible growth stimulation effect of fermented BJ using the multipurpose sweet basil as a plant model. The experimental design was the Randomized Complete Block design (RCB) with 15 replicates. A polyethylene pot (7 × 7 × 8 cm) was filled with white peat for young plants (Klassman-Deilmann TS 3 FINE type, Geeste, Germany). The characteristics of growth medium are as follows: fine structure, pH (H_2O) 6, N 140 mg L^{-1}, P (P_2O_5) 100 mg L^{-1}, K (K_2O) 180 mg L^{-1}, Mg 100 mg L^{-1}, S 150 mg L^{-1}. The sweet basil seeds were sown in the nursery substrate on 16 July 2018. The germinated seedlings were treated once a week with fermented BJ at different rates (i.e., 0.5, 1, and 2.5%) in the early stage of development (stage 1 BBCH) [27]. On 1 August 2018, the seedlings turned to stage 2 BBCH, all identical and healthy, were transferred to the pots. One pot contained one seedling and each treatment contained 15 pots. Fermented BJ was sprayed on the plants twice a week (on Tuesdays and Fridays) from 15 August till the end of the experiment (11 September) at rates of 0.5, 1.0, and 2.5%. Final application volume of 250 mL BJ was equally shared among all replicates of the same treatment (15 plants). The control plants (0.0%) were sprayed with the equivalent amounts of tap water. The experiment was terminated on 11 September (stage 5 BBCH) and plants were harvested, and samples were collected for the further biometric, physiological and anatomical analyses.

2.2.1. Analysis of Biometric Features of Basil Plants

At the end of the experiment, before flowering (BBCH stage 5), all plants in each treatment were harvested and the following vegetative parameters were measured: root and stem length (cm), root and stem volume (mL), root and stem fresh mass (g plant^{-1}), the number of leaves (pcs plant^{-1}) and leaf area. Roots were carefully removed from the pots and washed by tap water on a sieve to remove the adhered growth medium particles. Length was measured with a measuring ruler, while volume was determined by a graduated cylinder, fresh mass by OHAUS Pioneer PA214 analytical balance and the number of leaves by counting. The leaf area (cm^2) of sweet basil plants were measured by AreaMeter 350 (Opti-Sciences, Hudson, NY, USA). Six plants were analyzed and represented as a mean ± SD (n = 6). In the case of leaf area, we used the data of nine plants. All leaves of the examined plants were measured.

2.2.2. Determination of Chlorophyll Contents

At the end of the experimental period, the relative chlorophyll values were measured by SPAD 502 chlorophyll meter (Minolta, Japan) using the last fully developed leaves. The chlorophyll-a and -b contents were extracted based on the method of Moran and Porath [28] and determined by the method of Wellburn [29], Vidician and Cachita-Cosma [30]. We used the following formulas: "Chlorophyll a (mg·g^{-1}) = (11.65 a664–2.69 a647)" and "Chlorophyll b (mg·g^{-1}) = (20.81 a647–4.53 a664)". The samples were taken from the last fully expanded leaves and chlorophyll pigments were extracted by 5 mL *N*,*N*-dimethylformamide (DMF) added to 0.05 g leaf disc. The samples were soaked in this solvent for 48 h at room temperature in the dark. After 48 h, the discs were removed and the contents of chlorophyll-a and -b were measured by METEREKSP-830 spectrophotometer. All the measurements were repeated three times for each plant making up to nine measurements for each of the treatments.

2.2.3. Anatomical Analysis

We used three individual plants per treatment for the stem's histological examination. 15 different cross-sections per plant internodes (n = 45) were prepared as described in the following: each plant was cut into smaller pieces and the third internodes (from beneath) fixed separately in Strasburger-Flemming's solution [31], which is a mixture of glycerin:alcohol:water (1:1:1) for a week. Then, several cross-sections were prepared using blades, after clarification, they were stained with Toluidin-blue. Each analysis was performed under a light microscope (Zeiss Axioscope

2+; Zeiss International, Oberkochen, Ostalbkreis, Germany) with a compatible camera, and Scope Photo software (Scopetek, München, Germany) was used for processing the images. The measured parameters were: thickness of the epidermis, primary cortex, pith including primary and secondary vascular tissues.

2.3. Statistical Analysis

Results of the experiments were subjected to one-way ANOVA by SigmaPlot 12.0 and IBM SPSS Statistics 24 software and the means were compared by Tukey Test [32] at $p \leq 0.05$. Before the ANOVA test, in SPSS the Levene's Test for Equality of Variances was performed in the case of anatomical data. SigmaPlot 12.0 automatically ran a check test for Equality of Variance. The Equality of Variance test for different variables at the four treatments of BJ (i.e., 0.0, 0.5, 1.0, and 2.5%) were negative, $p \leq 0.05$, and the variances showed homogeneity.

3. Results

3.1. Plant Biometric Features

Results of root and stem length of sweet basil are presented in Figure 1. The treated plants with fermented BJ had taller stems and roots compared to the control plants (0.0%). Most importantly, fermented BJ enhanced plant development resulting in higher values of both root and stem. Despite all treated plants showed a taller stem than control plants (0.0%), increasing the rate of applied fermented BJ slightly reduced the stem length. However, the differences in stem length between treatments of fermented BJ were not significant. Treated plants with 0.5% fermented BJ showed an increase of 50.4% in stem length compared to control (0.0%) as the highest measured increase. Similarly, root length showed a relationship as root length gradually increased with increasing the rate of the applied fermented BJ. The application of fermented BJ significantly increased root length compared to the control (0.0%); however, no significant differences in root length were reported between different rates of fermented BJ (Figure 1).

Stem volume of sweet basil plants follows a dose-relationship response to fermented BJ; increasing rates of fermented BJ significantly and gradually caused an increase in stem volume. The highest increase in stem volume (38.6%) compared to untreated control plants (0.0%) was recorded by 2.5% BJ (Figure 2). Similar effect was noticed for root volume, as it increased upon spraying plants with increased rates of fermented BJ. Nevertheless, the highest rate of fermented BJ (2.5%) showed the same value as for treatment of 0.5% fermented BJ, which was two-fold higher than the control plants (0.0%). The highest root volume was 1.16 mL and was found when plants received 1.0% fermented BJ.

Results of stem and root fresh mass are presented in Figure 3. The fresh mass of stem gradually increased as a result of increased rates of fermented BJ. Significantly, all fermented BJ increased the stem fresh mass. While control plants (0.0%) possessed a stem fresh mass of 0.73 g plant^{-1}, treated plants with 2.5% fermented BJ showed a stem fresh mass of 1.52 g plant^{-1} as the highest recorded value. The differences between fermented BJ treatments were not significant. In contrast to stem fresh mass, root fresh mass increased as rate of fermented BJ increased up to 1.0%, then decreased at 2.5% fermented BJ. However, all treatments showed significant increases in the fresh mass of root system in comparison to control plants (0.0%) (Figure 3). Plants received 1.0% of fermented BJ, which was almost three-fold higher in their root fresh mass than control plants (0.0%).

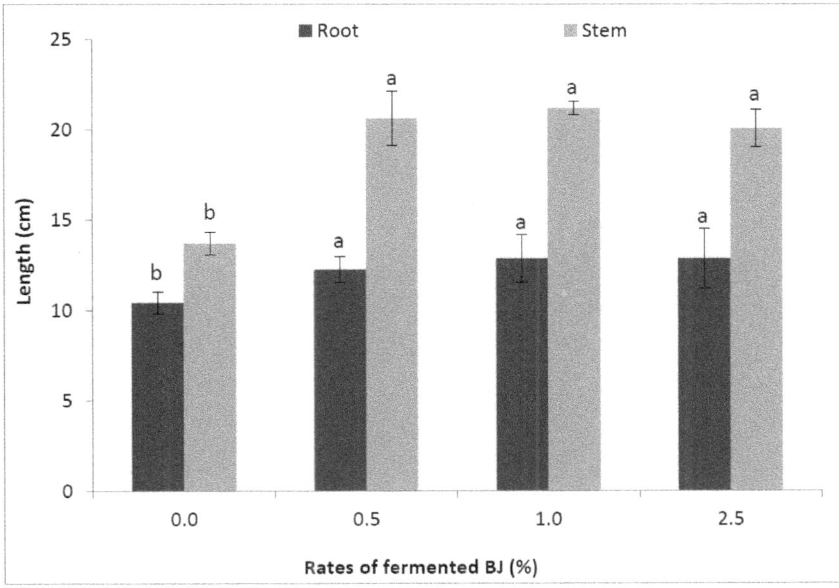

Figure 1. Root and stem length of sweet basil plants treated with different rates of fermented BJ. Sample size $n = 6$ (mean ± SD). Different letters above the same columns show significant differences at the level of $p \leq 0.05$.

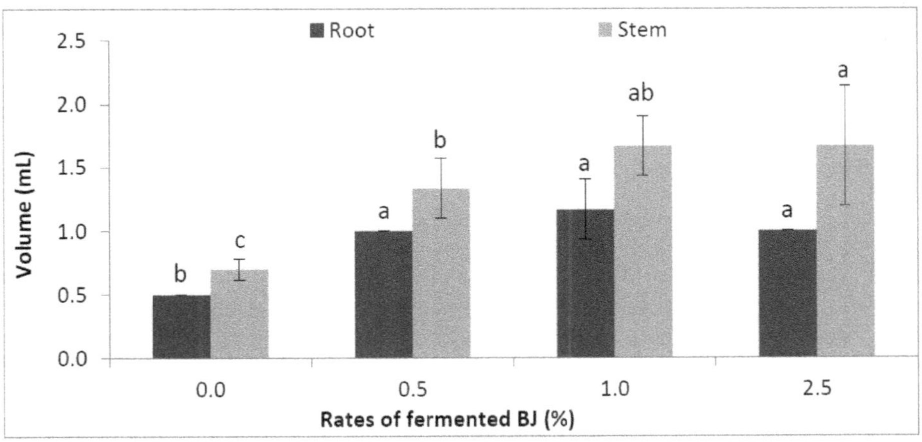

Figure 2. Stem and root volume of sweet basil plants treated with different rates of fermented BJ. Sample size $n = 6$ (mean ± SD). Different letters above the same columns show significant differences at the level of $p \leq 0.05$.

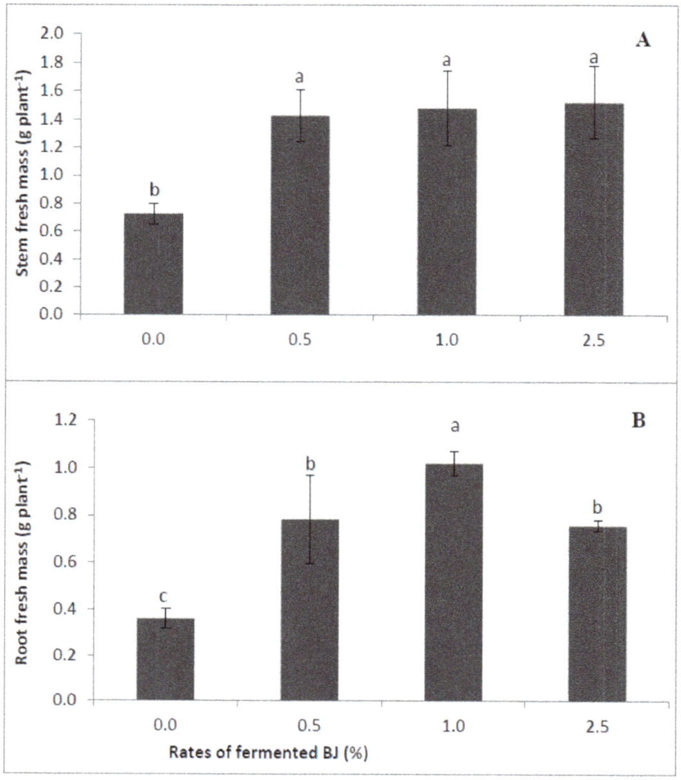

Figure 3. Fresh mass of stem (**A**) and root (**B**) of sweet basil plants treated with different rates of fermented BJ. Sample size $n = 6$ (mean ± SD). Different letters above the same columns show significant differences at the level of $p \leq 0.05$.

Figure 4 shows the results of the leaves fresh mass, number of leaves and leaf area of sweet basil plants after treating them with different rates of fermented BJ extracted from alfalfa biomass. Results showed that fermented BJ significantly improved plant growth as the leaves fresh mass of treated plants was higher than those untreated (control, 0.0%). All fermented BJ rates resulted in significantly higher values of leaves fresh mass than the control (0.0%). However, the highest applied rate of fermented BJ (2.5%) showed a lower fresh mass value than that measured at 0.5 and 1.0% of fermented BJ. The highest leaves fresh mass was recorded at treatment with 0.5 and 1.0% fermented BJ (Figure 4A). Contrarily, the number of leaves per plant gradually reduced with increasing rate of applied fermented BJ (Figure 4B). Control plants (0.0%) possessed the highest number of leaves (22 leaves per plant), while the plants sprayed with 2.5% fermented BJ had 11.7 leaves per plant. The differences among all treatments including the control (0.0%) were statistically significant. Despite control plants (0.0%) possessing a higher number of leaves than the treated plants with fermented BJ, the leaf area of plants sprayed with different rates of fermented BJ was larger than the control plants (0.0%). While the control plants (0.0%) had a leaf area of 71.4 cm^2, the plants of the treatment of 0.5% fermented BJ displayed a leaf area of 131.5 cm^2 (Figure 4C). Moreover, other treatments exhibited significantly higher leaf areas than control plants (0.0%) as well.

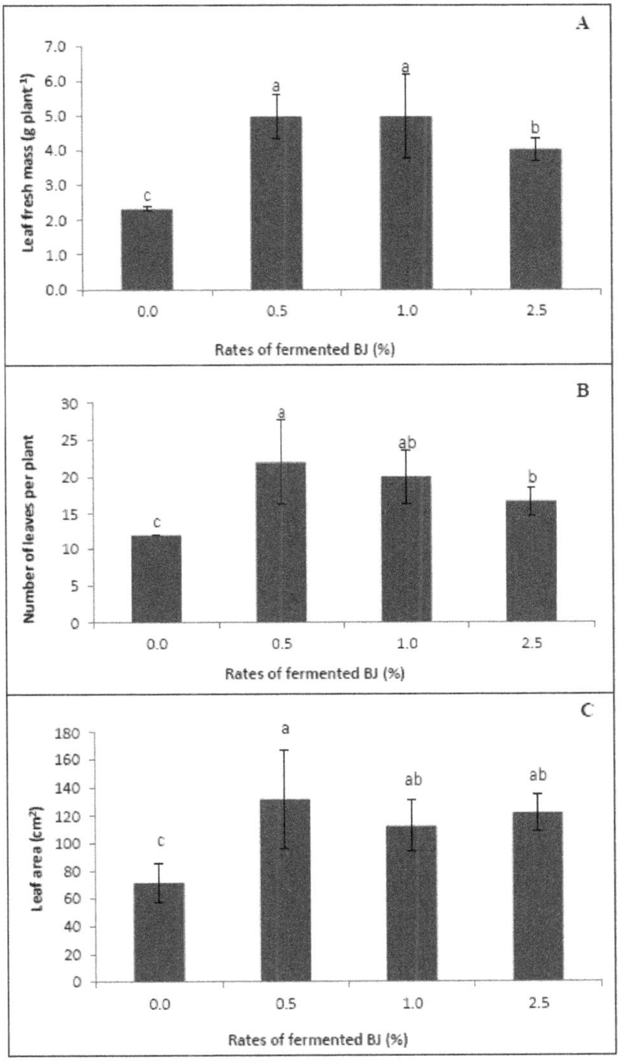

Figure 4. (**A**) Leaves fresh mass; (**B**) number of leaves; (**C**) leaf area of sweet basil plants treated with different rates of fermented BJ. Sample size $n = 6$ (mean ± SD). Different letters above the same columns show significant differences at the level of $p \leq 0.05$.

3.2. Contents of Photosynthetic Pigment

The results showed that the content of chlorophylls increased upon spraying sweet basil plants with fermented BJ (Table 1). The contents of chlorophyll-a (chl a) and chlorophyll-b (chl b) in the control (0.0%) leaves were 6.90 and 1.95 mg g^{-1}, respectively, and significantly increased to 8.04 (chl a) and 2.66 (chl b) mg g^{-1} after treating plants with 0.5% (chl a) and 2.5% (chl b) fermented BJ, respectively. The total content of chlorophylls ranged from 8.86 to 10.60 mg g^{-1}, 2.5% BJ application increased the chlorophyll content by 19% compared to the control (0.0%). Chlorophyll a/b ratio in treated plants with fermented BJ ranged from 2.97 to 3.50; while the control plants (0.0%) possessed a chlorophyll a/b ratio of 3.52 (Table 1). The SPAD value showed a significant, 28% increase when leaves were treated with 0.5% concentration of fermented alfalfa BJ in comparison to the control (0.0%).

Table 1. Content of chlorophyll in sweet basil leaves (mg g^{-1}) and their relative changes (%) compared to the control plants (0.0%) ($n = 9$).

	Chlorophyll-a	Chlorophyll-b	Chlorophyll-a/b Ratio	Total Chlorophyll	SPAD Value
0.0%	6.91 b	1.96 b	3.52 a	8.86 a	27.00 b
0.5%	8.05 a (16.52%)	2.46 ab (25.78%)	3.26 ab	10.51 ab (18.57%)	34.60 a (28.15%)
1.0%	6.94 ab (0.56%)	2.00 ab (2.26%)	3.47 ab	8.95 ab (0.93%)	32.40 ab (20.00%)
2.5%	7.93 ab (14.86%)	2.67 a (36.12%)	2.97 b	10.60 b (19.56%)	31.60 ab (17.04%)

Different letters (a, b, ab) in the same columns show significant differences at the level of $p \leq 0.05$.

3.3. Anatomical Traits of Sweet Basil

The cortex is made up of angular collenchyma (two to four cells thick) and typical parenchymatous cells. The pith consists of mainly parenchymatous cells, but the ratio of primary and secondary vascular tissues is dependent on the treatments. Treatments 0.5% and 1.0% resulted in well-developed secondary vascular tissues (Figure 5).

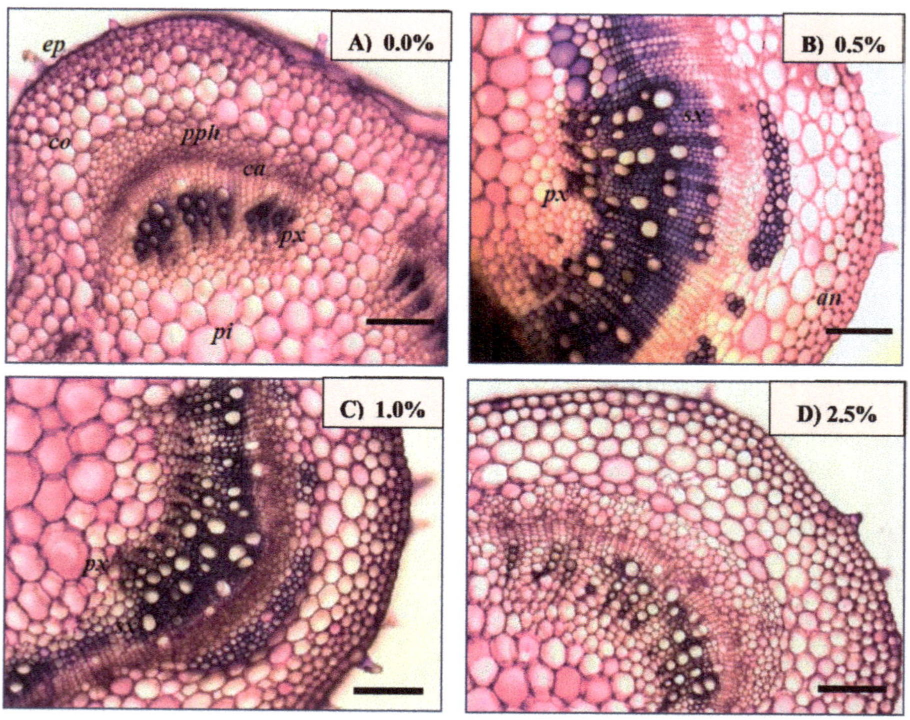

Figure 5. Anatomical sections of sweet basil stem. Subfigures show samples treated with different rates of fermented BJ such as (**A**) 0.0%, (**B**) 0.5%, (**C**) 1.0%, (**D**) 2.5%. ***ep*** epidermis, ***co*** cortex, ***ang*** angular collenchyma, ***pi*** pith, ***pph*** primary phloem ***ca*** cambium ***px*** primary xylem ***sx*** secondary xylem, after treating with different rates of fermented BJ (i.e., control (0.0%), 0.5%, 1.0%, 2.5%). Scale bar is 200 μm.

Differences which are visible in the tissues caused by the treatments were proven true by our measurements. All levels of concentration increased the thickness of the epidermis, however only two tests were significant. The impact of all treatments was a decrease in the thickness of the primary cortex.

Treatments 0.5% and 1.0% increased both the extension of the pith and the proportion of vascular tissue in it, which were proven statistically significant. Growth of the secondary vascular tissue was the highest at the 0.5% treatment. It can be concluded that a greater brown juice concentration, which was 2.5% in this analysis, may block the development of secondary vascular tissues (Table 2).

Table 2. Anatomical traits of basil stem tissue (epidermis, cortex, pith involved primary and secondary vascular tissues) (μm) after treatment with fermented BJ. Sample size $n = 45$ (Mean ± SD).

BJ Rate	Epidermis	Cortex	Pith	Vascular Tissue
0.0%	23.75 ± 5.55‡ b	273.55 ± 48.18 a	2202.89 ± 239.13 c	234.41 ± 53.75 c
0.5%	27.92 ± 7.81 a	259.39 ± 73.72 a	2529.03 ± 280.56 b	548.68 ± 107.39 a
1.0%	27.37 ± 8.22 ab	260.97 ± 51.36 a	2684.10 ± 259.04 a	539.41 ± 96.54 a
2.5%	28.88 ± 7.54 a	242.16 ± 56.36 a	2101.83 ± 162.30 c	290.38 ± 59.46 b

Different letters (a, b, ab) in the same columns show significant differences at the level of $p \leq 0.05$.

Consequently, the use of an incremental concentration of brown juice can increase the thickness of the stem and the vascular tissue in it but only to a certain extent.

4. Discussion

The BJ is largely produced during the isolation of leaf protein from several green leafy crops. The storage of fresh BJ in room temperature is the main concern due to the high content of carbohydrates particularly in the form of monosaccharaides. Lactic acid bacteria showed a considerable effect on reducing BJ pH as a result of organic acids production under an anaerobic condition and consequently increased the stability and handling of BJ at a pH of 4.5–4.8 [7].

Fermented BJ can be exploited as an organic fertilizer or growth stimulator due to its richness in free amino acids, soluble sugars, vitamins, organic acids and other nutrients. In addition, lactic acid bacteria—as plant growth promoting bacteria—represent an additional benefit for using fermented BJ as a fertilizer [7,33]. In the present study, fermented BJ proved its efficiency as a promising plant growth stimulator as it caused a significant increase in photosynthetic pigments including chl a, chl b, total chlorophyll and relative chlorophyll (SPAD value). Similar findings have been previously reported by Bergstrand et al. [34] who proved in their study on nitrogen speciation in pot experiment of sweet basil fertilized by different organic manures, i.e., blood meal + Baralith®Enslow and poultry manure, that the plant-based organic fertilizer treatment induced the chlorophyll content. Moreover, applying biofertilizers, i.e., Nitrajin (including Azotobacter, Azospirillum and Pseudomonase), increased the amount of photosynthetic pigments and the leaf area of sweet basil [35]. Additionally, Ertani et al. [36] mentioned that the extract prepared from alfalfa biomass using enzymes contained growth stimulant compounds like triacontanol and indole-3-acetic acid, which significantly improved the relative chlorophyll and growth of maize plants under salt-stress conditions.

Noticeably, all rates of BJ resulted in higher values of stem and root length (Figure 1), stem and root volume, stem and root fresh mass, number of leaves and leaf area. This was in agreement with those documented earlier by El-Ziat et al. [37]. They cited that organic fertilization and humic acid application improved growth parameters of 'Red Rubin' basil plant, i.e., fresh weight, plant height and leaf area, in a greenhouse experiment. Organic fertilization of basil using organic NPK fertilizer (4–3–4) (Organic Fertilizer, Mighty Grow®Fruitdale, AL) resulted in changes in both fresh and dry weight, and in nutrient uptake as well [24]. Onofrei et al. [18] stated that different organic foliar applications, i.e., Fylo®, Geolino Plants&Flowers®, Cropmax®, Fitokondi®, stimulated the content of total phenolic compounds contributing to healthier vegetable production of *Ocimum basilicum* L.

Unique positive changes in secondary vascular tissues observed as a result of 0.5% and 1.0% BJ treatments (Figure 5 and Table 2), which according to our best knowledge have not been published before. However, similar effects were reported on plumed cockscomb (*Celosia argantea* var. plumosa 'Arrabona') plants by Bákonyi et al. [7] who cited significant changes in histological parameters after treating plants with fermented BJ. The tissue structure of the stem of sweet basil analyzed is a typical structure of a plant at the age of 10–12 weeks. Stems were covered by the epidermis (single row) with a thin layer of cuticle. Contrary to Venkateshappa and Sreenath [38], more types of the trichomes are identified on the surface of the stem, e.g., non-glandular uniseriate hair (composed of three cells) and glandular capitate hairs, which supports the findings of Werker et al. [39] and Nasssar et al. [40].

In a few cases, the application of fermented BJ led to a slight reduction in some measured parameters. However, all rates of fermented BJ were better than the control plants (0.0%, untreated). This beneficial role of fermented BJ is owed to its high content of phytoavailable nutrients such as N, P, K, Ca, Mg, S, Mn, Fe, Cu, Zn, and Mo [7]. Similar findings were reported earlier by Ream et al. [41]. They revealed that BJ at the rate of 1.25 cm enhanced the yield and growth of corn, alfalfa and bromegrass. However, they also stated that higher rates of BJ (2.5 cm) caused a reduction in development and yields of all crops. Another study cited similar results where the application of BJ obtained from alfalfa biomass at low rates positively improved the germination of many crops, i.e., cowpea, mung bean and groundnut. Negative impacts were reported when BJ was used at higher rates, above 10% [33]. The detrimental effect of high rates of BJ may be attributed to the existence of some phytotoxic organic compounds in BJ [6]. Moreover, the reduction in plant growth at a high applied rate of fermented BJ could be due to the high electrical conductivity of BJ [7]. Additionally, spraying plants with acidic solutions (low pH as at the high rate of fermented BJ) is known to lessen the stomatal conductance [42]; therefore, a reduction in plant growth is expected. This hypothesis was supported by the earlier findings of Long et al. [43].

The results obtained from the present study alongside with our previous published work [7] strongly support the hypothesis about prospective studies on BJ either fermented or fresh to improve soil properties through the soil application technique. In alkaline, salt-affected soils, and sandy soils the expected benefits of using fermented BJ would be more due to the high content of macro and microelements, free amino acids [7,42,44]. Additionally, the soluble sugars will facilitate and improve the growth of rhizosphere bacteria supporting them against the unflavored growth conditions [45]. Another substantial result for using fermented BJ via soil application is increasing the soil aggregates due to a high sugar content in BJ. Moreover, the low pH of fermented BJ is an advantage of using fermented BJ as a soil conditioner since it will partially modify the local soil pH around the root system facilitating the uptake of especially microelements by plant roots [43]. However, in our next studies, we are going to work on these hypotheses.

5. Conclusions

This study demonstrates the potentiality of fermented alfalfa BJ to enhance both quantity and quality of one of the most important multipurpose crops, i.e., sweet basil, which is cultivated on large areas in many countries. Our results showed a significant increase in photosynthetic pigments leading to better vegetative growth; and consequently, enhancement in important medical phytochemicals in sweet basil could occur. For a plant such as sweet basil (used as spice and fresh vegetable, and culinary herb), larger leaf area, number of leaves, and thickness of stem (including vascular tissues) as well as fresh weight are very important decisive traits. The results also indicated that concentrations of 0.5 and 1.0% fermented BJ were the most valuable and have a considerable biostimulator effect on leaves as these treatments were more effective improving leaf parameters in comparison to control plants (0.0%).

Additionally, we came to the conclusion based on the results that fermented alfalfa BJ has great potential as a biofertilizer and plant growth promoter. Future experiments are needed to justify more of these findings with other horticultural and agricultural crops.

Author Contributions: Conceptualization, N.B., S.K. (Szilvia Kisvarga) and M.G.F.; methodology, S.K. (Szilvia Kisvarga), D.B. and I.O.T.; software, D.B.; validation, D.B.; formal analysis, G.C. and S.K. (Szilvia Kovács); investigation, S.V. and P.M.; resources, N.B. data curation, T.A.; writing—original draft preparation, N.B., T.A. and D.B.; writing—review and editing, N.B., T.A., S.V. and D.B.; visualization, G.C. and S.K. (Szilvia Kovács); supervision, M.G.F.; project administration, N.B.; funding acquisition, N.B. and M.G.F. All authors have read and agreed to the published version of the manuscript.

Funding: The research was financed by the Higher Education Institutional Excellence Programme (NKFIH-1150-6/2019) of the Ministry of Innovation and Technology in Hungary, within the framework of the Biotechnology thematic programme of the University of Debrecen.

Acknowledgments: We acknowledge the financial support of the followings: "Complex Rural Economic and Sustainable Development, Elaboration of its Service Networks in the Carpathian Basin (Project ID: EFOP-3.6.2-16-2017-00001, Hungary)" research project and Széchenyi 2020 under the GINOP-2.2.1-15-2017-00051.

Conflicts of Interest: The authors declare no conflicts of interest.

References

1. de Boer, J.; Aiking, H. Strategies towards healthy and sustainable protein consumption: A transition framework at the levels of diets, dishes, and dish ingredients. *Food Qual. Prefer.* **2019**, *73*, 171–181. [CrossRef]
2. Tenerio, A.T.; Kyriakopoulou, K.E.; Suarez-Garcia, E.; van den Berg, C.; van der Goot, A.J. Understanding differences in protein fractionation from conventional crops, and herbaceous and aquatic biomass—Consequences for industrial use. *Trends Food Sci. Technol.* **2018**, *71*, 235–245. [CrossRef]
3. Fári, M.G.; Domokos-Szabolcsy, E. Method for Producing Plant Protein Coagulum. Hun Patent WO2019150144, 8 August 2019.
4. Gaweł, E.; Grzelak, M.; Janyszek, M. Lucerne (*Medicago sativa* L.) in the human diet—Case reports and short reports. *J. Herb. Med.* **2017**, *10*, 8–16. [CrossRef]
5. Tripathi, A.D.; Mishra, R.; Maurya, K.K.; Singh, R.B.; Wilson, D.W. Estimates for world population and global food availability for global health. *Role Funct. Food Secur. Glob. Health* **2019**, 3–24. [CrossRef]
6. Pirie, N.W. *Leaf Protein and Other Aspects of Fodder Fractionation*; Cambridge University Press: Cambridge, UK, 1987; p. 209.
7. Bákonyi, N.; Kisvarga, S.; Barna, D.; Tóth, I.O.; El-Ramady, H.; Abdalla, N.; Kovács, S.; Rozbach, M.; Fehér, C.; Elhawat, N.; et al. Chemical Traits of Fermented Alfalfa Brown Juice: Its Implications on Physiological, Biochemical, Anatomical, and Growth Parameters of Celosia. *Agronomy* **2020**, *10*, 247. [CrossRef]
8. Jadhav, R.K.; Chhaya, G. Influence of deproteinised foliage fluid liquid biofertilizer on nitrate reductase activity of Eleusine coracana plants. *Curr. Bot.* **2018**, 33–36. [CrossRef]
9. Thomsen, M.H.; Bech, D.; Kiel, P. Manufacturing of Stabilised Brown Juice for L-lysine Production—From University Lab Scale over Pilot Scale to Industrial Production. *Chem. Biochem. Eng. Q.* **2003**, *18*, 37–46.
10. Bákonyi, N.; Tóth, I.O.; Barna, D.; Fári, M.G. *Preliminary Experiments on Preserving Alfalfa Brown Juice for Bio-Industrial Use (Előkísérletek a Lucerna Barnalé Tartósítására Bioipari Hasznosítás Céljából)*; XXXVII, Óvári Tudományos Napok; Széchenyi István University: Mosonmagyaróvár, Hungary, 2018.
11. Manwatkar, W.G.; Gogle, D.P. The Effect of Deproteinised Juice (DPJ) on Seed Germination and Seedling Growth of Different Plants (by Paper Towel Method). *Int. J. Life Sci.* **2014**, *A2*, 65–68.
12. Lamont, J.R.; Wilkins, O.; Bywater-Ekegärd, M.; Smith, D.L. From yogurt to yield: Potential applications of lactic acid bacteria in plant production. *Soil Biol. Biochem.* **2017**, *111*, 1–9. [CrossRef]
13. Cheynier, V.; Comte, G.; Davies, K.M.; Lattanzio, V.; Martens, S. Plant phenolics: Recent advances on their biosynthesis, genetics, and ecophysiology. *Plant Physiol. Biochem.* **2013**, *72*, 1–20. [CrossRef]
14. Shende, G.C.; Gogle, D.P. To study the effect of various concentration of deproteinized leaf juice (DPJ) of selected plants on growth of Aspergillus niger. *Int. J. Life Sci.* **2016**, *A6*, 186–188.
15. Jadhav, R.K.; Kadam, R.; Joshi, D. Physiology of Gramineae Crops by Deproteinised Foliage Extract and its Influence on Photosynthetic Chemistry. *J. Phytochem.* **2019**, *I*, 149–163.
16. Zanin, V. *A New Nutritional Idea for Man: Lucerne Leaf Concentrate*; APEF, Association Pour la Promotion des Extraits Foliaires en Nutrition: Paris, France, 1998.
17. Smith, D.L.; Subramanian, S.; Lamont, J.R.; Bywater-Ekegärd, M. Signaling in the phytomicrobiome: Breadth and potential. *Front. Plant Sci.* **2015**, *6*, 921. [CrossRef] [PubMed]
18. Onofrei, V.; Burducea, M.; Lobiuc, A.; Teliban, G.-C.; Ranghiuc, G.; Robu, T. Influence of Organic Foliar Fertilization on Antioxidant Activity and Content of Polyphenols in *Ocimum Basilicum* L. *Acta Pol. Pharm.* **2017**, *74*, 611–615.
19. Alishah, H.M.; Heidari, R.; Hassani, H.; Dizaji, A.A. Effect of Water Stress on Some Morphological and Biochemical Characteristics of Purple Basil (*Ocimum basilicum*). *J. Biol. Sci.* **2006**, *6*, 763–767. [CrossRef]

20. Nurzyńska-Wierdak, R.; Rożek, E.; Dzida, K.; Borowski, B. Growth response to nitrogen and potassium fertilization of common basil (*Ocimum basilicum* L.) plants. *Hortorum Cultus* **2012**, *11*, 275–288.
21. Gülçin, I.; Elmastaş, M.; Aboul-Enein, H.Y. Determination of antioxidant and radical scavenging activity of Basil (*Ocimum basilicum* L. Family Lamiaceae) assayed by different methodologies. *Phytother. Res.* **2007**, *21*, 354–361. [CrossRef]
22. Labra, M.; Miele, M.; Ledda, B.; Grassi, F.; Mazzei, M.; Sala, F. Morphological characterization, essential oil composition and DNA genotyping of *Ocimum basilicum* L. cultivars. *Plant Sci.* **2004**, *167*, 725–731. [CrossRef]
23. Hikmasari, I. *Handbook of Herbs and Spices*, 2nd ed.; Woodhead Publishing Limited: Cambridge, UK, 2012; Volume 1.
24. Bufalo, J.; Cantrell, C.; Astatkie, T.; Jeliazkov, V.; Gawde, A.; Boaro, C.S.F. Organic versus conventional fertilization effects on sweet basil (*Ocimum basilicum* L.) growth in a greenhouse system. *Ind. Crop. Prod.* **2015**, *74*, 249–254. [CrossRef]
25. Chiang, L.-C.; Ng, L.-T.; Cheng, P.-W.; Chiang, W.; Lin, C.-C. Antiviral activities of extracts and selected pure constituents of *Ocimum basilicum*. *Clin. Exp. Pharm. Physiol.* **2005**, *32*, 811–816. [CrossRef]
26. Phippen, W.B.; Simon, J.E. Anthocyanins in Basil (*Ocimum basilicum* L.). *J. Agric. Food Chem.* **1998**, *46*, 1734–1738. [CrossRef]
27. Hess, M.; Barralis, G.; Bleiholder, H.; Buhr, L.; Eggers, T.; Hack, H.; Stauss, R. Use of the extended BBCH scale—General for the descriptions of the growth stages of mono; and dicotyledonous weed species. *Weed Res.* **1997**, *37*, 433–441. [CrossRef]
28. Moran, R.; Porath, D. Chlorophyll Determination in Intact Tissues Using *N,N*-Dimethylformamide. *Plant Physiol.* **1980**, *65*, 478–479. [CrossRef]
29. Wellburn, A.R. The Spectral Determination of Chlorophylls a and b, as well as Total Carotenoids, Using Various Solvents with Spectrophotometers of Different Resolution. *J. Plant Physiol.* **1994**, *144*, 307–313. [CrossRef]
30. Vidican, T.I.; Cachita-Cosma, D. Determination of assimilator pigment content in cladodes of Opuntia fragilis var. Fragilis exposed to light of different colors emitted by LEDs. *Studia Univ. Vasile Goldis Arad* **2010**, *20*, 49–54.
31. Kálmán, K.; Nyakas, A.; Mihalik, E.; Nagy, E. Plantanatomical practicality. In *Növényanatómiai Praktikum*; JATE Press: Budapest, Hungary, 2019.
32. Tukey, J.W. Comparing Individual Means in the Analysis of Variance. *Biometrics* **1949**, *5*, 99–114. [CrossRef]
33. Reddy, G.U.; Deshmukh, V.R.; Joshi, R.N.; Kayama, R. Utilization of Alfalfa (*Mediacago sativa* L.) Whey as a Fertilizer in Irrigation. *Jap. J. Grassl. Sci.* **1987**, *33*, 32–37. [CrossRef]
34. Bergstrand, K.-J.; Löfkvist, K.; Asp, H. Dynamics of nitrogen availability in pot grown crops with organic fertilization. *Biol. Agric. Hortic.* **2018**, *35*, 143–150. [CrossRef]
35. Larimi, S.B.; Shakiba, M.; Mohammadinasab, A.D.; Vahed, M.M. Changes in nitrogen and chlorophyll density and leaf area of sweet basil (*Ocimum basilicum* L.) affected by biofertilizer and nitrogen application. *Int. J. Biosci.* **2014**, *5*, 256–265. [CrossRef]
36. Ertani, A.; Schiavon, M.; Muscolo, A.; Nardi, S. Alfalfa plant-derived biostimulant stimulate short-term growth of salt stressed *Zea mays* L. plants. *Plant Soil* **2013**, *364*, 145–158. [CrossRef]
37. El-Ziat, R.A.; Swaefy, H.M.; Esmail, S.E.A. The Response of Red Rubin Basil Plant to Organic Fertilizer and Humic Acid versus Chemical Fertilizers. *Middle East J. Agric. Res.* **2018**, *7*, 740–751.
38. Venkateshappa, S.M.; Sreenath, K.P. Some Species of Lamiaceae- Comparative Anatomical Studies. *Indo Am. J. Pharm. Res.* **2013**, *3*, 6.
39. Werker, E.; Putievsky, E.; Ravid, U.; Dudai, N.; Katzir, I. Glandular hairs and essential oil in developing leaves of *Ocimum basilicum* L. (Lamiaceae). *Ann. Bot.* **1993**, *71*, 43–50. [CrossRef]
40. Nassar, M.A.; El-Sagai, M.U.; Azoz, S.N. Anatomical and Phytochemical studies on *Ocimum basilicum* L. plant (Lamiaceae). *Int. J. Adv. Res.* **2014**, *2*, 204–226.
41. Ream, H.W.; Smith, D.; Walgenbach, R.P. Effects of Deproteinized Alfalfa Juice Applied to Alfalfa—Bromegrass, Bromegrass, and Corn1. *Agron. J.* **1977**, *69*, 685–689. [CrossRef]
42. Anugoolprasert, O.; Kinoshita, S.; Naito, H.; Shimizu, M.; Ehara, H. Effect of low pH on the growth, physiological characteristics and nutrient absorption of sago palm in a hydroponic system. *Plant Prod. Sci.* **2012**, *15*, 125–131. [CrossRef]

43. Long, A.; Zhang, J.; Yang, L.T.; Ye, X.; Lai, N.W.; Tan, L.L.; Lin, D.; Chen, L.S. Effects of low pH on photosynthesis, related physiological parameters, and nutrient profiles of citrus. *Front. Plant Sci.* **2017**, *8*, 185. [CrossRef]
44. Porra, R.J.; Thompson, W.A.; Kriedemann, P.E. Determination of accurate extinction coefficients and simultaneous equations for assaying chlorophylls a and b extracted with four different solvents: Verification of the concentration of chlorophyll standards by atomic absorption spectroscopy. *Biochim. Biophys. Acta* **1989**, *975*, 384–394. [CrossRef]
45. Ferreira, A.S.; Santos, M.A.D.; Corrêa, G.F. Soil microbial response to glucose and phosphorus addition under agricultural systems in the Brazilian Cerrado. *An. Acad. Bras. Ciênc.* **2013**, *85*, 395–403. [CrossRef]

MDPI
St. Alban-Anlage 66
4052 Basel
Switzerland
Tel. +41 61 683 77 34
Fax +41 61 302 89 18
www.mdpi.com

Agronomy Editorial Office
E-mail: agronomy@mdpi.com
www.mdpi.com/journal/agronomy

www.ingramcontent.com/pod-product-compliance
Lightning Source LLC
Chambersburg PA
CBHW061446220926
51850CB00004B/25
9783036500287